Internal Flow

This book describes the analysis and behavior of internal flows encountered in propulsion systems, fluid machinery (compressors, turbines, and pumps) and ducts (diffusers, nozzles and combustion chambers). The focus is on phenomena that are important in setting the performance of a broad range of fluid devices.

The authors show that even for complex processes one can learn a great deal about the behavior of such devices from a clear understanding and rigorous use of basic principles. Throughout the book they illustrate theoretical principles by reference to technological applications. The strong emphasis on fundamentals, however, means that the ideas presented can be applied beyond internal flow to other types of fluid motion.

The book equips students and practising engineers with a range of analytical tools, which offer enhanced interpretation and application of both experimental measurements and the computational procedures that characterize modern fluids engineering.

Edward M. Greitzer received his Ph.D. from Harvard University and is the H. N. Slater Professor of Aeronautics and Astronautics at the Massachusetts Institute of Technology. He spent ten years with United Technologies Corporation, at Pratt & Whitney and United Technologies Research Center. He has been a member of the US Air Force Scientific Advisory Board, the NASA Aeronautics Advisory Committee, and Director of the MIT Gas Turbine Laboratory. He is a three-time recipient of the ASME Gas Turbine Award, an ASME Freeman Scholar in Fluids Engineering, a fellow of AIAA and ASME, and a member of the National Academy of Engineering.

Choon Sooi Tan received his Ph.D. from the Massachusetts Institute of Technology and is currently a Senior Research Engineer in the Gas Turbine Laboratory at MIT.

Martin B. Graf received his Ph.D. from the Massachusetts Institute of Technology and is currently a Project Manager at the consulting firm Mars & Company. Before joining Mars he was with the Pratt & Whitney Division of United Technologies Corporation.

Internal Flow

Concepts and Applications

E. M. Greitzer
H. N. Slater Professor of Aeronautics and Astronautics
Massachusetts Institute of Technology

C. S. Tan
Massachusetts Institute of Technology

and

M. B. Graf
Mars & Company

PUBLISHED BY THE PRESS SYNDICATE OF THE UNIVERSITY OF CAMBRIDGE
The Pitt Building, Trumpington Street, Cambridge, United Kingdom

CAMBRIDGE UNIVERSITY PRESS
The Edinburgh Building, Cambridge, CB2 2RU, UK
40 West 20th Street, New York, NY 10011–4211, USA
477 Williamstown Road, Port Melbourne, VIC 3207, Australia
Ruiz de Alarcón 13, 28014 Madrid, Spain
Dock House, The Waterfront, Cape Town 8001, South Africa

http://www.cambridge.org

First published 2004

Printed in the United Kingdom at the University Press, Cambridge

Typefaces Times 10/13 pt. and Helvetica *System* LaTeX 2ε [TB]

A catalog record for this book is available from the British Library

Library of Congress Cataloging in Publication data
Greitzer, E. M., 1941–
Internal flow: concepts and applications / E. M. Greitzer, C. S. Tan, M. B. Graf.
p. cm. (Cambridge engine technology series ; 3)
Includes bibliographical references and index.
ISBN 0 521 34393 3
1. Fluid mechanics. I. Tan, Choon Sooi. II. Graf, M. B., 1968– III. Title. IV. Series.
TA357.G719 2003
620.1′064 – dc21 2003055357

ISBN 0 521 34393 3 hardback

Contents

10 Compressible internal flow 506

Preface

There are a number of excellent texts on fluid mechanics which focus on external flow, flows typified by those around aircraft, ships, and automobiles. For many fluid devices of engineering importance, however, the motion is appropriately characterized as an internal flow. Examples include jet engines or other propulsion systems, fluid machinery such as compressors, turbines, and pumps, and duct flows, including nozzles, diffusers, and combustors. These provide the focus for the present book.

Internal flow exhibits a rich array of fluid dynamic behavior not encountered in external flow. Further, much of the information about internal flow is dispersed in the technical literature and does not appear in a connected treatment that is accessible to students as well as to professional engineers. Our aim in writing this book is to provide such a treatment.

A theme of the book is that one can learn a great deal about the behavior of fluid components and systems through rigorous use of basic principles (the *concepts*). A direct way to make this point is to present illustrations of technologically important flows in which it is true (the *applications*). This link between the two is shown in a range of internal flow examples, many of which appear for the first time in a textbook.

The experience of the authors spans dealing with internal flow in an industrial environment, teaching the topic to engineers in industry and government, and teaching it to students at MIT. The perspective and selection of material reflects (and addresses) this span. The book is also written with the view that computational procedures for three-dimensional steady and unsteady flow are now common tools in the study of fluid motion. Our observation is that the concepts presented enable increased insight into the large amount of information given by computational simulations, and hence allow their more effective utilization.

The structure of the book is as follows. The first two chapters provide basic material, namely a description of the laws that determine the motion (Chapter 1) and the introduction of a number of useful concepts (Chapter 2). Among the latter are qualitative features of pressure fields and fluid accelerations, fundamentals of compressible channel flow, introduction to boundary layers, and applications of the integral forms of the conservation laws. Chapter 3 presents, and applies, the concepts of vorticity and circulation. These provide both a compact framework for describing the three-dimensional and unsteady fluid motions that characterize fluid devices and a route to increased physical insight concerning these motions. Chapter 4 discusses boundary layers and shear layers in the context of analysis of viscous effects on fluid component performance. Chapter 5 then gives an in-depth treatment of loss sources and loss accounting as a basis for the rigorous assessment of fluid component and system performance.

The remaining chapters are organized in terms of different phenomena that affect internal flow behavior. Chapter 6 deals with unsteadiness, including waves, oscillations, and criteria for instability in fluid systems. Chapter 7 treats flow in rotating passages and ducts, such as those in a turbomachine.

Swirling flow, including the increased potential for upstream influence, the behavior of vortex cores, boundary layers and jets in swirling flow, and vortex breakdown, is described in Chapter 8. Chapter 9 discusses the three-dimensional motions associated with embedded streamwise vorticity. Examples are 'secondary flows', which are inherent in non-uniform flow in curved passages, and the effects of streamwise vorticity on mixing. Chapter 10 addresses compressible flow including streams with mass, momentum, and energy (both work and heat) addition, with swirl, and with spatially varying stagnation conditions, all of which are encountered in fluid machinery operation. Effects of heat addition on fluid motions, described in Chapter 11, include an introduction to ramjet and scramjet propulsion systems and the interaction between swirl and heat addition. The final chapter (12) provides a broad view of non-uniform flow in fluid components such as contractions, screens, diffusers, and compressors, as well as the resulting interactions between the components. These chapters address different topics, but a shared paradigm is the creation of a rotational flow by non-uniform energy addition, external forces, or viscous forces and the consequent response to the pressure field (the dominant influence for the flows of interest) and wall shear stress associated with a bounding geometry.

In terms of accessibility, the material in the first two chapters underpins much of the material in the rest of the book. Sections 3.1–3.4, 3.8, 3.9, 3.14 and 4.1–4.3 are also often made use of in later chapters. Apart from these, however, the chapters (and to a large extent the sections) in the book can be read independently of the preceding material.

The text has been used in a one-semester MIT graduate course, generally taken after the student has had either an advanced undergraduate, or first year graduate, course in fluid dynamics. The lectures cover phenomena in which compressibility does not play a major role and include material in Chapters 2 (not including the compressible flow sections), Chapter 3, much of Chapter 5, and roughly half the material in Chapters 6, 7, 8, and 9. The text has also been used, along with a supplementary compressible flow reference, for a graduate compressible flow course that covers internal and external flow applications. In this latter context the material used is the development and application of the energy equation in Chapter 1 (which we find that many students need to review), the compressible flow sections in Chapter 2, Chapter 10, and roughly half of Chapter 11.

Many individuals have helped in the writing of this book and it is a pleasure to acknowledge this. Foremost among these are T. P. Hynes of Cambridge University and N. A. Cumpsty, formerly of Cambridge, now Chief Technologist of Rolls-Royce. Dr. Hynes was initially a coauthor, and provided the first versions of several chapters. Although the press of other work caused him to resign from coauthorship, he has been kind enough to provide information, answer many questions, and review (and much improve) several aspects of the work in progress. Dr. Cumpsty reviewed a number of aspects in different stages of the project. His high standards for clarity of exposition and selection of material have been extremely helpful in forming the final product. We also greatly appreciate the incisive comments on a number of the chapters by L. H. Smith of General Electric Aircraft Engines, especially his perspective and strong stance on what was, and was not, clear.

We are grateful for the feedback on different chapters that we have received from E. E. Covert (as well as for his trenchant comments on strategies for completion), D. L. Darmofal, M. Drela, D. R. Kirk, B. T. Sirakov, Z. S. Spakovszky, and I. A. Waitz of MIT; W. H. Heiser of Air Force Academy; J. S. Simon of Emhart Glass Research; A. J. Strazisar of NASA Glenn Research Center; Y. Dong, A. Prasad, D. Prasad, and J. S. Sabnis of Pratt & Whitney; M. V. Casey of Sulzer Innotec; C. N. Nett of United Technologies Research Center; and M. Brear of the University of Melbourne.

We also acknowledge material received from J. D. Denton, R. L. E. Fearn, E. F. Hasselbrink, A. R. Karagozian, A. Khalak, H. S. Khesgi, M. G. Mungal, and D. E. VanZante. In addition, we thank the several classes of graduate students who used portions of the manuscript as their text and made their way through arguments that were sometimes not as complete (or as coherent) as one had hoped. Input from all the above has resulted in considerable revision and the book is the better for it. For the parts of the book in which the exposition is still unclear, the authors are directly responsible.

It is difficult if not impossible for us to envision more effective help and creative solutions to editorial issues in the manuscript preparation than that rendered by Ms D. I. Park. We would also like to thank Ms R. Palazzolo for help in this regard. Much of our knowledge of internal flow has resulted from our research on propulsion system fluid dynamics, and we wish to thank long-time sponsors Air Force Office of Scientific Research, General Electric Aircraft Engines, NASA Glenn Research Center, and Pratt & Whitney. Our knowledge, and our research, have benefited in a major way from the keen insights that Professor F. E. Marble of Caltech has shared with us on many visits. It is also a great pleasure to acknowledge the faculty, staff, and students of the Gas Turbine Laboratory for the stimulating atmosphere in which this research was carried out.

Finally, E. M. Greitzer would like to acknowledge the financial support provided by the H. N. Slater Professorship and the Department of Aeronautics and Astronautics at MIT, E. F. Crawley, Department Head, H. L. Gallant, Administrative Officer, as well as the support of many kinds rendered by H. M. Greitzer during this lengthy process.

Acknowledgements

We wish to thank the following for permission to use figures and other materials: **American Institute of Aeronautics and Astronautics**: Figures 6.39 (Carta, 1967), 4.5 (Drela, 1998), 8.41 (Favaloro *et al.*, 1991), 9.43, and 9.45 (Fearn and Weston, 1974), 2.25 (Hawthorne, 1957), 11.6, and 11.7 (Heiser and Pratt, 1994), 4.26 (Kline *et al.*, 1983), 10.18 (Lin *et al.*, 1991), 6.38 (Lyrio and Ferziger, 1983), 5.26 (Patterson and Weingold, 1985), 6.40 (Smith, 1993), and 6.37 (Telionis and Romaniuk, 1978), © AIAA, reprinted with permission; **American Institute of Physics**: Figures 7.17–7.23 (Kheshgi and Scriven, 1985); **American Society of Mechanical Engineers**: Table 10.3 and Figure10.20 (Anderson *et al.*, 1970), Figures 12.24 (Barber and Weingold, 1978), 10.31, 10.37, and 10.38 (Bernstein *et al.*, 1967), 7.24 (Bo *et al.*, 1995), 8.36–8.38 (Chigier and Chervinsky, 1967), 10.13 (Chima and Strazisar, 1983), 8.30 and 8.31 (Daily and Nece, 1960), 5.6, 5.7, 5.9, 5.12, 5.14, and 5.28 (Denton, 1993), 8.25 and 8.26 (Dou and Mizuki, 1998), 12.27 and 12.28 (Graf *et al.*, 1998), 12.38 (Greitzer *et al.*, 1978), 12.48–12.50 (Greitzer and Strand, 1978), 6.27 (Hansen *et al.*, 1981), 4.36–4.39 (Hill *et al.*, 1963), 8.33–8.35 (Johnson *et al.*, 1990), 9.18 (Johnson, 1978), 9.12 (Langston, 1980), 12.29 (Longley *et al.*, 1996), 4.9 and 4.10 (Mayle, 1991), 7.28 (Moore, 1973a), 9.25 and 9.26 (Prasad and Hendricks, 2000), 4.3 (Reneau *et al.*, 1967), 5.29 and 5.30 (Roberts and Denton, 1996), 7.31 and 7.32 (Rothe and Johnston, 1976), 8.27 (Senoo *et al.*, 1977), 6.40 (Smith, 1966b), 12.19 (Stenning, 1980), 6.41 (Van Zante *et al.*, 2002), 12.39, 12.41, and 12.42 (Wolf and Johnston, 1969), and 11.12 and 11.13 (Young, 1995), permission granted by ASME; **Annual Reviews, Inc.**: Figures 8.2 (Escudier, 1987) and 8.9 (Hall, 1972), with permission from the *Annual Review of Fluid Mechanics*; **Cambridge University Press**: Figures 2.37 and 8.11 (Batchelor, 1967), 8.42 (Beran and Culick, 1992), 4.34 (Brown and Roshko, 1974), 8.16–8.22 (Darmofal *et al.*, 2001), 12.3(a), (b) (Davis, 1957), 12.8 (Elder, 1959), 9.46 (Hasselbrink and Mungal, 2001), 9.5 (Humphrey *et al.*, 1977), 9.6 (Humphrey *et al.*, 1981), 3.26 (Jacobs, 1992), 7.13 (Johnston *et al.*, 1972), 7.14 and 7.15 (Kristoffersen and Andersson, 1993), 6.10 and 6.11 (Kurosaka *et al.*, 1987), 4.33 (Lau, 1981), 7.25 and 7.26 (MacFarlane *et al.*, 1998), 3.44–3.46 (Nitsche and Krasny, 1994), 4.35 (Ricou and Spalding, 1961), 9.44 (Sykes *et al.*, 1986), 7.8 (Tatro and Mollo-Christensen, 1967), and 3.24 and 3.25 (Yang *et al.*, 1994); **Canadian Aero and Space Institute**: Figure 10.24 (Millar, 1971); **Concepts ETI Press**: Figures 4.12 (Johnston, 1986), and 4.2 and 4.4 (Kline and Johnston, 1986); **Dover Publications**: Figures 2.14, 2.26, 10.41, and 10.43 (Liepmann and Roshko, 1957), reprinted with permission from Dover Publications; **Educational Development Center**: Figures 3.34 (Abernathy, 1972), 2.5 (Shapiro, 1972), and 9.8 (Taylor, 1972); **Elsevier**: Figures 6.16 and 6.17 (Betchov and Criminale, 1967), 11.1 (Broadbent, 1976), 12.30 (Chue *et al.*, 1989), 4.14 (Clauser, 1956), 8.10 (Hall, 1966), 6.14 (Krasny, 1986), 6.29 and 6.30 (Marble and Candel, 1977), 4.29 and 4.32 (Roshko, 1993a), and 9.35, 9.36, 9.38, 9.39 (Waitz *et al.*, 1977), © Elsevier, reprinted with permission from Elsevier; **Institute of Mechanical Engineers**: Figures 4.11 (Abu-Ghannam and Shaw, 1980), 12.31, 12.32, 12.34, and

12.35 (Greitzer and Griswold, 1976), and 5.19 (Hall and Orme, 1955); **Janes Information Group Ltd**: Figure 10.1 (Gunston, 1999), reprinted with permission from Jane's Informaton Group – Jane's Aero-Engines; **McGraw-Hill**: Figures 4.6 and 4.13 (Cebeci and Bradshaw, 1977), 4.17, 4.28, 4.30, 4.31, and 8.28 (Schlichting, 1979), 5.8 (Schlichting, 1968), and 4.7, 4.8, 4.15, 4.16, and 6.15 (White, 1991), reprinted by permission of the McGraw-Hill Companies; **MIT Press**: Figures 10.9, 10.11, and 10.12 (Kerrebrock, 1992), and 4.19 (Tennekes and Lumley, 1972), © MIT Press, reprinted with permission from the MIT Press; **Oxford University Press**: Figures 9.20 (Lighthill, 1963), 3.36 (Thwaites, 1960), and 2.40 (Ward-Smith, 1980), reprinted by permission of the Oxford University Press; **Pearson Education**: Figures 10.7 (Hill and Peterson, 1992), 1.1 (Lee and Sears, 1963), and 10.42 (Sabersky *et al.*, 1989), reprinted with permission from Pearson Education; **Princeton University Press**: Figures 10.10, 10.16 and 10.17 (Crocco, 1958); © 1958 reprinted with permission of Princeton University Press; **Research Studies Press**: Figure 8.32 (Owen and Rogers, 1989); **The Royal Aeronautical Society (UK)**: Figures 9.9, 9.10, and 9.11 (Bansod and Bradshaw, 1971), and 6.5, 6.6, 6.7, 6.8, and 6.9 (Preston, 1961); **The Royal Society of London**: Figures 11.20, 11.21, 11.22, 11.24, 11.25, 11.27, 11.28, and 11.29 (Greitzer *et al.*, 1985), 9.19 (Hawthorne, 1951), and 6.37(b) (Patel, 1975), reprinted with permission of the Royal Society of London; **RTO/NATO**: Figures 8.43 and 8.44 (Cary and Darmofal, 2001), originally published by RTO/NATO in Meeting MP-069(I), March 2003; **SAE International**: Figure 5.11 (Denton, 1990), reprinted with permission from SAE SP-846 © 1990 SAE International; **Springer-Verlag**: Figures 6.12 (Eckert, 1987), 2.27 and 7.12 (Johnston, 1978), and 7.11 (Tritton and Davies, 1981), © Springer-Verlag Gmbh and Co. KG, reprinted with permission from Springer-Verlag; **United Technologies Corporation**: Figures 11.23 and 11.33 (Presz and Greitzer, 1988), 11.30 (Simonich and Schlinker, 1983), and 9.31 (Tillman *et al.*, 1992), © United Technologies Corp.; **von Karman Institute**: Figure 12.21 (Cumpsty, 1989, from C. Freeman in VKI Lecture Series 1985–05); **Wiley and Sons**: Figures 7.9 (Bark, 1996), 2.13, 2.15, 10.8, 10.15, and Table 10.1 (Shapiro, 1953), and 1.2 (Sonntag *et al.*, 1998), © reprinted by permission of Wiley and Sons, Inc.; **Individual authors**: Beer, J.M., Figures 8.39 and 8.40 (Beer and Chigier, 1972); Cumpsty, N.A., Figures 5.40, 10.13, and 12.21 (Cumpsty, 1989); Denton, J.D., Figure 10.13 (in Cumpsty, 1989); Drela, M., Figure 4.5 (Drela, 1998); Eckert, E., Figures 1.12(a), and 1.12(b) (Eckert and Drake, 1972), and 6.12 (Eckert, 1987); Fabri, J., Figure 9.13 (Gostelow, 1984); Ferziger, J.H., Figures 4.22–4.24 (Lyrio, Ferziger, and Kline, 1981); Heiser, W.H., Figures 11.5, 11.8, 11.9, and 11.10 (Pratt and Heiser, 1993); Johnston, J.P., Figures 12.43 and 12.44 (Wolf and Johnston, 1966); Lumley, J.L., Figure 4.18 (Lumley, 1967); McCormick, D., Figure 9.32 (McCormick, 1992); Prasad, D., Figures 9.27 and 9.28 (Prasad, 1998); Waitz, I., Tables 11.2 and 11.3 and Figure 11.16 (Underwood, Waitz, and Greitzer, 2000).

Conventions and nomenclature

Conventions

1. Vector quantities are shown in bold (**u**).
2. The task of integrating nomenclature from different fields has been a daunting one; not only is the terminology often not consistent, it is sometimes directly opposed. Our strategy has been, where possible, to keep to nomenclature in widespread use rather than inventing new symbols. This means that some symbols are used for two (or more!) quantities, for example h for the heat transfer coefficient and specific enthalpy, θ for momentum thickness, diffuser half-angle, and the circumferential coordinate, and W for work and for channel and diffuser width.
3. Several conventions have been used for station numbers. These are generally numerical: 0, 1, 2, 3, etc. Situations in which there is reference to inlet and exit conditions are denoted by i and e; these are noted where used. The subscripts i and o are used to denote inner and outer radii, and, again, the specific notation is defined where needed. The subscript E denotes the part of the stream which is outside ("external to") the viscous layer (boundary layer) adjacent to a solid surface.

   Far upstream and far downstream stations are denoted by $-\infty$ and ∞ respectively. In some cases two or more streams exist and these are denoted by 1, 2, etc. In situations in which there are two or more streams at different stations the convention used is that the first subscript denotes the stream and the second the station. As an example u_{1_i} denotes stream 1 at the inlet station.
4. In two dimensions the Cartesian coordinate system is defined such that x is along the mainstream direction and y is normal to it. Generally this implies that x is parallel to a boundary surface and y is normal to the boundary; for example y_E is the distance to just outside the edge of the boundary layer. For three dimensions, x and y maintain these conventions and z is defined as the third axis in a right-handed coordinate system.

   For axisymmetric geometries the x-coordinate direction is used as the axis of symmetry because the overall (bulk) flow motion is aligned with the axis of the machine in many devices.

   For rotating coordinate systems (Chapters 3 and 7) the z-axis is used as axis of rotation so the x-direction maintains the convention of being the main flow direction for a rotating passage.

Nomenclature

Letters

a	(1) Speed of sound
	(2) Vortex core radius
A	Area or surface

A_{port}	Area of ports (inlet and outlet) of a control volume
AR	Diffuser or nozzle area ratio (exit area/inlet area)
B_i	Components of a vector
$\mathbf{B}$	Vector
c_p	Specific heat at constant pressure
c_v	Specific heat at constant volume
C_c	Contraction coefficient (Eq. (2.10.3))
C_f	Skin friction parameter $(\tau_w/(\rho u_E^2/2))$
C_p	Pressure rise coefficient $((p_2 - p_1)/(\rho u_1^2/2))$
C_d	Dissipation coefficient (Eq. (5.4.10))
C_D	Drag coefficient
d	Diameter
$d(\ )$	Differential quantity
d_H	Hydraulic diameter (4A/perimeter)
$D(M)$	Compressible flow function (Eq. (2.5.3))
$\dot{D}$	Rate of mechanical energy dissipation per unit area in the boundary layer (Eq. (4.3.11))
d	Small amount of work or heat
D/Dt	Convective derivative
e	Internal energy per unit mass
e_t	Stagnation energy per unit mass $(e + u^2 / 2)$
$\mathbf{e}_r, \mathbf{e}_\theta, \mathbf{e}_x$	Unit vectors in r-, θ-, x-directions
E	Internal energy
E_t	Total energy of a thermodynamic system
$\mathbf{F}_{ext}, \mathbf{F}_{visc}$	External force, viscous force per unit mass
$\mathcal{F}_D$	Drag force in addition to wall shear stress (Eq. (10.3.4))
$\mathcal{F}_i, \mathcal{F}_x, \mathcal{F}_y$	Component of force
h	(1) Enthalpy per unit mass (2) Heat transfer coefficient (3) Separation parameter $((H-1)/H)$
h_t	Stagnation enthalpy per unit mass $(h + u^2/2)$
H	(1) Boundary layer or wake shape factor (δ^*/θ) (2) Non-dimensional enthalpy, $(c_pT/c_pT_{t_i})$ (3) Height of annular diffuser
$\mathcal{I}$	Fluid impulse
$\mathbf{I}$	Fluid impulse per unit mass
$\mathcal{I}_R, \mathcal{I}_{RS}$	Inertia parameter for rotors (R), rotors plus stators (RS)
J	Jet momentum flux
k	(1) Number of Fourier component (2) Thermal conductivity
K	(1) Acceleration parameter (Section 4.5) (2) Circulation/2π in an axisymmetric flow (ru_θ) (3) Non-dimensional kinetic energy $(u^2/2c_pT_{t_i})$
K	Screen pressure drop coefficient $[(\Delta P/(\rho u^2/2)]_{\text{screen}})$

l	Streamwise coordinate
$\mathbf{l}$	Unit vector in streamwise direction
$d\ell$	line element magnitude
$d\boldsymbol{\ell}$	line element vector
ℓ_{mix}	Mixing length in turbulent boundary layer (Eq. (4.6.12))
L	(1) Characteristic length scale
	(2) Duct length
m	Meridional coordinate
$\dot{m}$	Mass flow rate
M	(1) Mach number (u/a)
	(2) Molecular weight
M_Ω	Rotational Mach number ($\Omega r/a$)
M_E	Free-stream Mach number
M_c	Convective Mach number (Eq. (4.8.18))
n	Coordinate normal to streamline
n_i	Component of normal
$\mathbf{n}$	Outward pointing normal unit vector
N	(1) Diffuser length
	(2) Flow non-uniformity parameter (Eq. (5.6.17))
p	Pressure
p'	Perturbation (or disturbance) pressure
p_B	Back pressure in compressible channel flow
p_t	Stagnation (or "total") pressure
Pr	Prandtl number ($\mu c_p/k$)
q	Heat addition per unit mass
q_i	Component of heat flux vector
q_x, q_y	Heat flux in x-, y-direction
q_w	Wall heat flux
$\mathbf{q}$	Heat flux vector
Q	Heat addition
$\dot{Q}$	Rate of heat addition per unit mass
(r, θ, x)	Cylindrical coordinates
r	Radius
r_c	Radius of curvature
r_m	Mean radius
$\mathbf{r}$	Position vector
Δr	Annulus height ($r_o - r_I$)
$\mathcal{R}$	Universal gas constant
R	Gas constant $= \mathcal{R}/M$
Re	Reynolds number
Re_x, Re_θ, Re_{δ^*}	Reynolds numbers based on x-distance, momentum thickness, displacement thickness
s	Entropy per unit mass
S	Entropy

St	Stanton number (Section 11.1)
t	Time
T	Temperature
T_t	Stagnation (or "total") temperature ($T + u^2/2c_p$)
u	Velocity magnitude
u_i	Velocity component
$\overline{u}$	Mean or background velocity
u_τ	Friction velocity ($\sqrt{\tau_w/\rho}$)
u^+	Non-dimensional velocity (u/u_τ)
u_E	External, or free-stream, velocity
$\overline{u}$	Mean or background velocity
$\mathbf{u}$	vector velocity
(u_x, u_y, u_z),	Velocity components in Cartesian corrdinates
(u_r, u_θ, u_x)	Velocity components in cylindrical corrdinates
U	Reference velocity or characteristic velocity
v	Specific volume (volume per unit mass)
V	(1) Volume
	(2) Axial velocity ratio, external flow to vortex core
w	Work per unit mass
w_{loss}	Lost work per unit mass (Eq. (5.2.10))
w_{shaft}	Shaft work per unit mass
$\mathbf{w}$	Relative velocity
W	(1) Channel, diffuser width; blade, vortex pair spacing
	(2) Work
W_{eff}	Effective width of channel
W_{non-p}	Work over and above flow work done by inlet and exit pressures
W_{shaft}	Shaft work
(x, y, z) $(\mathbf{i}, \mathbf{j}, \mathbf{k})$	Cartesian coordinates and unit vectors
$\mathbf{x}$	Coordinate vector
X_i	Components of body forces
$\mathbf{X}$	Body force per unit mass
y_E	y-value at edge of boundary layer
y^+	Non-dimensional boundary layer coordinate (yu_τ/ν)

Symbols

$\Im$	Impulse function ($pA + \rho u^2 A$)
α	Flow angle measured from reference direction
β	(1) Reduced frequency ($\omega L/U$)
	(2) Shock angle
γ	(1) Specific heat ratio ($\gamma = c_p/c_v$)
	(2) Circulation per unit length

Γ	Circulation
Γ_{rel}	Relative circulation
δ	Boundary layer thickness
δ_{ij}	Kronecker delta
δ^*	Boundary layer or wake displacement thickness
Δ	Difference or change, e.g. Δp, Δh
ε	(1) Strain rate
	(2) Non-dimensional compressor tip clearance
	(3) Fraction of free-stream velocity
η	(1) Screen refraction coefficient (Eq. (12.2.17))
	(2) Amplitude of perturbation in vortex sheet position
θ	(1) Boundary layer or wake momentum thickness
	(2) Circumferential coordinate
	(3) Angle of flow deflection in bend
	(4) Planar diffuser half-angle
λ	Wavelength
μ	Viscosity
ν	Kinematic viscosity
ρ	Density
σ	(1) Normal stress
	(2) Fractional area of one stream in multiple stream flow
Ψ	Compressor or pump pressure rise coefficient
ψ	(1) Stream function
	(2) Force potential
	(3) Perturbation in compressor or pump pressure rise coefficient
τ, τ_{ij}	Shear stress
Φ	(1) Dissipation function (Section 1.10)
	(2) Axial velocity coefficient in compressor or pump
	(3) Non-dimensional impulse function (Eq. (11.4.2))
ϕ	Perturbation in axial velocity coefficient
φ	Velocity potential ($\mathbf{u} = \nabla\varphi$)
ω	(1) Radian frequency ($2\pi f$)
	(2) Vorticity magnitude
ω_n	Normal vorticity component
ω_s	Streamwise vorticity component
$\boldsymbol{\omega}$	Vorticity
$\boldsymbol{\Omega}$	Angular velocity (rotating coordinate system, fluid)
Ω	Magnitude of angular velocity ($\lvert\boldsymbol{\Omega}\rvert$)

Subscripts

av	Average
$body$	Body (as in body force)

B	Back (as in back pressure)
c	(1) Core
	(2) Contraction
CV	From control volume analysis
d	Flow field downstream of component
D	(1) Drag (as in drag force)
	(2) Duct (as in duct area)
E	External to boundary layer, edge of boundary layer
e	Exit station
eff	Effective
far	Denotes value in far field
i	(1) Inlet station
	(2) Inner radius station (as r_i)
inj	Properties of injected flow
$irrev$	Denotes an irreversible process
k	Fourier component number
m	(1) Mean
	(2) Meridional component
max	Maximum value
n	Normal coordinate, direction, or component
o	(1) Outer radius
	(2) Denotes uniform value of vorticity in vortex tube
p	Primary stream in ejector
$port$	Relating to the inlet and outlet ports of a control volume
r	Radial component
ref	Reference condition
rel	Relative frame
rev	Denotes a reversible process
s	(1) Streamwise component
	(2) Denotes process at constant entropy
	(3) Secondary stream in ejector
$shaft$	Due to rotating machinery or deforming control volume
$syst$	For a system
$surf$	For a surface
tan	Tangential to shock
T	Translation
TH	Station at channel or duct throat
$turb$	Denotes value due to turbulence
u	Denotes flow field upstream of component
$visc$	Denotes force from viscous (or turbulent) shear stress
vm	Vector mean
w	Evaluated at wall (bounding solid surface)
x, y, z	Components in x, y, z directions

θ	Component in circumferential direction
0	Reference station
0, 1, 2, etc.	(1) Station numbers
	(2) Numbers denoting different (e.g. initial, final) states
	(3) Component numbers
	(4) Numbers denoting different streams in multiple stream flow
∞	(1) Far downstream
	(2) Far away from wall or axis of rotation
$-\infty$	Far upstream

Superscripts and overbar symbols

$\sim$ (e.g. $\tilde{u}$)	Non-dimensional quantity
$\wedge$ (e.g. $\hat{u}$)	Non-dimensional quantity
$\overline{\ }$ (e.g. $\overline{u}$)	Mean or background flow variable
$(\ )^*$	Sonic condition (or critical swirl condition in Chapter 8)
$(\ )'$	Perturbation quantity
$+$	Pertains to normalized value in *BL*

1 Equations of motion

1.1 Introduction

This is a book about the fluid motions which set the performance of devices such as propulsion systems and their components, fluid machinery, ducts, and channels. The flows addressed can be broadly characterized as follows:

(1) There is often work or heat transfer. Further, this energy addition can vary between streamlines, with the result that there is no "uniform free stream". Stagnation conditions therefore have a spatial (and sometimes a temporal) variation which must be captured in descriptions of the component behavior.

(2) There are often large changes in direction and in velocity. For example, deflections of over 90° are common in fluid machinery, with no one obvious reference direction or velocity. Concepts of lift and drag, which are central to external aerodynamics, are thus much less useful than ideas of loss and flow deflection in describing internal flow component performance. Deflection of the non-uniform flows mentioned in (1) also creates (three-dimensional) motions normal to the mean flow direction which transport mass, momentum, and energy across ducts and channels.

(3) There is often strong swirl, with consequent phenomena that are different than for flow without swirl. For example, static pressure rise can be associated almost entirely with the circumferential (swirl) velocity component and thus essentially independent of whether the flow is forward (radially outward) or separated (radially inward). In addition the upstream influence of a fluid component, and hence the interaction between fluid components in a given system, can be qualitatively different than that in a flow with no swirl.

(4) The motions are often unsteady. Unsteadiness is necessary for work exchange in turbomachines. Waves, oscillations, and self-excited unsteadiness (instability) not only affect system behavior, but can sometimes be a limiting factor on operational regimes.

(5) A rotating reference frame is a natural vantage point from which to examine flow in rotating machinery. Such a reference frame, however, is a non-inertial coordinate system in which effects of Coriolis and centrifugal accelerations have a major role in determining the fluid motions.

(6) Perhaps the most important features of internal flows, however, are the constraints imposed because the flow is bounded within a duct or channel. This influence is felt in all flow regimes, but it is especially marked when compressibility is involved, as in many practical applications. If the effects of wall friction, losses in the duct, or energy addition or extraction are not assessed correctly, serious adverse effects on mass flow capacity and performance can result.

In the succeeding chapters we will see *when* these different effects are important, *why* they are important, and *how* to define and analyze the magnitude of their influence on a given fluid motion.

In this chapter we present a summary of the basic equations and boundary conditions needed to describe the motion of a fluid. The discussion given is self-contained, although it is deliberately brief because there are many excellent sources, with extended discussions of the topics covered; these are referred to where appropriate.

1.2 Properties of a fluid and the continuum assumption

For the applications in this book, we define a fluid as an isotropic substance which continues to deform in any way which leaves the volume unchanged as long as stresses are applied (Batchelor, 1967). In most engineering devices, except those that work at pressures several orders of magnitude below standard atmosphere or are of very small scale, the characteristic length scale of the motion in a gas will be many times the size of the mean free path (the mean distance between collisions for a molecule). This is not a very restrictive condition since the mean free path in a gas at standard temperature and pressure is approximately 10^{-7} m. In such situations we can ignore the detailed molecular structure and discuss the properties "at a point" as if the fluid were a continuous substance or *continuum*. In this context, we will use the term *fluid particle*, which can be defined as the smallest element of material having sufficient molecules to allow the continuum interpretation. For a liquid the corresponding condition is that the particle be much larger than the molecular size, which is of order 10^{-9} m for water (Lighthill, 1986a), again this is most typically the case.[1] In summary, at pressures, temperatures, and device dimensions commonly encountered, variations due to fluctuations on the molecular scale can be ignored and the fluid treated as a continuum.

1.3 Dynamic and thermodynamic principles

The principles that define the motion of a fluid may be expressed in a number of ways, but can be stated as follows: conservation of mass, conservation of momentum (Newton's second law of motion), and the first and second laws of thermodynamics. These must also be supplemented by the equation of state of the fluid, a relation between the thermodynamic properties, generally derived from observation. These conservation and thermodynamic laws are statements about *systems*, or *control masses*, which are defined here as collections of material of fixed identity. For example, conservation of mass is a statement that the mass of a fluid particle remains constant no matter how it is deformed. Newton's second law, force equals rate of change of momentum, also applies to a particle or to a given collection of particles.

In general, however, interest is not in fixed mass systems but rather in what happens in a fixed volume or at a particular position in space. For this reason, we wish to cast the equations for a system into a form which applies to a *control volume*, V, of arbitrary shape, bounded by a control surface, A,

[1] As an example, in a cube of air which is 10^{-3} mm (1 μm) on a side there are roughly 3×10^7 molecules at standard conditions. For water in a cube of these dimensions there are roughly 10^{10} molecules.

i.e. to transform the system (control mass) laws into control volume laws.[2] We will carry out these transformations in several steps. The concept of differentiation following a fluid particle, or sum of particles, is first introduced. This is then employed to express the conservation laws explicitly in a form tied to volumes and surfaces moving with the fluid. We then derive the relation between changes that occur in a volume moving with the fluid and changes in a volume fixed in an arbitrary coordinate system. This leads to expressions for the equations of motion in integral (control volume) as well as differential form.

1.3.1 The rate of change of quantities following a fluid particle

To describe what happens at a fixed volume or point in space we must inquire how the time rate of change for a particle can be described in a fixed coordinate system. For definiteness we take Cartesian coordinates x, y, z, and fluid velocity components u_x, u_y, and u_z. Suppose that c is some property of the fluid and we visualize a field of values of c continuously distributed throughout space. For small arbitrary and independent increments dx, dy, dz, and time, dt, the change in property c is

$$dc = \frac{\partial c}{\partial x}dx + \frac{\partial c}{\partial y}dy + \frac{\partial c}{\partial z}dz + \frac{\partial c}{\partial t}dt. \tag{1.3.1}$$

For a given particle, the increments dx, dy, and dz are related to the local instantaneous velocity components and the time increment, dt, by:

$$dx = u_x dt, \quad dy = u_y dt, \quad dz = u_z dt, \tag{1.3.2}$$

where u_x, u_y, and u_z are velocity components in the three spatial directions. Dividing each term by dt, the rate of change of c following a fluid particle can be written as

$$\text{rate of change of } c \text{ following a fluid particle} = \frac{Dc}{Dt} = u_x\frac{\partial c}{\partial x} + u_y\frac{\partial c}{\partial y} + u_z\frac{\partial c}{\partial z} + \frac{\partial c}{\partial t}. \tag{1.3.3}$$

In (1.3.3), the notation $D(\ )/Dt$ has been used to indicate a derivative defined following the fluid particle. This notation is conventional, and the quantity $D(\ )/Dt$, which occurs throughout the description of fluid motion, is known variously as the substantial derivative, the material derivative, or the convective derivative. Noting that in Cartesian coordinates the first three terms of the derivative are formally equivalent to $\mathbf{u} \cdot \nabla c$, the substantial derivative can be written more compactly as

$$\frac{Dc}{Dt} = \frac{\partial c}{\partial t} + (\mathbf{u} \cdot \nabla)c = \frac{\partial c}{\partial t} + u_i\frac{\partial c}{\partial x_i}. \tag{1.3.4}$$

In (1.3.4), and throughout the book, we use the convention that a repeated subscript implies summation over the appropriate indices. In (1.3.4),

$$u_i\frac{\partial c}{\partial x_i} = u_1\frac{\partial c}{\partial x_1} + u_2\frac{\partial c}{\partial x_2} + u_3\frac{\partial c}{\partial x_3}.$$

In this notation the derivative of the velocity following a fluid particle, which is the acceleration, is (for the ith component): $Du_i/Dt = \partial u_i/\partial t + u_j(\partial u_i/\partial x_j)$. In vector notation the acceleration is $D\mathbf{u}/Dt = \partial\mathbf{u}/\partial t + (\mathbf{u} \cdot \nabla)\mathbf{u}$.

[2] The terms *system* (or *control mass*) and *control volume* are used here in describing the two different viewpoints; these concepts are also referred to as *closed system* and *open system* respectively.

1.3.2 Mass and momentum conservation for a fluid system

We can use the derivative following a fluid particle to obtain expressions for the conservation laws, starting with the simplest, conservation of mass. If dm is the mass of a fluid particle, conservation of mass is obtained by taking c to be dm; i.e.

$$\frac{D}{Dt}(dm) = 0. \tag{1.3.5}$$

To obtain an expression valid for an assemblage of particles, i.e. a fluid system, we sum over the different particles in the system. In the continuum limit this can be represented by an integral over the masses:

$$\frac{D}{Dt}\int dm = 0. \tag{1.3.6}$$

In interpreting (1.3.6), it is important to keep in mind that the integral is taken over a fixed mass, which implies a volume fixed to fluid particles and moving with them.

Newton's second law can also be written for an assemblage of fluid particles as

$$\sum \mathcal{F}_{ext} = \frac{D}{Dt}\int \mathbf{u}\, dm. \tag{1.3.7}$$

In (1.3.7) $\mathcal{F}_{ext}$ represents the *external* forces acting on the particles and the summation includes all the forces that act on this mass. The forces can be body forces, which act throughout the mass, or can be surface forces exerted at the boundary of the system. Coriolis, gravity, and centrifugal forces are examples of the first of these; pressure and shear forces, which are exerted by the fluid or by bodies that bound the system, are examples of the second.

1.3.3 Thermodynamic states and state change processes for a fluid system

To describe the thermodynamics of fluid systems, we need to introduce the idea of a system state and define two classes of state change processes. The thermodynamic state of a system is defined by specifying the values of a small set of measured properties, such as pressure and temperature, which are sufficient to determine all other properties. In flow situations it is useful to express properties such as volume, V, or internal energy, E, which depend on the mass of the system, as a quantity per unit mass. The properties on this unit mass basis are referred to as specific properties and denoted here by lower case letters (v, e, for specific volume and specific internal energy respectively).

The state of a system in which properties have definite (unchanged) values as long as external conditions are unchanged is called an equilibrium state. Properties describe states only when the system is in equilibrium. For thermodynamic equilibrium of a system there needs to be: (i) mechanical equilibrium (no unbalanced forces), (ii) chemical equilibrium (no tendency to undergo a chemical reaction or a transfer of matter from one part of the system to another), and (iii) thermal equilibrium (all parts of the system at the same temperature, which is the same as that of the surroundings).

Fluid devices typically have quantities such as pressure which vary throughout, so that there is no single value that characterizes all the material within the device. If so the conditions for the three types of equilibrium to hold on a *global* basis (e.g. the absence of finite pressure differences or unbalanced forces) are not satisfied when we view the complete region of interest as a whole. To deal with this situation we can (conceptually) divide the flow field into a large number of small (differential) mass elements, over which the pressure, temperature, etc. have negligible variation, and consider each of these elements a different system with its own *local* properties.[3] In defining the behavior of the different systems the working assumption is that the local instantaneous relation between the thermodynamic properties of each element is the same as for a uniform system in equilibrium.[4]

Processes that change the state of a system can be classed as reversible or irreversible. Fluid process that are *irreversible* (also referred to as natural processes) include motions with friction, unrestrained expansion, heat transfer across a finite temperature difference, spontaneous chemical reaction, and mixing of matter of different composition or state. These processes have the common characteristic that they all take place spontaneously in nature. A further aspect is that "a cycle of changes A→B→A on a particular process, where A→B is a natural process, cannot be completed without leaving a change in some other part of the universe" (Denbigh, 1981).

A central role in thermodynamic analysis is played by *reversible* processes, defined as a process "whose direction can be reversed without leaving more than a vanishingly small change in any other system" (Denbigh, 1981). This means that the departures from thermodynamic equilibrium at any state in the process are also vanishingly small. In the case of forces, for instance, the internal forces exerted by the system must differ only infinitesimally from the external forces acting on the system. Similarly, for reversible heat transfer between surroundings and system, there can only be infinitesimal temperature differences between the two. A reversible process must also be *quasi-static*, i.e. slow enough that the time for the fluid to come to equilibrium when subjected to a change in conditions is much shorter than any time scale for the process, again so that the system essentially passes through a series of equilibrium states during the process. As with the continuum approximation this is not restrictive for the situations of interest: for example, equilibration times for air at room conditions are on the order of 10^{-9} seconds (Thompson, 1984).[5] All real fluid processes are in some measure irreversible although, as we will see, many processes can be analyzed to a high degree of accuracy assuming they are reversible.

Recognition of the irreversibility in a real process is vital in fluids engineering. A perspective on its effect is that "Irreversibility, or departure from the ideal condition of reversibility, reflects an increase in the amount of disorganized energy at the expense of organized energy" (Reynolds and Perkins, 1977). Organized energy is illustrated by a raised weight. Disorganized energy is represented by the random motions of the molecules in a gas (the internal energy of the gas). The importance of the distinction is that *all* the organized energy can, in principle, produce work, whereas a consequence of the second law of thermodynamics (Section 1.3.4) is that only a fraction of the disorganized energy is available to produce work. The transition from organized to disorganized energy brought about by irreversibility thus corresponds to a loss in opportunity to produce work (and hence power or propulsion) from a fluid device. In this connection Section 1.3.4 introduces the thermodynamic property

[3] A consequence is that the state definition requires specification of several functions rather than several variables. In addition, although we refer to the temperature and pressure *at a point*, the division into differential elements is made with the caveat expressed in Section 1.2.

[4] From a macroscopic point of view this assumption must be assessed by experience, which shows that its appropriateness is extremely well borne out for the flows of interest. The approximation made, referred to as the principle of local state, is discussed further by Kestin (1979) and Thompson (1984).

[5] For more complex molecules or temperatures much higher than room temperature, the equilibration time can be several orders of magnitude larger (times of 10^{-5} seconds are given by Thompson (1984) for gases at 3000 K). If so, the relaxation of the gas to the equilibrium state may need to be included. We do not examine these regimes.

entropy, which provides a quantitative measure of irreversibility; Section 1.10 discusses entropy generation in a flowing fluid; and Sections 5.1 and 5.2 examine the relation between irreversibility and the loss in capability for work production.

1.3.4 First and second laws of thermodynamics for a fluid system

The first law of thermodynamics can be expressed for a system as

$$\Delta E_t = Q - W \tag{1.3.8}$$

where ΔE_t is the change in the total energy of the system, Q is the heat received, and W is the work done by the system on the environment. In differential form (1.3.8) is

$$dE_t = đQ - đW. \tag{1.3.9}$$

The notations $d(\,)$ and $đ(\,)$ denote conceptual and physical differences between the terms in (1.3.9). The total energy, E_t, is a property. Changes in E_t (dE_t or its integral ΔE_t) represent state changes which do not depend on the path taken to achieve the change. Work and heat are not state variables and are only defined in terms of interactions with the system. For a specified change of state (specified initial and final states) ΔE_t is given, but the individual amounts of heat and work transfer to the system can vary, depending on the path by which the change is accomplished.[6] To emphasize the difference between the two types of quantities, we use $d(\,)$ for small changes in properties and $đ(\,)$ for the small amounts of heat and work transfer that bring these changes about.

For the systems we are concerned with, the total energy can be written as an integral, over the system mass, of the sum of the internal energy, e, per unit mass, and the kinetic energy, $u^2/2$, per unit mass. For flow situations the items of interest are generally the rates at which quantities change so it is useful to cast the first law as a rate equation:

$$\frac{DE_t}{Dt} = \frac{D}{Dt}\int \left[e + \frac{u^2}{2}\right] dm = \frac{đQ}{dt} - \frac{đW}{dt}. \tag{1.3.10}$$

In (1.3.10) $đQ/dt$ is the rate of heat transfer to the system and $đW/dt$ is the rate of work done by the system.

The second law of thermodynamics can be expressed in two parts.[7] The first part is a definition of the thermodynamic property entropy of the system, denoted as S. If $đQ_{rev}$ is the heat transferred to the system during a reversible incremental state transformation, and T is the temperature of the system,

$$dS = \frac{đQ_{rev}}{T}. \tag{1.3.11}$$

For a finite change from state 1 to state 2,

$$S_2 - S_1 = \int_1^2 \frac{đQ_{rev}}{T}. \tag{1.3.12}$$

[6] Discussion of this point is given in many texts. See, for example, Denbigh (1981), Kestin (1979), Reynolds and Perkins (1977) and Sonntag, Borgnakke, and Van Wylen (1998).

[7] See, for example, Abbott and Van Ness (1989), Denbigh (1981), and Kestin (1979) for additional discussion.

The second part of the second law states that for *any* process the change in entropy for the system is

$$dS \geq \frac{đQ}{T}. \tag{1.3.13}$$

The equality occurs only for a reversible process. A consequence of (1.3.13) for a system to which there is no heat transfer is

$$dS \geq 0 \quad \text{(for a system with } đQ = 0\text{)}. \tag{1.3.14}$$

Equation (1.3.13) can also be written as a rate equation in terms of the heat transfer rate and temperature of the fluid particles which comprise the system. With s the *specific entropy* or entropy per unit mass,

$$\frac{DS}{Dt} = \frac{D}{Dt}\int s dm \geq \sum \frac{1}{T}\frac{đQ}{dt}. \tag{1.3.15}$$

In (1.3.15), the summation is taken over all locations at which heat enters or leaves the system. Equation (1.3.15) will be developed in terms of fluid motions and temperature fields later in this chapter.

The fluids considered in this book are those described as simple compressible substances. The thermodynamic state of such fluids is specified when two independent intensive thermodynamic properties (pressure and temperature, for example) are given and the only reversible work mode is that associated with volume change (Reynolds and Perkins, 1977).

For incremental reversible processes in a simple compressible substance, the heat addition to the fluid is

$$đQ = TdS. \tag{1.3.16a}$$

If kinetic energy changes can be neglected (the change is in thermal energy only) the work done is

$$đW = pdV. \tag{1.3.16b}$$

Although the association of work with pdV and heat addition with TdS is only true for a reversible process, the sum of these, as expressed by the first law, is a relation between thermodynamic properties. For negligible kinetic energy changes, this relation is

$$de = Tds - pdv, \tag{1.3.17}$$

where s and v are the entropy and volume per unit mass. Equation (1.3.17), known as the Gibbs equation, can be regarded as a combined form of the first and second laws. It is a relation between thermodynamic properties and is not restricted to reversible processes.

A thermodynamic property which will be seen to occur naturally in flow processes is the *enthalpy*, denoted by h and defined as

$$h = e + p/\rho. \tag{1.3.18}$$

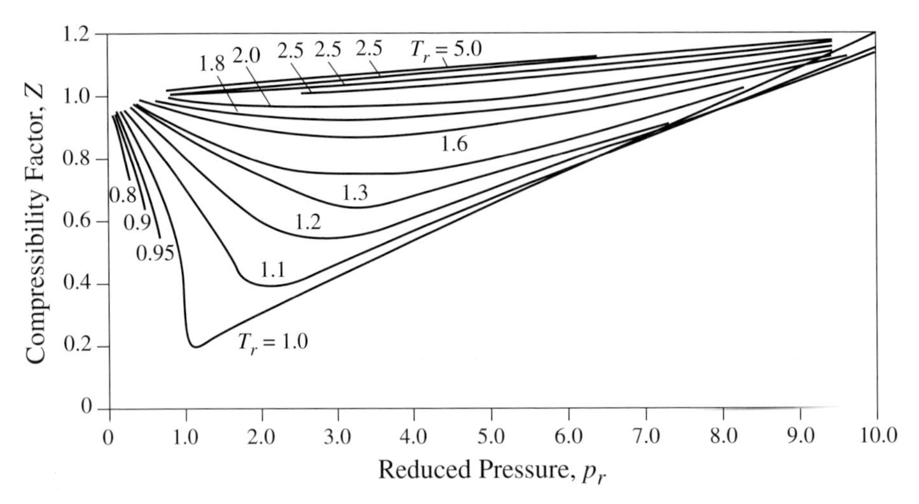

Figure 1.1: Compressibility factor $Z = p/\rho RT$, at low pressures; reduced temperature $T_r = T/T_c$, values of critical temperature. T_c, given in Table 1.1 (Lee and Sears, 1963).

A form of the Gibbs equation useful for flow processes can be written in terms of enthalpy changes, using the definition $v = 1/\rho$, as

$$dh = T ds + \frac{1}{\rho} dp. \tag{1.3.19}$$

As with (1.3.17), (1.3.19) is not restricted to reversible processes.

1.4 Behavior of the working fluid

1.4.1 Equations of state

The equations relating the intensive thermodynamic variables of a substance are called the equations of state. The flows examined in this book are very well represented using one of two equations of state. The first is for a *perfect* gas,

$$p = \rho RT, \tag{1.4.1}$$

where $R = \mathcal{R}/M$, with $\mathcal{R}$ the universal gas constant ($\mathcal{R} = 8.3145$ kJ/(kmol K))[8] and M the molecular weight of the gas. Equation (1.4.1) holds for air and other gases over a wide range of temperatures and pressures.

The ratio $p/\rho RT$ is called the compressibility factor, and its variation from unity gives a good measure of the applicability of (1.4.1). This quantity is plotted in Figure 1.1. The curves are averaged from experimental data on a number of monotonic and diatomic gases, plus hydrocarbons (Lee and Sears, 1963). The compressibility factor is given as a function of the reduced pressure,

[8] A kmol is a mass equal to the molecular weight of the gas in kilograms.

Table 1.1 *Critical pressures and temperatures for different gases (Lee and Sears, 1963)*

Substance	p_c (MPa)	T_c(K)
He	0.23	5.3
H_2	1.30	33.6
Air	3.77	132.7
O_2	5.04	154.5
CO_2	7.39	304.3
H_2O	22.1	647.4

defined as pressure/critical pressure[9] (p/p_c) for different reduced temperatures, T_r, defined as temperature/critical temperature (T/T_c). For reference, several values of p_c and T_c are listed in Table 1.1. For reduced temperatures between 1.6 and 5.0 and reduced pressures of less than approximately 3, the perfect gas approximation is valid to within 5%. For example, air at a pressure of 30 atmospheres and a temperature of 1650 K (conditions representative of the exit of the combustor in a gas turbine) corresponds to $p/p_c = 0.8$ and $T/T_c = 12.5$. Even at these conditions, the compressibility factor would be approximately 1.03.

The second equation of state that will be used is for an incompressible fluid, i.e. a fluid in which the volume of a given fluid mass (density) is constant. This is suitable for liquids. It is also a very good approximation for gases at low speeds. In Chapter 2 this statement is made more precise but, to give a numerical appreciation for the approximation, in air at standard temperatures the assumption of constant density holds within 3% for speeds of 100 m/s or less. Incompressible denotes that the volume of a fluid particle remains constant; it does not necessarily mean uniform density throughout the fluid.

1.4.2 Specific heats

Two important thermodynamic properties are the *specific heat at constant volume* and the *specific heat at constant pressure*. These quantities, denoted by c_v and c_p respectively for the values per unit mass, have a basic definition as derivatives of the internal energy and enthalpy. For a simple compressible substance, the energy difference between two states separated by small temperature and specific volume differences, dT and dv, can be expressed as

$$de = \left(\frac{\partial e}{\partial T}\right)_v dT + \left(\frac{\partial e}{\partial v}\right)_T dv. \tag{1.4.2}$$

The derivative $(\partial e/\partial T)_v$ is c_v. It is a function of state, and hence a thermodynamic property.

The name specific heat is somewhat of a misnomer because only in special circumstances is the derivative $(\partial e/\partial T)_v$ related to energy transfer as heat. For a constant volume reversible process, no work is done. Any energy increase is thus due only to energy transfer as heat, and c_v represents the

[9] The critical pressure and temperature correspond to p and T at the critical point, the highest pressure and temperature at which distinct liquid and gas phases of the fluid can coexist.

energy increase per unit of temperature and per unit of mass. In general, however, it is more useful to think of c_v in terms of the definition as a partial derivative, which is a thermodynamic property, rather than a quantity related to energy transfer as heat.

Just as c_v is related to a derivative of internal energy, c_p is related to a derivative of enthalpy. Writing the enthalpy as a function of T and p,

$$dh = \left(\frac{\partial h}{\partial T}\right)_p dT + \left(\frac{\partial h}{\partial p}\right)_T dp. \tag{1.4.3}$$

The derivative $(\partial h/\partial T)_p$ is called the specific heat at constant pressure and denoted by c_p. For reversible constant pressure heat addition, the amount of heat input per unit mass is given by $đq = c_p dT$.

Values of c_v and c_p are needed often enough that they have been determined for a large number of simple compressible substances. Numerical values of c_p for several gases are shown in Figure 1.2 (Sonntag, Borgnakke and Van Wylen, 1998).

For a perfect gas, the internal energy and enthalpy are defined to depend only upon temperature. Thus

$$de = c_v(T)dT, \tag{1.4.4a}$$

$$dh = c_p(T)dT, \tag{1.4.4b}$$

where c_v and c_p can depend on T. Further, $dh = de + d(pv) = c_v\, dT + RdT$. Hence, for a perfect gas (sometimes also referred to as an ideal gas (Reynolds and Perkins, 1977)),

$$c_v = c_p - R. \tag{1.4.5}$$

For other substances, e and h depend on pressure as well as temperature and, in this respect, the perfect gas is a special model.

Depending on the application, the variation in specific heat with temperature may be able to be neglected so that c_p and c_v can be treated as constant at an appropriate mean value. If so

$$e_2 - e_1 = c_v(T_2 - T_1), \tag{1.4.6a}$$

$$h_2 - h_1 = c_p(T_2 - T_1). \tag{1.4.6b}$$

Equations (1.4.6) hold only for a perfect gas with constant specific heats as do the relations that have been derived between changes in energy (or enthalpy) and temperature in (1.4.4).

For an *incompressible fluid*, the volume of a given fluid particle is constant and the internal energy is a function of a single thermodynamic variable, the temperature. The specific heat at constant volume is thus also a function of temperature but the change in internal energy of an incompressible fluid undergoing a temperature variation is

$$e_2 - e_1 = \int_{T_1}^{T_2} c_v(T)dT. \tag{1.4.7}$$

From the definition of enthalpy, $h = e + p/\rho$, the enthalpy change of an incompressible fluid for a specified pressure and temperature change is

$$h_2 - h_1 = e_2 - e_1 + \frac{1}{\rho}(p_2 - p_1). \tag{1.4.8}$$

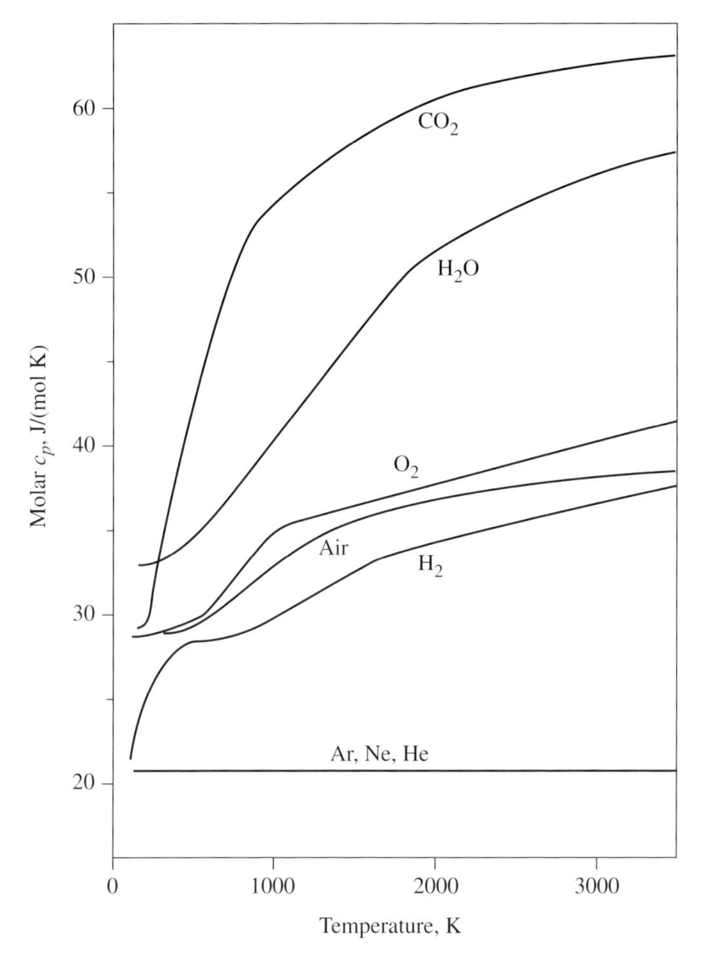

Figure 1.2: Constant-pressure specific heats for gases at zero pressure (Sonntag, Borgnakke, and Van Wylen, 1998).

Enthalpy changes for an incompressible fluid contain both thermodynamic (e) and mechanical (p) properties. From (1.4.7) and (1.4.8) and the definition of specific heat at constant pressure, we also have the relation

$$c_p = c_v = c \tag{1.4.9}$$

for an incompressible fluid.

1.5 Relation between changes in material and fixed volumes: Reynolds's Transport Theorem

The conservation statements in Section 1.3 are written in terms of material volumes, in other words volumes that move with the fluid particles. We wish to transform these statements to expressions

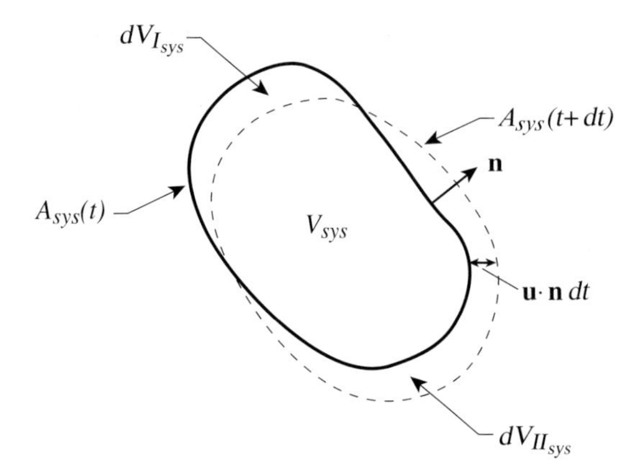

Figure 1.3: Relation between system volumes and surfaces and fixed control volumes and surfaces.

written in terms of volumes and surfaces which are fixed in space. This will provide an extremely useful way to view problems in fluid machinery. To start this transformation, consider the quantity c, which is a property per unit mass. For a finite mass:

$$C = \int c dm$$
$$= \int_{V_{sys}(t)} c\rho \, dV. \tag{1.5.1}$$

In (1.5.1) $V_{sys}(t)$, the system volume over which the integration is carried out, moves with the fluid.

Let us examine the volume V_{sys}, which is bounded by the surface $A_{sys}(t)$, at two times, t and $t + dt$, where dt is a small time increment. The volume is shown in Figure 1.3. The surface is a material surface (meaning that it is always made up of the same fluid particles) which moves and deforms with the fluid. At time, t, the material surface $A_{sys}(t)$ is taken to coincide with a fixed surface, A, which encloses the fixed volume, V, so the system is wholly inside the control surface. At the time, $t + dt$, the system has deformed to a volume $V_{sys}(t + dt)$, enclosed by the surface, $A_{sys}(t + dt)$, as indicated in Figure 1.3. With reference to the figure, the volumes at the two times are related by

$$V_{sys}(t + dt) = V_{sys}(t) + dV_{I_{sys}} + dV_{II_{sys}},$$

where $dV_{I_{sys}}$ and $dV_{II_{sys}}$ are defined in Figure 1.3. The change of the property C in time dt is thus

$$dt \frac{DC}{Dt} = \int_{V_{sys}(t+dt)} \rho c dV + \int_{\substack{dV_{II_s} \\ at\ t+dt}} \rho c dV + \int_{\substack{dV_{I_{sys}} \\ at\ t+dt}} \rho c dV - \int_{V_{sys}(t)} \rho c dV. \tag{1.5.2}$$

Referring to Figure 1.3, the sum of the volumes $dV_{I_{sys}}$ and $dV_{II_{sys}}$ is the volume swept out by the material surface as it deforms during the time, dt. Letting $dt \to 0$ and working to first order in dt, the volume swept out is $dt \int_{A_{sys}} u_i n_i$, where u_i and n_i represent the ith components of the velocity vector and the outward pointing normal respectively.[10]

The sum of $dV_{I_{sys}}$ and $dV_{II_{sys}}$ is a surface layer of local "thickness" (the word is in quotes since the value of the thickness can be negative) $u_i n_i dt$. Hence

$$dt\frac{DC}{Dt} = \int\limits_{V_{sys}(t+dt)} \rho c dV - \int\limits_{V_{sys}(t)} \rho c dV + \left[\int\limits_{A} \rho c u_i n_i dA\right] dt. \quad (1.5.3)$$

To first order in dt the first two terms on the right-hand side of (1.5.3) combine to give

$$\left[\int\limits_{V_{sys}(t)} \frac{\partial}{\partial t}(\rho c) dV\right] dt.$$

The control volume V and the material volume $V_{sys}(t)$ are initially coincident (at time t) so

$$\frac{DC}{Dt} = \underset{\substack{\text{(fixed}\\\text{volume)}}}{\int\limits_{V} \frac{\partial}{\partial t}(\rho) c dV} + \underset{\substack{\text{(fixed}\\\text{surface)}}}{\int\limits_{A} (\rho c) u_i n_i dA}, \quad (1.5.4)$$

or, from the definition of C,

$$\frac{D}{Dt}\int c dm = \frac{D}{Dt}\int\limits_{V_{sys}(t)} \rho c dV = \int\limits_{V} \frac{\partial}{\partial t}(\rho c) dV + \int\limits_{A} \rho c (u_i n_i) dA. \quad (1.5.5)$$

Equation (1.5.4) (or (1.5.5)) is a form of Reynolds's Transport Theorem (Aris, 1962). It relates the changes that occur in a system (mass of fixed identity) and in a fixed control volume bounded by a fixed control surface. The control volume formulation brings an additional term of the form $\int_A \rho c u_i n_i \, dA$, interpreted as a mass flux of property c in and/or out of the control volume, V, through its bounding surface, A.

1.6 Conservation laws for a fixed region (control volume)

Using the results of Section 1.5, the integral equations that describe the different conservation laws can be written for a fixed control volume by giving c various identities. If c is set equal to 1, we obtain the equation for conservation of mass:

$$\int\limits_{V} \frac{\partial \rho}{\partial t} dV + \int\limits_{A} \rho u_i n_i dA = 0. \quad (1.6.1)$$

[10] As mentioned previously, in the expression $u_i n_i$, and in what follows, the use of a repeated subscript implies that the index is summed over all values. The quantity $u_i n_i$ thus represents $u_1 n_1 + u_2 n_2 + u_3 n_3 = \mathbf{u} \cdot \mathbf{n}$, the scalar product of $\mathbf{u}$ and $\mathbf{n}$.

The common name for this equation is the continuity equation, not the conservation of mass, although we have used the latter principle to derive it. The issue here is physical continuity; the fluid stays as a continuum with no holes or gaps.

If c is taken as the specific volume, v, the statement

$$\frac{D}{Dt}\int c\,dm = 0 \tag{1.6.2}$$

becomes a statement that the specific volume of a fluid particle, in other words the density of the fluid particle, remains constant. This is the condition for an *incompressible fluid*. Use of (1.5.5) shows that the control volume form of the continuity equation for an incompressible fluid is

$$\int_A (u_i n_i)\,dA = 0. \tag{1.6.3}$$

If c is taken as the ith velocity component, u_i, the equation for conservation of momentum in the ith-direction becomes

$$\int_V \frac{\partial}{\partial t}(\rho u_i)\,dV + \int_A \rho u_i (u_j n_j) dA = \sum \mathcal{F}_{ext_i}. \tag{1.6.4}$$

The term $\sum \mathcal{F}_{ext_i}$ represents the ith component of the sum of all external forces acting on the fluid within the volume. Evaluation of this term generally involves surface or volume integrals.

In axisymmetric geometries such as turbomachines where there is a well-defined axis of rotation, it is often useful to consider changes in angular momentum. For a system, the rate of change of angular momentum is given by

$$\frac{D}{Dt}\int_A (\mathbf{u}\times\mathbf{r})_i\,dm = \sum (\mathcal{F}_{ext}\times\mathbf{r})_i, \tag{1.6.5}$$

where $\mathbf{r}$ is a position vector and where the notation $(\,)_i$ denotes the ith component of the cross-product. Setting c equal to $(\mathbf{u}\times\mathbf{r})_i$, an expression for the rate of change of angular momentum within a fixed control volume is obtained as

$$\int_V \frac{\partial}{\partial t}(\rho\mathbf{u}\times\mathbf{r})_i\,dV + \int_A (\rho\mathbf{u}\times\mathbf{r})_i\,u_j n_j dA = \sum (\mathcal{F}_{ext}\times\mathbf{r})_i. \tag{1.6.6}$$

Again, actual evaluation of the sum of the moments due to external forces generally involves integration over the volume V or the surface A.

To obtain the control volume form for the first law of thermodynamics, c is set equal to the energy per unit mass, $e + u^2/2$:

$$\int_V \frac{\partial}{\partial t}\left[\rho\left(e+\frac{u^2}{2}\right)\right]dV + \int_A \rho\left(e+\frac{u^2}{2}\right)u_i n_i dA = \frac{đQ}{dt} - \frac{đW}{dt}. \tag{1.6.7}$$

In (1.6.7), $đQ/dt$ and $đW/dt$ are the rate of heat transfer *to*, and the work, done *by*, the fluid in the volume. It is useful to separate work into a part due to the action of pressure forces at the inflow and outflow boundaries of the volume, and a part representing other work exchange. We discuss the reasons for this in detail later, but one basis on which to justify the separation is that the latter is the appropriate measure of energy added to a flowing stream by fluid machines and by external body forces.

The work done by pressure forces in time dt on a small element of surface dA is given by the product of the pressure force, pdA, which acts normal to the surface, times the displacement of the surface in the normal direction, $u_i n_i dt$. Integrating over the entire control surface yields the rate of work done by pressure forces on the surroundings external to the control volume:

$$\text{rate of work done by pressure forces} = \int_A p\, u_i n_i\, dA. \tag{1.6.8}$$

If $đW_{non\text{-}p}/dt$ is defined as the rate of work done by the fluid in the control volume, over and above that associated with pressure work at the inflow and outflow boundaries, (1.6.7) becomes

$$\int_V \frac{\partial}{\partial t}\left[\rho\left(e+\frac{u^2}{2}\right)\right] dV + \int_A \rho\left(e+\frac{p}{\rho}+\frac{u^2}{2}\right) u_i n_i dA = \frac{đQ}{dt} - \frac{đW_{non\text{-}p}}{dt}. \tag{1.6.9}$$

The quantity $e + (p/\rho)$ appears often in flow processes and is therefore defined as a separate specific property called enthalpy and denoted as h. Using this definition (1.6.9) is written more compactly as

$$\int_V \frac{\partial}{\partial t}\left[\rho\left(e+\frac{u^2}{2}\right)\right] dV + \int_A \rho\left(h+\frac{u^2}{2}\right) u_i n_i dA = \frac{đQ}{dt} - \frac{đW_{non\text{-}p}}{dt}. \tag{1.6.10}$$

1.7 Description of stress within a fluid

Equations (1.6.4), (1.6.6), and (1.6.10) are not yet in forms which can be directly applied in general because the force, work, and heat transfer terms are not linked to the other flow variables. In this section, expressions for these quantities are developed, starting with a description of the forces that can be exerted on the fluid within a control volume (see, e.g., Batchelor (1967), Landau and Lifschitz (1987)).

As mentioned in Section 1.3.2, forces on a fluid particle are of two types, body forces, which are forces per unit mass, and surface forces, which come about as the result of surface stresses exerted on a fluid particle either by other fluid particles or by adjacent solid surfaces. It is necessary to examine the state of stress in a fluid to describe these surface forces. To do this, we need to represent the force on a surface which is at an arbitrary angle to the coordinate axes, or more precisely, a surface defined by a normal at some arbitrary angle. As indicated in Figure 1.4, we consider the forces on a small, tetrahedron-shaped, fluid element with dimension dx_1, dx_2, dx_3 whose slant face has normal vector $\mathbf{n}$. The inertia and body forces acting on this tetrahedron are proportional to the volume, in

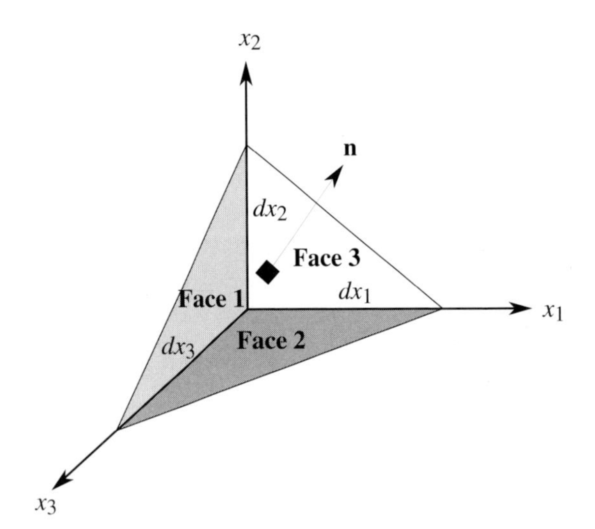

Figure 1.4: Tetrahedron-shaped fluid volume for examination of fluid stresses.

other words to dx^3, where dx is the characteristic dimension of the tetrahedron. The surface forces are proportional to the surface area and hence to dx^2. For equilibrium, as $dx \to 0$ the surface force on the slant face must balance the surface forces on the three sides which are perpendicular to the coordinate axes. This condition gives the relation needed to describe the force on the slanted surface.

The area of the slant face is denoted by dA. The areas of the other faces are dA_1, dA_2, dA_3, where the subscripts refer to the axis to which the face is perpendicular. On the face perpendicular to the x_1-axis, the tensile force per unit area in the x_1-direction is denoted by Π_{11}. The shear force per unit area (or shear stress) on this surface acting in the x_2-direction is Π_{12}, and that in the x_3-direction is Π_{13}, with similar notation for the other faces. Calling the force per unit area on the slant surface $\mathbf{F}$, with components F_i, a force balance gives

$$F_1 = \Pi_{11}\frac{dA_1}{dA} + \Pi_{21}\frac{dA_2}{dA} + \Pi_{31}\frac{dA_3}{dA} \tag{1.7.1}$$

with similar equations for the x_2- and x_3-directions. The ratios of the face areas, dA_1/dA, dA_2/dA, dA_3/dA, however, are just the three components of the direction cosines of the normal to the slant side. The expression for the surface forces per unit area (i.e. the surface stresses) on the element dA is thus:

$$F_1 = \Pi_{11}n_1 + \Pi_{21}n_2 + \Pi_{31}n_3, \tag{1.7.2a}$$

$$F_2 = \Pi_{12}n_1 + \Pi_{22}n_2 + \Pi_{32}n_3, \tag{1.7.2b}$$

$$F_3 = \Pi_{13}n_1 + \Pi_{23}n_2 + \Pi_{33}n_3. \tag{1.7.2c}$$

In general, to specify the surface stress nine numbers, Π_{ij}, would be needed because there are different components for different orientations of the plane. The nine quantities, however, are not all

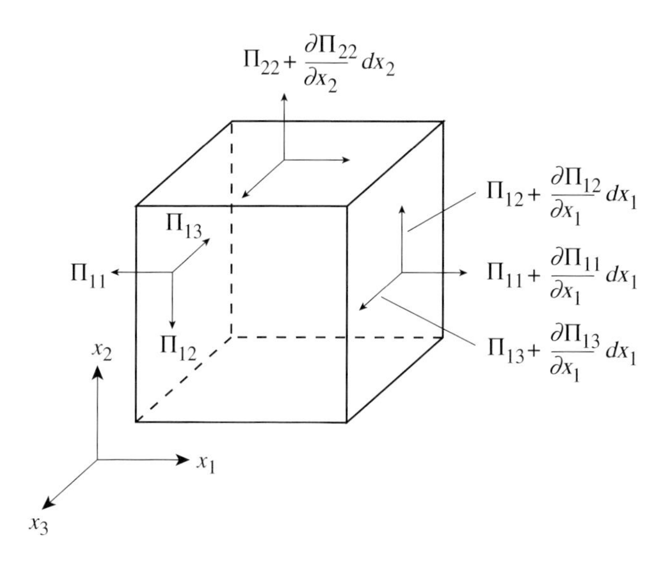

Figure 1.5: Stresses on fluid cube.

independent, as can be shown from examining the moment equilibrium of the small cube of Figure 1.5 about any axis, say, the x_3-axis. Moments due to shear stresses have contributions proportional to the third power of the dimension. (The shear force is proportional to the second power, and the moment arm to the first power.) Moments due to the body forces have contributions proportional to the fourth power of the dimension. (The body force is proportional to the third power, and the moment arm is proportional to the first power.) For equilibrium, the contributions proportional to dx^3 must therefore sum to zero which implies

$$\Pi_{12} = \Pi_{21}, \quad \Pi_{23} = \Pi_{32}, \quad \Pi_{13} = \Pi_{31}. \tag{1.7.3}$$

Only six stresses are thus independent. These form the components of a symmetric second order tensor,[11] the stress tensor, which is

$$\text{stress tensor} = \begin{bmatrix} \Pi_{11} & \Pi_{21} & \Pi_{31} \\ \Pi_{21} & \Pi_{22} & \Pi_{32} \\ \Pi_{31} & \Pi_{32} & \Pi_{33} \end{bmatrix}. \tag{1.7.4}$$

To better understand the relation of stress and force, and as a precursor of what is to come in the derivation of the differential forms of the equation of motion, it is helpful to examine the relationship between surface stresses and net forces on a fluid particle. To do this, consider the small cube of fluid of Figure 1.5 with sides parallel to the x_1-, x_2-, and x_3-axes. For clarity, not all the stresses are drawn, but there are three stress components acting on each of the six faces.

[11] The quantities Π_{ij} are "tensor components" because of the way the values of these quantities transform as we change reference from one coordinate system to another. Equations (1.7.2a)–(1.7.2c) state that when a coordinate change is made, the three sums $\Pi_{ij}n_i$ must transform as components of the vector $\mathbf{F}$. A set of nine quantities Π_{ij} which transform in this way is *by definition* a tensor of second rank. A tensor of first rank is a vector, whose three components transform so that the magnitude and direction remain invariant; a tensor of zeroth rank is a scalar (Aris, 1962; Goldstein, 1980).

The stresses vary throughout the fluid, and it is this variation that is responsible for the net surface forces on a fluid particle. This can be seen by summing up the stresses that act in one of the coordinate directions, for example the x_1-direction, working to lowest order in the cube dimension. The x_1-direction force is

$$\begin{aligned}
&\left[-\Pi_{11}+\left(\Pi_{11}+\frac{\partial \Pi_{11}}{\partial x_1}dx_1\right)\right]dx_2dx_3+\left[-\Pi_{21}+\left(\Pi_{21}+\frac{\partial \Pi_{21}}{\partial x_2}dx_2\right)\right]dx_1dx_3\\
&\quad+\left[-\Pi_{31}+\left(\Pi_{31}+\frac{\partial \Pi_{31}}{\partial x_3}dx_3\right)\right]dx_1dx_2\\
&\quad=\left(\frac{\partial \Pi_{11}}{\partial x_1}+\frac{\partial \Pi_{21}}{\partial x_2}+\frac{\partial \Pi_{31}}{\partial x_3}\right)dx_1dx_2dx_3\\
&\quad=\frac{\partial \Pi_{j1}}{\partial x_j}dx_1dx_2dx_3.
\end{aligned} \tag{1.7.5}$$

The first term comes from the stress on the two faces perpendicular to the x_1-direction, the second from the faces perpendicular to the x_2-direction, and the third from the faces perpendicular to the x_3-direction. The net force resulting from the stresses is proportional to the volume of the elementary cube; this must be the case if the surface forces are to balance the body and inertia forces.

Once surface forces are expressed in terms of stress tensor components, we are in a position to write the equations of motion in terms of surface stresses, which can then be related to various derivatives of the velocity. Before doing this, however, we make one change in notation, since it is customary (and helpful) to make a division into stresses due to fluid pressure (normal forces) and stresses due to viscous or shear forces, the stress tensor is written as

$$\Pi_{ij}=-p\delta_{ij}+\tau_{ij}. \tag{1.7.6}$$

In (1.7.6) τ_{ij} is the symmetric viscous stress tensor, and δ_{ij} is the Kronecker delta

$$\delta_{ij}=\begin{cases}0 & i\neq j\\ 1 & i=j\end{cases}.$$

The quantity p_M is defined as

$$p_M=-\tfrac{1}{3}\left(\Pi_{11}+\Pi_{22}+\Pi_{33}\right)=-\tfrac{1}{3}\Pi_{ii}, \tag{1.7.7}$$

which is the measurable mechanical pressure. For a compressible fluid at rest, the mechanical pressure, p_M, is equivalent to the thermodynamic pressure, $p=p(\rho,T)$. On the assumption that there is local thermodynamic equilibrium even when the fluid is in motion, plus the general conditions on fluid viscosity described in Section 1.13, this equivalence may be applied for a moving fluid. If the fluid is incompressible, the thermodynamic pressure is not defined and pressure must be taken as one of the fundamental dynamical variables. Based on (1.7.7), we define an inviscid fluid as one for which τ_{ij} is identically zero and only pressure forces are present.

1.8 Integral forms of the equations of motion

The expressions developed for surface forces and stresses can be applied to provide explicit forms of the control volume equations describing momentum and energy transfer to a flowing fluid (Liepmann and Roshko, 1957). Denoting the components of the body forces per unit mass by X_i, the momentum equation is

$$\int_V \frac{\partial}{\partial t}(\rho u_i)dV + \int_A \rho u_i(u_j n_j)dA = \int_V \rho X_i dV - \int_A p\delta_{ij}n_j dA + \int_A \tau_{ij}n_j dA. \tag{1.8.1}$$

The equation for angular moment (moment of momentum) is

$$\int_V \frac{\partial}{\partial t}(\rho e_{ijk}u_j r_k)dV + \int_A \rho e_{ijk}u_j r_k u_l n_l dA$$
$$= \int_V \rho e_{ijk}X_j r_k dV - \int_A e_{ijk}p\delta_{lj}n_l r_k dA + \int_A e_{ijk}\tau_{jl}n_l r_k dA. \tag{1.8.2}$$

In (1.8.2) the quantity e_{ijk} has been introduced to represent the vector product: e_{ijk} takes the value 1 if the subscripts are in cyclic order (i.e. $e_{123} = 1$), -1 if the subscripts are in anti-cyclic order ($e_{213} = -1$), and zero if any subscripts of e are repeated.

For the energy equation, the different effects that contribute to heat and work transfer need to be identified. Heat addition within the volume can take place due to internal heat sources with a rate of heat addition $\dot{Q}$ per unit mass. Heat can also be transferred via conduction, across the bounding surface. For an elementary area, dA, the net heat flux across the control surface is $q_i n_i dA$ where q_i is the ith component of the heat flux vector $\mathbf{q}$. The rate of work done within the volume by body forces is $\rho X_i u_i$ per unit volume. The rate done by the surface forces acting on the control surface, over and above the pressure work, is $\tau_{ij}n_j u_i$ per unit of surface area. Combining all these terms, the integral form of the energy equation becomes

$$\int_V \frac{\partial}{\partial t}\left[\rho\left(e + \frac{u^2}{2}\right)\right]dV + \int_A \rho\left(h + \frac{u^2}{2}\right)u_i n_i dA$$
$$= \int_V \rho\dot{Q}dV - \int_A q_i n_i dA + \int_V \rho X_i u_i dV + \int_A \tau_{ij}n_j u_i dA. \tag{1.8.3}$$

1.8.1 Force, torque, and energy exchange in fluid devices

An important application of the control volume equations arises in evaluating the performance of a device from the conditions of the fluid that enters and leaves, for example calculating the work put into a flowing stream by turbomachine blading and the force on a nozzle. To perform this type of analysis it is useful to choose a control surface that is coincident over some of its extent with the bounding surface(s) of the device. For the turbomachine this might be, depending on application, the hub and the casing of the annulus or the surface of the blading. For the nozzle the control surface would coincide with the nozzle wall. Use of such control surfaces aids in facilitating the analysis since there is typically no mass flux through these surfaces.

1.8.1.1 Force on a fluid in a control volume

The force exerted on the fluid is given by the integral of the surface forces over the device surface. In what follows we denote by A_{port} those parts of the control surface which do not coincide with the device surfaces; these are the ports for flow entering or leaving the control volume. If $\mathcal{F}_i$ are the components of the force exerted *by* the device *on* the fluid, from (1.8.1) the momentum equation is

$$\int_V \frac{\partial}{\partial t}(\rho u_i)dV + \int_A \rho u_i(u_j n_j)dA - \int_V \rho X_i dV + \int_{A_{port}} p\delta_{ij} n_j dA - \int_{A_{port}} \tau_{ij} n_j dA = \mathcal{F}_i. \tag{1.8.4}$$

Circumstances under which (1.8.4) is applied are often those of steady flow with negligible contributions from the shear forces at the inlet and exit stations. A common example is the inlet and outlet stations of a nozzle, with the exit and outlet control surfaces perpendicular to the flow. In this situation the components of the force exerted on the fluid are given by

$$\int_A \rho u_i u_j n_j dA - \int_V \rho X_i dV + \int_{A_{port}} p\delta_{ij} n_j dA = \mathcal{F}_i. \tag{1.8.5}$$

In (1.8.5) the integral of the momentum flux is taken over the whole surface A. If there is no flow through the part of the surface $A - A_{port}$ which coincides with the device surface, and no body forces, we can write (1.8.5) in terms of an integral over only the parts of the control surface at which fluid enters and exits (the inlet and exit stations),

$$\int_{A_{port}} (\rho u_i u_j + p\delta_{ij}) n_j dA = \mathcal{F}_i. \tag{1.8.6}$$

For unidirectional flow and uniform velocity and pressure at inlet and exit stations (or, as discussed in Chapter 5, if an appropriate average at these stations is defined) the magnitude of the force on the fluid between any two stations 1 and 2 with inflow and outflow areas A_1 and A_2 is given by[12]

$$[(\rho u^2 + p)A]_2 - [(\rho u^2 + p)A]_1 = \mathcal{F}. \tag{1.8.7}$$

1.8.1.2 Torque on a fluid in a control volume

Analyses similar to those for momentum can be carried out for the moment of momentum. We list here only the result for steady axisymmetric flow, negligible contributions of the shear stresses on

[12] It is hoped that the station notation subscripts will not be mixed with those used to indicate components in the velocity vector and stress tensor.

the surfaces A_{port}, and no body forces. (There is no contribution from the pressure because of the axisymmetry.)

$$\int_{A_{port}} \rho r u_\theta u_i n_i dA = \text{torque exerted on fluid.} \tag{1.8.8}$$

Equation (1.8.8) states that the torque exerted *on* the fluid *by* the device, about the axis of symmetry is the difference between the inlet and exit values of the mass-weighted integral of the angular momentum per unit mass, ru_θ.

1.8.1.3 Work and heat exchange with a fluid in a control volume

The total work exchange within the control volume consists of work done by the body forces and work done by surface forces. The latter, which is due to moving surfaces and encompasses the work associated with the presence of rotating turbomachinery blading, is commonly referred to as shaft work, denoted by W_{shaft}. We divide the rate of non-pressure work within the volume, $đW_{non\text{-}p}/dt$, into three parts to facilitate subsequent discussion of the role of fluid machinery shaft work:

$$\frac{đW_{non\text{-}p}}{dt} = \frac{đW_{shaft}}{dt} - \int_V \rho u_i X_i dV - \int_{A_{port}} \tau_{ij} u_i n_j dA. \tag{1.8.9}$$

Using the definition in (1.8.9), (1.6.10) becomes

$$\int_V \frac{\partial}{\partial t}\left[\rho\left(e + \frac{u^2}{2}\right)\right] dV + \int_A \rho\left(h + \frac{u^2}{2}\right) u_i n_i dA$$
$$- \int_V \rho X_i u_i dV - \int_{A_{port}} \tau_{ij} n_j u_i dA = \frac{đQ}{dt} - \frac{đW_{shaft}}{dt}. \tag{1.8.10}$$

Comparing (1.8.3) with (1.8.10), we see that the term $đQ/dt$ in (1.8.9) represents both heat flux across the control surface and heat generation within the volume.

1.8.1.4 The steady flow energy equation and the role of stagnation enthalpy

For steady flow with no body forces, no flow through the surface $A - A_{port}$, and negligible shear stress work on the surface A_{port}, (1.8.10) reduces to the "steady-flow energy equation" form of the first law for a control volume

$$\int_{A_{port}} \rho u_i n_i \left[h + \frac{u^2}{2}\right] dA = \frac{đQ}{dt} - \frac{đW_{shaft}}{dt}. \tag{1.8.11}$$

The integration is over the surface A_{port}, representing the locations of fluid entry and exit from the device so the fluid quantities evaluated are those at inlet and exit only.

The quantity $h + u^2/2$ in (1.8.10) and (1.8.11) occurs often in fluid flow problems. Consider the steady flow in a streamtube, defined as a tube of small cross-sectional area whose boundary is composed of streamlines so there is no flow across the streamtube boundary. With no body forces, if the net rate of work and heat transfer is zero across the boundary, (1.8.11) states that the quantity $h + u^2/2$ is invariant along the streamtube. We thus define a reference enthalpy corresponding to the stagnation state ($u = 0$) as the *stagnation enthalpy* (sometimes referred to as total enthalpy) denoted by h_t. Referring back to (1.8.11) and noting that $\rho u_i n_i dA$ is the mass flow rate $d\dot{m}$ through the element of surface area, dA, we obtain

$$\int_{A_{port}} h_t d\dot{m} = \frac{đQ}{dt} - \frac{đW_{shaft}}{dt}. \tag{1.8.12}$$

Steady flow through a control volume with heat and work transfer is a situation of such importance for fluid power and propulsion systems that it is worth obtaining the form of the first law for this case, (1.8.11), in an alternative (and simpler) manner. We thus examine the steady flow through the device of fixed volume in Figure 1.6, with a single stream at inlet and at outlet. Shaft work can be exchanged with the flow, for example by a turbomachine as depicted notionally in the figure, and heat added or extracted. The velocity and thermodynamic variables at inlet and exit are taken to be steady and to be uniform across the inlet and exit ports. The flow inside the control volume can be locally unsteady at a given point, but the overall quantities (defined as the integral over all the mass inside the control volume) do not change with time. We develop the appropriate form of the continuity equation first and then use this in the statement of the steady flow energy equation.

We examine the evolution of a system which initially consists of the fluid within the dashed lines. At time t a small mass, dm_{I}, which is part of the system, is outside the control volume boundaries in region I. The rest of the system is within the control volume. A short time dt later, the system has moved such that the small mass dm_{I} is inside the control volume and the small mass dm_{II}, which has different properties than dm_{I}, has emerged from the control volume into region II.

Denoting the mass between the stations $1'$ and 2 by m_{III} (see Figure 1.6), the system mass at times t and $t + dt$ can be written as

$$[m(t)]_{sys} = dm_{\mathrm{I}} + m(t)_{\mathrm{III}}, \tag{1.8.13a}$$

$$[m(t+dt)]_{sys} = dm_{\mathrm{II}} + m(t+dt)_{\mathrm{III}}. \tag{1.8.13b}$$

No time argument is indicated for dm_{I} and dm_{II} because these quantities are not changing with time. The mass of the system, m_{sys}, is constant. The mass m_{III} (the mass between stations $1'$ and 2) is also constant in time. From (1.8.13), $dm_{\mathrm{I}} = dm_{\mathrm{II}}$.

The masses dm_{I} and dm_{II} can be expressed in terms of stream properties at the inlet and exit stations as

$$dm_{\mathrm{I}} = \rho_1 A_1 u_1 dt; \qquad dm_{\mathrm{II}} = \rho_2 A_2 u_2 dt. \tag{1.8.14}$$

The quantity $\rho u A$ is the mass flow. Continuity thus implies that inlet and exit mass flows are the same:

$$\rho_1 A_1 u_1 = \dot{m}_1 = \rho_2 A_2 u_2 = \dot{m}_2 = \dot{m}. \tag{1.8.15}$$

The first law, (1.3.8), states that the change in total energy of a system, E_t (E_t is the the thermal and the kinetic energy summed over all the mass in the system) is equal to the heat received by the

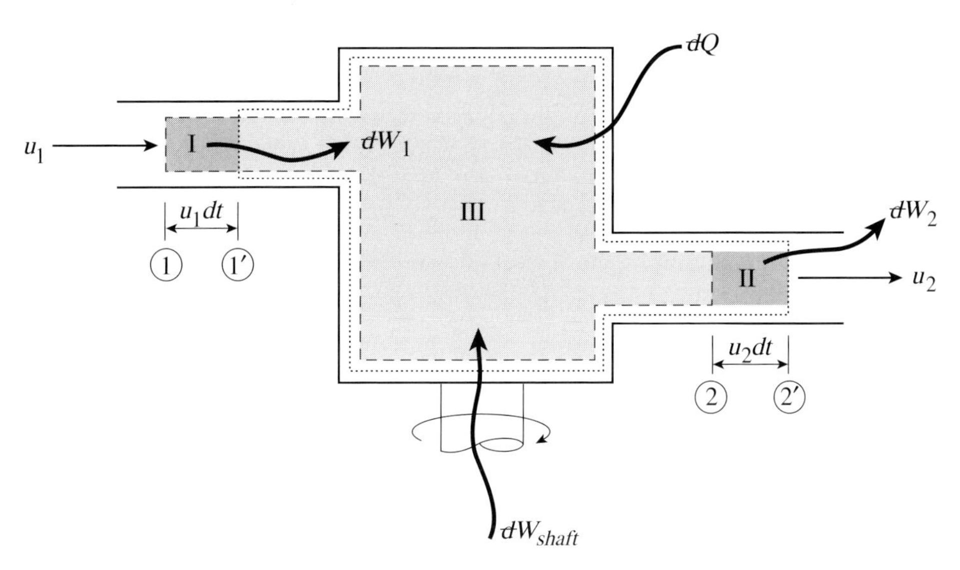

Figure 1.6: Steady flow through a fluid device (fixed control volume) with shaft work and heat transfer. Region I is between stations 1 and 1′, region II is between stations 2 and 2′, and region III between stations 1′ and 2.

system minus the work done by the system. For small changes the first law can be written as (with $\text{đ}Q$ and $\text{đ}W$ the transfers of heat and work)

$$[dE_t]_{sys} = \text{đ}Q - \text{đ}W. \tag{1.8.16}$$

For the fluid device in Figure 1.6 two types of work exist. One is the shaft work, denoted by $\text{đ}W_{shaft}$. The second is the work done by the fluid within the system on the external environment, in other words on the fluid outside of the system. This is indicated by the quantities $\text{đ}W_1$ and $\text{đ}W_2$ in the figure. During the time interval dt the net work done on the fluid external to the system is given by $\text{đ}W_2 - \text{đ}W_1$. At each station the force is pA and the distance moved is udt, so this quantity is

$$\text{net work on the fluid external to the system} = (p_2A_2)u_2dt - (p_1A_1)u_1dt. \tag{1.8.17}$$

The total energy change of the system during dt is

$$\begin{aligned} d\,[E_t]_{sys} &= [E_t\,(t+dt)]_{sys} - [E_t\,(t)]_{sys} \\ &= E_{t_{\text{III}}}\,(t+dt) - E_{t_{\text{III}}}\,(t) - \left(e_1 + \frac{u_1^2}{2}\right) m_{\text{I}} + \left(e_2 + \frac{u_2^2}{2}\right) m_{\text{II}} \\ &= E_{t_{\text{III}}}\,(t+dt) - E_{t_{\text{III}}}\,(t) - \left(e_1 + \frac{u_1^2}{2}\right) \rho_1 A_1 u_1 dt \\ &\quad + \left(e_2 + \frac{u_2^2}{2}\right) \rho_2 A_2 u_2 dt. \end{aligned} \tag{1.8.18}$$

Combining (1.8.16), (1.8.17), and (1.8.18), and using the fact that $E_{t_{III}}$ does not change with time, (1.8.18) becomes

$$\left(e_2 + \frac{u_2^2}{2} + \frac{p_2}{\rho_2}\right)\rho_2 A_2 u_2 - \left(e_1 + \frac{u_1^2}{2} + \frac{p_1}{\rho_1}\right)\rho_1 A_1 u_1 = \frac{đQ}{dt} - \frac{đW_{shaft}}{dt}. \tag{1.8.19}$$

The terms $đQ/dt$ and $đW_{shaft}/dt$ represent the rates of heat transfer to, and shaft work done by, the stream between control stations 1 and 2. Making use of the mass flow rate defined in (1.8.15) and the definition of stagnation enthalpy ($h_t = e + p/\rho + u^2/2 = h + u^2/2$), (1.8.19) can be written compactly as a relation between change in stagnation enthalpy, mass flow and rates of heat and work exchange:

$$\dot{m}\left[h_{t_2} - h_{t_1}\right] = \frac{đQ}{dt} - \frac{đW_{shaft}}{dt}. \tag{1.8.20}$$

Equation (1.8.20) can also be expressed in terms of heat transfer and shaft work per unit mass, q and w_{shaft}:

$$(h_{t_2} - h_{t_1}) = q - w_{shaft}. \tag{1.8.21}$$

Equations (1.8.20) and (1.8.21) show the key role of stagnation enthalpy as a measure of energy interactions in aerothermal devices.

1.9 Differential forms of the equations of motion

To develop the differential forms of the equations of motion, we begin with the integral forms and make use of the Divergence Theorem,

$$\int_V \frac{\partial B_i}{\partial x_i} dV = \int_A B_i n_i dA, \tag{1.9.1}$$

where B_i are the components of any vector $\mathbf{B}$ and the repeated subscript denotes summation over the indices. The Divergence Theorem is used to transform surface integrals into volume integrals so that all the terms in the various equations have the same domain of integration, a necessary step in obtaining the differential forms.

1.9.1 Conservation of mass

To illustrate the procedure to be followed, the Divergence Theorem is applied to the surface integral in the equation for mass conservation, (1.6.1), which becomes

$$\int_V \left[\frac{\partial \rho}{\partial t} + \frac{\partial}{\partial x_i}(\rho u_i)\right] dV = 0. \tag{1.9.2}$$

The volume V is arbitrary. For (1.9.2) to hold, therefore, the integrand must be zero everywhere, so

$$\frac{\partial \rho}{\partial t} + \frac{\partial}{\partial x_i}(\rho u_i) = 0 \quad \left(\frac{\partial \rho}{\partial t} + \boldsymbol{\nabla}\cdot(\rho \mathbf{u}) = 0, \text{ in vector notation}\right). \tag{1.9.3}$$

Equation (1.9.3) is the differential form of the mass conservation, or continuity, equation. It can also be expressed in terms of the substantial derivative of the density as

$$\frac{1}{\rho}\frac{D\rho}{Dt} + \frac{\partial u_i}{\partial x_i} = 0 \quad \left(\frac{1}{\rho}\frac{D\rho}{Dt} + \nabla \cdot \mathbf{u} = 0, \quad \text{in vector notation}\right). \tag{1.9.4}$$

The continuity equation for an incompressible fluid can be written as an explicit statement that the density of a fluid particle remains constant:

$$\frac{D\rho}{Dt} = 0. \tag{1.9.5}$$

Equation (1.9.5) implies that for an incompressible flow

$$\frac{\partial u_i}{\partial x_i} = 0 \quad (\text{or } \nabla \cdot \mathbf{u} = 0). \tag{1.9.6}$$

As mentioned in Section 1.6, this is a condition on the rate of change of fluid volume, as can be seen from the Divergence Theorem:

$$\int_V \frac{\partial u_i}{\partial x_i} dV = \int_A (u_i n_i)\, dA = 0. \tag{1.9.7}$$

The term $\int_A (u_i n_i) dA$ is the volume flux out of a closed surface (see (1.6.3)), and must be zero for an incompressible flow.

1.9.2 Conservation of momentum

The Divergence Theorem can be applied to each component of the momentum equation, (1.8.1), to obtain the differential statement of conservation of momentum. For example, transformation of the x_i component of the momentum flux term gives

$$\int_A \rho u_i u_j n_j dA = \int_V \frac{\partial}{\partial x_j}(\rho u_i u_j) dV. \tag{1.9.8}$$

Application of the Divergence Theorem to (1.8.1) gives, with some rearrangement,

$$\frac{\partial}{\partial t}(\rho u_i) + \frac{\partial}{\partial x_j}(\rho u_i u_j) = -\frac{\partial p}{\partial x_i} + \rho X_i + \frac{\partial \tau_{ij}}{\partial x_j}. \tag{1.9.9}$$

Equation (1.9.9) is often referred to as the "conservation form" of the momentum equation. Expanding the derivatives in the first two terms and using the continuity equation yields the more commonly encountered form

$$\frac{\partial u_i}{\partial t} + u_j \frac{\partial u_i}{\partial x_j} = -\frac{1}{\rho}\frac{\partial p}{\partial x_i} + X_i + \frac{1}{\rho}\frac{\partial \tau_{ij}}{\partial x_j}. \tag{1.9.10}$$

The shear forces now appear as derivatives of the surface stresses in the last term on the right-hand side of (1.9.10).

1.9.3 Conservation of energy

Using the same procedure as previously on (1.8.3), the energy equation in differential form is found as

$$\frac{\partial}{\partial t}\left[\rho\left(e+\frac{u^2}{2}\right)\right]+\frac{\partial}{\partial x_i}\left[\rho\left(h+\frac{u^2}{2}\right)u_i\right]$$
$$=\rho\dot{Q}-\frac{\partial q_i}{\partial x_i}+\rho X_i u_i+\frac{\partial}{\partial x_j}(\tau_{ij}u_i). \tag{1.9.11}$$

By expanding the derivatives and using the equation of continuity, (1.9.11) can be written in terms of substantial derivatives of stagnation energy or stagnation enthalpy

$$\frac{D}{Dt}\left(e+\frac{u^2}{2}\right)=\dot{Q}+u_i X_i-\frac{1}{\rho}\frac{\partial}{\partial x_i}(pu_i)-\frac{1}{\rho}\frac{\partial q_i}{\partial x_i}+\frac{1}{\rho}\frac{\partial}{\partial x_j}(\tau_{ij}u_i) \tag{1.9.12}$$

or

$$\frac{D}{Dt}\left(h+\frac{u^2}{2}\right)=\dot{Q}+u_i X_i+\frac{1}{\rho}\frac{\partial p}{\partial t}-\frac{1}{\rho}\frac{\partial q_i}{\partial x_i}+\frac{1}{\rho}\frac{\partial}{\partial x_j}(\tau_{ij}u_i). \tag{1.9.13}$$

Because of the convenient and natural role of the stagnation enthalpy in flow processes (1.9.13) is a form in which the energy equation is frequently used in internal flows. For inviscid flow, with no shear stresses and no heat transfer, (1.9.13) becomes

$$\frac{D}{Dt}\left(h+\frac{u^2}{2}\right)=\dot{Q}+u_i X_i+\frac{1}{\rho}\frac{\partial p}{\partial t}. \tag{1.9.14}$$

In such a flow the stagnation enthalpy of a fluid particle can be changed only by heat sources within the flow, the action of body forces, or unsteadiness, as reflected in the term $(1/\rho)\partial/\partial t$. We will see considerable application of this last term in Chapter 6.

1.10 Splitting the energy equation: entropy changes in a fluid

The equation given as (1.9.12) describes changes in thermal and mechanical energy together. It is instructive to look at each of these separately (Liepmann and Roshko, 1957), because this allows a direct connection with the second law of thermodynamics and the entropy production in the fluid. To begin, we multiply each ith component of the momentum equation by the corresponding ith velocity component and sum the resulting equations to obtain

$$\frac{D}{Dt}\left(\frac{u^2}{2}\right)=u_i X_i-\frac{1}{\rho}u_i\frac{\partial p}{\partial x_i}+\frac{1}{\rho}u_i\frac{\partial \tau_{ij}}{\partial x_j}. \tag{1.10.1}$$

Equation (1.10.1), which describes the changes in kinetic energy per unit mass for a fluid particle, can be subtracted from (1.9.12) or (1.9.13) to obtain an equation for the rate of change of the thermodynamic quantities' thermal energy or enthalpy:

$$\frac{De}{Dt}=\dot{Q}-\frac{p}{\rho}\frac{\partial u_i}{\partial x_i}-\frac{1}{\rho}\frac{\partial q_i}{\partial x_i}+\frac{1}{\rho}\tau_{ij}\frac{\partial u_i}{\partial x_j}, \tag{1.10.2}$$

or

$$\frac{Dh}{Dt} = \dot{Q} + \frac{1}{\rho}\frac{Dp}{Dt} - \frac{1}{\rho}\frac{\partial q_i}{\partial x_i} + \frac{1}{\rho}\tau_{ij}\frac{\partial u_i}{\partial x_j}. \tag{1.10.3}$$

The enthalpy form of the Gibbs equation, (1.3.19), $dh = Tds + (1/\rho)\partial p$, holds for all small changes. It can thus be written to express the entropy changes experienced by a fluid particle:

$$T\frac{Ds}{Dt} = \frac{Dh}{Dt} - \frac{1}{\rho}\frac{Dp}{Dt}. \tag{1.10.4}$$

Combining (1.10.4) with (1.10.3) gives an expression for the rate of change of entropy per unit mass:

$$T\frac{Ds}{Dt} = \dot{Q} - \frac{1}{\rho}\frac{\partial q_i}{\partial x_i} + \frac{1}{\rho}\tau_{ij}\frac{\partial u_i}{\partial x_j}. \tag{1.10.5}$$

The entropy of a fluid particle can be changed by heat addition, either from heat sources or heat flux (q_i), or by shear forces. Pressure forces and body forces have no effect. The product $\tau_{ij}(\partial u_i/\partial x_j)$ represents the heat generated per unit volume and time by the dissipation of mechanical energy; it is conventionally denoted as Φ and referred to as the dissipation function.

1.10.1 Heat transfer and entropy generation sources

Further insight into the content of (1.10.5) can be obtained if we use the relation between conduction heat flux and temperature distribution. Experiments show that the conduction heat flux is given by the same expression as that for heat transfer in solids, namely

$$q_i = -k\frac{\partial T}{\partial x_i}, \quad i = 1, 2, 3, \tag{1.10.6}$$

where k is the thermal conductivity. The thermal conductivity is often approximated as a constant but it can have a variation with fluid properties, most notably temperature.

We suppose that internal heat sources, $\dot{Q}$, can be neglected. Employing (1.10.6), dividing (1.10.5) by T, and integrating throughout the interior volume, $V_{sys}(t)$, of a closed surface, $A_{sys}(t)$, moving with the fluid, we obtain

$$\int_{V_{sys}} \rho\frac{Ds}{Dt}dV = \int_{V_{sys}} \frac{\Phi}{T}dV + \int_{V_{sys}} \frac{1}{T}\frac{\partial}{\partial x_i}\left(k\frac{\partial T}{\partial x_i}\right)dV.$$

Integration by parts yields

$$\int_{V_{sys}} \rho\frac{Ds}{Dt}dV = \int_{V_{sys}} \frac{\Phi}{T}dV + \int_{V_{sys}} \frac{k}{T^2}\left(\frac{\partial T}{\partial x_i}\right)^2 dV + \int_{A_{sys}} \frac{k}{T}\frac{\partial T}{\partial x_i}n_i dA. \tag{1.10.7}$$

The first two integrals on the right-hand side of (1.10.7) are positive definite. The third term represents heat transfer in and out of the volume and can be positive or negative. The entropy of a fluid particle can thus decrease only if there is heat conducted out of the particle. If the boundary is insulated so there is no heat transfer across it, the entropy can only increase.

The second and third terms, on the right-hand side of (1.10.7) connect entropy changes to temperature gradients. The third term, associated with heat transfer across the surface that bounds the fluid volume, represents the entropy change due to heat inflow or outflow. It can be either positive

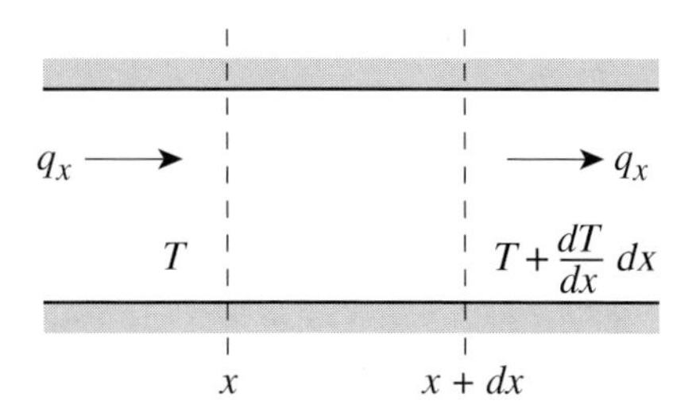

Figure 1.7: Entropy production in a solid bar; heat is flowing from left to right at constant rate q_x per unit area.

or negative. The second term, which is quadratic and always positive, is different in nature. It represents entropy production due to internal irreversibility. Its role can be understood by analogy with one-dimensional steady flow of heat in a solid bar of unit area, as shown in Figure 1.7. There is no heat transfer from the top or bottom, so the heat flux $q_x = -k(dT/dx)$ is uniform in the bar and has only an x-component. The small element, dx, gains entropy at a rate q_x/T, at its left-hand side. The entropy that flows out of the element at the right-hand side is

$$\frac{q_x}{\left(T + \dfrac{dT}{dx}dx\right)} \cong \frac{q_x}{T}\left(1 - \frac{1}{T}\frac{dT}{dx}dx\right).$$

Since dT/dx must be negative for heat to flow in the direction indicated, the entropy outflow is greater than the entropy inflow to the element. The net rate of entropy production in the element per unit volume is

$$\text{entropy production per unit volume} = \frac{k}{T^2}\left(\frac{dT}{dx}\right)^2. \tag{1.10.8}$$

The expression for entropy production in (1.10.8) has the same form as the quadratic temperature gradient term in (1.10.7). Both represent entropy production due to an irreversible process, heat flow across a finite temperature difference.

Equation (1.10.7) can now be interpreted as a statement that entropy changes are due to two causes, irreversibilities and heat transfer. For a unit mass, therefore,

$$ds = ds_{irrev} + \frac{đq}{T}. \tag{1.10.9}$$

The first term on the right-hand side of (1.10.9) represents the effect of irreversibility. As discussed in more depth in Chapter 5, understanding of the entropy change caused by irreversible processes plays a key role in addressing improvements in the efficiency of fluid devices.

1.11 Initial and boundary conditions

The solution to the general time-dependent equations for a particular flow situation requires the specification of an initial condition and boundary conditions. The flow field at any instant is determined by its initial state and the boundary conditions which may vary in time or be time-independent. If the boundary conditions are time-independent, the solution will often approach a time-independent asymptotic state. There are, however, situations in which, even for time-independent boundary conditions, self-excited fluid motions (instabilities) can occur. We will examine some examples of these in Chapters 6 and 12.

From a system perspective, the boundary conditions can be viewed as the forcing to which the flow must respond. The response is captured in the equations of motion. In the next two subsections, we discuss the imposition of boundary conditions on solid surfaces, boundary conditions on the far field, and the use of inflow and outflow boundary conditions as approximations to far field boundary conditions.

1.11.1 Boundary conditions at solid surfaces

At any point on a boundary formed by a solid impermeable surface, continuity requires that the velocity component normal to the surface be the same for the fluid and for the surface. This boundary condition is purely kinematic. If the solid boundary is stationary so the surface position is not changing with time and if we define $\mathbf{n}$ as the local normal to the surface, then $\mathbf{u} \cdot \mathbf{n} = 0$ on the surface. Two important cases in which the solid body is not stationary are uniform translation with velocity $\mathbf{v}_T$ (where we have used $\mathbf{v}$ to denote a velocity other than a fluid velocity) for which the boundary condition becomes

$$\mathbf{u} \cdot \mathbf{n} = \mathbf{v}_T \cdot \mathbf{n} \tag{1.11.1}$$

and rotation with angular velocity $\boldsymbol{\Omega}$, for which the condition takes the form

$$\mathbf{u} \cdot \mathbf{n} = (\boldsymbol{\Omega} \times \mathbf{r}) \cdot \mathbf{n}, \tag{1.11.2}$$

where $\mathbf{r}$ is a position vector from the axis of rotation.

A more complicated situation is encountered when a body is changing shape (deforming) with time, such as might be the case for flow about vibrating surfaces. Suppose the equation of the surface is $G(\mathbf{x}, t) = 0$. The components of the unit normal to the surface are given by

$$\mathbf{n} = \frac{\nabla G}{|\nabla G|}. \tag{1.11.3}$$

If $\mathbf{v}_{surf}$ is the velocity of a point $\mathbf{x}$ on the surface at time t, the equation for the surface at a small time later $t + dt$ is

$$G(\mathbf{x} + \mathbf{v}_{surf} dt, t + dt) = 0. \tag{1.11.4}$$

Equation (1.11.4) is equivalent to

$$\mathbf{v}_{surf} \cdot \nabla G + \frac{\partial G}{\partial t} = 0. \tag{1.11.5}$$

The component of $\mathbf{v}_{surf}$ along the normal is $\mathbf{v}_{surf} \cdot \mathbf{n}$. The gradient of G, ∇G, is also along the normal so that

$$\mathbf{v}_{surf} \cdot \mathbf{n} = \frac{-\left(\dfrac{\partial G}{\partial t}\right)}{|\nabla G|}. \tag{1.11.6}$$

However, the fluid velocity at the surface along the normal is equal to the instantaneous velocity of the surface in this direction:

$$\mathbf{u} \cdot \mathbf{n} = \mathbf{v}_{surf} \cdot \mathbf{n}. \tag{1.11.7}$$

The boundary condition on the fluid velocity at the deforming surface is therefore

$$\mathbf{u} \cdot \mathbf{n} = \frac{-\left(\dfrac{\partial G}{\partial t}\right)}{|\nabla G|}. \tag{1.11.8}$$

Using (1.11.3), we can write (1.11.8) in terms of the substantial derivative of G

$$\frac{\partial G}{\partial t} + \mathbf{u} \cdot \nabla G = \frac{DG}{Dt} = 0. \tag{1.11.9}$$

Equation (1.11.9) is called the kinematic surface condition. Its physical description is the statement that particles on the surface stay on the surface, because the velocity of a particle on the surface with respect to the surface is purely tangential or zero (Goldstein, 1960).

Situations also exist for which the solid surfaces are permeable, for example suction into, or blowing from, a surface. If the normal component of the suction velocity is known, then $\mathbf{u} \cdot \mathbf{n}$ is also known. In other cases, such as flow through a porous plate with a given pressure differential (which is actually a dynamic, rather than wholly kinematic, boundary condition), the normal velocity at the surface will not be known *a priori*, and will be part of the solution. In such cases there will be matching conditions on the normal velocity which need to be specified. Chapter 12 presents examples of this latter situation.

The boundary conditions described so far are kinematic and do not depend on the nature of the fluid. For a real, i.e. viscous, fluid, no matter how small the viscosity, there is an additional condition on the tangential velocity. For fluids at the pressures that are of interest here (essentially all situations excluding rarefied gases), the surface boundary condition for a viscous fluid is that there is no tangential velocity relative to the surface, i.e. no slip, at a solid boundary.

1.11.2 Inlet and outlet boundary conditions

In addition to surface conditions there are generally other boundary conditions that are needed in the description of a flow. For flow about an object in a duct, such as in Figure 1.8, conditions are needed on the object, on the duct walls, and also at the locations in the duct at which we wish to terminate the calculation domain, the "inlet" and "outlet". A condition often applied at the upstream location is that the static pressure is uniform, i.e. that the upstream influence of the disturbance due to the flow round the body is not felt at this station. As we will see in Chapter 2, this puts constraints on the location of the inlet and outlet stations with respect to the body position.

At the downstream station the situation is less straightforward because the flow conditions may be part of the solution, and thus unable to be precisely specified in advance. An assumption about the decay of pressure disturbances is often also made for the downstream station, and in many cases this is adequate. One way in which this can be implemented in a computation is to put a condition on derivatives in the streamwise directions. A constant static pressure boundary condition will be specified in many of the applications examined, but there are situations in which this must be modified. These will be discussed in Chapters 6 and 12.

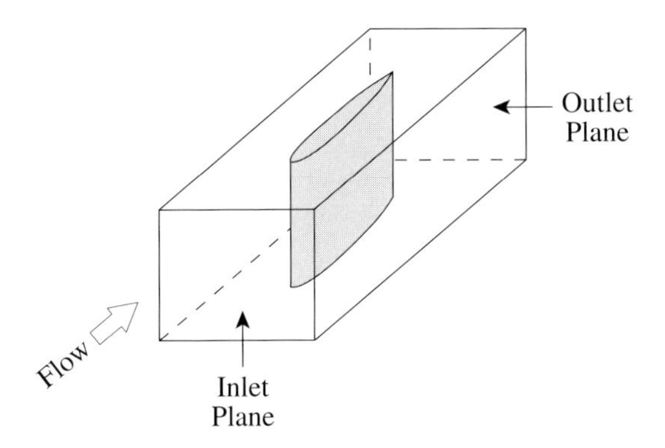

Figure 1.8: Flow about an obstacle in a duct.

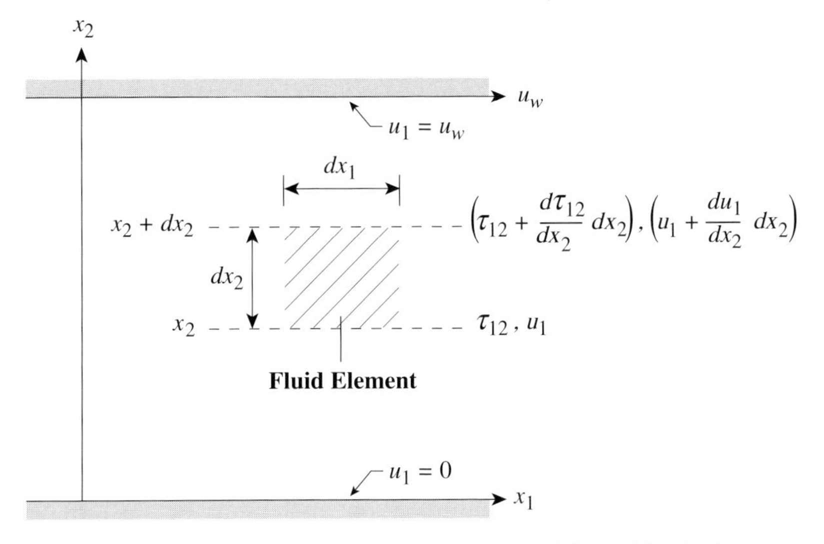

Figure 1.9: Shear stresses and velocities for unidirectional flow with velocity component $u_1 = u_1(x_2)$.

1.12 The rate of strain tensor and the form of the dissipation function

Various products of shear stresses and velocity derivatives have appeared in the different forms of the energy equation. In this section we introduce these terms from another viewpoint to give insight into the physical processes they represent.

To start, consider a fluid motion in which the only component of velocity is in the x_1-direction, with this component being a function of x_2 only. The situation is shown in Figure 1.9 which depicts flow in an infinite two-dimensional channel. The fluid motion is caused by the movement of the upper wall, with velocity u_w in the x_1-direction relative to a lower wall (at $x_2 = 0$) with zero velocity. There are no variations in the x_1- and x_3-directions and it is only the shear stresses on the top and bottom of a fluid element that have dynamical consequences. The net force on the element per unit depth into the page is $(d\tau_{12}/dx_2)dx_1dx_2$ or $d\tau_{12}/dx_2$ per unit area in the plane.

Net work is done on the element by the shear stresses. The rate of work per unit depth into the page on the bottom surface is $u_1\tau_{12}dx_1$. The rate of work on the top surface is $[u_1 + (du_1/dx_2)dx_2][\tau_{12} + (d\tau_{12}/dx_2)dx_2]dx_1$. To order dx_1dx_2, the net rate of work is

$$\text{net rate of work on element } dx_1dx_2 = \left(\tau_{12}\frac{du_1}{dx_2} + u_1\frac{d\tau_{12}}{dx_2}\right)dx_1dx_2$$

$$= \left[\frac{d}{dx_2}(u_1\tau_{12})\right]dx_1dx_2. \tag{1.12.1}$$

Equation (1.12.1) is a special case of the expression for shear work that appears on the right-hand side of the energy equation, (1.9.12) or (1.9.13).

The term $u_1(d\tau_{12}/dx_2)$, which has the form of a velocity times a force, appears in the equation for the rate of change of kinetic energy, (1.10.1). Its contribution is to the mechanical energy of the fluid element. The term $\tau_{12}(du_1/dx_2)$, which has the form of a shear stress times a velocity gradient, appears in the entropy production equation, (1.10.5). For the specific flow we are describing, the entropy production can be evaluated directly. The only terms in the momentum equation are due to shear forces so that (1.9.10) reduces to

$$\frac{\partial\tau_{ij}}{\partial x_j} = 0,$$

or $\tau_{12} =$ constant. The rate of entropy production, (1.10.5), is

$$\text{rate of entropy production per unit volume} = \frac{1}{T}\tau_{12}\frac{du_1}{dx_2}. \tag{1.12.2a}$$

Neglecting changes in temperature and integrating (1.12.2a) from $x_2 = 0$ to the upper wall yields

$$\text{rate of entropy production/unit length} = \frac{1}{T}\tau_{12}u_w. \tag{1.12.2b}$$

The rate of work done on the fluid per unit length of the wall is $\tau_{12}u_w$. From these arguments it can be seen that the quantity $(1/T)\tau_{ij}(\partial u_i/\partial x_j)$ can be regarded as an entropy source term which represents the dissipation of mechanical energy per unit volume.

Another basic situation is that of flow in the direction of the x_1-axis with variation in this direction only, as shown in Figure 1.10. Consider a streamtube of unit cross-section. The rate of work done on the left-hand side of the fluid element by shear stresses is $\tau_{11}u_1$. The rate of work per unit area on the right-hand face is $\tau_{11}u_1 + [d(\tau_{11}u_1)/dx_1]dx_1$, so the net rate of work done is $u_1(d\tau_{11}/dx_1) + \tau_{11}(du_1/dx_1)$ per unit volume. The work associated with shear stress can again be broken into two parts, one with the form of a velocity times a force, which contributes to changes in mechanical energy, and one with the form of the product of shear stress and velocity gradient, which contributes to entropy production.

For a general three-dimensional flow, additional terms appear in the expression for the net work done on an element. In the two examples just discussed, the velocity gradients were the fluid strain rates, and it thus seems reasonable to inquire whether this is also true for the three-dimensional situation. To answer this, we need to develop expressions for the rates of strain in three dimensions. The tensor $\partial u_i/\partial x_j$, which expresses the rate of deformation of a fluid element, is first broken into a

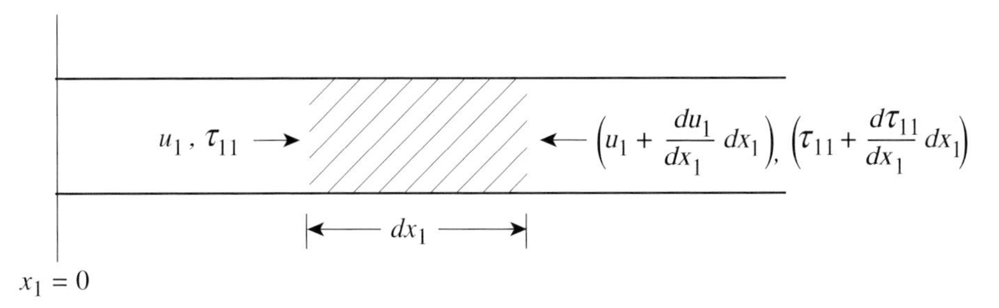

Figure 1.10: Shear stresses and velocities in a one-dimensional flow, $u_1 = u_1(x_1)$.

symmetric and an anti-symmetric part as follows:

$$
\frac{\partial u_i}{\partial x_j} = \begin{bmatrix} \frac{\partial u_1}{\partial x_1} & \frac{\partial u_1}{\partial x_2} & \frac{\partial u_1}{\partial x_3} \\ \frac{\partial u_2}{\partial x_1} & \frac{\partial u_2}{\partial x_2} & \frac{\partial u_2}{\partial x_3} \\ \frac{\partial u_3}{\partial x_1} & \frac{\partial u_3}{\partial x_2} & \frac{\partial u_3}{\partial x_3} \end{bmatrix}
$$

$$
= \begin{bmatrix} \frac{\partial u_1}{\partial x_1} & \frac{1}{2}\left(\frac{\partial u_1}{\partial x_2} + \frac{\partial u_2}{\partial x_1}\right) & \frac{1}{2}\left(\frac{\partial u_1}{\partial x_3} + \frac{\partial u_3}{\partial x_1}\right) \\ \frac{1}{2}\left(\frac{\partial u_1}{\partial x_2} + \frac{\partial u_2}{\partial x_1}\right) & \frac{\partial u_2}{\partial x_2} & \frac{1}{2}\left(\frac{\partial u_2}{\partial x_3} + \frac{\partial u_3}{\partial x_2}\right) \\ \frac{1}{2}\left(\frac{\partial u_1}{\partial x_3} + \frac{\partial u_3}{\partial x_1}\right) & \frac{1}{2}\left(\frac{\partial u_2}{\partial x_3} + \frac{\partial u_3}{\partial x_2}\right) & \frac{\partial u_3}{\partial x_3} \end{bmatrix}
$$

$$
- \begin{bmatrix} 0 & \frac{1}{2}\left(\frac{\partial u_2}{\partial x_1} - \frac{\partial u_1}{\partial x_2}\right) & -\frac{1}{2}\left(\frac{\partial u_1}{\partial x_3} - \frac{\partial u_3}{\partial x_1}\right) \\ -\frac{1}{2}\left(\frac{\partial u_2}{\partial x_1} - \frac{\partial u_1}{\partial x_2}\right) & 0 & \frac{1}{2}\left(\frac{\partial u_3}{\partial x_2} - \frac{\partial u_2}{\partial x_3}\right) \\ \frac{1}{2}\left(\frac{\partial u_1}{\partial x_3} - \frac{\partial u_3}{\partial x_1}\right) & -\frac{1}{2}\left(\frac{\partial u_3}{\partial x_2} - \frac{\partial u_2}{\partial x_3}\right) & 0 \end{bmatrix}. \tag{1.12.3}
$$

The splitting of the deformation tensor in this manner has physical significance, which can be seen by examining one of the three components of the anti-symmetric part, for example $\frac{1}{2}(\partial u_2/\partial x_1 - \partial u_1/\partial x_2)$, with respect to the two perpendicular fluid lines OA and OB depicted in Figure 1.11. At time t these lines are parallel to the x_1- and x_2-axes. At a slightly later time, $t + dt$, the points A and B have moved (relative to point O) to A′ and B′. The distances AA′ and BB′ are $(\partial u_1/\partial x_2)dx_2 dt$ and $(\partial u_2/\partial x_1)dx_1 dt$ respectively. The angles through which OA and OB have rotated in the counterclockwise direction are therefore $-(\partial u_1/\partial x_2)dt$ and $(\partial u_2/\partial x_1)dt$ and the average rate of rotation of the two perpendicular fluid lines about the x_3-axis is $\frac{1}{2}[(\partial u_2/\partial x_1) - (\partial u_1/\partial x_2)]$. A corresponding statement can be made about the other two components. These arguments show that the terms

$$
\frac{1}{2}\left(\frac{\partial u_3}{\partial x_2} - \frac{\partial u_2}{\partial x_3}\right), \quad \frac{1}{2}\left(\frac{\partial u_1}{\partial x_3} - \frac{\partial u_3}{\partial x_1}\right), \quad \frac{1}{2}\left(\frac{\partial u_2}{\partial x_1} - \frac{\partial u_1}{\partial x_2}\right)
$$

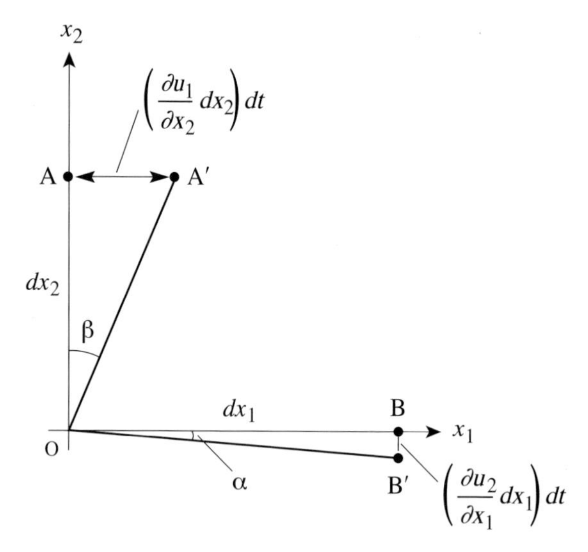

Figure 1.11: Rates of rotation of two perpendicular fluid lines.

which appear in the anti-symmetric part of the deformation tensor are the rates of angular rotation of the fluid element about axes through its center and parallel to the three coordinate axes. The angular velocity is a vector and the vector which is twice the angular velocity of a fluid element is known as the vorticity. Examination of the vorticity field provides considerable insight into fluid motion as discussed in Chapter 3.

Angular rotations do not strain the fluid element. For example, a rigid body rotation would be an extreme case for which no work at all is done by the shear stresses. All the strain must therefore be expressed by the symmetrical part of the deformation tensor. The quantities $\partial u_1/\partial x_1$, $\partial u_2/\partial x_2$, $\partial u_3/\partial x_3$ are tensile strain rates, as can be seen by considering the deformation in the coordinate directions of elements aligned with the three axes. The remaining quantities ($\frac{1}{2}$ $(\partial u_1/\partial x_2 + \partial u_2/\partial x_1)$, etc.), are the rate of shear strain. With reference to Figure 1.11, they represent the average rate at which the two originally perpendicular elements depart from a right angle orientation. If one writes out all the individual parts of the term $\tau_{ij}(\partial u_i/\partial x_j)$, it is seen that only the symmetric part of the deformation tensor contributes to this term.

1.13 Relationship between stress and rate of strain

The momentum equation for a fluid was given in Section 1.9 as

$$\frac{\partial u_i}{\partial t} + u_j \frac{\partial u_i}{\partial x_j} = -\frac{1}{\rho}\frac{\partial p}{\partial x_i} + X_i + \frac{1}{\rho}\frac{\partial \tau_{ij}}{\partial x_j}, \tag{1.9.10}$$

since

$$\Pi_{ij} = -p\delta_{ij} + \tau_{ij}. \tag{1.7.6}$$

In this section, we develop expressions for τ_{ij} (called the deviatoric stress tensor) in terms of the velocity gradients which represent the fluid strain rates.

In Section 1.12 the velocity gradient tensor was decomposed into symmetric and anti-symmetric parts, where

$$\frac{\partial u_i}{\partial x_j} = \underbrace{\frac{1}{2}\left(\frac{\partial u_i}{\partial x_j} + \frac{\partial u_j}{\partial x_i}\right)}_{\text{symmetric}} + \underbrace{\frac{1}{2}\left(\frac{\partial u_i}{\partial x_j} - \frac{\partial u_j}{\partial x_i}\right)}_{\text{anti-symmetric}}. \tag{1.12.3}$$

The anti-symmetric terms describe angular rotations of a fluid element which do not contribute to element deformation. Stresses in the fluid must therefore be generated by the remaining rate of strain terms, known as the strain rate tensor,

$$e_{ij} = \frac{1}{2}\left(\frac{\partial u_i}{\partial x_j} + \frac{\partial u_j}{\partial x_i}\right). \tag{1.13.1}$$

To relate the strain rate tensor to the deviatoric stress tensor, several properties of the stress and strain rate tensors will be used. The first is that, in accord with experimental findings, stress in a fluid is linearly proportional to strain rate. Based on this we can propose a relation between τ_{ij} and e_{ij} as

$$\tau_{ij} = k_{ijmn} e_{mn}, \tag{1.13.2}$$

where k_{ijmn} is a fourth order tensor with 81 components (Batchelor, 1967). This relationship is conceptually similar to Hooke's law from solid mechanics which assumes proportionality between stress and strain.

The second property we invoke is that the fluid of interest is an isotropic medium, i.e. the fluid has no preferred directional behavior. All gases are statistically isotropic as are most simple fluids. A consequence is that the stress–strain rate relationship for these substances is independent of rotation of the governing coordinate system. This invariance is only possible when k_{ijmn} is an isotropic tensor. Further, it is known that any isotropic tensor of even order can be expressed in terms of products of δ_{ij} (Aris, 1962). A fourth order isotropic tensor can be written as

$$k_{ijmn} = \lambda\delta_{ij}\delta_{mn} + \mu\delta_{im}\delta_{jn} + \zeta\delta_{in}\delta_{jm}, \tag{1.13.3}$$

where λ, μ, and ζ are scalars that are a function of the local thermodynamic state.

The third property invoked is that only the symmetric portion of the strain rate tensor imparts stress. This implies that the stress tensor must be symmetric. If τ_{ij} is a symmetric tensor, k_{ijmn} must also be a symmetric tensor and this is only true if (1.13.3) has

$$\zeta = \mu. \tag{1.13.4}$$

Combining (1.13.2), (1.13.3), and (1.13.4) gives an expression for the relationship between the deviatoric stress tensor and the strain rate tensor:

$$\begin{aligned}\tau_{ij} &= (\lambda\delta_{ij}\delta_{mn} + \mu\delta_{im}\delta_{jn} + \zeta\delta_{in}\delta_{jm})\, e_{mn} \\ &= \lambda e_{ij}\delta_{ij} + 2\mu e_{ij}.\end{aligned} \tag{1.13.5}$$

The complete stress tensor (1.7.6) can now be written as

$$\Pi_{ij} = -p\delta_{ij} + \lambda e_{ij}\delta_{ij} + 2\mu e_{ij}. \tag{1.13.6}$$

In (1.13.6)

$$e_{ij}\delta_{ij} = \nabla \cdot \mathbf{u} = \frac{\partial u_i}{\partial x_i}. \tag{1.13.7}$$

The two scalars μ and λ can be further related by setting $i = j$ and summing over the repeated index. From (1.13.6)

$$\begin{aligned}\Pi_{ii} &= -3p + (2\mu + 3\lambda)\, e_{ii} \\ &= -3p + (2\mu + 3\lambda)\, \frac{\partial u_i}{\partial x_i}.\end{aligned} \tag{1.13.8}$$

This allows the thermodynamic pressure to be defined as

$$p = -\frac{1}{3}\Pi_{ii} + \left(\frac{2}{3}\mu + \lambda\right)\frac{\partial u_i}{\partial x_i}. \tag{1.13.9}$$

where the mechanical pressure was given in Section 1.7 as

$$p_M = -\frac{1}{3}\Pi_{ii}. \tag{1.7.7}$$

The difference between the mechanical and thermodynamic pressures is thus

$$p - p_M = \left(\frac{2}{3}\mu + \lambda\right)\frac{\partial u_i}{\partial x_i}. \tag{1.13.10}$$

For an incompressible fluid, the mechanical and thermodynamic pressures are the same since $\partial u_i/\partial x_i$ is equal to zero. As seen from (1.13.10), λ plays no role in an incompressible flow. For a compressible fluid there are two different definitions of pressure. The assumption that the thermodynamic and mechanical pressures are equal is often referred to as Stokes's assumption (White, 1991) and implies

$$\lambda = -\frac{2}{3}\mu. \tag{1.13.11}$$

Equation (1.13.11) is supported by kinetic theory for a monatomic gas, although not for other fluids (Sherman, 1990). We adopt its use here, noting this proviso, but also noting that the impact of this assumption has been found to be small in flows in engineering applications.[13]

Combining the above results, the complete stress tensor for a compressible fluid can be obtained. Substitution of (1.13.7) and (1.13.11) into (1.13.6) gives

$$\Pi_{ij} = -p\delta_{ij} - \frac{2}{3}\mu\frac{\partial u_k}{\partial x_k}\delta_{ij} + 2\mu\, e_{ij}. \tag{1.13.12}$$

This linear relationship between stress and strain rate is consistent with the definition of the viscosity coefficient for parallel flows given by Newton, in which case (1.13.12) reduces to $\mu(\partial u_1/\partial x_2)$. Hence, fluids which obey this constitutive relationship and the underlying assumptions are called Newtonian. For the special case of an incompressible Newtonian fluid, (1.13.12) reduces to

$$\Pi_{ij} = -p\delta_{ij} + 2\mu e_{ij}, \tag{1.13.13}$$

where p is interpreted as the mean mechanical pressure.

[13] For additional discussion of this point, see Thompson (1984), Schlichting (1979), Sherman (1990), and White (1991).

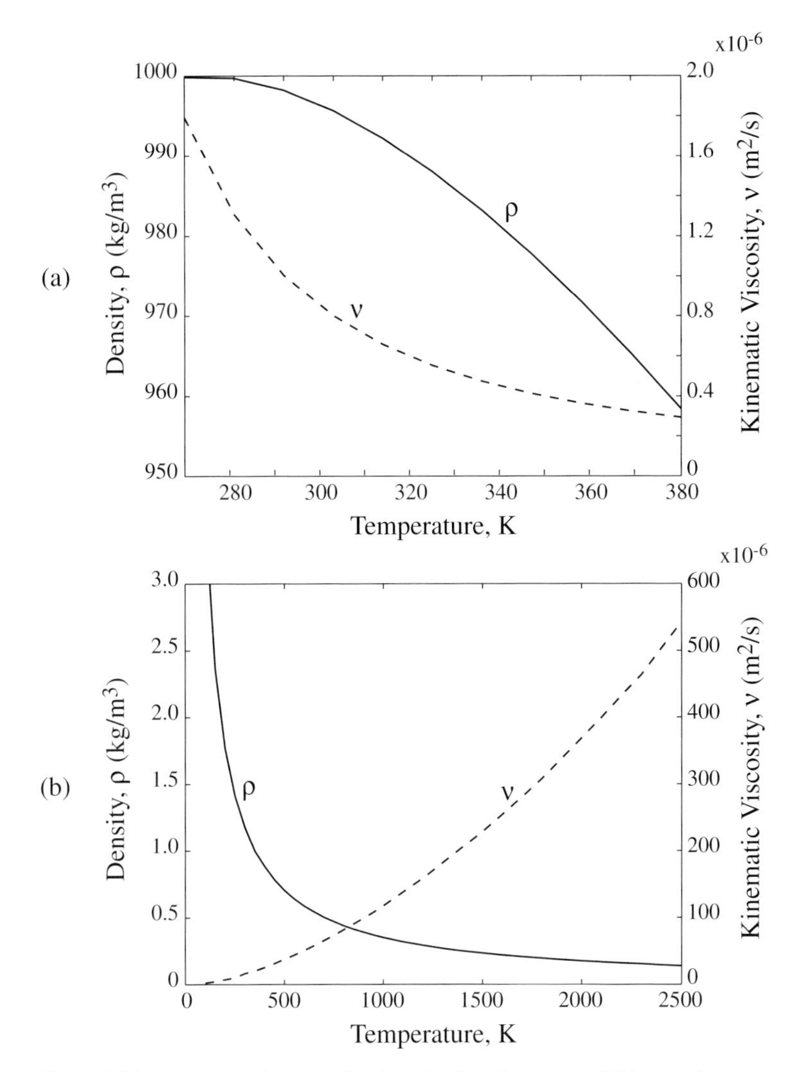

Figure 1.12: Density and kinematic viscosity for (a) water and (b) air at 1 atmosphere (Eckert and Drake, 1972).

1.14 The Navier–Stokes equations

The governing equation of motion for a Newtonian fluid can now be obtained by substituting the constitutive relationship for τ_{ij}, (1.13.12), into the momentum equation, (1.9.10), to yield

$$\frac{\partial u_i}{\partial t} + u_j \frac{\partial u_i}{\partial x_j} = -\frac{1}{\rho}\frac{\partial p}{\partial x_i} + X_i + \frac{1}{\rho}\frac{\partial}{\partial x_j}\left(2\mu e_{ij} - \frac{2}{3}\mu \frac{\partial u_k}{\partial x_k}\delta_{ij}\right). \tag{1.14.1}$$

This is known as the general form of the Navier–Stokes equation, the momentum equation for a compressible Newtonian fluid. The kinematic viscosity $\nu = \mu/\rho$ and the density for water and air at 1 atmosphere pressure as a function of temperature are shown in Figures 1.12(a) and 1.12(b). These are representative of the temperature dependence for other gases and liquids. Although viscosity is a function of thermodynamic state, there are many situations in which μ can be assumed constant. If so, the Navier–Stokes equations can be simplified to

$$\frac{\partial u_i}{\partial t}+u_j\frac{\partial u_i}{\partial x_j}=-\frac{1}{\rho}\frac{\partial p}{\partial x_i}+X_i+\frac{\mu}{\rho}\left[\frac{\partial}{\partial x_j}\left(\frac{\partial u_i}{\partial x_j}+\frac{\partial u_j}{\partial x_i}\right)-\frac{2}{3}\frac{\partial}{\partial x_i}\left(\frac{\partial u_i}{\partial x_i}\right)\right]$$

$$=-\frac{1}{\rho}\frac{\partial p}{\partial x_i}+X_i+\frac{\mu}{\rho}\left[\frac{\partial^2 u_i}{\partial x_j\partial x_j}+\frac{1}{3}\frac{\partial}{\partial x_i}\left(\frac{\partial u_i}{\partial x_i}\right)\right]. \tag{1.14.2}$$

For an incompressible flow with constant kinematic viscosity, ν, (1.14.2) further reduces to

$$\frac{\partial u_i}{\partial t}+u_j\frac{\partial u_i}{\partial x_j}=-\frac{1}{\rho}\frac{\partial p}{\partial x_i}+X_i+\nu\frac{\partial^2 u_i}{\partial x_j\partial x_j} \tag{1.14.3}$$

or, in vector notation,

$$\frac{D\mathbf{u}}{Dt}=-\frac{1}{\rho}\boldsymbol{\nabla}p+X_i+\nu\boldsymbol{\nabla}^2\mathbf{u}. \tag{1.14.4}$$

This is the Navier–Stokes equation for an incompressible flow.

For reference, the components of the momentum equation and the continuity equation are given below for two coordinate systems that are used often in this book: Cartesian coordinates and cylindrical coordinates.

1.14.1 Cartesian coordinates

$$\rho\left[\frac{\partial u_x}{\partial t}+u_x\frac{\partial u_x}{\partial x}+u_y\frac{\partial u_x}{\partial y}+u_z\frac{\partial u_x}{\partial z}\right]$$
$$=-\frac{\partial p}{\partial x}+X_x+\frac{\partial}{\partial x}\left[\mu\left(2\frac{\partial u_x}{\partial x}-\frac{2}{3}\boldsymbol{\nabla}\cdot\mathbf{u}\right)\right]$$
$$+\frac{\partial}{\partial y}\left[\mu\left(\frac{\partial u_x}{\partial y}+\frac{\partial u_y}{\partial x}\right)\right]+\frac{\partial}{\partial z}\left[\mu\left(\frac{\partial u_z}{\partial x}+\frac{\partial u_x}{\partial z}\right)\right], \tag{1.14.5a}$$

$$\rho\left[\frac{\partial u_y}{\partial t}+u_x\frac{\partial u_y}{\partial x}+u_y\frac{\partial u_y}{\partial y}+u_z\frac{\partial u_y}{\partial z}\right]$$
$$=-\frac{\partial p}{\partial y}+X_y+\frac{\partial}{\partial y}\left[\mu\left(2\frac{\partial u_y}{\partial y}-\frac{2}{3}\boldsymbol{\nabla}\cdot\mathbf{u}\right)\right]$$
$$+\frac{\partial}{\partial z}\left[\mu\left(\frac{\partial u_y}{\partial z}+\frac{\partial u_z}{\partial y}\right)\right]+\frac{\partial}{\partial x}\left[\mu\left(\frac{\partial u_x}{\partial y}+\frac{\partial u_y}{\partial x}\right)\right], \tag{1.14.5b}$$

$$\rho\left[\frac{\partial u_z}{\partial t}+u_x\frac{\partial u_z}{\partial x}+u_y\frac{\partial u_z}{\partial y}+u_z\frac{\partial u_z}{\partial z}\right]$$
$$=-\frac{\partial p}{\partial z}+X_z+\frac{\partial}{\partial z}\left[\mu\left(2\frac{\partial u_z}{\partial z}-\frac{2}{3}\boldsymbol{\nabla}\cdot\mathbf{u}\right)\right]$$
$$+\frac{\partial}{\partial x}\left[\mu\left(\frac{\partial u_z}{\partial x}+\frac{\partial u_x}{\partial z}\right)\right]+\frac{\partial}{\partial y}\left[\mu\left(\frac{\partial u_y}{\partial z}+\frac{\partial u_z}{\partial y}\right)\right]. \tag{1.14.5c}$$

The continuity equation is

$$\frac{\partial \rho}{\partial t} + \frac{\partial}{\partial x}(\rho u_x) + \frac{\partial}{\partial y}(\rho u_y) + \frac{\partial}{\partial z}(\rho u_z) = 0. \tag{1.14.6}$$

For incompressible flow with constant viscosity, (1.14.5) and (1.14.6) simplify to

$$\rho\left[\frac{\partial u_x}{\partial t} + u_x\frac{\partial u_x}{\partial x} + u_y\frac{\partial u_x}{\partial y} + u_z\frac{\partial u_x}{\partial z}\right]$$
$$= -\frac{\partial p}{\partial x} + X_x + \mu\left(\frac{\partial^2 u_x}{\partial x^2} + \frac{\partial^2 u_x}{\partial y^2} + \frac{\partial^2 u_x}{\partial z^2}\right), \tag{1.14.7a}$$

$$\rho\left[\frac{\partial u_y}{\partial t} + u_x\frac{\partial u_y}{\partial x} + u_y\frac{\partial u_y}{\partial y} + u_z\frac{\partial u_y}{\partial z}\right]$$
$$= -\frac{\partial p}{\partial y} + X_y + \mu\left(\frac{\partial^2 u_y}{\partial x^2} + \frac{\partial^2 u_y}{\partial y^2} + \frac{\partial^2 u_y}{\partial z^2}\right), \tag{1.14.7b}$$

$$\rho\left[\frac{\partial u_z}{\partial t} + u_x\frac{\partial u_z}{\partial x} + u_y\frac{\partial u_z}{\partial y} + u_z\frac{\partial u_z}{\partial z}\right]$$
$$= -\frac{\partial p}{\partial z} + X_z + \mu\left(\frac{\partial^2 u_z}{\partial x^2} + \frac{\partial^2 u_z}{\partial y^2} + \frac{\partial^2 u_z}{\partial z^2}\right), \tag{1.14.7c}$$

$$\frac{\partial u_x}{\partial x} + \frac{\partial u_y}{\partial y} + \frac{\partial u_z}{\partial z} = 0. \tag{1.14.8}$$

1.14.2 Cylindrical coordinates (*x*, axial; θ, circumferential; *r*, radial)

We list only the incompressible form of the equations for cylindrical coordinates:

$$\rho\left[\frac{\partial u_r}{\partial t} + u_r\frac{\partial u_r}{\partial r} + \frac{u_\theta}{r}\frac{\partial u_r}{\partial \theta} - \frac{u_\theta^2}{r} + u_x\frac{\partial u_r}{\partial x}\right]$$
$$= -\frac{\partial p}{\partial r} + X_r + \mu\left(\frac{\partial^2 u_r}{\partial r^2} + \frac{1}{r}\frac{\partial u_r}{\partial r} - \frac{u_r}{r^2} + \frac{1}{r^2}\frac{\partial^2 u_r}{\partial \theta^2} - \frac{2}{r^2}\frac{\partial u_\theta}{\partial \theta} + \frac{\partial^2 u_r}{\partial x^2}\right), \tag{1.14.9a}$$

$$\rho\left[\frac{\partial u_\theta}{\partial t} + u_r\frac{\partial u_\theta}{\partial r} + \frac{u_\theta}{r}\frac{\partial u_\theta}{\partial \theta} + \frac{u_r u_\theta}{r} + u_x\frac{\partial u_\theta}{\partial x}\right]$$
$$= -\frac{1}{r}\frac{\partial p}{\partial \theta} + X_\theta + \mu\left(\frac{\partial^2 u_\theta}{\partial r^2} + \frac{1}{r}\frac{\partial u_\theta}{\partial r} + \frac{1}{r^2}\frac{\partial^2 u_\theta}{\partial \theta^2} + \frac{\partial^2 u_\theta}{\partial x^2} + \frac{2}{r^2}\frac{\partial u_r}{\partial \theta} - \frac{u_\theta}{r^2}\right), \tag{1.14.9b}$$

$$\rho\left[\frac{\partial u_x}{\partial t} + u_r\frac{\partial u_x}{\partial r} + \frac{u_\theta}{r}\frac{\partial u_x}{\partial \theta} + u_x\frac{\partial u_x}{\partial x}\right]$$
$$= -\frac{\partial p}{\partial x} + X_x + \mu\left(\frac{\partial^2 u_x}{\partial r^2} + \frac{1}{r}\frac{\partial u_x}{\partial r} + \frac{1}{r^2}\frac{\partial^2 u_x}{\partial \theta^2} + \frac{\partial^2 u_x}{\partial x^2}\right). \tag{1.14.9c}$$

The continuity equation is

$$\frac{\partial u_r}{\partial r} + \frac{1}{r}u_r + \frac{1}{r}\frac{\partial u_\theta}{\partial \theta} + \frac{\partial u_x}{\partial x} = 0. \tag{1.14.10}$$

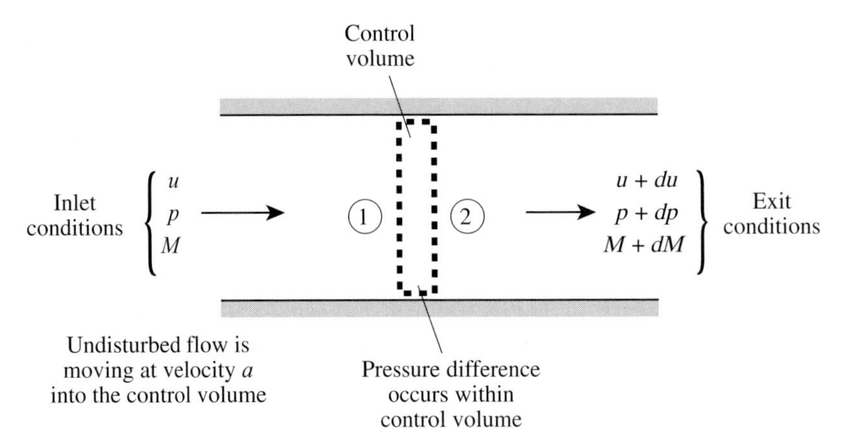

Figure 1.13: Control volume fixed to a propagating small disturbance in a compressible fluid.

1.15 Disturbance propagation in a compressible fluid: the speed of sound

A quantity which plays a major role in a number of the flows to be discussed is the speed at which small amplitude pressure disturbances propagate in a compressible medium. To find this we consider a disturbance propagating in a frictionless, non-heat-conducting, perfect gas in a channel of uniform area. As shown in Figure 1.13, we choose a control volume moving with the disturbance at a velocity, a, so that flow relative to the control volume is steady. The pressure at the left-hand side of the control volume where the disturbance has not yet arrived is p, the velocity is a, and the density is ρ. At the right-hand side of the control volume the pressure is $p + dp$, the velocity is $a + du$, and the density is $\rho + d\rho$. For small disturbances, the ratios of the disturbance quantities to the background flow variables (e.g. du/a, dp/p) will be much less than 1 so that products of these quantities can be neglected.

The continuity equation applied across the control volume in Figure 1.13 is

$$\rho a = (\rho + d\rho)(a + du),$$

or, to first order in the small disturbance terms,

$$a d\rho + \rho du = 0. \tag{1.15.1}$$

Application of the control volume form of the momentum equation in a similar manner plus use of (1.15.1) gives a relation between pressure and velocity changes across the control volume,

$$\frac{dp}{\rho} = -a du. \tag{1.15.2}$$

Combining (1.15.1) and (1.15.2) yields an expression for the disturbance speed, a, in terms of the ratio of changes in pressure and density:

$$a^2 = \frac{dp}{d\rho}. \tag{1.15.3}$$

To define the ratio given in (1.15.3) explicitly, we apply the energy equation to the control volume to provide a relation between enthalpy and velocity changes:

$$dh = -a du. \tag{1.15.4}$$

Comparison with (1.15.2) shows that, for the disturbances considered,

$$dh - dp/\rho = T ds = 0. \tag{1.15.5}$$

The relation between changes in density and pressure in (1.15.3) is therefore that existing in an isentropic process, p/ρ^{γ} = constant, and the speed of the small amplitude disturbances can be written as

$$a = \sqrt{\left(\frac{\partial p}{\partial \rho}\right)_s} \tag{1.15.6}$$

or, for a perfect gas with $p = \rho RT$,

$$a = \sqrt{\gamma RT} = \sqrt{\frac{\gamma p}{\rho}}. \tag{1.15.7}$$

Sound waves are small amplitude disturbances of this type, and the speed, a, is therefore referred to as the speed of sound. For air at room temperature and pressure a is roughly 340 m/s.

1.16 Stagnation and static quantities

The performance of internal flow devices is generally characterized by two attributes: the energy transfer and the losses (or efficiency) that are associated with the flow processes. This characterization is most naturally expressed in terms of changes in stagnation pressure and stagnation enthalpy, conditions associated with a zero velocity state, rather than the static temperature and pressure which are the state conditions associated with the local velocity.

The stagnation enthalpy has already been introduced as the enthalpy which would be attained by a fluid element brought to rest in a steady manner with no net heat and work transfer. If so, to recap the result from (1.8.11), all along a streamline

$$h_t = h + \frac{u^2}{2} = \text{constant}. \tag{1.16.1}$$

For a perfect gas with constant specific heats, (1.16.1) provides a relation between the static temperature, T, and the stagnation temperature, T_t:

$$T_t = T + \frac{u^2}{2c_p}. \tag{1.16.2}$$

In contrast to stagnation temperature, the conditions that define the stagnation pressure are more restrictive in that the deceleration must also be reversible and hence isentropic. For a perfect gas with constant specific heats, stagnation pressure can be related to static pressure, static temperature, and stagnation temperature through the isentropic relation

$$\frac{p_t}{p} = \left(\frac{T_t}{T}\right)^{\frac{\gamma}{\gamma-1}}. \tag{1.16.3}$$

Other stagnation quantities can also be defined but temperature, pressure, and entropy (which is the same as the static entropy) are those most frequently encountered.

Entropy changes between thermodynamic states can also be given in terms of stagnation quantities using (1.3.19):

$$T_t ds = dh_t - \frac{1}{\rho_t} dp_t. \tag{1.16.4}$$

For steady adiabatic flow with no shaft work the stagnation enthalpy is constant along a streamline, whether or not the flow is reversible. For a perfect gas, the entropy at two locations along a streamtube is therefore given by the integral of (1.16.4) with $dh_t = 0$:

$$s_2 - s_1 = -R \ln \frac{p_{t_2}}{p_{t_1}}, \tag{1.16.5}$$

where p_{t_1} and p_{t_2} refer to the stagnation pressure at locations (1) and (2) respectively. For adiabatic flows one can view the change in stagnation pressure as a measure of the change in entropy, and hence the irreversibility, between two stations. We will discuss the utility and application of (1.16.5) in Chapter 5.

Two points can be noted concerning stagnation pressure and temperature. First, stagnation (rather than static) quantities are generally most convenient to measure in internal flow devices, with the interpretation of changes in these quantities directly connected to experimental results. Second, the process by which the fluid is brought to the stagnation state need not be one that occurs in the actual flow. Even in situations with unsteadiness, heat transfer, or losses, therefore, one can still refer to local stagnation properties although there are a number of situations in which one or both of the stagnation temperature and pressure quantities remains constant along a streamline, so these quantities often furnish a useful reference level.

1.16.1 Relation of stagnation and static quantities in terms of Mach number

The ratio of the local velocity magnitude to the speed of sound, u/a, is a non-dimensional parameter known as the Mach number and denoted by M: $M = u/a$. For a perfect gas with constant specific heats, the ratio of the stagnation and static quantities can be presented in terms of Mach number, using (1.16.2) and (1.16.3), and the relations between c_p and R as

$$\frac{T_t}{T} = 1 + \frac{\gamma - 1}{2} M^2 \tag{1.16.6}$$

and

$$\frac{p_t}{p} = \left(1 + \frac{\gamma - 1}{2} M^2\right)^{\gamma/\gamma-1}. \tag{1.16.7}$$

1.17 Kinematic and dynamic flow field similarity

An important concept in fluid mechanics is similarity between flow fields. The specific question is under what conditions can information about one flow field be applied to another with different parameters. This issue is examined below, first for incompressible flow and then for the compressible flow regime.

1.17.1 Incompressible flow

An initial step in determining similarity is to cast the equations in a non-dimensional form where the parameters necessary for similarity are explicitly defined. The fluid motion considered has a constant density ρ, a coefficient of viscosity μ, a geometry with characteristic dimension L, a characteristic velocity U and a reference pressure[14] p_{ref}. If the flow is unsteady, a characteristic time over which there are appreciable changes can be defined as $1/\omega$, where ω is the radian frequency corresponding to the unsteadiness of interest. With no body forces, $X_i = 0$, the equations describing the flow become:

$$\frac{\partial u_i}{\partial x_i} = 0, \tag{1.9.6}$$

$$\frac{\partial u_i}{\partial t} + u_j \frac{\partial u_i}{\partial x_j} = -\frac{1}{\rho}\frac{\partial p}{\partial x_i} + \nu \frac{\partial^2 u_i}{\partial x_j \partial x_j}. \tag{1.14.3}$$

These equations can be put into a non-dimensional form by dividing length by L, velocities by U, pressure differences by ρU^2, and time by $1/\omega$. This amounts to adopting new measurement scales in which length is measured in units of L, velocity in units of U, pressure differences in units of ρU^2 and time in units of $1/\omega$. The variables measured in terms of these units will be denoted by a tilde ($\sim$)

$$\tilde{x}_i = \frac{x_i}{L}, \quad \tilde{t} = t\omega, \quad \tilde{u}_i = \frac{u_i}{U}, \quad \tilde{p} = \frac{p - p_{ref}}{\rho U^2}. \tag{1.17.1}$$

In incompressible flow, the absolute pressure level plays no role in determining the fluid motion. The non-dimensional pressure in (1.17.1) is therefore defined using the difference between local and reference pressures.

Equations (1.9.6) and (1.14.4) can be written in non-dimensional form

$$\frac{\partial \tilde{u}_i}{\partial \tilde{x}_i} = 0, \tag{1.17.2}$$

$$\frac{\omega L}{U}\frac{\partial \tilde{u}_i}{\partial \tilde{t}} + \tilde{u}_j \frac{\partial \tilde{u}_i}{\partial \tilde{x}_j} = -\frac{\partial \tilde{p}}{\partial \tilde{x}_i} + \frac{\nu}{UL}\frac{\partial^2 \tilde{u}_i}{\partial \tilde{x}_j \partial \tilde{x}_j}. \tag{1.17.3}$$

Equations (1.17.2) and (1.17.3) show the flow field depends on two non-dimensional parameters, UL/ν and $\omega L/U$, and the variables $\tilde{x}$ and $\tilde{t}$.

[14] The length, L, could represent the length or width of a duct, channel or blade passage, and the velocity, U, could represent the inlet velocity, the mean velocity across a duct, or the velocity at some other station. Similarly the reference pressure, p_{ref}, (as well as other reference quantities to be introduced later) could represent the pressure at inlet. The central point is that an appropriate quantity is one that figures prominently in *characterizing* (describing scales and features of) the motion.

1.17.2 Kinematic similarity

In defining two flows as similar, two sets of conditions must be met. The first is similarity in geometry. To scale the flow in a turbomachine to a smaller or larger machine, geometrical parameters such as blade profile, blade stagger angle, blade spacing/chord ratio, and hub/tip radius ratio must be kept the same. The normal velocity boundary conditions, which are set by the geometry, must also be the same. If one configuration has a condition of zero normal velocity, for example, the scaled configuration must also have this condition; it cannot have flow through the wall.

This set of conditions defines *kinematic similarity*. Kinematic similarity is necessary but not sufficient for full similarity, although for some applications kinematic similarity can be all that is needed to compare flow fields.

A class of motions for which kinematic similarity is all that is necessary is incompressible irrotational flow, which is described by a velocity potential whose gradient is the velocity

$$u_i = \frac{\partial \varphi}{\partial x_i}. \tag{1.17.4}$$

The above form of the velocity plus the continuity equation for incompressible flow leads to a single equation (Laplace's equation) for the velocity potential:

$$\frac{\partial u_i}{\partial x_i} = \frac{\partial^2 \varphi}{\partial x_i \partial x_i} = 0. \tag{1.17.5}$$

This equation plus the kinematic boundary conditions on normal velocity determine the velocity field. For this type of flow the momentum equation can be regarded as an auxiliary relation for determining the pressure.

An example is the static pressure difference from inlet to exit for steady incompressible flow in a converging channel. If the value of UL/ν is large enough, as we will see in Chapter 2, any viscous effects will be confined to thin layers near the walls and the flow over almost all of the channel will be described by (1.17.5). In this situation the pressure change will be determined essentially by kinematic considerations; all nozzles having the same shape will have the same non-dimensional pressure difference to within several percent.

1.17.3 Dynamic similarity

More generally, dynamic similarity is also needed. For a steady flow, dynamic similarity for geometrically similar bodies of different sizes requires the values of the free-stream velocity and the constitution of the fluid (ρ and μ or both) to be such that the value of the non-dimensional quantity UL/ν is the same for the two flows. For kinematically similar steady flows, the behavior thus depends only on this single parameter, $Re = UL/\nu$, known as the Reynolds number.

For unsteady flows, there is an additional non-dimensional parameter, $\omega L/U$, known as the *reduced frequency*, $\beta = \omega L/U$. Both reduced frequency and Reynolds number Re must have the same value in two flows for them to be dynamically similar.

It is generally desirable to process the results of measurements or computations using dimensionless parameters so the information can be applied to other situations with different ρ, U, ω, L, and

μ. Further, if it is shown that the influence of a non-dimensional parameter is small, the similarity can be applied over a range of conditions and not just at the exact comparison point.

1.17.4 Compressible flow

For compressible flow, variations in fluid properties (viscosity, thermal conductivity) due to temperature differences often need to be taken into account. In contrast to incompressible flow, the pressure enters both as a dynamical variable in the momentum equation ((1.9.10), (1.14.2)) and also as a thermodynamic variable in the energy equation ((1.9.13) or (1.10.3)) and the equation of state (1.4.1) (Lagerstrom, 1996). The implication is that when making the momentum equation dimensionless, a pressure difference referenced to $\rho_{ref}U^2$ should be used, while in the equation of state and the energy equation the normalizing variable is the reference pressure, p_{ref}. For a compressible flow, there are additional non-dimensional variables to those defined in Section 1.17.1:

$$\hat{p} = \frac{p}{p_{ref}}, \qquad \hat{\rho} = \frac{\rho}{\rho_{ref}}, \qquad \hat{h} = \hat{T} = \frac{T}{T_{ref}},$$
$$\tilde{\tau}_{ij} = \frac{\tau_{ij}L}{\mu_{ref}U}, \qquad \tilde{\mu} = \frac{\mu}{\mu_{ref}}. \tag{1.17.6}$$

(For convenience we use the shear stress here rather than writing out all the velocity derivatives.) In (1.17.6) the notation (ˆ) has been used to denote that the dimensionless quantity enters as a thermodynamic variable. The non-dimensional pressures are related by $\hat{p} = \gamma M^2 \tilde{p} + 1$. For a perfect gas with constant specific heats (c_p and c_v), no internal heat generation, and constant Prandtl number ($Pr = \mu c_p / k$), the equations of motion are:

$$\beta \frac{\partial \hat{\rho}}{\partial \tilde{t}} + \tilde{u}_j \frac{\partial \hat{\rho}}{\partial \tilde{x}_j} + \hat{\rho} \frac{\partial \tilde{u}_j}{\partial \tilde{x}_j} = 0, \tag{1.17.7}$$

$$\beta \frac{\partial \tilde{u}_i}{\partial \tilde{t}} + \tilde{u}_j \frac{\partial \tilde{u}_i}{\partial \tilde{x}_j} + \frac{1}{\hat{\rho}} \frac{\partial \tilde{p}}{\partial \tilde{x}_i} = \frac{1}{Re \hat{\rho}} \frac{\partial \tilde{\tau}_{ij}}{\partial \tilde{x}_j}, \tag{1.17.8}$$

$$\beta \frac{\partial \hat{h}}{\partial \tilde{t}} + \tilde{u}_j \frac{\partial \hat{h}}{\partial \tilde{x}_j} - \left(\frac{\gamma - 1}{\gamma} \right) \frac{1}{\hat{\rho}} \left[\beta \frac{\partial \hat{p}}{\partial \tilde{t}} + \tilde{u}_j \frac{\partial \hat{p}}{\partial \tilde{x}_j} \right]$$
$$= \frac{1}{RePr} \frac{1}{\hat{\rho}} \frac{\partial}{\partial \tilde{x}_i} \left(\tilde{\mu} \frac{\partial \hat{h}}{\partial \tilde{x}_j} \right) + \frac{(\gamma - 1) M^2}{Re \hat{\rho}} \tilde{\tau}_{ij} \frac{\partial \tilde{u}_i}{\partial \tilde{x}_j}, \tag{1.17.9}$$

$$\hat{p} = \hat{\rho} \hat{T} = \hat{\rho} \hat{h}. \tag{1.17.10}$$

In (1.17.8) and (1.17.9), the Reynolds number and the Mach number are defined based on the reference conditions.

For similarity, the non-dimensional surface heat flux $\tilde{q}_w$ $[= q_w L/(c_p T_{ref} U)]$ must be the same for two flows implying similarity in the non-dimensional surface temperature. This condition may be stated more conveniently as similarity in Stanton number, St, defined as

$$St(\tilde{x}_w, \tilde{t}) = \frac{q_w}{\rho_{ref} c_p U (T_w - T_{ref})} \tag{1.17.11a}$$

or Nusselt number, Nu, defined as

$$Nu(\tilde{x}_w, \tilde{t}) = \frac{q_w L}{k_{ref}(T_w - T_{ref})}. \tag{1.17.11b}$$

Complete dynamical similarity of compressible flows requires identical values of β, Re, Pr, M, γ, and also Nu or St for situations involving heat transfer. It also requires the same dependence of $\tilde{\mu}$ on temperature variation. There are thus many more non-dimensional parameters characterizing compressible flows than incompressible flows, although (fortunately!) often not all of these are important in a given problem.

1.17.5 Limiting forms for low Mach number

The distinction in the roles of pressure in the momentum equation and in the energy and state equations can be seen when examining the limiting form of the compressible equations for low Mach number. In terms of $\tilde{p}$, the equation of state (1.17.10) is

$$\hat{p} = \hat{\rho}\hat{T} = 1 + \gamma M^2 \tilde{p}. \tag{1.17.12}$$

Pressure enters the momentum equation as a dynamic variable. In the limit of $M \to 0$, as shown by (1.17.12), it has no other effect and should be made dimensionless with respect to $\rho_{ref}U^2$. Replacing $\hat{p}$ in (1.17.9) with $1 + \gamma M^2 \tilde{p}$ as $M \to 0$, (1.17.7) and (1.17.8) are unchanged but (1.17.9) and (1.17.10) are altered in form and the compressible flow equations now become:

$$\beta \frac{\partial \hat{\rho}}{\partial \tilde{t}} + \tilde{u}_j \frac{\partial \hat{\rho}}{\partial \tilde{x}_j} + \hat{\rho} \frac{\partial \tilde{u}_j}{\partial \tilde{x}_j} = 0, \tag{1.17.7}$$

$$\beta \frac{\partial \tilde{u}_i}{\partial \tilde{t}} + \tilde{u}_j \frac{\partial \tilde{u}_i}{\partial \tilde{x}_j} + \frac{1}{\hat{\rho}} \frac{\partial \tilde{p}}{\partial \tilde{x}_i} = \frac{1}{Re\hat{\rho}} \frac{\partial \tilde{\tau}_{ij}}{\partial \tilde{x}_j}, \tag{1.17.8}$$

$$\beta \frac{\partial \hat{T}}{\partial \tilde{t}} + \tilde{u}_j \frac{\partial \hat{T}}{\partial \tilde{x}_j} = \frac{1}{RePr} \frac{1}{\hat{\rho}} \frac{\partial}{\partial \tilde{x}_i} \left(\tilde{\mu} \frac{\partial \hat{T}}{\partial \tilde{x}_j} \right), \tag{1.17.13}$$

$$\hat{\rho}\hat{T} = 1. \tag{1.17.14}$$

For $\hat{p} = 1$, the equations of incompressible flow are recovered.

The low Mach number limit of (1.17.12) is used in Chapters 2 and 11 in describing flows with heat addition. It can be stated in a more physical manner starting from an estimate for the size of the static pressure variations in a steady flow. With U and ΔU the characteristic velocity and velocity variation of the motion, and L the characteristic length scale, the accelerations have magnitude $U\Delta U/L$ and the pressure variations along the stream, Δp, have magnitude $\rho U \Delta U$. The velocity variation will be the same size as the velocity, or less ($\Delta U \le U$), so a (crude but conservative) estimate for the bound on the ratio of pressure variations to the ambient pressure level is

$$\frac{\Delta p}{p} \approx \frac{U^2}{(p/\rho)} \approx \frac{U^2}{a^2} = M^2. \tag{1.17.15}$$

For Mach numbers much less than unity, pressure variations are much less than ambient pressure. Variations in temperature, however, which can be driven by combustion processes, are not necessarily small compared to ambient temperatures. For Mach numbers much smaller than unity the equation of state is

$$p = p_{ref}[1 + O(M^2)] = \rho T. \tag{1.17.16}$$

In such situations large changes in temperature must be closely balanced by large changes in density, and the equation of state can be approximated as (to order M^2)

$$\rho T = \rho_{ref} T_{ref} = \text{constant}. \tag{1.17.17}$$

Incompressible flow ($\rho = \rho_{ref} = \text{constant}$) is included as a condition described by this equation of state.

2 Some useful basic ideas

2.1 Introduction

This chapter introduces a variety of basic ideas encountered in analysis of internal flow problems. These concepts are not only useful in their own right but they also underpin material which appears later in the book.

The chapter starts with a discussion of conditions under which a given flow can be regarded as incompressible. If these conditions are met, the thermodynamics have no effect on the dynamics and significant simplifications occur in the description of the motion.

The nature and magnitude of upstream influence, i.e. the upstream effect of a downstream component in a fluid system, is next examined. A simple analysis is developed to determine the spatial extent of such influence and hence the conditions under which components in an internal flow system are strongly coupled.

Many flows of interest cannot be regarded as incompressible so that effects associated with compressibility must be addressed. We therefore introduce several compressible flow phenomena including one-dimensional channel flow, mass flow restriction ("choking") at a geometric throat, and shock waves. The last of these topics is developed first from a control volume perspective and then through a more detailed analysis of the internal shock structure to show how entropy creation occurs within the control volume.

The integral forms of the equations of motion, utilized in a control volume formulation, provide a powerful tool for obtaining an overall description of many internal flow configurations. A number of situations are analyzed to show their application. These examples also serve as modules for building descriptions of more complex devices.

The last sections of the chapter introduce two related topics which lead into more detailed discussions in later chapters. The first is the role of viscous effects, as manifested in the creation of wall boundary layers, and their effect on flow regimes in channels and ducts. The second is the irreversibility of real (i.e. viscous) fluid motions, namely the fore and aft asymmetry of flow over bodies and through ducts, a key concept in understanding the behavior of flow devices.

2.2 The assumption of incompressible flow

Simplification in the analysis of fluid motions occurs when one can consider the density of a fluid particle to be invariant. If so, the continuity equation reduces to $\nabla \cdot \mathbf{u} = 0$ so the velocity field is solenoidal. Flows with this character are referred to as incompressible. The motion is defined by **u** and p and is independent of the thermodynamics. We examine under what conditions this approximation is valid, first for steady flow and then for unsteady flow.

2.2.1 Steady flow

The starting point in the assessment of whether a flow can be considered incompressible is the continuity equation (1.9.4):

$$\nabla \cdot \mathbf{u} = -\frac{1}{\rho}\frac{D\rho}{Dt}. \tag{1.9.4}$$

If velocity changes in the flow are of magnitude ΔU and occur over a length L, the sizes of the individual terms on the left-hand side of (1.9.4) are $\Delta U/L$. The term on the right-hand side will be of order $(U/L)(\Delta\rho/\rho)$, where U and ρ are representative magnitudes of the velocity and density. The task is to assess under what situations the term on the right-hand side will be much smaller than the individual terms on the left, i.e. when the ratio $(\Delta\rho/\rho)/(\Delta U/U)$ is much less than unity.

The equation of state for a perfect gas implies that small changes in density scale approximately as

$$\frac{\Delta\rho}{\rho} \sim \frac{\Delta p}{p} - \frac{\Delta T}{T}. \tag{2.2.1}$$

Density changes can occur due to variations in pressure or temperature. In general, there are three sources of pressure differences for a flowing fluid: (i) fluid accelerations (inertial forces), (ii) body forces, represented here by centrifugal force, and (iii) fluid friction. Heat addition or extraction can change temperature. These four effects, and their impact on density changes, are now discussed in turn.

(i) For a steady flow with characteristic velocity magnitude U and velocity change ΔU, the pressure differences along the stream have magnitude $\Delta p \sim \rho U \Delta U$ (Section 1.17). Thus

$$\frac{\Delta U}{U} \sim \frac{\Delta p}{\rho U^2}. \tag{2.2.2}$$

For situations without externally imposed temperature differences, the quantities $(\Delta\rho/\rho)$ and $(\Delta p/p)$ in (2.2.1) have similar magnitudes. The ratio $(\Delta\rho/\rho)/(\Delta U/U)$ can thus be estimated as

$$\left[\frac{\Delta\rho}{\rho}\right] \Big/ \left[\frac{\Delta U}{U}\right] \sim \frac{\rho U^2}{p} \sim \frac{U^2}{a^2} \sim M^2. \tag{2.2.3}$$

The criterion for a flow to be viewed as incompressible is thus $M^2 \ll 1$. If this criterion is met, the expression for the stagnation pressure, (1.16.7), can be expanded as a power series in M^2, the first two terms of which yield

$$p_t = p + \frac{1}{2}\rho u^2. \tag{2.2.4}$$

Equation (2.2.4) is the definition of stagnation pressure used for incompressible flow. It can also serve as one guide to when flow can be regarded as incompressible through examining the ratio $\frac{1}{2}\rho u^2/(p_t - p)$ for a compressible flow. This ratio differs from unity by about 2% at $M = 0.3$ and by less than 5% for $M < 0.4$ so that, depending on the accuracy required, the incompressible flow assumption can be used even up to these values. A somewhat more conservative guide is to ensure that the density ratio $\Delta\rho/\rho_t$ is much less than unity, say less than 5%. This implies that the Mach number is limited to roughly 0.3. The two results are quoted because a point to note is that the applicability of the approximation depends on the specific usage in mind.

(ii) In a rotating environment such as a turbomachinery impeller, pressure changes can occur due to centrifugal forces. Consider the balance between pressure difference and centrifugal force for fluid at rest in a radial channel rotating about an axis with rotation speed Ω. Over a small length Δr in the radial direction

$$\Delta p \sim \rho\Omega^2 r^2 \frac{\Delta r}{r}. \tag{2.2.5}$$

As in (i), we set the condition under which we can neglect effects of compressibility as $\Delta\rho/\rho(\sim \Delta p/p) \ll 1$. Applying this to (2.2.5) and defining a rotational Mach number $M_\Omega = \Omega r/a$, we find the condition as

$$M_\Omega^2 \frac{\Delta r}{r} \ll 1. \tag{2.2.6}$$

The quantities Δr and r are often not greatly different and thus $M_\Omega^2 \ll 1$ gives a conservative criterion.

(iii) Departures from incompressible flow can also arise due to viscous effects. An example is furnished by fully-developed flow in a constant area duct of length L. For this situation, the pressure drop can be represented in terms of the skin friction coefficient, C_f (= wall shear stress$/\frac{1}{2}\rho\overline{u}^2$, where $\overline{u}$ denotes the mean velocity in the duct) and the ratio, L/d_H, length to hydraulic diameter (4 times the cross-sectional area divided by the wetted perimeter)[1] as

$$\Delta p = \frac{1}{2}\rho\overline{u}^2 \cdot (4C_f)\frac{L}{d_H}. \tag{2.2.7}$$

Departure from incompressible flow occurs when the ratio $\Delta\overline{u}/\overline{u}$, (hence $\Delta\rho/\rho$) becomes appreciable compared to unity. Friction-dominated flow can be regarded as incompressible when $\Delta p/p$ is much less than unity or when $C_f M^2\ (L/d_H) \ll 1$, with $M = \overline{u}/a$.

(iv) Even with the Mach number much less than unity, departures from incompressible behavior can occur when external heating or cooling is imposed or when internal heat sources, such as combustion, are present. In this situation, the pressure changes due to dynamical effects will be (as described just above) of order M^2 compared to ambient pressure. Changes in density can thus be expressed as

$$\frac{\Delta\rho}{\rho} \sim \frac{\Delta T_{imposed}}{T_{ref}} + O(M^2), \tag{2.2.8}$$

where $\Delta T_{imposed}$ is a representative imposed temperature difference (for example, between the wall and free stream or between the inlet and exit of a combustor) and T_{ref} is a reference temperature (e.g. ambient temperature or combustor inlet temperature). For example, temperature changes can be of the same (or larger) magnitude as the ambient temperature in combustion or in mixing of streams of non-uniform temperature. If so, density changes can have magnitudes comparable to the initial density whatever the Mach number. Thus, $\Delta T_{imposed}/T_{ref}$ must be much less than unity for density changes to be neglected.

[1] The concept of hydraulic diameter is often used as a means to correlate friction factor data for turbulent flow in pipes of different cross-section. Discussion of the hydraulic diameter, as well as data for pipes of non-circular cross-section, can be found in the work by Schlichting (1979).

2.2.2 Unsteady flow

Departures from incompressible behavior can also be caused by flow unsteadiness. Following Lighthill (1963), to assess such departures we compare the sizes of terms on the left-hand and right-hand sides of the continuity equation for a situation where the flow is periodic with radian frequency, ω, the application of most interest in fluid machinery. The magnitude of the density fluctuations is

$$\frac{1}{\rho}\frac{D\rho}{Dt} \sim \omega\frac{\Delta\rho}{\rho} \sim \omega\frac{\Delta p}{p}, \tag{2.2.9}$$

where $\Delta\rho$ and Δp are the perturbations in density and pressure associated with fluctuations at the frequency ω, and ρ and p are the mean or ambient levels of these quantities. If ΔU is the magnitude of a typical fluctuation in velocity and L is the relevant length of the device, balancing the local (unsteady) fluid accelerations with pressure differences in the momentum equation leads to

$$\Delta p \sim \rho\omega L \Delta U. \tag{2.2.10}$$

There may also be terms of order $\rho U \Delta U$ contributing to Δp, but if $M^2 \ll 1$, these will not invalidate the conditions under which the flow can be regarded as incompressible.

The above estimate of the pressure fluctuations shows the term $(1/\rho)\,(D\rho/Dt)$ in the continuity equation (1.9.4), is of magnitude $\omega^2 L\Delta U/a^2$, whereas the magnitude of the individual terms in $\nabla \cdot \mathbf{u}$ are $\Delta U/L$. The criterion for the flow to be regarded as incompressible is therefore $\omega^2 L^2/a^2 \ll 1$. An interpretation of this criterion is that L must be small compared to the "radian wavelength", a/ω, of a sound wave of frequency ω. This condition can also be expressed in terms of the reduced frequency β $(= \omega L/U)$, which was defined in Chapter 1, as $\beta^2 M^2 \ll 1$.

To summarize, a flow can be considered incompressible under the following circumstances:

(a) The square of the Mach number is small compared to unity ($M^2 \ll 1$).

(b) In a rotating environment

$$\left(\frac{\Omega r}{a}\right)^2 \frac{\Delta r}{r} = M_\Omega^2 \frac{\Delta r}{r} \ll 1.$$

(c) In a duct flow involving friction, $C_f M^2\,(L/d_H) \ll 1$.

(d) In flows involving imposed heat addition from external or internal sources, $\Delta T_{imposed}/T_{ref} \ll 1$.

(e) For unsteady flow, $(\omega L/a)^2 \ll 1$ or, equivalently $(\beta M)^2 \ll 1$.

2.3 Upstream influence

A question often encountered with fluid machinery is when components should be considered aerodynamically coupled, in the sense that there is significant interaction between them. One aspect of this concerns the spacing needed for mixing of wakes from upstream components before the flow enters the downstream component. Another, and very different, consideration, however, is that of upstream influence. By this is meant the axial extent of the upstream non-uniformity in pressure and velocity which is created by a downstream component or geometrical feature such as a bend or row of struts. This impacts not only upstream component behavior but also the choice of measurement locations

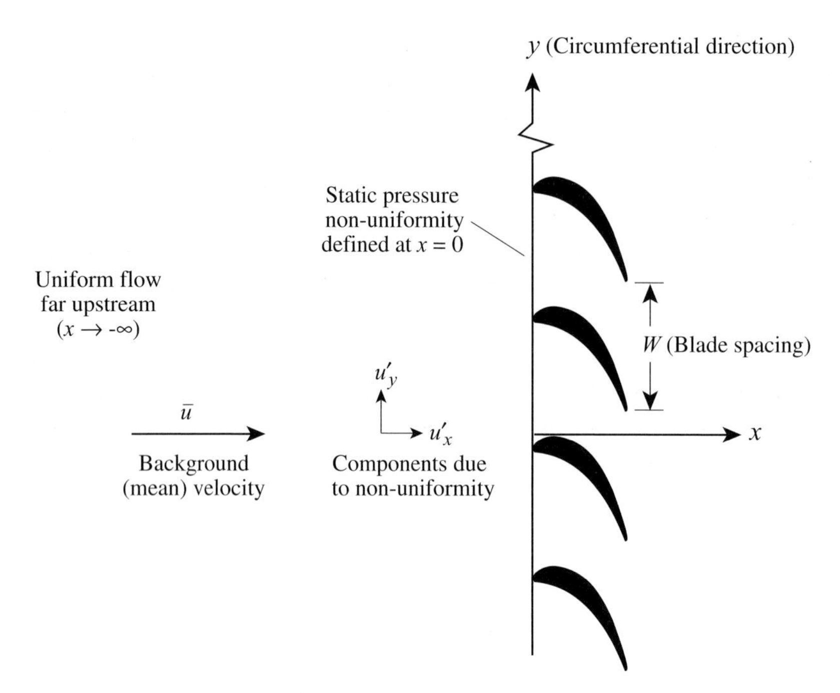

Figure 2.1: Flow domain used in the estimation of the upstream influence region for a periodic array (turbine blade row); the region of interest is $x < 0$.

to obtain accurate performance representations as well as selection of boundaries for computational domains. Upstream influence is examined in several contexts in the book. In this introduction to the topic we concentrate on the development of basic scaling rules which allow estimates of the magnitude of the effect in many situations.

2.3.1 Upstream influence of a circumferentially periodic non-uniformity

We proceed by example, starting with the upstream effect of a circumferentially periodic flow non-uniformity, such as that presented by a turbomachinery blade row. A two-dimensional representation of this is sketched in Figure 2.1, which shows a row of turbine airfoils with spacing W; the figure can be taken as representative of a blade row in an annular region of high hub/tip radius ratio. The x-coordinate represents axial distance and the y-coordinate represents distance in the circumferential direction around the turbomachine annulus. The aerodynamic loading on the blading causes the static pressure to vary circumferentially, with period W, upstream of the blade row, and the specific issue is how this static pressure variation attenuates with upstream axial distance.

The length scale in the problem which characterizes the non-uniformity in the y-direction is the spacing, W. If this is the relevant length scale over which the flow quantities vary upstream of the blades, for high Reynolds number flow an order of magnitude analysis shows viscous forces are much smaller than inertial forces[2] in this upstream region and an inviscid description of the

[2] If the characteristic velocity has magnitude U the inertial and viscous forces have magnitudes $\rho U^2/W$ and $\mu U/W^2$, respectively, in the upstream region. The ratio of the two is ν/UW or $1/$(Reynolds number).

pressure field will suffice. Further, while the ratio of the non-uniformities in pressure or velocity (for example, the variation in static pressure about the mean compared to the dynamic pressure based on average axial velocity) *near* the blades may be of order unity, the question of interest concerns the upstream decay of these variations. Over much of the region of interest flow non-uniformities will be small, in a non-dimensional sense, with the implication being that a linearized description is appropriate. The problem can thus be posed as determining the upstream pressure variations about a uniform inviscid flow due to the presence of the blade row shown in Figure 2.1. The treatment below is for steady incompressible flow, but comments on the extension to the compressible case will be given.

As implied by the figure, the background flow, which can be thought of as that existing in the absence of the blading, is axial. The velocity components and pressure field for this mean or background flow are: $u_x = \overline{u} =$ constant; $u_y = 0$; $p = \overline{p} =$ constant, where $\overline{p}$ is the static pressure far upstream of the blades. The flow field can be represented as

$$u_x = \overline{u} + u'_x, \quad u_y = u'_y, \quad p = \overline{p} + p', \tag{2.3.1}$$

where $(u'_x/\overline{u})$, $(u'_y/\overline{u})$, and $(p'/(\frac{1}{2}\rho\overline{u}^2))$ are all taken to be much less than unity.

Substituting (2.3.1) into the continuity and momentum equations and (based on the assumption of small non-uniformities) neglecting terms which are products of the disturbance velocities yields a set of linearized equations for the two velocity components and the pressure:

$$\overline{u}\frac{\partial u'_x}{\partial x} = -\frac{1}{\rho}\frac{\partial p'}{\partial x}, \tag{2.3.2a}$$

$$\overline{u}\frac{\partial u'_y}{\partial x} = -\frac{1}{\rho}\frac{\partial p'}{\partial y}, \tag{2.3.2b}$$

$$\frac{\partial u'_x}{\partial x} + \frac{\partial u'_y}{\partial y} = 0. \tag{2.3.2c}$$

Differentiation of (2.3.2a) with respect to x and (2.3.2b) with respect to y and use of (2.3.2c), gives Laplace's equation for the disturbance pressure field p' $(= p - \overline{p})$:

$$\nabla^2 p' = \frac{\partial^2 p'}{\partial x^2} + \frac{\partial^2 p'}{\partial y^2} = 0. \tag{2.3.3}$$

An immediate conclusion about upstream influence can be drawn from the structure of (2.3.3). Laplace's equation has no intrinsic length scale. If a length scale, W, is specified in the y-direction, as is the case for a blade row of spacing W, the length scale in the x-direction, which is essentially the extent of the upstream influence, must also be W. This idea is basic in understanding upstream influence in the situations addressed, and we now proceed to make it more quantitative.

Regardless of the loading on the blades, any periodic pressure distribution at $x = 0$ can be represented as a Fourier series in y:

$$p'|_{x=0} = \sum_{\substack{k=-\infty \\ k\neq 0}}^{\infty} b_k e^{(2\pi i k y/W)}. \tag{2.3.4}$$

To match this boundary condition, the solution for p' must also be of this form:

$$p' = \sum_{\substack{k=-\infty \\ k\neq 0}}^{\infty} f_k(x)\, b_k e^{(2\pi i k y/W)}. \tag{2.3.5}$$

Substituting (2.3.5) into (2.3.3) yields a form for $f_k(x)$ which has exponentials $e^{2\pi kx/W}$ and $e^{-2\pi kx/W}$. The solutions of physical interest decay with upstream distance and must be bounded at $x = -\infty$, so the form for p' is

$$p' = \sum_{\substack{k=-\infty \\ k\neq 0}}^{\infty} b_k e^{(2\pi |k| x/W)} e^{(2\pi i k y/W)}. \tag{2.3.6}$$

Equation (2.3.6) exhibits several generic features of the upstream pressure field. First, the upstream decay distance, say the distance at which the non-uniformity is reduced to some given percentage of its value at $x = 0$, is proportional to the y-direction length scale. For a disturbance with wavelength W in the y-direction (the longest wavelength disturbances in this situation) at a location a distance $W/2$ upstream of the blade row the non-uniformity is 4% of the value at $x = 0$. Second, the lowest Fourier component ($|k| = 1$) has the greatest upstream influence. Higher spatial harmonic components have an upstream influence with an axial extent smaller by a factor of $1/|k|$, where k is the harmonic number. Unless the pressure profile is skewed strongly to higher harmonics, the first Fourier component is the most important in setting the upstream influence. Third, although nonlinearities will alter the quantitative rate of decay near the blades, we are dealing with non-uniformities which are small over most of the region of interest, and nonlinear effects will not appreciably affect either the extent or which harmonic components are most important. Fourth, although the example shown is for a non-uniformity with a length scale equal to the blade spacing, it is applicable to any periodic non-uniformity. For instance, the non-uniformity associated with an inlet distortion in a compressor can have a y-direction length scale of the circumference of the machine, implying a correspondingly large extent of upstream influence. Finally, for computations, the upstream boundary of the domain should be far enough away so that the flow at this location is unaffected by downstream non-uniformities. The specific requirement thus depends on the circumferential length scale in the problem of interest, and this is also true for the question of when components can be considered aerodynamically coupled.

2.3.2 Upstream influence of a radial non-uniformity in an annulus

A second example concerns the radially non-uniform flow in an annular region. Figure 2.2 shows an annulus with inner radius, r_i, and outer radius r_o. At an axial location $x = 0$, there is a non-uniform pressure or velocity field, as would occur with a downstream geometry such as a blade row or duct curvature. The question again is how far upstream will the influence of the non-uniformity extend. Following the discussion in Section 2.3.1, it suffices to develop a linearized, inviscid, steady description of the variations in static pressure and velocity about a uniform axial background flow.

The interest here is in radial variations so the non-uniformities about the background state of $u_x = \overline{u} =$ constant and $p = \overline{p} =$ constant are taken as axisymmetric, i.e. $\partial/\partial\theta = 0$, with $u_\theta = 0$. Using cylindrical coordinates, the linearized equations which describe the non-uniformities u'_r, u'_x, and p' are the r- and x-components of the inviscid momentum equation and the incompressible flow

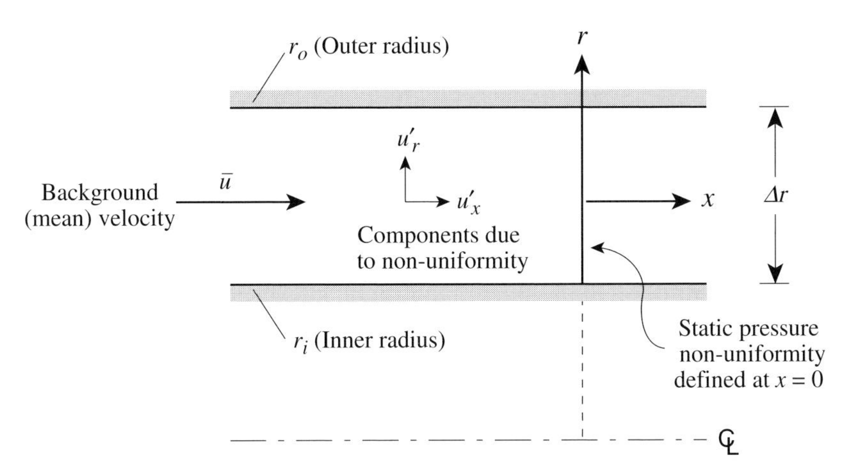

Figure 2.2: Annular flow geometry used in the estimation of the upstream influence region for axisymmetric flow; the region of interest is $x < 0$.

form of the continuity equation (see Section 1.14)

$$\bar{u}\frac{\partial u'_x}{\partial x} = -\frac{1}{\rho}\frac{\partial p'}{\partial x}, \tag{2.3.7a}$$

$$\bar{u}\frac{\partial u'_r}{\partial x} = -\frac{1}{\rho}\frac{\partial p'}{\partial r}, \tag{2.3.7b}$$

$$\frac{\partial u'_r}{\partial r} + \frac{u'_r}{r} + \frac{\partial u'_x}{\partial x} = 0. \tag{2.3.7c}$$

Differentiating (2.3.7a) with respect to x and (2.3.7b) with respect to r, and invoking the continuity equation leads to Laplace's equation for p' in cylindrical coordinates:

$$\nabla^2 p' = \frac{\partial^2 p'}{\partial r^2} + \frac{1}{r}\frac{\partial p'}{\partial r} + \frac{\partial^2 p'}{\partial x^2} = 0. \tag{2.3.8}$$

Further simplification of (2.3.8) is possible for annular regions of high hub/tip radius ratio. The non-uniformities of interest have a radial variation with length scale $\Delta r = r_o - r_i$ (or less). The ratio of the second term in (2.3.8) to the first is of order $(\Delta r/r_m)$, where r_m is the annulus mean radius. For annuli of high hub/tip radius ratio, where $(\Delta r/r_m)$ is much less than unity, this term can be neglected, and (2.3.8) becomes

$$\frac{\partial^2 p'}{\partial r^2} + \frac{\partial^2 p'}{\partial x^2} = 0. \tag{2.3.9}$$

This is the same equation that was derived in Section 2.3.1, although the two coordinates are here x and r (axial and radial), compared with x and y (axial and circumferential) in Section 2.3.1.

The boundary conditions for solution of (2.3.9) are different than for a periodic geometry. Appropriate conditions are the specification of the radial static pressure non-uniformity at $x = 0$ and the imposition of no normal velocity at the inner and outer radii, $u_r = 0$ at $r = r_i$ and $r = r_o$, for any value of x. From (2.3.7b), this is equivalent to the condition that the radial derivative of the static pressure non-uniformity is zero at the inner and outer radii: $\partial p'/\partial r = 0$ at $r = r_i$ and $r = r_o$.

Solutions to (2.3.9) can again be written as a Fourier series. From Section 2.3.1, however, we know that the first Fourier component, which has the largest length scale, sets the maximum extent of upstream influence. We thus need to consider only this component. Using similar arguments as those in Section 2.3.1, the solution for p' can be written as

$$p' = b_1 e^{(\pi x/\Delta r)} \cos\left[\frac{\pi(r - r_i)}{\Delta r}\right]. \tag{2.3.10}$$

The upstream radial static pressure field in the annulus has exponential decay similar to the periodic disturbance, although the quantitative features are different. The previous comments concerning upstream influence thus capture the basic scaling and also apply to this second example.

The discussion up to now has addressed incompressible flow. To extend the ideas to compressible flow for moderate subsonic Mach numbers ($M_x < 0.6$, say, where M_x is the axial Mach number associated with the mean flow) one can use the Prandtl–Glauert transformation (Liepmann and Roshko, 1957; Sabersky *et al.*, 1989) to convert the incompressible solutions to compressible form. For a subsonic compressible flow the first Fourier component of the radial non-uniformity in the upstream pressure field has the form

$$p' = b_1 e^{\pi x/(\Delta r\sqrt{1-M_x^2})} \cos\left[\frac{\pi(r - r_i)}{\Delta r}\right]. \tag{2.3.11}$$

The axial extent of the upstream influence is thus reduced as the axial Mach number increases.

2.4 Pressure fields and streamline curvature: equations of motion in natural coordinates

2.4.1 Normal and streamwise accelerations and pressure gradients

The momentum equation for inviscid steady flow is

$$(\mathbf{u} \cdot \boldsymbol{\nabla})\mathbf{u} = -\boldsymbol{\nabla} p/\rho \tag{2.4.1}$$

for incompressible and compressible fluids. With **u** as the magnitude of the velocity, l as the distance along a streamline,[3] $\mathbf{l}$ as a unit vector tangent to the streamline, n as the outward distance along the principal normal to the streamline, and $\mathbf{n}$ as an outward-pointing unit vector normal to the streamline, (2.4.1) can be written in terms of changes along and normal to the streamlines as

$$u\frac{\partial(u\mathbf{l})}{\partial l} = \mathbf{l}u\frac{\partial u}{\partial l} + u^2\frac{\partial \mathbf{l}}{\partial l} = -\frac{1}{\rho}\left(\mathbf{l}\frac{\partial p}{\partial l} + \mathbf{n}\frac{\partial p}{\partial n}\right). \tag{2.4.2}$$

There is a component of fluid acceleration along the streamlines and a component normal to the streamlines. The former is a consequence of changes in the velocity magnitude and is related to the

[3] Some notes on nomenclature and conventions: The definition of the unit normal vector, $\mathbf{n}$, as pointing in the direction outward from the center of curvature of a streamline is opposite to the usual convention for the principal normal in the description of a space curve. It is adopted, however, to be consistent both with the definition of the "n-direction" for natural coordinates and with the use of a positive outward pointing normal in the description of control volumes. The variable, l, is used for streamwise distance instead of the perhaps more mnemonic s to avoid use of s for both entropy and streamwise distance. (We would otherwise encounter the quantity $\partial s/\partial s$ later in the chapter!) To help reinforce this convention, $\mathbf{l}$ is used to denote the unit vector in the streamwise direction.

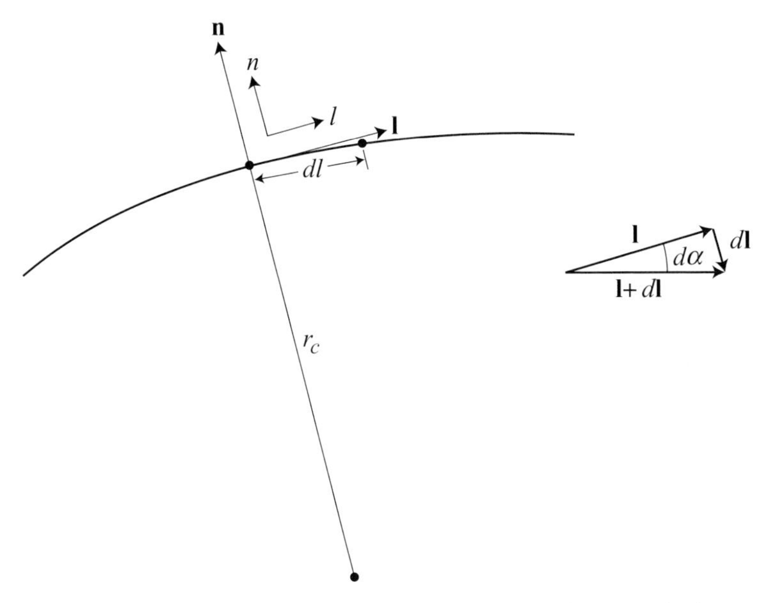

Figure 2.3: Normal and streamwise coordinates and rate of change of unit vector, **l**, in streamwise direction.

component of the pressure gradient in the streamwise direction:

$$u \frac{\partial u}{\partial l} = -\frac{1}{\rho} \frac{\partial p}{\partial l}. \tag{2.4.3}$$

The second is a consequence of changes in the direction of the velocity. The unit vector **l** cannot have changes in magnitude, so its changes must be in the normal direction. As indicated in Figure 2.3 the change in **l** is given by $\mathbf{n}d\alpha$, where $d\alpha$ is the change in angle of the streamline over a distance l. With r_c denoting the local radius of curvature of the streamline

$$\frac{\partial \mathbf{l}}{\partial l} = -\frac{\mathbf{n}}{r_c}. \tag{2.4.4}$$

The minus sign means that the acceleration is in the direction towards the local center of curvature. The component of the pressure gradient normal to the streamline is therefore

$$\rho \frac{u^2}{r_c} = \frac{\partial p}{\partial n}. \tag{2.4.5}$$

The quantity u^2/r_c is the centripetal acceleration familiar from particle dynamics. Equation (2.4.5) states that, in a steady flow, streamline curvature is associated with a component of the pressure gradient force normal to the streamlines and pointing toward the local center of curvature.

2.4.2 Other expressions for streamline curvature

Equation (2.4.5) can be derived in another manner which further illustrates the l, n coordinate system. Consider the steady, inviscid, two-dimensional flow through the control surface of Figure 2.4. The upper and lower parts of the control surface (AB and DC) are along streamlines and the left and right

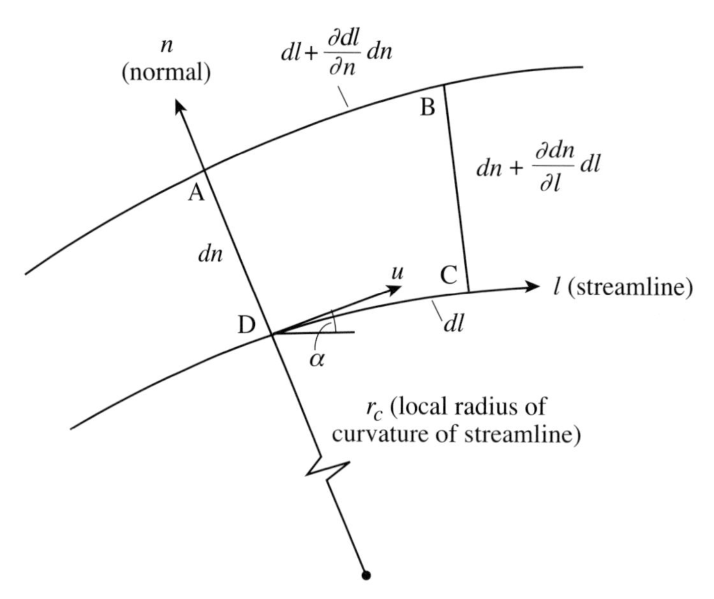

Figure 2.4: Natural coordinates: u, α are functions of l and n.

hand parts (DA and BC) are normal to the streamlines. The streamlines and their normals define a natural coordinate system (l, n) with n measured normal to streamlines and l the distance along the streamline. The local radius of curvature of the streamline is r_c and α is the local angle of the streamline with respect to a reference direction.

The flux of momentum in the n-direction out of the control volume is equal to the net force on the control surface. The only forces are pressure forces. The net momentum flux in the n-direction is $-\rho u^2 d\alpha dn$ (the difference between the momentum flux out and the momentum flux in), plus higher order terms in the quantities dn and $d\alpha$. The net pressure force in the n-direction, along the radius of curvature, is $(-\partial p/\partial n)dndl$, plus higher order terms. Equating the net momentum flux to the force on the element, using the relation between changes in streamline angle, $d\alpha$, the distance along the streamline, and the local radius of curvature ($d\alpha = dl/r_c$,), and taking the limit as dn and dl become vanishingly small, yields (2.4.5).

The l- and n-directions are referred to as natural, or intrinsic, coordinates. In addition to l and n components of the momentum equation ((2.4.3) and (2.4.5)) the other necessary equations for a two-dimensional inviscid, adiabatic flow are:

Continuity: $\rho u dn =$ constant (2.4.6)

Energy (constant entropy along a streamline): $\dfrac{\partial s}{\partial l} = 0.$ (2.4.7)

These plus the equation of state and the boundary conditions describe the flow field. (For a three-dimensional flow there would be a third direction, perpendicular to both the streamline and the normal (Tsien, 1958).)

It is often helpful to cast these natural coordinates in terms of the angle, α, which the streamlines make with a reference direction, as indicated in Figure 2.4 (Liepmann and Roshko, 1957). This allows

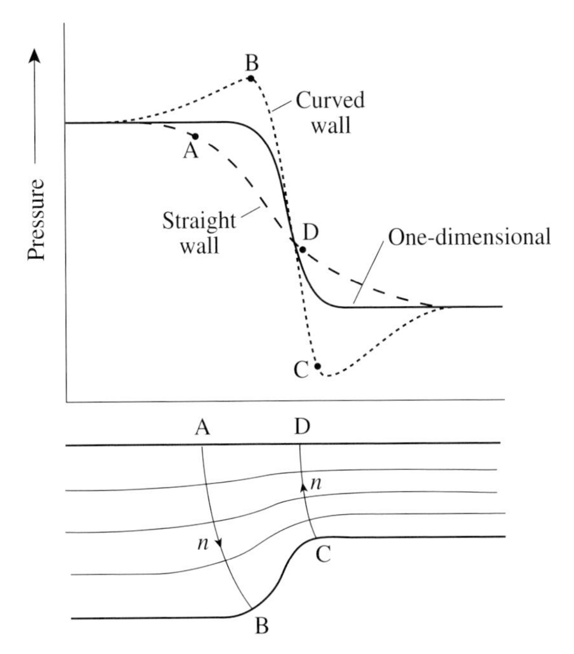

Figure 2.5: Streamlines and wall static pressure distributions for two-dimensional contractions (Shapiro, 1972).

another interpretation of the normal equation of motion. The local radius of curvature is related to the flow angle by $1/r_c = \partial\alpha/\partial l$ so that (2.4.5) can be written in terms of the flow angle as

$$\frac{\partial p}{\partial n} = \rho u^2 \frac{\partial \alpha}{\partial l}. \tag{2.4.8}$$

Equation (2.4.8) states that a normal component of the pressure gradient exists if the velocity vector changes direction along a streamline.

Streamline curvature is a feature of essentially all flows of technological interest, although (depending on the magnitude of the curvature) the pressure difference normal to the streamline may or may not have substantial impact on the effect being studied. A flow which is uni-directional in an overall sense, but in which streamline curvature can be important, is a contraction in a two-dimensional asymmetric channel, as shown in Figure 2.5. The streamlines (taken from flow visualization pictures) and the measured pressure distributions on each of the walls of the channel are indicated (Shapiro, 1972). The sense of the normal component of the pressure gradient is also sketched. The streamline curvature has one sign in the upstream part of the contraction and another sign at the downstream part, because the radius of curvature points one way near the start of the contraction and the other way towards the end. The quasi-one-dimensional pressure distribution, based on the local flow through area, is also indicated. For this particular geometry the differences in pressure are a substantial fraction of the dynamic pressure. Depending on the objective, inclusion of the pressure differences in the normal direction in the problem description could be important.

The ideas concerning streamline curvature and normal components of the pressure gradient can be related to the results of Section 2.3, where linearized forms of the momentum equation were used to derive upstream static pressure variations. Within the approximation made, the x-direction was the streamwise direction and the y-direction the normal direction in the first example, while the r-direction was the normal direction in the second. If the departures from uniform flow ($u_x = \overline{u} =$ constant) are small such that products of terms representing the non-uniformity can be neglected, the angle the flow makes with the x-axis, α, is given by (for the example in Section 2.3.1)

$$\tan\alpha \approx \alpha \approx \frac{u'_y}{\overline{u}}. \tag{2.4.9a}$$

For the axisymmetric situation of Section 2.3.2 the corresponding expression is

$$\tan\alpha \approx \alpha \approx \frac{u'_r}{\overline{u}}. \tag{2.4.9b}$$

For small departures from uniformity (so that $x \approx l$ and $u^2 \approx \overline{u}^2$) (2.4.8) becomes (using a prime to denote the perturbation from uniform flow)

$$\frac{\partial p'}{\partial y} = -\rho\overline{u}^2\frac{\partial \alpha'}{\partial x}. \tag{2.4.10}$$

Using (2.4.9a),

$$\frac{\partial p'}{\partial y} = -\rho\overline{u}\frac{\partial u'_y}{\partial x}, \tag{2.4.11a}$$

which is the expression given for a two-dimensional flow in Section 2.3.1. The corresponding term for the axisymmetric flow of Section 2.3.2 is

$$\frac{\partial p'}{\partial r} = -\rho\overline{u}\frac{\partial u'_r}{\partial x}. \tag{2.4.11b}$$

Equations (2.4.11a) and (2.4.11b) can be interpreted as linearized forms of the expression relating streamline curvature and the normal component of pressure gradient.

To summarize Sections 2.3 and 2.4, in many of the flows to be examined there are regions in which the motion can be viewed in terms of a balance between pressure and inertial forces. The connection between streamline curvature, fluid accelerations, and pressure fields, shown compactly in (2.4.5) and (2.4.8) is an important key in understanding such flows.

2.5 Quasi-one-dimensional steady compressible flow

When the conditions given in Section 2.2 are not met, the motion cannot be considered incompressible and the coupling of thermodynamics and dynamics which occurs in a compressible flow must be addressed. In this section we describe an approach for analyzing compressible flow which is particularly helpful in internal flow configurations. Geometries encountered in fluid machinery and propulsion systems can often be viewed as duct- or channel-like because the length which characterizes changes in the geometry along the flow direction is much larger than the channel width. Under such conditions, perhaps to a surprising degree when these conditions are only partially met, a quasi-one-dimensional description of the flow has considerable utility and, as a result, has found wide application for analysis of fluid devices. Nozzles are a prime example of such geometries, but turbomachinery blading can also be approached in this manner. The phrase "quasi-one-dimensional" means here that flow properties are functions of one variable only, for example the distance along the channel or, for isentropic flow, the local channel area.

The quasi-one-dimensional approach assumes: (i) the channels have small divergence (or convergence), (ii) curved channels have a large radius of curvature compared to their width, and (iii) the velocity and temperature are uniform across the channel. A consequence of (i) is that the velocity components at a given station along the channel are nearly parallel. If so, the velocity components normal to the mean direction of the channel are small compared to the velocity components along the mean direction and the transverse accelerations thus also small compared to some measure of streamwise accelerations (say, the dynamic pressure). The consequence of (ii) is that the static pressure difference across the channel due to streamline curvature is small. As developed in Section 2.4, the pressure difference across (normal to) the channel, Δp_n, is roughly

$$\Delta p_n \approx \frac{\partial p}{\partial n} W = \rho u^2 \frac{W}{r_c},$$

where r_c is a representative value of the radius of curvature of the channel. Taking the pressure difference along the channel, Δp_l, to be some appreciable fraction of the dynamic pressure $\rho u^2/2$, as in many cases of interest, the ratio of the normal pressure difference to the pressure difference along the channel thus scales as (dropping the numerical factors)

$$\frac{\Delta p_n}{\Delta p_l} \propto \frac{W}{r_c}.$$

The inference is that, if both (i) and (ii) hold, static pressure differences across the channel can be neglected and the pressure regarded as a function of the streamwise coordinate only. Further, the velocities need not be distinguished from the components along the mean direction of the channel. The above arguments also imply that the quasi-one-dimensional treatment applies locally to the behavior of a given slender streamtube even if large cross-stream variations in static pressure exist.

For inviscid flows the assumption of velocity uniformity (iii) can be quite a good approximation, but this cannot hold across the whole channel for a viscous fluid, which has zero velocity at the wall. Effects of viscosity and heat conduction, however, can be taken into account in an approximate manner within the one-dimensional approach. Further, within the framework of the theory effects of velocity and temperature non-uniformities can be accounted for by using appropriate average values. We present only a summary of the methodology; detailed exposition can be found in a number of texts, for example Shapiro (1953), Crocco (1958), Anderson (1990), and Hill and Peterson (1992).

2.5.1 Corrected flow per unit area

On a one-dimensional basis, if ρ and u are the density and velocity at a given station, the mass flow through the area, A, at that station can be written as

$$\dot{m} = \rho u A. \tag{2.5.1}$$

We wish to cast this in terms of stagnation quantities p_t and T_t, which serve as useful references. The first step is to use the perfect gas equation of state to give

$$\dot{m} = \frac{p}{p_t} \frac{T_t}{T} \frac{p_t}{RT_t} u A. \tag{2.5.2}$$

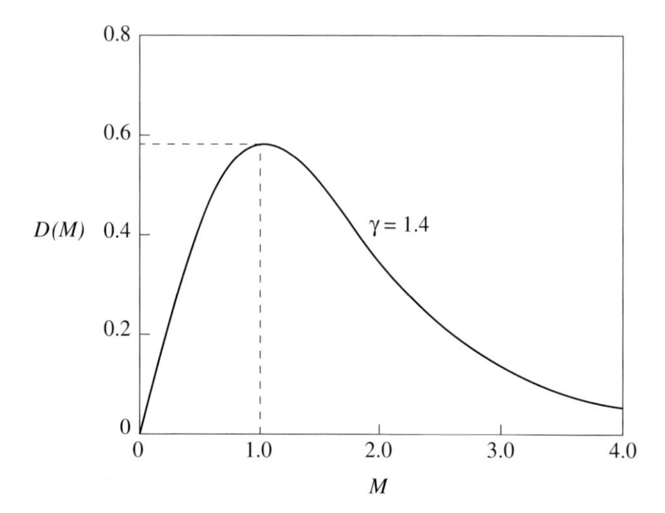

Figure 2.6: Corrected flow function, $D(M)$ versus M; $\gamma = 1.4$.

Introducing the relations between stagnation and static quantities in terms of Mach number ((1.16.6) and (1.16.7), $T_t/T = 1 + (\gamma - 1)M^2/2$, $p_t/p = [1 + (\gamma - 1)M^2/2]^{\gamma/\gamma-1}$) and writing the velocity in terms of the Mach number and the speed of sound provides a relation for the non-dimensional variable sometimes referred to as *corrected flow per unit area*. For a given gas (given value of R and specific heat, γ), the corrected flow per unit area (the quantity on the left-hand side of (2.5.3)) is a function of Mach number only:

$$\frac{\dot{m}\sqrt{RT_t}}{Ap_t\sqrt{\gamma}} = \frac{M}{\left(1 + \frac{\gamma - 1}{2}M^2\right)^{\frac{1}{2}\left(\frac{\gamma+1}{\gamma-1}\right)}} = D(M). \tag{2.5.3}$$

The corrected flow function, $D(M)$, is plotted in Figure 2.6 for $\gamma = 1.4$.

Examination of (2.5.3) and Figure 2.6 shows several important features. For a given Mach number, the physical mass flow per unit area ($\dot{m}/A$, in kg/(s m^2)) is proportional to the stagnation pressure and inversely proportional to the square root of the stagnation temperature, with the stagnation pressure and temperature interpreted as local values. Figure 2.6 shows that corrected flow per unit area rises as the Mach number increases for $M < 1$, falls as the Mach number increases for $M > 1$, and has a maximum at $M = 1$. The value of the maximum depends on γ and is 0.579 for $\gamma = 1.4$. For air at room conditions (20 °C, 0.1013 MPa), the dimensional maximum flow per unit area, $\dot{m}/A$, is 239 kg/(s m^2). In terms of fluid component and system performance, similarity of operating regimes implies similar Mach numbers and thus similar corrected flows per unit area.

The corrected flow function, $D(M)$, can also be viewed in a complementary fashion. For steady isentropic flow in a channel, stagnation quantities and mass flow are constant, so that the product DA is also. Denoting sonic conditions ($M = 1$) by $(\ )^*$,

$$\frac{D^*}{D(M)} = \frac{A(M)}{A^*}. \tag{2.5.4}$$

The sonic condition occurs with D a maximum at $D(1) = D^*$ and the area, A, a minimum at A^*. The quantity A/A^* provides a useful measure of how much area margin one has to allow to pass a desired

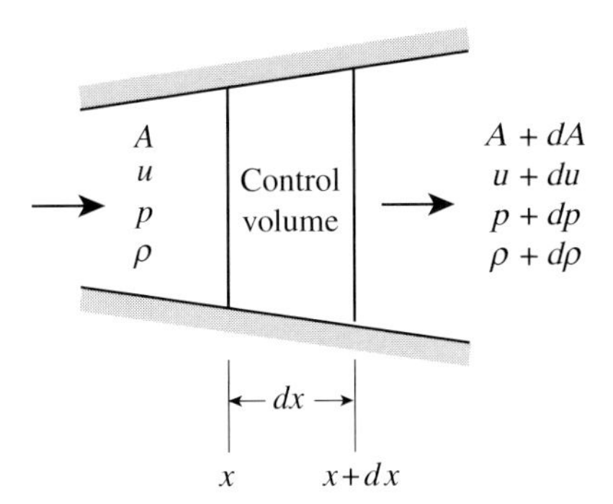

Figure 2.7: Elementary control volume for analysis of quasi-one-dimensional channel flow.

flow. The value of A/A^* is 1.09 for $M = 0.7$ and drops to 1.009 at $M = 0.9$ so devices that operate with Mach numbers near unity can exhibit substantial changes in Mach number for small changes in area.

The use of corrected flow allows direct interpretation of the effects of friction and heat transfer. Equation (2.5.3) and the form of Figure 2.6 show that processes which either increase the stagnation temperature of a steady flow (for example, heat addition) or decrease the stagnation pressure (friction) increase D. When such processes are present, in both subsonic and supersonic regimes, the Mach number is pushed closer to unity from a given initial state. Further, suppose changes in stagnation temperature or pressure exist between stations 1 and 2. The relation between the sonic areas at the two locations is

$$\frac{A_2^*}{A_1^*} = \frac{p_{t1}}{p_{t2}}\sqrt{\frac{T_{t2}}{T_{t1}}}. \tag{2.5.5}$$

Equation (2.5.5) shows that processes which increase the stagnation temperature or decrease the stagnation pressure increase the area needed to pass a given physical mass flow.

2.5.2 Differential relations between area and flow variables for steady isentropic one-dimensional flow[4]

The one-dimensional approach allows a simple derivation of the relation between changes in flow variables along a channel or streamtube and variations in geometry. We confine attention here to frictionless steady flow with no heat transfer and no body forces. Using the control volume shown in Figure 2.7, which is bounded by the channel walls and the control surfaces at x and $x + dx$ a small distance away, the quasi-one-dimensional forms of the continuity and momentum equations are

$$\frac{du}{u} + \frac{d\rho}{\rho} + \frac{dA}{A} = 0, \tag{2.5.6}$$

$$u\,du + \frac{dp}{\rho} = 0. \tag{2.5.7}$$

[4] This term *one-dimensional* is the one in general use, and we will employ it from now on, rather than the more cumbersome "quasi-one-dimensional flow".

The energy equation can be expressed as $s =$ constant or, equivalently for this situation,

$$c_p dT + udu = 0. \tag{2.5.8}$$

Equations (2.5.6)–(2.5.8) can be combined with the definition of the speed of sound to relate variations in local flow properties and variations in channel area. As an illustration, the expression for velocity is:

$$\frac{du}{u} = \frac{-\dfrac{dA}{A}}{1 - M^2}. \tag{2.5.9}$$

Equation (2.5.9) shows several important features of compressible channel flow:

(1) For Mach numbers less than unity an increase in area gives a decrease in velocity. The behavior in this regime is qualitatively similar to the behavior for incompressible ($M = 0$) flow.
(2) For Mach numbers greater than unity, an increase in area gives an increase in velocity. At supersonic conditions the density decreases more rapidly than the velocity decreases, and an increase in area is necessary to maintain conservation of mass.
(3) At the condition $M = 1$, the area variation is zero, and the area is a minimum, as seen in the discussion of corrected flow. The existence of a minimum area at $M = 1$ means that to isentropically accelerate a flow from subsonic to supersonic a converging–diverging nozzle must be used. The conditions at the throat are that the Mach number is equal to unity.

The transition to sonic flow, which occurs at a throat, is known as choking. This phenomenon plays a key role in compressible channel flow. To gain further insight into the conditions associated with flow at a throat, we use the isentropic relation between density and pressure to write the momentum equation (2.5.7) in the form (Coles, 1972)

$$\frac{d\rho}{\rho} + M^2 \frac{du}{u} = 0. \tag{2.5.10}$$

At a throat the area has a minimum, $dA = 0$. The continuity (2.5.6) thus becomes

$$\frac{d\rho}{\rho} + \frac{du}{u} = 0. \tag{2.5.11}$$

Equations (2.5.10) and (2.5.11) are two homogeneous algebraic equations for the quantities $d\rho/\rho$ and du/u at the throat. If the Mach number at the throat is not equal to 1, the two equations can be satisfied only if $d\rho/\rho$ and du/u are zero. This means that changes in the density and velocity (and consequently pressure) have either a maximum or minimum at the throat with the flow having local symmetry about the throat conditions.

If the Mach number at the throat is equal to unity, however, (2.5.10) and (2.5.11) become identical. If so, $d\rho/\rho$ and du/u cannot both be determined from a single equation and there is no longer a requirement for them to be zero. The velocity, density, and pressure can increase or decrease continuously through a sonic throat and the flow does not need to be symmetric about the throat conditions.

The equations for the differential changes in flow variables can be numerically integrated to find the properties corresponding to any area, but useful information can often be obtained from the values of the coefficient differentials themselves. For example, (2.5.9) shows that in both subsonic and supersonic flows, the effect of a small change in area on the velocity becomes much more significant as Mach numbers approach unity. In addition, although the relation between area and velocity changes sign at $M = 1$, (2.5.7) shows that increases in velocity always correspond to decreases in static pressure.

2.5.3 Steady isentropic one-dimensional channel flow

For isentropic flow the relation between the Mach number, the stagnation pressure, and the static pressure ((1.16.7), see also Section 2.5.1) can be written as an expression for the Mach number as a function of the ratio of stagnation to static pressure, p_t/p:

$$M = \sqrt{\left(\frac{2}{\gamma - 1}\right)\left[\left(\frac{p_t}{p}\right)^{(\gamma-1)/\gamma} - 1\right]}. \tag{2.5.12}$$

Equation (2.5.12) applies to non-isentropic flow as well as isentropic flow provided the stagnation pressure is interpreted as $p_t(x)$, the value that actually exists at the location of interest.

For steady isentropic flow the stagnation pressure is constant along the channel and equal to the inlet value, p_{t_i}. The Mach number at any location x along the channel, $M(x)$, is therefore defined by the local ratio of static to inlet stagnation pressure, $p(x)/p_{t_i}$:

$$M(x) = \sqrt{\left(\frac{2}{\gamma - 1}\right)\left[\left(\frac{p_{t_i}}{p(x)}\right)^{(\gamma-1)/\gamma} - 1\right]}. \tag{2.5.13}$$

For a given value of $p_{t_i}/p(x)$ the Mach number is determined as is the value of $A(x)/A^*$. In fact any one of T_{t_i}/T, p_{t_i}/p, A/A^*, or M, together with the inlet stagnation pressure and temperature, is enough to determine the velocity and the thermodynamic states at any station in the channel.

In steady isentropic flow the ratio of exit pressure to inlet stagnation pressure, p_{exit}/p_{t_i}, determines the channel exit Mach number and hence the corrected flow per unit area. Because we know the inlet stagnation states, quantities such as the physical flow rate per unit area, the static temperature and density, and the exit velocity can be determined. For situations in which the flow can be approximated as isentropic the capability to obtain flow properties from knowledge of only inlet stagnation conditions and exit static pressure is extremely useful. We return to the general topic of one-dimensional channel flow in Chapter 10.

2.6 Shock waves

Flow compressibility is associated with the existence of propagating disturbances or waves such as the small amplitude motions examined in Section 1.15. The behavior of finite amplitude disturbances, or "shock waves", is also of interest because they can have a large effect on performance of fluid components.

We work with a control volume moving with the disturbance and consider the one-dimensional situation. For finite amplitude disturbances, in contrast to the discussion of Section 1.15, terms which arise from products of the disturbance quantities cannot be neglected. With reference to stations 1

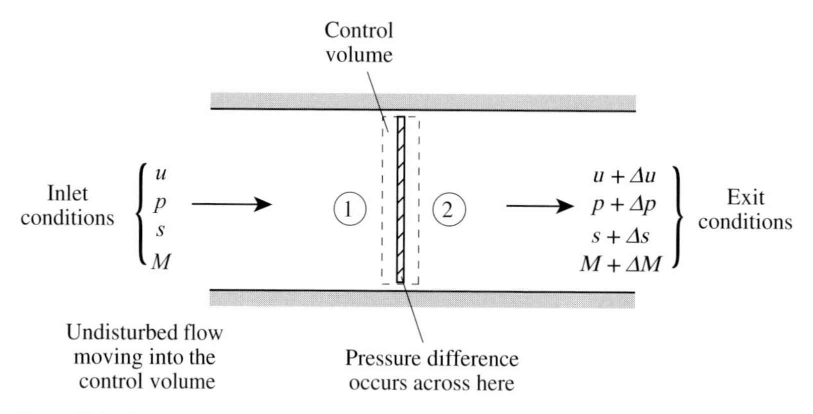

Figure 2.8: Control volume fixed to shockwave.

and 2 of the control volume shown in Figure 2.8, the equations for conservation of mass, momentum and energy across a shock wave normal to the flow are:

$$\rho_1 u_1 = \rho_2 u_2 = \dot{m}, \tag{2.6.1}$$

$$p_1 + \rho_2 u_1^2 = p_2 + \rho_2 u_2^2, \tag{2.6.2}$$

$$h_1 + \frac{u_1^2}{2} = h_2 + \frac{u_2^2}{2} = h_t. \tag{2.6.3}$$

In (2.6.1) $\dot{m}$ denotes the mass flow per unit area.

The numerical solution of (2.6.1) and (2.6.3) can be expressed non-dimensionally as functions of the upstream Mach number, M_1 as in Figure 2.9 in which the ratios of stagnation pressure (p_{t2}/p_{t1}), static pressure (p_2/p_1), Mach number (M_2/M_1), and entropy rise $T(s_2 - s_1)/u_1^2$ across the shock wave are presented. The solutions are for compressive disturbances. There is also a trivial solution, with no change in the flow variables, and a solution in which the flow undergoes a finite amplitude rarefaction from subsonic to supersonic. As seen in Figure 2.9, entropy increases in the compression. It would decrease in the rarefaction which, for this adiabatic flow, is a violation of the second law. Only the compression is thus physically possible. The non-dimensional entropy increase across a shock wave is small for Mach numbers up to roughly 1.25, after which it rises rapidly. Shock waves at Mach numbers below this are efficient ways to diffuse the flow, and the presence of weak shock waves can be a desirable feature in devices where one wishes to diffuse in a short distance.

2.6.1 The entropy rise across a normal shock

We can understand the way in which the increase of entropy across a shock scales with Mach number by using the conservation equations to derive an expression for the change in stagnation pressure (Liepmann and Roshko, 1957). In this, it is useful to work in terms of the ratio $(p_2 - p_1)/p_1 = \Delta p/p_1$, where Δp is the pressure rise across the shock and p_1 is the upstream pressure; $\Delta p/p_1$ gives a measure of shock strength. From (2.6.2) the pressure difference can be written as

$$\frac{\Delta p}{p_1} = \frac{p_2}{p_1} - 1 = \gamma M_1^2 \left(1 - \frac{u_2}{u_1}\right). \tag{2.6.4}$$

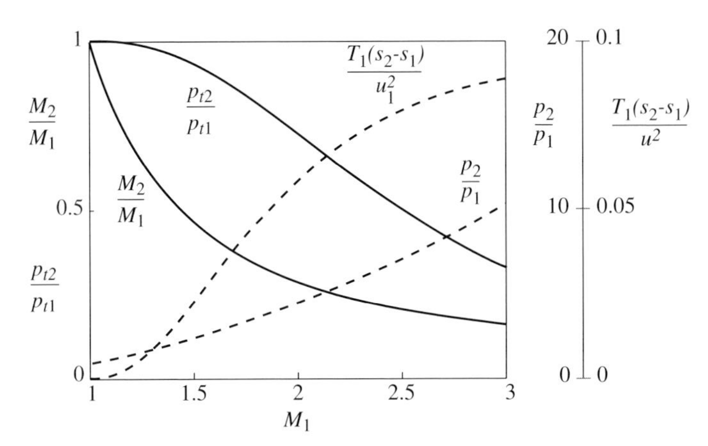

Figure 2.9: Changes in stagnation pressure, static pressure, Mach number, and entropy across a shock wave as functions of upstream Mach number, $\gamma = 1.4$.

We can find the ratio u_2/u_1 in terms of Mach number as follows. Equation (2.6.3) implies, with a^* denoting the speed of sound at sonic conditions and $M^* = u/a^*$,

$$\frac{u^2}{2} + \frac{a^2}{\gamma - 1} = \frac{1}{2}\left(\frac{\gamma + 1}{\gamma - 1}\right) a^{*2} \tag{2.6.5a}$$

or

$$M^{*2} = \frac{(\gamma + 1)M^2}{2 + (\gamma - 1)M^2}. \tag{2.6.5b}$$

Use of the relation $a^2 = \gamma p/\rho$ to eliminate p and ρ from (2.6.1) and (2.6.2) results in

$$\frac{a_1^2}{\gamma u_1} + u_1 = \frac{a_2^2}{\gamma u_2} + u_2. \tag{2.6.6}$$

Equation (2.6.6) can be combined with (2.6.5a) to obtain an expression for u_1, u_2, and a^{*2} through elimination of a_1^2 and a_2^2. Upon simplifying the expression, we obtain

$$M_1^* M_2^* = \frac{u_1}{a^*}\frac{u_2}{a^*} = 1, \tag{2.6.7}$$

where the subscript on a^* has been omitted since $a_1^* = a_2^*$. Equation (2.6.7) allows the ratio u_2/u_1 to be expressed as

$$\frac{u_2}{u_1} = \frac{u_1 u_2}{u_1^2} = \frac{1}{M_1^{*2}}. \tag{2.6.8}$$

Equations (2.6.5b) and (2.6.8) can now be used to rewrite (2.6.4) in terms of M_1 as

$$\frac{\Delta p}{p_1} = \frac{p_2}{p_1} - 1 = \frac{2\gamma}{\gamma + 1}\left(M_1^2 - 1\right). \tag{2.6.9}$$

The shock strength thus scales as $M_1^2 - 1$.

Use of (2.6.5b) and (2.6.7) also gives M_2 in terms of M_1 as

$$M_2^2 = \frac{2 + (\gamma - 1)\, M_1^2}{2\gamma M_1^2 - (\gamma - 1)}, \tag{2.6.10}$$

which allows the stagnation pressure ratio across the shock to be expressed in terms of M_1 as

$$\frac{p_{t_2}}{p_{t_1}} = \left(\frac{p_2}{p_1}\right)\frac{\left(\frac{p_{t_2}}{p_2}\right)}{\left(\frac{p_{t_1}}{p_1}\right)} = \frac{\left[\frac{(\gamma+1)\, M_1^2}{2 + (\gamma - 1)\, M_1^2}\right]^{\gamma/\gamma-1}}{\left[1 + \frac{2\gamma}{\gamma+1}\left(M_1^2 - 1\right)\right]^{1/\gamma-1}}. \tag{2.6.11}$$

Substituting the shock strength $\Delta p/p_1$ for the Mach number in (2.6.11) and expanding the resulting expression in a Taylor series about zero shock strength ($M_1 = 1$), it is found that the terms that are linear and quadratic in $\Delta p/p_1$ are both zero. For moderate shock strengths, therefore the change in stagnation pressure scales as the third power of the shock strength:

$$1 - \frac{p_{t_2}}{p_{t_1}} = \frac{(\gamma+1)}{12\gamma^2}\left(\frac{\Delta p}{p_1}\right)^3 + 0\left[\left(\frac{\Delta p}{p_1}\right)^4\right]$$

or

$$\frac{\Delta p_t}{p_{t_1}} \cong -\frac{2\gamma}{3(\gamma+1)^2}\left(M_1^2 - 1\right)^3 = -\frac{(\gamma+1)}{12\gamma^2}\left(\frac{\Delta p}{p_1}\right)^3 + \cdots. \tag{2.6.12}$$

There is no change in stagnation temperature across the shock wave and the entropy change is thus

$$\frac{(s_2 - s_1)}{R} \cong \frac{(\gamma+1)}{12\gamma^2}\left(\frac{\Delta p}{p_1}\right)^3 = \frac{2}{3}\frac{\gamma}{(\gamma+1)^2}\left(M_1^2 - 1\right)^3 \tag{2.6.13}$$

plus terms which are higher order in $\Delta p/p_1$. Equations (2.6.12) and (2.6.13) show the scaling of the entropy change in terms of shock strength.

2.6.2 Shock structure and entropy generation processes

The approach to shock waves in the preceding section is global, in that the shock is treated as a control volume and the details of flow within the shock are not dealt with. For insight into the mechanisms by which the entropy change is produced, we need to look into the structure of the flow within the shock, i.e. within the control volume that contains the shock. This procedure is carried out below for a purely one-dimensional flow with a normal planar shock wave, but the analysis is a useful model problem for more complex configurations because the shock radius of curvature is almost always (unless the pressure is very low or the device length scale is small) much larger than the length scales within the shock which characterize the viscous and heat transfer processes.

For one-dimensional flow, the variables depend only on a single coordinate (x). The total entropy rise is a function of the end states only and is independent of the viscous stresses and heat transfer occurring within the shock. As seen, the total entropy rise can be derived using control volume arguments, but we wish here to examine entropy generation within the shock. The discussion that follows is based largely on that given in Liepmann and Roshko (1957).

For one-dimensional flow through a steady shock wave, the continuity, momentum, and energy equations are (where in this one-dimensional flow we again omit the subscript on the velocity)

$$\frac{d}{dx}(\rho u) = 0, \tag{2.6.14}$$

$$\frac{d\rho u^2}{dx} = -\frac{dp}{dx} + \frac{d\tau_{xx}}{dx}, \tag{2.6.15}$$

$$\frac{d\rho u h_t}{dx} = \frac{d}{dx}(\tau_{xx} u - q_x). \tag{2.6.16}$$

In (2.6.15) and (2.6.16), the viscous stress τ_{xx} is τ_{11} in terms of the equations in Chapter 1. The rate of heat transfer in the x-direction per unit area is denoted by q_x. Equations (2.6.14)–(2.6.16) can be integrated to yield[5]

$$\rho u = \dot{m} = \text{constant}, \tag{2.6.17}$$

$$\rho u^2 + p - \tau_{xx} = \rho_1 u_1^2 + p_1, \tag{2.6.18}$$

$$\rho u h_t - \tau_{xx} u + q_x = \dot{m} h_{t_1}. \tag{2.6.19}$$

The subscript 1 denotes the conditions upstream of the shock, where shear stress and heat transfer vanish. If the integration is carried to a far downstream station where the stress (τ_{xx}) and heat transfer rate (q_x) also vanish, the jump conditions at a normal shock given previously in this section are obtained:

$$\rho_1 u_1 = \rho_2 u_2, \tag{2.6.1}$$

$$p_1 + \rho_1 u_1^2 = p_2 + \rho_2 u_2^2, \tag{2.6.2}$$

$$h_{t_1} = h_{t_2}. \tag{2.6.3}$$

The heat flux and stresses do not influence the downstream state but they are directly linked to the rate of entropy rise. The latter can be evaluated using the combined first and second law ((1.10.4)), in the form

$$T\frac{Ds}{Dt} = \frac{Dh}{Dt} - \frac{1}{\rho}\frac{Dp}{Dt}. \tag{2.6.20}$$

From (2.6.15), (2.6.16), and (2.6.17) the local rate of change of entropy is

$$\dot{m}\frac{ds}{dx} = \frac{\tau_{xx}}{T}\frac{du}{dx} - \frac{1}{T}\frac{dq_x}{dx}. \tag{2.6.21}$$

Integrating from station 1 (upstream of the shock) to a given location x',

$$\dot{m}(s - s_1) = \int_1^{x'} \frac{\tau_{xx}}{T}\frac{du}{dx}dx - \int_1^{x'} \frac{1}{T}\frac{dq_x}{dx}\,dx. \tag{2.6.22}$$

[5] Again, in this section $\dot{m}$ denotes the mass flow per unit area.

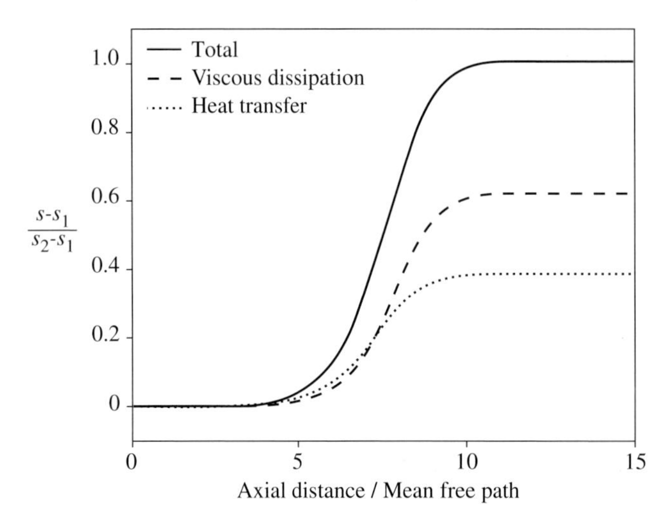

Figure 2.10: Normalized entropy distribution across a shock; $M_1 = 1.5$, upstream entropy taken as 0, downstream value = 1.0 (Teeple, 1995).

As discussed in Sections 1.7 and 1.13, the stress and heat flux are related to derivatives of the velocity and temperature:

$$\tau_{xx} = (2\mu + \lambda)\frac{du}{dx}, \quad q_x = -k\frac{dT}{dx}, \tag{2.6.23}$$

where λ is the second coefficient of viscosity. The overall entropy change from upstream to downstream of the shock is

$$\dot{m}(s_2 - s_1) = \int_1^2 (2\mu + \lambda)\left(\frac{du}{dx}\right)^2 dx + \int_1^2 \frac{1}{T}\frac{d}{dx}\left(k\frac{dT}{dx}\right)dx. \tag{2.6.24}$$

The second integral can be integrated by parts to yield

$$\dot{m}(s_2 - s_1) = \int_1^2 (2\mu + \lambda)\left(\frac{du}{dx}\right)^2 dx + \int_1^2 \frac{k}{T^2}\left(\frac{dT}{dx}\right)^2 dx. \tag{2.6.25}$$

The two terms in (2.6.25), respectively, represent dissipation (irreversible conversion of mechanical energy to internal energy due to viscous stress) and production of entropy due to heat transfer across a temperature difference, as illustrated in Section 1.10. The quantities $(2\mu + \lambda)$ and k are both positive as are both integrals. The stagnation enthalpy is the same far upstream and far downstream but it is not uniform throughout the region in which viscous stresses and heat transfer are non-zero. The non-uniformity, however, has no effect on the flow field external to this region.

The results of a numerical integration of the one-dimensional equations are shown in Figure 2.10 (Teeple, 1995). The temperature dependence of viscosity is modeled using Stokes's assumption of $\lambda = -2/3\mu$ plus the behavior $\mu/\mu_o = (T/T_o)^{0.77}$ (based on measurements in air), the Prandtl number is 0.71, and the upstream Mach number (M_1) is 1.5. The abscissa in Figure 2.10 is the shock thickness

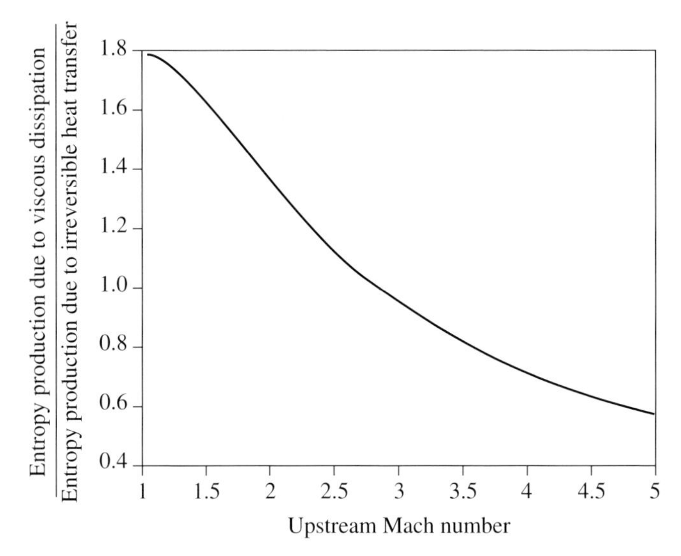

Figure 2.11: Relative sources of entropy production across a shock wave, $\gamma = 1.4$, $Pr = 0.71$ (Teeple, 1995).

in terms of the mean free path corresponding to upstream conditions and the ordinate is the entropy rise normalized by the total entropy rise from far upstream to far downstream. For the parameters shown the entropy rise is monotonic and occurs over roughly five mean free paths, i.e. over a distance of order 10^{-6} m at standard conditions, so the shock is indeed thin in comparison to representative dimensions of fluids engineering devices.

Figure 2.10 shows that at $M = 1.5$, the contribution of viscous dissipation to the entropy rise is more important than the effect of heat transfer. This proportion drops as the Mach number increases, as shown in Figure 2.11, which gives the ratio of the overall entropy increase due to viscous dissipation to that due to transfer of heat across the temperature difference. The two are roughly equal at an upstream Mach number of 3, with the heat transfer dominating at higher Mach numbers than this.

2.7 Effect of exit conditions on steady, isentropic, one-dimensional compressible channel flow

The material in Sections 2.5 and 2.6 provides the basis for a general description of the effect of exit conditions on flow regimes in compressible channel flow. The ratio of static pressure to stagnation pressure at any location determines the local Mach number. For isentropic (i.e. frictionless, adiabatic) flow, if the Mach number is known at one location in a channel of specified area variation, the conditions everywhere in the channel are defined. It is often the case that the static pressure at the exit of a nozzle, diffuser, or turbomachine is a known and controlled variable. An important issue, therefore, is the behavior of the flow in a channel as the ratio of the exit static to stagnation pressure is altered. We examine this first for a converging nozzle and then for a converging–diverging nozzle, following Shapiro (1953).

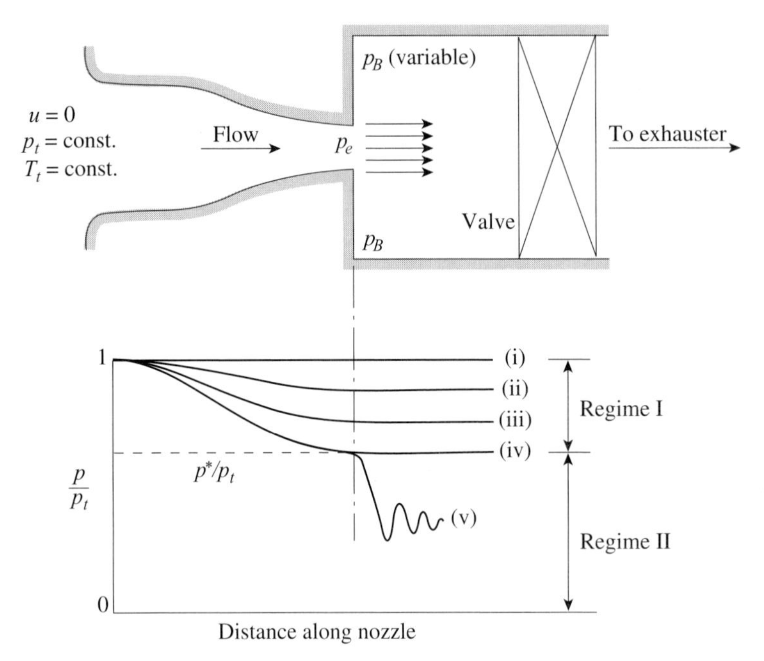

Figure 2.12: Operation of a converging nozzle at different back pressures.

2.7.1 Flow regimes for a converging nozzle

The discussion can be given in terms of the configuration in Figure 2.12 which shows a converging nozzle fed from a large reservoir (e.g. the atmosphere) at constant stagnation pressure and temperature, p_t and T_t. The nozzle discharges into a chamber, whose pressure can be controlled through the combination of an exhauster and a valve, as sketched in the top part of the figure. The chamber pressure, denoted by p_B, is commonly referred to as the *back pressure* and we adopt this nomenclature here. The flow is isentropic from the inlet to the nozzle exit. We address the behavior of the mass flow and nozzle exit pressure as the ratio of back pressure to stagnation pressure, p_B/p_t, is reduced from an initial value of unity.[6]

At $p_B/p_t = 1$ there is no flow in the channel, as indicated by curve (i) in the lower part of Figure 2.12. If p_B/p_t is reduced to a value slightly below unity, the flow in the nozzle will be subsonic everywhere, with a pressure that decreases along the channel as indicated by curve (ii). In the subsonic regime the pressure at the nozzle exit, p_e, is essentially equal to the back pressure, p_B. The argument for this can be seen if we suppose the exit pressure, p_e, to be substantially different from p_B, say higher. If so, there would be streamline curvature with the stream expanding laterally on leaving the nozzle (see Section 2.4). However, this would cause the stream pressure downstream to be even higher than at the nozzle exit. Since the back pressure is the pressure which the stream must eventually attain in the exhaust chamber, this situation cannot occur and the exit pressure cannot be higher than the back pressure. A similar argument can be made to rule out an exit pressure lower than the back pressure.

[6] The ratio between stagnation and static pressure is reported in the literature both as p/p_t, as in this section, and as p_t/p, often referred to as the expansion ratio, as in Section 2.3 and Chapter 10. We will make use of both conventions in this text, depending on context.

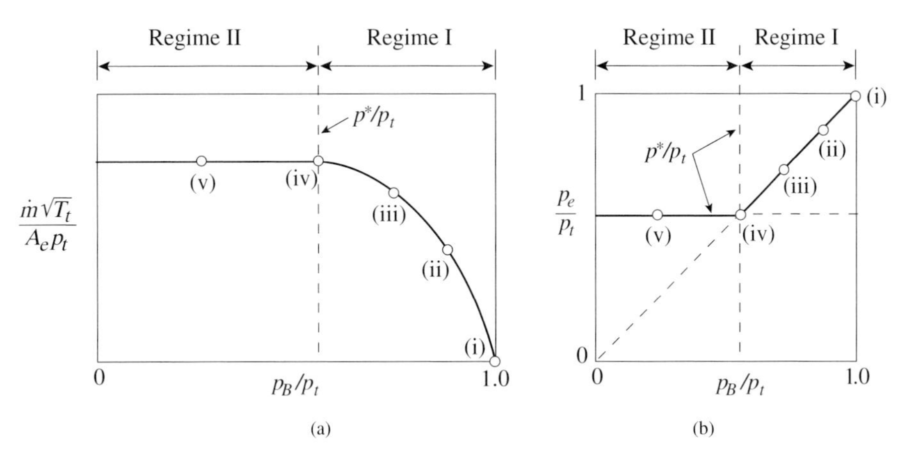

Figure 2.13: Corrected flow per unit area (a) and nozzle exit pressure (b) as a function of the back pressure ratio (p_B/p_t) for a converging nozzle (Shapiro, 1953).

For subsonic flow the conclusion is thus that the exit pressure and the back pressure are the same, $p_e = p_B$. Curve (ii) in the figure is thus extended at a constant level from the nozzle exit into the chamber.

If the back pressure is reduced further, to a value representing curve (iii) the Mach number everywhere in the channel increases. The highest value is still at the exit, with this value less than unity and the flow subsonic everywhere. There is no qualitative change in behavior from that seen along curve (ii).

Similar conditions apply until the back pressure reaches the critical pressure p^* ($p_B/p_t = p^*/p_t$) indicated by curve (iv). At this condition the Mach number at the exit of the channel, M_e, is equal to unity and the corrected flow through the nozzle has its maximum possible value. Further reduction of the back pressure cannot increase the corrected flow and thus cannot alter any of the flow quantities upstream of the exit. At any value of p_B/p_t lower than the critical value, represented by curve (v), the pressure distribution within the channel, the value of p_e/p_t, and the flow rate are all identical with the corresponding quantities for condition (iv). The pressure distribution outside the channel cannot be described within a one-dimensional framework and is indicated only notionally by a wavy line. The critical pressure ratio, p^*/p_t, can be regarded as the boundary between Regime I (unchoked) and Regime II (choked) depicted in Figure 2.12.

The behavior of the nozzle corrected flow per unit exit area ($\dot{m}\sqrt{T_t}/A_e p_t$) and the ratio of nozzle exit static pressure to stagnation pressure (p_e/p_t) are shown in Figures 2.13(a) and 2.13(b) as functions of the back pressure ratio (p_B/p_t). Exit conditions corresponding to curves (i)–(v) are indicated in both plots, which can be described with reference to two regimes separated by the critical pressure ratio $p_B/p_t = p^*/p_t$.

In Regime I the exit corrected flow per unit area increases as the back pressure decreases. It reaches a maximum, with the exit Mach number equal to unity, when the back pressure ratio drops to p^*/p_t. Further decreases in back pressure which occur in Regime II have no effect on exit corrected flow or nozzle exit Mach number. The exit pressure is equal to the back pressure in Regime I until the latter drops to p^*/p_t, after which, in Regime II, it remains constant.

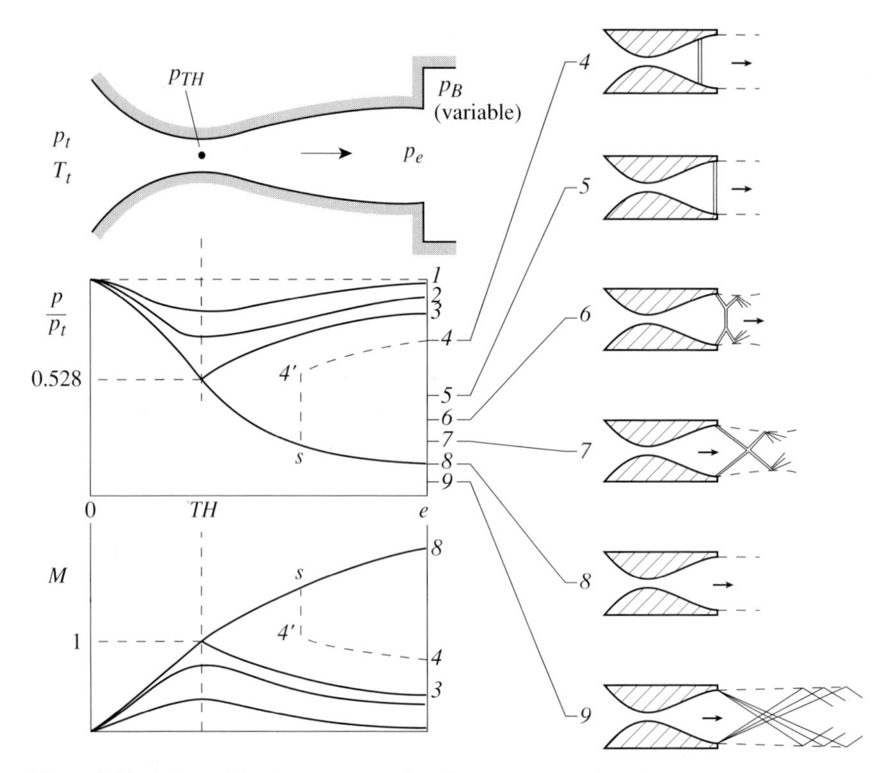

Figure 2.14: Effect of back pressure on flow in converging-diverging nozzle ($p^*/p_t = 0.528$ for $\gamma = 1.4$), *TH* denotes nozzle throat location (Liepmann and Roshko, 1957).

2.7.2 Flow regimes for a converging–diverging nozzle

There is no supersonic region within a converging nozzle, whatever the back pressure ratio. We thus now examine converging–diverging nozzles, in which supersonic regions exist at low back pressures. A plot of the static pressure along a converging–diverging nozzle discharging into a chamber is given in Figure 2.14 for different back pressures, p_B. For back pressures such as p_1 and p_2, which are above the value corresponding to $M = 1$ at the throat, the static pressure first decreases along the channel and then increases, with a corresponding increase and decrease in velocity. For frictionless adiabatic flow, solutions for this regime of operation are subsonic, continuous, and isentropic, and exhibit the local symmetry about the throat mentioned in Section 2.5.

As the back pressure is decreased to a value p_3, the Mach number reaches unity at the throat. For all back pressures below this, the conditions at the throat also correspond to $M = 1$. The flow upstream of the throat is subsonic, but its conditions are fixed because $M = 1$ at the throat and pressure information from downstream of this location cannot travel upstream.

With the throat Mach number equal to unity, two continuous solutions are possible. In one the exit flow is subsonic, with back pressure corresponding to $p_B = p_3$. In the other the flow downstream of the throat is supersonic with back pressure $p_B = p_8$. These correspond to the two points at which a horizontal line intersects the curve of $D(M)$ versus M in Figure 2.6 with one intersection for $M < 1$ and the other for $M > 1$.

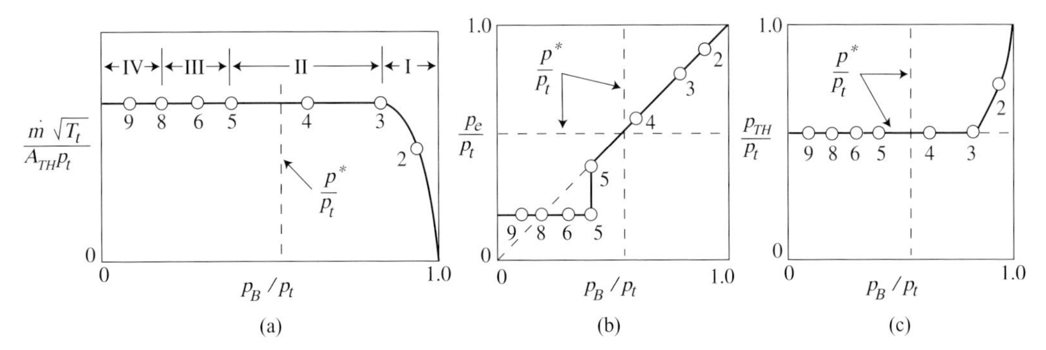

Figure 2.15: Performance of a converging–diverging nozzle with various ratios of back pressure to inlet stagnation pressure. Numbers correspond to conditions in Figure 2.14 (not all numbers shown for clarity). (a) Flow regimes and corrected flow per unit area versus (p_B/p_t); (b) exit-plane pressure (p_e/p_t) versus ratio of back pressure to inlet stagnation pressure (p_B/p_t); (c) throat pressure (p_{TH}/p_t) versus (p_B/p_t) (Shapiro, 1953).

The two solution curves in Figure 2.14 corresponding to p_3 and p_8 are the only possibilities for isentropic, one-dimensional steady flow. To describe the behavior at other levels of back pressure, the constraint of isentropic flow must be relaxed. In the range of back pressures (more precisely back pressure ratios) between p_3 and p_8 the pressure and velocity in the nozzle are discontinuous. Between p_3 and p_5 there is a region of supersonic flow downstream of the throat, followed by a normal shock and then a region of subsonic flow. Because the exit flow is subsonic the exit pressure is equal to the back pressure. This condition sets the strength of the shock. Lowering the back pressure means the shock strength increases (see Figure 2.9) and the shock occurs at a higher value of Mach number which corresponds to a location further downstream in the diverging section of the nozzle.

At a back pressure level of p_5, the normal shock stands at the nozzle exit and the flow is supersonic from the throat to the nozzle exit. No additional change inside the nozzle can occur as the pressure is lowered from this point. Adjustment between the nozzle exit and downstream for back pressures between p_5 and p_8 does not take place in a one-dimensional manner but rather through a series of oblique shock waves as sketched. For back pressures between p_5 and p_8, the flow is referred to as overexpanded.

Decreasing the back pressure beyond p_8 means the flow at the exit is at a higher pressure than the surroundings. Adjustment to a final state with a pressure equal to the back pressure then occurs through a series of expansion waves. For back pressures lower than p_8, the flow is said to be underexpanded.

The behavior can also be portrayed in terms of the relation of: (a) the corrected mass flow per unit area at the throat $(\dot{m}\sqrt{T_t}/A_{TH}\,p_t)$, (b) the non-dimensional exit-plane pressure, (p_e/p_t), and (c) throat pressure (p_{TH}/p_t), to the back pressure ratio (p_B/p_t). These are depicted in Figure 2.15, where the numbers correspond to the flow fields of Figure 2.14. (Figures 2.15(a) and 2.15(b) can be compared with Figure 2.13 for the converging nozzle.) Four regimes can be identified (in Figure 2.15 for clarity not all the conditions in Figure 2.14 are marked). Regime I has entirely subsonic flow, with the corrected flow sensitive to the level of back pressure. The dividing line between Regimes I and II occurs at $p_B = p_3$ with the Mach number unity at the throat and the throat pressure, $p_{TH} = p^*$. Regime II has a shock standing in the diverging section of the channel, with subsonic deceleration after the shock. In this regime exit pressure and back pressure are essentially the same, but the corrected flow per unit area in the channel is not affected by back pressure level.

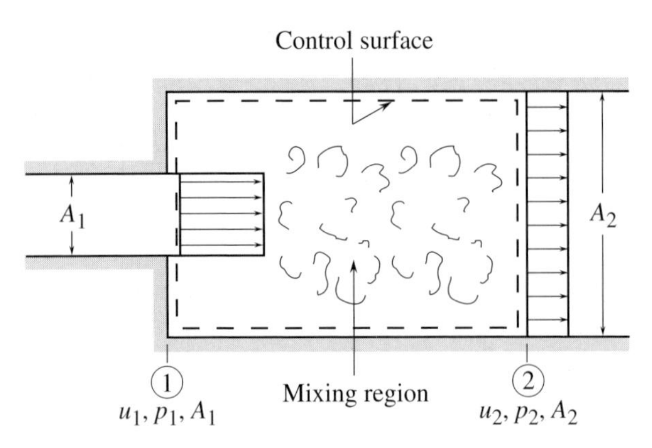

Figure 2.16: Sudden expansion in a pipe.

In Regime III, corresponding to back pressures between p_5 and p_8 the exit-plane pressure is lower than the back pressure. Compression from p_e to p_B occurs, as indicated in Figure 2.14, through oblique shock waves (see Section 2.8.7) outside the channel. At condition 8, the boundary between Regimes III and IV, the exit plane pressure is equal to the back pressure; at this condition the nozzle is referred to as ideally expanded. In Regime IV the expansion from exit-plane pressure to back pressure occurs outside the nozzle through oblique expansion waves. In Regimes III and IV the flow pattern within the entire nozzle is independent of back pressure and corresponds to the flow pattern at the "design condition" for which the exit pressure is equal to the back pressure.

2.8 Applications of the integral forms of the equations of motion

The integral forms of the equations of motion developed in Chapter 1 provide powerful tools for analysis of flow problems in which the details of the motion within a control volume are not needed. This use is illustrated in this section, starting with a constant density, unidirectional flow situation, and working up to more complex configurations. To show the applications with a minimum of algebraic complexity, the flows examined have inlet and exit states which are characterized by a single value of velocity, pressure, or temperature, but it is emphasized that this approximation is not necessary to apply control volume approaches.

2.8.1 Pressure rise and mixing loss at a sudden expansion

The first example is the pressure rise and mixing loss at a sudden expansion as indicated in Figure 2.16, where the steady flow from a duct of area A_1 exits into a larger duct of area A_2. The stream emerging from the smaller pipe at station 1 mixes with the surrounding fluid and, at some further downstream location, 2, becomes essentially uniform with velocity, u_2. For simplicity the flow is taken here as incompressible, but the approach is generalized to include compressibility in Chapter 5.

The integral forms of the continuity and momentum equations applied to the control surface shown as a dashed line in Figure 2.16 provide the means to calculate conditions at station 2 without reference to flow details inside the surface. In the figure, the jet is indicated as entering the large area duct with the area and velocity it had in the smaller duct, in other words, the flow separates from the bounding surface geometry at the exit corner of the smaller duct. In Section 2.10, we discuss this behavior in more detail, and for now we state as an experimental observation that fluid motions in configurations with a sharp edge (such as a sudden expansion or a nozzle exit) are observed not to follow the geometry, but rather occur as roughly parallel jets having area and velocity equal to that just upstream of the duct or nozzle exit.

In the fluid surrounding the jet, near the start of the expansion in the large pipe, the velocities are low, and the static pressure is thus nearly uniform and equal to that in the jet. The pressure at station 1 can therefore be taken as if it were uniform across the duct, with the pressure on the left-hand wall approximated as equal to that in the entering jet.[7] The continuity equation gives

$$A_2 u_2 = A_1 u_1 \tag{2.8.1}$$

in this one-dimensional treatment. Neglecting any contribution from friction forces on the walls of the pipe, the momentum equation in the flow direction is

$$A_2(p_2 - p_1) = \rho A_1 u_1^2 - \rho A_2 u_2^2. \tag{2.8.2}$$

Combining (2.8.1) and (2.8.2), the static pressure rise in the mixing process can be expressed in terms of the dynamic pressure of the incoming stream and the expansion area ratio $AR = A_2/A_1$ as a pressure rise coefficient, C_p,

$$C_{p_{\substack{sudden \\ expansion}}} = \frac{p_2 - p_1}{\frac{1}{2}\rho u_1^2} = \frac{2}{AR}\left(1 - \frac{1}{AR}\right). \tag{2.8.3}$$

The non-dimensional loss in stagnation pressure as a result of the mixing is

$$\frac{p_{t_1} - p_{t_2}}{\frac{1}{2}\rho u_1^2} = \left(1 - \frac{1}{AR}\right)^2. \tag{2.8.4}$$

The static pressure rise and the stagnation pressure loss given by (2.8.3) and (2.8.4) are shown in Figure 2.17. As the area ratio of the expansion is increased from unity, the static pressure rise increases to a maximum $(0.5 \times \frac{1}{2}\rho u_1^2)$ at $AR = 2$. It then drops to zero at high values of area ratio as the loss in stagnation pressure dominates the static pressure increase associated with fluid deceleration. If the expansion were lossless the stagnation pressure would be constant along a streamline. From the definition of stagnation pressure in an incompressible flow ((2.2.4)) this means

$$p + \tfrac{1}{2}\rho u^2 = p_t = \text{constant along a streamline.} \tag{2.8.5}$$

The statement that $p + \frac{1}{2}\rho u^2$ is constant along a streamline is known as *Bernoulli's equation*. It can be combined with the continuity equation to give an expression for the static pressure rise for incompressible flow in a reversible (or lossless) expansion (no change in stagnation pressure),

$$C_{p_{rev}} = \frac{p_2 - p_1}{\frac{1}{2}\rho u_1^2} = 1 - \frac{1}{AR^2}, \tag{2.8.6}$$

[7] This is an analogous argument to that given in Section 2.7 for the nozzle exit pressure in subsonic flow.

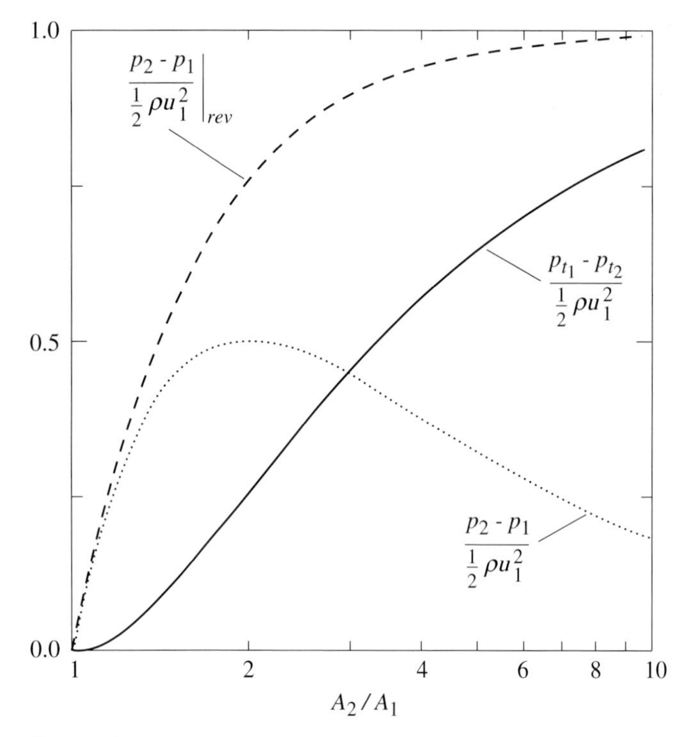

Figure 2.17: Static pressure rise and stagnation pressure loss for sudden expansion and static pressure rise for reversible (lossless) expansion.

shown by the dashed line in Figure 2.17. For the reversible expansion the static pressure coefficient increases monotonically to unity as $AR \to \infty$.

The control volume analysis of the static pressure rise coefficient at a sudden expansion is compared with experiment in Figure 2.18, where the ratios of measured to calculated static pressure rise versus distance downstream of an expansion in a circular duct are shown for a range of values of A_2/A_1. The measured maximum pressure rise coefficient agrees to within roughly 5% with the control volume analysis, showing that neglect of skin friction in the mixing region is a good approximation. In terms of the static pressure rise, mixing is effectively complete by roughly five diameters downstream of the expansion, and even a crude estimate (see Section 2.9) shows that frictional effects over this short distance are small compared to the pressure and momentum flux terms in the overall momentum balance expressed in (2.8.2).

2.8.2 Ejector performance

The sudden expansion analysis serves as part of the description of ejectors, or mixing tubes, which are used to pump fluid. A representative configuration, shown in Figure 2.19, has a high pressure primary stream of stagnation pressure p_{t_p} exiting into a constant area mixing tube at station 1. The initial area of the high pressure stream is a fraction, σ, of the mixing tube area, A. The secondary stream enters the tube with a lower stagnation pressure, for example from the atmosphere as pictured.

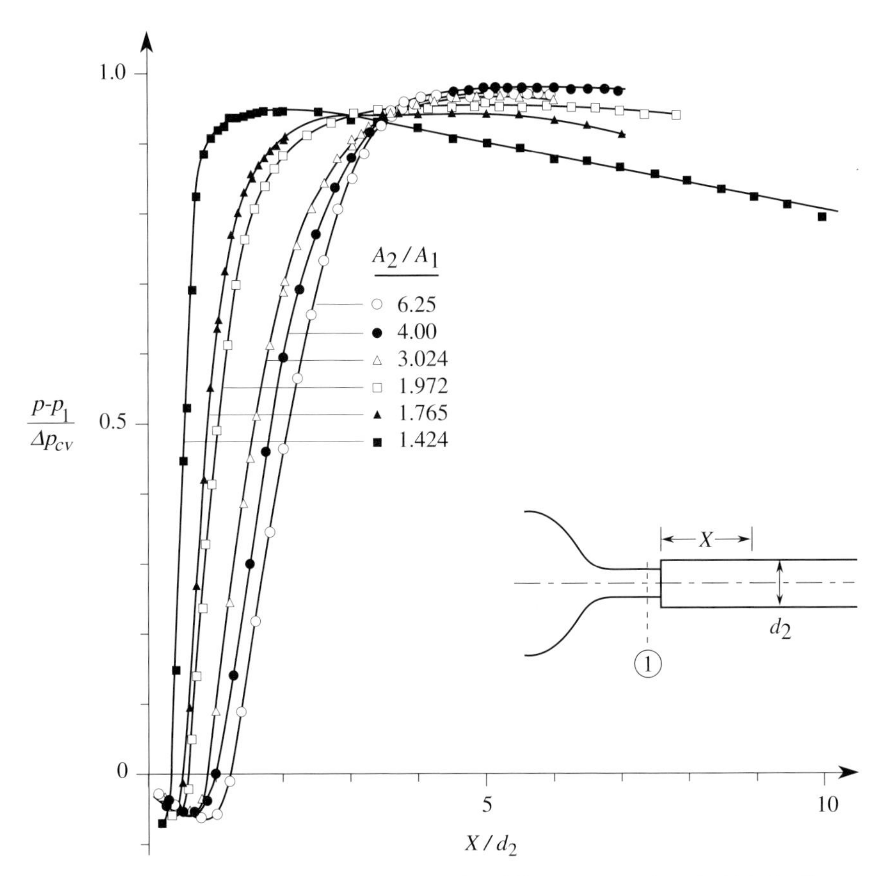

Figure 2.18: Sudden expansion pressure rise for different area ratios (A_2/A_1) (wall pressure measurements) (Ackeret, 1967).

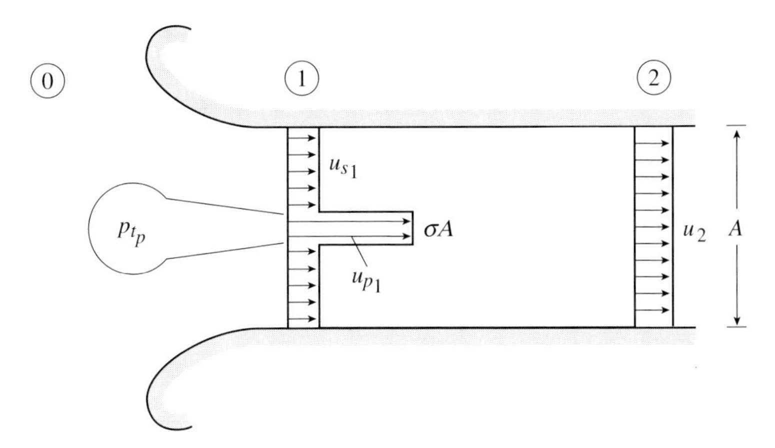

Figure 2.19: Schematic of an ejector showing the locations used in the analysis.

The mixing tube is long enough such that the exit flow can be taken as fully mixed with uniform velocity, u_2. The discharge is to atmosphere. We wish to determine the total amount of fluid pumped by the ejector, regarding the process as incompressible and constant density.

The results of Section 2.8.1 can be used here with the area ratio equal to $1/\sigma$ and the velocity u_1 replaced by the difference between primary and secondary velocities $(u_{p_1} - u_{s_1})$. Thus,

$$p_2 - p_1 = \rho\,[(u_{p_1} - u_{s_1})^2]\,\sigma(1-\sigma). \tag{2.8.7}$$

The velocity in the primary stream at station 1, u_{p_1}, is related to the reservoir stagnation pressure p_{t_p} by

$$p_{t_p} - p_1 = \tfrac{1}{2}\rho\, u_{p_1}^2. \tag{2.8.8}$$

Two other statements about the flow are needed. First, the secondary stream from ambient conditions (zero velocity, pressure $= p_0$) to the start of the mixing plane is assumed lossless:

$$p_0 = p_1 + \tfrac{1}{2}\rho\, u_{s_1}^2. \tag{2.8.9}$$

Second, the flow at station 2 exits the tube as a jet with the static pressure constant across the exit jet and equal to the ambient pressure outside the jet:

$$p_0 = p_2. \tag{2.8.10}$$

Equations (2.8.7)–(2.8.10) can be combined into a quadratic equation for $\dot{m}_s/\dot{m}_p$, the ratio of mass flow pumped by the ejector to mass flow through the primary stream:

$$\left(\frac{\dot{m}_s}{\dot{m}_p}\right)^2\left[\left(\frac{\sigma}{1-\sigma}\right)^2 + 1\right] + 4\frac{\dot{m}_s}{\dot{m}_p} - 2\left(\frac{1-\sigma}{\sigma}\right) = 0. \tag{2.8.11}$$

The only parameter that enters into (2.8.11) is the fractional area occupied by the primary stream, σ. The level of stagnation pressure in the reservoir has no effect on the ratio $\dot{m}_s/\dot{m}_p$ (or the ratio u_{s_1}/u_{p_1}). As with the sudden expansion, kinematic similarity is all that is needed for similarity in $\dot{m}_s/\dot{m}_p$ because any dependence on Reynolds number has been neglected. However, the stagnation pressure (or rather $p_{t_p} - p_0$, the driving pressure difference for the flow) does determine the physical quantity of fluid pumped; all the velocities in the problem scale with $\sqrt{(p_{t_p} - p_0)/\rho}$.

2.8.3 Fluid force on turbomachinery blading

The control volume formulation also enables derivation of the force on a row of turbomachine airfoils, or blades, in steady flow. Figure 2.20 shows the blade row and defines a coordinate system fixed to the blades. The flow is treated as incompressible and inviscid.

At the stations far enough in front of, and behind, the blades, the velocity is uniform. If W is the spacing between the blades, the continuity equation is

$$u_{x_1}W = u_{x_2}W = u_x W, \quad \text{or} \quad u_{x_1} = u_{x_2} = u_x, \tag{2.8.12}$$

where u_x is the axial velocity component far away from the blades.

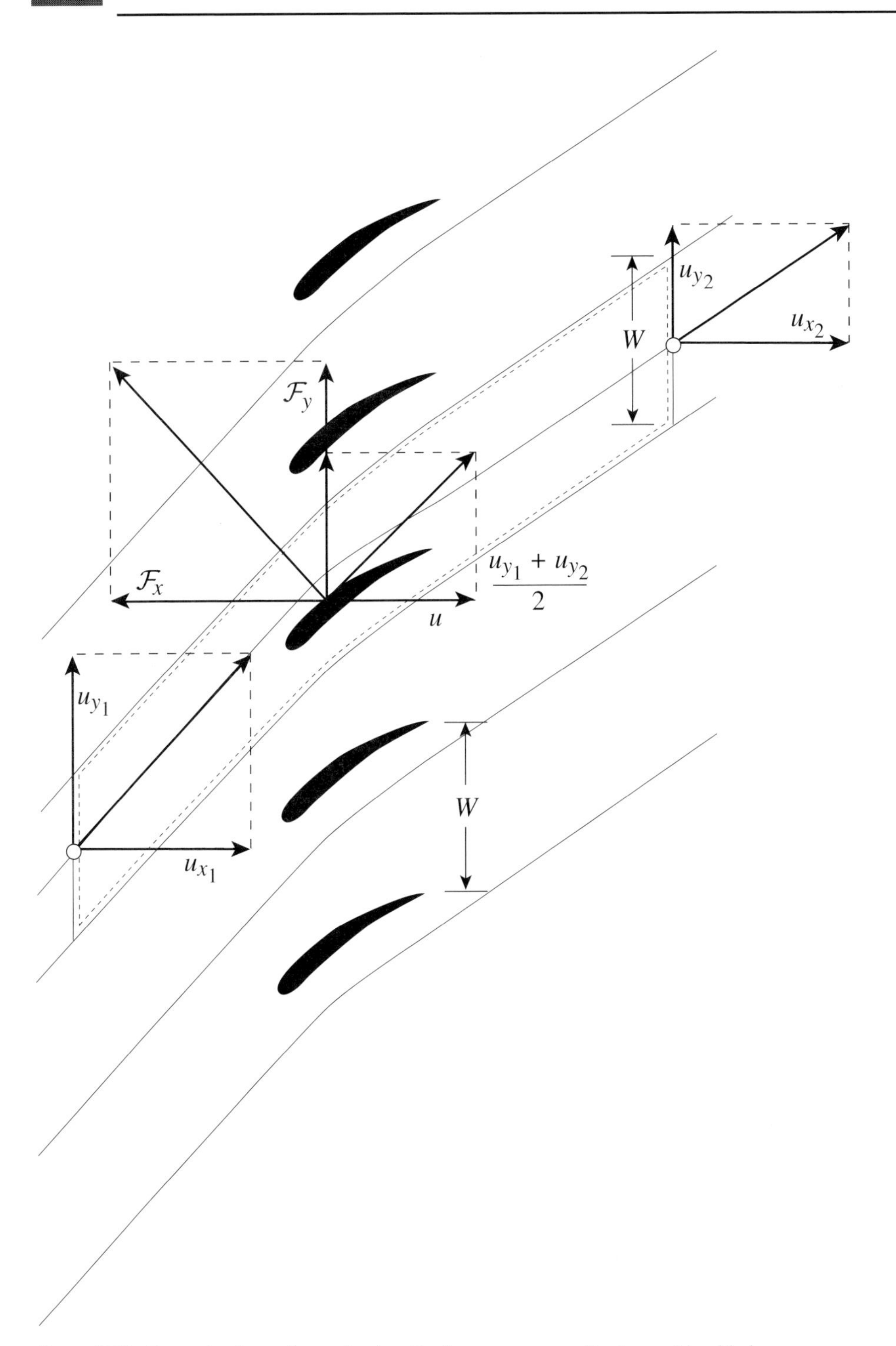

Figure 2.20: Control volume for evaluating the force on a row of turbomachine blades.

From the condition of constant stagnation pressure along a streamline the static pressure difference across the blades is related to the velocities by

$$p_2 - p_1 = \frac{\rho}{2}\left(u_{x_1}^2 + u_{y_1}^2 - u_{x_2}^2 - u_{y_2}^2\right). \tag{2.8.13}$$

Because the axial velocity is the same at locations 1 and 2

$$p_2 - p_1 = \frac{\rho}{2}\left(u_{y_1}^2 - u_{y_2}^2\right). \tag{2.8.14}$$

To apply the momentum theorem, we use the control volume indicated by the dashed lines in Figure 2.20. The bounding surfaces are two streamlines a distance apart equal to the blade spacing, W, and two vertical lines parallel to the plane of the blade row which are far upstream and far downstream respectively. The depth of all faces of the control surface can be taken as unity.

There is no flow through the two streamline surfaces. Further, because conditions are the same in each blade passage, the sum of the net force on these two surfaces is zero. The momentum flux and pressure force contributions from the upstream and downstream vertical surfaces are thus all that need to be found. The axial (x) velocity is the same at the upstream and downstream locations, so there is no net flux of axial momentum out of the control volume and the axial component of the force on the blade is given by

$$\mathcal{F}_x = W(p_1 - p_2). \tag{2.8.15}$$

There is no component of pressure force in the y-direction, but there is a net flux of y-momentum out of the control volume. Equating this to the force on the blade yields

$$\mathcal{F}_y = \rho u_x W(u_{y_1} - u_{y_2}). \tag{2.8.16}$$

The quantity $(W(u_{y_2} - u_{y_1}))$ is referred to as the circulation and denoted by Γ. As will be seen in Chapter 3, this quantity is of considerable interest; for now it is simply noted as a property of the flow field through the blades.

Using (2.8.14), the x-component of the force on the blade is given by

$$\mathcal{F}_x = \rho\Gamma(u_{y_1} + u_{y_2})/2. \tag{2.8.17}$$

Using (2.8.16), the y-component is

$$\mathcal{F}_y = \rho\Gamma u_x. \tag{2.8.18}$$

The ratio $\mathcal{F}_x/\mathcal{F}_y$ is $(u_{y_1} + u_{y_2})/2u_x$. The resultant of $\mathcal{F}_x$ and $\mathcal{F}_y$ is therefore at right angles to the resultant velocity formed from the axial velocity u_x and the mean of the upstream and downstream y-velocities, $(u_{y_1} + u_{y_2})/2$. Denoting the magnitude of this resultant force by $\mathcal{F}$, and defining a vector mean velocity u_{vm}, with components u_x and $(u_{y_1} + u_{y_2})/2$, leads to an expression relating the magnitudes of the resultant force, the circulation, and the vector mean velocity:

$$\mathcal{F} = \rho|\Gamma|u_{vm}. \tag{2.8.19}$$

Equation (2.8.19) has a form similar to the Kutta–Jukowski relation for the lift of an isolated airfoil.

The limiting case of large blade spacing is the isolated airfoil. Increasing W, the distance between neighboring blades, while holding the circulation around a blade constant, means the difference $(u_{y_1} - u_{y_2})$ shrinks inversely with the spacing. As W approaches infinity, the velocity difference

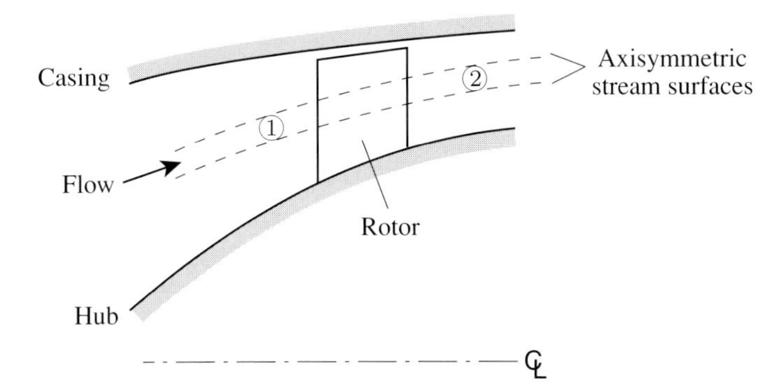

Figure 2.21: Axisymmetric stream surfaces used for an annular control volume.

approaches zero, and the velocities in front of and behind the one blade left at a finite position approach one another, provided the distance from the blade is large enough. The vector mean velocity can thus be represented by the velocity far from the blade row, u_∞, which is the same on either side. In this limiting case, the Kutta–Jukowski result for the magnitude of the force on an isolated airfoil is recovered:

$$\mathcal{F} = \rho|\Gamma|u_\infty. \tag{2.8.20}$$

2.8.4 The Euler turbine equation

Equation (1.8.8) provides a relation between the torque (the moment of the forces) exerted within a control volume and the net outflux of angular momentum. Figure 2.21 shows a control volume consisting of the region between two axisymmetric stream surfaces in a turbomachine. The flow enters at radius r_1 with a circumferential velocity u_{θ_1} and leaves at radius r_2 with circumferential velocity u_{θ_2}. The mass flow between the stream surfaces is given by

$$d\dot{m} = 2\pi\rho_1 r_1 u_{x_1} dr_1 = 2\pi\rho_2 r_2 u_{x_2} dr_2, \tag{2.8.21}$$

where dr is the radial distance between stream surfaces. The difference in angular momentum flux between stations 1 and 2 for the axisymmetric streamtube has magnitude $d\dot{m}(r_2 u_{\theta_2} - r_1 u_{\theta_1})$ and is equal to the torque exerted by the blades over the region bounded by the two stream surfaces.

Integrating over the total mass flow gives the total torque exerted by the blade row on the fluid as

$$\text{torque} = \left[\left(\int r u_\theta d\dot{m}\right)_2 - \left(\int r u_\theta d\dot{m}\right)_1\right]. \tag{2.8.22}$$

An average value of the angular momentum per unit mass, ru_θ, at each axial station can be defined as

$$(ru_\theta)_{av} = \frac{\int r u_\theta \, d\dot{m}}{\int d\dot{m}} = \frac{\int r u_\theta \, d\dot{m}}{\dot{m}}. \tag{2.8.23}$$

The total torque can now be written in terms of the conditions at the inlet and exit as

$$\text{torque exerted by the blade row} = \dot{m}[(r_2 u_{\theta_2})_{av} - (r_1 u_{\theta_1})_{av}] \tag{2.8.24}$$

For a rotating blade row, or rotor, with angular velocity Ω, the power needed by the blade row is related to the torque by

$$\text{power needed} = -\Omega \times \text{torque}. \tag{2.8.25}$$

The kinematic quantities (velocities) can now be related to the thermodynamic states at the inlet and exit. The steady-flow energy equation (1.8.10) states that for an adiabatic flow the *power output* is equal to the rate of stagnation enthalpy decrease of the fluid:

$$(h_{t_1} - h_{t_2})\dot{m} = -\Omega \times \text{torque}. \tag{2.8.26}$$

In (2.8.26) the convention for torque is defined as in (2.8.24). Using (2.8.24) and taking the flow to be uniform at stations 1 and 2,

$$h_{t_2} - h_{t_1} = \Omega(r_2 u_{\theta_2} - r_1 u_{\theta_1}). \tag{2.8.27}$$

Equation (2.8.27) is known as the Euler turbine equation and applies to both compressible and incompressible flow.

For constant density, adiabatic, and lossless flow ($ds = 0$), $dh = (1/\rho)dp$, and the Euler turbine equation becomes

$$\Delta\left(\frac{p_t}{\rho}\right) = \Omega\Delta(r u_\theta). \tag{2.8.28}$$

2.8.5 Thrust force on an inlet

Two other examples of the use of control volumes are related to the axial force on an inlet (which can be a large fraction of the net thrust of a propulsion system) and the production of thrust through heat addition. The streamline pattern for an inlet varies as a function of the ratio of the velocity in the inlet to the onset, or ambient, velocity, as shown schematically in Figure 2.22 for subsonic flow. Figure 2.22(a) represents near static (take-off) conditions for a jet engine and Figure 2.22(b) represents cruise-type conditions (Küchemann, 1978). A control volume approach allows computation of the axial force exerted on the inlet without detailed reference to the streamline pattern. The control volume used, shown in Figure 2.23, is axisymmetric. The inlet is approximated as being a constant section from some given distance behind the lip and the discussion here is restricted to incompressible, constant density flow.

The axial (x-direction) velocity at a station 0 far upstream is denoted by u_0 and the pressure by p_0. Quantities at the station inside the inlet control volume are denoted by 1. The integral momentum equation applied to the control volume in the figure is

$$\rho u_0^2 A_0 + p_0 A_0 - \rho u_{x_1}^2 A_1 - p_1 A_1 - \left(\int_{A_N} p\, dA_N\right)$$
$$- p_0(A_0 - A_1 - A_N) - (\rho u_0 A_0 - \rho u_{x_1} A_1)u_0 = 0. \tag{2.8.29}$$

As described by Küchemann and Weber (1953) the first two terms in (2.8.29) represent the flux of x-momentum of the mass flow, $\rho u_0 A_0$, through the forward surface A_0 of the control volume and the pressure force which acts on that surface. The two terms following are the corresponding quantities for the flow through the internal duct. The fifth term is the integral of the static pressure p over the

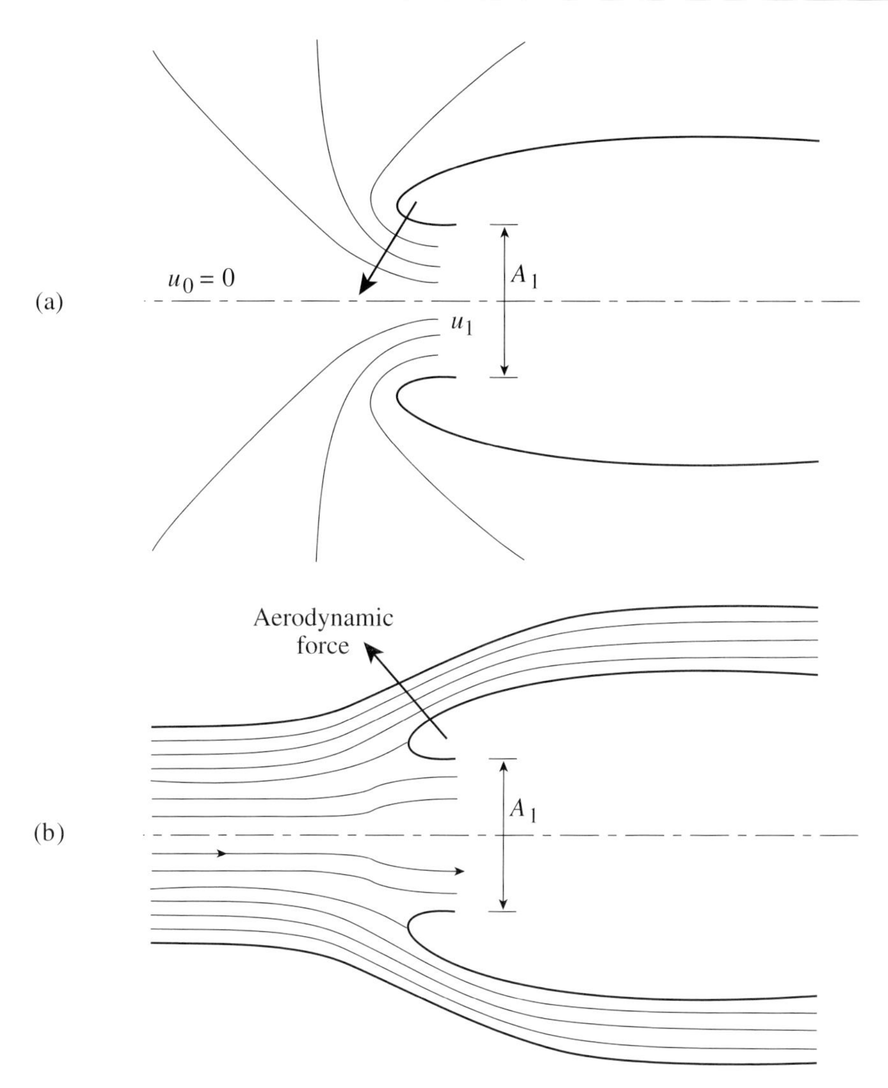

Figure 2.22: Streamline patterns upstream of a subsonic inlet: (a) u_1/u_0 much larger than unity (near take-off conditions); (b) u_1/u_0 less than unity (cruise-type conditions).

surface of the intake, with dA_N a surface element normal to the mean flow (x) direction. The next term is the force on the base of the control surface outside the intake, with the streamlines assumed to be straight and the pressure thus equal to the far upstream value. The last term is the momentum of the flow through the base of the control volume and the curved (cylindrical) part of the control surface, with the control cylinder large enough so the axial velocity at the control surface can be taken as u_0 in evaluating this term.

Cancelling terms in (2.8.29) allows the equation to be simplified to

$$\int_{A_N} (p - p_0)\, dA_N = \rho u_{x_1} A_1 (u_0 - u_{x_1}) - (p_1 - p_0) A_1. \tag{2.8.30}$$

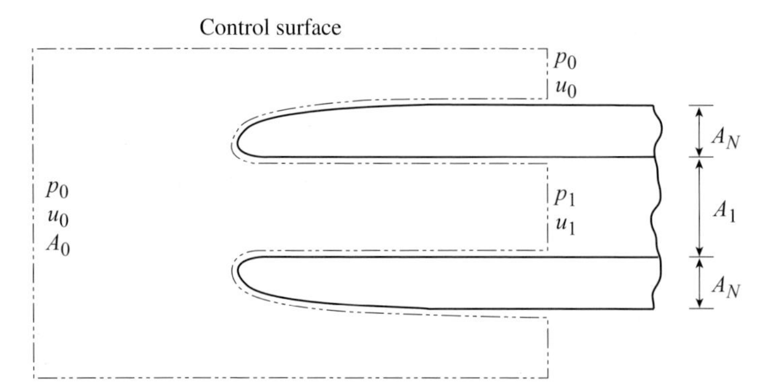

Figure 2.23: Control surface round inlet lip for the application of the momentum theorem.

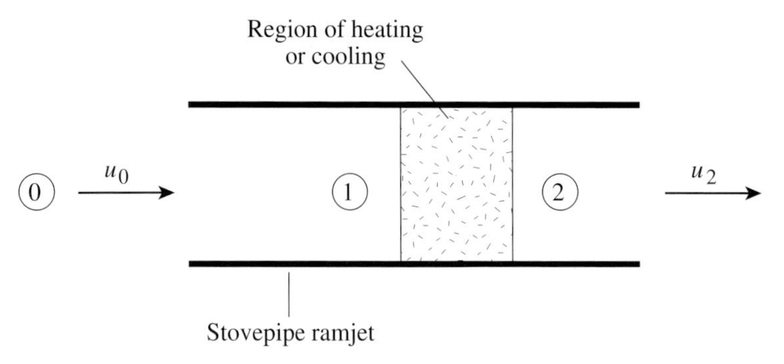

Figure 2.24: Tube with heating or cooling (idealized ramjet).

Applying Bernoulli's equation ($p_0 + \frac{1}{2}\rho u_0^2 = p_1 + \frac{1}{2}\rho u_1^2$) in (2.8.30) between far upstream and the station inside the inlet yields a relation for the upstream pointing force on the inlet, $\mathcal{F}_I$, in terms of the inlet area and the velocities at stations 0 and 1:

$$\frac{\mathcal{F}_I}{\frac{1}{2}\rho u_{x_1}^2 A_1} = \frac{\int\limits_{A_N} (p_0 - p)\, dA_N}{\frac{1}{2}\rho u_{x_1}^2 A_1} = \left(\frac{u_0}{u_{x_1}} - 1\right)^2. \tag{2.8.31}$$

The force $\mathcal{F}_I$ represents the difference between the pressure force on the curved part of the inlet (the lip) and the force due to a pressure p_0 acting on the cross-sectional area of the straight section of the inlet, i.e. the force is referenced to a condition with p_0 acting on the rear of the inlet cross-section, A_N. The force on the inlet, as thus defined, is positive (in other words is a thrust) for all mass flow conditions except $u_{x_1} = u_0$, independent of the outer shape and cross-section of the inlet.

2.8.6 Thrust of a cylindrical tube with heating or cooling (idealized ramjet)

The inlet thrust result can be used in an analysis of a basic "stovepipe" ramjet consisting of a hollow thin tube of uniform cross-section, with a region of frictionless heat addition or extraction, as shown in Figure 2.24. The ideas can be illustrated with reference to low Mach number flow. For $M^2 \ll 1$, the equation of state can be approximated as (see Section 1.17)

$$\rho T = \text{constant}. \tag{1.17.17}$$

The increase in temperature between stations 1 and 2 means a decrease in density and hence, from continuity, an increase in velocity between the stations. There is consequently a pressure drop across the region of heat addition:

$$p_1 - p_2 = \rho_2 u_2^2 - \rho_1 u_1^2. \tag{2.8.32}$$

Ahead of the region of heat addition, the density can be taken as constant and the stagnation pressure is uniform. We can therefore set ρ_1 equal to ρ_0 in (2.8.32) and use Bernoulli's equation to relate p_0 and p_1:

$$p_0 - p_1 = \tfrac{1}{2}\left(\rho_0 u_1^2 - \rho_0 u_0^2\right). \tag{2.8.33}$$

The streamlines at the trailing edge (station 2) exit tangentially to the tube wall (i.e. axially) and the pressure at this station is equal to the ambient pressure, p_0:

$$p_2 = p_0. \tag{2.8.34}$$

Equations (2.8.32)–(2.8.34) describe the flow from upstream to the ramjet exit. They can be combined to yield an expression for the velocity in the tube upstream of the region of heat transfer,

$$\frac{u_1}{u_0} = \left[\frac{1}{2(\rho_0/\rho_2) - 1}\right]^{1/2} = \left[\frac{1}{2(T_2/T_1) - 1}\right]^{1/2}. \tag{2.8.35}$$

For the idealized ramjet, all the surfaces other than the inlet lip have zero projection in the axial direction. The thrust can therefore only be due to the flow round the inlet lip.[8] The expression for inlet thrust given previously, which did not depend on the details of the lip geometry, can be applied here. Values of u_0/u_1 and u_0/u_2 are plotted in Figure 2.25 as functions of the density ratio (or temperature ratio) across the heat transfer zone along with streamline patterns for $\rho_2 < \rho_1$ and $\rho_2 > \rho_1$. The thrust is zero only for a density ratio of unity; at any other condition, either heating or cooling, thrust is generated.

2.8.7 Oblique shock waves

In the description of shock waves presented in Section 2.6 the shocks were normal to the flow. In general, however, we need to consider configurations in which shock waves are not normal but rather oblique to the incoming velocity. The last example of control volume analysis is thus a derivation of the relation between the upstream and downstream quantities for such an oblique shock wave.

Figure 2.26 shows a typical geometry in which oblique shock waves would be encountered, a so-called compression ramp which creates an oblique shock at an angle β to the incoming flow. The figure also indicates the control volume for developing the relations between upstream and downstream conditions across the shock.

[8] If the tube is infinitely thin the thrust must be developed by an infinite negative pressure at the leading edge, similar to the infinite negative pressure at the leading edge of an infinitely thin wing, since nowhere else can thrust be sustained.

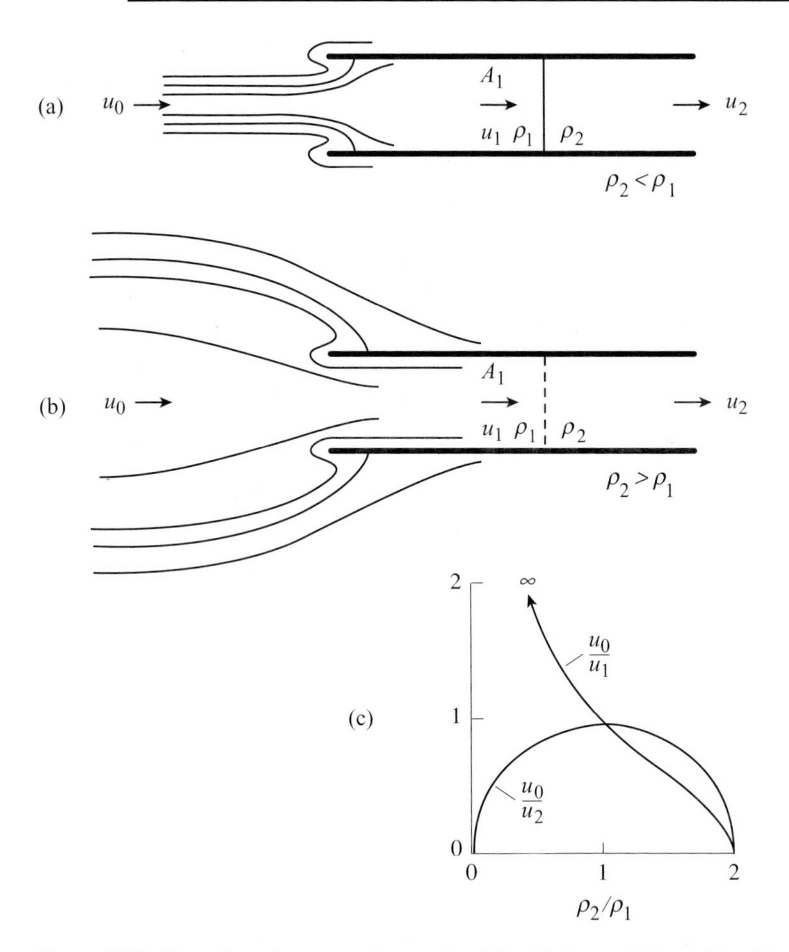

Figure 2.25: Flow through a stovepipe ramjet: (a) with heat addition; (b) with cooling; (c) the effect of the density ratio (Hawthorne, 1957).

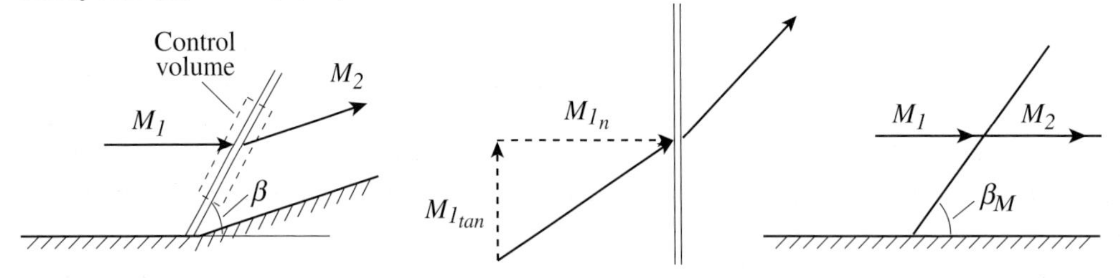

Figure 2.26: Flow through an oblique shock wave (Liepmann and Roshko, 1957).

We resolve the incoming Mach number into components normal and tangential to the shock, M_{1_n} and $M_{1_{tan}}$. The mass flow per unit area into and out of the control volume is the same and is equal to the product of upstream density and upstream component of velocity normal to the shock. Consider the flux of tangential momentum in and out of the control volume. There is no net force in the tangential direction so the tangential velocity component must be the same upstream and downstream of the shock. In consequence the changes in pressure, stagnation pressure, and in fact in all the flow quantities, must be set by the upstream normal Mach number. Another way to argue this is to view the flow through a normal shock from a coordinate system traveling with a constant velocity $u_{1_{tan}}$ along the shock. In such a frame of reference, the perceived velocity is oblique to the shock. Since no flow processes are altered by adoption of this constant velocity, the shock properties must depend on the normal component of the upstream Mach number only.

Oblique shock properties can be found using the three conservation laws given as (2.6.1)–(2.6.3) applied to the normal Mach number, plus the condition of unchanged tangential velocity across the shock. The results are described in detail in many texts (e.g. Liepmann and Roshko (1957), Kerrebrock (1992), Sabersky, Acosta, and Hauptmann (1989) and Hill and Peterson (1992)) and we mention here only three further aspects. First, because the tangential velocity remains the same but the normal velocity decreases, the flow angle will change, i.e. the flow will be deflected through the shock as indicated in Figure 2.26. Second, for a given upstream Mach number, solution of the equations yields two solutions, a weak oblique shock, with supersonic flow downstream of the shock, and a strong oblique shock. The solution that occurs depends on the conditions downstream of the shock. Third, the minimum angle for an oblique shock occurs when the normal Mach number drops to unity. At this condition, the shock becomes an oblique compression wave, called a Mach wave or, more appropriately, a Mach line,[9] analogous to the small disturbance examined in Section 1.15. The flow angle at which this occurs is related to the upstream Mach number by

$$\sin(\beta_M) = \frac{a_1}{u_1} = \frac{1}{M_1}. \tag{2.8.36}$$

The angle β_M referred to as the Mach angle, is shown in Figure 2.26.[10]

2.9 Boundary layers

A useful tactic in the analysis of fluid motions is the partitioning, at least conceptually, of the flow into zones in which different effects play a major role. This provides help in the definition of relevant mechanisms. An illustration of this approach is seen in the treatment of the viscous layers which occur adjacent to solid surfaces and which are referred to as *boundary layers*. In these thin layers the velocity rises from zero at the wall, because of the zero velocity condition at the solid surface, to the free-stream value and viscous effects are important. The part of the flow external to the viscous layers, which is referred to by such (roughly equivalent) terms as inviscid core, external flow, and free-stream flow can often be treated as if it behaves inviscidly. There is a well-developed methodology for calculating the properties of boundary layers which is discussed in some depth in Chapter 4. The purpose in this chapter is to introduce the concept, to show the behavior in a qualitative way, and to point out some of the links between boundary layer behavior and the overall performance of fluid devices.

2.9.1 Features of boundary layers in ducts

Some features of the way in which the boundary layers and the core flow interact can be seen in Figure 2.27, which is a sketch of the flow through an inlet bellmouth into a constant width two-dimensional

[9] At any point in a two-dimensional flow there are two families of Mach lines intersecting the streamline at the angle θ_M. They are also referred to as the characteristics.

[10] The well-accepted notation for shock angle is β; it should not be confused with the use of β to denote reduced frequency, also another well-accepted notation!

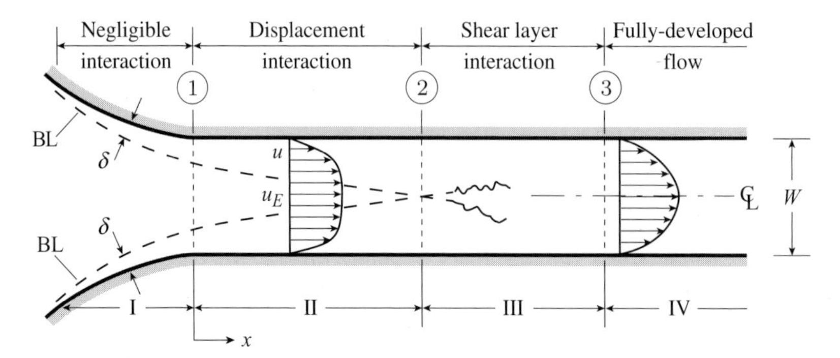

Figure 2.27: Effects of viscous forces on flow regimes in a channel (Johnston, 1978, 1986).

duct. Four regions are indicated and described below in sequence for incompressible flow (Johnston, 1978, 1986).

In Region I, almost all of the duct is occupied by flow that behaves in an inviscid manner, except for the thin boundary layers near the wall, denoted by BL in Figure 2.27. We can estimate the thickness of the boundary layers in order to assess their influence in representative situations. If viscous effects are significant, they must be of the same magnitude as inertial forces. If the length scale in the direction of flow is L, the inertial forces, represented by terms such as $\rho u_x(\partial u_x/\partial x)$ in the momentum equation, will be of order $\rho U^2/L$, where U is a characteristic velocity, say the average velocity. The largest gradients in velocity occur normal to the surface. For laminar flow, the viscous forces represented by terms such as $\mu(\partial^2 u/\partial y^2)$ will thus be of order $\mu U/\delta^2$, where δ is the thickness of the boundary layer. These two forces will be of the same magnitude if

$$\frac{\delta}{L} \sim \sqrt{\frac{\nu}{UL}} = \frac{1}{\sqrt{Re_L}}, \tag{2.9.1}$$

where Re_L is the Reynolds number based on length. The balance between viscous and inertial forces thus leads to the estimate of boundary layer thickness, δ, given in (2.9.1).

Reynolds numbers for many industrial internal flow devices (turbomachines, diffusers, nozzles) are 10^5 or higher,[11] so that boundary layers are much smaller than channel heights in many cases of interest. If the streamwise length scale and the channel height, W, are roughly the same, as in Region I, (2.9.1) shows that the boundary layer thickness is two orders of magnitude smaller than the channel height for a Reynolds number of 10^5. Under these conditions a description of the inviscid core flow based on geometry and inviscid flow analysis provides a good estimate of the static pressure distribution. Note that there is no sharp transition between boundary layer and core flow and the quantity δ is generally specified as a location at which the velocity has come to some specified fraction of the core velocity, say 0.99.

It is of interest to examine the relationship of the velocity components along the wall (x-direction) and normal to the wall (y-direction), and the pressure difference across the boundary layer. The

[11] The length Reynolds number for an air flow with a velocity of 100 m/s is 6×10^6 per meter.

continuity equation for two-dimensional incompressible flow provides a scaling for the first of these:

$$\frac{\partial u_x}{\partial x} + \frac{\partial u_y}{\partial y} = 0. \tag{2.9.2}$$

The y-distance in which the velocity normal to the wall reaches the value outside the boundary layer is the boundary layer thickness, δ, and an estimate for $\partial u_y/\partial y$ is u_y/δ. This must be of the same magnitude as the rate of change in x-velocity along the direction of the stream, $\partial u_x/\partial x$ which is U/L. The magnitude of the ratio u_y/u_x is therefore δ/L, or $1/\sqrt{Re_L}$; for high Reynolds numbers, velocities normal to the wall are much smaller than velocities along the wall.

Using this scaling in the y-momentum equation allows estimation of the pressure difference across the boundary layer. The y-momentum equation is given as (2.9.3), with the magnitude of the different terms shown below it:

$$u_x \frac{\partial u_y}{\partial x} + u_y \frac{\partial u_y}{\partial y} = -\frac{1}{\rho}\frac{\partial p}{\partial y} + \nu\left(\frac{\partial^2 u_y}{\partial x^2} + \frac{\partial^2 u_y}{\partial y^2}\right). \tag{2.9.3}$$

$$\frac{u_x u_y}{L} \qquad \frac{u_y^2}{\delta} \qquad\qquad \frac{\Delta p_y}{\rho\delta} \qquad \nu\left(\frac{u_y}{L^2} \qquad \frac{u_y}{\delta^2}\right)$$

In (2.9.3) Δp_y denotes the magnitude of the change in pressure across the boundary layer. The two terms on the left-hand side and the last term on the right-hand side are of the same magnitude, from the arguments presented above. The term $\partial^2 u_y/\partial x^2$ is $(\delta/L)^2$ smaller than these. The change in pressure across the boundary layer is thus $\Delta p_y \sim \rho u_y^2 \sim \rho u_x^2\ (\delta/L)^2 = \rho u_x^2(1/Re_L)$. For the Reynolds numbers that characterize fluid machinery, unless there are large curvature effects (see Chapter 4), the pressure can be regarded as uniform across the boundary layer and equal to the pressure outside the boundary layer.

2.9.2 The influence of boundary layers on the flow outside the viscous region

Equation (2.9.1) shows that the thickness of the viscous layer grows with the square root of the length scale in the streamwise direction, in this case the streamwise distance from the start of the channel. At some location, denoted by the start of Region II, the boundary layers have grown enough so their influence on the inviscid region can no longer be neglected. The effect on the velocity in the inviscid region, u_E, can be described with reference to a two-dimensional control volume bounded by the wall, a surface a distance y_{CV} from the wall, and two surfaces, 1 and 2, perpendicular to the wall, as in Figure 2.28. At the upstream face of the control volume (station 1) we suppose the boundary layer thickness to be much less than y_{CV}, so the volume flow through the face is approximately $u_{E_1} y_{CV}$, where u_{E_1} is the velocity external to the boundary layers at station 1. At the downstream face, the boundary layer has grown so δ is larger than y_{CV}. The volume flow is consequently less than $u_{E_1} y_{CV}$ and the streamlines diverge from the wall, with a corresponding convergence of streamlines in the core. The effect is similar to that which would occur if the flow were inviscid and the geometric area decreased in the direction of flow. We can thus view the presence of the boundary layer as creating an effective channel area which is smaller than the geometric area.

This idea can be made more quantitative as follows, where, for simplicity, we consider a symmetric channel. We introduce the effective height, W_{eff}, as the height that would be needed to carry the channel

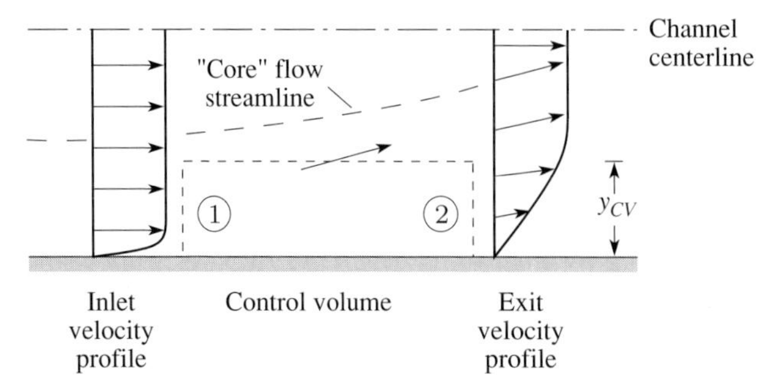

Figure 2.28: Convergence of streamlines in the inviscid (core) region due to boundary layer growth (not to scale).

volume flow if it were all at the inviscid region axial velocity, u_E:

$$u_E W_{eff} = \int_0^W u_x dy$$

$$= W u_E - \int_0^W (u_E - u_x) dy. \tag{2.9.4}$$

Dividing both sides by u_E provides an expression for W_{eff} in terms of a boundary layer parameter, δ^*, referred to as the *displacement thickness*. For a situation in which the boundary layers on the two walls are the same, the displacement thickness is given by

$$W_{eff} = W - 2\int_0^{W/2} \left(1 - \frac{u}{u_E}\right) dy = W - 2\delta^*. \tag{2.9.5}$$

In the integral in (2.9.5), the velocity is equal to the velocity in the free stream for values of y greater than δ and the integrand is zero in this range. When the profiles of u/u_E are similar along the channel, the displacement thickness δ^* and the boundary layer thickness δ are proportional; for a constant pressure laminar boundary layer the proportionality is approximately $\delta^* \sim \delta/3$.

The name displacement thickness derives from external flow applications, for which one interpretation of δ^* is the amount by which a streamline outside the boundary layer is displaced in the direction normal to the boundary. For internal flow applications, the most important characteristic is the effect of the displacement thickness on the core flow, which can be regarded as the flow "blockage" illustrated in Figure 2.29. The representation on the right has the same core velocity and volume flow but occurs in a channel of reduced height, W_{eff}, compared to the actual geometry. The displacement thickness is equal to the blocked height for the lower part of the channel shown.

A relation between changes in blockage and changes in static pressure can be derived from the incompressible form of the incompressible channel flow equations applied to the core flow. For the two-dimensional channel with the boundary layers the same on both walls, the continuity equation is

$$\frac{du_E}{u_E} = -\frac{dW_{eff}}{W_{eff}} = \frac{-d(W - 2\delta^*)}{W - 2\delta^*}. \tag{2.9.6}$$

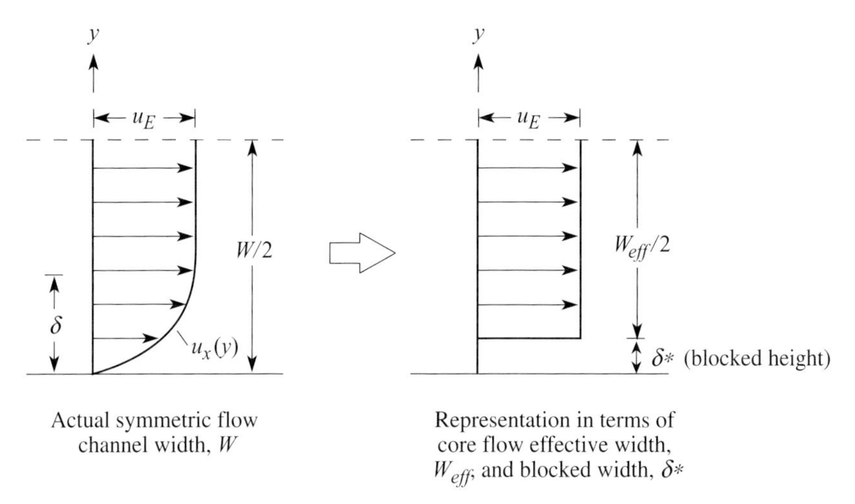

Figure 2.29: Interpretation of displacement thickness in terms of flow blockage.

Substituting this into the momentum equation for the core gives

$$\frac{dp}{\rho u_E^2} = \frac{dW - 2d\delta^*}{W - 2\delta^*}. \tag{2.9.7}$$

Static pressure changes due to boundary layer growth alone (constant channel width) are thus

$$\left(\frac{dp}{\rho u_E^2}\right)_{\substack{boundary \\ layer\ growth}} = -\frac{2d\delta^*}{W - 2\delta^*} \tag{2.9.8}$$

or, for $\delta^*/W \ll 1$,

$$\left(\frac{dp}{\rho u_E^2}\right)_{\substack{boundary \\ layer\ growth}} \approx -2d\left(\frac{\delta^*}{W}\right). \tag{2.9.9}$$

A further implication of (2.9.7) is that if the displacement thickness grows rapidly enough so $2d\delta^* > dW$, increases in geometrical area result in decreases in static pressure.

Because of the connection between displacement thickness and static pressure, a critical part of the problem of finding pressure distribution in a channel or passage often hinges on accurate assessment of the boundary layer displacement thickness. In Chapter 4 we describe techniques for the quantitative prediction of boundary layers focusing on this aspect.

In Region III the boundary layers start to overlap and there is no streamline for which the stagnation pressure is equal to the initial value. For sufficiently long ducts, Region IV can be reached in which the flow obtains a fully developed state so the velocity profiles no longer change with streamwise coordinate. In this region, for incompressible flow, the static and stagnation pressure decrease linearly with x.

2.9.3 Turbulent boundary layers

In fluid machinery, Reynolds numbers can often be high enough that the flow is turbulent rather than laminar. In turbulent flow, the velocity components and pressure can be viewed as composed of an average or mean part plus a fluctuating part. Turbulent boundary layers are examined in Chapter 4, and for now we only mention some properties which differentiate them from laminar boundary layers. The fluctuating velocities in turbulent flow greatly increase the transfer of momentum and energy. Because of this, turbulent shear stresses are much higher than those due to viscous effects alone. For example, for a zero pressure gradient boundary layer at a Reynolds number of 10^6, the skin friction coefficient, $C_f = [\tau_w/(\frac{1}{2}\rho u_E^2)]$, where τ_w is the wall shear stress and u_E is the velocity external to the boundary layer, is approximately seven times higher for a turbulent boundary layer than for a laminar one (0.0047 versus 0.00067). The region of retarded flow produced by the increased shear stresses is also larger so turbulent boundary layers are thicker than laminar boundary layers. For a 0.3 meter long duct at a velocity of 50 m/s (Reynolds number of 10^6), the thicknesses of the laminar and turbulent boundary layers are approximately 1.5 mm and 7 mm, respectively.

Even with the differences between laminar and turbulent flow, the classification of flow regimes is still applicable. Rough guidelines for turbulent boundary layers might be $x/W \sim 15$–25 to the start of Region III and $x/W > 40$ for Region IV although these depend on factors such as turbulence level, Reynolds numbers, and surface roughness. Internal flow devices tend to be designed to be compact so values of x/W are such that operation is often in Region I or II.

One final point concerns operation in the region where the boundary layers have merged. If the flow changes that take place occur in a length short compared to the length needed to merge the boundary layers, the flow can often be treated as inviscid but non-uniform. In other words, for changes that occur over length scales short compared with those required for viscous effects to penetrate to the midst of the channel the influence of viscous forces can be small. In the succeeding chapters we will see a number of situations in which viscous effects, acting over a long distance, have created a non-uniform flow which then undergoes some alteration in a comparatively short distance. In this situation, an inviscid description can be of great use.

2.10 Inflow and outflow in fluid devices: separation and the asymmetry of real fluid motions

2.10.1 Qualitative considerations concerning flow separation from solid surfaces

The inlet and exit flows for the geometries in Section 2.8 have been represented as having a fundamental front-to-rear asymmetry. In Figure 2.22(a) streamlines which *enter* the inlet are shown originating from essentially all directions of the flow domain. In contrast, flow which exits the ejector (Figure 2.19) or the ramjet (Figure 2.25) is described as a parallel jet with velocity in the direction of the exit nozzle. To emphasize the point Figure 2.30 is a sketch of flow into and out of a pipe in a quiescent fluid. For inflow to the pipe (Figure 2.30(a)) the streamlines have approximately spherical symmetry and the pipe entrance appears from afar as a "point sink". For outflow from the pipe (Figure 2.30(b)) the fluid leaves as a jet, similar to the situation at the ramjet and ejector exits. This

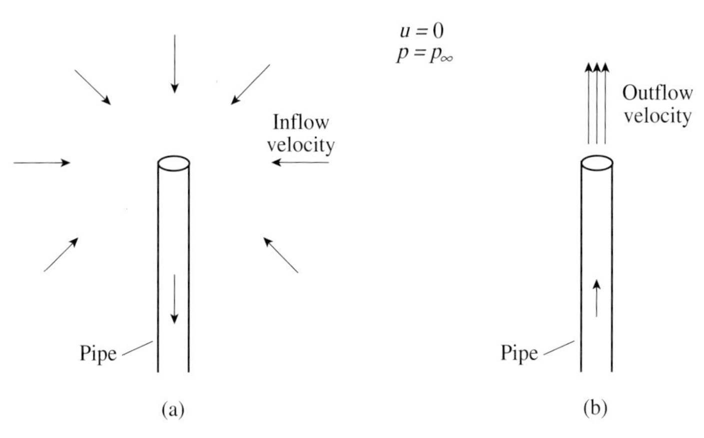

Figure 2.30: Flow into (a) and out of (b) a pipe in a quiescent fluid; $u = 0$, $p = p_\infty$ far away.

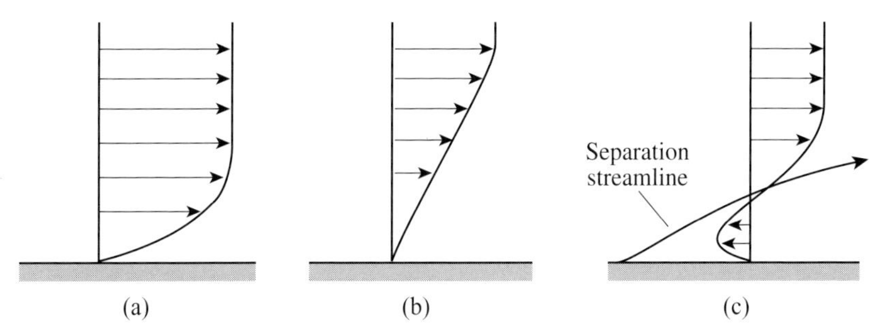

Figure 2.31: Velocity profiles in a boundary layer subjected to a pressure rise: (a) start of pressure rise; (b) after small pressure rise; (c) after separation.

asymmetry, which is a feature of all real (i.e. viscous) flows, is implicit in the control volume analysis of these devices and it is thus worthwhile to examine the rationale behind its use.

The reason for the asymmetry is associated with the no-slip condition at a solid surface in a viscous fluid and the consequent presence of a boundary layer adjacent to the surface, which has lower velocity than the free stream (Section 2.9). For high Reynolds numbers and thin boundary layers the pressure field is set by the flow outside the boundary layer which behaves in an inviscid manner. If u_E is the free-stream (or "external") velocity the maximum pressure rise which can be achieved by the free stream is $\frac{1}{2}\rho u_E^2$. Fluid in the boundary layer, however, has been retarded by viscous forces and has a lower velocity than the free stream. As a result, the pressure rise at which the velocity of boundary layer fluid particles falls to zero is less than $\frac{1}{2}\rho u_E^2$, in other words less than that which the free stream could attain.

The evolution of a boundary layer subjected to a pressure rise is sketched notionally in Figure 2.31. Figure 2.31(a) shows the boundary layer at the start of the pressure rise and Figure 2.31(b) shows the situation after some increase in static pressure. For larger (or more sudden) increases in pressure the result can be reversed flow and a breaking away, or *separation*, of the wall streamline from the solid surface as illustrated in Figure 2.31(c). Quantitative definitions of "larger" and "more sudden" will be given in Chapter 4; for now we combine these qualitative considerations concerning separation

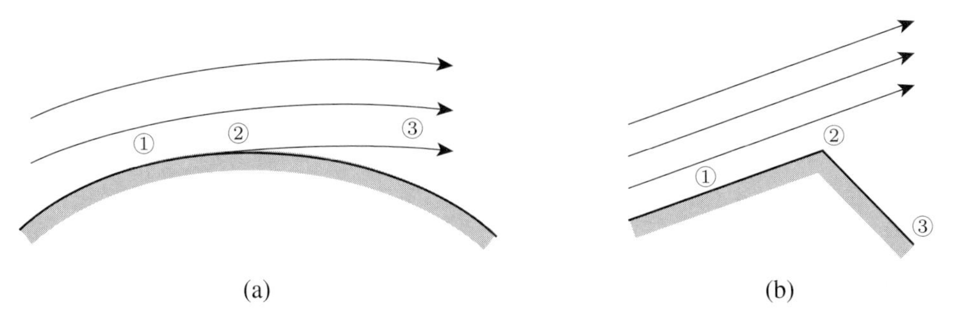

Figure 2.32: Flow separation from a surface: (a) a smooth body; (b) a salient edge (after Batchelor, 1967).

with a description of the static pressure field near the entrance of the pipe to provide a conceptual picture of the observed asymmetry.

There is one further aspect of separation that needs to be introduced, namely the difference between separation from a smooth body and separation from a body with a salient edge. The difference is indicated in Figure 2.32 from Batchelor (1967). For the smooth body (Figure 2.32(a)), the streamlines leaving the surface are tangential to the body. If this were not the case, and a non-zero angle existed between the separation streamline and the body (i.e. a non-zero angle 123 where 1, 2, and 3 are points on the separation streamline) the inviscid flow outside the boundary layer would have a stagnation point at location 2. The fluid in the boundary layer would not be able to negotiate such a pressure rise, and separation would occur upstream of point 2.

For a salient edge with discontinuity in slope (Figure 2.32(b)), inviscid streamlines that followed the geometry would have infinite curvature (zero radius of curvature) and an infinitely low pressure at the discontinuity (point 2). Although engineering devices do not have slope discontinuities when viewed at close range, the point is that, as suggested by the inviscid flow arguments, high curvatures lead to large decreases in pressure and hence severe adverse pressure gradients downstream of the region of high curvature. A viscous fluid will thus separate from a salient edge, as indicated in Figure 2.32(b) with the streamlines leaving tangential to the upstream part of the body. In such cases (e.g. at the pipe exit in Figure 2.30) the velocity of the flow outside the boundary layer does not decrease as the separation point is approached.

2.10.2 The contrast between flow in and out of a pipe

With Section 2.10.1 as background, we can now describe flow in and out of the pipe. Inflow streamlines in the vicinity of the entrance are sketched in Figure 2.33 for a high Reynolds number flow with thin boundary layers. From 1 to 2 there is a favorable pressure gradient with acceleration of the fluid in the boundary layer and thus no tendency for separation. From Section 2.4, location 2 at the entrance lip would be expected to be at low pressure because of the sharp curvature of the streamlines around the lip. The static pressure along the streamline rises from 2 to 3, where the flow outside the boundary layers becomes uniform across the pipe. From 2 to 3 there is some overall streamline convergence (the area normal to the streamlines at 2 is larger than that at 3) which lessens the severity of the adverse pressure gradient. Further, the entrance lip can be shaped to minimize the

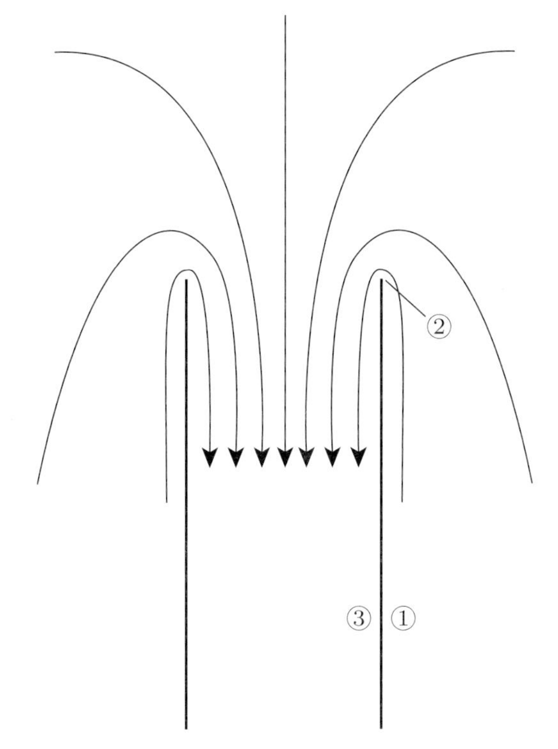

Figure 2.33: Inflow from a quiescent fluid into a pipe: flow near the pipe entrance.

pressure rise, or rather to make it mild enough so that separation does not occur; this is one of the requirements for good inlet design. For high Reynolds numbers the streamlines entering the pipe will thus follow the geometry and look generally similar to those for inviscid flow.

If we ask whether the outflow from the pipe will have a streamline configuration that looks like that of the inlet, however, the answer is no. For this to occur the exiting fluid would have to flow round the pipe entrance and negotiate a pressure rise to stagnation conditions; there is a pressure rise associated not only with the streamline curvature round the lip, but also with the increase in overall streamtube area. Fluid in the boundary layer on the pipe wall cannot do this because of its low velocity (compared to the free stream) and separation will occur.

There is a further difference between outflow and inflow. The function of the exit nozzle is to ensure the flow leaves in a certain direction, rather than flowing round the nozzle lip. This can readily be achieved in practice since it is essentially the case of separation at a sharp edge (in fact it is hard not to have happen). With flow that exits the pipe, therefore, the direction of the velocity is along the line of the pipe, the static pressure and velocity are not altered as the fluid approaches the lip, and the exit configuration is a parallel jet along the axis of the pipe. The static pressure in the exit jet is the same as that of the surrounding environment for a subsonic flow, as argued in Section 2.5.

The asymmetry in streamline configurations which has been described occurs due to the presence of viscosity. Viscous motions are not thermodynamically reversible and generally not kinematically reversible (i.e. changing the direction of the flow does not mean that the streamlines will retain

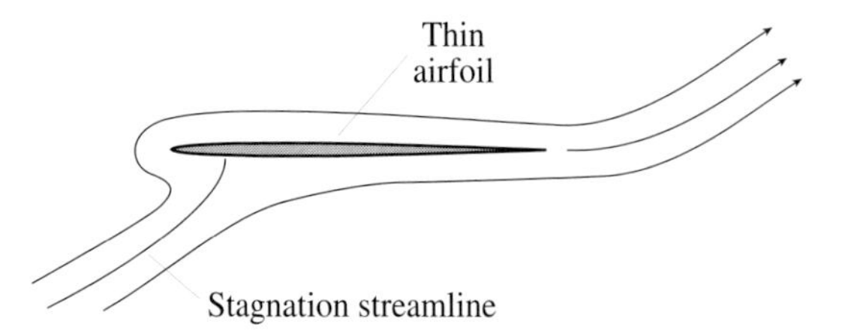

Figure 2.34: Flow round a thin airfoil at an angle of attack.

their form).[12] A well-known example of this is the flow round a thin wing sketched in Figure 2.34. Classical thin airfoil theory describes a flow which curves round the leading edge (with a locally infinite velocity for a thin flat plate), and leaves the trailing edge tangential to the airfoil, as simulated in the Kutta–Jukowski condition. There is a direct analogy with the flow entering and exiting the ramjet. Describing the flow leaving a straight nozzle as a jet parallel to the nozzle axis is similar to the Kutta–Jukowski condition for the airfoil in that it is an assumption that allows us to capture features of the viscous flow with an inviscid description. This assumption can also be used to describe the flow leaving a cascade of closely spaced turbine or compressor blades, where the idealization is also a sharp trailing edge. In that situation the leaving angle of the flow depends little on the angle at which the flow enters the cascade and can be regarded as constant over a range of inlet conditions.

2.10.3 Flow through a bent tube as an illustration of the principles

An example that incorporates many of the above ideas is given by the constant density flow through a bent tube of uniform area A, as in Figure 2.35. We examine two situations, first flow exiting the tube through the two areas at the ends of the tube (e and e′) and second flow entering the tube through these areas. In the former situation the fluid enters at the center at O and exits through the two bent parts of the tube at stations e and e′. With the tube free to rotate around O and the velocity through the tube u_1, we wish to know the rate of rotation. This can be found by considering the angular momentum flux through a cylindrical control surface centered on O with a radius greater than the tube radius. The fluid enters at the center of rotation with very small radius and thus no angular momentum about O. With the tube free to rotate, no torque is applied and the fluid also leaves with no angular momentum. The angular momentum flux across the outer control surface is zero, and this can only occur if the tangential velocity is zero. For this to occur the velocity at which the fluid exits the bent tube, relative to the tube, must therefore be equal and opposite to the tangential velocity of the tube end so their sum is zero. The rate of rotation, Ω, is thus given by the condition $\Omega r_{tube} = u_1$, or $\Omega = u_1/r_{tube}$. This result can also be derived viewed from a coordinate system rotating with the tube by balancing the Coriolis forces on the radial part of the tube with the pressure forces in the bend that turn the flow into the tangential direction.

From another perspective if the tube is held stationary, the exit flux of angular momentum around O is $\rho u_1 A \Omega r_{tube}$, so there must be a torque about point O. A stationary tube which is not restrained will (in the absence of friction) therefore increase its rotation rate, Ω, until it attains the value u_1/r_{tube}.

[12] At Reynolds numbers (UL/ν) much less than unity, when inertial forces are much less than viscous forces, fluid motions do exhibit kinematic reversibility (Taylor, 1972).

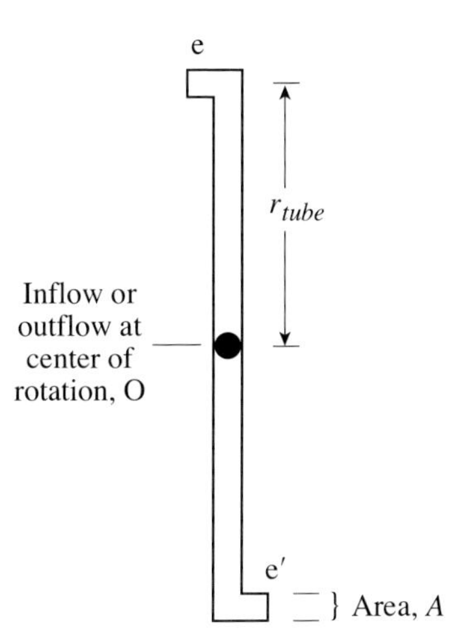

Figure 2.35: Freely rotating bent tube. Outflow or inflow at tube ends e and e′; velocity through the tube is u_1.

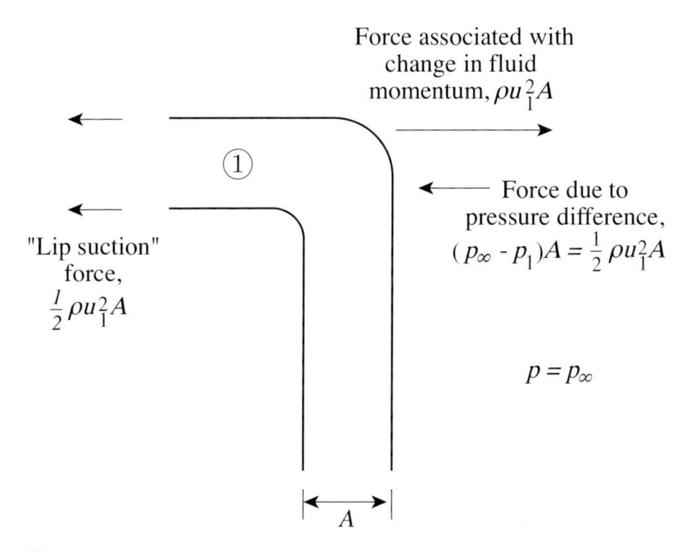

Figure 2.36: Forces on bent tube with inflow; $u = 0$, $p = p_\infty$ far from tube.

Suppose now, as recounted in graphic terms by Feynman (1985), the direction in which the fluid is pumped is reversed, so that fluid is sucked into the tube at e and e′, and exits at O. What is the rate of rotation in this situation? If the surrounding fluid is without rotation, as it would be if the tube were fed from a still atmosphere, the flux of angular momentum across the outer cylindrical control surface is zero. The flux of angular momentum out at O is also essentially zero. These two statements imply no torque on the tube. If the tube is at rest, it will remain at rest, contrary to the first case.

It is helpful to see why this occurs from a different viewpoint through examination of the tangential forces that act on the tube. These are indicated in Figure 2.36 for the condition in which the tube is stationary. The discussions in Section 2.8 imply there is a "lip suction" force of magnitude $\frac{1}{2}\rho u_1^2 A$ pointing forward. (We assume the section of the tube perpendicular to the radius is short enough so it

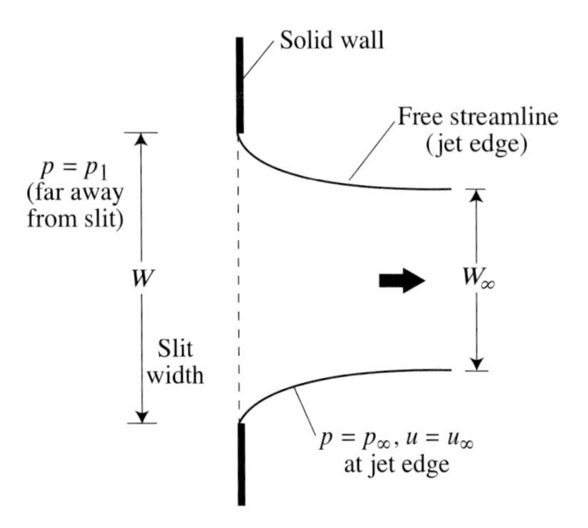

Figure 2.37: Calculated inviscid steady flow through a two-dimensional slit to a uniform pressure region (Batchelor, 1967).

can be taken as pointing in the tangential direction.) The force due to the pressure difference between the inside and outside of the bent tube is $(p_\infty - p_1)A$, where p_1 is the static pressure at station 1 inside the tube and p_∞ is the pressure of the still fluid far from the tube. From the Bernoulli equation this force is equal to $\frac{1}{2}\rho u_1^2 A$ and points in the same direction as the lip suction force. Finally, the force associated with the change in direction of the velocity (i.e. with the momentum change) as the fluid is turned in the bend has magnitude $\rho u_1^2 A$ and points in the direction opposite to the other two. As shown in the figure, therefore, the sum of the three contributions is zero.

2.10.4 Flow through a sharp edged orifice

Separation at a sharp edge or corner must be accounted for in descriptions of the flow through orifices and grids such as perforated plates (e.g. plates with sharp edged circular holes). The basic behavior can be seen in the model problem of inviscid, constant density, steady flow through a two-dimensional slit in a wall between a reservoir at a pressure p_1 and an ambient pressure, p_∞, as shown in Figure 2.37. If the inviscid flow is to capture the basic features of the actual (viscous fluid) situation the stream that emerges from the reservoir should separate at the termination of the solid wall, with the velocity at the edge of the resulting jet tangent to the wall at the separation location. Far downstream the jet velocity is uniform, parallel, and perpendicular to the plate. Although the term "far downstream" is used here to denote the asymptotic form of the jet, the considerations of length scales in Section 2.3 imply that the distance in which this condition is achieved is roughly one slit width.

The downstream jet width is less than the width of the slit, W, and this contraction between initial and asymptotic jet areas is common to flow through sharp edge orifices. The general features and streamline pattern in such configurations are essentially unchanged for values of Reynolds numbers (based on an appropriate length scale of the orifice) above roughly a thousand.

In Figure 2.37 the "free streamline" that bounds the jet once it leaves the solid wall is subjected to ambient pressure p_∞ all along its length. (We use the subscript ∞ for consistency with, and in the

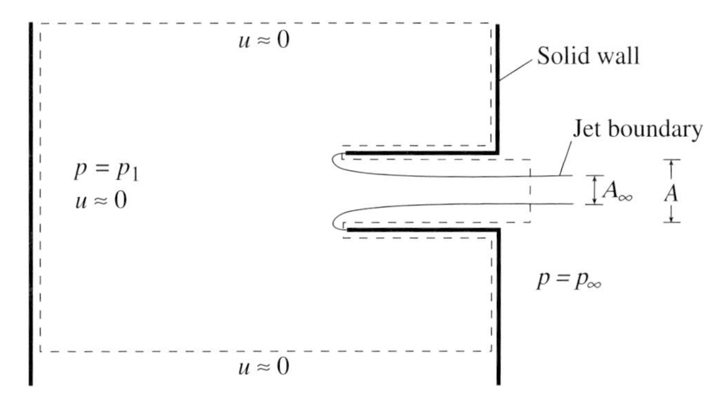

Figure 2.38: Separated flow from a reservoir through a reentrant channel (Borda's mouthpiece).

same sense as, the term far downstream.) The velocity on this free streamline is thus constant. In the vicinity of the plane of the slit, there is streamline curvature in the jet associated with the pressure gradient force; there is a higher pressure at the jet centerline than at the edge of the jet.

From the Bernoulli equation, with p_1 the stagnation pressure, the far downstream jet velocity is

$$u_\infty = \sqrt{\frac{2(p_1 - p_\infty)}{\rho}}. \tag{2.10.1}$$

The ratio of the actual jet flow to a reference flow rate based on the velocity u_∞ and the slit width is often referred to as the discharge coefficient. For the two-dimensional problem the discharge coefficient is given from the free streamline analysis as $W_\infty/W = \pi/(\pi + 2) = 0.611$ (Batchelor, 1967), a result which is close to the experimental value.

The above arguments imply that to increase the discharge coefficient the exit should be shaped so the stream leaves the solid surface with a velocity parallel to the far downstream direction. For a well-designed nozzle, for example, discharge coefficients are close to unity. In contrast a reentrant geometry such as in Figure 2.38, in which the direction of the velocity at separation is opposite to the far downstream jet direction, would be expected to have a discharge coefficient lower than that for a slit or orifice in a plane wall. Discharge coefficients for a number of two- and three-dimensional geometries are given by Miller (1990) and Ward-Smith (1980), but the discharge coefficient for the configuration in Figure 2.38 can be found using control volume concepts.

The flow round the sharp edge of the reentrant channel separates from the channel wall as drawn in Figure 2.38. If the channel is short enough so the flow does not reattach to the channel wall (from Section 2.8 this means the length must be less than four or five channel widths) the pressure on the free streamline at the edge of the jet is ambient throughout its length. For the control surface in Figure 2.38 the force exerted on the fluid in the control volume is $(p_1 - p_\infty)A$, where A is the channel area. Equating this force to the outflow of momentum at the far downstream station, where the jet has achieved its final area and velocity, yields

$$(p_1 - p_\infty)A = \rho u_\infty^2 A_\infty. \tag{2.10.2}$$

Substituting the expression for the far downstream velocity, u_∞, from (2.10.1) into (2.10.2) we obtain the ratio of areas as $A_\infty/A = 1/2$, a result that applies whether the channel is two- or three-dimensional.

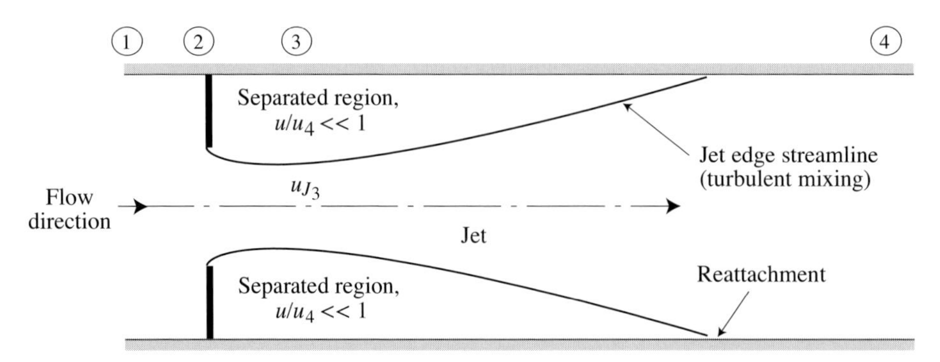

Figure 2.39: Flow through a sharp edged orifice in a duct: jet and reattachment; the jet edge turbulent region is the mixing layer (not to scale); free-streamline theory applies from station 2 to station 3, station 3 is location of minimum jet area.

If the channel is long enough that the jet flow through the orifice or slit reattaches to the channel wall, as shown in Figure 2.39, there is a pressure rise associated with the mixing and reattachment process. The pressure to which the jet discharges is therefore lower than ambient and the mass flow is increased. This situation can be analyzed by combining the results for the sudden expansion (Section 2.8) with the ideas introduced concerning the flow downstream of sharp edged orifices, as done by Ward-Smith (1980) for a circular orifice in a cylindrical duct.

The stations used in the analysis are given in Figure 2.39. At stations 1 and 4 the velocity and static pressure are taken as uniform. At station 3 the jet area has reached its minimum value and the jet velocity is denoted by u_{J_3}. Denoting the contraction coefficient between the jet minimum area and orifice (or slit) area, A_3/A_2, as C_c the equations that describe the flow are:

$$u_1 = C_c \frac{A_2}{A_1} u_{J_3} = u_4, \tag{2.10.3}$$

$$p_1 + \frac{\rho}{2} u_1^2 = p_3 + \frac{\rho}{2} (u_{J_3})^2, \tag{2.10.4}$$

$$p_3 + \rho C_c \frac{A_2}{A_1} (u_{J_3})^2 = p_4 + \rho u_4^2. \tag{2.10.5}$$

Equations (2.10.3)–(2.10.5) can be combined to give a relation for the stagnation pressure (or, equivalently, static pressure) drop between stations 1 and 4 in terms of the orifice area to duct area ratio, A_2/A_1, and the contraction coefficient as

$$\frac{p_{t_1} - p_{t_4}}{\rho u_1^2/2} = \frac{p_1 - p_4}{\rho u_1^2/2} = \left[\left(\frac{A_1}{A_2} \right) \left(\frac{1}{C_c} \right) - 1 \right]^2. \tag{2.10.6}$$

Measurements of pressure drop then allow one to find the relation between contraction coefficient and the ratio of orifice area to duct area, A_2/A_1, as plotted in Figure 2.40. This information can also be applied to the behavior of perforated plates (see also Cornell (1958)).

In the flows illustrated in Figures 2.37–2.39 a common phenomenon is that the jet downstream of the obstacle or plate has a smaller area, and therefore a larger velocity, than that inferred based on the open area in the channel (the total area minus the geometric blocked area). The resulting

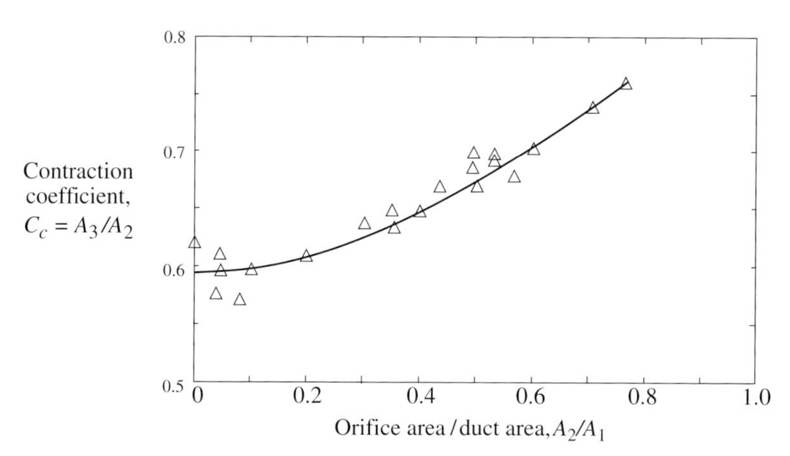

Figure 2.40: Variation of contraction coefficient, C_c, as a function of orifice area/duct area (see Figure 2.39) for orifice plates with square edges, constant density flow in a circular pipe (Ward-Smith, 1980).

static, and stagnation, pressure drop for flow past sharp edged geometries is thus typically several times (or more) larger than that based on purely one-dimensional geometric area versus velocity considerations. Information on the numerical values for pressure drop in a variety of internal flow configurations involving separations from sharp edges or corners (as well as in configurations with no sharp edges) are given by Ward-Smith (1980), Fried and Idelchik (1989), and Miller (1990).

Finally, it is worth noting that an analogous situation concerning separation occurs for external flow past bluff bodies with salient edges (e.g. a thin flat plate normal to a stream) in which the wake width is considerably larger than the lateral dimension of the body. Roshko (1993b, 1993c) presents insightful discussions of such configurations.

3 Vorticity and circulation

3.1 Introduction

In many internal flows there are only limited regions in which the velocity can be considered irrotational; i.e. in which the motion is such that particles travel without local rotation. In an irrotational, or potential, flow the velocity can be expressed as the gradient of a scalar function. This condition allows great simplification and, where it can be employed, is of enormous utility. Although we have given examples of its use, potential flow theory has a narrower scope in internal flow than in external flow and the description and analysis of non-potential, or rotational, motions plays a larger role in the former than in the latter. One reason for this difference is the greater presence of bounding solid surfaces and the accompanying greater opportunity for viscous shear forces to act. Even in those internal flow configurations in which the flow can be considered inviscid, however, different streamtubes can receive different amounts of energy (from fluid machinery, for example), resulting in velocity distributions which do not generally correspond to potential flows. Because of this, we now examine two key fluid dynamic concepts associated with rotational flows: *vorticity*, which has to do with the local rate of rotation of a fluid particle, and *circulation*, a related, but more global, quantity.

Before formally introducing these concepts, it is appropriate to give some discussion concerning the motivation for working with them, rather than velocity and pressure fields only. The equations of motion for a fluid contain expressions of forces and acceleration, derived from Newton's laws. On one level there is no need to introduce concepts relating to the angular rotation rate of a fluid particle explicitly. The idea of introducing local fluid rotation can be motivated, however, by analogy with rigid body dynamics. There, in addition to dealing with forces and linear velocity and momentum, use of the concepts of moment of force (torque), angular velocity, and angular momentum gives rise to additional, very effective, tools for examining problems involving rotation.

Ideas of vorticity and circulation are introduced in a similar context; it is not the necessity of describing fluid mechanics in terms of these concepts that gives rise to their wide application, but rather the demonstrated utility. A goal of this chapter, therefore, is to demonstrate that focus on these concepts provides a useful framework for the physical interpretation and qualitative understanding of fluid phenomena, particularly where three-dimensional or unsteady effects are concerned.

The plan and scope of the material to be covered stem from our observation that, although the algebraic manipulations needed to derive the equations describing the evolution of vorticity and circulation present little difficulty, there is often uneasiness about the physical content, the question of why one considers vorticity, and the point of recasting the equations of motion in this form. We thus illustrate with physical examples how one can use these concepts in situations of practical interest, as well as make connections between this material and more familiar areas of dynamics.

Discussions of vorticity and circulation are presented along parallel paths, so that the relation between changes in the two quantities can be seen and overall ideas concerning fluid rotation reinforced. Both concepts are developed in stages, starting with constant density, inviscid flow and then incorporating the complicating factors of viscosity and compressibility one at a time, so that the role of each is apparent. The initial discussion addresses changes of vorticity and circulation and what this implies about the evolution of the flow features. The last part of the chapter describes the relationship between a general distribution of vorticity and the velocity field, and shows how this relation can be exploited in computing fluid motions.

3.2 Vorticity kinematics

The vorticity, $\boldsymbol{\omega}$, is formally defined as

$$\omega = \nabla \times \mathbf{u}. \tag{3.2.1}$$

To tie this to a specific example, consider a plane flow in which there is a small cylinder of fluid rotating with local angular velocity $\boldsymbol{\Omega}$ within this flow. The magnitude of the average vorticity over the area, A, of the cylinder is then given by

$$\omega_{av} = \frac{1}{A}\iint \nabla \times \mathbf{u} \cdot \mathbf{n}\, dA, \tag{3.2.2}$$

where the unit vector $\mathbf{n}$ is normal to the planar area A. If the cylinder is small enough in cross-section for the angular velocity to be considered constant over the area of the cylinder, ω_{av} becomes the local value, ω. Using Stokes's Theorem, the above expression can be written as an integral over the line elements $d\boldsymbol{\ell}$ of a contour C that bounds the cylinder area. As the area shrinks to zero, this becomes an expression for the magnitude of the vorticity

$$\omega = \frac{1}{A}\oint_C \mathbf{u} \cdot d\boldsymbol{\ell}, \qquad \text{as } A \to 0. \tag{3.2.3}$$

Another way to define the vorticity is thus as the line integral round the contour that bounds the small area. For a circular cylinder of radius r rotating with angular velocity $\boldsymbol{\Omega}$, as shown in Figure 3.1, the value of the integral is $2\pi r u_\theta = 2\pi r^2 \Omega$ and the magnitude of the vorticity is

$$\omega = 2\Omega. \tag{3.2.4}$$

As defined in (3.2.1), the sense of the vorticity is positive if the rotation is anti-clockwise as seen from above and negative if clockwise. The fluid element in Figure 3.1 therefore has positive vorticity.

In the planar configuration just examined the magnitude of the vorticity was shown to be twice the local rate of fluid rotation. However, the flow does not have to be planar for this result to hold. For a fluid particle small enough that the rotation rate can be regarded as constant over the area of integration, we can carry out similar operations with reference to the three component directions.

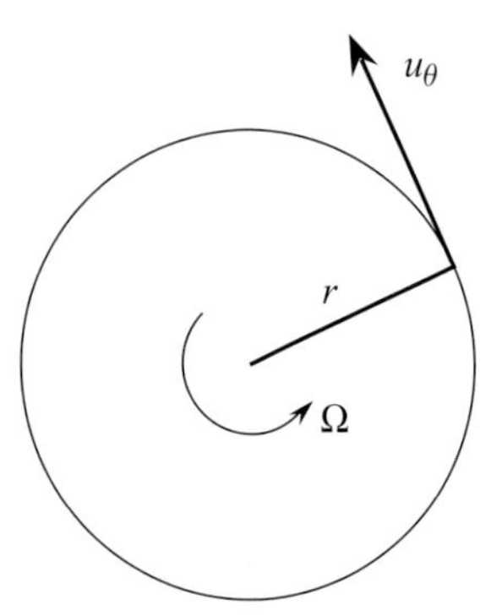

Figure 3.1: Circumferential velocity (u_θ) and angular velocity (Ω) for a small cylindrical fluid element; $u_\theta = \Omega r$.

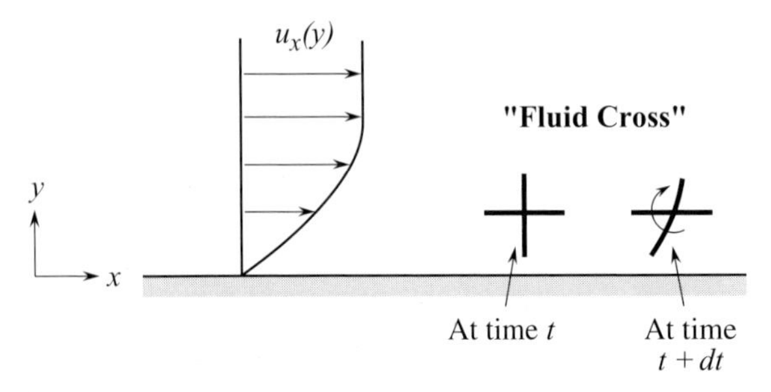

Figure 3.2: Rotation of fluid element in a uni-directional shear flow.

The vorticity vector, $\boldsymbol{\omega}$, is therefore related to the local angular velocity of the fluid, $\boldsymbol{\Omega}$, by

$$\boldsymbol{\omega} = 2\boldsymbol{\Omega}. \tag{3.2.5}$$

A physical interpretation of (3.2.5) is that if a small sphere of fluid were instantaneously solidified with no change in angular momentum, the local vorticity would be twice the local angular velocity of the sphere. The rotation convention is such that there is a "right-hand rule" between velocity and vorticity directions.

As with angular velocity, vorticity is a vector. On a component by component basis, the components of the vorticity vector are the sum of the rotation rate of two mutually perpendicular fluid lines. For example consider the planar uni-directional flow shown in Figure 3.2. The velocity $\mathbf{u}$ is given by $u_x(y)\mathbf{i}$ (with $\mathbf{i}$ the unit vector in the x-direction) and the streamlines are parallel. Examination of the components of $\nabla \times \mathbf{u}$ shows that $\omega_x = \omega_y = 0$, but the z-component of $\boldsymbol{\omega}$ is non-zero:

$$\omega_z = -\frac{du_x}{dy}. \tag{3.2.6}$$

The quantity $(-du_x/dy)$ is the clockwise rotation rate of the fluid line initially parallel to the y-axis. Because the fluid line parallel to the x-axis does not rotate, the average rotation rate is $\frac{1}{2}(du_x/dy)$ and the vorticity is as given in (3.2.6).

The general planar case is depicted in Figure 3.3, which shows the rotations of the lines OP and OQ about point O, the center of a fluid particle. The two lines, of lengths dx and dy respectively, are initially perpendicular. After a short time, dt, they have moved to positions OP′ and OQ′with

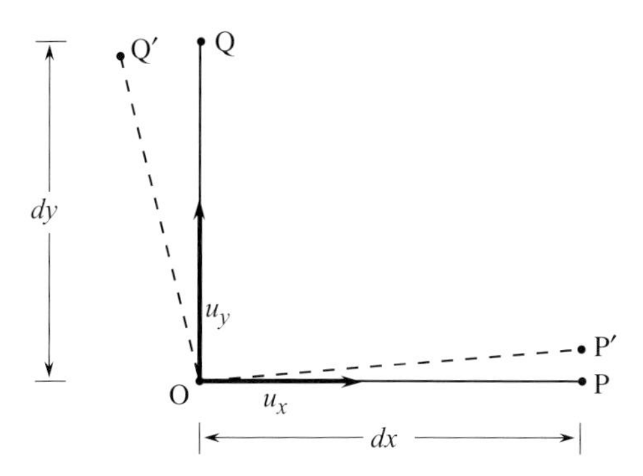

Figure 3.3: Rotation of two initially perpendicular fluid lines, OP and OQ, during a short time dt; u_x and u_y are velocity components at point O.

reference to point O, as shown by the dashed lines. If u_x and u_y are the velocity components at point O, the rate of counterclockwise rotation of OP is $(\partial u_y/\partial x)$ and that of OQ $(-\partial u_x/\partial y)$. The average rate of rotation is one-half the sum of these two quantities so the vorticity is $[(\partial u_y/\partial x) - (\partial u_x/\partial y)]$. For the x–y planar flow illustrated, this would be the magnitude of the z-component of vorticity. For a three-dimensional velocity field, the two other (y–z and z–x) components of the vorticity vector could be obtained by carrying out these operations for their respective planes. Note that in Figure 3.3, the orientation of the x–y coordinate system was arbitrary with respect to the flow field; the mean angular rotation at a given location, and thus the vorticity, has the same value independent of coordinate orientation.

3.2.1 Vortex lines and vortex tubes

Applications of vorticity concepts are often connected to an overall, rather than just local, description of flow fields. To link the local definition given in (3.2.1) and the overall field, we introduce the idea of *vortex lines*, which are lines in the fluid tangent to the local vorticity vector. A general result for all vector fields is that the divergence of a curl is identically zero, so that for a vector **B**

$$\nabla \cdot [\nabla \times \mathbf{B}] = 0. \tag{3.2.7}$$

Thus, since $\boldsymbol{\omega} = \nabla \times \mathbf{u}$,

$$\nabla \cdot \boldsymbol{\omega} = 0. \tag{3.2.8}$$

Equation (3.2.8) is purely kinematic and holds for any flow. A vector whose divergence is zero is referred to as *solenoidal* and (3.2.8) is often referred to as stating that the vorticity field is solenoidal. This is a strong constraint about the behavior of vortex lines, as described in the next several paragraphs.

Applying the Divergence Theorem to (3.2.8), we obtain a statement about the vortex lines that thread through a closed surface as

$$\oiint \boldsymbol{\omega} \cdot \mathbf{n}\, dA = 0. \tag{3.2.9}$$

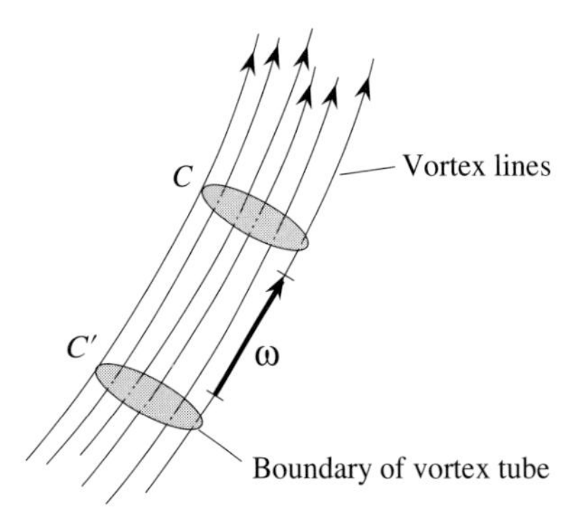

Figure 3.4: Individual vortex lines and a vortex tube.

Equation (3.2.9) states that the integral of the normal component of vorticity is zero over any closed surface. The vortex lines that enter the surface must therefore also leave it (else the integral would not be zero) so that *vortex lines cannot end in a fluid*. The vorticity field obeys the same continuity equation as an incompressible velocity field ($\nabla \cdot \mathbf{u} = 0$), for which: (1) streamlines (lines tangent to the local velocity vector) cannot end in the fluid, and (2) concentrations of the streamlines occur where the velocity is high. Similarly, vortex lines are closely spaced in regions of high vorticity and sparse where the vorticity is small.

The analogy can be taken a step further by introducing the concept of a *vortex tube* as a tube with boundaries formed by vortex lines which intersect a closed curve, as in Figure 3.4. The vorticity, which is everywhere parallel to the vortex lines only penetrates surfaces which cut the tube such as those bounded by the curves C and C'. Equation (3.2.9) shows that the total (integrated) vorticity, $\oiint \boldsymbol{\omega} \cdot \mathbf{n}\, dA$, threading through both of these two surfaces, or through any other two surfaces which completely cut the vortex tube, will be the same. The flux of vorticity ($\iint \boldsymbol{\omega} \cdot \mathbf{n}\, dA$) is analogous to the volume flow along a stream tube (a tube composed of streamlines through a closed curve) in an incompressible fluid. A streamline is a curve locally tangent to the velocity, so that no fluid leaves the stream tube through its sides. The volume flow $\iint \mathbf{u} \cdot \mathbf{n}\, dA$ must be the same at any location along the streamtube and, when the streamtube area decreases, the velocity increases. Similarly, the quantity $\iint \boldsymbol{\omega} \cdot \mathbf{n}\, dA$, which is often referred to as the strength of the vortex tube, is constant along the length of the vortex tube. When the vortex tube area decreases, the local vorticity magnitude increases. In addition, since the individual vortex lines within the vortex tube cannot end in the fluid, vortex tubes also cannot end in the fluid.

The concept of a vortex tube is especially applicable when there are regions of concentrated vorticity, and in situations of this type it is possible to deduce features of the velocity field from vorticity considerations. A basic example is an infinite, straight vortex tube of radius a in an unbounded flow which is irrotational outside the tube,[1] such as is shown in Figure 3.5. The tube is specified to have vorticity of uniform magnitude ω_o, with no vorticity outside. The strength of vorticity, which is the strength of the vortex tube, is $\pi a^2 \omega_o$. This is also the total (integrated) vorticity through any circular area of radius $r > a$ centered on the tube axis and normal to it. The total vorticity that threads through an area, however, can also be expressed as a line integral round the contour that bounds the region,

[1] See Convention 3 in the Nomenclature section concerning the use of the same variable for two different quantities.

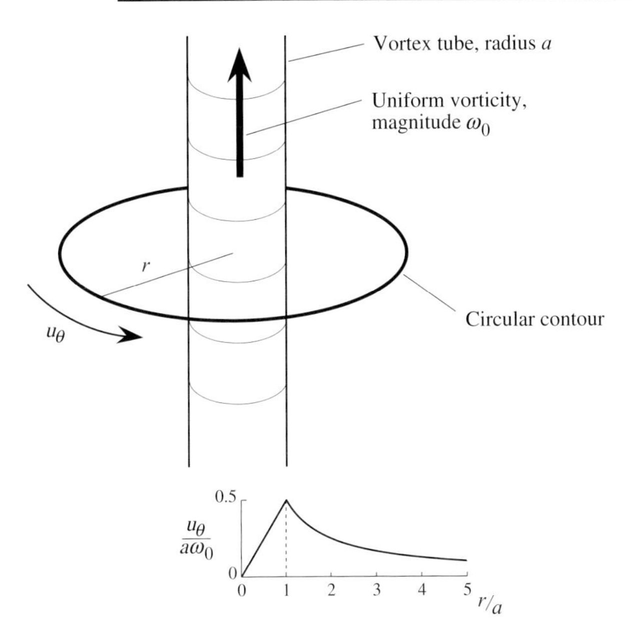

Figure 3.5: Velocity field associated with a straight vortex tube.

as discussed previously, and is given by

$$\int \mathbf{u} \cdot d\mathbf{l} = \pi a^2 \omega_o. \tag{3.2.10}$$

The scalar quantity defined by $\int \mathbf{u} \cdot d\mathbf{l}$, which represents an integral property of the vortex tube, is called the *circulation* and will be discussed at length in Section 3.8.

From symmetry, in the region outside of the vortex tube the only component of velocity is axisymmetric in the circumferential (θ) direction. The circulation around the vortex tube is constant for $r > a$ and so the θ-component of velocity is given by

$$u_\theta = \frac{a^2 \omega_o}{2r}, \quad r > a. \tag{3.2.11}$$

In the irrotational region outside the vortex tube, the θ-component of velocity varies inversely with radius. For radii less than a, the flux of vorticity depends on radius and at any radius, $r \leq a$,

$$u_\theta = \frac{\omega_o r}{2}, \quad r \leq a. \tag{3.2.12}$$

At $r = a$, the two velocity distributions are continuous. The corresponding θ-velocity distribution is also sketched in Figure 3.5.

The infinite straight vortex tube is not in any strict sense a representation of flows of engineering interest, but it does give qualitative guidelines about the velocity field in more complex configurations, for example, the curved vortex tube in Figure 3.6. If the tube has a diameter small compared to the radius of curvature of the tube axis, then the predominant motion will locally resemble that of the infinite tube, i.e. a swirl round the tube, with the resulting velocity as sketched: downwards on the outside of the loop and upwards on the inside. We will explore the connection between the velocity and vorticity field in greater depth later in this chapter as a means for not only qualitative, but quantitative, flow descriptions.

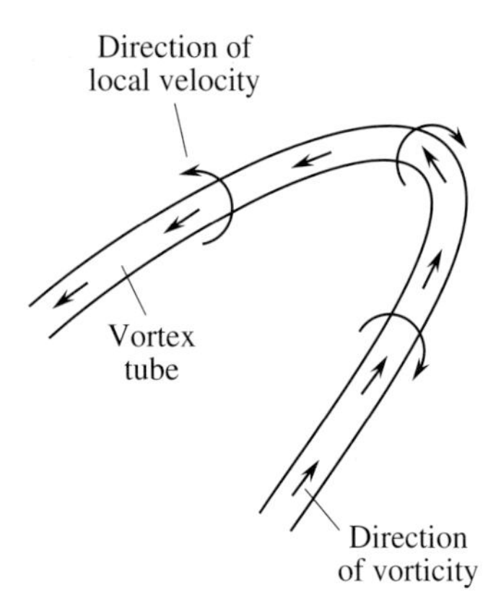

Figure 3.6: Velocity field associated with a curved vortex tube.

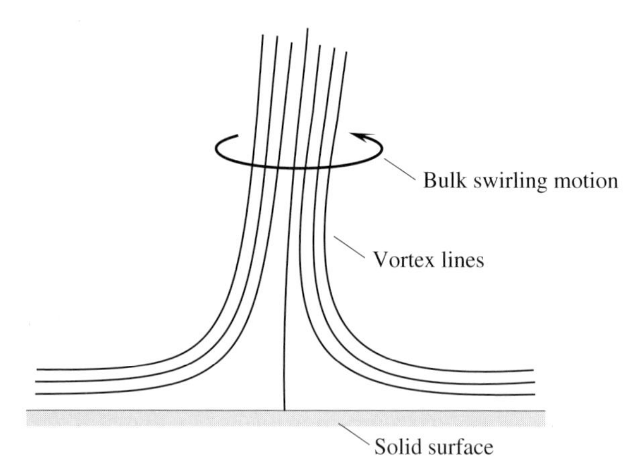

Figure 3.7: Behavior of vortex lines at a solid surface; vorticity must be tangential except for isolated vortex lines (with zero circulation).

3.2.2 Behavior of vortex lines at a solid surface

We conclude this section with a description of the behavior of vortex lines at a solid surface. For a stationary boundary, the no-slip condition requires that the velocity of the fluid at the surface be zero. The circulation round any contour drawn in the solid surface is therefore also zero. This means that there are no vortex lines threading through such a contour and hence no normal component of vorticity. At stationary solid surfaces, the vortex lines must be tangential, except possibly for isolated vortex lines (with zero circulation) similar to dividing streamlines. In contrast, for a rotating surface, there is a normal component of vorticity at the surface, with a magnitude twice the surface angular velocity. Thus vortex lines can terminate on rotating surfaces. This implies that for a flow with stationary boundaries, vortex lines must either form closed loops or "go to infinity"; except for isolated instances they cannot end on the solid boundary. A sketch of such a configuration is given in Figure 3.7, which shows vortex lines associated with a swirling flow over a stationary solid surface.

3.3 Vorticity dynamics

The foregoing has been purely kinematic, and the results are applicable to viscous and inviscid, compressible as well as incompressible, flows. To make real use of the vorticity as an aid in developing physical understanding, it is necessary to consider the dynamical aspects, in particular, to address how the vorticity distribution evolves in a general flow field.

The starting point for this is the momentum equation, (1.9.10), written in the form

$$\frac{D\mathbf{u}}{Dt} = \frac{\partial \mathbf{u}}{\partial t} + \mathbf{u}\cdot\nabla\mathbf{u} = -\frac{1}{\rho}\nabla p + \mathbf{X} + \mathbf{F}_{visc}. \tag{3.3.1}$$

The forces acting on the fluid are represented as three types: pressure forces per unit mass ($\nabla p/\rho$), body forces per unit mass ($\mathbf{X}$), and viscous forces per unit mass ($\mathbf{F}_{visc}$), allowing the effect of each to be examined separately. Using the vector identity

$$(\mathbf{u}\cdot\nabla)\,\mathbf{u} \equiv \nabla\left(\frac{u^2}{2}\right) - \mathbf{u}\times(\nabla\times\mathbf{u}) = \nabla\left(\frac{u^2}{2}\right) - \mathbf{u}\times\boldsymbol{\omega}, \tag{3.3.2}$$

(3.3.1) can be written

$$\frac{\partial \mathbf{u}}{\partial t} + \nabla\left(\frac{u^2}{2}\right) - \mathbf{u}\times\boldsymbol{\omega} = -\frac{1}{\rho}\nabla p + \mathbf{X} + \mathbf{F}_{visc}. \tag{3.3.3}$$

An equation for the rate of change of vorticity is obtained by taking the curl of (3.3.3):[2]

$$\frac{D\boldsymbol{\omega}}{Dt} = (\boldsymbol{\omega}\cdot\nabla)\,\mathbf{u} - \boldsymbol{\omega}(\nabla\cdot\mathbf{u}) - \nabla\times\left(\frac{1}{\rho}\nabla p\right) + \nabla\times\mathbf{X} + \nabla\times\mathbf{F}_{visc}. \tag{3.3.4}$$

Equation (3.3.4) describes changes in vorticity for a fluid particle. Rather than examine the general form immediately, it is helpful to build up the different effects from several simpler situations. We thus examine the following classes of fluid motions:

(1) incompressible ($\nabla\cdot\mathbf{u} = 0$), uniform density, inviscid ($\mathbf{F}_{visc} = 0$) flow with conservative body forces ($\nabla\times\mathbf{X} = 0$);
(2) incompressible, non-uniform density, inviscid flow with conservative body forces;
(3) uniform density, viscous flow with conservative body forces;
(4) compressible, inviscid flow with conservative body forces.

For the phenomena considered in this book, the most important non-conservative body force is the Coriolis force, which is encountered when describing flows in rotating machinery. The effects of Coriolis forces will be examined in depth in Chapter 7.

[2] Vector identities used in obtaining (3.3.4) are $\nabla\times(\mathbf{u}\times\boldsymbol{\omega}) \equiv (\boldsymbol{\omega}\cdot\nabla)\mathbf{u} - (\mathbf{u}\cdot\nabla)\boldsymbol{\omega} - \boldsymbol{\omega}(\nabla\cdot\mathbf{u})$ and $\nabla\times\nabla(u^2/2) \equiv 0$.

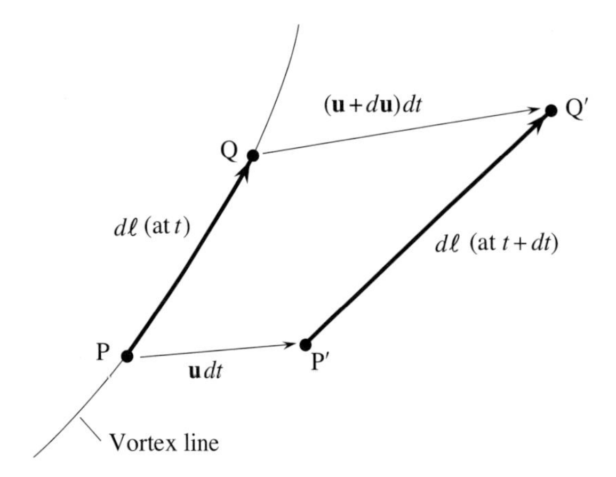

Figure 3.8: Change in length and orientation of vortex line element PQ during a short time interval, dt.

3.4 Vorticity changes in an incompressible, uniform density, inviscid flow with conservative body force

For an incompressible uniform density flow with conservative body forces, the terms $\nabla \cdot \mathbf{u}$, $(1/\rho \nabla p)$, and $\nabla \times \mathbf{X}$ are all equal to zero. Equation (3.3.4) thus becomes

$$\frac{D\boldsymbol{\omega}}{Dt} = (\boldsymbol{\omega} \cdot \nabla)\, \mathbf{u}. \tag{3.4.1}$$

The term on the right-hand side of (3.4.1) is the magnitude of the vorticity times the rate of change of the velocity with respect to distance along the vortex line. Its meaning can be interpreted with regard to Figure 3.8, where P and Q are points a short distance $d\boldsymbol{\ell}$ apart on a vortex line. The rate of change of velocity along the direction of the vortex line is $\partial \mathbf{u}/\partial \ell$, where $\partial/\partial \ell$ denotes differentiation in the direction of $d\boldsymbol{\ell}$. The term $(\boldsymbol{\omega} \cdot \nabla)\mathbf{u}$ in (3.4.1) can thus be represented by $\omega(\partial \mathbf{u}/\partial \ell)$:

$$\frac{D\boldsymbol{\omega}}{Dt} = \omega\left(\frac{\partial \mathbf{u}}{\partial \ell}\right). \tag{3.4.2}$$

The physical content of (3.4.2) can be seen by examining the change in the element $d\boldsymbol{\ell}$, which moves with the fluid during a short time interval dt. At time t, this line element extends from point P to point Q so that $d\boldsymbol{\ell}(\mathrm{t}) = \mathbf{r}(\mathrm{Q}) - \mathbf{r}(\mathrm{P})$, where $\mathbf{r}$ denotes the distance of a point from the origin. At time dt later, the ends of the line element have moved to P′ and Q′ so that $d\boldsymbol{\ell}(t + dt) = \mathbf{r}(\mathrm{Q}') - \mathbf{r}(\mathrm{P}')$. During this interval, the velocity of point P is $\mathbf{u}\,[\mathbf{r}(\mathrm{P})]$ and the velocity of point Q is $\mathbf{u}\,[\mathbf{r}(\mathrm{Q})] = \mathbf{u}\,[\mathbf{r}(\mathrm{P}) + d\boldsymbol{\ell}]$, or $\mathbf{u} + (d\boldsymbol{\ell} \cdot \nabla \mathbf{u}) = \mathbf{u} + (\partial \mathbf{u}/\partial \ell) d\ell$ for small $d\boldsymbol{\ell}$. The velocity of point Q with respect to point P, $d\mathbf{u}$, is thus given by $(\partial \mathbf{u}/\partial \ell) d\ell$. Likewise, the change in the vector $d\boldsymbol{\ell}$ in the time interval dt is given by

$$\begin{aligned} d\boldsymbol{\ell}(t + dt) - d\boldsymbol{\ell}(t) &= [\mathbf{r}(Q') - \mathbf{r}(Q)] - [\mathbf{r}(P') - \mathbf{r}(P)] \\ &= (\mathbf{u} + d\mathbf{u})\; dt - \mathbf{u} dt \\ &= \left[\left(\frac{\partial \mathbf{u}}{\partial \ell}\right) d\ell\right] dt. \end{aligned} \tag{3.4.3}$$

In the small time interval, dt, then, the fractional rate of change in the element $d\boldsymbol{\ell}$ is

$$\frac{1}{d\ell}\frac{D(d\ell)}{Dt} = \left(\frac{\partial \mathbf{u}}{\partial \ell}\right). \tag{3.4.4}$$

The notation D/Dt is appropriate because the change is evaluated following the same fluid particles. Comparing (3.4.2) and (3.4.4), we have

$$\frac{1}{d\ell}\frac{D(d\ell)}{Dt} = \frac{1}{\omega}\left(\frac{D\omega}{Dt}\right). \tag{3.4.5}$$

The relation between vorticity and the length of a vortex line element satisfying (3.4.5) is a direct proportionality:

$$\boldsymbol{\omega} = C\, d\boldsymbol{\ell}, \tag{3.4.6}$$

where C is a constant. The magnitudes of $\boldsymbol{\omega}$ and $d\boldsymbol{\ell}$ are thus related by

$$\frac{|\omega|}{d\ell} = \text{constant}. \tag{3.4.7}$$

Equations (3.4.5) and (3.4.7) show that the behavior of vortex lines and of material lines (lines composed of the same fluid particles at all times) is identical. In an inviscid, uniform density fluid, tilting or stretching of the material lines to alter orientation or length affects vortex lines in precisely the same manner. Another way to state this is that *the vortex lines move with the fluid*, or equivalently, that vortex lines can be regarded as "locked" to the fluid particles; fluid once possessing vorticity will do so forever. In a three-dimensional flow, where different parts of a vortex line move with the local fluid particles at different convection rates, the vorticity vector will change in both orientation and magnitude. Equation (3.4.2) expresses this change as a function of the vorticity and the velocity derivatives.

Because phenomena associated with the alteration of the components of vorticity due to the stretching and tipping of vortex lines are so important, it is worthwhile to examine the consequences of (3.4.1) on a component by component basis. We do this with reference to Figure 3.9, which shows a flow in which the x-component of velocity, u_x, varies with y, and in which, at some given position, there is a component of vorticity in the y-direction. The x-component of (3.4.1) is

$$\frac{D\omega_x}{Dt} = \omega_x\frac{\partial u_x}{\partial x} + \omega_y\frac{\partial u_x}{\partial y} + \omega_z\frac{\partial u_x}{\partial z}. \tag{3.4.8}$$

The term $\omega_y(\partial u_x/\partial y)$ is non-zero so there will be a change in ω_x as the flow evolves. For the velocity field shown, the term $(\partial u_x/\partial y)$ is positive and a positive x-component of vorticity will be created. Figure 3.9 shows that as the vorticity initially in the y-direction moves with the fluid it is tipped into the x-direction.

We can also note the implication of (3.4.1) for a planar two-dimensional flow (velocity components which depend on two coordinates, say x and y, and $u_z = 0$). In this situation, the vorticity has only a component in the z-direction, and $(\boldsymbol{\omega}\cdot\boldsymbol{\nabla})\mathbf{u}$ is identically zero. For a two-dimensional, constant density, inviscid incompressible flow, (3.4.1) reduces to the statement that the magnitude, ω, of the z-component of vorticity is invariant:

$$\frac{D\omega}{Dt} = 0;\ \text{planar, two-dimensional, inviscid, uniform density, incompressible flow.} \tag{3.4.9}$$

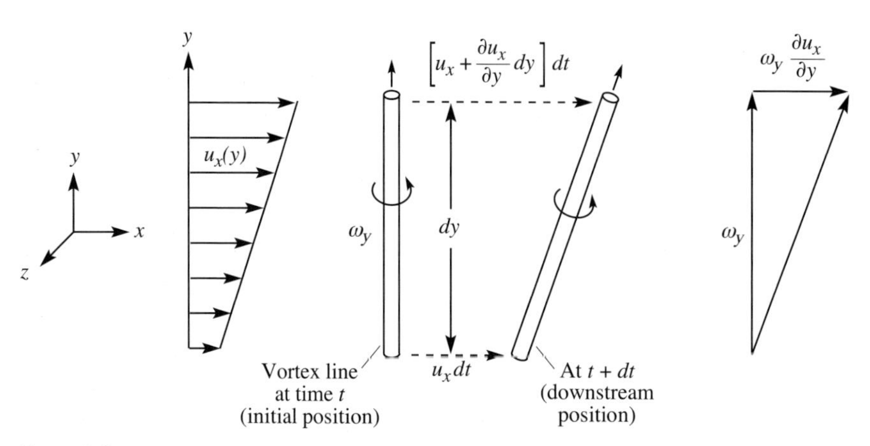

Figure 3.9: Creation of the x-component of vorticity by tipping of the element of the vortex line initially in the y-direction into the x-direction due to differential convection.

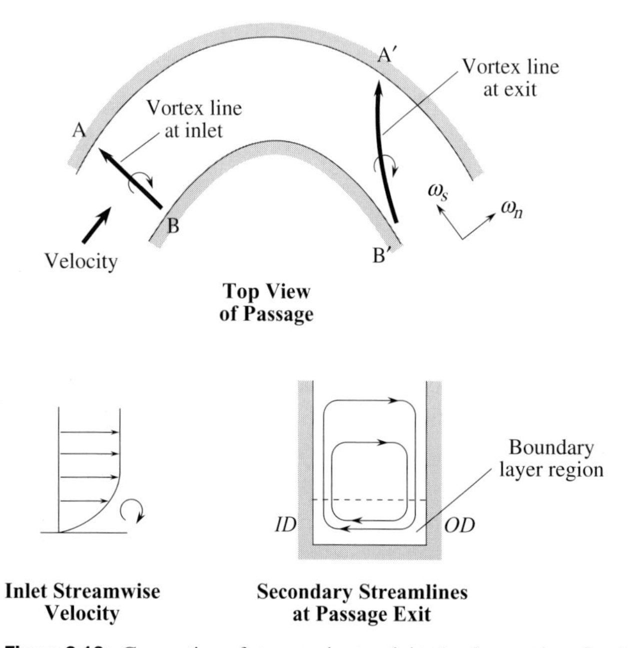

Figure 3.10: Generation of streamwise vorticity (and secondary flow) from the convection of vortex lines through a bend.

3.4.1 Examples: Secondary flow in a bend, horseshoe vortices upstream of struts

An example showing the creation of vorticity components due to the non-uniform convection rate of different parts of a vortex line is the so-called secondary flow that occurs in flow round a bend or in a turbomachinery passage. The topic will be addressed further in Chapter 9 but Figure 3.10, which shows flow in a channel, illustrates the basic situation. At the inlet, suppose there is a boundary layer on the floor of the passage and that the free-stream velocity can be considered approximately uniform in a direction across the passage. The vortex lines run across the channel normal to the inlet velocity, as indicated by the arrow AB and are located near the channel floor where the flow has non-uniform velocity. We can view this situation approximately as a distribution of vortex lines which are convected by an irrotational background or "primary" flow. The evolution of the vorticity distribution produced then leads to a "secondary" motion normal to the primary flow streamlines.

As the flow proceeds round the bend, the fluid near the inner wall will have a higher velocity than that near the outer wall. Particles on the outside wall also have farther to travel. The net result is that a line of particles AB, initially normal to the mean flow, ends up oriented as A′B′, at the passage exit. Because vortex lines and material lines behave the same way, the vortex lines at the exit will also be "tipped" and stretched into the streamwise direction. The result is a component of streamwise vorticity at the exit giving a secondary circulation as indicated in the channel cross-section shown in Figure 3.10. This secondary flow generates an inward motion of fluid in the floor boundary layer.

It was stated in Section 3.1 that characterization of flow patterns in terms of vorticity forms a complement to the use of pressure and fluid accelerations, but that the two viewpoints embody the same dynamical concepts. Which view is more attractive in terms of furnishing insight depends on the specific problem to be attacked; for example, the illustration given above of the secondary flow in a bend can also be described in terms of the pressure field. As discussed in Section 2.4, in the free stream above the boundary layer on the floor of the bend, there is a pressure gradient normal to the streamlines ($\partial p/\partial n$) which balances the normal acceleration of the fluid particles moving round the bend with velocity u_E and streamline radius of curvature r_c:

$$\frac{\partial p}{\partial n} = \rho \frac{u_E^2}{r_c}. \tag{3.4.10}$$

The fluid in the boundary layer on the floor of the channel also experiences the same pressure gradient, but has a lower velocity. The boundary layer streamlines must therefore have a smaller radius of curvature than the free stream, so the boundary layer fluid is swept towards the inner radius of the bend.

Another aspect of the behavior of the vorticity field is the possibility of amplification due to stretching of vortex lines. As given explicitly in (3.4.7), if a material line is stretched, the component of vorticity along that line is stretched in the same proportions. This can intensify weak swirling motions into concentrated vortices with high swirl velocities.

A frequently encountered example of such intensification occurs in the flow of a boundary layer round a strut or other obstacle that protrudes through it, as sketched in Figure 3.11. Far upstream, vortex lines in the boundary layer are straight and normal to the velocity vectors (line AA′). As they approach the obstacle, vortex lines are bent round the obstacle (line BB′), because fluid particles on the plane of symmetry are slowed down (approaching the stagnation point), whereas those away from this plane speed up. Further, particles in the plane of symmetry must remain at the front of the obstacle, whereas those that are off this plane eventually move downstream (line CC′). As a result, the material lines and hence the vortex lines are stretched and the vorticity increases. The strongest stretching occurs on the plane of symmetry, with the vorticity and the associated swirl velocity being greatest there. Portions of different vortex lines near the plane of symmetry will rotate about each other faster than those which are off to the sides, so they twist round one another like the strands of a rope and a strong vortex can be formed on the upstream side of an obstacle. Figure 3.12 is a

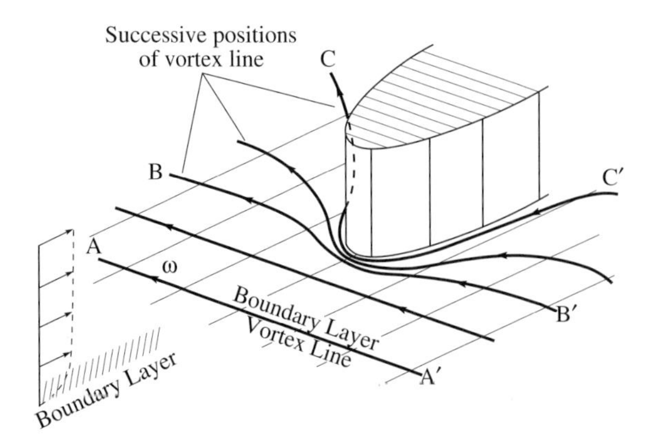

Figure 3.11: Boundary layer vortex lines wrapping round an obstacle.

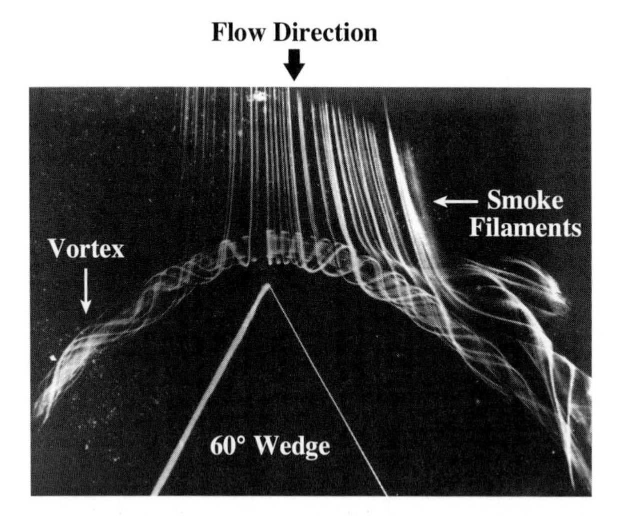

Figure 3.12: Smoke flow: a visualization of a horseshoe vortex upstream of a 60° wedge in a channel; vortex on the bottom floor of channel, view from top of channel (Schwind, 1962).

visualization of such a vortex, located upstream of a 60° wedge in a channel (Schwind, 1962), where smoke flow streaklines have been used to indicate the nature of the flow. The view is from the top, looking down parallel to the sides of the wedge. The increase of the swirl velocities is more naturally described here in terms of the intensification of vorticity; arguments in terms of the pressure field are more difficult to apply, and this appears to be generally true for flows in which there is strong swirl.

In a real (viscous) flow beneath a highly swirling structure such as that shown in Figure 3.12, the shear stress and heat transfer can be an order of magnitude larger than far upstream. A natural manifestation of this effect is shown in Figure 3.13, which is a photograph of steady flow round a log. The scouring of the snow in front of, and on the sides of, the log can be plainly seen; these regions mark the trace of the vortex. The schematic in Figure 3.13 shows a cross-section of the flow process. Vortices generated by an obstacle in a flow are often referred to as horseshoe vortices because of the general U-shaped configuration they form, and are widespread in fluid engineering situations.

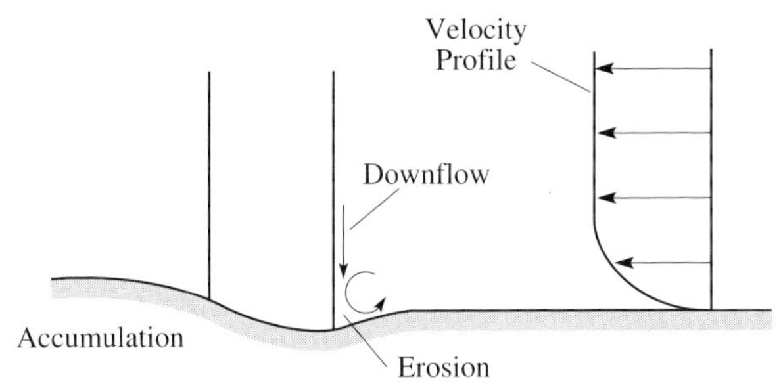

Figure 3.13: Erosion caused by a high scouring rate due to a horseshoe vortex; flow round a log.

3.4.2 Vorticity changes and angular momentum changes

Upon encountering vorticity dynamics for the first time, there is a natural tendency to try to link the concepts with material encountered previously concerning three-dimensional dynamics. In doing this, there can be confusion in the interpretation of precisely what (3.4.1) describes ($D\boldsymbol{\omega}/Dt = (\boldsymbol{\omega} \cdot \boldsymbol{\nabla})\mathbf{u}$). This equation is a statement about the way in which the local *angular velocity* of a fluid particle changes, not a statement about angular momentum. To see this, consider the changes in vorticity in a small incompressible fluid sphere of radius r undergoing a pure straining motion, or a motion without shear, as shown in Figure 3.14(a). The strain rate is $\partial u_y/\partial y = \varepsilon$ in the y-direction and $\partial u_x/\partial x = \partial u_z/\partial z = -\varepsilon/2$ in the x- and z-directions. The sum of the strain rates is zero because the fluid is incompressible.

Suppose the vorticity vector at time t has magnitude ω_0 and is in the plane of the paper pointing at 45° to the x-axis so $\omega_x = \omega_y = \omega_0/\sqrt{2}$, $\omega_z = 0$. After a short interval dt, the spherical particle will have the form of an ellipsoid of revolution, as indicated in Figure 3.14(b). The y-axis of the ellipsoid has a length that is $(1 + \varepsilon dt)$ of the initial length, and the x-component of vorticity will be increased in just this proportion as expressed in (3.4.5) or (3.4.7). Similarly, the x-dimension of the ellipsoid will be $(1 - (\varepsilon/2)dt)$ of the original length with the y-component of vorticity decreased by this factor. The vorticity vector will thus undergo a net increase in magnitude and a reorientation into the y-direction, as shown by the heavy arrow in the right-hand side of the figure.

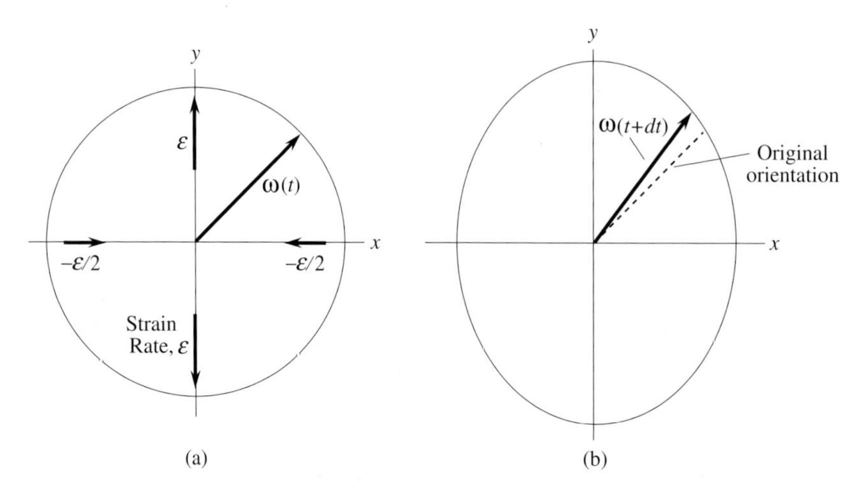

Figure 3.14: Spherical fluid particle with radius r and vorticity vector ω:
(a) at time, t: $\partial u_y/\partial y = \varepsilon$; $\partial u_x/\partial x = \partial u_z/\partial z = -\varepsilon/2$;
(b) at time, $t + dt$: particle deformed and vorticity vector rotated and stretched.

Now consider the angular momentum of the fluid particle during the time dt. The rate of change of angular momentum about the center of mass of the spherical particle is proportional to the net torque about this center. The only forces acting on the particle, however, are pressure forces, which act normal to the spherical surface and do not exert a torque. The angular momentum of the particle is thus unchanged during the interval, even though the vorticity varies.

Examining the differences between changes in vorticity and angular momentum using the tools of three-dimensional dynamics also gives a different perspective from which to view the differences between fluid and rigid body dynamics. Let us calculate the angular momentum of the particle about its center at initial and final times separated by the interval dt. The initial angular momentum of the sphere, $\mathbf{H}(t)$, can be written in terms of the inertia tensor and the vorticity components (recalling that the vorticity is twice the angular velocity) as

$$\mathbf{H}(t) = I_{xx}\left(\frac{\omega_x}{2}\right)\mathbf{i} + I_{yy}\left(\frac{\omega_y}{2}\right)\mathbf{j} = I_{xx}\left(\frac{\omega_0}{2\sqrt{2}}\right)\mathbf{i} + I_{yy}\left(\frac{\omega_0}{2\sqrt{2}}\right)\mathbf{j}. \tag{3.4.11}$$

The terms I_{xx} and I_{yy} are the elements of the inertia tensor for the fluid sphere and have the values $I_{xx} = I_{yy} = I_0 = \frac{2}{5}mr^2$, where m is the mass of the fluid particle and r is the radius.

At time dt later, the particle is an ellipsoid of revolution with semi-major axis $[r(1 + \varepsilon dt)]$ and semi-minor axis $[r(1 - (\varepsilon/2)dt)]$. Moments of inertia about the x- and y-axes for an ellipsoid of revolution with semi-major and semi-minor axes a and b are

$$I_{yy} = \frac{2}{5}mb^2 \quad \text{and} \quad I_{xx} = \frac{1}{5}m(a^2 + b^2). \tag{3.4.12}$$

To first order in dt, the moments of inertia of the fluid particle at time $t + dt$ are thus

$$I_{yy}(t + dt) = I_0(1 - \varepsilon dt) \quad \text{and} \quad I_{xx} = I_0\left(1 + \frac{\varepsilon}{2}dt\right). \tag{3.4.13}$$

The two components of angular momentum at $t + dt$ are:

$$H_x(t + dt) = I_0\left(1 + \frac{\varepsilon}{2}dt\right)\frac{\omega_0}{2\sqrt{2}}\left(1 - \frac{\varepsilon}{2}dt\right) \cong \frac{I_0\omega_0}{2\sqrt{2}}, \tag{3.4.14a}$$

$$H_y(t + dt) = I_0(1 - \varepsilon dt)\frac{\omega_0}{2\sqrt{2}}(1 + \varepsilon dt) \cong \frac{I_0\omega_0}{2\sqrt{2}}. \tag{3.4.14b}$$

The moments of inertia have altered so the angular momentum about the center of mass of the particle remains constant, even though the vorticity (the angular velocity) has changed. This example demonstrates the central message of this section: vorticity is a measure of local angular velocity not angular momentum, and (3.4.1) describes the evolution of this angular velocity.

3.5 Vorticity changes in an incompressible, non-uniform density, inviscid flow

We next examine inviscid flows in which the density is non-uniform but still incompressible, because changes in pressure are insufficient to produce a significant variation of the density of a given fluid particle. The density field is therefore described by

$$\frac{D\rho}{Dt} = 0 \tag{3.5.1}$$

and the velocity field is solenoidal ($\nabla \cdot \mathbf{u} = 0$). One situation of this type is a thermally stratified flow at low Mach number. For this case, (3.3.4) becomes (again with conservative body forces)

$$\frac{D\boldsymbol{\omega}}{Dt} = (\boldsymbol{\omega} \cdot \nabla)\mathbf{u} - \nabla \times \left(\frac{1}{\rho}\nabla p\right) \tag{3.5.2}$$

or, since $\nabla \times \nabla \mathrm{p} \equiv 0$,

$$\frac{D\boldsymbol{\omega}}{Dt} = (\boldsymbol{\omega} \cdot \nabla)\mathbf{u} + \frac{1}{\rho^2}(\nabla\rho \times \nabla p). \tag{3.5.3}$$

The second term on the right-hand side of (3.5.3) shows that changes in vorticity occur whenever the surfaces of constant density and constant pressure are not aligned so $\nabla\rho \times \nabla p$ is non-zero. This is illustrated in the sketch of a cylindrical fluid particle of radius r_0 with a non-uniform density in Figure 3.15. The lines of constant density are shown dashed, and the density distribution is such that $\rho_3 > \rho_2 > \rho_1$. The center of mass of the particle is at C, which does not coincide with the center (O) but is displaced from it by η_c. If there are no body forces, the only force that acts in an inviscid fluid is pressure. The variation in magnitude of this force around the cylinder is indicated by the arrows. The resultant of the pressure force will act through the geometric center (O) so that there will be a net torque about the center of mass, and a consequent angular acceleration.

The creation of vorticity in a fluid with non-uniform density can also be derived from classical dynamics arguments by analyzing the behavior of the small cylinder of fluid in Figure 3.15. For purposes of the argument, it is sufficient to consider two-dimensional flow, for which the first term on the right-hand side of (3.5.3) is zero and the only agency for changing the vorticity is the interaction

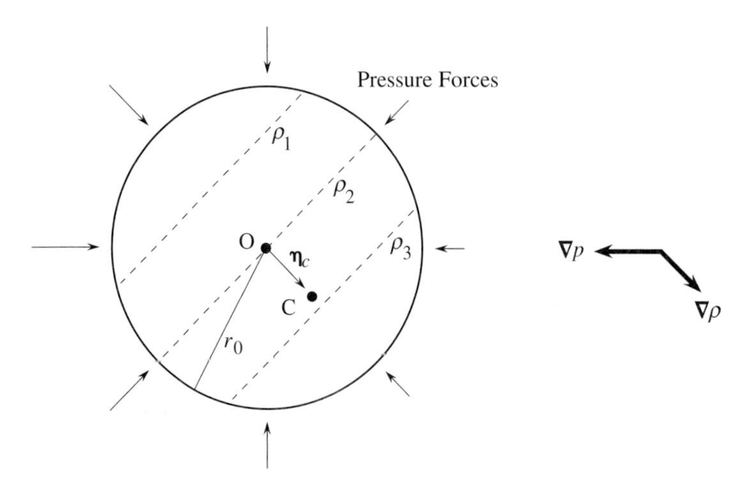

Figure 3.15: Generation of vorticity due to the interaction of pressure and density gradients: pressure force torque about the center of mass of a cylindrical fluid particle of radius r_0 with a non-uniform density and center of mass at C.

of pressure and density gradients. The rate of change of angular velocity of the cylinder is

$$\frac{d\Omega}{dt} = \frac{d(\omega/2)}{dt} = \frac{\text{torque about the center of mass}}{\text{moment of inertia about the center of mass}}. \tag{3.5.4}$$

The pressure forces are of magnitude $|\nabla p|$ per unit volume and act through the geometric center of the cylinder. The torque (per unit depth) about the center of mass is

$$\text{torque} = \boldsymbol{\eta}_c \times (-\nabla p)(\pi r_0^2) \tag{3.5.5}$$

where the vector $\boldsymbol{\eta}_c$ is the distance from the geometric center, O, to the center of mass, C, and r_0 is the radius of the cylinder. For a linear variation of density, $\boldsymbol{\eta}_c$ is

$$\boldsymbol{\eta}_c = -\frac{\nabla \rho}{4\rho_0} r_0^2. \tag{3.5.6}$$

The moment of inertia of the cylinder about its center of mass is

$$I = \rho_0 \pi \frac{r_0^4}{2} \left[1 - \frac{1}{\rho_0^2} \left(\frac{d\rho}{d\eta} \right)^2 \frac{r_0^2}{8} \right], \tag{3.5.7}$$

where $d\rho/d\eta$ denotes the derivative of density in the direction of $\boldsymbol{\eta}_c$. If the cylinder radius is small compared to the characteristic length over which density changes, then $(r_0/\rho)(d\rho/d\eta) \ll 1$, and the inertia can be approximated as

$$I = \rho_0 \pi \frac{r_0^4}{2} \tag{3.5.8}$$

and ∇p can be taken as uniform over the cylinder. Substituting (3.5.5) and (3.5.8) into (3.5.4) yields an expression for rate of change of angular velocity of the cylinder:

$$\frac{1}{\rho_0^2} \nabla \rho \times \nabla p = \frac{d\omega}{dt}, \tag{3.5.9}$$

which is the two-dimensional form of (3.5.3).

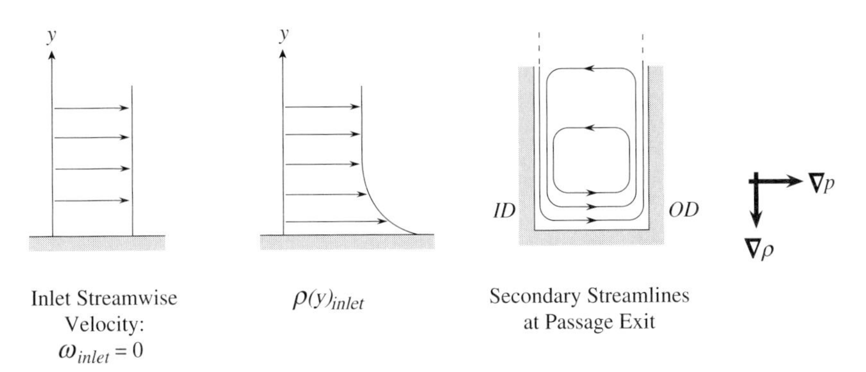

Figure 3.16: Generation of streamwise vorticity (and secondary flow) due to the interaction of the pressure and density gradients.

3.5.1 Examples of vorticity creation due to density non-uniformity

An example of vorticity creation associated with a density non-uniformity occurs in flow round a bend. The geometry is similar to that in Section 3.4.1, but the fluid now has *uniform velocity* upstream (so $\omega = 0$), and non-uniform density. Assuming y is the coordinate perpendicular to the channel floor, the inlet conditions are shown on the left of Figure 3.16. We can view this as a layer of cool fluid, in which the density is larger toward the lower part of the channel, so the density gradient $(d\rho/dy)$ is negative (i.e. pointing toward the bottom of the channel). The pressure gradient in the bend is approximately normal to the free-stream streamlines and points radially outward. The product $(1/\rho^2)\nabla\rho \times \nabla p$ is thus in the streamwise direction and at the bend exit there will be a component of streamwise vorticity and a secondary circulation as shown.

This secondary flow can also be described in terms of pressure forces. The argument is similar to that in Section 3.4.1 except that the fluid in the layer near the wall now has a value of ρu^2 higher than the free stream because of its increased density. The pressure gradient, however, is still set up by the free-stream flow. The radius of curvature for the streamlines containing the higher density, larger inertia fluid particles is thus larger than that of the free-stream flow, resulting in these particles moving outwards as they pass through the bend.

Another instance in which vorticity is created by the interaction of pressure and density gradients is in the flow of a stratified fluid from a reservoir through a nozzle or from a duct of large area through a contraction, as illustrated in Figure 3.17. In the reservoir or large area part of the channel, the lines of constant density are horizontal, the pressure (at station i, say) is approximately uniform, and the velocity variation is small. At the exit of the contraction, station e, the pressure is again uniform across the duct but the velocity is non-uniform so that vorticity has been produced. The physical argument associated with the generation of vorticity is that the two streams (high and low density) have the same pressure difference acting on them; the acceleration and hence the velocity at exit will be larger for the lower density fluid. Flows such as this occur in turbine vanes in gas turbine engines because the combustor exit typically has a non-uniform temperature and density distribution.

The velocity variation at the channel exit can be found using Bernoulli's equation. Assuming that the area at station i is large enough so we can neglect the dynamic pressure there, the duct exit velocity field is given by

$$p_i - p_e = \tfrac{1}{2}\rho(y)[u_x(y)]_e^2 .$$

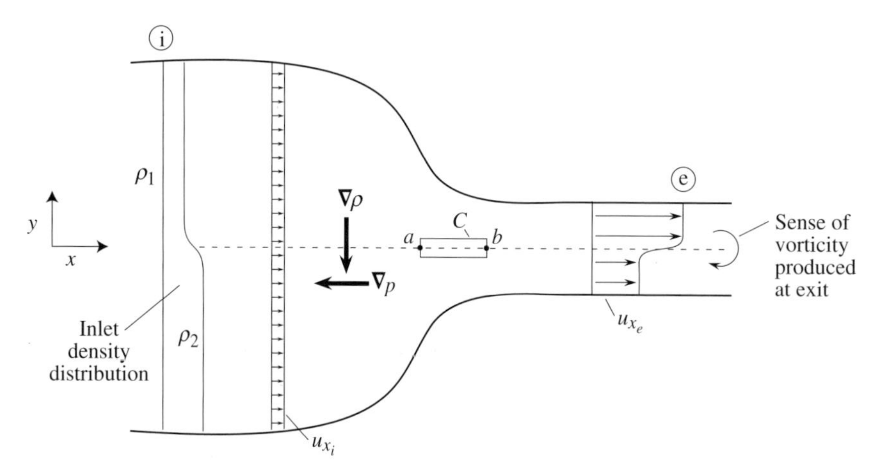

Figure 3.17: Vorticity production in a fluid of non-uniform density; two-dimensional nozzle.

For the streamline at the exit with the mean exit density ρ_m

$$p_i - p_e = \frac{1}{2}\rho_m[u_x(y_m)]_e^2,$$

where y_m refers to the level at which this streamline exits. The velocity at any location y, with density $\rho(y)$, is thus

$$\left[\frac{u_x(y)}{u_x(y_m)}\right]_e = \sqrt{\frac{\rho_m}{\rho(y)}}. \tag{3.5.10}$$

Figure 3.17 shows a sharp change in density to illustrate the concepts but suppose, as is closer to the case in practice, that the exit density distribution can be approximated as linear across the exit channel width, W,

$$\frac{\rho(y)}{\rho_m} = 1 + \frac{1}{\rho_m}\left(\frac{d\rho}{dy}\right)y. \tag{3.5.11}$$

If the quantity $(W/\rho_m)(d\rho/dy)$ is much less than unity, we can expand the square root in (3.5.10) to yield the approximate form

$$\left[\frac{u_x(y)}{u_x(y_m)}\right]_e \cong 1 - \frac{1}{2\rho_m}\left(\frac{d\rho}{dy}\right)y. \tag{3.5.12}$$

The sense of rotation associated with the vorticity is as shown in Figure 3.17.

3.6 Vorticity changes in a uniform density, viscous flow with conservative body forces

For an incompressible, constant property, viscous flow with conservative body forces, the general form of the equation for changes in vorticity can be obtained from (3.3.4) as

$$\frac{D\boldsymbol{\omega}}{Dt} = (\boldsymbol{\omega}\cdot\boldsymbol{\nabla})\mathbf{u} + \boldsymbol{\nabla}\times\mathbf{F}_{visc}. \tag{3.6.1}$$

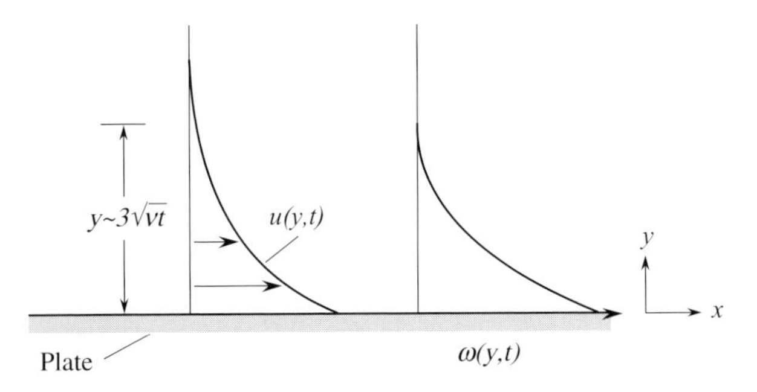

Figure 3.18: Generation of vorticity due to the action of viscous forces; impulsively started plate: $u(0, t) = 0$, $t < 0$; $u(0, t) = u_w$, $t \geq 0$; $u/u_w \sim 0.01$ at $y = 3\sqrt{\nu t}$.

Because the flow is incompressible, the viscous force per unit mass, $\mathbf{F}_{visc}$, is (Section 1.14)

$$\mathbf{F}_{visc} = \nu(\nabla^2 \mathbf{u}). \tag{1.14.4}$$

Applying the vector identity

$$\nabla^2 \mathbf{B} = \nabla(\nabla \cdot \mathbf{B}) - \nabla \times (\nabla \times \mathbf{B}) \tag{3.6.2}$$

and using the continuity equation allows representation of the viscous force per unit mass in terms of the curl of the vorticity:

$$\mathbf{F}_{visc} = -\nu[\nabla \times \omega]. \tag{3.6.3}$$

Equation (3.6.1) can thus be recast as:

$$\begin{aligned} \frac{D\omega}{Dt} &= (\omega \cdot \nabla)\mathbf{u} - \nabla \times [\nu(\nabla \times \omega)] \\ &= (\omega \cdot \nabla)\mathbf{u} + \nu \nabla^2 \omega. \end{aligned} \tag{3.6.4}$$

The term $-\nabla \times \nu(\nabla \times \omega)$ $(= \nu\nabla^2\omega)$, which is discussed in this section, represents the effect of viscosity in spreading, or diffusing, vorticity.

To gain familiarity with this effect, we begin by considering the two-dimensional flow adjacent to an infinite plate, which is impulsively given a velocity, u_w, in its own plane at time $t = 0$. The domain of interest is the semi-infinite region shown in Figure 3.18.

The boundary conditions and geometry are independent of the distance along the plate (x), so $\partial u_x/\partial x = 0$ and from the continuity equation $\partial u_y/\partial y = 0$ everywhere. The condition of zero normal velocity at $y = 0$ means that the y-component of velocity is zero throughout the flow field. The only non-trivial component of the momentum equation is the x-component, which reduces to

$$\frac{\partial u_x}{\partial t} = \nu \frac{\partial^2 u_x}{\partial y^2}. \tag{3.6.5}$$

The boundary conditions are $u_x(0, t) = u_w$, $u_x(\infty, t) = 0$, $u_x(y, 0) = 0$; $y > 0$. Equation (3.6.5) is the one-dimensional diffusion equation, which has the solution

$$\frac{u_x}{u_w} = 1 - \frac{2}{\sqrt{\pi}} \int_0^{y/2\sqrt{\nu t}} e^{-\xi^2}\, d\xi. \tag{3.6.6}$$

Equation (3.6.6), which is an exact solution of the Navier–Stokes equations, shows that u_x/u_w is only a function of the non-dimensional distance from the wall, $y/2\sqrt{\nu t}$.

The equation for the rate of change of vorticity is obtained by taking $\partial/\partial y$ of (3.6.5) to give

$$\frac{\partial \omega}{\partial t} = \nu \frac{\partial^2 \omega}{\partial y^2}, \tag{3.6.7}$$

where ω $(= -\partial u_x/\partial y)$ is the z-component of vorticity. Equation (3.6.7) has solution

$$\frac{\omega\sqrt{\nu t}}{u_w} = \frac{1}{\sqrt{\pi}} e^{-y^2/4\nu t}. \tag{3.6.8}$$

Since (3.6.5) and (3.6.7) are of the same form as that governing the time-dependent heat diffusion in a solid body, an analogy is often drawn between heat conduction and the diffusion of vorticity. This is helpful in understanding how changes in vorticity are produced by viscous effects, but the analogy is only strictly appropriate for two-dimensional flows, as there is no counterpart in the energy equation to the term $(\boldsymbol{\omega} \cdot \boldsymbol{\nabla})\mathbf{u}$ which occurs in three-dimensional flow.

Several features are shown by the solution (3.6.8) of (3.6.7). If we integrate the vorticity in y through the viscous layer to get the total vorticity per unit length at the plate, what is obtained is just the velocity difference

$$u(\infty, t) - u(0, t) = u_w.$$

Anticipating the results of Section 3.8, this is the circulation per unit length along the plate

$$\int_0^\infty \omega dy = \int_0^\infty \left[-\left(\frac{\partial u}{\partial y}\right) dy\right] = u(\infty, t) - u(0, t).$$

All the vorticity in the flow was created at time $t = 0$ by the motion of the wall and no additional vorticity is introduced as long as u_w is held constant.

From (3.6.8) the characteristic magnitude of the maximum vorticity can be shown to be $u_w/\sqrt{\nu t}$. The distance over which the vorticity has diffused, or the thickness of the viscous layer in which the vorticity is appreciably different from zero, is thus of order $\sqrt{\nu t}$, with the rate of vorticity diffusion also scaling as $\sqrt{\nu t}$. The concept of a characteristic time for diffusion of vorticity can be applied not only in unsteady flows but wherever one can form a time scale from a characteristic length and velocity. For example, in a steady flow with characteristic length, L, and velocity, U, the time scale is L/U. The thickness of the layer in which diffusion is able to spread appreciable vorticity is thus $\sqrt{\nu L/U}$. In this context, the thickness of a laminar boundary layer can be interpreted as being set by the diffusion of vorticity for a (convection) time equal to L/U.

3.6.1 Vorticity changes and viscous torques

Changes in vorticity from viscous effects can also be developed by examining the balance of torque and changes in angular momentum if one chooses a situation in which angular momentum and angular velocity are aligned. As an example, consider a square element of fluid with $dx = dy$, in a two-dimensional flow, as in Figure 3.19 (Hornung, 1988; Sherman, 1990). The stress components on the different faces are illustrated; τ is the shear stress, σ_x and σ_y are the normal stresses. Only variations in σ_x are shown, but the other stress components are also functions of x and y. Expanding

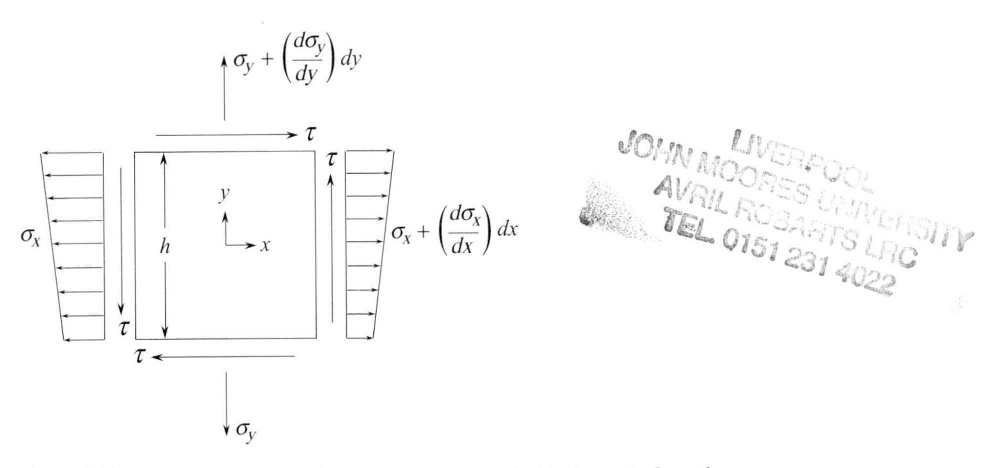

Figure 3.19: Viscous stresses and torques on a square fluid element; $dx = dy$, $\tau = \tau_{xy}$.

the stresses in a Taylor series in x and y about the center of the square and integrating to get the total contribution, the torque about the center of the square is

$$\text{magnitude of clockwise torque} = \frac{(dx)^4}{12}\left[\frac{\partial^2\sigma_x}{\partial x\partial y} - \frac{\partial^2\sigma_y}{\partial x\partial y} + \frac{\partial^2\tau}{\partial y^2} - \frac{\partial^2\tau}{\partial x^2}\right]. \tag{3.6.9}$$

The moment of inertia of the square fluid element per unit depth normal to the page is $(dx)^4/6$. The angular velocity of the element is equal to half the vorticity, $\omega = 2\Omega$, so the equation for the rate of vorticity is

$$\frac{(dx)^4}{12}\cdot\frac{d\omega}{dt} = \text{torque}. \tag{3.6.10}$$

With reference to (3.3.4), the term in square brackets on the right-hand side of (3.6.9) can be seen to be the two-dimensional version of $(\boldsymbol{\nabla} \times \mathbf{F}_{visc})$ so (3.6.9) and (3.6.10) are equivalent to the expression for the rate of change of vorticity due to viscous forces given in (3.3.4) derived in a quite different manner.

3.6.2 Diffusion and intensification of vorticity in a viscous vortex

The examples so far have dealt with one effect at a time, and it is instructive to examine a flow in which viscous forces, which tend to reduce vorticity magnitude through diffusion, and vortex stretching, which increases the vorticity, are both present. The specific configuration is the steady state of a straight axisymmetric vortex, which is stretched along its axis at constant strain rate, ε, where $\partial u_x/\partial x = \varepsilon$ everywhere (Batchelor, 1967). This allows an exact solution of the Navier–Stokes equations as well as furnishing insight into the balance between vortex stretching and diffusion, which sets the radius of vortex cores in many flows.

We adopt a cylindrical coordinate system, with the x-axis aligned with the axis of the vortex, r the distance normal to the x-axis, and θ the circumferential coordinate. For strain rate ε, with $u_x(0, r) = 0$, the axial velocity is $u_x = \varepsilon x$. The continuity equation is

$$\frac{\partial u_x}{\partial x} + \frac{1}{r}\frac{\partial}{\partial r}(r u_r) = 0, \tag{3.6.11}$$

which, with the condition that $u_r = 0$ at $r = 0$, requires that the radial velocity be given by

$$u_r = -\frac{\varepsilon r}{2}. \tag{3.6.12}$$

Because strain rate, ε, is invariant with x, the radial and circumferential velocities must also be independent of x. Furthermore, the only component of vorticity is parallel to the vortex axis (x-axis) and is obtained from the x-component of the cylindrical coordinate form of (3.6.4),

$$u_r\frac{\partial \omega_x}{\partial r} = \omega_x\frac{\partial u_x}{\partial x} + \nu\left[\frac{1}{r}\frac{\partial}{\partial r}\left(r\frac{\partial \omega_x}{\partial r}\right)\right]. \tag{3.6.13}$$

The three terms in (3.6.13) represent, respectively, convection of vorticity inward by the radial velocity, production of vorticity due to vortex stretching, and diffusion of vorticity by viscous stresses.

The expressions for $\partial u_x/\partial x (= \varepsilon)$ and $u_r (= -\varepsilon r/2)$ can be substituted in (3.6.13) to yield an ordinary differential equation for ω_x:

$$-\frac{\varepsilon}{2}\frac{d}{dr}(\omega_x r^2) = \nu\frac{d}{dr}\left(r\frac{\partial \omega_x}{\partial r}\right). \tag{3.6.14}$$

Integrating once:

$$-\frac{\varepsilon}{2}\omega_x r^2 = \nu r\frac{d\omega_x}{dr} + \text{constant}. \tag{3.6.15}$$

If ω_x is finite at $r = 0$, the constant term must be zero, and (3.6.15) can be integrated again to give the radial distribution of axial vorticity:

$$\omega_x(r) = \frac{\Gamma}{\pi}e^{-\varepsilon r^2/(4\nu)}. \tag{3.6.16}$$

The constant Γ/π is determined by the conditions that existed prior to the steady state.

In this flow, the region in which vorticity is appreciable (say greater than 1% of the value on the axis) is confined to radii less than approximately $4\sqrt{\nu/\varepsilon}$. The vortex core radius is set by the strain rate, ε, i.e. the rate of stretching; the higher this rate, the thinner the vortex core. The three-way balance between convection, production, and diffusion of vorticity, represented by (3.6.13), is illustrative of the processes that occur in more complex flows.

The circumferential velocity can now be found from the definition of the x-component of vorticity in an axisymmetric flow:

$$\omega_x = \frac{1}{r}\frac{d}{dr}(r u_\theta), \tag{3.6.17}$$

leading to

$$u_\theta = \frac{\Gamma}{2\pi r}\left[1 - e^{-\varepsilon r^2/(4\nu)}\right]. \tag{3.6.18}$$

The term $e^{-\varepsilon r^2/(4\nu)}$ is less than 0.01 for $r > 4.5\sqrt{\nu/\varepsilon}$. For values of r larger than this, the second term in the brackets is negligible compared to unity and the circumferential velocity has the $1/r$

dependence derived in Section 3.2 for the infinite vortex tube of constant vorticity. In other words, for radii far outside the vortex the internal structure within the vortex has no effect.

Finally, because the flow is axisymmetric and the angular momentum and angular velocity have the same orientation and axis, we can use statements about the conservation of angular momentum to describe this flow in terms familiar from dynamics. A cylindrical fluid element will have a radius that is contracting because of the axial strain. If no torque were exerted, the angular velocity would increase as the radius fell because the angular momentum is constant. Viscous stresses, however, exert a torque in a direction to decrease the angular momentum and hence limit the angular velocity.

3.6.3 Changes of vorticity in a fixed volume

The discussion so far has been of the changes of vorticity of a fluid element, but it is sometimes useful to examine the changes of vorticity that occur in a volume of fixed identity. The starting point is obtained from (3.6.4), written as (for uniform density and conservative body forces)

$$\frac{\partial \boldsymbol{\omega}}{\partial t} = -(\mathbf{u} \cdot \boldsymbol{\nabla})\boldsymbol{\omega} + (\boldsymbol{\omega} \cdot \boldsymbol{\nabla})\mathbf{u} - \nu \boldsymbol{\nabla} \times (\boldsymbol{\nabla} \times \boldsymbol{\omega}). \tag{3.6.19}$$

This is integrated over a fixed volume V, bounded by a surface A, making use of the vector identity

$$\int_V \boldsymbol{\nabla} \times \mathbf{B}\, dV = \int_A \mathbf{n} \times \mathbf{B}\, dA, \tag{3.6.20}$$

where $\mathbf{B}$ is any vector and $\mathbf{n}$ is the unit normal to the surface A. The expression for the vector triple product is also used to write several of the terms in the resulting equation as integrals over a surface:

$$\frac{\partial}{\partial t}\int_V \boldsymbol{\omega}\, dV = \underbrace{\int_V (\boldsymbol{\omega} \cdot \boldsymbol{\nabla})\mathbf{u}\, dV}_{\text{(i)}} - \underbrace{\int_A (\mathbf{n} \cdot \mathbf{u})\boldsymbol{\omega}\, dA}_{\text{(ii)}} - \underbrace{\nu \int_A \mathbf{n} \times (\boldsymbol{\nabla} \times \boldsymbol{\omega})\, dA}_{\text{(iii)}}. \tag{3.6.21}$$

The rate of change of vorticity inside the volume can be regarded as due to three different effects. Term (i) represents the production of vorticity within the volume from vortex stretching. Term (ii) arises because the volume considered is fixed in space rather than moving with the fluid, representing the convection of vorticity through the bounding surface. Lastly, term (iii) represents the component of the viscous forces exerted tangential to the bounding surface.

The application of (3.6.21) can be illustrated with reference to the steady-state vortex stretched along its axis as described in Section 3.6.2. Figure 3.20 shows a cylindrical control volume whose radius is taken at a location where viscous stresses are negligible, say $r > 10\sqrt{\nu/\varepsilon}$. At this location the vorticity is also negligible (see (3.6.16)). The integrals of $\mathbf{n} \times (\boldsymbol{\nabla} \times \boldsymbol{\omega})$ over the top and bottom of the cylinder sum to zero so (3.6.21) reduces to

$$\int_V (\boldsymbol{\omega} \cdot \boldsymbol{\nabla})\mathbf{u}\, dV = \int_A (\mathbf{n} \cdot \mathbf{u})\boldsymbol{\omega}\, dA. \tag{3.6.22}$$

This is an explicit balance between vorticity production within the volume due to vortex stretching, and the net flux of vorticity out of the control volume through the top and bottom surfaces of the cylindrical volume.

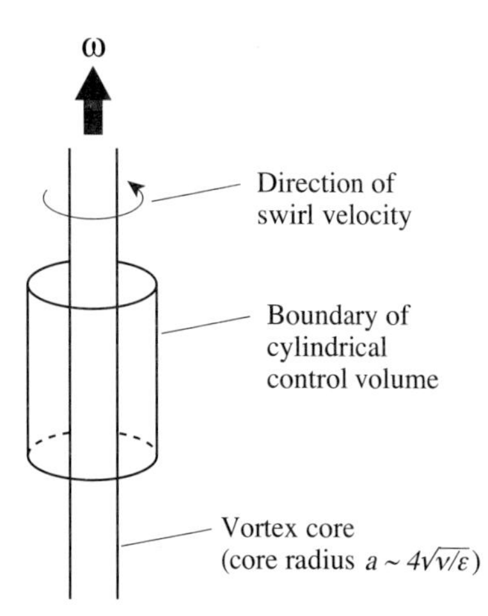

Figure 3.20: Vortex core and cylindrical control volume.

3.6.4 Summary of vorticity evolution in an incompressible flow

To recap, for incompressible flow the equation for the rate of change of vorticity of a fluid particle is

$$\frac{D\boldsymbol{\omega}}{Dt} = \underset{\text{(i)}}{(\boldsymbol{\omega}\cdot\nabla)\mathbf{u}} + \underset{\text{(ii)}}{\frac{\nabla\rho\times\nabla p}{\rho^2}} + \underset{\text{(iii)}}{\nabla\times\mathbf{X}} + \underset{\text{(iv)}}{\nabla\times\mathbf{F}_{visc}}. \tag{3.6.23}$$

Terms (i)–(iv) represent the effects of: (i) reorientation or stretching of vortex filaments (Section 3.4); (ii) creation of vorticity when density and pressure gradients are not aligned (Section 3.5); (iii) torques due to non-conservative body forces (to be addressed in Chapter 7); and (iv) diffusion of vorticity associated with viscous torque (Section 3.6).

3.7 Vorticity changes in a compressible inviscid flow

For compressible flows, the roles of viscous and body forces are similar to those in incompressible flow, although the expression for the viscous forces is more complicated. We thus consider only inviscid compressible flows with conservative body forces. The starting point is again (3.3.4). From continuity we can substitute $(-1/\rho)\,(D\rho/Dt)$ for $\nabla\cdot\mathbf{u}$ in the term $\boldsymbol{\omega}(\nabla\cdot\mathbf{u})$ so that (3.3.4) can be written as

$$\frac{D}{Dt}\left(\frac{\boldsymbol{\omega}}{\rho}\right) = \left(\frac{\boldsymbol{\omega}}{\rho}\cdot\nabla\right)\mathbf{u} - \frac{1}{\rho}\nabla\times\left(\frac{1}{\rho}\nabla p\right). \tag{3.7.1}$$

Comparison of (3.7.1) with (3.5.2) shows that the quantity $\boldsymbol{\omega}/\rho$ in a compressible flow behaves similarly to $\boldsymbol{\omega}$ for incompressible flow.

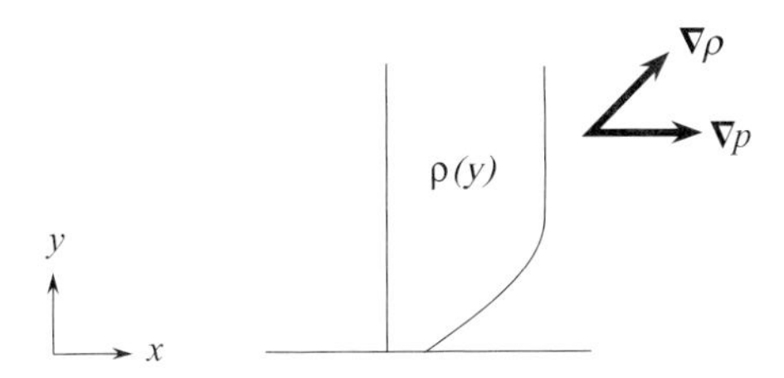

Figure 3.21: Density and pressure gradients in a high speed boundary layer with an adiabatic wall and an adverse pressure gradient.

An alternative form of (3.7.1) involving gradients of temperature and entropy, which is often useful, can be obtained as follows. The Gibbs equation (1.3.19) can be written in terms of gradients in the thermodynamic quantities as

$$T\nabla s = \nabla h - \frac{1}{\rho}\nabla p, \tag{3.7.2}$$

allowing (3.7.1) to be expressed as

$$\frac{D}{Dt}\left(\frac{\omega}{\rho}\right) = \left(\frac{\omega}{\rho}\cdot\nabla\right)\mathbf{u} + \frac{1}{\rho}\nabla T \times \nabla s. \tag{3.7.3}$$

For a compressible fluid, ω/ρ can be changed whenever the density, ρ, is not a function of pressure only ($\rho \neq \rho(p)$) or, equivalently, the entropy is not only a function of temperature. Such conditions occur, for example, at the exit of a gas turbine combustor, where the flow has approximately constant pressure but non-uniform temperature. They also occur behind turbomachines which typically have radial variations in stagnation temperature due to radially non-uniform work input.

Flows in which the density depends on pressure only are called *barotropic*, while those in which the density is not only a function of pressure are called *baroclinic*. The production of vorticity through the interaction of pressure and density fields is thus often referred to as the production of vorticity through baroclinic torque.

Even if both terms on the right of (3.7.1) or (3.7.3) are zero, the vorticity of a fluid particle can change in a compressible flow if density changes. For example, in a two-dimensional isentropic flow with incoming vorticity in an accelerating passage such as a nozzle, the exit density is lower than at the inlet, and the vorticity is therefore also lower, since (ω/ρ) remains constant.

An example of vorticity generation due to the density gradient–pressure gradient interaction represented by the second term in (3.7.1) occurs in a high speed boundary layer subjected to a pressure gradient along the bounding wall. If the boundary is adiabatic, the static temperature increases towards the wall and the density decreases. The density gradient will have components both normal and parallel to the wall, although only the former is effective in producing vorticity. For an adverse pressure gradient, the relation of ∇p and $\nabla\rho$ is as shown in Figure 3.21. The vorticity produced by this effect points into the paper and has a clockwise sense. Production of vorticity of this sign means that the boundary layer velocity at a given y location will be reduced due to the $\nabla\rho \times \nabla p$ term and the boundary layer consequently thickened.

3.8 Circulation

A quantity closely linked to the vorticity is the circulation, which is defined as the integral of the velocity around a closed contour, C:

$$\Gamma = \int_C \mathbf{u} \cdot d\boldsymbol{\ell}. \tag{3.8.1}$$

The relation between circulation and vorticity can be seen by applying Stokes's Theorem to this definition resulting in

$$\Gamma = \iint_A \boldsymbol{\omega} \cdot \mathbf{n} \, dA, \tag{3.8.2}$$

where A is a surface bounded by the contour C and $\mathbf{n}$ is the normal to that surface. The circulation is a scalar measure of the strength of all the vortex tubes threading through the area enclosed by C or, equivalently, the net flux of vorticity through the surface A, enclosed by contour C.

3.8.1 Kelvin's Theorem

The description of changes in circulation can provide considerable insight into fluid motions. We begin by examining the evolution of the circulation around a *closed fluid contour of fixed identity*, or a curve that consists always of the same fluid particles.

The rate of change of circulation, Γ, for C is given by

$$\frac{D\Gamma}{Dt} = \frac{D}{Dt} \oint_C \mathbf{u} \cdot d\boldsymbol{\ell}. \tag{3.8.3}$$

The convective operator can be taken inside the integral because we are examining a group of fluid particles of fixed identity.[3]

$$\frac{D\Gamma}{Dt} = \oint_C \frac{D\mathbf{u}}{Dt} \cdot d\boldsymbol{\ell} + \oint_C \mathbf{u} \cdot \frac{D}{Dt} d\boldsymbol{\ell}. \tag{3.8.4}$$

Interpretation of the second term on the right can be made by referring to Figure 3.22, which shows an element $d\boldsymbol{\ell}$ of the fluid (or material) contour, C. At time t, the ends of the element are at P and Q. A short time, dt, later, point P has been displaced by $\mathbf{u}\,dt$ to P′, point Q by an additional $(\partial \mathbf{u}/\partial \ell)\, d\ell \, dt$

[3] Another way to think of this is to consider the term $D\Gamma/Dt$ as the sum over many small fluid line elements that comprise the curve C:

$$\frac{D\Gamma}{Dt} = \frac{D}{Dt} \sum_j \mathbf{u}_j \, d\boldsymbol{\ell}_j.$$

The operation D/Dt is carried out for fixed fluid elements, so

$$\frac{D\Gamma}{Dt} = \frac{D}{Dt} \sum_j \mathbf{u}_j \cdot d\boldsymbol{\ell} = \sum_j \frac{D}{Dt}(\mathbf{u}_j \cdot d\boldsymbol{\ell}) = \sum_j \frac{D\mathbf{u}_j}{Dt} \cdot d\boldsymbol{\ell} + \mathbf{u}_j \cdot \frac{Dd\boldsymbol{\ell}_j}{Dt}.$$

Taking the limiting case of infinitesimal elements gives the integral form, (3.8.4).

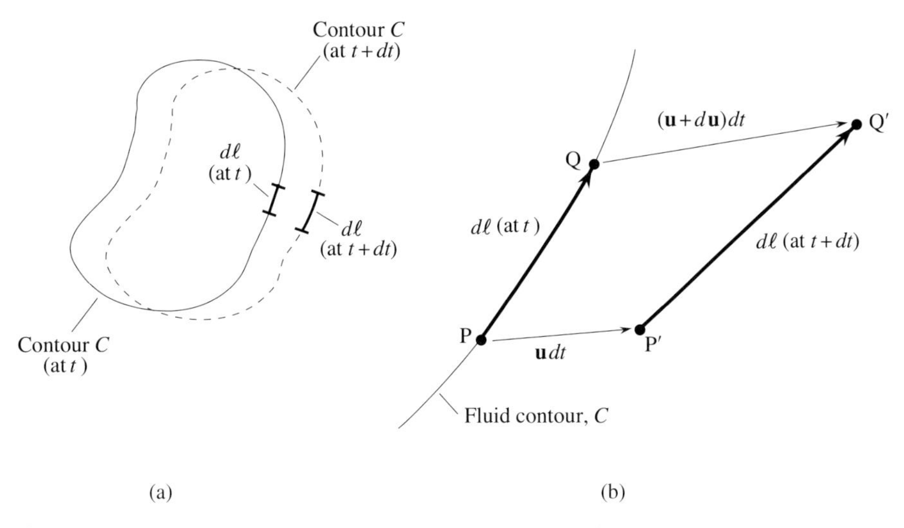

Figure 3.22: Change in length and orientation of an element $d\boldsymbol{\ell}$ (b) of fluid contour, C (a).

to Q′, and the line element is now the vector $d\boldsymbol{\ell} + (\partial\mathbf{u}/\partial\ell)\,d\ell\,dt$. As discussed in Section 3.4, the rate of change of the fluid contour element $d\boldsymbol{\ell}$ is given by

$$\frac{Dd\boldsymbol{\ell}}{Dt} = \frac{\partial\mathbf{u}}{\partial\ell}d\ell = d\mathbf{u}. \tag{3.8.5}$$

The second term on the right-hand side of (3.8.4) now becomes

$$\oint_C \mathbf{u}\cdot\frac{Dd\boldsymbol{\ell}}{Dt} = \oint_C \mathbf{u}\cdot d\mathbf{u} = \oint_C d\left(\frac{u^2}{2}\right) = 0, \tag{3.8.6}$$

because it is an exact differential integrated around a closed contour. The expression for the rate of change of circulation round a fluid contour is therefore

$$\frac{D\Gamma}{Dt} = \oint_C \frac{D\mathbf{u}}{Dt}\cdot d\boldsymbol{\ell}, \tag{3.8.7}$$

or, using the momentum equation,

$$\frac{D\Gamma}{Dt} = \oint_C \left(-\frac{1}{\rho}\nabla p + \mathbf{X} + \mathbf{F}_{visc}\right)\cdot d\boldsymbol{\ell}. \tag{3.8.8}$$

Equation (3.8.8) shows several mechanisms for changing circulation.

For the case of inviscid flow and conservative body forces (for which $\oint_C \mathbf{X}\cdot d\boldsymbol{\ell} = 0$, since $\mathbf{X}$ is the gradient of a potential), (3.8.8) takes the form

$$\frac{D\Gamma}{Dt} = -\oint_C \frac{\nabla p}{\rho}\cdot d\boldsymbol{\ell}. \tag{3.8.9}$$

Equation (3.8.9) is an important result known as *Kelvin's Theorem*. We now examine the consequences of (3.8.8) and (3.8.9) in different types of flows.

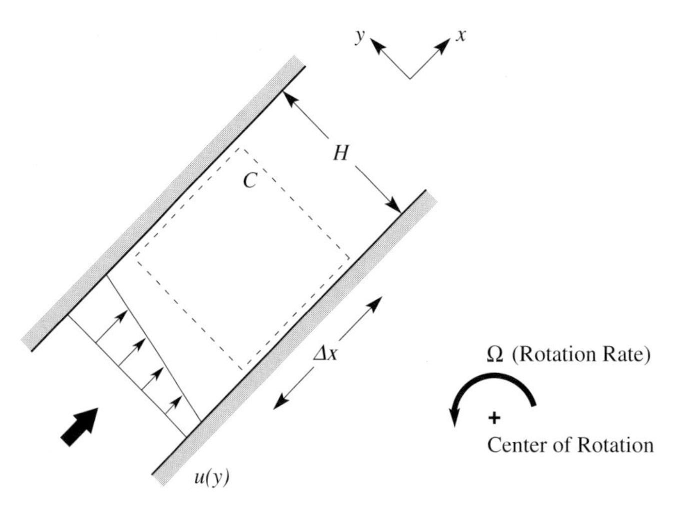

Figure 3.23: Relative velocity distribution in a rotating straight channel.

3.9 Circulation behavior in an incompressible flow

3.9.1 Uniform density inviscid flow with conservative body forces

Under the above conditions, the third term on the right-hand side of (3.8.8) is zero. The pressure gradient term is also zero since it is an exact differential:

$$\frac{1}{\rho}\oint_C \nabla p \cdot d\boldsymbol{\ell} = \frac{1}{\rho}\oint_C dp = 0. \tag{3.9.1}$$

Since a conservative force can be expressed as the gradient of a potential, the second term integrates to zero round a closed contour. Equation (3.8.8) reduces to

$$\frac{D\Gamma}{Dt} = 0. \tag{3.9.2}$$

Equation (3.9.2) is for inviscid, incompressible, uniform density flow with conservative body forces and finds wide applicability in a number of areas. An important special case is a flow without circulation at some given time. The circulation about any arbitrary contour will remain zero, and the flow will have zero vorticity. An example is a flow started from rest or from a very large reservoir with $\mathbf{u} \approx 0$, so that Γ is initially zero. The resulting velocity field will have $\boldsymbol{\nabla} \times \mathbf{u} = 0$ throughout so that $\mathbf{u}$ can be expressed as the gradient of a potential, greatly simplifying analysis. Methods based on potential flow have been applied in many areas of fluids engineering for which inviscid analysis is an appropriate approximation.

Another example occurs in a rotating passage, such as the outer part of a centrifugal compressor impeller. A simplified geometry is shown in Figure 3.23, where the z-axis is the axis of rotation and the x-axis is in the direction of flow. Fluid machinery is often fed from a reservoir where the velocity, and hence the circulation, are essentially zero. Provided viscous effects are negligible in the absolute (stationary) coordinate system, the circulation will remain zero as the fluid flows through the passage. With Γ and $\mathbf{u}$ denoting the circulation and velocity in the absolute coordinate system, therefore,

$$\Gamma = \oint_C \mathbf{u} \cdot d\boldsymbol{\ell} = 0. \tag{3.9.3}$$

The absolute velocity is related to the relative velocity $\mathbf{w}$ by

$$\mathbf{u} = \mathbf{w} + \boldsymbol{\Omega} \times \mathbf{r}, \tag{3.9.4}$$

where $\mathbf{w}$ is the velocity seen by an observer at $\mathbf{r}$ rotating with the channel at angular velocity $\boldsymbol{\Omega}$. Defining

$$\Gamma_{rel} = \oint_C \mathbf{w} \cdot d\boldsymbol{\ell}, \tag{3.9.5}$$

it follows, since $D\Gamma/Dt = 0$ and $\Gamma = 0$, that

$$\Gamma_{rel} = -\oint_C (\boldsymbol{\Omega} \times \mathbf{r}) \cdot d\boldsymbol{\ell}. \tag{3.9.6}$$

Applying Stokes's Theorem,

$$\Gamma_{rel} = -\iint_{A_c} 2\boldsymbol{\Omega} \cdot \mathbf{n} dA_c, \tag{3.9.7}$$

where dA_c is an element of area enclosed by C and $\mathbf{n}$ is the normal to this area. For the contour C shown in Figure 3.23, the relative circulation is thus

$$\Gamma_{rel} = -2\Omega A_c, \tag{3.9.8}$$

where A_c is the area enclosed by the contour. Equation (3.9.8) shows that the magnitude of the relative vorticity is

$$(\omega_z)_{rel} = -2\Omega. \tag{3.9.9}$$

If the channel geometry is such that changes in the y-direction are small, then the relative vorticity can be approximated as

$$(\omega_z)_{rel} = -\frac{dw_x}{dy}. \tag{3.9.10}$$

The velocity profile is as sketched in Figure 3.23, with the inviscid flow in the rotating channel possessing a non-uniform velocity and relative vorticity. The phenomenon of relative vorticity generated in this manner is often referred to as the "relative eddy" and is seen to be a kinematic consequence of Kelvin's Theorem. We will examine this in more depth in Chapter 7.

Kelvin's Theorem also provides an explanation for the observation of "prewhirl", or the axisymmetric swirling of flow in the direction of rotor rotation sometimes seen upstream of a turbomachine. Such swirling motions can be encountered upstream of a pump or compressor at conditions of high aerodynamic loading, and they can occupy a significant fraction of the annulus. A circular fluid contour in the swirling region, centered on the machine axis of rotation, would have a net circulation given by $\Gamma = 2\pi r V_\theta$. Far upstream, however, the circulation is typically zero because the flow is usually drawn from a large chamber or still atmosphere. From Kelvin's Theorem (or more precisely, (3.8.8)), finite circulation can only arise because of viscous forces, which are associated with fluid that has passed through the rotor and then undergone reversed flow. One can thus state that the prewhirl (when the rotor is the first airfoil row to be encountered) must be associated with local flow reversal in the turbomachine; indications of upstream swirl are therefore identical to indications of reverse flow in some portion of the turbomachine.

3.9.2 Incompressible, non-uniform density, inviscid flow with conservative body forces

When the density is non-uniform, the term $\nabla p/\rho$ is no longer generally an exact differential and the circulation of a fluid contour can change with time. The rate of change of circulation for an inviscid flow is given from (3.8.9) as

$$\frac{D\Gamma}{Dt} = -\oint_C \frac{\nabla p}{\rho} \cdot d\boldsymbol{\ell}. \tag{3.8.9}$$

This can be put into a more familiar form by using Stokes's Theorem to yield an integral over the surface, A, bounded by the curve, C:

$$\frac{D\Gamma}{Dt} = \iint_A \frac{\nabla \rho \times \nabla p}{\rho^2} \cdot \mathbf{n}\, dA. \tag{3.9.11}$$

Like vorticity, circulation is produced when density gradients are not aligned with the pressure gradients. This mechanism was introduced in Section 3.5 in the context of vorticity production, and is applied here in a more global fashion.

Such circulation production occurs when fluids of different densities are taken through converging or diverging channels as shown in Figure 3.17, which we now examine with regard to changes in circulation. The density at the inlet varies as indicated while the inlet velocity is uniform.

Consider the contour C which straddles the density difference. Since the flow is in a converging passage, the pressure gradient will point upstream. Across the density interface, the pressure remains continuous and the term ∇p will have essentially the same values on both horizontal legs of contour C. The term $\oint(\nabla p/\rho) \cdot d\boldsymbol{\ell}$ in (3.8.9) can thus be approximated as

$$-\oint_C \frac{\nabla p}{\rho} \cdot d\boldsymbol{\ell} \cong \left(\frac{1}{\rho_2} - \frac{1}{\rho_1}\right) \int_a^b \nabla p \cdot d\boldsymbol{\ell}, \tag{3.9.12}$$

where the integral is taken from one end of the contour to the other along the horizontal direction. The rate of change of circulation for the contour becomes

$$\frac{D\Gamma}{Dt} \cong \left(\frac{1}{\rho_2} - \frac{1}{\rho_1}\right) \Delta p, \tag{3.9.13}$$

where Δp is the change in pressure from one end of the contour to the other. When $\rho_1 < \rho_2$, this term has a negative value and circulation of a clockwise sense is produced around the contour C, leading to the exit velocity profile indicated in Figure 3.17.

3.9.3 Uniform density viscous flow with conservative body forces

For this situation, (3.8.8) takes the form

$$\frac{D\Gamma}{Dt} = \oint_C \mathbf{F}_{visc} \cdot d\boldsymbol{\ell} = -\nu \oint_C \nabla \times \boldsymbol{\omega} \cdot d\boldsymbol{\ell}, \tag{3.9.14}$$

which shows that changes in circulation can also result from the action of viscous forces along the contour.

3.10 Circulation behavior in a compressible inviscid flow

In the derivation of the expression for the rate of change of circulation for a fluid contour, (3.8.9), there was no restriction to incompressible flow. For an inviscid compressible flow, Kelvin's Theorem has the same form as that for incompressible flow

$$\frac{D\Gamma}{Dt} = -\oint_C \frac{\nabla p}{\rho} \cdot d\boldsymbol{\ell} \tag{3.8.9}$$

or

$$\frac{D\Gamma}{Dt} = \iint \frac{\nabla \rho \times \nabla p}{\rho^2} \cdot \mathbf{n}\, dA. \tag{3.10.1}$$

Using the relation $\nabla p/\rho = \nabla h - T\nabla s$, and noting that $\oint \nabla h = 0$, (3.10.1) can be put into a form involving gradients in entropy and temperature,

$$\frac{D\Gamma}{Dt} = \iint \nabla T \times \nabla s \cdot \mathbf{n}\, dA. \tag{3.10.2}$$

If the flow is such that the density, ρ, is only a function of pressure, p, (as it would be, for example, if the entropy were constant) or the entropy, s, is only a function of temperature, T, then the circulation round a closed fluid contour is constant.

An example in which this occurs is compressible isentropic flow, where p/ρ^γ = constant. In this situation, $\nabla p/\rho$ takes the form $\nabla p/\rho(p)$, which yields an exact differential. Thus, the conclusions derived for incompressible flow, for example the persistence of irrotational flow, the relative eddy, and the origin of prewhirl, carry over directly into the compressible regime provided that the flow is isentropic.

3.10.1 Circulation generation due to shock motion in a non-homogeneous medium

An example of circulation generation in compressible flow occurs in the passage of a shock wave through a non-homogeneous fluid, a phenomenon with application to mixing augmentation at high speed. A configuration of interest is the two-dimensional unsteady flow in Figure 3.24, where a cylinder of low density gas sits in a heavier medium through which a shock is passing. The density gradient is radially outward from the center of the cylinder and the pressure gradient is normal to the shock wave. Around the periphery of the cylinder, except at the front and rear, the two gradients are not parallel. Equation (3.10.1) applied to a thin contour which sits on both sides of the density

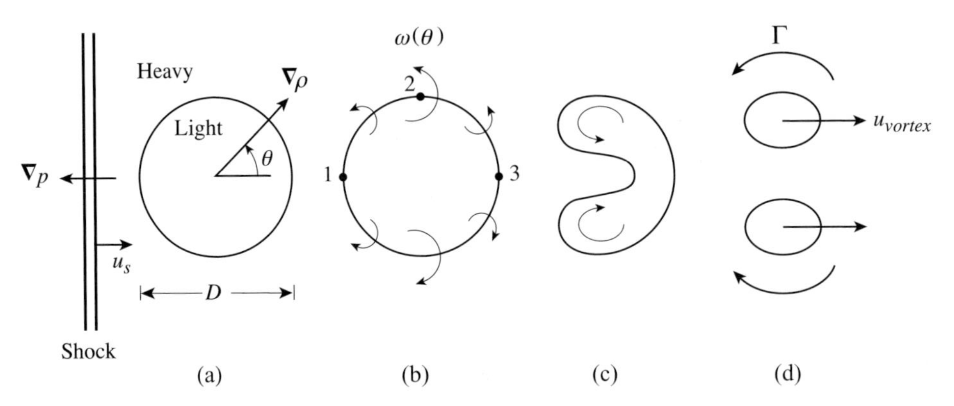

Figure 3.24: Schematic of a two-dimensional unsteady shock-induced vortical flow: (a) before interaction, (b) vorticity distribution immediately after interaction, (c) roll up, (d) steady-state vortex pair (Yang, Kubota, and Zukoski, 1994).

discontinuity gives an appreciation for the flow evolution. When pressure and density gradients are not aligned, the cross-product has a finite value (i.e. $\nabla\rho \times \nabla p \neq 0$) and circulation is generated; the rate of generation is maximum when the two gradients are perpendicular. The angle between the two gradient vectors increases from zero at (1) (Figure 3.24) to a maximum of 90° at (2), and the rate of generation thus varies from zero at (1) and (3) to a maximum at (2). After the passage of the shock the pressure gradient is removed, but the circulation on the interface remains and leads to a deformation of the interface, as shown in Figure 3.24.

The circulation generation occurs over a time interval of order d/u_s, where d is the diameter of the cylinder of light gas and u_s is the mean propagation velocity of the shock across the region. Equation (3.10.1) can be integrated to give the circulation for a half-plane of the flow field as

$$\Gamma = \oiint_C dxdy\left[\int_0^\infty \frac{dt(\nabla\rho \times \nabla p)}{\rho^2}\right], \tag{3.10.3}$$

where C is a contour that encloses all the vorticity in the half-plane of the flow field. Assuming the shock is weak enough so that, while it passes through the cylinder, the interface does not deform appreciably, an estimate for the circulation is (with $\delta(\ )$ denoting the Dirac delta function)

$$\Gamma = \int_0^\pi \sin\theta\, d\theta\, \Delta p\, \Delta\left(\frac{1}{\rho}\right)\left\{\int_0^\infty r\delta\left(r - \frac{d}{2}\right) dr\right\}\left\{\int_0^\infty \delta(x - u_s t)\, dt\right\}. \tag{3.10.4}$$

Thus

$$\Gamma \propto \frac{d}{u_s}\Delta p\, \Delta\left(\frac{1}{\rho}\right), \tag{3.10.5}$$

where $p_2 - p_1 = \Delta p$ is the static pressure rise across the shock, and $\Delta\rho$ is the density difference between the heavy medium and the light cylinder gas. In (3.10.4), the two Dirac delta functions denote the interface at $r = d/2$ and the shock location at time t so that ∇p and $\nabla(1/\rho)$ can be written as $\Delta p\delta(x - u_s t)$ and $\Delta(1/\rho)\,\delta(r - d/2)$ respectively, with $\Delta(\ )$ denoting the change in flow variable

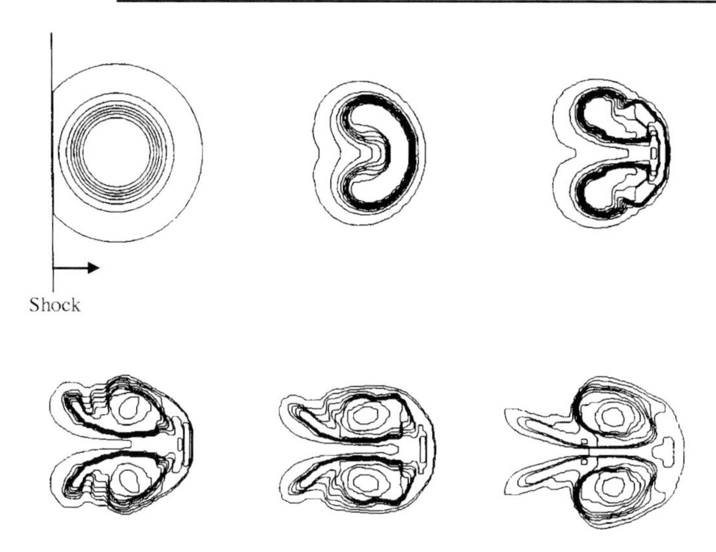

Figure 3.25: Computed density contour plots at $\tilde{t} = ta/d = 0, 10, 20, 40, 50, 70$, $M_s = 1.1$, density ratio (light gas/heavy gas) $= 0.14$ (Yang *et al.* 1994).

across the shock. The approximation embodied in (3.10.4) and (3.10.5) is valid for flow situations where the shock can be considered weak and $\Delta\rho/\overline{\rho} \ll 1$; in this case, $\overline{\rho}$ can be taken to be ρ_2.

Calculations demonstrating the evolution of the cylinder of low density gas are shown in Figure 3.25 at different non-dimensional times, $\tilde{t} = ta/d$, where a is the speed of sound. The initially cylindrical shape is deformed into a vortex pair-like structure. This can also be seen in the flow visualization, from experiments carried out with a cylinder of helium in air, in Figure 3.26.

3.11 Rate of change of circulation for a fixed contour

The expressions derived have been for the rate of change of circulation round a contour moving with the fluid. A complement to this is the rate of change of circulation for a contour fixed in space. This finds most application for two-dimensional flows. The development below is for a uniform density fluid with conservative body forces, but extensions to other cases follow along similar lines.

The scalar product of the momentum equation (3.3.1) with a line element $d\boldsymbol{\ell}$ integrated along a curve AB, yields an equation for the time rate of change of circulation on the curve AB:

$$\frac{\partial \Gamma_{AB}}{\partial t} = \frac{p_{t_A} - p_{t_B}}{\rho} + \int_A^B \mathbf{u} \times \boldsymbol{\omega} \cdot d\boldsymbol{\ell} + \int_A^B \mathbf{X} \cdot d\boldsymbol{\ell} + \int_A^B \mathbf{F}_{visc} \cdot d\boldsymbol{\ell}. \tag{3.11.1}$$

Substituting the form of $\mathbf{F}_{visc}$ for an incompressible constant viscosity fluid and noting that only the component of velocity normal to the contour, u_n, contributes to the second term, we obtain

$$\frac{\partial \Gamma_{AB}}{\partial t} = \frac{p_{t_A} - p_{t_B}}{\rho} - \int_A^B \omega u_n d\ell + \int_A^B \mathbf{X} \cdot d\boldsymbol{\ell} + \nu \int_A^B \frac{\partial \omega}{\partial n} d\ell. \tag{3.11.2}$$

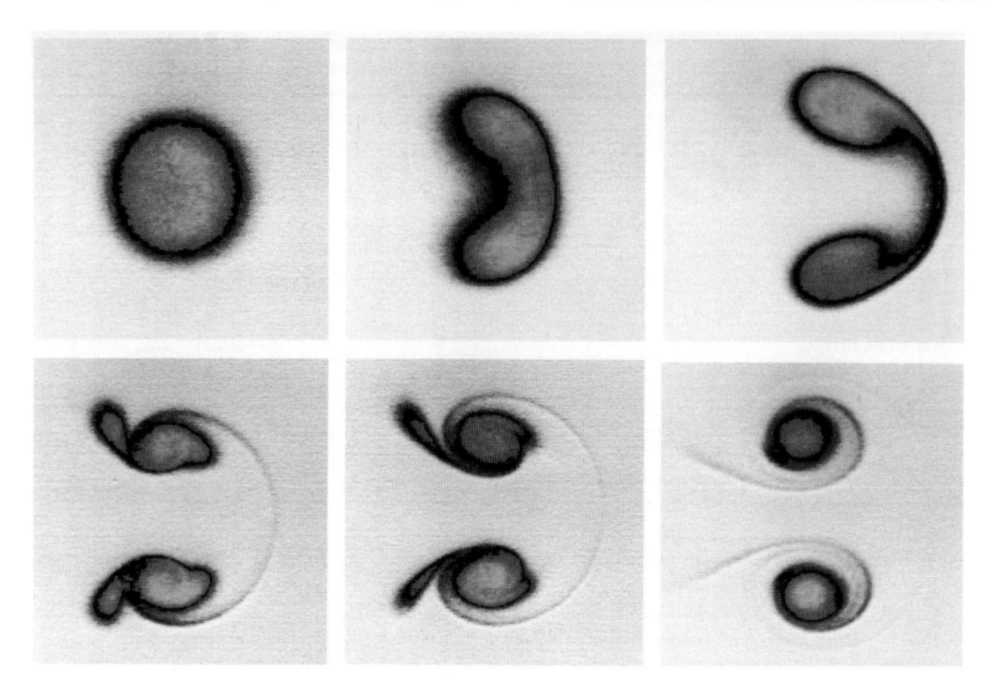

Figure 3.26: Flow visualization showing the evolution of light gas following shock passage, $M_s = 1.1$, density ratio (light gas/heavy gas) = 0.14 (Jacobs, 1992).

In (3.11.2), $\partial\omega/\partial n$ is the derivative of the vorticity in the direction of the outward pointing normal to the contour. For a closed contour, the first and third terms on the right-hand side of (3.11.2) are zero, so

$$\frac{\partial \Gamma}{\partial t} = -\oint \omega u_n d\ell + \nu \oint \frac{\partial \omega}{\partial n} d\ell. \tag{3.11.3}$$

Equation (3.11.3) expresses the change in circulation around a contour fixed in space as due to the difference between the net convection and diffusion of vorticity across the contour. For a steady flow (circulation round the fixed contour constant), the rate of convection of vorticity into the contour is equal to the rate at which vorticity is diffused across it. For the vortex stretching example given in Section 3.6.2, if we examine a circular contour within the core, the radial velocity convects axial vorticity inwards at a rate that balances the outwards diffusion across the contour with the circulation constant.

3.12 Rotational flow descriptions in terms of vorticity and circulation

In many situations, a useful approximation is to regard the flow as inviscid, with density a function of pressure $\rho = \rho(p)$. With no non-conservative body forces acting, the circulation round a given fluid contour remains invariant. This type of flow, which occurs in many engineering problems, is a good arena to illustrate the concepts.

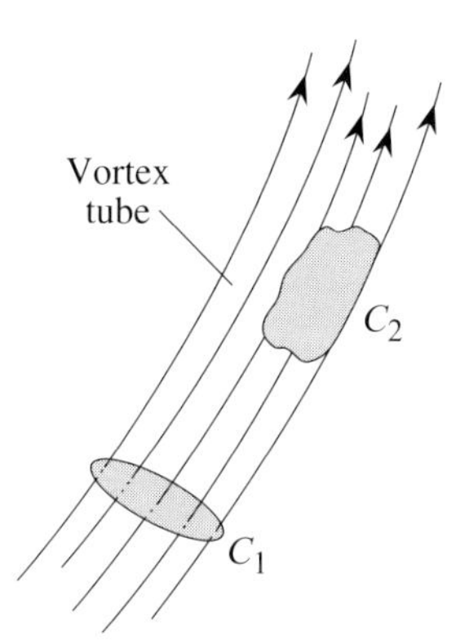

Figure 3.27: Vortex tube showing contour C_1, which encloses all vortex lines in the tube, and contour C_2, which has zero circulation.

For this class of flows, the laws of vortex motion can be brought together and summarized as:

(1) Vortex lines never end in the fluid. The circulation is the same for every contour enclosing the vortex line. (This result is purely kinematic and always true.)
(2) Vortex lines are fluid or material lines; a fluid or material line which at any one time coincides with a vortex line will coincide with it forever.
(3) For a vortex tube of fixed identity, $\omega/\rho d\ell = \text{constant}$, where $d\ell$ is a small length element along the vortex tube. If the vortex tube is stretched, the vorticity increases.

3.12.1 Behavior of vortex tubes when $D\Gamma/Dt = 0$

The behavior of vortex tubes furnishes an introductory application of Kelvin's Theorem to obtain (3) above. Figure 3.27 shows two fluid contours on a vortex tube, one which encloses all the vortex lines in the vortex tube, and is denoted as C_1, and another which lies on the surface of the vortex tube, denoted as C_2. As the vortex tube moves, the circulation around these contours is constant; all the vortex lines will remain enclosed by C_1, and C_2 will stay on the surface of the vortex tube maintaining zero circulation. Because the vortex tube can be made arbitrarily small, this is another view of the statement that vortex lines move with the fluid. If $D\Gamma/Dt = 0$, a fluid, or material, line, which is a vortex line at some time, is always a vortex line.

Conservation of mass for an element of the vortex tube, as shown in Figure 3.28, can be written as

$$\rho \, dA \, d\ell = \text{constant for a fluid element.} \tag{3.12.1}$$

If we take the vortex tube small enough for the vorticity to be considered uniform over the area then

$$\omega \, dA = \text{constant.} \tag{3.12.2}$$

Combining (3.12.1) and (3.12.2) yields

$$\frac{\omega}{\rho d\ell} = \text{constant for a fluid element.} \tag{3.12.3}$$

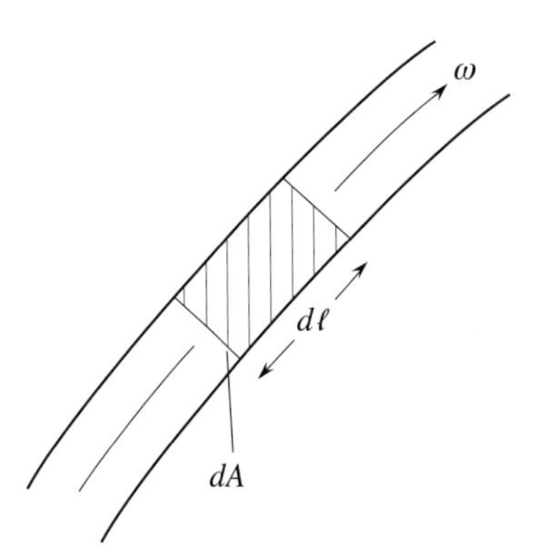

Figure 3.28: Fluid element in a vortex tube; mass $= \rho\, dA\, d\ell$.

If the density is uniform throughout the flow, this reduces to

$$\frac{\omega}{d\ell} = \text{constant}. \tag{3.12.4}$$

Equations (3.12.3) and (3.12.4) again show the relation between vortex stretching and changes in vorticity seen in Sections 3.4 and 3.7, as well as the correspondence between ω in incompressible flow and ω/ρ in compressible flow. Equation (3.12.4) is a statement involving only kinematic quantities, because the force relationships are contained within the derivation of Kelvin's Theorem.

3.12.2 Evolution of a non-uniform flow through a diffuser or nozzle

Equation (3.12.3), or for simplicity its incompressible form (3.12.4), can be applied to describe the evolution of a flow non-uniformity through a diffuser or a nozzle, as illustrated in Figures 3.29(a) and 3.29(b). Figure 3.29(a) shows flow through a nozzle, with a component of vorticity in the streamwise direction. In Figure 3.29(b) the vorticity is in the transverse direction. In discussing these examples, we make the approximation (as has been done several times before) that the vortex lines can be considered to be carried along by a mean flow which is known, in other words, that the three-dimensional flow associated with the vorticity field is weak enough to be approximated as a superposition on a known background or primary flow.

In Figure 3.29(a), the streamwise component of vorticity implies velocity components in directions normal to the primary stream. Along a streamline from the inlet (station i) to the exit (station e) the mean velocity increases. From continuity, the length of an incompressible fluid element increases in proportion to velocity and fluid elements at the inlet and exit are sketched in the figure showing this relationship. The ratio of the streamwise vorticity at the nozzle inlet and the exit of the nozzle is thus

$$\frac{\omega_{x_e}}{\omega_{x_i}} = \frac{\overline{u}_{x_e}}{\overline{u}_{x_i}}, \tag{3.12.5}$$

where $\overline{u}_{x_i}$ and $\overline{u}_{x_e}$ are the background velocities at the inlet and exit. The streamwise vorticity and the maximum swirl velocity are therefore both increased.

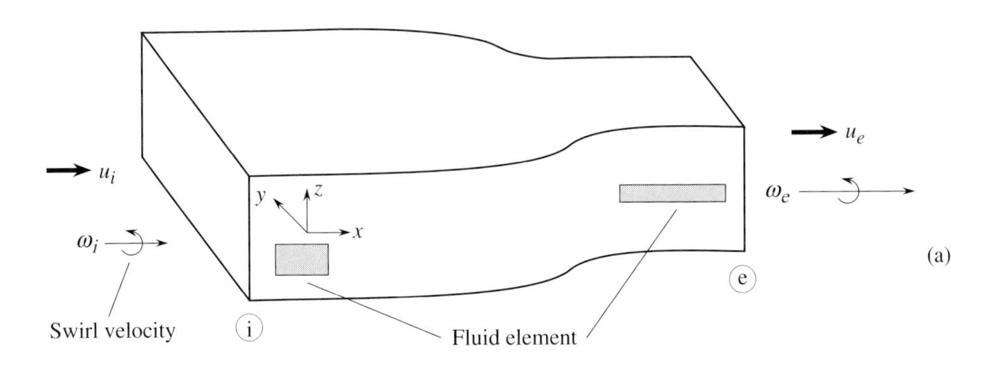

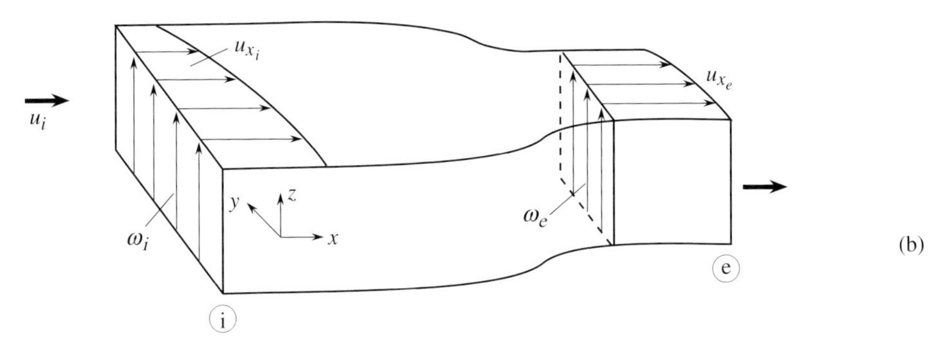

Figure 3.29: Non-uniform rotational flow in a nozzle. (a) streamwise vorticity, $\omega_i \sim \omega_x \mathbf{i}$; (b) normal or transverse vorticity, $\omega_i \sim \omega_z \mathbf{k}$ ($\mathbf{i}$, $\mathbf{k}$ are unit vectors in the x-, z-directions).

Often, what is of most interest is the relative uniformity of a flow. A better measure of this than swirl velocity alone is swirl angle, α, given by

$$\tan\alpha \sim \frac{\text{swirl velocity}}{\text{axial velocity}}. \tag{3.12.6}$$

For a circular vortex tube of radius r, the upstream swirl angle can be approximated as

$$\alpha_i \sim \frac{\omega_{x_i} r_i}{2\overline{u}_{x_i}}. \tag{3.12.7}$$

A vortex tube in this flow is approximately a streamtube and the relation between the streamtube radius and the velocity can be taken as $r^2\overline{u}_x$ = constant. The inlet and exit swirl angles are thus related by

$$\frac{\alpha_e}{\alpha_i} \sim \frac{r_e}{r_i} \sim \sqrt{\text{area ratio}}. \tag{3.12.8}$$

Equation (3.12.8) shows that nozzles tend to increase the uniformity of the flow with regard to swirl angularity, while diffusers tend to worsen it.

In Figure 3.29(b), the vorticity is in the z-direction ($\omega_z = -\partial u_x/\partial y$) and is associated with a non-uniformity in streamwise (x) velocity, u_x. In the constant area straight sections at the inlet and exit, the streamlines will be parallel and the y- and z-components of velocity zero. Thus

$$\omega_{z_i} = \frac{du_{x_i}}{dy} = \omega_{z_e}. \tag{3.12.9}$$

The local velocity gradient remains the same, but as the channel width decreases the level of velocity non-uniformity across the channel, Δu_x, decreases in the ratio

$$\frac{\Delta u_{x_e}}{\Delta u_{x_i}} = \text{area ratio}. \tag{3.12.10}$$

As before, what is generally of most interest are the normalized quantities, in this case the fractional velocity non-uniformity, $\Delta u_x/\overline{u}_x$, which is given by

$$\frac{\Delta u_{x_e}/\overline{u}_{x_e}}{\Delta u_{x_i}/\overline{u}_{x_i}} = (\text{area ratio})^2. \tag{3.12.11}$$

As a third example, consider the same geometry as in Figures 3.29(a) and 3.29(b) but an inlet velocity distribution having vorticity in the y-direction only. Vortex filaments in the y-direction will be compressed in length in proportion to the decrease in channel width, so the vorticity will decrease in proportion to the area ratio. The velocity non-uniformity across the channel height is reduced as before, in this case because of a reduced velocity gradient over a constant height, and the same decrease in Δu_x is obtained as with the vorticity in the z-direction.

3.12.3 Trailing vorticity and trailing vortices

The requirement that vortex lines do not end in the flow has implications for flow downstream of bodies with circulation, such as turbomachine blades. The no-slip conditions at solid surfaces mean that in a viscous fluid all the vorticity that comprises what we view as the circulation round a body is actually contained in the boundary layers on the body. To expand on this point we can make a comparison with classical inviscid analysis of the flow round an airfoil. For this example the airfoil is modeled as a flat plate at an angle of attack with "bound vorticity", $\gamma_b(\ell)$, as shown in Figure 3.30(a). To extend to three-dimensional motions, we must connect this model more directly with real fluid behavior by assessing the situation from the perspective of the viscous boundary layers and their vorticity, as shown in Figure 3.30(b). Doing so leads from arguments concerning the kinematics of vorticity in Section 3.2 to the concept of trailing vorticity discussed below.

The situation of interest is that of a three-dimensional body, for example a turbomachine blade with a tip clearance between the blade tip and the outer casing. At the end of the blade the vortex lines, which thread through the boundary layer on the blade surface and are roughly radial, cannot end in the fluid. The no-slip condition on the velocity means there is zero circulation in any contour on the casing over the tip. Vortex lines therefore cannot end on the casing but must leave the blade surface and trail downstream.

Figures 3.31(a) and 3.31(b) show this situation for a rotor blade with tip clearance. The net circulation round the blade row has the sense of the vorticity in the boundary layer on the suction surface of the blade. Vorticity from the pressure and suction sides of the blade leaves at or near the tip as shown in Figure 3.31(b). The net effect is a vortex layer (or shear layer) with circulation of the same sense as that in the suction surface boundary layer.

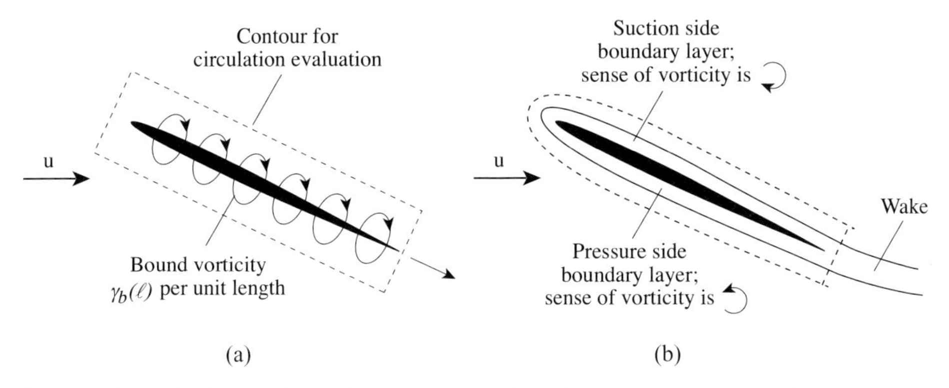

Figure 3.30: (a) Inviscid analysis of flow past a flat plate airfoil using bound vorticity, $\gamma_b(\ell)$; $\Gamma = \int_0^{chord} \gamma_b(\ell)d\ell$. (b) View of airfoil circulation as contained in boundary layer vorticity. Circulation evaluated around a contour just outside the boundary layers and perpendicular to wake.

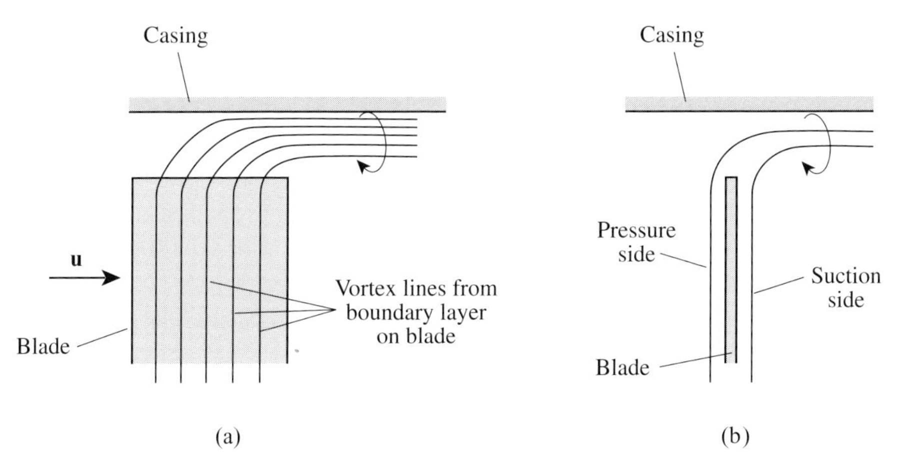

Figure 3.31: Sketch of vortex lines in a turbomachinery tip clearance: (a) view looking normal to blade; (b) view looking upstream at blade edge.

Trailing circulation also occurs at the ends of a blade when there is no tip clearance, for example at the hub of a rotor. The circulation around the blade, evaluated on the hub surface, is zero, so there is a change in circulation round the blade with radius. The vortex lines associated with the circulation round the blade away from the hub must turn tangentially to the hub and trail off in the downstream direction.

In summary, trailing vorticity occurs whenever there is a non-uniform distribution of circulation round a body. The occurrence of trailing vorticity is a kinematic result associated with the fact that the vorticity distribution is solenoidal ($\boldsymbol{\nabla} \cdot \boldsymbol{\omega} = 0$) and applies to all flow regimes.

An often seen consequence of trailing vorticity is a downstream region containing discrete vortices which are compact in scale. A qualitative rationale for this can be given with respect to Figure 3.32, analogous to the situation found behind a finite wing.

Figure 3.32 shows an idealized view of the vortex layer shed from the blade tip at a given axial location. The direction of the vorticity is into the page. As indicated, we can consider the vortex

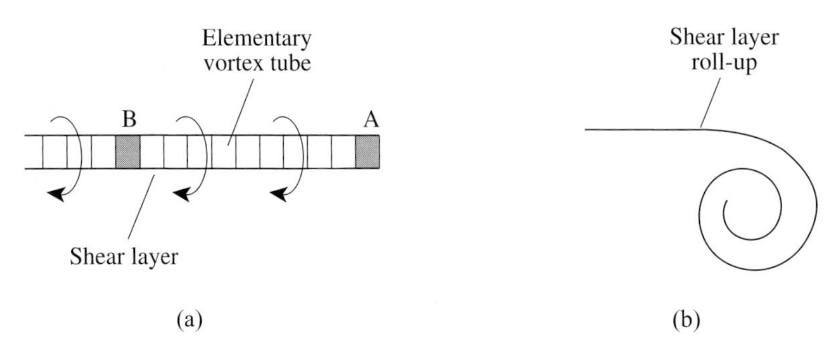

Figure 3.32: Tip clearance shear layer modeled as an array of elementary vortex tubes all with circulation in the sense shown. The velocity at A is the sum of contributions all of one sign. The velocity at B is sum of contributions of opposite sign. (a) Sketch of the initial configuration showing downward velocity near the edge; (b) roll-up of the shear layer.

layer to be made of elementary vortex tubes. Although all of the tubes do not necessarily have the same strength, they have the same sense of circulation. Let us examine the velocity field associated with the shed vorticity at two locations on the sheet, say a station A near the edge and B far from the edge. If we regard the velocity associated with each elementary vortex tube as roughly that of a straight vortex with the local strength, we see that the velocity at the edge of the sheet is that due to the summation of a number of small contributions, weighted with respect to the local strength and distance from the various tubes (falling off as 1/distance), but all with the same sign. If we consider the situation at B the behavior is different. At B there are both positive and negative contributions (both upward and downward velocities). The downward velocity of A in the plane of the page is thus greater than that at B and the layer will have a tendency to roll up into a discrete vortical structure. This behavior, which we have only qualitatively described, implies that flow downstream of devices with a non-uniform circulation distribution along the body (wings, turbomachinery blading, forced mixer lobes) can often contain embedded discrete vortical structures. Quantitative results illustrating this phenomenon are presented in Section 3.15.

Even without roll up and formation of vortices, the presence of trailing vorticity means that the flow downstream of the device will be rotational. Depending on the scale of the information one wishes to extract and the strength and distribution of the trailing vorticity, there are situations in which it is appropriate to view the entire downstream region as filled with trailing vorticity. Examples are the axisymmetric representation of flow in a turbomachine annulus, in which the downstream vorticity field is essentially a "smeared out" representation of the trailing vorticity which originates on the solid surfaces that make up the individual blades and the hub and casing, and the secondary flow type of representation shown in Figure 3.10 and described in Chapter 9.

3.13 Generation of vorticity at solid surfaces

We have not yet considered in any depth the question of how vorticity and circulation are introduced into a flow at solid surfaces. Answering this is necessary because the equations that have been developed contain no mechanism for the production of circulation in a fluid of uniform density or in which $\rho = \rho(p)$. While vortex filaments can be turned and stretched, creating changes in vorticity magnitude and direction, this is basically processing of existing vorticity in a manner to conserve circulation. The viscous forces within the flow modify this processing, but they serve only to redistribute the existing vorticity. In contrast we address here the generation of vorticity, in other words the addition of "local positive or negative circulation to the flow" (Fric and Roshko, 1994), which occurs at solid surfaces.

3.13.1 Generation of vorticity in a two-dimensional flow

We describe the generation of vorticity at a stationary solid surface in a constant density fluid, first for two-dimensional flow and then for three dimensions. A starting point is the momentum equation evaluated at the solid surface. Because the velocity is zero, this reduces to

$$\left(\frac{1}{\rho}\nabla p = \nu\frac{\partial^2 \mathbf{u}}{\partial n^2}\right)_{surface}, \tag{3.13.1}$$

where n is the normal to the surface. For two-dimensional flow with the surface as the plane $y = 0$, use of the continuity equation and the zero velocity condition allows us to write (3.13.1) in terms of the derivative of the vorticity as

$$\left(\frac{1}{\rho}\frac{dp}{dx} = \nu\frac{\partial^2 u_x}{\partial y^2} = -\nu\frac{\partial \omega}{\partial y}\right)_{y=0}. \tag{3.13.2}$$

Equation (3.13.2) shows that whenever a pressure gradient exists along a solid boundary, there is a gradient of tangential vorticity at the surface in the wall-normal direction and hence a diffusion of vorticity into the fluid. This is interpreted as a flux of vorticity from the solid surface at a rate of ν times the gradient of the vorticity along the normal to the surface (Lighthill, 1963). The entering vorticity can be of either sense depending on the sign of the pressure gradient. For cases in which the pressure increases in the flow direction ($dp/dx > 0$), positive, or counterclockwise, vorticity enters the flow.

For a boundary layer, where the pressure gradient is determined by the inviscid flow in the free stream,[4] (3.13.2) can be cast in terms of the spatial and temporal variations in free-stream or "external" velocity, u_E:

$$\frac{\partial u_E}{\partial t} + u_E\frac{\partial u_E}{\partial x} = \nu\left(\frac{\partial \omega}{\partial y}\right)_{y=0}. \tag{3.13.3}$$

These arguments can be given from another viewpoint by computing the circulation round the rectangular contour, ABCD, in Figure 3.33, which encloses a section of a boundary layer on a solid surface. The bottom of the contour is on the solid surface, while the upper edge is just outside the boundary layer in the free stream, and the two vertical legs are perpendicular to the solid surface. The velocity on the upper edge has the free-stream value, u_E, and if the contour is of length dx, the counterclockwise circulation is $(-u_E)dx$ plus the contributions due to the two vertical legs. With the boundary layer of thickness δ, the net contribution of these vertical legs is approximately

[4] We use this term to denote the flow external to the boundary layer.

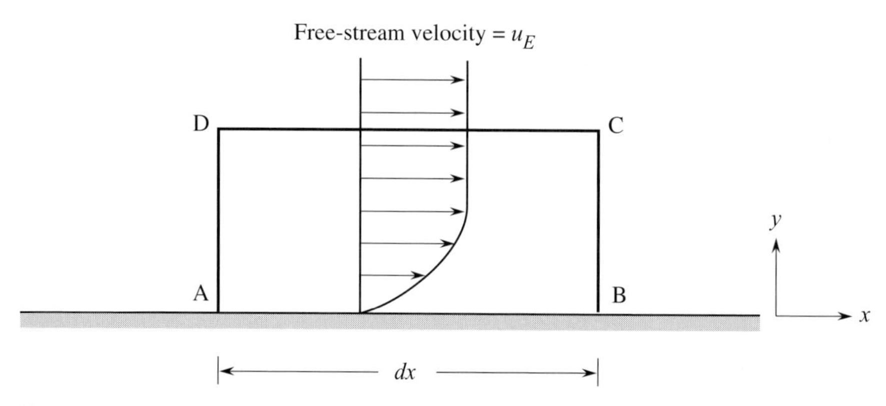

Figure 3.33: Contour used for evaluation of circulation in boundary layer; $\Gamma_{ABCD} = -u_E$.

$(d/dx)/(u_y\delta)dx$ and the ratio of this contribution to that of the upper surface is

$$\frac{\frac{d}{dx}(u_y\delta)}{u_E} \sim \frac{u_y\delta}{u_E L}, \tag{3.13.4}$$

where L is a representative length scale in the streamwise direction. As described in Section 2.9, the ratio of velocity components is $u_y/u_E \sim \delta/L$, so the net contribution of the vertical legs compared to that of the upper leg is of order $(\delta/L)^2$, much smaller than unity for both laminar and turbulent boundary layers. To a very good approximation, the counterclockwise circulation round the contour per unit length, or the net strength of all the vortex tubes threading through the contour, is thus given by

$$\text{circulation per unit length} = -u_E = -[\text{free-stream velocity}]. \tag{3.13.5}$$

We now apply these ideas to a steady boundary layer in a region where the velocity is increasing in the flow direction, such as in a contraction. The free-stream velocity and the circulation per unit length in the boundary layer increase in the downstream direction. This can only occur if vorticity diffuses into the flow from the solid wall. Equation (3.13.3) shows that this is the case, because there is diffusion of clockwise vorticity (the same sign as the existing vorticity) into the fluid. A surface over which the free-stream velocity is increasing (and the pressure decreasing) can thus be regarded as being covered with sources of vorticity of clockwise sense, whereas if the free-stream velocity decreases (and the pressure increases), the sources will be of opposite sign. The strength of these sources is given by (3.13.2) or (3.13.3).

A further aspect concerning vorticity diffusion is illustrated in Figures 3.34(a) and 3.34(b), which are drawn from experimental measurements in a 2:1 contraction (Abernathy, 1972). The streamline distance from the surface is h. Figure 3.34(b) indicates that the boundary layer at station 2 is thinner than that at station 1, not only because of the decrease in channel height, but also because of a decrease in the ratio of boundary layer thickness, δ, to the distance to the streamline in the free stream, h. This can be understood in terms of vorticity diffusion. There is additional vorticity added between stations 1 and 2, and this vorticity has less time to diffuse away from the wall than the vorticity which was already present at station 1. At station 2, a larger percentage of the total vorticity in the

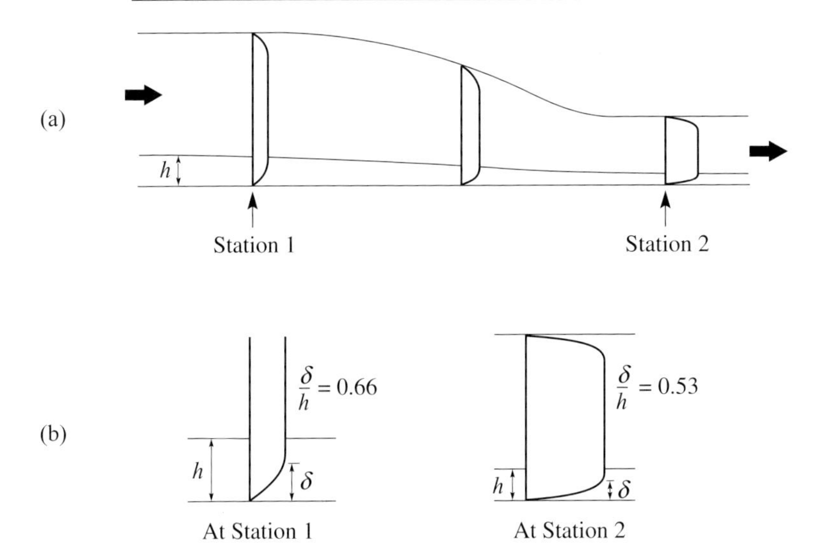

Figure 3.34: Flow in a 2:1 contraction; h is the distance to a streamline outside the boundary layer: (a) overall velocity profiles; (b) blowup of (a) at stations 1 and 2. Tracing of hydrogen bubble flow visualization (Abernathy, 1972).

boundary layer is near the wall than at station 1, so the velocity at a given fraction of the boundary layer thickness will be higher at 2 than at 1.

As before, an alternative explanation can be given in terms of forces and fluid accelerations. The low velocity fluid within the boundary layer will experience a larger velocity change for a given drop in static pressure than the fluid in the free stream. This can be seen from the one-dimensional form of the inviscid momentum equation $du = -dp/\rho u$, where the lower the velocity the larger the velocity increment for a given dp. The boundary layer will therefore be made thinner relative to the free stream as shown in Figure 3.34.

Diffusion of vorticity can also be described in reference to the horseshoe vortex, mentioned in Section 3.4.1, which forms upstream of a strut or obstacle. In Figure 3.35, a contour ABCD is shown on the plane of symmetry of a strut, which protrudes through a boundary layer. Vortex lines from far upstream (with clockwise sense) are continually convected downstream and swept into the left-hand leg (DA) of the contour, and then wrap around the strut. Because the vortex lines cannot be cut, and thus cannot leave the contour, it might seem that the net vorticity inside the contour would continually increase and a steady state would never be obtained. This clearly contradicts experience, so we know that vorticity of the opposite sign must also be entering the contour, and this is provided by the vorticity sources which exist on side AB of the contour. If the free-stream pressure distribution can be regarded as being impressed on the wall, the wall static pressure on the symmetry plane will increase from far upstream to the strut as a result of its upstream influence. An adverse pressure gradient at the wall means that counterclockwise vorticity (opposite sign to that convected in) will be diffused into the contour. The steady state can be viewed as a balance between the two processes, convection and diffusion.

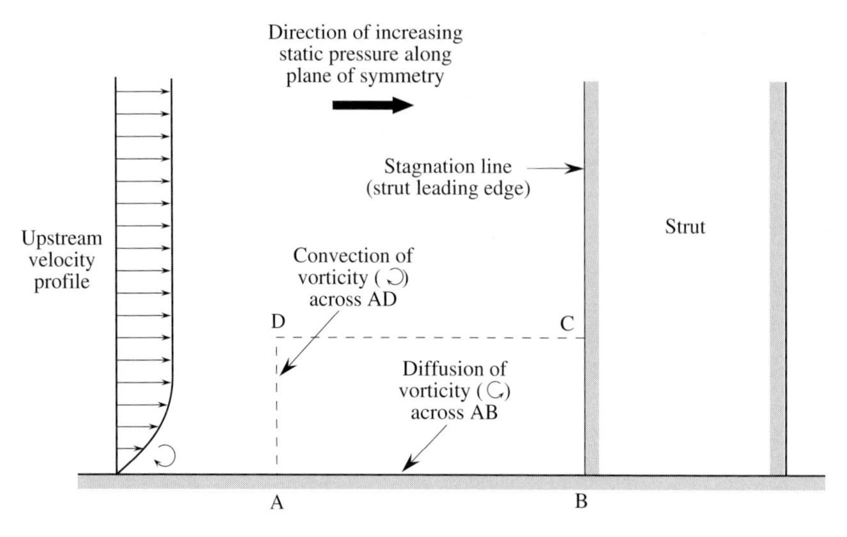

Figure 3.35: Convection and diffusion of vorticity into contour ABCD on the plane of symmetry upstream of a strut.

Referring back to Figure 3.33, we now examine the situation for unsteady flow. If the free-stream velocity changes with time, (3.13.3) implies that the circulation around a contour such as that in Figure 3.33 also changes with time because of the gradient of vorticity at the wall. If the contour were at a station where the free-stream flow was not varying with x, the free-stream momentum equation would be

$$\frac{\partial u_E}{\partial t} = -\frac{1}{\rho}\frac{dp}{dx} = \nu\left(\frac{\partial \omega}{\partial y}\right)_{y=0}. \tag{3.13.6}$$

Integrating (3.13.6) over a time interval during which the free-stream velocity changes by Δu_E,

$$\Delta u_E = \int_{t_{initial}}^{t_{final}} \nu\left(\frac{\partial \omega}{\partial y}\right)_{y=0} dt. \tag{3.13.7}$$

The total vorticity diffused into the contour during the interval is equal to the change in circulation round the contour (which is Δu_E per unit length along the surface). Equation (3.13.7) gives an explicit statement of the link between changes in circulation and vorticity generation at the solid boundary.

The foregoing considerations lead to an interesting interpretation of vorticity generation in a constant pressure boundary layer on a flat plate. The circulation per unit length is constant all along the plate since u_E is constant. The gradient of tangential vorticity at the surface is also zero. All the vorticity in the boundary layer is put into the flow at the leading edge of the plate.

Finally, we look at generation of vorticity in situations in which the *surfaces* are moving. A situation described previously is the infinite flat plate given an impulsive velocity, u_w, at time $t = 0$, with this velocity subsequently maintained constant. For this flow, all the vorticity is introduced at time $t = 0$, when the plate is accelerated. Once the acceleration is completed, the circulation per unit length remains constant at u_w, and no further vorticity enters, although there is a redistribution of the existing vorticity through diffusion to greater distances from the plate.

3.13.2 Vorticity flux in thin shear layers (boundary layers and free shear layers)

Vorticity generated at solid surfaces is subsequently convected away and the resulting vorticity flux past a given station becomes important in considerations of unsteady flow round objects and in the discussion of conditions at trailing edges. For a two-dimensional thin shear layer in which the velocity in the x-direction (which is roughly aligned with the streamwise direction) is much larger than that in the y-direction, the counterclockwise vorticity can be represented by $\omega \cong -(\partial u_x/\partial y)$. The expression for the flux of vorticity past a streamwise station is then

$$\text{flux of vorticity past a given station} = \int_{y_L}^{y_U} u_x \omega \, dy = \int_{y_L}^{y_U} -\left(u_x \frac{\partial u_x}{\partial y}\right) dy = \frac{-u^2(y_U) + u^2(y_L)}{2}. \tag{3.13.8}$$

The integral is carried from y_L to y_U, where y_U and y_L denote the upper and lower boundaries of the shear or boundary layer. For a boundary layer on a stationary surface, y_L coincides with the surface, $u_x(y_L) = 0$, and $u_x(y_U) = u_E$, the free-stream velocity. The vorticity flux is thus $u_E^2/2$.

The mean convection velocity for the vorticity is defined as the net vorticity flux divided by the net amount of vorticity in a unit length of the layer:

$$\text{mean convection velocity of vorticity} = \frac{\int_{y_L}^{y_U} u_x \omega \, dy}{\int_{y_L}^{y_U} \omega \, dy} = \frac{u_x(y_U) + u_x(y_L)}{2}. \tag{3.13.9}$$

For either a laminar or a turbulent boundary layer, the local mean convective velocity of vorticity is therefore half the free-stream velocity.

For the contour in Figure 3.33, the difference in the flux of vorticity across the left and right vertical surfaces is $(d/dx)(u_E^2/2)$ or $(u_E du_E/dx)$. From (3.13.3) this is the rate of diffusion of vorticity across the lower surface of the contour (AB) in steady flow. This again shows the direct connection between changes in the flux of vorticity in the streamwise direction and vorticity diffusion into the flow from the solid wall.

The ideas about vorticity flux can also be used to make a statement about conditions at the trailing edge of a body in a viscous flow following Thwaites (1960). Figure 3.36 shows a fixed contour round a two-dimensional body with flow separation occurring at locations S_U and S_L. Part UAL of the contour is outside the rotational part of the flow, parts US_U and LS_L are perpendicular to the local velocity in the boundary layer, and S_UTS_L is on the surface downstream of the separation locations. The vorticity is thus zero along UAL, and there is no convection of vorticity across S_UTS_L. The convection of vorticity across S_LL and S_UU is given by (3.13.8). In the separated part of the flow, the velocity gradients can be taken to be small adjacent to the body, so diffusion of vorticity can be

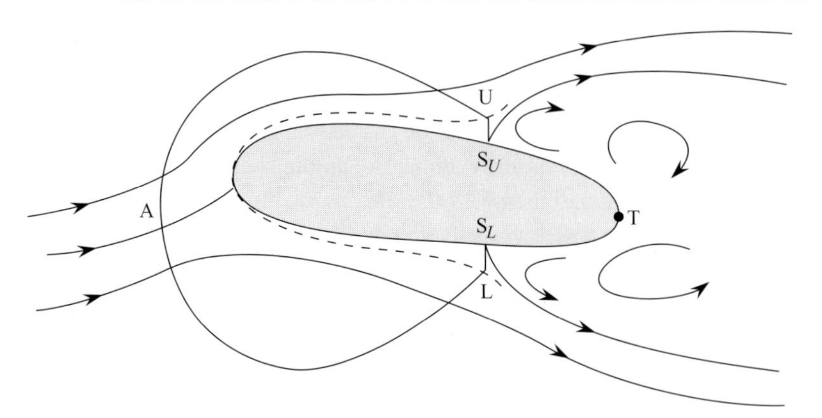

Figure 3.36: Contour used for computation of circulation and vorticity flux for a body with separation (after Thwaites, (1960)).

neglected on S_UTS_L. Diffusion of vorticity in the streamwise direction across S_LL and S_UU is also neglected compared with convection.

In steady flow, the circulation round the body on the contour does not change with time. The net vorticity flux from the body into the wake must be zero, because there is no diffusion across the contour. Vorticity leaves the body in two layers, one from the point of separation of the flow on the upper side of the body and one from the point of separation on the lower part with vorticity fluxes of $u_U^2/2$ and $u_L^2/2$, respectively, where u_U and u_L are the free-stream velocities at the separation points. Because the net vorticity flux is zero, the free-stream velocities and hence the static pressures must be equal at these points. The static pressure between S_U and S_L will be essentially uniform because the fluid velocities are low in the separated region.[5] The condition of no net vorticity flux can therefore be regarded as determining the location of the separation points and the overall circulation round the body.

For unsteady flow, it is no longer necessary that there be zero net vorticity flux into the wake, because the circulation around the body can change. If the location of the separation points is fixed, as it might be if there were a sharp corner or salient edge on the body, the net flux of vorticity into the wake at any given time is $u_U^2/2 - u_L^2/2$ which is equal to the net rate of change of circulation round the body. Evaluating the circulation round a fixed contour from S_L to S_U, from (3.11.2)

$$\frac{\partial \Gamma_{LU}}{\partial t} + \frac{u_U^2}{2} - \frac{u_L^2}{2} + \frac{p_U - p_L}{\rho} = 0. \tag{3.13.10}$$

If diffusion of vorticity in the separated region is negligible, the sum of the first three terms must be zero. In an unsteady flow, the static pressure is thus also approximately uniform at the rear of the body between S_U and S_L.

The difference in velocities at the two separation points, $(u_U^2 - u_L^2)/2$, can be written in a manner that directly exhibits the net vorticity flux into the downstream wake. The flux of vorticity into the wake is given by $\overline{u}\gamma$, where the average velocity $\overline{u}$ is given by $\overline{u} = (u_U + u_L)/2$ and $\gamma = u_U - u_L$,

[5] As discussed in Chapter 5, however, the static pressure in this base region is generally not equal to (and lower than) the free-stream value.

the circulation per unit length of the wake. If the flow leaves the body at the trailing edge, (3.13.10) becomes

$$\frac{\partial \Gamma}{\partial t} = -\{\bar{u}\gamma\}_{trailing\, edge}. \tag{3.13.11}$$

3.13.3 Vorticity generation at a plane surface in a three-dimensional flow

In three dimensions we again examine the gradient of vorticity at the solid surface to develop an expression for the vorticity flux. The gradient of a vector, **B**, is defined in Cartesian coordinates by

$$\nabla \mathbf{B} = \mathbf{i}\frac{\partial \mathbf{B}}{\partial x} + \mathbf{j}\frac{\partial \mathbf{B}}{\partial y} + \mathbf{k}\frac{\partial \mathbf{B}}{\partial z}, \tag{3.13.12}$$

where **i**, **j**, **k**, are unit vectors in the x-, y-, z-directions respectively (Morse and Feshbach, 1953; Gibbs, 1901). We are interested in the gradient in the wall-normal direction, here the y-direction.[6]

The term of interest here corresponds to $\mathbf{j}\partial\boldsymbol{\omega}/\partial y$, which is a vector with three components: $(\partial\omega_x/\partial y)$, $(\partial\omega_y/\partial y)$, $(\partial\omega_z/\partial y)$. Writing out the vorticity components in terms of velocity components, and using the continuity equation to infer that both $\partial^2 u_y/\partial x\partial y$ *and* $\partial^2 u_y/\partial z\partial y$ are zero at the surface ($y = 0$) yields:

$$\left(\nu\frac{\partial \omega_x}{\partial y} = \nu\frac{\partial^2 u_z}{\partial y^2} = \frac{1}{\rho}\frac{\partial p}{\partial z}\right)\Bigg|_{y=0}, \tag{3.13.13a}$$

$$\left(\nu\frac{\partial \omega_z}{\partial y} = -\nu\frac{\partial^2 u_x}{\partial y^2} = -\frac{1}{\rho}\frac{\partial p}{\partial x}\right)\Bigg|_{y=0}. \tag{3.13.13b}$$

The derivative $\partial\omega_y/\partial y$ can be written, using the condition of zero velocity at the solid surface, as

$$\left(\nu\frac{\partial \omega_y}{\partial y} = \frac{1}{\rho}\left[\frac{\partial \tau_{xy}}{\partial z} - \frac{\partial \tau_{zy}}{\partial x}\right]\right)\Bigg|_{y=0}. \tag{3.13.13c}$$

Equations (3.13.13) are the three components of the vorticity flux in the wall-normal direction at a plane solid surface:

$$\text{vorticity flux in the wall-normal } (y) \text{ direction} = -\mathbf{j}\times(\nabla p)|_{y=0} - \mathbf{j}[\mathbf{j}\cdot(\nabla\times\boldsymbol{\tau}_w)]. \tag{3.13.14}$$

In (3.13.14) the term $(\nabla p)|_{y=0}$ is the pressure gradient term evaluated at the wall and $\boldsymbol{\tau}_w$ is the vector with components equal to the wall shear stresses.

The first term on the right-hand side of (3.13.14) is the vorticity source term due to a wall pressure gradient, analogous to the description in Section 3.13.1 for a two-dimensional flow. The flux of vorticity produced by this is tangent to the wall. The second term, which has a torque-like quality, accounts for the gradient of wall-normal vorticity. The vorticity at the wall must be tangential, so the normal component at the wall is zero. However, there can be a flux of normal vorticity and, immediately above the wall, a component of normal vorticity can exist.

[6] As described by Fric and Roshko (1994) the vorticity flux out of the wall can be interpreted as $\mathbf{n}\cdot\mathbf{J}_0$, where $\mathbf{J}_0 = -\nu(\nabla\boldsymbol{\omega})|_w$ is the vorticity flux tensor at the solid surface and **n** is the wall-normal unit vector. See also Panton (1984) for a useful discussion of this topic.

For attached viscous flows (more specifically, for flows in which the viscous layer thickness is much smaller than the x or z length scales) the pressure gradient term is dominant and the shear stress contribution can be neglected. (The latter is zero for two-dimensional flow.) For example, the vorticity flux in three-dimensional attached boundary layers is well described as a flux of tangential vorticity only. For those three-dimensional separations, however, where the length scales in the x- and z-directions (along the wall) become comparable to the relevant length scales normal to the wall, the flux of wall-normal vorticity associated with the $\nabla \times \boldsymbol{\tau}_w$ term can be important. As pointed out by Fric and Roshko (1994), one situation of this type occurs on a solid surface underneath the spiral flow in a "tornado-like" motion.

3.14 Relation between kinematic and thermodynamic properties in an inviscid, non-heat-conducting fluid: Crocco's Theorem

The equations of motion can be written in several forms which involve the vorticity and relate the kinematic and thermodynamic properties of the flow. These are especially useful when effects of viscosity and thermal conductivity can be neglected and so the development is presented for this situation only. To begin, we substitute the Gibbs equation (1.3.19) into the inviscid momentum equation ((3.3.3) with viscous forces set equal to zero). The momentum equation becomes

$$-\frac{\partial \mathbf{u}}{\partial t} + (\mathbf{u} \times \boldsymbol{\omega}) = \nabla h - T\nabla s + \frac{1}{2}\nabla(u^2) - \mathbf{X}. \tag{3.14.1}$$

If the body force is conservative, it can be represented by a potential function: $\mathbf{X} = -\nabla\psi$. Therefore

$$-\frac{\partial \mathbf{u}}{\partial t} - (\mathbf{u} \times \boldsymbol{\omega}) = \nabla\left(h + \frac{1}{2}u^2 + \psi\right) - T\nabla s$$

or, in terms of the stagnation enthalpy,

$$-\frac{\partial \mathbf{u}}{\partial t} + (\mathbf{u} \times \boldsymbol{\omega}) = \nabla(h_t + \psi) - T\nabla s. \tag{3.14.2}$$

For steady flow, (3.14.2) reduces to

$$\mathbf{u} \times \boldsymbol{\omega} = \nabla(h_t + \psi) - T\nabla s. \tag{3.14.3}$$

Equations (3.14.2) and (3.14.3) imply:

(1) In a steady irrotational flow ($\boldsymbol{\omega} = 0$), either (i) the entropy or temperature must be uniform because all the other terms in (3.14.3) are pure gradients, or (ii) the variations in h_t, ψ, and s are such that the gradients exactly cancel (Smith, 2001); this can occur in a parallel flow only.

(2) In a steady flow, if the entropy and the quantity $(h_t + \psi)$ are uniform throughout, the velocity field is either irrotational or the velocity and vorticity are parallel. If $\mathbf{u}$ and $\boldsymbol{\omega}$ are parallel, $\mathbf{u} \times \boldsymbol{\omega} = 0$: this is known as a Beltrami flow.

(3) In steady flow with no body forces, the relation between variations in the thermodynamic properties and the kinematic quantities (vorticity and velocity) is

$$(\mathbf{u} \times \boldsymbol{\omega}) = \nabla h_t - T\nabla s. \tag{3.14.4}$$

Equation (3.14.4) is known as Crocco's Theorem.

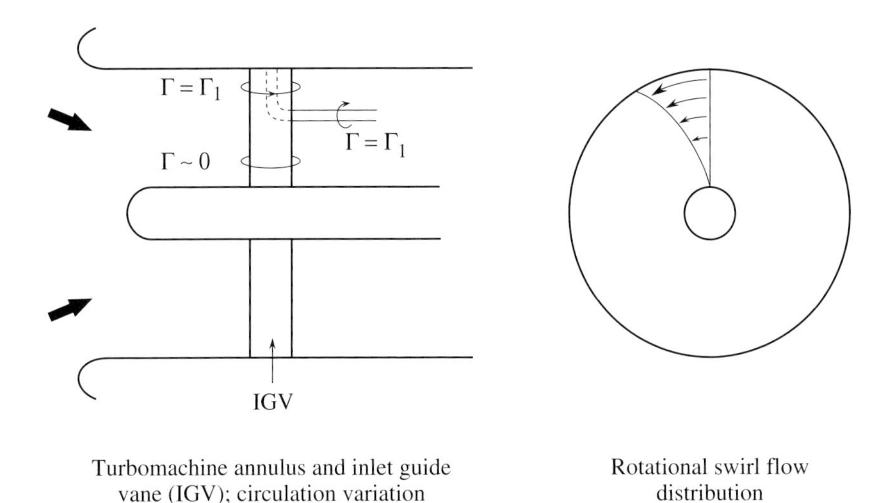

Figure 3.37: Trailing vorticity downstream of an inlet guide vane.

(4) For an irrotational flow with no body forces, the stagnation enthalpy can only vary if the flow is unsteady.

An important subset of the above flows is those with no body forces and in which the fluid can be regarded as incompressible and uniform density. The relation corresponding to (3.14.4) for that situation is

$$-\frac{\partial \mathbf{u}}{\partial t} + (\mathbf{u} \times \boldsymbol{\omega}) = \frac{\nabla p_t}{\rho}. \tag{3.14.5}$$

For steady flow this becomes

$$\mathbf{u} \times \boldsymbol{\omega} = \frac{\nabla p_t}{\rho}. \tag{3.14.6}$$

Under these conditions, if the stagnation pressure is constant, either the flow is irrotational or the vorticity is parallel to the velocity. Further, for an irrotational flow the stagnation pressure can only change if the flow is unsteady.

3.14.1 Applications of Crocco's Theorem

Crocco's Theorem provides a useful description for a number of types of rotational flows encountered in practice. We present three illustrations.

3.14.1.1 Flow downstream of an inlet guide vane (stationary blade row) in a turbomachine

Even in ideal or lossless turbomachines, the flow is not necessarily irrotational. As an example, we examine the inlet guide vane row (or IGV) shown in Figure 3.37. This is typically the first row of blades in a turbomachine and is used to direct the flow, considered here as entering from a large reservoir at uniform stagnation conditions. For a steady reversible flow, the entropy and the stagnation enthalpy downstream of the vane row will be uniform and equal to the upstream values, and the right-hand side of (3.14.4) will be zero. Crocco's Theorem therefore tells us that the vorticity must be parallel to the velocity vector.

Suppose the guide vane row is designed to create a radially non-uniform deflection of the flow or, as has sometimes been the case, to produce swirl in one direction at one radius and in another direction at another radius. At any spanwise location, the circulation around the vane will be the product of the difference in the inlet and exit circumferential velocities and the blade-to-blade spacing. The vortex lines associated with circulation round the IGV cannot end in the fluid and since the circulation varies with radius, the vortex lines must trail off the vane as sketched on the left-hand side of Figure 3.37. The vortex lines are parallel to the velocity vectors, like the trailing vorticity behind a finite wing. For an invsicid steady flow all the downstream vortex lines are contained in discrete vortex sheets, which leave the trailing edge of each vane. The circumferentially averaged effect of these sheets is an axisymmetric swirling flow such as that sketched on the right-hand side of Figure 3.37.

3.14.1.2 Flow downstream of a rotor (moving blade row) in a turbomachine

The radial distribution of blade circulation is also generally non-uniform for the rotating blades in a turbomachine. The stagnation enthalpy change across the moving blade row is given by the Euler turbine equation (2.8.27):

$$h_{t_2} - h_{t_1} = \Omega(r_2 u_{\theta_2} - r_1 u_{\theta_1}), \tag{2.8.27}$$

where Ω is the rotational speed and where 1 and 2 denote stations at the inlet and exit of the blade row. If fluid particles enter and exit the blade row at the same radius,

$$h_{t_2} - h_{t_1} = \Omega r(u_{\theta_2} - u_{\theta_1}). \tag{3.14.7}$$

The velocity difference $(u_{\theta_2} - u_{\theta_1})$ is not generally proportional to $1/r$ so there is a radial variation of stagnation enthalpy. Similar to the IGV discussed above, the circulation around the blade at a particular radius is given by $(u_{\theta_2} - u_{\theta_1})W$, where W is the blade spacing. Equation (3.14.7) can therefore be written in terms of the blade circulation $\Gamma_{blade}(r)$ as

$$h_{t_2} - h_{t_1} = \frac{\Omega r \Gamma_{blade}}{W}. \tag{3.14.8}$$

Since stagnation enthalpy gradients typically exist downstream of the rotor blade rows, the exit flow field will generally have non-zero vorticity.

3.14.1.3 Flow downstream of a non-uniform strength shock wave

Across a shock wave, stagnation enthalpy is conserved and entropy increases. If a shock is curved, or if the Mach number upstream of the shock varies, the shock strength and the entropy rise will vary along the shock and, in accord with (3.14.4), the flow downstream of the shock will be rotational. An illustration of this occurs in the supersonic flow round the leading edge of an airfoil or bluff body. As discussed in Chapter 2, the entropy rise across a shock is small for Mach numbers of 1.3 or less (the non-dimensional change in entropy, $T_2(s_2 - s_1)/u_1^2 = 0.012$ for $M_1 = 1.3$), so the influence of shock curvature on vorticity creation does not become appreciable until higher Mach numbers. To illustrate

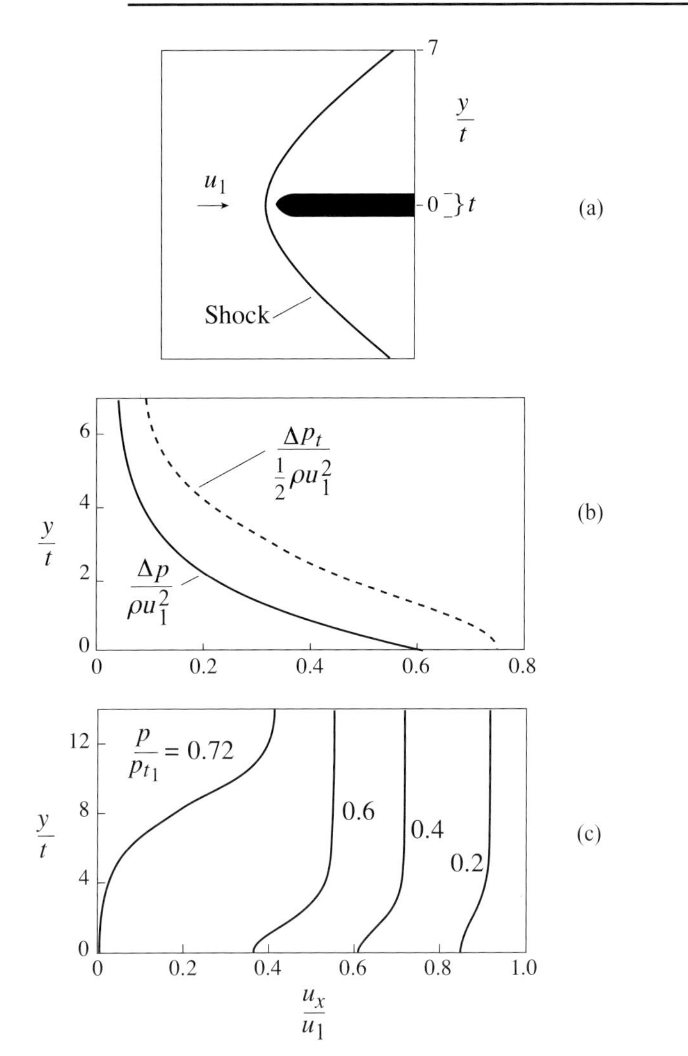

Figure 3.38: Rotational flow downstream of a curved shock, upstream Mach number = 2.0, t is plate thickness: (a) geometry and shock configuration; (b) static pressure rise and stagnation pressure decrease across shock, $\Delta p/\frac{1}{2}\rho u_1^2$, $\Delta p_t/\frac{1}{2}\rho u_1^2$, versus vertical distance from plate center; (c) axial velocity profiles u_x/u_1 for different levels of downstream static pressure, p/p_{t_1}.

the effect, Figures 3.38(a), (b), and (c) present computational results for the two-dimensional inviscid flow past a cascade of flat plates, at a Mach number of 2.0. The airfoils have a 10% thickness to chord ratio and elliptical leading edges. The blade spacing to thickness ratio is 30 so that there is only a small effect of the neighboring blade, and the local flow behavior is close to what it would be with an isolated airfoil.

Figure 3.38(a) shows the computed configuration of the shock and Figure 3.38(b) indicates the static pressure rise across the shock and the stagnation pressure decrease downstream of the shock normalized by the upstream dynamic pressure as a function of the vertical distance from the center of the plate in units of blade thickness. As described in Chapter 2, the decrease in stagnation pressure is directly reflected in the entropy rise ($(s_2 - s_1)/R = \ln(p_{t_1}/p_{t_2})$). On the line of symmetry the shock is normal to the upstream flow, and the stagnation pressure change corresponds to the value for a normal shock at a Mach number of 2.0 (Figure 2.9). Away from the airfoil the shock is inclined to the flow. As discussed in Section 2.8, the stagnation pressure change is associated with the Mach number normal to the shock. For streamlines in which the shock is more inclined to the upstream flow, the magnitudes of the stagnation pressure drop and the entropy rise are decreased and an entropy gradient exists downstream of the shock.

Because of the non-uniform entropy (or stagnation pressure), the flow downstream of a curved shock is rotational. This can also be seen by considering the region far downstream of the shock where the flow is parallel. The discussion of Section 2.3 implies that the only velocity component is in the x-direction. Equation (3.14.4) can therefore be written as an explicit relation between vorticity and the gradient of stagnation pressure normal to the flow:

$$\omega = \frac{T}{u_x}\frac{\partial s}{\partial y} = -\frac{1}{\rho u_x}\frac{\partial p_t}{\partial y}. \tag{3.14.9}$$

Figure 3.38(c) depicts a consequence of the non-uniformity associated with rotationality. The figure shows the velocity profiles (assuming parallel flow in the x-direction) corresponding to different levels of downstream static pressure. These would represent a situation where the flow downstream of the shock is subject to further pressure change. The profiles are plotted for overall pressure levels from $p/p_{t_1} = 0.2$ to $p/p_{t_1} = 0.72$ which is close to the limit at which the flow at the plate will reverse. The scale is extended twice as far as in (a) or (b) to indicate the changes in profile. As a reference, the level of pressure just downstream of a normal shock at $M_1 = 2.0$ is $0.575p_{t_1}$. In terms of pressures and fluid accelerations, particles with the lowest stagnation pressure downstream of the shock also have the lowest velocity and density and are thus decelerated the most for a given pressure rise. This effect results in the observed thickening of the low stagnation pressure region with increased pressure rise.

Finally, the evolution of the vorticity distribution over and above what might occur in a uniform density situation can also be commented on using the arguments given in Section 3.7. First, as the pressure rises the density of a fluid particle increases so that the vorticity also increases. Second, the static temperature, and hence the density, in the downstream flow is non-uniform. For a pressure distribution which increases in the direction of flow, the torque associated with the $\nabla p \times \nabla \rho$ effect creates additional clockwise vorticity. Both of these effects enhance the velocity defect and drive the flow towards reversal.

3.15 The velocity field associated with a vorticity distribution

We have used the concepts of vorticity and circulation to provide physical insight into a number of different situations. Another role these ideas can play in dealing with fluid motions is to provide a route to quantitative descriptions as applied in various types of "vortex methods" (see Section 3.15.5). To illustrate this aspect, we now address the question of defining the velocity field associated with a given distribution of vorticity.

The starting point for the process is a general result from vector analysis known as Helmholtz's Decomposition Theorem, which we apply to the velocity vector $\mathbf{u}$. The theorem states that any vector, here represented by the velocity $\mathbf{u}$, can be defined as the sum of two simpler vectors, $\mathbf{u}_1$ and $\mathbf{u}_2$. The vector $\mathbf{u}_1$ is solenoidal, $\nabla \cdot \mathbf{u}_1 = 0$, and the vector $\mathbf{u}_2$ is the gradient of a potential, $\mathbf{u}_2 = \nabla\varphi$. From the vector identity $\nabla \times \nabla\varphi \equiv 0$, we infer that $\nabla \times \mathbf{u}_2 \equiv 0$, so $\mathbf{u}_2$ must be irrotational. Derivation of the theorem is given in a number of texts, for example Aris (1962), Sommerfeld (1964), or Plonsey and Collin (1961).

From what has been said so far concerning $\mathbf{u}_1$ and $\mathbf{u}_2$, the representation is not unique, because we could choose any potential field and subtract it from $\mathbf{u}$ to get the same $\mathbf{u}_2$. A unique decomposition can, however, be made by choosing $\mathbf{u}_1$ and $\mathbf{u}_2$ to be the velocity fields associated with the distribution of vorticity, $\nabla \times \mathbf{u}$, and the distribution of $\nabla \cdot \mathbf{u}$ throughout the flow field, as described below. The former term is the vorticity, $\boldsymbol{\omega}$, while the latter term represents the departure from a solenoidal velocity distribution due to compressibility or volume addition for example from heat addition or phase change.

For a velocity field which is defined everywhere in space and vanishes at infinity, $\mathbf{u}_1$ and $\mathbf{u}_2$ are given by volume and surface integrals of the vorticity and source distributions:

$$\mathbf{u}(\mathbf{x}) = \frac{1}{4\pi}\int_V \frac{\boldsymbol{\omega}(\mathbf{x}') \times \mathbf{r}}{r^3} dV' + \frac{1}{4\pi}\int_V \frac{\nabla' \cdot \mathbf{u}(\mathbf{x}')\,\mathbf{r}}{r^3} dV'. \tag{3.15.1}$$

In (3.15.1) $\mathbf{r} = (\mathbf{x} - \mathbf{x}')$ and is the radius vector from the source or element of vorticity ($\mathbf{x}'$) to the location of interest ($\mathbf{x}$). The notation ∇' signifies that the operator is defined with respect to $\mathbf{x}'$, and the notation V' that the integration is carried out over $\mathbf{x}'$.

For an incompressible fluid with $\nabla \cdot \mathbf{u} = 0$ the velocity field is related directly to the vorticity distribution by

$$\mathbf{u}(\mathbf{x}) = \int_V \frac{\boldsymbol{\omega}(\mathbf{x}') \times \mathbf{r}}{4\pi r^3} dV', \tag{3.15.2}$$

where, again, (3.15.2) implies that $\mathbf{u}$ is defined everywhere in space and vanishes at infinity. Equation (3.15.2) is known as the Biot–Savart law.

In general, the velocity is not defined everywhere in space because of bounding surfaces (exterior boundaries) or solid bodies (interior boundaries). Equation (3.15.2) must therefore be supplemented with suitable boundary conditions. This can be accomplished by extending the Decomposition Theorem to include surface distributions of vorticity and surface sources. A physical example of the former is a thin boundary layer (a region of concentrated vorticity) on the surface of a body and an example of the latter is suction or blowing normal to a solid surface. With this extension a relation between the vorticity and the velocity known as the Representation Theorem is obtained,

$$\begin{aligned}\mathbf{u}(\mathbf{x}) = &\int_V \frac{\boldsymbol{\omega}(\mathbf{x}') \times \mathbf{r}}{4\pi r^3} dV' + \int_A \frac{[\mathbf{u}(\mathbf{x}') \times \mathbf{n}] \times \mathbf{r}}{4\pi r^3} dA' \\ &+ \int_V \frac{\nabla' \cdot \mathbf{u}(\mathbf{x}')\,\mathbf{r}}{4\pi r^3} dV' + \int_A \frac{[\mathbf{u}(\mathbf{x}') \cdot \mathbf{n}]\,\mathbf{r}}{4\pi r^3} dA'.\end{aligned} \tag{3.15.3}$$

Equation (3.15.3) is a kinematic result which is valid for steady and unsteady flow.

For incompressible flow with no surface sources a general relation for the velocity field is

$$\mathbf{u}(\mathbf{x}) = \int_V \frac{\boldsymbol{\omega}(\mathbf{x}') \times \mathbf{r}}{4\pi r^3}\, dV' + \int_A \frac{[\mathbf{u}(\mathbf{x}') \times \mathbf{n}] \times \mathbf{r}}{4\pi r^3}\, dA'. \tag{3.15.4}$$

Equation (3.15.4) provides a complete description of the velocity field for incompressible flow. We show below that the surface integral in this equation describes the vorticity on the surface of a body. The physical interpretation of (3.15.4) can be summed up in the statement by Saffman (1981) that "all the problems of such flows can be posed as questions about the strength and location of the vorticity".

3.15.1 Application of the velocity representation to vortex tubes

There are many situations in which the only vorticity present is confined to tubes of small cross-sectional area. If so, and the tube radius is small compared to $\mathbf{r}$, the variation of $\mathbf{r}/4\pi r^3$ in (3.15.1) over the tube can be neglected and the volume integration performed by first integrating over the cross-sectional area, and then along the length of the tube:

$$\mathbf{u}(\mathbf{x}) = \int_V \boldsymbol{\omega} \times \frac{\mathbf{r}}{4\pi r^3}\, dV' = \int_{\substack{tube\\length}} \left(\int_A \boldsymbol{\omega}\, dA \right) \times \frac{\mathbf{r}}{4\pi r^3} d\ell. \tag{3.15.5}$$

The integral of $\boldsymbol{\omega}$ over the cross-sectional area of the tube is constant along its length and equal numerically to the circulation around the tube:

$$\int_A \boldsymbol{\omega}\, dA = \Gamma \mathbf{m},$$

where $\mathbf{m}$ is a unit vector along the tube. Equation (3.15.5) then becomes

$$\mathbf{u}(\mathbf{x}) = \Gamma \int d\boldsymbol{\ell}(\mathbf{x}') \times \frac{\mathbf{r}}{4\pi r^3}. \tag{3.15.6}$$

As shown in Section 3.2 using Stokes's Theorem and symmetry, the velocity field outside a straight vortex tube is in the θ-direction and is inversely proportional to r. This result also comes directly from (3.15.6).

In this case, $\mathbf{x} = r\mathbf{e}_r + x\mathbf{e}_x$ and the expression for $\mathbf{u}$ given in (3.15.6) becomes

$$\begin{aligned}\mathbf{u} &= \Gamma \int_{x'=-\infty}^{\infty} \frac{dx'\mathbf{e}_x \times (r\mathbf{e}_r + x\mathbf{e}_x - x'\mathbf{e}_x)}{4\pi (r^2 + (x - x')^2)^{3/2}} \\ &= \frac{\Gamma r \mathbf{e}_\theta}{4\pi} \int_{x'=-\infty}^{\infty} \frac{dx'}{[r^2 + (x - x')^2]^{3/2}} \\ &= \frac{\Gamma r \mathbf{e}_\theta}{4\pi} \left[\frac{x' - x}{r^2 (r^2 + (x' - x)^2)^{1/2}} \right]_{-\infty}^{\infty} = \frac{\Gamma \mathbf{e}_\theta}{2\pi r},\end{aligned} \tag{3.15.7}$$

in agreement with that in the earlier section.

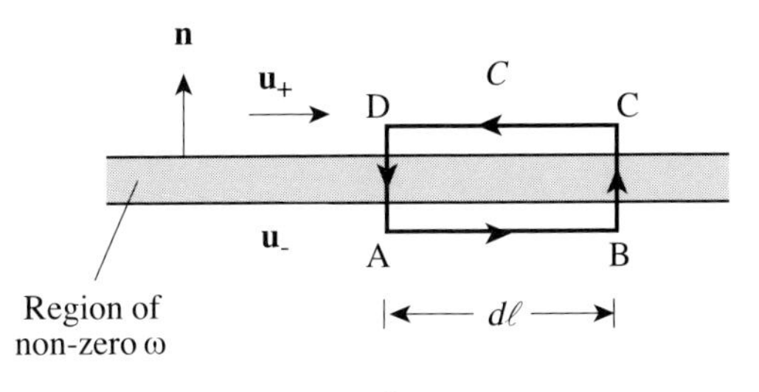

Figure 3.39: Vortex layer and curve C used for deriving vortex sheet jump conditions.

3.15.2 Application to two-dimensional flow

For two-dimensional flow, the Representation Theorem and the general ideas about the relationship between the velocity and vorticity field can be simplified. Two of the three components of the vorticity vanish identically, with the remaining non-zero component being perpendicular to the plane in which motion takes place. All boundaries and vortex lines are independent of the coordinate perpendicular to the plane of the motion and the volume and surface integrals in the Representation Theorem of (3.15.3) can be integrated in this direction to give surface integrals over the region occupied by fluid and line integrals around boundaries. The result is

$$\mathbf{u}(\mathbf{x}) = \int_A \frac{\boldsymbol{\omega}(x') \times \mathbf{r}}{2\pi r^2} dA' + \int_A \frac{[\nabla' \cdot \mathbf{u}(x')]\mathbf{r}}{2\pi r^2} dA' + \oint_C \frac{[\mathbf{u}(x') \times \mathbf{n}] \times \mathbf{r}}{2\pi r^2} d\ell' + \oint_C \frac{[\mathbf{u}(x') \cdot \mathbf{n}]\,\mathbf{r}}{2\pi r^2} d\ell'. \tag{3.15.8}$$

3.15.3 Surface distributions of vorticity

To understand the surface integral in (3.15.4) we apply it to describing the flow associated with thin sheets of vorticity, for example boundary layers on solid surfaces. Consider the curve, C, shown in Figure 3.39 which passes either side of such a thin vortex layer. The application of Stokes's Theorem to this curve and to the surface A it encloses gives

$$\oint_C \mathbf{u} \cdot d\boldsymbol{\ell} = \oint_A \boldsymbol{\omega} \cdot \mathbf{m}\, dA, \tag{3.15.9}$$

where $\mathbf{m}$ is a unit vector out of the page. The contributions to the line integral from the portions of the curve which cross the sheet, BC and DA, can be made vanishingly small by letting the lengths BC and DA tend to zero. On AB and CD, $d\boldsymbol{\ell}$ can be written $|d\boldsymbol{\ell}|\mathbf{n} \times \mathbf{m}$ and $-\,|d\boldsymbol{\ell}|\mathbf{n} \times \mathbf{m}$ respectively. The integrand on the left-hand side of (3.15.9) can be written as

$$\begin{aligned}\mathbf{u} \cdot d\boldsymbol{\ell} &= \mathbf{u}_+ \cdot \mathbf{n} \times \mathbf{m}\, d\ell - \mathbf{u}_- \cdot \mathbf{n} \times \mathbf{m}\, d\ell \\ &= \mathbf{m} \cdot [\mathbf{u}] \times \mathbf{n}\, d\ell,\end{aligned} \tag{3.15.10}$$

with $[\mathbf{u}]$ denoting the change in $\mathbf{u}$ across the sheet.

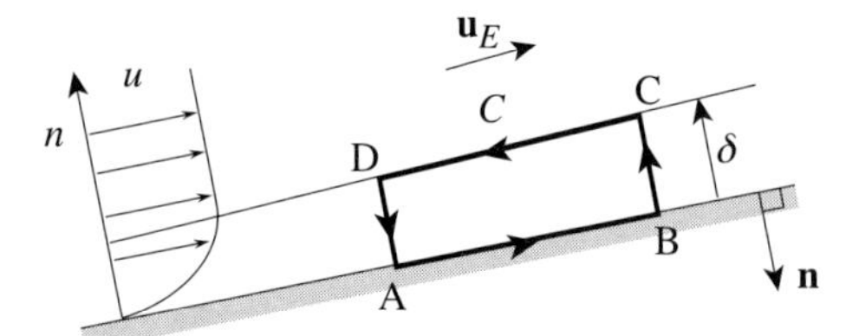

Figure 3.40: Boundary layer and curve C used to derive $\int_0^\delta \omega dn = \mathbf{u}_E \times \mathbf{n}$.

Taking the limit, AB = CD= $d\ell$ and BC and DA tending to zero, the right-hand side of (3.15.9) becomes

$$\int_A \boldsymbol{\omega} \cdot \mathbf{m}\, dA = \int \boldsymbol{\gamma} \cdot \mathbf{m}\, d\ell, \tag{3.15.11}$$

where $\boldsymbol{\gamma}$ is defined as the strength of the sheet and the integral on the right is carried out along the surface.

Equating the expressions in (3.15.10) and (3.15.11) implies that $\mathbf{m} \cdot \boldsymbol{\gamma} = \mathbf{m} \cdot [\mathbf{u}] \times \mathbf{n}$. Because the curve C can be reoriented such that $\mathbf{m}$ takes an arbitrary direction within the sheet, the difference in velocity across the sheet must satisfy

$$\boldsymbol{\gamma} = [\mathbf{u}] \times \mathbf{n}. \tag{3.15.12}$$

Further, the difference in velocity must in fact be a difference in the tangential components only, because a jump in the normal component is not consistent with satisfying continuity. This result can be put in context for viscous flow by considering a boundary layer on a surface. If we take a curve similar to that shown in Figure 3.40, but now with segment AB lying on the surface of the body and segment CD just outside the boundary layer, to the level of approximation used in boundary layer theory, (3.15.12) becomes

$$\boldsymbol{\gamma} = \int_0^\delta \boldsymbol{\omega}\, dn = \mathbf{u}_E \times \mathbf{n}, \tag{3.15.13}$$

where $\mathbf{u}_E$ is the free-stream velocity. The integral of the vorticity in the boundary layer, per unit length, has the value of the free-stream velocity and is the vorticity needed to bring the flow to rest at the surface. This is the strength of the surface vortex sheet that would be needed to approximate the boundary layer in an equivalent inviscid flow.

In summary, for a viscous fluid vortex lines cannot end in the fluid or at non-rotating boundaries, but must turn tangentially to the surface as the boundary is approached. In an inviscid fluid, if we imagine that the velocity field is extended (as zero) into the interior of the solid boundary, the vortex lines turn into the surface and are viewed as part of the equivalent surface vorticity distribution.

3.15.4 Some specific velocity fields associated with vortex structures

The velocity–vorticity relation enables considerable insight into fluid motions, particularly the overall structures of flows with concentrated vorticity. An illustration seen in Section 3.4 is the horseshoe

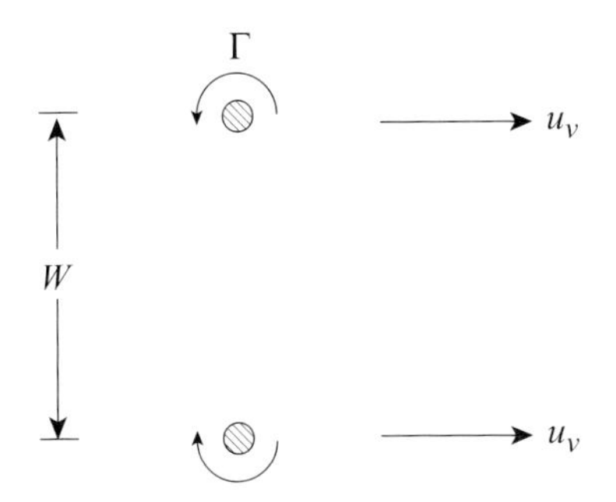

Figure 3.41: Vortex pair with circulation Γ and spacing W; the velocity of the vortex pair is equal to the velocity u_v.

vortex, but there are other generic structures whose velocity field can be readily inferred from the vorticity distribution.

One example is the motion of a two-dimensional vortex pair, with vortices of equal strength, Γ, and opposite sign, such as occurs in the starting flow through a slot or past a symmetric bluff body. The configuration to be analyzed is shown in Figure 3.41. The fluid is taken as inviscid and of uniform density. The velocity field associated with the presence of the upper vortex, evaluated at the location of the lower vortex, and the corresponding velocity associated with the lower vortex, evaluated at the location of the upper vortex, are indicated. The two velocities are equal so that the two vortices move on parallel trajectories, and with uniform velocity, to the right. The speed of the motion can be found from application of the expression for the velocity field associated with a straight vortex. If the magnitude of the circulation around each vortex is Γ and the spacing between them is W, the velocity of the pair is (refer to Figure 3.41)

$$u_v = u_{vortex\ pair} = \frac{\Gamma}{2\pi W}. \tag{3.15.14}$$

Equation (3.15.14) is strictly applicable only to the behavior of two line vortices (radius $= 0$), since elements of vorticity in each vortex tube of finite radius a have different contributions to the motion of the other vortex tube. However, the velocity field is found to deviate from (3.15.14) to order (a/W) and, if the vortex tube radius is much less than the distance apart, this expression will provide a good description.

An extension of this application is to the motion of two vortices of the same sign. In this situation, the tendency will be for the vortices to spiral around their vortex "center of gravity", which is determined by the strength of the two elements. For two elements of equal strength and distance W apart, their motion will be circular around the midpoint of the line between them with angular velocity $\Gamma/(\pi W^2)$.

A second example is the motion of a vortex ring, such as that formed in the starting flow out of a tube or through an orifice, as well as in the coherent vortex structures in the shear layers that surround an axisymmetric jet. Consider the sense of the velocity of a given element of the ring in a direction parallel to the ring axis of symmetry. At any location on the ring all the vorticity elements in the remainder of the ring are associated with an induced velocity along this axis. (The velocity–vorticity relationship is linear so that the contributions of different vortex elements are additive.) If Figure 3.41 is taken as a cut through the ring, a ring which has vorticity with the sense of that shown would move to the right. The distinct structure associated with vortex ring motion has been described strikingly

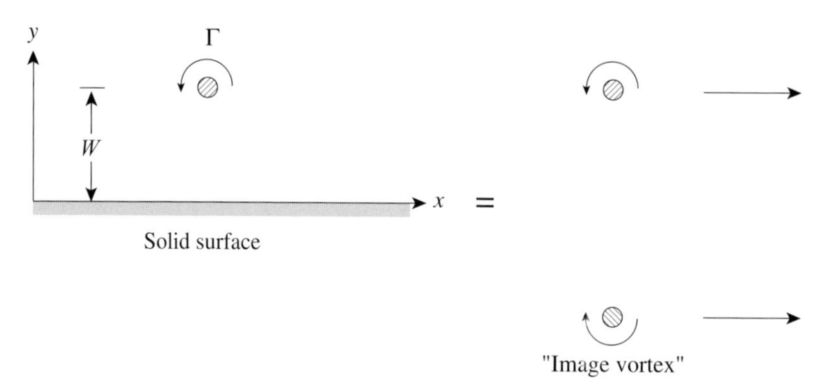

Figure 3.42: Kinematic equality between the vortex and the infinite plane surface and a vortex pair (original vortex plus image vortex).

by Lighthill (1963) as the reason why one can blow out a candle (through creation of a vortex ring, and of a consequent large fluid velocity at the candle, when one blows through one's lips) but cannot suck one out (inhaling creates a sink, which ingests fluid from all directions so that the velocities at the candle will be much lower; see Section 2.10).

Some further comments can also be made concerning vortex rings. For the same vortex tube thickness and circulation, the larger the ring diameter the lower the ring-induced velocity. Consider two rings having the same axis of symmetry, which start out with the same diameter. The induced velocity field of the two rings is such that the rear one will shrink in diameter and the forward one increase. The rear ring can thus catch up and move through the initially forward ring, with the two rings interchanging roles and the process then starting again. References to the experimental demonstration of this so-called "leapfrogging" process are given by Saffman (1992).

Two final examples are provided by the behavior of a vortex, or vortices near a surface. First, consider a two-dimensional vortex at a given distance, say $W/2$, from an infinite solid surface. If an inviscid description is appropriate for the situation of interest, the necessary boundary condition is the purely kinematic one of zero flow normal to the wall. This can be achieved if we imagine the wall removed and a fictitious *image vortex* placed an equivalent distance below the surface, as indicated schematically in Figure 3.42. The velocity field is that associated with the original vortex plus that associated with the image, and on the symmetry plane there is no normal velocity. The velocity field at values of y greater than 0 in Figure 3.42 is therefore kinematically the same as that for the vortex and the infinite wall. As inferred from (3.15.14) a vortex of strength Γ a distance W from a plane surface moves parallel to the surface with a velocity equal to $\Gamma/4\pi W$.

These considerations can also be used to explain the behavior of vortex pairs or rings approaching a plane surface, as shown in Figure 3.43, which gives the actual configuration and the kinematically similar image representation. The discussion above implies that the motion of the vortex pair (or ring) will be towards the surface. To obtain the symmetry condition of no normal velocity at the surface, an image vortex pair is needed. As the vortex pair (or ring) approaches the wall, the velocity field associated with the image vortex pair leads to trajectories of the type shown in Figure 3.43. The vortices which originally make up the vortex pair move in opposite directions along the wall with

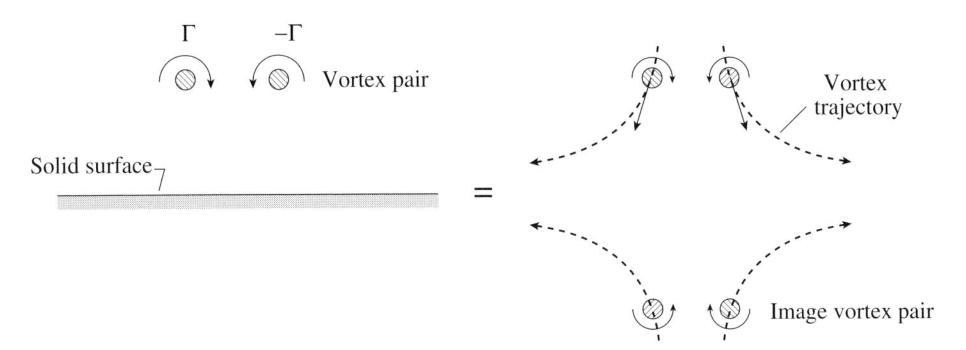

Figure 3.43: Kinematic equality between a vortex pair approaching an infinite plane surface and a vortex pair plus image pair. The trajectory of vortices is shown as a dashed line.

the magnitude of their asymptotic velocity equal to the far upstream velocity with which the pair originally approached the wall.

3.15.5 Numerical methods based on the distribution of vorticity

A number of numerical methods have been developed which are based on the representation of the velocity field in terms of vorticity and/or source distributions. These include the large class of numerical calculation procedures for inviscid flow referred to as "panel methods", which make use of distributions of surface singularities (either vortex elements or, equivalently, distributions of dipoles) which are discretized on surface panels. Panel methods have been effective in describing flows over complex geometries, such as aircraft. The overall procedure is to solve for the distribution of discrete vortex elements which produce, for an inviscid flow, the desired normal velocity (generally zero) at a point on each panel. For a two-dimensional geometry, if these methods can be applied, the problem is reduced to the one-dimensional problem of specifying the elements around a curve. Similarly for a three-dimensional geometry the problem becomes a two-dimensional one involving, for incompressible flow, only values of the elements on one or more surfaces. The gridding and computational requirements are thus generally much less than for methods in which the entire domain must be analyzed. Surface vorticity and panel methods are described in detail by Kerwin *et al.* (1987), Lewis (1991), and Katz and Plotkin (2001).

Vortex methods have also been used to examine unsteady flows, in which one must account for the effect of vorticity shed into the region downstream of the body, so that the location of the wake vorticity and its strength can be found. This is typically done by tracking the shed vortex elements and thus, in addition to the kinematic statements, there must be a description of the motion of these elements once they leave the body. An advantage, however, is that the computation need only deal with the sections of the flow in which there is appreciable vorticity, such as on the surface of a body or in a wake (Sarpkaya, 1988, 1994; Leonard, 1985).

An example of a vortex method computation is given in Figures 3.44–3.46, which show the unsteady exit flow from a tube. In this situation, the cylindrical vortex sheet, which leaves at the exit of the tube, rolls up to form a vortex ring. In the computation, elements of vorticity are released

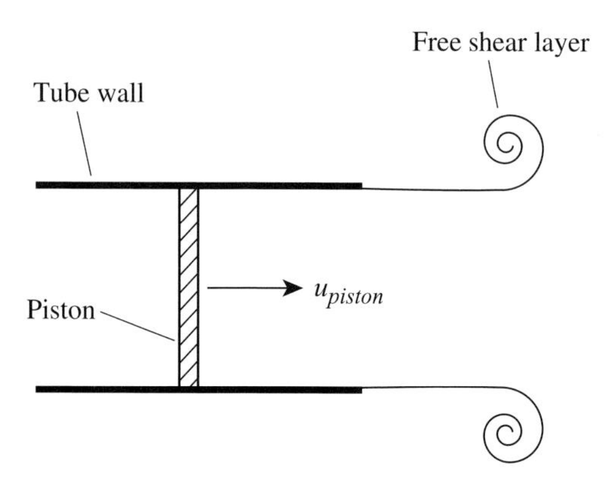

Figure 3.44: Schematic diagram of the experiment showing piston, tube wall, and free shear layer (Nitsche and Krasny, 1994).

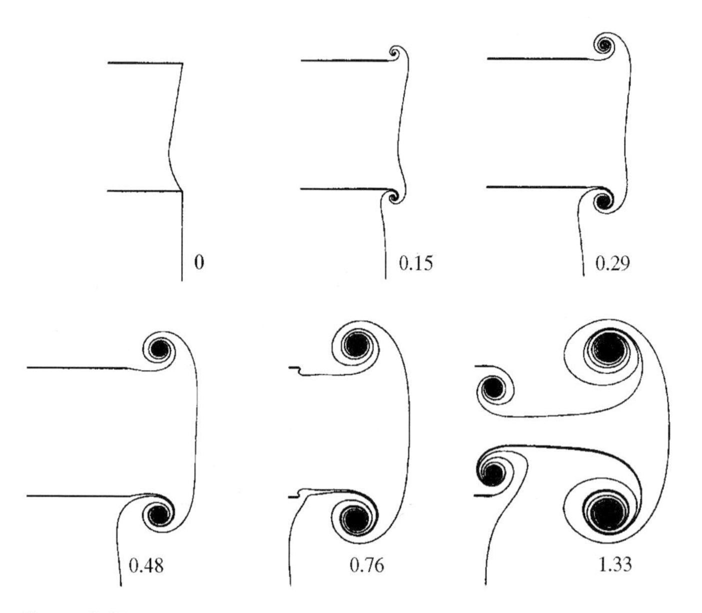

Figure 3.45: Vortex method computation showing vortex-ring formation; numbers refer to non-dimensional times, $\tilde{t} = t \times$ piston/tube length (Nitsche and Krasny, 1994).

at the end of the tube and are convected by the resulting flow. Kelvin's Theorem implies that the position of the vortices, which is known at any time, can be updated by tracking the fluid particles to which the vortex lines are locked. The kinematic vorticity–velocity relationship in (3.15.1) can then be used at any time step to find the velocity, which is then used for the next convection step.

Figure 3.44 shows the basic experimental configuration. In Figure 3.45 computations of a marked line of particles are shown at several different times, depicting the different stages of the roll up process in some detail. Figure 3.46 shows the corresponding experimental flow visualization. The vortex method captures the features of the experiment well, although it is to be noted that there are a number of computational subtleties which need to be taken into account and which we have by

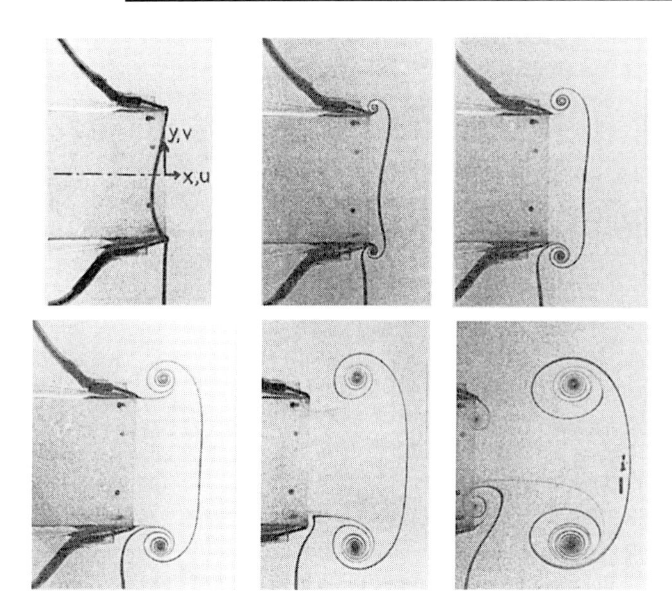

Figure 3.46: Flow visualization showing vortex-ring formation; times correspond to Figure 3.45 (Didden, 1979, as given by Nitsche and Krasny, 1994).

no means addressed. These methods are, again, most effective when the vorticity is concentrated on thin sheets or surfaces.

Vortex method computations have also been used in flows where the location at which vorticity leaves the body surface is not known *a priori*. In this situation, there needs to be some description, such as a boundary layer computation (see Chapter 4), of the processes that set the separation point. With this proviso, however, vortex methods have been applied to bluff body flows and also to the stalled flow around airfoils. For a description of these applications see Lewis (1991). In summary, a number of methods exist for computing flows based on the velocity–vorticity relationship given in Section 3.15, many of which have application to the geometries of interest for internal flows.

4 Boundary layers and free shear layers

4.1 Introduction

In this chapter, we discuss the types of thin shear layers that occur in flows in which the Reynolds number is large. The first of these is the boundary layer, or region near a solid boundary where viscous effects have reduced the velocity below the free-stream value. The reduced velocity in the boundary layer implies, as mentioned in Chapter 2, a decrease in the capacity of a channel or duct to carry flow and one effect of the boundary layer is that it acts as a blockage in the channel. Calculation of the magnitude of this blockage and the influence on the flow external to the boundary layer is one issue addressed in this chapter. Boundary layer flows are also associated with a dissipation of mechanical energy which manifests itself as a loss or inefficiency of the fluid process. Estimation of these losses is a focus of Chapter 5. The role of boundary layer blockage and loss in fluid machinery performance is critical; for a compressor or pump, for example, blockage is directly related to pressure rise capability and boundary layer losses are a determinant of peak efficiency that can be obtained.

Another type of shear layer is the *free shear layer* or *mixing layer*, which forms the transition region between two streams of differing velocity. Examples are jet or nozzle exhausts, mixing ducts in a jet engine, sudden expansions, and ejectors. In such applications the streams are often parallel so the static pressure can be regarded as uniform, but the velocity varies in the direction normal to the stream. For mixing layers a central problem is to assess the rate at which the two streams transfer momentum and energy, because this can affect how downstream components are designed to achieve the desired performance. Wakes and jets are another type of free shear layer where it is of interest to determine how rapidly mixing occurs, and, in the case of the wake, what the effect of the blockage on the free-stream flow is.

Boundary layers and free shear layers are subjects in which there has been an enormous amount of research. The objectives of this chapter are to give an introduction to these aspects of particular interest in internal flows, to provide tools for estimating the principal effects in engineering situations, and to guide further exploration into the extensive literature in this area.

Several main ideas thread through the chapter. First, as mentioned in Section 2.9, a high Reynolds number flow can be conceptually and usefully partitioned into regions in which viscous effects are important and regions in which they can be neglected and the flow behaves as if it were inviscid. Second, the regions in which viscous effects must be addressed are thin, in the sense that the characteristic length scale for velocity variations in a direction normal to the stream is much less than in the streamwise direction. Third, this difference in scale allows the development of a reduced form of the Navier–Stokes equations, referred to as boundary layer or thin shear layer equations,

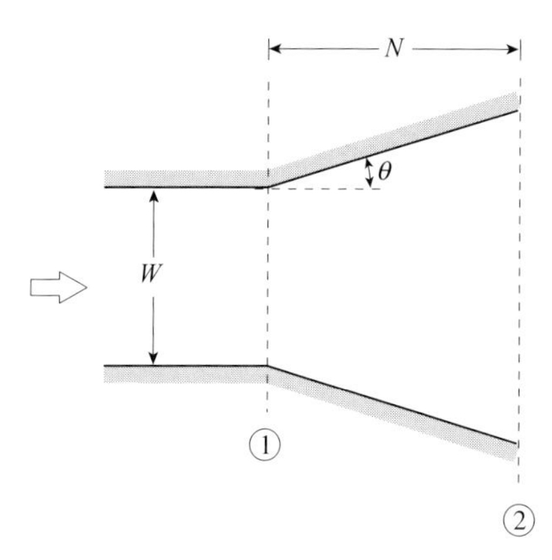

Figure 4.1: Nomenclature for a two-dimensional straight channel diffuser; area ratio, $AR = W_2/W_1$.

which describe the flow in these shear layers very well and are much simpler to solve. Finally, the effect of the viscous regions on the inviscid-like flow outside these regions can be captured through coupling the former, through the behavior of a small number of overall, or integral, boundary layer parameters, with the latter. This coupling allows a consistent description of both regions and hence of the flow as a whole.

In Section 4.1.1 we use the performance of a basic internal flow device, the diffuser, to illustrate one role of boundary layer behavior and its linkage with the flow outside the viscous region. The boundary layer form of the equations of motion is then developed, first for laminar flow and then for turbulent flow (which is the more common occurrence in fluid machinery applications), along with descriptions of solution procedures and the circumstances in which "transition" occurs from the laminar to the turbulent state. Definitions of the relevant integral quantities used to couple the boundary layer behavior to the flow outside the boundary layer are also given. These concepts are then used together to examine diffuser behavior in more depth as a vehicle for the discussion of interactions between the boundary layer and the inviscid-like region. The last several sections describe free shear layers including rates of mixing and behavior in pressure gradients.

4.1.1 Boundary layer behavior and device performance

The role that boundary layers play in determining fluid component performance can be made more definite by examining the behavior of a two-dimensional straight channel diffuser. This simple geometry incorporates many of the issues addressed in Chapter 4 and the description of its behavior illustrates the aspects of shear layers which typically need to be captured by predictive techniques. Diffusers are used as the central application of the chapter to focus the discussion on specific items of interest in the context of fluid machinery.

A two-dimensional straight diffuser is shown in Figure 4.1. The functions of a diffuser are to change a major fraction of the kinetic energy of the entering flow into static pressure and to decrease the velocity magnitude. From Figure 4.1 the diffuser area ratio, AR is W_2/W_1, the non-dimensional length is N/W_1, and the diffuser opening angle θ is given by $\tan\theta = (AR - 1)/[2(N/W_1)]$. For an

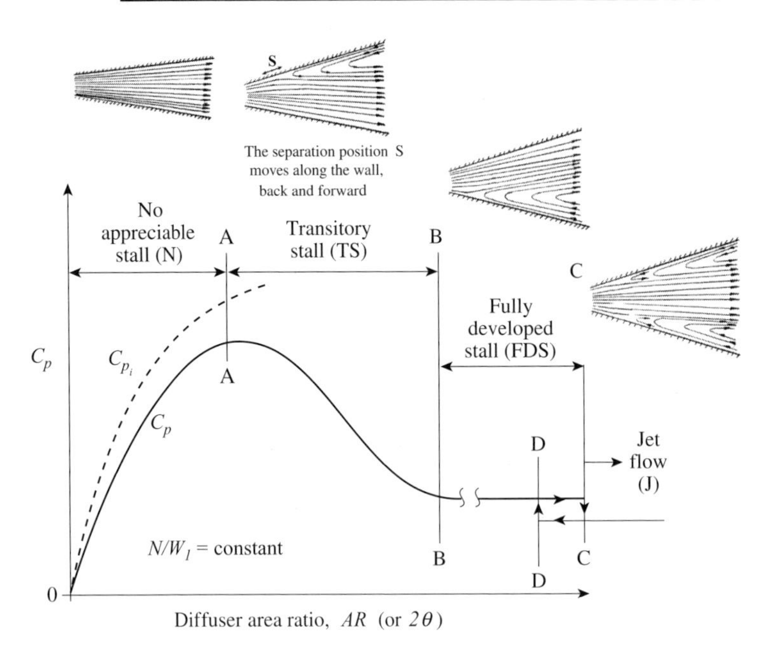

Figure 4.2: Relation of C_p to diffuser flow regimes (after Kline and Johnston (1986)).

ideal flow, from the one-dimensional form of the continuity equation and Bernoulli's equation, the pressure rise coefficient, C_p, is given in terms of area ratio by

$$C_p = \frac{p_2 - p_1}{\frac{1}{2}\rho u_1^2} = 1 - \frac{1}{AR^2}. \tag{4.1.1}$$

Figure 4.2 shows a sketch of measured diffuser pressure rise versus area ratio, AR, for diffusers of high enough aspect ratio to be considered two-dimensional. The pressure rise coefficient for ideal one-dimensional flow is denoted by C_{p_i}. For a range of area ratios the measured C_p generally follows the ideal curve, although at a lower value, but it peaks and then decreases for larger area ratios while the ideal curve monotonically increases. The labels in the figure describe flow regimes encountered as the area ratio is increased. Only for area ratios below the line AA can the flow be said to follow the geometry in that the streamlines diverge and the velocity drops, in qualitative accord with the ideal one-dimensional picture. At area ratios above AA, the streamline pattern does not reflect the divergence of the boundaries and the flow does not look even qualitatively like the ideal case. As the area ratio increases still further the pressure rise coefficient decreases.

Sketches of streamlines in the different regimes (no appreciable stall, transitory stall, fully developed stall, and jet flow) taken from measurement and flow visualization, are also included in Figure 4.2. In the region of "no appreciable stall", the boundary layers are thin and the effective area of the channel and the geometrical area both grow similarly. "Transitory stall" defines a regime in which there are large amplitude fluctuations, with a repeated build up and wash out of regions of reversed velocity along the walls of the diffuser. In "fully developed stall" there is a region of back flow (generally on one wall) and a free shear layer penetrates substantially into the interior of the channel. In the "jet flow" regime both boundary layers have separated from the wall, there is a large region of

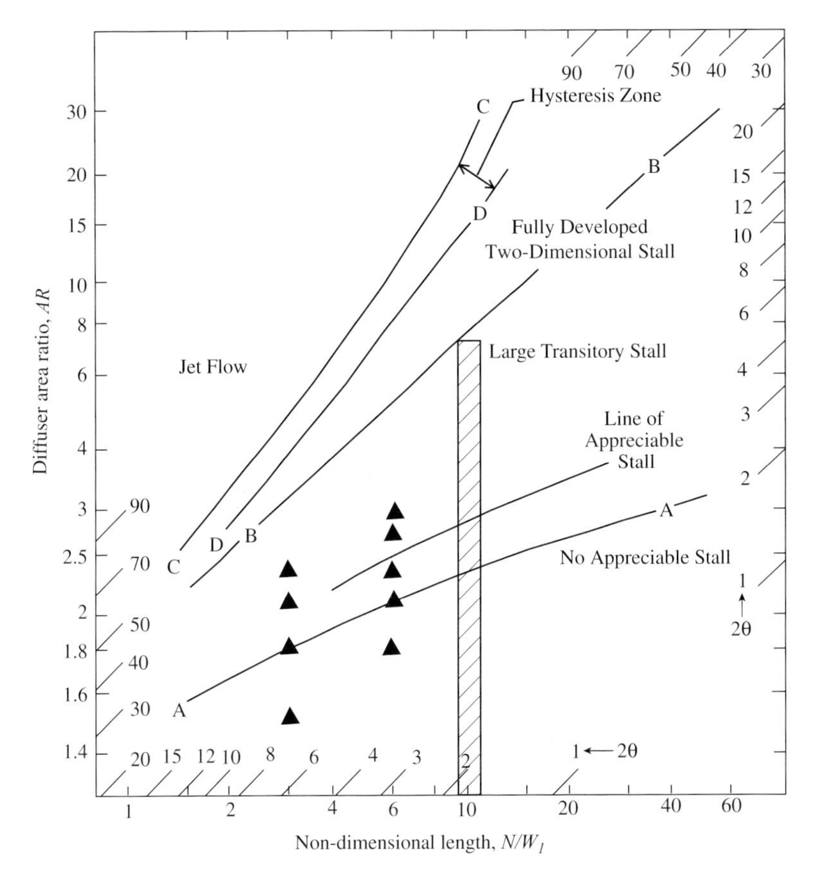

Figure 4.3: Two-dimensional diffuser flow regime as established by Reneau, Johnston, and Kline (1967). Solid symbols and shaded area are geometries whose performance is described in Section 4.7.

back flow on each side, and the effective area for the core flow is not much larger than the diffuser inlet area.

Figure 4.3 shows a measured diffuser flow regime map expressed in terms of area ratio, AR, and aspect ratio, N/W_1, with included angles referenced. For a given area ratio, changing the non-dimensional length moves the operation through different flow regimes. For example, changing the length from 3 to 10, at an area ratio of 2.5, results in moving from a stalled to an unstalled regime and, although not shown in the figure, an increase in pressure rise.

Viewing this overall behavior in terms of a boundary layer parameter, the displacement thickness (defined in Section 2.9 and interpreted there as a flow blockage) provides a perspective on those items we wish to evaluate. The relation of the displacement thickness to the effective flow area for the diffuser is shown in Figure 4.4. For equal boundary layer displacement thicknesses on the two walls, the effective channel height for the inviscid-like core flow is $W - 2\delta^*$.

To illustrate the way in which the displacement thickness affects the pressure rise as the diffuser area ratio changes we substitute the effective area ratio into (4.1.1) and differentiate the pressure rise coefficient with respect to geometric area ratio, AR. The behavior of interest is associated with displacement thickness growth at station 2. As such we assume the displacement thickness at the

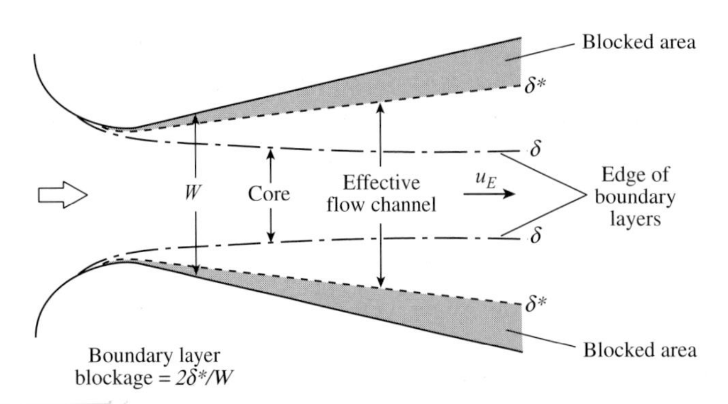

Figure 4.4: View of displacement thickness as a boundary layer blockage (Kline and Johnston, 1986).

inlet (station 1) is small enough so it, and its changes, can be neglected and the inlet area taken as the geometric area. Under these conditions the rate of change of the diffuser pressure rise coefficient is

$$\frac{dC_p}{d(AR)} = \frac{2(1 - C_p)}{AR}\left\{1 + \frac{[d(1 - 2\delta_2^*/W_2)]/(1 - 2\delta_2^*/W_2)}{d(AR)/AR}\right\}. \tag{4.1.2}$$

The quantity $W_2 - 2\delta_2^*$ represents the effective width of the channel at the exit and the term $1 - 2\delta_2^*/W_2$ in (4.1.2) is therefore the fractional effective width. Equation (4.1.2) indicates that the rate of change of the pressure rise coefficient with the geometric area ratio can be positive or negative depending on the rate of variation of this effective fraction, and hence of the exit blockage ($2\delta_2^*/W_2$).

The variation in the diffuser flow regime versus length in Figure 4.3 shows a different feature of the phenomena of interest, the rate dependence of the relevant processes. There is a competition between pressure forces, which decelerate the slow moving wall layers more than the free-stream fluid, and mixing processes which can transfer momentum to the lowest velocity parts of the boundary layer and inhibit separation. The effect of the latter depends on the length over which they are able to act.

This chapter will provide tools for estimating, and understanding, the manner in which the geometry of internal flow devices affects displacement thickness and hence pressure change and mass flow capacity. Another important issue is the viscous loss associated with dissipation of mechanical energy in the boundary layers. As discussed in Section 4.3, there is a different boundary layer thickness parameter which reflects this loss and which the methods described will enable us to find.

4.2 The boundary layer equations for plane and curved surfaces

4.2.1 Plane surfaces

As described in Section 4.1, the central approximation of boundary layer theory is that rates of change at high Reynolds number of the velocity components and their derivatives, or the temperature and its derivatives, in the direction normal to the bounding surface are much larger than the corresponding rates of change along the surface, allowing simplification of the expressions for viscous forces and heat transfer rates. The equations that describe the behavior of boundary layers were introduced in

Section 2.9. We now examine them in more depth to enable their use in a wider range of situations. For a compressible fluid, there are not only velocity boundary layers, but also thermal boundary layers, in which the temperature changes from that of the boundary to that of the free-stream outside. For values of the Prandtl number ($\mu c_p/k$) of order unity, the thicknesses of the viscous and thermal boundary layers are comparable.

For the purposes here a two-dimensional treatment of the steady-flow situation with no body forces is sufficient; extensions to three dimensions and the inclusion of body forces can be found in the texts by White (1991), Cebeci and Bradshaw (1977), and Schlichting (1979) and discussion of aspects due to flow unsteadiness are given in Chapter 6. The boundary layer approximation implies that δ_u and δ_T, the thicknesses of the velocity and temperature boundary layers, and thus the characteristic scales in the direction *normal* to the main flow, are small compared with the length scale *along* the channel or body. In the viscous boundary layer the velocity increases from zero at the wall to the free-stream value u_E, and in the thermal boundary layer the temperature changes from the value T_w at the wall to the value T_E in the free-stream.

We begin by examining the momentum equation for two-dimensional steady flow (1.9.10) in component form, where the coordinate normal to the surface is y and that along the surface is x.

$$\rho\left(u_x\frac{\partial u_x}{\partial x}+u_y\frac{\partial u_x}{\partial y}\right)=-\frac{\partial p}{\partial x}+\left(\frac{\partial \tau_{xx}}{\partial x}+\frac{\partial \tau_{xy}}{\partial y}\right), \tag{4.2.1a}$$

$$\rho\left(u_x\frac{\partial u_y}{\partial x}+u_y\frac{\partial u_y}{\partial y}\right)=-\frac{\partial p}{\partial y}+\left(\frac{\partial \tau_{xy}}{\partial x}+\frac{\partial \tau_{yy}}{\partial y}\right). \tag{4.2.1b}$$

The continuity equation (1.9.4) written out is

$$\frac{\partial}{\partial x}(\rho u_x)+\frac{\partial}{\partial y}(\rho u_y)=0. \tag{4.2.2}$$

The basic arguments for reducing (4.2.1) to boundary layer form are as follows:[1]

(a) From the continuity equation (4.2.2) the velocity components in the layer scale as

$$\frac{\partial u_x}{\partial x}\sim\frac{u_x}{L}\quad \frac{\partial u_y}{\partial y}\sim\frac{u_y}{\delta},$$

where L is a characteristic length scale in the x (streamwise) direction and δ is the boundary layer thickness. Therefore,

$$\frac{u_y}{u_x}\sim\frac{\delta}{L}.$$

(b) From (a) and the constitutive relations between the shear stress and the rate of strain given in Section 1.13, the ratio of viscous forces in (4.2.1a) is

$$\frac{\left(\dfrac{\partial \tau_{xx}}{\partial x}\right)}{\left(\dfrac{\partial \tau_{xy}}{\partial y}\right)}\sim\frac{\mu\left(\dfrac{\partial^2 u_x}{\partial x^2}\right)}{\mu\left(\dfrac{\partial^2 u_x}{\partial y^2}\right)}\sim\left(\frac{\delta}{L}\right)^2,$$

so that $\partial\tau_{xx}/\partial x$ can be neglected in (4.2.1a).

[1] See also Section 2.9.

(c) In the boundary layer, there is a balance between fluid accelerations and viscous forces (and possibly pressure forces) so that the first two quantities are of the same magnitude.

From (a) and (b) the magnitude of the terms on the left-hand side of (4.2.1a) is $\rho U^2/L$, where U is a representative velocity magnitude. Dividing by this quantity to normalize and non-dimensionalize all the terms, the magnitudes of the pressure gradient and the viscous force are unity and $(L^2/\delta^2)(1/Re)$, where Re is the Reynolds number UL/ν. For (c) to be valid, δ/L must scale as $1/\sqrt{Re}$, which is small for the devices of interest; a gas turbine compressor airfoil with a chord of 0.03 m and blade speed 300 m/s has a Reynolds number of 6×10^5.

Using the information on the magnitude of δ/L we can estimate the magnitude of the pressure difference across the boundary layer, Δp_n, from (4.2.1b):

$$\Delta p_n \sim \rho U^2(\delta/L)^2.$$

The estimate shows that the pressure difference across the boundary layer can be neglected and the pressure through the boundary layer taken as equal to the free-stream value, p_E. The momentum equation in the direction along the surface thus becomes

$$\rho\left(u_x\frac{\partial u_x}{\partial x} + u_y\frac{\partial u_x}{\partial y}\right) = -\frac{dp_E}{dx} + \frac{\partial \tau}{\partial y}. \tag{4.2.3}$$

In (4.2.3) p_E is a function of the distance along the surface and we have dropped the subscript on τ_{xy} because this is the only viscous stress that is retained.

Using similar arguments the energy equation (1.10.3) takes the form

$$\rho\left(u_x\frac{\partial c_pT}{\partial x} + u_y\frac{\partial c_pT}{\partial y}\right) = u_x\frac{dp_E}{dx} - \frac{\partial q_y}{\partial y} + \tau\frac{\partial u_x}{\partial y}. \tag{4.2.4}$$

Equations (4.2.2), (4.2.3), and (4.2.4) are known as boundary layer or thin shear layer equations. Comparing the magnitudes of the various terms shows that the ratio of the thermal and viscous boundary layer thicknesses, δ_u and δ_T, is

$$\frac{\delta_u}{\delta_T} \cong \sqrt{\frac{\mu c_p}{k}} = \sqrt{Pr}. \tag{4.2.5}$$

The assumption that the two thicknesses are of the same order is thus equivalent to the assumption that the Prandtl number is of order unity. For air the Prandtl number is roughly 0.7 and varies by approximately 5% over temperatures from 200 to 2000 K, so this assumption is well borne out, as it is for a number of other gases. For liquids the Prandtl number varies over a much larger range, from 10^3 for engine oils at room temperature to 10^{-2}–10^{-3} for liquid metals, and the assumption is not justified. For information concerning these latter situations see, for example, Incropera and De Witt (1996) or Schlichting (1979).

An alternative form of the boundary layer energy equation, in terms of the stagnation enthalpy, can be obtained by multiplying the momentum equation (4.2.3) by u_x and adding it to (4.2.4) or by applying the boundary layer approximations to (1.9.13) for the rate of change of h_t. Carrying out either yields

$$\rho u_x\frac{\partial h_t}{\partial x} + \rho u_y\frac{\partial h_t}{\partial y} = -\frac{dq_y}{dy} + \frac{\partial}{\partial y}(u_x\tau). \tag{4.2.6}$$

The first term on the right-hand side is the heat transfer to a given streamtube and the second is the net work done by shear stresses on the streamtube.

Equations (4.2.2)–(4.2.4), or (4.2.6), describe the velocity and temperature field within the boundary layer only. As such, the boundary conditions differ from those for the Navier–Stokes equations. Conditions at the surface are the same as those given in Chapter 1, namely that for an impermeable surface both components of the velocity are zero and either the wall temperature or the heat flux (or some combination) is specified. At the outer edge of the boundary layer, however, what is required is that the boundary layer velocity and temperature match the distribution ($u = u_E$, $T = T_E$) in the flow outside the boundary layer. Because of the smooth transition, defining the "edge" or thickness of the boundary layer, δ, is somewhat arbitrary, although one convention is to locate it[2] at $u/u_E = 0.99$.

4.2.2 Extension to curved surfaces

The equations for two-dimensional boundary layers on surfaces with radius of curvature, r_c, can be inferred from (1.14.9) for flow in cylindrical coordinates. In particular, for situations in which δ/r_c is small, the normal or radial momentum gradient becomes (neglecting terms of first order or higher in δ/r_c)

$$\frac{\partial p}{\partial r} = \frac{\rho u^2}{r_c}, \tag{4.2.7}$$

with the pressure gradient normal to the wall balancing the centrifugal force. There is a pressure difference across the boundary layer, Δp_n of order $\rho u^2(\delta/r_c)$ which can be neglected if $\delta/r_c \ll 1$. (Since we take $\delta_u/\delta_T \sim 0(1)$ the subscript on δ has been dropped.) For flow along curved surfaces, examination of the different terms in (1.14.9) shows that to order δ/r_c the form of the momentum and energy equation is unmodified from that for a plane surface so that, to this order, the boundary layer equations remain the same as for a plane surface.

4.3 Boundary layer integral quantities and the equations that describe them

4.3.1 Boundary layer integral thicknesses

Three definitions of boundary layer thickness based on integral properties have found useful application in describing the overall effect of the layer on the external flow. The first of these is the *displacement thickness*, δ^*, defined as

$$\delta^* = \int_0^{y_E} \left(1 - \frac{\rho u_x}{\rho_E u_E}\right) dy. \tag{4.3.1}$$

The incompressible form of this quantity was introduced in Section 2.9. The integration is taken to a value of y slightly larger than the "edge" of the boundary layer; the precise value does not matter because the contribution to the integral is essentially zero outside $y = \delta$.

[2] For a constant pressure laminar boundary layer the value of $[(\delta/x)\sqrt{Re}]$ based on this is roughly 5.

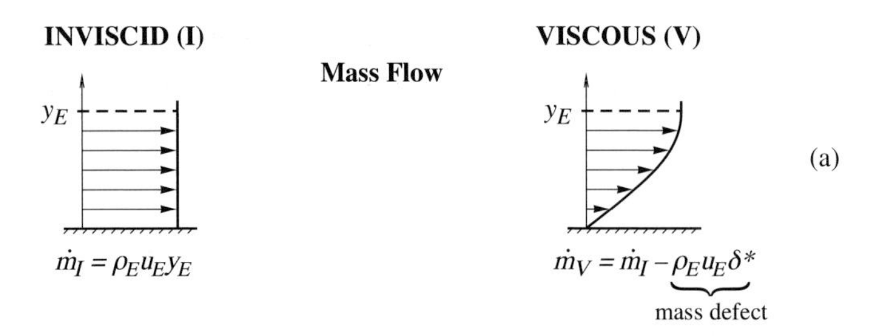

Momentum Flow (force, F): Comparison of F_I and F_V done with same $\dot{m}$

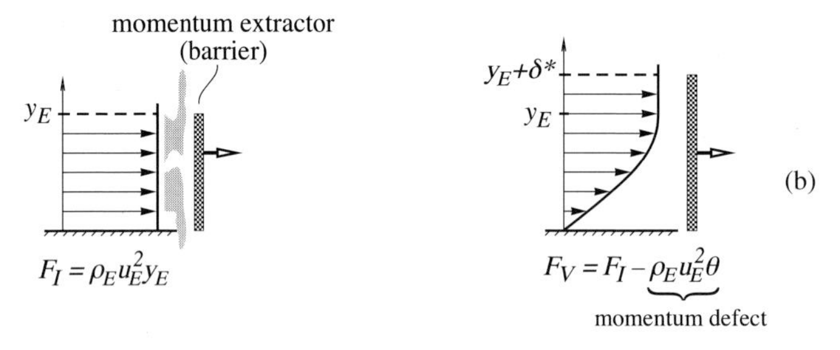

Kinetic Energy Flow (power, P): Comparison of P_I and P_V done with same $\dot{m}$

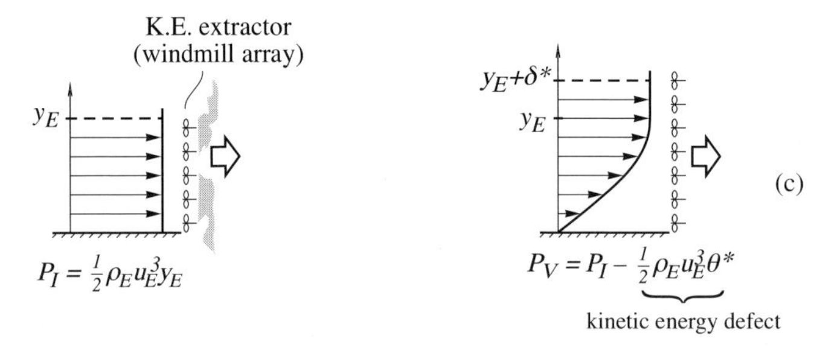

Figure 4.5: Interpretation of boundary layer integral thicknesses (Drela, 2000; see also Drela, 1998).

A physical interpretation of the displacement thickness is given by considering the mass flow rate that would occur in an inviscid fluid which has velocity u_E and density ρ_E, and comparing this to the actual, viscous, situation. This is shown schematically in Figure 4.5(a), where $\rho_E u_E \delta^*$ is the defect in mass flow due to the flow retardation in the boundary layer. The effect on the flow outside the boundary layer is therefore equivalent to displacing the surface outwards, in the normal direction, a distance δ^*. For a two-dimensional channel aligned in the x-direction, with boundary layers on upper and lower surfaces, the mass flow is

$$\dot{m} = \int_{\substack{lower\\surface}}^{\substack{upper\\surface}} \rho u_x dy = \rho_E u_E [W - (\delta^*_{lower} + \delta^*_{upper})].$$

For a given $\rho_E u_E$, the effective width of a two-dimensional channel is thus reduced by the sum of δ^*_{upper} and δ^*_{lower}.

For incompressible flow, the definition of displacement thickness can also be given an interpretation in terms of the total vorticity in the boundary layer (Lighthill, 1958). The displacement thickness in an incompressible flow is

$$\delta^* = \int_0^{y_E} \left(1 - \frac{u_x}{u_E}\right) dy. \tag{4.3.2}$$

The expression for the average distance at which the boundary layer vorticity resides is

$$\text{average distance of vorticity} = \frac{1}{u_E} \int_0^{y_E} y \frac{\partial u_x}{\partial y}\, dy, \tag{4.3.3}$$

where the small term $\partial u_y/\partial x$ has been neglected consistent with the boundary layer approximation. Integrating (4.3.3) by parts,

$$\text{average distance} = \int_0^{y_E} \left(1 - \frac{u_x}{u_E}\right) dy = \delta^*.$$

In this view the displacement thickness is the distance from the wall at which a vortex sheet, having local circulation per unit length equal to that of the boundary layer, would be located. Within the layer of thickness δ^* there is zero flow, consistent with the displacement thickness representing an equivalent blockage next to the boundary.

The *momentum thickness*, θ, is defined as

$$\theta = \int_0^{y_E} \left(1 - \frac{u_x}{u_E}\right) \frac{\rho u_x}{\rho_E u_E}\, dy. \tag{4.3.4}$$

Referring to Figure 4.5(b) the quantity $\rho_E u_E^2 \theta$ represents the defect in streamwise momentum flux between the actual flow and a uniform flow having the density ρ_E and velocity u_E outside the boundary layer. It can be regarded as being produced by extraction of flow momentum and is thus related to drag.

The third quantity is the *kinetic energy thickness*, θ^*, which measures the defect between the flux of kinetic energy (or mechanical power) in the actual flow and that in a uniform flow with u_E and ρ_E the same as outside the boundary layer. The kinetic energy thickness, portrayed in Figure 4.5(c), is defined as

$$\theta^* = \int_0^{y_E} \left(1 - \frac{u_x^2}{u_E^2}\right) \frac{\rho u_x}{\rho_E u_E}\, dy. \tag{4.3.5}$$

This defect can be regarded as being produced by the extraction of kinetic energy. The power extracted is linked to device losses, and the kinetic energy thickness is a key quantity in characterizing losses in internal flow devices.

In summary, the parameters δ^*, θ, and θ^* provide measures of the defects in mass, momentum, and kinetic energy attributable to the boundary layer. They can be computed for any flow, whether compressible or incompressible, laminar or turbulent. Further, since the transverse direction variations have been integrated out, the thickness parameters are only functions of the primary flow direction.

4.3.2 Integral forms of the boundary layer equations

The integral boundary layer thicknesses "wash out" the details of the flow within the boundary layer, and it is consistent to examine their evolution using a set of equations which have this same level of information. Such an approach is provided by the integral forms of the boundary layer equations. To derive these, we integrate the boundary layer equations in y from the wall to y_E, the edge of the boundary layer. Doing so transforms the partial differential boundary layer equations (in x and y) into ordinary differential equations (in x) for the different thicknesses. The two integral forms derived below are for momentum and kinetic energy thicknesses. There is not a separate equation expressing continuity because this condition enters through its application in the derivation of the integral forms.

To obtain the two-dimensional steady flow integral momentum equation we begin here[3] by multiplying the continuity equation by $(u_E - u_x)$ and adding it to the momentum equation, also making use of the free-stream relation

$$u_E \frac{du_E}{dx} = -\frac{1}{\rho_E}\frac{dp_E}{dx}.$$

Performing these operations yields

$$\frac{\partial}{\partial x}[(u_E - u_x)\rho u_x] + \frac{\partial}{\partial y}[(u_E - u_x)\rho u_y] = -(\rho_E u_E - \rho u_x)\frac{du_E}{dx} - \frac{\partial \tau}{\partial y}. \tag{4.3.6}$$

Integrating (4.3.6) term by term, and invoking the definition of the displacement and momentum thicknesses, we obtain, with τ_w denoting the wall shear stress,

$$\frac{d}{dx}\left(\rho_E u_E^2 \theta\right) + \rho_E u_E \delta^* \frac{du_E}{dx} = \tau_w. \tag{4.3.7}$$

In non-dimensional form, (4.3.7) becomes

$$\frac{d\theta}{dx} + \left(H + 2 - M_E^2\right)\frac{\theta}{u_E}\frac{du_E}{dx} = \frac{C_f}{2}, \tag{4.3.8}$$

where $C_f (= \tau_w/(\frac{1}{2}\rho u_E^2))$ is the skin friction coefficient and $H\ (= \delta^*/\theta)$ is the *boundary layer shape parameter*.

For incompressible flow, (4.3.8) reduces to

$$\frac{d\theta}{dx} + (H + 2)\frac{\theta}{u_E}\frac{du_E}{dx} = \frac{C_f}{2}. \tag{4.3.9}$$

In the above discussion, as well as in the derivations of the integral equations for the kinetic energy deficit and the stagnation enthalpy below, the forms of the wall shear stress, τ_w, and wall heat flux, q_w, have not been explicitly specified. The equations obtained are thus applicable to the time mean quantities in turbulent flow as well as to laminar flow, as described further in Section 4.6.

To obtain the equation for the kinetic energy thickness, we multiply the continuity equation by $(u_x^2 - u_E^2)$ and add it to the product of $2u_x$ multiplied by the momentum equation. After integrating,

[3] The integral momentum equation can also be obtained by setting up the overall momentum balance for an element, dx, of the boundary layer (see Young (1989) and Schlichting (1979)).

the result is (Young, 1989; White, 1991; Schlichting, 1979):

$$\frac{d}{dx}\left(\rho_E u_E^3 \theta^*\right) = -\int_0^{y_E} 2u_x \frac{d\tau}{dy} dy = -2u_x\tau\Big|_0^{y_E} + 2\int_0^{y_E} \tau \frac{\partial u_x}{\partial y} dy. \tag{4.3.10}$$

The term $u_x\tau$ is zero at both $y = y_E$ and $y = 0$, while the term $\int_0^{y_E} \tau(\partial u_x/\partial y)dy$, henceforth denoted by $\dot{D}$, represents the rate of dissipation of mechanical energy in the boundary layer, per unit surface area.

The non-dimensional form of the kinetic energy equation is

$$\frac{d\theta^*}{dx} + \left(3 - M_E^2\right)\frac{\theta^*}{u_E}\frac{du_E}{dx} = \frac{2\dot{D}}{\rho_E u_E^3} = 2C_d, \tag{4.3.11}$$

where C_d is referred to as the dissipation coefficient. For incompressible flow this reduces to

$$\rho\frac{d}{dx}\left(\theta^* u_E^3\right) = 2\dot{D}. \tag{4.3.12}$$

Equations (4.3.11) and (4.3.12) find considerable application in the estimation of losses described in Chapter 5.

A third integral equation which relates to the thermal energy in the flow is that for the stagnation enthalpy. It is obtained by integrating (4.2.6) in y and using the continuity equation

$$\frac{d}{dx}\left[\int_0^{y_E} \rho u_x (h_t - h_{t_E}) dy\right] = -q_w. \tag{4.3.13}$$

Equation (4.3.13) equates the rate of change of the flux of stagnation enthalpy difference between the boundary layer and the free stream to the rate of heat transfer to the fluid at the surface. For an adiabatic surface this is zero. There is no work term because no work is done by the stationary surface at $y = 0$ and there is no shear stress at $y = y_E$.

4.4 Laminar boundary layers

4.4.1 Laminar boundary layer behavior in favorable and adverse pressure gradients

Procedures for computations of laminar boundary layers are well described in depth elsewhere (e.g. Schlichting (1979), Sherman (1990), and White (1991)), and we thus present a short description only of boundary layer behavior in response to different types of pressure gradient. The simplest (and historically the most prominent) situation, the constant pressure laminar boundary layer, is not addressed as a separate topic, but is rather recovered as a special case of the boundary layer with a pressure gradient.

To exhibit the generic features of laminar boundary layers in adverse and favorable pressure gradients we examine a family of self-similar boundary layer solutions (Cebeci and Bradshaw, 1977). Non-similar solutions can also readily be computed, but the qualitative features do not differ from those shown, and similarity allows compact display of the overall results. The solutions are the

Falkner–Skan velocity profiles for incompressible flow which apply to free-stream velocities of the form

$$u_E = cx^m, \tag{4.4.1}$$

where c is a constant. The solution family represents boundary layers in both adverse ($m < 0$) and favorable ($m > 0$) pressure gradients.

The existence of the similarity variables can be made plausible by noting that if the streamwise length scale is x, a normal length scale, δ_n, of the same form as that for the constant pressure boundary layer discussed in Section 2.9 is given by $\delta_n/x = 1/\sqrt{u_E x/\nu}$, or $\delta_n = \sqrt{\nu x/u_E}$. An appropriate non-dimensional boundary layer coordinate is thus

$$\eta = \frac{y}{\delta_n} = y\sqrt{\frac{u_E}{x\nu}}. \tag{4.4.2}$$

For two-dimensional flow a stream function, ψ, can be defined so that

$$u_x = \frac{\partial \psi}{\partial y}, \quad u_y = -\frac{\partial \psi}{\partial x}. \tag{4.4.3}$$

The stream function ψ automatically satisfies the continuity equation. A natural scaling for the stream function is $u_E\delta_n$, so that a non-dimensional form of the stream function can be taken as

$$F(\eta) = \frac{\psi}{\sqrt{u_E \nu x}}. \tag{4.4.4}$$

Using (4.4.2) and (4.4.4) in (4.2.3) yields a non-linear ordinary differential equation for the function $F(\eta)$. With the prime denoting differentiation with respect to η:

$$F''' + \frac{(m+1)}{2} F F'' + m[1 - (F')^2] = 0. \tag{4.4.5}$$

The solutions of (4.4.5) are independent of x if the boundary conditions are also. Suitable boundary conditions for describing this class of flows are

$$\eta = 0: \qquad F = \text{constant}, \quad F' = 0,$$

corresponding to $u_x = u_y = 0$ on the boundary, and

$$\eta \to \infty: \qquad F' = 1$$

corresponding to $u_x = u_E$ as $\eta \to \infty$.

Numerical solutions of (4.4.5) (known as the Falkner–Skan equation) giving the velocity u/u_E as a function of η are shown in Figure 4.6 for different values of m. Profiles corresponding to favorable pressure gradients, $m > 0$, are fuller than for adverse pressure gradients, $m < 0$. The profiles for $m < 0$ become S-shaped and the skin friction coefficient at the wall falls as m decreases. The condition at which the wall shear stress = 0 and separation occurs is $m = -0.0904$. The condition $m = 0$ corresponds to the Blasius constant pressure boundary layer solution for which (4.4.5) takes the form

$$F''' + \frac{F F''}{2} = 0.$$

Table 4.1 *Behavior of Falkner–Skan-type boundary Layers; free stream has* $u_E = cx^m$ *(Cebeci and Bradshaw, 1977)*

	m	$C_f Re_x^{1/2}$	$(\delta^*/x)\, Re_x^{1/2}$	$H = \delta^*/\theta$
$\frac{dp_E}{dx} < 0$	1	2.465	0.648	2.216
	1/3	1.515	0.985	2.297
	0.1	0.903	1.348	2.422
$\frac{dp_E}{dx} = 0$	0	0.664	1.721	2.591
$\frac{dp_E}{dx} > 0$	−0.01	0.632	1.780	2.622
	−0.05	0.427	2.117	2.818
	−0.0904	0	3.428	3.949

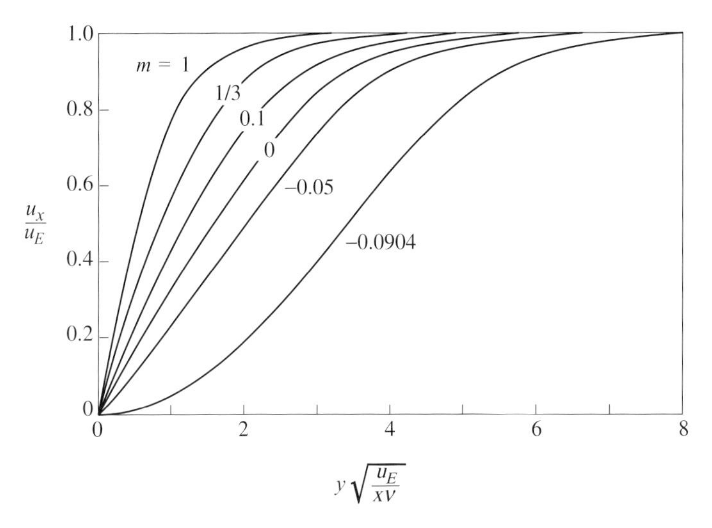

Figure 4.6: Boundary layer velocity profiles in favorable and adverse pressure gradients – solutions of the Falkner–Skan equations with free-stream flow $u_E = cx^m$ (Cebeci and Bradshaw, 1977).

Results for non-dimensional wall shear stress and boundary layer integral parameters are given in Table 4.1. As the pressure gradient is made more adverse, the skin friction falls and the shape parameter increases.

4.4.2 Laminar boundary layer separation

The pressure rise that the boundary layer can withstand without separating from a surface is a quantity of great interest. A simple and useful estimate of this pressure rise for laminar boundary layers is given by a method due to Thwaites (White, 1991). This starts with the momentum integral equation for incompressible flow multiplied by $u_E\theta/\nu$ and written in the form

$$\frac{\tau_w\theta}{\mu u_E} = \frac{u_E\theta}{\nu}\frac{d\theta}{dx} + \frac{\theta^2}{\nu}\frac{du_E}{dx}(2+H). \tag{4.4.6}$$

It was observed from examination of boundary layer solutions that the shape parameter, H, and the skin friction coefficient, $\tau_w\theta/\mu u_E$, can both be regarded to good approximation as functions of a single parameter, $\lambda = (\theta^2/\nu)(du_E/dx)$, so that

$$\frac{\tau_w\theta}{\mu u_E} \approx S(\lambda), \tag{4.4.7b}$$

$$H = \frac{\delta^*}{\theta} \approx H(\lambda). \tag{4.4.7c}$$

Equation (4.4.6) can then be expressed as

$$u_E\frac{d}{dx}\left[\lambda\Big/\left(\frac{du_E}{dx}\right)\right] \approx 2[S(\lambda) - \lambda(2+H)] = F(\lambda). \tag{4.4.8}$$

Thwaites noted that the known analytic and experimental results were well fitted by the function

$$F(\lambda) = 0.45 - 6\lambda. \tag{4.4.9}$$

If we substitute (4.4.9) into (4.4.8) and multiply the resulting equation by u_E^5 we obtain an exact differential which then allows a closed form solution of (4.4.6):

$$\frac{1}{\nu}\frac{d}{dx}\left(\theta^2 u_E^6\right) = 0.45u_E^5. \tag{4.4.10}$$

Integrating (4.4.10) from an initial location (0) to x gives

$$\frac{\theta^2 u_E^6}{\nu} = 0.45\int_0^x u_E^5 dx' + \left(\frac{\theta^2 u_E^6}{\nu}\right)_0. \tag{4.4.11}$$

Equation (4.4.11) allows the momentum thickness to be found for any distribution $u_E(x)$. With this established, the parameter λ can be found and thus the skin friction and displacement thickness from Figure 4.7, or from the tabulated values of $H(\lambda)$ and $S(\lambda)$ given by White (1991), who presents an example of the application of Thwaites's method to a linearly decelerating flow, $u_E(x) = u_{E_0}(1 - x/L)$. Figure 4.8 shows the results and a comparison with a finite difference solution. Figure 4.8 also implies that the pressure rise which can be tolerated by a laminar boundary layer is roughly 20% of the initial free-stream dynamic pressure.[4] As we will see, turbulent boundary layers can withstand several times this value.

One consequence of a laminar separation is the formation of a laminar free shear layer which can become unstable, evolve to a turbulent shear layer, and reattach as a turbulent boundary layer. Even without separation, however, if the Reynolds number is high enough, laminar layers will naturally undergo transition to turbulence. As a prelude to discussion of turbulent boundary layers which are much more common in fluid machinery than laminar boundary layers, in the next section we describe some features of transition from laminar to turbulent flow.

[4] While this gives a general guideline, the specifics of the conclusion depend strongly on the shape of the $u_E(x)$ distribution. As discussed in the preceding section, a similarity boundary layer can be decelerated to $u_E \approx 0$. Rapid deceleration after a long constant pressure flow, however, will cause separation with only a small percentage decrease in u_E.

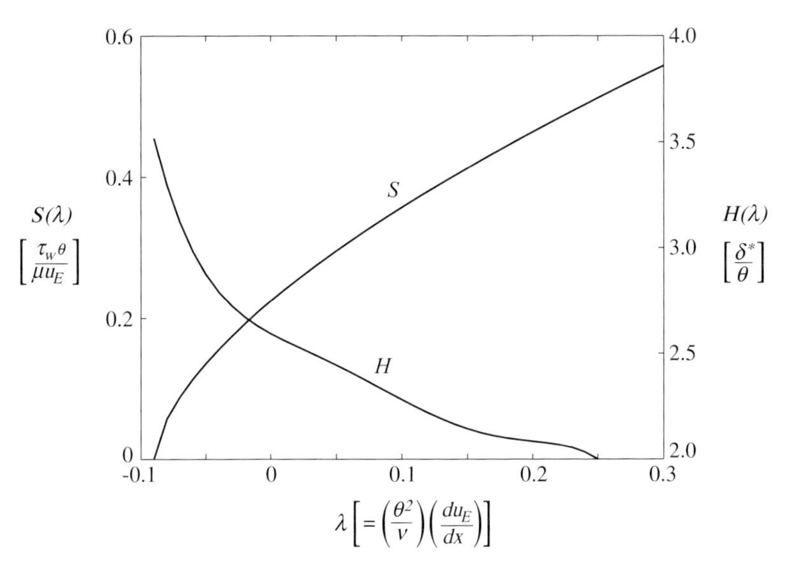

Figure 4.7: Laminar boundary layer correlation functions suggested by White (1991).

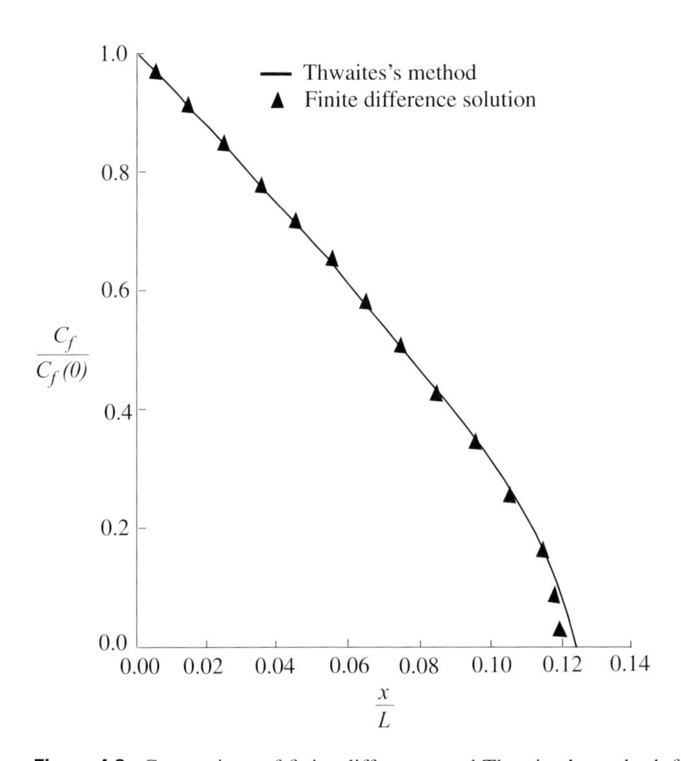

Figure 4.8: Comparison of finite-difference and Thwaites's method, for wall friction in a linearly decelerating flow, $u_E/u_{E_0} = 1 - x/L$ (White, 1991).

4.5 Laminar–turbulent boundary layer transition

Transition from laminar to turbulent flow can have several stages and generally take place over a three-dimensional space. The mechanisms for transition can be classified into *natural transition*, in which the first stage of the process is the growth of small amplitude disturbances in the boundary layer, and *bypass transition*, in which the level of free-stream turbulence is high enough to bypass the initial stages of the natural process and cause the onset of turbulent flow. This is typically the mode observed in multistage turbomachinery, for example, where wakes from the upstream blading impinge on the boundary layer.

In this discussion we present information to allow estimates of the conditions under which transition occurs. Figure 4.9 ((Mayle, 1991) from whom much of the discussion of transition given here is taken) shows the topology of the different modes of transition plotted in a momentum thickness Reynolds number (Re_θ) versus acceleration parameter K ($= (\nu/u_E^2)(du_E/dx)$) format, with both parameters evaluated at the beginning of transition. Lines of constant turbulence level represent the value of the momentum thickness Reynolds number at which transition begins for that value of turbulence level and acceleration parameter. The line marked "stability criterion" is the line above which boundary layer instability, the self-excited amplification of small disturbances within the boundary layer, is possible. The line marked "separation criterion" is the calculated laminar boundary layer separation limit, defined by Thwaites (1960) as $Re_\theta^2 K = -0.082$.

Figure 4.9 illustrates the large effect of the free-stream pressure gradient (manifested through changes in the value of boundary layer shape parameter, H) on the start of transition. Favorable pressure gradients require much higher values of Re_θ for transition than adverse gradients. Strong favorable pressure gradients, such as occur in nozzles of large contraction, or turbines, can even cause turbulent boundary layers to re-laminarize. For strong adverse pressure gradients, on the other hand, the momentum thickness Reynolds numbers for transition are much reduced from the value for zero pressure gradients.

Natural transition involves several stages: (1) at a critical value of the momentum thickness Reynolds number the laminar boundary layer becomes unstable to small disturbances; (2) the instability amplifies to a point where three-dimensional disturbances grow and develop into loop-shaped vortices; (3) the fluctuating portions of the flow develop into turbulent spots, localized regions of turbulent flow, which grow as they convect downstream, until they coalesce into a turbulent boundary layer. These stages occur over a finite length and it is appropriate to describe transition as a process rather than an event occurring at a point (White, 1991; Sherman, 1990; Schlichting, 1979).

A special type of natural transition occurs when a laminar boundary layer separates. If this occurs, the growth of instability is much more rapid in the resulting free shear layer, promoting transition to turbulence and reattachment as a turbulent boundary layer. A laminar separation/turbulent reattachment "bubble" thus exists on the surface. The bubble length depends on the transition process within the free shear layer and can involve all of the stages listed above. The process is depicted schematically in Figure 4.10, which indicates an upstream region of nearly constant pressure and a downstream region with pressure recovery.

Bypass transition occurs when there is a high level of free-stream turbulence. The first two stages of the natural transition process can be completely bypassed so that turbulent spots are produced directly.

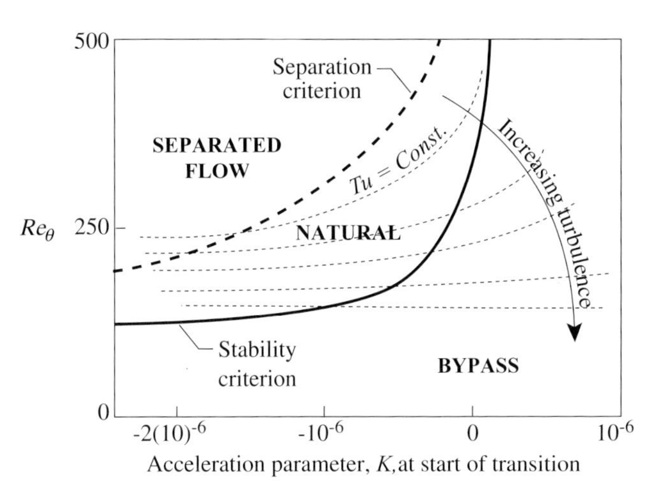

Figure 4.9: Topology of the different modes of transition in a Reynolds number, acceleration parameter K $(= (\nu/u_E^2)(du_E/dx))$; Tu is turbulence intensity $(= (\overline{u'^2}/3)^{1/2}/u_E)$ (Mayle, 1991).

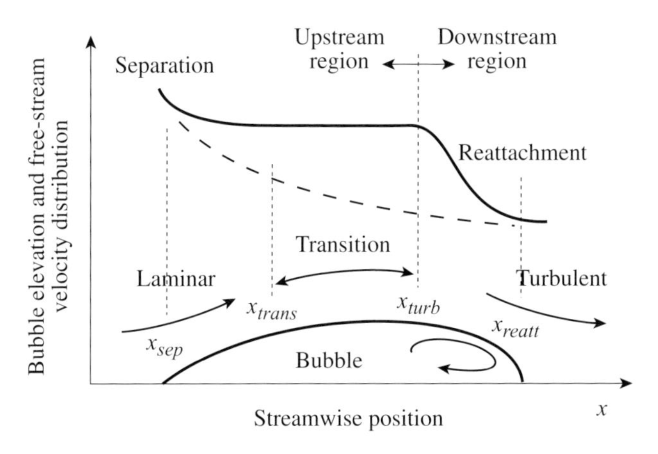

Figure 4.10: Flow around a separation bubble and the corresponding pressure distribution (Mayle, 1991).

In this case the linear instability mechanism associated with natural transition is not appropriate, and in fact Figure 4.9 shows that for high levels of turbulence and a favorable pressure gradient, transition can occur before the stability criterion is reached.

Detailed coverage of transition is beyond the scope of this text, but Figure 4.11 is presented to make quantitative some of the points that have been discussed. The figure gives momentum thickness Reynolds number at the start and the end of transition for a constant pressure boundary layer as a function of the free-stream turbulence level. As the turbulence level increases, the momentum thickness Reynolds number at which transition can start decreases but there is a minimum value (given in Abu-Ghannam and Shaw (1980) as 163) below which transition cannot occur. Although

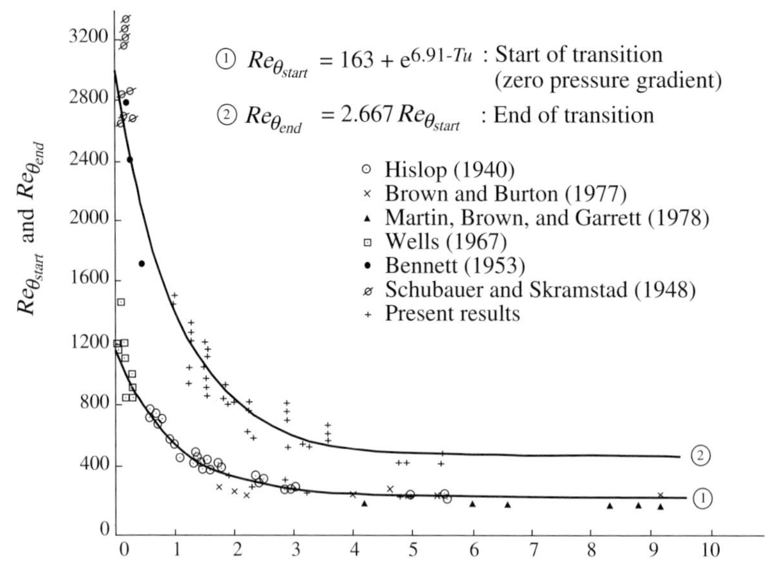

Figure 4.11: Momentum thickness Reynolds number at the start and end of transition for zero pressure gradient (Abu-Ghannam and Shaw, 1980).

the length of the transition region is not shown explicitly, the figure implies, and measurements show, the finite spatial extent.

4.6 Turbulent boundary layers

4.6.1 The time mean equations for turbulent boundary layers

Turbulent flow is characterized by flow property fluctuations about the time mean values. Associated with these fluctuations is a greatly increased transfer rate of mass, momentum, and energy compared to laminar flow. To introduce ideas concerning turbulent boundary layers we resolve variables into time mean quantities and fluctuations about the mean. For example the time mean velocity is

$$\overline{\mathbf{u}}(\mathbf{x}) = \frac{1}{t_{int}} \int_0^{t_{int}} \mathbf{u}(\mathbf{x}, t)dt, \tag{4.6.1}$$

where the integration time t_{int} is large compared to the fluctuation period. Denoting the fluctuating quantities by the curved overbar (e.g. $\breve{u}$), for a two-dimensional boundary layer the velocity components and the pressure are

$$u_x = \overline{u}_x + \breve{u}_x, \tag{4.6.2a}$$

$$u_y = \overline{u}_y + \breve{u}_y, \tag{4.6.2b}$$

$$p = \overline{p} + \breve{p}. \tag{4.6.2c}$$

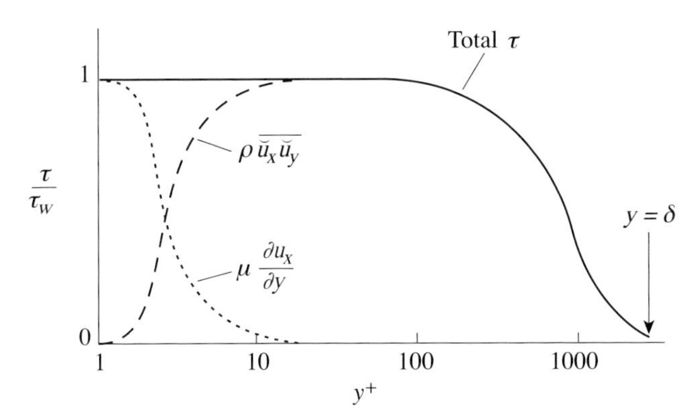

Figure 4.12: Shear stress in a turbulent boundary layer as a function of the non-dimensional distance from the wall; $y^+ = yu_\tau/\nu$, and friction velocity $u_\tau = \sqrt{(\tau_w/\rho)}$ (Johnston, 1986).

The discussion here is confined to the incompressible case. For compressible flows there would also be fluctuations in temperature and density.

We now apply the averaging procedure defined by (4.6.1) to the boundary layer equations to develop equations for turbulent flow. The continuity equation is linear in the velocity components, so that time averaging does not change the form from that in the laminar case, hence:

$$\text{time mean:} \quad \frac{\partial \overline{u}_x}{\partial x} + \frac{\partial \overline{u}_y}{\partial y} = 0, \tag{4.6.3a}$$

$$\text{fluctuations:} \quad \frac{\partial \breve{u}_x}{\partial x} + \frac{\partial \breve{u}_y}{\partial y} = 0. \tag{4.6.3b}$$

A different situation occurs for the momentum equation, which is quadratic in the velocity components. Expressing the velocity and pressure as in (4.6.2), substituting into the x-component of the momentum equation, and taking the time average yields

$$\overline{u}_x \frac{\partial \overline{u}_x}{\partial x} + \overline{u}_y \frac{\partial \overline{u}_x}{\partial y} = -\frac{1}{\rho}\frac{d\overline{p}}{dx} + \nu\left(\frac{\partial^2 \overline{u}_x}{\partial y^2}\right) - \frac{\partial}{\partial x}\left(\overline{\breve{u}_x\breve{u}_x}\right) - \frac{\partial}{\partial y}\left(\overline{\breve{u}_x\breve{u}_y}\right). \tag{4.6.4}$$

There are now additional terms in the time mean momentum equation compared with laminar flow. These terms involve products of the turbulent fluctuations. The product terms are not known *a priori* and we cannot find them from the time mean equations, because information has been lost through the averaging process. Equations additional to continuity and momentum are thus needed to close the problem.

The quadratic fluctuation terms in (4.6.4) function as additional stresses. This can be seen by considering the flux of x-momentum across a control plane at a constant value of y. If the fluctuations in $\breve{u}_x$ and $\breve{u}_y$ are correlated so that the product $(\overline{\breve{u}_x\breve{u}_y})$ is positive, there is transport of fluid particles with positive x-momentum upwards across the plane and transport of fluid particles having negative x-momentum downwards. The result is a net upwards transfer of x-momentum, of magnitude $\rho\overline{\breve{u}_x\breve{u}_y}$ per unit area and unit time. Terms of this type are known as *Reynolds stresses*, and the total stress in a time mean turbulent flow is the sum of the viscous and Reynolds stresses. Figure 4.12 shows a sketch of the stresses in a turbulent boundary layer, plotted versus the non-dimensional distance from the wall. Over most of the turbulent boundary layer, except near the wall, the Reynolds stresses are much larger than the viscous stresses.

Modeling of the stress terms (or of similar terms in equations which define the evolution of the stresses) is the central problem in turbulent flow. We do not address techniques for doing this in any detail and rather present basic approaches for calculating the overall properties of the time mean flow. These are more appropriately regarded as *scaling* arguments concerning mean flow behavior (Roshko, 1993a) rather than theories of turbulent shear flow, but they have proved useful in helping organize the large amount of empirical information about this complex subject.

The arguments used in deriving the laminar boundary layer equations must be modified for turbulent flow. As before, the situations to be considered for the time mean flow are those for which the characteristic length scale normal to the bounding surface (the boundary layer thickness) is much less than the length scale along the surface. We cannot, however, state that this is true for the fluctuating velocities. Experiments show that the instantaneous x- and y-velocity fluctuations are comparable as are the x- and y-length scales associated with the fluctuations. The approximation made is thus that derivatives of the time mean quantities vary much less in the streamwise direction than in the normal direction. In what follows, the overbars will be dropped so that u_x, for example, will represent the time mean x-velocity component. The x-momentum equation is approximated as

$$u_x \frac{\partial u_x}{\partial x} + u_y \frac{\partial u_x}{\partial y} = -\frac{1}{\rho}\frac{dp}{dx} + \nu\left(\frac{\partial^2 u_x}{\partial y^2}\right) + \frac{\partial}{\partial y}\left(-\overline{\breve{u}_x \breve{u}_y}\right)$$

$$= -\frac{1}{\rho}\frac{\partial p}{\partial x} + \frac{\partial \tau_{viscous}}{\partial y} + \frac{\partial \tau_{turbulence}}{\partial y}. \tag{4.6.5}$$

(As for the discussion of laminar boundary layers τ denotes τ_{xy}.) The dominant forces due to Reynolds stresses in a two-dimensional turbulent boundary layer arise from the y-derivative of the $(\overline{\breve{u}_x \breve{u}_y})$ term, and this is the only one we consider.

Using the above arguments, the y-momentum equation becomes

$$\frac{1}{\rho}\frac{\partial \overline{p}}{\partial y} = \frac{\partial}{\partial y}\left(\overline{\breve{u}_y^2}\right). \tag{4.6.6}$$

Equation (4.6.6) can be integrated across the boundary layer to give the normal pressure difference as

$$\Delta p_n = \rho\left(\overline{\breve{u}_y^2}\right). \tag{4.6.7}$$

The variation of pressure across a turbulent shear layer is from one to several percent of the dynamic pressure based on the free-stream velocity. The lower value is for a boundary layer, the higher value for a jet, based on the maximum jet velocity. This pressure difference can generally be neglected in computations of turbulent boundary layer behavior. In summary, the equations that describe two-dimensional turbulent boundary layers in incompressible flow are (4.6.3) and (4.6.5) plus specification of the pressure gradient imposed on the layer. For compressible flow there are additional terms due to the correlations between fluctuating density and velocity: for these equations see White (1991) or Cebeci and Bradshaw (1977).

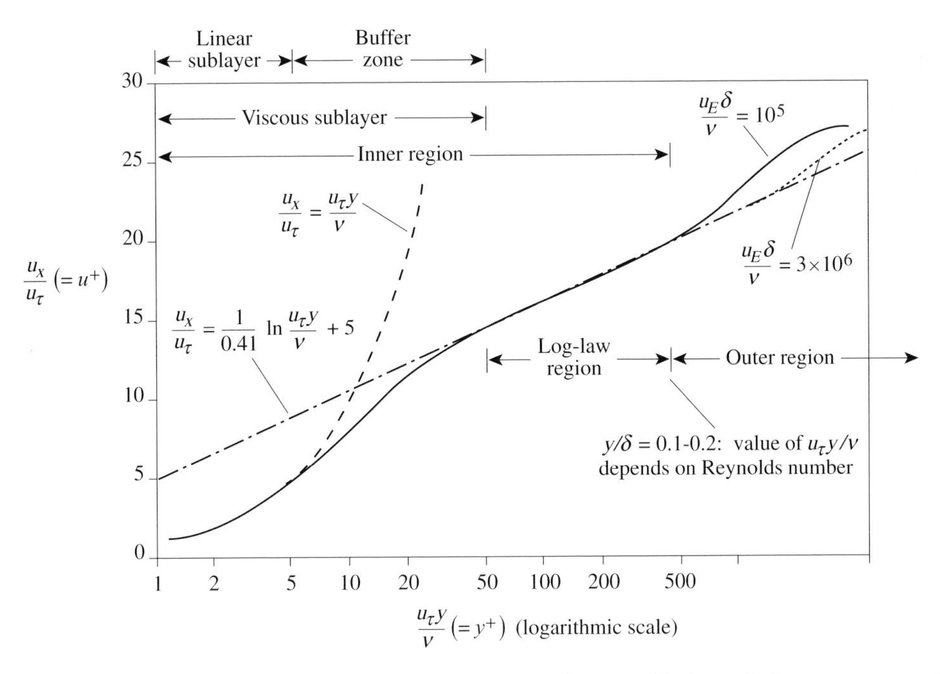

Figure 4.13: Regions of a turbulent boundary layer. Outer-layer profile shown is for u_E = constant (Cebeci and Bradshaw, 1977).

4.6.2 The composite nature of a turbulent boundary layer

An important feature of a turbulent boundary layer is the difference in the behavior of the region near the surface (the inner region) and the rest of the boundary layer (the outer region). This is illustrated by examining a constant pressure flow. As we have seen, for a laminar boundary layer a dimensionless normal coordinate can be defined which represents the velocity profile at any x-station. For a turbulent boundary layer, however, this is not the case. The reason is that the velocity profile in the inner region of the boundary layer is dependent on viscosity, while that in the outer region depends on the Reynolds stresses. The scaling of the two regions is thus quite different.

In the inner region the relevant quantities are wall shear stress, density, kinematic viscosity, and distance from the wall, y. It is helpful to make use of the friction velocity defined as $u_\tau = \sqrt{\tau_w/\rho}$, where τ_w is the wall shear stress. From dimensional analysis an appropriate non-dimensional grouping for the velocity dependence is

$$\frac{u_x}{u_\tau} = f\left(\frac{yu_\tau}{\nu}\right). \tag{4.6.8}$$

The conventional notation is to define $u^+ = u_x/u_\tau$ and $y^+ = yu_\tau/\nu$ so (4.6.8) can be written

$$u^+ = f(y^+). \tag{4.6.9}$$

Figure 4.13 is a plot of non-dimensional velocity u^+ versus y^+; the logarithmic scale should be noted. The region up to roughly $y^+ = 10$, where viscous stresses dominate, is known as the linear sublayer. Further away from the wall, say $y^+ \approx 50$, the stress is still close to τ_w but the stress and rate

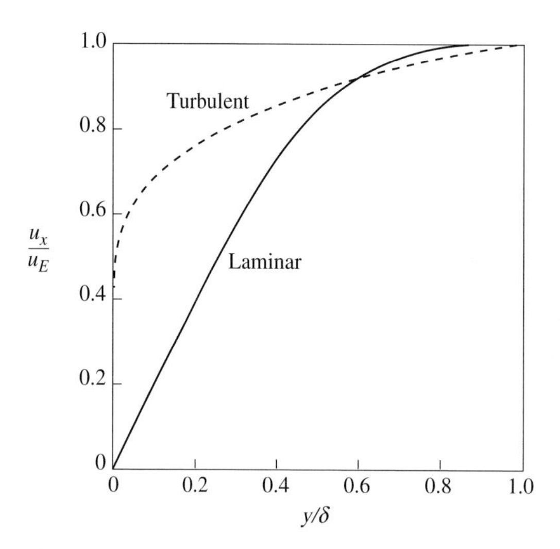

Figure 4.14: Comparison of the shapes of laminar and turbulent boundary layers (Clauser, 1956).

of strain no longer depend on viscosity. If so, the only dimensionally correct relationship is (Cebeci and Bradshaw, 1977),

$$\frac{\partial u_x}{\partial y} = \frac{u_\tau}{\kappa y}. \tag{4.6.10}$$

The non-dimensional constant κ has been found experimentally to be 0.41. Equation (4.6.10) can be integrated to give the form of the velocity profile outside the linear sublayer, but still in the inner region, as

$$u^+ = \frac{1}{\kappa}\ln y^+ + C, \tag{4.6.11}$$

where C is found experimentally to be 5.0. Equation (4.6.11) is known as the "law of the wall".

Figure 4.13 illustrates the regions of the turbulent boundary layer. The inner region can be plotted as a single curve using u^+ and y^+ for all Reynolds numbers. In the outer region whose extent depends on the Reynolds number, velocity profiles for different Reynolds numbers will not collapse in this manner, even for a constant pressure flow. The inner region typically occupies 10–20% of the overall boundary layer thickness, δ. In the outer region the velocity profile does not depend directly on the viscosity, and an appropriate choice of variables is to scale the velocity defect with the friction velocity.

A general view of time–mean turbulent boundary layer velocity profiles is provided by Figures 4.14 (Clauser, 1956) and 4.15 (White, 1991). The first shows the velocity distribution (u_x/u_E) as a function of (y/δ) for constant pressure laminar and turbulent boundary layers. The latter has much steeper velocity gradients near the wall than the former, even allowing for the fact that the boundary layer thickness is larger for the turbulent boundary layer at the same Reynolds number. The difference in transport mechanisms in the inner and outer regions also implies a difference in characteristic length scales between these regions. This is seen in the turbulent velocity profile, which shows (see Figure 4.14) large differences in the local slope across the layer. For the laminar boundary layer,

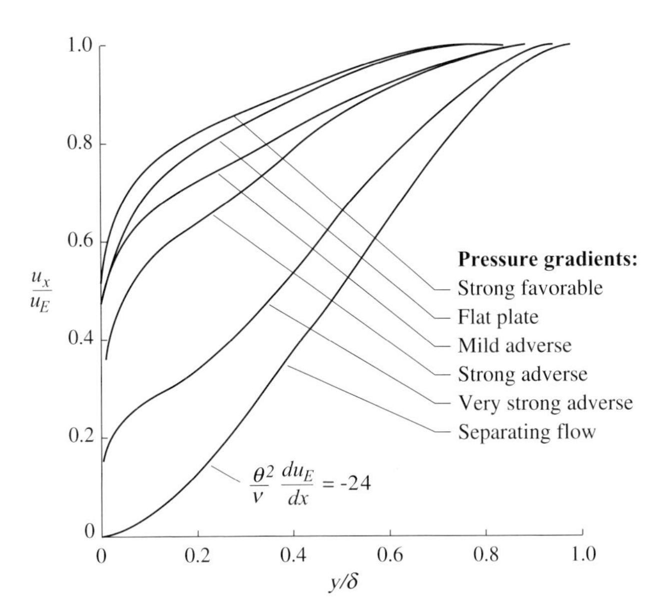

Figure 4.15: Experimental turbulent boundary layer velocity profiles for various pressure gradients. From White (1991), after data of Coles and Hirst (1968).

the differences in local slope are much less, consistent with the existence of a single characteristic length.

The steep velocity gradient near the wall of a turbulent boundary layer can be viewed as associated with the behavior of the effective turbulent viscosity as one moves away from the wall (see Section 4.6.3). In the outer layer the effective viscosity can be two or more orders of magnitude larger than the actual viscosity, resulting in a much higher velocity gradient near the wall than in the outer region. This provides the near wall flow an enhanced capability (compared to that of a laminar boundary layer) to resist separation in adverse pressure gradients because of the increased momentum transfer from faster moving fluid.

Figure 4.15 presents time mean turbulent velocity distributions for a range of favorable and adverse pressure gradients, also in terms of (u_x/u_E) versus (y/δ), which show similar features to the constant pressure situation.

4.6.3 Introductory discussion of turbulent shear stress

To close the problem of analyzing turbulent boundary layers a relation is needed to link the turbulent (or Reynolds) stress and the rate of strain. Approaches for supplying this via the definition of a turbulent momentum diffusivity, or *eddy diffusivity*, range from dimensional analysis coupled with experiment, to computational procedures in which the eddy diffusivity is calculated from other turbulent quantities (Bradshaw, 1996). The difficulty is that the transport coefficient is not a property of the fluid, as for laminar flow, but rather a property of the flow field itself.

Over the past century or more, a number of approaches have been pursued to address this closure. Initial attempts were aimed at connecting the eddy diffusivity to features of the time mean flow field. These have been reasonably successful in providing estimates of turbulent boundary layer development, although they must be used with caution in cases far from previous experience. A basic proposal concerning turbulent shear stress (due to Prandtl (White, 1991)) is that the fluctuating velocity is related to a mixing length scale and the velocity gradient. If so, the Reynolds stress is given by

$$\overline{\left(\breve{u}_x \breve{u}_y\right)} \propto \left[\ell_{mix}\left(\frac{\partial u_x}{\partial y}\right)\right]^2, \tag{4.6.12}$$

where ℓ_{mix}, the mixing length, is to be defined. From (4.6.12) an eddy diffusivity can be defined such that

$$\nu_{turb}\frac{\partial \overline{u}_x}{\partial y} = -\overline{\left(\breve{u}_x \breve{u}_y\right)} = (\ell_{mix})^2 \left|\frac{\partial u_x}{\partial y}\right|^2. \tag{4.6.13}$$

To make (4.6.13) useful, we need a way to connect the length scale to the flow conditions. Because of the composite nature of the turbulent boundary layer, this needs to be done in two parts (White, 1991). For the inner region the necessary relation is provided by the empirical expression

$$(\ell_{mix})_{inner\ region} \approx \kappa y\left[1 - e^{-y^+/Y}\right]. \tag{4.6.14}$$

The quantity in the square bracket is a damping factor that accounts for the decrease in turbulent transport properties very near the wall. For a flat plate boundary layer the non-dimensional parameter Y is approximately 26 and at a value of $y^+ = 60$, the exponential quantity is only 0.1. Over most of the logarithmic region therefore, the mixing length can be taken to be proportional to the distance from the wall, y. In the outer region, measurements imply the mixing length scales with boundary layer thickness:

$$(\ell_{mix})_{outer\ region} \sim 0.09\delta. \tag{4.6.15}$$

Relations such as the above do not reveal any fundamental information concerning the turbulent flow, and they are perhaps best viewed as correlations of data which, coupled with the appropriate forms of the time mean equations of motion, allow estimates of the time mean velocity and pressure fields. The eddy viscosity, μ_{turb} varies across the turbulent boundary layer, but it is roughly constant in the outer region and can be scaled as

$$\mu_{turb} \propto \rho u_E \delta^*$$

or

$$\frac{\mu_{turb}}{\mu} \approx 0.016 Re_{\delta^*}. \tag{4.6.16}$$

Figure 4.16 shows computations of eddy viscosity across a turbulent boundary layer for three different values of Re_{δ^*}. The straight lines, which go from the origin to the constant values, represent the behavior in the inner layer. The dashed lines are modifications to the estimation based on the fact that the outer portion of the boundary layer contains fluid which is not turbulent (i.e. patches of irrotational fluid from the free stream). The fraction of the time a probe might see turbulent fluid varies from near unity at $y/\delta \approx 0.5$ to close to zero at $y/\delta \approx 1$, with a consequent fall off in magnitude of the turbulent transport properties.

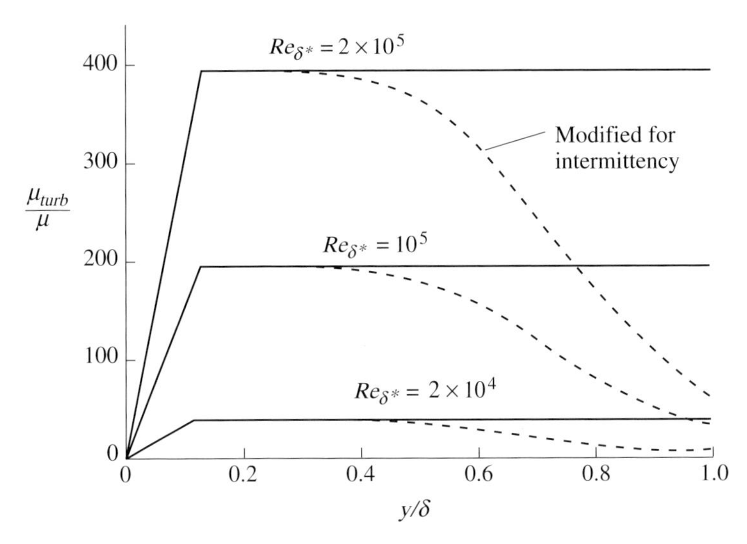

Figure 4.16: Eddy viscosity distribution in a turbulent boundary layer computed from the inner law and outer law (White, 1991).

As a closing note to this section, it may be worthwhile to comment on the current state of turbulent flow computations. It is now common that standard computational procedures employ turbulence models with two auxiliary equations for the evolution of the turbulent kinetic energy and for dissipation of turbulence energy. The local eddy viscosity is scaled with the square of the former divided by the latter, with the proportionality factor for the scaling obtained from experiment. For such models there are also other proportionality factors which must be supplied, and it has been found that the values of these are not universal for all flows. There is considerable research on large eddy simulations, in which the larger eddies are computed and only the smaller ones represented by empirical expressions, and on direct simulation of the Navier–Stokes equations, although these are not yet standard industry tools (Moin, 2002; Moin and Mahesh, 1998).

4.6.4 Boundary layer thickness and wall shear stress in laminar and turbulent flow

It is useful to compare some of the overall properties of laminar and turbulent boundary layers. We examine two aspects, the wall shear stress and the boundary layer thickness for a constant pressure incompressible flow. For the laminar boundary layer, the boundary layer thickness obtained, δ, is calculated to be

$$\frac{\delta}{x} \approx \frac{5}{\sqrt{\dfrac{u_E x}{\nu}}} \approx \frac{5}{\sqrt{Re_x}}. \tag{4.6.17}$$

The wall shear stress is

$$\frac{\tau_w(x)}{\frac{1}{2}\rho u_E^2} = C_f = \frac{0.664}{\sqrt{Re_x}}. \tag{4.6.18}$$

Integrating (4.6.18) from $x = 0$ to $x = L$, the total frictional force per unit width, F_w, on a plate of length L is

$$\frac{F_w}{\frac{1}{2}\rho L u_E^2} = \frac{1.33}{\sqrt{Re_L}}. \tag{4.6.19}$$

For the turbulent boundary layer we make use of the integral momentum equation (4.3.9) in the form

$$\tau_w = \rho u_E^2 \frac{d\theta}{dx}. \tag{4.6.20a}$$

This can be integrated to give

$$F_w = \int_0^x \tau_w(x')dx' = \rho u_E^2 \int_0^x \frac{d\theta}{dx}dx = \rho u_E^2 \theta. \tag{4.6.20b}$$

The local wall shear stress is related to the derivative of the momentum thickness and the non-dimensional force is just the value of momentum thickness at the exit station.

To proceed further, we need a link between the wall shear stress and the boundary layer parameters. A simple relation of this type is provided by the empirical expression (Schlichting, 1979)

$$\frac{\tau_w}{\frac{1}{2}\rho u_E^2} = 0.045\left(\frac{\nu}{u_E \delta}\right)^{1/4}. \tag{4.6.21}$$

To relate momentum thickness to boundary layer thickness, we also need a suitable velocity profile which can be used in the definition of the former, (4.3.4). An appropriate representation for this purpose, valid for Reynolds numbers from 10^5 to 10^7, has been found to be[5]

$$\frac{u_x}{u_E} = \left(\frac{y}{\delta}\right)^{1/7}. \tag{4.6.22}$$

Substituting (4.6.22) in the definition of momentum and displacement thickness, (4.3.4) and (4.3.2), yields

$$\theta = \frac{7}{72}\delta, \quad \delta^* = \frac{\delta}{8}. \tag{4.6.23}$$

Using (4.6.21) and (4.6.23) in (4.6.20a) yields an expression for the growth of the boundary layer in x:

$$0.023\left(\frac{\nu}{u_E \delta}\right)^{\frac{1}{4}} = \frac{7}{72}\frac{d\delta}{dx}. \tag{4.6.24}$$

Finally, integrating (4.6.24) from the starting conditions (taken here as $\delta = 0$ at $x = 0$) gives an expression for the boundary layer thickness, δ, as a function of x:

$$\delta(x) = \frac{0.37x}{(Re_x)^{1/5}}. \tag{4.6.25}$$

[5] The distribution in (4.6.22) gives good representation of the *overall* shape of the turbulent boundary layer velocity profile in Figure 4.14, although it cannot be valid in the near-wall region because the derivative is unbounded at the wall. However, local details of the velocity field such as this (which are not captured) are unimportant for the estimation of integral properties.

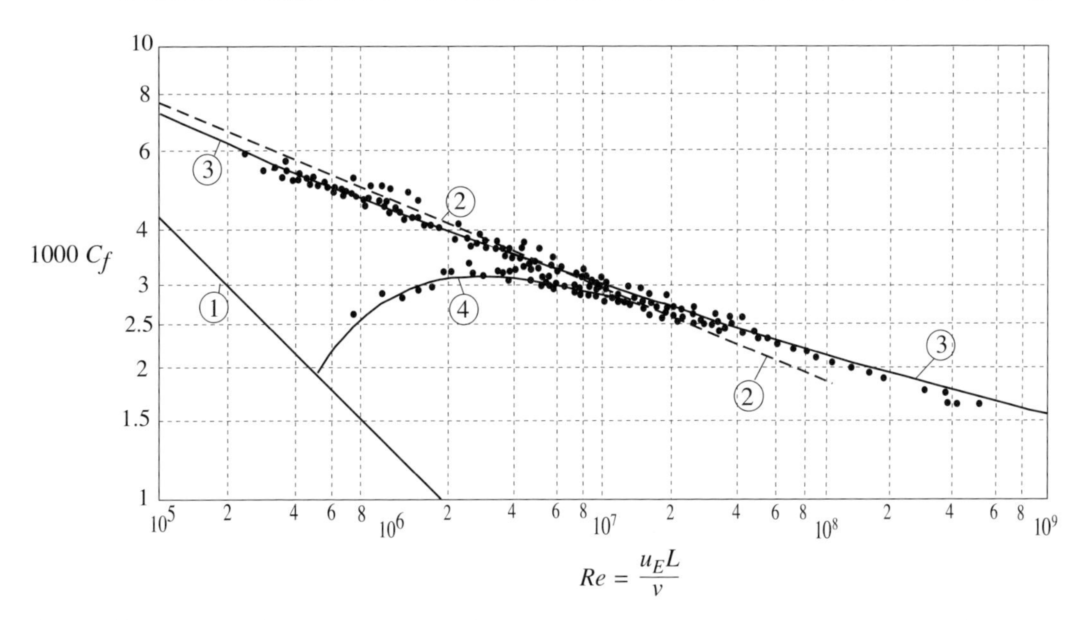

Figure 4.17: Resistance formulas for a smooth flat plate, theory and measurement: curve 1 for a laminar layer (4.6.18); curve 2 is based on (4.6.26); curve 3 is the data fit $C_f = 0.455/(\log Re)^{2.58}$; curve 4 represents the laminar–turbulent transition regime (Schlichting, 1979).

The momentum thickness is proportional to $\delta(x)$ and is

$$\theta(x) = \frac{0.036x}{(Re_x)^{1/5}}. \tag{4.6.26}$$

The boundary layer thickness is found to increase as $x^{4/5}$ in turbulent flow compared with (4.6.17), which shows a thickness growth as $x^{1/2}$ in laminar flow. For a length Reynolds number of 10^6 the boundary layer thicknesses are $\delta/x = 0.005$ and $\delta/x = 0.023$ for laminar and turbulent flow respectively. From (4.6.25) the rate of growth of a turbulent layer at a Reynolds number of 10^6 is approximately 1 in 50. These numbers emphasize the relative thinness of constant pressure boundary layers.

Figure 4.17 shows a plot of the wall shear stress on a smooth flat plate versus length Reynolds numbers. The curve marked 1 is for a laminar layer and is (4.6.18). The curve marked 2 is based on (4.6.26). Data for flat plate turbulent boundary layers are also shown, and it is seen that use of (4.6.26) gives a reasonable estimate for Reynolds numbers of 10^5–10^7. Schlichting (1979) and White (1991) describe other approaches for estimating skin friction which give improved agreement at higher Re. The curve marked 3 is an empirical fit to the data, $C_f = 0.455/(\log R_L)^{2.58}$. The curve marked 4 represents the regime of laminar to turbulent transition.

4.6.5 Vorticity and velocity fluctuations in turbulent flow

Several features of turbulent flows can be connected in an instructive way with the concepts concerning vorticity that were developed in Chapter 3. One property of turbulence is an overall transfer of kinetic energy from larger to smaller length scales across a broad spectrum of motions. At one end of the spectrum are motions with length scales on order of the boundary layer thickness. At the

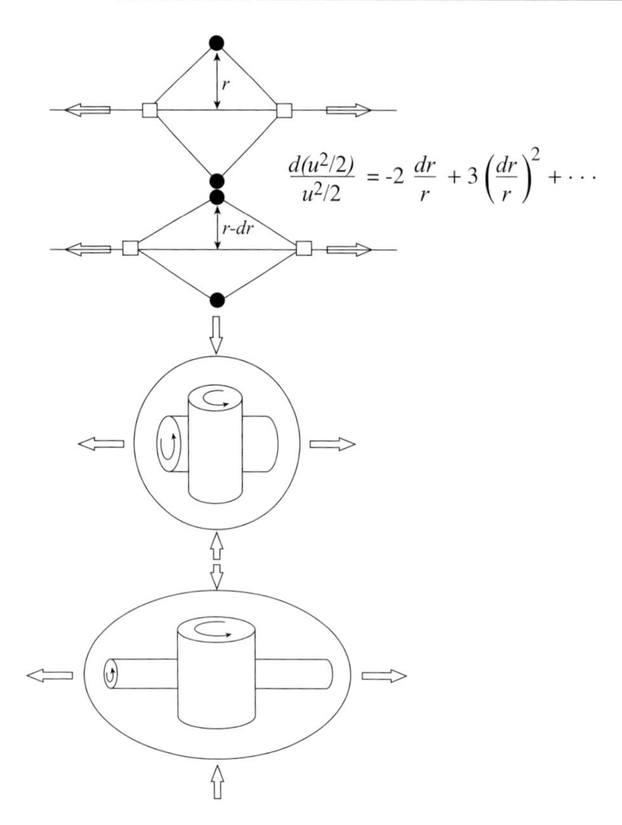

Figure 4.18: Inertial transfer in turbulent flow by the interaction of a strain-rate field and vorticity; kinetic energy per unit mass $u^2/2$ (Lumley, 1967).

other are motions with the smallest length scales in the flow, namely those for which length scale Reynolds numbers are small enough so that viscous effects dominate. Although we have described turbulent flow in a two-dimensional manner, if we look in more depth, vortex stretching, which is an inherently three-dimensional phenomenon, is at the heart of this evolution from larger to smaller scale motions.

A view of this "energy cascade" process is shown schematically in Figure 4.18. The figure depicts two vortex elements in a strain-rate field. As the vortex elements are strained, the lengthened vortex gains more energy than the shortened one loses. With kinetic energy per unit mass, $u^2/2$, for a strain dr/r there is a change $d(u^2/2)/(u^2/2)$, as given in the figure, with energy being removed from the large scale strain-rate field and put into the smaller scale vortex motion.

A second three-dimensional aspect concerns the Reynolds stresses. A general flow field consists of a time mean flow field plus a fluctuation:

$$\mathbf{u} = \overline{\mathbf{u}} + \breve{\mathbf{u}}. \tag{4.6.27}$$

The time mean momentum equation can be written for an incompressible flow as

$$\overline{\mathbf{u}} \times \overline{\boldsymbol{\omega}} = \boldsymbol{\nabla}\left[\frac{\overline{p}_t}{\rho} - \overline{(\breve{\mathbf{u}}^2)}\right] - \overline{\breve{\boldsymbol{\omega}} \times \breve{\mathbf{u}}} - \nu\boldsymbol{\nabla}^2\overline{\mathbf{u}}. \tag{4.6.28}$$

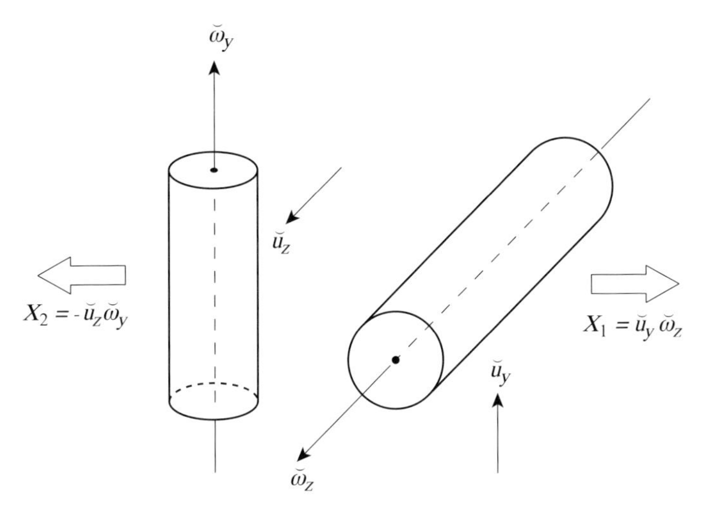

Figure 4.19: The vorticity–velocity cross-product generates effective body forces (per unit mass) X_1 and X_2 (Tennekes and Lumley, 1972).

The terms in the square brackets are normal stress terms. The contribution of the turbulence to these normal stresses is not significant because $\overline{u^2} \ll \overline{u}^2$ (Tennekes and Lumley, 1972). To show the effect of the other terms, consider a two-dimensional time mean flow and apply the boundary layer approximations. The equation for $\overline{u}_x$ can be written as

$$\overline{u}_x\frac{\partial\overline{u}_x}{\partial x}+\overline{u}_y\frac{\partial\overline{u}_x}{\partial y}=-\frac{1}{\rho}\frac{\partial\overline{p}}{\partial x}+(\overline{\breve{u}_y\breve{\omega}_z}-\overline{\breve{u}_z\breve{\omega}_y})+\nu\frac{\partial^2\overline{u}_x}{\partial y^2}. \tag{4.6.29}$$

Comparison with (4.6.5) shows that the vortex terms represent the cross-stream derivative of the Reynolds stress $-(\overline{\breve{u}_x\breve{u}_y})$:

$$\frac{\partial}{\partial y}(-\overline{\breve{u}_x\breve{u}_y})=(\overline{\breve{u}_y\breve{\omega}_z}-\overline{\breve{u}_z\breve{\omega}_y}). \tag{4.6.30}$$

The Reynolds stress term may be given an interpretation as shown in Figure 4.19. The vorticity–velocity cross-product generates an effective body force per unit mass, which can be regarded as a generalization of the result for lift due to the flow past an airfoil with vorticity aligned along its span (Lighthill, 1962).

4.7 Applications of boundary layer analysis: viscous–inviscid interaction in a diffuser

Chapter 2 introduced the idea that the presence of a boundary layer creates flow blockage and makes the effective flow area of a channel or duct less than the geometric area, decreasing the mass flow for a given total-to-static pressure ratio. To calculate this effect in a general situation requires addressing the interaction of the inviscid-like flow outside the boundary layer and the viscous layer.[6] Historically, the method initially developed to deal with this problem was one of successive approximations in which the flow external to the boundary layer was first calculated neglecting the presence of the boundary layer with the resulting pressure distribution used in a boundary layer calculation. Computing the displacement thickness and using it to modify the body shape, one could recalculate the external flow, obtain an improved pressure distribution and then recompute the boundary layer. This procedure works well if the boundary layers are thin (in an appropriate non-dimensional sense) but the inherently uni-directional passing of information does not capture situations in which there are substantial viscous–inviscid interactions and it fails in regions of flow separation. For these flows in which there is strong coupling between boundary layers and the inviscid region, a different approach is needed. A method for attacking the problem which is well suited to many internal flow situations is that of interactive boundary layer theory in which the boundary layer and the flow external to it are essentially computed simultaneously. This method also provides insight into the effects which drive the behavior of interest.

We illustrate the procedure here for a quasi-one-dimensional channel flow. Extensions to more general situations are described by Drela and Giles (1987), Strawn, Ferziger, and Kline (1984), and Tannehill, Anderson, and Pletcher (1997). As stated earlier in the chapter, it is the overall effect of the boundary layer (for example the displacement thickness) which is often of most interest, so that the analysis is given in terms of an integral boundary layer computation. This is also the simplest viscous–inviscid approach to implement, although there is no fundamental limit to posing the problem in terms of a differential computation for the boundary layer.

The specific configuration to be investigated is similar to that sketched in Figure 4.4, a symmetrical diffusing duct of length-to-width ratio such that a quasi-one-dimensional description of the inviscid-like core flow, which has velocity u_E, can be used. The core is bounded by viscous layers on the top and bottom walls.

The evolution of the momentum thickness is given by the integral form of the momentum equation, (4.3.9). Interaction between the core and the boundary layer is captured by the global continuity equation for the channel which is of local width, W:

$$u_E[W - 2\delta^*] = \text{constant}, \tag{4.7.1a}$$

or

$$\frac{du_E}{dx}[W - 2\delta^*] + \left(\frac{dW}{dx} - \frac{2d\delta^*}{dx}\right) u_E = 0. \tag{4.7.1b}$$

In (4.3.9) and (4.7.1) there are five unknowns: (i) core velocity, u_E, (ii) displacement thickness, δ^*, (iii) momentum thickness, θ, (iv) wall friction coefficient, C_f, and (v) boundary layer shape parameter, H. The integral momentum equation, the continuity equation (4.7.1), and the definition of H $(= \delta^*/\theta)$ provide three relations connecting these quantities, so that two additional relations are necessary to close the problem. The selection of conditions for closure for general two-dimensional boundary layers is discussed further in Section 4.7.2. In the next section we present a simple illustration of features of the flow to be expected using an idealized model of the boundary layer.

[6] The term "inviscid flow" or sometimes core flow is often used to describe the region outside the boundary layer. This does not mean that the fluid is inviscid, but rather that the velocity gradients are small enough so that shear stresses can be neglected and the flow treated *as if* it were inviscid.

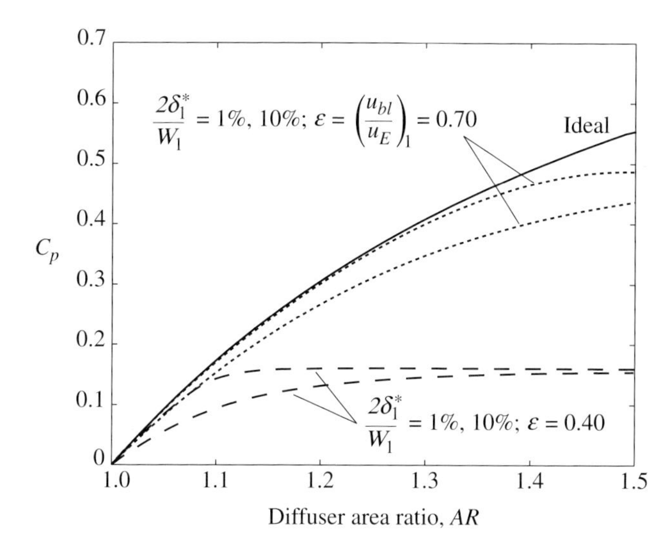

Figure 4.20: Effect of the initial boundary layer thickness, shape parameter $H_1 = 1/\varepsilon$, and area ratio on the diffuser pressure rise; station 1 is inlet, station 2 exit.

4.7.1 Qualitative description of viscous–inviscid interaction

To qualitatively illustrate the features of viscous–inviscid interactions we take the boundary layer to be represented by an inviscid stream with uniform velocity ε times the local free-stream value ($u_{bl} = \varepsilon u_E$, $\varepsilon < 1$). The displacement and momentum thicknesses are then given by $\delta^*/\delta = (1 - \varepsilon)$ and $\theta/\delta = [\varepsilon(1 - \varepsilon)]$ respectively, with shape parameter, $H = 1/\varepsilon$. The momentum integral equation, the continuity equation for the channel, and the continuity equation for the core flow furnish three coupled differential equations for the core flow velocity, u_E (or, non-dimensionally, the ratio of core flow velocity at a given station x to the core flow velocity at the initial station, u_E/u_{E_1}), the parameter ε, and the local boundary layer thickness, δ.

For this idealized example the integration of the equations can be carried out explicitly. The conditions of constant stagnation pressure in the core flow and in the boundary layer, continuity for the core and boundary layer, and the condition that the core and boundary layer stream heights add up to the channel height generate four coupled algebraic equations for the velocities and stream heights of the core and boundary layers at the inlet (station 1) and exit (station 2) locations. Solution of these shows that the effects of the boundary layer shape parameter and boundary layer blockage influence the overall pressure rise differently. The core velocity corresponding to the maximum pressure rise for a given boundary layer velocity parameter ε_1 occurs when $u_{bl} = 0$ or

$$\varepsilon_1^2 = 1 - \left(\frac{u_{E_2}}{u_{E_1}}\right)^2 = \frac{\Delta p_{max}}{\frac{1}{2}\rho u_{E_1}^2} = C_{p_{max}}. \tag{4.7.2}$$

The initial boundary layer thickness does not affect the maximum pressure rise that can be obtained as the area ratio is varied but it does determine, for a given geometry, the maximum pressure rise.

Figure 4.20 shows the effect of the initial boundary layer blockage, δ_1^*/W_1, and the boundary layer shape parameter, H_1, on the pressure rise coefficient as a function of diffuser area ratio. (For

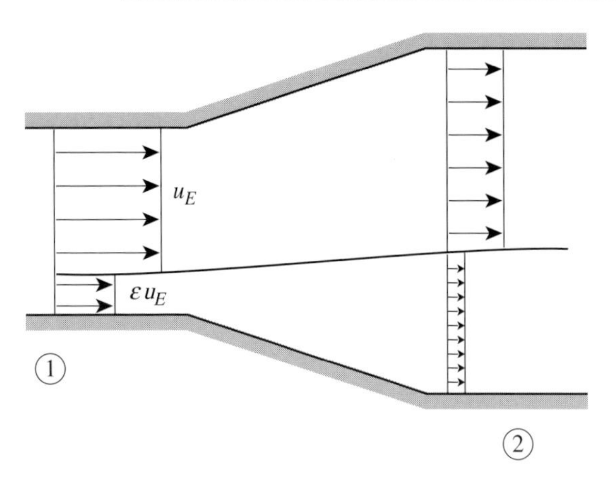

Figure 4.21: Relative growth of the low velocity region due to pressure rise ($p_2 > p_1$).

reference, shape parameters for constant pressure laminar and turbulent boundary layers are roughly 2.5 and 1.4 respectively, corresponding to ε_1 of 0.4 and 0.7.) The area ratio is taken only to a value of 1.5 because this simple representation of the boundary layer cannot capture the actual separation process, but the figure shows features seen in experiment such as the reduced pressure rise as either the inlet blockage or inlet boundary layer shape parameter increases.

The decrease in pressure rise compared to the ideal behavior based on geometry occurs because of the growth of the low velocity region, as shown in Figure 4.21. Along any streamline, the relative change in velocity magnitude is

$$\frac{du}{u} = -\frac{dp}{\rho u^2}.$$

Boundary layer and core experience the same pressure rise so that the former, which has lower velocity, has a larger relative deceleration. As indicated in Figure 4.21, the effective area ratio for the core flow is less than the geometrical area ratio. Although the arguments are strictly correct for inviscid flow only, the general trend applies to boundary and free shear layers.

4.7.2 Quantitative description of viscous–inviscid interaction

As mentioned, in addition to the momentum integral equation, the equation expressing global continuity across the channel, and the definition of the boundary layer shape parameter, two remaining relations are needed. There is no unique approach to such closure for turbulent boundary layers and a number of approaches exist in the literature. Examples are the use of an equation describing the rate at which free-stream fluid is brought, or entrained, into the boundary layer and an equation for the rate of change of kinetic energy defect. Because of the averaging process the latter has a different content than the momentum equation and can be used as a separate piece of information. We describe one approach below as representative, but we emphasize that a number of methods exist, consisting of a set of ordinary differential equations which can be integrated along the channel or duct, plus supplementary empirical algebraic relations between parameters which close the problem (Drela and Giles, 1987; Strawn *et al.*, 1984; White, 1991; Johnston, 1997).

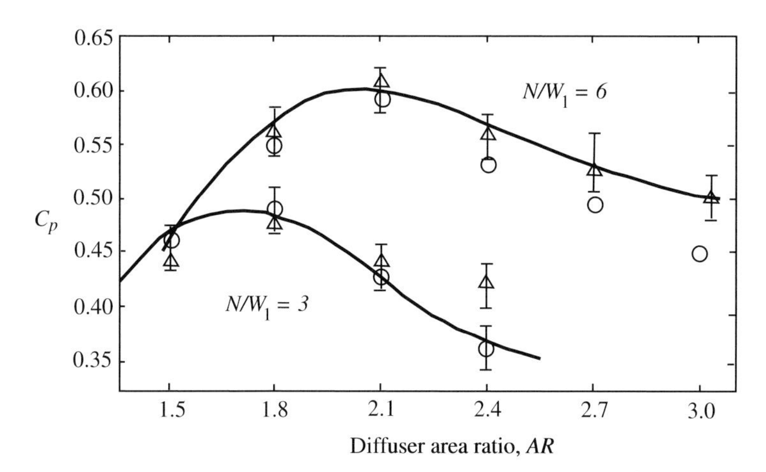

Figure 4.22: Diffuser pressure rise coefficient. Solid lines are integral boundary layer calculations, symbols are data (Lyrio *et al.*, 1981, based on data of Carlson and Johnston, 1965).

In the method of Drela and Giles (1987) the differential equations employed are: (i) the momentum integral equation, (ii) an expression for the variation in shape parameter, *H*, derived from the kinetic energy equation, and (iii) overall mass conservation for boundary layers and the core flow. These are

$$\frac{d\theta}{dx} = f_1(\theta, H, u_E), \tag{4.7.3a}$$

$$\frac{dH}{dx} = f_2(\theta, H, u_E), \tag{4.7.3b}$$

$$\frac{du_E}{dx} = f_3(\theta, H, u_E), \tag{4.7.3c}$$

where the displacement thicknesses on both surfaces have been taken to be the same.

Equations (4.7.3) need to be supplemented by empirical relations linking H^* (H^* = kinetic energy thickness/momentum thickness, θ^*/θ), C_f (the skin friction coefficient, $\tau_w/\frac{1}{2}\rho_E u_E^2$) and C_d (the dissipation coefficient, dissipation per unit length/$\rho_E u_E^3$) to the variables θ, H, and u_E. Equations (4.7.3) provide information about the evolution of a characteristic length scale, boundary layer shape parameter, and velocity. Integrating them along the channel with a given $W(x)$ yields a solution in which $u_E(x)$ is computed (rather than specified), supplying the desired interaction between core flow and the boundary layer. Because the computation includes this interaction, procedures of this type are suitable for attached, separating, and reattaching flows.

To illustrate the results of an integral boundary layer approach to viscous–inviscid interaction, as well as to show some quantitative features of internal flows in adverse pressure gradients, we return to the theme of computing diffuser pressure rise behavior. Figure 4.22 shows integral boundary layer calculations and measurements for two channel diffusers, $N/W_1 = 3$ and 6, at area ratios from 1.4 to 3.1 (Lyrio, Ferziger, and Kline, 1981). The data span regimes from operation with no appreciable stall, through the peak value of C_p, to well into transitory stall (see Figure 4.2). The inlet ratio of displacement thickness to width (δ^*/W_1) is 0.03.

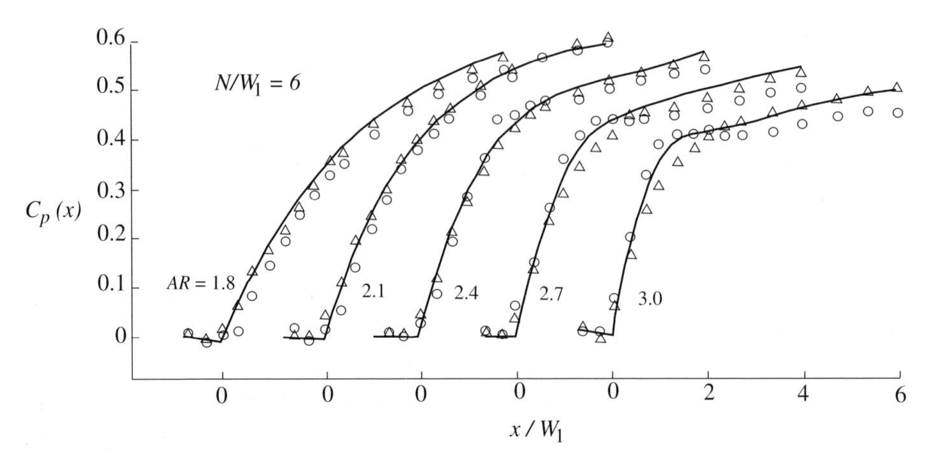

Figure 4.23: Pressure rise coefficient along $N/W_1 = 6$ diffuser wall (Lyrio *et al.*, 1981). Lines are integral boundary layer computation; symbols are data from Carlson and Johnston (1967).

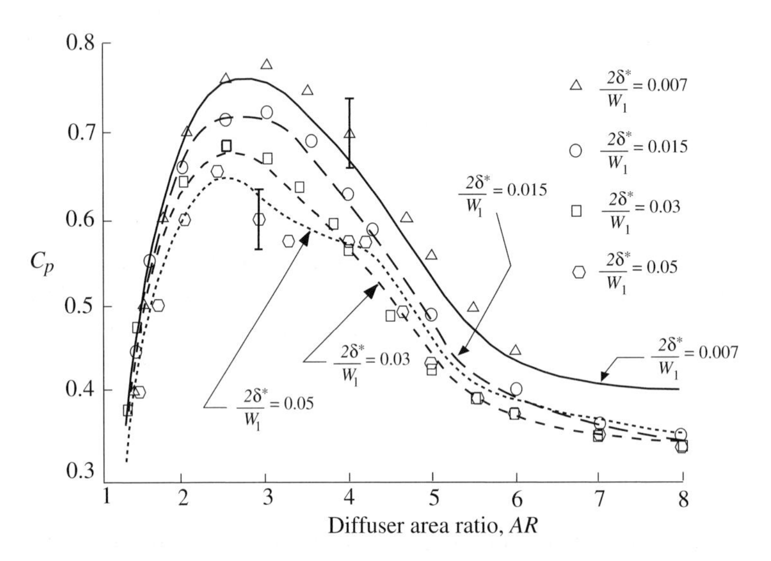

Figure 4.24: Effect of inlet blockage on pressure rise at $N/W_1 = 10$ (Lyrio *et al.*, 1981). Lines are integral boundary layer computation; symbols are data from Reneau *et al.* (1967).

The static pressure coefficient along the diffuser is given in Figure 4.23 for the $N/W_1 = 6$ case, for area ratios both less and greater than that corresponding to the peak pressure rise. In the regime with no appreciable stall the pressure rises smoothly along the entire diffuser, although the rate of rise decreases with distance. For area ratios larger than 2.1, however, the diffuser is operating in transitory stall and the pressure distribution shows a marked flattening.

A third feature is captured in Figure 4.24, which shows the effect of the initial displacement thickness on pressure rise for a family of diffusers of constant non-dimensional length, $N/W_1 =$ 10. The flow regimes extend from unstalled to fully stalled. As implied by the flow regime map in

Figure 4.3, the onset of the different regimes is little affected by inlet blockage but the pressure rise at any set of geometric parameters does depend on this parameter.

4.7.3 Extensions of interactive boundary layer theory to other situations

4.7.3.1 Non-one-dimensional flow

Although the geometries addressed so far were those in which the inviscid flow could be considered quasi-one-dimensional, the approaches described are also applicable to situations with strong streamline curvature. In such a case the normal component of the inviscid momentum equation shows (see Section 2.4) that the boundary layers on the two walls of the passage are subjected to different pressure gradients. Normal pressure gradients in the channel must thus be obtained as part of the solution procedure with the core described by a suitable inviscid model. (For an incompressible irrotational core, for example, Laplace's equation is appropriate.) Viscous–inviscid approaches are able to capture the strong interaction of boundary layer and free stream in cases that include this effect, compressibility, and flow rotationality outside the boundary layer as described in the above references. Even with symmetric geometries, regions of separation and back flow often occur asymmetrically so that the streamline curvature is produced not by the physical geometry but by the need for the flow to detour around large regions of nearly stagnant or reverse flow. The inviscid portion of the flow field must also be treated in a two-dimensional manner in these situations.

4.7.3.2 Boundary layers on rough walls

Discussion in this chapter has been for boundary layers on smooth walls, but considerable data and methodology exist to allow one to estimate the behavior of boundary layers on rough walls. This includes guidelines for the characterization of the roughness, data on the increase in skin friction with roughness, and methods to include the effects of roughness in boundary layer calculations. For discussion see White (1991) or Schlichting (1979).

4.7.4 Turbulent boundary layer separation

The reader can by now infer that a critical issue concerning boundary layer behavior in adverse pressure gradients is when and where the boundary layer will separate from the surface. Historically, the approach used for estimating this in fluid machines has been through correlations that connect the limits of pressure capability to appropriate overall geometric parameters for the device of interest. This method works well for a range of geometries of similar type and has been used with success for diffusers (e.g. the flow regime map of Figure 4.3) including two-dimensional straight channel, curved, conical and annular geometries (Kline and Johnston, 1986; Sovran and Klomp, 1967). It has also been widely used for estimates of separation limits in turbomachinery (Cumpsty, 1989; Casey, 1994; Kerrebrock, 1992).

At a less empirical level, the conditions at which separation occurs can be linked to local boundary layer integral properties. Turbulent separation or "detachment" (as many workers refer to it) is more realistically viewed as a process rather than as a discontinuous change. This process can be described in terms of a parameter, ξ, the percentage of instantaneous forward flow in the viscous sublayer

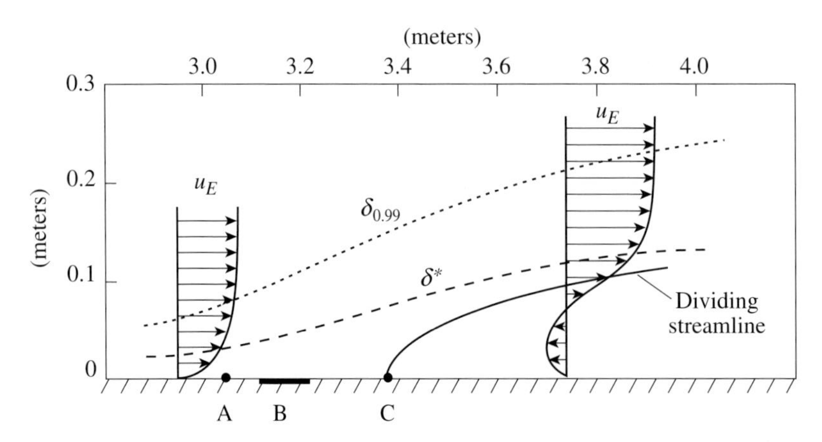

Figure 4.25: Schematic of two-dimensional detachment (not to scale) (Kline *et al.*, 1983).

(Kline, Bardina, and Strawn, 1983; Simpson, 1996). For a two-dimensional laminar boundary layer the behavior of ξ would be (at least conceptually) a step function, with a sudden shift from 100% to 0% forward flow. For a turbulent boundary layer, measurements show that as one goes from a location at which the flow is fully attached to one where it is fully detached, the flow near the wall fluctuates, with the percentage of the time the velocity is in the upstream direction increasing with streamwise distance. The parameter ξ is thus one metric for the degree of detachment. As sketched in Figure 4.25 at station A measurable backflows near the wall are first observed with $\xi > 0.5\%$. In zone B, appreciable backflow occurs with ξ between 5 and 50% and this region is denoted as incipient detachment. Point C is the location of full detachment, where $\xi = 50\%$ and $\tau_w = 0$. The detachment process occurs over a length which can be several boundary layer heights, with the boundary layer shape parameter changing from a value associated with attached flow to one reflecting detached flow.

The conditions that characterize separation are portrayed in Figure 4.26 in a plane based on the non-dimensional parameters h and χ, where

$$h = \frac{\delta^* - \theta}{\delta^*} = \frac{H - 1}{H} \quad \text{and} \quad \chi = \frac{\delta^*}{\delta}. \tag{4.7.4}$$

Use of the parameters h and χ has several advantages. First, the relation between h and χ is approximately linear for turbulent boundary layers near separation and depends only weakly on the Reynolds number. Two fits to experimental data are shown in the figure, one for $Re_{\delta^*} = 10^3$ and one for $Re_{\delta^*} = 10^6$, and it is seen that the differences are small. For high Reynolds numbers, a good approximation, which is essentially the trajectory of the boundary layer state as conditions near detachment, is $h = 1.5\chi$. Experiments show that the conditions for intermittent detachment occur at $h = 0.63$ and full detachment at $h = 0.75$ ($\chi = 0.5$) (indicated in Figure 4.26).

4.8 Free turbulent flows

4.8.1 Similarity solutions for incompressible uniform density free shear layers

In this section we describe some basic features of constant density, incompressible, free shear layers: mixing layers, jets, and wakes. The discussion is restricted to situations in which the flows have

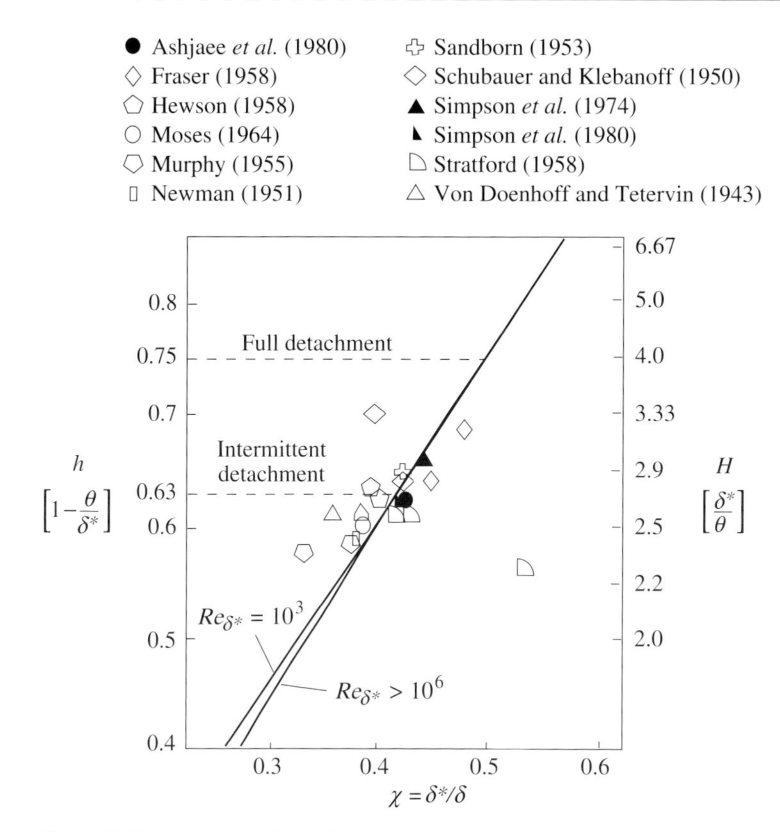

Figure 4.26: Correlation for turbulent boundary layer detachment (Kline *et al.*, 1983).

self-similarity, i.e. the region of interest is far enough downstream so that velocity and shear stress are functions of a similarity parameter. For these important cases, useful scaling information can be obtained without solving the equations of motion. Consider first a constant pressure, two-dimensional or plane turbulent jet.[7] The equations for the time mean velocity are

$$u_x \frac{\partial u_x}{\partial x} + u_y \frac{\partial u_y}{\partial y} = \frac{1}{\rho}\frac{\partial \tau_{xy}}{\partial y}, \tag{4.8.1a}$$

$$\frac{\partial u_x}{\partial x} + \frac{\partial u_y}{\partial y} = 0, \tag{4.8.1b}$$

where τ_{xy} is the turbulent stress. Because the jet spreads in a constant pressure environment, the jet momentum, J, remains invariant with axial distance and at any axial station,

$$\int_{-\infty}^{\infty} u_x^2 dy = \frac{J}{\rho} = \text{constant}. \tag{4.8.2}$$

[7] In free shear layers transition to turbulence occurs at much lower Reynolds numbers than in boundary layers (see Chapter 6). We thus consider only the turbulent case.

With b the local width of the jet, and u_{cl} the local centerline velocity, if the time mean velocity field is similar at different axial locations, then

$$\frac{u_x}{u_{cl}} = f_1\left(\frac{y}{b}\right), \tag{4.8.3a}$$

$$\frac{\tau}{\rho u_{cl}^2} = f_2\left(\frac{y}{b}\right). \tag{4.8.3b}$$

In (4.8.3), b is a characteristic jet width, say the width for the location where the mean velocity is half the centerline value. In (4.8.3), f_1 and f_2 are functions whose form can remain unknown. If we look for similarity of the form $b \sim x^p$ and $u_{cl} = x^{-q}$ respectively, the terms in the equation of motion have the behavior

$$u_x \frac{\partial u_x}{\partial x} \sim x^{-2q-1}; \quad u_y \frac{\partial u_x}{\partial y} \sim x^{-2q-p}; \quad \frac{\partial \tau}{\partial y} \sim x^{-2q-p}. \tag{4.8.4}$$

To have the equation independent of x, in other words to have the profiles exhibit similarity, requires that

$$2q + 1 = 2q + p \quad \text{or} \quad p = 1.$$

The invariance of momentum flux expressed in (4.8.2) implies that x^{-2q+p} must be constant so that $q = 1/2$. The plane jet thus spreads linearly with x and the centerline velocity, u_{cl}, decreases as $11\sqrt{x}$.

A similar analysis can be applied to the circular jet, which has equations

$$u_x \frac{\partial u_x}{\partial x} + u_r \frac{\partial u_x}{\partial r} = \frac{1}{\rho r}\frac{\partial}{\partial r}(r\tau_{rx}), \tag{4.8.5}$$

$$\frac{\partial}{\partial x}(ru_x) + \frac{\partial}{\partial r}(ru_r) = 0. \tag{4.8.6}$$

Jet momentum invariance is given by

$$2\pi \int_{-\infty}^{\infty} u_x^2 r dr = \frac{J}{\rho} = \text{constant}. \tag{4.8.7}$$

The results are a jet width which increases linearly with x and a centerline velocity which decreases as $1/x$.

Like arguments can also be made for wakes. The conditions for similarity to apply are that the locations are far enough downstream so the velocity variation in the wake, Δu, obeys $\Delta u = u_E - u_x \ll u_E$. In this case, the momentum equation for a plane wake can be approximated as that for a uni-directional flow:

$$u_E \frac{\partial \Delta u}{\partial x} = \frac{1}{\rho}\frac{\partial \tau_{xy}}{\partial y}. \tag{4.8.8}$$

Consistent with this approximation conservation of momentum is

$$u_E \int_{-\infty}^{\infty} \Delta u dy = \frac{J}{\rho} = \text{constant}. \tag{4.8.9}$$

Table 4.2 *Power laws for the increase in width and decrease in centerline velocity in terms of distance x for free turbulent shear layers (Schlichting, 1979)*

	Width, b	Centerline velocity u_{cl} *or velocity defect* Δu_{cl}
Mixing layer (free jet boundary)	x	x^0
Two-dimensional jet	x	$x^{-1/2}$
Circular jet	x	x^{-1}
Two-dimensional wake	$x^{1/2}$	$x^{-1/2}$
Circular wake	$x^{1/3}$	$x^{-2/3}$

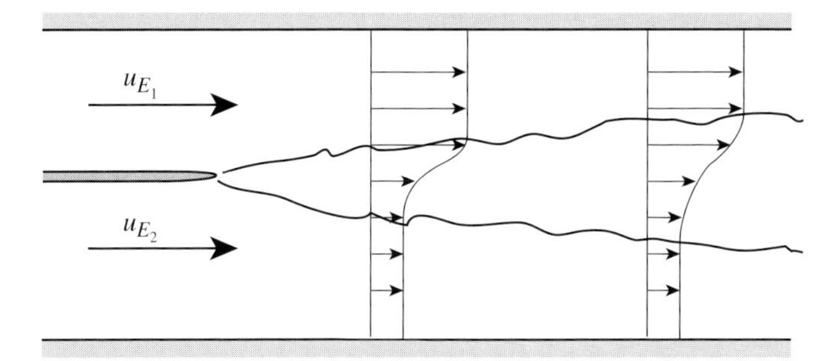

Figure 4.27: Schematic of a mixing layer between parallel streams of differing velocity.

The results for the wake width and wake velocity defect are: wake half-width, $b \propto x^{1/2}$, centerline velocity defect, $\Delta u_{cl} \propto 1/\sqrt{x}$.

Table 4.2 summarizes the similarity scaling for width and centerline velocity for different free shear layers.

To determine the time mean velocity profile in these flows, we can use the similarity to infer the behavior of the eddy viscosity. With the shear stress given by $\tau_{xy} = \mu_{turb}\, \partial u_x/\partial y$, the eddy viscosity, μ_{turb}, scales as x^{p-q}. From Table 4.2, μ_{turb} is constant for the round jet and the plane wake, implying that the spreading behavior should be similar to a laminar flow with a much higher viscosity than the actual value.

4.8.2 The mixing layer between two streams

An often encountered situation is the smoothing out of a velocity discontinuity between two streams at u_{E_1} and u_{E_2} as sketched in Figure 4.27. For this mixing layer flow the similarity considerations show that the eddy viscosity scales as x. Since the width of the mixing layer also scales with x, the eddy viscosity is proportional to the shear layer width, b (this is also consistent with the approximation of a uniform eddy viscosity in the outer part of a boundary layer). The characteristic velocity is the velocity difference between the two streams so the eddy viscosity is given by

$$\mu_{turb} = \text{constant} \cdot \rho x (u_{E_1} - u_{E_2}). \tag{4.8.10}$$

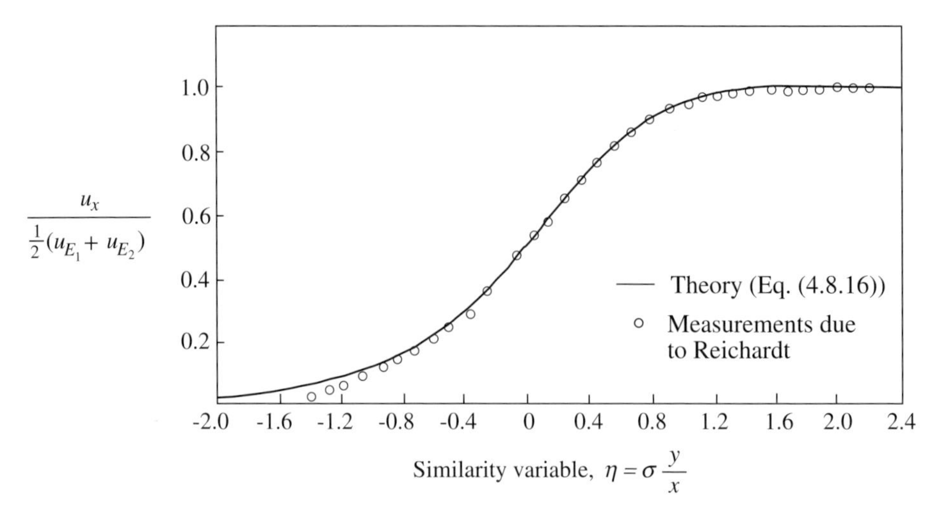

Figure 4.28: Velocity distribution in the mixing zone of a jet; $\sigma = 13.5$ (Schlichting, 1979).

The equations describing incompressible constant pressure mixing layer evolution are thus

$$u_x \frac{\partial u_x}{\partial x} + u_y \frac{\partial u_x}{\partial y} = kx \frac{\partial^2 u_x}{\partial y^2}, \tag{4.8.11}$$

where k is a constant, and

$$\frac{\partial u_x}{\partial x} + \frac{\partial u_y}{\partial y} = 0. \tag{4.8.12}$$

If the approximation is made that $(u_{E_1} - u_{E_2})/(u_{E_1} + u_{E_2})$ is much less than unity, the equations allow an analytical solution (Schlichting, 1979). Using a similarity variable η of the form $\eta = \sigma(y/x)$, where σ is a constant, a stream function can be defined as $\psi = x(u_{E_1} + u_{E_2})\, F(\eta)$, with the axial velocity given as

$$u_x = \left(\frac{u_{E_1} + u_{E_2}}{2}\right) \sigma F'(\eta). \tag{4.8.13}$$

Substituting into the momentum equation (4.8.11) leads to an ordinary differential equation for F:

$$F''' + 2\eta F'' = 0, \tag{4.8.14}$$

with boundary conditions $F'(\eta) = \pm 1$ at $\eta = \pm\infty$. The solution is

$$F'(\eta) = \mathrm{erf}(\eta) = \frac{2}{\sqrt{\pi}} \int_0^{\eta} e^{-z^2} dz \tag{4.8.15}$$

or

$$u_x = \frac{u_{E_1} + u_{E_2}}{2} \left[1 + \left(\frac{u_{E_1} - u_{E_2}}{u_{E_1} + u_{E_2}}\right) \mathrm{erf}(\eta)\right]. \tag{4.8.16}$$

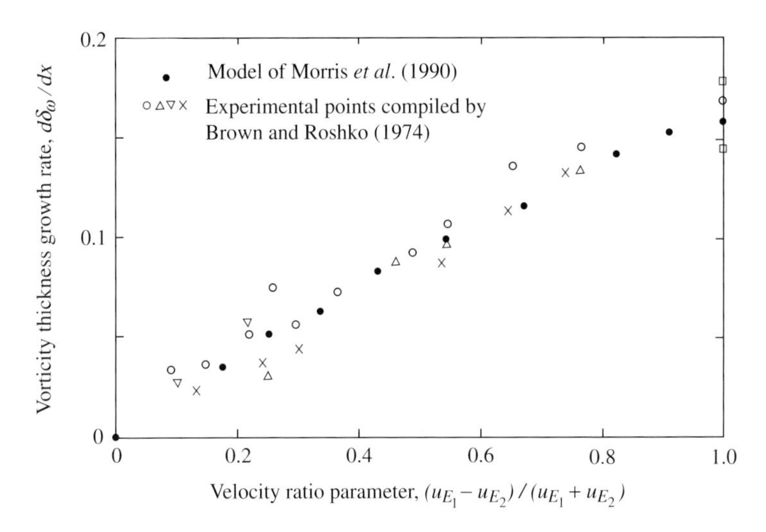

Figure 4.29: Growth rate of the free shear layer; dependence on velocity difference (Roshko, 1993a).

A comparison of (4.8.16) with data is given in Figure 4.28. The parameter σ, which must be found from experiment, has been determined to be 13.5. The rate of spreading of the edge of a shear layer with $u_{E_2} = 0$ is thus roughly $1/10$; this can be compared with the $1/50$ rate of growth of a turbulent boundary layer. The calculated eddy viscosity is $\mu_{turb} = 0.014\rho u_1$, independent of the Reynolds number.

The scaling implied by (4.8.16) can also be compared against experimental results for different values of the velocity ratio parameter $(u_{E_1} - u_{E_2})/(u_{E_1} + u_{E_2})$ in Figure 4.29 (Roshko, 1993a; see also Brown and Roshko, 1974). The growth rate used in the figure is the derivative of the vorticity thickness, δ_ω, defined as

$$\delta_\omega = \frac{1}{|\omega|_{max}} \int_{-\infty}^{\infty} |\omega| dy,$$

where $\omega = -\partial u_x/\partial y$. The vorticity thickness is appropriate, because modern theories of turbulent shear layers view their growth as "basically the kinematic problem of the unstable motion induced by the vorticity" (Brown and Roshko, 1974). The shear layer grows linearly with x and the derivative of the vorticity thickness is given by

$$\frac{d\delta_\omega}{dx} = \frac{\delta_\omega}{x - x_{vo}}, \tag{4.8.17}$$

where x_{vo} is the virtual origin of the mixing layer. The derivative of the vorticity thickness can be related to the spreading parameter σ, when the profile is fitted by an error function, as $\sigma d\delta_\omega/dx = \sqrt{\pi}$. The best-fit line for the data has the equation $d\delta_\omega/dx = 0.18(u_{E_1} - u_{E_2})/(u_{E_1} + u_{E_2})$. Also included in the figure are the results of computations by Morris *et al.* (1990), based on a model of the vortical structure in the shear layer, which have no empirical constants.

Schlichting (1979) shows a number of examples using approximate analyses such as that described above. Figure 4.30 also taken from that reference shows the velocity profile in the wake behind a

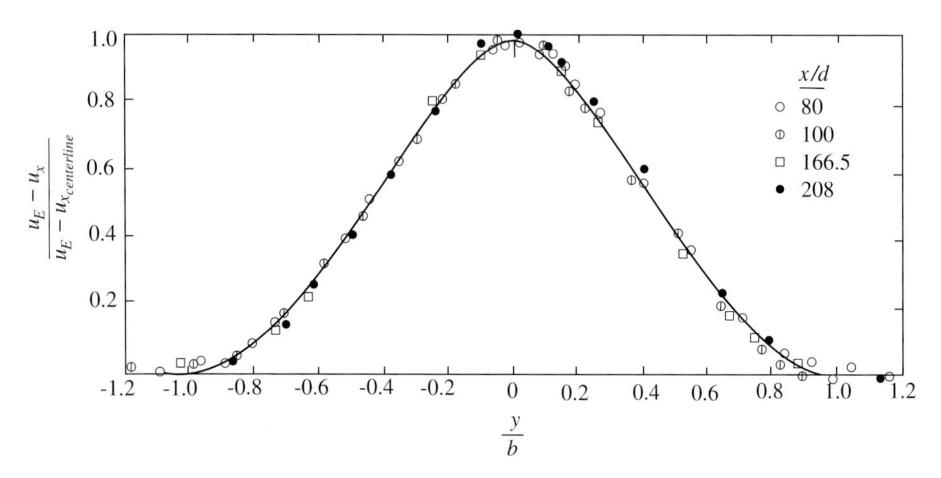

Figure 4.30: Velocity distribution in a two-dimensional wake of half-width b behind a circular cylinder of diameter d. Comparison between theory and measurement after Schlichting (1979).

two-dimensional cylinder as a function of the similarity variable $\eta = y/\sqrt{xC_D d}$, where d is the cylinder diameter and C_D is the drag coefficient. The theoretical expression is the solid curve and the symbols show measurements. The scaling with downstream distance is shown in Figure 4.31. The wake width measured to the half-velocity points, $b_{1/2}$, is given by $b_{1/2} = \frac{1}{4}(xC_D d)^{1/2}$.

4.8.3 The effects of compressibility on free shear layer mixing

The analysis and experiments presented have all been for incompressible flow. It is well documented that the spreading rate of a two-dimensional shear layer decreases as the flow Mach number increases. It has been suggested (Papamoschou and Roshko, 1988) that to a large extent the effects of compressibility can be viewed as a function of the convective Mach numbers of the large scale vortical structures which are found in shear layers. The convective Mach numbers measure the relative free-stream Mach numbers as seen from a frame of reference translating with these structures. For streams of velocities u_{E_1} and u_{E_2}, speeds of sound a_1 and a_2 respectively, and a velocity of the large scale structures equal to u_c, the convective Mach numbers, M_c, of the two streams are

$$M_{c_1} = \frac{u_{E_1} - u_c}{a_1}, \quad M_{c_2} = \frac{u_c - u_{E_2}}{a_2}. \tag{4.8.18}$$

A connection with the theory of shear layer instability has also been made in that, as described by Roshko (1993a), the strong effect of compressibility in decreasing growth rate correlates with the corresponding effect on the amplification rate of small disturbances within the shear layer. Figure 4.32 shows both these points. In the figure the derivative of the shear layer vorticity thickness $d\delta_\omega/dx$ and the disturbance growth rate, both normalized by their respective values at Mach number $= 0$, are plotted versus the convective Mach number. There is a decrease in both of roughly a factor of 5 in going from a convective Mach number of zero to unity. Discussion of this effect, including the development of the arguments for the use of convective Mach number are given in Papamoschou and Roshko (1988), Dimotakis (1991), and Coles (1985).

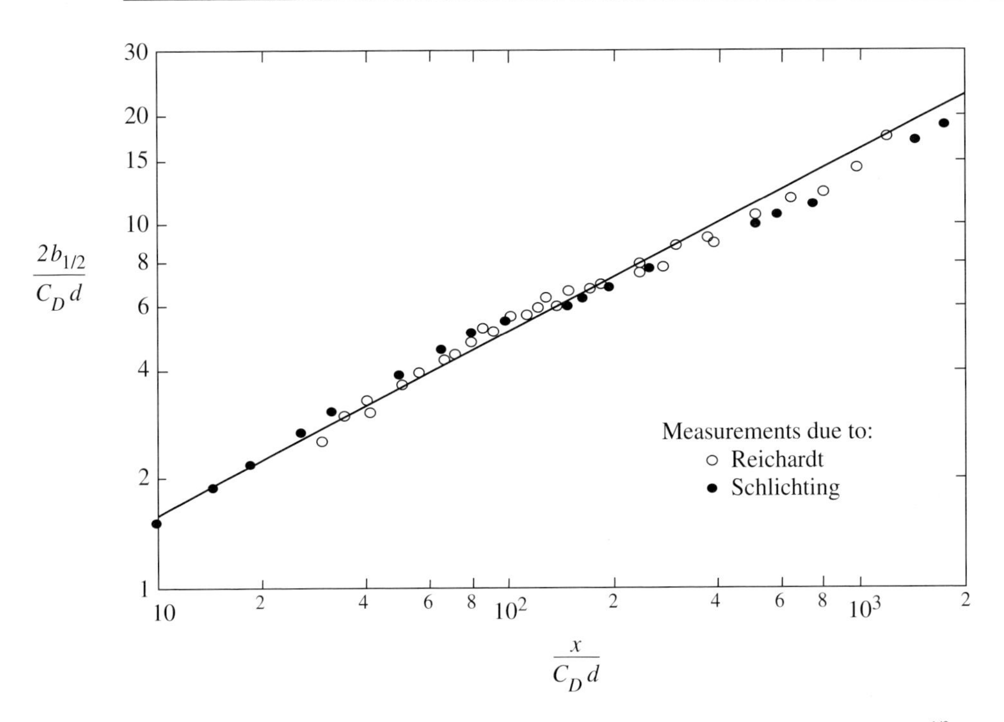

Figure 4.31: Increase of wake width behind circular cylinder. The straight line is $b_{1/2} = 1/4\,(xC_D d)^{1/2}$ (Schlichting, 1979).

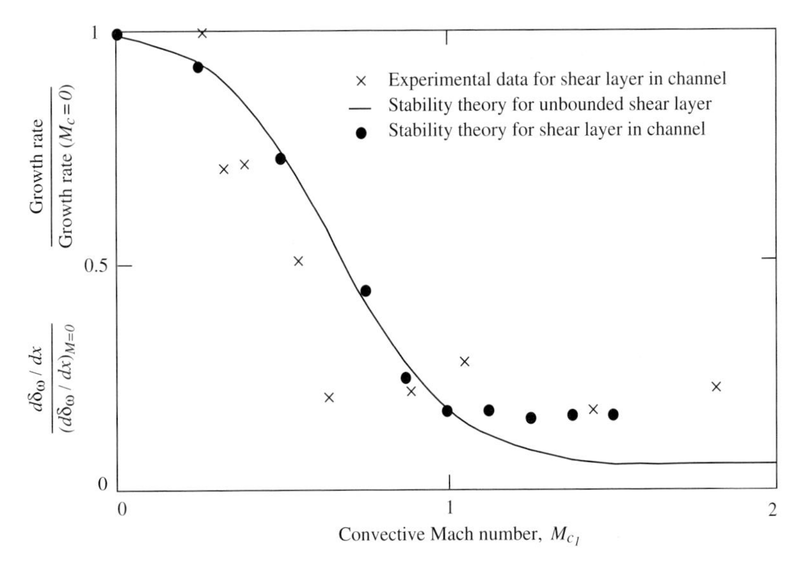

Figure 4.32: Effect of compressibility on turbulent free shear layers (Roshko, 1993a). All data normalized at $M_c = 0$. Growth rates from experiment. Amplification rates from linear stability theory (Papamoschou and Roshko, 1988; Zhuang, Kubota, and Dimotakis, 1990).

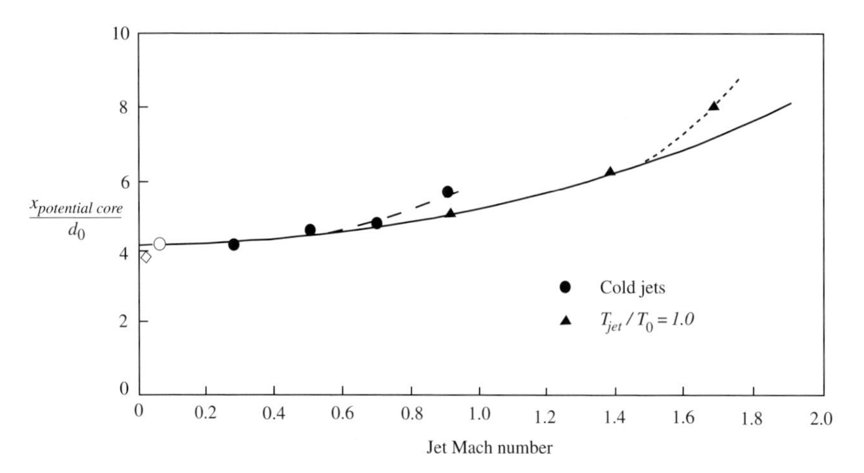

Figure 4.33: Variation of the round jet potential-core length ($x_{potential\ core}/d_0$) with Mach number; d_0 is the initial jet diameter (Lau, 1981).

4.8.4 Appropriateness of the similarity solutions

We will not explore non-similar free shear layers in any depth, but it is worthwhile to describe the conditions over which the similarity holds. We do this in the context of a round jet, which we can consider as the flow exiting from a nozzle into a still atmosphere. At the exit, the shear layers which separate the jet from the surroundings are thin compared to the jet diameter, and the jet is composed of a potential core with an axisymmetric shear layer bounding it. For a constant pressure jet, the centerline velocity in this potential core does not vary with axial distance. As one moves further downstream, the shear layers thicken, with the potential core disappearing when they merge. Figure 4.33 (Lau (1981); see also Schetz (1980)) shows the length of the potential core region measured on the centerline, in units of initial jet diameter, d_0, versus jet Mach number. There is an increase in this length as the Mach number increases, consistent with the decreased growth of the shear layers shown in Figure 4.32. The conditions for similarity are not reached until sometime after the disappearance of the potential core, say $x/d \approx 6-8$, which can be taken as a rough guideline for the situation with zero free-stream velocity.

There is a large body of work on organized structures in turbulent free shear layers. On a time-resolved basis, the shear layer has been found to consist of discrete vortical structures as in the flow visualization results of Figure 4.34 (Roshko, 1976). The increasing length scale of the vortices, and the consequent growth of the shear layer with downstream distance, can be noted. Time-resolved data show that growth of the shear layer is associated with vortex pairing. Dimotakis (1986) has used this idea to develop a model for shear layer growth which does not rely on the eddy viscosity concept, and which contains the basic processes shown by the experiments. Papamoschou and Roshko (1988) have extended the analysis to compressible mixing layers using a similar approach. Direct computational simulations of the mixing layer are also being carried out which are able to capture the overall structure as well as provide additional details of the mixing layer (see, for example, Sandham and Reynolds (1990)).

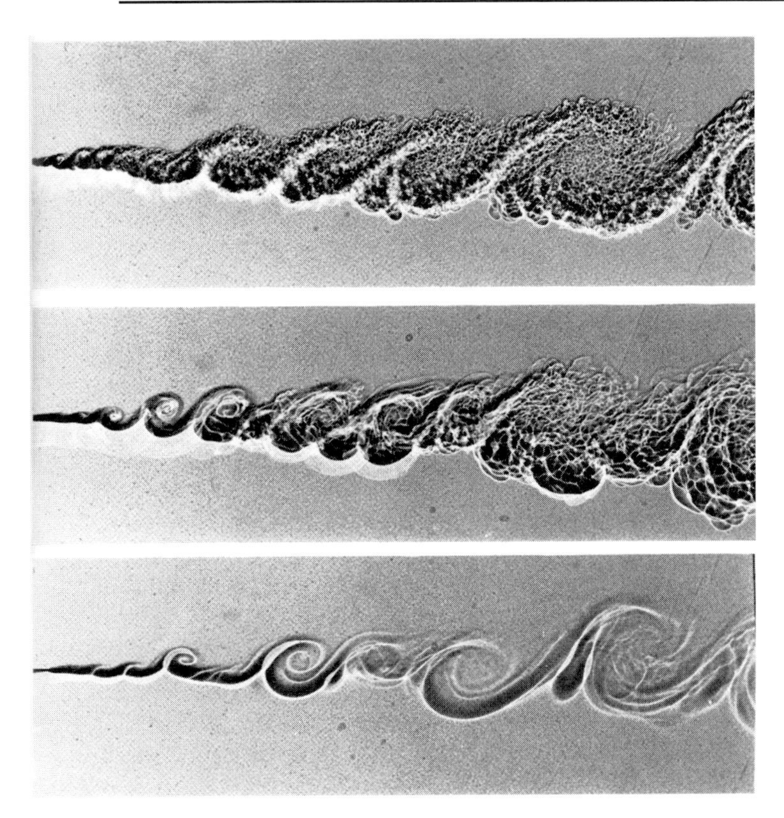

Figure 4.34: Mixing layer between helium and nitrogen $u_2/u_1 = 0.38$; $\rho_2/\rho_1 = 7$; $u_1 L/\mu_1 = 1.2$, 0.6, and 0.3×10^5, respectively, from top to bottom (L is the width of the picture) (Brown and Roshko, 1974).

4.9 Turbulent entrainment

Shear layers entrain fluid from the free stream, so there is a net flow into the layer. This *entrainment* is connected with the shear layer's ability to reattach and is also a key feature in the performance of devices such as ejectors. Turbulent entrainment can be illustrated by the behavior of a high Reynolds number circular jet issuing from a nozzle of diameter d into a still atmosphere. As described in Section 4.8, the momentum flux of the jet is constant with downstream distance. For locations far enough downstream so the similarity description applies, the jet width grows with x (see Table 4.2) and the centerline velocity decays as $1/x$, so the jet mass flux grows with x. Dimensional analysis for a jet with momentum flux J, issuing into a still atmosphere with density ρ_1, shows that mass flow in the jet, $\dot{m}$, scales as[8]

$$\frac{\dot{m}}{x J^{1/2} \rho_1^{1/2}} = C, \tag{4.9.1}$$

[8] The relevant parameters are jet mass flow and momentum flux, ambient density, and downstream distance. Thus $f(\dot{m}, J, \rho_1, x) = 0$. The non-dimensional parameter that can be made from these four quantities is that given in (4.9.1).

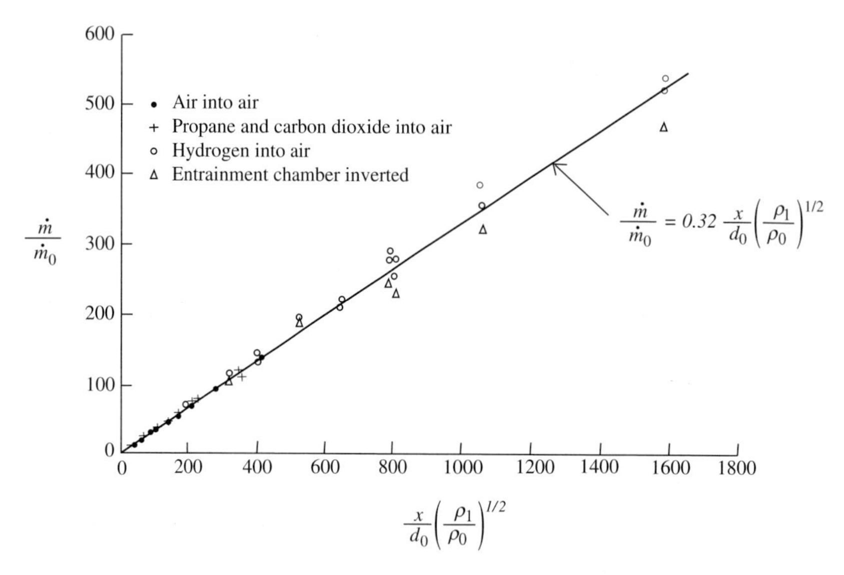

Figure 4.35: Entrainment rate for isothermal jets of density ρ_0 discharging into a still fluid of density ρ_1; $\dot{m}_0$ is the mass flow at the jet nozzle exit, d_0 is the nozzle diameter (Ricou and Spalding, 1961).

where C is a constant. The momentum flux, J, can be evaluated at the location where the jet issues (station 0). If the velocity is uniform at the nozzle exit with diameter d_0,

$$J = J_0 = \frac{\pi}{4} d_0^2 \rho_0 u_0^2. \tag{4.9.2}$$

The mass flux at the initial station is

$$\dot{m}_0 = \frac{\pi}{4} d_0^2 \rho_0 u_0. \tag{4.9.3}$$

Equations (4.9.2) and (4.9.3) can be combined to yield an expression for the local mass flux of a jet of density ρ_0 discharging into another gas of density ρ_1:

$$\frac{\dot{m}}{\dot{m}_0} = 0.32 \frac{x}{d_0} \left(\frac{\rho_1}{\rho_0}\right)^{1/2}. \tag{4.9.4}$$

The constant (0.32) in (4.9.4) has been determined from data the (Ricou and Spalding, 1961; see also Sforza and Mons, 1978) shown in Figure 4.35. The data represent a range of injected jet densities of over a factor of 20. In the case shown, nearly all the mass flux in the jet is from the surroundings, but all the momentum flux is put in through the initial jet fluid. Additional information concerning shear layer entrainment is given by Dimotakis (1986) and Turner (1986).

4.10 Jets and wakes in pressure gradients

There are many configurations in which jets and wakes are subjected to streamwise pressure gradients. Examination of this situation is not only of interest for these applications but it also provides an instructive view of the competition between (turbulent) shear forces and pressure fields which is inherent in the behavior of viscous layers in pressure gradients. The central issue is indicated by

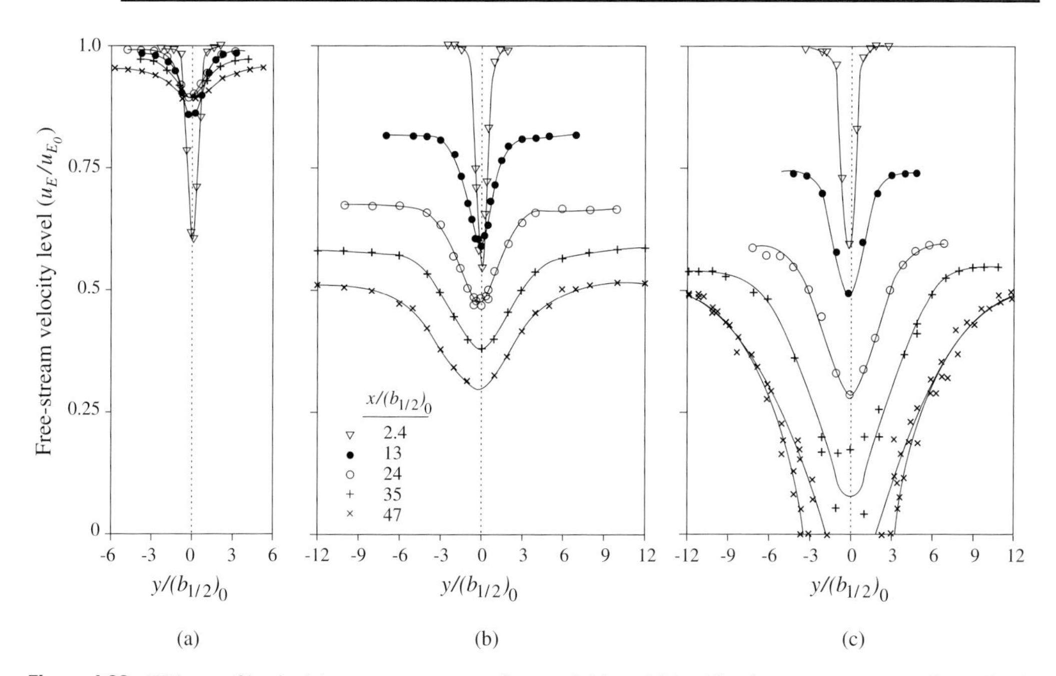

Figure 4.36: Wake profiles in (a) constant pressure flow and (b) and (c) with adverse pressure gradient; $(b_{1/2})_0$ is initial wake half-width (data of Hill *et al.*, 1963).

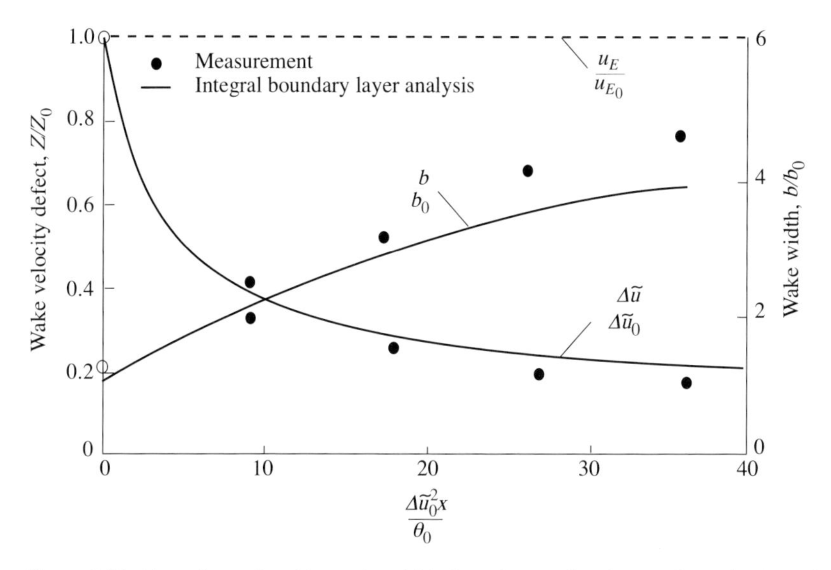

Figure 4.37: Two-dimensional jet wake width, b, and normalized centerline velocity defect, $\Delta\tilde{u} = (u_E - u_{cl})/u_E$, as a function of downstream distance at constant pressure; $\Delta\tilde{u}_0 = 0.4$ (Hill *et al.*, 1963).

Figures 4.36(a)–(c) (Hill, Schaub, and Senoo, 1963) which show measured velocity profiles of the wake of a two-dimensional plate at different downstream locations for three different streamwise pressure gradients. Figure 4.36(a) is essentially constant static pressure, Figure 4.36(b) an adverse gradient, and Figure 4.36(c) a stronger adverse gradient. The wake defect decays less rapidly in the presence of an adverse gradient. If the pressure rise is large and rapid enough, the wake can

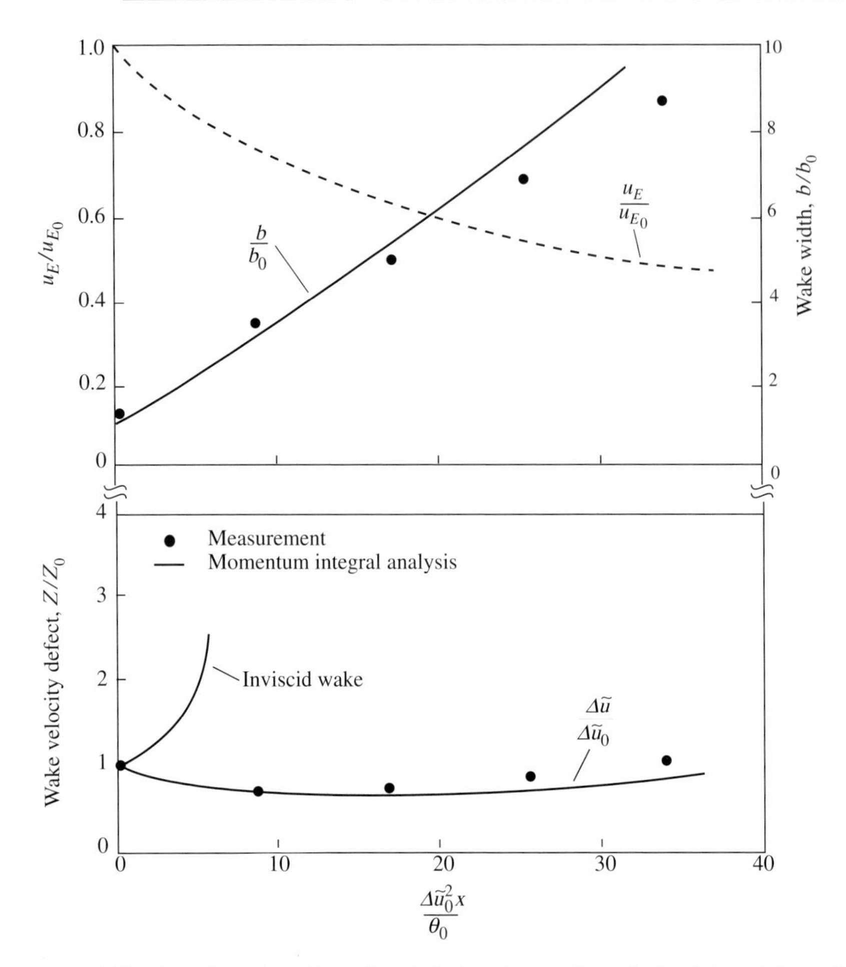

Figure 4.38: One-dimensional jet wake width, b, and centerline velocity defect, $\Delta\tilde{u}$, as a function of downstream distance in adverse pressure gradient; $\Delta\tilde{u}_0 = 0.4$ (Hill *et al.*, 1963).

stagnate or reverse in direction because of the proportionally larger deceleration than in the free stream. The increase in wake width is due to the response of the low stagnation pressure region to the static pressure field and is basically an inviscid effect. This mechanism, described in Section 4.7, underpins many phenomena that occur in flows with non-uniform stagnation pressure.

There are two competing effects in the wake. Turbulent shear forces tend to accelerate the wake fluid while pressure forces decelerate it. The general trend is that situations in which a given pressure rise occurs over a longer distance provide more opportunity for the shear forces to act. If the pressure rise occurs over a short distance, the role of the shear forces is diminished and the pressure forces thus have a greater relative effect.

The wake response can be analyzed using the momentum integral equation. The integration is across the whole wake and the shear stress is zero at both wake edges so the momentum integral takes the form

$$\frac{1}{\theta}\frac{d\theta}{dx} + (2 + H)\frac{1}{u_E}\frac{du_E}{dx} = 0. \tag{4.10.1}$$

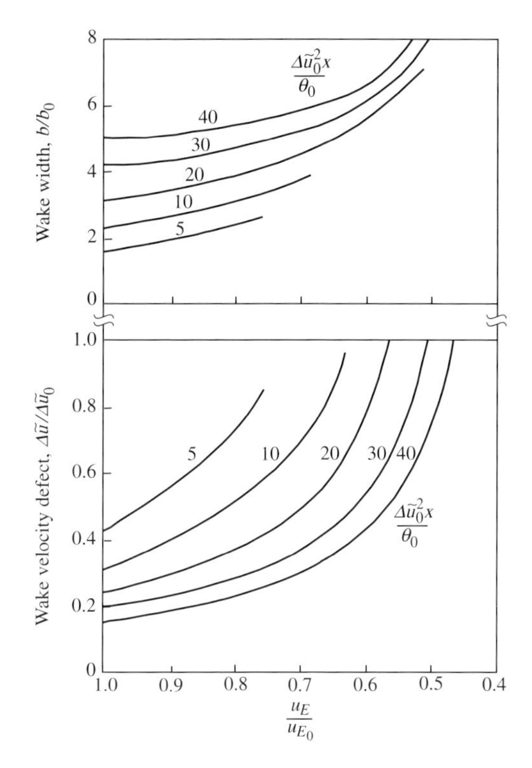

Figure 4.39: Effect of the streamwise length scale/wake thickness on the two-dimensional jet wake width and centerline velocity defect as a function of the downstream velocity level, u_E (Hill *et al.*, 1963).

For the profiles depicted in Figure 4.36 the boundary layer shape parameter, H, is approximately constant and near unity. If H is taken to be $1 + \zeta$, where ζ is a small positive constant, (4.10.1) can be integrated to yield an expression relating the momentum thickness between two levels of free-stream velocity:

$$\frac{\theta_2}{\theta_1} = \left(\frac{u_{E_1}}{u_{E_2}}\right)^{3+\zeta}. \tag{4.10.2}$$

Schlichting (1979) suggests that if (4.10.2) is used starting from an airfoil trailing edge an appropriate value for ζ is 0.2. For far downstream conditions where the wakes have $u_E/u_{E_0} \ll 1$, $H \rightarrow 1$ and $\zeta \rightarrow 0$, so the momentum thickness growth is proportional to the cube of the free-stream velocity ratio.

Figures 4.37 (for constant pressure) and 4.38 (for an adverse pressure gradient) show wake behavior as a function of the non-dimensional parameter $\Delta\tilde{u}_0^2 x/\theta_0$, where $\Delta\tilde{u}$ is the normalized centerline velocity defect, $\Delta\tilde{u} = (u_E - u_{cl})/u_E$. Station 0 denotes the initial station for the measurements. The solid curves are the result of an integral boundary layer calculation (Hill *et al.*, 1963). The figures show the wake half-width divided by the initial half-width, and the velocity defect parameter, as a function of the non-dimensional parameter $\Delta\tilde{u}_0^2 x/\theta_0$. For reference the behavior of an inviscid stream with the initial velocity defect is also indicated in Figure 4.38. This reaches zero velocity at a value of $u_E/u_{E_0} = 0.8$ for the conditions indicated, showing the role of the shear stresses in enabling the actual wake to negotiate the pressure rise.

Figure 4.39 presents a different view of the effect of streamwise distance in enabling a wake to undergo an adverse pressure gradient. The wake width and the velocity defect parameter are given as functions of pressure rise (as reflected by the free-stream velocity ratio) for different values of the parameter $\Delta\tilde{u}_0^2 x/\theta_0$. For a given level of u_E/u_{E_0} and initial wake defect $\Delta\tilde{u}_0$, the longer the non-dimensional distance over which the pressure rise occurs (x/θ), the lower the resultant wake velocity defect.

5 Loss sources and loss accounting

5.1 Introduction

Efficiency can be the most important parameter for many fluid machines and characterizing the losses which determine the efficiency is a critical aspect in the analysis of these devices. This chapter describes basic mechanisms for loss creation in fluid flows, defines the different measures developed for assessing loss, and examines their applicability in various situations.

In external aerodynamics, drag on an aircraft or vehicle is most frequently the measure of performance loss. The product of drag and forward velocity represents the power that has to be supplied to drive the vehicle. Defining drag, however, requires defining the direction in which it acts and determining the power expended requires specification of an appropriate velocity. The choice of direction is clear for most external flows but it is less evident in internal flows. Within gas turbine engines, for example, there are situations in which viscous forces can be nearly perpendicular to the mean stream direction or in which the mean stream direction changes by as much as 180°, as in a reverse flow combustor. There is also some ambiguity in the choice of an appropriate reference velocity for power input, even in simple internal flow configurations such as nozzles or diffusers where the velocity changes from inlet to outlet.

Because of this, the most useful indicator of performance loss and inefficiency in internal flows is the entropy generated due to irreversibility. The arguments that underpin this statement are presented in the first part of the chapter to illustrate quantitatively the connection between entropy rise and work lost through an irreversible process. Different entropy generation phenomena in internal flow devices are then addressed to define ways to characterize the losses and levels of efficiency in situations of interest.

Fluid flows within real devices are generally non-uniform. A basic question thus concerns the representation of the thermodynamic state, and hence loss, by a single number, i.e. the approximation of a non-uniform flow by a uniform flow with suitable average values of fluid dynamic and thermodynamic variables. This concept is discussed in some depth along with methods for arriving at an appropriate loss metric. The conditions under which one can notionally construct such averages, for example letting the flow fully mix at constant area or at constant pressure, however, are often not met. A consequence, as will be seen, is that the overall level of loss depends on the processes downstream of loss generating components as well as the flow through the component.

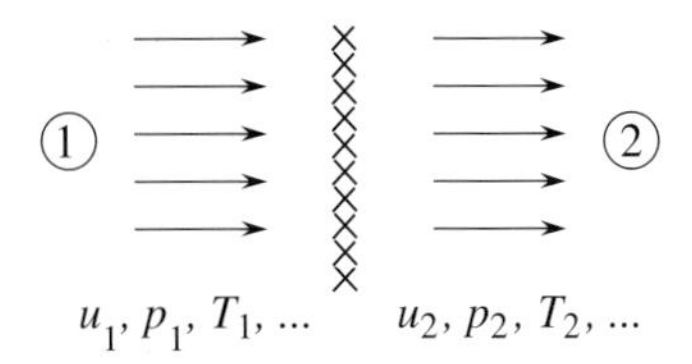

Figure 5.1: Flow through a uniform screen.

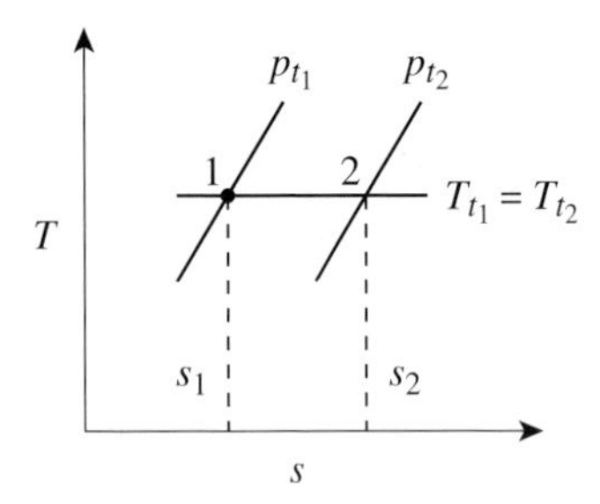

Figure 5.2: Thermodynamic states for flow through a screen.

5.2 Losses and entropy change

5.2.1 Losses in a spatially uniform flow through a screen or porous plate

We introduce the ideas through analysis of a model problem, the steady flow of a perfect gas through a uniform screen or porous plate, as sketched in Figure 5.1 (Taylor, 1971). This is a representation of a generic adiabatic throttling process. The upstream flow is uniform in space. The downstream station is taken far enough from the screen so that velocity non-uniformities arising in connection with the local details of the flow through the screen have decayed and the flow can again be considered uniform. In the irreversible state transition from upstream to downstream of the screen no shaft work is done and no heat is exchanged.

The steady flow energy equation for a flow with no work due to body forces (1.8.21) relates the stagnation enthalpy change per unit mass to the difference between heat addition and shaft work, both per unit mass,

$$\left(h_{t_2} - h_{t_1}\right) = q - w_{shaft}. \tag{1.8.21}$$

If there is no shaft work or heat exchange, the stagnation enthalpy and, for constant specific heat, the stagnation temperature, is the same at stations 1 and 2. Viscous processes, associated with flow through the screen and downstream mixing before the flow comes to a uniform state at station 2, cause an increase in entropy. The states at stations 1 and 2 can be represented as in Figure 5.2 with T_t the stagnation temperature and p_t the stagnation pressure. With a constant stagnation temperature the entropy rise in this adiabatic process is characterized only by the change in stagnation pressure. The relation between loss and entropy change can be seen by defining an ideal reversible process to restore the medium at the downstream stagnation state 2 to the initial stagnation state 1. For a perfect gas, the internal energy per unit mass, e, is a function of temperature only, and the internal energy corresponding to stagnation states 1 and 2 is the same, $e_{t_1} = e_{t_2}$. For any process between these states, the first law ((1.3.8)) reduces to

$$q = w, \tag{5.2.1}$$

with q the heat received, and w the work done, per unit mass of fluid.

For a reversible process, the heat received per unit mass is related to the change in entropy per unit mass, s, as[1]

$$đq_{rev} = Tds. \tag{5.2.2}$$

The heat received, and therefore the work done, is in general path-dependent because it is a function of the temperature at which any reversible heat exchanges occur. For flow through a screen, the stagnation temperature is constant and provides a useful reference. Equation (5.2.2) can thus be integrated to give

$$q_{rev} = T_{t_1}(s_2 - s_1) = T_{t_1}\Delta s. \tag{5.2.3}$$

From (5.2.1), therefore, the reversible work per unit mass to restore the fluid to the initial state is directly proportional to the entropy change:

$$w_{rev} = T_{t1}\Delta s. \tag{5.2.4}$$

This representation of entropy changes as the amount of work that would have to be supplied to restore the fluid to the initial state provides one view of what entropy changes represent. It also makes it plausible that the quantity Tds, where T is an appropriate temperature characterizing the process, is a basic metric for loss. The question of what temperature to use for a more general process, when the stagnation temperature is not constant, still remains to be resolved. In Section 5.2.3 we return to this topic to address this issue for the general situation.

As given in Section 1.16, in terms of stagnation states the entropy change for a perfect gas with constant specific heats is

$$ds = \frac{c_p dT_t}{T_t} - \frac{dp_t}{\rho_t T_t}. \tag{5.2.5}$$

For a process with constant stagnation temperature ($dT_t = 0$), integration of (5.2.5) yields

$$s_1 - s_2 = -R \ln \frac{p_{t_2}}{p_{t_1}}. \tag{1.16.5}$$

Substituting this in (5.2.4) we find the work per unit mass of fluid needed to restore the medium to its initial state as

$$w_{rev} = -RT_{t_1} \ln\left(\frac{p_{t_2}}{p_{t_1}}\right). \tag{5.2.6}$$

Equation (5.2.6) connects the work to restore the fluid to the original condition to the decrease in stagnation pressure due to passage through the screen.

[1] As described in Chapter 1 the notation $đ$ indicates that $đq$ is not the differential of a property but rather represents a small amount of heat.

If $(p_{t_1} - p_{t_2})/p_{t_1} \ll 1$, the logarithm in (5.2.6) can be approximated by the first term in its power series expansion to give

$$w_{rev} \cong \frac{RT_{t_1}}{p_{t_1}}\left(p_{t_1} - p_{t_2}\right)$$
$$= \frac{\left(p_{t_1} - p_{t_2}\right)}{\rho_{t_1}}. \quad (5.2.7)$$

5.2.2 Irreversibility, entropy generation, and lost work

The connections between the entropy rise, the lack of reversibility, and the development of appropriate measures of loss can be given more applicability by examining a general process which takes a system of unit mass from state a to state b (Kestin, 1979). Consider two processes, one ideal, or reversible, and the other irreversible. In both cases the system is allowed to exchange heat with a reservoir. For the reversible process the first law (for unit mass) states

$$de_t = đq_{rev} - đw_{rev}. \quad (5.2.8)$$

For the irreversible process the actual heat and work transfers are related by

$$de_t = đq - đw. \quad (5.2.9)$$

The energy, e_t, is a state variable. Because both processes are defined to be between the same end states, the state change, de_t, is the same in the two cases. A comparison of (5.2.8) and (5.2.9) thus yields

$$đw_{rev} - đw = đq_{rev} - đq = đw_{loss}. \quad (5.2.10)$$

The difference in the work done for the two processes, $đw_{rev} - đw$, can be regarded as work "lost", "dissipated", or "made unavailable", owing to the irreversibility. The difference represents work which could have been obtained ideally, but which has been lost to us. This *lost work*, which will be related to entropy changes in the following, is a rigorous measure of "loss".

Because the reversible and irreversible processes have the same initial and final states the change in entropy is the same for both. The entropy change of the system can be written for the reversible process as

$$\text{reversible process: } ds = \frac{đq_{rev}}{T}. \quad (5.2.11)$$

Using (5.2.11) in (5.2.10) gives, for the irreversible process,

$$\text{irreversible process: } ds = \frac{đw_{rev}}{T} - \frac{đw}{T} + \frac{đq}{T} = \frac{đw_{loss}}{T} + \frac{đq}{T}. \quad (5.2.12)$$

The second law (Section 1.3.3) enables us to make a statement about the sign of the lost work. Assume for simplicity that the reservoir is at temperature T, the system temperature, in both reversible and irreversible processes.[2] The second law states that the total entropy change, system plus reservoir,

[2] This must be the case for the reversible process, although not for the irreversible process, but the arguments can be generalized to account for this situation (Kestin, 1979).

is either zero (for the reversible process) or positive. The total entropy change is given by the right-hand side of (5.2.12) plus the entropy change $đq_{reservoir}/T$. The heat lost (or gained) by the reservoir is equal and opposite to the heat gained (or lost) by the system, so the total entropy change is

$$ds_{total} = \frac{đw_{loss}}{T} \geq 0. \tag{5.2.13}$$

The quantity

$$ds_{irrev} = \frac{đw_{loss}}{T} \tag{5.2.14}$$

is the entropy produced or generated by the irreversible process. Equations (5.2.14) and (5.2.10) taken together show that the reversible process is the "best we can do" in terms of maximizing work received (or minimizing work input) for the specified system state change.

Equations (5.2.12) and (5.2.13) show that system entropy changes can be grouped into two types: changes associated with heat transfer $đq$ and changes due to irreversibility. The entropy change due to heat transfer ($đq/T$) can be positive or negative. The entropy change represented by $đw_{loss}/T = ds_{irrev}$ is equal to or greater than zero: zero for a reversible process and positive if the process is irreversible.

Equation (5.2.12) can also be written as a rate equation

$$\begin{aligned} \frac{ds}{dt} &= \frac{1}{T}\frac{đq}{dt} + \frac{1}{T}\frac{đw_{loss}}{dt} \\ &= \frac{1}{T}\frac{đq}{dt} + \frac{ds_{irrev}}{dt}. \end{aligned} \tag{5.2.15}$$

Equation (5.2.15) gives the rate of entropy change for a system as due to a flow of entropy per unit mass into or out of the system from heat transfer, $(đq/dt)/T$, plus an additional entropy generation associated with irreversibility.

For an adiabatic process, (5.2.15) reduces to

$$\frac{ds}{dt} = \frac{1}{T}\frac{đw_{loss}}{dt}. \tag{5.2.16}$$

In this situation the rate of entropy production in a system is only associated with irreversibility. One such example is the flow through the screen in Section 5.2.1. Another is the flow through a turbine, shown by the thermodynamic representation in Figure 5.3. The abscissa and the ordinate are the entropy (s) and enthalpy (h) per unit mass, respectively. If the change in kinetic energy from the inlet and the outlet is negligible the shaft work per unit mass produced by the turbine as the fluid passes from an inlet pressure p_1 to an exit pressure p_2 is equal to the change in enthalpy. For an isentropic (adiabatic, reversible) process the shaft work per unit mass is $h_1 - h_{2_{rev}}$. For the actual process, in which the fluid exits at the same pressure as in the ideal situation but at a higher entropy, the work is $h_1 - h_2$. If the difference in work done per unit mass between the reversible and the actual processes is much smaller than the actual work done (as is generally the case), the difference between the work in the actual and reversible processes can be approximated as

$$\left(h_2 - h_{2_{rev}}\right) = \left(\frac{\partial h}{\partial s}\right)_{p_2}\left(s_2 - s_{2_{rev}}\right). \tag{5.2.17}$$

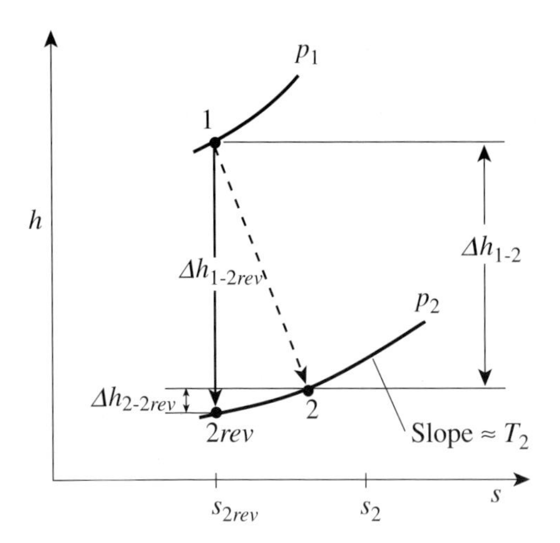

Figure 5.3: Turbine expansion process on an h–s diagram; the slope of the p_2 isobar $\approx T_2$ because $[(h_2 - h_{2rev})/\Delta h_{1-2}] \ll 1$.

Since the slope $(\partial h/\partial s)_{p_2} = T_2$, the difference between the actual and reversible turbine work, $\Delta h_{2-2_{rev}}$, is

$$\Delta h_{2-2_{rev}} = T_2\left(s_2 - s_{2_{rev}}\right). \tag{5.2.18}$$

The turbine component efficiency (generally referred to as adiabatic efficiency) is the ratio of actual to ideal work,[3] or

$$\text{turbine component efficiency} = \frac{h_1 - h_2}{h_1 - h_{2_{rev}}} \approx 1 - \frac{T_2\left(s_2 - s_{2_{rev}}\right)}{h_1 - h_2}. \tag{5.2.19}$$

The preceding discussion has served to connect entropy, loss, irreversibility, and the component efficiency. On a fundamental level, local irreversibility in a fluid flow can always be represented by the two quadratic terms in the integrals in (1.10.7). It is useful, however, to categorize the important sources of irreversibility in a more operational manner in terms of flow processes as:

(a) viscous dissipation;
(b) mixing of mass, momentum, and energy;
(c) heat transfer across a finite temperature difference;
(d) shocks (Section 2.6 showed that this is really a combination of (a) and (c)).

5.2.3 Lost work accounting in fluid components and systems

There are two issues connected with loss accounting which we now need to resolve. The first concerns the relation between the entropy change due to irreversibility and the lost work. Three examples have been presented in which expressions for lost work were developed: adiabatic flow through a screen, an incremental general process in which heat was exchanged with a reservoir at temperature, T, and

[3] Ideal here means work that would be received in a reversible process. This is the maximum work the turbine could produce.

adiabatic flow through a turbine. In all of these the lost work was represented by the product of the change in entropy due to irreversibility and a temperature. Three different temperatures, however, were used in the different examples. The link between lost work and entropy change thus needs to be further defined.

The second issue concerns different perspectives for loss measurement that can be adopted. The discussion so far has been on losses as seen in the context of assessing fluid component performance. Such components typically operate as a part of a more complex fluid system, for example an engine. An important question is the relation between the (local) loss measures for components and loss measures based on global system (i.e. thermodynamic cycle) considerations.

These two issues can be addressed employing the concept of *flow availability*. Flow availability is a property whose change measures the maximum useful work (i.e. work over and above flow work done on the surroundings) obtained for a given state change. The concept is developed in depth by, for example, Bejan (1988, 1996), Horlock (1992), Sonntag *et al.* (1998), and we present only an introduction here.

Consider a steady-flow device, which can exchange heat and shaft work with the surroundings. The first law for a control volume, (1.8.11), states that the shaft work per unit mass obtained from a stream which passes from an initial state 1 to a subsequent state 2 is

$$w_{shaft} = q + \left(h_{t_1} - h_{t_2}\right). \tag{1.8.11}$$

The convention is that q, the heat addition per unit mass, is positive for heat addition *to* the stream. For given initial and final states the change in stagnation enthalpy is specified. The first law gives no information concerning the magnitude of the heat addition, q, and (1.8.11) shows that the larger the heat addition the larger the shaft work. The second law, however, puts a bound on the maximum heat addition and thus the maximum work that can be obtained for a given state change.

This upper limit can be determined by examining a situation in which the stream exchanges heat only with a reservoir at temperature T_0. For purposes of illustration the reservoir is regarded as the atmosphere, since that is the environment in which most fluid systems operate and to which heat is eventually rejected, but it is to be emphasized that this is not necessarily the case for the arguments that follow.[4] For a unit mass of the stream that undergoes the given state change the entropy change of the reservoir is $\Delta s = -q/T_0$. From the second law the entropy change of the stream between inlet and exit must be such as to make the total entropy changes occurring in the device plus the environment equal to or greater than zero. Any difference from zero represents the departure from reversibility. The second law, applied to a unit mass of fluid which passes from state 1 to state 2, is

$$(s_2 - s_1) - \frac{q}{T_0} = \Delta s_{irrev} \geq 0. \tag{5.2.20}$$

The quantity Δs_{irrev} is the entropy generated per unit mass as a result of irreversible processes. Combining (1.8.11) and (5.2.20),

$$w_{shaft} = \left(h_{t_1} - T_0 s_1\right) - \left(h_{t_2} - T_0 s_2\right) - T_0 \Delta s_{irrev}. \tag{5.2.21}$$

[4] The results can also be extended to situations in which heat is interchanged with any number of reservoirs (in addition to the atmosphere) at different temperatures (Bejan, 1988; Horlock, 1992; Sonntag *et al.* 1998).

The entropy change Δs_{irrev} is equal to zero or positive. The maximum shaft work that can be obtained for a state change from 1 to 2 is therefore the difference in the quantity $(h_t - T_0 s)$,

$$[w_{shaft}]_{maximum} = \left(h_{t_1} - T_0 s_1\right) - \left(h_{t_2} - T_0 s_2\right). \tag{5.2.22}$$

Comparison of (5.2.21) and (5.2.22) shows that for the given state change the difference between the maximum shaft work and the shaft work actually obtained is $T_0 \Delta s_{irrev}$ which is the lost work for the process.

The quantity $(h_t - T_0 s)$ is known as the steady-flow availability function (Horlock, 1992) or, more simply, the flow availability (Bejan, 1988). It is a composite property which depends on both the state of the fluid and the temperature of the environment. By tracing work received and availability changes one can determine the locations in a system which provide the largest potential for improvements in overall performance.

With this background the difference between the three situations can be described. In the example of flow through the screen the temperature at which heat is seen as being exchanged between system and surroundings is the stagnation temperature, so the quantity T_0 in (5.2.21) and (5.2.22) in the evaluation of $(h_t - T_0 s)$ would be replaced by T_t. The difference in this quantity between states 1 and 2 is thus $T_t \Delta s_{irrev}$ and, because there is no shaft work, this is also the lost work per unit mass. For the second process (the incremental state change) the system is in equilibrium with the heat reservoir and the temperature at which heat is exchanged with the surroundings is the local system temperature, T. The lost work is thus computed from analysis of the changes in $(h - Ts)$ as $T\Delta s_{irrev}$, consistent with the direct evaluation of this quantity in (5.2.14). Finally for the turbine, the "reservoir temperature" which the example corresponds to is the turbine exit temperature, T_2.

Expressions for lost work in terms of Δs_{irrev} are seen to be, just as is the availability, composite quantities which depend on the properties of both the system and the temperature of the surrounding medium with which heat is exchanged. The point is succinctly expressed by Cravalho and Smith (1981) who state "the irreversibility cannot be related to the loss of useful work until a specification is given for the final location (specifically the temperature) of the entropy which has been generated."

The different expressions for lost work have a fundamental connection with each other which can be seen through a comparison of the metrics for fluid device loss considered as an isolated component and as a part of an overall system which exchanges heat with the atmosphere at T_{atm}. We illustrate the point using the adiabatic throttling process across the screen; the analysis also applies directly to an adiabatic duct or blade row.

From the component perspective (considering the flow across the screen by itself) the lost work per unit mass for a given state change was given as $T_t \Delta s_{irrev}$ in Section 5.2.1. Considering the screen or duct as a part of a more complex system which exchanges heat with the atmosphere, (5.2.21) shows that the lost work for the same state changes is $T_{atm} \Delta s_{irrev}$. The difference between the two, $(T_t - T_{atm})\Delta s_{irrev}$, is equal to the work per unit mass, w_C, that *could* be obtained by a Carnot cycle, operating between T_t and T_{atm} with an entropy change Δs_{irrev}. The quantity w_C represents an opportunity for doing useful work. However, if *none* of the work represented by the hypothetical Carnot cycle is realized, $T_t \Delta s_{irrev}$ is also the lost work for the system. Both situations are found in practice. For example blade row inefficiencies in multistage turbines mean that the work output of the succeeding blade rows is higher than if the upstream rows were isentropic. For an exhaust nozzle, in contrast, there is no chance to recover additional work from a stream that emerges at a temperature greater than T_{atm}. The difference between these processes arises because "useful work can be realized during

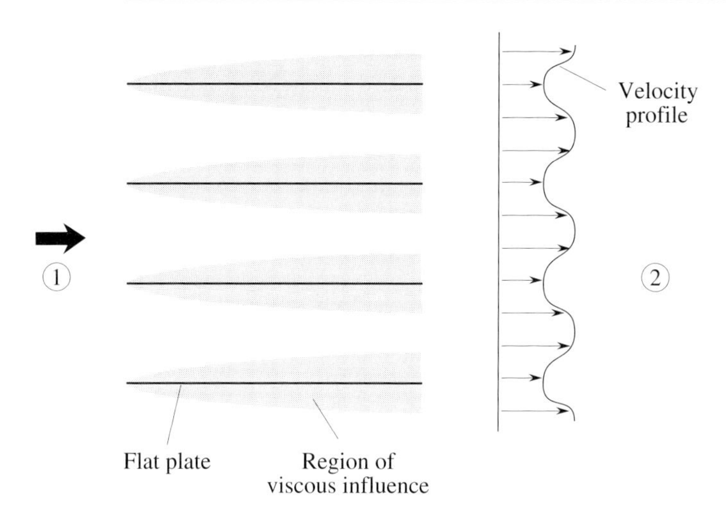

Figure 5.4: Flat plate cascade and downstream velocity distribution.

the series of processes that transfer the generated entropy from the high temperature to the entropy sink at ambient temperature" (Cravalho and Smith, 1981).

Local loss measures for fluid components (e.g. boundary layers, compressor blade rows) do not explicitly account for the possibility that some fraction $(1 - T_{atm}/Tt)$ of the energy dissipated by irreversible processes *might* be converted to useful work because of the difference between the stagnation temperature and the temperature of the environment. Whether this occurs or not depends on the configuration of the specific system in which the component is embedded. One can relate the two measures of loss (component and system) using the ideas just described. In the rest of the chapter we therefore focus on the component metrics, which are the basic building blocks for developing a description of complex system performance.

5.3 Loss accounting and mixing in spatially non-uniform flows

We now consider a more general situation in which the velocity and static temperature downstream of a device vary spatially. Specifically, let the screen used in the previous section be replaced by an array of plates parallel to the stream as in Figure 5.4, which can be regarded as a model of a turbomachinery cascade. Station 2 represents a location at which the velocity defects due to the plate boundary layers have not yet mixed out. To develop an expression for the loss at this station, we compute the increase in entropy flux through stations 1, at which the flow is uniform, and 2, at which it is not,

$$\Delta\left[\text{specific entropy flux}\right] = \frac{\int_{\dot{m}} (s_2 - s_1)\, d\dot{m}}{\int d\dot{m}}. \tag{5.3.1}$$

In (5.3.1) the integral is taken over a passage.

We wish to proceed as in the previous section. On an overall basis, no work is done and no heat is transferred so the quantity $\int_{\dot{m}} c_p T_t d\dot{m}$ remains invariant. Although we cannot say the local stagnation enthalpy is uniform, this is a very good approximation in adiabatic steady flows of this type, not only on a global basis but along a streamline. (Invariance of the stagnation temperature along a streamline is equivalent to the statement that the non-pressure work done by a given streamtube on the flow external to the streamtube and the heat transfer to the streamtube are in balance.)

The power expended to restore the flow of station 2 to its original state, per passage, is, with A_2 the area occupied by the flow from a single passage at station 2,

$$\text{power} = T_{t_1} \int_{\dot{m}} (s_2 - s_1) d\dot{m} = T_{t_1} \int_{A_2} (s_2 - s_1)\rho_2 u_{x_2} dA_2. \tag{5.3.2}$$

Equation (5.3.2) can be written in terms of the mass flow rate, $\dot{m}$, for a single passage as

$$\text{power} = T_{t_1} \dot{m} \left(\overline{s}_2^M - s_1\right). \tag{5.3.3}$$

Equations (5.3.2) and (5.3.3) introduce the *mass average* specific entropy, $\overline{s}^M$, defined as

$$\overline{s}^M = \frac{\int_{\dot{m}} s d\dot{m}}{\dot{m}}. \tag{5.3.4}$$

The power needed to restore the flow to its original state can also be related to the stagnation pressure distribution at station 2 by making use of (1.16.5),

$$\text{power} = -RT_{t_1} \int_{\dot{m}} \ln \frac{p_{t_2}}{p_{t_1}} d\dot{m} \tag{5.3.5}$$

If $(p_{t_1} - p_{t_2})/p_{t_1} \ll 1$, (5.3.5) can be approximated as

$$\text{power} = \frac{\dot{m}}{\rho_{t_1}} \left(p_{t_1} - \overline{p}_{t_2}^M\right), \tag{5.3.6}$$

where $\overline{p}_{t_2}^M$ is the mass average total pressure. Equation (5.3.6), which is a relevant description[5] of many flow processes, finds wide use as a measure of loss.

The location of station 2 has not been specified except to say it was downstream of the cascade. Mixing occurs continuously from the trailing edge of the plates with the entropy flux increasing to a final value at the far downstream, fully mixed state. In general, one cannot say that $(s_{far\ downstream} - s_2) \ll (s_2 - s_1)$ because the mixing losses depend on both the nature of the loss creating device and the nature of the downstream flow. As described below, a downstream pressure increase (such as in flow through a diffuser) increases mixing loss whereas a downstream static pressure decrease (as in flow through a nozzle) decreases it.

The preceding discussion highlights several issues in developing a procedure for assessing loss. One is the development of the means to estimate rates of entropy production in order to determine loss generation within a component or fluid element. A second is the characterization of flows downstream of the component, particularly where the device length is not sufficient to allow complete mixing to occur. In this situation the flow will be non-uniform, and an appropriate methodology is needed to describe the state of the flow. A third is that mixing does not always occur at constant area and we need to be able to account for the effect of downstream flow processes on the overall loss levels. These issues are addressed in this chapter.

[5] The limitations on the use of stagnation pressure as a measure of loss are given in Section 5.5.

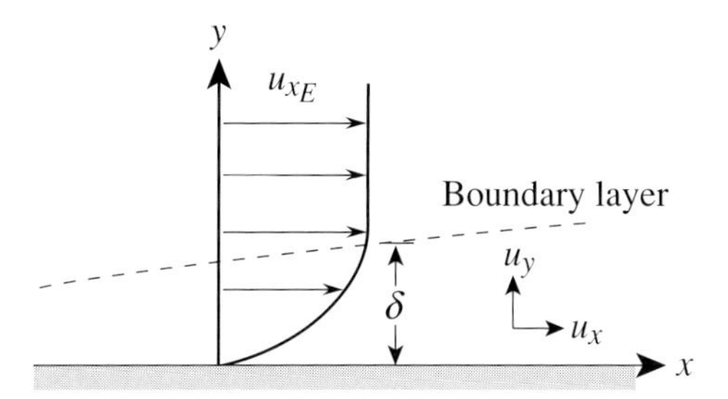

Figure 5.5: Notation for a two-dimensional boundary layer.

5.4 Boundary layer losses

5.4.1 Entropy generation in boundary layers on adiabatic walls

A major source of loss is entropy generation in boundary layers on solid surfaces (Denton, 1993). To exhibit this process, we derive an expression for entropy production in the steady two-dimensional boundary layer sketched in Figure 5.5. The starting point is (1.10.5), which gives the rate of change of entropy for a fluid particle:

$$T\frac{Ds}{Dt} = \dot{Q} - \frac{1}{\rho}\frac{\partial q_i}{\partial x_i} + \frac{1}{\rho}\tau_{ij}\frac{\partial u_i}{\partial x_j}. \tag{1.10.5}$$

For the situation shown the mainstream flow is in the x-direction. As discussed in Chapter 4, to describe the boundary layer we retain only the shear stress term τ_{xy} and the derivative of the heat flux in the y-direction. For a flow without heat sources ($\dot{Q} = 0$) the boundary layer form of (1.10.5) is

$$\rho T\left(u_x\frac{\partial \Delta s}{\partial x} + u_y\frac{\partial \Delta s}{\partial y}\right) = -\frac{\partial q_y}{\partial y} + \tau_{xy}\frac{\partial u_x}{\partial y}. \tag{5.4.1}$$

In (5.4.1) Δs is the specific entropy difference between the local value and that outside of the boundary layer, with the latter taken as uniform in the y-direction.

A case of interest is that of an adiabatic wall with $q_y(y=0) = 0$. In this situation variations in static temperature and density through the boundary layer are of order M_E^2 compared to the absolute temperature, where M_E is the Mach number at the edge of the boundary layer. For low Mach number flows the temperature and density can thus be taken as constant in application of (5.4.1).[6]

With no heat transfer from the wall to the fluid, the change in entropy flux between two stations at different x-locations results only from entropy generation associated with irreversibility. The rate

[6] The rationale for this approximation is as follows. Variations in entropy, temperature, and density all scale as u_E^2, but the three quantities appear in (1.10.5) and (5.4.1) in different ways. For low Mach number, the temperature and density enter as a quantity, say T_E, which has fractional variations of order M_E^2, which can be neglected. For the entropy, however, it is the *variations alone* that are of interest. Put another way, the effects that are captured scale as M_E^2 (i.e. $\Delta s/c_p \propto M_E^2$). Inclusion of the variations in temperature and density would have an effect on this quantity of order M_E^4. The temperature and density anywhere in the flow field therefore can be chosen as the reference value.

of change of entropy flux along the surface, per unit depth, is found by integrating from $y = 0$ to y_E, the edge of the boundary layer, as

$$\begin{aligned} \dot{S}_{irrev} &= \frac{d}{dx} \int_0^{y_E} \rho u_x (s - s_E)\, dy \\ &= \frac{d\delta}{dx} \left[\rho u_x (s - s_E) \right] \Big|_{y=y_E} + \int_0^{y_E} \left\{ \frac{\partial}{\partial x} \left[\rho u_x (s - s_E) \right] \right\} dy, \end{aligned} \tag{5.4.2}$$

using differentiation under the integral sign. We denote the rate of change of entropy flux per unit depth by $\dot{S}$, which is also interpreted as the entropy production in the boundary layer per unit area of surface. The first term on the right-hand side of (5.4.2) is zero because the entropy at the edge of the boundary layer is just the free-stream entropy, $s(x, y_E) = s_E$. The second term can be written as

$$\int_0^{y_E} \left\{ \frac{\partial}{\partial x} \left[\rho u_x (s - s_E) \right] \right\} dy = \int_0^{y_E} \left\{ (s - s_E) \frac{\partial}{\partial x} (\rho u_x) \right\} dy + \int_0^{y_E} \left\{ \rho u_x \frac{\partial}{\partial x} (s - s_E) \right\} dy. \tag{5.4.3}$$

Using the continuity equation to replace $[\partial/\partial x\, (\rho u_x)]$ in the first term on the right-hand side of (5.4.3), integrating by parts, and rearranging gives

$$\begin{aligned} \dot{S}_{irrev} &= \frac{d}{dx} \int_0^{y_E} \rho u_x (s - s_E)\, dy \\ &= \int_0^{y_E} \left\{ \rho u_x \frac{\partial}{\partial x} (s - s_E) + \rho u_y \frac{\partial}{\partial y} (s - s_E) \right\} dy. \end{aligned} \tag{5.4.4}$$

Comparing the integrand in (5.4.4) with (5.4.1), the expression for the rate of change of entropy flux along the surface is

$$T\dot{S}_{irrev} = \frac{d}{dx} \int_0^{y_E} \rho u_x \left[T (s - s_E) \right] dy = \int_0^{y_E} \left(\tau_{xy} \frac{\partial u_x}{\partial y} \right) dy, \tag{5.4.5}$$

where the conditions of an adiabatic wall and no heat flux at the edge of the boundary layer mean that the integral of the heat transfer term $\partial q_y / \partial y$ is zero. Equation (5.4.5) is an expression for the rate of entropy production, from conversion into heat of work done by viscous shear stresses, per unit length along the wall and unit depth (i.e. into the page in Figure 5.5).

Comparison with (4.3.10) and the discussion just thereafter shows that the quantity $T\dot{S}$ is the dissipation term labeled as $\dot{D}$ in Section 4.3. For incompressible flow the total dissipation per unit depth can be linked to the kinetic energy thickness parameter, θ^*, using (5.4.5) as

$$\rho \frac{d}{dx} \left(u_E^3 \theta^* \right) = 2\dot{D} = 2T\dot{S}. \tag{5.4.6}$$

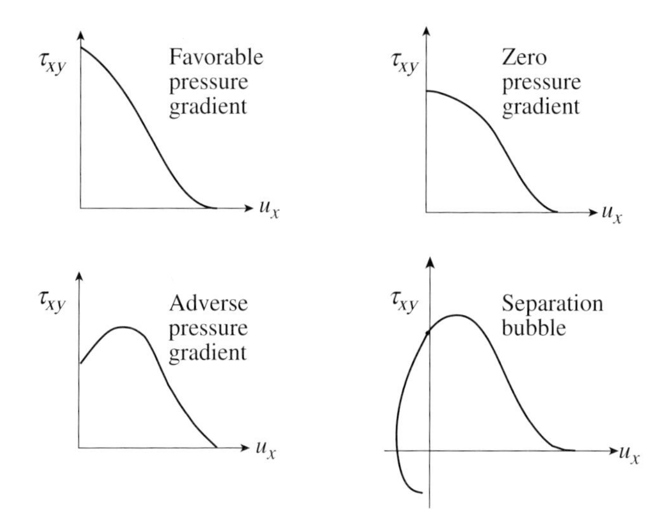

Figure 5.6: Sketch of shear stress (τ_{xy}) versus velocity (u_x) in different boundary layer regimes: $T\dot{S}_{irrev} = \int_0^{u_E} \tau_{xy} du_x$ (Denton, 1993).

Integration of (5.4.6) along the length of the surface from an initial location at $x = 0$ to an arbitrary station, x, yields

$$\left(\rho u_E^3 \theta^*\right)\Big|_0^x = 2\int_0^x T\dot{S}dx'. \tag{5.4.7}$$

If the kinetic energy thickness is negligible at $x = 0$, (5.4.7) reduces to

$$\theta^* = \frac{2}{\rho u_E^3}\int_0^x T\dot{S}dx', \tag{5.4.8}$$

where the free-stream velocity, u_E, is evaluated at the station x. The kinetic energy thickness at this location is thus proportional to the cumulative rate of dissipation per unit depth in the boundary layer, up to that station.

For laminar boundary layers the entropy production can be computed directly from the equations of motion with no additional hypotheses (White, 1991; Sherman, 1990; Bejan, 1996). In contrast, for turbulent boundary layers which are more often encountered in practice, this is not the case. We thus focus on the latter.

Equation (5.4.5) can be given a graphical interpretation if we express the entropy production term, $\int_0^{y_E} \tau_{xy}\partial u_x/\partial y dy$ as an integral over the velocity, $\int_0^{u_E} \tau_{xy} du_x$ (Denton, 1993),

$$T\dot{S}_{irrev} = \frac{d}{dx}\int_0^{y_E} \rho u_x[T(s - s_E)]dy = \int_0^{u_E} \tau_{xy} du_x. \tag{5.4.9}$$

Representative curves of shear stress as a function of velocity are sketched in Figure 5.6 for different types of boundary layers, ranging from accelerating flow to a situation with a region of reversed flow near the wall. The shear stress integral in (5.4.9) gives the area under the curve. For turbulent

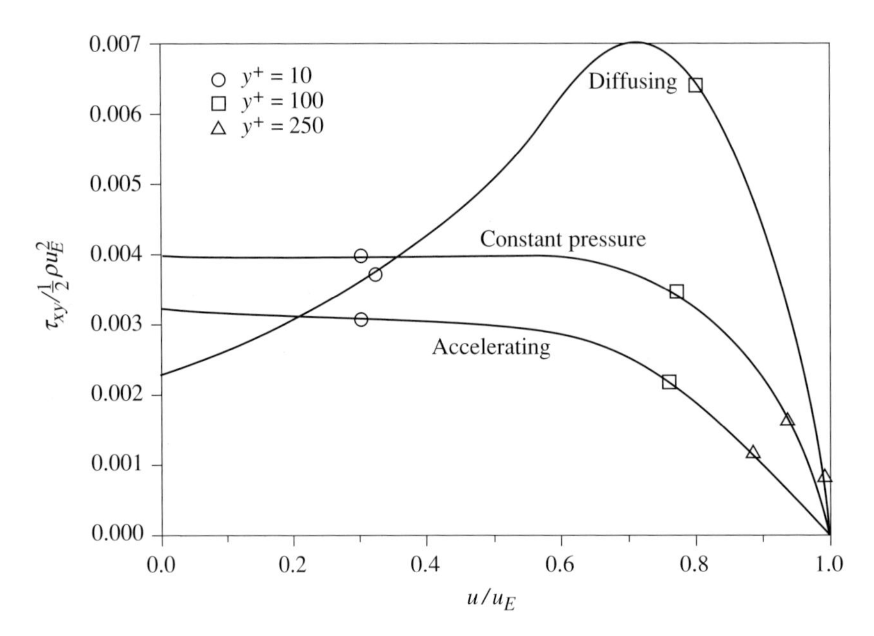

Figure 5.7: Variation of shear stress with velocity through boundary layers with $Re_\theta = 1000$ (Denton, 1993).

flow, the velocity in the boundary layer changes most rapidly near the surface, and most of the entropy generation occurs in this region rather than in the outer parts of the boundary layer. The figure is a sketch only but, as it suggests, for a given external velocity the overall dissipation in a turbulent boundary layer is found to depend only weakly on the state of the boundary layer (Denton, 1993). This result will be seen to allow a simple and useful estimate to be made for overall entropy production. Calculations of the variation of shear stress with velocity through turbulent boundary layers are given in Figure 5.7, with values of the non-dimensional boundary layer inner region variable y^+ indicated on the figure. The outer part of the boundary layer ($y^+ > 250$) is most affected by the streamwise pressure gradient, but in this region there is little shear stress and, as a result, little entropy generation.

5.4.2 The boundary layer dissipation coefficient

To explore the applicability of the ideas in the previous section, it is useful to turn the entropy production rate into a dimensionless boundary layer dissipation coefficient defined by

$$C_d = \frac{T\dot{S}_{irrev}}{\rho u_E^3}, \tag{5.4.10}$$

where u_E is the velocity at the edge of the boundary layer. For turbulent flow, the value of the dissipation coefficient cannot yet be calculated from first principles and we need to have recourse to experimental findings. Figure 5.8 shows values of the dissipation coefficient, C_d, and the skin friction coefficient, C_f, for momentum thickness Reynolds numbers from 10^3 to 10^5. Information is given for a range of shape factors from 1.2 to 2.0 for C_d and from 1.2 to 2.4 for C_f. A striking result is that the dissipation coefficient is much less dependent on the shape factor than the more familiar skin friction

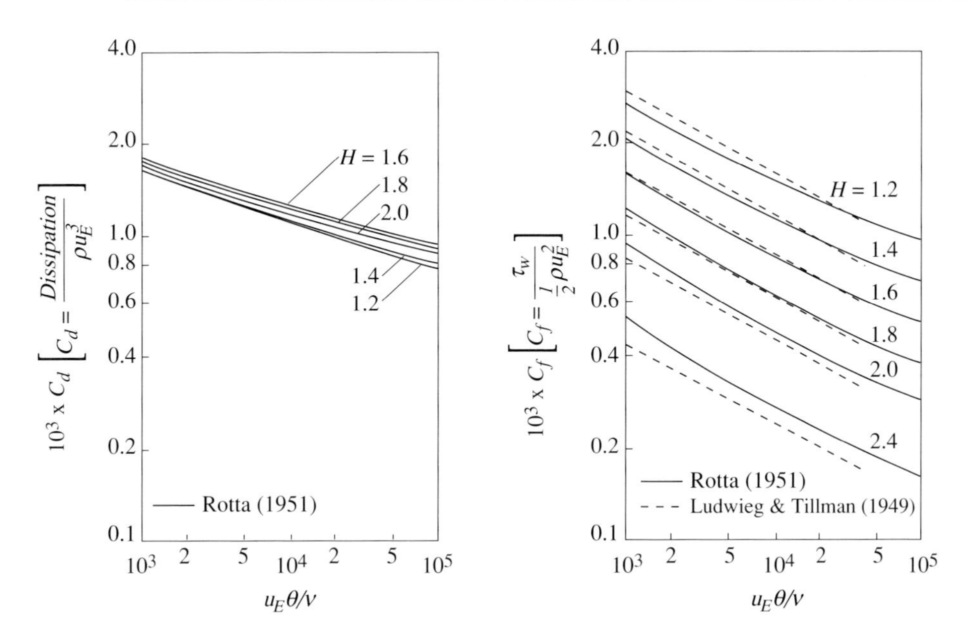

Figure 5.8: Turbulent boundary layer properties (Schlichting, 1968).

coefficient. Although the turbulent skin friction coefficient decreases by a factor of roughly 3 as the shape factor increases from 1.2 to 2.0, the dissipation coefficient varies by less than 10% over this range. Further, the dependence on Re_θ is weak. Based on the data in Figure 5.8, Schlichting (1979) suggests a curve fit for C_d as

$$C_d = 0.0056\,(Re_\theta)^{-1/6}. \tag{5.4.11}$$

For laminar boundary layers the dissipation coefficient depends more strongly on Re_θ, with an $(Re_\theta)^{-1}$ dependence (see Schlichting (1979)) as described by Truckenbrodt (1952). Even for laminar boundary layers, however, calculations carried out by Denton (1993) suggest little dependence on the state of the boundary layer. The variation of the dissipation coefficient with Re_θ is shown in Figure 5.9 for a range in which both laminar and turbulent boundary layers could exist, say $300 < Re_\theta < 1000$. The dissipation coefficient for the laminar boundary layer is lower by a factor of between 2 and 3 than for the turbulent boundary layer at the same momentum thickness Reynolds number.

The above results are based on, and apply strictly to, low Mach number flow. There are few data for the effect of Mach number on dissipation coefficient. However, since there is only a 20% decrease in the skin friction coefficient over the range, $0 < M_E < 2$, it may be reasonable to use the low speed results as a useful approximation. The temperature can no longer be considered constant if M^2 is not small compared to unity but, because the majority of the entropy production takes place near the surface, a suitable modification might be to use the recovery or adiabatic wall temperature, T_{rf}, as the appropriate parameter in defining C_d. An approximation for the recovery temperature is given by

$$\frac{T_{rf}}{T_E} = 1 + r\,\frac{(\gamma - 1)}{2}\,M_E^2, \tag{5.4.12}$$

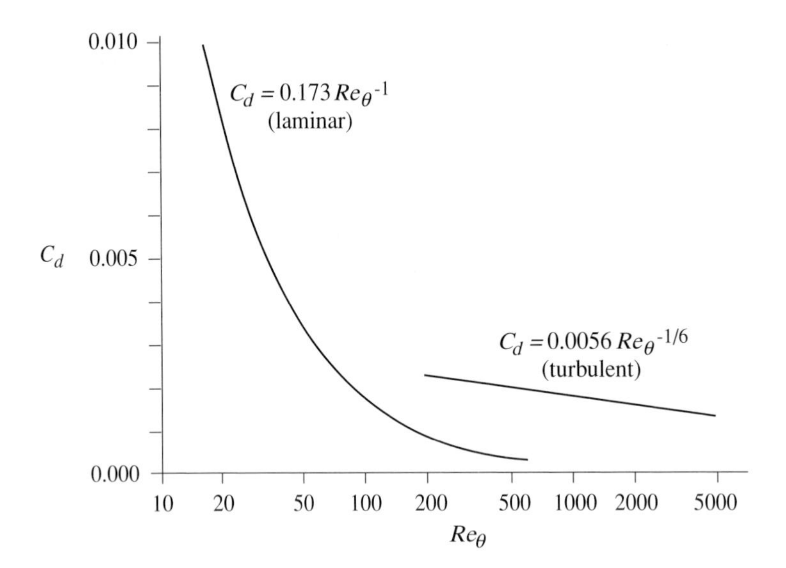

Figure 5.9: Dissipation coefficients for laminar and turbulent boundary layers (Truckenbrodt (1952) as reported in Denton (1993)).

where $r = \sqrt{Pr}$ for laminar flow and $\sqrt[3]{Pr}$ in turbulent flow, where Pr is the Prandtl number (Schlichting, 1979). This is the surface temperature in a boundary layer along an insulated wall.

For the estimation of entropy production, the weak variation of the dissipation coefficient with Re_θ implies that a useful approximation is to take the dissipation coefficient, C_d, as constant at some representative value of Re_θ, say $C_d = 0.002$ for turbomachinery blading (Denton, 1993). For flow through a two-dimensional passage, the total rate of boundary layer entropy generation per unit depth can then be estimated by integrating the expression (5.4.10) for $\dot{S}_{irrev}$ over the length of the solid surface:

$$T\dot{S}_{total} = C_d \sum_{\substack{all \\ surfaces}} \rho L U^3 \int_0^{x_{final}} \left(\frac{u_E}{U}\right)^3 d\left(\frac{x}{L}\right). \tag{5.4.13}$$

In (5.4.13), L is a reference length (say airfoil chord or duct axial length), x is the distance measured along the solid surface, U is a reference velocity, and $\dot{S}_{total}$ is the rate of entropy production per unit depth in the boundary layer from the initial ($x = 0$) to the final station. The dissipation scales as the *cube* of the free-stream velocity, so that regions of locally high free-stream velocity contribute strongly to entropy generation.

The entropy generation in the blade passages can also be related to commonly used loss coefficients for fluid machinery. The mass-averaged entropy change per unit depth at a given downstream station is related to $\dot{S}_{total}$ by

$$\dot{m}(\overline{s}^M - s_1) = \dot{S}_{total}. \tag{5.4.14}$$

For low Mach number adiabatic flows,

$$\dot{m}\frac{\overline{\Delta p_t}^M}{\rho} = T\dot{S}_{total}. \tag{5.4.15}$$

From (5.4.15), a non-dimensionalized mass-averaged loss coefficient $(\overline{\Delta p_t}^M/(\frac{1}{2}\rho U^2))$ can be related to the entropy production by

$$\frac{\overline{\Delta p_t}^M}{\frac{1}{2}\rho U^2} = \frac{T\dot{S}_{total}}{\frac{1}{2}\dot{m}U^2}. \tag{5.4.16}$$

If U is taken as the average velocity at the inlet, as it might be for a diffuser or a compressor blade row, and W is the height of the passage at the inlet station, the loss coefficient in (5.4.16) can be calculated from

$$\frac{\overline{\Delta p_t}^M}{\frac{1}{2}\rho u_1^2} = 2C_d \frac{L}{W} \sum_{\substack{all \\ surfaces}} \int_0^{x_{final}} \left(\frac{u_E}{u_1}\right)^3 d\left(\frac{x}{L}\right). \tag{5.4.17}$$

5.4.3 Estimation of turbomachinery blade profile losses

To illustrate the way in which (5.4.17) enables insight into features of fluid machinery performance we give an example drawn from axial turbine behavior. If the turbine blade surface velocity distribution and variation of the dissipation coefficient C_d are known, (5.4.17) allows estimation of the blade boundary layer or "profile loss" coefficient. The difference in values of C_d for laminar and turbulent flows implies that the boundary layers should be kept laminar as long as practical, although at the high values of turbulence intensity in turbomachines transition is likely to occur in the range $Re_\theta \approx$ 200–500. Because of the weak variation of the dissipation coefficient in a turbulent flow, we can take it to be constant, with a value of 0.002, over the range of momentum thickness Reynolds numbers representative of those encountered in gas turbine blading. While such an approximation cannot give precise quantitative results, it does allow the development of systematic trends for variation in loss with turbine blade characteristics.

One aspect to be addressed is the existence of an optimum value of the blade space/chord ratio (Denton, 1993). Consider an idealized rectangular velocity distribution around the blade, with high velocity on the suction surface and low velocity on the pressure surface, as sketched in Figure 5.10. Using now the velocity at the exit (station 2) for the reference velocity, as is conventional for turbines, the integral in (5.4.17) can be evaluated as (see Figure 5.10 for notation)

$$\text{loss coefficient} = \frac{\overline{\Delta p_t}^M}{\frac{1}{2}\rho u_2^2} = 2C_d \frac{\text{blade length}}{\text{spacing}}\left[2\left(\frac{\overline{u}}{u_2}\right)^3 + 6\left(\frac{\overline{u}}{u_2}\right)\left(\frac{\Delta u}{u_2}\right)^2\right]. \tag{5.4.18}$$

The circulation round the blade is the product of the length along the blade (approximated here by the chord) and the average velocity difference between suction and pressure sides of the blade. As shown in Section 2.8, the circulation is also the product of the difference between inlet and exit circumferential velocities and the blade-to-blade spacing, with the circumferential (u_y) and axial (u_x) velocities related by the flow angle α: $u_y = u_x \tan\alpha$. Combining these statements the loss coefficient based on mean velocity $\overline{u}$ is

$$\text{loss coefficient} = C_d\left(\frac{2\overline{u}}{\Delta u} + 6\frac{\Delta u}{\overline{u}}\right)(\tan\alpha_2 - \tan\alpha_1). \tag{5.4.19}$$

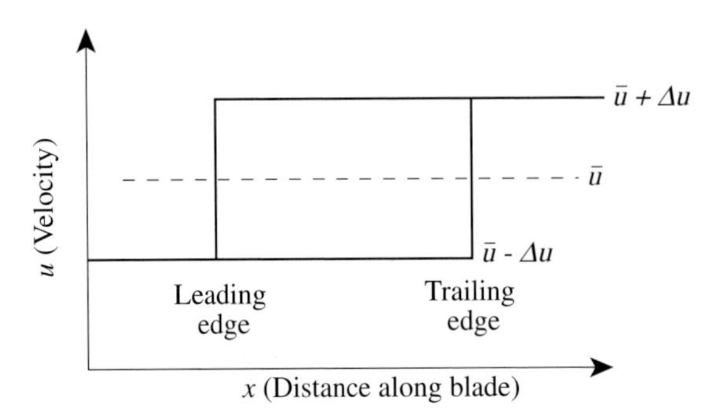

Figure 5.10: Idealized blade surface velocity distribution on a turbine blade.

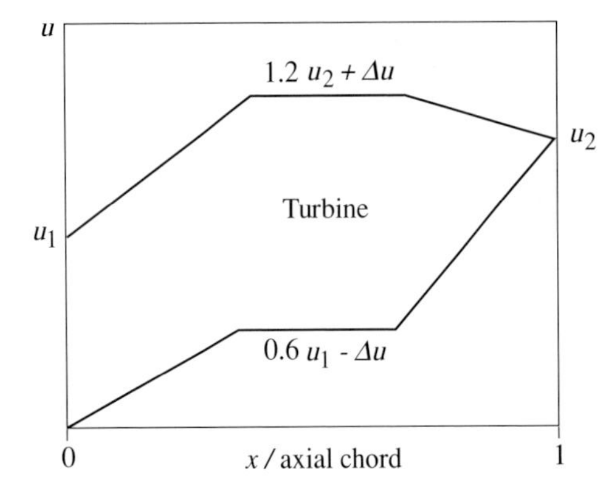

Figure 5.11: Generic surface velocity distributions for turbine blades (Denton, 1990).

This has a minimum value corresponding to the optimum value of blade space/chord when $(\Delta u/\overline{u}) = (1/\sqrt{3})$. With $C_d = 0.002$, representative blade profile losses can be found using this method.

Denton (1993) has employed this idea, with the (more realistic) family of generic turbine velocity distributions shown in Figure 5.11, to generate optimum blade space/chord ratios and blade loss coefficients for turbine blade rows over a range of inlet and exit angles. Figure 5.12 shows profile loss coefficients as a function of the inlet and exit flow angles for blade rows which have the calculated optimum space/chord ratio; the loss estimates generated agree fairly well with measurements.

5.5 Mixing losses

5.5.1 Mixing of two streams with non-uniform stagnation pressure and/or temperature

A common situation in fluid machinery and propulsion systems is the mixing of two coflowing streams with different stagnation conditions. A model of such a configuration is the constant area mixing of two streams of different stagnation temperatures and pressures as in Figure 5.13. The mixing process can be analyzed using a control volume approach so that details of the mixing need not be addressed. The stagnation pressure and temperature at the inlet of the mixing region, station "i", and the initial area of each stream are specified, as is the static pressure at this location. (The

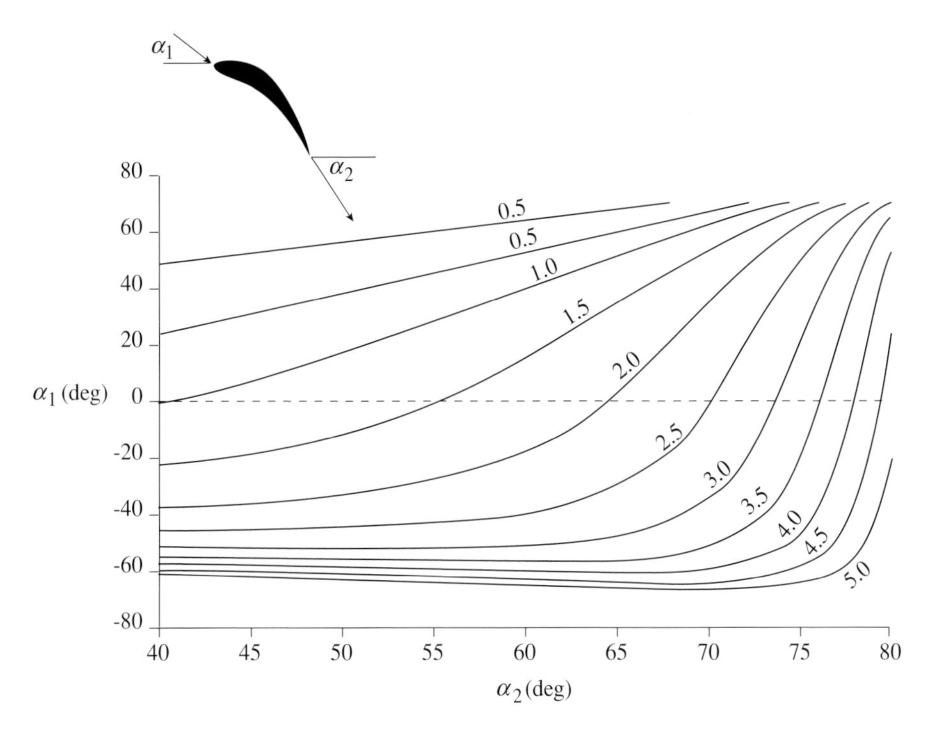

Figure 5.12: Turbine blade profile loss coefficients, $\Delta\overline{p}_t^M/\frac{1}{2}\rho u_2^2$, at optimum pitch/chord ratio estimated using velocity cubed approach (in % loss) (Denton, 1993).

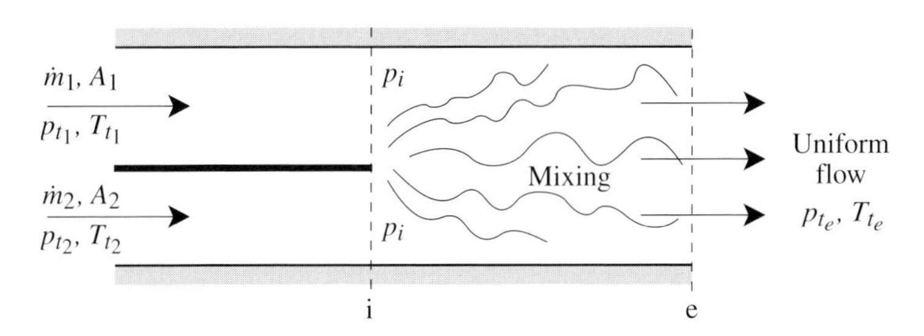

Figure 5.13: Mixing of two streams in a constant area duct.

latter can be thought of as being controlled by opening or closing a throttle at the exit of a chamber into which the mixing duct discharges.) As described in Section 2.8 wall shear stresses are neglected and the walls are taken as adiabatic. The mixing proceeds from the specified inlet state to a uniform (fully mixed out) state at the exit of the control volume.

The calculation of mixed out conditions follows from application of conservation of mass, momentum, and energy, plus the equation of state. For a specified pressure p_i at the inlet station, the ratios $(p_{t_{1_i}}/p_i)$ and $(p_{t_{2_i}}/p_i)$, and hence the inlet Mach numbers of streams 1 and 2 are known (see Section 2.5). Mass flows and velocities in each stream can thus be found. Mass conservation between the inlet and the exit of the control volume is

$$\dot{m}_e = \dot{m}_{1_i} + \dot{m}_{2_i}, \tag{5.5.1}$$

where the subscript "e" denotes the fully mixed out location. Making use of the impulse function, $\Im$, defined as $\Im = pA + \rho A u_x^2 = pA + \dot{m}u_x$, the equation for conservation of momentum in the constant area duct is

$$\Im_e = \Im_{1_i} + \Im_{2_i}. \tag{5.5.2}$$

For a perfect gas with constant specific heats the steady flow energy equation gives

$$T_{t_e} = \frac{\dot{m}_{1_i} T_{t_{1_i}} + \dot{m}_{2_i} T_{t_{2_i}}}{\left(\dot{m}_{1_i} + \dot{m}_{2_i}\right)}. \tag{5.5.3}$$

A non-dimensional form of the impulse function can be defined as

$$\tilde{\Im}_e = \frac{\Im_e}{\left(\dot{m}_{1_i} + \dot{m}_{2_i}\right)\sqrt{c_p T_{t_e}}}. \tag{5.5.4}$$

The impulse function at mixed out conditions is a function of Mach number and γ, given by

$$\tilde{\Im}_e = \left(\frac{\sqrt{\gamma - 1}}{\gamma M_e}\right)\left[\frac{1 + \gamma M_e^2}{\sqrt{1 + \frac{1}{2}(\gamma - 1)M_e^2}}\right]. \tag{5.5.5}$$

Equation (5.5.5) is an implicit expression for the exit (mixed out) Mach number. With given Mach number, stagnation temperature, and duct area, all other mixed out flow properties can be found. There are two possible values of Mach number which satisfy (5.5.5), one subsonic and one supersonic. If both entering flows are subsonic, only the subsonic solution is compatible with an increase in entropy flux. If one or both of the entering streams are supersonic, both subsonic and supersonic solutions are possible.

Figure 5.14 presents contours of an entropy rise coefficient, defined as (with s_i the inlet mass average entropy)

$$\text{entropy rise coefficient} = \frac{\overline{T}_t\left(s_e - s_i\right)}{\frac{1}{2}\overline{u}_i^2} \tag{5.5.6}$$

for the constant area mixing of two streams with equal areas at the start of mixing. The inlet stagnation pressure of one stream is $\overline{p}_t + \Delta p_t$ and that of the other is $\overline{p}_t - \Delta p_t$. The stagnation temperatures are similarly specified as $T_t = \overline{T}_t \pm \Delta T_t$. The inlet static pressure has a value which would produce a Mach number of 0.5 if the inlet stagnation pressures and temperatures were uniform; this is representative of conditions in aeropropulsion components at which a number of mixing processes occur.

The calculated loss coefficients shown in Figure 5.14 are roughly symmetric about both axes indicating that, although the entropy increase depends on both the stagnation temperature and pressure differences, the increase of entropy due to an initial stagnation pressure difference is almost independent of the initial difference in stagnation temperature, and vice-versa.[7] The entropy changes are due to heat transfer across a finite temperature difference (primarily associated with the stagnation temperature difference) and the dissipation of mechanical energy (mainly associated with the stagnation pressure difference).

[7] If the contours *were*, for example, ellipses symmetric about the horizontal and vertical axes, each would be described by $1 = [(P_{t_1} - P_{t_2})_i / A_{\Delta S}]^2 + [(T_{t_1} - T_{t_2})/B_{\Delta S}]^2$, where $A_{\Delta s}$ and $B_{\Delta s}$ are (dimensional) quantities representing the semi-major and semi-minor axes of an ellipse corresponding to a given entropy rise. The greater the entropy rise, the larger $A_{\Delta s}$ and $B_{\Delta s}$. From the form of the equation it can be seen that the entropy change in mixing associated with an initial stagnation pressure difference is not affected by the initial stagnation temperature difference, and vice-versa.

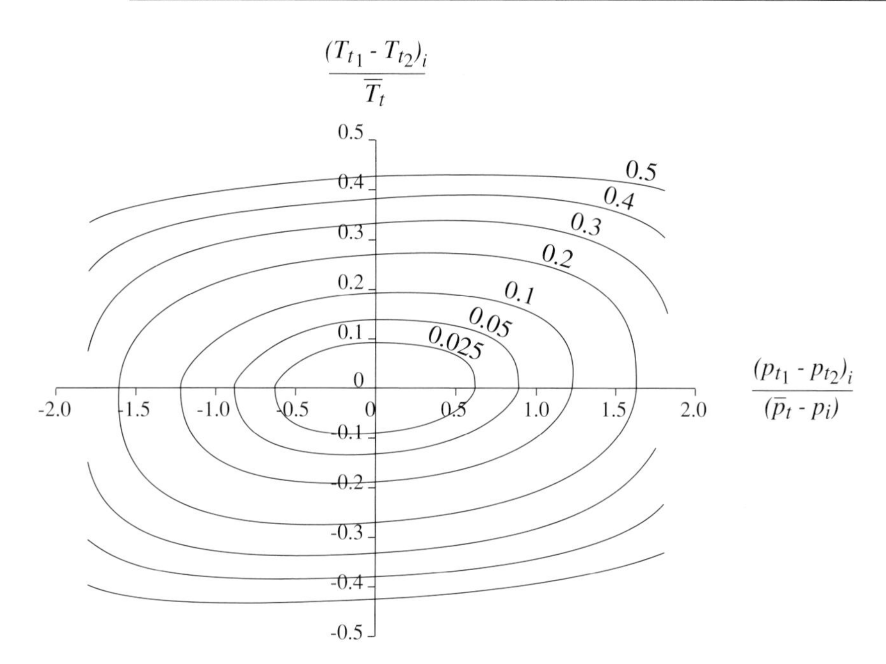

Figure 5.14: Entropy rise coefficient (defined in (5.5.6)) for the constant area mixing of two equal area streams at different stagnation pressures and temperatures; $p_{t_i} = \overline{p}_t \pm \Delta p_t$, $T_{t_i} = \overline{T}_t \pm \Delta T_t$ (Denton, 1993).

The non-dimensionalization for the stagnation pressure and temperature is not the same. The denominator for the former is the quantity $(\overline{p}_t - p_i)$. This reduces to the inlet dynamic pressure, $\frac{1}{2}\rho u^2$, as $M \rightarrow 0$. The quantity $(p_t - p)$ is more appropriate for compressible flow because it represents the pressure rise achievable from reversible adiabatic deceleration to the stagnation state. The denominator for the latter is the mean stagnation temperature. The reason for the different treatment of stagnation pressure and temperature is the topic of much of the next two subsections.

5.5.2 The limiting case of low Mach number $M^2 \ll 1$ mixing

Numerical results for mixed out quantities can readily be generated for arbitrary Mach number but it is useful to examine the case of low Mach number for several reasons. First, for mixing of streams with non-uniform stagnation temperature the connection between changes in stagnation pressure and component (or system) loss is different than for adiabatic flow. At low Mach numbers the analytic solution which exists can be used to demonstrate explicitly the role and behavior of changes in stagnation pressure and entropy as loss metrics. Discussion of this limit also reinforces, from a different perspective than in Section 2.2, what is meant by stating that a flow is incompressible. Finally, the resulting expressions, although strictly applicable only for $M^2 \ll 1$, give useful guidelines[8] concerning the behavior to be expected for Mach numbers up to 0.5–0.6.

For low Mach number flow the equation of state can be (see Section 1.17) approximated as $\rho T = \text{constant} + O(M^2)$, i.e. the effect on density or temperature due to pressure changes (which

[8] The limits of the approximation can be seen in Greitzer *et al.* (1985).

are of order Mach number squared) can be neglected. Differences between the stagnation and static temperature also have an impact of order Mach number squared and can be neglected. The relations appropriate to low Mach number mixing are thus (5.5.1), (5.5.2), plus the low Mach number form of (5.5.3)

$$T_e = \frac{\dot{m}_{1_i} T_{1_i} + \dot{m}_{2_i} T_{2_i}}{\left(\dot{m}_{1_i} + \dot{m}_{2_i}\right)}, \tag{5.5.7}$$

and the equation of state: $\rho_{1_i} T_{1_i} = \rho_{2_i} T_{2_i} = \rho_e T_e$.

The inlet non-uniformities in stagnation pressure and temperature can be characterized as the stream-to-stream temperature ratio, $TR = T_{2_i}/T_{1_i}$, and the stream-to-stream stagnation pressure difference, $\chi = (p_{t_1} - p_{t_{2_i}})/(\frac{1}{2}\rho_{1_i} u_{1_i}^2)$. The geometry is specified by the ratio of stream 1 area at inlet to total area, $\sigma = A_{1_i}/A$.

From dimensional considerations the stagnation pressure difference (between the inlet value in stream 1 or 2 and the mixed out value) scales with a representative inlet dynamic pressure and is a function of the quantities TR, χ, and σ, independent of Mach number, i.e.

$$\frac{p_{t_{1_i}} - p_{t_e}}{\frac{1}{2}\rho_{1_i} u_{1_i}^2} = \text{function}\,(TR, \chi, \sigma). \tag{5.5.8}$$

As an example, for two streams with equal areas ($\sigma = \frac{1}{2}$) and equal stagnation pressures at inlet ($\chi = 0$), neglecting wall shear stress and heat transfer, the expression for stagnation pressure change from inlet to mixed out conditions is[9]

$$\frac{p_{t_i} - p_{t_e}}{\frac{1}{2}\rho_{1_i} u_{1_i}^2} = \left[\sqrt{TR} + \frac{1}{\sqrt{TR}} - 2\right]. \tag{5.5.9}$$

The mixed out quantities of most interest are an entropy rise coefficient (analogous to that defined above) and the stagnation pressure difference, from inlet to mixed out conditions. For simplicity in dealing with the latter we define the reference state to be stream 1 at the inlet.

For compressible two-stream mixing with a uniform stagnation pressure at the inlet the entropy rise is given by

$$\frac{(s_e - s_i)}{c_p} = \left[\left(\frac{\dot{m}_{1_i}}{\dot{m}_e}\right)\ln\left(\frac{T_{t_e}}{T_{t_{1_i}}}\right) + \left(\frac{\dot{m}_{2_i}}{\dot{m}_e}\right)\ln\left(\frac{T_{t_e}}{T_{t_{2_i}}}\right)\right] - \frac{(\gamma - 1)}{\gamma}\ln\left(\frac{p_{t_e}}{p_{t_i}}\right). \tag{5.5.10}$$

For $M^2 \ll 1$, the entropy rise coefficient, normalized by the inlet velocity in stream 1, can be expressed as

$$\frac{\overline{T}_t (s_e - s_i)}{u_{1_i}^2} = \left[\frac{1}{(\gamma - 1) M_{1_i}^2}\right]\left[\left(\frac{\dot{m}_{1_i}}{\dot{m}_e}\right)\ln\left(\frac{T_{t_e}}{T_{t_{1_i}}}\right) + \left(\frac{\dot{m}_{2_i}}{\dot{m}_e}\right)\ln\left(\frac{T_{t_e}}{T_{t_{2_i}}}\right)\right] + \frac{\left(p_{t_i} - p_{t_e}\right)}{\rho u_1^2}. \tag{5.5.11}$$

[9] The explicit form of the function is not needed in the arguments that follow, although it can readily be found by application of the low Mach number form of the conservation laws for the control volume (see Section 11.7). Equation (5.5.7) yields the exit temperature and thus density. Conservation of mass then gives the mixed out velocity. Conservation of momentum gives the mixed out static pressure thus allowing calculation of the stagnation pressure.

In (5.5.11) $\overline{T}_t$ is the average stagnation temperature as defined in Section 5.5.1. Although the first term in (5.5.11) appears to become unbounded as Mach number decreases, this is an artifact of the normalization. From the low Mach number form of (5.5.10) (or multiplying (5.5.11) by $M_{1_i}^2$), we recover

$$\frac{(s_e - s_i)}{c_p} = \left[\left(\frac{\dot{m}_{1_i}}{\dot{m}_e}\right)\ln\left(\frac{T_e}{T_{1_i}}\right) + \left(\frac{\dot{m}_{2_i}}{\dot{m}_e}\right)\ln\left(\frac{T_e}{T_{2_i}}\right)\right] + (\gamma - 1)\, M_{1_i}^2 \left[\frac{\left(p_{t_i} - p_{t_e}\right)}{\rho u_1^2}\right]. \tag{5.5.12}$$

The effect of temperature equilibration on the entropy change does not depend on the inlet Mach number. At low Mach number, therefore, the specific entropy rise associated with a non-uniform inlet stagnation temperature approaches a constant value which depends on temperature and inlet area ratios. The contribution of the stagnation pressure decrease to the entropy change scales with the dynamic pressure and is proportional to M^2.

5.5.3 Comments on loss metrics for flows with non-uniform temperatures

Equation (5.5.12) shows the qualitative difference in the behavior of entropy and stagnation pressure in flows with non-uniform stagnation temperatures. In the low Mach number limit the change in p_t is linked to the change in mechanical energy per unit volume (as discussed later in this section). The change in entropy measures not only this effect but also the lost work associated with the thermal mixing of the two streams. If there is thermal mixing, the physical effects connected with entropy change and stagnation pressure change do not correspond as they did in flows with uniform stagnation temperature. As far as changes in stagnation pressure are concerned, the mixing process could have been regarded as a purely mechanical event with two streams of densities ρ_1 and ρ_2, both at the same temperature. In that case the same equations would be used to describe the process except T would be replaced by (constant/ρ) in (5.5.7) and the result interpreted as conservation of volume flow for incompressible mixing.

Two implications can be drawn from the above. First, for steady flow with $M^2 \ll 1$ the thermodynamics do not affect the dynamics. This is another statement of what constitutes the incompressible flow approximation. Second, the loss metric depends on whether one is interested only in the degradation of the mechanical energy within a fluid component or in the overall system losses. In the latter case the entropy change associated with heat transfer across a finite temperature difference must be accounted for: someone has paid to have one fluid heated or cooled, and a comprehensive system accounting must include this.

5.5.4 Mixing losses from fluid injection into a stream

In many applications two or more streams initially at an angle to one another are brought together to mix. A sketch of a typical configuration in which one stream is injected into a primary flow is given in Figure 5.15. The situation can be analyzed in a simple manner for arbitrary Mach number when the flow rate of the injected stream is small compared to the mainstream flow. If so, the differential

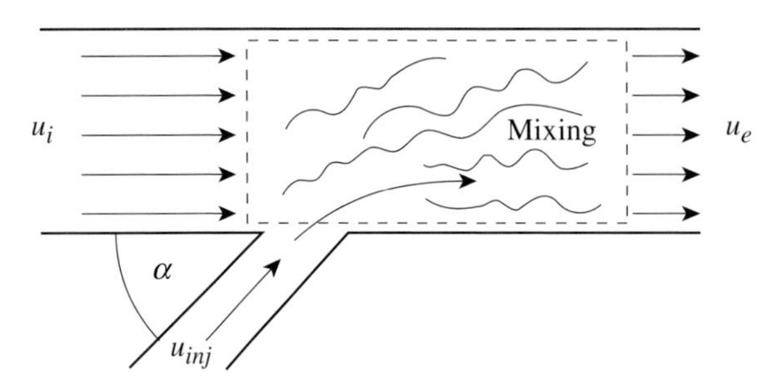

Figure 5.15: Mixing of injected flow with a mainstream flow at a different velocity, temperature, angle; injected flow quantities $d\dot{m}$, u_{inj}, T_{inj}, etc.

expressions for mass, momentum, and energy conservation across the control volume can be written to first order in $d\dot{m}/\dot{m}$, the ratio of injected to mainstream flow as (Shapiro, 1953)

$$\frac{d\rho}{\rho} + \frac{du_x}{u_x} = \frac{d\dot{m}}{\dot{m}}, \tag{5.5.13}$$

$$\frac{dp}{p} + \gamma M^2 \frac{du_x}{u_x} = \frac{d\dot{m}}{\dot{m}}\left[\gamma M^2\left(\frac{u_{x_{inj}}}{u_x} - 1\right)\right], \tag{5.5.14}$$

$$\frac{dT_t}{T_t} = \frac{d\dot{m}}{\dot{m}}\left(\frac{T_{t_{inj}}}{T_t} - 1\right). \tag{5.5.15}$$

The subscript "*inj*" denotes properties of the injected fluid, with the other variables denoting the mainstream quantities. All mainstream velocities in this section are in the x-direction. Equations (5.5.13), (5.5.14), and (5.5.15) must be supplemented by the differential forms of the perfect gas equation of state (5.5.16), the definitions of stagnation enthalpy (5.5.17) and stagnation pressure (5.5.18), and the Gibbs equation for entropy changes (1.3.19) in a form appropriate for a perfect gas with constant specific heats (referred to below as (1.3.19a)):

$$\frac{dp}{p} - \frac{d\rho}{\rho} - \frac{dT}{T} = 0, \tag{5.5.16}$$

$$\frac{dT_t}{T_t} - \frac{1}{\left(1 + \frac{\gamma - 1}{2} M^2\right)} \frac{dT}{T} - \frac{(\gamma - 1) M^2}{\left(1 + \frac{\gamma - 1}{2} M^2\right)} \frac{du_x}{u_x} = 0, \tag{5.5.17}$$

$$\frac{dp_t}{p_t} - \frac{\gamma}{\gamma - 1}\left(\frac{dT_t}{T_t} - \frac{ds}{c_p}\right) = 0, \tag{5.5.18}$$

$$\frac{ds}{c_p} - \frac{dT}{T} + \frac{\gamma - 1}{\gamma}\frac{dp}{p} = 0. \tag{1.3.19a}$$

In (5.5.13)–(5.5.18) and (1.3.19a), the known quantities that drive the changes, namely the non-dimensional mass, x-momentum, and energy added to the mainstream, appear on the right-hand side, and the seven unknowns: du_x/u_x, $d\rho/\rho$, dT/T, dp/p, ds/c_p, dT_t/T_t, and dp_t/p_t on the left. These equations can be combined to yield expressions for changes in two quantities of interest concerning loss, specific entropy, and stagnation pressure:

$$\frac{ds}{c_p} = \frac{d\dot{m}}{\dot{m}}\left[\left(1+\frac{\gamma-1}{2}M^2\right)\left(\frac{T_{t_{inj}}}{T_t}-1\right)+(\gamma-1)\,M^2\left(1-\frac{u_{x_{inj}}}{u_x}\right)\right], \tag{5.5.19}$$

$$\frac{dp_t}{p_t} = \frac{d\dot{m}}{\dot{m}}\left[-\frac{\gamma M^2}{2}\left(\frac{T_{t_{inj}}}{T_t}-1\right)-\gamma\,M^2\left(1-\frac{u_{x_{inj}}}{u_x}\right)\right]. \tag{5.5.20}$$

For $M^2 \ll 1$ (5.5.19) and (5.5.20) reduce to

$$\frac{Tds}{u_x^2} = \frac{d\dot{m}}{\dot{m}}\left[\frac{1}{M^2(\gamma-1)}\left(\frac{T_{inj}}{T}-1\right)+\left(1-\frac{u_{x_{inj}}}{u_x}\right)\right], \tag{5.5.21}$$

$$\frac{dp_t}{\rho u_x^2} = \frac{d\dot{m}}{\dot{m}}\left[-\frac{1}{2}\left(\frac{T_{inj}}{T}-1\right)-\left(1-\frac{u_{x_{inj}}}{u_x}\right)\right]. \tag{5.5.22}$$

These are changes in the mainstream quantities and do not include the entropy change of the injected flow (Denton, 1993). As described in Section 5.5.2, the relation between entropy and stagnation pressure changes in flows with non-uniform stagnation temperatures is qualitatively different from the correspondence between the two that occurs with uniform stagnation temperature.

For low Mach number flow, changes in stagnation pressure can be interpreted in terms of mechanical energy as follows. The equation of state in differential form is

$$\frac{d\rho}{\rho}+\frac{dT}{T}=0. \tag{5.5.23}$$

For $M^2 \ll 1$ the conservation equations are

$$\frac{d\rho}{\rho}+\frac{du_x}{u_x}=\frac{d\dot{m}}{\dot{m}}, \tag{5.5.24}$$

$$\frac{dp}{\rho u_x^2}+\frac{du_x}{u_x}=\frac{d\dot{m}}{\dot{m}}\left(\frac{u_{x_{inj}}}{u_x}-1\right), \tag{5.5.25}$$

$$\frac{d\rho}{\rho}=\frac{d\dot{m}}{\dot{m}}\left(1-\frac{\rho}{\rho_{inj}}\right). \tag{5.5.26}$$

Equations (5.5.24)–(5.5.26) describe the mixing of streams of non-uniform density at constant temperature which is a purely mechanical process. From these

$$\frac{dp_t}{\rho u_x^2} = \frac{d\dot{m}}{\dot{m}}\left[-\frac{1}{2}\left(\frac{\rho}{\rho_{inj}}-1\right)-\left(1-\frac{u_{x_{inj}}}{u_x}\right)\right], \tag{5.5.27}$$

which is equivalent to (5.5.22) for mixing of different temperature fluids.

5.5.5 Irreversibility in mixing

The previous two subsections have described the differences between the behavior of stagnation pressure changes and entropy changes.[10] As discussed in Section 5.2, a direct measure of loss is the

[10] This topic is addressed further, for flows with heat addition, in Section 11.3.

entropy creation due to irreversible processes. It is therefore important to develop a framework for understanding entropy creation in mixing processes. In this we follow the illuminating discussion of Young and Wilcock (2001), based on the example of the fluid injection into a stream.

The entropy created within a control volume such as that in Figure 5.15 is the difference between the leaving and entering entropy flux. The entropy flux leaving the control volume is $(\dot{m} + d\dot{m})(s + ds)$, where s is the entropy in the main stream at the inlet station of the control volume. The entering entropy flux is the mainstream entropy flux plus the entropy flux from the injected fluid, or $\dot{m}s + d\dot{m}s_{inj}$. The difference between the two represents entropy created because of irreversibilities. In terms of the entropy creation per unit mass this is (to first order in the small changes across the control volume)

$$ds_{irrev} = ds - (s_{inj} - s)\frac{d\dot{m}}{\dot{m}}. \tag{5.5.28}$$

The entropy change ds is given by (5.5.19). The difference $(s_{inj} - s)$ can be written (because the injected flow enters the control volume at the mainstream static pressure) as

$$(s_{inj} - s) = c_p \int_T^{T_{inj}} \frac{d\hat{T}}{\hat{T}}, \tag{5.5.29}$$

where $\hat{T}$ denotes here a dummy variable of integration. Using (5.5.19) and (5.5.28) in (5.5.29), and writing the stagnation temperatures in terms of static temperatures and velocities ($T_t = T + u_x^2/2c_p$; $T_{t_{inj}} = T_{inj} + ((u_{x_{inj}})^2 + (u_{y_{inj}})^2)/2c_p$), the entropy creation per unit mass within the control volume is found to be

$$\frac{ds_{irrev}}{c_p} = \frac{d\dot{m}}{\dot{m}} \left\{ \left[\frac{\left(u_x - u_{x_{inj}}\right)^2 + \left(u_{y_{inj}}\right)^2}{2c_pT} \right] + \left[\int_T^{T_{inj}} \left(\frac{1}{T} - \frac{1}{\hat{T}}\right) d\hat{T} \right] \right\}. \tag{5.5.30}$$

Equation (5.5.30) gives considerable insight into entropy creation during mixing. The first square bracket represents the entropy change from mixing of two streams at different velocities, i.e. the dissipation of bulk kinetic energy as mainstream and injection velocities mix to a uniform state. The first quadratic term in the bracket refers to velocity equilibration in the mainstream (x) direction. The second shows that in the mixing process all kinetic energy associated with injection normal to the mainstream also appears in the entropy rise. The second square bracket is the entropy change associated with thermal mixing of the injected flow and the mainstream to a uniform temperature. This term, multiplied by $\dot{m}c_pT$, is the power that could theoretically be obtained from a Carnot engine coupled between the mainstream flow at constant temperature T and the injected flow as the temperature of the latter changes from T_{inj} to T (Young and Wilcock, 2001).

5.5.6 A caveat: smoothing out of a flow non-uniformity does not always imply loss

Although a number of illustrations of losses caused by mixing out of flow non-uniformities have been presented, it should not be assumed that the presence of a non-uniformity always implies an increase in entropy (or decrease in stagnation pressure) as the flow comes to a final uniform state. A counterexample is furnished by the steady, two-dimensional, frictionless irrotational flow downstream of a obstacle or row of obstacles, for example a row of turbomachine blades.

Far downstream of the blade row, the flow is uniform and parallel with velocity components u_{x_∞} and u_{y_∞} in the x- and y-directions respectively. Near the blade row, the velocity field is non-uniform and can be described using a disturbance velocity potential, φ, whose gradients are the disturbance velocity components denoted by u'_x and u'_y. For Mach numbers low enough that the flow can be considered incompressible, the equation satisfied by φ is Laplace's equation, $\nabla^2\varphi = 0$. The velocity components are:

$$u_x = u_{x_\infty} + \frac{\partial \varphi}{\partial x} = u_{x_\infty} + u'_x, \tag{5.5.31a}$$

$$u_y = u_{y_\infty} + \frac{\partial \varphi}{\partial y} = u_{y_\infty} + u'_y. \tag{5.5.31b}$$

Similar to the description in Section 2.3, in the region downstream of the blades the solution for φ is periodic with the blade spacing W and decays with distance from the blade row. The form of φ which meets these conditions, as can be verified by direct substitution in Laplace's equation, is

$$\varphi = \sum_{k=1}^{\infty} \left[\left(A_k \sin \frac{2\pi k y}{W} + B_k \cos \frac{2\pi k y}{W} \right) e^{-2\pi k x/W} \right]. \tag{5.5.32}$$

If either the axial or the tangential velocity distribution is specified at a given x-location, which we can take as $x = 0$, the coefficients A_k and B_k can be found.

Because the flow is irrotational and steady, the stagnation pressure is everywhere constant throughout the downstream region, *whatever* the velocity variation at $x = 0$. It is of interest to examine the use of a control volume analysis with the objective of showing why the presence of the axial velocity non-uniformity here does not lead to a decrease in stagnation pressure. The reason is seen by considering the static pressure,

$$\begin{aligned} p &= p_t - \tfrac{1}{2}\rho\left(u_x^2 + u_y^2\right) \\ &= \underline{p_t - \tfrac{1}{2}\rho\left(u_{x_\infty}^2 + u_{y_\infty}^2\right)} - \tfrac{1}{2}\rho\left[2u_{x_\infty}u'_x + 2u_{y_\infty}u'_y + (u'_x)^2 + (u'_y)^2\right]. \end{aligned} \tag{5.5.33}$$

The underlined group of terms is a constant equal to p_∞, the static pressure far downstream. Equation (5.5.33) can therefore be written as

$$C_p = \frac{p - p_\infty}{\frac{1}{2}\rho u_{x_\infty}^2} = \left[2\frac{u'_x}{u_{x_\infty}} + 2\frac{u_{y_\infty}}{u_{x_\infty}}\frac{u'_y}{u_{x_\infty}} + \frac{(u'_x)^2}{u_{x_\infty}^2} + \frac{(u'_y)^2}{u_{x_\infty}^2} \right]. \tag{5.5.34}$$

Equation (5.5.34) shows that the static pressure along the control surface at $x = 0$ is not uniform in y, in contrast to the other cases we have examined so far. The variation in pressure implies streamline curvature at station 1 and consequently streamtube convergence and divergence downstream of this station. The average static pressure at $x = 0$ is lower than at $x \to \infty$ because the axial momentum flux is higher at $x = 0$, but the change in momentum is brought about solely by pressure forces. The forms of the velocity components given above can be used in the x-momentum control volume equation to see the consistency between constant stagnation pressure and the attenuation of the axial velocity non-uniformity.

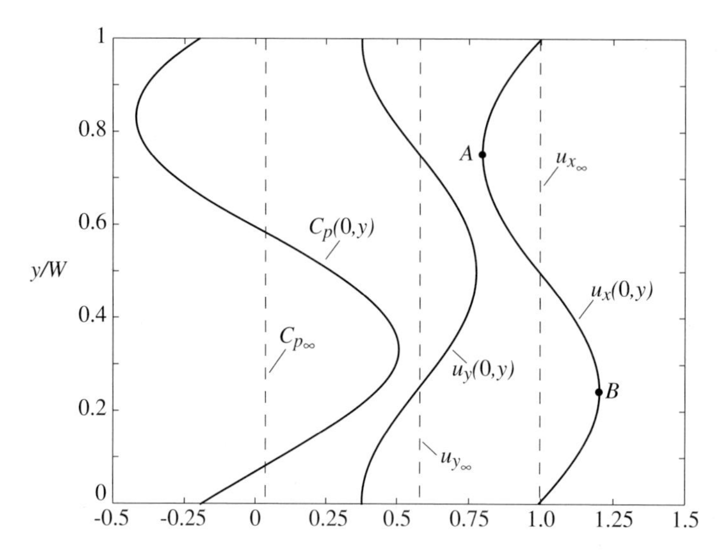

Figure 5.16: Velocity components and static pressure in a periodic irrotational flow. Mean exit flow angle = $\tan^{-1}(u_{y_\infty}/u_{x_\infty}) = 30^\circ$ (subscript ∞ denotes conditions far downstream).

Figure 5.16 shows the x- and y-velocity components and the static pressure for a single sinusoidal component of the disturbance velocity potential,

$$\varphi = A_1 \sin \frac{2\pi y}{W} e^{-2\pi x/W}. \tag{5.5.35}$$

The axial velocity variation is $0.20u_{x_\infty}$ at $x = 0$ and the exit angle from the blade row, based on u_{x_∞} and u_{y_∞}, is 30° from axial. The convergence of the streamlines will be such as to increase the velocity from location A to the downstream location and to decrease the velocity from point B to downstream.

The point to note is that there are situations in which static pressure variations over the inlet (or outlet) stations of a control volume must be addressed. The assumption that the static pressure is uniform is just that – an assumption – and is not always appropriate.

5.6 Averaging in non-uniform flows: the average stagnation pressure

5.6.1 Representation of a non-uniform flow by equivalent average quantities

Loss generation processes typically create a non-uniform flow, with subsequent mixing downstream. Measurement stations must often be placed at locations in which mixing is not complete, for example in multistage turbomachinery where the performance of one blade row is desired but the presence of downstream blading means the instrumentation is at a location with incomplete mixing. A specific issue we need to address in more depth, therefore, is how one accounts for losses in a flow in which the properties have a spatial variation, i.e. how one defines an appropriate average value for a flow property in a non-uniform stream.

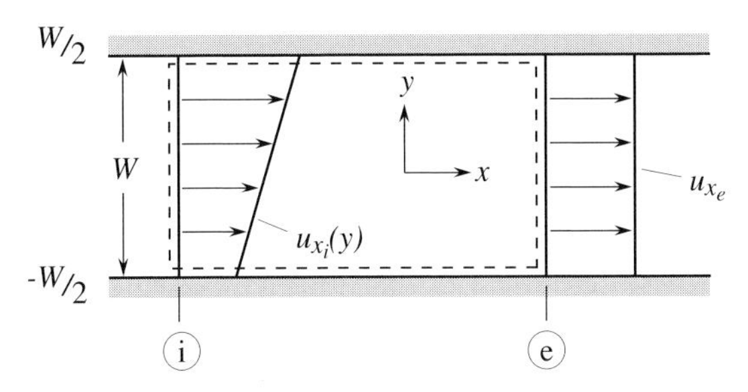

Figure 5.17: System and control volume used for mixing analysis; inlet station i: non-uniform velocity; exit station e: uniform ("mixed out") velocity.

This is only one aspect of a much broader question concerning the representation of a non-uniform flow with an "equivalent" average uniform flow, namely what general procedure is appropriate for capturing the behavior of a non-uniform flow using average values of the flow variables? Unfortunately, there is no unique answer to the question as posed. More precisely, as stated in Pianko and Wazelt (1983): "No uniform flow exists which simultaneously matches all the significant stream fluxes, aerothermodynamic and geometric parameters of a non-uniform flow." A main purpose of Section 5.6 is thus to sensitize the reader to the choices to be made, and methodology to be used, in developing useful approaches to averaging. In this context we develop several basic procedures and show their parametric behavior, first for constant density fluid motions and then for compressible flow. Quantitative information is also presented about the differences that exist between various averages. The final subsection takes up the specific question of how one chooses an appropriate method for obtaining an average value in a particular situation. Discussion and examples are given to show the way in which this depends on the application for which the average is to be used.

5.6.2 Averaging procedures in an incompressible uniform density flow

We turn first to the basic features of the averaging process in connection with the question of defining an average stagnation pressure. Three definitions of average stagnation pressure in common use are examined: area average, mass average, and mixed out average. To illustrate the behavior we work through the implications of each for incompressible uniform density flow (Sections 5.6.2 and 5.6.3) and then for compressible flow (Section 5.6.4).

The incompressible analysis both serves as an introduction and provides a framework to view results for compressible flow. The formulation is general, but it is helpful to cast the discussion in terms of a specific situation, steady flow in a two-dimensional channel of width W with a linearly varying velocity, as shown in Figure 5.17. The velocity at inlet station i has an x-component only with distribution

$$u_{x_i}(y) = \overline{u}\left(1 + \Lambda \frac{y}{W}\right), \tag{5.6.1}$$

where $\overline{u}$ is the mean velocity ($\overline{u} = (u_{max} + u_{min})/2$). The maximum velocity non-uniformity is thus $|\Delta u_x|_{max} = \Lambda \overline{u}$. The average stagnation pressure at station 1 will be found using each of the three averaging procedures.

5.6.2.1 Area average ($\overline{p}_t^A$)

The area average stagnation pressure is defined as

$$\overline{p}_t^A = \frac{1}{A}\int_A p_t dA \tag{5.6.2}$$

at any station in the duct. The static pressure is constant across the duct for a parallel flow so

$$\overline{p}_{t_i}^A - p_i = \frac{\rho}{2W}\int_{-W/2}^{W/2} u_x^2(y)\,dy. \tag{5.6.3}$$

Using the velocity distribution of (5.6.1),

$$\overline{p}_{t_i}^A - p_i = \frac{1}{2}\rho\overline{u}^2\left(1 + \frac{\Lambda^2}{12}\right). \tag{5.6.4}$$

The area average is presented first because of its simplicity, but this is essentially its only merit. In contrast to the other stagnation pressure averages to be introduced, the area average stagnation pressure is not associated with application of any conservation law and there is no fundamental reason for its use.

5.6.2.2 Mass average ($\overline{p}_t^M$)

To obtain the mass average for any quantity the area elements are weighted by the mass flow per unit area, with the integral taken over the channel mass flow. The mass average stagnation pressure is defined as

$$\overline{p}_t^M = \frac{1}{\dot{m}}\int_{\dot{m}} p_t\,d\dot{m} = \frac{\int_{-W/2}^{W/2}\left(p + \frac{1}{2}\rho u_x^2\right)\rho u\,dy}{\int_{-W/2}^{W/2}\rho u_x dy}. \tag{5.6.5}$$

For the velocity distribution of (5.6.1),

$$\overline{p}_{t_i}^M - p_i = \frac{1}{2}\rho\overline{u}^2\left(1 + \frac{\Lambda^2}{4}\right). \tag{5.6.6}$$

The mass average was previously encountered during the discussion of entropy flux in Section 5.3. It was shown there that, for uniform stagnation enthalpy and changes in stagnation pressure small compared to the (upstream) reference value, the mass average stagnation pressure at a given location represents the entropy flux at that station.

5.6.2.3 Mixed out average ($\overline{p}_t^X$)

The mixed out average stagnation pressure[11] is defined as the stagnation pressure that would exist after full mixing at constant area. To find this value we apply conservation of mass and momentum to the non-uniform profile, using the constant area control volume in Figure 5.17 and neglecting frictional forces on the top and bottom walls of the channel.

The flow is uniform at the exit station, e ($u_{x_e} = \bar{u}$), and the continuity equation is

$$\int_{-W/2}^{W/2} u_{x_i} dy = \overline{u} W. \tag{5.6.7}$$

The momentum equation is

$$\int_{-W/2}^{W/2} \left(p_i + \rho u_{x_i}^2\right) dy = \left(p_e + \rho u_{x_e}^2\right) W = \left(p_e + \rho \overline{u}^2\right) W. \tag{5.6.8}$$

Using (5.6.1) in (5.6.8) gives the static pressure rise associated with mixing:

$$p_e - p_i = \rho \overline{u}^2 \left(\frac{\Lambda^2}{12}\right). \tag{5.6.9}$$

The mixed out average stagnation pressure at the exit station is

$$\overline{p}_t^X = p_e + \rho \frac{\overline{u}^2}{2}. \tag{5.6.10}$$

Combining (5.6.9) and (5.6.10) yields

$$\overline{p}_t^X - p_i = \frac{1}{2} \rho \overline{u}^2 \left(1 + \frac{\Lambda^2}{6}\right). \tag{5.6.11}$$

For averaging processes that make use of a mixing analysis, the manner in which the mixing occurs must be specified. For example, instead of constant area the mixing process might occur at constant pressure. In this case the exit area at station e would not be the same as that at station i. For the linear inlet velocity distribution of (5.6.1), conservation of mass and momentum applied to mixing within a control volume with uniform pressure, p_i, on the bounding surfaces gives

$$u_{x_e} W_e = \int_{-W_i/2}^{W_i/2} u_{x_i} \, dy \quad \text{and} \quad u_{x_e}^2 W_e = \int_{-W_i/2}^{W_i/2} u_{x_i}^2 \, dy. \tag{5.6.12}$$

The ratio of stream areas for constant pressure mixing is

$$\frac{W_e}{W_i} = \frac{A_e}{A_i} = \frac{1}{\left(1 + \dfrac{\Lambda^2}{12}\right)}. \tag{5.6.13}$$

[11] This term and nomenclature were suggested by Smith (2001).

The mixed out stagnation pressure for constant pressure mixing is

$$\overline{p}_t^X\Big|_{\substack{constant\\ pressure}} - p_i = \frac{1}{2}\rho\overline{u}^2\left(1 + \frac{\Lambda^2}{12}\right)^2. \tag{5.6.14}$$

Constant pressure mixing is less commonly used as a model than is constant area mixing, but it is also a consistent way to look at mixing and may be the most pertinent in some situations. While general mixing processes tend to be neither precisely constant area nor constant pressure, these two situations furnish useful reference cases from which to view overall mixing behavior.

Several inferences can be drawn from the results of the three averaging processes. One is that there are different plausible ways to define an average flow quantity in a non-uniform flow. The example here is stagnation pressure but the comment applies to other variables as well.

The relative placement of the levels of the three average quantities is a general result for constant density flow. The mass average value is the highest of the three, because the higher stagnation pressure part of the stream is more heavily weighted. The area average is the lowest since it weights all parts equally. As mentioned the mass average stagnation pressure is directly related to the loss generated up to the averaging plane. Mixing generates further losses and the mass average stagnation pressure falls. The mixed out average, which can be regarded as a mass average at the final uniform state, is thus lower than the mass average but higher than the area average at the upstream station i.

The losses due to non-uniform flow are quadratic in the non-uniformity in that all three average total pressures involve Λ^2. We can connect this to the discussion in Section 2.8 by adopting a coordinate system moving with the lowest velocity in the flow. The loss due to mixing is unaltered, since the entropy rise is invariant with a change of reference frame. In the moving coordinate system, however, some part of the flow has zero velocity so the situation is similar to mixing in a sudden expansion where the stagnation pressure loss, and indeed all pressure changes, scale as the square of the velocity.

5.6.3 Effect of velocity distribution on average stagnation pressure (incompressible uniform density flow)

The linear variation in velocity is only one type of non-uniformity encountered, and the range of velocity distributions seen in practice includes boundary layers, wakes, and step-type profiles. It is thus relevant to assess the effect of velocity profile on average stagnation pressure. To address this we compare results for the linear profile with those derived for a very different velocity distribution, the step-type profile shown in the inset of Figure 5.18, which has two parallel streams with velocities u_E and εu_E. Denoting the fractional area occupied by the low velocity stream as σ, the average velocity is

$$\overline{u} = [\sigma\varepsilon + (1-\sigma)]u_E. \tag{5.6.15}$$

For constant density flow the stagnation pressure averages are formed as defined in the preceding section. For example the mass average stagnation pressure, normalized by the dynamic pressure based on the average velocity, is

$$\frac{\overline{p}_t^M - p}{\rho\overline{u}^2/2} = \frac{[\sigma\varepsilon^3 + (1-\sigma)]}{[\sigma\varepsilon + (1-\sigma)]^3}. \tag{5.6.16}$$

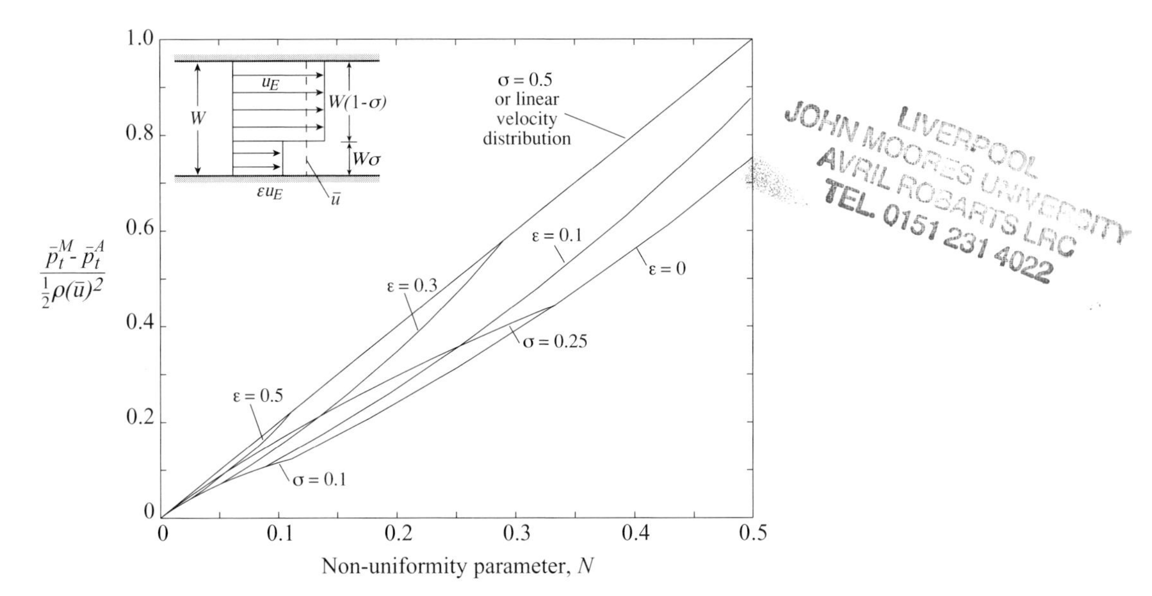

Figure 5.18: Difference between mass average and area average stagnation pressure as a function of non-uniformity parameter, N (5.6.17), for step-type (see inset) profiles and for linear velocity distribution; constant density flow.

The differences between the three averages for stagnation pressure depend on both the velocity non-uniformity parameter, ε, and the proportion of the duct occupied by the low and high speed flows, σ. For a given value of σ the differences increase as ε decreases from 1 to 0. The behavior with ε is more complicated: for a given value of ε the difference between averages increases as σ increases from 0 to 0.5 but can either increase or decrease for values of σ above this.

A simple quadratic measure of non-uniformity that captures the dependence on both parameters is the ratio of the average of the square of the velocity to the square of the average velocity, which we incorporate in a non-uniformity parameter, N, as

$$N = \frac{\int u_x^2 d(y/W)}{\left[\int u_x d(y/W)\right]^2} - 1 = \frac{\sigma\varepsilon^2 + (1-\sigma)}{[\sigma\varepsilon + (1-\sigma)]^2} - 1. \tag{5.6.17}$$

The parameter N goes to 0 when σ goes to 0 and 1 and when ε goes to 1. From (5.6.4) N is $\Lambda^2/12$ for the two-dimensional linear velocity distribution of (5.6.1).

The upper bound on differences between the stagnation pressure averages is that between mass average and area average. Presenting this upper bound as a function of N enables a general view of the trends in its magnitude, not only for different values of σ and ε but also for different velocity profiles. Figure 5.18 thus shows the difference between mass average and area average stagnation pressures, normalized by the dynamic pressure based on average velocity, as a function of non-uniformity parameter. (This normalization convention has been adopted to allow direct comparison with the results of Section 5.6.2.) Results are given for velocity non-uniformity (ε) from 0.5 to 0 for three values of σ (0.1, 0.25, 0.5) as well as for the linear velocity distribution in (5.6.1). Traversing a curve of constant σ in the direction of increasing N corresponds to increasing the velocity non-uniformity (decreasing ε) while holding the fractional area of low and high velocity streams constant. Contours of constant ε are also indicated: the curves for the different values of σ terminate at $\varepsilon = 0$, the condition of zero velocity in the low velocity stream. For the linear velocity distribution the difference between mass average and area average stagnation pressure[12] is $2N$ (i.e. $\Lambda^2/6$) which coincides with the line corresponding to $\sigma = 0.5$.

The principal trend in Figure 5.18 is a monotonic increase in the difference between mass average and area average stagnation pressure as N is increased. Although the differences between averages do not collapse to a single curve as a function of N, the parameter provides a guide to when effects of non-uniformities are likely to be important in loss or performance accounting. A 1% change in N implies (again, for $\sigma \leq 0.5$) a maximum difference between the stagnation pressure averages of 2% of the dynamic pressure based on the average velocity and thus a difference between mass average and mixed out average of 1% or less.

5.6.4 Averaging procedures in a compressible flow

In extending the averaging procedures to compressible flow the definition of an area average remains unchanged. The mass average, however, now includes the density variation

$$\overline{p}_t^M = \frac{\int\limits_A p_t d\dot{m}}{\int\limits_A d\dot{m}} = \frac{\int\limits_A p_t \rho u_x dA}{\int\limits_A \rho u_x dA}. \tag{5.6.18}$$

The definition of the mixed out average is based on a mixing process that implies the use of the conservation equations. For compressible flow an additional equation describing energy conservation is needed. If we specify no mass, momentum, heat, or work transfer to the stream from the duct walls, the three conservation equations defining the mixed out state in the duct are:

$$\text{conservation of mass:} \left(\int\limits_A \rho u_x dA\right)\Bigg|_{at\ (i)} = \rho_e u_{x_e} A = \dot{m}, \tag{5.6.19}$$

$$\text{conservation of momentum:} \ p_e A - p_i A = \left(\int\limits_A \rho u_x^2 dA\right)\Bigg|_{at\ (i)} - \dot{m} u_{x_e}, \tag{5.6.20}$$

$$\text{conservation of energy:} \ \frac{1}{\dot{m}}\left(\int\limits_A \rho u_x h_t dA\right)\Bigg|_{at\ (i)} = h_{t_e} = \overline{h}_t^M. \tag{5.6.21a}$$

[12] For values of σ greater than 0.5 and ε near 0, differences in the non-dimensional average stagnation pressure as defined above (and used in the figure) increase rapidly. For values of σ near unity the flow is essentially a narrow high speed jet in a much wider slowly moving stream and the non-dimensionalization used is not appropriate. The basic issue is one of choosing the relevant dynamic pressure for the context of the problem. For a constant density flow in which the mean velocity is not greatly different than the maximum, it can be argued that the dynamic pressure based on mean velocity is a, if not the, relevant form. In contrast, for a flow which has a narrow region with a velocity much greater than the mean, it is generally more useful to base the dynamic pressure on the velocity in the high speed stream. An example is the sudden expansion in Section 2.8, where the reference dynamic pressure is that of the stream entering into the larger duct. (If the difference in velocities, $(1-\varepsilon)u_E$, is substituted for the inlet velocity in a sudden expansion of area ratio $1/(1-\sigma)$, the results for static pressure rise and stagnation pressure decrease due to mixing can be applied directly.)

Neither of the choices for non-dimensionalization is *incorrect* and it is rather a question of which is more helpful as a measure of the behavior of interest; the objective here is to make a comparison of two profile families in a consistent and general way. Had we used a dynamic pressure based on the high speed flow we would find a difference in non-dimensional average stagnation pressures which varied between 0 and 1 for all σ and ε.

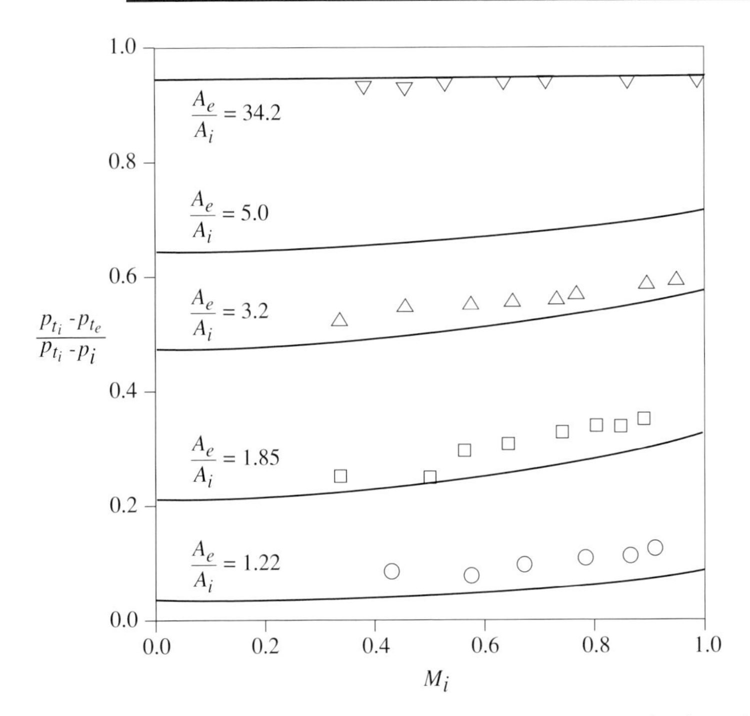

Figure 5.19: Stagnation pressure decrease across a sudden expansion in a pipe (experimental data from Hall and Orme (1955)).

For a perfect gas with constant specific heats, which is the case treated here, $c_p T_t$ can be substituted for h_t in (5.6.21a),

$$\frac{1}{\dot{m}}\left(\int_A \rho u_x T_t dA\right)\Bigg|_{at\ (i)} = T_{t_e} = \overline{T}_t^M. \tag{5.6.21b}$$

The effect of the Mach number level on mixed out stagnation pressure in a sudden expansion from A_i to A_e is shown in Figure 5.19 which gives the stagnation pressure decrease across the expansion as a function of the inlet Mach number. The different curves, which are derived from a compressible control volume analysis, correspond to different area ratios. The stagnation pressure decrease is non-dimensionalized by the difference between the inlet stagnation and static pressure, $(p_{t_i} - p_i)$.

There is a gradual rise in non-dimensional stagnation pressure drop as the upstream Mach number increases. Values of the stagnation pressure decrease for $M_i = 1$ are roughly 50% above those for $M_i = 0$ for the lower area ratios but as the area ratio of the expansion increases, this effect reduces. The control volume mixing analysis is seen to give a good estimate for the stagnation pressure changes.

5.6.4.1 Effects of inlet entropy and/or stagnation temperature non-uniformity

In defining averages for a compressible flow an inlet property additional to those specified for constant density flow must be given. Two choices for this, which model conditions found in practice, are uniform stagnation temperature and uniform entropy. The processes represented are quite different. The former corresponds to a non-uniformity created by losses whose magnitudes vary across the

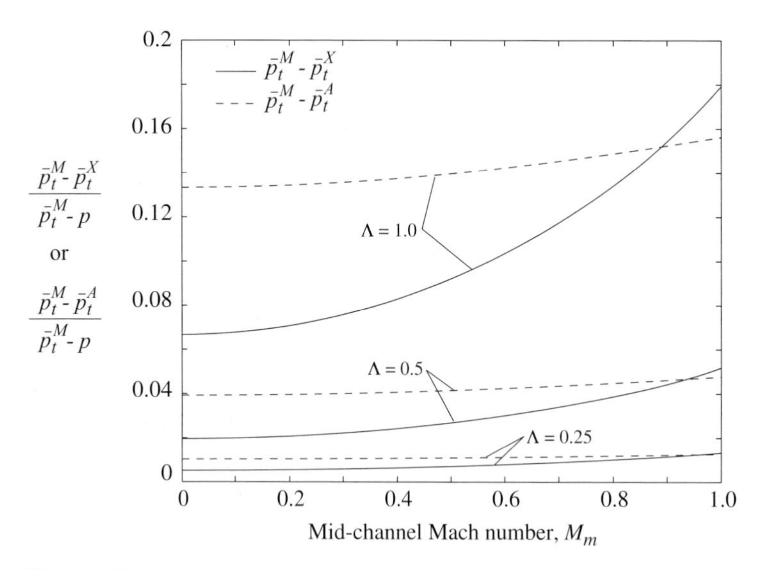

Figure 5.20: Difference between mass average and mixed out average *or* mass average and area average stagnation pressures, normalized by $\overline{p}_t^M - p$, versus mid-channel Mach number; two-dimensional channel of width, W, $u_x(y) = \overline{u}\,(1 + \Lambda y/W)$, uniform entropy at inlet.

flow, for example stationary obstacles (fences, screens) that block part of the channel. The latter might represent the conditions downstream of a compressor stage designed for non-uniform work input where the entropy change does not vary along the blade height.

For inlet conditions of uniform stagnation temperature, over the range of parameters shown there is no qualitative change relative to the constant density situation, and results for this case are therefore not shown. There is a quantitative change in that the non-dimensional difference between the averages increases from the constant density results as Mach number increases, in a manner roughly similar to that in Figure 5.19.

For uniform entropy at the inlet there is a qualitative change in the behavior of the average stagnation pressure compared to the constant density situation. Figure 5.20 shows this information for a two-dimensional straight channel. The figure presents the differences between: (i) mass average and mixed out stagnation pressures and (ii) mass average and area average stagnation pressures, normalized by $\overline{p}_t^M - p$. The initial velocity is the linear variation of (5.6.1): $u_x = \overline{u}\,(1 + \Lambda y/W)$. The differences in stagnation pressure[13] are given as a function of channel midheight Mach number, M_m, for three values of the velocity variation parameter Λ.

For $\Lambda = 0.5$ and 1.0 the value of $(\overline{p}_t^M - \overline{p}_t^X)$ (the solid curves) is larger than the value of $(\overline{p}_t^M - \overline{p}_t^A)$ (the dashed curves) for Mach numbers M_m near unity, which means that the mixed out stagnation pressure is lower than the area average stagnation pressure. This effect is not directly dependent on compressibility in that similar behavior occurs at low Mach number in a flow with uniform inlet stagnation pressure but non-uniform stagnation temperature. In that situation the mass average and

[13] The choice of which reference stagnation pressure should be used in these comparisons is not without some arbitrariness. The mass average stagnation pressure, however, is familiar, is defined using only inlet quantities, and is linked (with the qualifications expressed above) to the entropy flux. Its use also allows us to present the comparisons in Figure 5.21 in terms of the two stagnation pressures and the static pressure, without the necessity for the definition of an additional quantity.

the area average stagnation pressures are equal, with the mixed out stagnation pressure lower than both (Greitzer, Paterson, and Tan, 1985). Mixed out stagnation pressures can therefore be lower than area averages in a compressible flow and a non-constant density incompressible flow.

Figure 5.20 also indicates that for small values of Λ there is little difference in the three averages, and this is also true with uniform stagnation temperature at inlet. For the Mach number range in the figure, at a value of $\Lambda = 0.25$, the maximum differences are roughly 1% of $(\overline{p}_t^M - p)$.

For compressible flow (or incompressible flow with non-uniform density) the behavior of the averages is parametrically complex and Figure 5.20 should be interpreted as indicating trends only over the range of Mach number and non-uniformity shown. For example the mass average stagnation pressure is larger than the area average stagnation pressure for both uniform inlet stagnation temperature and uniform inlet entropy over the range of parameters in Figure 5.20, but this is not true under all circumstances. If the density variation in the flow is large enough, the portion of the stream with higher stagnation pressure can be weighted less by mass averaging than by area averaging, resulting in a mass average stagnation pressure which is lower than the area average value.

5.6.5 Appropriate average values for stagnation quantities in a non-uniform flow

We are now equipped to address the question posed at the beginning of the section, namely which procedure is most appropriate to represent "the" average quantities in a given non-uniform compressible flow (bearing in mind the overall caveat concerning representation of a non-uniform flow by an average uniform flow). A starting premise is that the mass and stagnation enthalpy fluxes, which together define the heat and shaft work exchanges with a fluid system, are quantities that should be the same in the average and the actual non-uniform flow. From the steady-flow energy equation the natural representation of the stagnation enthalpy flux is the mass average stagnation enthalpy.

To define other quantities such as the average stagnation pressure, however, additional considerations are needed. It is worthwhile to state explicitly what is desired of the average quantity because there are a number of ways to proceed. A useful approach is through the idea that for any given situation we wish to define average values corresponding to a uniform flow which retains the "essence of the action of the machine" (Smith, 2001) when compared to the actual flow in the situation of interest. One procedure for achieving this is to enforce the condition that fluxes of mass, linear momentum, stagnation enthalpy, and entropy are to be the same in the actual and the averaged flows. This provides a route to the definition of an average stagnation pressure.[14]

5.6.5.1 Definition and application of the entropy flux average (availability average) stagnation pressure

The entropy flux and the mass average entropy are related by

$$\int_A \left(s - s_{ref}\right) \rho u_x dA = \int_{\dot{m}} \left(s - s_{ref}\right) d\dot{m} = (\overline{s}^M - s_{ref})\dot{m}. \tag{5.6.22}$$

[14] If the discussion is extended to annular swirling flow, there is an additional variable, the circumferential velocity component that needs to be averaged. It is appropriate to use the mass average, because it is the difference in mass flux of angular momentum which is equal to the torque exerted on the fluid.

In (5.6.22) the subscript "*ref*" denotes an appropriate reference state, for example the region of the stream outside boundary layers or wakes. From (5.2.5), for a perfect gas with constant specific heats, the entropy change between any (stagnation) state and an initial reference state is

$$\frac{s - s_{ref}}{c_p} = \ln\left(\frac{T_t}{T_{ref}}\right) - \frac{\gamma - 1}{\gamma}\ln\left(\frac{p_t}{p_{ref}}\right). \tag{5.6.23}$$

Equation (5.6.23) can be integrated over the mass flow to find the entropy flux. The requirement for the averaged flow to have the same stagnation enthalpy flux as the actual flow yields the condition for equality of entropy flux between the actual and the averaged flow as

$$\begin{aligned}\frac{\overline{s}^M - s_{ref}}{c_p} &= \left(\frac{1}{\dot{m}}\right)\left\{\int_{\dot{m}} \ln\left[\left(\frac{T_t}{T_{ref}}\right)\left(\frac{p_{ref}}{p_t}\right)^{\frac{\gamma-1}{\gamma}}\right] d\dot{m}\right\} \\ &= \ln\left[\left(\frac{\overline{T}_t^M}{T_{ref}}\right)\left(\frac{p_{ref}}{\overline{p}_t^S}\right)^{\frac{\gamma-1}{\gamma}}\right].\end{aligned} \tag{5.6.24}$$

Equation (5.6.24) defines an average stagnation pressure, $\overline{p}_t^S$, based on equality of entropy flux between actual and average flows, as

$$\frac{\overline{p}_t^S}{p_{ref}} = \left\{\frac{\overline{T}_t^M}{T_{ref}}\right\}^{\frac{\gamma}{\gamma-1}} \left(\exp\left\{\frac{\gamma}{(\gamma-1)}\frac{1}{\dot{m}}\int_{\dot{m}} \ln\left[\left(\frac{T_{ref}}{T_t}\right)\left(\frac{p_t}{p_{ref}}\right)^{\frac{\gamma-1}{\gamma}}\right] d\dot{m}\right\}\right). \tag{5.6.25}$$

The definition maintains the same steady-flow availability function, $h_t - T_0 s$ (see Section 5.2), for the actual and averaged flows, and the stagnation pressure derived in this manner is thus sometimes referred to as the availability average stagnation pressure. An attribute of this definition is that we correctly account not only for the total energy input between any two states, or locations (through matching the mass flux of stagnation enthalpy) but also for the potential for shaft work resulting from a transformation between the two states (through matching the flux of flow availability function) (Cumpsty and Horlock, 1999).

Figure 5.21 shows the differences between the entropy flux average stagnation pressure, $\overline{p}_t^S$, and the mass average stagnation pressure, $\overline{p}_t^M$, for a two-dimensional straight channel with uniform inlet stagnation temperature and a velocity that varies linearly across the channel. As in Figure 5.20 the abscissa is the Mach number at the channel midheight location, the stagnation pressure differences are normalized by the quantity $\overline{p}_t^M - p$, and the curves are for different values of the velocity variation parameter, Λ. In the limit of low Mach number, for uniform inlet stagnation temperature, $\overline{p}_t^S$ reduces to the mass averaged stagnation pressure, $\overline{p}_t^M$, as mentioned in Section 5.3. Further, over a substantial parameter regime the availability average and mass average stagnation pressures are close and there may be little difference in practice in which is employed.

For a uniform inlet stagnation temperature $\overline{p}_t^M$ is larger than $\overline{p}_t^S$, although this is not always true for a non-uniform stagnation temperature. The relation of the two stagnation pressures can be seen using the example of a stream with step-type profiles in either stagnation pressure or temperature and a uniform value of the other property. Applying (5.6.25) to a uniform stagnation temperature stream, with the reference temperature corresponding to the uniform value and the reference pressure to the

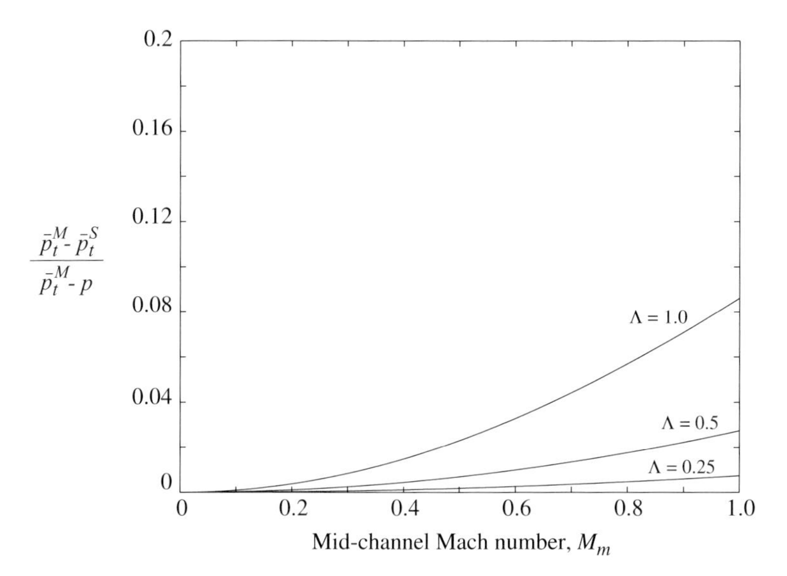

Figure 5.21: Difference between mass average and entropy flux average (availability average) stagnation pressures, normalized by $\overline{p}_t^M - p$, versus midchannel Mach number; two-dimensional channel of width, W, $u_x(y) = \overline{u}\,(1 + \Lambda y/W)$, uniform stagnation temperature at inlet.

mass average value yields

$$\frac{\overline{p}_t^S}{\overline{p}_t^M} = \exp\left\{\left(\frac{1}{\dot{m}}\right)\left[\int_{\dot{m}} \ln\left(\frac{p_t}{\overline{p}_t^M}\right) d\dot{m}\right]\right\}. \tag{5.6.26}$$

For a two-stream step-type profile with mass flows and stagnation pressures $\dot{m}_1, \dot{m}_2, p_{t_1}, p_{t_2}$, the integration gives

$$\frac{\overline{p}_t^S}{\overline{p}_t^M} = \left(\frac{p_{t_1}}{\overline{p}_t^M}\right)^{\left(\frac{\dot{m}_1}{\dot{m}_1+\dot{m}_2}\right)} \left(\frac{p_{t_2}}{\overline{p}_t^M}\right)^{\left(\frac{\dot{m}_2}{\dot{m}_1+\dot{m}_2}\right)}. \tag{5.6.27}$$

For $\dot{m}_1 = \dot{m}_2$ (5.6.27) simplifies to

$$\frac{\overline{p}_t^S}{\overline{p}_t^M} = \frac{2\sqrt{p_{t_1} p_{t_2}}}{p_{t_1} + p_{t_2}}, \tag{5.6.28}$$

a ratio which is always less than unity.

For a two-stream profile with uniform stagnation pressure, a non-uniform stagnation temperature, and $\dot{m}_1 = \dot{m}_2$, a similar analysis gives the ratio of the entropy flux average stagnation pressure to mass average stagnation pressure (which is also the actual uniform value) as

$$\frac{\overline{p}_t^S}{\overline{p}_t^M} = \left(\frac{T_{t_1} + T_{t_2}}{2\sqrt{T_{t_1} T_{t_2}}}\right)^{\frac{\gamma}{\gamma-1}}. \tag{5.6.29}$$

The ratio in (5.6.29) is larger than unity so, in this case, the average stagnation pressure derived from matching the entropy flux is larger than the actual (uniform) stagnation pressure.

5.6.5.2 Some general principles concerning averaging of non-uniform flows

From the above discussion several general principles that relate to averaging of non-uniform flows can be inferred. The first and most important follows from the statement at the start of this section concerning the inability to represent all attributes of a non-uniform flow by an average flow; the methodology and approach for defining an "equivalent" uniform flow must be developed within the context of the problem of interest. For example, if averaging is carried out at the exit of a given component, matching the entropy flux (in addition to the stagnation enthalpy and mass fluxes) debits the upstream component with the loss produced only up to the averaging station. Use of a mixed out average, in contrast, includes additional loss due to mixing that occurs downstream. Which is preferred, or even whether some other definition should be used, is the basic question faced in choosing an averaging scheme.

The nature of the application must be considered in addressing this question, as described by Smith (2001), who gives several examples that point to different choices for averaging. With reference to the propelling nozzle performance, for instance, it is suggested that thrust is the relevant metric and an appropriate average stagnation pressure might be based on matching the thrust of the actual flow to that of a uniform stream with the same mass flow.

Smith (2001) also mentions the different considerations that arise in defining average *inlet* properties for components when the stagnation pressure is uniform but the stagnation temperatures are non-uniform, a circumstance representative of turbine entry conditions in a gas turbine engine. The averaging constraints encountered in such situations, can be illustrated by examination of the question of defining a suitable average for non-uniform one-dimensional flow through a choked nozzle. We take the non-uniformity to be a step-type (two-stream) profile with uniform stagnation pressure and non-uniform stagnation temperature. The attribute we desire for the average is that the mass flow is well represented.

For a choked nozzle of given area we compare the mass flows based on two sets of average properties with the actual mass flow. The mass flows based on average conditions are: (i) $\dot{m}_M$, based on the mass average stagnation temperature and mass average stagnation pressure, and (ii) $\dot{m}_S$, based on the mass average stagnation temperature and entropy flux average stagnation pressure. If we require the behaviors of the average and the actual flow to be similar, the mass flows through the nozzle obey the choked flow relation (see Section 2.5),

$$\frac{\dot{m}_S\sqrt{\overline{T}_t^M}}{\overline{p}_t^S} = \frac{\dot{m}_M\sqrt{\overline{T}_t^M}}{\overline{p}_t^M} = \left(\frac{\dot{m}_{ref}\sqrt{T_{t_{ref}}}}{p_{t_{ref}}}\right) = \text{constant}. \tag{5.6.30}$$

In (5.6.30) the subscript "*ref*" denotes reference values of the quantities in a uniform one-dimensional choked flow.

Two questions can be asked about the mass flows based on average properties. First, for all average flows with the same mass average stagnation temperature as the actual flow (in other words for a mass average stagnation temperature equal to $T_{t_{ref}}$), what is the ratio of actual mass flow to mass flow based on the average stagnation pressures? Second, what is the mass flow ratio for arbitrary variations in stagnation temperature of the two streams, in other words for arbitrary variation in the ratio of the mass average stagnation temperature to $T_{t_{ref}}$?

Figure 5.22 provides answers to these questions. The figure shows the ratios of actual nozzle mass flow, $\dot{m}_{actual}$, to calculated mass flow based on average properties. The latter is derived using (5.6.30),

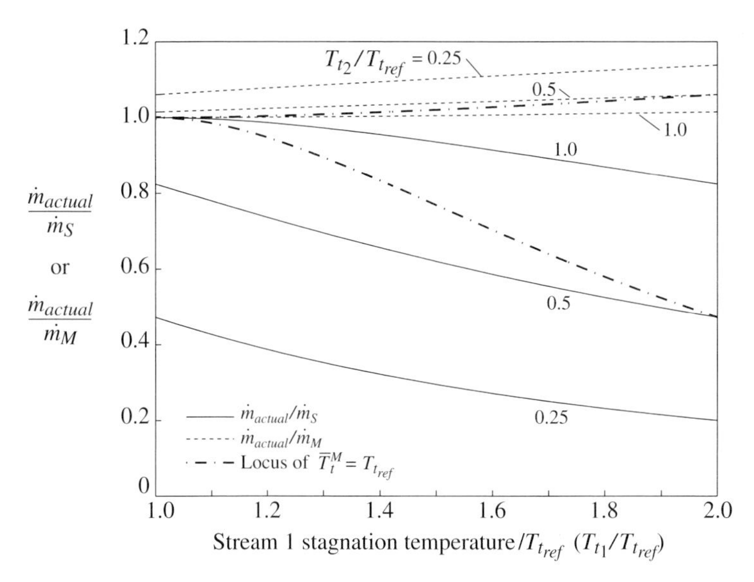

Figure 5.22: Ratio of actual mass flow in a choked nozzle to mass flows $\dot{m}_S$ and $\dot{m}_M$ defined using entropy flux average and mass average stagnation pressures respectively. Two-stream step-type profile with equal stream areas, uniform stagnation pressure.

an average stagnation pressure, and the mass average stagnation temperature. The mass flow ratios are shown as a function of the ratio of the stagnation temperature in the higher temperature stream, T_{t_1}, to the reference (uniform flow) stagnation temperature. The different curves correspond to different ratios of stagnation temperature in the lower stagnation temperature stream to the reference temperature. The locus of constant mass average stagnation temperature, $T_t^M / T_{t_{ref}} = 1$ (mass average stagnation temperature equal to stagnation temperature in the reference uniform flow), is also indicated.

For any T_{t_1} and T_{t_2} different than $T_{t_{ref}}$ the nozzle mass flow based on either average stagnation pressure is different from the actual flow. Use of the mass average stagnation pressure, however, provides a much better estimate for nozzle flow than use of the entropy flux average, with almost an order of magnitude difference for many conditions. Equation (5.6.29) shows the entropy flux average stagnation pressure is considerably higher than the actual pressure for large stream-to-stream stagnation temperature differences, leading to the poor estimate of mass flow in these conditions.

Figure 5.22 also illustrates a second aspect of flow averaging, namely that the attempt to represent a non-uniform flow by an "equivalent" average flow means that some properties will have different values than those in the actual flow. For the choked nozzle if we wish the mass flow to be well represented (defined here as having the averaged flow obey the one-dimensional choked nozzle relationship), the entropy flux must be different from the value in the actual flow. Another example is provided in comparing two channel flows, one uniform and one non-uniform, which have the same mass flux, stagnation enthalpy flux, entropy flux, and linear momentum; the calculated static pressure is different in these two flows. Discussion of this point, as well as of some other aspects of averaging procedures, is given by Pianko and Wazelt (1983).

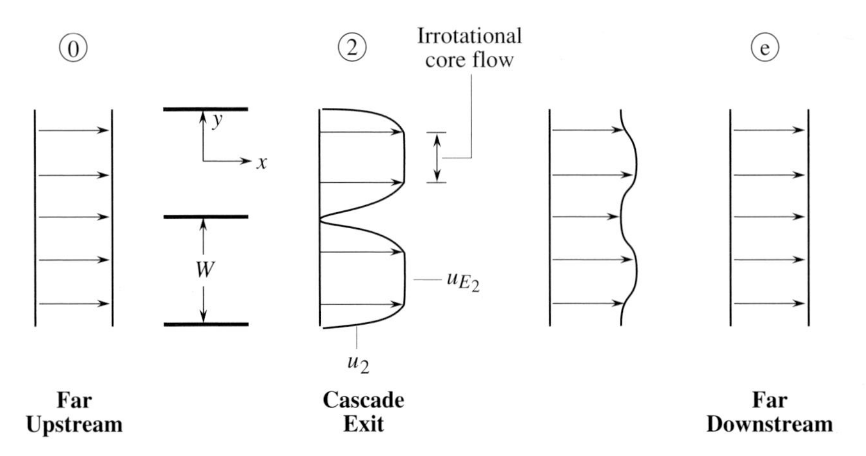

Figure 5.23: Stations used in analysis of flow losses.

The third, and final, aspect is that although the focus of Section 5.6 has been on stagnation pressure the ideas pertain more generally to the issue of averaging the equations of motion to give a reduced dimensionality (e.g. axisymmetric or one-dimensional) set of equations. Averaging the equations of motion in a formal manner leads to the appearance of Reynolds stress-like terms which are spatial averages of the products of various non-uniformities.[15] Discussions of the forms of these terms, their magnitudes, and some methodologies for including them, are given for non-uniform flow in ducts by Crocco (1958), Livesey and Hugh (1966), Livesey (1972), and Pianko and Wazelt (1983) and for turbomachinery flows by Smith (1966a), Köppel *et al.* (1999), and Adamczyk (2000).

5.7 Streamwise evolution of losses in fluid devices

We now return to the relation between loss produced inside a device and loss which occurs downstream. The topic is discussed in the context of incompressible constant density flow through the cascade of thin flat plate airfoils shown in Figure 5.23. We show how the different measures of average stagnation pressure at the exit of the cascade are linked to integral boundary layer properties and how they relate to the far downstream mixed out state (Mayle, 1973).

5.7.1 Stagnation pressure averages and integral boundary layer parameters

The mass average stagnation pressure at station 2, the trailing edge of the cascade, is given by

$$p_{t_0} - \overline{p}_{t_2}^M = \frac{\displaystyle\int_{-W/2}^{W/2} \rho u_0 p_{t_0}\, dy - \left[\int_{-W/2}^{W/2} \rho u_x p_t\, dy\right]_{station\ 2}}{\rho u_0 W}. \tag{5.7.1}$$

[15] Such terms always occur in a non-uniform flow because of the quadratic nature of the momentum flux. A simple example is the mixing out of a non-uniform constant density flow in a straight duct discussed in Section 5.6.1. As given in (5.6.9), the difference between the static pressure at the inlet and exit of the duct is the average of a term which is quadratic in the velocity non-uniformity.

In (5.7.1) the uniform far upstream velocity is denoted by u_0. Viscous effects are confined to thin boundary layers at the exit of the cascade, and the static pressure, p_2, is approximated as independent of y. The stagnation pressure in the free-stream region between the boundary layers, with cascade exit velocity, u_{E_2}, is equal to the upstream stagnation pressure:

$$p_0 + \frac{1}{2}\rho u_0^2 = p_{E_2} + \frac{1}{2}\rho u_{E_2}^2. \tag{5.7.2}$$

Carrying out the integration in (5.7.1) and using mass conservation, the change in mass average stagnation pressure between upstream and the cascade exit can be written in terms of the cascade exit velocity distribution as

$$p_{t_0} - \overline{p}_{t_2}^M = \rho \frac{u_{E_2}^3}{2Wu_0} \left[\int_{-W/2}^{W/2} \frac{u_x}{u_{E_2}} \left(1 - \frac{u_x^2}{u_{E_2}^2}\right) dy \right]_{\text{station 2}}. \tag{5.7.3}$$

The integral on the right-hand side of (5.7.3) is the kinetic energy thickness, θ^* (Sections 4.3 and 5.4), referenced to the local free-stream conditions.

To non-dimensionalize the stagnation pressure change by the far upstream velocity, which is a more convenient reference, we need to relate u_0 to u_{E_2}. From mass conservation for a passage,

$$\frac{u_0}{u_{E_2}} = 1 - \frac{1}{W} \int_{-W/2}^{W/2} \left(1 - \frac{u_x}{u_{E_2}}\right) dy = 1 - \frac{\delta_2^*}{W}, \tag{5.7.4}$$

where δ_2^* is the displacement thickness (Sections 2.9 and 4.3). The mass average stagnation pressure loss coefficient can now be expressed in terms of the kinetic energy thickness and the displacement thickness as

$$\frac{p_{t_0} - \overline{p}_{t_2}^M}{\frac{1}{2}\rho u_0^2} = \frac{\theta_2^*}{W} \frac{1}{\left(1 - \frac{\delta_2^*}{W}\right)^3}. \tag{5.7.5}$$

For viscous regions which are thin compared to the spacing between the blades ($\delta^*/W \ll 1$), (5.7.5) can be approximated as

$$\frac{p_{t_0} - \overline{p}_{t_2}^M}{\frac{1}{2}\rho u_0^2} \cong \frac{\theta_2^*}{W}. \tag{5.7.6}$$

If the flow at the cascade exit were taken to a fully mixed state at constant area, the mixed out average stagnation pressure, $\overline{p}_t^X = p_{t_e}$, would be obtained. This quantity can be found by applying the integral form of the mass and momentum conservation equations to a rectangular control volume with the upstream side at station 2, the downstream side at station e, and the top and bottom at $y = \pm W/2$. Doing this and forming the downstream stagnation pressure yields

$$\frac{p_{t_0} - \overline{p}_{t_2}^X}{\frac{1}{2}\rho u_0^2} = \frac{p_{t_0} - p_{t_e}}{\frac{1}{2}\rho u_0^2} = \frac{\left(\frac{\delta_2^*}{W}\right)^2 + \frac{2\theta_2}{W}}{\left(1 - \frac{\delta_2^*}{W}\right)^2}. \tag{5.7.7}$$

In (5.7.7) θ_2 is the momentum thickness at cascade exit. For $\delta^*/W \ll 1$, an approximate form of (5.7.7) is

$$\frac{p_{t_0} - p_{t_e}}{\frac{1}{2}\rho u_0^2} \cong \frac{2\theta_2}{W}. \tag{5.7.8}$$

The magnitude and direction of the far upstream and downstream velocities are equal so $(p_{t_0} - p_{t_e}) = (p_0 - p_e)$. Equation (5.7.8) therefore provides an expression for the drag of the cascade.

The area average stagnation pressure is given by

$$\frac{p_{t_0} - \overline{p}_{t_2}^A}{\frac{1}{2}\rho u_0^2} = \frac{\dfrac{\delta_2^*}{W} + \dfrac{\theta_2}{W}}{\left(1 - \dfrac{\delta_2^*}{W}\right)^2}. \tag{5.7.9}$$

For $\delta_2^*/W \ll 1$,

$$\frac{p_{t_0} - \overline{p}_{t_2}^A}{\frac{1}{2}\rho u_0^2} \cong \frac{\delta_2^*}{W} + \frac{\theta_2}{W}. \tag{5.7.10}$$

The area average and mixed out average stagnation pressure loss coefficients can be compared using the boundary layer shape parameter, $H = \delta^*/\theta$. The range of H is from 1.0 for a wake with a small fractional velocity defect to roughly 1.4 for a constant pressure turbulent boundary layer, to 2.5–3 for turbulent boundary layers near separation. The area average stagnation pressure loss coefficient for the cascade is, using (5.7.8) and (5.7.10),

$$\frac{p_{t_0} - \overline{p}_{t_2}^A}{\frac{1}{2}\rho u_0^2} = \left(\frac{1 + H_2}{2}\right)\left(\frac{p_{t_0} - p_{t_e}}{\frac{1}{2}\rho u_0^2}\right). \tag{5.7.11}$$

Equation (5.7.11) shows that the area average stagnation pressure at the cascade exit is lower than the mixed out average.

To give some reference for the magnitudes of the quantities defined above, the area average, mass average, and mixed out average stagnation pressure loss coefficients at the trailing edge for a single boundary layer with $\delta = 10\%$ of the passage and profile $(u_{x_2}/u_{E_2}) = (y/\delta)^{1/7}$ $(H = 1.29)$ are 0.023, 0.018, and 0.020 respectively. For a triangular exit velocity profile ($H = 3$), representative of the exit of a highly loaded compressor blade row, and the same δ, the three values are 0.074, 0.029, and 0.040.

The ratio of stagnation pressure loss between upstream (station 0) and the cascade exit (station 2) to that between upstream and the far downstream (station e) can also be put in terms of boundary layer parameters as

$$\frac{\text{loss in cascade}}{\text{overall loss}} = \frac{p_{t_0} - \overline{p}_{t_2}^M}{p_{t_0} - p_{t_e}} = \frac{\left(\dfrac{\theta_2^*}{W}\right)}{\left(\dfrac{\delta_2^*}{W}\right)\left(\dfrac{2}{H_2} + \dfrac{\delta_2^*}{W}\right)\left(1 - \dfrac{\delta_2^*}{W}\right)}$$

$$\cong \frac{H_2\theta_2^*}{2\delta_2^*} = \frac{\theta_2^*}{2\theta_2}. \tag{5.7.12}$$

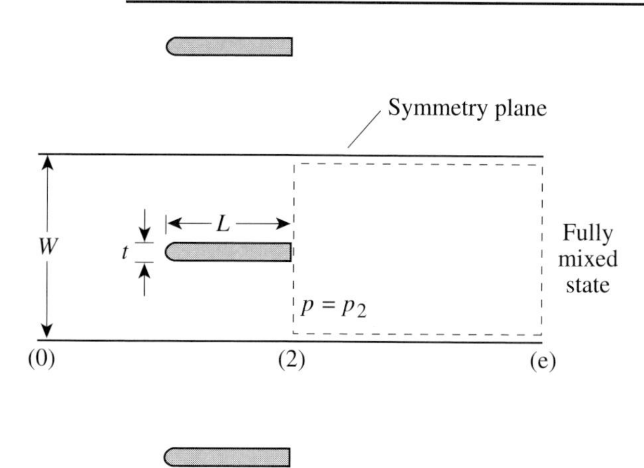

Figure 5.24: System and control volume for analysis of boundary layer and mixing loss for flow through an array of struts.

5.7.2 Comparison of losses within a device to losses from downstream mixing

As summarized by (5.7.12), the extent to which the loss can be regarded as occurring within the device depends on the form of the exit velocity profile. The examples above had most of the loss occurring within the device, but this is not always the case. More specifically the applications described so far have been mainly boundary layers on thin flat plates. Mixing situations also include wakes from bluff bodies and bodies with trailing edges thick compared to the boundary layer. In such cases losses generated from downstream mixing are important and even dominant. A case in point is the loss at the sudden expansion, discussed in Section 2.8, where the contribution of the losses in the boundary layers in the smaller diameter pipe could be neglected. When this approximation is appropriate, the mass flux of entropy (relative to an upstream station) at the beginning of the large diameter pipe is zero, and it is *only* downstream mixing that is responsible for the entropy generation.

The split between losses created within a component and losses due to mixing downstream of the component is illustrated by considering the flow past a periodic array of symmetric struts of non-zero thickness.[16] The control volume used to analyze the mixing process is given in Figure 5.24. The struts have a blunt trailing edge from which the flow separates. The static pressure is taken as uniform across the channel at the trailing edge, station 2.

Figure 5.25 shows the ratio of the loss occurring between station 0 and station 2 (from far upstream to trailing edge) to the overall loss, from station 0 to far downstream (station e), for three arrays of struts having thicknesses 0, 5, and 10% of chord. The chord/spacing ratio for the array is unity. The boundary layer loss was computed with an interactive boundary layer analysis (Drela and Giles, 1987) assuming fully turbulent flow. For the zero thickness strut, roughly 90% of the loss is incurred by the trailing edge location. For the 10% thick strut, the ratio drops to approximately 45% even though the boundary layer loss slightly increases.

[16] The periodic configuration is equivalent to a single strut in a constant area straight channel with width equal to the strut spacing.

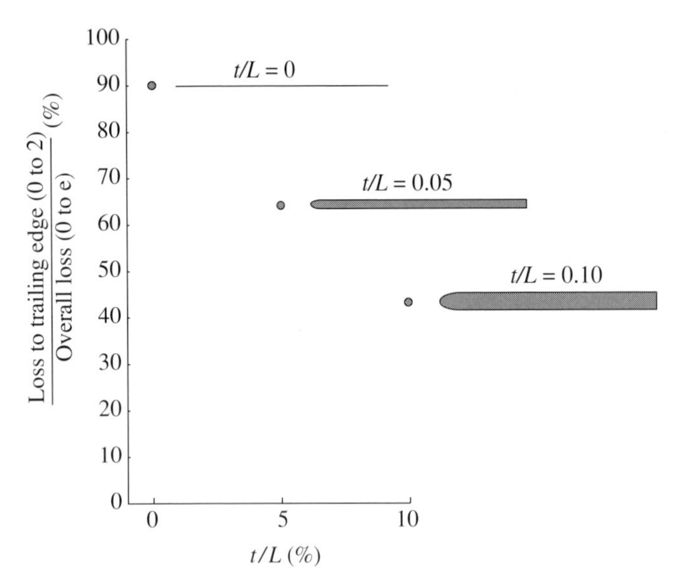

Figure 5.25: Ratio of losses for a cascade of symmetric struts, $L/W = 1.0$ (station numbers refer to those in Figure 5.24).

5.8 Effect of base pressure on mixing losses

The flow behind a bluff body or airfoil with a finite thickness trailing edge contains another feature affecting loss, referred to as the base pressure defect. Experiments show that the static pressure at the rear of such bodies is lower than the free-stream value. An example is given in Figure 5.26, which shows the pressure near the rear of a flat plate with a blunt trailing edge (Paterson and Weingold, 1985). The phenomena that determine base pressure are outside the scope of this discussion except to mention that unsteady flow associated with vortex shedding at the trailing edge is an important part of the process for subsonic flow.[17] For present purposes, it suffices to note that rough magnitudes of the base pressure coefficient, defined here as $C_{p_B} = (p_B - p_E)/\frac{1}{2}\rho u_E^2$, are from -0.1 to -0.2 for trailing edges which are thick compared to the surface boundary layers (Denton, 1993).[18]

We can carry out an approximate analysis to estimate the effect of base pressure on loss generation for the array of struts examined earlier. With reference again to Figure 5.24, the assumption about uniformity of pressure at station 2 is now dropped and a pressure p_B, different than the free-stream pressure, is taken to exist on the trailing edge of the body. This cannot be strictly correct because

[17] Suppressing vortex shedding through use of a trailing edge splitter plate reduces the magnitude of the base pressure coefficient by nearly a factor of 2 (Roshko, 1954). Conversely if vortex shedding is enhanced, the magnitude of the base pressure coefficient increases (by approximately 30% in the experiments of Kurosaka *et al.* (1987)).

[18] Base pressure coefficients quoted for bluff bodies in an external flow, such as cylinders or wedges of large included angle, are defined as $(p_B - p_0)/\frac{1}{2}\rho u_0^2$. The values are roughly 4–6 times the values shown in Figure 5.26. Aside from the difference in reference pressure, a large part of this disparity lies in the dynamic pressure used in defining the coefficient. For bluff bodies, the far upstream dynamic pressure is used, while for the trailing edge the local free-stream dynamic pressure is employed, and the free-stream dynamic pressure at separation for a bluff body is from 2 to 3 times the far upstream value. This does not completely resolve the difference, but it does give substantial reconciliation between the two values (Paterson and Weingold, 1982).

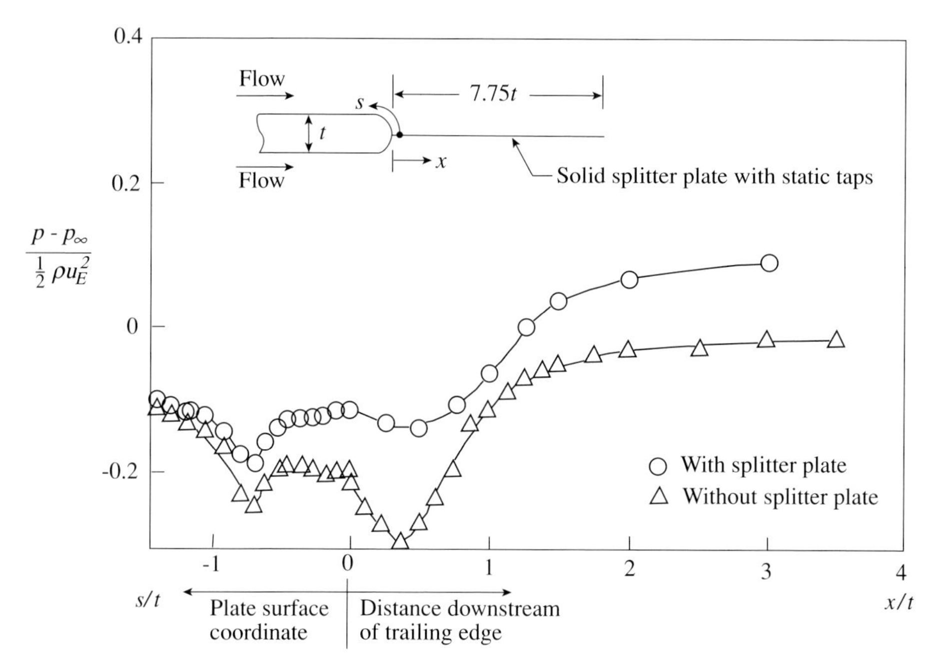

Figure 5.26: Static pressure coefficient for blunt trailing edge, $\delta^*/t = 0.18$, $u_E t/\nu = 56 \times 10^3$ (Paterson and Weingold, 1985).

the static pressure is not discontinuous in a subsonic flow, but the approach allows a useful parameterization of losses due to base pressure (Denton, 1993). The continuity and momentum equations applied to the control volume in Figure 5.24 are:

$$\dot{m} = \rho u_{E_2}(W - t - \delta_2^*) = \rho u_{x_e} W, \tag{5.8.1}$$

$$(W - t)\, p_2 + t p_B + \dot{m} u_{E_2} - \rho u_{E_2}^2 \theta_2 = W p_2 + \dot{m} u_{x_2}. \tag{5.8.2}$$

In (5.8.1) the notation u_{E_2} denotes the free-stream velocity at station 2. The resulting expression for the stagnation pressure decrease between far upstream and far downstream is

$$\frac{p_{t_0} - p_{t_e}}{\frac{1}{2}\rho u_{E_2}^2} = -C_{p_B}\left(\frac{t}{W}\right) + \frac{2\theta_2}{W} + \left(\frac{\delta_2^* + t}{W}\right)^2. \tag{5.8.3}$$

Equation (5.8.3) reduces to the expressions given in Section 5.7 (see (5.7.8)) when both C_{pB} and t/W are 0. If δ_2^*/W and $t/W \ll 1$, (5.8.3) becomes

$$\frac{p_{t_0} - p_{t_e}}{\frac{1}{2}\rho u_{E_2}^2} = -C_{p_B}\left(\frac{t}{W}\right) + \frac{2\theta_2}{W}. \tag{5.8.4}$$

To illustrate the effect of base pressure on loss level, as well as to provide comparison with more detailed methods for the assessment of this point, Figure 5.27 presents the local loss coefficient, based on the mass average stagnation pressure, $(p_{t_0} - \overline{p}_t^M(x))/(\frac{1}{2}\rho u_0^2)$, and the mixed out loss coefficient (from upstream to far downstream) for a 10% thickness periodic strut array, with a chord/spacing ratio of unity. The results are from an interactive boundary layer computation using a semi-empirical

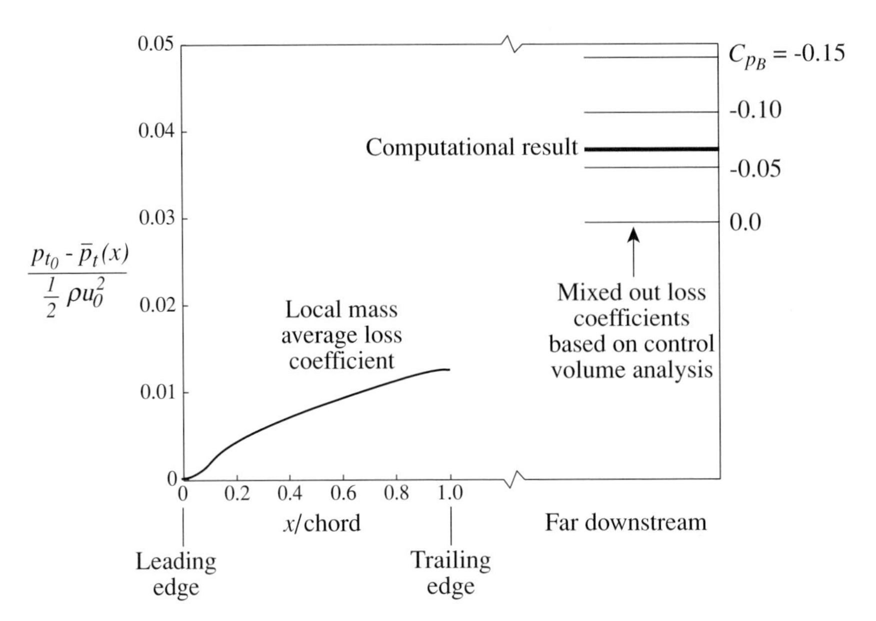

Figure 5.27: Loss generated within and downstream of a cascade of symmetric airfoils for different back pressure coefficients, $t/L = 0.10$, $L/W = 1.0$.

wake closure model for the base region (Drela, 1989). Values of the mixed out loss coefficient from the control volume analysis (5.8.3) are indicated for different values of the base pressure coefficient, C_{pB}. The mixing losses given by the computations correspond to a C_{p_B} of roughly -0.06; the wake closure model assumes boundary layers are thick relative to trailing edges and thus does not fully capture blunt trailing edge behavior.

Figure 5.28 shows results from a compressible control volume analysis for the entropy rise coefficient of a cascade of finite thickness flat plates as a function of Mach number. The conditions of the calculations are that there is no boundary layer and the trailing edge thickness is 10% of the spacing. The different curves correspond to the specified values of the base pressure coefficient. There is a substantial increase with Mach number, in accord with the experimental finding that trailing edge losses increase rapidly as the downstream Mach number approaches unity (Denton, 1993).

Measurements of the evolution of loss in the wake of an airfoil are given in Figure 5.29. The airfoil had a trailing edge thickness 2% of chord and was subjected to a representative turbine blade pressure distribution through contouring the bounding passage walls. Mach numbers were much less than unity and the boundary layers at the trailing edge were turbulent. Two types of loss coefficient are shown which are slightly different in definition, but analogous, to those described above. The first, shown by the symbols, is an overall loss coefficient based on a constant area mixing process using the measured velocity and stagnation pressure profiles as the upstream conditions for the control volume. It is defined as

$$\text{overall loss coefficient} = \frac{\displaystyle\int_\delta (p_e - p)\,dy}{\frac{1}{2}\rho u_{E_e}^2 \left(\dfrac{u_{E_e}}{u_E}\right) t} + \frac{\displaystyle\int_\delta \frac{u_x}{u_E}\left(1 - \frac{u_x}{u_E}\right) dy}{\frac{1}{2}\left(\dfrac{u_{E_e}}{u_E}\right)^2 t}. \tag{5.8.5}$$

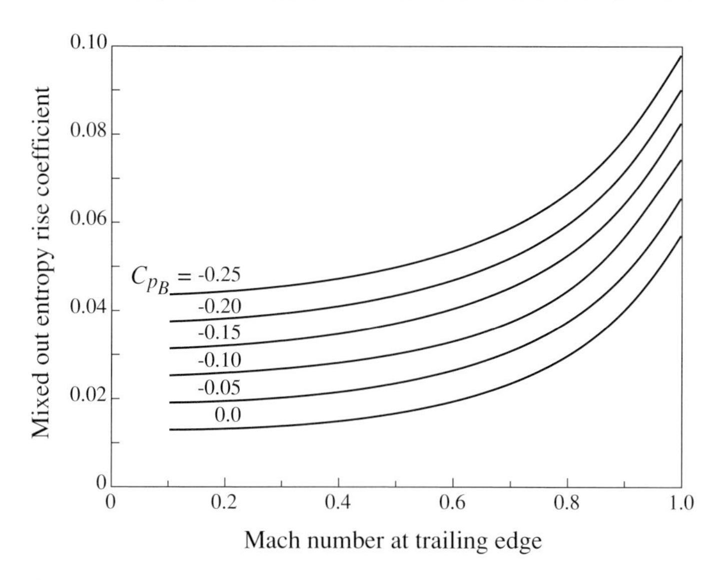

Figure 5.28: Variation of the trailing edge loss coefficient based on the entropy rise $(T_i(s_e - s_i)/\frac{1}{2}\rho u_i^2)$ with base pressure coefficient and Mach number for a 10% thick body with zero boundary layer thickness; control volume analysis of Denton (1993).

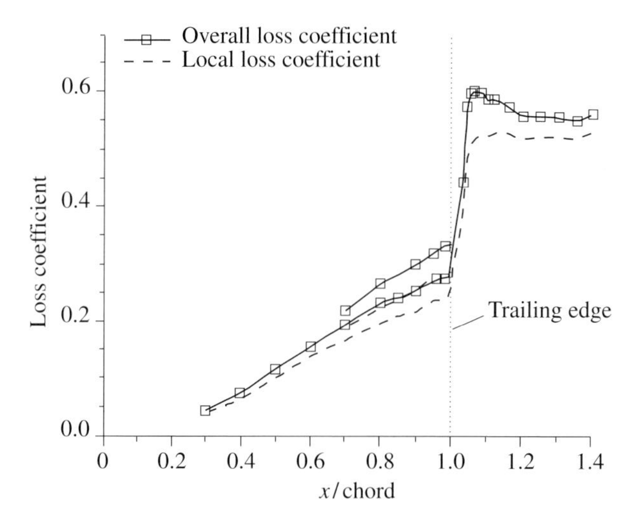

Figure 5.29: Streamwise evolution of loss coefficients on an airfoil with representative turbine pressure distribution. Suction surface measurements start from 0.3 (x/chord), both surfaces from 0.7 chord (Roberts and Denton, 1996).

In (5.8.5) the integration is carried out across the boundary layer or wake, depending on the station examined. The reference velocity used is the free-stream velocity at the exit station of the channel, U_{E_e}. Overall loss coefficients associated with the suction surface boundary layer are plotted from the 0.3 chord station and data including both surfaces are given from 0.7 chord.

If static pressure variations are negligible over the integration domain, and the free-stream velocity does not change between the local station and the exit station, (5.8.5) reduces to twice the momentum

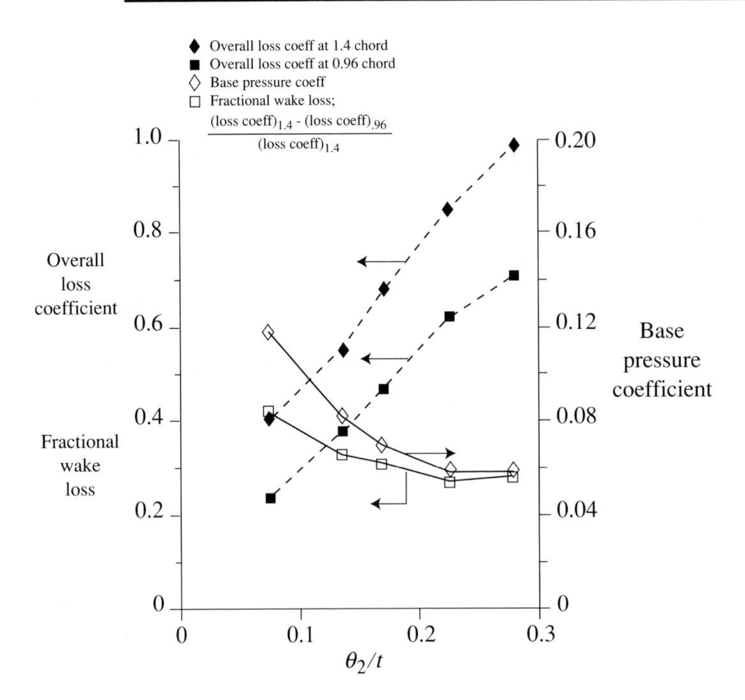

Figure 5.30: Overall loss coefficients, fractional wake loss, and base pressure coefficient versus suction surface momentum thickness/traiting edge thickness; airfoil with representative turbine pressure distribution; θ_2 denotes momentum thickness at 0.96 chord location (Roberts and Denton, 1996).

thickness divided by the trailing edge thickness, $2\theta/t$. Multiplying this limiting value by the ratio of thickness to passage spacing, t/W, yields the mixed out loss coefficient defined previously in (5.7.8).

The second loss coefficient is based on the entropy created up to the station indicated, which, for $M^2 \ll 1$, is equal to the mass average stagnation pressure defect at that location. The definition is

$$\text{local loss coefficient} = \frac{\int_{\delta} u_x(p_{t_e} - p_t)dy}{\frac{1}{2}\rho(u_{E_e})^3 t}. \tag{5.8.6}$$

The behavior of the local loss coefficient is given by the dashed line in Figure 5.29. For the limiting conditions of uniform static pressure at the station of integration and no change in external velocity to the exit station, (5.8.6) reduces to the kinetic energy thickness divided by the thickness, θ^*/t. For a passage this corresponds to the mass average loss coefficient defined in (5.7.6).

The non-dimensionalizations in (5.8.5) and (5.8.6) are in terms of trailing edge thickness because interest is in loss per trailing edge. As mentioned, to connect with previous results in terms of passage width the loss coefficients in Figure 5.29 (and Figure 5.30) should be multiplied by the trailing edge thickness ratio (t/W); for comparison with the 10% thick symmetric airfoil results in Figure 5.27 this means division of loss numbers by 5.

An evident feature in Figure 5.29 is the rapid increase in loss within 0.05 chord length (2.5 trailing edge thicknesses) downstream of the trailing edge. A substantial portion of the total loss is seen to be associated with processes that take place downstream of the trailing edge.[19] In this context a distinction can be made between *all* the processes which occur downstream of the trailing edge and those which may be more properly defined as *wake loss*. The argument is that "if the boundary layers mix out at the local flow area, the associated loss is independent of the nature of the wake flowfield and should not be included in the definition of wake loss." (Roberts and Denton, 1996). Wake loss is thus defined as the difference in overall loss coefficients evaluated at the downstream and upstream stations. On this basis there is a distinction between the wake loss and the difference between overall and entropy flux loss coefficients. The two quantities were measured to be 33% and 41% of the downstream overall loss respectively.

The measurements can also be related to the approximate expression for overall loss given by (5.8.3). Using the measured momentum thickness, displacement thickness, and base pressure coefficients, the calculated overall loss is approximately 10% below the actual value. The three terms in (5.8.3), $2\theta_2/t$, C_{PB}, and $(\delta_2^* + t)^2/(tW)$ (where the evaluation is done at the 0.96 chord station), had values of 66%, 18%, and 15% of the total respectively.

Figure 5.30 shows the overall loss coefficient (5.8.5) at 0.96 chord and at the farthest downstream station (1.4 chord), the fractional wake loss, and the base pressure coefficient, all as functions of suction surface momentum thickness θ_2/t. The increase in magnitude of the base pressure coefficient as the momentum thickness decreases is associated with an observed increase in vortex shedding intensity (i.e. an increase in *rms* velocity fluctuation) of roughly 50%.

Figures 5.27, 5.29, and 5.30 provide quantitative information about the ratio of loss produced in a device compared to that produced far downstream. In accord with trends mentioned earlier, an increase in the ratio of trailing edge thickness to boundary layer thickness is associated with an increase in the fraction of overall loss that occurs downstream of the device.

5.9 Effect of pressure level on average properties and mixing losses

In many configurations static pressure increases or decreases occur downstream of fluid components. Such changes in pressure level impact mixing loss. To give insight into this behavior three examples are presented for a constant density incompressible flow: an introductory discussion of the effect of pressure level on two-stream mixing losses; an extension of the analysis of Section 5.6 for linear velocity variation to include the effect of pressure level; and a description of pressure level effects on wake mixing loss.

5.9.1 Two-stream mixing

Consider two streams of constant density fluid in adjacent ducts, as sketched in Figure 5.31. Stream 1 comes from a reservoir at stagnation pressure p_{t_1} and stream 2 from a reservoir at p_{t_2}. The combined mass flow of the two streams is $\dot{m}$, with fraction f in stream 1, so $\dot{m}_l = f\dot{m}$ and $\dot{m}_2 = (1 - f)\dot{m}$. These mass flow fractions will be held fixed in the analysis to follow. In addition, because we are assessing

[19] The slight fall in the measured overall loss gives an indication that there is some error in the measurements, but this is small enough that it does not affect the conclusions.

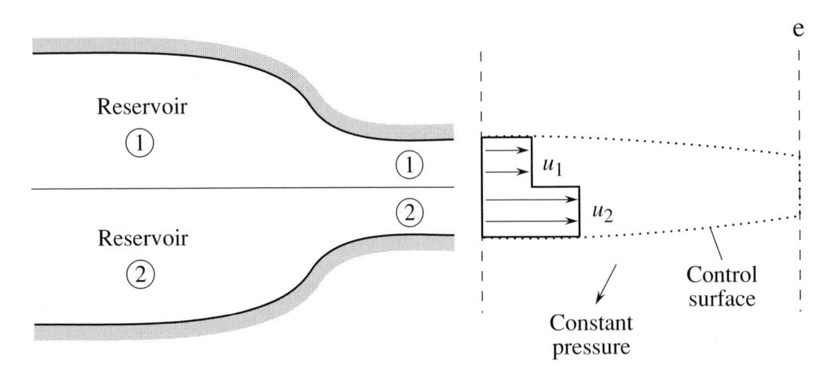

Figure 5.31: Two-stream constant pressure mixing.

the effect of pressure level, the mixing is taken to occur at constant pressure. The conclusions do not depend on this assumption but it allows for a more straightforward interpretation (Taylor, 1971).

For the constant pressure control surface in Figure 5.31, the one-dimensional form of the momentum equation is

$$f u_{x_1} + (1 - f)\, u_{x_2} = u_{x_e}. \tag{5.9.1}$$

In (5.9.1) the subscripts 1 and 2 denote the two streams and the subscript "*e*" denotes the fully mixed state at the exit of the control volume.

Suppose the static pressure of the reservoir into which the streams are discharged is altered by dp, but f and the reservoir stagnation pressures are held constant. (To keep f constant, the ratio of exit flow areas would need to be changed.) From the definition of stagnation pressure the change in p_{t_e} that results is

$$dp_{t_e} = dp + \rho u_{x_e} du_{x_e}. \tag{5.9.2}$$

From (5.9.1) the velocity changes associated with the static pressure change are related by

$$f du_{x_1} + (1 - f) du_{x_2} = du_{x_e}. \tag{5.9.3}$$

The reservoir pressures p_{t_1} and p_{t_2} are fixed so that du_{x_1} and du_{x_2} are related only to the change in pressure, dp, ($dp = -\rho u_{x_j}\, du_{x_j}$ for $j = 1, 2$), as is du_{x_e} through (5.9.3). Substitution in (5.9.2) yields an expression for the dependence of the mixed out stagnation pressure on the static pressure level:

$$\left(\frac{\partial p_{t_e}}{\partial p}\right)_f = [(f - 1)\, f] \left[\frac{u_{x_1}}{u_{x_2}} + \frac{u_{x_2}}{u_{x_1}} - 2\right]. \tag{5.9.4}$$

The second square bracket in (5.9.4) can be rewritten as $(\sqrt{u_{x_1}/u_{x_2}} - \sqrt{u_{x_2}/u_{x_1}})^2$, which is positive whatever the values of u_{x_1}/u_{x_2}. Since $f < 1$, the right-hand side of (5.9.4) is negative and

$$\left(\frac{\partial p_{t_e}}{\partial p}\right)_f < 0. \tag{5.9.5}$$

The interpretation of (5.9.5) is that increasing the level of static pressure at which mixing occurs decreases the mixed out stagnation pressure, while decreasing the static pressure increases the mixed out stagnation pressure. This is due to the effect of pressure level on the velocity differences between the streams; as discussed previously, the mixing losses scale with the square of this difference. For a small change in static pressure, the ratio of the velocity change in stream 2 to that in stream 1 is

$$\frac{du_{x_2}}{du_{x_1}} = \frac{u_{x_1}}{u_{x_2}}. \tag{5.9.6}$$

If u_{x_1} is larger than u_{x_2}, stream 2 experiences a larger velocity change than stream 1. If the static pressure drops, the velocities u_{x_1} and u_{x_2} will draw closer together; if it rises, they become farther apart.

These conclusions can be extended to finite changes in the static pressure level. The decrease in mass average total pressure during mixing is

$$\overline{p}_t^M - p_{t_e} = f\, p_{t_1} + (1-f)\, p_{t_2} - p_{t_e}. \tag{5.9.7}$$

Because mixing occurs at constant pressure, (5.9.7) can be written as

$$\overline{p}_t^M - p_{t_e} = \frac{\rho}{2}\left[f\, u_{x_1}^2 + (1-f)\, u_{x_2}^2 - u_{x_e}^2\right]. \tag{5.9.8}$$

Eliminating the downstream mixed out velocity, u_{x_e}, by using (5.9.1) yields

$$\overline{p}_t^M - p_{t_e} = \frac{\rho}{2} f(1-f)(u_{x_1} - u_{x_2})^2. \tag{5.9.9}$$

For a fixed value of f, the mixing loss is proportional to the square of the velocity difference between the two streams which, in turn, is set by the level of static pressure at which the mixing takes place. Defining Δp_t $(= p_{t_2} - p_{t_1})$ as the difference in stagnation pressure between the two streams, the non-dimensional velocity difference prior to mixing is

$$\frac{u_{x_1} - u_{x_2}}{\sqrt{\dfrac{2\Delta p_t}{\rho}}} = \sqrt{\frac{p_{t_1} - p}{\Delta p_t}} - \sqrt{\frac{p_{t_1} - p}{\Delta p_t} - 1}. \tag{5.9.10}$$

Figure 5.32 shows the effects of the pressure level on velocity difference and stagnation pressure decrease in constant pressure two-stream mixing. The abscissa is the static pressure level, referenced to the stagnation pressure of the high velocity stream, non-dimensionalized by the stagnation pressure difference between the two streams. The ordinates are the non-dimensional velocity difference between the streams at the start of mixing (the solid line corresponding to the scale on the left) and the stagnation pressure decrease due to mixing (the dashed lines corresponding to the scale on the right). The non-dimensional velocity difference is independent of f but $(\overline{p}_t^M - p_{t_e})/\Delta p_t$ depends on f as well as the static pressure level. The highest static pressure on the abscissa corresponds to a value of the pressure coefficient $(p_{t_1} - p)/\Delta p_t$ of 1.0; at this value, the low stagnation pressure stream has zero velocity. As $(p_{t_1} - p)/\Delta p_t$ is increased, the static pressure drops, the velocity difference at the start of mixing decreases, and the overall mixing losses reduce.

5.9.2 Mixing of a linear shear flow in a diffuser or nozzle

Another example of the effects of pressure level on both mixing loss and average values of stagnation pressure is provided by the constant density linear velocity variation of Section 5.6 taken through a diffuser or a nozzle with no mixing and then mixed or averaged as shown in Figure 5.33. Three

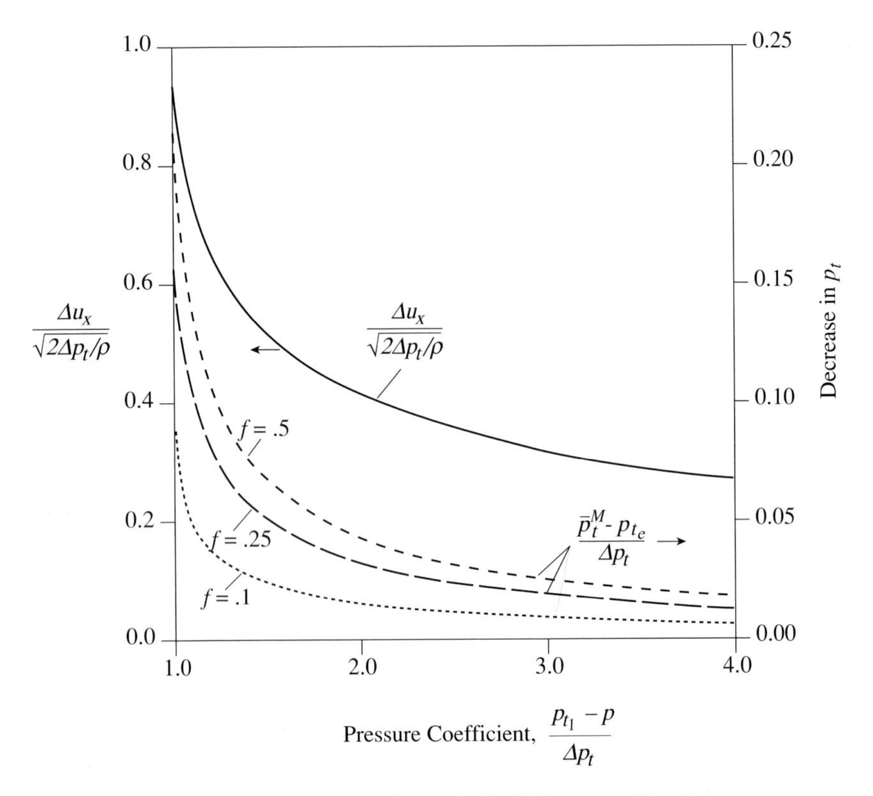

Figure 5.32: The effect of pressure level on velocity difference and loss in constant pressure two-stream mixing (see Figure 5.31); $f =$ mass fraction of lower stagnation pressure fluid.

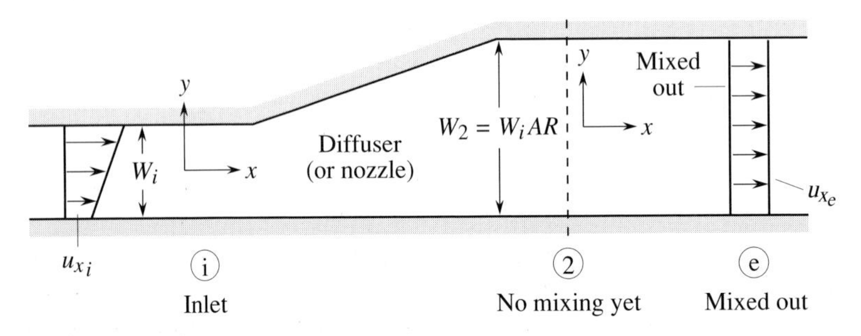

Figure 5.33: Diffuser (or nozzle) with non-uniform inlet flow.

stations are shown in the figure: (i), at which the profile is defined; (2), after the diffuser (or nozzle); and (e), after constant area mixing from (2). The height ratio or area ratio for a two-dimensional flow, from station i to 2, W_2/W_i, is denoted by AR and there is no mixing between i and 2.

The velocity field at 2 can be found from the equation describing the vorticity in the region between i and 2:

$$\frac{D\omega}{Dt} = 0. \tag{5.9.11}$$

With a straight section at i or 2, so the streamlines are parallel and $u_y = 0$, the vorticity is related to the x-component of velocity only:

$$\omega = -\frac{du_x}{dy}. \tag{5.9.12}$$

From (5.9.11), (5.9.12) and the definition of the velocity profile given in (5.6.1),

$$\omega_2 = \omega_i = -\Lambda\frac{\overline{u}}{W_i}. \tag{5.9.13}$$

The vorticity is uniform across the duct at both stations.

Substituting (5.9.12) into (5.9.13), we can integrate to find u_{x_2}:

$$u_{x_2} = \int \Lambda\frac{\overline{u}}{W_i}dy_2 + C. \tag{5.9.14}$$

In (5.9.14) C is a constant of integration, obtained from continuity, giving the result

$$u_{x_2} = \overline{u}\left(\frac{1}{AR} + \Lambda AR\frac{y_2}{W_2}\right). \tag{5.9.15}$$

The velocity gradient at station 2 is the same as that at station i, but the duct width is different ($W_i AR$ versus W_i). Velocity differences at station 2 are greater than at station i for a diffuser ($AR > 1$) and less than station i for a nozzle ($AR < 1$).

The results for diffusers are confined to the situation in which there is forward flow at all locations so that the connection between the vorticity at stations i and 2 can be made. If reverse flow were to occur, we would have to know the vorticity of particles coming from downstream. Explicitly, the constraint is that $u_{x_2} \geq 0$ at the bottom wall, $y_2 = -AR(W/2)$; particles with the lowest stagnation pressure are initially at $y_i = -W/2$ and these move along the bottom wall. From the form of u_{x_2} given in (5.9.15), this implies that the area ratio for which the description is applicable is $0 \leq AR \leq \sqrt{2/\Lambda}$.

We now compute the three average stagnation pressures defined in Section 5.6 beginning with the area average. Integration of (5.6.2) over the area at station 2 yields

$$\frac{\overline{p}_{t_2}^A - p_2}{\frac{1}{2}\rho\overline{u}^2} = \frac{1}{AR^2} + \frac{\Lambda^2(AR)^2}{12}. \tag{5.9.16}$$

This stagnation pressure is referred to p_2, rather than p_i. To make a comparison with the reference quantities at station i, the difference $p_2 - p_i$ must be found. The static pressure difference is the same along any streamline, and those at the top or bottom of the channel, where the properties are known, can be used to find the static pressure difference $p_2 - p_i$. From Bernoulli's equation and continuity,

$$\frac{p_2 - p_i}{\frac{1}{2}\rho\overline{u}^2} = \left(1 - \frac{1}{AR^2}\right) - \left[\frac{\Lambda^2}{4}(AR^2 - 1)\right]. \tag{5.9.17}$$

The term in the first parentheses, $(1 - 1/AR^2)$, is the result obtained for a uniform flow; the rest of the expression represents the decrease in static pressure rise associated with the non-uniformity. From (5.9.16) and (5.9.17) $\overline{p}_{t_2}^A - p_i$ is

$$\frac{\overline{p}_{t_2}^A - p_i}{\frac{1}{2}\rho\overline{u}^2} = 1 + \frac{\Lambda^2}{12}[3 - 2(AR)^2]. \tag{5.9.18}$$

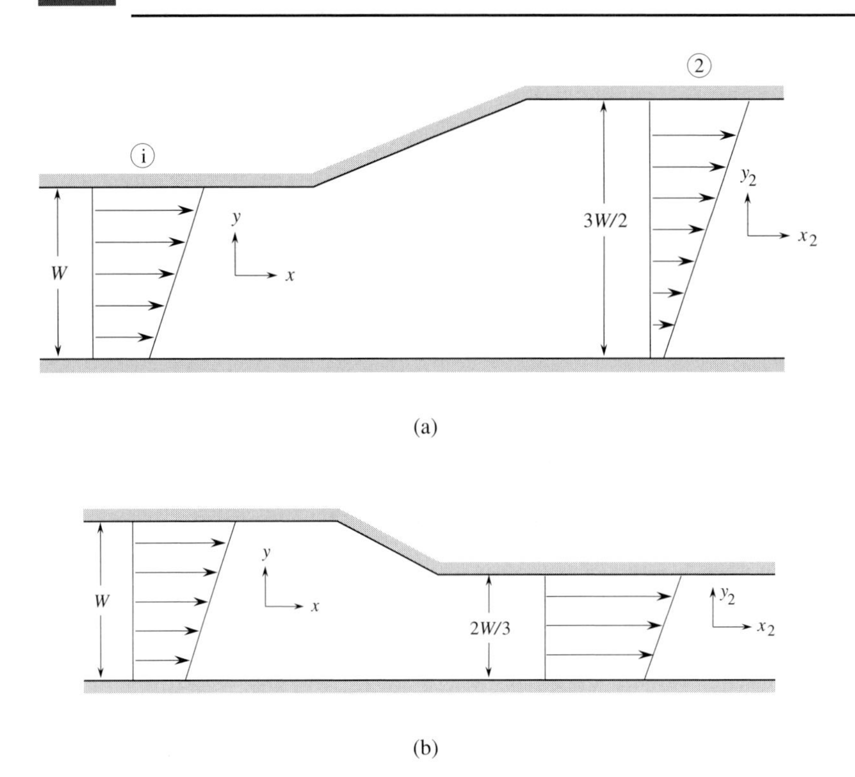

Figure 5.34: Effect of pressure level on velocity non-uniformity, $u_{x_i} = \bar{u}[1 + \frac{2}{3}(y_i/W)]$ (profiles drawn to scale): (a) diffuser: $AR = 3/2$, $u_{x_2} = (2\bar{u}/3)(1 + y_z/W)$; (b) nozzle: $AR = 2/3$, $u_{x_2} = (3\bar{u}/2)[1 + \frac{1}{2}(y_1/W)]$.

Comparing the right-hand side of (5.9.18) with (5.6.4) (which is for $AR = 1$) shows that the area average stagnation pressure is lowered in a diffusing flow ($AR > 1$) and raised in a nozzle ($AR < 1$).

The difference in area average stagnation pressure is due to the inviscid distortion of the velocity profile as it is subjected to a pressure increase or decrease before mixing. Figure 5.34 gives velocity profiles for two cases: (a) a diffuser of area ratio $3/2$ and (b) a nozzle of area ratio $2/3$, both of which have initially linear profiles

$$u_{x_i} = \bar{u}\left(1 + \frac{2}{3}\frac{y}{W}\right). \tag{5.9.19}$$

The velocity non-uniformity is increased in the diffuser and decreased in the nozzle.

To see the results in another way, Figure 5.35 presents normalized duct velocity profiles. The abscissa is velocity divided by mean velocity at that station, and the ordinate is percentage of the local channel height. All the profiles intersect at 50% height with $u_x/\bar{u} = 1.0$. Increasing pressure level shows up as an increase in the normalized velocity distortion. The creation of the increased velocity non-uniformity can be understood from the arguments in Section 4.7 relating to growth of a low stagnation pressure region in an adverse pressure gradient.

Mass average and mixed out average stagnation pressures referenced to p_i are also of interest. Using (5.6.5), the mass average stagnation pressure at station 2 is

$$\frac{\overline{p}_{t_2}^M - p_i}{\frac{1}{2}\rho\bar{u}^2} = 1 + \frac{\Lambda^2}{4}. \tag{5.9.20}$$

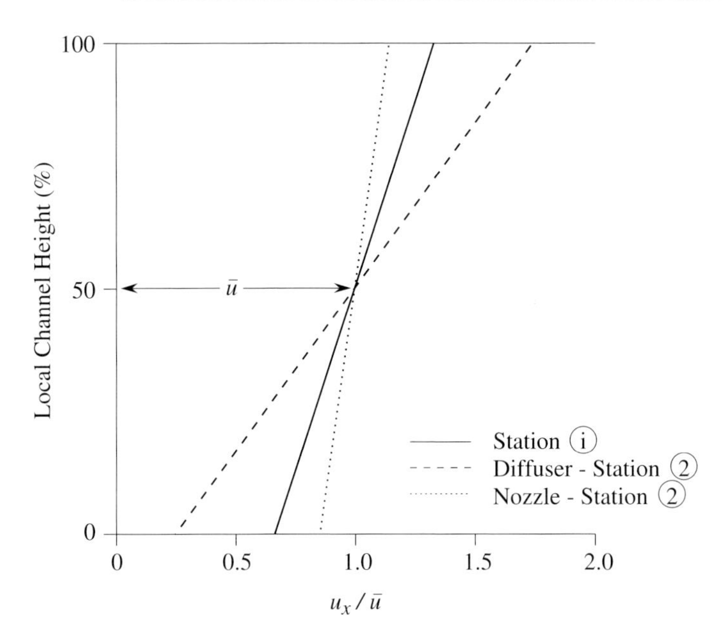

Figure 5.35: Normalized velocity profiles at stations i and 2 for geometry of Figure 5.34.

This is the same result as for the constant area, constant pressure situation; the mass average stagnation pressure is not changed by pressure level. This conclusion applies to any inviscid adiabatic steady flow. For any streamtube the mass flow and stagnation pressure do not change between stations i and 2.

The mixed out average stagnation pressure obtained by mixing the flow to a uniform condition is given by

$$\frac{\overline{p}^X_{t_2} - p_i}{\frac{1}{2}\rho\bar{u}^2} = 1 + \frac{\Lambda^2}{12}(3 - AR^2). \tag{5.9.21}$$

This result lies between the mass average and area average values.

The three averages are shown in Figure 5.36 for the initial velocity distribution of Figure 5.34. The abscissa is $(AR - 1)$ and the ordinate is the non-dimensional average stagnation pressure. The curves are drawn from $(AR - 1) \rightarrow -1$, which represents a nozzle with a very large contraction ratio, up to the forward flow limit for the parameters used. The three averages converge as the contraction ratio increases (and the velocity difference in the duct decreases) and diverge as the pressure rise increases.

5.9.3 Wake mixing

A third example of the effect of pressure level on mixing losses is the loss due to wake mixing (Denton, 1993). Figures 5.37 and 5.38 show, respectively, a schematic of the geometry and the results for a square wake with an initial wake velocity defect Δu_i, which is taken through a change in free-stream velocity with no mixing and then allowed to mix. Application of Bernoulli's equation to the free-stream and the wake provides the change in wake velocity for a given free-stream velocity ratio, u_2 / u_i. Acceleration before mixing reduces the stagnation pressure loss, because the free-stream and wake velocities are brought closer together, whereas deceleration increases mixing losses.

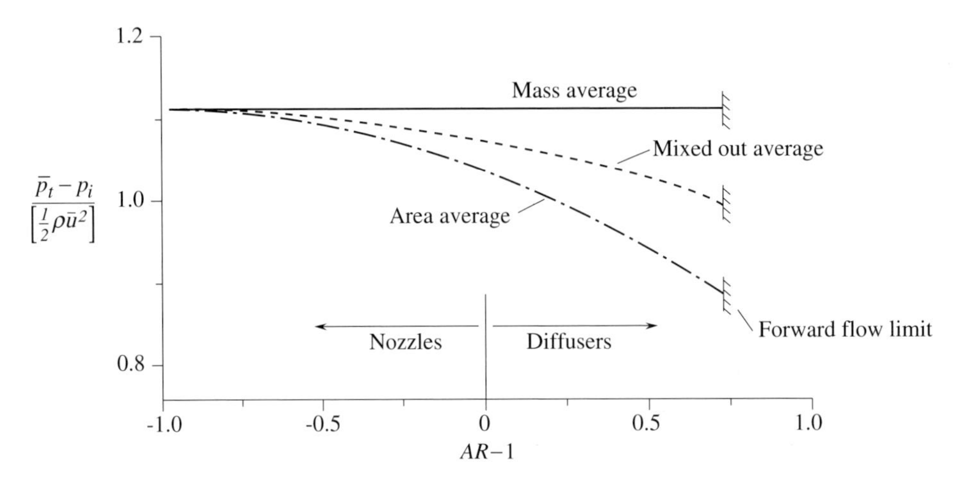

Figure 5.36: Effect of exit pressure level on average total pressure; linear inlet velocity profile, $u_{x_i} = \overline{u}[1 + \frac{2}{3}(y_i/W)]$.

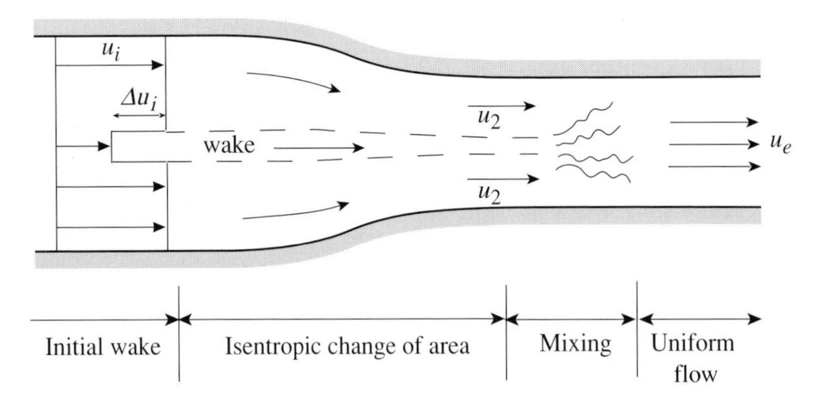

Figure 5.37: Wake mixing through a duct.

5.10 Losses in turbomachinery cascades

The ideas developed thus far can be extended to more general situations. An example is mixing downstream of a two-dimensional cascade of turbomachine blades, as shown in Figure 5.39. At exit, station 2, the velocity and static pressure distributions are specified and we wish to find the quantities at the mixed out conditions denoted by station e.

The flow is taken to be constant density and steady. As in the initial analysis of wakes and boundary layers, at any x-location downstream of the cascade the static pressure is assumed uniform in y, $\partial p/\partial y = 0$. For the thin wakes characteristic of cascades operating near design, a reasonable approximation is also to take the flow angle at the trailing edge, α_2, as constant across the passage. Denoting the magnitude of the velocity at the exit as u_2, conservation of mass and the x- and

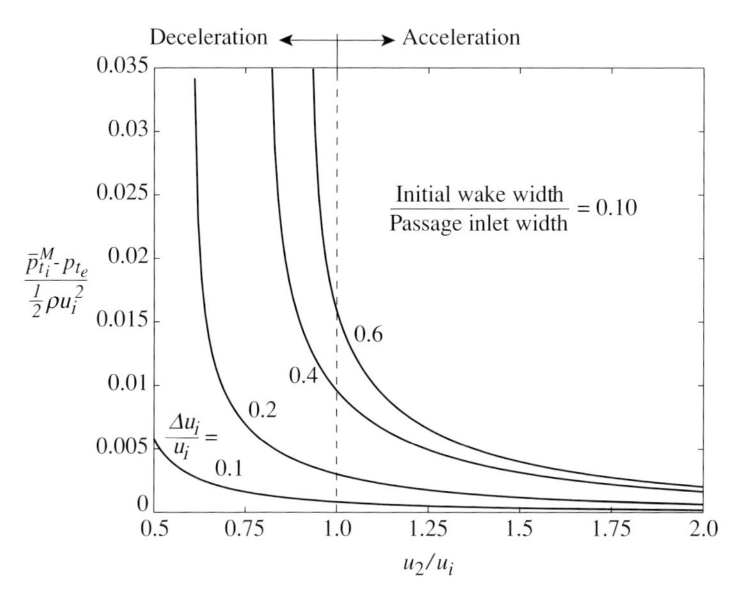

Figure 5.38: Effect of wake acceleration or deceleration on mixing loss.

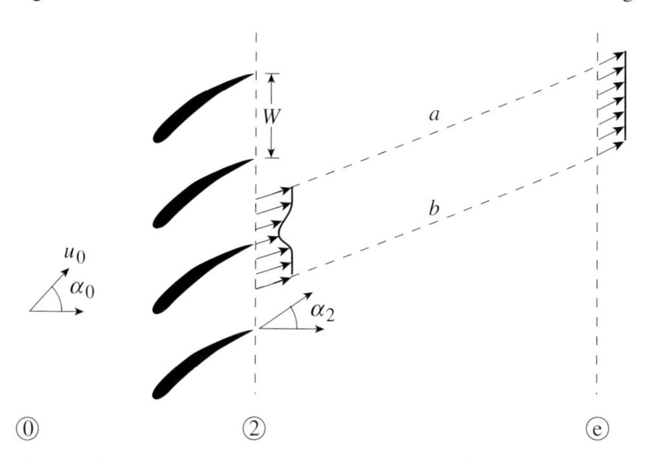

Figure 5.39: Mixing out of wakes downstream of a cascade of turbomachine blades.

y-momentum equations provide the relations needed to obtain u_e, α_e, and $p_e - p_2$:

continuity:
$$\int_{-W/2}^{W/2} u_2 \cos\alpha_2 dy = W\, u_e \cos\alpha_e; \tag{5.10.1}$$

y-momentum:
$$\int_{-W/2}^{W/2} u_2^2 \cos\alpha_2 \sin\alpha_2 dy = W u_e^2 \sin\alpha_e \cos\alpha_e; \tag{5.10.2}$$

x-momentum:
$$W(p_e - p_2) = \rho \int_{-W/2}^{W/2} u_2^2 \cos^2\alpha_2 dy - \rho W u_e^2 \cos^2\alpha_e. \tag{5.10.3}$$

In writing (5.10.2) and (5.10.3), periodicity of the cascade has been invoked so that there are no net forces on sides a and b of the control surface shown in Figure 5.39.

Carrying out the integrations and making use of the integral boundary layer parameters yields:

$$\left(1-\frac{\delta_2^*}{W}\right)u_2\cos\alpha_2 = u_e\cos\alpha_e, \tag{5.10.4}$$

$$(p_e - p_2) = -\rho u_2^2\left[\frac{\theta_2}{W} - \left(1-\frac{\delta_2^*}{W}\right)\right]\cos^2\alpha_2 - \rho u_e^2\cos^2\alpha_e, \tag{5.10.5}$$

$$u_2^2\left[\left(1-\frac{\delta_2^*}{W}\right) - \frac{\theta_2}{W}\right]\cos\alpha_2\sin\alpha_2 = u_e^2\cos\alpha_e\sin\alpha_e. \tag{5.10.6}$$

Solution of (5.10.4)–(5.10.6) provides the mixed-out conditions u_e, p_e, and α_e. The stagnation pressure loss from far upstream to far downstream can then be obtained by relating the conditions at the cascade exit to those far upstream:

$$\frac{p_{t_0} - p_{t_e}}{\frac{1}{2}\rho u_{0_x}^2} = \frac{\left(\frac{\delta_2^*}{W}\right)^2 + \frac{2\theta_2}{W} - 1}{\left(1-\frac{\delta_2^*}{W}\right)^2} + \frac{1}{\left(1-\frac{\delta_2^*}{W}\right)^2\cos^2\alpha_2} - \frac{\left(1-\frac{\delta_2^*}{W}-\frac{\theta_2}{W}\right)^2}{\left(1-\frac{\delta_2^*}{W}\right)^4}\tan^2\alpha_2. \tag{5.10.7}$$

For $\delta_2^*/W \ll 1$, (5.10.7) can be approximated as

$$\frac{p_{t_0} - p_{t_e}}{\frac{1}{2}\rho u_{0_x}^2} = \left(\frac{2\theta_2}{W} - 1\right) + \frac{1}{\cos^2\alpha_2} - \left(1-\frac{\theta_2}{W}\right)^2\tan^2\alpha_2. \tag{5.10.8}$$

For $\alpha_2 = 0$ these results reduce to those given for the flat plate cascade, (5.7.7) and (5.7.8) respectively.

Figure 5.40 shows the calculated loss after mixing for an idealized wake of width δ with representative compressor exit conditions. The dashed line indicates a typical magnitude of profile loss with fully attached boundary layers. The wake needs to extend over roughly an eighth of the passage width before the mixing loss becomes larger than the boundary layer losses in the cascade, but the mixing loss rises rapidly as the wake thickness becomes larger than this value. The ratio of the tangent of the far downstream angle to that of the exit angle is

$$\frac{\tan\alpha_e}{\tan\alpha_2} = \frac{1-\frac{\delta_2^*}{W}-\frac{\theta_2}{W}}{\left(1-\frac{\delta_2^*}{W}\right)^2}. \tag{5.10.9}$$

For well-designed cascades at or near design operation, the quantities δ_2^*/W and θ_2/W will be much less than unity and (5.10.9) can be expanded to yield the approximate form

$$\alpha_e - \alpha_2 = \left[\frac{\delta_2^*}{W} - \frac{\theta_2}{W}\right]\sin\alpha_2\cos\alpha_2 \tag{5.10.10}$$

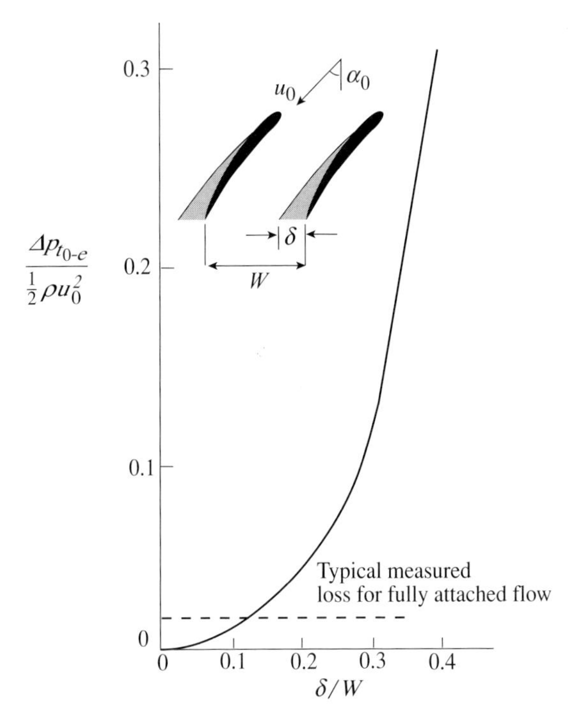

Figure 5.40: Calculated loss for fully mixed out incompressible flow with an idealized wake of width δ having zero velocity. Inlet flow angle $\alpha_0 = 35°$, outlet flow angle $\alpha_E = 0°$, trailing edge thickness $= 0$ (Cumpsty, 1989).

or

$$\Delta\alpha = \left[\frac{\delta_2^*}{2W}\left(\frac{H_2 - 1}{H_2}\right)\right]\sin(2\alpha_2). \tag{5.10.11}$$

Equation (5.10.11) shows that $\Delta\alpha > 0$, and that the flow is generally turned towards tangential due to wake mixing. Using the conventions customarily adopted for blade rows, compressor cascades thus lose turning because the far downstream flow angles will be larger than the exit flow angles, whereas turbines gain turning. This effect is typically less than a degree unless the wake thickness is larger than 10% of the passage width.

5.11 Summary concerning loss generation and characterization

There have been a number of concepts introduced in Chapter 5 concerning loss generation and characterization. These are summarized below:

(1) The appropriate metric for loss is the change in entropy due to irreversibility. This measures the "lost work", i.e. the loss of the opportunity to obtain work.

(2) For steady flows with a uniform stagnation temperature the entropy rise, and thus the losses, can be related to changes in the stagnation pressure. For a non-uniform stagnation temperature this correspondence is not valid.
(3) A useful way in which to characterize losses associated with boundary layers is through the rate of dissipation per unit area of surface. The dissipation scales with the cube of the local free-stream velocity so local regions of high velocity contribute strongly to entropy production.
(4) The ratio of loss measured at a given location to the overall loss from far upstream to fully mixed out conditions depends on the configuration. In general, bodies with a trailing edge geometric thickness much larger than the trailing edge boundary layer thickness (and hence a wake thickness much larger than the trailing edge boundary layer thickness) have a substantial fraction of the entropy rise generated downstream of the body whereas bodies with trailing edges thin compared to boundary layers have most of the losses generated upstream of the trailing edge.
(5) Principles that underpin the averaging of flow quantities in a non-uniform flow, or characterizing a non-uniform flow by an equivalent (uniform) average flow, have been developed. No uniform flow can simultaneously match all significant stream fluxes and properties of a non-uniform flow. There is thus no unique average, in other words no representation of the latter by an equivalent average, which is suitable in all situations. As such, the choice of which averaging procedure is most appropriate depends on the application of interest. The concepts presented in this chapter enable the user to make this choice in an informed manner.
(6) Different definitions for average stagnation pressure have been given that capture such features as the irreversibility creation up to the plane at which averaging is carried out and the downstream losses associated with the evolution to fully mixed conditions. The material presented also gives a background from which to guide the decision on which of these, or other, averaging procedures is to be used in a given situation.
(7) The magnitude of the overall loss for a given fluid dynamic device depends not only on the process within the device, but also on the downstream flow process, and in particular, on the level of pressure at which mixing occurs. An increase in static pressure level, such as would be obtained in a diffuser, increases the velocity non-uniformity. Mixing losses scale quadratically with the magnitude of the non-uniformity and are thus increased. Flow through a nozzle, which has a decrease in static pressure, creates a more uniform velocity profile and a decrease in mixing loss.
(8) Non-uniform velocity does not necessarily lead to loss. Velocity uniformity can be achieved reversibly through pressure forces, as well as irreversibly through mixing.
(9) The concepts introduced, which have been for geometrically simple configurations, can be extended to assess losses in more complex configurations as well as to include other phenomena such as swirl, non-two-dimensional behavior, and wake mixing in flow machinery that is predominantly radial rather than axial.

6 Unsteady flow

6.1 Introduction

Unsteady flow phenomena are important in fluid systems for several reasons. First is the capability for changes in the stagnation pressure and temperature of a fluid particle; the primary work interaction in a turbomachine is due to the presence of unsteady pressure fluctuations associated with the moving blades. A second reason for interest is associated with wave-like or oscillatory behavior, which enables a greatly increased influence of upstream interaction and component coupling through propagation of disturbances. The amplitude of these oscillations, which is set by the unsteady response of the fluid system to imposed disturbances, can be a limiting factor in defining operational regimes for many devices. A final reason is the potential for fluid instability, or self-excited oscillatory motion, either on a local (component) or global (fluid system) scale. Investigation of the conditions for which instability can occur is inherently an unsteady flow problem.

Unsteady flows have features quite different than those encountered in steady fluid motions. To address them Chapter 6 develops concepts and tools for unsteady flow problems.

6.2 The inherent unsteadiness of fluid machinery

To introduce the role unsteadiness plays in fluid machinery, consider flow through an adiabatic, frictionless turbomachine, as shown in Figure 6.1 (Dean, 1959). At the inlet and outlet of the device, and at the location where the work is transferred (by means of a shaft, say), conditions are such that the flow can be regarded as steady. We also restrict discussion to situations in which the average state of the fluid within the control volume is not changing with time. Under these conditions, the energy equation for steady flow, (1.8.11), states that the relation between the inlet and outlet stagnation enthalpies (h_t) and the work done per unit mass is

$$h_{t_i} - h_{t_e} = \frac{\text{work done by turbomachine}}{\text{per unit mass}}.$$

Suppose now that we analyze the internal workings of this device using the steady form of the momentum equation. Along a representative streamline through the machine (shown dashed in the figure) the pressure, velocity and density are related by

$$-\frac{1}{\rho}dp = u\,du. \qquad (2.5.7)$$

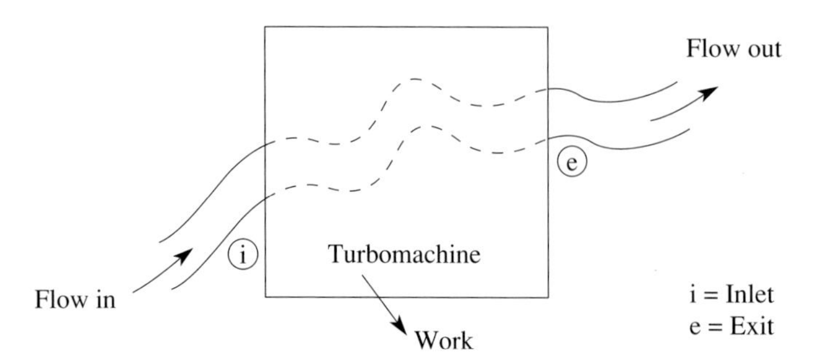

Figure 6.1: Flow through a frictionless, adiabatic turbomachine.

For small changes in state

$$dh = Tds + \frac{1}{\rho}dp. \tag{1.3.19}$$

Since the turbomachine is adiabatic and frictionless, the entropy change along a streamline is zero. Combining (2.5.7) and (1.3.19) we obtain

$$dh = -udu. \tag{6.2.1}$$

Equation (6.2.1) can be integrated to yield

$h + \frac{1}{2}u^2 = h_t =$ constant along a streamline.

Hence, from inlet to exit $h_{t_e} = h_{t_i}$ and the turbomachine does no work.

This conclusion, which is contrary to intuition and experience, motivates the question of where the source of the apparent inconsistency lies. A step on the way to the conclusion was use of the steady-flow form of the momentum equation *through* the machine. In fact, the flow inside the device is unsteady, and we are not justified in neglecting the effects of this unsteadiness. We now thus reexamine the problem including the unsteady terms. For inviscid flow with no body forces the momentum equation is (3.3.3) with $\mathbf{F}_{visc} = \mathbf{X} = 0$,

$$\begin{aligned} \frac{\partial \mathbf{u}}{\partial t} - \mathbf{u} \times \omega + \nabla\left(\frac{u^2}{2}\right) &= -\frac{1}{\rho}\nabla p \\ &= -\nabla h + T\nabla s. \end{aligned} \tag{6.2.2}$$

Taking the scalar product of $\mathbf{u}$ with (6.2.2) and making use of the fact that entropy is constant for a fluid particle yields

$$\begin{aligned} \frac{\partial}{\partial t}\left(\frac{u^2}{2}\right) + \mathbf{u}\cdot\nabla\left(\frac{u^2}{2}\right) &= -\mathbf{u}\cdot\nabla h - T\frac{\partial s}{\partial t} \\ &= -\mathbf{u}\cdot\nabla h - \frac{\partial h}{\partial t} + \frac{1}{\rho}\frac{\partial p}{\partial t}. \end{aligned}$$

Combining terms into the stagnation enthalpy, h_t, allows a compact statement concerning the rate of change of stagnation enthalpy for a fluid particle:

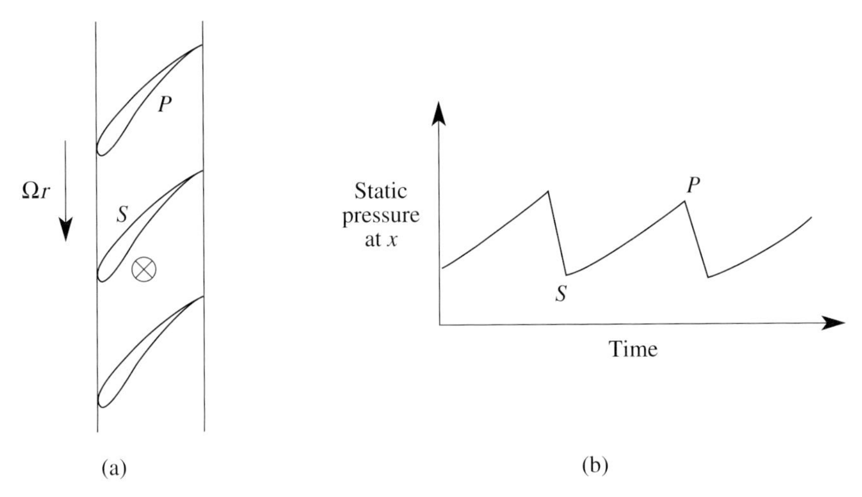

Figure 6.2: Time-resolved pressure over an axial compressor rotor: (a) axial compressor rotor showing location x; (b) unsteady pressure as measured at ⊗.

$$\frac{Dh_t}{Dt} = \frac{1}{\rho}\frac{\partial p}{\partial t}. \tag{6.2.3}$$

Equation (6.2.3) is not restricted to situations with constant entropy throughout the flow. It refers to the broader class of isentropic flows where the entropy of a given fluid particle is constant, but the entropy can vary from fluid particle to particle. In these situations, (6.2.3) shows that the stagnation enthalpy of a fluid particle can change only if the flow is unsteady.

An illustration of this point is furnished by the axial compressor rotor with radius r sketched in Figure 6.2(a). The pressure field of the blades, which has pressure increasing from the suction surface (S) to the pressure surface (P), moves with the blades. An observer sitting at the fixed point ($\times$) on the casing would measure a pressure variation with time as in Figure 6.2(b). Particles passing through the rotor see positive $\partial p/\partial t$ and hence experience positive values of Dh_t/Dt. For a turbine the variations in pressure are opposite and the change in stagnation enthalpy of a particle is negative. Unsteady effects are therefore essential for the changes in stagnation enthalpy and pressure achieved by fluid machinery.

For situations in which the density can be regarded as constant and the stagnation pressure given by $p_t = p + \frac{1}{2}\rho u^2$, (6.2.3) reduces to

$$\frac{Dp_t}{Dt} = \frac{\partial p}{\partial t} \tag{6.2.4}$$

for inviscid, adiabatic flow.

6.3 The reduced frequency

The non-dimensional parameter that characterizes the importance of unsteadiness in a given situation is known as the reduced frequency. It was introduced in Section 1.17. To develop this parameter in a

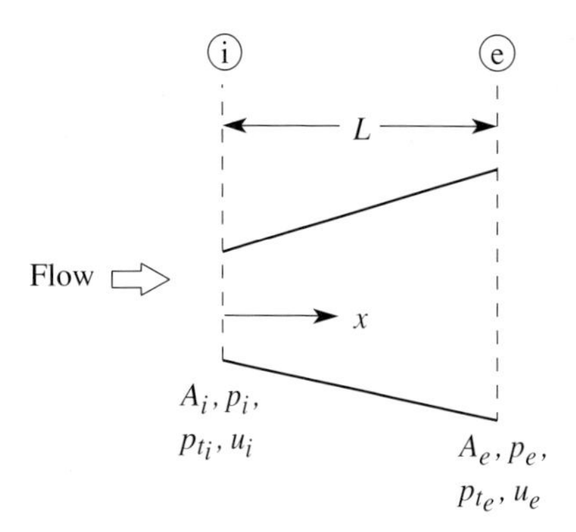

Figure 6.3: Unsteady flow in a diffuser passage; fluctuation in stagnation pressure p_{t_i} specified at inlet, constant static pressure at exit.

more specific context, consider a fluid device (an airfoil, a diffuser, a turbomachine blade passage, etc.) which experiences a time varying flow of the form $e^{i\omega t}$. The time scale associated with the unsteadiness is $1/\omega$, with significant changes occurring in a time of the order of $1/\omega$.

There is another time scale in the problem, the time for fluid particle transport through the device. If the length of the device is L and a characteristic throughflow velocity is U, this time is L/U. The change in local flow quantities during the passage of the particle depends on the ratio of the two times, or $\omega L/U$, which is the reduced frequency, β. Small values of β imply that fluid particles experience little change due to unsteadiness, while large values imply a substantial variation during the transit time. The magnitude of the reduced frequency is therefore a measure of the relative importance of unsteady effects:

$\beta \ll 1$ unsteady effects small – quasi-steady flow;
$\beta \gg 1$ unsteady effects dominate;
$\beta \sim 1$ both unsteady and quasi-steady effects important.

Many fluid machinery situations are characterized by values of β of order unity.

6.3.1 An example of the role of reduced frequency: unsteady flow in a channel

The manner in which the reduced frequency can enter into the description of an unsteady flow is illustrated by analysis of one-dimensional, inviscid, uniform density, incompressible flow in a channel subjected to a time varying inlet stagnation pressure. This can be considered an elementary model of a turbomachine rotor blade passage moving through a spatially non-uniform stagnation pressure.

The configuration of interest is shown in Figure 6.3, where the channel is drawn as a diffuser. Station i corresponds to the inlet and station e to the exit. The coordinate x measures distance along the diffuser and L is the diffuser length. The inlet perturbation in stagnation pressure is taken to be of the form $e^{i\omega t}$. At the exit the static pressure is constant, as would be the case if the diffuser discharged into a large volume.

All flow quantities will be expressed as a time mean value plus an unsteady perturbation which has small enough amplitude that a linearized description can be adopted. Denoting the time mean quantities by overbars ($\overline{}$) and the perturbations by primes ($'$), the inlet stagnation pressure, for example, can be written as

$$p_{t_i} = \overline{p}_{t_i} + p'_{t_i} = \overline{p}_{t_i} + \varepsilon\, e^{i\omega t},$$

where ε is the amplitude of the perturbation.

The one-dimensional form of the momentum equation is

$$\frac{\partial u}{\partial t} + u\frac{\partial u}{\partial x} = -\frac{1}{\rho}\frac{\partial p}{\partial x}. \tag{6.3.1}$$

Integrating (6.3.1) from inlet to exit yields

$$\int_i^e \frac{\partial u}{\partial t}dx = -\left(\frac{p}{\rho} + \frac{u^2}{2}\right)\Bigg|_i^e = \frac{p_{t_i}}{\rho} - \frac{p_{t_e}}{\rho}. \tag{6.3.2}$$

Equation (6.3.2) shows that differences in stagnation pressure along the diffuser are created only through unsteadiness.

The one-dimensional continuity equation for the passage is

$$uA = \text{constant} = u_i A_i, \tag{6.3.3}$$

where A, the local area, is a function of distance along the passage and A_i is the area at the inlet. Using (6.3.3), the time derivative in (6.3.2) can be written as

$$\begin{aligned}\int_i^e \frac{\partial u}{\partial t}dx &= \frac{du_i}{dt}\int_i^e \frac{A_i}{A}dx \\ &= \mathcal{L}\frac{du_i}{dt}.\end{aligned} \tag{6.3.4}$$

Equation (6.3.4) defines the quantity $\mathcal{L}$, an "effective length" of the diffuser, which is a function of diffuser geometry only. An example is a linear area variation with length $A = A_i + (A_e - A_i)\,(x/L)$ which gives, upon substitution into (6.3.4),

$$\mathcal{L} = \frac{\ln\dfrac{A_e}{A_i}}{\left(\dfrac{A_e}{A_i} - 1\right)}. \tag{6.3.5}$$

With the definition of $\mathcal{L}$, the integral of the momentum equation in (6.3.2) takes the form

$$\mathcal{L}\frac{du_i}{dt} = \frac{p_{t_i}}{\rho} - \frac{p_{t_e}}{\rho}. \tag{6.3.6}$$

We now use the idea that the unsteady perturbations have small amplitude compared to the mean flow quantities and linearize, neglecting products of perturbation quantities. Equation (6.3.6) becomes

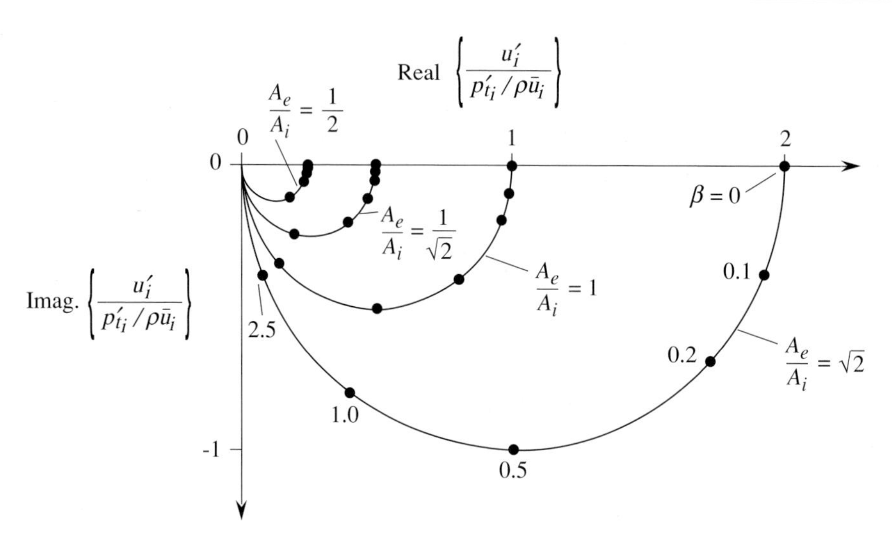

Figure 6.4: Channel inlet velocity perturbation as a function of reduced frequency, $\beta = \omega L/\overline{u}_i$; inlet stagnation pressure fluctuation $p'_{t_i} \propto e^{i\omega t}$.

$$\mathcal{L}\frac{du_i'}{dt} = \frac{\overline{p}_{t_i} + p'_{t_i} - (\overline{p}_e + p'_e)}{\rho} - \frac{\overline{u}_e^2}{2} - \overline{u}_e u'_e. \tag{6.3.7}$$

In (6.3.7), the stagnation pressure at the exit is separated into static and dynamic pressures because the boundary condition involves the exit static pressure, p_e. For the time mean flow the stagnation pressure is the same at the inlet and the exit. This, plus the prescribed condition of constant static pressure at the exit, $p'_e = 0$, allows the equation for the perturbation quantities to be written as

$$\mathcal{L}\frac{du_i'}{dt} = \frac{p'_{t_i}}{\rho} - \overline{u}_e u'_e. \tag{6.3.8}$$

The inlet velocity perturbation, u'_i, is the quantity sought. The continuity equation (6.3.3) can be used to eliminate the exit velocity, and the resultant expression solved to obtain u'_i in terms of the imposed inlet stagnation pressure non-uniformity, p'_{t_i}. Defining the reduced frequency, β, as $\omega\mathcal{L}/\overline{u}_i$, this is

$$\frac{u_i'}{(p'_{t_i}/\rho\overline{u}_i)} = \frac{\dfrac{1}{(A_e/A_i)^2} - i\beta}{\dfrac{1}{(A_e/A_i)^4} + \beta^2}. \tag{6.3.9}$$

Equation (6.3.9) is plotted in Figure 6.4, which shows the real and imaginary parts of $u'_i/(p'_{t_i}/\rho\overline{u}_i)$ as a function of reduced frequency, β, for different values of A_e/A_i, the exit/inlet area ratio. The values range from $A_e/A_i = \sqrt{2}$, representative of an axial compressor, to 1.0 for a straight channel, to $1/\sqrt{2}$ and $1/2$ which are representative of a turbine. For any value of β, a vector drawn from the origin to the curve represents the quantity $u'_i/(p'_{t_i}/\rho\overline{u}_i)$ in magnitude and phase. All the plots are semi-circles and can be collapsed into a single curve if one plots $\{[u'_i(A_i/A_e)^2]/(p'_{t_i}/\rho\overline{u}_i)\}^2$ versus $\beta(A_e/A_i)^2$; this has not been done in order to exhibit both the role of the reduced frequency and the effect of the area ratio.

Several general features are shown in Figure 6.4:

(1) At low reduced frequency ($\beta \ll 1$), the non-dimensional velocity perturbation is close to the steady-state values (2.0 for the diffuser, 1.0 for the straight channel, 0.5 and 0.25 for the nozzle) and there is little difference in phase between velocity and stagnation pressure perturbations.
(2) At high reduced frequency ($\beta \gg 1$), there is a phase difference of close to $\pi/2$ between velocity and stagnation pressure perturbations and a greatly reduced amplitude of the velocity non-uniformity. In this situation, the local accelerations dominate the convective acceleration terms.
(3) Diffusing passages respond more strongly to perturbations than do nozzles.

For rotating machinery, periodic disturbances are often associated with a spatially non-uniform flow through which the moving blade rows pass. Common occurrences are wakes of an upstream stationary blade row, inlet separation or flow distortions produced by upstream ducting, or downstream obstacles such as struts. In this situation, a radian frequency, ω, for the unsteadiness seen by the rotor can be related to a characteristic wavelength, λ, of the stationary non-uniformity by

$$\omega = \frac{2\pi\Omega r_m}{\lambda},$$

where r_m is the mean radius of the blade row and Ω is the rotational velocity. With U and L the characteristic through-flow velocity and length respectively, a reduced frequency can thus be defined as

$$\beta = \frac{2\pi\Omega r_m \cdot L}{\lambda U}. \tag{6.3.10}$$

For many fluid devices, Ωr_m and U are roughly comparable. If so, the reduced frequency scales as

$$\beta \propto 2\pi\frac{L}{\lambda}$$

with the proportionality constant of order unity. This is an interpretation of reduced frequency in terms of the ratio of the wavelength of the imposed flow non-uniformity to the characteristic length of the device, L. For disturbance wavelengths which are long compared to L the device can be considered to be embedded in a slowly varying flow, with a response close to quasi-steady. For disturbances with wavelength of order L or shorter, the reduced frequency will be roughly 2π or higher and unsteadiness will be important. In rotating machinery, λ is an integer fraction of the circumference. If so, $\lambda = 2\pi r_m/n$, where n is the number of "lobes" of the disturbance, and the reduced frequency is given by

$$\beta \propto n\frac{L}{r_m}.$$

A third view of reduced frequency is provided by direct examination of (6.3.1). Suppose the temporal and spatial variations of the velocity have the same magnitude, ΔU. With L the characteristic length and ω the radian frequency, the relative magnitudes of the two acceleration terms on the left-hand side of (6.3.1) are $\omega L/U$ and unity. In this context the reduced frequency can be regarded as a measure of the contribution of unsteadiness to the static pressure changes in the flow.

6.4 Examples of unsteady flows

6.4.1 Stagnation pressure changes in an irrotational incompressible flow

The relation between flow unsteadiness and stagnation pressure takes a compact and useful form in a constant density, inviscid, irrotational flow. For this condition the momentum equation is

$$\frac{\partial \mathbf{u}}{\partial t} + \nabla\left(\frac{u^2}{2} + \frac{p}{\rho}\right) = 0. \tag{6.4.1}$$

Because the flow is irrotational, $\mathbf{u}$ can be defined as the gradient of a velocity potential φ, $\mathbf{u} = \nabla\varphi$ and $(\partial\mathbf{u}/\partial t) = (\partial/\partial t)\nabla\varphi$. The operations $\partial/\partial t$ and ∇ commute and (6.4.1) can be integrated to yield

$$\frac{\partial \varphi}{\partial t} + \frac{p_t}{\rho} = f(t). \tag{6.4.2}$$

The term on the right of (6.4.2) is purely a function of time which is determined if its value at any location in the flow field is known. Consider a situation where the unsteadiness is caused by an object moving through the flow, so that regions at large distances from the object are undisturbed by its movement. Then $f(t)$ is constant and (6.4.2) becomes

$$\frac{\partial \varphi}{\partial t} + \frac{p_t}{\rho} = \frac{P_0}{\rho} = \text{constant}. \tag{6.4.3}$$

The value of the constant has no effect on the flow pattern and can be absorbed into the definition[1] of φ. Equation (6.4.3) will be made much use of in what follows.

6.4.2 The starting transient for incompressible flow exiting a tank

An example which shows a number of features of interest is furnished by the flow of an inviscid, incompressible fluid from a pressurized tank (Preston, 1961). Figure 6.5 shows a large tank containing an incompressible fluid, which can exit through a pipe of length L and diameter d, with $L/d \gg 1$. A closed valve on the end of the pipe is opened at time $t = 0$ and the liquid starts to leave the tank. The pressure difference between the tank and the exit is maintained constant at Δp_0. The question to be addressed is how the velocity and stagnation pressure evolve in time during the approach to steady state.

We make use of (6.4.3), which holds throughout the flow domain. The velocity in the tank is much less than in the pipe so that P_0 is equal to Δp_0. In the pipe, the velocity is uniform in x so the velocity potential has the form

$$\varphi = U(t)x. \tag{6.4.4}$$

[1] Even if $f(t)$ were not constant, it could still be absorbed into the definition of φ by defining a new velocity potential, φ_I, as

$$\varphi_I = \varphi + \int_{-\infty}^{t} f(\xi)d\xi.$$

This would make no difference to the velocity field which is determined only by the spatial derivatives of φ.

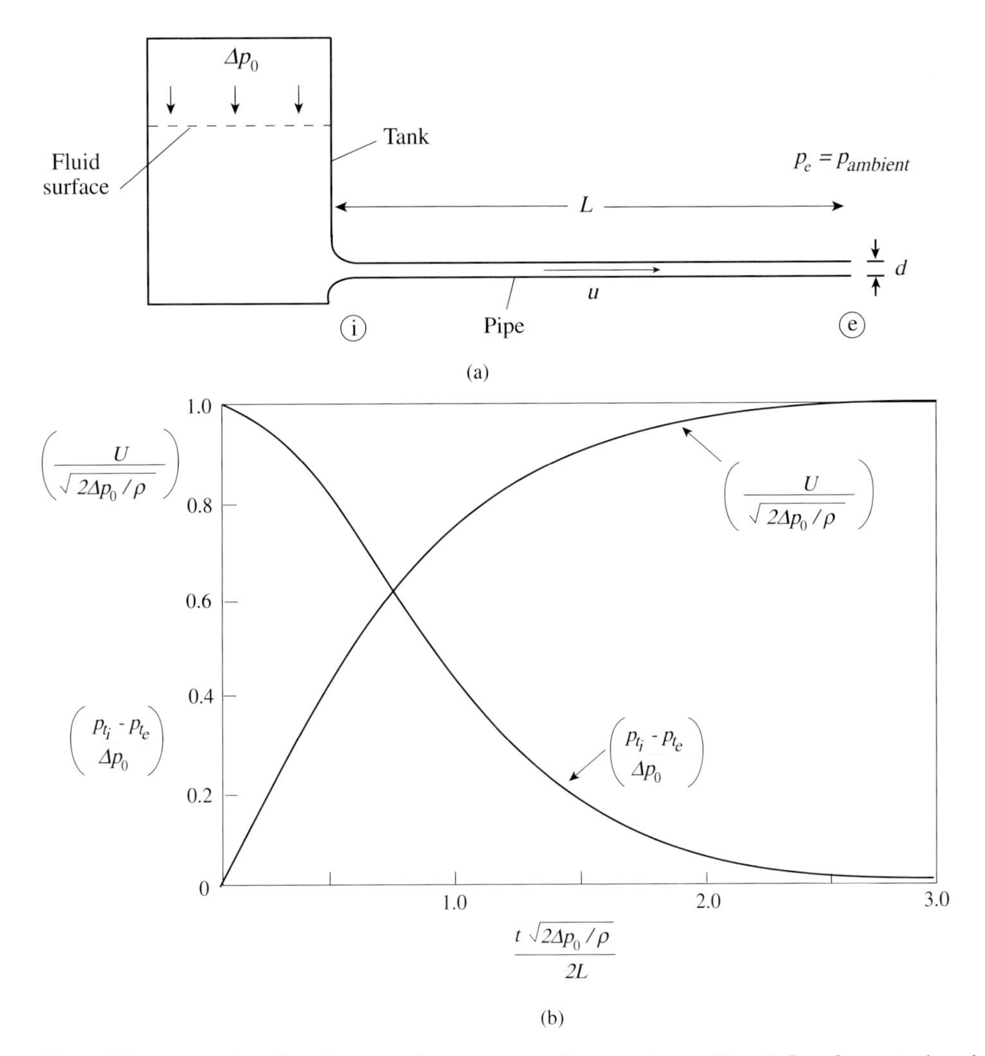

Figure 6.5: (a) Transient flow from a tank: geometry and nomenclature; (b) exit flow from a tank: velocity and stagnation pressure variation with time (Preston, 1961).

This velocity potential is defined with $\varphi = 0$ at station i just inside the pipe. Application of (6.4.3) and (6.4.4) between stations 1 and 2, plus continuity in the form $u_i = u_e = U$, yields

$$p_i - p_e = \rho L \frac{dU}{dt}. \tag{6.4.5}$$

Between station i and the surface of the incompressible fluid, the velocity varies with position, but we can employ a simplified flow description in this region because the tank area is much larger than the pipe area. From continuity, the velocity magnitude in the region of the tank near the pipe inlet will be similar to that of a "sink" so that

$$u \approx U \frac{(d/2)^2}{r^2}, \tag{6.4.6}$$

where r is the distance from the virtual location of the sink (roughly a radius into the pipe). The velocity potential in the tank thus has the form

$$\varphi \approx -U\frac{(d/2)^2}{r}. \tag{6.4.7}$$

From (6.4.7) the difference in the value of $\partial\varphi/\partial t$ from station i to the upper surface is of order $Ud/\Delta t$, where Δt is the characteristic time scale over which the transient occurs. Comparison with (6.4.5) shows that if $L/d \gg 1$, the contribution to the variation in stagnation pressure from motion in the tank is much less than the contribution from the unsteady flow in the pipe and the former can be neglected. Another way of stating this is that the reduced frequency associated with the entrance region flow into the pipe is small and the inlet region behavior can be considered quasi-steady. The reduced frequency of the flow in the pipe (inlet to exit), however, is such that unsteady effects must be taken into account. The concept of treating some regions of a flow field as quasi-steady, while accounting for unsteadiness in other regions, as we do here, is a significant simplifying feature for a number of applications.

From the arguments in the preceding paragraph we can connect conditions at the pipe entry and the surface by

$$p_i + \tfrac{1}{2}\rho U^2 = \Delta p_0. \tag{6.4.8}$$

Combining (6.4.8) with (6.4.5) gives

$$\Delta p_0 - \tfrac{1}{2}\rho U^2 = \rho L\frac{dU}{dt},$$

or

$$dt = 2L\frac{dU}{\dfrac{2\Delta p_0}{\rho} - U^2}. \tag{6.4.9}$$

Using the initial condition of $U = 0$ at time $t = 0$, (6.4.9) can be integrated as

$$\frac{U}{\sqrt{2\Delta p_0/\rho}} = \tanh\left(\frac{t\sqrt{2\Delta p_0/\rho}}{2L}\right). \tag{6.4.10}$$

This solution is shown in Figure 6.5(b).

The stagnation pressure is found from (6.4.8) as

$$\frac{p_{t_i} - p_{t_e}}{\Delta p_0} = 1 - \frac{U^2}{2\dfrac{\Delta p_0}{\rho}} = \operatorname{sech}^2\left(\frac{t\sqrt{2\Delta p_0/\rho}}{2L}\right), \tag{6.4.11}$$

and this is also shown in Figure 6.5(b). At time $t = 0$ the available pressure difference Δp_0 is all used to accelerate the fluid in the pipe. At times large compared with $2L/\sqrt{2\Delta p_0/\rho}$ there is no stagnation pressure difference and Δp_0 is manifest as the dynamic pressure at the exit of the pipe.

6.4.3 Stagnation pressure variations due to the motion of an isolated airfoil

A source of unsteadiness in fluid machinery is the presence of moving airfoils. We examine the resulting flow in the stationary system which is set up by airfoil motion, starting with a basic model

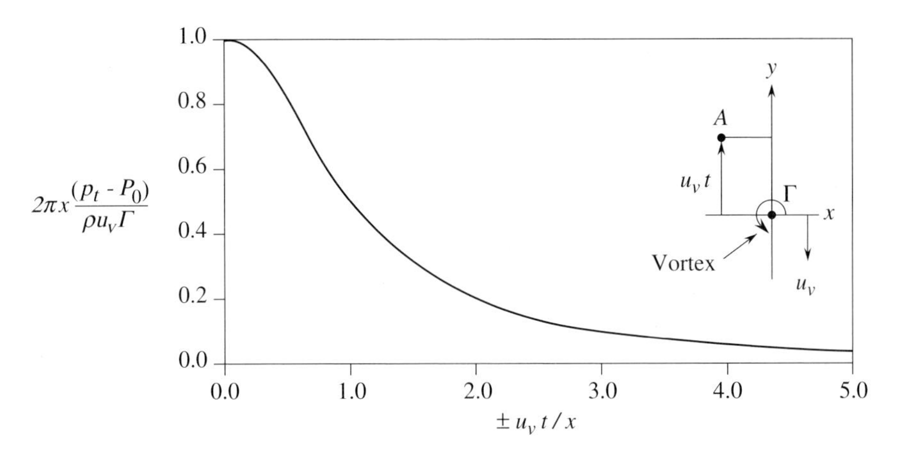

Figure 6.6: Uniform motion of vortex past a fixed point; change of stagnation pressure with time (Preston, 1961).

for a single airfoil, or blade, moving past a fixed observer and then developing the concepts for a row of moving blades (Preston, 1961).

The model for the blade is a bound vortex of circulation Γ, representing the circulation round the airfoil as sketched in Figure 6.6. The flow is assumed two-dimensional, constant density, inviscid, and irrotational. Because the velocity can be derived from a velocity potential, φ, application of the continuity equation, $\nabla \cdot \mathbf{u} = 0$, means the potential satisfies Laplace's equation

$$\frac{\partial^2 \varphi}{\partial x^2} + \frac{\partial^2 \varphi}{\partial y^2} = 0. \tag{6.4.12}$$

There are well-known solutions of this equation for configurations such as fluid sources, vortices, doublets, etc. and we make use of the solution for a stationary vortex. The velocity potential associated with a vortex at the origin is

$$\varphi = \frac{\Gamma}{2\pi}\theta, \tag{6.4.13a}$$

where

$$\theta = \tan^{-1}\left(\frac{y}{x}\right) \tag{6.4.13b}$$

and y and x are the vertical and horizontal coordinates shown in Figure 6.6. Equations (6.4.13) apply everywhere outside the origin.

The velocity components are obtained from differentiation of (6.4.13):

$$u_x = -\frac{\Gamma}{2\pi}\left(\frac{y}{x^2 + y^2}\right), \tag{6.4.14a}$$

$$u_y = \frac{\Gamma}{2\pi}\left(\frac{x}{x^2 + y^2}\right). \tag{6.4.14b}$$

Equations (6.4.14) describe a circular flow about the origin with velocity magnitude $\Gamma/(2\pi\sqrt{x^2 + y^2})$.

If the vortex is in steady motion with negative (downward) velocity u_v parallel to the y-axis, as indicated in Figure 6.6, the coordinates of a fixed point, A, relative to the vortex, are $y = u_v t$ and x. As seen by an observer at point A,

$$\left.\frac{\partial \varphi}{\partial t}\right|_{\substack{\text{as seen in}\\ \text{stationary}\\ \text{frame}}} = u_v \left.\frac{\partial \varphi}{\partial y}\right|_{\substack{\text{as seen in}\\ \text{moving}\\ \text{frame}}} = u_v u_y\Big|_{\substack{\text{as seen in}\\ \text{moving}\\ \text{frame}}}. \tag{6.4.15}$$

In (6.4.15) $\partial\varphi/\partial y$ and u_y are evaluated in the coordinate system fixed to the moving vortex. Choosing the time origin so point A has its y-coordinate equal to zero at time $t = 0$, the velocity components at point A seen by an observer in the vortex (moving) system at time t are:

$$u_x = -\frac{\Gamma}{2\pi}\frac{u_v t}{(u_v t)^2 + x^2}, \tag{6.4.16a}$$

$$u_y = \frac{\Gamma}{2\pi}\frac{x}{(u_v t)^2 + x^2}. \tag{6.4.16b}$$

From (6.4.3), the variation in stagnation pressure seen by the stationary observer at any x-location is

$$\frac{p_t - P_0}{\rho} = \frac{u_v \Gamma}{2\pi}\frac{x}{(u_v t)^2 + x^2} \tag{6.4.17}$$

or, non-dimensionally,

$$2\pi x\left(\frac{p_t - P_0}{\rho u_v \Gamma}\right) = \frac{1}{1 + \left(\dfrac{u_v t}{x}\right)^2}. \tag{6.4.18}$$

The stagnation pressure variation with time for a moving vortex is shown in Figure 6.6. Appreciable fluctuations in stagnation pressure occur when the vortex is "near" the observer, say, for times $|t| < 2x/u_v$.

6.4.4 Moving blade row (moving row of bound vortices)

The above ideas can be extended to situations more representative of those in fluid machinery by considering the flow due to a *moving row* of bound vortices, a model for a rotor blade row moving relative to a stationary observer.[2] The configuration is illustrated in Figure 6.7, which shows a row of bound vortices representing the circulation around the blades of a turbomachine rotor. The vortices have a circulation of Γ in the counterclockwise direction, a spacing W, and move in the negative y-direction (downward) with velocity u_v.

The velocity potential, obtained by summing up the potentials for an infinite row of vortices, is (Lamb, 1945):

$$\phi = \frac{\Gamma}{2\pi}\tan^{-1}\left[\tan\left(\frac{\pi y}{W}\right)\coth\left(\frac{\pi x}{W}\right)\right]. \tag{6.4.19}$$

[2] As will be seen subsequently, this model is also of help in understanding features of the unsteady behavior of wakes.

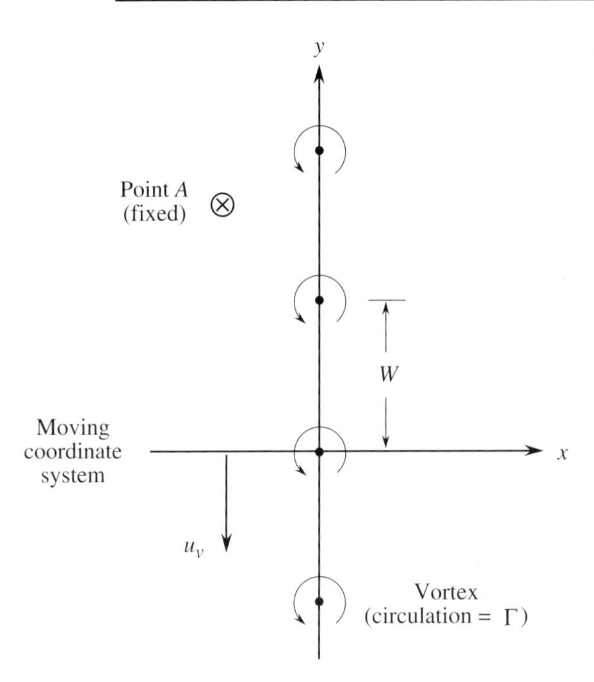

Figure 6.7: Moving row of vortices and fixed observation point A (Preston, 1961).

In (6.4.19) x and y are in a coordinate system attached to the moving row. The velocity components in the moving system are:

$$u_x = -\frac{\Gamma}{2W}\left[\frac{\sin(2\pi y/W)}{\cosh(2\pi x/W) - \cos(2\pi y/W)}\right], \tag{6.4.20a}$$

$$u_y = \frac{\Gamma}{2W}\left[\frac{\sinh(2\pi x/W)}{\cosh(2\pi x/W) - \cos(2\pi y/W)}\right]. \tag{6.4.20b}$$

The transformation from spatial derivatives in the moving system to time derivatives in the stationary system is as described in the previous section. Substituting the expressions for the velocity components (6.4.20) into Eq. (6.4.3) yields

$$\frac{p_t - P_0}{\rho u_v \Gamma/W} = \frac{\sinh(2\pi x/W)}{\cosh(2\pi x/W) - \cos(2\pi u_v t/W)}. \tag{6.4.21}$$

Equation (6.4.21) is an expression for the instantaneous stagnation pressure as measured by a stationary observer at point A who has coordinates x, $y = (u_v t)$ in the moving system.

Several features are to be noted concerning the form of (6.4.21). First, at $x = -\infty$ (far upstream) and $+\infty$ (far downstream) respectively, the stagnation pressures are:

$$p_{t-\infty} = P_0 - \rho\frac{u_v \Gamma}{2W}, \tag{6.4.22a}$$

$$p_{t+\infty} = P_0 + \rho\frac{u_v \Gamma}{2W}. \tag{6.4.22b}$$

The change in stagnation pressure from far upstream to far downstream is

$$\Delta p_t = \rho \frac{u_v \Gamma}{W}. \tag{6.4.23}$$

The change in "tangential" velocity (y-velocity) from $-\infty$ to $+\infty$ is Γ/W, so (6.4.23) expresses the change in p_t given by the Euler turbine equation (see Section 2.8) applied to this incompressible inviscid flow.

We also examine the average stagnation pressure over a "cycle", the passage of one vortex, $0 < u_v t/W < 1.0$. The time mean stagnation pressure (denoted by an overbar) is:

$$-(\overline{p_t - P_0}) = \rho \overline{\frac{\partial \varphi}{\partial t}}$$

$$= \rho \left(\frac{u_v}{W}\right) \int_0^{W/u_v} \frac{\partial \varphi}{\partial t} dt. \tag{6.4.24}$$

Therefore,

$$-\left(\overline{p_t - P_0}\right) = \rho \frac{u_v}{W} (\varphi|_{t=W/u_v} - \varphi|_{t=0}). \tag{6.4.25}$$

Referring to the expression for φ in (6.4.19), we find that for any positive value of x (downstream), $\overline{p}_t = P_0 + \rho u_v \Gamma/(2W)$, whereas for any negative value (upstream) $\overline{p}_t = P_0 - \rho u_v \Gamma/(2W)$. The time mean stagnation pressure is independent of x on either side of the vortex row and changes discontinuously across the row by $\rho \Gamma u_v/W$.

The variations in stagnation pressure seen in the stationary frame are shown in two different ways in Figures 6.8 and 6.9. In Figure 6.8, the variations have been plotted versus the horizontal location of point A in units of x/W, for different times during the passage of the row of vortices. The time taken for the row to move one vortex spacing is W/u_v and this has been used to make the time non-dimensional.

The unsteady stagnation pressure variations near the vortex row are a substantial fraction of the time mean stagnation pressure change across the row. As one moves away from the blade row to a distance of $x/W = 0.5$, however, the fluctuations decrease to roughly 10% of the stagnation pressure change across the row and at $x/W = 1.0$ they are less than 1%. A pressure probe would see appreciable fluctuations in stagnation pressure if placed in close enough proximity to this row, but the fluctuations would be negligible if it were a blade spacing away.

Figure 6.9 is a cross-plot of Figure 6.8 showing instantaneous stagnation pressure versus time. The different curves, which correspond to different positions of the observer, again in terms of x/W, give another picture of the rate of decay of the unsteady variations.

6.4.5 Unsteady wake structure and energy separation

The analysis of the flow associated with a moving row of vortices can be extended qualitatively to describe compressible flows. If the flow is irrotational, the inviscid momentum equation can be

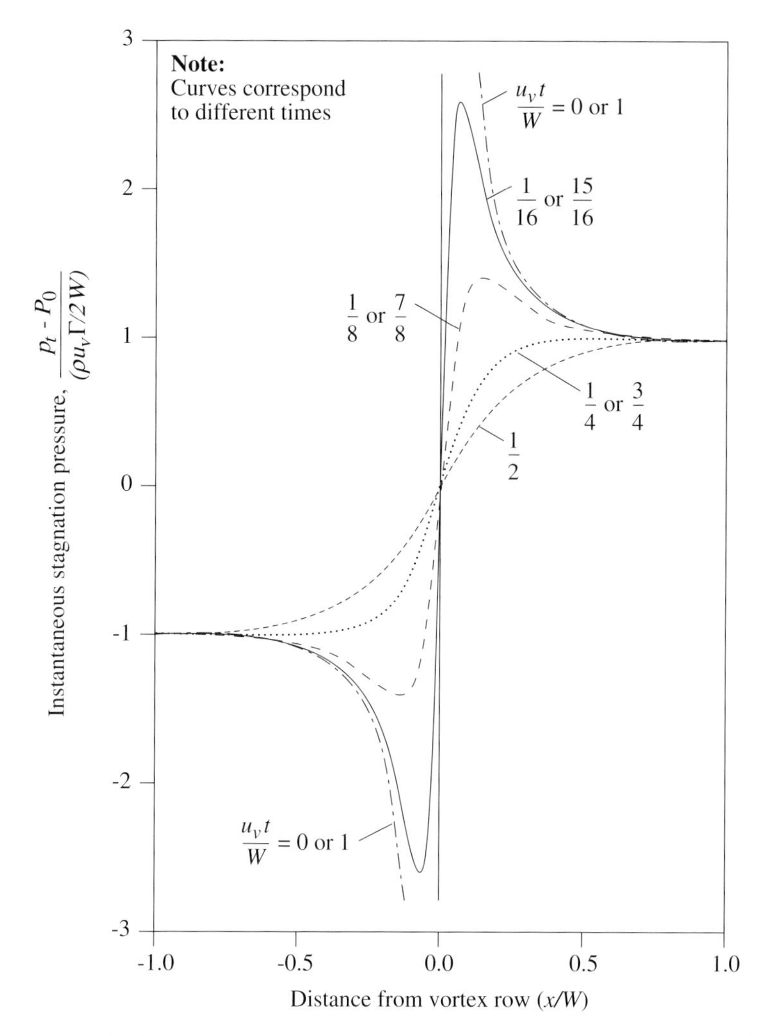

Figure 6.8: Instantaneous stagnation pressure versus time for a moving row of vortices (Preston, 1961).

written as

$$\frac{\partial \mathbf{u}}{\partial t} + \nabla\left(h + \frac{u^2}{2}\right) = 0, \tag{6.4.26}$$

or, integrating,

$$\frac{\partial \varphi}{\partial t} + h_t = f(t). \tag{6.4.27}$$

In compressible flow the link is between unsteadiness in the velocity potential and stagnation enthalpy, rather than the stagnation pressure as in the incompressible case. Because of the coupling between density and velocity, the velocity field due to a row of vortices in a compressible flow is not the same

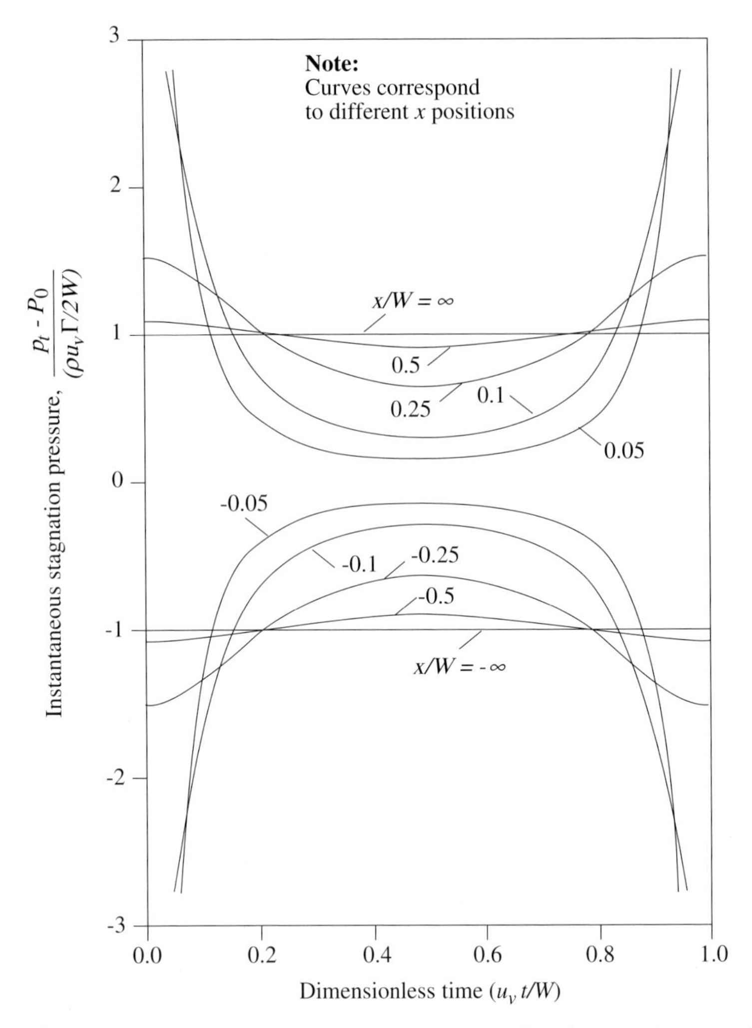

Figure 6.9: Instantaneous stagnation pressure versus time for a moving row of vortices (Preston, 1961).

as in incompressible flow. For flows in which all velocities are subsonic, however, the behavior will be qualitatively similar, and the ideas of Section 6.4.5 can be used to examine the phenomenon of energy separation in wakes.

Discussions of wakes (including the earlier sections of this text) generally portray them as steady constant pressure regions with a lower velocity than the free stream and a roughly equal uniform stagnation temperature. As mentioned in Chapter 4, the shear layers that form the wakes have an unsteady vortical structure. An observer in the stationary (fixed) system downstream of a body sees an unsteady flow with two rows of vortices of opposite sign convecting past. The wake structure actually evolves spatially, but we can approximate the situation as two infinite rows of counterrotating vortices and apply the ideas developed in the previous section for the single row of moving vortices.

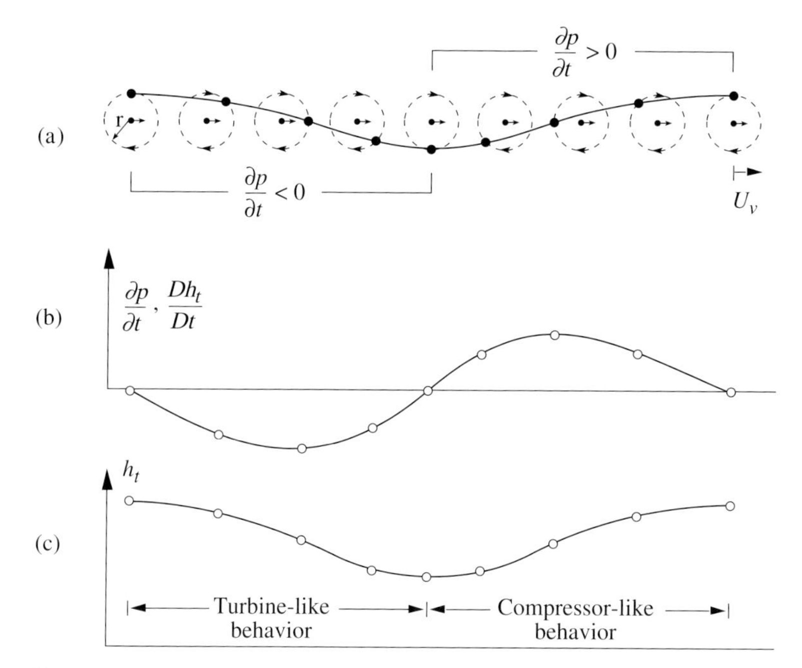

Figure 6.10: Variation of stagnation enthalpy h_t along a pathline for a moving row of vortices: (a) streamlines in vortex fixed system (dashed) and pathline in a stationary system; (b) behavior of static pressure; and (c) behavior of stagnation enthalpy (Kurosaka *et al.*, 1987).

The result of principal interest, namely that the stagnation temperature is lower in the wake than in the free stream, can be motivated using the following argument (Kurosaka *et al.*, 1987). Close to any one of the vortices in the moving rows the streamlines in the vortex fixed system are approximately circular. This is sketched in Figure 6.10, which shows these circular streamlines (as seen in the vortex fixed system) around a vortex moving to the right with velocity u_v. On the circular streamlines, the velocity magnitude and the static pressure will be approximately constant with the static pressure decreasing toward the vortex center.

Let us now examine the points indicated by the small solid symbols in Figure 6.10(a) from the perspective of observers in the stationary system. The observers located in the first half of the cycle all see a negative value of $\partial p/\partial t$ (and hence Dh_t/Dt), whereas observers at the points in the second half of the cycle see positive values. All the observers, however, correspond to points on particle paths (as seen in the stationary system) which are indicated by the solid line in Figure 6.10(a). Fluid particles thus have their stagnation temperature (and stagnation pressure) decreased in the first part of the cycle and increased in the second half, as illustrated in parts (b) and (c) of the figure.

A kinematic argument can also be made for this result. Fluid particles on the circular streamlines have approximately constant static temperature. For a stationary observer, fluid on the outside of the wake (above the moving row) has a higher velocity than that below because the velocity associated with the vortex adds to the convection velocity for the former and subtracts for the latter. Particles

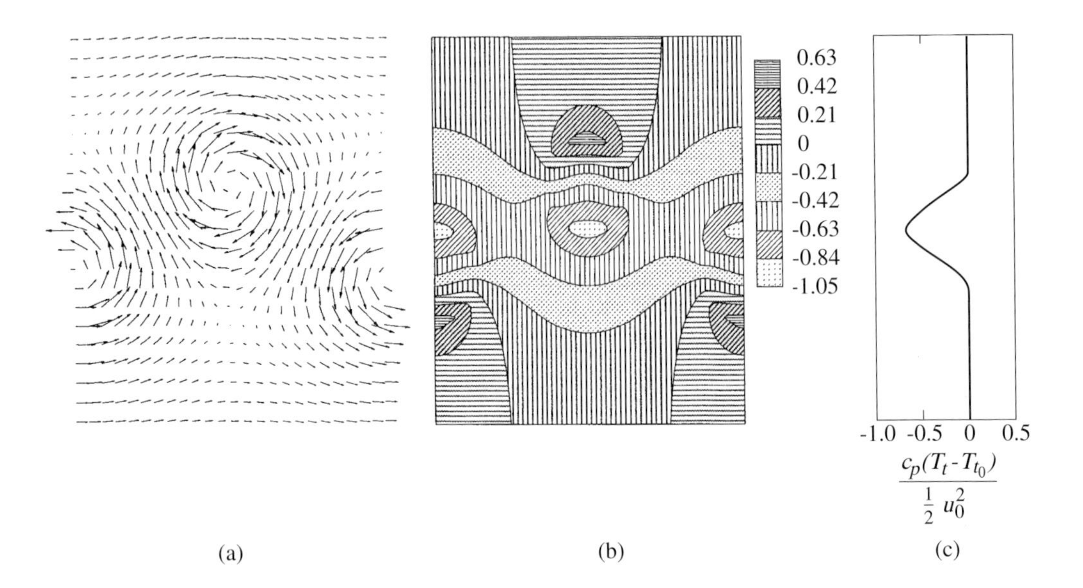

Figure 6.11: Computed stagnation temperature field in a vortex wake (a) vortex wake; (b) instantaneous stagnation temperature; (c) time mean stagnation temperature (Kurosaka *et al.*, 1987).

outside the moving rows therefore have a higher stagnation temperature than those inside. The terms "turbine" and "compressor" given in Figure 6.10 are appropriate because kinematically the vortices act as turbines for the passage of fluid across the row from the upper side and as compressors for the passage from below to above the row.

The analogy between the moving vortices and turbomachine blading can now be applied to the two vortex rows which bound a wake as shown in Figure 6.11. From free stream to wake, the effect is analogous to that across a turbine blade row, a drop in stagnation temperature (consider the circulation of the vortices shown in Figure 6.10). The drop in stagnation temperature seen in the stationary system is roughly $u_v\Gamma/W$, where u_v is the vortex velocity, Γ is the circulation, and W the vortex spacing.

Computations of this effect have been carried out with the vortices modeled as having finite cores of uniform vorticity, rather than as point vortices (Kurosaka *et al.*, 1987). Results are given in Figure 6.11, which is based on parameters representative of the wake of an axial compressor blade. Figure 6.11(a) shows velocity vectors as seen in the coordinate system moving with the vortices, Figure 6.11(b) the instantaneous stagnation temperature field, and Figure 6.11(c) the time mean stagnation temperature. With a finite core there is no discontinuity in stagnation quantities across the vortex row.

The arguments concerning the stagnation temperature non-uniformity take no account of viscosity and heat transfer effects, which decrease the effects that we have been discussing. Energy separation in a wake should therefore be most evident when the wake width is large compared to regions directly affected by viscous forces, as occurs in bluff body flows. Figure 6.12 shows an example, the stagnation temperature distribution in the wake of a cylinder. The shape and magnitude of the distribution are similar to those in Figure 6.11.

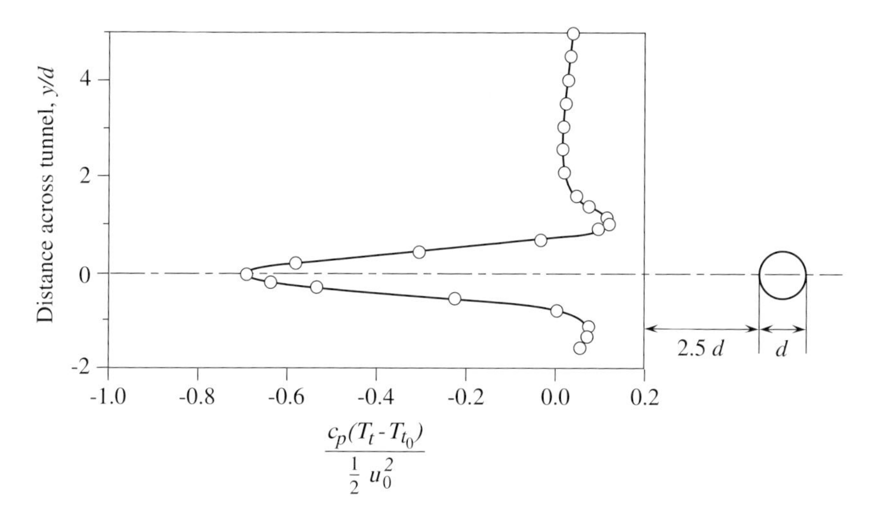

Figure 6.12: Measured stagnation temperature distribution through a cylinder wake; T_{t_0} and u_0 are the temperature and velocity far upstream of the cylinder (data of Ryan (1951) as cited in Eckert (1987)).

From the considerations given, the stagnation temperature difference can be estimated as

$$\frac{c_p(T_t - T_{t_0})}{\frac{1}{2}u_0^2} = \frac{\left(\frac{u_v \Gamma}{W}\right)}{\frac{1}{2}u_0^2} = \frac{2\Gamma}{u_0 W}\left(\frac{u_v}{u_0}\right). \tag{6.4.28}$$

The quantities T_0 and u_0 in (6.4.28) represent the temperature and velocity at a location far upstream of the cylinder.

The velocity field in the wake of a cylinder has been extensively investigated, and the measured values of $\Gamma/(u_0 W) \approx 0.55$, $u_v/u_0 \approx 0.75$ can be used in (6.4.28). Substituting these gives $c_p(T_t - T_{t_0})/(\frac{1}{2}u_0^2) = 0.84$, in rough agreement with the experimental value of 0.74 and nearly an order of magnitude larger than the stagnation temperature difference across a boundary layer on an adiabatic wall (see Eckert (1984), Schlichting (1979)). In the boundary layer, non-uniformities in stagnation temperature are due to imbalances between heat transfer and viscous work for a given streamtube; these have a net impact much less than that due to the basically inviscid unsteady effects in the bluff body wake.

6.5 Shear layer instability

The stability of fluid motions can be an important criterion for fluid device performance. Illustrations of the qualitative changes associated with instability already encountered in Chapter 4 were transition to turbulence and the genesis of large scale vortex structure in free shear layers.

Stability can be operationally defined as "the quality of being immune to small disturbances" (Betchov and Criminale, 1967). The issue is whether a given system, which is operating in some equilibrium state and which is given a small disturbance, will change by a proportionally small amount or whether the disturbance will be amplified and the state undergo evolution to a new equilibrium

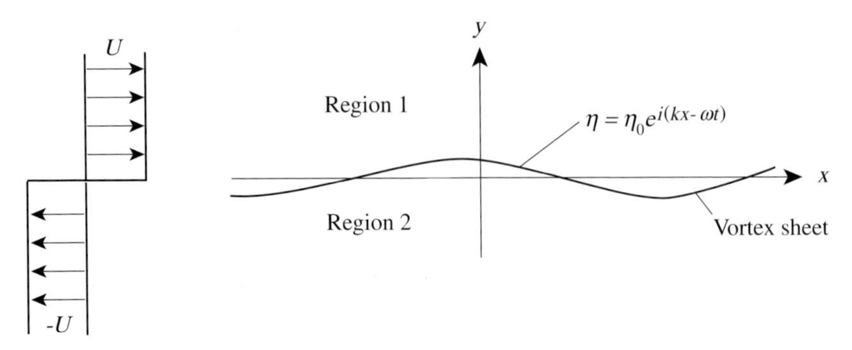

Figure 6.13: Kelvin–Helmholtz instability of a vortex sheet.

quite different from the initial one. A number of different self-excited fluid motions exist, many of which are described in the texts by Drazin and Reid (1981), Betchov and Criminale (1967), and Tritton (1988). In this section we address instability phenomena associated with parallel shear flow with application to boundary and free shear layer behavior.

6.5.1 Instability of a vortex sheet (Kelvin–Helmholtz instability)

As an introduction to the topic of shear layer instability consider the instability of a two-dimensional plane vortex sheet in an incompressible, inviscid, constant density fluid. This is known as Kelvin–Helmholtz instability. The geometry is presented in Figure 6.13. The time mean flow is a parallel shear flow with a vortex sheet of zero thickness at $y = 0$, separating streams of velocities u_1 and u_2 with $u_1 > u_2$. Adopting a coordinate system moving with average velocity $(u_1 + u_2)/2$, the flow looks as drawn in the figure, with the magnitude of the velocity U given by $U = (u_1 - u_2)/2$. To determine stability, we inquire into the transient behavior when the interface between the two streams is subjected to a small displacement, $\eta(x, t)$ as in Figure 6.13. Any such small displacement can be analyzed as a sum of Fourier components with the displacement taken to be of the form $\eta(x, t) = \eta_0 e^{i(kx-\omega t)}$, where k is the wave number ($k = 2\pi$/disturbance wavelength) and $\omega = 2\pi \times$ frequency. The disturbance is a propagating wave with phase velocity $c = \omega/k$.

From Kelvin's Theorem (Sections 3.8 and 3.9) disturbing the interface will not change the circulation around any contour outside the vortex sheet, and the flow remains irrotational everywhere except within the sheet. We cast the problem in terms of two *disturbance velocity potentials*, φ_1 and φ_2, with the former applying to the region above the sheet and the latter to the region below. Using appropriate matching conditions across the interface, the two disturbance potentials can be connected to give a description of the motion which is valid throughout.

To analyze the unsteady small amplitude behavior, a linearized flow field description, which includes only quantities that are first order in the small disturbances, is appropriate. Since the sheet displacement η is proportional to $e^{i(kx-\omega t)}$, all the disturbance quantities will have this form, where the real part of the complex quantity is implied. For the disturbance potentials we seek a solution to Laplace's equation with a spatial periodicity of the disturbance wavelength. Such a solution has already been derived in the context of the periodic pressure field analyzed in Section 2.3. With that development as reference, and the requirement that the velocities are bounded at $y = \pm\infty$, the forms for φ_1 and φ_2 are given by

$$\varphi_1 = Ae^{-ky+i(kx-\omega t)} \quad \text{and} \quad \varphi_2 = Be^{+ky+i(kx-\omega t)}. \tag{6.5.1}$$

The two necessary matching conditions across the vortex sheet are that pressure and displacement are continuous across the sheet. The pressure can be evaluated using the linearized form of the x-momentum equation. Writing the velocity as a time mean, denoted by U plus a small disturbance, denoted by a prime ($'$), the linearized form of the x-momentum equation in the region above the sheet is

$$\frac{\partial u'_{1x}}{\partial t} + U\frac{\partial u'_{1x}}{\partial x} = -\frac{1}{\rho}\frac{\partial p'_1}{\partial x}. \tag{6.5.2}$$

Equation (6.5.2) can be written in terms of the velocity potential as

$$(\omega - kU)i\varphi_1 = \frac{p'_1}{\rho}. \tag{6.5.3}$$

A corresponding relation holds for Region 2. Continuity of pressure across the vortex sheet implies

$$\left[\left(\frac{\omega}{k} - U\right)\varphi_1 = \left(\frac{\omega}{k} + U\right)\varphi_2\right]_{y=0}. \tag{6.5.4}$$

To implement the second matching condition we make use of the kinematic boundary condition developed in Section 1.11 to relate the y-component of velocity and the sheet displacement. The linearized form of the kinematic surface condition for the upper region is

$$u_{1y}(x, 0, t) = \frac{\partial \eta}{\partial t} + U\frac{\partial \eta}{\partial x}. \tag{6.5.5a}$$

Similarly, for the lower region,

$$u_{2y}(x, 0, t) = \frac{\partial \eta}{\partial t} - U\frac{\partial \eta}{\partial x}. \tag{6.5.5b}$$

Substituting $u_y = \partial\varphi/\partial y$ and combining (6.5.5a) and (6.5.5b) gives a second relation between the velocity potentials in the upper and lower regions:

$$\left[\left(\frac{\omega}{k} + U\right)\varphi_1 = \left(-\frac{\omega}{k} + U\right)\varphi_2\right]_{y=0}. \tag{6.5.6}$$

Equations (6.5.4) and (6.5.6) are two homogeneous equations linking the two unknown constants A and B defined in (6.5.1). For these to have a non-trivial solution, the coefficient determinant for the two-equation system must be zero. Imposition of this condition provides an equation for the frequency ω, the imaginary part of which is the growth rate of the disturbance:

$$\frac{\omega}{k} = \pm iU = \pm\frac{i}{2}(u_1 - u_2). \tag{6.5.7}$$

All wavelengths are unstable and the growth rate (ω) is linear with wave number, k.

This linearized analysis only describes the initial stages of the vortex sheet instability, but nonlinear numerical computations using vortex methods (Krasny, 1986) can be used to track the evolution to the final state. Figure 6.14 shows the growth of sinusoidal disturbances and the formation of discrete vortices, similar to the flow visualization of a shear layer in Section 4.8.

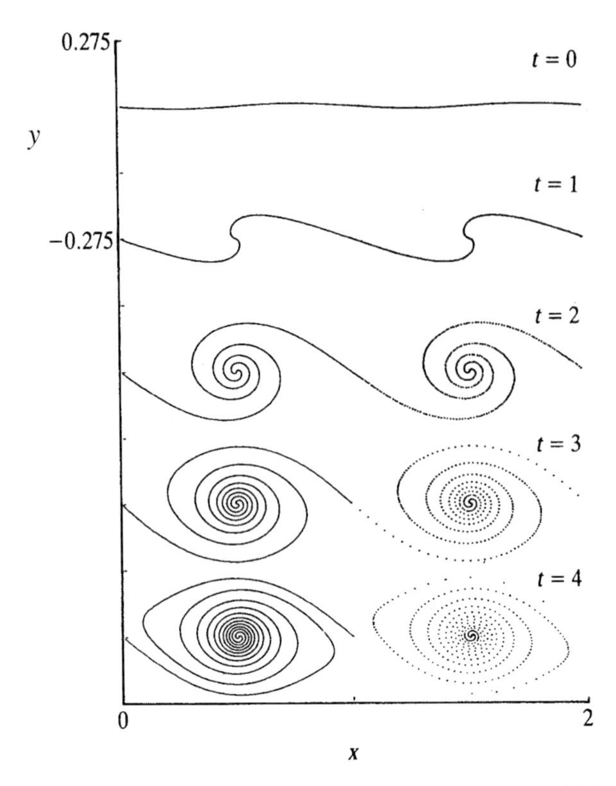

Figure 6.14: Nonlinear rollup of a vortex sheet (Krasny, 1986).

6.5.2 General features of parallel shear layer instability

While vortex sheet evolution demonstrates features of shear layer instability, the vortex sheet is a special example and we need to explore a broader class of instability problems. Of particular interest are questions such as what types of velocity profiles are most sensitive to instability and what differences exist between wall bounded and free shear flows. To address these we derive a set of linear equations that describe the behavior of small disturbances in a general inviscid, constant density parallel shear flow. The two-dimensional continuity and momentum equations, linearized to first order in the disturbance quantities, yield the required set of equations, where $\mathbf{u} = (\overline{u}_x + u'_x, u'_y)$ and $p = \overline{p} + p'$:

$$\frac{\partial u'_x}{\partial x} + \frac{\partial u'_y}{\partial y} = 0, \tag{6.5.8a}$$

$$\frac{\partial u'_x}{\partial t} + \overline{u}_x \frac{\partial u'_x}{\partial x} + u'_y \left(\frac{d\overline{u}_x}{dy} \right) = -\frac{1}{\rho} \frac{\partial p'}{\partial x}, \tag{6.5.8b}$$

$$\frac{\partial u'_y}{\partial t} + \overline{u}_x \frac{\partial u'_y}{\partial x} = -\frac{1}{\rho} \frac{\partial p'}{\partial y}. \tag{6.5.8c}$$

We again take the disturbances to be of the form $e^{i(kx-\omega t)}$, where k is real, and consider a single component of a Fourier series in x. For a general shear flow we cannot make use of a velocity potential because the flow is not irrotational. We can, however, introduce a disturbance stream function which identically satisfies continuity:

$$\psi(x, y, t) = f(y)e^{i(kx-\omega t)} \quad \text{and} \quad u'_x = \frac{\partial \psi}{\partial y}, \quad u'_y = -\frac{\partial \psi}{\partial x}. \tag{6.5.9}$$

Substituting (6.5.9) into (6.5.8b) and (6.5.8c) and cross-differentiating to eliminate the pressure yields a second order equation for the function $f(y)$, known as Rayleigh's equation:

$$(\overline{u} - c)\left(\frac{d^2 f}{dy^2} - k^2 f\right) - \left(\frac{d^2\overline{u}}{dy^2}\right) f = 0. \tag{6.5.10}$$

In (6.5.10) the quantity $c = \omega/k$ is the phase velocity of the disturbance. The boundary conditions that are appropriate vary depending on the specific geometry investigated, but if the flow is bounded by walls at upper and lower locations y_U and y_L, where $u'_y = 0$, then $f(y_U) = f(y_L) = 0$.

Using (6.5.10) we can make a strong statement about the conditions on the type of time mean profiles that lead to instability (Betchov and Criminale, 1967; Sherman, 1990). To see this we multiply the equation by f^*, the complex conjugate of f, divide by $(\overline{u} - c)$ and integrate the result between the limits y_U and y_L. This yields, after some rearrangement of terms,

$$\int_{y_L}^{y_U} \left[\frac{d}{dy}\left(f^* \frac{df}{dy}\right) - \left(\frac{df^*}{dy}\frac{df}{dy}\right) - k^2 f^* f\right] dy = \int_{y_L}^{y_U} \left[\left(\frac{d^2\overline{u}}{dy^2}\right)\left(\frac{f^* f}{\overline{u} - c}\right)\right] dy. \tag{6.5.11}$$

The first term on the left of (6.5.11) can be integrated as

$$\int_{y_L}^{y_U} \frac{d}{dy}\left(f^* \frac{df}{dy}\right) dy = \left[f^* \frac{df}{dy}\right]_{y_L}^{y_U}. \tag{6.5.12}$$

The boundary condition on f means that both real and imaginary parts of f vanish at the limits so the integral in (6.5.12) is zero. The two other terms in the integral on the left in (6.5.11) both have the form $()^*()$ (a quantity times its conjugate) so they are positive definite. The value of the integral is thus equal to $-\Upsilon^2$, where Υ is a constant. This means that (6.5.11) can be written as

$$-\Upsilon^2 = \int_{y_L}^{y_U} \left[\left(\frac{d^2\overline{u}}{dy^2}\right)\left(\frac{f^* f}{\overline{u} - c}\right)\right] dy. \tag{6.5.13}$$

The phase speed, c, is now expressed in terms of real and imaginary parts:

$$c = c_R + ic_I. \tag{6.5.14}$$

Substituting (6.5.14) into (6.5.13) and examining the imaginary part of the result we obtain

$$c_I \left(\int_{y_L}^{y_U} \left[\frac{(d^2\overline{u}/dy^2)(f^* f)}{(\overline{u} - c_R)^2 + c_I^2}\right] dy\right) = 0. \tag{6.5.15}$$

Equation (6.5.15) means that either c_I is zero, in which case the disturbance wave is not growing or decaying and the flow is neutrally stable, or the integral vanishes. If the disturbances are to grow, the integral must be zero, but every term in the integrand is positive except possibly the second derivative of the time mean velocity profile. Further, the integral can only be zero if the second derivative is positive over some part of the interval in y and negative over the rest of the interval, implying that $(d^2\overline{u}/dy^2)$ passes through 0 at one or more values of y. A necessary condition for disturbances in the shear layer to grow, therefore, is that the time mean velocity profile must possess a point of inflection $(d^2\overline{u}/dy^2 = 0)$. This theorem was first proved by Rayleigh over a hundred years ago. Since then others have extended it to show that a growing wave can only exist in a parallel shear flow if the time mean vorticity, $(-d\overline{u}/dy)$, has a maximum (see Sherman (1990)).

Rayleigh's Theorem provides an important qualitative distinction between flows with an inflection point in the velocity profile, such as jets and free shear layers, and flows without an inflection point, such as the constant pressure boundary layer and Poiseuille flow in a channel. The instability mechanism in the former type of shear layer is much more powerful. Shear layers with an inflection point[3] are unstable in the inviscid limit and can be stabilized by viscosity at low enough Reynolds number but the values needed are on the order of 10–100. For profiles without an inflection point, instability occurs only at much higher Reynolds numbers when viscosity has the "remarkable destabilizing influence" described by Betchov and Criminale (1967).

Further, from the conditions at a solid surface developed in Section 3.13, we see that boundary layers with an adverse pressure gradient have an inflection point in the velocity (and a maximum value of the vorticity) away from the wall. (The constant pressure boundary layer has its second derivative equal to 0 at the wall: $(\partial^2\overline{u}/\partial y^2) = 0$ at $y = 0$.) This provides insight into why, as mentioned in the discussion of natural transition in Section 4.5, instability of boundary layers in adverse pressure gradients occurs at much lower Reynolds numbers than with favorable pressure gradients. Adverse pressure gradients increase the boundary layer shape parameter, H, and, as shown in Figure 6.15 (White, 1991), the critical Reynolds number, Re_{δ^*}, at which disturbance waves will grow decreases sharply with H.

Other features of shear layer instability can be seen from the numerical solution of (6.5.10) for the shear layer profile $\overline{u}(y) = U\tanh(y/W)$, where W is the half-width of the shear layer in Figure 6.16 (Betchov and Criminale, 1967; see also Lucas *et al.*, 1997). The abscissa is the non-dimensional wave number, kW. Two quantities are shown on the ordinate, c_I/U, the disturbance growth rate, and $\omega_I W/U$. The value of $\omega_I W/U$ for the Kelvin–Helmholtz results is also indicated. For disturbances with wavelengths large compared to the shear layer thickness ($kW \ll 1$), the finite thickness shear layer behavior is similar to that of a vortex sheet. As the wave number increases, the growth rate for the finite thickness layer peaks and falls to 0 at a disturbance wavelength of $2\pi W$. For very short wavelengths, the behavior can be viewed as similar to disturbance waves in a uniform shear, a flow which does not have a point of inflection.

Figure 6.17 shows the growth rates for an unbounded shear layer and for a shear layer with a wall $3W$ from the zero of velocity. Long wavelength disturbances (say, wavelengths larger than the distance of the point of inflection to the wall) "feel" the effect of the wall and are stabilized. Shorter wavelength disturbances do not and exhibit a behavior similar to that in the unbounded shear layer.

To summarize this section, three aspects of shear layer instability have been discussed. The first is the role of an inflection point in the velocity profile as a qualitative indicator of the tendency for

[3] If we think of the vortex sheet as the limiting case of a continuous velocity distribution across a symmetric shear layer, there is an inflection point at the midpoint of the layer.

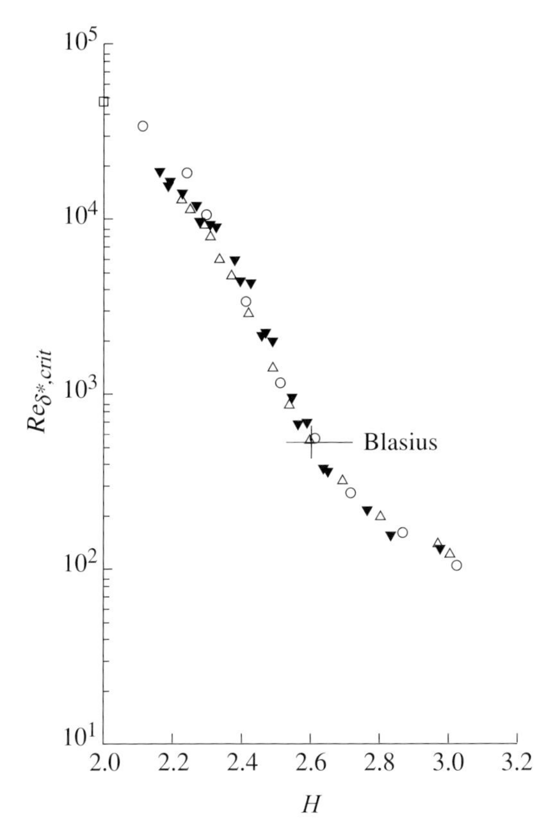

Figure 6.15: Computations of critical Reynolds number ($u_E\delta^*/\nu$) for instability versus boundary layer shape (after Wazzan *et al.*, as presented by White (1991)).

shear layer instability. The second is the different disturbance behavior depending on the ratio of wavelength to shear layer thickness. The third is the increased stability associated with the presence of a wall. An example in which these factors conspire to promote an accelerated growth of disturbance waves is the separated shear layer, with the result being a rapid transition to turbulence in the shear layer.

6.6 Waves and oscillations in fluid systems: system instabilities

Another important class of instabilities arise in the context of overall system unsteadiness. This, as well as the response of systems to external forcing, belongs to the general topic of waves and oscillations in flow systems (Lighthill, 1978). The features of this type of self-excited motion, particularly the dynamic coupling between the components in a fluid system, will be addressed from the perspective of unsteady one-dimensional small disturbances to an inviscid compressible fluid.

We begin with the linearized one-dimensional continuity and momentum equations:

$$\frac{\partial \rho'}{\partial t} + \overline{u}\frac{\partial \rho'}{\partial x} + \overline{\rho}\frac{\partial u'}{\partial x} = 0, \tag{6.6.1a}$$

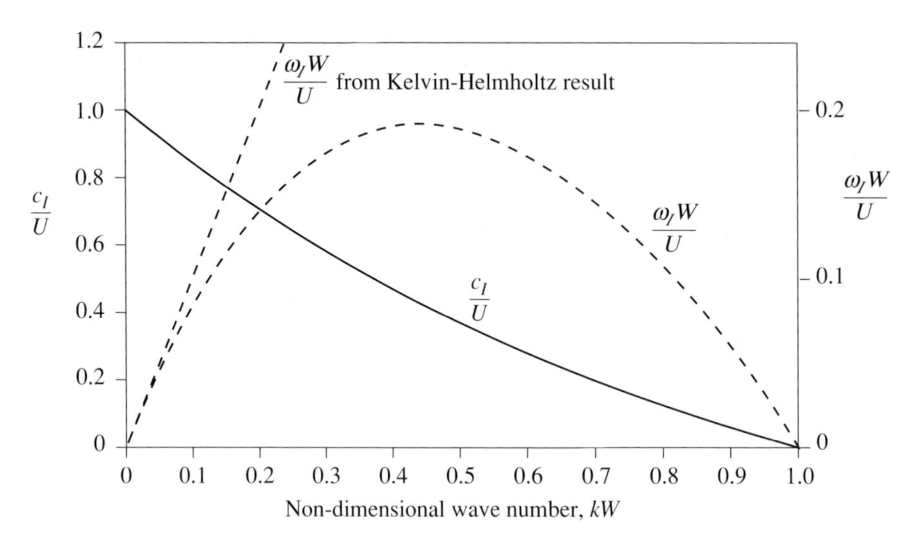

Figure 6.16: Disturbance growth rate for a shear layer with time mean velocity, $\overline{u}(y) = U\tanh(y/W)$ (Betchov and Criminale, 1967).

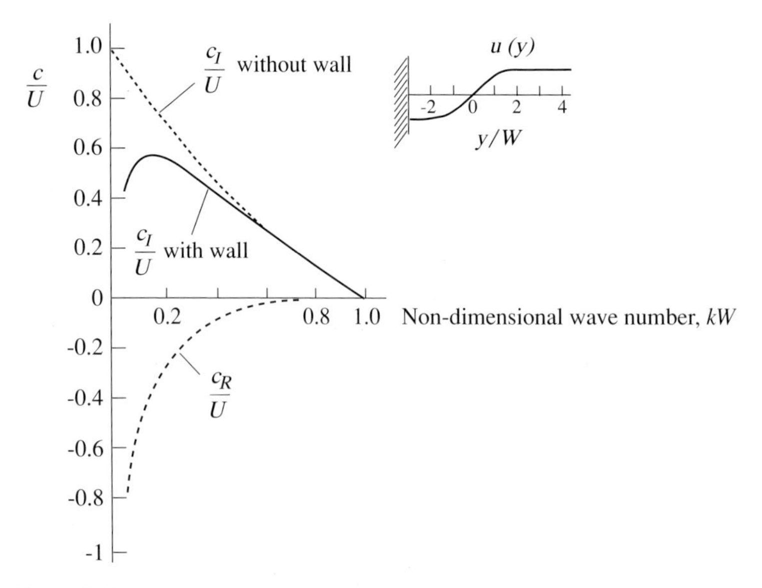

Figure 6.17: Eigenvalues for a shear layer in the vicinity of a wall; $\overline{u}(y) = U\tanh(y/W)$ (Betchov and Criminale, 1967).

$$\frac{\partial u'}{\partial t} + \overline{u}\frac{\partial u'}{\partial x} + \left(\frac{1}{\overline{\rho}}\right)\frac{\partial p'}{\partial x} = 0. \tag{6.6.1b}$$

For the motions considered the relation between small changes in density and pressure is isentropic:

$$\frac{p'}{\overline{p}} = \frac{\gamma\rho'}{\overline{\rho}}. \tag{6.6.2}$$

In (6.6.1) the subscript x has been dropped because the only velocity component is in the x-direction. Combining (6.6.1) and (6.6.2) yields equations for the disturbance pressure, $p'(x, t)$, and the velocity $u'(x, t)$:

$$\left(\frac{1}{\overline{a}}\frac{\partial}{\partial t}+\overline{M}\frac{\partial}{\partial x}\right)^2\begin{bmatrix}u'\\p'\end{bmatrix}-\frac{\partial^2}{\partial x^2}\begin{bmatrix}u'\\p'\end{bmatrix}=0. \tag{6.6.3}$$

In (6.6.3) the matrix notation implies that the same operators apply to both pressure and velocity disturbances. The variable $\overline{a}$ is the speed of sound (Section 1.15), which is equal to $\sqrt{\gamma\overline{p}/\rho}$.

If we confine the discussion to periodic disturbances of the form $e^{i\omega t}$, the solutions to (6.6.3) can be seen by substitution to have the form

$$u' = Ae^{i(\omega t-k_+x)} + Be^{i(\omega t+k_-x)}, \tag{6.6.4a}$$

$$p' = A\overline{\rho}\,\overline{a}e^{i(\omega t-k_+x)} - B\overline{\rho}\,\overline{a}e^{i(\omega t+k_-x)}. \tag{6.6.4b}$$

In (6.6.4) A and B are constants determined by the boundary conditions. The wave numbers k_+ and k_- are given by

$$k_+ = \left(\frac{\omega}{\overline{a}}\right)\frac{1}{1+M}, \quad k_- = \left(\frac{\omega}{\overline{a}}\right)\frac{1}{1-M}. \tag{6.6.5}$$

The two wave numbers represent waves traveling downstream and upstream respectively, at the speed of sound relative to the mean flow.

In situations where the mean Mach number is much less than unity ($\overline{M}^2 \ll 1$) (6.6.3) reduce to the acoustic wave equations:

$$\frac{1}{\overline{a}^2}\frac{\partial^2}{\partial t^2}\begin{bmatrix}u'\\p'\end{bmatrix}-\frac{\partial^2}{\partial x^2}\begin{bmatrix}u'\\p'\end{bmatrix}=0. \tag{6.6.6}$$

An application of (6.6.6) is to determine the form of the acoustic disturbance (the "acoustic mode") in a duct of length L which, for example, is open at one end, $x = 0$, and closed at the other, $x = L$. At the open end the pressure is constant, so $p(0, t) = 0$. At the closed end the velocity must be 0, so that $u(L, t) = 0$. For periodic disturbances of the forms $e^{i\omega t}$ the pressure and velocity therefore have the forms

$$u'(x,t) = -A\cos\left(\frac{\pi x}{2L}\right)e^{i\omega t}, \tag{6.6.7a}$$

$$p'(x,t) = -A\overline{\rho}\,\overline{a}\sin\left(\frac{\pi x}{2L}\right)e^{i\omega t}. \tag{6.6.7b}$$

Velocity fluctuations are maximum at $x = 0$ and pressure fluctuations are maximum at $x = L$.

6.6.1 Transfer matrices (transmission matrices) for fluid components

It is of considerable interest to be able to couple different fluid elements to describe unsteady disturbances in general fluid systems. For simple systems the most direct approach is to work from the conservation equations for each of the components. As the number of components increases, however, it is helpful to have a more formal procedure to build up the system model. Transfer matrices (also referred to as transmission matrices) provide such a methodology for dynamically coupling fluid components (Brennen, 1994; Lucas *et al.*, 1997; Munjal, 1987). The idea is that (in a one-dimensional sense) for any component the pressure and velocity at the inlet can be written in terms of the pressure and velocity at the exit as follows:

$$\begin{bmatrix} p' \\ \overline{\rho}\,\overline{a}u' \end{bmatrix}_i = \begin{bmatrix} 2\times 2 \text{ transfer matrix} \\ \text{for the element} \end{bmatrix} \begin{bmatrix} p' \\ \overline{\rho}\,\overline{a}u' \end{bmatrix}_e . \tag{6.6.8}$$

In (6.6.8) the quantity $\overline{\rho}\,\overline{a}$ has been introduced as a multiplier for the disturbance velocity u' so the matrix elements are non-dimensional.

6.6.1.1 The transfer matrix for a duct

We develop the transfer matrices for some common fluid system components, starting with the constant area duct. Using the forms of pressure and velocity given in (6.6.4) and substituting the values at $x = -L$ (inlet) and $x = 0$ (exit), the transfer matrix for a constant area duct of length L can be represented as

$$\begin{bmatrix} p' \\ \overline{\rho}\,\overline{a}u' \end{bmatrix}_{x=-L} = \begin{bmatrix} \frac{1}{2}\left(e^{ik_+L} + e^{-ik_-L}\right) & \frac{1}{2}\left(e^{ik_+L} - e^{-ik_-L}\right) \\ \frac{1}{2}\left(e^{ik_+L} - e^{-ik_-L}\right) & \frac{1}{2}\left(e^{ik_+L} + e^{-ik_-L}\right) \end{bmatrix} \begin{bmatrix} p' \\ \overline{\rho}\,\overline{a}u' \end{bmatrix}_{x=0} . \tag{6.6.9}$$

For situations in which the Mach numbers are small enough so the effect of the mean velocity can be neglected, (6.6.9) takes the form

$$\begin{bmatrix} p' \\ \overline{\rho}\,\overline{a}u' \end{bmatrix}_{x=-L} = \begin{bmatrix} \cos kL & i\sin kL \\ i\sin kL & \cos kL \end{bmatrix} \begin{bmatrix} p' \\ \overline{\rho}\,\overline{a}u' \end{bmatrix}_{x=0} \tag{6.6.10}$$

with $k = 2\pi/\text{disturbance wavelength} = \omega/\overline{a}$.

An important simplification of the duct transfer matrix occurs when the duct length and disturbance wavelength are such that $(kL)^2 = (\omega L/\overline{a})^2 \ll 1$. If so (see Section 2.2.2), the flow in the duct can be considered incompressible, and, for a constant area duct, the inlet and exit velocities are the same. There can, however, be a difference between the (inlet and exit) pressure perturbations across the duct. As seen below in connection with fluid system behavior, this pressure difference needs to be included in describing the phenomena of interest.

When $(\omega L/\overline{a})^2 \ll 1$ the transfer matrix for a duct can be derived using the incompressible form of the one-dimensional continuity equation and the one-dimensional momentum equation, applied to periodic disturbances. For a constant area duct these are:

$$u'_e = u'_i = u',$$
$$p'_e - p'_i = -\overline{\rho}L\frac{\partial u'}{\partial t} = -i\omega\overline{\rho}Lu'.$$

The transfer matrix for incompressible flow in a constant area duct thus has the form

$$\begin{bmatrix} p' \\ \overline{\rho}\,\overline{a}u' \end{bmatrix}_i = \begin{bmatrix} 1 & (iL\omega/\overline{a}) \\ 0 & 1 \end{bmatrix} \begin{bmatrix} p' \\ \overline{\rho}\,\overline{a}u' \end{bmatrix}_e . \tag{6.6.11}$$

Under these conditions the duct has only inertance and no mass storage capability.

6.6.1.2 The transfer matrix for a plenum (chamber of large cross-section)

Another useful component model is a plenum or chamber of large cross-sectional area such that velocities inside are small compared to the values in the inlet and outlet ports. The only attribute of this type of element is the mass storage capability, or capacitance. The pressures at the chamber inlet and exit are the same, but the velocity at the exit can differ from that at the inlet because of transient mass storage. The transfer matrix for a capacitance has the form

$$\begin{bmatrix} p' \\ \bar{\rho}\bar{a}u' \end{bmatrix}_i = \begin{bmatrix} 1 & 0 \\ (i\omega V/(\bar{a}A)) & 1 \end{bmatrix} \begin{bmatrix} p' \\ \bar{\rho}\bar{a}u' \end{bmatrix}_e, \tag{6.6.12}$$

where V is the chamber volume and A is the inlet and exit port area.

6.6.1.3 The transfer matrix for a contraction or expansion

Contractions or expansions, such as those that occur in nozzles and diffusers, can also be handled through a transfer matrix approach. If the element is such that $(\omega L/a)^2 \ll 1$, where L is the relevant length scale, there is no mass storage and the volume flow is the same at the inlet and exit. In addition, if the reduced frequency is low enough that convective accelerations dominate over local accelerations, the Bernoulli equation (or the momentum equation in the case of the sudden expansion) can be used in a quasi-steady manner to link the velocities and pressures at the inlet and outlet of the device. To derive the transfer matrices for contractions or expansions at low mean Mach number, we linearize the steady-state relation between pressure change and flow velocity about the operating condition of interest. For a nozzle with $AR = A_e/A_i$ the transfer matrix is

$$\begin{bmatrix} 1 & \dfrac{M_i}{AR}(1 - AR^2) \\ 0 & AR \end{bmatrix}. \tag{6.6.13a}$$

For a sudden expansion the result is

$$\begin{bmatrix} 1 & 2M_i(1 - AR) \\ 0 & AR \end{bmatrix}. \tag{6.6.13b}$$

6.6.1.4 The transfer matrix for a screen, perforated plate, or throttle

For low Mach number flows through screens or perforated plates, the pressure drop across the screen is found to be related to velocity by

$$\Delta p = \mathcal{K}\tfrac{1}{2}\rho u^2. \tag{6.6.14}$$

In (6.6.14) $\mathcal{K}$ is a constant whose value depends on screen solidity, or blocked area.[4] Viewed on the scale of the screen mesh elements, the screen is a contraction (through the area between the individual screen wires) followed by sudden expansion and mixing out. The pressure changes in both of these processes scale with the dynamic pressure of the entering flow. Unless one is in a regime in which

[4] The value of $\mathcal{K}$ for a round wire screen of 50% solidity is roughly 2. For other values of solidity, $\mathcal{K}$ can be estimated as $\mathcal{K} = 0.8s/(1 - s)^2$, where s is the solidity (Cornell, 1958).

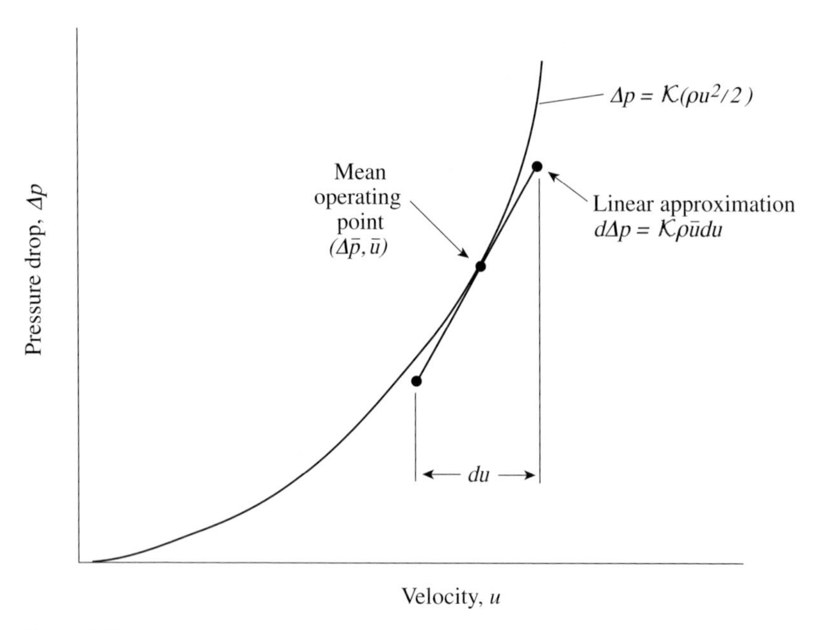

Figure 6.18: Linearized relation between screen pressure drop and velocity.

there are strong effects of Reynolds number (wire diameter Reynolds number much less than 10^3), the scaling for pressure drop versus flow rate is quadratic.

For small disturbances about a time mean velocity, the screen pressure drop can be linearized about the time mean, as shown in Figure 6.18. The transfer matrix relating the disturbance quantities is

$$\begin{bmatrix} p' \\ \overline{\rho}\,\overline{a}u' \end{bmatrix}_i = \begin{bmatrix} 1 & \Xi \\ 0 & 1 \end{bmatrix} \begin{bmatrix} p' \\ \overline{\rho}\,\overline{a}u' \end{bmatrix}_e, \tag{6.6.15}$$

where Ξ is the non-dimensional *slope* of the screen pressure drop versus screen mass flow per unit area curve, given by

$$\Xi = \mathcal{K}M.$$

This quantity is also known as the "acoustic throttle slope". In deriving (6.6.15) the screen pressure drop is taken as small compared to the ambient pressure level so density, and hence velocity, is the same on both sides of the screen. There is an entropy increase across the screen and in the regions of wake mixing (which occurs in roughly ten mesh spacings or less), but this can be lumped into the description of screen loss and the flow is well approximated using the isentropic equations outside the screen.

Fluid elements such as junctions or throttles can also be treated by transfer matrix methods. At a junction the sum of all the volume flows is the same upstream and downstream of the junction. Throttles are essentially resistances and are treated in a similar fashion to a screen, although their mean operating point, and hence equivalent value of $\mathcal{K}$, is a function of system operating point rather than fixed by geometry as with a screen.

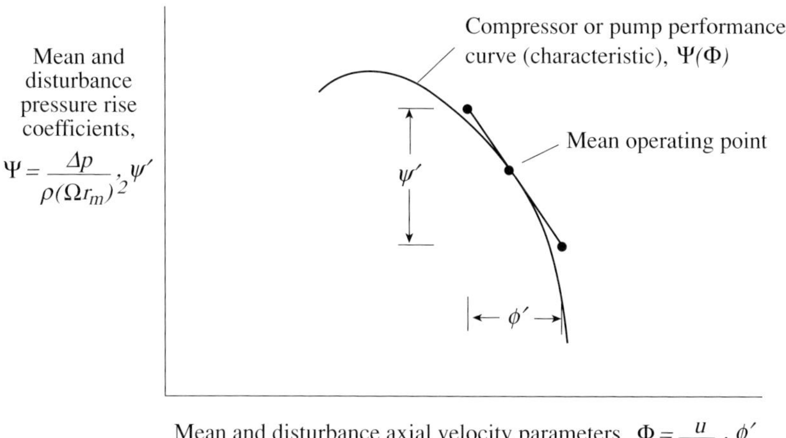

Figure 6.19: Linear approximation for compressor or pump performance curve.

6.6.1.5 The transfer matrix for a compressor or pump

Compressors or pumps are elements of many fluid systems. These devices differ in kind from the components described so far because they are active, in the sense of being able to add mechanical energy into the system. Steady-state performance of a compressor or pump is often presented as pressure rise versus mass flow or axial velocity for a constant rotational speed, Ωr_m, where r_m is the mean radius of the rotating blade row. In non-dimensional terms we define the pressure rise coefficient, $\Psi = \Delta p/[\rho(\Omega r_m)^2]$, as a function of the axial velocity parameter, $\Phi = u/\Omega r_m$ (essentially a non-dimensional mass flow): $\Psi = \Psi(\Phi)$. For low reduced frequency the compressor operating point can be viewed as tracking quasi-steadily along the steady-state (Ψ, Φ) curve, or "compressor characteristic". For small disturbances, the excursions can be approximated as linear about the time mean operating condition. The quasi-steady linear approximation to the pressure rise versus flow relation is $\psi' = (d\Psi/d\Phi)\phi'$, as shown in Figure 6.19, where ψ' and ϕ' are the departures from the time mean condition and $(d\Psi/d\Phi)$ is evaluated at this time mean condition.

From the above the transfer matrix for a compressor or pump with pressure rise small compared to the ambient level is

$$\begin{bmatrix} p \\ \overline{\rho}\,\overline{a}u \end{bmatrix}_i = \begin{bmatrix} 1 & -\Pi \\ 0 & 1 \end{bmatrix} \begin{bmatrix} p \\ \overline{\rho}\,\overline{a}u \end{bmatrix}_e . \tag{6.6.16}$$

In (6.6.16) Π is the "acoustic compressor slope" (Gysling, 1993), defined as

$$\Pi = \frac{\Omega r_m}{2\overline{a}} \frac{\partial \Psi_{t/t}}{\partial \Phi}, \tag{6.6.17}$$

where $\Psi_{t/t}$ is the stagnation pressure rise characteristic for the machine. Equation (6.6.16) is based on treatment of the pump or compressor as an element with $(\omega L/a)^2 \ll 1$, so the flow within the element is taken as incompressible.

A compressor or pump in a fluid system is often followed by a plenum in which there is essentially no static pressure rise. Under these conditions the stagnation pressure rise from inlet to plenum pressure is actually the inlet stagnation pressure to exit static pressure rise, $\Psi_{t/s}$, and the appropriate slope is $\partial\Psi_{t/s}/\partial\Phi$. The quantity $\Psi_{t/s}$ will be used as the relevant compressor pressure rise (denoted by Ψ_C) in what follows. Although not dealt with here, it can be mentioned that more complicated pumping devices can be modeled using this type of approach, for example cavitating turbopumps, in which there can also be mass flow storage (Ng and Brennen, 1978; Greitzer, 1981).

6.6.2 Examples of unsteady behavior in fluid systems

Transfer matrices have been employed in the description of many complex fluid systems, particularly with respect to the acoustics of these devices (Munjal, 1987; Poinsot *et al.*, 1987; Lucas *et al.*, 1997). The discussion here is confined to several examples which show both how the methodology is used and illustrate features of the dynamic behavior of fluid machinery.

6.6.2.1 The Helmholtz resonator

The Helmholtz resonator is a compliance plus an inertance in series, with the compliance closed at the downstream side (Dowling and Ffowcs Williams, 1983). The properties of this system can be worked out by multiplying the matrices from (6.6.11) and (6.6.12)

$$\begin{bmatrix} p' \\ \bar{\rho}\bar{a}u' \end{bmatrix}_i = \begin{bmatrix} 1 & (iL\omega/\bar{a}) \\ 0 & 1 \end{bmatrix} \begin{bmatrix} 1 & 0 \\ (iV\omega/(\bar{a}A)) & 1 \end{bmatrix} \begin{bmatrix} p' \\ \bar{\rho}\bar{a}u' \end{bmatrix}_e . \tag{6.6.18}$$

The two boundary conditions that apply are a pressure fluctuation equal to 0 at the inlet and a velocity fluctuation equal to 0 at the exit. Carrying out the matrix multiplication and imposing the boundary conditions leads to an equation for the frequency of the oscillation, ω, which is an eigenvalue of the system

$$\omega = \bar{a}\sqrt{\frac{A}{VL}}. \tag{6.6.19}$$

6.6.2.2 A model for gas turbine engine system instability

A slightly more complex example is shown in Figure 6.20, which models a gas turbine engine system. There are four elements: (i) a duct, (ii) a compressor, (iii) a plenum or chamber (typically the combustion chamber), and (iv) a throttle (or turbine nozzle). The transmission matrices for this arrangement are (with the assumption that both M^2 and $(\omega L/a)^2 \ll 1$)

$$\begin{bmatrix} p' \\ \bar{\rho}\bar{a}u' \end{bmatrix}_i = \begin{bmatrix} 1 & iL\omega/\bar{a} \\ 0 & 1 \end{bmatrix} \begin{bmatrix} 1 & -\Pi \\ 0 & 1 \end{bmatrix} \begin{bmatrix} 1 & 0 \\ iV\omega/(\bar{a}A) & 1 \end{bmatrix} \begin{bmatrix} 1 & \Xi \\ 0 & 1 \end{bmatrix} \begin{bmatrix} p' \\ \bar{\rho}\bar{a}u' \end{bmatrix}_e . \tag{6.6.20}$$

The boundary conditions are no pressure fluctuations at the inlet of the duct or the exit of the throttle. Carrying out the matrix multiplications leads to an eigenvalue equation for the frequency:

$$\omega^2 + i\bar{a}\left(\frac{\Pi}{L} - \frac{A}{V\Xi}\right)\omega + \frac{\bar{a}^2 A}{VLT}(\Pi - \Xi) = 0. \tag{6.6.21}$$

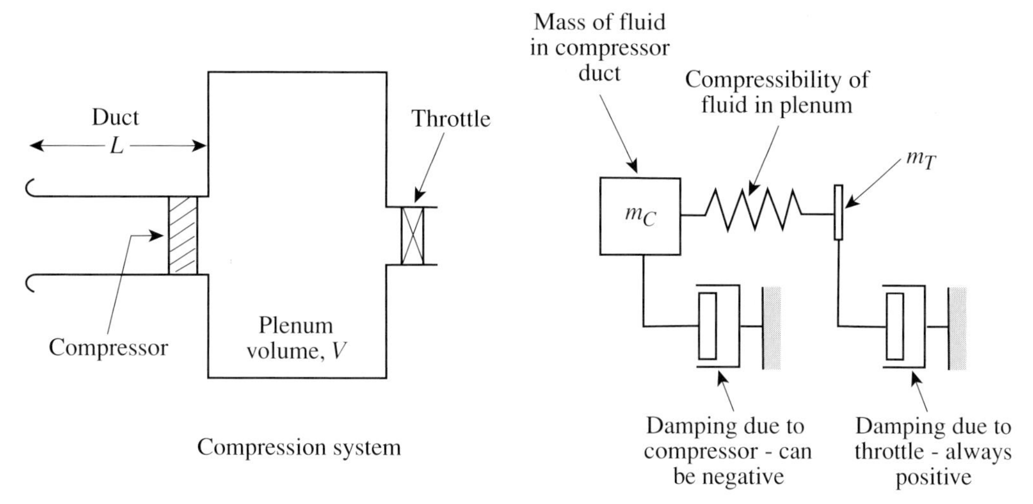

Figure 6.20: Compression system and mass-spring-damper mechanical analogue.

Equation (6.6.21) has complex roots in general:

$$\omega = -i\frac{\overline{a}}{2}\left(\frac{\Pi}{L} - \frac{A}{V\Xi}\right) \pm i\frac{\overline{a}}{2}\sqrt{\left(\frac{\Pi}{L} - \frac{A}{V\Xi}\right)^2 + \frac{4A}{VL\Xi}(\Pi - \Xi)}. \tag{6.6.22}$$

6.6.2.3 Static and dynamic instability

The imaginary part of the roots in (6.6.22) define the growth or decay of oscillations and hence the stability or instability of the system to small disturbances. There are two criteria corresponding to static and dynamic stability respectively:[5]

$$\Xi - \Pi < 0 \quad \text{or} \quad \Pi > \Xi, \quad \text{static instability,} \tag{6.6.23}$$

and

$$\frac{A}{V\Xi} - \frac{\Pi}{L} < 0 \quad \text{or} \quad \Pi\Xi > \frac{LA}{V}, \quad \text{dynamic instability.} \tag{6.6.24}$$

[5] The terms dynamic and static instability can be made more quantitative in the context of a second order system described by the equation

$$\frac{d^2x}{dt^2} + 2\alpha\frac{dx}{dt} + \beta x = 0,$$

where α and β are constants. The transient response to an initial perturbation is given by

$$x = A\exp\left[(-\alpha + \sqrt{a^2 - \beta})t\right] + B\exp\left[(-\alpha + \sqrt{a^2 - \beta})t\right],$$

where A and B are determined by the initial conditions. If $\beta > \alpha^2$, the condition for instability is simply $\alpha < 0$, which corresponds to oscillations of exponentially growing amplitude. Instability will also occur if $\beta < 0$, independent of the value of α; however, in this case the exponential growth is non-oscillatory. It is usual to denote these two types of instability as dynamic and static respectively. Static stability ($\beta > 0$) is a necessary but not sufficient condition for dynamic stability.

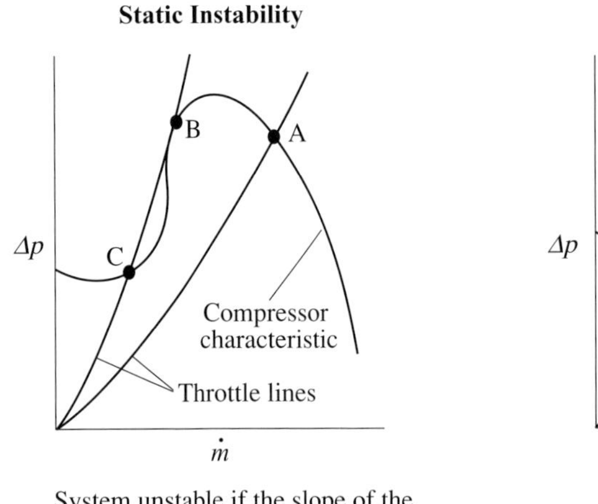

System unstable if the slope of the compressor characteristic is greater than the slope of the throttle line (point B)

Even if statically stable, the system can be dynamically unstable (point D)

Figure 6.21: Static and dynamic compression system instability.

In terms of the compressor and throttle characteristic curves, Ψ_C and Ψ_T, the criteria are:

$$\frac{\partial \Psi_C}{\partial \Phi} > \frac{\partial \Psi_T}{\partial \Phi}, \quad \text{static instability} \tag{6.6.25}$$

and

$$\frac{\partial \Psi_C}{\partial \Phi} \cdot \frac{\partial \Psi_T}{\partial \Phi} > \frac{1}{B^2}, \quad \text{dynamic instability.} \tag{6.6.26}$$

In (6.6.26) the parameter B is defined as

$$B = \frac{\Omega r_m}{2a}\sqrt{\frac{V}{AL}}.$$

The throttle characteristic, Ψ_T, is given by $\Psi_T = \Delta p_{throttle}/[\frac{1}{2}\rho(\Omega r_m)^2] = \chi\Phi^2$.

The static stability criterion in (6.6.25) indicates the system is unstable if the slope of the compressor pumping characteristic is steeper than the slope of the throttle pressure drop curve. Static stability can be assessed from the steady-state attributes of a system, which in this case are the slopes of the compressor and throttle characteristics. For a mass-spring-damper system, static instability corresponds to a "negative spring constant", with a pure exponential divergence from the initial equilibrium position (Greitzer, 1981).

The left-hand side of Figure 6.21 shows a sketch of a pressure rise versus mass flow compressor characteristic as well as two throttle lines (pressure drop versus mass flow for the throttle) to illustrate the situation for static stability. The steady-state operating point of the compressor is at a condition where the compressor and throttle flows are equal and the pressure rise across the compressor is the same as the pressure drop across the throttle. This occurs at the intersection of the compressor and throttle curves. Points A, B, and C are three such points. Inspection of the pressure changes that occur in the throttle and the compressor in response to a small mass flow decrease from the steady-state operating point shows that A and C are statically stable, because a pressure imbalance will be set

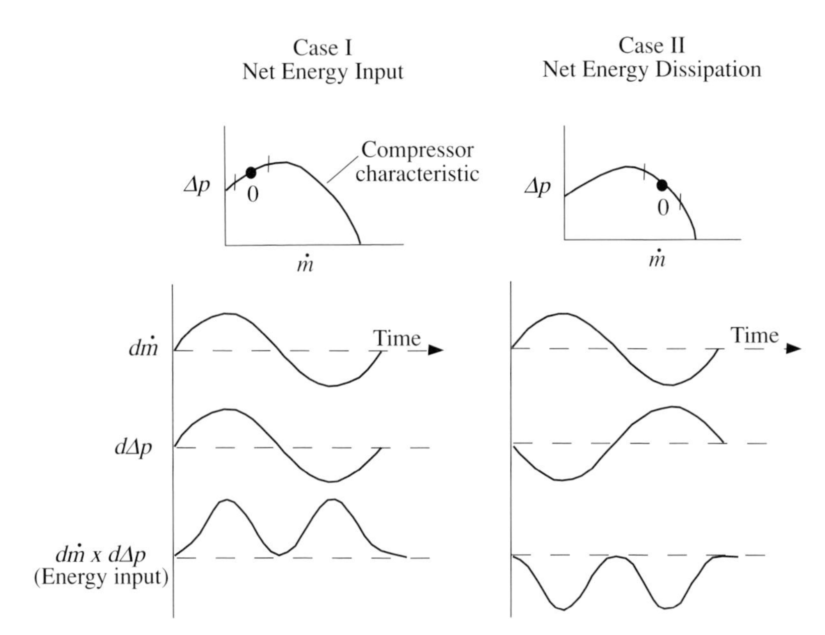

Figure 6.22: Physical mechanism for compression system dynamic instability. Point 0 on the compressor characteristic is the mean operating condition and the two short vertical lines denote a nominal oscillation amplitude.

up to return the system to the initial operating point. For point B, however, at which the throttle line is tangent to the compressor characteristic, the pressure imbalances move the operating point away from the initial value. The system is therefore statically unstable.

It is the dynamic stability criterion, represented by (6.6.26), which is most important in practice. Dynamic instability can occur even if the system is statically stable. Criteria for dynamic instability depend on the unsteady behavior of the system and thus cannot be found from knowledge of steady-state attributes. In terms of the analogy between the compression system and the mass-spring-damper system of Figure 6.20, dynamic instability corresponds to "negative damping".

6.6.2.4 Mechanism for dynamic compression system instability

The mechanism of dynamic instability for the compression system is associated with operation on the positively sloped part of the compressor characteristic. For this condition fluctuations in compressor mass flow have the effect of providing negative mechanical damping. This can be seen in Figure 6.22, which presents sketches of compressor characteristics, instantaneous disturbances in mass flow and pressure rise, and their product; the product is the instantaneous flux of mechanical energy out of the compressor over and above the steady-state value. For operation on the positively sloped part of the compressor curve, high mass flow and high pressure rise occur together, giving rise to a net flux of disturbance mechanical energy. For operation on the negatively sloped part of the compressor curve, high mass flow occurs at the same time as the low pressure rise and the net effect is to extract energy from the oscillations. Whether instability occurs in a specific system depends on the balance between mechanical energy fed into the oscillations by the compressor and that extracted by the throttle (in which dissipation occurs) but the above description shows how the compressor

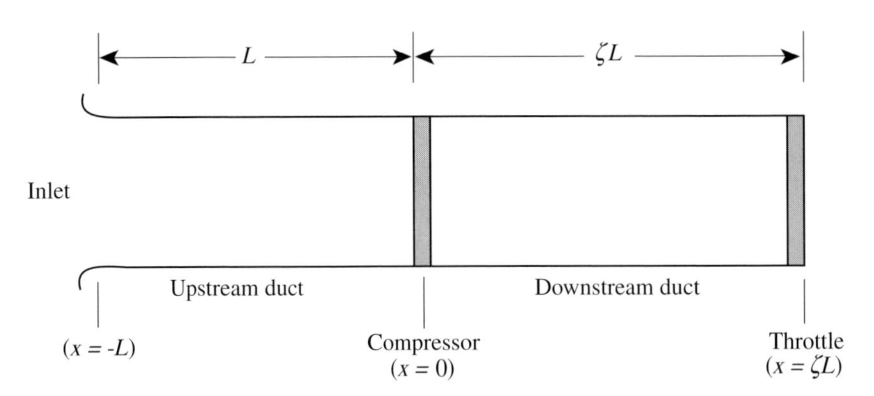

Figure 6.23: One-dimensional model of compression system with distributed inertance and capacitance.

is able to feed mechanical energy to grow the oscillation amplitude. For situations in which the downstream volume is large in a non-dimensional sense (more precisely, if the non-dimensional parameter $B = (\Omega r_m/2a)\sqrt{V/AL}$ is large (Greitzer, 1981)) the criterion for the onset of dynamic instability becomes a statement that instability occurs when the compressor operating point passes the peak of the pressure rise curve. We return to this point in Section 6.6.3.

6.6.2.5 Instability in distributed (non-lumped parameter) fluid systems

The above examples are all in the context of lumped parameter descriptions of a fluid system, but there are many situations in which disturbance spatial structure influences both frequency response and stability. Figure 6.23 illustrates a compressor/throttle combination, in which these two components sit at different stations in a constant area duct. In this situation, as indicated schematically in Figure 6.24, closing the throttle changes the behavior of the system from one similar to an open-duct mode to one that is nearly a closed/open mode.

Analysis of this system can be carried out with four transfer matrices: in the latter there is distributed capacitance and inertance. The representation in terms of transfer matrices is

$$\begin{bmatrix} p' \\ \overline{\rho}\,\overline{a}u' \end{bmatrix}_{x=-L} = \begin{bmatrix} Z_1 & Z_2 \\ Z_2 & Z_1 \end{bmatrix} \begin{bmatrix} 1 & -\Pi \\ 0 & 1 \end{bmatrix} \begin{bmatrix} Z_3 & Z_4 \\ Z_4 & Z_3 \end{bmatrix} \begin{bmatrix} 1 & \Xi \\ 0 & 1 \end{bmatrix} \begin{bmatrix} p' \\ \overline{\rho}\,\overline{a}u' \end{bmatrix}_{x=\zeta L}. \tag{6.6.27}$$

The matrix elements Z_i in the two duct transfer matrices in (6.6.27) correspond to the terms for constant area ducts given in (6.6.9). Applying boundary conditions of no pressure disturbances downstream of the throttle and at the upstream end of the inlet duct leads to an eigenvalue equation:

$$\Xi\,[Z_1Z_3 + Z_4(Z_2 - Z_1\Pi)] + Z_1Z_4 + Z_3(Z_2 - Z_1\Pi) = 0. \tag{6.6.28}$$

Solutions of (6.6.28) for the damping ratio and frequency as functions of acoustic throttle slope are shown in Figure 6.25. The upper part of the figure gives the critical damping ratio (damping/value of damping for which oscillatory motion does not occur) as a function of acoustic throttle slope for three compressor operating points, one on the negatively sloped part of the characteristic, one at the peak pressure rise (zero slope) and one having a positive slope of pressure rise versus flow characteristic. The increase in throttle slope causes a stable operating system to become unstable. The bottom part

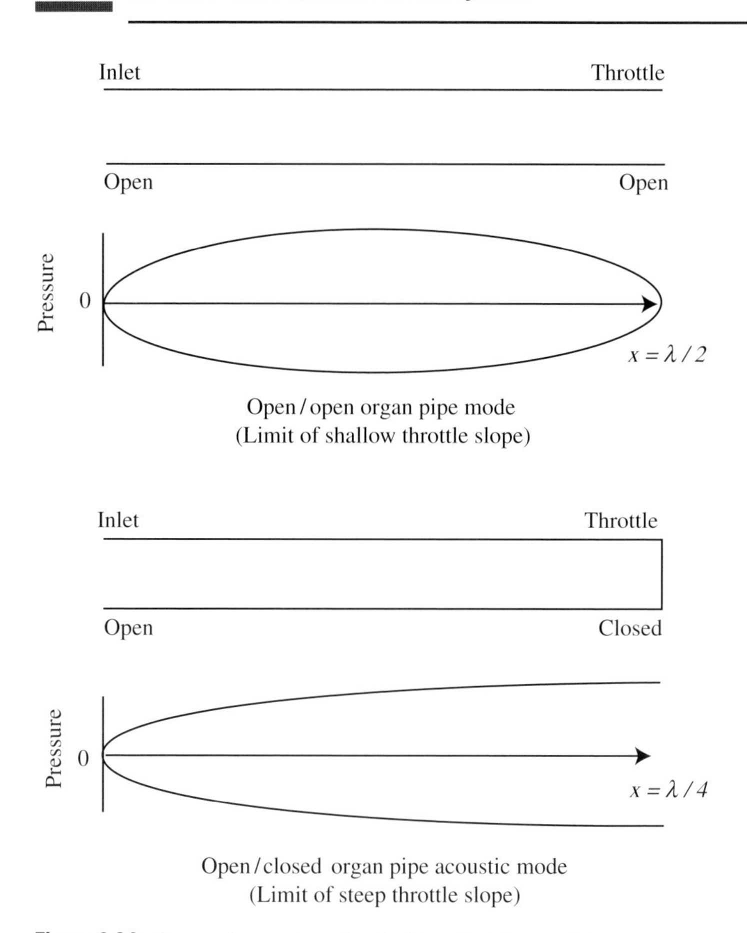

Figure 6.24: Organ pipe analogy for limiting throttle conditions.

of the figure indicates a decrease in frequency of close to a factor of 2, corresponding to the modal behavior evolving from open duct to closed/open.

6.6.3 Nonlinear oscillations in fluid systems

In addition to the identification of conditions for the onset of system instability, behavior subsequent to the onset, such as the amplitude and eventual form of the disturbance, is also of interest. This question is beyond the scope of linear analysis. To answer it we need to address nonlinear oscillations in these non-conservative systems. For nonlinear oscillations the behavior depends on conditions in a possibly large region surrounding the initial operating point, rather than just at the initial operating point as in Section 6.6.2, so the motions have a global, rather than local, character.

The basic compression system model of Section 6.6.2 consisting of the compressor duct, the representation of the compressor by its pumping characteristic, the plenum or collector, and the throttle can again be employed. Now, however, the compressor pressure rise is not linearized about an initial operating point but rather is specified as a nonlinear function of compressor mass flow, $\Psi_C = \Psi_C(\Phi)$. The throttle mass flow and plenum pressure are related by $\Phi_T = \Phi_T(\Psi)$, also nonlinear. The

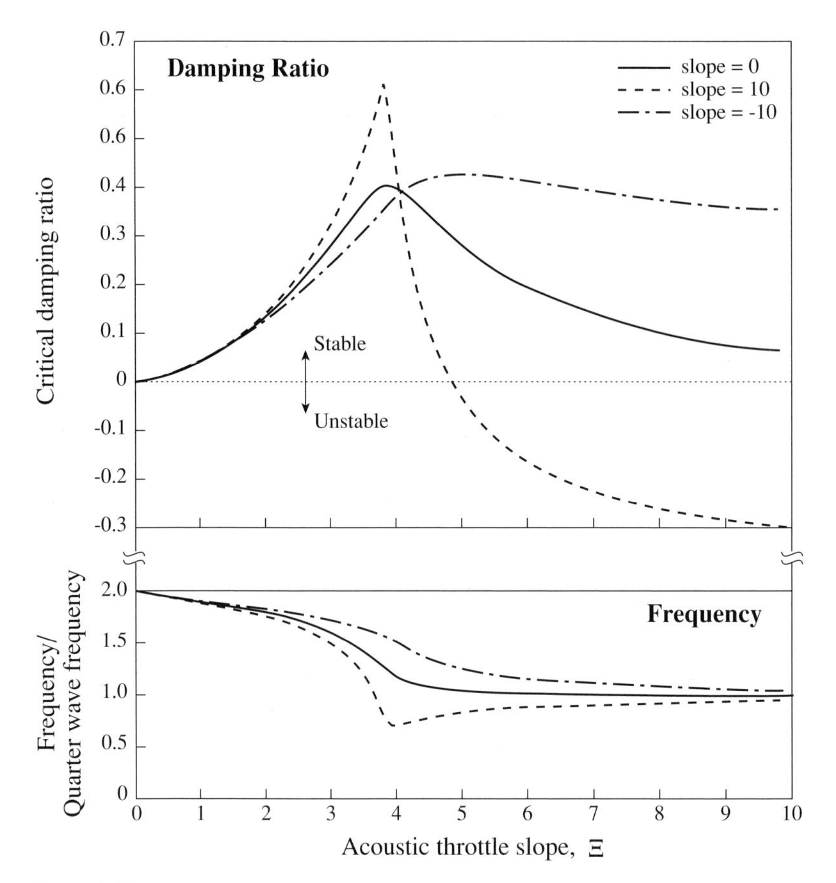

Figure 6.25: One-dimensional modes of a compression system as a function of acoustic throttle slope; results for compressor slopes of −10, 0, 10 (Gysling, 1993).

quantities $\Psi_C(\Phi)$ and $\Phi_T(\Psi)$, the steady-state curves of compressor pressure rise as a function of mass flow and throttle mass flow as a function of plenum pressure, are both applied here in a quasi-steady manner.

Because we are interested in the time domain behavior (rather than the eigenvalues as in Section 6.6.2) it is useful to express the system response in terms of the evolution of appropriate state variables. Using the Helmholtz resonator model of the system described in Section 6.6.2[6] and applying conservation of momentum to the fluid in the compressor duct and conservation of mass to the plenum, the representation of the compression system dynamics in non-dimensional form is:

$$\frac{d\phi}{d\tilde{t}} = B[\Psi_C(\phi) - \psi], \quad \text{momentum equation for the compressor duct,} \tag{6.6.29}$$

$$\frac{d\psi}{d\tilde{t}} = \frac{1}{B}[\phi - \Phi_T(\psi)], \quad \text{conservation of mass in plenum.} \tag{6.6.30}$$

[6] This implies that the system pressure rise is much less than the ambient level so that in the plenum $p' = (\gamma\overline{p}/\overline{\rho})\rho'$, with $\overline{p}$ and $\overline{\rho}$ the mean values, is still a good approximation.

The two system state variables are instantaneous (non-dimensional) compressor mass flow, ϕ, and plenum pressure, ψ. The non-dimensional time variable in (6.6.29) and (6.6.30) is $\tilde{t} = \omega_H t$, where ω_H is the Helmholtz resonator frequency. The other quantities are defined in Section 6.6.2. (There should be no confusion with the velocity potential φ used in Sections 6.3 and 6.4 or the streamfunction ψ of Section 6.5.)

6.6.3.1 Limit cycle oscillations

Numerical solutions of (6.6.29) and (6.6.30) are available elsewhere (Cumpsty, 1989; Fink, Cumpsty and Greitzer, 1992) and we concentrate here on the qualitative features of the oscillation which can be discussed with reference to the mechanical energy input over different parts of the cycle. The dynamic system represented by (6.6.29) and (6.6.30) exhibits a widely encountered behavior known as limit cycle oscillations. Limit cycles are an inherently nonlinear motion of non-conservative systems in which energy is fed into the oscillation over part of the cycle and extracted over the rest, with the amplitude of the resulting motion determined by the balance between energy input and dissipation (Ogata, 1997; Strogatz, 1994). To derive conditions under which periodic motions exist we thus examine a quadratic quantity analogous to mechanical energy[7] and determine under what conditions periodic motion (rather than growth or decay) will occur.

In the discussion it is convenient to transform the state equations to a coordinate system in which the origin is at the system initial operating point (ϕ_0, ψ_0). This is the intersection of the steady-state throttle and compressor characteristics and is an equilibrium point for the system. The transformation is implemented through the substitutions:

$$\hat{\phi} = \phi - \phi_0, \tag{6.6.31a}$$

$$\hat{\psi} = \psi - \psi_0, \tag{6.6.31b}$$

$$\hat{\Psi} = \Psi_C(\hat{\phi} + \phi_0) - \Psi_C(\phi_0), \tag{6.6.31c}$$

$$\hat{\Phi}_T = \Phi_T(\hat{\psi} + \psi_0) - \Phi_T(\psi_0). \tag{6.6.31d}$$

The resulting representation of the transformed origin is shown in Figure 6.26, where the compressor and throttle curves are also indicated. The operating point shown is on the positively sloped part of the compressor characteristic, where, from the arguments in Section 6.6.2, linear instability might be expected.

The solution behavior can be given in terms of ϕ and ψ as functions of time, but it is often more useful to plot the solution trajectory in the ϕ–ψ plane with time as a parameter. A sketch of such a

[7] The concept can be readily illustrated for a second order differential equation corresponding to a mass-spring damping system with a nonlinear frictional force of the form $\varepsilon(x^2 - 1)\dot{x}$, where $\dot{x}$ denotes dx/dt. The non-dimensional differential equation for the system is $\ddot{x} + \varepsilon(x^2 - 1)\dot{x} + x = 0$, known as the Van der Pol oscillator (Stoker, 1950; Morse and Ingard, 1968; Strogatz, 1994). Multiplying the differential equation by $\dot{x}$ and integrating over a cycle leads to an expression for the change in mechanical energy $\Delta(\dot{x}^2 + x^2)/2$ over the cycle. This increases, decreases, or remains constant depending on whether the integral $\int(1 - x^2)\dot{x}^2 dt$ is positive, negative, or zero. For small amplitude oscillations (amplitude < 1) the integral, which represents the mechanical power input associated with the damping force, is positive. For larger amplitudes (amplitude > 1), however, the power input is negative. The eventual amplitude of the motion is set when the oscillation grows to a level at which the integral is zero and the power input over one part of a cycle is balanced by the dissipation over the rest of the cycle.

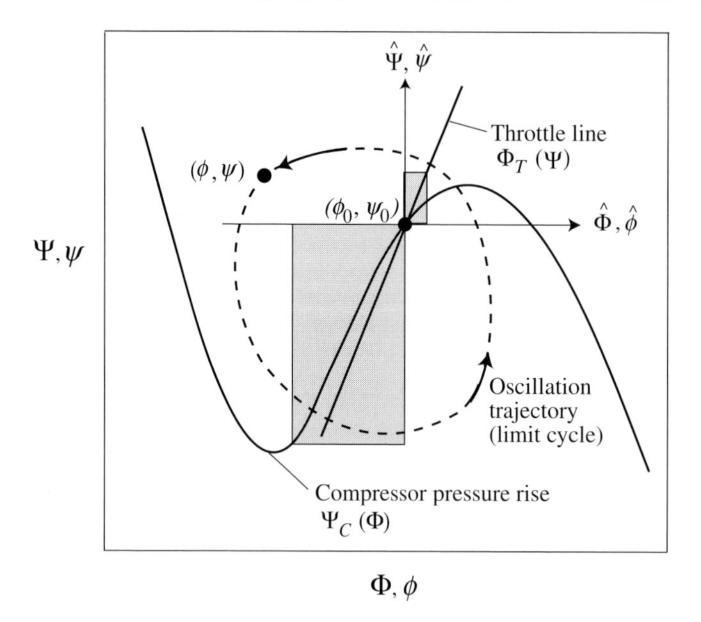

Figure 6.26: Representation of the system characteristic with the origin at the system initial operating point (ϕ_0, ψ_0).

trajectory is shown as the dashed line in the figure. We employ this description of the motion in the discussion that follows.

6.6.3.2 Liapunov function description of nonlinear fluid system oscillations

A general approach for determining the overall behavior of nonlinear oscillations in a given system is to examine an energy-like function, or "Liapunov function" (Ogata, 1997; Strogatz, 1994), and establish under what conditions the energy-like quantity grows or decays. Although this method does not provide details of the trajectory, it allows assessment of stability and information about the existence and qualitative character of limit cycles. The choice of Liapunov function, denoted here by V, is not unique, but an appropriate candidate for the compression system is (Simon and Valavani, 1991)

$$V(\hat{\phi}, \hat{\psi}) = \frac{1}{2}\left(\frac{1}{B}\hat{\phi}^2 + B\hat{\psi}^2\right). \tag{6.6.32}$$

The first term on the right-hand side of (6.6.32) can be viewed as representing the incremental kinetic energy of the gas in the compressor duct and the second the incremental potential energy stored through compression of the gas in the plenum. Curves of constant V are nested ellipses around the origin of the new coordinates, with increasing V corresponding to increasing energy in the motion. The shapes of the ellipses are dependent on the B-parameter with larger values of B leading to elongation of the ellipses in the horizontal direction (larger mass flow fluctuations).

Differentiating (6.6.32) with respect to time, and substituting the system equations in the resulting expression (this amounts to taking the derivative along a trajectory) gives

$$\frac{dV}{d\tilde{t}} = \hat{\phi} \cdot \hat{\Psi}_C(\hat{\phi}) - \hat{\phi}_T(\hat{\psi}) \cdot \hat{\psi}. \tag{6.6.33a}$$

If (6.6.33a) is integrated over a time interval $\Delta t = t_{final} - t_{initial}$, the change in V is found as

$$\Delta V|_{initial}^{final} = \int [\hat{\phi} \cdot \hat{\Psi}_C(\hat{\phi}) - \hat{\phi}_T(\hat{\psi}) \cdot \hat{\psi}]d\tilde{t}. \tag{6.6.33b}$$

The change in the energy-like quantity, V, thus depends on the instantaneous mass flow and pressure rise and the shape of the resistive-like elements (the curves of compressor and throttle pressure change versus mass flow). For a limit cycle ΔV over a period of the cycle must be 0.

The two terms on the right-hand side of (6.6.33) are products of pressure and mass flow and thus power-like. The first, which can be interpreted as incremental mechanical power production due to the oscillating flow through the compressor, is positive or negative depending on the operating point and amplitude. The second can be interpreted as incremental mechanical power dissipation in the throttle due to the oscillations and is always a positive quantity.

For small amplitude motions around operating points on the positively sloped part of the compressor characteristic the power production term is positive, as discussed in Section 6.6.2. This situation is sketched in Figure 6.26, where the shaded rectangles represent the values of the two terms in the quantity $dV/d\tilde{t}$ at one particular point on a cycle. For large enough amplitude oscillations it can be seen that there are times during the cycle, when the flow is either large and positive or large and negative, that the compressor acts to dissipate the mechanical energy associated with the oscillation. At such conditions the product $\hat{\phi} \cdot \hat{\Psi}_C(\hat{\phi})$, which represents an energy source, is negative. The amplitude of the limit cycle (although the term "amplitude" is used, oscillations associated with nonlinear systems are non-sinusoidal) is set by the balance between power production and dissipation in both compressor and throttle.

The B-parameter does not occur explicitly in (6.6.33) but it has a role through the dynamic equations in determining the relation between the compressor and throttle mass flow excursions and thus the relative sizes of the power production and dissipation terms. Larger values of B mean larger compressor mass flow variations for a given throttle mass flow fluctuation, and hence a trend towards more vigorous oscillatory motion.

An example of a limit cycle oscillation is given in Figure 6.27 which shows the measured and computed transient behavior of a compression system with a centrifugal turbocharger (Hansen, Jorgensen, and Larsen, 1981). The axes are non-dimensional mass flow and pressure rise. The solid lines are the measured compressor pressure rise curve and the throttle line, which have similar shapes to those sketched in Figure 6.26. The solid points are the measurements, and the dashed line is the computed trajectory given by a model which is a slightly extended version of that described by (6.6.29) and (6.6.30). In line with the arguments presented, the compressor characteristic is such as to make $dV/d\tilde{t}$ positive for the region near the initial operating point, and negative at values of mass flow away from this region (i.e. at large positive, or negative, flows). The measurements of mass flow, made with a hot wire, have some scatter especially in the reverse flow region, but the limit cycle nature of the oscillation, which is known as compressor surge, is evident. Further

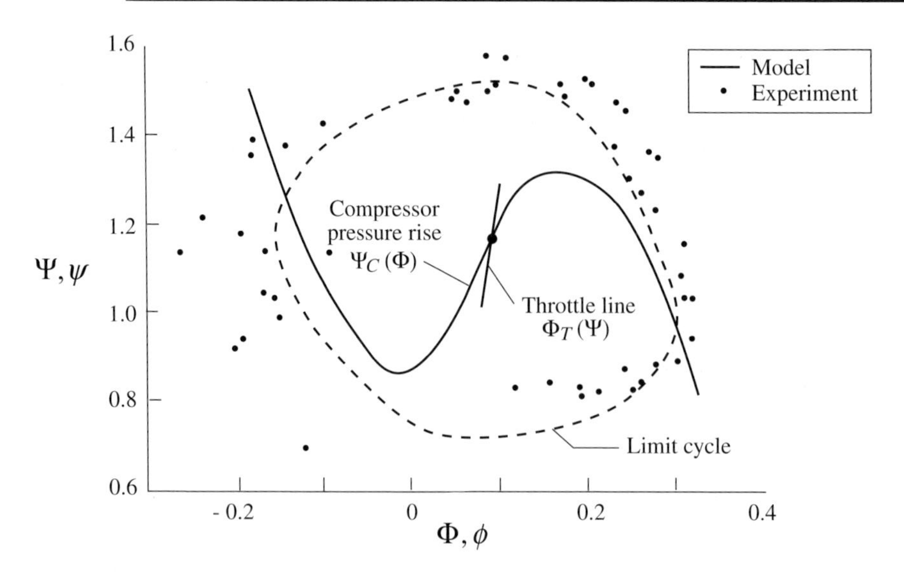

Figure 6.27: Surge limit cycle in a centrifugal compression system, $B = 0.55$ (Hansen *et al.*, 1981).

discussion of surge is given by Cumpsty (1989), Stetson (1984), Greitzer (1981), and Fink *et al.* (1992).

6.6.3.3 An energy approach to instability onset

Finally, we can connect the approach based on energy considerations to the discussions of instability onset in Section 6.6.2.2. For assessment of instability onset it is sufficient to consider *small perturbations* in mass flow and plenum pressure, ϕ' and ψ'. Equations (6.6.29) and (6.6.30) thus take the *linearized* form (again with $\tilde{t} = \omega_H t$)

$$\frac{d\phi'}{d\tilde{t}} = B\left[\frac{d\Psi_C}{d\Phi}\phi' - \psi'\right], \tag{6.6.34}$$

$$\frac{d\psi'}{d\tilde{t}} = \frac{1}{B}\left[\phi' - \left(\frac{d\Phi_T}{d\Psi}\right)\psi'\right] = \frac{1}{B}\left[\phi' - \frac{\psi'}{\left(\dfrac{d\Psi_T}{d\Phi}\right)}\right]. \tag{6.6.35}$$

In (6.6.34) and (6.6.35) the derivatives of the pressure versus mass flow characteristics are evaluated at the equilibrium point. Equations (6.6.34) and (6.6.35) can be combined into a single equation for ϕ' or ψ', which is (6.6.21) in another guise:

$$\frac{\partial^2\psi'}{\partial\tilde{t}^2} + \left[\frac{1}{B\left(\dfrac{d\Psi_T}{d\Phi}\right)} - B\left(\frac{d\Psi_C}{d\Phi}\right)\right]\frac{\partial\psi'}{\partial\tilde{t}} + \left[1 - \frac{\left(\dfrac{d\Psi_C}{d\Phi}\right)}{\left(\dfrac{d\Psi_T}{d\Phi}\right)}\right]\psi' = 0. \tag{6.6.36}$$

From (6.6.36) the requirement for stability to small disturbances (i.e. the requirement that all perturbations of the form $e^{s\tilde{t}}$ have a negative real part) is that both quantities in square brackets are positive.

The first of these is a resistance-like term. The condition $(1 - B^2(d\Psi_C/d\Phi)(d\Psi_T/d\Phi) = 0)$ marks the point at which system damping goes from positive to negative. Larger values of B, more positive compressor slopes, and steeper throttle lines all promote instability. From the form of the term it can be seen that for either very steep throttle lines, $(d\Psi_T/d\Phi) \to \infty$, or very large B, instability occurs at the peak of the compressor characteristic, $(d\Psi_C/d\Phi) = 0$.

Examination from an energy perspective using the Liapunov function gives further insight into this behavior. For the linearized system the quantity dV/dt is

$$\frac{dV}{d\tilde{t}} = \left(\frac{d\Psi_C}{d\Phi}\phi'\right)\phi' - \left(\frac{\psi'}{\dfrac{d\Psi_T}{d\Phi}}\right)\psi' = \left(\frac{d\Psi_C}{d\Phi}\right)(\phi')^2 - \frac{(\psi')^2}{\left(\dfrac{d\Psi_T}{d\Phi}\right)}. \tag{6.6.37}$$

Integrating (6.6.37) over a cycle yields

$$\Delta V_{cycle} = \frac{d\Psi_C}{d\Phi}\langle(\phi')^2\rangle - \frac{\langle(\psi')^2\rangle}{\dfrac{d\Psi_T}{d\Phi}}. \tag{6.6.38}$$

The quantities $\langle(\phi')^2\rangle$ and $\langle(\psi')^2\rangle$ are the mean square values of perturbations in compressor mass flow and plenum pressure over the cycle and are positive definite. The value of ΔV_{cycle} depends on the ratios of these quantities and the slopes of the compressor and throttle characteristic curves. Equation (6.6.38) is analogous to a net mechanical energy input to the oscillations and extends the qualitative arguments of Section 6.6.2 to include dissipation in the throttle.

Substituting the values of $\langle(\phi')^2\rangle$ and $\langle(\psi')^2\rangle$ from solution of (6.6.34) and (6.6.35) and the condition $B^2(d\Psi_C/d\Phi)(d\Psi_T/d\Phi) = 1$ (which holds at the stability boundary) into (6.6.38) reveals that the condition $\Delta V_{cycle} = 0$ corresponds to instability onset. For a given compressor operating condition, (6.6.38) implies that as the throttle line is steepened the dissipation in the system associated with the perturbations decreases relative to the energy production, and the tendency towards instability is increased. For the infinitely steep (vertical) throttle line, there is no dissipation in the throttle (because the mass flow perturbations in the throttle are zero) so any positive slope of the compressor characteristic is enough to cause instability. Because throttle slopes are generally steep, operation on the positive slope is to be avoided for compressors and pumps. Equation (6.6.36) and the subsequent discussion also highlight the point that dynamic instability associated with negative damping, rather than static instability, is the more severe problem in practice.

6.7 Multi-dimensional unsteady disturbances in a compressible inviscid flow

We now describe the general unsteady small disturbances which can exist in an inviscid compressible flow. The velocity and thermodynamic quantities are once again decomposed into a steady, uniform part, denoted by $\overline{\mathbf{u}}, \overline{p}, \overline{\rho}$, etc, and a small disturbance denoted by $\mathbf{u}', p', \rho'$. The latter have amplitudes such that terms involving products of disturbance quantities can be neglected and a linearized version of the equations of motion serves to describe the behavior of the disturbances. The disturbance equations are thus:

$$\frac{\partial \rho'}{\partial t} + \overline{\mathbf{u}} \cdot \nabla \rho' + \overline{\rho} \nabla \cdot \mathbf{u}' = 0, \tag{6.7.1a}$$

$$\frac{\partial \mathbf{u}'}{\partial t} + \overline{\mathbf{u}} \cdot \nabla \mathbf{u}' + \frac{1}{\overline{\rho}} \nabla p' = 0, \tag{6.7.1b}$$

$$\frac{\partial s'}{\partial t} + \overline{\mathbf{u}} \cdot \nabla s' = 0. \tag{6.7.1c}$$

Equations (6.7.1) are supplemented by the linearized form of the equation of state for a perfect gas with constant specific heats.

Because (6.7.1) are linear, a general solution can be constructed by superposition of particular solutions. We exploit this fact, choosing solutions which each emphasize one particular aspect of the properties of the general solutions.

We start by taking the curl of (6.7.1b) to obtain, using $D(\ )/Dt = (\partial/\partial t + \overline{\mathbf{u}} \cdot \nabla)(\,)$,

$$\frac{D\omega'}{Dt} = 0. \tag{6.7.2}$$

Equation (6.7.2) states that vorticity disturbances are convected without alteration by the uniform background flow. We can thus consider solutions to (6.7.2) which have constant density and which have the velocity field associated with ω' also convected unchanged by the background flow. No acceleration of a fluid particle is associated with these rotational disturbances, and there are correspondingly no pressure perturbations.

Equation (6.7.1c) describes the behavior of entropy variations. Solutions to (6.7.1c) have variations in density but no associated variations in pressure and satisfy

$$\frac{Ds'}{Dt} = \frac{D\rho'_{ent}}{Dt} = 0, \tag{6.7.3}$$

where ρ'_{ent} are the density fluctuations associated with entropy non-uniformities. The entropic density disturbances, like the vorticity disturbances, are convected unchanged by the background flow, and there are no variations in velocity associated with ρ'_{ent}.

The two types of perturbation discussed satisfy requirements for small disturbances of vorticity and entropy. To obtain a complete solution to (6.7.1), we now seek disturbances which are irrotational and which have uniform entropy so that

$$\mathbf{u}'_{irrot} = \nabla \varphi \quad \text{and} \quad s'_{irrot} = 0, \tag{6.7.4}$$

where the subscript "*irrot*" signifies that these disturbances are irrotational. For these disturbances (6.7.1a) and (6.7.1b) can be written as

$$\frac{1}{\overline{\rho}} \frac{D\rho'_{irrot}}{Dt} + \nabla^2 \varphi = 0 \tag{6.7.5a}$$

and

$$\frac{D}{Dt} \nabla \varphi + \frac{1}{\overline{\rho}} \nabla p'_{irrot} = 0. \tag{6.7.5b}$$

Eliminating ρ' and ρ' from (6.7.5) yields an equation for the disturbance velocity potential (or, equivalently, for the static pressure disturbances) as

$$\nabla^2\varphi - \frac{1}{\bar{a}^2}\frac{D^2\varphi}{Dt^2} = 0. \tag{6.7.6}$$

Disturbances in velocity potential are propagated at the local speed of sound *relative to the background flow*. These irrotational (or acoustic) disturbances have an associated static pressure variation which also propagates at the local speed of sound relative to the background flow.

To review, there are three types of small amplitude disturbances which can be superposed on a uniform, steady, compressible background flow: an irrotational velocity perturbation, which carries the static pressure information, a vorticity perturbation (or equivalently a rotational velocity perturbation), and an entropy perturbation. Any solution to (6.7.1) can be written as a combination of these as

$$\mathbf{u}' = \mathbf{u}'_{rot} + \nabla\varphi, \tag{6.7.7a}$$

$$s' = -c_p\frac{\rho'_{ent}}{\bar{\rho}}, \tag{6.7.7b}$$

which have the three independent disturbances. Other disturbance quantities such as

$$\rho' = \rho'_{irrot} + \rho'_{ent}, \tag{6.7.8a}$$

$$p' = p'_{irrot}, \tag{6.7.8b}$$

can then be derived from (6.7.7).

With a uniform background flow, the three types of disturbance do not interact. The irrotational velocity disturbances propagate at the speed of sound relative to the background flow, while the rotational velocity disturbance and the entropy disturbance are convected without change at the velocity of the background flow. Coupling between disturbances arises, as shown below, either through boundary conditions or when the background flow is non-uniform.

If compressibility effects are negligible, a simpler form of the equations is obtained. In this case, all the variation in density must come from the entropy perturbations (from local heating or cooling). The form of (6.7.7) for situations in which compressibility is not important is

$$\frac{D\rho'}{Dt} = 0. \tag{6.7.9}$$

The equation for the velocity potential under these conditions is Laplace's equation:

$$\nabla^2\phi = 0. \tag{6.7.10}$$

Equations (6.7.9) and (6.7.10), plus (6.7.2) which is unaltered, describe the behavior of small disturbances to a uniform flow in an incompressible, non-uniform density, fluid. If the density is constant, only (6.7.10) and (6.7.2) are needed.

6.8 Examples of fluid component response to unsteady disturbances

The flow disturbances described are independent if the background flow is uniform[8] which, for an internal flow, can only occur in a uniform duct. Disturbance interaction (or coupling) is therefore associated with boundary conditions including variations in geometry along the flow direction or the presence of a screen or device such as a turbomachine. Convection of a vorticity perturbation into a screen or turbomachine, for instance, generally results in the modification of the original disturbance, the creation of pressure disturbances (both upstream and downstream), and the creation of entropy disturbances on the downstream side of the device.

In the following sections we present examples of the behavior of unsteady small amplitude disturbances in a compressible flow. Two main aspects are illustrated. First is the coupling of disturbances, shown for a nozzle and a turbomachine blade row. Second is the change in component behavior, in other words the dynamic response of the device, as the reduced frequency varies. This topic was introduced in Section 6.3 and the present section builds on the concepts developed there.

6.8.1 Interaction of entropy and pressure disturbances

6.8.1.1 Density waves in an incompressible flow

We begin with one-dimensional flows in which the only disturbances are entropy and pressure. The results to be expected can be motivated in a qualitative manner through the model problem of constant velocity convection of an incompressible, non-uniform density fluid through a nozzle. As sketched in Figure 6.28 the density variation we impose has a wavelength in the flow direction which is long compared to the nozzle length. The reduced frequency of the unsteady flow in the nozzle is therefore much less than unity and the nozzle response quasi-steady. The pressure difference across the nozzle is thus given by $p_i - p_e = \frac{1}{2}\rho\overline{u}^2[(1/AR^2) - 1]$, where p_i and p_e are the values just upstream and downstream of the nozzle and $\rho = \overline{\rho} + [\rho'(x - \overline{u}t)]$. If p_e is constant, as would be the case if the nozzle discharged to a large reservoir, there is a pressure fluctuation upstream of the nozzle:

$$p_i' = \frac{1}{2}(\rho')\overline{u}^2\left(\frac{1}{AR^2} - 1\right). \tag{6.8.1}$$

Equation (6.8.1) illustrates density and pressure disturbance coupling. Pressure disturbances from this mechanism are important for density wave generation in two-phase flow (Greitzer, 1981).

6.8.1.2 Passage of an entropy disturbance through a choked nozzle

A compressible flow example concerns the pressure disturbances due to the passage of an entropy variation through a choked nozzle with supersonic exit flow. The geometry is similar to that shown in Figure 6.28, but the flow in the duct now has a non-zero Mach number. We describe first the behavior

[8] This is not the case with a non-uniform background flow on which the disturbances are superposed, even for small amplitude perturbations. An example of this is the parallel shear flow discussed in Section 6.5, another is the presence of mean swirl addressed in Chapter 12.

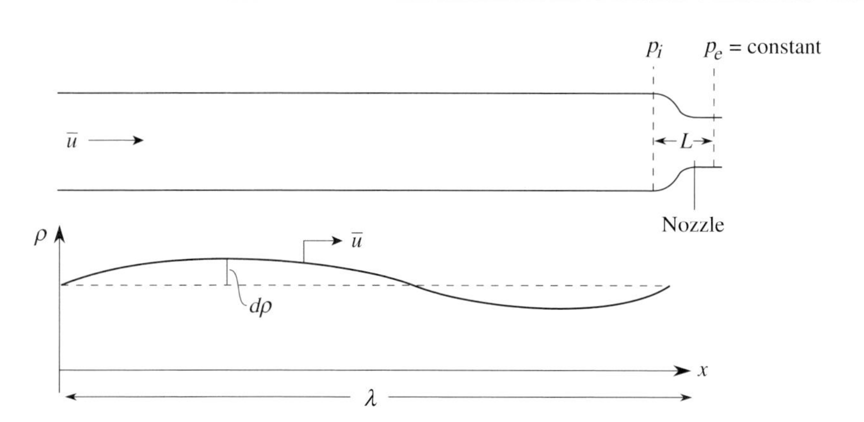

Figure 6.28: Pressure fluctuations at a nozzle inlet due to the passage of a convected density wave through the nozzle; constant upstream velocity $\overline{u}$, $\rho = \overline{\rho} + [\rho'(x - \overline{u}t)]$.

when the nozzle is short enough that the response is quasi-steady and then consider the effect of finite reduced frequency.

If the nozzle length is such that $(\omega L/u)^2 \ll 1$, the flow within the nozzle can be modeled as quasi-steady, with no mass storage within the nozzle and stagnation enthalpy the same at the inlet and exit. The nozzle geometry is represented by inlet–outlet matching conditions derived from the steady-flow performance of the device. The corrected flow per unit area (see Section 2.5) into the nozzle is a function of the inlet Mach number M_i, denoted as $D(M_i)$

$$\frac{\dot{m}\sqrt{RT_t}}{Ap_t\sqrt{\gamma}} = D(M_i)\,. \tag{2.5.3}$$

Using $\dot{m}/A = \rho u$ and the condition that for a choked nozzle $D(M_i)$ is constant, we obtain

$$\frac{\rho'}{\overline{\rho}} + \frac{u'}{\overline{u}} + \frac{T'}{2\overline{T}} - \frac{p'}{\overline{p}} = 0. \tag{6.8.2}$$

All the quantities in (6.8.2) can be separated into irrotational (or acoustic) and entropic disturbances. For perturbations with frequency ω, the former are of the form

$$\frac{\rho'_{irrot}}{\overline{\rho}},\ \frac{T'_{irrot}}{\overline{T}},\ \frac{p'}{\overline{p}},\ \frac{u'}{\overline{u}} \propto e^{i\omega[t - x/(\overline{u} \pm \overline{a})]}. \tag{6.8.3}$$

The latter are

$$\frac{\rho'_{ent}}{\overline{\rho}},\ \frac{T'_{ent}}{\overline{T}} \propto e^{i\omega[t - x/\overline{u}]}. \tag{6.8.4}$$

From the momentum equation, the relation between the velocity and pressure disturbances is

$$\frac{u'}{\overline{u}} = -\frac{1}{\gamma\overline{M}}\frac{p'_{irrot}}{\overline{p}}. \tag{6.8.5}$$

For the entropy disturbances,

$$\frac{s'}{c_p} = \frac{T'_{ent}}{\overline{T}} \quad \text{(since } p'_{ent} \equiv 0\text{)}. \tag{6.8.6}$$

Substituting these expressions for disturbances into (6.8.2) shows that a convected entropy disturbance into a choked nozzle results in upstream propagating pressure waves from the nozzle with strength

$$\frac{p'}{\overline{p}} = \left\{ \frac{-\gamma \dfrac{\overline{M}_i}{2}}{1 + \frac{1}{2}(\gamma - 1)\,\overline{M}_i} \right\} \frac{s'}{c_p}, \quad \text{upstream waves.} \tag{6.8.7}$$

The entropy disturbance also causes pressure waves at the nozzle exit. In a coordinate system moving with the flow these disturbances propagate upstream and downstream with the speed of sound, $\overline{a}$, but in the absolute (nozzle fixed) reference frame the disturbances move downstream (since $\overline{u} > \overline{a}$) and have the form

$$\frac{p'_+}{\overline{p}} = A e^{i\omega[t - x/(\overline{u}+\overline{a})]}, \tag{6.8.8a}$$

$$\frac{p'_-}{\overline{p}} = B e^{i\omega[t - x/(\overline{u}-\overline{a})]}. \tag{6.8.8b}$$

where A and B are constants.

The velocity disturbance waves can be directly related to the pressure disturbances in the two directions since each wave is independent. In other words, to have matching spatial and temporal behavior, the upstream moving pressure disturbances must be linked to upstream moving velocity disturbances only and similarly for the downstream waves. Substitution in the momentum equation yields the form of the velocity disturbances downstream of the nozzle:

$$\frac{u'_+}{\overline{u}} = A \frac{1}{\gamma \overline{M}} e^{i\omega[t - x/(\overline{u}+\overline{a})]}, \tag{6.8.9a}$$

$$\frac{u'_-}{\overline{u}} = -B \frac{1}{\gamma \overline{M}} e^{i\omega[t - x/(\overline{u}-\overline{a})]}. \tag{6.8.9b}$$

The result in (6.8.9) can be used to derive expressions for the nozzle exit pressure perturbations generated by an entropy perturbation convected into the nozzle inlet:

$$\frac{p'_+}{\overline{p}} = \frac{\gamma(\overline{M}_e - \overline{M}_i)}{4} \left[\frac{\dfrac{s'}{c_p}}{1 + \dfrac{(\gamma - 1)}{2}\overline{M}_i} \right], \tag{6.8.10a}$$

$$\frac{p'_-}{\overline{p}} = -\frac{\gamma(\overline{M}_e + \overline{M}_i)}{4} \left[\frac{\dfrac{s'}{c_p}}{1 + \dfrac{(\gamma - 1)}{2}\overline{M}_i} \right]. \tag{6.8.10b}$$

All the preceding results refer to the situation in which the nozzle length is very short compared to disturbance wavelength, i.e. to the low reduced frequency limit. We now wish to assess the effect of reduced frequency on unsteady nozzle response. The nozzle geometry must be specified to carry out the calculations, and the example chosen has a linearly varying velocity with the Mach number

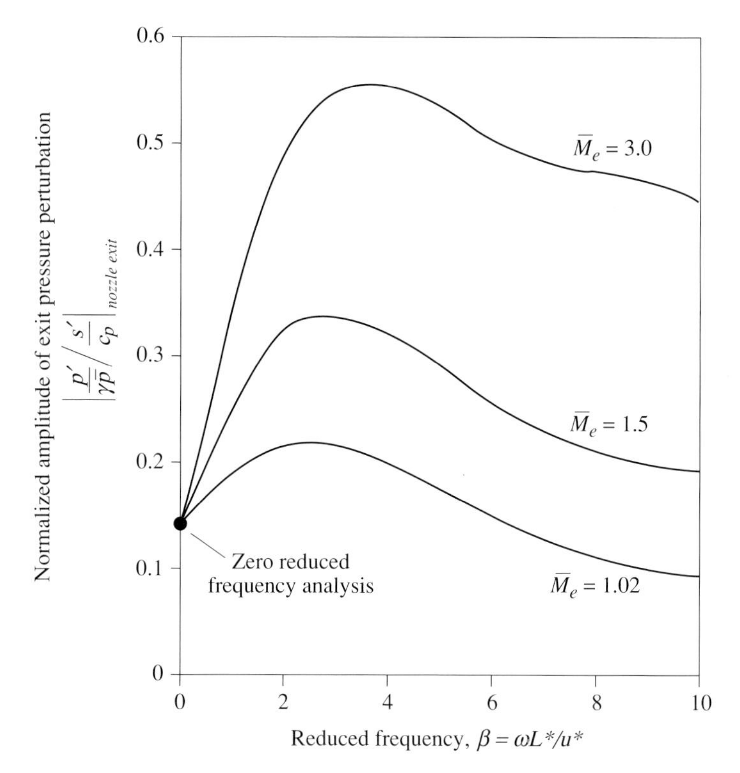

Figure 6.29: Dependence of the nozzle exit pressure amplitude on reduced frequency for entropy perturbations in the nozzle; nozzle inlet Mach number = 0.29 (u^* is the sonic velocity at the throat, L^* is the distance from the nozzle inlet to the throat) (Marble and Candel, 1977).

subsonic at the inlet and supersonic at exit. The reference velocity used in the definition of reduced frequency is the sonic speed at the throat, u^*. The reference length is the distance from the nozzle inlet to the throat, L^*: $\beta = \omega L^*/u^*$.

Figure 6.29 shows the normalized amplitude of the pressure disturbance at the nozzle exit as a function of reduced frequency. The curves are for an upstream Mach number of 0.29 and three exit Mach numbers. The short nozzle (or long wavelength) limiting case results correspond to zero reduced frequency and are independent of exit Mach number.

The magnitude of the pressure amplitude at the nozzle exit exhibits an initial rise with reduced frequency then a fall-off. Examination of the amplitude and phase relationships of the p'_+ and p'_- pressure waves shows that this behavior is associated with the phasing of these two waves. At low reduced frequency the magnitude of each individual wave is large, but the waves are 180° out of phase at the nozzle exit and their combination has a small resultant. As the reduced frequency increases, the magnitude of the exit pressure waves decreases, but the angle between them shifts so their resultant is larger than for zero reduced frequency. Figure 6.30 gives a phase diagram of the composition of the exit pressure fluctuation at an exit Mach number $\overline{M}_e = 3$ to illustrate this relationship.

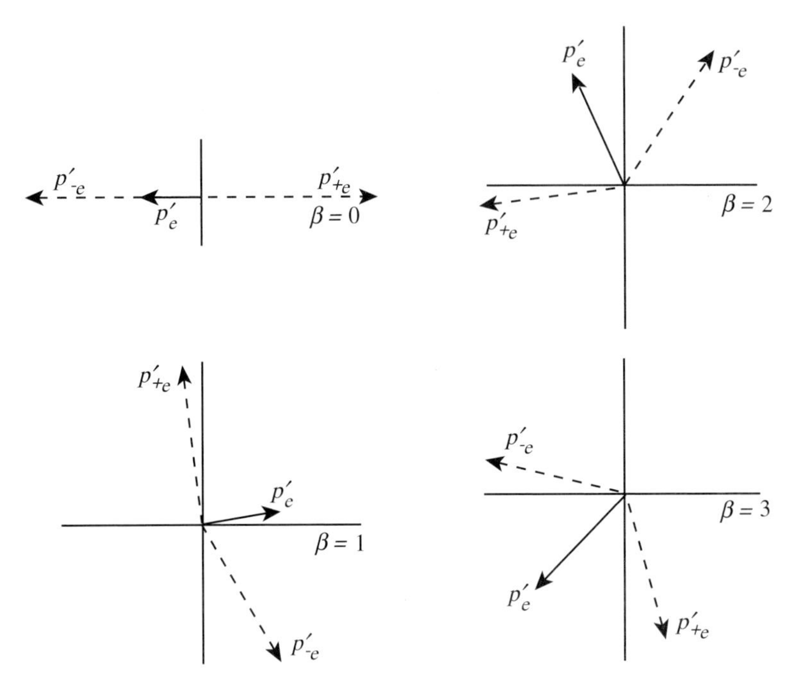

Figure 6.30: Phase diagrams showing the composition of p'_{+e} and p'_{-e} waves to form pressure fluctuation p'_e at the nozzle exit; $\overline{M}_i = 0.29$, $\overline{M}_e = 3.0$; reduced frequencies of 0, 1, 2, 3 (Marble and Candel, 1977).

6.8.2 Interaction of vorticity and pressure disturbances

Although situations with three different types of disturbances can readily be addressed, the features of disturbance coupling are seen more clearly when only two types interact. The next example thus concerns coupling of vorticity and pressure disturbances. Two problems are discussed related to small amplitude disturbances incident on a two-dimensional cascade (blade row) of flat plate airfoils in a subsonic flow. The first is a vorticity (rotational velocity) disturbance and the second is a pressure disturbance from downstream which propagates upstream into the cascade.

6.8.2.1 A vorticity disturbance entering a blade row in an incompressible flow

The geometry for this example is shown in Figure 6.31. There is no time mean aerodynamic loading, hence no time mean change of flow direction across the flat plate cascade. The velocity field associated with the rotational disturbance, which is convected from far upstream to the cascade, has the form

$$u'_{x_{rot}} = 0, \quad u'_{y_{rot}} = u'_{y_0}\, e^{i\omega[t-(x/\overline{u}_x)]}, \tag{6.8.11}$$

where $\overline{u}_x$ is the x-component of the background velocity. No pressure disturbances are associated with this incoming velocity field which is a pure shear disturbance. To restrict discussion to pressure and vorticity disturbances the flow through the blades is taken as lossless and the entropy uniform throughout.

To show the overall features of the disturbance field in a simple manner we initially take the blade chord length, b, such that the reduced frequency, $\omega b/\overline{u}$, is much less than unity, returning later to

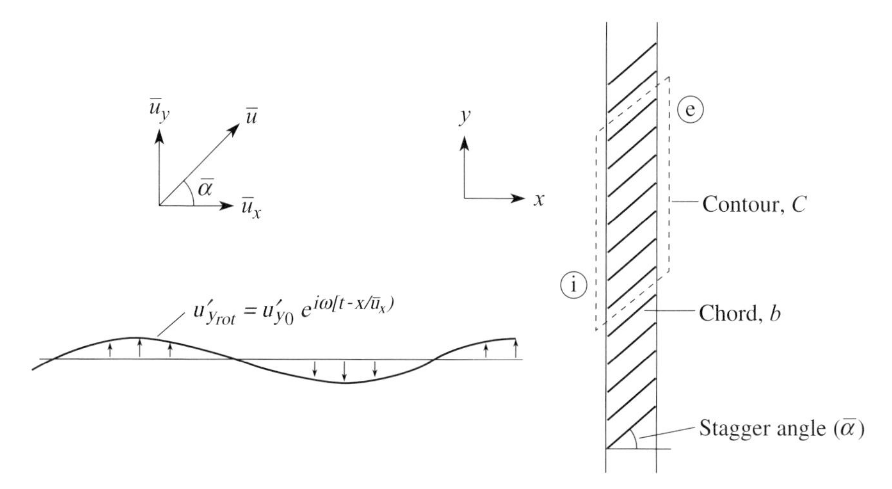

Figure 6.31: Vorticity disturbance incident on a two-dimensional cascade of flat plate airfoils.

examine the effect of reduced frequency on cascade response. For $\omega b/\overline{u}$ much less than unity the cascade can be described as quasi-steadily responding to the local instantaneous conditions. This approximation for blade row response is known as the *actuator disk* representation. We also assume the blades are closely spaced (small circumferential spacing/blade chord) so the exit flow is well guided and the angle at the exit of the cascade, α_e, is constant and equal to the stagger angle (the angle between the chord-line and the axial direction). This is the time mean flow angle throughout $(\overline{\alpha}_i = \overline{\alpha}_e = \overline{\alpha})$.

Before looking at specific numerical results, some features can be extracted from consideration of the incompressible flow case. From the continuity equation, the form of the imposed velocity disturbance, and the fact that any irrotational velocity disturbance must have the same argument, one can infer that

$$\frac{\partial u'_x}{\partial x} = 0. \tag{6.8.12}$$

From (6.8.11) the axial velocity disturbance is thus zero throughout the flow field.

The incoming vorticity disturbance corresponds to a cascade airfoil angle of incidence fluctuation of $\alpha'_i = (\cos^2 \overline{\alpha}) u'_{y_{rot}}/\overline{u}_x$ and a variation of $\rho \overline{u}_y u'_{y_{rot}}$ in the incident dynamic pressure. The pressure difference across the cascade is obtained from the linearized form of the quasi-steady Bernoulli equation as

$$(p_e - p_i)' = \rho \overline{u}_y u'_{y_{rot}}. \tag{6.8.13}$$

The pressure difference across the cascade is related to the lift fluctuation on the blade. If we consider the contour C shown in Figure 6.31, the cascade circulation per unit length in the y-direction is the difference in the y-velocity component on the two vertical sides of the contour. The condition of constant leaving angle plus the fact that there are no axial velocity perturbations mean that downstream of the cascade there are no y-velocity perturbations. Hence the circulation per unit length along the cascade is just the incoming rotational perturbation, $u'_{y_{rot}}$, evaluated at the leading edge of the cascade. With $\overline{u}$ $(= \sqrt{\overline{u}_x^2 + \overline{u}_y^2})$ the magnitude of the time mean velocity and

Γ' the perturbation in cascade circulation per unit length in the y-direction, the lift fluctuation per unit length is given by the steady-state Kutta–Joukowski expression (see Section 2.8.3):

$$\text{lift per unit length of the cascade} = \rho\bar{u}\Gamma'. \tag{6.8.14}$$

Noting that $\Gamma' = u'_{y_{rot}}$, the pressure difference can be seen to be the x-component of the lift, as derived in Section 2.8.3 for steady flow.

Because there is no downstream y-component of perturbation velocity there is no vorticity in the flow downstream of the cascade. For a two-dimensional, inviscid, incompressible flow, vorticity is convected with fluid particles. The vorticity flux into the upstream side of the cascade must therefore be cancelled by vorticity shed by the blade. Applying the concepts developed in Section 3.11, concerning vorticity changes associated with a fixed contour, to curve C in Figure 6.31, the rate of change in cascade circulation per unit length in the y-direction is

$$\frac{\partial \Gamma'_{unit\ length}}{\partial t} = \frac{\partial u'_{y_{rot}}(0, y, t)}{\partial t} = i\omega u'_{y_0} e^{i\omega(t - x/\bar{u}_x)}. \tag{6.8.15}$$

Associated with this change is the vorticity shed by the blades which is equal and opposite to that convected through the cascade, creating zero velocity disturbance in the downstream region. The production of shed vorticity in this inviscid flow is connected with the imposition of a constant leaving angle, a constraint which is analogous to the application of the Kutta condition at the trailing edge of an airfoil. Both of these are inviscid models for the viscous (boundary layer) processes that cause the actual flow to leave the trailing edge smoothly. The change in circulation of the blades arises from the ability of the leaving angle condition to capture (to a good approximation) the effect of viscous processes on the flow external to the blade boundary layer and wake.

6.8.2.2 Vorticity and pressure disturbances entering a blade row in a compressible subsonic flow

The approach for the compressible problem is similar to that for unsteady flow through the nozzle and still in the low reduced frequency (actuator disk) limit. We develop equations for the disturbances upstream and downstream of the cascade using the control volume shown in Figure 6.31 and match them across the cascade to obtain a solution which is applicable for the whole domain (Horlock, 1978). The matching conditions are:

$$\text{conservation of mass:} \quad \rho_i u_{x_i} = \rho_e u_{x_e} \tag{6.8.16a}$$

$$\text{constant exit angle:} \quad \tan\alpha_e = \tan\bar{\alpha} = \frac{u_{y_e}(0, y, t)}{u_{x_e}(0, y, t)} = \text{constant} \tag{6.8.16b}$$

conservation of energy (stagnation enthalpy constant across the cascade):

$$c_p T_i + u_{x_i}^2 + u_{y_i}^2 = c_p T_e + u_{x_e}^2 + u_{y_e}^2. \tag{6.8.16c}$$

These, plus the condition of no entropy change across the cascade, are the required matching relations. In (6.8.16), the subscript i denotes the conditions on the upstream side of the cascade and e the conditions on the downstream side, with both quantities being evaluated at $x = 0$.

Linearizing (6.8.16) we obtain,

$$\frac{\rho_i'}{\overline{\rho}} + \frac{u'_{x_i}}{\overline{u}_x} = \frac{\rho_e'}{\overline{\rho}} + \frac{u'_{x_e}}{\overline{u}_x}, \tag{6.8.17a}$$

$$u'_{x_e} \tan\overline{\alpha} = u'_{y_e}, \tag{6.8.17b}$$

$$c_p T_i' + \overline{u}_x u'_{x_i} + \overline{u}_y u'_{y_i} = c_p T_e' + \overline{u}_x u'_{x_e} + \overline{u}_y u'_{y_e}. \tag{6.8.17c}$$

Subscripts on the time mean quantities have been omitted because there is no change through the cascade. Equations (6.8.17) and the linearized forms of the governing field equations can be solved in terms of the incident rotational velocity u'_{y_0} to give the upstream and downstream disturbance fields. For example, the propagating pressure disturbances and the downstream convecting vorticity disturbance are:

$$\frac{p'_{-_i}}{\overline{\rho}\,\overline{u}u'_{y_0}} = -\left(\frac{1+\overline{M}\cos\overline{\alpha}}{1-\overline{M}\cos\overline{\alpha}}\right)\frac{\sin\overline{\alpha}}{[2+\overline{M}\cos\overline{\alpha}(2+\tan^2\overline{\alpha})]}e^{i\omega[t-x/(\overline{u}_x-\overline{a})]},$$

upstream pressure, (6.8.18a)

$$\frac{p'_{+_e}}{\overline{\rho}\,\overline{u}u'_{y_0}} = \frac{\sin\overline{\alpha}}{[2+\overline{M}\cos\overline{\alpha}(2+\tan^2\overline{\alpha})]}e^{i\omega[t-x/(\overline{u}_x+\overline{a})]}, \text{ downstream pressure;} \tag{6.8.18b}$$

$$\frac{u'_{y_{rote}}}{u'_{y_0}} = \frac{\overline{M}\sin\overline{\alpha}\tan\overline{\alpha}}{[2+\overline{M}\cos\overline{\alpha}(2+\tan^2\overline{\alpha})]}e^{i\omega[t-x/\overline{u}_x]}, \text{ downstream vorticity disturbance.} \tag{6.8.18c}$$

Figure 6.32 shows the amplitudes of the upstream and downstream pressure disturbances and the axial velocity disturbance, due to a vortical perturbation incident on the cascade, as a function of the cascade stagger angle, $\overline{\alpha}$, for several time-mean Mach numbers, $\overline{M}$. The pressure disturbances are zero at zero stagger angle because there is no component of blade force normal to the cascade plane. They again approach zero at 90° because the incidence fluctuations approach zero. The response at Mach number of 0.01 is similar to that in incompressible flow where the axial velocity disturbances are zero, but as the Mach number increases, the axial velocity becomes non-zero and the pressure response alters.

Figure 6.33 presents upstream and downstream pressure disturbances and downstream rotational velocity disturbance for Mach number $\overline{M} = 0.5$, as a function of blade stagger angle. Results are given for a convecting vortical disturbance (Figure 6.33(a)) and for a pressure wave from downstream (Figure 6.33(b)). In the latter situation the magnitudes of the upstream and downstream pressure disturbances are the acoustic reflection and transmission coefficients.[9] The behavior changes from

[9] The reflection and transmission coefficients for the cascade are:

$$\text{reflection coefficient} = \left|\frac{p'_{+_e}}{p'_{-_e}}\right| = \frac{\overline{M}\sin\overline{\alpha}\tan\overline{\alpha}}{[2+\overline{M}\cos\overline{\alpha}(2+\tan^2\overline{\alpha})]}$$

$$\text{transmission coefficient} = \left|\frac{p'_{-_i}}{p'_{-_e}}\right| = \frac{2(1-\overline{M}^2)}{(1-\overline{M}\cos\overline{\alpha})[2+\overline{M}\cos\overline{\alpha}(2+\tan^2\overline{\alpha})]}.$$

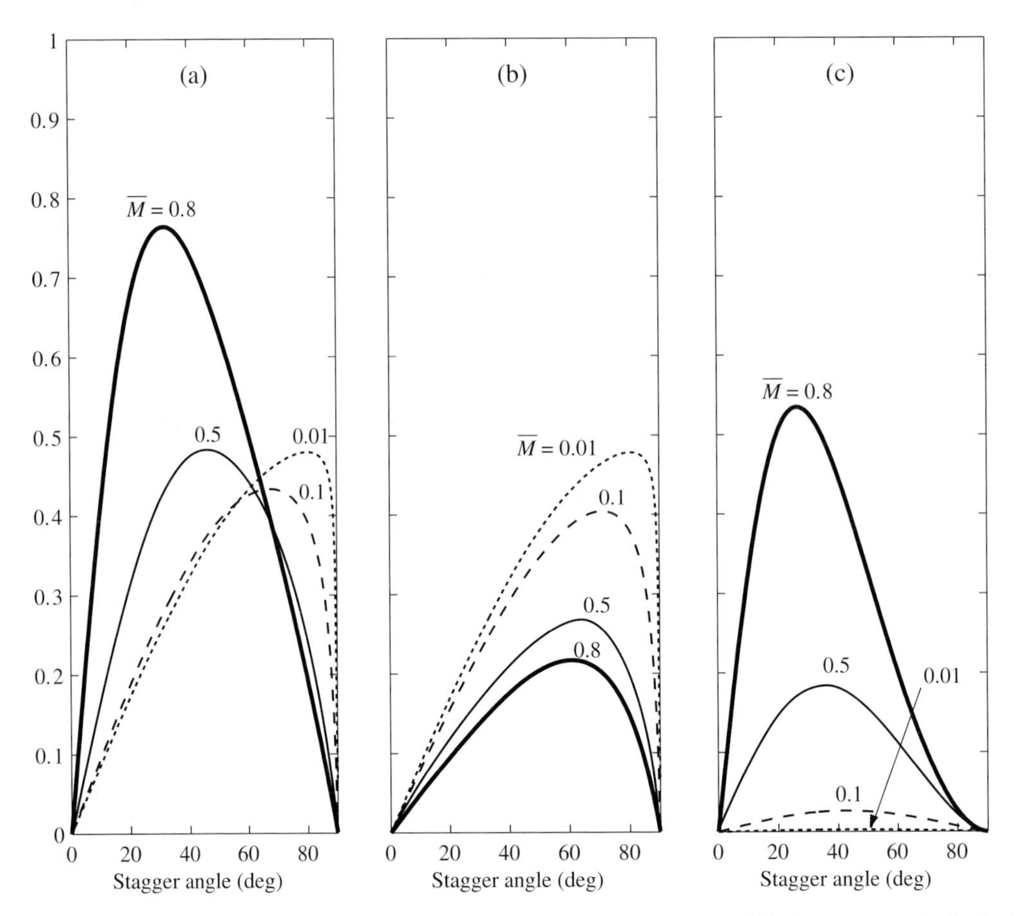

Figure 6.32: Disturbance amplitudes for a flat plate cascade as a function of blade stagger angle for incident vortical disturbance at different Mach numbers, reduced frequency, $\beta = \omega b/\overline{u} \ll 1$: (a) upstream pressure disturbance, $|p'_{-_i}/\overline{\rho}\,\overline{u}u'_{y0}|$; (b) downstream pressure disturbance, $|p'_{+_e}/\overline{\rho}\,\overline{u}u'_{y0}|$; (c) upstream axial velocity disturbance, $|u'_{x_i}/u'_{y0}|$.

zero reflection for zero stagger (the blades are parallel to the direction of wave propagation and the transmission is 100%) to zero transmission for 90° stagger when the blades are normal to the direction of wave propagation. For both the vortical and pressure incident disturbances the response is not only modification of the incoming disturbance by the cascade but creation of the other type of disturbance; pressure disturbances fed into the cascade cause the generation of vorticity disturbances and vorticity disturbances generate pressure disturbances.

It is worthwhile to note that these results are for cascades with semi-infinite upstream and downstream domains. With different upstream and downstream geometry the upstream and downstream pressure and velocity disturbances, although not the relations between incident conditions and changes across the cascade, will be different. A simple illustration showing this is a cascade in incompressible flow with the exit boundary condition of $p'_e = 0$ (as would be the case if the cascade discharged into a large chamber). In this situation (analogous to the nozzle example in the previous

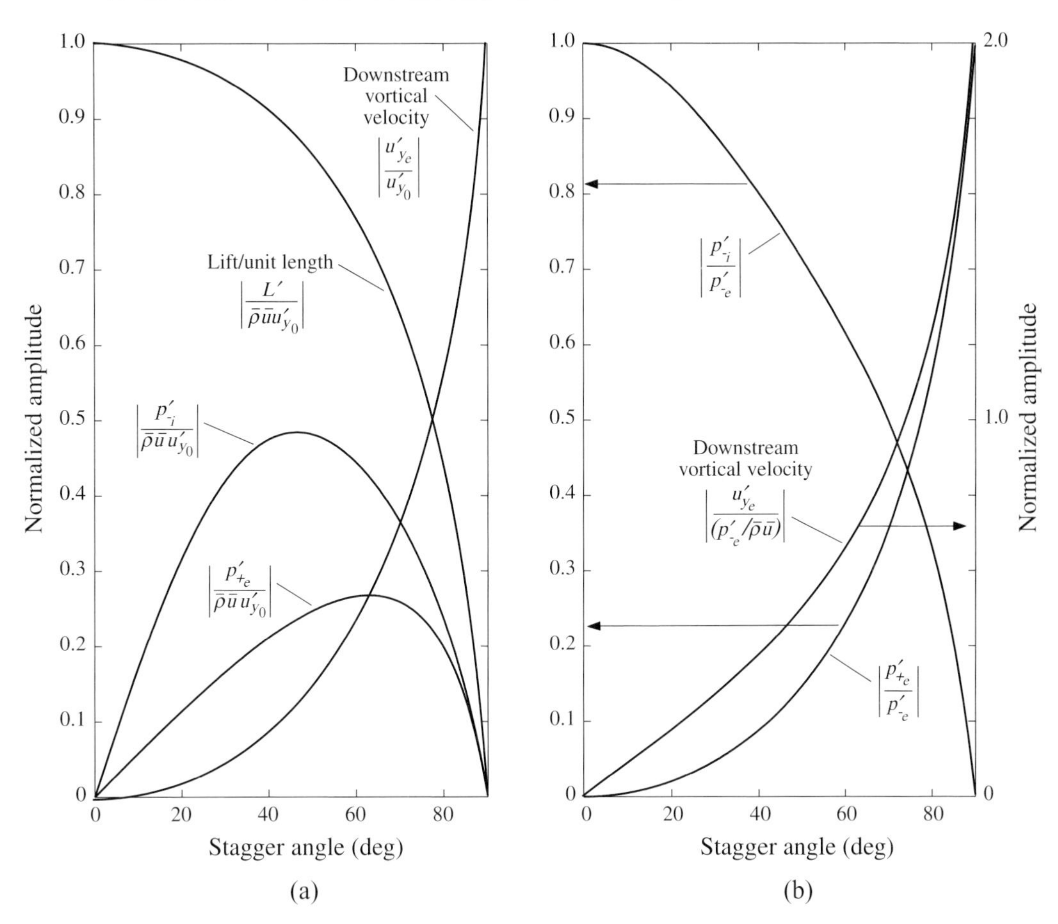

Figure 6.33: Disturbance amplitudes for flat plate cascades as a function of blade stagger angle at $\overline{M} = 0.5$, $\beta = \omega b/\overline{u} \ll 1$: (a) incident vortical disturbance; (b) pressure wave from downstream.

section) all pressure disturbances occur upstream of the cascade. The point is that components such as cascades are generally part of a fluid system; one needs to consider the coupling to other components to completely define the overall disturbance response.

The above results are based on a low reduced frequency approximation and, as in the nozzle example, it is of interest to see when the quasi-steady approach is valid. Figure 6.34 thus shows the magnitude and phase of the unsteady lift fluctuation for a cascade of flat plate airfoils of 60° stagger angle as a function of reduced frequency, at a Mach number of 0.5 (Khalak, 2000). The zero reduced frequency result is essentially that for the actuator disk (without the restriction to constant leaving angle) and the value from the actuator disk analysis is indicated on the figure. As the reduced frequency is increased, the magnitude of the lift decreases. At the highest reduced frequency shown more than a wavelength of the disturbance is within the blade passage, and the lift has decayed to roughly a third of the quasi-steady value. The phase between the lift fluctuation and the incident disturbance at the cascade leading edge is also shown in the figure. This is zero at the low reduced

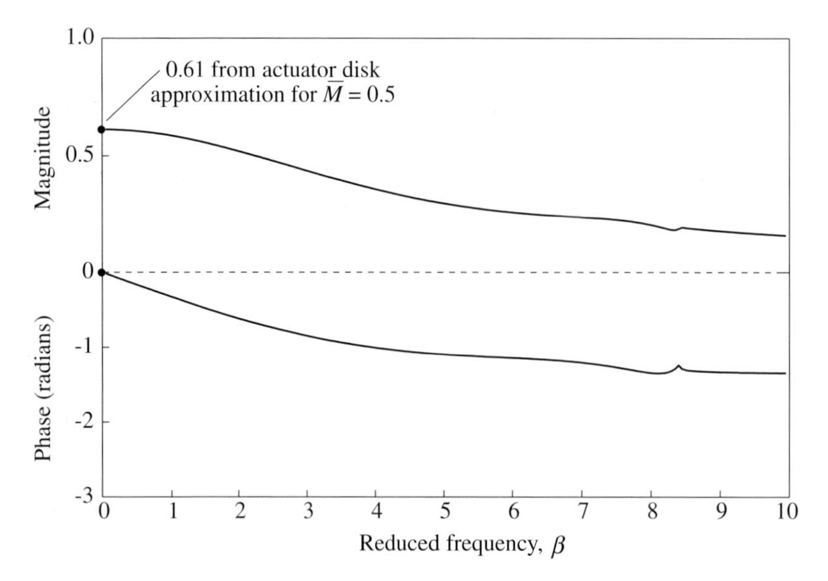

Figure 6.34: Lift response of a cascade of flat plate airfoils: stagger angle $= 60°$, space/chord ratio $= 0.8$, $\overline{M} = 0.5$.

frequency (actuator disk) limit but increases to close to $\pi/2$ at a reduced frequency of 10. We will see in Section 6.9 that stronger departures from quasi-steady behavior can occur for unsteady viscous flows.

6.8.3 Disturbance interaction caused by shock waves

Shock-wave/disturbance interaction also couples flow disturbances and, in general, passage of any one type of disturbance across a shock will create the other two. A problem examined by a number of authors (see e.g. Mahesh, Lele, and Moin (1997) and Andreopoulis, Agui, and Briassulis (2000)) concerns pressure perturbations generated by vorticity disturbances that convect through the shock. This is of interest in connection with noise generation by high speed machinery and aircraft. For details regarding matching conditions and numerical results the above references can be consulted.

6.8.4 Irrotational disturbances and upstream influence in a compressible flow

In this section, we examine the effect of compressibility on upstream influence, specifically the upstream effect of a moving two-dimensional periodic array as a model for a turbomachinery blade row. For the situation in which the background velocity is in the x-direction the equation for the disturbance velocity potential, (6.7.6), takes the form

$$\frac{1}{\overline{a}^2}\left(\frac{\partial}{\partial t} + \overline{u}_x \frac{\partial}{\partial x}\right)^2 \varphi - \left(\frac{\partial^2 \varphi}{\partial x^2} + \frac{\partial^2 \varphi}{\partial y^2}\right) = 0. \tag{6.8.19}$$

Equation (6.8.19) describes a disturbance which propagates at speed $\overline{a}$ with reference to a coordinate system traveling in the x-direction at background velocity $\overline{u}_x$. In a compressible flow we expect there is a possibility for waves, rather than only upstream decaying solutions as were seen in Section 6.4 for incompressible flow.

The disturbance is caused by, and moves with, the airfoils, at velocity Ωr_m (r_m can be interpreted as representing conditions at a mean radius in this two-dimensional treatment) in the negative y-direction. The disturbance must also have a wavelength equal to the blade spacing, W. The axial velocity perturbation at an axial location which we may take as $x = 0$ is therefore of the form (for the first Fourier harmonic)

$$u_x = u_0 \exp\left[2\pi i\left(\frac{y}{W} + \frac{\Omega r_m t}{W}\right)\right]. \tag{6.8.20}$$

Equation (6.8.19) is a linear differential equation with constant coefficients and its solution must have the same dependence on y and t as the impressed disturbance. The velocity perturbation, φ, is therefore

$$\varphi = f(x)\exp\left[2\pi i\left(\frac{y}{W} + \frac{\Omega r_m t}{W}\right)\right]. \tag{6.8.21}$$

Defining the axial and blade Mach numbers as $M_x = \overline{u}_x/\overline{a}$ and $M_B = \Omega r_m/\overline{a}$, and substituting (6.8.21) into (6.8.19) yields a second order differential equation for $f(x)$

$$\left(1 - M_x^2\right)\frac{d^2 f}{dx^2} - \frac{4\pi i}{W}M_x M_B \frac{df}{dx} + \left(\frac{2\pi}{W}\right)^2\left(M_B^2 - 1\right) = 0. \tag{6.8.22}$$

There are two solutions

$$f(x) = C_\pm \exp\left\{\frac{2\pi x}{W}\left[\frac{\pm\left(1 - M_x^2 - M_B^2\right)^{1/2} + iM_x M_B}{\left(1 - M_x^2\right)}\right]\right\} \tag{6.8.23}$$

The constants C_+ and C_- are set by the specific boundary conditions, but the most important aspect is the form of the exponential term. For $M_x^2 + M_B^2 < 1$, the exponent has a real part, implying either growth or decay with x. The former is not acceptable on physical grounds so $C_- = 0$. For $M_x^2 + M_B^2 \geq 1$, the exponent is purely imaginary, implying wave-like solutions (i.e. solutions for φ of the form $\exp[i(k_x x + k_y y - \omega t)]$, where k_x and k_y are wave numbers in the x- and y-directions). In this situation, the boundary condition far upstream is that the waves are outgoing, or radiating from the moving blades.

To explore the rate at which disturbances die away with upstream distance, we examine the behavior of the exponent in (6.8.23) as the blade Mach number M_B increases from zero, holding the ratio of M_x to M_B constant at $M_x/M_B = 0.5$, a value roughly representative of aeroengine axial compressors. Increasing M_B thus implies increasing blade speed while keeping the relative flow angle constant. Holding M_x/M_B constant is also equivalent to keeping the reduced frequency, based on a length W and the mean relative velocity, $\sqrt{\overline{u}_x^2 + (\Omega r_m)^2}$, constant. Although the reduced frequency is invariant with blade speed, the product of reduced frequency and Mach number, βM, which is a descriptor of the impact of compressibility (see Section 2.2), scales with Mach number.

The decay of the axial velocity disturbance amplitude is illustrated in Figure 6.35 for several values of M_B. The vertical axis is the amplitude of the axial velocity non-uniformity, normalized by the value at $x = 0$, and the horizontal axis is the upstream position, non-dimensionalized by the blade spacing, W, which is the disturbance wavelength.

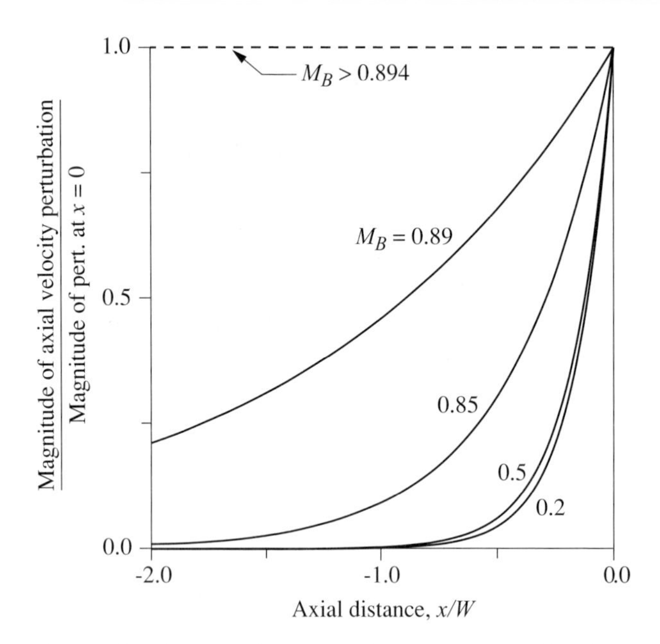

Figure 6.35: Upstream decay of axial velocity perturbation due to a rotor ($M_x = 0.5\, M_B$).

For low Mach numbers ($M_B \leq 0.5$), the extent of the upstream influence is similar to incompressible flow (see Figure 6.8). However, as blade Mach numbers increase past roughly 0.8, the extent of upstream influence rapidly increases. For high enough blade Mach numbers ($M_B \geq 2/\sqrt{5} = 0.894$ in this case), there is no decay of the upstream velocity and pressure perturbations with distance, and disturbances propagate upstream. This occurs when the quantity in the square root in (6.8.23) becomes negative. It marks the condition at which waves are no longer "cut off" but can propagate upstream, with the implication that acoustic pressure disturbances will propagate rather than being attenuated.

Viewed in another way, the condition at which propagating waves occur is that at which the relative Mach number seen by an observer traveling with the disturbance is unity, i.e. $M_{relative} = \sqrt{M_B^2 + M_x^2} = 1$. In a coordinate system traveling with the rotor the flow is steady, the relative velocity has x- and y-components, $\overline{u}_x$ and Ωr_m respectively, and the equation for φ becomes,

$$\left(1 - M_x^2\right)\frac{\partial^2 \varphi}{\partial x^2} + \left(1 - M_y^2\right)\frac{\partial^2 \varphi}{\partial y^2} + 2M_x M_y \frac{\partial^2 \varphi}{\partial x \partial y} = 0. \tag{6.8.24}$$

In (6.8.24) $M_y = \Omega r_m/\overline{a}$ is the y-component of the Mach number seen by an observer moving with the rotor. For $M_x = 0$, (6.8.24) reduces to the result for flow along a wavy wall (Liepmann and Roshko, 1957), where the condition for propagating disturbances is that the Mach number of the flow along the wall is supersonic.

6.8.5 Summary concerning small amplitude unsteady disturbances

We conclude the discussion of small disturbances in a compressible flow with some remarks concerning the overall applicability of the results. The description of the different types of disturbances has been developed under the idealization that the background flow is uniform. This is a useful approximation in many circumstances, and even when not quantitatively correct often provides qualitative insight into overall flow features.

For disturbances of amplitudes large enough such that nonlinear effects need to be accounted for, the independence of the different disturbances described here does not hold. An example is a vortex in an infinite stationary fluid, where the associated static pressure field has a magnitude proportional to the square of the circulation. Another example is pressure disturbances in an incompressible, uniform density, inviscid flow. Taking the divergence of the momentum equation and invoking the continuity equation gives, to first order in the disturbance strength, an equation for the pressure, p', as $\nabla^2 p' = 0$. If second order terms are included, the equation for pressure is

$$\nabla^2 p = \rho\left(\tfrac{1}{2}\omega^2 - e^2\right), \tag{6.8.25}$$

where ω^2 is the square of the magnitude of the vorticity vector and e^2 ($= e_{ij}e_{ij}$, where e_{ij} is the strain rate tensor, see (1.13.1)) is the sum of the squares of the principal rates of strain associated with the disturbance flow. In summary, nonlinear effects couple disturbances so that pressure disturbances depend on vorticity and velocity perturbations (Bradshaw and Koh, 1981).

Finally, although we have divided the different types of disturbances into irrotational velocity perturbations, vorticity or rotational velocity perturbations, and entropy perturbations, it should be noted that there are other equivalent sets of independent flow disturbances that can be employed (Goldstein, 1978).

6.9 Some features of unsteady viscous flows

We now turn to features of unsteady viscous flows. Two exact solutions of the Navier–Stokes equations for an incompressible fluid are of interest as a means of illustrating some of the important concepts: the flow due to an oscillating plane boundary and the flow in a channel with a periodic pressure gradient. Unsteady boundary layer behavior is also discussed.

6.9.1 Flow due to an oscillating boundary

We first examine the viscous flow due to an oscillating infinite plane boundary in a semi-infinite fluid region, referred to as Stokes's second problem. The x- and y-coordinates are parallel and perpendicular to the boundary motion. There is no variation of any flow variable in the x-direction and the continuity equation plus the condition of zero x-velocity at the plate requires that the y-component of velocity be zero throughout the flow. The momentum equation thus reduces to

$$\frac{\partial u_x}{\partial t} = \nu \frac{\partial^2 u_x}{\partial y^2}. \tag{3.6.5}$$

The x-velocity boundary condition at the wall, $y = 0$, is that u_x must match the boundary velocity. If the latter is harmonic with amplitude u_w and frequency ω,

$$u_x(x, 0, t) = u_w e^{i\omega t}. \tag{6.9.1}$$

The final boundary condition is that u_x goes to zero as $y \to \infty$.

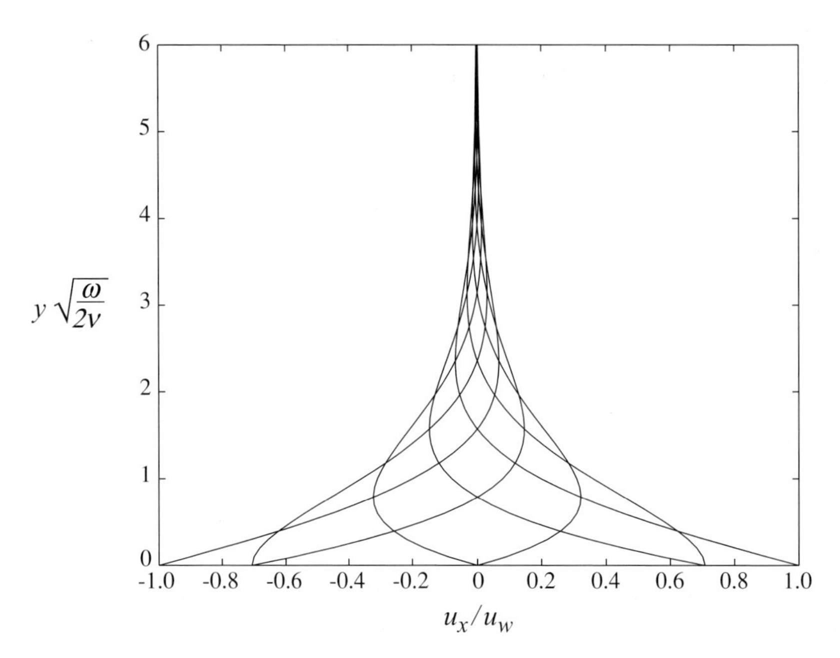

Figure 6.36: Velocity profiles for a flat plate oscillating in a viscous fluid at rest at $y \to \infty$. Oscillation is of the form $u_x(x, 0, t) = u_w e^{i\omega t}$. Profiles are at intervals of $\omega t = \pi/4$ for $0 \leq \omega t \leq 2\pi$.

For the linear equation (3.6.5), with the boundary condition (6.9.1), u_x must be of the form $f(y)\, e^{i\omega t}$. Substituting this form into the momentum equation and solving yields

$$\frac{u_x}{u_w} = \exp\left\{-i\left(\frac{y}{\sqrt{2\nu/\omega}} - \omega t\right) - \frac{y}{\sqrt{2\nu/\omega}}\right\}. \tag{6.9.2}$$

Equation (6.9.2) is a harmonic oscillation which is damped in the y-direction. The amplitude of the velocity variation, u_x/u_w, at any value of y is $e^{-y/\sqrt{2\nu/\omega}}$. In addition there is a phase lag between different values of y. Figure 6.36 gives velocity profiles, u_x/u_w, at different times in the period of oscillation, $2\pi/\omega$. Analogous to the impulsively started plate (Section 3.6) where the effective depth of penetration of the velocity was of order $\sqrt{\nu t}$, the velocity penetration depth here is of order $\sqrt{\nu/\omega}$. We can view the unsteady flow as due to the diffusion of vorticity from the wall, with $\sqrt{\nu/\omega}$ the effective diffusion distance. This result carries over qualitatively to unsteady boundary layers where effects of unsteadiness are "felt" to a depth of order $\sqrt{\nu t}$ or $\sqrt{\nu/\omega}$.

6.9.2 Oscillating channel flow

Another example illustrating the concept of penetration depth is the flow due to an oscillating pressure gradient in a two-dimensional channel of width W. The pressure gradient is uniform with x and varies with t as

$$-\frac{1}{\rho}\frac{dp}{dx} = Ce^{i\omega t}, \tag{6.9.3}$$

where C is a constant. With this pressure gradient the velocity is a function of y and t only and there is only one velocity component, u_x. The x-momentum equation is

$$\frac{\partial u_x}{\partial t} = -\frac{1}{\rho}\frac{dp}{dx} + \nu\frac{\partial^2 u_x}{\partial y^2}. \tag{6.9.4}$$

The boundary conditions are

$$u_x\left(x, \frac{-W}{2}, t\right) = u_x\left(x, \frac{W}{2}, t\right) = 0.$$

Substituting the form of the pressure gradient in (6.9.4) and noting that the velocity must also be of the form $e^{i\omega t}$, we obtain

$$u_x = -i\frac{Ce^{i\omega t}}{\omega}\left[1 - \frac{\cosh\left(\sqrt{\frac{i\omega}{\nu}}\frac{W}{2}\right)\frac{2y}{W}}{\cosh\left(\sqrt{\frac{i\omega}{\nu}}\frac{W}{2}\right)}\right]. \tag{6.9.5}$$

The non-dimensional parameter that characterizes the behavior of the solution in (6.9.5) is $\sqrt{\omega/\nu}(W/2)$, which can be regarded as the ratio of the channel half-height to the penetration depth of the vorticity generated at the wall. For values of this parameter large compared to unity, viscous effects are confined to a thin layer of thickness $\sqrt{\nu/\omega}$ near the walls, frequently referred to as a Stokes layer. For values of $\sqrt{\omega/\nu}(W/2)$ much smaller than unity, viscous effects are felt throughout the channel.

The limiting forms of the solutions for high and low values of the parameter $\sqrt{\omega/\nu}(W/2)$ show this behavior explicitly. For low frequency, $\sqrt{\omega/\nu}(W/2) \ll 1$, (6.9.5) becomes

$$u_x = -\frac{1}{2\mu}\frac{dp}{dx}\frac{W^2}{4}\left(1 - \frac{4y^2}{W^2}\right). \tag{6.9.6}$$

Equation (6.9.6) describes quasi-steady Poiseulle flow, with the velocity field and the pressure gradient in phase. The velocity distribution is the same as that for fully developed laminar flow at the instantaneous value of the pressure gradient.

For high frequency, $\sqrt{\omega/\nu}(W/2) \gg 1$, we use the approximation that $\cosh\zeta \to e^{\zeta}/2$ for $\zeta \gg 1$ and find

$$u_x = \underset{\text{(I)}}{\frac{-iC}{\omega}e^{i\omega t}}\left\{1 - \underset{\text{(II)}}{\exp\left[(1+i)\sqrt{\frac{\omega}{2\nu}}\frac{W}{2}\left(\frac{2y}{W} - 1\right)\right]} - \underset{\text{(III)}}{\exp\left[-(1+i)\sqrt{\frac{\omega}{2\nu}}\frac{W}{2}\left(\frac{2y}{W} + 1\right)\right]}\right\} \tag{6.9.7}$$

The form of the velocity distribution, which is quite different from the quasi-steady case, is usefully viewed as the sum of three different parts. The first term (I) is the unsteady response associated with the inertia of the fluid in the channel and resulting from the inviscid effects described in the unsteady diffuser example of Section 6.3, with $A_e/A_i \to 1$. The velocity associated with I is constant across the channel and has a phase of $-\pi/2$ with respect to the driving pressure force per unit mass.

Terms II and III represent viscous layers near the two walls at $y = \pm W/2$. (Term II gives the behavior near $y = W/2$ and term III corresponds to $y = -W/2$.) The thickness of these viscous layers is of order $\sqrt{\nu/\omega}$. The velocity field described by terms II and III has similarities with that for the previous section, with a phase difference in velocity across the layer. The wall shear stress lags the pressure force per unit mass $(-1/\rho)(dp/dx)$ by $\pi/4$. The phase difference between the velocity in the inviscid-like region between the two viscous layers and the wall shear stress is thus $\pi/4$ (a phase lead of the shear stress). We will find this same behavior in the unsteady response of laminar boundary layers at high frequencies described in the next section.

6.9.3 Unsteady boundary layers

The ideas of the previous section are helpful in extending the discussion to unsteady boundary layers, although only a short introduction to this general topic can be given. We wish to define the regimes in which boundary layer unsteadiness is important, and describe some features of these unsteady motions. Situations where unsteady boundary layers occur include the generation of flows on solid surfaces starting from rest, effects due to unsteadiness in the free-stream velocity or pressure, and unsteady flow associated with motion or deformation of a body. Periodic motions are most common in fluid machines and we thus focus on these.

To develop a framework for characterizing the flow regimes consider an unsteady laminar boundary layer having a characteristic frequency ω, in which the unsteadiness can be regarded as a perturbation to the steady flow. An analogy can be drawn between the boundary layer thickness, δ, and the channel height in the oscillating flow in Section 6.9.2, although this is meant more to motivate what follows than to be an exact comparison. To describe "how unsteady" the boundary layer flow is, an appropriate non-dimensional parameter is $\delta\sqrt{\omega/\nu}$, the ratio of steady-state boundary layer thickness at a given location to the penetration depth of the unsteady viscous layer. The steady-state thickness scales as $\delta \propto \sqrt{\nu x/\overline{u}_E}$, where $\overline{u}_E$ characterizes the time-mean free-stream velocity, so the ratio is

$$\delta\sqrt{\frac{\omega}{\nu}} \propto \sqrt{\frac{\omega x}{\overline{u}_E}}. \tag{6.9.8}$$

The parameter $\sqrt{\omega x/\overline{u}_E}$, or $\omega x/\overline{u}_E$ as generally written, gives a measure of the spatial influence of unsteadiness in a boundary layer with an impressed periodic disturbance. Small values imply close to quasi-steady response. Large values mean the unsteady viscous effects occupy a small fraction of the boundary layer and can be regarded as a secondary boundary layer (Stokes layer) located next to the wall. For large values of $\omega x/\overline{u}_E$ the inertial forces are dominated by local rather than convective accelerations and the oscillations are essentially independent of the mean flow.

We can also develop the parameter in (6.9.8) from consideration of the physical processes associated with the development of viscous flow over a solid surface (Stuart, 1963). There are three processes of interest: (1) the rate of vorticity convection by $\overline{u}_E$ over a length scale x, (2) the rate of vorticity diffusion through a distance δ (normal to the surface), and (3) the rate of vorticity diffusion through a distance that scales with frequency as $\sqrt{\nu/\omega}$. In steady flow the boundary layer thickness δ is set by the balance between the convection of vorticity over a distance x in the flow direction (process 1) and diffusion of vorticity through a distance δ normal to the surface (process 2), giving the laminar flow result $\delta \propto \sqrt{\nu x/\overline{u}_E}$ (Section 2.9). In an unsteady flow the rate of vorticity diffusion is ω and the ratio of this to the rate of vorticity convection by $\overline{u}_E$ over distance x $(\overline{u}_E/x)$ is $\omega x/\overline{u}_E$. Although the discussion has been based on laminar flow, $\omega x/\overline{u}_E$ is used to characterize turbulent unsteady boundary layers, and results of calculations and experiments on unsteady boundary layers are often presented with $\omega x/\overline{u}_E$ as the independent variable.

The unsteady boundary layer equations can be developed using the arguments presented in Chapter 4, with the local acceleration terms now included. For incompressible flow the continuity equation remains the same and the x-component of the momentum equation becomes

$$\frac{\partial u_x}{\partial t} + u_x \frac{\partial u_x}{\partial x} + u_y \frac{\partial u_x}{\partial y} = -\frac{1}{\rho}\frac{\partial p}{\partial x} + \frac{\partial \tau}{\partial y}. \tag{6.9.9}$$

The relation between the free-stream velocity and the pressure gradient also now includes an unsteady term:

$$\frac{\partial u_E}{\partial t} + u_E \frac{\partial u_E}{\partial x} = -\frac{1}{\rho}\frac{\partial p}{\partial x}. \tag{6.9.10}$$

We can write (6.9.9) and (6.9.10) in non-dimensional forms using x, $\overline{u}_E$, and $1/\omega$ as the characteristic length, velocity and time scale. Following the procedure used in Section 1.17 the corresponding non-dimensional form of the equations with the dimensionless parameters $(\omega x/\overline{u}_E)$ and $(\overline{u}_E x/\nu)$ appearing explicitly can be written as:

$$\left(\frac{\omega x}{\overline{u}_E}\right)\frac{\partial \tilde{u}_x}{\partial \tilde{t}} + \tilde{u}_x \frac{\partial \tilde{u}_x}{\partial \tilde{x}} + \tilde{u}_y \frac{\partial \tilde{u}_x}{\partial \tilde{y}} = -\frac{\partial \tilde{p}}{\partial \tilde{x}} + \left(\frac{\nu}{\overline{u}_E x}\right)\frac{\partial^2 \tilde{u}_x}{\partial \tilde{y}^2} \tag{6.9.11}$$

and

$$-\frac{\partial \tilde{p}}{\partial \tilde{x}} = \left(\frac{\omega x}{\overline{u}_E}\right)\frac{\partial \tilde{u}_E}{\partial \tilde{t}} + \tilde{u}_E \frac{\partial \tilde{u}_E}{\partial \tilde{x}}, \tag{6.9.12}$$

where $(\tilde{\ })$ denotes a dimensionless variable.

Equation (6.9.11) along with the continuity equation can be solved numerically for any value of $\omega x/\bar{u}_E$ but it is instructive to describe the limiting cases of $\omega x/\overline{u}_E \ll 1$ (low frequency) and $\omega x/\overline{u}_E \gg 1$ (high frequency). In the former situation, as shown by Lighthill (see Rosenhead, (1963), Chapter VII), for small amplitude unsteady fluctuations the magnitude of the departure from quasi-steady behavior can be expressed as a quantity which is linear in the reduced frequency. In the latter case, for large values of $\omega x/\overline{u}_E$, convective accelerations can be neglected and the boundary layer equation reduced to

$$\left(\frac{\omega x}{\overline{u}_E}\right)\frac{\partial \tilde{u}_x}{\partial \tilde{t}} = -\frac{\partial \tilde{p}}{\partial \tilde{x}} + \left(\frac{\nu}{\overline{u}_E x}\right)\frac{\partial^2 \tilde{u}_x}{\partial \tilde{y}^2}. \tag{6.9.13}$$

The free-stream momentum equation in this case is

$$\left(\frac{\omega x}{\overline{u}_E}\right)\frac{\partial \tilde{u}_E}{\partial \tilde{t}} = -\frac{\partial \tilde{p}}{\partial \tilde{x}}. \tag{6.9.14}$$

In the high frequency limit the equations are similar to those for the oscillating channel flow of Section 6.9.2 and the unsteady boundary layer is independent of the time mean velocity profile.

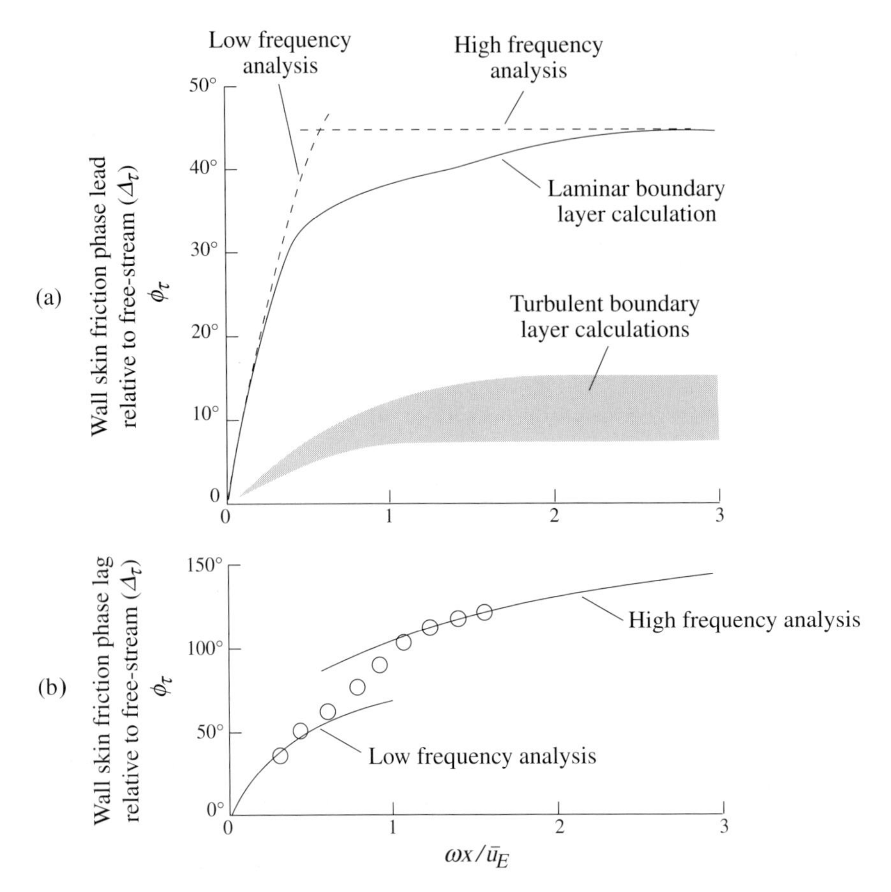

Figure 6.37: Unsteady boundary layer skin friction phase angle with respect to free-stream velocity, ϕ_τ, as a function of frequency. (a) Oscillating free stream, $u_E = \overline{u}_E + u_{unst} \cos \omega t$, the solid line is the laminar boundary layer calculation by Telionis and Romaniuk (1978), the dashed lines are high and low frequency analyses by Lighthill (1954), the turbulent results are as given in Lyrio and Ferziger (1983), ϕ_τ, denotes phase lead. (b) Travelling wave imposed on a laminar boundary layer $u_E = \overline{u}_E + u_{unst} \cos \omega[t - (x/u_{wave})]$ with $u_{wave} = 0.77$ $\overline{u}_E$ (Patel, 1975), symbols are experimental results, solid lines are high and low frequency analyses.

Figure 6.37(a) shows the phase of the skin friction fluctuation compared to the free-stream velocity perturbation, as a function of $\omega x/\overline{u}_E$ for an unsteady boundary layer. For a developing boundary layer on a device, at small x there can be regions in which the response is quasi-steady whereas further back on the device, at large x, there can be large departures from quasi-steady behavior.

The dashed curves labeled low frequency and high frequency in Figure 6.37(a) are from analyses by Lighthill (1954) based on approximations for these regimes. The numerical result of Telionis and Romaniuk (1978), shown as the solid line, indicates the transition from low frequency to high frequency regimes. In the high frequency limit (6.9.13) shows that the boundary layer response is a balance between pressure gradient, viscous force, and local accelerations. There is a phase shift between the free-stream velocity fluctuation and a skin friction of $\pi/4$ (phase lead of the shear stress), similar to that for the oscillating channel flow in the high frequency limit. The figure also

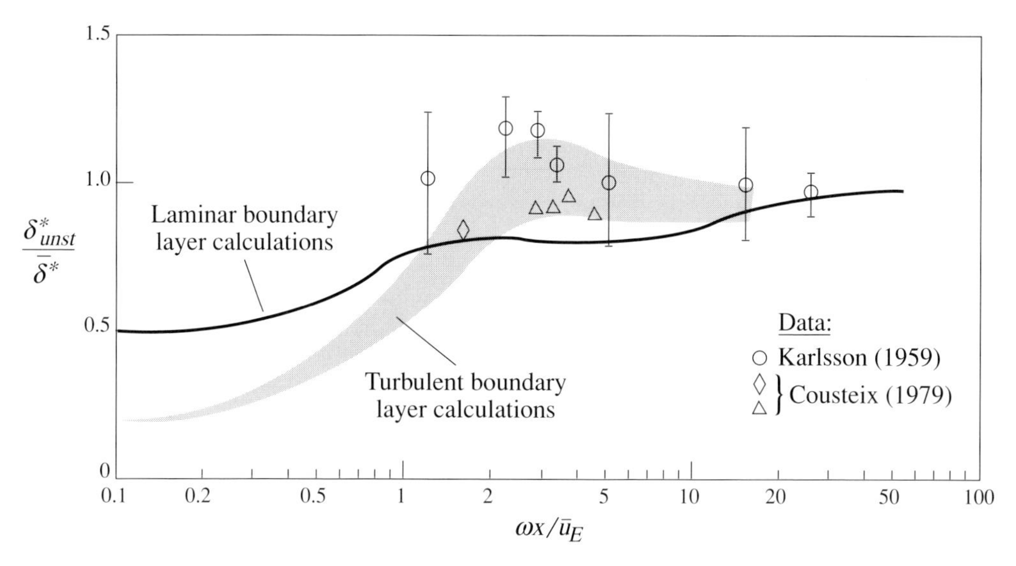

Figure 6.38: Amplitude of displacement thickness for an unsteady boundary layer; $u_E = \bar{u}_E(1 + 0.125 \sin \omega t)$, $\delta^* = \bar{\delta}^* + 0.125\delta^*_{unst} \sin(\omega t + \pi + \Delta_t)$. Laminar boundary layer calculations from McCroskey and Philippe (1975); turbulent boundary layer calculations, and data for turbulent boundary layers are as given in Lyrio and Ferziger (1983).

gives information on the phase of the skin friction from computations of turbulent boundary layers. There is a range of values, depending on the particular turbulence model used, but the skin friction phase shift is much less than with laminar flow.

Figure 6.37(b) shows the response to an impressed unsteadiness of the form $\cos \omega[t - (x/u_{wave})]$, a traveling disturbance with velocity u_{wave}, a situation more representative of turbomachines. The value of u_{wave} used is $0.77\, \bar{u}_E$. The high frequency limit here is not the same as that for Figure 6.37(a) because for a constant phase speed the wave number of the unsteady disturbance increases with frequency and convective accelerations remain important.

Figure 6.38 gives the computed magnitude of the displacement thickness variation for an unsteady turbulent boundary layer, along with experimental data. The change in response as $\omega x/\bar{u}_E$ is increased is more marked with the turbulent layer than with the laminar layer; in the latter it also depends on Reynolds number.

For unsteady laminar boundary layers, numerical methods exist that well capture the observed behavior (McCroskey, 1977; Telionis, 1979). With unsteady turbulent flow, the bands shown in the figures, representing a range of several results given in the literature, reflect different approaches to closure of the turbulent boundary layer equations (Section 4.6).

6.9.4 Dynamic stall

Dynamic stall is a phenomenon in which large effects of unsteadiness occur even at relatively low values of reduced frequency. On an oscillating airfoil whose incidence is increasing rapidly, the onset of stall can be delayed to incidence angles considerably in excess of the angle at which stall occurs under steady-state conditions. Associated with this delay are values of lift which can be up to 30%

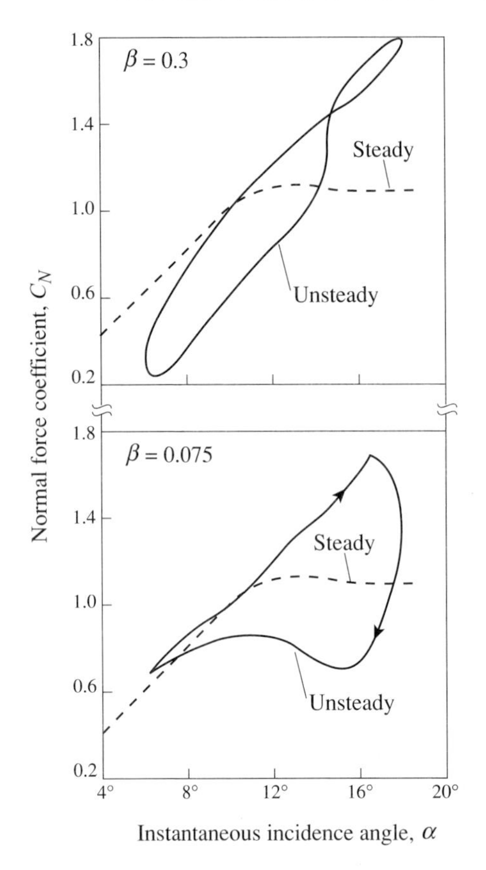

Figure 6.39: Unsteady normal force for the NACA 0012 airfoil oscillated in pitch about the quarter-chord; $\alpha = 12^\circ + 6^\circ \sin \omega t$, and Mach number = 0.3 (Carta, 1967).

greater than the peak steady-state value and which have a finite hysteresis as the angle of incidence is varied. Figure 6.39 shows the measured unsteady lift (shown as the normal force coefficient) for an airfoil pitching about an axis at the quarter-chord, for two values of reduced frequency (Carta, 1967). The time-dependent behavior in the dynamic stall regime is characterized by the shedding of a large scale vortical disturbance from the leading edge region (McCroskey and Pucci, 1982; Ekaterinaris and Platzer, 1997). Local low pressures from the passage of this vortex over the upper surface of the airfoil are associated with the observed increase in lift. Dynamic stall is a striking example of the differences between steady-state and unsteady behavior.

6.9.5 Turbomachinery wake behavior in an unsteady environment

The discussions of wake response to pressure fields in Chapters 4 and 5 refer to steady flow. Wake passage through a pressure rise was seen to result in wake growth (as measured by momentum thickness, for example) and an increase in mixing losses. In an unsteady environment the wake behavior can be qualitatively different and wake passage through a pressure rise can result in a decrease in wake size.

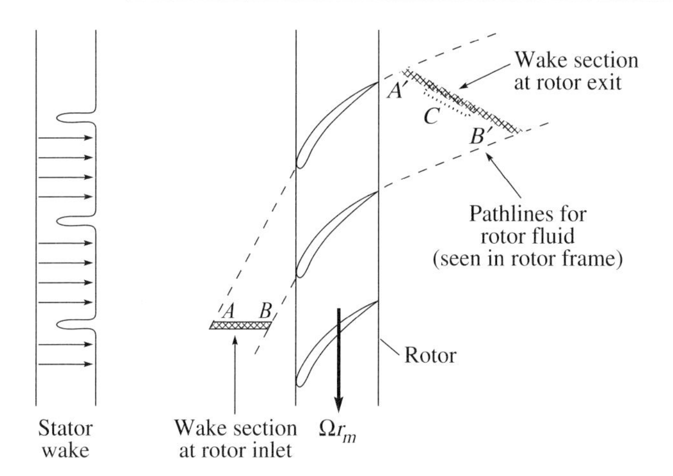

Figure 6.40: Passage of stator wake through a rotor (after Smith (1966b, 1993)).

This effect is present in turbomachines which have multiple closely spaced blade rows so that wakes are not fully mixed when they enter the succeeding row. Experiments in multistage axial compressors have shown efficiency increases of up to several percent as the axial spacing between the rows is decreased (Smith, 1970). An explanation for one contribution to this effect, based on wake behavior in an unsteady flow, is sketched in Figure 6.40 (Smith, 1966b, 1993). The figure is a two-dimensional representation of a stationary blade row (stator) wake being transported through a rotating blade row (rotor). The physical mechanism can be introduced by viewing the wake as an inviscid velocity defect. For a constant density inviscid fluid Kelvin's Theorem states that the circulation around contour C is constant as the wake moves through the rotor. Because of: (i) the streamtube divergence in the rotor and (ii) the difference in convection time for particles on the suction and pressure surfaces (due to the circulation around the blades), the wake length increases from rotor inlet to exit, with a commensurate increase in the length of contour C. Since the circulation round the contour is equal to the product of the velocity difference (between the free stream and the wake) and the contour length, the velocity difference decreases if the wake length increases. The loss due to mixing is thus lower than if the wake had fully mixed before entering the rotor.

The process can also be viewed through examination of stagnation pressure changes for particles in the free stream and in the wake as they move through the downstream row. The stagnation pressure change for an inviscid constant density fluid is given by

$$\frac{Dp_t}{Dt} = \frac{\partial p}{\partial t}. \tag{6.2.4}$$

Particles in the wake have a lower axial velocity than particles in the free stream, a longer residence time in the rotor passage, and hence, from (6.2.4), a larger increase in stagnation pressure than those in the free stream. The difference in stagnation pressure, and hence velocity magnitude, between the wake and the free stream is therefore lessened.

The figure and the arguments refer to the passage of a stator wake through a rotor, but the same mechanism applies to attenuation of rotor wakes passing through stators. Figure 6.41 shows analyses and measurements of the evolution of axial compressor rotor wake depth in a downstream stator. for two operating conditions: peak efficiency and peak pressure rise. The solid and dashed lines are

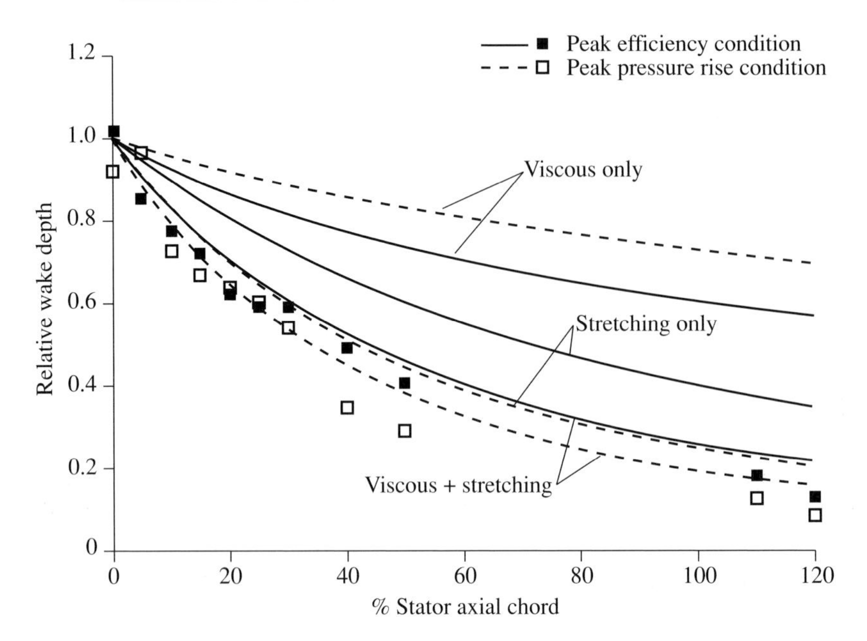

Figure 6.41: Evolution of a compressor rotor wake through a stator passage; lines refer to analysis, symbols to data (Van Zante *et al.*, 2002).

results from approximate analyses of the decrease in wake depth (Van Zante *et al.*, 2002). The three curves for each condition indicate the effects of viscous decay alone (based on a steady wake at constant pressure), from wake stretching alone, and from the two in combination. The symbols are laser anemometer measurement results. At the peak pressure rise condition there is an increase in wake stretching associated with the higher aerodynamic loading. Two-dimensional unsteady Navier–Stokes computations of wake evolution bear out the ideas and show that the magnitudes of the effect are in overall agreement with the approximate analyses (Valkov and Tan, 1999).

7 Flow in rotating passages

7.1 Introduction

In the analysis of fluid machinery behavior, it is often advantageous to view the flow from a coordinate system fixed to the rotating parts. Adopting such a coordinate system allows one to work with fluid motions which are steady, but there is a price to be paid because the rotating system is not *inertial*. In an inertial coordinate system, Newton's laws are applicable and the acceleration on a particle of mass m is directly related to the vector sum of forces through $\mathbf{F} = m\mathbf{a}$. In a rotating coordinate system, the perceived accelerations also include the Coriolis and centrifugal accelerations which must be accounted for if we wish to write Newton's second law with reference to the rotating system.

In this chapter we examine flows in rotating passages (ducts, pipes, diffusers, and nozzles). These typically operate in a regime where rotation has an effect on device performance but does not dominate the behavior to the extent found in the geophysical applications which are considered in much of the literature (e.g. Greenspan (1968)). The objectives are to develop criteria for when phenomena associated with rotation are likely to be important and to illustrate the influence of rotation on overall flow patterns. A derivation of the equations of motion in a rotating frame of reference is first presented to show the origin of the Coriolis and centrifugal accelerations, with illustrations provided of the differences between flow as seen in fixed (often called absolute) and rotating (often called relative) systems. Quantities that are conserved in a steady rotating flow are then discussed, because these find frequent use in fluid machinery. A brief description of fluid motion when the effects of rotation dominate is also given, because phenomena exist which are strikingly different from those situations without rotation. The last four sections focus on specific attributes of inviscid and viscous flows in rotating passages.

7.1.1 Equations of motion in a rotating coordinate system

The relation between the relative velocity, $\mathbf{w}$, seen in the rotating coordinate system and the absolute velocity, $\mathbf{u}$, seen in the stationary, or inertial, coordinate system, is

$$\mathbf{u} = \mathbf{w} + (\boldsymbol{\Omega} \times \mathbf{r}), \tag{7.1.1}$$

where $\boldsymbol{\Omega}$ is the angular velocity of the rotating system and $\mathbf{r}$ is a position vector from the origin of rotation to the point of interest. Equation (7.1.1) is an illustration of the general transformation between derivatives of vectors in rotating and stationary systems: for any vector **B**

$$\left(\frac{d\mathbf{B}}{dt}\right)_{stationary} = \left(\frac{d\mathbf{B}}{dt}\right)_{rotating} + \boldsymbol{\Omega} \times \mathbf{B}. \tag{7.1.2}$$

The term on the left is the derivative as observed in the stationary system and the first term on the right is the derivative as observed in the rotating system. If **B** is set equal to the position vector **r** of a fluid particle, (7.1.1) is recovered. For application to fluid flows the differentiation is interpreted as the rate of change experienced by a fluid particle, or substantial derivative (Section 1.3.1), and (7.1.2) assumes the form

$$\left(\frac{D\mathbf{B}}{Dt}\right)_{stationary} = \left(\frac{D\mathbf{B}}{Dt}\right)_{rotating} + \boldsymbol{\Omega} \times \mathbf{B} \tag{7.1.3}$$

for the transformation between derivatives as observed in the rotating and stationary systems.

For *scalar* quantities such as density or entropy, the substantial derivative is the same in the rotating and the stationary systems:

$$\left(\frac{D[\text{scalar}]}{Dt}\right)_{stationary} = \left(\frac{D[\text{scalar}]}{Dt}\right)_{rotating}. \tag{7.1.4}$$

Spatial derivatives, which are taken at fixed time, are also the same in rotating and stationary systems:

$$\boldsymbol{\nabla}_{stationary} = \boldsymbol{\nabla}_{rotating}. \tag{7.1.5}$$

The equations describing fluid motion in the absolute frame can be transformed to the rotating frame by using (7.1.3), (7.1.4), and (7.1.5). From (1.9.4) the continuity equation can be written as

$$\left(\frac{D\rho}{Dt}\right)_{stationary} + \rho\boldsymbol{\nabla}\cdot\mathbf{u} = \left(\frac{D\rho}{Dt}\right)_{rotating} + \rho\boldsymbol{\nabla}\cdot\mathbf{w} + \rho\boldsymbol{\nabla}\cdot(\boldsymbol{\Omega}\times\mathbf{r}) = 0.$$

The term $\boldsymbol{\nabla}\cdot(\boldsymbol{\Omega}\times\mathbf{r})$ is zero since it represents a rigid body rotation with no change of volume. The continuity equation therefore has the same form in the rotating and stationary systems:

$$\frac{1}{\rho}\left(\frac{D\rho}{Dt}\right)_{rotating} + \boldsymbol{\nabla}\cdot\mathbf{w} = 0. \tag{7.1.6}$$

This is also seen by considering mass conservation for a control volume fixed in the rotating frame.

To relate the acceleration as seen in the stationary system to the acceleration in the rotating system, we apply (7.1.3) to the velocity **u** given by (7.1.1):

$$\left(\frac{D\mathbf{u}}{Dt}\right)_{stationary} = \left(\frac{D[\mathbf{w}+\boldsymbol{\Omega}\times\mathbf{r}]}{Dt}\right)_{rotating} + \boldsymbol{\Omega}\times[\mathbf{w}+\boldsymbol{\Omega}\times\mathbf{r}]. \tag{7.1.7}$$

In (7.1.7) the velocity observed in the stationary system is denoted by **u**, the velocity observed in the rotating system by **w**, and the subscripts indicate to which coordinate system the derivatives are referred. Carrying out the differentiations and restricting the development to constant angular velocity, the situation of most interest, leads to

$$\left(\frac{D\mathbf{u}}{Dt}\right)_{stationary} = \left(\frac{D\mathbf{w}}{Dt}\right)_{rotating} + \boldsymbol{\Omega} \times (\boldsymbol{\Omega} \times \mathbf{r}) + 2\boldsymbol{\Omega} \times \mathbf{w}. \tag{7.1.8}$$

The angular velocity $\boldsymbol{\Omega}$ of the rotating system is also taken to be constant in the rest of the chapter.

The momentum equation can be written in terms of relative (rotating) frame accelerations as (neglecting external body forces)

$$\left(\frac{D\mathbf{w}}{Dt}\right)_{rotating} = \frac{\partial \mathbf{w}}{\partial t} + (\mathbf{w} \cdot \nabla)\,\mathbf{w} = -\frac{1}{\rho}\nabla p + \mathbf{F}_{visc} - \boldsymbol{\Omega} \times (\boldsymbol{\Omega} \times \mathbf{r}) - 2\boldsymbol{\Omega} \times \mathbf{w}. \tag{7.1.9}$$

The interpretation of (7.1.9) is that the real forces felt in the inertial system must be modified by the presence of reaction terms, or "fictitious forces", which are a consequence of observing the motion from an accelerated reference frame.

Using (7.1.1) in the expression for viscous stresses given in Section 1.13 shows that $\mathbf{F}_{visc}$ takes the same form as in a stationary system with $\mathbf{w}$ replacing $\mathbf{u}$ and with the spatial derivatives evaluated in the rotating frame. This is because a rigid body rotation leads to no local strain and hence no stress.

The momentum equation is changed because of the presence of the last two terms in (7.1.9), known as centrifugal and Coriolis accelerations respectively. Regarding these terms as fictitious forces per unit mass allows the momentum equation in the rotating system to have a similar form to that in the stationary system. It should be kept in mind, however, that these two terms do not represent actual forces but are rather kinematic consequences of viewing the motion from a rotating coordinate system.

7.1.2 Rotating coordinate systems and Coriolis accelerations

The expressions for Coriolis accelerations were developed in a formal manner, and it is useful to derive the result from another perspective which brings out the physical significance more directly (Den Hartog, 1948). We begin by considering one-dimensional incompressible flow in a constant area channel rotating with angular velocity, $\boldsymbol{\Omega}$, around an axis at 0, as drawn in Figure 7.1. The particles in the channel move radially outwards with a constant radial velocity, w_r.

The absolute[1] acceleration of a fluid particle can be calculated by examining the absolute velocity at two instants a short time, dt, apart, when the particle is at positions 1 and $2'$. In the absolute system, the path of the particle is a spiral. The absolute velocity at point 1, at radius r, is the vector sum of the radial velocity, w_r, and the circumferential velocity of the channel at that point, Ωr. The vector addition is similar for point $2'$ at $r + dr$, but the channel velocity at $r + dr$ is $\Omega(r + dr) = \Omega(r + w_r dt)$. The absolute acceleration is the difference between the two absolute velocities divided by the time interval, dt. The components used in calculating the velocity difference are referred to the directions parallel to, and perpendicular to, the line 0–1–2. For the small time interval the terms

[1] The terms "absolute" and "relative" are in common use in the fluid machinery community to denote the velocities and accelerations in the stationary (inertial) and rotating frames of reference. We adopt this usage from here on. The substantial derivative in the rotating system is thus denoted as $[D/Dt]_{rel}$.

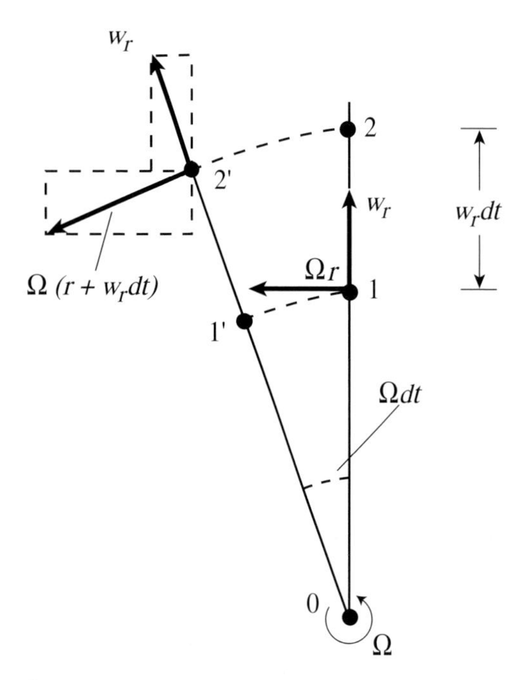

Figure 7.1: Fluid particle motion in a rotating straight channel as seen in the stationary system; w_r (radial velocity) = constant.

sin $\Omega\, dt$ and cos $\Omega\, dt$ which appear in writing the two components can be approximated by Ωdt and 1 respectively. In the direction parallel to 0–1–2, therefore, working to first order in dt,

$$du = [w_r - \Omega\,(r + w_r dt)\,\Omega dt] - w_r = -\Omega^2 r dt,$$

or

$$\text{absolute acceleration in the radial direction} = \left[\frac{du}{dt}\right]_{radial} = -\Omega^2 r. \tag{7.1.10}$$

In the direction perpendicular to 0–1–2, the velocity change is

$$du = [\Omega\,(r + w_r dt) + w_r \Omega dt] - \Omega r = 2\Omega w_r dt,$$

or

$$\text{absolute acceleration in the circumferential direction} = \left[\frac{du}{dt}\right]_{circumferential} = 2\Omega w_r. \tag{7.1.11}$$

The absolute acceleration consists of two components, one radial, $-\Omega^2 r$, and one circumferential and to the left, $2\Omega w_r$. The former can be referred to as the rotating frame acceleration (the acceleration of the channel at the particular location of interest). The latter is the Coriolis acceleration. This nomenclature provides a useful statement of the different "pieces" that make up the absolute acceleration, which can be described as the vector sum of three components: the relative acceleration, the rotating frame acceleration, and the Coriolis acceleration. The Coriolis acceleration is perpendicular

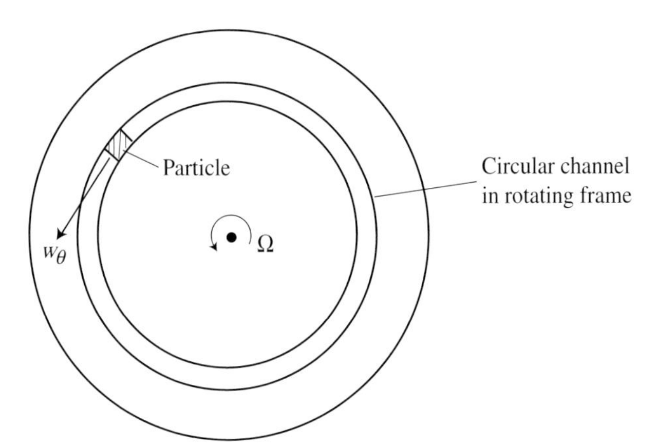

Figure 7.2: Particle motion on a concentric circular channel in a rotating frame.

to the relative velocity and to the angular velocity vector and has the magnitude $2\Omega w_\perp$, where $w_\perp$ is the component of the relative velocity perpendicular to the axis of rotation. This statement is seen to be true for the radial velocity, and we show below its application in general.

A second demonstration of the statement is steady motion, with relative velocity w_θ in the circumferential direction, in a thin circular channel rotating around the axis of symmetry, as in Figure 7.2. The absolute velocity of the fluid is $w_\theta + \Omega r$ and its path is a circle of radius r, so the acceleration in the inertial frame is in the radial direction with magnitude given by

$$\begin{array}{l}\text{magnitude of} \\ \text{acceleration} \\ \text{in inertial frame}\end{array} \quad \nabla = \frac{(w_\theta + \Omega r)^2}{r} = \underset{\text{(a)}}{\frac{w_\theta^2}{r}} + \underset{\text{(b)}}{2\Omega w_\theta} + \underset{\text{(c)}}{\Omega^2 r}.$$

As before, the absolute acceleration can be separated into three parts: (a) the relative acceleration, which is the acceleration seen in the rotating coordinate system; (b) the Coriolis acceleration; and (c) the rotating frame or centripetal acceleration. All are radially inward and there is a corresponding radial pressure gradient:

$$\frac{dp}{dr} = \rho\left(\frac{w_\theta^2}{r} + 2\Omega w_\theta + \Omega^2 r\right). \tag{7.1.12}$$

In terms of an observer in the rotating system, the perception is that Coriolis and centrifugal forces act to oppose this pressure gradient so the only acceleration seen is w_θ^2/r. For the relative frame (7.1.12) would therefore be rearranged as

$$\frac{w_\theta^2}{r} = \frac{1}{\rho}\frac{dp}{dr} - 2\Omega w_\theta - \Omega^2 r. \tag{7.1.13}$$

Equation (7.1.13) demonstrates how Coriolis and centripetal accelerations enter the momentum equation as apparent forces per unit mass.

The last case considered is the relative velocity parallel to the axis of rotation, as in Figure 7.3. The absolute velocity of the particle in space has a component parallel to the axis of rotation and a

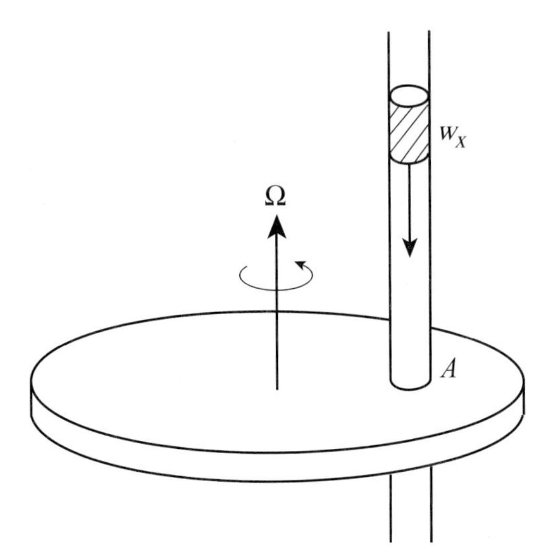

Figure 7.3: Particle motion with relative velocity parallel to the axis of rotation.

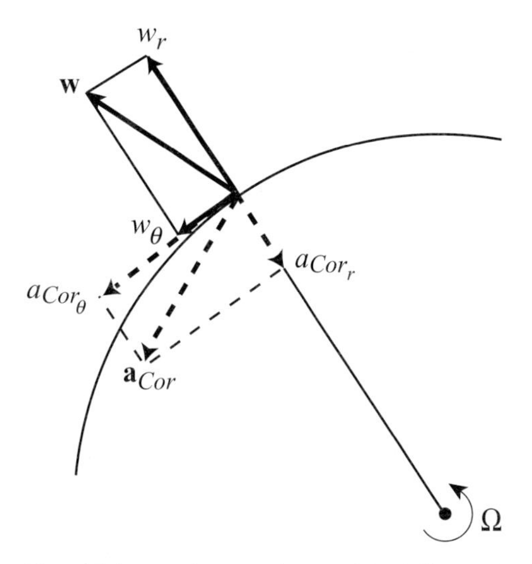

Figure 7.4: Relative velocities and Coriolis accelerations.

circumferential component Ωr. The absolute acceleration is equal to the rotating frame acceleration and there is no Coriolis acceleration.

We now extend the above three special cases to particle motion with all three velocity components (axial, radial, and circumferential). As just described, the axial component does not contribute to the Coriolis acceleration. The other two components lie in the plane of rotation so the resulting Coriolis acceleration is also in that plane. Figure 7.4 shows relative velocities, indicated by the solid lines, and Coriolis accelerations, indicated by dashed lines. The resultant Coriolis acceleration is perpendicular to the resultant relative velocity vector and proportional to it, in accordance with the general statement.

7.1.3 Centrifugal accelerations in a uniform density fluid: the reduced static pressure

The term $\mathbf{\Omega} \times (\mathbf{\Omega} \times \mathbf{r})$, which occurs in the momentum equation, (7.1.9), can be written as $-\nabla(\Omega^2 r^2/2)$, where r represents, the distance from the axis of rotation.[2] For a fluid of uniform density, this term, which is identified with the centrifugal force, can be combined with the static pressure to form the reduced static pressure, $p - \frac{1}{2}\rho\Omega^2 r^2$. Working in terms of the reduced static pressure is similar to the procedure of subtracting out the hydrostatic pressure to eliminate the (non-dynamical) effects of gravitational forces in a uniform fluid; as seen from (7.1.9), it is gradients in reduced static pressure that cause accelerations in the relative system. An illustration is a fluid in solid-body rotation, i.e. no relative motion. For this case, the pressure field is $p - p_{axis} = \frac{1}{2}\rho\Omega^2 r^2$, the pressure gradient is $\nabla p = \rho\Omega^2 \mathbf{r}$, and the reduced static pressure is constant throughout the fluid. For a uniform density fluid, provided none of the boundary conditions involve static pressure, it is useful to work in terms of reduced static pressure.

The reduced static pressure can also be interpreted in terms of a measurement in rotating machinery (Moore, 1973a). Suppose that static pressure taps are located on the blades of a turbomachine at a radial location r, but the pressure is recorded by a transducer located on the axis. The fluid in the tubing connecting the axis to the pressure tap at r is in hydrostatic equilibrium (due to the pressure gradient dp/dr and the centrifugal force $\rho\Omega^2 r$) so the pressure difference between the tap and the axis is $\rho\Omega^2 r^2/2$. The reduced static pressure can therefore be viewed as the pressure one would obtain from a measuring device located on the axis of rotation.

7.2 Illustrations of Coriolis and centrifugal forces in a rotating coordinate system

The role played by Coriolis and centrifugal forces is sometimes difficult to see clearly in flows that are geometrically complex. To demonstrate the origin of these forces, we present a situation in which the flow can be simply examined in both stationary and rotating frames of reference. The specific configuration addressed is inviscid, constant density, two-dimensional flow due to a combined source and vortex at the origin. The velocity field is axisymmetric, and the velocity components in the stationary system are

$$u_r = \frac{Q_V}{2\pi r}, \tag{7.2.1a}$$

where Q_V is the volume flow rate per unit height, and

$$u_\theta = \frac{\Gamma}{2\pi r}, \tag{7.2.1b}$$

[2] Although $\mathbf{r}$ was introduced as a position vector from the origin, the component parallel to the axis has zero contribution to $(\mathbf{\Omega} \times \mathbf{r})$. We can thus interpret $\mathbf{r}$ in the term $\mathbf{\Omega} \times (\mathbf{\Omega} \times \mathbf{r})$ as marking distance from the axis of rotation. Using the vector identity $\mathbf{A} \times (\mathbf{B} \times \mathbf{C}) = \mathbf{B}(\mathbf{A} \cdot \mathbf{C}) - \mathbf{C}(\mathbf{A} \cdot \mathbf{B})$, the quantity $\mathbf{\Omega} \times (\mathbf{\Omega} \times \mathbf{r}) = -r\Omega^2 \mathbf{e}_r$, where $\mathbf{e}_r$ is the unit vector in the r-direction. The gradient of a scalar in cylindrical coordinates is

$$\nabla = \frac{\partial}{\partial r}\mathbf{e}_r + \frac{1}{r}\frac{\partial}{\partial \theta}\mathbf{e}_\theta + \frac{\partial}{\partial z}\mathbf{e}_z \text{ and } \nabla(-\Omega^2 r^2/2) = -r\Omega^2 \mathbf{e}_r$$

which is equal to $\mathbf{\Omega} \times (\mathbf{\Omega} \times \mathbf{r})$.

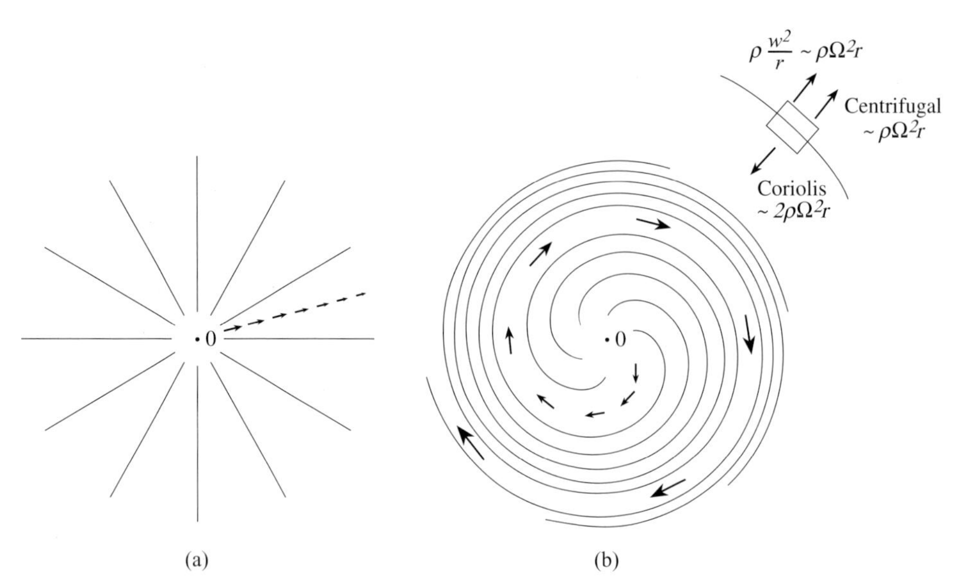

Figure 7.5: Source flow viewed from stationary and rotating coordinate systems: (a) stationary system; (b) rotating system.

where Γ is the circulation. In the rotating coordinate system, the radial velocity is the same but the circumferential velocity is given by

$$w_\theta = \frac{\Gamma}{2\pi r} - \Omega r. \tag{7.2.1c}$$

Consider first the case $\Gamma = 0$. Streamlines and velocity vectors in the stationary system are given in Figure 7.5(a). The solid lines illustrate streamlines with the length of the arrows proportional to the magnitude of the velocity vectors. In the stationary system, the streamlines extend radially outward from the axis at 0.

The flow seen in the rotating system is shown in Figure 7.5(b). The streamlines are now spirals curving to the right as the flow moves radially outward. The relative velocity vectors (each of which represents the velocity at the midpoint of the arrow) increase in magnitude with radius. The relative streamlines are strongly curved; from the viewpoint of an observer in the rotating system, it is the Coriolis forces that cause the streamline curvature.

As the radius increases, the relative velocity inclines more and more towards the circumferential direction. Equations (7.2.1) show that at large radii ($r \gg u_r/\Omega$) the absolute velocity is small compared to the relative velocity, implying that the static pressure gradient is small compared to the Coriolis and centrifugal forces and the streamlines in the relative frame are nearly concentric circles. In these regions the normal momentum equation in the relative system is essentially a balance between accelerations due to streamline curvature, $\rho w_\theta^2/r$, centrifugal forces, $\rho\Omega^2 r$, and Coriolis forces, $2\rho\Omega w_\theta$, with magnitudes and directions as indicated by Figure 7.5(b). The figure emphasizes again that the centrifugal force and the Coriolis force arise as kinematic consequences of describing the motion in a rotating system.

An example closer to a practical flow geometry is shown in the stationary and rotating system velocity fields of Figures 7.6(a) and (b). The flow in the stationary system now has a substantial

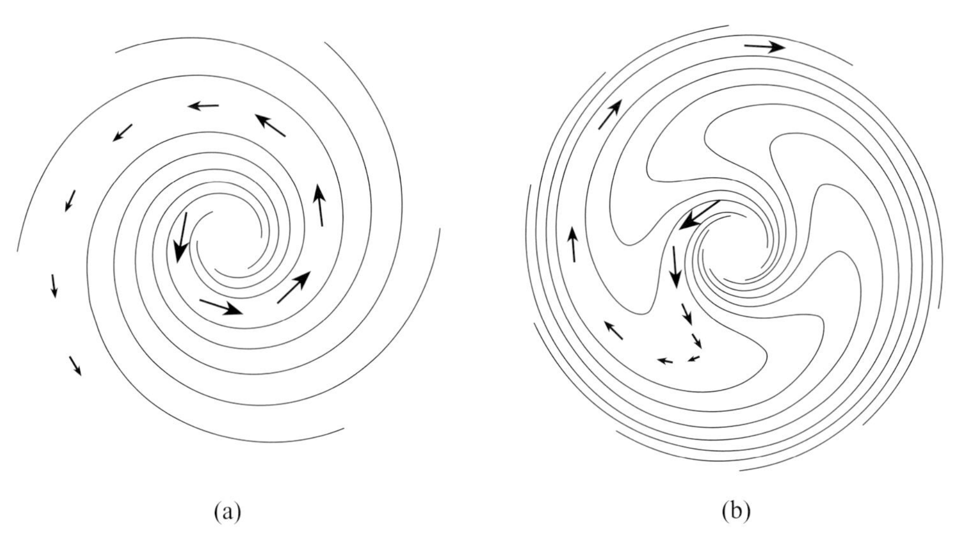

Figure 7.6: Swirling flow (combined vortex/source with $\Gamma/Q_V = 5$) viewed from stationary and rotating coordinate systems: (a) stationary system; (b) rotating system.

swirl velocity, $\Gamma/Q_V = 5$, or $u_\theta/u_r = 5$, as might be representative of the flow leaving a radial impeller.

In the stationary system, the streamlines are spirals having constant angle with the radial direction. (Both u_r and u_θ are inversely proportional to the radius so their ratio is invariant with radius.) In the relative system, the curvature of the streamlines is initially concave to the left in the region close to the inner radius of the picture, because the radial pressure gradient, which is the only "actual" force, is important. As the radius increases, the influence of the pressure gradient decreases, while that of the Coriolis and centrifugal forces increases. The curvature of the streamlines therefore becomes concave to the right and the direction of motion of the particle changes. At large radius, the balance is between relative frame streamline curvature and Coriolis and centrifugal forces, as in the previous example.

7.3 Conserved quantities in a steady rotating flow

For steady adiabatic flow in a stationary system, with no work transfer between streamlines, the stagnation enthalpy is constant along a streamline (Section 1.8). If the flow can be considered frictionless, the stagnation pressure is also constant along the streamline. Analogous conserved flow quantities exist in a steady rotating flow and serve as useful constraints in analyzing fluid motions in rotating systems.

To derive the conserved quantities we take the scalar product of the momentum equation (7.1.9) with $\mathbf{w}$ to yield an equation for the change in mechanical energy of a fluid particle seen in the rotating (relative) frame:

$$\rho\left(\frac{D}{Dt}\frac{w^2}{2}\right)_{rel} = -\mathbf{w}\cdot\boldsymbol{\nabla}p + \rho\mathbf{w}\cdot\boldsymbol{\nabla}\left(\frac{\Omega^2 r^2}{2}\right) + w_i\frac{\partial \tau_{ij}}{\partial x_j}. \tag{7.3.1}$$

As mentioned previously $(D/Dt)_{rel}$ means the substantial derivative following a particle in the relative (rotating) frame. The Coriolis force acts perpendicularly to $\mathbf{w}$ and makes no contribution to the change of mechanical energy of a fluid particle.

The internal energy equation, (1.10.2), can be written in the rotating system as

$$\rho\left(\frac{De}{Dt}\right)_{rel} = -p\nabla\cdot\mathbf{w} - \frac{\partial q_i}{\partial x_i} + \tau_{ij}\frac{\partial w_i}{\partial x_j} + \dot{Q}. \tag{7.3.2}$$

Combining (7.3.1) and (7.3.2) gives

$$\rho\left[\frac{D}{Dt}\left(e + \frac{w^2}{2} - \frac{\Omega^2 r^2}{2}\right)\right]_{rel} = -\nabla\cdot p\mathbf{w} - \frac{\partial q_i}{\partial x_i} + \frac{\partial(w_i\tau_{ij})}{\partial x_j} + \dot{Q}. \tag{7.3.3}$$

Use of the continuity equation allows (7.3.3) to be rewritten in terms of the quantity we seek:

$$\rho\left[\frac{D}{Dt}\left(e + \frac{p}{\rho} + \frac{w^2}{2} - \frac{\Omega^2 r^2}{2}\right)\right]_{rel} =$$

$$\rho\left(\frac{DI_t}{Dt}\right)_{rel} = \frac{\partial p}{\partial t} - \frac{\partial q_i}{\partial x_i} + \frac{\partial(w_i\tau_{ij})}{\partial x_j} + \dot{Q}. \tag{7.3.4}$$

The quantity I_t is termed *rothalpy*. It appears often in problems involving rotating machinery and is defined as

$$I_t = h + \frac{w^2}{2} - \frac{\Omega^2 r^2}{2} = (h_t)_{rel} - \frac{\Omega^2 r^2}{2}. \tag{7.3.5}$$

In (7.3.5) $(h_t)_{rel}$ is the stagnation enthalpy $(h + \frac{1}{2}w^2)$ as measured in the rotating system. Equation (7.3.4) implies that a change in rothalpy for a fluid particle can result from flow unsteadiness, heat transfer, work done by viscous stresses (or real body forces, which are not considered here), or internal heat sources.

For an adiabatic steady rotating flow with no work transfer, or for the less restrictive situation in which the sum of shear work on, and the heat transfer to, a given streamline is zero, (7.3.4) reduces to

$$\mathbf{w}\cdot\nabla I_t = 0. \tag{7.3.6}$$

Equation (7.3.6) is a statement that rothalpy is conserved along a relative streamline. This is true as long as there is no net energy transfer between the streamtube and its surroundings, even if the flow is irreversible. If the flow on the streamline of interest can be considered frictionless with no heat transfer, entropy is also conserved along a relative streamline. Rothalpy in a rotating system thus plays an analogous role to stagnation enthalpy in a stationary system.

One can use conservation of rothalpy to derive the Euler turbine equation ((2.8.27), $h_{t_2} - h_{t_1} = \Omega(r_2 u_{\theta_2} - r_1 u_{\theta_1})$) from a different point of view than given in Section 2.8. The steps in the procedure are to set the inlet and exit rothalpy equal, split the rothalpy into enthalpy and kinematic quantities, and then write out the velocity components and use the relation between relative and absolute circumferential velocity $(u_\theta = w_\theta + \Omega r)$ to relate the change in stagnation enthalpy of a fluid particle to the change in the tangential component of the absolute velocity. This gives a complementary view of the approximations made (steady relative flow, no net energy transfer to the relative streamtube) in applying the Euler turbine equation.

For incompressible flow, the analogous quantity is the reduced stagnation pressure[3] $p_{t_{red}}$:

$$p_{t_{red}} = (p_t)_{rel} - \frac{\rho\Omega^2 r^2}{2} = p_t - \rho \mathbf{u} \cdot (\boldsymbol{\Omega} \times \mathbf{r}), \tag{7.3.7}$$

where $(p_t)_{rel}$ is the stagnation pressure, $p + \frac{1}{2}\rho w^2$, as measured in the rotating system. For inviscid flow $p_{t_{red}}$ is conserved along a relative streamline and the Euler turbine equation becomes (2.8.28), $p_{t_2} - p_{t_1} = \rho\Omega(r_2 u_{\theta_2} - r_1 u_{\theta_1})$.

7.4 Phenomena in flows where rotation dominates

7.4.1 Non-dimensional parameters: the Rossby and Ekman numbers

When effects of rotation become dominant, fluid motions exhibit properties quite different from those with no rotation. To define this regime it is necessary to develop a measure of the importance of rotation in a given situation. For a uniform density fluid the momentum equation can be written in terms of reduced pressure so the centrifugal force does not explicitly appear. For steady flow (7.1.9) is thus

$$(\mathbf{w} \cdot \nabla)\,\mathbf{w} = -\frac{1}{\rho}\nabla p_{red} - 2\boldsymbol{\Omega} \times \mathbf{w} + \nu\nabla^2\mathbf{w}. \tag{7.4.1}$$

If w_{ref} and L are representative velocity and length scales for the flow of interest, (7.4.1) can be put in non-dimensional form as

$$\left[\frac{w_{ref}}{\Omega L}\right](\tilde{\mathbf{w}} \cdot \nabla)\,\tilde{\mathbf{w}} = -\nabla \tilde{p}_{red} - 2\mathbf{k} \times \tilde{\mathbf{w}} + \left[\frac{\nu}{\Omega L^2}\right]\nabla^2\tilde{\mathbf{w}}, \tag{7.4.2}$$

where the tilde (˜) denotes non-dimensional variables and where $\mathbf{k}$ is the unit vector in the direction of the axis of rotation. The two terms in the square brackets are non-dimensional parameters which characterize the importance of rotation and of viscous effects respectively.

The parameter $w_{ref}/(\Omega L)$ gives a measure of the ratio of relative flow accelerations to Coriolis accelerations (or, equivalently, relative frame inertia forces to Coriolis forces). It is known as the Rossby number, *Ro*. Flows in which rotation dominates have Rossby numbers much less than unity. In flows with Rossby numbers much larger than unity effects of rotation are not likely to be significant. Turbomachinery tends to have Rossby numbers of order unity (generally the relative velocity has comparable magnitude to the wheel speed) so both Coriolis and relative accelerations can be important.

One application in which low Rossby number phenomena are important is meteorological flows in which the length scales are hundreds or thousands of kilometers and, even with the small value of the Earth's rotation, the Rossby number can still be much less than unity. For example, if the relative fluid velocity is 20 m/s (which is a strong wind) and the length scale is 10^3 km, at a latitude of 45° the Rossby number is less than 0.4. Effects of rotation are important for this choice of parameters. For larger scale weather patterns or lower wind speeds they dominate the flow pattern.

The term $\nu/(\Omega L^2)$, referred to as the Ekman number, *Ek*, represents a ratio between viscous and Coriolis forces. For a small Ekman number we expect thin viscous layers, whereas for a large Ekman

[3] In some treatments this is referred to as the rotary stagnation pressure.

number viscous effects are felt throughout the flow domain. The Reynolds number ($Re = w_{ref}L/\nu$) is related to the Rossby and Ekman numbers by ($Re = Ro/Ek$) so that any two of the three parameters *Ro*, *Re*, and *Ek* (plus the geometry and boundary conditions) characterize the flow.

7.4.2 Inviscid flow at low Rossby number: the Taylor–Proudman Theorem

For steady flow at low Rossby number, the term $(\mathbf{w} \cdot \nabla)\mathbf{w}$ in (7.4.1) is negligible compared to the Coriolis and pressure gradient terms. With the z-axis as the axis of rotation the components of the inviscid ($Ek = 0$) momentum equation are:

$$-2\Omega w_x = -\frac{1}{\rho}\frac{\partial p_{red}}{\partial y}, \tag{7.4.3a}$$

$$2\Omega w_y = -\frac{1}{\rho}\frac{\partial p_{red}}{\partial x}, \tag{7.4.3b}$$

$$0 = -\frac{1}{\rho}\frac{\partial p}{\partial z}. \tag{7.4.3c}$$

Taking the x-derivative of (7.4.3a) and the y-derivative of (7.4.3b) yields

$$\frac{\partial w_x}{\partial x} + \frac{\partial w_y}{\partial y} = 0. \tag{7.4.4}$$

Comparing (7.4.4) with the continuity equation for an incompressible flow,

$$\frac{\partial w_x}{\partial x} + \frac{\partial w_y}{\partial y} + \frac{\partial w_z}{\partial z} = 0, \tag{1.9.6}$$

leads to the result

$$\frac{\partial w_z}{\partial z} = 0. \tag{7.4.5}$$

For low Rossby number flows, the physical interpretation of (7.4.3) and (7.4.5) is that w_x, w_y, and w_z are functions of x and y only and the velocity and pressure fields are the same at any station along the z-direction (the axis of rotation). Further, if the boundary condition is that w_z is zero on any plane perpendicular to the axis of rotation, it is zero throughout the flow field. These remarkable results, which are known as the Taylor–Proudman Theorem, are often expressed in the statement that slow steady inviscid motion of a rotating incompressible fluid must be two-dimensional (Batchelor, 1967; Tritton, 1988).

From (7.4.3) a further consequence of a low Rossby number can be inferred namely that the relative velocity, $\mathbf{w}$, is perpendicular to the gradient of reduced static pressure, ∇p_{red}, i.e. the relative velocity is parallel to lines of constant reduced static pressure. As illustration, Figure 7.7 shows sketches of streamlines and isobars (lines of constant static pressure) for a two-dimensional channel. The pictures on the left correspond to stationary ($Ro \to \infty$) inviscid motion and those on the right to low Rossby number inviscid motion in a rotating system. The streamline pattern, shown in the upper two figures, is sketched as roughly similar in both cases, but the contours of constant static pressure (for stationary flow) and constant reduced pressure (for rotating flow), and hence the pressure gradients, are quite

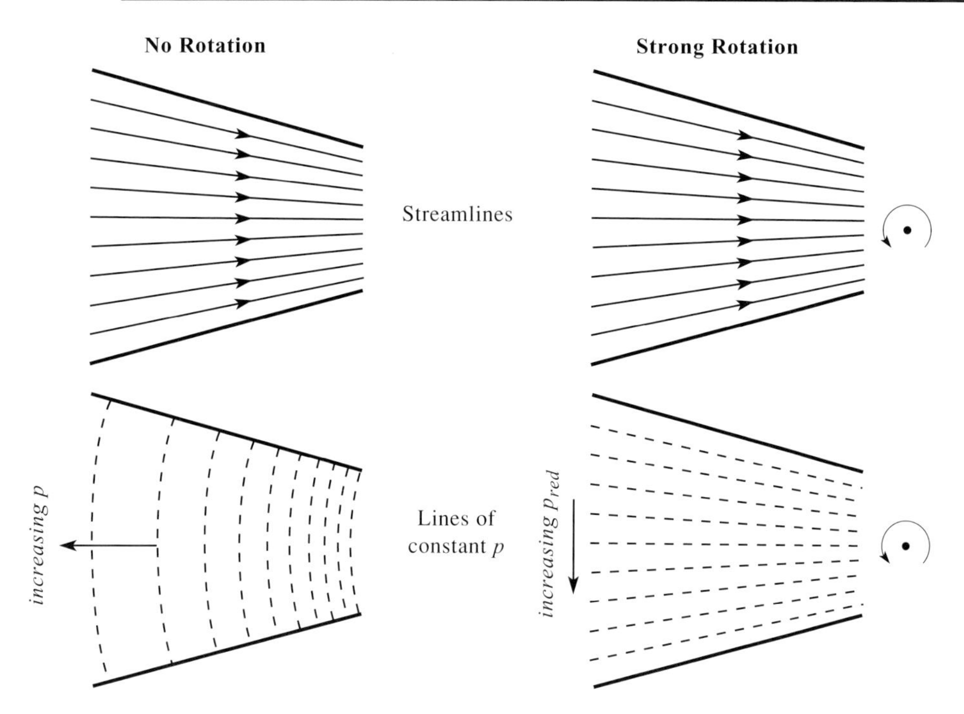

Figure 7.7: Streamlines and isobars in a converging channel with no rotation and with strong rotation (low Rossby number).

different. In the stationary case the isobars are perpendicular to the streamlines. At low Rossby number in the rotating flow the isobars of reduced pressure are aligned with the streamlines. For the stationary channel, the direction of the pressure gradient is independent of the direction of flow but for the rotating channel if the direction of flow is reversed so is the sense of the reduced pressure gradient.

7.4.3 Viscous flow at low Rossby number: Ekman layers

For steady viscous flow at low Rossby numbers, (7.4.1) takes the form

$$0 = -\frac{1}{\rho}\nabla p_{red} - 2\boldsymbol{\Omega} \times \mathbf{w} + \nu\nabla^2\mathbf{w}. \tag{7.4.6}$$

Equation (7.4.6) is linear, with components

$$-2\Omega w_y = -\frac{1}{\rho}\frac{\partial p_{red}}{\partial x} + \nu\frac{\partial^2 w_x}{\partial z^2}, \tag{7.4.7a}$$

$$2\Omega w_x = -\frac{1}{\rho}\frac{\partial p_{red}}{\partial y} + \nu\frac{\partial^2 w_y}{\partial z^2}, \tag{7.4.7b}$$

$$0 = \frac{\partial p_{red}}{\partial z}. \tag{7.4.7c}$$

For a uniform free-stream over a plane surface (which we set at $z = 0$) perpendicular to the axis of rotation an analytic solution to the low Rossby number equations exists. The x-axis is taken to be aligned with the free-stream flow. Away from the region where viscosity is important the flow is therefore uniform in the x-direction, of magnitude w_{Ex}, and

$$\frac{\partial p_{red}}{\partial z} = \frac{\partial p_{red}}{\partial x} = 0, \tag{7.4.8a}$$

$$\frac{\partial p_{red}}{\partial y} = -2\rho\Omega w_{E_x}. \tag{7.4.8b}$$

Using (7.4.8b), (7.4.7b) may be put in the form

$$2\Omega(w_x - w_{E_x}) = \nu\frac{\partial^2 w_y}{\partial z^2}. \tag{7.4.9}$$

Equations (7.4.9) and (7.4.7a) (with $\partial p_{red}/\partial x = 0$) are two coupled equations for w_x and w_y. The boundary conditions on the velocity components are

$$w_x = w_y = 0 \qquad \text{at } z = 0, \tag{7.4.10a}$$

$$\left\{\begin{matrix} w_x \to w_{E_x} \\ w_y \to 0 \end{matrix}\right\} \qquad \text{as } z \to \infty. \tag{7.4.10b}$$

Eliminating w_y from (7.4.7a) and (7.4.9) yields a fourth order linear equation for w_x:

$$\frac{\partial^4}{\partial z^4}(w_x - w_{E_x}) + \frac{4\Omega^2}{\nu^2}(w_x - w_{E_x}) = 0. \tag{7.4.11}$$

Defining $\Delta = \sqrt{\nu/\Omega}$ as a viscous length scale for this problem, solutions of (7.4.11) (or the coupled (7.4.7a) and (7.4.9)) which satisfy the boundary conditions are:

$$w_x = w_{E_x}\left[1 - e^{-z/\Delta}\cos\left(\frac{z}{\Delta}\right)\right], \tag{7.4.12a}$$

$$w_y = w_{E_x}\left[e^{-z/\Delta}\sin\left(\frac{z}{\Delta}\right)\right]. \tag{7.4.12b}$$

The velocity distribution of (7.4.12) is referred to as an Ekman layer. It is independent of both x and y. Using the continuity equation for incompressible flow, which has the form $\partial w_z/\partial z = 0$, and the normal velocity boundary condition ($w_z = 0$) at the wall it is seen that w_z is zero throughout the flow.

The Ekman layer profile is depicted in Figure 7.8, which shows theoretical and measured velocities and flow angles at a Rossby number of 0.125. Features of this viscous layer which differ from the non-rotating situation are: flow angle variation through the layer, from 0° at the edge (i.e. aligned with the main stream) to 45° at the wall; invariance of layer thickness with x and y position; and a velocity magnitude within the layer which is higher than in the free stream. In a non-rotating flow, for example a zero pressure gradient boundary layer, shear forces continually decrease the momentum of the flow and the boundary layer thickness grows with downstream distance. In the Ekman layer, there is a component of the Coriolis force opposite to the viscous forces in both the x- and y-directions, which is sufficient to maintain the thickness at a constant level.

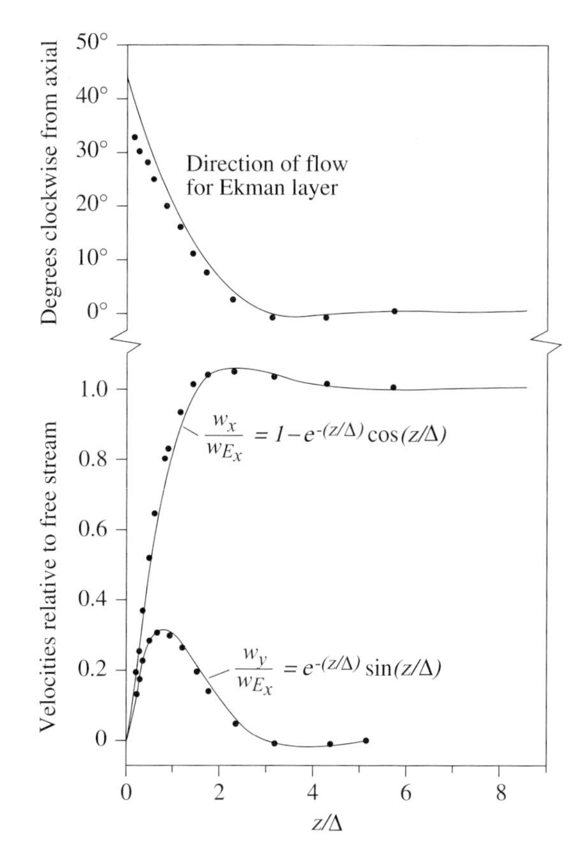

Figure 7.8: Velocity profiles for a laminar Ekman layer; Rossby number = 0.125; $\Delta = \sqrt{\nu/\Omega}$ (data of Tatro and Mollo-Christensen (1967)).

The length scale, $\sqrt{\nu/\Omega}$, which characterizes the thickness of the region in which viscous effects are significant can also be obtained from estimates of viscous and Coriolis forces. If the viscous layer is influenced by Coriolis forces, the two are the same order of magnitude. The Ekman number based on the boundary layer thickness is thus of order unity, and if δ is the thickness of the layer:

$$\delta \approx \sqrt{\frac{\nu}{\Omega}}. \tag{7.4.13}$$

Another solution of (7.4.7), which is more relevant for discussion of internal flows, is the rotating system analog of plane Poiseuille flow. This is flow between two parallel walls a distance H apart in a system rotating at velocity Ω, under the influence of a constant reduced pressure gradient (Bark, 1996),

$$\frac{\partial p_{red}}{\partial x} = \text{constant}. \tag{7.4.14}$$

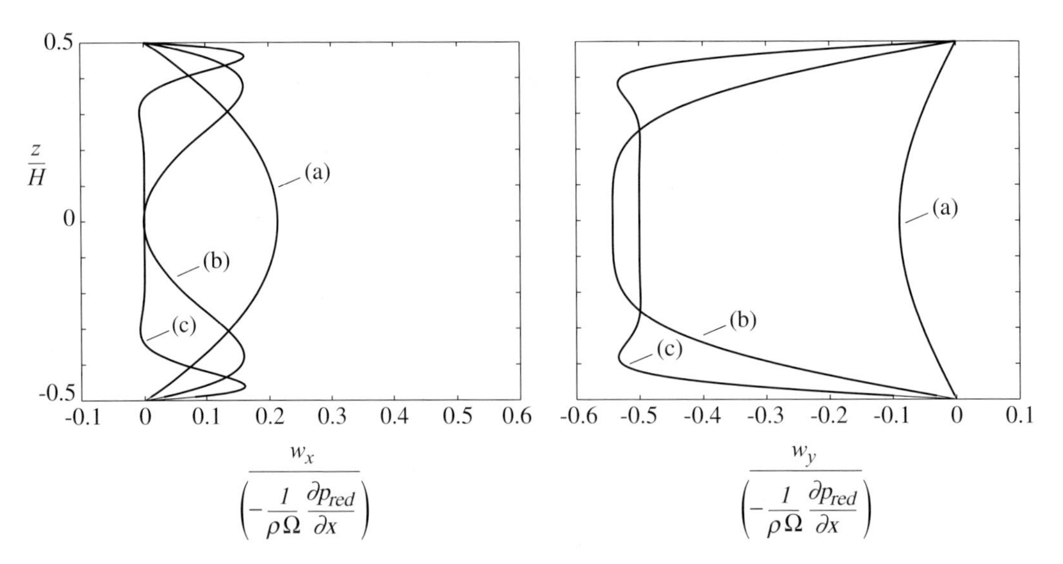

Figure 7.9: Velocity profiles in a rotation-modified plane Poiseuille flow: (a) $Ek = 1$, (b) $Ek = 0.1$, (c) $Ek = 0.01$ (Bark, 1996).

The walls are perpendicular to the axis of rotation and the boundary conditions are that the velocity is zero on the walls, $\mathbf{w} = 0$ at $z = \pm H/2$. The solution to (7.4.7) is given compactly as

$$\frac{w_x - iw_y}{\left(-\frac{1}{\rho\Omega}\frac{\partial p_{red}}{\partial x}\right)} = \frac{i}{2}\left\{1 - \frac{\cosh\left[\frac{(1-i)}{\sqrt{Ek}}\left(\frac{2z}{H}\right)\right]}{\cosh\left[\frac{(1-i)}{\sqrt{Ek}}\right]}\right\}, \tag{7.4.15a}$$

$$w_z = 0. \tag{7.4.15b}$$

The character of the solution is indicated in Figure 7.9, which shows plots of w_x and w_y for three values of the Ekman number ($\nu/(\Omega H^2)$). For large Ekman number the solution resembles that for the non-rotating situation, with the balance being basically between pressure gradient and viscous forces. For small values of the Ekman number, Ek, the solution has the asymptotic form

$$\frac{w_x - iw_y}{\left(-\frac{1}{\rho\Omega}\frac{\partial p_{red}}{\partial x}\right)} \approx \frac{i}{2}; \quad 1 \pm \left(\frac{2z}{H}\right) = O(1), \tag{7.4.16a}$$

$$\frac{w_x - iw_y}{\left(-\frac{1}{\rho\Omega}\frac{\partial p_{red}}{\partial x}\right)} \approx \frac{i}{2}\left\{1 - \exp\left[\frac{(i-1)}{\sqrt{Ek}}\left(1 \pm \frac{2z}{H}\right)\right]\right\}; \quad 1 \pm \frac{2z}{H} = O\left(\sqrt{\mathrm{Ek}}\right). \tag{7.4.16b}$$

The form of (7.4.16) is similar to that described in Section 6.9.2 for the high frequency limit of unsteady Poiseuille flow, with an inviscid core and two thin viscous layers near the walls, except here the thin layers are Ekman layers rather than Stokes layers. The free stream is a region in which Coriolis and pressure forces balance, and the velocity is perpendicular to the reduced pressure gradient.

7.5 Changes in vorticity and circulation in a rotating flow

As underpinning for discussion of three-dimensional flows in rotating systems it is useful to have reference to expressions for vorticity and circulation changes in a rotating flow. The relevant development is outlined below for uniform density incompressible flow. Taking the curl of the relation between absolute and relative velocities, (7.1.1), yields a relationship between the absolute vorticity ($\boldsymbol{\omega} = \boldsymbol{\nabla} \times \mathbf{u}$) and the vorticity observed in a rotating frame ($\boldsymbol{\omega}_{rel} = \boldsymbol{\nabla} \times \mathbf{w}$):

$$\boldsymbol{\omega} = \boldsymbol{\omega}_{rel} + 2\boldsymbol{\Omega}. \tag{7.5.1}$$

(Equation (7.5.1) should be no surprise if one recalls that vorticity is twice the local fluid angular velocity.)

To derive the equation for changes in $\boldsymbol{\omega}_{rel}$ we take the curl ($\boldsymbol{\nabla} \times [\ ]$) of the momentum equation, (7.1.9). The curl of the centrifugal acceleration term is zero since it is the curl of the gradient of a scalar. The curl of the Coriolis acceleration term is

$$\boldsymbol{\nabla} \times (2\boldsymbol{\Omega} \times \mathbf{w}) = -(2\boldsymbol{\Omega} \cdot \boldsymbol{\nabla})\,\mathbf{w}. \tag{7.5.2}$$

The rate of change of relative vorticity for a uniform density incompressible fluid is thus[4]

$$\frac{D\boldsymbol{\omega}_{rel}}{Dt} = (\boldsymbol{\omega}_{rel} \cdot \boldsymbol{\nabla})\,\mathbf{w} + (2\boldsymbol{\Omega} \cdot \boldsymbol{\nabla})\mathbf{w} + \nu\nabla^2\boldsymbol{\omega}_{rel}. \tag{7.5.3}$$

The term $(2\boldsymbol{\Omega} \cdot \boldsymbol{\nabla})\mathbf{w}$ in (7.5.3) does not appear for a stationary coordinate system. The consequence of its appearance is that in an inviscid rotating flow, relative vortex lines do not move with the fluid and the relative circulation about a material curve need not remain constant.

Reexamination of the Taylor–Proudman Theorem introduced in Section 7.4.2 provides an application of the concepts of relative vorticity and relative circulation and illustrates the behavior of these quantities in a flow with strong rotation (Tritton, 1988). We interpret this theorem from two different perspectives, first using the vorticity equation and then using the expression for the rate of change of circulation.

For inviscid flow (7.5.3) reduces to

$$\frac{D\boldsymbol{\omega}_{rel}}{Dt} - (\boldsymbol{\omega}_{rel} \cdot \boldsymbol{\nabla})\mathbf{w} = -2\Omega\frac{\partial \mathbf{w}}{\partial z} \tag{7.5.4}$$

with the axis of rotation along the z-direction. The two terms on the left-hand side of (7.5.4) represent the variations in the relative vorticity. The term on the right-hand side describes the change in magnitude and direction of the background vorticity ($2\boldsymbol{\Omega}$) associated with variations of the relative velocity field along the direction of the axis of rotation. If L is the length scale for the flow variation

[4] Equation (7.5.3) should be compared with the general expression for the rate of change of vorticity in a constant density fluid, (3.6.23) with $\boldsymbol{\nabla}\rho = 0$. Writing this in a rotating coordinate system with $\mathbf{X}$ representing external body forces,

$$\frac{D\boldsymbol{\omega}_{rel}}{dt} = (\boldsymbol{\omega}_{rel} \cdot \boldsymbol{\nabla})\mathbf{w} + \boldsymbol{\nabla} \times \mathbf{X} + \nu\nabla^2\boldsymbol{\omega}_{ref}.$$

The term involving the Coriolis acceleration in (7.5.3) appears as a (non-conservative) body force whose effect on the rate of relative vorticity production is equal to $\boldsymbol{\nabla} \times \mathbf{X}$.

along the axis of rotation and w_{ref} is a characteristic velocity magnitude, the term on the right-hand side has magnitude $\Omega w_{ref}/L$. The two terms on the left-hand side have magnitudes

$$(\boldsymbol{\omega}_{rel} \cdot \nabla \mathbf{w}), \quad (\mathbf{w} \cdot \nabla \boldsymbol{\omega}_{rel}) \approx \frac{w_{ref}^2}{L^2}.$$

The Rossby number can thus be interpreted as

$$Ro = \frac{w_{ref}}{\Omega L} \approx \frac{|\mathbf{w} \cdot \nabla \boldsymbol{\omega}_{rel}|}{2\Omega \dfrac{\partial \mathbf{w}}{\partial z}} \approx \frac{|\boldsymbol{\omega}_{rel}|}{\Omega}. \tag{7.5.5}$$

For Rossby numbers small compared to unity no "slow convection of small relative vorticity" (Lighthill, 1966) can balance the change in the large background vorticity associated with a variation in the velocity in the direction of the axis of rotation. More directly, at low Rossby numbers the inviscid vorticity equation reduces to

$$2\Omega \frac{\partial \mathbf{w}}{\partial z} \approx 0. \tag{7.5.6}$$

The relative velocity field cannot vary in the direction of the rotation axis and the flow is two-dimensional in planes perpendicular to the rotation axis. The (absolute) vortex tubes tend to remain parallel to the axis of rotation and resist bending, shrinking, or stretching.

The absolute circulation can be written in terms of the relative velocity and the angular velocity of rotation, $\boldsymbol{\Omega}$, as,

$$\Gamma = \oint_C \mathbf{w} \cdot d\boldsymbol{\ell} + \oint_C (\boldsymbol{\Omega} \times \mathbf{r}) \cdot d\boldsymbol{\ell}. \tag{7.5.7}$$

Using Stokes's Theorem, (7.5.7) becomes

$$\Gamma = \Gamma_{rel} + 2\Omega \iint dA_n, \tag{7.5.8}$$

where Γ_{rel} is the circulation seen in the relative frame, and A_n is the projection of the area enclosed by the contour onto a plane normal to the axis of rotation. For a constant density inviscid fluid with no external body force, $D\Gamma/Dt = 0$ (Sections 3.8 and 3.9) so

$$\frac{D\Gamma_{rel}}{Dt} = -2\Omega \frac{DA_n}{Dt}. \tag{7.5.9}$$

Equation (7.5.9) states that circulation round a fluid contour, as measured in the rotating system, alters when the area enclosed by the contour changes. An illustration of this concept is the radially outward flow described in Section 7.2. The absolute circulation round any contour of radius r in the stationary system is zero and the relative circulation at any radius is $\Gamma_{rel} = -2\pi r^2 \Omega$. The agent for the change in relative circulation as particles move outward is the non-conservative Coriolis force.

Equation (7.5.9) provides a further look at the Taylor–Proudman Theorem. Over a given time interval, the magnitude of changes in area and in circulation are related by

$$\frac{\Delta\Gamma_{rel}}{2A_n\Omega} \approx \frac{\Delta A_n}{A_n}. \tag{7.5.10}$$

Changes in relative circulation will be of order Lw_{ref} (or less). The left-hand side of (7.5.10) thus represents the ratio between the magnitude of the relative vorticity, w_{ref}/L, and the angular velocity of rotation, Ω, which is the Rossby number:

$$\frac{w_{ref}}{\Omega L} \approx \frac{\Delta A_n}{A_n}.$$

For small Rossby number, fractional changes in the area enclosed by any contour on a plane normal to the axis of rotation will be small and the area enclosed essentially constant. Applying this constraint to contours both with and without projections on planes normal to the axis of rotation leads to the conclusion that flows in which the projected areas remain constant must be two-dimensional.

7.6 Flow in two-dimensional rotating straight channels

7.6.1 Inviscid flow

Inviscid uniform density flow in a two-dimensional straight channel illustrates a number of features relevant to fluid machinery components. The channel has width W, and rotates around the z-axis with angular velocity of magnitude Ω, as shown in Figure 7.10 (Prandtl, 1952). The supply to the channel is from a reservoir in which the fluid is irrotational in the absolute (stationary) system. Such a configuration represents an approximation to flow in the radial section of a centrifugal impeller, into which irrotational flow is drawn from the atmosphere. The length/width ratio of the channel is taken as large enough that variations along the channel can be neglected compared to those across the channel. This carries with it the assumption that we are an appropriate distance from the inlet or

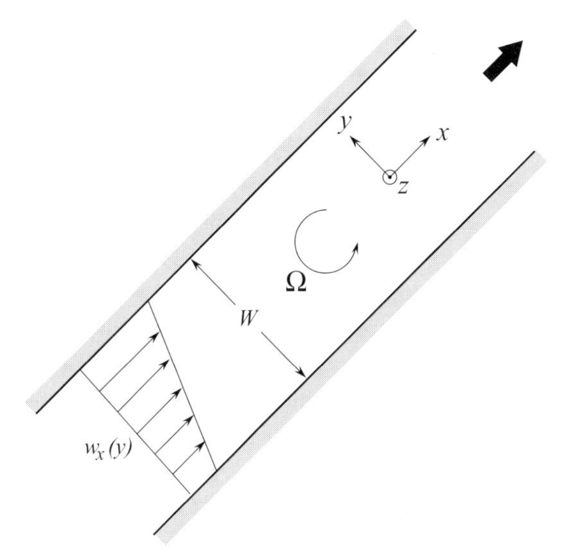

Figure 7.10: Two-dimensional inviscid flow in a rotating channel (x and y denote coordinates fixed in the rotating system); flow is irrotational in the absolute system.

exit of the channel, as described in more detail in Section 7.8. In terms of the relative frame x–y–z coordinate system sketched in Figure 7.10, the approximation made is that $\partial/\partial x = 0$.

The two-dimensional form of the continuity equation, plus the condition $\partial/\partial x = 0$, means that $\partial w_y/\partial y$ is zero. Because the y-component of velocity is zero at the channel wall, it is zero everywhere, and the only velocity component is w_x.

From Kelvin's Theorem the absolute flow remains irrotational. The relative vorticity, $\boldsymbol{\omega}_{rel}$, is given by

$$\boldsymbol{\omega}_{rel} = \boldsymbol{\nabla} \times \mathbf{w} = -2\boldsymbol{\Omega}. \tag{7.6.1}$$

The relative vorticity is in the z-direction, along the axis of rotation, with the value

$$(\omega_z)_{rel} = -\frac{dw_x}{dy} = -2\Omega. \tag{7.6.2}$$

If the flow rate per unit depth of the channel is $\overline{w}_x W$, the solution of (7.6.2) for w_x is

$$w_x = 2\Omega y + \overline{w}_x. \tag{7.6.3}$$

In (7.6.3) the channel spans from $y = -W/2$ to $y = +W/2$. The relative velocity field is composed of a uniform throughflow with velocity $\overline{w}_x$, plus a uniform shear of 2Ω. This shear, which is equal and opposite to the angular rotation, is often referred to as (one manifestation of) the relative eddy.

We now discuss the pressure field. There are no fluid accelerations seen by an observer in the relative system ($D\mathbf{w}/Dt = 0$), and the momentum equation represents a balance between the reduced static pressure gradient and the Coriolis force. The components of the momentum equation can be written in terms of the reduced pressure as

$$\frac{\partial p_{red}}{\partial x} = 0, \tag{7.6.4a}$$

$$\frac{\partial p_{red}}{\partial y} = -2\rho w_x \Omega = -2\rho\Omega\overline{w}_x - 4\rho\Omega^2 y. \tag{7.6.4b}$$

From (7.6.4a) and (7.6.4b), the form of the reduced pressure is

$$p_{red} = -2\rho\Omega\overline{w}_x y - 2\rho\Omega^2 y^2 + \text{constant}. \tag{7.6.5}$$

The absolute level of static pressure has no effect in an incompressible flow and the constant can be taken as equal to zero. (This amounts to choosing the y-location at which $p_{red} = 0$.) There is a pressure difference across the channel $\Delta p_{red} = 2\rho\Omega\overline{w}_x W$, although the relative flow streamlines are straight; in a rotating flow, curvature of the relative streamlines is not necessary to have normal pressure gradients. As the fluid moves radially along the channel, its absolute angular momentum about the axis is changing, and the torque necessary for the change is associated with the gradient of reduced pressure.

The actual (as opposed to reduced) static pressure can be found by substituting the value of the radius, $\sqrt{x^2 + y^2}$, into the definition of p_{red}:

$$p = \text{constant} + \frac{\rho}{2}\Omega^2(x^2 + y^2) - 2\rho\Omega\overline{w}_x y - 2\rho\Omega^2 y^2. \tag{7.6.6}$$

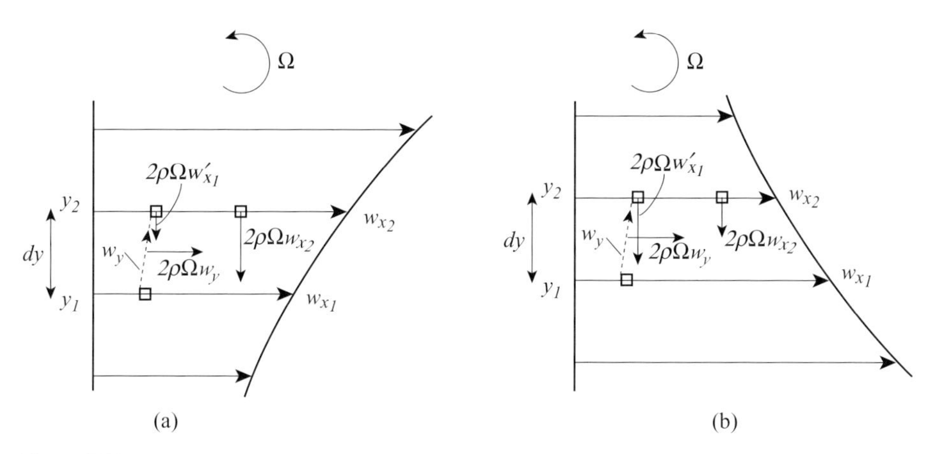

Figure 7.11: Coriolis forces on particles in a rotating flow: (a) relative vorticity and background rotation with opposite senses; Coriolis forces are destabilizing if the shear is large enough; (b) relative vorticity and background rotation with same sense; Coriolis forces are stabilizing (Tritton and Davies, 1981).

Taking the gradient of the difference between the actual and the reduced static pressure gives

$$\frac{\partial}{\partial x}(p - p_{red}) = \rho\Omega^2 x, \tag{7.6.7a}$$

$$\frac{\partial}{\partial y}(p - p_{red}) = \rho\Omega^2 y. \tag{7.6.7b}$$

Equations (7.6.7a) and (7.6.7b) denote the x- and y-components of the centrifugal force. These play no role in creating fluid accelerations in the relative system.

7.6.2 Coriolis effects on boundary layer mixing and stability

Viscous flows in two-dimensional channels exhibit substantial alterations in behavior as a function of rotation. The mixing processes in turbulent boundary and shear layers are modified due to rotation, as are the stability and transition characteristics of laminar boundary layers.

The mechanism that leads to this alteration in behavior can be described following Tritton and Davies (1981) by examining the forces on particles that are displaced from their initial position in a rotating two-dimensional parallel shear flow. The Coriolis force associated with the velocity component in the x-direction (velocity along the channel) is normal to the channel walls. Figure 7.11(a) shows the Coriolis force $2\rho\Omega w_{x_2}$ acting on an undisplaced particle at y_2 $(= y_1 + dy)$ and the force $2\rho\Omega w'_{x_1}$ acting on a particle displaced a distance dy from its initial position y_1 where its velocity was w_{x_1}. The same reduced pressure gradient in the y-direction is acting on both of these particles, and the displaced particle will be further displaced (a condition of static instability) if $w'_{x_1} < w_{x_2}$.

The velocity w'_{x_1} is different from the original velocity of the particle, w_{x_1}, because Coriolis forces have acted during its displacement. The change in velocity is

$$w'_{x_1} - w_{x_1} = 2\Omega w_y dt = 2\Omega dy. \tag{7.6.8}$$

This velocity difference must be compared with the difference in the undisturbed velocities at y_1 and y_2:

$$w_{x_2} - w_{x_1} = \left(\frac{dw_x}{dy}\right) dy. \tag{7.6.9}$$

Hence $w'_{x_1} < w_{x_2}$ only if $dw_x/dy > 2\Omega$.

Two general cases can be defined, as shown in Figure 7.11. In case (a) the relative vorticity and the background rotation have opposite senses and Coriolis effects are destabilizing if the shear is large enough. In case (b), the relative vorticity and the background rotation have the same sense and Coriolis effects are stabilizing. Destabilization i.e. enhancement of the initial displacement) can thus occur when the absolute vorticity $(2\Omega - dw_x/dy)$ has the opposite sign than the background vorticity, 2Ω. For a given shear dw_x/dy (taken positive), rotation is destabilizing if 2Ω lies in the range $0 < 2\Omega < dw_x/dy$ and stabilizing otherwise. From consideration of velocity profiles in the viscous two-dimensional channel flow, case (a) corresponds to conditions on the high pressure ("pressure") side of the channel while case (b) corresponds to the low pressure ("suction") side.

A non-dimensional parameter which captures the above arguments has been introduced by Bradshaw (1969) as

$$\text{rotating flow stability parameter} = -\frac{2\Omega(dw_x/dy - 2\Omega)}{(dw_x/dy)^2}. \tag{7.6.10}$$

Small values of this parameter imply little change in stability compared to a non-rotating flow. Negative values indicate the tendency towards destabilization.

A qualitative analogy exists between the effect just described and the centrifugal instability that occurs on concave surfaces in a stationary frame of reference. In the latter case the balance is between pressure gradients normal to the surface and centrifugal forces. The arguments concerning the enhancement or suppression of particle motions on the inner and outer walls of a curved passage, however, are similar to those given for the rotating channel, as sketched in Figure 7.12 (Johnston, 1978). Further, for laminar boundary layers instability in a rotating channel takes the form of streamwise vortices, analogous to the Gortler vortices (see e.g. Schlichting (1979) for a description of these) seen in flow over a concave surface (Lezius and Johnston, 1976; Yang and Kim, 1991). The presence of these vortices enhances momentum transfer and shear stress along the surface.

For turbulent boundary layers, the mechanism described can be regarded as either damping or encouraging motions that already exist in a direction normal to the wall. Momentum transfer, for example, is increased on the high pressure side of the channel and decreased on the low pressure side. Because of this there is an asymmetry to the rotating channel boundary layer behavior and velocity profile. A sketch of a channel flow geometry is shown in Figure 7.13 with the regions of stability and instability indicated.

Figure 7.14 shows velocity profiles across a rotating two-dimensional channel derived from direct simulations of the Navier–Stokes equations for fully developed turbulent flow (Kristofferson and Andersson, 1993). Time mean velocity profiles are given for different values of Rossby number based on the average velocity, $w/\Omega W$. As the Rossby number decreases, the asymmetry in wall layer behavior becomes increasingly evident, with the flow away from the walls tending towards the inviscid description given in Section 7.6.1. Figure 7.15 shows the velocity near the wall in wall layer coordinates (Section 4.6) for the conditions of Figure 7.14. As the Rossby number decreases the velocity profiles depart further and further from the law of the wall relationship obtained in stationary

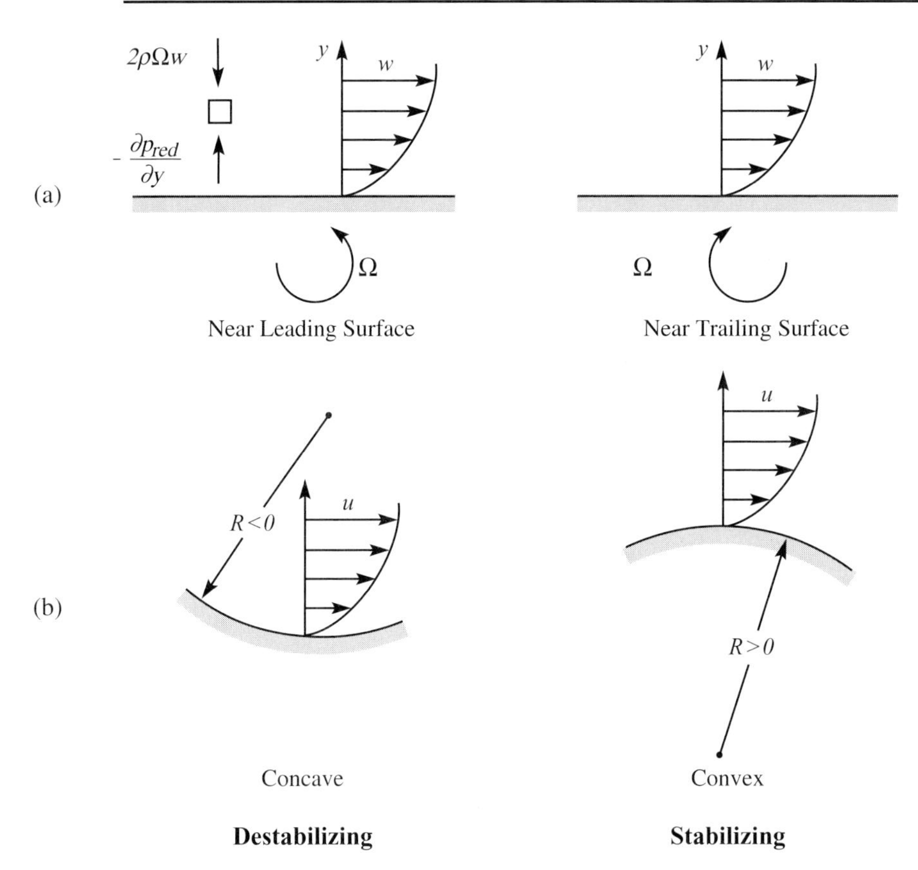

Figure 7.12: Schematic of the effects of rotation and wall curvature on local instability in boundary layers: (a) effects of system rotation, (b) effects of wall curvature (after Johnston (1978)).

flows which is indicated by the dashed line. The simulations, and the experiments of Johnston *et al.* (1972), show a cellular structure in the unstable regions of the channel.

The mechanism described has implications for boundary layer behavior in adverse pressure gradients. As mentioned, destabilization means that momentum interchange is increased (compared to the situation with no rotation) and this increases the resistance of the boundary layer to separation. Stabilization, with an associated decrease in momentum interchange, has the opposite effect. For a rotating passage with adverse reduced pressure gradients, boundary layers in the destabilized region will thus be more resistant to separation than those in the stabilized region. We will see evidence of this trend in Section 7.8.

7.7 Three-dimensional flow in rotating paassages

7.7.1 Generation of cross-plane circulation in a rotating passage

We discuss three-dimensional flows in rotating passages in several steps, starting with a description of the overall concepts in order to provide a framework for viewing the phenomena. Numerical

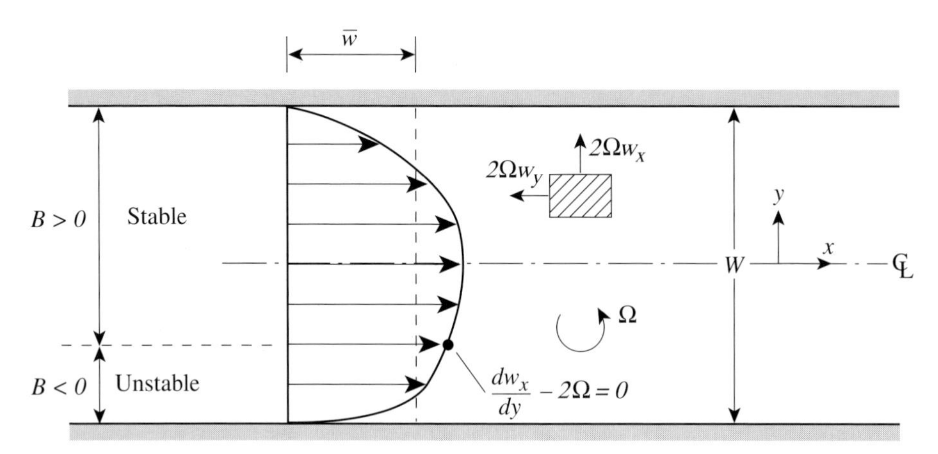

Figure 7.13: Sketch of time average velocity and regions of stability in fully-developed two-dimensional rotating channel flow (Johnston, Halleen, and Lezius. 1972).

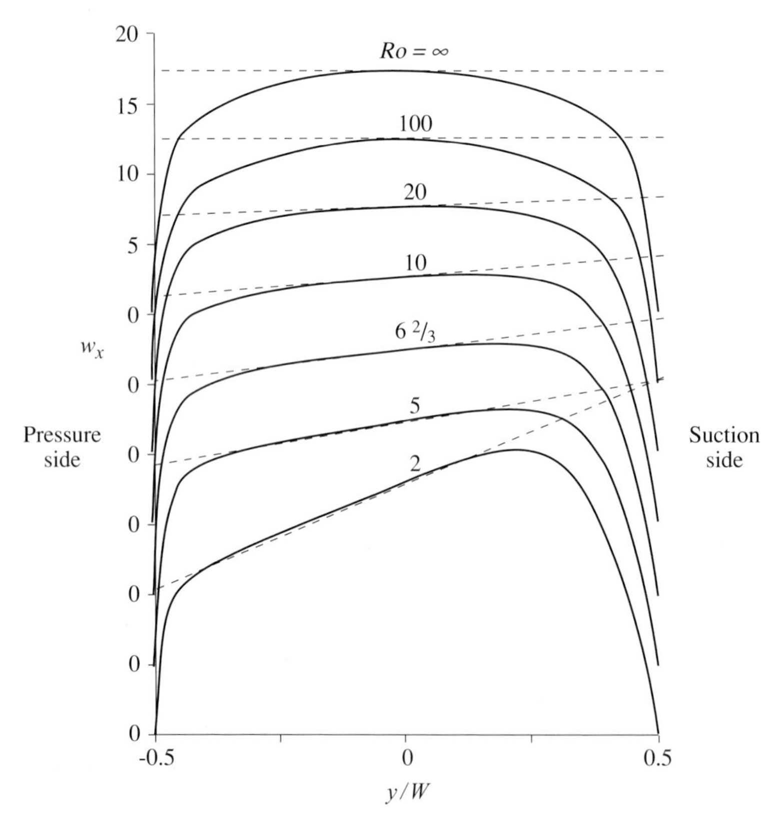

Figure 7.14: Time mean turbulent velocity profiles across a rotating two-dimensional channel for different Rossby numbers. The coordinate system is the same orientation as in Figure 7.13. The straight lines have slope 2Ω. Direct numerical solution of Kristoffersen and Andersson (1993) at Reynolds number of 2900 based on mean velocity and channel width.

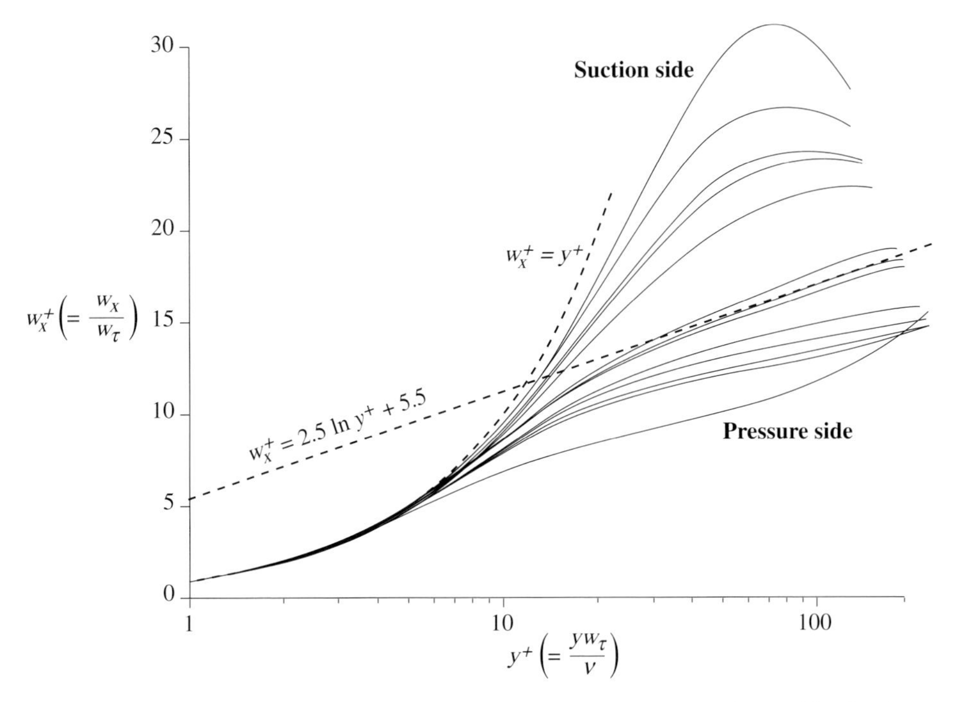

Figure 7.15: Rotating two-dimensional channel time mean near wall velocity profiles for the conditions of Figure 7.14. Direct numerical solution of Kristoffersen and Andersson (1993).

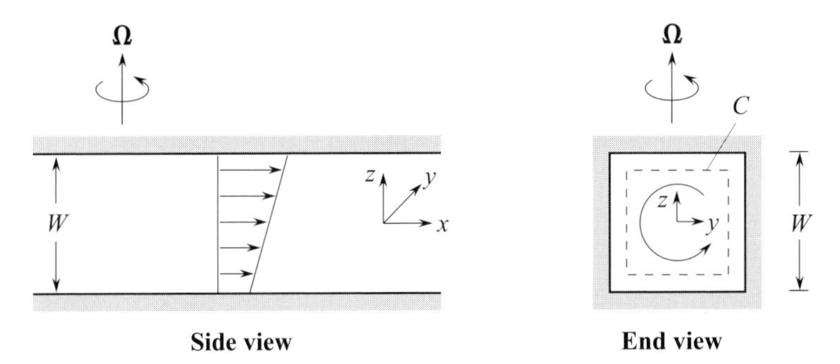

Figure 7.16: Generation of three-dimensional motion in a rotating square passage. The end view looking upstream (velocity out of page) shows the direction of secondary circulation.

and experimental results are then presented for laminar and turbulent flows in rotating passages to illustrate specific features and parametric dependence.

Non-conservative Coriolis forces which can change the relative circulation around a fluid contour are a source for generation of three-dimensional motions in rotating channels. Figure 7.16 depicts a straight square channel of side W, rotating about the z-axis. Suppose at a given station along the channel there is a two-dimensional inviscid shear flow which has a velocity component only in the x-direction, is uniform in y, and has a velocity that increases with z. A basic question is whether such a two-dimensional shear flow can persist (as it would in a stationary system) or whether a three-dimensional motion, with streamwise vorticity, will arise.

To answer the question we examine a contour C, indicated by the dashed line in the end view of the channel in Figure 7.16. For an inviscid uniform density fluid the rate of change of relative circulation for this contour is obtained from (3.8.8) as

$$\frac{D\Gamma_{rel}}{Dt} = \oint \mathbf{F}_{Coriolis} \cdot d\boldsymbol{\ell}.$$

The Coriolis force is

$$\mathbf{F}_{Coriolis} = -2\Omega w_x \mathbf{j}, \tag{7.7.1}$$

with $\mathbf{j}$ a unit vector in the y-direction. If the velocity varies linearly from $w_{x_{max}}$ at the top of the channel to $w_{x_{min}}$ at the bottom of the channel,

$$\frac{D\Gamma_{rel}}{Dt} = \oint \mathbf{F}_{Coriolis} \cdot d\boldsymbol{\ell} = 2(w_{x_{max}} - w_{x_{min}})\Omega W. \tag{7.7.2}$$

Equation (7.7.2) indicates that a circulatory motion with spiral streamlines will be generated in the y–z plane, with the sense of rotation indicated in Figure 7.16.

A rough estimate for the initial rate of generation of the cross-plane velocity (valid for streamwise locations such that the x-velocity variation is not appreciably affected by the three-dimensional flow) can be obtained from (7.7.2). We approximate the relative circulation in the cross-plane (y–z plane) as $\Gamma_{cross} \approx 4Ww_{cross}$, where w_{cross} is an average value of the cross-plane velocity along the dashed contour. The rate of change of circulation, $D\Gamma_{cross}/Dt$, is estimated as $\overline{w}_x(d\Gamma_{cross}/dx)$, where $\overline{w}_x$ is the mean velocity along the channel. Substituting into (7.7.2) yields

$$\frac{dw_{cross}}{dx} \approx \frac{\Delta w_x \Omega}{2\overline{w}_x}, \tag{7.7.3}$$

with $\Delta w_x = w_{x_{max}} - w_{x_{min}}$. With no cross-plane flow at an initial station $x = x_i$, integration of (7.7.3) yields

$$\frac{w_{cross}}{\overline{w}_x} \approx \left(\frac{\Delta w_x}{2\overline{w}_x}\right)\left(\frac{\Omega W}{\overline{w}_x}\right)\frac{(x - x_i)}{W}. \tag{7.7.4}$$

The initial generation of cross-plane circulation thus scales with three factors: the velocity difference along the direction of the axis of rotation (which determines the imbalance in Coriolis forces), the inverse of the Rossby number (which measures the importance of rotation), and the distance over which the Coriolis forces act. The example is based on a shear profile occupying the whole channel, but similar arguments apply to shear regions of more limited extent such as boundary layers on the top or bottom walls of a rotating passage (MacFarlane, Joubert, and Nickels, 1998).

We can also interpret the generation of cross-plane circulation from the perspective of an inertial coordinate system. Even in rectilinear relative motion, fluid particles undergo an absolute acceleration as they move along the passage. The magnitude of the acceleration is greater for fluid with a larger x-velocity. The cross-plane circulation can be viewed as a consequence of the fluid in the upper part of the duct, which has a larger value of w_x, having larger resistance to taking a curved path than the fluid in the lower part. We return to the topic of streamwise vorticity generation in rotating passages in Chapter 9.

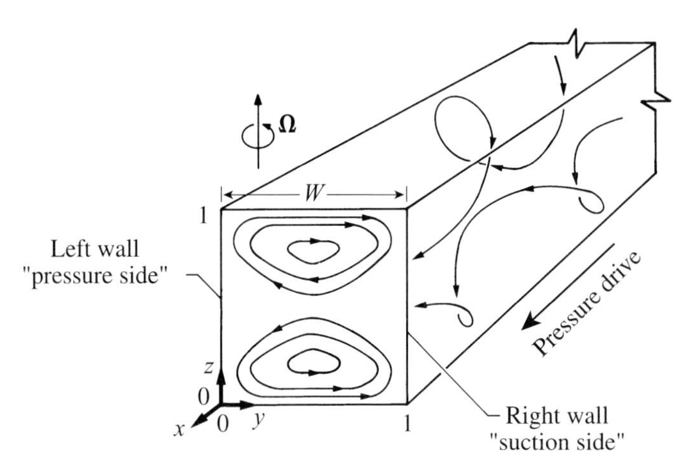

Figure 7.17: Fully developed flow in a rotating straight passage (Kheshgi and Scriven, 1985).

7.7.2 Fully developed viscous flow in a rotating square duct

Fully developed, pressure-driven, laminar viscous flow in a rotating square passage provides a useful vehicle for introducing the features and parametric behavior of three-dimensional viscous motion in rotating systems. In contrast to the situation in a stationary system, fully developed flow in a rotating straight passage is not rectilinear. One type of flow pattern that exists is sketched in Figure 7.17. (This, and the next several figures, are from Kheshgi and Scriven (1985).) Figure 7.17 depicts the direction of the overall driving pressure force (with reduced pressure gradient $\partial p_{red}/\partial x$ = constant) and the cross-plane circulatory motion produced by Coriolis forces, in which the higher velocity fluid moves towards the high reduced pressure side ("pressure side") of the passage.

Fully developed viscous flow in a rotating passage is characterized by the two non-dimensional quantities introduced in Section 7.4: the Ekman number (*Ek*) and the Rossby number (*Ro*). For the fully developed case there is no dependence on x, and the axial and cross-plane velocities are functions of y and z only. Even with this simplification, the parametric behavior is complex because different force balances can occur depending on the values of *Ro* and *Ek*.

At zero Rossby number (the rapidly rotating passage limit) the balance is between Coriolis, pressure, and viscous forces. Figure 7.18 shows the computed cross-plane velocities and the axial profile at $Ro = 0$ and $Ek = 0.01$. As pictured in Figure 7.17, the cross-plane motion consists of two cells. The axial velocity is symmetric across the passage about $y/W = 0.5$. The cross-plane flow near the top and bottom walls at $Ro = 0$ exhibits behavior like that of an Ekman layer with an overshoot in the axial velocity. This can be seen in more detail in Figure 7.19 which shows the axial velocity, normalized with the pressure gradient and angular velocity of rotation $((\rho\Omega w_x/|\partial p_{red}/\partial x|)\sqrt{Ek})$ at a location midway across the passage ($y/W = 0.5$) versus the Ekman layer coordinate $(z/(\sqrt{Ek}W))$. The curves correspond to Ekman numbers of 0.01 and 0.001 and to an asymptotic solution representing the limit of low Ekman number. In this limit the flow in the interior of the passage (away from the walls) is uniform with a non-dimensional velocity of $(1/(2\sqrt{Ek}))$ and represents a balance between pressure and Coriolis forces. Flows that exhibit such a balance are known as *geostrophic* and appear often in meteorological applications. The Ekman-like layers on the top and bottom

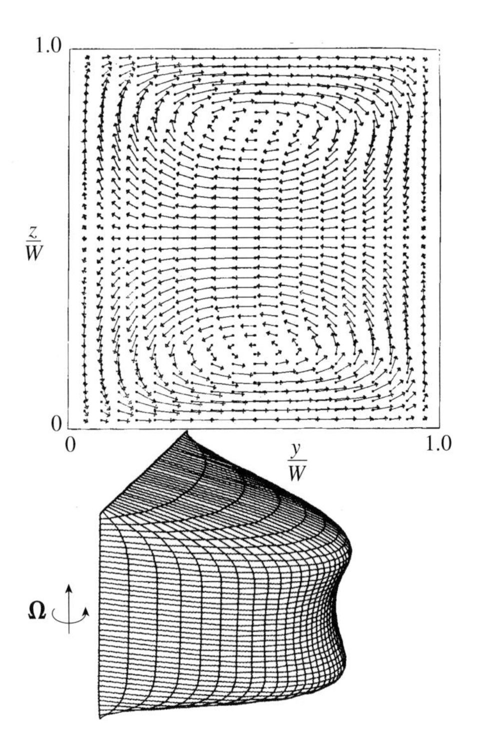

Figure 7.18: Fully developed flow in a rotating straight passage; cross-plane velocity vectors (top) and axial velocity distribution (bottom); $Ro = 0$, $Ek = 0.01$; computations of Kheshgi and Scriven (1985).

walls (as well as the more complex "double layer" viscous regions on the side walls whose extents scale as $Ek^{1/4}$; see e.g. (Bark 1996)) represent a balance between these two and the viscous forces.

Figure 7.20 shows the dependence of the non-dimensional volume flow rate on Ekman number, for zero Rossby number. The volume flow rate is defined as

$$\text{non-dimensional volume flow rate,} \quad \dot{V} \equiv -\frac{\rho\Omega Ek}{(\partial p_{red}/\partial x)} \int_0^1 \int_0^1 w_x d\left(\frac{y}{W}\right) d\left(\frac{z}{W}\right). \tag{7.7.5}$$

For large Ekman numbers (viscous forces important throughout the passage) the flow is similar to that in a stationary channel and the motion is rectilinear. For small Ekman numbers, the viscous effects are confined to thin regions, the velocity away from the walls is uniform, and the non-dimensional volume flow rate approaches a limiting form given by

$$\dot{V} \to \frac{\sqrt{Ek}}{2} \quad \text{as} \quad Ek \to 0. \tag{7.7.6}$$

For non-zero Rossby numbers inertial forces play a role in the force balance. At low Ekman numbers the flow in the interior of the passage, away from the thin viscous layers on the walls is

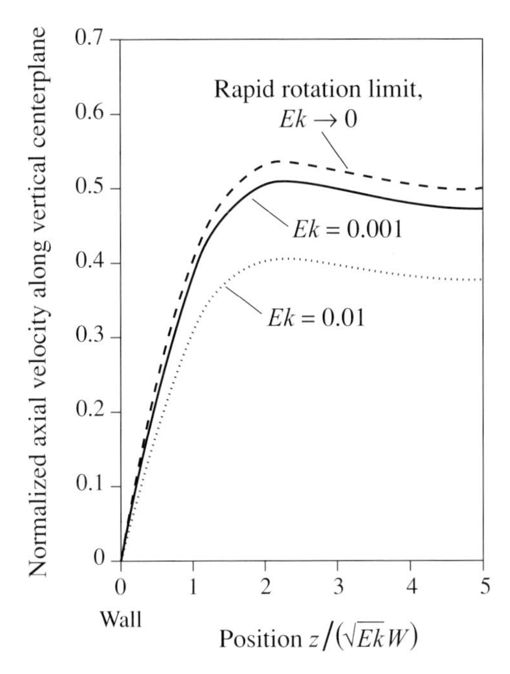

Figure 7.19: Fully developed flow in a rotating straight passage: axial velocity along centerplane ($y/W = 0.5$) for $Ro = 0$; computations of Kheshgi and Scriven (1985).

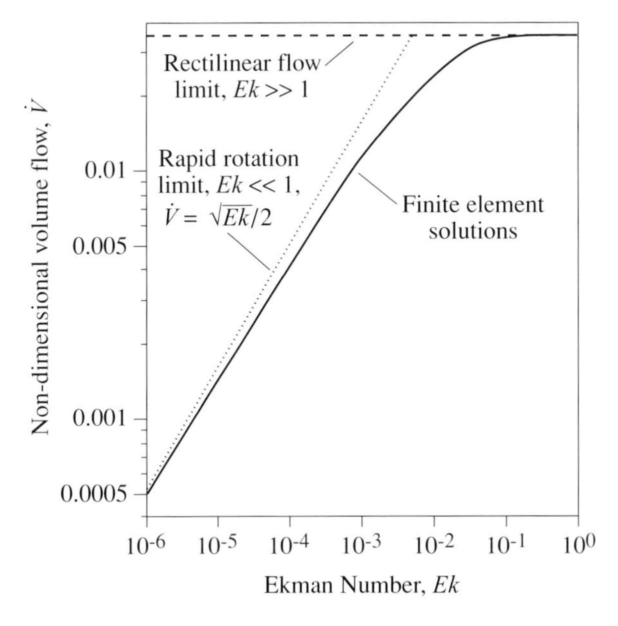

Figure 7.20: Fully developed flow in a rotating straight passage: non-dimensional volume flux $\dot{V}$ (defined in (7.7.5)) as a function of Ekman number; $Ro = 0$; computations of Kheshgi and Scriven (1985).

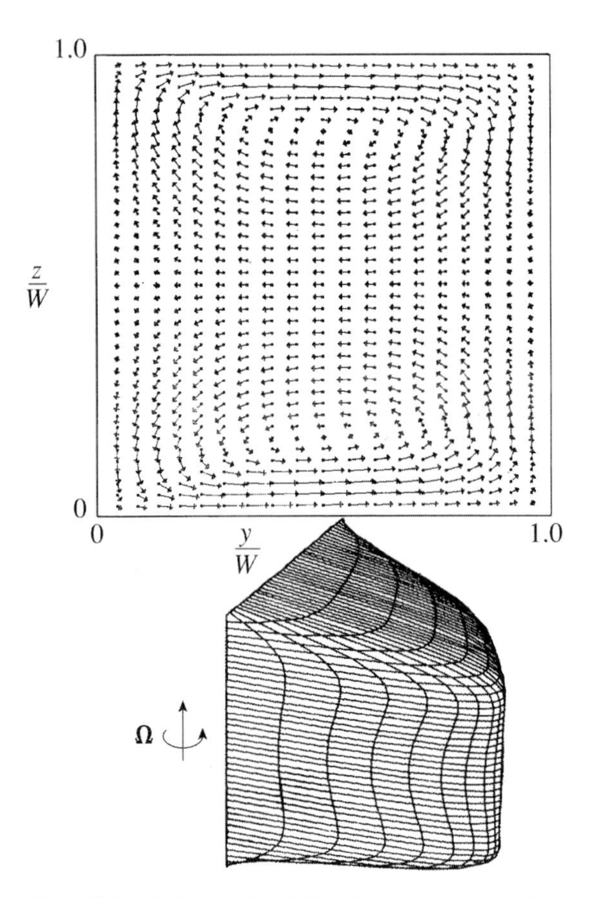

Figure 7.21: Fully developed flow in a rotating straight passage; cross-plane velocity vectors (top) and axial velocity distribution (bottom); $Ro = 1.5$, $Ek = 0.01$; computations of Kheshgi and Scriven (1985).

primarily a balance between inertia, Coriolis, and pressure forces. Figure 7.21 shows the cross-plane flows and the overall form of the axial velocity for an Ekman number of 0.01 (the same value of Ek as in Figure 7.18) but a Rossby number of 1.5. The axial velocity varies approximately linearly with y across the interior of the passage. In addition for this three-dimensional motion the variation is in the *opposite* sense to the inviscid, two-dimensional flow pictured in Figure 7.14, with the highest axial velocity now nearer to the high reduced pressure side of the passage.

The origin of this feature is the cross-plane fluid transport associated with the Coriolis forces. A qualitative argument for the shape of the velocity profile can be given as follows. For no rotation (infinite Rossby number) the flow in the duct is rectilinear with the largest velocity in the center of the passage. For purposes of illustration consider the three-dimensional motion at finite Rossby number as a perturbation on this rectilinear flow. Based on the x-component velocity field for no rotation, larger y-components of Coriolis forces (which depend on the local velocity) exist in the central region of the duct than near the walls. The differential between the y-component of Coriolis forces at different z-coordinate locations is equivalent to a torque whose sense is to cause the type

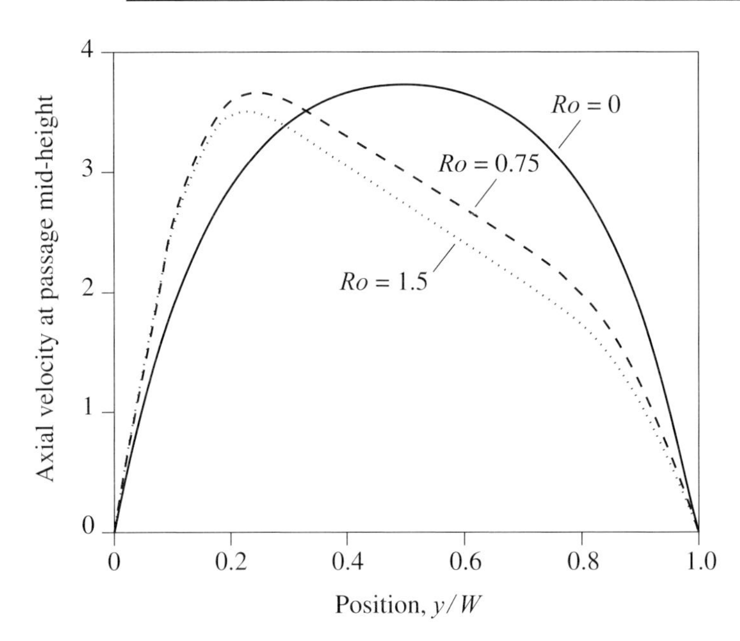

Figure 7.22: Fully developed flow in a rotating straight passage; non-dimensional axial velocity at midheight ($z/W = 0.5$); computations of Khesghi and Scriven (1985).

of cross-plane motion sketched. The consequence of high x-velocity flow convected towards the high pressure surface and low x-velocity flow moved towards the low pressure surface is a shift in the region of highest streamwise velocity from the center of the duct towards the high pressure surface.[5]

The axial velocity across the passage at the midheight station is presented in more detail in Figure 7.22 for an Ekman number of 0.01 and three values of Rossby number: 0, 0.75, and 1.5. The figure shows the change in profile shape with Rossby number, as well as the extent of the nearly linear velocity distribution at the two higher values of Ro.

For larger values of Rossby number the two-cell structure is unstable. One result is that other pairs of vortices appear on the pressure side of the channel, as indicated in Figure 7.23; this behavior is also seen in computations of fully developed turbulent flow, as described below.

The fully-developed flow examined is a square passage, because this geometry is similar to that in turbomachinery, but computations in other geometries such as circular pipes show many of the same features; see for example Lei and Hsu (1990) and Lei *et al.* (1994). Additional information concerning fully developed flow in rotating passages can be found in Morris (1981).

[5] There is an analogy between this situation and that of fully developed flow in a curved duct. For inviscid, irrotational, two-dimensional flow in a curved passage the highest velocity is near the inner wall. For fully developed flow in a curved pipe, however, (see for example Berger, Talbot, and Yao (1983)) the region of highest velocity is between the center of the pipe and the outer wall. The position of the high velocity region is associated with the local imbalance between centrifugal forces and pressure differences across the pipe, which cause the high velocity fluid to move outwards and the low velocity fluid near the walls to move inwards. Detailed computational results concerning the analogy between fully developed rotating flow in a straight circular pipe and in a curved stationary circular pipe are given by Ishigaki (1994, 1996).

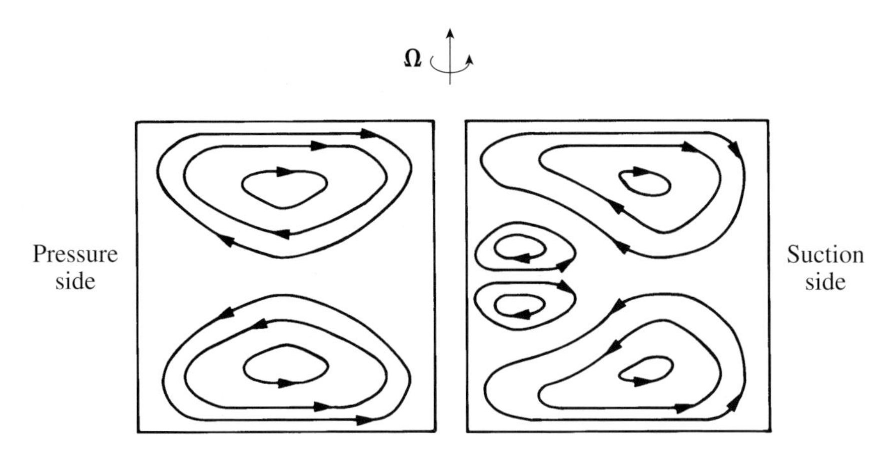

Figure 7.23: Rotating square passage looking upstream: sketch of cross-plane flow patterns for two-vortex and four-vortex configurations in laminar fully developed flow (Kheshgi and Scriven, 1985).

7.7.3 Comments on viscous flow development in rotating passages

The previous description of fully developed rotating passage flow provides a basis for discussing situations in which the flow is evolving in the streamwise direction. An illustration is furnished by the computations of the developing turbulent flow in a rotating square passage, (Bo, Iacovides, and Launder, 1995) from which Figure 7.24 is taken. Cross-plane velocity vectors and contours of axial velocity magnitude are shown at two streamwise locations, $x/W = 3$ and $x/W = 9$. The conditions of the computation are given in the figure. Migration of the high velocity fluid to the pressure side, under the influence of the Coriolis force, can be seen, similar to the laminar computations. The downstream station velocity vectors show a four- vortex configuration, with an additional developing pair of vortices near the pressure surface.

Another example of a developing flow is shown schematically in Figure 7.25, based on measurements of the cross-plane streamline pattern for developing turbulent boundary layers in a rotating passage (MacFarlane *et al.*, 1998). Measurements corresponding to this situation are given in Figure 7.26. Features of the cross-plane structure seen in the fully developed flow are also represented in the developing flow, with two major cells driven by Coriolis forces and a weaker pair of vortices near the pressure surface. There is a strong cross-passage motion in the boundary layers with a return flow taking place over most of the rest of the channel height.

Two other points concerning the development of boundary layers in rotating passages can be noted. The first concerns the pressure rise that can be achieved in a rotating passage. It is the gradient of *reduced* static pressure, $p_{red} = p - \rho(\Omega^2 r^2/2)$, which affects the boundary layer behavior, rather than the gradient of the actual static pressure. This means that it is possible to achieve a large static pressure rise in a radially outward flow with little adverse effect on the boundary layers.

The second point concerns a difference between the Ekman layer type of behavior described earlier in the chapter and the boundary layers on the top and bottom of passages in which the flow external to the boundary layers is irrotational in the absolute coordinate system (Moore, 1973b). This would occur for example in fluid components which are short enough such that direct effects of viscosity do not have time to spread throughout the duct.

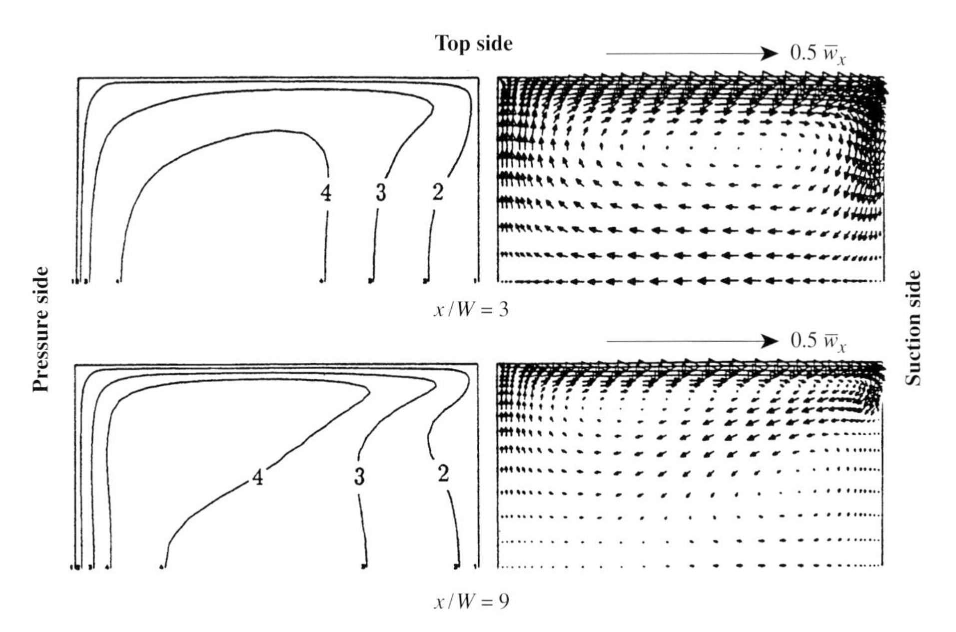

Figure 7.24: Developing turbulent flow in a rotating square passage; contours of velocity and cross-plane velocity vectors, Rossby number $\overline{w}_x/(\Omega W) = 4.2$, $Ek = 1.7 \times 10^{-4}$ ($Re = \overline{w}_x W/\nu = 25,000$). Fully developed stationary flow velocity distribution applied as the inlet condition. Computations of Bo *et al.* (1995).

We motivate the difference as follows. For thin boundary layers in a constant-area rotating channel, the reduced pressure gradient in the streamwise (x-direction) can be neglected and the component of the momentum equation that describes the evolution of the axial velocity, w_x, is

$$w_x \frac{\partial w_x}{\partial x} + w_y \frac{\partial w_x}{\partial y} + w_z \frac{\partial w_x}{\partial z} = 2\Omega w_y + (F_{visc})_x. \tag{7.7.7}$$

As described in Section 7.6.1 the axial velocity external to the boundary layer, w_{E_x}, obeys

$$\frac{dw_{E_x}}{dy} = 2\Omega. \tag{7.7.8}$$

Combining (7.7.8) with (7.7.7) gives

$$w_x \frac{\partial w_x}{\partial x} + w_y \left[\frac{\partial}{\partial y} \left(w_x - w_{E_x} \right) \right] + w_z \frac{\partial w_x}{\partial z} = (F_{visc})_x. \tag{7.7.9}$$

The terms w_x and w_{E_x} can be comparable over much of the boundary layer. If so, and the term in square brackets in (7.7.9) is small, the statement concerning the evolution of w_x is approximately

$$w_x \frac{\partial w_x}{\partial x} + w_z \frac{\partial w_x}{\partial z} \approx (F_{visc})_x. \tag{7.7.10}$$

Equation (7.7.10) implies that (away from the side walls) the top and bottom wall boundary layers in a rotating passage with irrotational absolute flow exhibit streamwise evolution which is qualitatively more akin to the boundary layer in a stationary passage than to an Ekman layer.

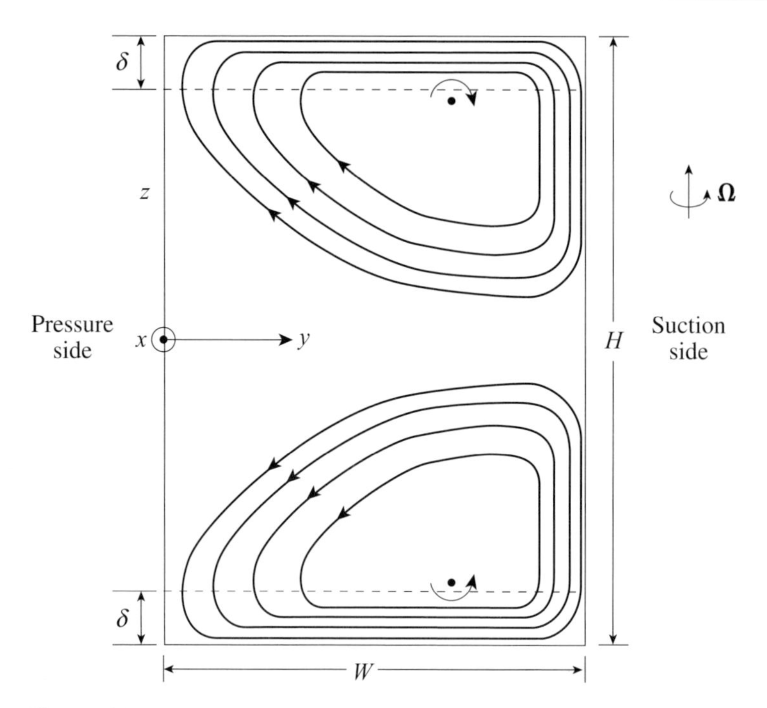

Figure 7.25: Rotating rectangular passage: schematic diagram of the cross-stream flow pattern for a developing turbulent flow in a rotating rectangular passage looking upstream (MacFarlane *et al.*, 1998).

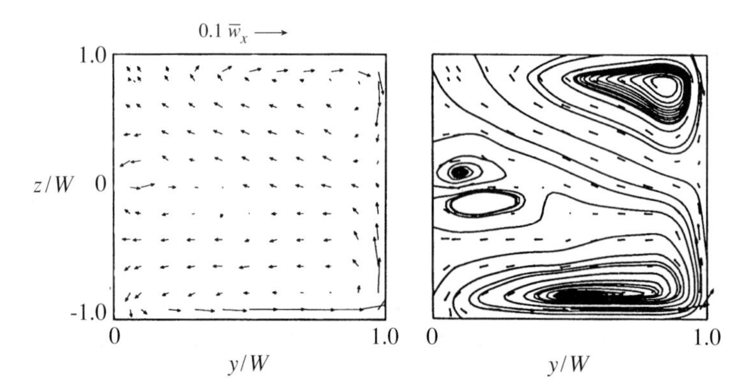

Figure 7.26: Turbulent flow in a rotating straight passage: cross-plane velocities and streamlines, $Ro = \overline{w}_x/(\Omega W) = 10.6$, $Ek = 1 \times 10^{-4}$ ($Re = \overline{w}_x W/\nu = 1 \times 10^5$); measurements of MacFarlane *et al.* (1998).

7.8 Two-dimensional flow in rotating diffusing passages

7.8.1 Quasi-one-dimensional approximation: irrotational absolute flow

We now turn our attention to rotating diffusing passages and introduce the overall behavior using the channel flow solution given in Section 7.6. A radial diffusing passage is sketched in Figure 7.27. The appropriate (relative frame) coordinate system is polar coordinates, r and θ, with velocity components

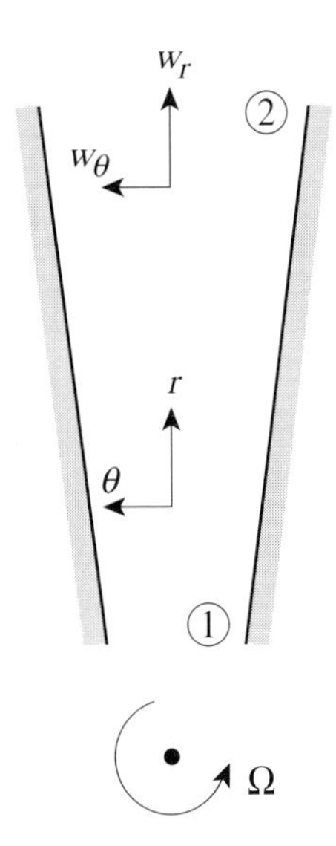

Figure 7.27: Rotating channel and coordinates.

w_r and w_θ. If the passage is slender enough that $w_\theta/w_r \ll 1$, an approximation for the relative vorticity is (Moore, 1973a)

$$\frac{1}{r}\frac{\partial w_r}{\partial \theta} = 2\Omega. \tag{7.8.1}$$

The radial velocity is obtained by integrating (7.8.1). Denoting the average radial velocity across the passage at a given radius by $\overline{w}_r$ and the passage width by W,

$$w_r \cong 2\Omega\theta r + \frac{\overline{w}_r W}{2\theta_w r}. \tag{7.8.2}$$

In (7.8.2) the angle θ is measured from the midpoint of the channel and θ_w is the half-angle of the channel.

To assess the utility of this description of the velocity field, we examine the effect of circumferential velocity on the pressure difference across the channel and compare it with that due to radial velocity (through Coriolis forces) alone. The continuity equation allows an estimate of w_θ to be made:

$$\frac{1}{r}\frac{\partial}{\partial r}(r w_r) + \frac{1}{r}\frac{\partial w_\theta}{\partial \theta} = 0. \tag{7.8.3}$$

Introducing the radial velocity from (7.8.2) and applying the boundary condition that w_θ must vanish on the channel walls at $\theta = \pm\theta_w$ yields

$$w_\theta = 2\Omega r(\theta_w^2 - \theta^2). \tag{7.8.4}$$

The pressure difference associated with w_θ is found from the θ-component of the momentum equation to be of order $\overline{w}_r\, w_\theta/r$ (the average value of w_θ across the channel is $\frac{4}{3}\Omega r\theta_w^2$). The pressure difference due to the radial velocity is approximately $2\Omega\overline{w}_r$ so their ratio is

$$\frac{\overline{w}_r w_\theta/r}{2\Omega\overline{w}_r} = \frac{w_\theta}{2\Omega r} \approx \theta_w^2. \tag{7.8.5}$$

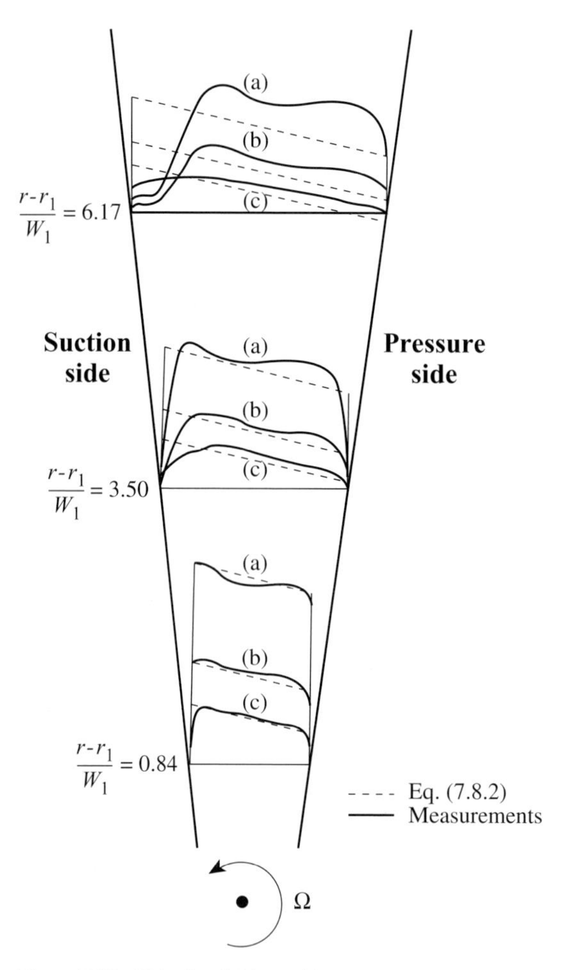

Figure 7.28: Velocity field at midpassage height in a rotating diffusing passage, ratio of passage height/width at inlet (station 1) = 1.0: (a) $(\overline{w}_{r_1}/(\Omega W_1)) = 21$; (b) $(\overline{w}_{r_1}/(\Omega W_1)) = 10$; (c) $(\overline{w}_{r_1}/(\Omega W_1)) = 5.3$ (Moore, 1973a).

For a representative geometry, say a centrifugal impeller with twelve blades, $\theta_w = 15^\circ$, the magnitude of the ratio is roughly 0.05, and the channel flow treatment is appropriate.

Equation (7.8.2) also shows that, for a given volume flow rate, the velocity on the pressure side decreases as the radius increases, becoming zero at a radius $\sqrt{w_r W/\Omega/2\theta_w}$, with the implication that there is a region of backflow at larger radii. As the volume flow rate is lowered, this zero velocity point occurs at a smaller radius. To illustrate this effect, Figure 7.28 shows the measured radial velocity in a rotating diffusing passage at three different flow rates. Away from the walls (7.8.2) gives a description which captures the general features, although viscous effects, in particular blockage due to the wall boundary layers, become important in the outer parts of the passage.

7.8.2 Two-dimensional inviscid flow in a rotating diffusing blade passage

The description in the previous section applies within the passage, but we also need to examine the behavior near the passage inlet and outlet where the channel flow approximation does not apply.

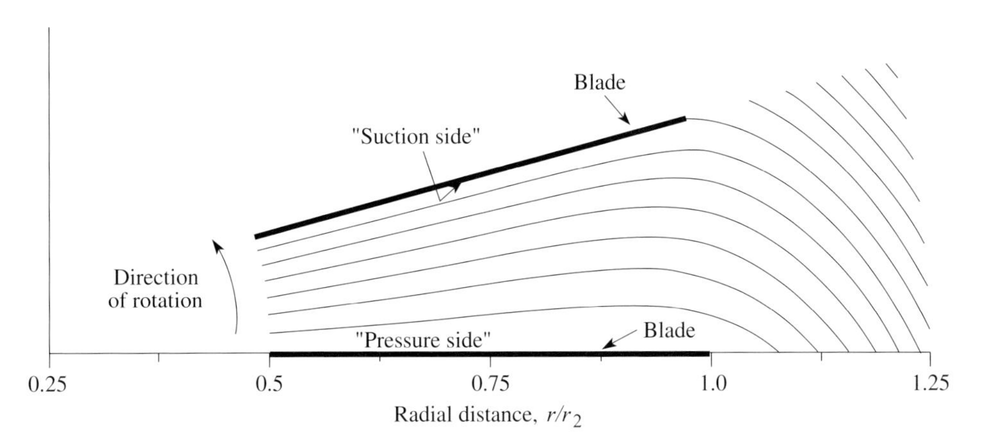

Figure 7.29: Two-dimensional inviscid flow streamlines in a radial blade diffusing rotating passage; streamlines of equal increments, $\overline{w}_r/(\Omega W_1) = 1.3$, $r_1/r_2 = 0.5$.

Suppose the passage is formed by two radial blades in an impeller (rotor) of a centrifugal compressor as shown in Figure 7.29. At the outer end of the blade there is no pressure difference between the suction and pressure surfaces of the blade. An adjustment must thus take place near the end to decrease the pressure difference across the channel. Because of the decreased pressure difference, the Coriolis forces in this region drive the flow towards the pressure side resulting in a relative velocity with a direction opposite to the rotation.

The adjustment process can be analyzed in terms of a two-dimensional inviscid flow in the channel. We define a stream function, ψ, for the relative velocity field as

$$w_r = \frac{1}{r}\frac{\partial \psi}{\partial \theta}, \tag{7.8.6a}$$

$$w_\theta = -\frac{\partial \psi}{\partial r}. \tag{7.8.6b}$$

With this definition, the continuity equation is satisfied identically. If the absolute flow is irrotational, the relative vorticity satisfies (7.6.1),

$$-\frac{1}{r}\frac{\partial w_r}{\partial \theta} + \frac{1}{r}\frac{\partial}{\partial r}(r w_\theta) = -2\Omega. \tag{7.8.7}$$

Combining (7.8.6) and (7.8.7) yields a Poisson equation for ψ:

$$\nabla^2 \psi = \frac{\partial^2 \psi}{\partial r^2} + \frac{1}{r}\frac{\partial \psi}{\partial r} + \frac{1}{r^2}\frac{\partial^2 \psi}{\partial \theta^2} = 2\Omega. \tag{7.8.8}$$

Equation (7.8.8) has boundary conditions of zero normal relative velocity at the blades (θ-component), which extend from a non-dimensional radius $r/r_2 = 0.5$ to $r/r_2 = 1.0$. The additional boundary conditions imposed here are that at the outer radius ($r/r_2 \gg 1$) the flow is axisymmetric, at the inner radius the radial velocity of Section 7.8.1 is prescribed, and periodicity is imposed on radial lines that are extensions of the blades to the outer boundary.

Computed streamlines from (7.8.8) plotted in Figure 7.29 show flow adjustment from radial outward to axisymmetric occurring in a radial distance on the scale of the passage width. This is also seen in the accompanying Figure 7.30, which gives contours of relative velocity magnitude. In the channel, up to a radius of roughly $0.9r_2$, the flow is similar to that described by (7.8.2), while

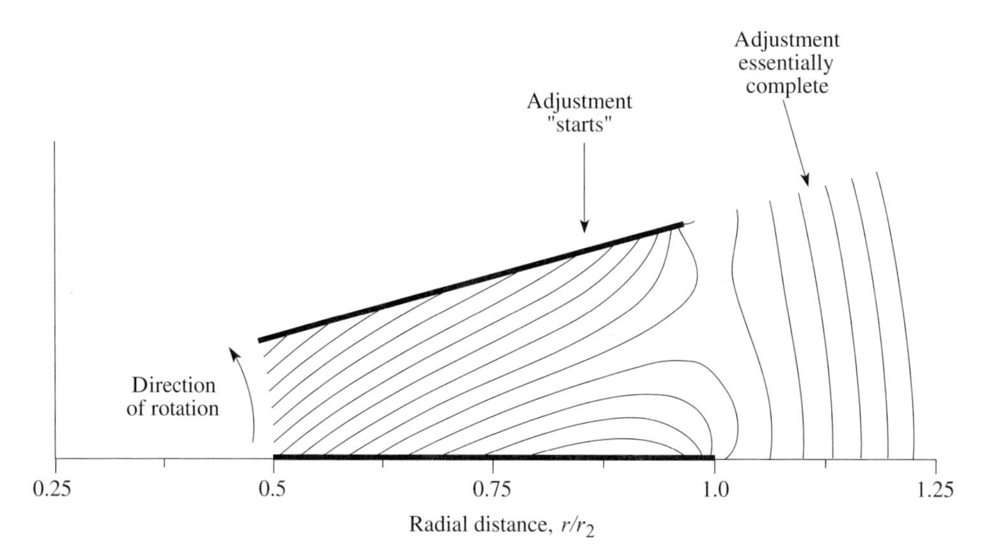

Figure 7.30: Contours of relative velocity magnitude in a two-dimensional diffusing rotating passage (inviscid flow) showing the adjustment zone; $\overline{w}_r/(\Omega W_1) = 1.3$, $r_1/r_2 = 0.5$, $w_{max}/\Omega r_2 = 0.8$, contours are $0.03\ \Omega r_2$.

from approximately $r/r_2 = 0.8$ to $r/r_2 = 1.1$, a transition occurs to axisymmetric swirling flow. As in other irrotational flows we have examined, the two length scales in the problem, the passage width and the radial length scale for flow adjustment, are linked by the condition of irrotationality.

7.8.3 Effects of rotation on diffuser performance

In the description of boundary layers in rotating systems in Section 7.6.2 it was stated that the ability of a turbulent boundary layer to negotiate a given pressure rise is decreased on the suction side of a rotating channel, because of the suppression of mixing and shear stress, and increased on the pressure side, because of the mixing enhancement. The performance of a rotating diffuser will therefore be adversely affected by rotation. Experiments addressing this point have been carried out by Rothe and Johnston (1976).

The flow patterns relating to stall (see Section 4.1 for a description of diffuser flow regimes) were altered in several ways by system rotation. Stall appeared only on the suction side of the rotating diffuser, in contrast to the bi-stable stall regime encountered in stationary diffusers. In addition, the separated flow patterns were more two-dimensional, and more steady, than for stationary diffusers, with no regime of large transitory stall. Rothe and Johnston point out the connection of this two-dimensionality with the Taylor–Proudman Theorem, applied to the nearly stagnant (and hence nearly solid-body rotation) fluid in the stalled region.

Figure 7.31 shows a flow regime map for a rotating two-dimensional diffuser. The abscissa is $1/Ro$ (or $\Omega W_1/\overline{w}_1$, which is often referred to as the "rotation number") and the ordinate is the diffuser area ratio. As the area ratio is increased at constant Rossby number, stall is first encountered in the suction corners of the diffuser. Further increase leads to the build up of a two-dimensional region of low speed reverse flow on the suction side of the passage. Flow separation occurs at a wall position which depends on the area ratio and Rossby number. An additional area ratio increase, or Rossby

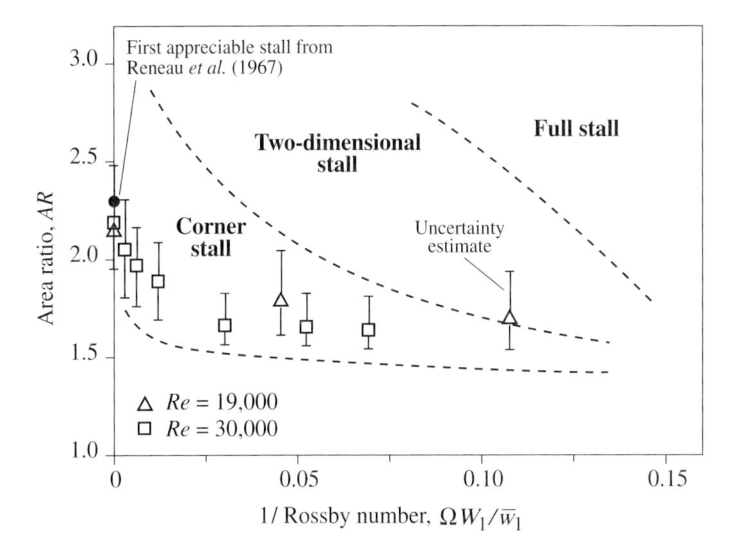

Figure 7.31: Comparison of pressure recovery data with a stall regime map in a rotating diffuser, length/width ratio $N/W_1 = 6$; $\overline{w}_1$ is mean velocity at the inlet, inlet flow is fully developed. Symbols represent the area ratio at which peak pressure recovery occurred at fixed Rossby number (Rothe and Johnston, 1976).

number decrease, moves the separation line upstream; the condition at which the separation line is two inlet widths from the throat is referred to as full stall. The onset of first appreciable stall for this geometry in a stationary diffuser is denoted by the solid circle on the vertical axis. Also shown in the figure are the area ratios at which peak pressure recovery occurred for a given value of Rossby number.

The effect of rotation on the peak diffuser pressure rise coefficient, C_p (based on the reduced static pressure), is shown in Figure 7.32. There is a roughly 30% decrease in this quantity at the highest rotation speed (which is several times smaller than the non-dimensional rotational speed typically encountered in centrifugal turbomachinery).

7.9 Features of the relative flow in axial turbomachine passages

The relative eddy mentioned in Section 7.6 in connection with irrotational absolute flow in a rotating frame is seen in several ways in axial turbomachines. (Axial turbomachines have the predominant direction of flow along the axis of rotation rather than perpendicular to it as with centrifugal machines.) To introduce the topic in a basic manner, consider a rotating passage in a constant density inviscid flow, in which the blades are radial and axial so the trace of the blade remains at the same θ location for any axial station.[6] If the blade passage is sufficiently long, the flow away from the ends will have $\partial/\partial x = 0$ and a stream function for the tangential and radial velocities can be defined which obeys the two-dimensional Poisson equation in r and θ similar to (7.8.8).

[6] Such blades are referred to as having zero stagger angle, the stagger angle being defined as the angle between the blade chord and the axial direction.

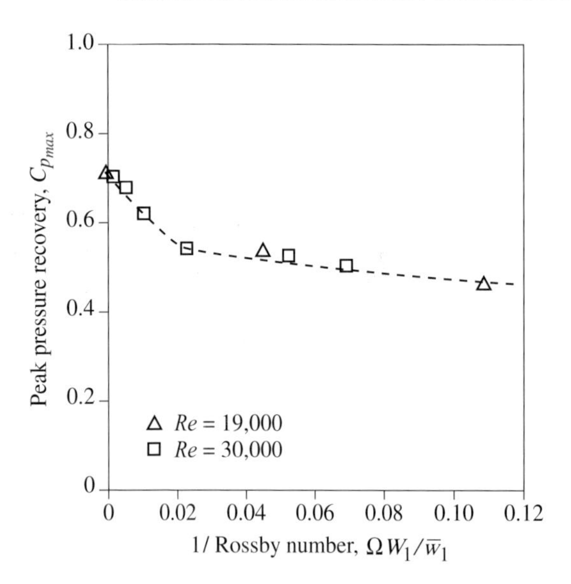

Figure 7.32: Pressure recovery peak in a rotating diffuser, length/width ratio $N/W_1 = 6$; $\overline{w}_1$ is the mean velocity at inlet, inlet flow is fully developed (area ratio varied at fixed Rossby number) (Rothe and Johnston, 1976).

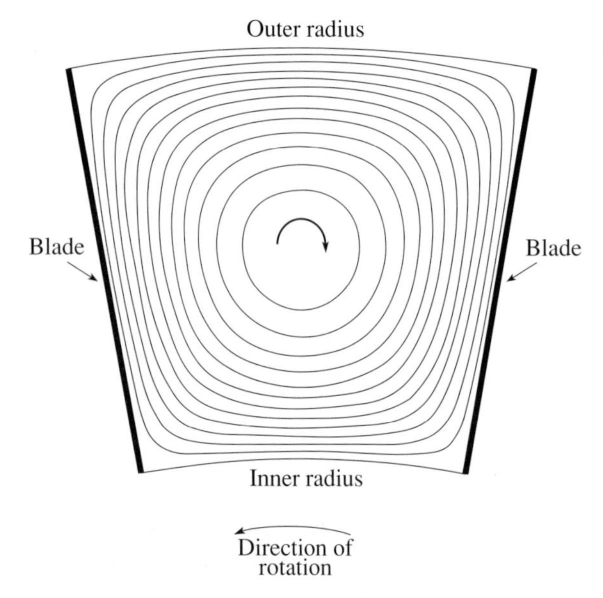

Figure 7.33: Streamlines of a relative eddy in an axial turbomachinery passage with zero stagger blades; contours of equal increments in relative streamfunction ψ.

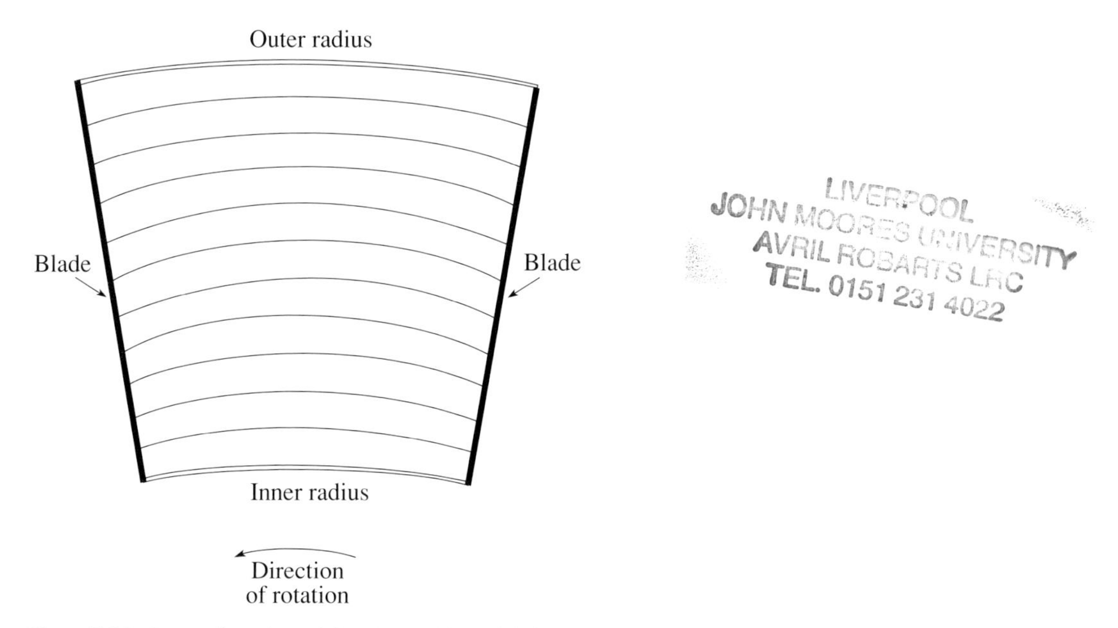

Figure 7.34: Streamlines for axial turbomachine with 60° blade stagger; contours of equal increments in ψ.

Computed streamlines for this flow are shown in Figure 7.33. The blade passage has an inner/outer (hub/tip) radius ratio of 0.7, and the annulus height is equal to the midspan passage width. A relative rotation of the flow, with sense opposite to that of the passage rotation, is evident.

Comparing Figure 7.33 with Figures 7.5 or 7.6 we can infer that the circulatory motion is due to the presence of the radial walls. The stream function solution for this two-dimensional flow consists of a complementary and a homogeneous solution, $\psi = \psi_C + \psi_H$. The former obeys the Poisson equation $\nabla^2\psi_C = -2\Omega$ and is associated with a solid-body flow of vorticity -2Ω with streamlines that are concentric circles centered on the axis of rotation. This part of the solution obeys the boundary conditions of zero normal velocity on inner and outer radii but has a non-zero normal (θ) relative velocity at the location of the radial blades. To obtain zero normal velocity at the blades, the homogeneous solution is needed. The homogeneous solution is irrotational in the relative frame, since $\nabla^2\psi_H = 0$. Although numerical solution is carried out for the combined stream function, the conceptual breakup into homogeneous and complementary solutions helps illustrate that the irrotational relative flow, associated with the presence of non-circular boundaries, carries the information about non-axisymmetric variations in velocity and pressure.

The streamline pattern is quite different for blades with a non-zero stagger angle. Instead of the blades being axial suppose now the blades are inclined to the axial direction by 60° (blade stagger angle 60°). Relative flow streamlines for such a blade passage in an r–θ plane are shown in Figure 7.34. The axial component of relative vorticity is the same as in the previous example. The streamlines do not show an obvious relative eddy because a large relative tangential velocity is superposed on the circulatory flow. For non-zero stagger w_θ is not equal to zero at the blade surface because the boundary condition is $w_\theta/w_x = \tan\gamma$ at the blade surface where γ is the stagger angle. The relative eddy seen in Figure 7.33 thus has a mean circumferential velocity superposed on it which overwhelms the circulatory pattern. Subtracting an appropriate w_θ from the overall velocity field would, however, allow us to extract the circulating flow. The observance of streamline features thus depends on the background velocity field, or coordinate system, from which the observations are made.

An additional feature concerning the relative eddy has been pointed out by Smith (2001). The existence of a relative eddy is connected with a jump in radial (spanwise) velocity across the blade. For blading designed for free vortex flow ($u_\theta \propto 1/r$), i.e. absolute circulation constant with radius, no vortex lines can thread downstream from the blade, there will hence be no such jump, and thus no relative eddy.

8 Swirling flow

8.1 Introduction

Many fluid machinery applications involve swirling flow. Devices in which swirl phenomena have a strong influence include combustion chambers, turbomachines and their associated ducting, and cyclone separators. In this chapter, we examine five aspects of swirling flows: (i) an introductory description of pressure and velocity fields in these types of motion; (ii) the increased capability for downstream conditions to affect upstream flow; (iii) instabilities and propagating waves on vortex cores; (iv) the behavior of vortex cores in pressure gradients; and (v) viscous swirling flow, specifically the influence of swirl on boundary layers, jets, mixing, and recirculation. The behavior of vortex cores ((iii) and (iv)) is described in some depth because this type of embedded structure features in a number of fluid devices. Further, much of the focus is on inviscid flow because the dominant effects of swirl are inertial in nature.

In the discussion it is necessary to modify some of the concepts developed for non-swirling flow. For example, there can be a large variation in static pressure through a vortex core at the center of a swirling flow, in contrast to the essentially uniform static pressure across a thin shear layer or boundary layer in a flow with no swirl. This pressure variation affects the vortex core evolution. The length scales which characterize the upstream influence of a fluid component are also altered when swirl exists.

Different parameters exist in the literature for representing the swirl level in a given flow. These have been developed to enable the definition of flow regimes and behavior. To characterize overall swirl level, we will use the circumferential velocity divided by the axial velocity, u_θ/u_x, denoted by S and referred to as the *swirl parameter*. This can be evaluated in several ways, for example based on the peak u_θ for a vortex core, the conditions at the mean radius, or some average condition; the specific usage will be defined where necessary. The swirl parameter has the advantage that it appears explicitly in the equations describing many of the different phenomena to be examined, facilitating insight into parametric dependence. We note, however, that it is not only the swirl level that is important but also the swirl distribution or, equivalently, the distributions of stagnation pressure and vorticity. Therefore, although a single parameter serves to define different regimes for one specific type of swirling flow (for example solid-body rotation) no one parameter by itself suffices across all different types of swirl distribution. This point is discussed further in Section 8.4.

Chapter 8 deals with uniform density incompressible swirling fluid only. Chapters 10 and 11 address: (i) compressible swirling flow and (ii) the effect of swirl in flows with density variation due to heat addition.

8.2 Incompressible, uniform density, inviscid swirling flows in simple radial equilibrium

The simplest class of swirling flows can be referred to as cylindrical, or *simple radial equilibrium*, flows. The latter term is in common usage and will be adopted here. A simple radial equilibrium flow is defined as one which: (a) is steady; (b) is axisymmetric; (c) has radial velocity, u_r, zero everywhere; and (d), as a consequence of (a), (b), and (c), has axial and circumferential velocity components, u_x and u_θ, and pressure, p, which are only functions of the radius r.

With this set of conditions the continuity and axial and circumferential momentum equations are automatically satisfied. The only non-trivial component of the momentum equation is the radial component, which expresses a balance between radial and centripetal acceleration:

$$\frac{1}{\rho}\frac{dp}{dr} = \frac{u_\theta^2}{r} = \frac{K^2}{r^3}. \tag{8.2.1}$$

In (8.2.1), the quantity K, which can be a function of r, is defined as $K = ru_\theta$, the circulation round a circular contour at radius r, divided by 2π. In defining a simple radial equilibrium flow we are free to specify the radial distribution of u_x and any one of the variables p, u_θ, or K. The circumferential and axial vorticity components are given by[1]

$$\omega_x = \frac{1}{r}\frac{d}{dr}(ru_\theta) = \frac{1}{r}\frac{dK}{dr}, \tag{8.2.2a}$$

$$\omega_\theta = -\frac{du_x}{dr}. \tag{8.2.2b}$$

For a uniform density flow, if the circulation distribution is specified, (8.2.1) can be integrated to find the static pressure distribution. With r_{ref} a reference radius at which a reference pressure, p_{ref}, is specified,

$$\frac{p - p_{ref}}{\rho} = \int_{r_{ref}}^{r} \frac{K^2}{r'^3} dr'. \tag{8.2.3}$$

The stagnation pressure is then

$$\frac{p_t - p_{ref}}{\rho} = \frac{1}{2}\left(u_x^2 + u_\theta^2\right) + \int_{r_{ref}}^{r} \frac{K^2}{r'^3} dr', \tag{8.2.4}$$

or, since $u_\theta = K/r$,

$$\frac{p_t - p_{ref}}{\rho} = \frac{1}{2}u_x^2 + \frac{1}{2}u_{\theta_{ref}}^2 + \int_{r_{ref}}^{r} \frac{K}{r'^2}\frac{dK}{dr'} dr'. \tag{8.2.5}$$

[1] In this chapter the axis of symmetry is denoted by x whereas the axis of rotation in Chapter 7 was denoted by z. The coordinate choice (which it is hoped will cause no confusion!) has been made based on the convention that the overall flow direction has been identified throughout with the x-direction. Chapter 7 described passage flows in the rotating system, with the x-coordinate along the passage. In contrast, Chapter 8 addresses mainly swirling flows with the throughflow direction parallel to the axis of symmetry.

It is generally more relevant to define the flow as having a given p_t and flow angle, α ($\tan\alpha = u_\theta/u_x$), distributions, since these are quantities one can set, rather than p and u_x. Equation (8.2.4) can thus be regarded as a prescription for the variation in velocity magnitude compatible with the requirements for given p_t, flow angle, and the condition of simple radial equilibrium. Equations (8.2.3) and (8.2.4) are alternative statements for radial equilibrium; if one is satisfied, so is the other.

8.2.1 Examples of simple radial equilibrium flows

Several simple radial equilibrium flows represent most of the range of practical interest. We shall see that an important feature of a swirling flow is whether it contains axial vorticity. The first flow to be examined is therefore the irrotational swirling motion referred to as *free vortex flow*. It has u_x and K constant, say u_{x_0} and K_0, and the circumferential velocity is given by $u_\theta = K_0/r$. Equation (8.2.5) shows that p_t is constant and that

$$\frac{p - p_{ref}}{\rho} = \frac{K_0^2}{2}\left(\frac{1}{r_{ref}^2} - \frac{1}{r^2}\right). \tag{8.2.6}$$

Free vortex flow is irrotational everywhere except at $r = 0$.

A simple rotational swirling flow is *solid-body rotation*, also known as *forced vortex flow* (we use the latter term to include the general situation with u_x a function of r). This is often, but not necessarily, specified as having a constant axial velocity, $u_x = u_{x_0} =$ constant. The circumferential velocity is proportional to r (as if the fluid were a "solid body"):

$$u_\theta = \Omega r; \quad \Omega = \text{constant}, \tag{8.2.7}$$

so that $K = \Omega r^2$. Taking conditions on the axis as the reference, the static and stagnation pressures for uniform axial velocity are:

$$\frac{p - p(0)}{\rho} = \frac{\Omega^2 r^2}{2}, \tag{8.2.8a}$$

$$\frac{p_t - p(0)}{\rho} = \frac{1}{2}u_{x_0}^2 + \Omega^2 r^2. \tag{8.2.8b}$$

Forced vortex flow has an axial component of vorticity given by $\omega_x = 2\Omega$. If the axial velocity is non-uniform there is also a circumferential component of vorticity $\omega_\theta = -du_x/dr$.

A third elementary swirl flow, also rotational, is *constant circumferential velocity*, $u_\theta = u_{\theta_0}$, and constant axial velocity, u_{x_0}. For this flow, the stagnation pressure distribution is

$$\frac{p_t - p_{ref}}{\rho} = \frac{1}{2}u_{x_0}^2 + u_{\theta_0}^2 \ln\left(\frac{r}{r_{ref}}\right) + \frac{u_{\theta_0}^2}{2}. \tag{8.2.9}$$

Constant circumferential velocity flow has an axial component of vorticity:

$$\omega_x = \frac{u_{\theta_0}}{r}. \tag{8.2.10}$$

As with the forced vortex, there could also be a circumferential component of vorticity.

Circumferential velocity and stagnation pressure distributions are shown in Figure 8.1 for these three flows in an annulus of inner/outer radius ratio, $r_i/r_o = 0.5$. The normalized circumferential velocity is set to be the same for the three flows at $r/r_o = 0.75$.

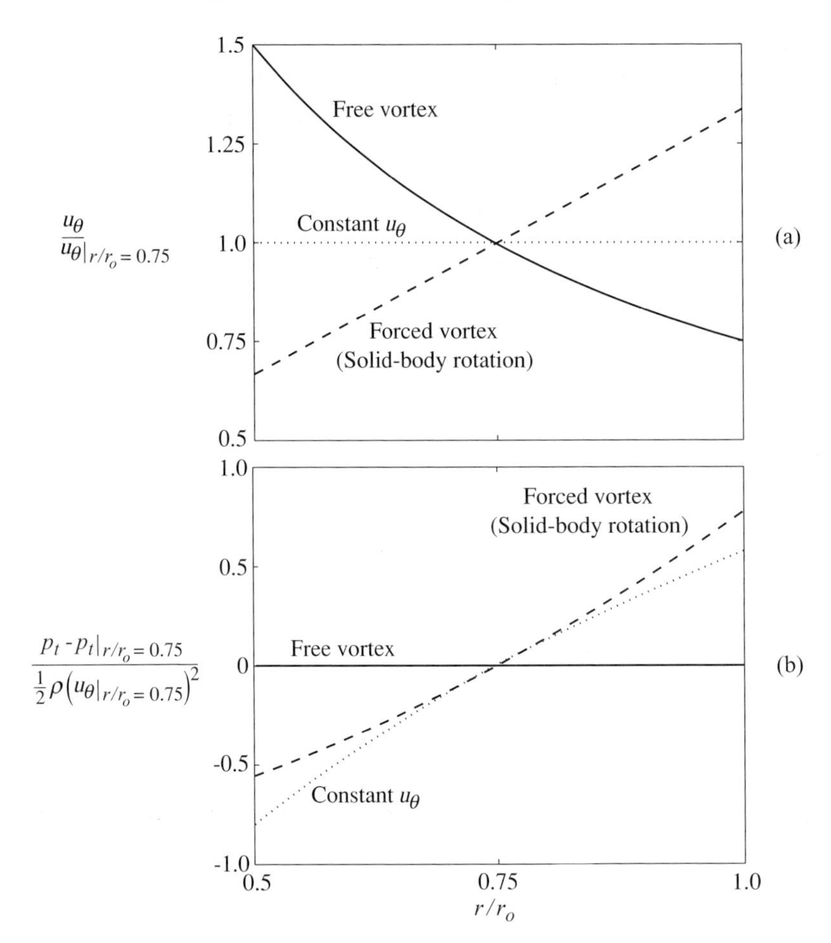

Figure 8.1: Circumferential velocity and stagnation pressure distributions for free vortex, forced vortex, and constant swirl velocity annular flow; inner/outer radius ratio $r_i/r_o = 0.5$, u_θ matched at $r/r_o = 0.75$: (a) circumferential velocity; (b) stagnation pressure based on uniform axial velocity.

Another simple radial equilibrium flow of interest arises in turbomachinery applications where fluid is drawn from a large reservoir into a row of swirl vanes. The fluid entering the vanes has uniform stagnation pressure. If the vanes are designed so that loss variations are small along the span of the vanes, the swirling flow at the vane exit can also be regarded as having uniform stagnation pressure. Differentiating (8.2.5) then gives a constraint on the axial velocity distribution as

$$\frac{d}{dr}\left(u_x^2\right) = -\frac{1}{r^2}\frac{d}{dr}(K^2). \tag{8.2.11}$$

Equation (8.2.11) can also be obtained by substituting the expressions for the vorticity components into Crocco's equation for an incompressible, uniform density fluid, (3.14.6). For the case of uniform stagnation pressure this becomes

$$\mathbf{u} \times \boldsymbol{\omega} = 0. \tag{8.2.12}$$

Except for the free vortex the axial velocity is not constant in a uniform stagnation pressure swirling flow because the variation in axial velocity is linked to the variation in circulation (or circumferential velocity) distribution. For example, with forced vortex swirl, $K = \Omega r^2$, the equation for the axial velocity is

$$\frac{d}{dr}\left(u_x^2\right) = -4\Omega^2 r. \tag{8.2.13}$$

If the axial velocity is u_{x_a} at a radius $r = a$, the axial velocity distribution is

$$\frac{u_x(r)}{u_{x_a}} = \sqrt{1 + \frac{2\Omega^2 a^2}{u_{x_a}^2}\left(1 - \frac{r^2}{a^2}\right)}. \tag{8.2.14}$$

8.2.2 Rankine vortex flow

The free vortex furnishes a useful description in an annular region, but the velocity becomes infinite as the radius, r, approaches zero. In reality viscous effects limit the velocity gradients. A description which reflects this behavior and can be used in a cylindrical duct is a combination of two of the above velocity fields, a *core* of solid-body rotation embedded in a free vortex flow. Denoting the radius of the core by $r = a$, the configuration, often termed a *Rankine vortex*, has the velocity distribution

$$u_\theta = \Omega r, \quad u_x = u_x(r) \quad r \leq a, \tag{8.2.15a}$$

$$u_\theta = \Omega a^2/r, \quad u_x = u_{x0} \quad r \geq a. \tag{8.2.15b}$$

The axial velocity for $r \leq a$ is left as a function of radius in (8.2.15b) because there can be a velocity variation within vortex cores.

The static pressure variation in the Rankine vortex is

$$\frac{p - p(0)}{\rho} = \frac{\Omega^2 r^2}{2} \quad r \leq a, \tag{8.2.16a}$$

$$\frac{p - p(0)}{\rho} = \Omega^2 a^2 - \frac{\Omega^2 a^4}{2r^2} \quad r \geq a. \tag{8.2.16b}$$

Half the overall pressure drop from a station at large radius ($r \to \infty$) to the axis occurs in the irrotational region and half occurs within the solid-body core from $r = a$ to $r = 0$, independent of core radius. The magnitude of the overall static pressure variation, however, is dependent upon the core radius:

$$\frac{\Delta p_{overall}}{\rho} = \Omega^2 a^2. \tag{8.2.17}$$

The axial and circumferential components of vorticity within the core ($r \leq a$) are 2Ω and $-du_x/dr$. The vorticity is zero outside the core. The stagnation pressure distribution is

$$\frac{p_t - p(0)}{\rho} = \frac{[u_x(r)]^2}{2} + \Omega^2 r^2; \quad r \leq a, \tag{8.2.18a}$$

$$\frac{p_t - p(0)}{\rho} = \frac{u_{x0}^2}{2} + \Omega a^2 = \text{constant}; \quad r > a. \tag{8.2.18b}$$

The relation between K and Ω is

$$K = \Omega r^2; \quad r \leq a, \tag{8.2.19a}$$

$$K = \Omega a^2; \quad r > 0. \tag{8.2.19b}$$

8.3 Upstream influence in a swirling flow

Flows with swirl exhibit a much enhanced potential for upstream influence, defined here as the ability to cause a change in the structure of the upstream velocity profile or streamline distribution, compared to non-swirling flow. An example is seen in Figure 8.2, which shows experimentally visualized streamlines in a cylindrical duct downstream of a simulated combustor geometry. Figures 8.2(a) and 8.2(b) correspond to a lower swirl parameter than Figures 8.2(c) and 8.2(d). For these lower swirl conditions, placing a contraction on the downstream end of the duct has little effect on the streamline pattern (compare Figures 8.2(a) and 8.2(b)). With the higher swirl in Figures 8.2(c) and 8.2(d) a substantial change in the streamlines is seen, and the effect of the exit contraction is felt more than three diameters upstream of the duct exit. This behavior is different from the upstream influence with no swirl in which (as will be seen in the next section) pressure disturbances have upstream exponential decay over a length scale of roughly a duct radius.

The increased upstream influence means that, for swirling flow, the guidelines for assuming no coupling between fluid components or for the selection of the type of boundary conditions needed in computational studies are different than for non-swirling flow. Upstream influence will be addressed on several levels. An approximate analysis is given in this section to introduce the topic and provide some general guidelines concerning the impact of swirl. In the following section the topic is explored in more depth to determine parametric dependencies.

We emphasize again it is not swirl level alone which is relevant. If the flow is irrotational (free vortex), no matter what the swirl magnitude the upstream axial and radial velocities are derivable from a potential that (for incompressible flow) obeys Laplace's equation and is independent of the swirl. Upstream influence in an incompressible flow is not altered by free vortex swirl.

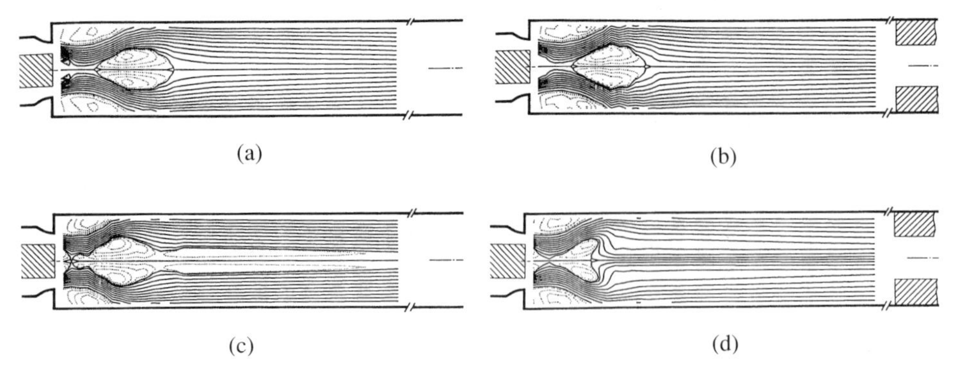

Figure 8.2: Influence of an exit contraction on measured streamlines in a swirling flow in a combustor geometry, $Re = 10,600$: (a) no exit contraction; $S = 5.2$; (b) with exit contraction; $S = 5.2$ (54.5% diameter reduction); (c) no exit contraction; $S = 22.4$; (d) with exit contraction; $S = 22.4$ (Escudier, 1987).

The presence of rotationality in the swirl distribution, particularly the presence of axial vorticity, is the key to the change in upstream influence and is the focus of the present section. The basic phenomena are brought out by examining the behavior of steady axisymmetric disturbances, or perturbations, superposed on a background flow composed of a forced vortex with angular velocity Ω and a uniform axial velocity $\overline{u}_x$. The perturbed motion has velocity components $(\overline{u}_x + u'_x, \Omega r + u'_\theta, u'_r)$. The equations that describe this axisymmetric flow are (Section 1.14):

$$\frac{\partial(u'_x)}{\partial x} + \frac{1}{r}\frac{\partial}{\partial r}(r u'_r) = 0, \tag{8.3.1a}$$

$$\frac{D}{Dt}(\overline{u}_x + u'_x) = -\frac{1}{\rho}\frac{\partial(\overline{p} + p')}{\partial x}, \tag{8.3.1b}$$

$$\frac{D}{Dt}(\Omega r + u'_\theta) + u'_r\frac{(\Omega r + u'_\theta)}{r} = 0, \tag{8.3.1c}$$

$$\frac{Du'_r}{Dt} - \frac{(\Omega r + u'_\theta)^2}{r} = -\frac{1}{\rho}\frac{\partial(\overline{p} + p')}{\partial r}, \tag{8.3.1d}$$

where, for steady flow,

$$\frac{D}{Dt} = (\overline{u}_x + u'_x)\frac{\partial}{\partial x} + u'_r\frac{\partial}{\partial r}. \tag{8.3.2}$$

For small amplitude disturbances squares and products of the perturbation terms can be neglected, resulting in linearized momentum equations for the perturbations:

$$\overline{u}_x\frac{\partial u'_x}{\partial x} = -\frac{1}{\rho}\frac{\partial p'}{\partial x}, \tag{8.3.3a}$$

$$\overline{u}_x\frac{\partial u'_\theta}{\partial x} + 2\Omega u'_r = 0, \tag{8.3.3b}$$

$$\overline{u}_x\frac{\partial u'_r}{\partial x} - 2\Omega u'_\theta = -\frac{1}{\rho}\frac{\partial p'}{\partial r}. \tag{8.3.3c}$$

Eliminating the pressure between (8.3.3a) and (8.3.3c) gives

$$\overline{u}_x\frac{\partial}{\partial x}\left(\frac{\partial u'_r}{\partial x} - \frac{\partial u'_x}{\partial r}\right) - 2\Omega\frac{\partial u'_\theta}{\partial x} = 0. \tag{8.3.4}$$

To obtain a solution of these equations, a perturbation stream function, ψ, can be introduced which satisfies the continuity equation identically:

$$u'_x = \frac{1}{r}\frac{\partial \psi}{\partial r}, \quad u'_r = -\frac{1}{r}\frac{\partial \psi}{\partial x}. \tag{8.3.5}$$

Substituting (8.3.5) into (8.3.4), and eliminating u'_θ using (8.3.3b), yields an equation for the disturbance stream function, ψ:

$$\frac{\partial}{\partial x}\left[\frac{\partial^2 \psi}{\partial x^2} + \frac{\partial^2 \psi}{\partial r^2} - \frac{1}{r}\frac{\partial \psi}{\partial r} + \left(\frac{2\Omega}{\overline{u}_x}\right)^2 \psi\right] = 0. \tag{8.3.6}$$

To demonstrate in a simple manner the effect of swirl on upstream influence, we confine attention (for now) to annular regions of high inner/outer radius ratio, i.e. r_i/r_o near unity. In this situation order

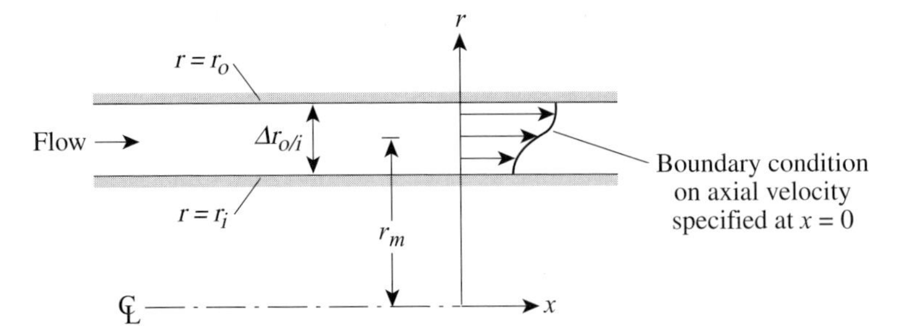

Figure 8.3: Geometry for the analysis of upstream influence in an annular swirling flow; the domain is the annular region upstream of $x = 0$.

of magnitude arguments can be used to eliminate a term in (8.3.6). The length scale for variations in ψ in the radial direction is of order $\Delta r_{o/i}$, where $\Delta r_{o/i}$ is the annulus height, $r_o - r_i$. The ratio of the two r-derivative terms, $(1/r)(\partial\psi/\partial r)$ and $(\partial^2\psi/\partial r^2)$, is thus roughly $\Delta r_{o/i}/r_m$, where r_m is the mean radius. For high inner/outer radius ratio ($\Delta r_{o/i}/r_m \ll 1$) the first r-derivative term can be neglected compared to the second and (8.3.6) reduced to

$$\frac{\partial}{\partial x}\left[\frac{\partial^2\psi}{\partial x^2} + \frac{\partial^2\psi}{\partial r^2} + \left(\frac{2\Omega}{\overline{u}_x}\right)^2 \psi\right] = 0. \tag{8.3.7}$$

Equation (8.3.7) describes the steady axisymmetric disturbance flow field in a high hub/tip radius ratio annulus.

To close the problem specification we take the flow to have an axial velocity distribution that varies with radius at the station $x = 0$ (see Figure 8.3) and ask how far upstream the influence of this non-uniformity will be felt. For definiteness the axial velocity perturbation at $x = 0$ is given by

$$u'_x(0, r) = \varepsilon \overline{u}_x \sin\frac{\pi(r - r_m)}{\Delta r_{o/i}}. \tag{8.3.8}$$

The disturbance stream function must give an axial velocity consistent with the boundary condition at $x = 0$ and obey the condition of no normal velocity along the inner and outer walls of the annulus or "hub" and "tip" ($r = r_m \pm \Delta r_{o/i}/2$). Therefore,

$$\frac{1}{r}\frac{\partial\psi}{\partial r}(0, r) = u'_x(0, r), \tag{8.3.9}$$

$$\frac{\partial\psi}{\partial x}\left(x, r_m + \frac{\Delta r_{o/i}}{2}\right) = \frac{\partial\psi}{\partial x}\left(x, r_m - \frac{\Delta r_{o/i}}{2}\right) = 0. \tag{8.3.10}$$

The disturbance must also be bounded far upstream.

As can be verified by direct substitution, a suitable form of ψ satisfying the boundary conditions given in (8.3.9) and (8.3.10) is[2]

$$\psi = \frac{-\varepsilon \overline{u}_x r_m \Delta r_{o/i}}{\pi} f(x) \cos\frac{\pi(r - r_m)}{\Delta r_{o/i}}, \tag{8.3.11}$$

[2] In (8.3.11) we have replaced r by r_m in the coefficient of the stream function, consistent with the approximation made previously in dropping the term $(1/r)(\partial\psi/\partial r)$.

where $f(x)$, which describes the axial variation, is to be determined. Substituting (8.3.11) into (8.3.7) yields an equation for $f(x)$:

$$\frac{d}{dx}\left\{\frac{d^2 f}{dx^2} + \left[\left(\frac{2\Omega}{\overline{u}_x}\right)^2 - \frac{\pi^2}{(\Delta r_{o/i})^2}\right] f\right\} = 0. \tag{8.3.12}$$

The solution of (8.3.12) which decays upstream ($x < 0$) has the form

$$f(x) \propto e^{(\pi x/\Delta r_{o/i})\sqrt{1-[\Omega r_m/\overline{u}_x]^2[2\Delta r_{o/i}/(\pi r_m)]^2}}.$$

The term inside the square root has been written in terms of the swirl parameter at the mean radius, $\Omega r_m/\overline{u}_x$, and a term $2\Delta r_{o/i}/\pi r_m$ representing the inner/outer radius ratio of the annulus. The form of the disturbance stream function, ψ, is

$$\psi = \frac{-\varepsilon \overline{u}_x r_m \Delta r_{o/i}}{\pi} \cos\frac{\pi(r - r_m)}{\Delta r_{o/i}} e^{(\pi x/\Delta r_{o/i})\sqrt{1-[\Omega r_m/\overline{u}_x]^2[2\Delta r_{o/i}/(\pi r_m)]^2}}. \tag{8.3.13}$$

The exponential decay sets the extent of upstream influence. Without swirl the exponent would be $\pi x/\Delta r_{o/i}$ (Section 2.3). As the swirl parameter $\Omega r_m/\overline{u}_x$ is increased, the decay with upstream distance decreases. At swirl parameters equal to, or greater than, $\pi r_m/(2\Delta r_{o/i})$, the exponent is zero or imaginary and disturbances do not decay upstream. The solutions then have a wave-like, rather than decaying, structure and different boundary conditions need to be applied that take this into account.

The lengthened upstream distance over which a disturbance can be felt in a swirling flow compared to the no-swirl situation is sometimes referred to as the stiffening effect of vortex lines. It is essentially the same phenomenon we encountered in rotating flows (Section 7.4), namely that for large values of background axial vorticity, $\Omega r_m/\overline{u}_x$, the flow exhibits strong tendencies towards motions which do not vary along the axis of rotation.

8.4 Effects of circulation and stagnation pressure distributions on upstream influence

The previous section introduced qualitative features of upstream influence in a swirling flow. We now make the conclusions more quantitative and demonstrate how radial distributions of circulation (swirl) and stagnation pressure affect the extent over which a downstream non-uniformity impacts the upstream motion. The approach is to derive an equation relating the stream function to the radial distributions of circulation and stagnation pressure. Solution of this equation defines the upstream decay rate of a velocity variation with radius specified at a given axial station.

The effects of interest are described in the context of steady, axisymmetric, inviscid flow. For this situation circulation and stagnation pressure are conserved along streamlines so that $K = \Gamma/2\pi = K(\psi)$ and $p_t = p_t(\psi)$. From the definition of the axisymmetric stream function, (8.3.5), the circumferential component of vorticity, ω_θ, is

$$\omega_\theta = \frac{\partial u_r}{\partial x} - \frac{\partial u_x}{\partial r} = -\frac{1}{r}\left(\frac{\partial^2 \psi}{\partial x^2} - \frac{1}{r}\frac{\partial \psi}{\partial r} + \frac{\partial^2 \psi}{\partial r^2}\right). \tag{8.4.1}$$

The x-component of the Crocco form of the momentum equation allows us to link ω_θ to K and p_t. The Crocco equation is

$$\mathbf{u} \times \boldsymbol{\omega} = \frac{\nabla p_t}{\rho}. \tag{3.14.6}$$

The x-component is

$$u_r\omega_\theta - u_\theta\omega_r = \frac{1}{\rho}\frac{\partial p_t}{\partial x}. \tag{8.4.2}$$

To write (8.4.2) in terms of ψ, K and p_t note that the radial component of vorticity, ω_r, is given by

$$\omega_r = -\frac{\partial u_\theta}{\partial x} = -\frac{\partial}{\partial x}\left(\frac{K}{r}\right). \tag{8.4.3}$$

Because K is a function of ψ only,

$$\frac{\partial}{\partial x}K(\psi) = r\frac{\partial u_\theta}{\partial x} = \frac{dK}{d\psi}\frac{\partial \psi}{\partial x},$$

yielding the radial component of vorticity as

$$\omega_r = u_r\frac{dK}{d\psi}. \tag{8.4.4}$$

The axial variation of the stagnation pressure can also be written in terms of the stream function as

$$\frac{\partial p_t}{\partial x} = \frac{dp_t}{d\psi}\frac{\partial \psi}{\partial x}. \tag{8.4.5}$$

Substituting (8.4.3)–(8.4.5) into (8.4.2) produces the desired equation for the stream function in terms of derivatives of stagnation pressure and circulation:

$$\frac{\partial^2 \psi}{\partial x^2} - \frac{1}{r}\frac{\partial \psi}{\partial r} + \frac{\partial^2 \psi}{\partial r^2} = r^2\frac{d(p_t/\rho)}{d\psi} - K\frac{dK}{d\psi}. \tag{8.4.6}$$

Equation (8.4.6), which is due to Bragg and Hawthorne (1950) (see also Batchelor (1967), Leibovich and Kribus (1990)), explicitly links the stagnation pressure and circulation distributions to the stream function behavior.

As a first example of the use of (8.4.6), we reexamine in more depth the problem considered in Section 8.3, upstream influence in an annulus with far upstream forced vortex swirl and uniform axial velocity. At the far upstream location[3]

$$u_x(-\infty, r) = \frac{1}{r}\left(\frac{\partial \psi}{\partial r}\right)\bigg|_{x=-\infty} = \overline{u}_x \tag{8.4.7a}$$

[3] We use the notation $(-\infty)$ to emphasize that the station is distant enough not to see any upstream influence as well as to distinguish from the subscript that denotes conditions at the outer radius (e.g. r_o, p_o).

or

$$\psi|_{x=-\infty} = \overline{u}_x \frac{r^2}{2}, \tag{8.4.7b}$$

$$p_t(-\infty, r) = p_t(-\infty, r_i) + \rho\left(r^2\Omega^2 - r_i^2\Omega^2\right) = -\rho r_i^2\Omega^2 + 2\rho\frac{\Omega^2\psi}{\overline{u}_x}, \tag{8.4.7c}$$

$$K = \Omega r^2 = \frac{2\Omega\psi}{\overline{u}_x}. \tag{8.4.7d}$$

Because K and p_t are functions of ψ only, the derivatives with respect to ψ have the same value at any axial station and (8.4.6) takes the form

$$\frac{\partial^2\psi}{\partial r^2} - \frac{1}{r}\frac{\partial\psi}{\partial r} + \frac{\partial^2\psi}{\partial x^2} = \frac{2\Omega^2 r^2}{\overline{u}_x} - \frac{4\Omega^2\psi}{\overline{u}_x^2}. \tag{8.4.8}$$

The stream function ψ can be defined in two parts as

$$\psi = \frac{1}{2}\overline{u}_x r^2 + \psi_{up}. \tag{8.4.9}$$

The first term represents a forced vortex, uniform axial velocity flow, undisturbed by any downstream boundary conditions. The second term, ψ_{up}, which expresses the departure from the far upstream forced vortex flow, defines the upstream influence. Substituting (8.4.9) into (8.4.8) provides the equation for ψ_{up}:

$$\frac{\partial^2\psi_{up}}{\partial r^2} - \frac{1}{r}\frac{\partial\psi_{up}}{\partial r} + \frac{\partial^2\psi_{up}}{\partial x^2} + \left(\frac{2\Omega}{\overline{u}_x}\right)^2\psi_{up} = 0. \tag{8.4.10}$$

To assess the upstream influence, as in Section 8.3, we examine the upstream decay of a velocity non-uniformity at a specified axial location. To do this it is not necessary to define the solution to (8.4.10) in detail. If we separate variables and write the stream function as

$$\psi_{up} = R(r)X(x) \tag{8.4.11}$$

the x-dependence is found to be of the form $X = e^{\lambda x/\Delta r_{o/i}}$. The non-dimensional quantity in the exponent, λ, is determined by solving (8.4.10), imposing the boundary conditions of no normal (or radial) velocity at $r = r_o$ and $r = r_i$. The value of $1/\lambda$ gives an indication of the upstream distance over which downstream disturbances attenuate and hence of the extent of upstream influence.

Figure 8.4 shows λ versus the non-dimensional parameter $\Omega\Delta r_{o/i}/\overline{u}_x$, for four different values of inner/outer radius ratio, r_i/r_o. The dashed line is the approximate solution of Section 8.3,[4] for which λ is equal to π at $\Omega\Delta r_{o/i}/\overline{u}_x = 0$ (the upstream influence result from Section 2.3) and falls to zero at $\Omega\Delta r_{o/i}/\overline{u}_x = \pi/2$. For forced vortex flow the effect of swirl on the upstream extent over which disturbances are felt can be seen by the fact that all the curves drop to 0 as $\Omega\Delta r_{o/i}/\overline{u}_x$ increases to between $\pi/2$ and 1.9. For values of $\Omega\Delta r_{o/i}/\overline{u}_x$ in excess of those for which λ is 0, there is no decay with upstream distance.

Although the initial consideration of upstream influence was focused on the forced vortex velocity distribution because it provides a clear example of the effects of interest, the ideas are readily extended

[4] Note that $\overline{u}_x$ in Section 8.3 has been replaced by $u_{x_{-\infty}}$ in the more general treatment in Section 8.4.

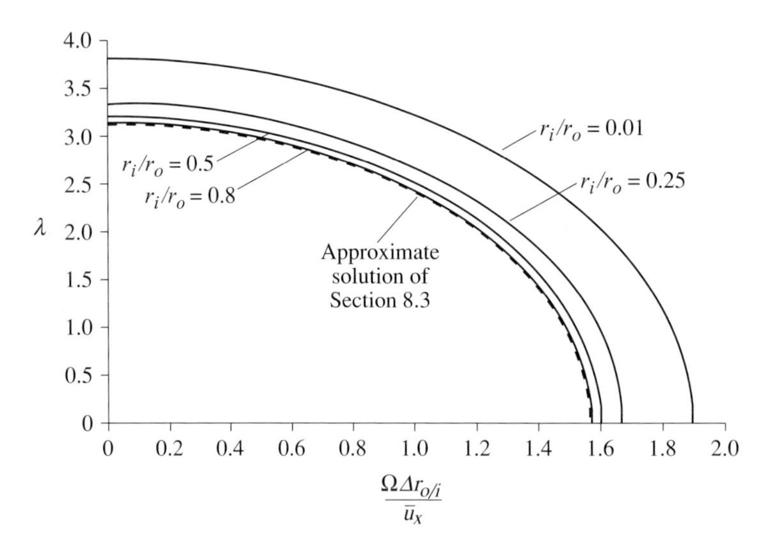

Figure 8.4: Upstream decay exponent for a forced vortex flow ($u_\theta = \Omega r$, where Ω is a constant) in an annulus with r_o = outer radius, r_i = inner radius, $\Delta r_{o/i} = r_o - r_i$; upstream disturbance velocity decay $\propto e^{\lambda x/\Delta r_{o/i}}$.

to more general swirl and axial velocity distributions. The problem can be posed as in the previous section. At a given axial station, $x = 0$, there is a radially non-uniform axial velocity, $u_x(r, 0)$. This could result from duct geometry (e.g. a radius increase in an annular duct or the presence of a nozzle) or the influence of turbomachinery. For a given far upstream distribution of swirl (K) and stagnation pressure (p_t) we wish to determine the upstream distance over which there is an appreciable effect of this imposed downstream axial velocity distribution.

To proceed further specific statements must be made about the configuration to be studied. Two geometries are considered, an annular region with an inner/outer radius ratio of 0.5 and a cylindrical duct. The former primarily illustrates the effect of the circulation distribution, the latter the effect of the stagnation pressure distribution. At $x = 0$ the axial velocity is taken to have the form (with the far upstream axial velocity, $u_{x_{-\infty}}$, no longer restricted to be uniform)

$$u_x(0, r) = \left[1 + \varepsilon \sin\frac{\pi(r - r_i)}{\Delta r_{o/i}}\right] u_{x_{-\infty}}. \tag{8.4.12}$$

One further approximation will be made to simplify (8.4.6). For ε small compared with unity the disturbances considered (for example the disturbance in axial velocity) are small compared to the mean values of these quantities over *much* of the region of interest, and in many situations over *all* of this region. We can take advantage of this and solve a linearized form of (8.4.6) without affecting the overall conclusions concerning the extent of the upstream influence. The linearization is that local quantities on the right-hand side of (8.4.6) are replaced by their value at the far upstream condition, denoted by the subscript "$-\infty$". A physical statement of this approximation is that stagnation pressure and circulation are regarded as convected along the undisturbed streamlines, which are helices of constant radius, rather than along the actual streamlines, which have a radius change. If $\Delta u_x/u_{x_{-\infty}}$ is

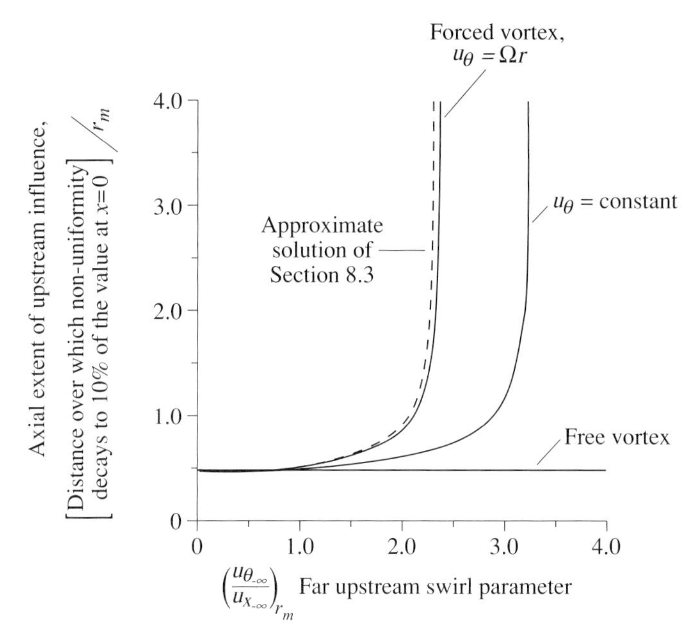

Figure 8.5: Upstream influence for different swirl distributions; annular flow, $r_i/r_o = 0.5$ (subscript "r_m" denotes conditions at radius $r_m = 0.75r_o$).

everywhere small compared to unity, the linearized solution will be a good quantitative descriptor but, even if $\Delta u_x/u_{x_{-\infty}}$ is not small compared to unity, as long as there is no reverse flow the description will be qualitatively useful.

With the above proviso, the equation for the disturbance stream function, ψ_{up}, associated with the departure from far upstream conditions, is

$$\frac{\partial^2 \psi_{up}}{\partial r^2} - \frac{1}{r}\frac{\partial \psi_{up}}{\partial r} + \frac{\partial^2 \psi_{up}}{\partial x^2} = \left\{ r^2 \left[\frac{d(p_t/\rho)}{d\psi}\right]_{x=-\infty} - \left[K\left(\frac{dK}{d\psi}\right)\right]_{x=-\infty} \right\} \psi_{up}. \tag{8.4.13}$$

In (8.4.13) the square-bracketed terms are functions of radius.

Assigning a numerical value to the extent of upstream influence has some degree of arbitrariness, but a metric which illustrates the point is the axial distance at which the magnitude of the axial velocity non-uniformity has decreased to 10% of the value at the downstream boundary where the non-uniformity is imposed. Figure 8.5 shows this "upstream influence distance", normalized by the mean radius of the duct, r_m, as a function of the far upstream swirl parameter $[u_{\theta_{-\infty}}/u_{x_{-\infty}} = K/(ru_{x_{-\infty}})]$ evaluated at the mean radius. Results from solution of (8.4.13) are presented for three different circulation distributions: free vortex ($K = ru_\theta$ = constant), constant circumferential velocity, and forced vortex (K proportional to r^2). Results from the approximate solution of Section 8.3 are also indicated. For all these the far upstream axial velocity is uniform. The far upstream values of axial vorticity at the mean radius for the three cases are $\omega_{x_{-\infty}} r_m/u_{x_{-\infty}} = 0$, $(u_{\theta_{-\infty}}/u_{x_{-\infty}})_{r_m}$, and $2(u_{\theta_{-\infty}}/u_{x_{-\infty}})_{r_m}$ for the free vortex, uniform u_θ, and forced vortex flows respectively.

For irrotational steady flow, K and p_t are uniform and (8.4.6) and (8.4.13) reduce to an equation in which the swirl level does not appear:

$$\frac{\partial^2 \psi_{up}}{\partial r^2} - \frac{1}{r}\frac{\partial \psi_{up}}{\partial r} + \frac{\partial^2 \psi_{up}}{\partial x^2} = 0. \tag{8.4.14}$$

For irrotational flow upstream influence does not depend on $u_{\theta_{-\infty}} / u_{x_{-\infty}}$.

For uniform u_θ and forced vortex distributions, the behavior is different. Figure 8.5 indicates that the region of upstream influence increases as the parameter $(u_{\theta_{-\infty}}/u_{x_{-\infty}})_{r_m}$ is increased. Further, as described in Section 8.3, there is a value of swirl parameter above which axial velocity disturbances do not decay.

The discussion so far has been in terms of differences in circulation distribution. The stagnation pressure distribution is also different for the two rotational flows and this affects upstream influence. To exhibit the trends with the stagnation pressure profile, we examine a Rankine vortex swirling flow in a cylindrical duct in which the far upstream flow has a forced vortex distribution over the inner part of the duct, from $r = 0$ to $r = 0.5r_o$, and constant circulation at radii greater than $r = 0.5r_o$. Calculations have been carried out using (8.4.13) for three families of far upstream axial velocity profiles: (1) axial velocity $(u_{x_{-\infty}})$ uniform with radius, (2) axial velocity having a linear decrease or increase with radius in the *inner* part of the duct (denoted by *ID*), and (3) axial velocity having a linear decrease or increase with radius in the *outer* part of the duct (denoted by *OD*). The downstream boundary condition for the disturbance flow in all cases is

$$\frac{1}{r}\frac{\partial \psi_{up}}{\partial r}(0, r) = u_x(0, r) - u_{x_{-\infty}}(0) = \varepsilon (u_{x_{-\infty}})_{r_m} \sin(\pi r / r_o). \tag{8.4.15}$$

Figure 8.6 shows the results. Because of the interacting parameters, a range of cases has been included. Figure 8.6(a) illustrates the circumferential velocity distribution far upstream while Figures 8.6(b) and 8.6(c) show the far upstream axial velocity distributions. Figures 8.6(d)–(g) portray the upstream stagnation pressure distributions (referenced to the static pressure on the centerline, p_{cl} ($= p(-\infty, 0)$)) corresponding to Figures 8.6(b) and 8.6(c), for two levels of swirl parameter. Figures 8.6(d) and 8.6(f) correspond to the axial velocity profiles in Figure 8.6(b), while Figures 8.6(e) and 8.6(g) correspond to the axial velocity profiles in Figure 8.6(c). The curves in Figures 8.6(d) and 8.6(e) correspond to $u_{\theta_{-\infty}}/u_{x_{-\infty}} = 0.5$ at the mean radius, r_m, and those in Figures 8.6(f) and 8.6(g) to $u_{\theta_{-\infty}}/u_{x_{-\infty}} = 1.0$ at the mean radius. The nomenclature for the axial velocity is that I-1, I-2, and so on correspond to profiles 1, 2, etc. with axial velocity variation in the inner region of the duct, and O-1, O-2, etc. correspond to profiles with axial velocity variation in the outer part.

The results of the calculations are summarized in Figure 8.7, which shows the extent of upstream influence versus the far upstream swirl parameter evaluated at the mean radius, $(u_{\theta_{-\infty}}/u_{x_{-\infty}})_{r_m}$. The figure illustrates that the form of the stagnation pressure distribution has a major impact on upstream influence. In particular a decrease in stagnation pressure in the inner part of the duct (where the stagnation pressure is low even with uniform axial velocity) has a stronger effect than a decrease in the outer part of the duct. The spread in the values of the swirl parameter at which the upstream influence increases rapidly is more than a factor of 10 larger for the I-1 to I-5 profiles than for the O-1 to O-5 profiles. Figure 8.7 shows it is not only the axial velocity distribution that is important,

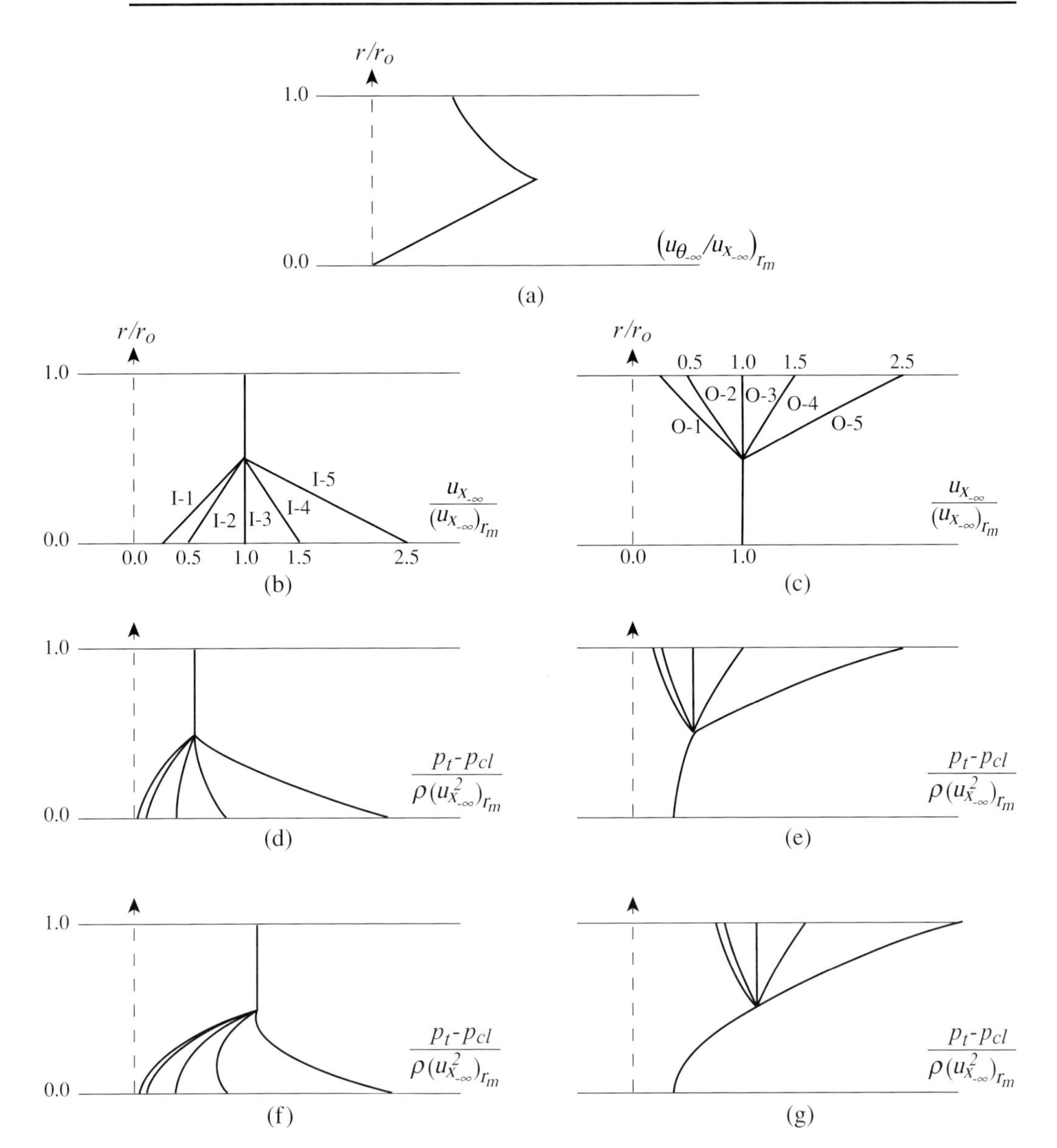

Figure 8.6: Far upstream circumferential and axial velocities and stagnation pressure distributions used to illustrate the parametric behavior of upstream influence for swirling flow in a cylindrical duct. *ID* and *OD* denote axial velocity variation in the inner and outer parts of the duct, respectively. Far upstream velocities: $u_{\theta_{-\infty}} = \Omega r, r \le 0.5\ r_o$, $u_{\theta_{-\infty}} = 0.25\ \Omega r_o^2/r, r > 0.5\ r_o$, $u_{x_{-\infty}}$ as shown in (b) and (c); subscript "r_m" denotes value at $r = 0.5\ r_o$ (duct mean radius); $p_t = p_t(-\infty, r)$, $p_{cl} = p(-\infty, 0)$: (a) Far upstream swirl distribution; (b) axial velocity for *ID* velocity variations; (c) axial velocity for *OD* velocity variations; (d) stagnation pressure distribution corresponding to (b), $(u_{\theta_{-\infty}}/u_{x_{-\infty}})_{r_m} = 0.5$; (e) stagnation pressure distribution corresponding to (c), $(u_{\theta_{-\infty}}/u_{x_{-\infty}})_{r_m} = 0.5$; (f) stagnation pressure distribution corresponding to (b), $(u_{\theta_{-\infty}}/u_{x_{-\infty}})_{r_m} = 1.0$; (g) stagnation pressure distribution corresponding to (c), $(u_{\theta_{-\infty}}/u_{x_{-\infty}})_{r_m} = 1.0$.

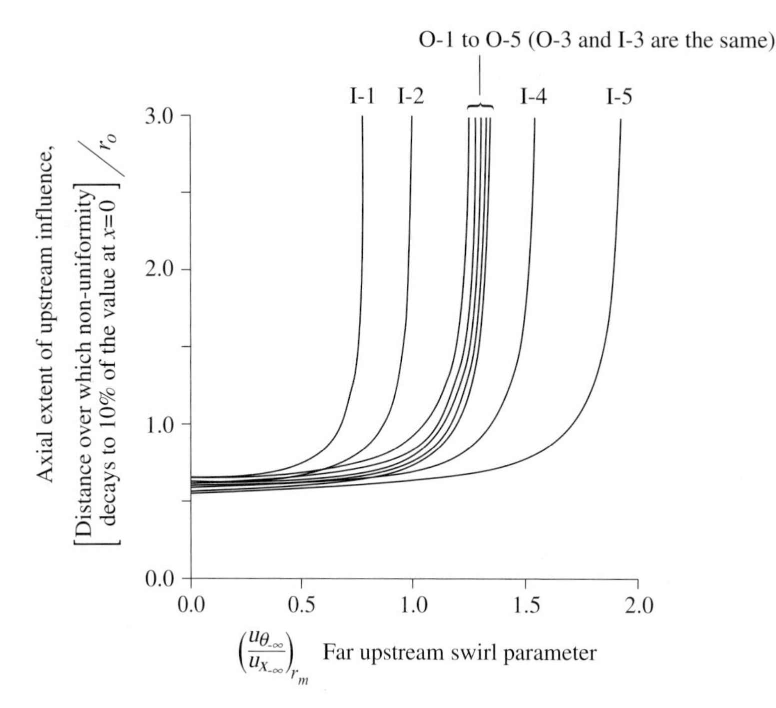

Figure 8.7: Upstream influence for different stagnation pressure distributions, flow in a cylindrical duct; see Figure 8.6 for the key to axial velocity and stagnation pressure distributions.

since flows with the same axial velocity but different stagnation pressures (as in Figures 8.6(d) and 8.6(f) for example) exhibit very different behaviors with regard to upstream influence.

8.5 Instability in swirling flow

Flows with swirl exhibit a variety of unsteady phenomena. In this section a basic instability associated with swirl is described. In Section 8.6 two additional aspects of unsteady behavior are addressed, wave propagation on vortex cores and the stabilizing effect of swirl on shear layer (Kelvin–Helmholtz) instability.

The instability associated with the presence of swirl means that some circumferential velocity distributions consistent with simple radial equilibrium are unstable and not achievable in practice. To assess stability (as described in Chapter 6) one subjects a steady flow to a small amplitude unsteady perturbation and determines the subsequent dynamic behavior of such perturbations, in particular whether they grow or decay. For axisymmetric disturbances in an inviscid, uniform density, incompressible fluid, this question can be settled without formally solving the equations using an argument originally given by Rayleigh (see, for example, Howard (1963), Tritton (1988)).

One form of this argument is as follows (Howard, 1963). From Section 1.14 the equations of inviscid axisymmetric flow in cylindrical coordinates are:

$$\frac{1}{r}\frac{\partial}{\partial x}(ru_r)+\frac{\partial u_x}{\partial x}=0, \tag{8.5.1a}$$

$$\frac{\partial u_r}{\partial t}+u_r\frac{\partial u_r}{\partial r}+u_x\frac{\partial u_r}{\partial x}=-\frac{1}{\rho}\frac{\partial p}{\partial r}+\frac{(ru_\theta)^2}{r^3}, \tag{8.5.1b}$$

$$\frac{\partial u_x}{\partial t}+u_r\frac{\partial u_x}{\partial r}+u_x\frac{\partial u_x}{\partial x}=-\frac{1}{\rho}\frac{\partial p}{\partial x}, \tag{8.5.1c}$$

$$\frac{\partial}{\partial t}(ru_\theta)+u_r\frac{\partial}{\partial r}(ru_\theta)+u_x\frac{\partial}{\partial x}(ru_\theta)=0. \tag{8.5.1d}$$

Equation (8.5.1d) implies that the quantity ru_θ is constant following a fluid particle. Equations (8.5.1a)–(8.5.1c) show that the motion described is as if the only velocity components were u_x and u_r but the fluid were subjected to a body force $(ru_\theta)^2/r^3$ in the outward radial direction. This can be viewed as the force due to an equivalent radial gravitational field of strength $1/r^3$ acting on a density distribution proportional to $(ru_\theta)^2$. The interpretation of $(ru_\theta)^2$ as a density is appropriate because $(ru_\theta)^2$ is constant following a particle. An analogy can therefore be drawn between an axisymmetric swirling flow of a uniform density fluid and the axisymmetric, non-swirling flow of a non-homogeneous incompressible fluid with density proportional to $(ru_\theta)^2$ in a radial gravitational field of strength $1/r^3$.

The condition for stability of a steady simple radial equilibrium flow with $u_x = u_r = 0$, $u_\theta = u_\theta(r)$ follows from this analogy. The flow will be stable if $(ru_\theta)^2$ increases outwards and unstable if $(ru_\theta)^2$ decreases; the analogy is stability when denser fluid is outside less dense fluid. In summary, Rayleigh's criterion is that a swirling flow is stable to axisymmetric perturbations if the square of the circulation increases with radius.

Free vortex flow, with ru_θ constant, defines the neutral stability condition. Swirling flows in which the circumferential velocity drops off more rapidly with radius than a free vortex are unstable. Forced vortex swirl, with $(ru_\theta)^2=\Omega^2r^4$ which is increasing outwards, and constant circumferential velocity swirl, are examples of stable swirling flows.

Rayleigh's criterion can also be derived by considering two thin rings of fluid, one at r_1 and one at r_2, where $r_1 < r_2$. Suppose the rings are interchanged. Initially each was in equilibrium such that

$$\frac{1}{\rho}\frac{\partial p}{\partial r}=\frac{u_\theta^2}{r}=\frac{(ru_\theta)^2}{r^3}. \tag{8.5.2}$$

During the displacement, both rings keep their initial value of ru_θ. When $r=r_2$, for the ring initially at r_1

$$\frac{u_{\theta 2}^2}{r_2}=\frac{(r_1u_{\theta 1})^2}{r_2^3}. \tag{8.5.3}$$

The radial pressure gradient is set by the conditions outside the ring and is equal to $(r_2u_{\theta 2})^2/r_2^3$ at $r=r_2$. If the pressure gradient is greater than the centripetal acceleration, a radial motion will be created to return the ring to its initial radius. This requires $(r_2 u_{\theta 2})^2 > (r_1 u_{\theta 1})^2$, in other words, that the circulation increases outwards, as was derived above. The arguments developed are for the case $u_x = 0$, but they apply to $u_x = $ constant also, because this is just equivalent to changing the frame of reference of the observer.

8.6 Waves on vortex cores

Vortex cores are a feature of many flows. Examples are the clearance vortices found in turbomachines, the vortices on the centerline of swirl flow chambers, and the vortices that form at the inlet to gas turbine engines. The geometry in which these vortex cores are created is often non-axisymmetric, but if the core thickness is small compared to the characteristic scale of the region in which they are embedded the vortex structure can be approximated as axisymmetric, as in the treatment here.

In this section we examine the characteristics of axisymmetric wave motions in vortex cores. The discussion in Section 8.5 implies that swirl distributions in which the circumferential velocity decreases more slowly than $1/r$ exhibit a restoring force to return fluid particles to their original positions when radially displaced. This situation is one in which wave motions would be expected. We will see in Section 8.7 that the wave propagation speed obtained from the analysis is also helpful as a guide to the flow regimes expected for steady vortex cores in pressure gradients. In particular, this speed will be seen to play a role analogous to the speed of sound in one-dimensional compressible flow.

8.6.1 Control volume equations for a vortex core

We use the Rankine vortex model of Section 8.2 consisting of a forced vortex with core of radius a, surrounded by an irrotational swirling flow. The core center is aligned with the x-axis. The core radius and axial velocity, u_x (taken here as uniform across the core), are both functions of the axial coordinate, x, and the time, t, as indicated in Figure 8.8. The circulation of the core, denoted by K_c,

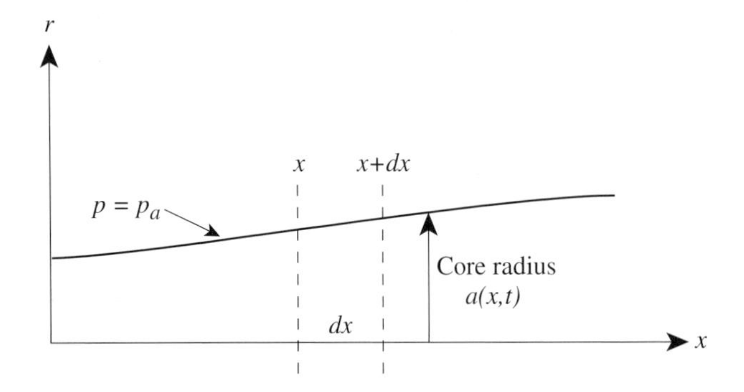

Figure 8.8: Schematic of a quasi-one-dimensional model showing a vortex core of radius $a(x,t)$ with control volume. The pressure force at x is $\int_o^a 2\pi p r dr = p_a A - (\rho \pi K_c^2/4)$; $K_c = a u_{\theta max}$.

is a constant of the motion.[5] At any axial location there is a Rankine distribution of circumferential velocity:

$$u_\theta(r,x,t) = \begin{cases} \dfrac{K_c r}{a^2}, & r \le a(x,t) \\ \dfrac{K_c}{r}, & r > a(x,t) \end{cases}, \quad K_c = \text{constant}. \tag{8.6.1}$$

The maximum swirl velocity $u_\theta = K_c/a$ occurs at the core edge $r = a$. The swirl parameter for the vortex core, S_c, is defined in terms of the core velocity components and radius as

$$S_c = \frac{u_{\theta_{\max}}}{u_x} = \frac{K_c}{a u_x}. \tag{8.6.2}$$

With the approximation that radial velocities are negligible, the radial momentum equation reduces to simple radial equilibrium, applied locally in x,

$$\frac{\partial p}{\partial r} = \rho \frac{u_\theta^2}{r}. \tag{8.6.3}$$

Equations (8.6.1) and (8.6.3), along with the assumption that the flow outside the vortex core is irrotational, imply the axial velocity outside the core is uniform in r, although its value need not be the same as in the core.

An expression for the static pressure is obtained by integrating (8.6.3) with the specified circumferential velocity distribution of (8.6.1). Using the notation p_a for the core edge pressure, $p(a, x, t)$, this is

$$p(r,x,t) - p_a(x,t) = \begin{cases} -\dfrac{1}{2}\rho\left(\dfrac{K_c}{a}\right)^2\left[1-\left(\dfrac{r}{a}\right)^2\right], & r \le a \\ \dfrac{1}{2}\rho\left(\dfrac{K_c}{a}\right)^2\left[1-\left(\dfrac{a}{r}\right)^2\right], & r > a. \end{cases} \tag{8.6.4}$$

With reference to the control volume of Figure 8.8 we assume the core boundary is a streamline. This plus integration of (8.6.4) across the core to find the pressure force enables derivation of the conservation equations for the core. Denoting the local core area, πa^2, as A, these are:

$$\text{conservation of mass: } \frac{\partial}{\partial t}(A) + \frac{\partial}{\partial x}(A u_x) = 0, \tag{8.6.5}$$

$$\text{conservation of momentum: } \frac{\partial}{\partial t}(A u_x) + \frac{\partial}{\partial x}\left(A u_x^2\right) = -\frac{A}{\rho}\frac{\partial p_a}{\partial x}. \tag{8.6.6}$$

Equations (8.6.5) and (8.6.6) are two equations for three unknowns, A, u_x, and p_a. To close the problem the variation in core edge pressure must be either specified through imposition of the far field pressure (in the case of an unconfined vortex flow) or linked to A and u_x through a description of the bounding geometry in a confined flow.

[5] As previously, K_c is used rather than the actual core circulation, Γ_c, to avoid having to bookkeep the factor of 2π in the equation.

For an unconfined geometry the expression for pressure in (8.6.4) can be used to cast (8.6.6) in terms of changes in the far field ($r \gg a$) pressure, p_{far}, as

$$\frac{\partial}{\partial t}(Au_x) + \frac{\partial}{\partial x}\left(Au_x^2 + \frac{\pi K_c^2}{2}\ln A\right) = -\frac{A}{\rho}\frac{dp_{far}}{dx}. \tag{8.6.7}$$

For vortex cores in confined geometries the duct shape is given in terms of a specified area $A_D(x) = \pi[r_D(x)]^2$. The core occupies the region $r = 0$ to $r = a(x, t)$ with irrotational flow between $r = a(x, t)$ and $r = r_D(x)$. Conservation of mass and momentum in the outer region close the problem. With U_x the axial velocity in the outer flow, the two statements are:

conservation of mass:

$$\frac{\partial}{\partial t}(A_D - A) + \frac{\partial}{\partial x}[(A_D - A)U_x] = 0, \tag{8.6.8}$$

conservation of momentum:

$$\frac{\partial}{\partial t}[(A_D - A)U_x] + \frac{\partial}{\partial x}\left[(A_D - A)U_x^2\right]$$
$$= -\left(\frac{A_D - A}{\rho}\right)\frac{\partial p_a}{\partial x} + \frac{\pi K_c^2}{2}\left(\frac{A_D}{A} - 1\right)\frac{1}{A}\frac{\partial A}{\partial x}. \tag{8.6.9}$$

Equations (8.6.5), (8.6.6), (8.6.8), and (8.6.9) describe the evolution of A, u_x, U_x, and p_a for confined vortex cores.

8.6.2 Wave propagation in unconfined geometries

To examine small amplitude wave propagation along the core we linearize the conservation equations by taking the velocity, core area, and pressure to be composed of a mean state, uniform in x and denoted by an overbar, plus a small perturbation denoted by a prime. The simplest configuration exhibiting wave propagation is a vortex core in an unconfined geometry with far field pressure, p_{far}, uniform in x, for which the motion is described by the appropriate linearized forms of (8.6.5) and (8.6.7). Making use of (8.6.5) in (8.6.7) the wave equations for the vortex core are:

$$\frac{\partial A'}{\partial t} + \bar{u}_x\frac{\partial A'}{\partial x} + \bar{A}\frac{\partial u_x'}{\partial x} = 0, \tag{8.6.10a}$$

$$\frac{\partial u_x'}{\partial t} + \bar{u}_x\frac{\partial u_x'}{\partial x} + \left(\frac{\pi K_c^2}{2\bar{A}^2}\right)\frac{\partial A'}{\partial x} = 0. \tag{8.6.10b}$$

Equations (8.6.10) provide a "long wavelength" (i.e. a wavelength long compared to the core diameter) approximate description of wave propagation on the vortex core. The waves are taken to be of the form $e^{i(kx-\omega t)}$, where k is the wave number in the x-direction and ω is the radian frequency:

$$\begin{bmatrix} u_x' \\ A' \end{bmatrix} = \begin{bmatrix} u_{x0} \\ A_0 \end{bmatrix} e^{i(kx-\omega t)}. \tag{8.6.11}$$

In (8.6.11) u_0 and A_0 are (possibly complex) constants relating the amplitude and relative phase of the velocity and area perturbations.

Substituting (8.6.11) into (8.6.10) leads to two algebraic equations for u_0 and A_0. For these to have a non-trivial solution, the determination of the coefficient matrix must be zero, giving an eigenvalue relation for the wave phase speed, ω/k:

$$\frac{\omega}{k} = \overline{u}_x \pm \frac{K_c}{\sqrt{2}\overline{a}} = \overline{u}_x \pm \frac{\Omega \overline{a}}{\sqrt{2}}. \tag{8.6.12}$$

Equation (8.6.12) shows that waves on the core propagate upstream and downstream with a velocity of $K_c/(\sqrt{2}\overline{a})$ relative to the core fluid.

An analogy exists between these waves and waves in a compressible fluid. From Section 6.6 the equations that describe the propagation of one-dimensional isentropic small disturbances in a uniform compressible fluid are:

$$\frac{\partial \rho'}{\partial t} + \overline{u}_x \frac{\partial \rho'}{\partial x} + \overline{\rho} \frac{\partial u'_x}{\partial x} = 0, \tag{8.6.13a}$$

$$\frac{\partial u'_x}{\partial t} + \overline{u}_x \frac{\partial u'_x}{\partial x} + \frac{1}{\overline{\rho}} \left(\frac{\gamma \overline{p}}{\overline{\rho}} \right) \frac{\partial \rho'}{\partial x} = 0. \tag{8.6.13b}$$

There is a direct correspondence between (8.6.13) and (8.6.10) with the core area playing the role of fluid density and $K_c/(\sqrt{2}\overline{a})$ corresponding to the speed of sound, $\sqrt{\gamma \overline{p}/\overline{\rho}}$. The waves described by (8.6.13) depend on fluid compressibility as the restoring force or "elasticity" responsible for the ability to support waves. In a vortex core the increase in circulation with radius means that if a ring of particles in the core is displaced the resulting pressure imbalance creates a restoring force to return the ring to the initial position.

We can build on the analogy further. Similar to the way the Mach number appears in a compressible flow, it is useful to work in terms of the ratio of the axial velocity $\overline{u}_x$ to the speed of propagation of small amplitude waves, $K_c/(\sqrt{2}\overline{a})$, to characterize the state of the vortex core. We thus define a dimensionless criticality parameter, D, which depends on the mean core properties, as

$$D = \frac{\sqrt{2}\overline{a}\,\overline{u}_x}{K_c}. \tag{8.6.14}$$

The parameter D is related to the reciprocal of the core swirl parameter (8.6.2) by $D = \sqrt{2}/S_c$.

Situations in which $D > 1$, so that the core velocity is larger than the wave propagation velocity and waves do not travel upstream, are called supercritical. Flows in which $D < 1$ are referred to as subcritical. The condition $D = 1$ corresponds to a core swirl parameter equal to $\sqrt{2}$, i.e. to a velocity ratio $(\overline{u}_\theta/\overline{u}_x)$ at the vortex core edge equal to $\sqrt{2}$. In Section 8.8 we will see that passage through this value is associated with large core expansion.

A number of assumptions have been made in developing the approximate description. As one assessment of their impact the result for the critical swirl parameter ($S_{c_{crit}} = \sqrt{2}$) from the quasi-one-dimensional analysis can be compared with the value of roughly 1.2 obtained from solutions of the exact inviscid small disturbance equations (Maxworthy, 1988; Marshall, 1993) for a Rankine vortex. The quasi-one-dimensional model provides a useful estimate for the critical swirl parameter. In the following it will be seen to enable insight into other trends in vortex core behavior.

8.6.3 Wave propagation and flow regimes in confined geometries: swirl stabilization of Kelvin–Helmholtz instability

A similar analysis to that in Section 8.6.2 can be carried out for waves on a vortex core in a confined geometry by linearizing (8.6.5), (8.6.6), (8.6.8), and (8.6.9). This leads to the following set of equations:

$$\frac{\partial A'}{\partial t} + \overline{u}_x \frac{\partial A'}{\partial x} + \overline{A}\frac{\partial u'_x}{\partial x} = 0, \tag{8.6.15a}$$

$$\frac{\partial u'_x}{\partial t} + \overline{u}_x \frac{\partial u'_x}{\partial x} + \frac{1}{\rho}\frac{\partial p'_a}{\partial x} = 0, \tag{8.6.15b}$$

$$\frac{\partial A'}{\partial t} + \overline{U}_x \frac{\partial A'}{\partial x} - (A_D - \overline{A})\frac{\partial U'_x}{\partial x} = 0, \tag{8.6.15c}$$

$$\frac{\partial U'_x}{\partial t} + \overline{U}_x \frac{\partial U'_x}{\partial x} + \frac{1}{\rho}\frac{\partial p'_a}{\partial x} - \left(\frac{\pi K_c^2}{2\overline{A}^2}\right)\frac{\partial A'}{\partial x} = 0. \tag{8.6.15d}$$

Substituting the disturbance form $e^{i(kx-\omega t)}$ into (8.6.15) and evaluating the determinant of the coefficient matrix of the resulting algebraic equations gives an eigenvalue equation for the wave speed. With $c = K_c/(\sqrt{2\overline{a}})$ this is

$$\frac{\omega}{k} = \frac{(A_D - \overline{A})\overline{u}_x + A\overline{U}_x}{A_D} \pm \sqrt{\left(1 - \frac{\overline{A}}{A_D}\right)\left[c^2 - \frac{\overline{A}}{A_D}(\overline{U}_x - \overline{u}_x)^2\right]}. \tag{8.6.16}$$

If we define effective convective and wave speeds as

$$\overline{u}_{x_{eff}} = \frac{(A_D - \overline{A})\overline{u}_x + \overline{A}\,\overline{U}_x}{A_D}, \tag{8.6.17a}$$

$$c_{eff}^2 = \left(1 - \frac{\overline{A}}{A_D}\right)\left[c^2 - \frac{\overline{A}}{A_D}(\overline{U}_x - \overline{u}_x)^2\right], \tag{8.6.17b}$$

the eigenvalues can be written as $\lambda_\pm = \overline{u}_{x_{eff}} \pm c_{eff}$.

For a confined flow the eigenvalues can be complex, corresponding to a long wavelength Kelvin–Helmholtz type instability (Section 6.5) of the cylindrical vortex sheet between the core and the external flow. For stability the eigenvalues must be real so that c_{eff}^2 is equal to or greater than 0, or,

$$\frac{\pi K_c^2}{2\overline{A}} \geq \frac{\overline{A}}{A_D}(\overline{U}_x - \overline{u}_x)^2 \quad \text{for stability.} \tag{8.6.18}$$

Equation (8.6.18) explicitly shows the stabilizing effect of swirl. The larger the difference in axial velocity between the core and outer regions, the greater the swirl needed for stability. An analogy that applies is that between the stability of an axisymmetric swirling flow and that of a stratified shear layer in a gravitational field. The connection can be inferred from the similarity of equations for the two situations (see for example Section 232 in Lamb (1945)). For the stratified shear layer, stabilization arises from the difference in fluid density.

For a confined vortex core it is helpful to define an effective criticality parameter as

$$D_{eff} = \frac{\overline{u}_{x_{eff}}}{c_{eff}}, \tag{8.6.19}$$

with values of $D_{eff} > 1$ indicating supercritical behavior. For given $\overline{A}/A_D$ and $\overline{u}_x/\overline{U}_x$ the core swirl parameter for which $D_{eff} = 1$ is

$$S_{C_{crit}} = \sqrt{2\left[1 + \frac{\left(\overline{U}_x/\overline{u}_x\right)^2}{A_D/\overline{A} - 1}\right]}. \tag{8.6.20}$$

The parameter $S_{C_{crit}}$ marks the division between subcritical and supercritical flows. It will also be seen to be useful as an indicator of conditions at which rapid expansion can occur for vortex cores in confined geometries.

8.7 Features of steady vortex core flows

8.7.1 Pressure gradients along a vortex core centerline

Although the static pressure within a boundary layer can be taken to be the same as in the free stream just outside of the layer, this is not true for a vortex core. The pressure variation within the core has important effects on the velocity at the core centerlines. Assuming that the rate of velocity variation along the core is much less than the rate of variation across the core, we apply the simple radial equilibrium equation to give an estimate of these effects (Hall, 1972). With the core edge taken to be a streamline, and the axial variation along this core edge streamline denoted by (dp_a/dx), the difference between the axial pressure gradients along the core outer radius, a, and the centerline is

$$\frac{dp_a}{dx} - \left.\frac{\partial p}{\partial x}\right|_{r=0} = \frac{d}{dx}\left(\int_0^a \frac{\rho u_\theta^2}{r} dr\right). \tag{8.7.1}$$

For a forced vortex core with circulation K_c and circumferential velocity $u_\theta = K_c r/a^2$

$$\frac{dp_a}{dx} - \left.\frac{\partial p}{\partial x}\right|_{r=0} = \frac{d}{dx}\left(\rho \frac{K_c^2}{2a^2}\right). \tag{8.7.2}$$

The core circulation is constant and the term on the right-hand side of (8.7.2) is non-zero only because of changes in core radius. Carrying out the differentiation yields an expression for the difference in rates of change of pressure with x in terms of da/dx, the half-angle of the core streamtube divergence:

$$\left.\frac{\partial p}{\partial x}\right|_{r=0} - \frac{dp_a}{dx} = \frac{\rho K_c^2}{a^3}\left(\frac{da}{dx}\right). \tag{8.7.3}$$

Equation (8.7.3), while strictly applicable only to forced vortex rotation, provides a useful guide for the general case. It shows that when the core area increases, the pressure gradient along the axis is larger than that along the core outer radius by an amount proportional to the square of the circulation. Changes in axial velocity on the axis are thus more pronounced than outside the core. This is seen in Figure 8.9, which shows calculated axial velocity and pressure variations at the core radius and at the centerline for inviscid flow in a cylindrical duct with the initial radial distributions of swirl and axial velocity shown in the inset. The amplification of pressure and velocity differences on the axis compared to those on the outer radius is evident.

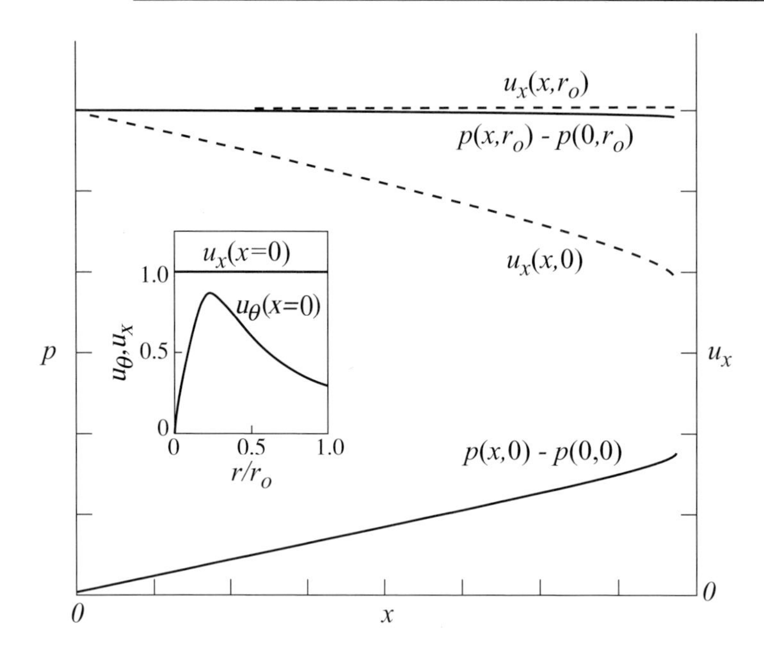

Figure 8.9: Calculated variations of pressure p and axial velocity u_x along the axis ($r = 0$) and along the outer radius ($r = r_o$) for inviscid swirling flow in a cylindrical duct with the initial velocity distribution shown in inset (Hall, 1972).

The evolution of the centerline velocity can be expressed even more simply for a situation with large swirl. The pressure gradient on the axis is much greater than that at $r = a$. For small changes in core radius, therefore,

$$\Delta p|_{r=0} \approx \rho \frac{K_c^2}{a^2} \frac{\Delta a}{a}. \tag{8.7.4}$$

From the x-component of the inviscid momentum equation, the corresponding change in u_x on the axis is

$$\left.\frac{\Delta u_x}{u_x}\right|_{r=0} \approx -\left(\frac{K_c^2}{u_x^2 a^2}\right) \frac{\Delta a}{a}. \tag{8.7.5}$$

Small changes in vortex core area can lead to large changes in centerline axial velocity.

Figure 8.10 gives the centerline velocity (computed using the full axisymmetric inviscid equations) as a function of initial swirl parameter for a vortex core taken from initial radius a_i at axial station, x_i, to radius $a_i(1 + E)$. For small swirl parameters, the relation between velocity and area changes for one-dimensional flow in a circular streamtube ($du_x/u_x \approx -2dr/r$) is recovered, but for initial swirl parameters which are not small compared to unity the effect of area change on axial velocity is strongly amplified.

The core centerline axial velocity behavior can be interpreted in terms of vorticity kinematics. Suppose the core and free stream have equal axial velocity far upstream and there is only an axial component of vorticity, ω_x, so the vortex lines are parallel to the x-axis. The fluid particles along the vortex lines spiral about the axis of symmetry. If the core undergoes a radius increase at some

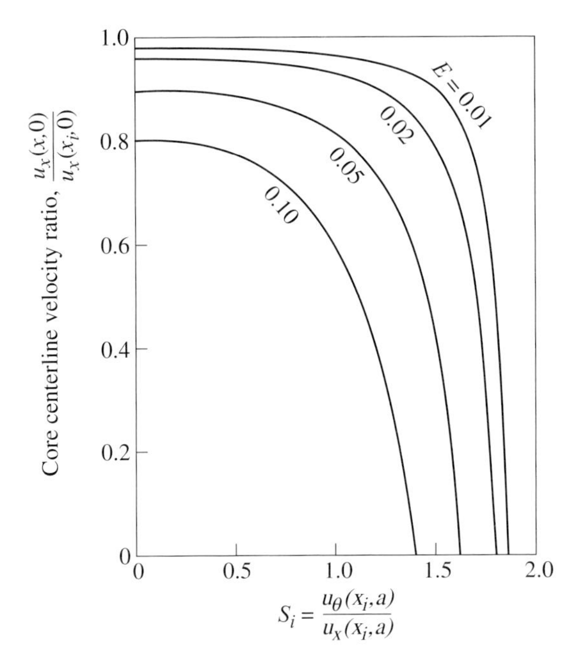

Figure 8.10: Effect of initial swirl parameter S_i and core expansion E (E defined by $a = a_i(1 + E)$) on axial velocity decrease along the vortex core axis; axisymmetric inviscid flow (Hall, 1966).

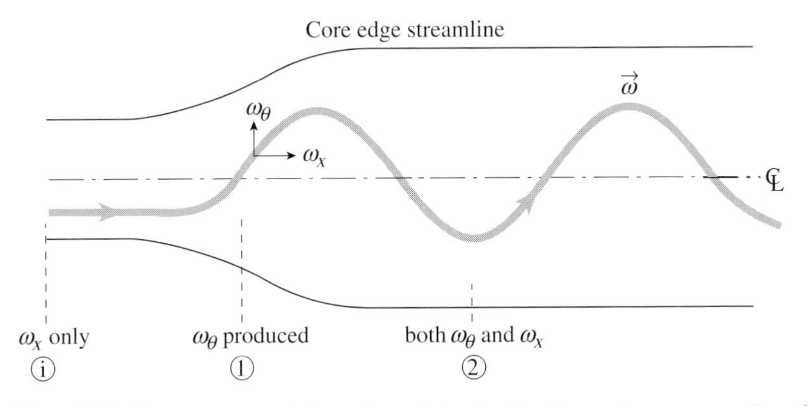

Figure 8.11: Downstream evolution of an axial vortex line in a vortex core: creation of circumferential vorticity (after Batchelor (1967)). Stations i (initial), 1, 2 denote regions of differing behavior.

downstream location, the angular velocity of a particle about the axis decreases. Because the vortex lines are continuous they must therefore tip into the circumferential direction, creating a θ-component of vorticity, ω_θ, as sketched in Figure 8.11 (Batchelor, 1967; see also Brown and Lopez, 1990).

This creation of ω_θ can also be seen from the vorticity equation. For small area change,

$$\frac{D\omega_\theta}{Dt} = (\boldsymbol{\omega} \cdot \boldsymbol{\nabla})\, u_\theta \cong \omega_x \frac{\partial u_\theta}{\partial x}. \tag{8.7.6}$$

For an increase in radius and hence a decrease in u_θ with x, circumferential vorticity, ω_θ, is created, with the sense indicated in Figure 8.11. If $u_r \ll u_x$, ω_θ can be written as

$$\omega_\theta \cong -\frac{\partial u_x}{\partial r}. \tag{8.7.7}$$

Equations (8.7.6) and (8.7.7) show that core growth is linked to generation of circumferential vorticity and that there is a greater reduction in axial velocity near the axis than in the outer parts of the core. The initial axial vorticity is critical to this process; without it the creation of circumferential vorticity does not occur.

8.7.2 Axial and circumferential velocity distributions in vortex cores

The variation in static pressure in a vortex core means that the axial velocity distribution is typically different from that in a boundary layer (Batchelor, 1964). For example consider a trailing vortex downstream of a wing. All streamlines in the vortex core originate (far upstream) in a region of uniform static pressure, $p_{-\infty}$, and uniform velocity with components $(u_{x_{-\infty}}, 0, 0)$. In the core at a given downstream station,

$$\frac{p}{\rho} + \frac{1}{2}\left(u_x^2 + u_\theta^2 + u_r^2\right) = \frac{p_{-\infty}}{\rho} + \frac{u_{x_{-\infty}}^2}{2} - \Delta p_t, \tag{8.7.8}$$

where Δp_t is the change in stagnation pressure between far upstream and the given station. Application of simple radial equilibrium for the pressure in the core then yields

$$u_x^2 = u_{x_{-\infty}}^2 + \int_r^\infty \frac{1}{r^2}\frac{d\left[(ru_\theta)^2\right]}{dr}dr - \frac{2\Delta p_t}{\rho}, \tag{8.7.9}$$

where the pressure at $r \to \infty$ has been taken equal to $p_{-\infty}$.

For a core tangential velocity distribution with $u_\theta = \Omega r$ and a stagnation pressure loss coefficient, $C_{p_t}(= \Delta p_t / \frac{1}{2}\rho u_{x_{-\infty}}^2)$, the axial velocity in the core is given by

$$\frac{u_x}{u_{x_{-\infty}}} = \left[1 - C_{p_t} + \frac{2\Omega^2 a^2}{u_{x_{-\infty}}^2}\left(1 - \frac{r^2}{a^2}\right)\right]^{1/2}. \tag{8.7.10}$$

Equation (8.7.10) is plotted in Figure 8.12 for different swirl parameters $\Omega a / u_{x_{-\infty}}$ and a loss coefficient distribution of the form $C_{p_t} = [1 - (r/a)^2]$. As the swirl parameter increases, the axial velocity in the vortex core changes from wake-like to jet-like behavior, and the axial velocity on the centerline exceeds that of the free stream for swirl parameters greater than $0.707 = 1/\sqrt{2}$. Other distributions of C_{p_t} give different quantitative results, but the main point is that the axial velocity in a vortex core can be appreciably larger than that outside the core. This is typically the case for isolated wing tip vortices (Green, 1995) with the converse (a velocity defect) generally existing for compressor blade tip clearance vortices (e.g., Khalid *et al.* (1999)).

8.7.3 Applicability of the Rankine vortex model

In a number of examples in this chapter the vortex core circumferential velocity distribution has been represented by a Rankine vortex, and it is worthwhile to address how this approximation

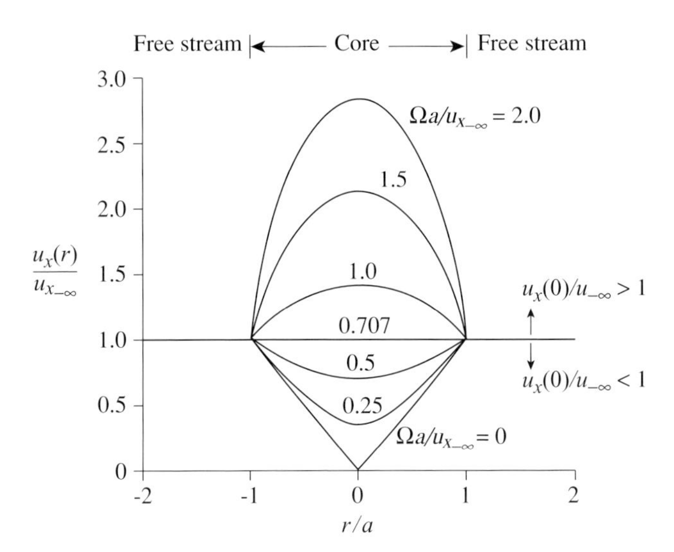

Figure 8.12: Axial velocity distribution in a Rankine vortex for different values of swirl parameter, $\Omega a/u_{x_{-\infty}}$(or $K_c/au_{x_{-\infty}}$); stagnation pressure loss distribution $C_{p_t} = [p_t(a) - p_t(0)]/(\frac{1}{2}\rho u_{x_{-\infty}}^2) = 1 - (r/a)^2$.

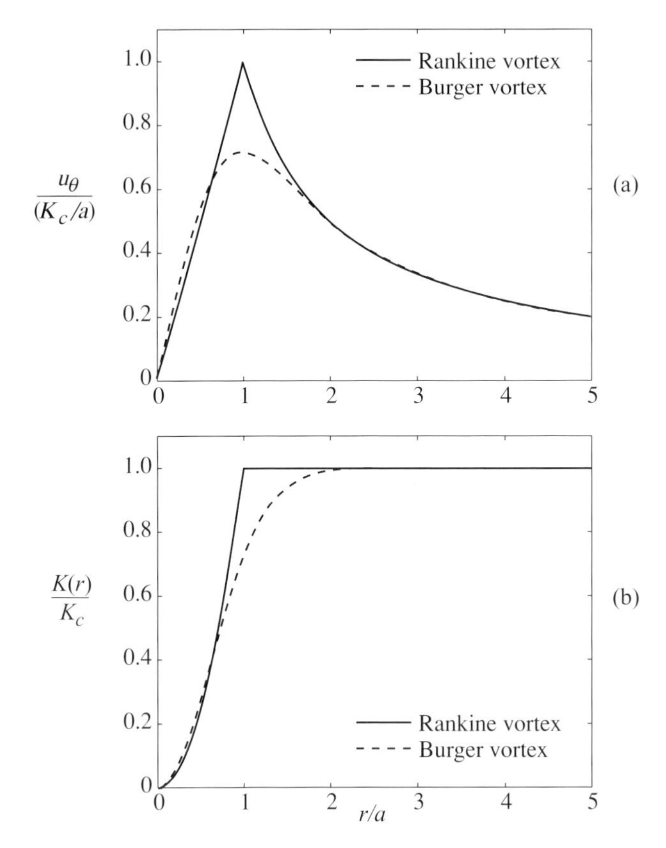

Figure 8.13: Circumferential velocity (a) and circulation (b) in Rankine ($u_\theta = \Omega r$) and Burger vortex (u_θ given by (8.7.11)) models. K = circulation/2π, K_c is vortex core circulation.

characterizes an actual flow. A circumferential velocity profile which represents experimental data well is the Burger vortex (or q-vortex) (Delery, 1994), with the form

$$u_\theta = \frac{K_c}{r}\left\{1 - \exp\left[-1.26\left(\frac{r}{a}\right)^2\right]\right\}. \tag{8.7.11}$$

In (8.7.11) K_c is interpreted as (circulation$/2\pi$) at locations far away from the axis and a is interpreted as the location at which the circumferential velocity is the maximum. Figure 8.13 shows circumferential velocity and K/K_c as functions of r/a, for a Rankine vortex and for (8.7.11). For the same circulation the Rankine vortex has a larger maximum swirl velocity than the Burger vortex. The pressure difference between the core edge and the axis is thus somewhat larger, as is (for a given initial axial velocity distribution) the response of streamtubes on the axis to changes in the external flow. The Rankine approximation, however, captures the observed parametric trends and we make further use of it below to describe vortex core behavior.

8.8 Vortex core response to external conditions

8.8.1 Unconfined geometries (steady vortex cores with specified external pressure variation)

Conditions under which a large growth in vortex core area occurs are perhaps the most important technological issue associated with vortex core flows. In this section we use the Rankine vortex model to describe the response of a steady vortex core to external conditions in unconfined and confined geometries.

The mass average core stagnation pressure plays an important role in phenomena associated with vortex core growth. The behavior of the mass average core stagnation pressure is seen by combining the steady-state form of (8.6.5) and (8.6.6) to give (noting that u_x is modeled as uniform across the core at any axial station)

$$\frac{\partial}{\partial x}\left(p_a + \frac{1}{2}\rho u_x^2\right) = 0. \tag{8.8.1}$$

The quantity $p_a + \frac{1}{2}\rho u_x^2$ is the core mass average stagnation pressure, denoted by $\overline{p}_{t_c}^M$ and defined as

$$\overline{p}_{t_c}^M \equiv \frac{2\pi}{Au_x}\int_0^a \left[p + \frac{1}{2}\rho\left(u_\theta^2 + u_x^2\right)\right]u_x r\,dr = p_a + \frac{1}{2}\rho u_x^2. \tag{8.8.2}$$

Equation (8.8.1), which states that the mass average core stagnation pressure is constant along the core, can be regarded as a quasi-Bernoulli relation between core edge pressure and core velocity. Invoking continuity, it can be written in a form that connects changes in core area and core edge static pressure from an initial station i as

$$\frac{p_a - p_{a_i}}{\frac{1}{2}\rho u_{x_i}^2} = \frac{\Delta p_a}{\frac{1}{2}\rho u_{x_i}^2} = 1 - \left(\frac{A_i}{A}\right)^2. \tag{8.8.3}$$

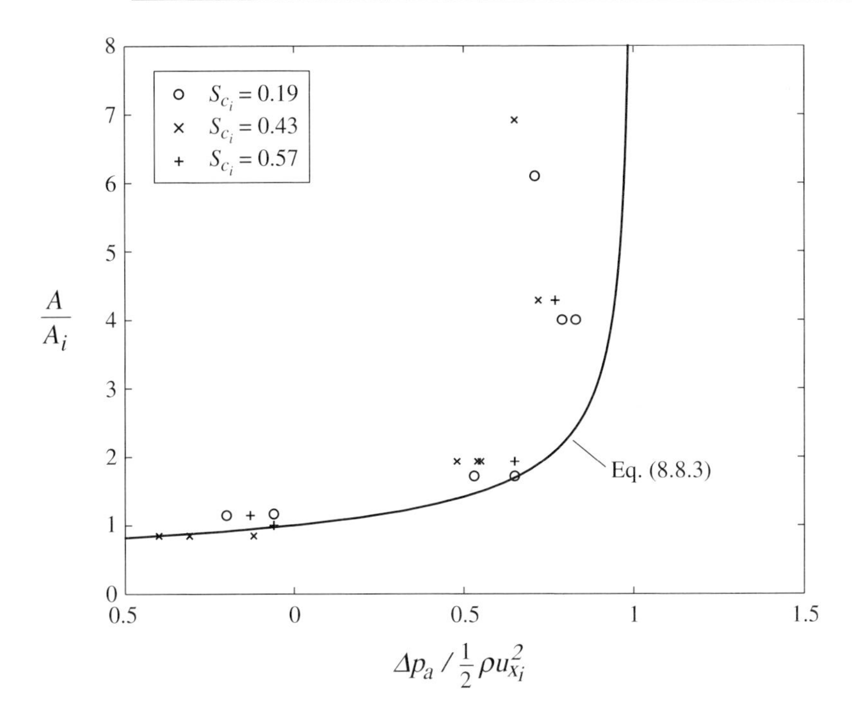

Figure 8.14: Vortex core expansion A/A_i versus core edge pressure rise $\Delta p_a/(\frac{1}{2}\rho u_{x_i}^2)$. Data for inlet core swirl parameters $S_{c_i} = 0.19, 0.43$ and 0.57 (Cho, 1995).

In (8.8.3) A_i and u_{x_i} are the initial core area and axial velocity and Δp_a ($= p_a - p_{a_i}$) is the core edge pressure rise from the initial to the current location. Equation (8.8.3) applies to both confined and unconfined geometries. Measurements of core area variation as a function of core edge pressure rise, Δp_a, given in Figure 8.14, show that (8.8.3) provides a guide to the value of Δp_a at which large core growth occurs, although the one-dimensional theory cannot accurately describe the core area variation in these situations because the radial velocities become comparable to the axial velocities.

For an unconfined vortex core, the effect of external conditions is expressed by the far field pressure distribution, $p_{far}(x)$, the pressure at large radius, $r/a \gg 1$. The far field pressure is related to the core stagnation pressure, core radius, core circulation, and criticality parameter, $D\,(= \sqrt{2}/S_c = \sqrt{2}au_x/K_c)$, by

$$\frac{\overline{p}_{t_c}^M - p_{far}}{\frac{1}{2}\rho\left(\dfrac{K_c^2}{2a^2u_x}\right)^2} = D^2(D^2 - 2). \tag{8.8.4}$$

For steady continuous flow, $[K_c^2/(2a^2u_x)]$ and $\overline{p}_{t_c}^M$ are invariant. Equation (8.8.4) thus provides the relation between far field pressure and the criticality parameter illustrated in Figure 8.15. Increases in far field pressure drive D towards unity for any initial value of D. The difference between the mass average stagnation pressure and the far field pressure reaches a minimum when $D = 1$. At this

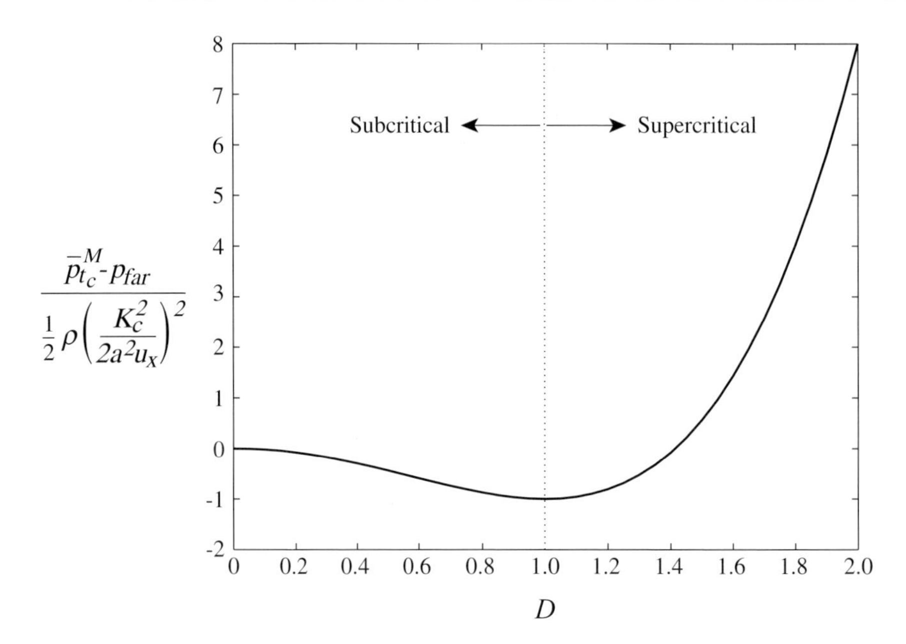

Figure 8.15: Relationship of the stagnation pressure, $\overline{p}_{t_c}^M$, and the far field pressure, p_{far}, for a steady flow.

condition, denoted by $(\ )^*$, the far field pressure is given by

$$\overline{p}_{t_c}^M - p_{far}^* = \left(\overline{p}_{t_c}^M - p_{far}\right)\Big|_{D=1} = -\frac{1}{2}\rho\left(\frac{K_c^2}{2a^2u_x}\right)^2. \tag{8.8.5}$$

If the far field pressure is greater than p_{far}^*, the core cannot remain isentropic. Time-resolved computations of vortex flows show that a discontinuity (analogous to a shock in compressible flow) develops and propagates upstream (Darmofal *et al.*, 2001).

Another view of critical conditions ($D = 1$ or equivalently $S_c = \sqrt{2}$) is seen by combining the steady form of (8.6.5), (8.6.6), plus (8.6.2) to yield an expression for differential change in core area:

$$\frac{dA}{A} = \left(\frac{1}{1 - S_c^2/2}\right)\left(\frac{dp_{far}}{\rho u_x^2}\right) = \left(\frac{D^2}{D^2 - 1}\right)\left(\frac{dp_{far}}{\rho u_x^2}\right). \tag{8.8.6}$$

Equation (8.8.6) implies that critical conditions correspond to a maximum of p_{far}.

Experiments reported by Pagan (1989) and Delery (1994), shown in Figure 8.16, support the idea of a maximum pressure rise. The flow regimes are mapped in terms of swirl parameter versus pressure rise and the figure indicates a limiting curve above which rapid core expansion (or vortex breakdown) occurs. For low swirl the behavior is similar to that of a wake in a pressure gradient (Section 4.10) but as S_c increases the effects of swirl play a dominant role in the dynamics. The quasi-one-dimensional description shows trends similar to the data with the maximum pressure rise increasing as the swirl decreases, although breakdown occurs in the experiments at a swirl parameter approximately 25% below the critical conditions given by the Rankine vortex model.

The dependence of core area behavior on the initial swirl parameter S_{c_i} and the far field pressure rise Δp_{far} $(= p_{far} - p_{far_i})$ can be brought out from the quasi-Bernoulli equation (8.8.1). Using this

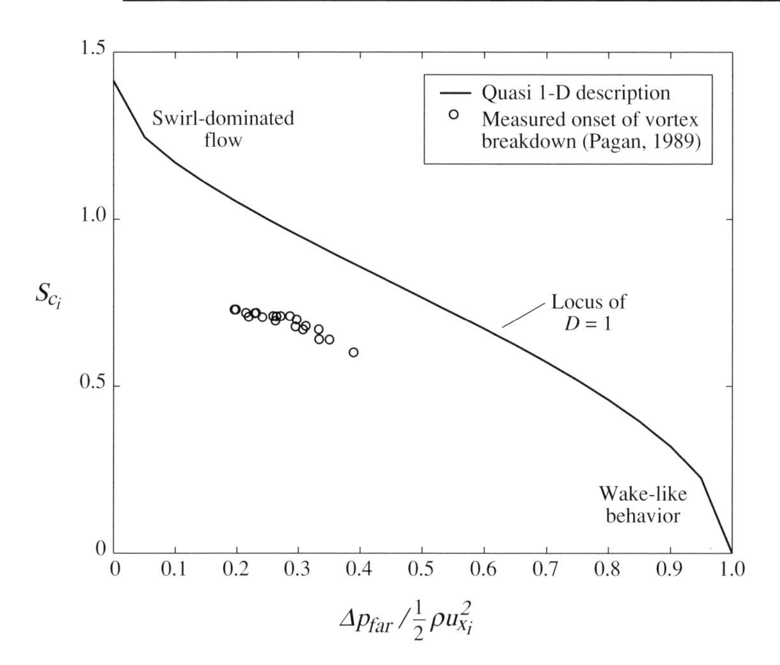

Figure 8.16: Initial core swirl parameter $S_{c_i}(= u_{\theta_{max}}/u_x)_i$ versus far field pressure rise $\Delta p_{far}/(\frac{1}{2}\rho u_{x_i}^2)$ showing the limiting curve and vortex breakdown data (Darmofal *et al.*, 2001).

together with the continuity equation for the core gives an expression for the ratio of core area to inlet core area:

$$\frac{S_{c_i}^2}{A/A_i} - \left(\frac{1}{A/A_i}\right)^2 - \frac{\Delta p_{far}}{\frac{1}{2}\rho u_{x_i}^2} = S_{c_i}^2 - 1. \tag{8.8.7}$$

Figure 8.17 shows the variation of A/A_i with far field pressure difference, for initial core swirl parameter S_{c_i} from 0 to 2.5. The variation of core area with respect to far field pressure changes sign when the flow switches from subcritical to supercritical. For supercritical vortices increases in far field pressure create decreases in the core axial velocity and increases in core area, so $dA/dp_{far} > 0$, qualitatively similar to behavior in non-swirling flows. Similarly, decreases in far field pressure produce decreases in core area. For subcritical vortices the situation is reversed and increases in far field pressure are associated with decreases in core area, so $dA/dp_{far} < 0$. The relationship between changes in p_a and area, however, is independent of flow regime; it is the relation between changes in p_a and p_{far} which switches sign at critical conditions.

The maximum pressure rise for each value of inlet swirl parameter, S_{c_i}, in Figure 8.17 coincides with the vortex becoming locally critical (local swirl ratio of $\sqrt{2}$). At this condition, as can be found by differentiating (8.8.7), the local core area A/A_i is $2/S_{c_i}^2$. The maximum pressure rise that can be achieved (which occurs at the critical conditions) is a function of the initial swirl parameter

$$\left(\frac{\Delta p_{far}}{\frac{1}{2}\rho u_{x_i}^2}\right)^* = 1 - S_{c_i}^2 + \frac{S_{c_i}^4}{4}. \tag{8.8.8}$$

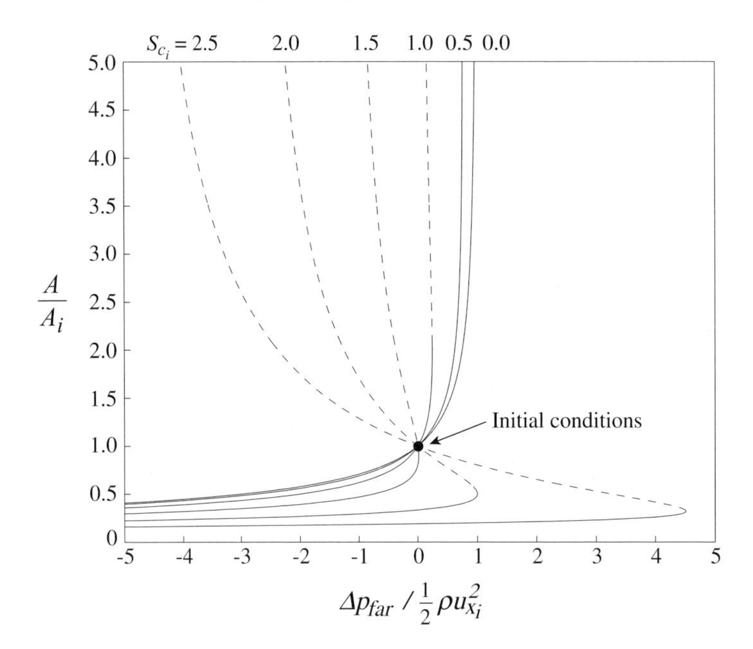

Figure 8.17: Dependence of A/A_i on far field pressure rise for steady continuous flows. Initial swirl parameter, $S_{c_i} = 0.0, 0.5, 1.0, 1.5, 2.0$, and 2.5. Solid lines represent supercritical flows, dashed lines subcritical flows. Steady continuous solutions which change isentropically from supercritical to subcritical flow are unstable (Darmofal *et al.*, 2001).

8.8.2 Confined geometries (steady vortex cores in ducts with specified area variation)

For a confined flow, the quantities needed to define the behavior are the duct area variation and three non-dimensional parameters that characterize the inlet state. One choice of these is the inlet axial velocity ratio, $V_i = U_{x_i}/u_{x_i}$, the inlet core/duct area ratio, $\sigma_i = A_i/A_{D_i}$, and the inlet core swirl parameter, $S_{c_i} = u_{\theta max_i}/u_{x_i}$. Application of continuity and $\overline{p}_{t_c}^M$ invariance in the core and outer flow leads to an equation for the core area in terms of the initial conditions and the duct area ratio as

$$\frac{S_{c_i}^2}{A/A_i} - \left(\frac{1}{A/A_i}\right)^2 + \left[\frac{V_i(1-\sigma_i)}{A_D/A_{D_i} - \sigma_i A/A_i}\right]^2 = S_{c_i}^2 - 1 + V_i^2. \tag{8.8.9}$$

The relation between local changes in core and duct areas is found by differentiating (8.8.9) as

$$\frac{dA}{A} = 2\left[\frac{(U_x/u_x)^2}{\left(S_{c_{crit}}^2 - S_c^2\right)(1 - A/A_D)}\right]\frac{dA_D}{A_D}. \tag{8.8.10}$$

In (8.8.10) $S_{c_{crit}}$ is the critical swirl number for confined flows defined in (8.6.20). Equations (8.8.10) and (8.8.6) show the critical swirl condition has similar roles in unconfined and confined flows. Equation (8.8.10) also implies that continuous behavior, with a finite value of dA/A at the critical swirl ratio, can only exist if there is a geometric throat. For geometries in which a throat does

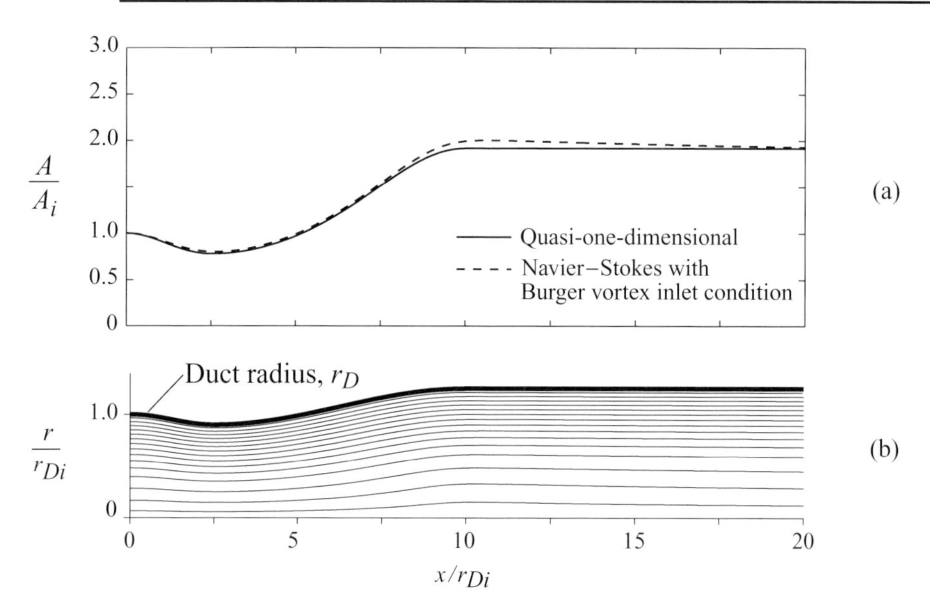

Figure 8.18: Vortex core area ratio and streamlines for inlet core swirl parameter $S_{c_i} = 0.56$: (a) core area variation for confined vortex flow in a converging–diverging pipe, $V_i = 1.09$ (wake) and $\sigma_i = 0.30$; (b) stream surfaces for confined vortex flow in a converging–diverging pipe, $V_i = 1.09$ (wake) and $\sigma_i = 0.30$ (Darmofal *et al.*, 2001).

not exist, for example a monotonic increase in area from one value to another, the transition between supercritical and subcritical conditions will be discontinuous.[6]

Figures 8.18 and 8.19 show area ratios (a) and streamline plots (b) for two different swirl conditions from the quasi-one-dimensional description and from axisymmetric laminar Navier–Stokes computations. At the lower initial core swirl, $S_{c_i} = 0.56$, the flow is nearly columnar with no reversed flow, whereas at $S_{c_i} = 0.78$, a large recirculation bubble forms. The ratio of final to initial area is captured by the one-dimensional analysis although the existence of the reverse flow region is not.

Confined vortex flow is parametrically complex, and a relevant question is which choice of non-dimensional parameters yields the most direct insight into the trends. It is emphasized that no one parameter can capture all of the behavior variation. The calculations reported by Darmofal *et al.* (2001), however, show that use of the core mass average stagnation pressure defect and the swirl parameter (rather than the axial velocity and the swirl parameter, for example) does allow some aspects to be viewed in terms of a dominant dependence of one parameter. The three parameters, S_{c_i}, V_i, and the core stagnation defect coefficient, $C_{p_{t,c}}$ (the difference between core and outer stream stagnation pressure), are related by:

$$C_{p_{t,c}} \equiv \frac{\overline{p}_{t_c}^M - p_{t_{outer}}}{\frac{1}{2}\rho\left(u_{\theta_{max_i}}^2 + U_{x_i}^2\right)} = \frac{1}{S_{c_i}^2 + V_i^2} - 1. \tag{8.8.11}$$

[6] Calculations by Darmofal (2002) suggest such flows are unstable. Although steady solutions can be constructed which go from supercritical to subcritical at a throat, in practice a steady continuous decrease through the critical value of D will not be observed; an analogy exists with the unstable transition from supersonic to subsonic conditions at a throat in a compressible flow (see Section 10.5).

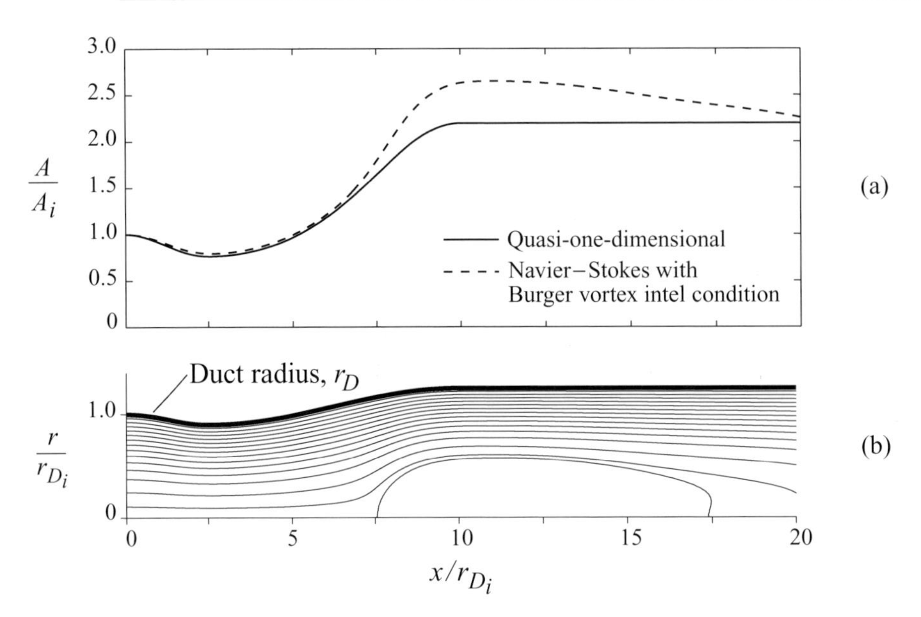

Figure 8.19: Vortex core area ratio and streamlines for inlet core swirl parameter $S_{c_i} = 0.78$: (a) core area variation for confined vortex flow in a converging–diverging pipe, $V_i = 1.09$ (wake) and $\sigma_i = 0.30$; (b) stream surfaces for confined vortex flow in a converging–diverging pipe, $V_i = 1.09$ (wake) and $\sigma_i = 0.30$ (Darmofal *et al.*, 2001).

For $C_{p_{t,c}} < 0$, the core has a mass average stagnation pressure deficit relative to the outer stream. For the range of conditions investigated by Darmofal *et al.* (2001) two general trends were found: (i) proportionally small increases in core area occur as the duct area increases for vortex cores with low stagnation pressure defect ($C_{p_{t,c}}$ much less than unity) and large increases in core area occur if $C_{p_{t,c}}$ is an appreciable fraction of unity; (ii) these results are weakly affected by the swirl parameter up to values of the latter of unity. Again, however, these should be regarded as rough guidelines only; no single parameter can completely characterize the behavior.

8.8.3 Discontinuous vortex core behavior

The conservation equations derived in Section 8.6 admit both continuous (smooth) and discontinuous (jump) solutions depending on the boundary conditions. The discontinuous jump solutions do not have a constant flux of stagnation pressure in the core and hence can be considered as "non-isentropic", in analogy with shocks in compressible flow.[7]

We can analyze such motions without the need to describe the flow within the region of stagnation pressure loss by considering the states that must exist on the two sides of a stationary discontinuity in stagnation pressure, velocity, or area. The relationships satisfied by these two end states are described below, with the initial and final states denoted by the subscripts 1 and 2, respectively. The core jump

[7] The analogy does not hold for all aspects; the axial length over which the transition takes place is generally one or more core diameters compared to the very thin transition region for a shock.

conditions are expressed in terms of the jump brackets, $[[f]] = f_2 - f_1$. In the vortex core, conservation of mass across the jump is

$$[[Au_x]] = 0. \tag{8.8.12}$$

Conservation of momentum is

$$\left[\!\!\left[Au_x^2 + \frac{1}{\rho} p_a A + \frac{\pi K_c^2}{2} ln A \right]\!\!\right] - \left(\frac{1}{\rho} p_{a_1} + \frac{\pi K_c^2}{2A_1} \right) [\![A]\!] = 0. \tag{8.8.13}$$

For unconfined vortex cores, substitution of (8.6.4) into (8.8.13) and solution of the resulting nonlinear system of equations yields an implicit relationship for the right (downstream) value of D (D_2) in terms of the upstream value (D_1) (Landahl and Widnall, 1971; Marshall, 1991) as

$$D_1^2 = \frac{2ln\,(D_2/D_1)}{(D_2/D_1)^2 - 1}. \tag{8.8.14}$$

Equation (8.8.14) admits "shocks" which increase as well as decrease D, but only the latter are allowed because the mass average core stagnation pressure must decrease through a jump. From (8.8.4), (8.8.13) and (8.8.14), the jump in $\overline{p}_{t_c}^M$ ($[\![\overline{p}_{t_c}^M]\!] = \overline{p}_{t_{c_2}}^M - \overline{p}_{t_{c_1}}^M$) across a steady discontinuity can be expressed in terms of the ratio D_r/D_l as

$$\frac{\Delta \overline{p}_{t_c}^M}{\frac{1}{2}\rho u_{x_1}^2} \equiv \frac{[\![\overline{p}_{t_c}^M]\!]}{\frac{1}{2}\rho u_{x_1}^2} = \frac{2}{D_1^2} \left\{ \left[\left(\frac{D_2}{D_1} \right)^2 + 1 \right] ln \frac{D_2}{D_1} - \left(\frac{D_2}{D_1} \right)^2 + 1 \right\}. \tag{8.8.15}$$

For values of D_1 near unity, the change in $\overline{p}_{t_c}^M$ can be approximated as

$$\frac{\Delta \overline{p}_{t_c}^M}{\frac{1}{2}\rho u_{x_1}^2} \approx -\frac{32}{3} (D_1 - 1)^3. \tag{8.8.16}$$

The decrease in stagnation pressure across a discontinuous vortex jump thus scales with $(D_1 - 1)^3$, analogous to the dependence of entropy rise across a shock with upstream Mach number (Section 2.6).

The change in core edge pressure across the jump (Δp_a) is given as

$$\frac{\Delta p_a}{\frac{1}{2}\rho u_{x_1}^2} = 2 \left(\frac{D_1^2 - D_2^2}{D_1^4} \right). \tag{8.8.17}$$

The core area ratio (or equivalently the axial velocity ratio) across the jump is

$$\frac{A_1}{A_2} = \frac{u_{x_2}}{u_{x_1}} = \left(\frac{D_2}{D_1} \right)^2. \tag{8.8.18}$$

Equations (8.8.14)–(8.8.18) show that the properties of the discontinuous vortex core solution are set by the criticality parameter D_1. We thus now examine the dependence of u_{x_2}/u_{x_1}, edge pressure jump Δp_a, and mass average core stagnation pressure decrease $\Delta \overline{p}_{t_c}^M$, on this parameter.

Figure 8.20 shows four quantities: D_2, A_1/A_2, the edge pressure jump, and the mass average stagnation pressure decrease, as functions of D_1, the upstream value of D. The core edge pressure rise across a jump has a maximum near $D_1 = 1.3$. This behavior can be motivated by the following physical considerations. From (8.8.2), the pressure jump is

$$\Delta p_a = \Delta \overline{p}_{t_c}^M - \Delta\left(\tfrac{1}{2}\rho u_x^2\right).$$

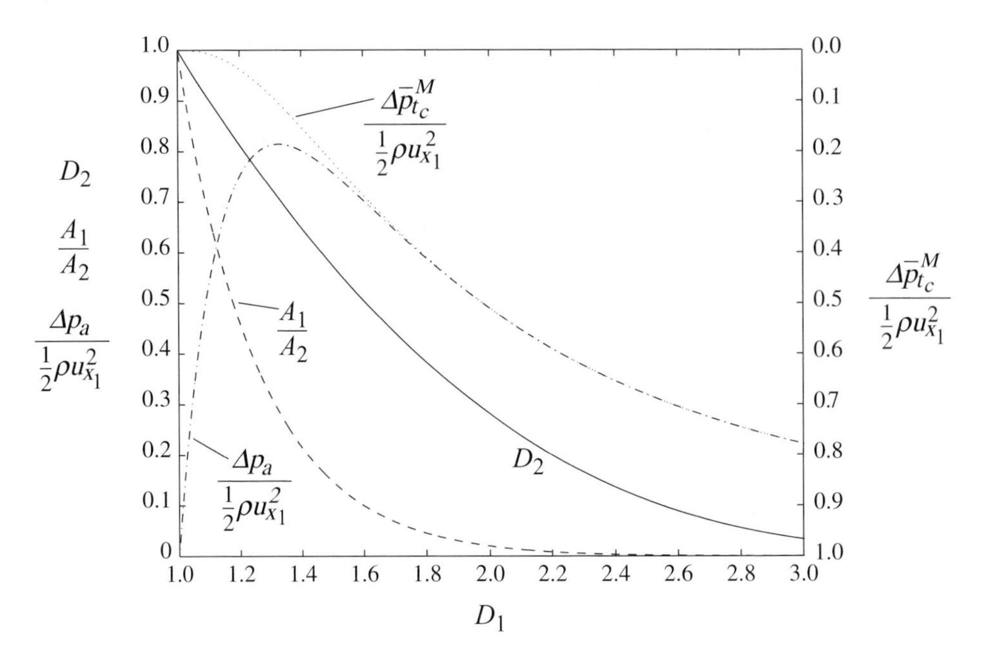

Figure 8.20: Core property changes across a discontinuous vortex jump: D_2, A_1/A_2, $\Delta p_a/(\frac{1}{2}\rho u_{x_1}^2)$, and $\Delta \overline{p}_{t_c}^M/(\frac{1}{2}\rho u_{x_1}^2)$ as a function of D_1; upstream and downstream states denoted by 1 and 2 respectively (Darmofal *et al.*, 2001)

For weak discontinuities the change in core stagnation pressure can be neglected (for $D_1 < 1.3$, $\Delta \overline{p}_{t_c}^M/\frac{1}{2}\rho u_{x_1}^2 < 0.1$), and the pressure rise approximated as

$$\frac{\Delta p_a}{\frac{1}{2}\rho u_{x_1}^2} \approx 1 - \left(\frac{u_{x_2}}{u_{x_1}}\right)^2. \tag{8.8.19}$$

For strong discontinuities, the right state is near stagnation, and the pressure rise can be approximated as

$$\frac{\Delta p_a}{\frac{1}{2}\rho u_{x_1}^2} \approx \frac{\Delta \overline{p}_{t_c}^M}{\frac{1}{2}\rho u_{x_1}^2} + 1. \tag{8.8.20}$$

The maximum core edge pressure rise marks a transition between nearly lossless discontinuities, in which the core edge pressure rise increases with upstream D, and discontinuities with large losses, in which the pressure rise decreases with upstream D. Figure 8.21 shows the weak and strong discontinuity approximations for edge pressure rise compared to the computed value. The value of A_1/A_2 drops sharply with upstream D, and for jumps with D_1 larger than 1.6 there is more than a ten-fold increase in vortex core area through the jump.

The jump conditions may also be superimposed on Figure 8.15 as connections between supercritical and subcritical states. This is done in Figure 8.22, which shows the admissible jump states as the end points of the dashed lines. The arrows indicate that jumps can only occur from supercritical to subcritical states. Figure 8.22 is the key to the construction of continuous and discontinuous steady vortex flow solutions. If the flow is continuous, D varies along the solid line in accordance with

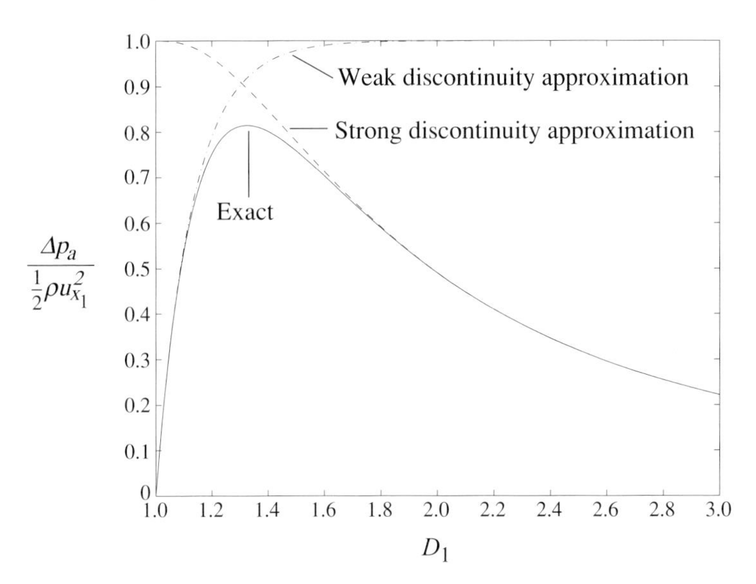

Figure 8.21: Core edge pressure change across a discontinuous vortex jump (Darmofal *et al.*, 2001).

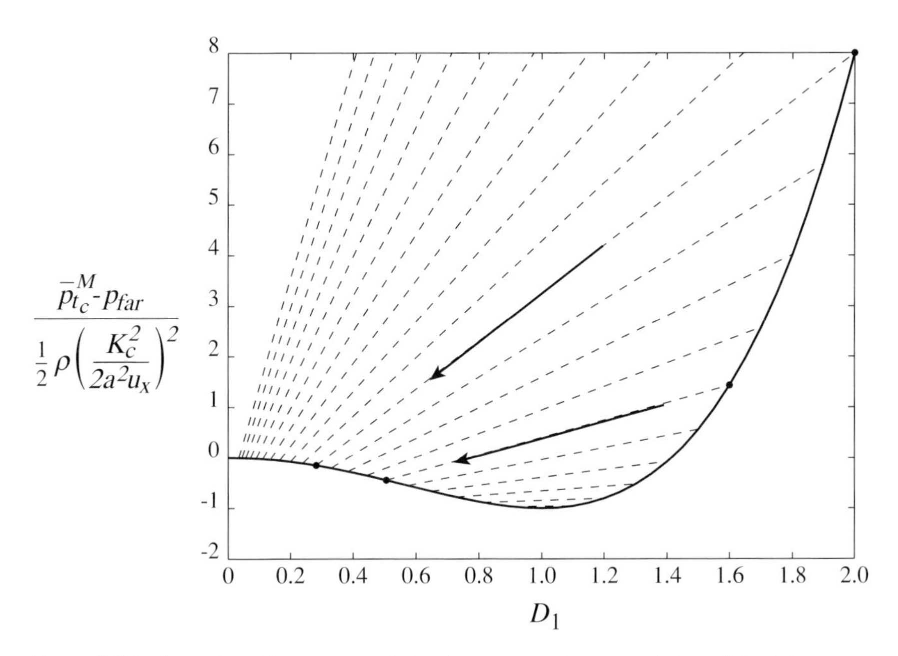

Figure 8.22: Core stagnation pressure behavior for a vortex core. Admissible states are connected by dashed lines with arrows indicating that discontinuous jumps can only occur from $D_1 > 1$ to $D_2 < 1$ (Darmofal *et al.*, 2001).

changes in the far field pressure. At a discontinuous transition, the vortex jumps from a supercritical state with large D_1 and small area to a subcritical state with small D_2 and large area, which are the end points of the dashed lines in Figure 8.22. The stagnation pressure loss across the jump is the difference in the value of the ordinate between the two states.

8.9 Swirling flow boundary layers

8.9.1 Swirling flow boundary layers on stationary surfaces and separation in swirling flow

Swirling flow boundary layers exhibit different features than the two-dimensional motions described in Chapter 4. Cross-flows can occur with velocity components at right angles to the local free-stream direction. The mechanism of separation is also different than for a two-dimensional boundary layer.

The contrasts between separation in two-dimensional flows, as discussed in Chapter 4, and in swirling flows (or general three-dimensional flows) have been described by Lighthill (1963) in terms of the behavior of the vorticity at the wall. (See also Tobak and Peake (1982) or Delery (2001).) Lighthill posed the issue in terms of asking under what conditions does the thickness of a steady boundary layer remain small. For a two-dimensional boundary layer at a small distance from the surface the velocity has velocity components $\omega_w y$ parallel to the surface and $-(d\omega_w/dl)y^2/2$ normal to the surface, where ω_w is the vorticity evaluated at the wall and where the derivative with respect to l denotes differentiation along the surface. Streamlines close to the surface remain so unless ω_w goes to zero. For volume flow $\dot{V}$ per unit depth, $\omega_w y^2/2 = \dot{V}$, and the distance of the streamline from the surface varies like $(\omega_w)^{-1/2}$. Points on the surface where $\omega_w = 0$ (and shear stress consequently zero) and $d\omega_w/dl < 0$, so there is a normal velocity component away from the surface, are separation points.

For a swirling (or three-dimensional) boundary layer the situation is different. To see this take Cartesian coordinates x, y, z, with x and z parallel to the wall and y again perpendicular to it. A small distance from the surface, the velocity is approximately $\mathbf{u} = \boldsymbol{\beta}_w y$, where $\boldsymbol{\beta}_w = \boldsymbol{\omega}_w \times \mathbf{n} = \boldsymbol{\tau}_w/\mu$, with $\mathbf{n}$ the outward normal to the surface and $\boldsymbol{\tau}_w/\mu$ the skin friction vector. Streamlines very near the surface, which are often referred to as limiting streamlines, are essentially along skin friction lines (curves to which $\boldsymbol{\beta}_w$ and thus $\boldsymbol{\tau}_w/\mu$ are everywhere tangential). With $\dot{V}$ the volume flow along a streamtube of rectangular section the base of which is the portion of the surface Δz_w between two limiting streamlines (skin friction lines), the height of the streamtube, y, is given by

$$\omega_w y^2 \Delta z_w/2 = \dot{V}. \tag{8.9.1}$$

The distance of the limiting streamlines from the surface therefore varies as $(\omega_w \Delta z_w)^{-1/2}$. Streamlines can greatly increase their distance from the surface not only when the vorticity at the wall, ω_w, goes to zero, as in the two-dimensional situation, but also when Δz_w does, in other words where limiting streamlines at the surface converge. This latter mechanism of separation is seen with swirling flow boundary layers.

As in Chapter 4 the boundary layer behavior is introduced in the context of a fluid device. The swirl flow analog of the straight channel diffuser of Chapter 4 is a radial vaneless diffuser. A vaneless diffuser is shown in two views in Figure 8.23. Figure 8.23(b) indicates the inviscid region and

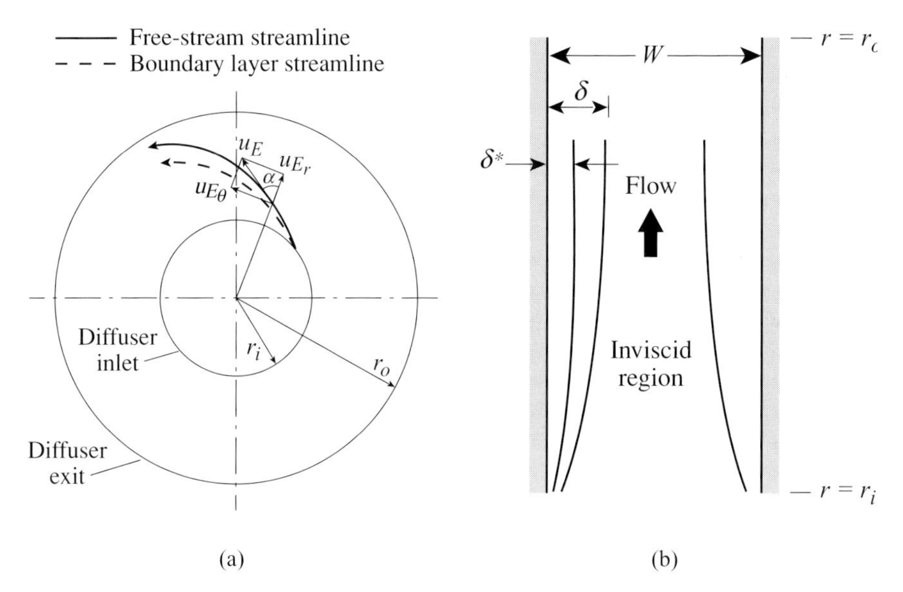

Figure 8.23: Vaneless diffuser geometry; (a) axial view (r–θ plane); (b) view on r–x plane.

the wall boundary layers. With the boundary layers on both walls assumed the same, the effective width of the diffuser at any radial location is $W_{eff} = W - 2\delta^*$.

The equations describing incompressible, steady, axisymmetric flow in the inviscid region are (using the subscript "E" to denote conditions external to the boundary layers):

$$\frac{\partial u_{E_r}}{\partial r} + \frac{u_{E_r}}{r} + \frac{u_{E_r}}{W_{eff}}\frac{dW_{eff}}{dr} = 0, \tag{8.9.2a}$$

$$u_{E_r}\frac{\partial u_{E_r}}{\partial r} - \frac{u_{E_\theta}^2}{r} = -\frac{1}{\rho}\frac{dp}{dr}, \tag{8.9.2b}$$

$$u_{E_r}\frac{\partial (r u_{E_\theta})}{\partial r} = 0. \tag{8.9.2c}$$

The boundary layer equations are:

$$\frac{\partial u_r}{\partial r} + \frac{u_r}{r} + \frac{\partial u_x}{\partial x} = 0, \tag{8.9.3a}$$

$$u_r\frac{\partial u_r}{\partial r} - \frac{u_\theta^2}{r} + u_x\frac{\partial u_r}{\partial x} = -\frac{1}{\rho}\frac{dp}{dr} + \frac{1}{\rho}\frac{\partial \tau_{rx}}{\partial x}, \tag{8.9.3b}$$

$$u_r\frac{\partial u_\theta}{\partial r} + u_x\frac{\partial u_\theta}{\partial x} + \frac{u_r u_\theta}{r} = \frac{1}{\rho}\frac{\partial \tau_{\theta x}}{\partial x}. \tag{8.9.3c}$$

Two shear stress components are now included in the boundary layer description.

Computation of viscous swirling flow can be carried out by procedures similar to those for the two-dimensional situation and both boundary layer and Navier–Stokes techniques are in use. We thus address the qualitative illustration of the features mentioned in the first paragraph of the section

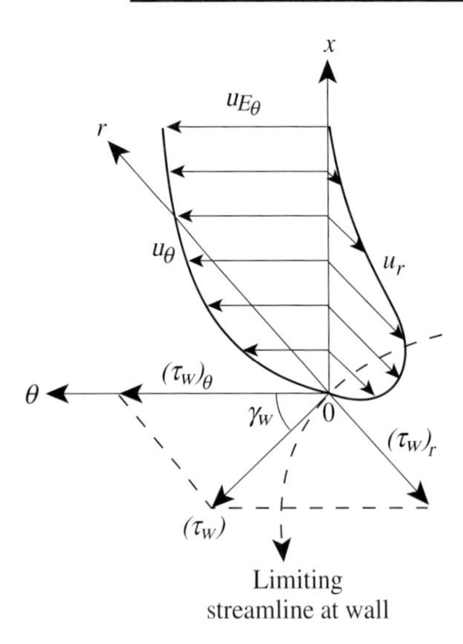

Figure 8.24: Velocity profile in a swirling flow boundary layer with a purely circumferential free stream; γ_w is the angle between u_E and the limiting streamlines at the wall, u_r is inwards in boundary layer.

with reference to the difficulty of maintaining radial outflow in a boundary layer with high values of swirl.

The problem can be framed with reference to Figure 8.23(a), which shows streamlines in a swirling flow boundary layer. Consider a vortex line in the boundary layer which we can view as convected with a local velocity proportional to that in the free stream. The circumferential velocity in the free stream is larger at smaller radius, and the vortex line will thus be increasingly tipped into the circumferential direction as it moves outward. This implies cross-flow generation in the boundary layer and a tendency for inward flow. The mechanism for cross-flow generation can also be described in terms of pressure fields (see also Section 3.4). The radial pressure gradient is set by the free stream. Because the fluid in the boundary layer has a lower velocity, the radius of curvature of boundary layer streamlines must be smaller than that for the free stream, resulting in streamline trajectories such as sketched in Figure 8.23(a). For high swirl angles (large values of u_θ/u_r) the streamlines are close to tangential and it takes little deviation of the boundary layer to produce inward flow. For a flow with a purely circumferential free stream the boundary layer profiles would be as sketched in Figure 8.24.

We can amplify these arguments using an approach similar to that in Section 4.7, with the boundary layer regarded as an inviscid one-dimensional layer with a velocity magnitude ε times the free-stream value ($\varepsilon < 1$). The local flow angle, α, with respect to the radial direction is given by $\tan\alpha = u_\theta/u_r$. We wish to determine changes in the angle between free-stream and boundary layer streamlines as the fluid undergoes a small change in radius. For such changes,

$$d(\tan\alpha) = \left[\frac{du_\theta}{u_\theta} - \frac{du_r}{u_r}\right]\tan\alpha. \tag{8.9.4}$$

Applying (8.9.4) and the inviscid form of (8.9.3) to changes in u_r and u_θ in the boundary layer and the free stream (the circumferential velocity obeys $ru_\theta = \text{constant}$ in this approximation), and taking

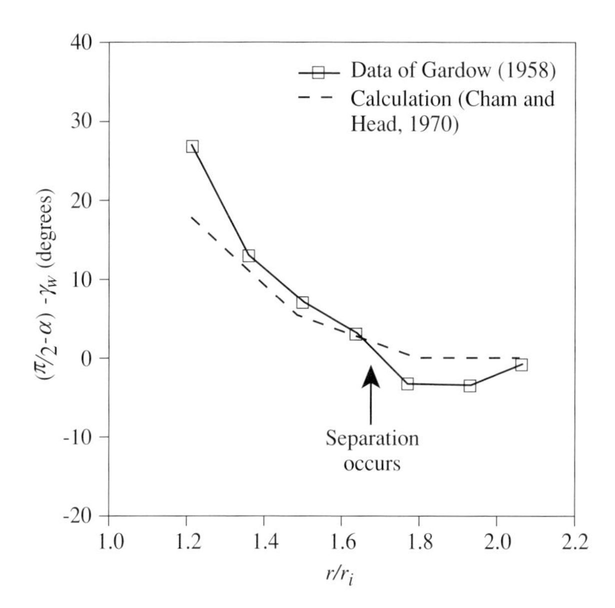

Figure 8.25: Flow angle at the wall versus the radius ratio in a vaneless diffuser; $\alpha_i = 35.5^\circ$, γ_w is defined in Figure 8.24, Reynolds number $= 3.8 \times 10^5$ (after Dou and Mizuki, 1998).

the initial flow angles in the boundary layer and free stream to be the same yields

$$d(\alpha - \alpha_E) = \left(1 - \frac{1}{\varepsilon^2}\right)\left(\frac{du_{E_r}}{u_{E_r}} - \tan^2\alpha\frac{dr}{r}\right)\left(\frac{\sin 2\alpha}{2}\right). \tag{8.9.5}$$

Combining (8.9.5) and (8.9.2a) provides an expression for the difference between boundary layer and free-stream flow angles in terms of effective area and radius change as

$$d(\alpha - \alpha_E) = \left(\frac{1}{\varepsilon^2} - 1\right)\left[\tan\alpha\frac{dr}{r} + \left(\frac{\sin 2\alpha}{2}\right)\frac{dW_{eff}}{W_{eff}}\right]. \tag{8.9.6}$$

If there is negligible variation in the effective width of the diffuser, (8.9.6) becomes

$$d(\alpha - \alpha_e) = \left(\frac{1}{\varepsilon^2} - 1\right)\tan\alpha\left(\frac{dr}{r}\right). \tag{8.9.7}$$

For large swirl angles ($\alpha \to 90^\circ$) (8.9.7) shows that substantial angle changes can be produced between the boundary layer and free stream, causing inward flow (regions of reversed radial velocity flow) and consequent large blockage in the passage.

Figure 8.25 shows measurements and calculations of the difference between the angles at the wall and at the free stream, as a function of radius, in a vaneless diffuser with parallel walls (Dou and Mizuki, 1998). As the radius increases, the limiting streamlines (the streamlines at the walls) become increasingly circumferential and more closely spaced. At a non-dimensional radius of 1.7, for this diffuser, the angle at the wall is such that wall flow is circumferential, the limiting wall streamlines converge as sketched in Figure 8.26 and the flow separates. The flow angle data in Figure 8.25 indicate that the region of separation closes at larger radius.

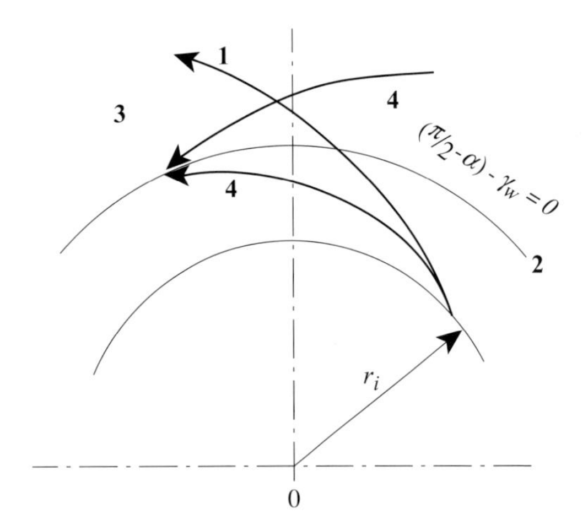

Figure 8.26: Flow pattern in radial vaneless diffusers: 1 – main flow streamline, 2 – separation line, 3 – reverse flow region, 4 – wall limiting streamline (Dou and Mizuki, 1998).

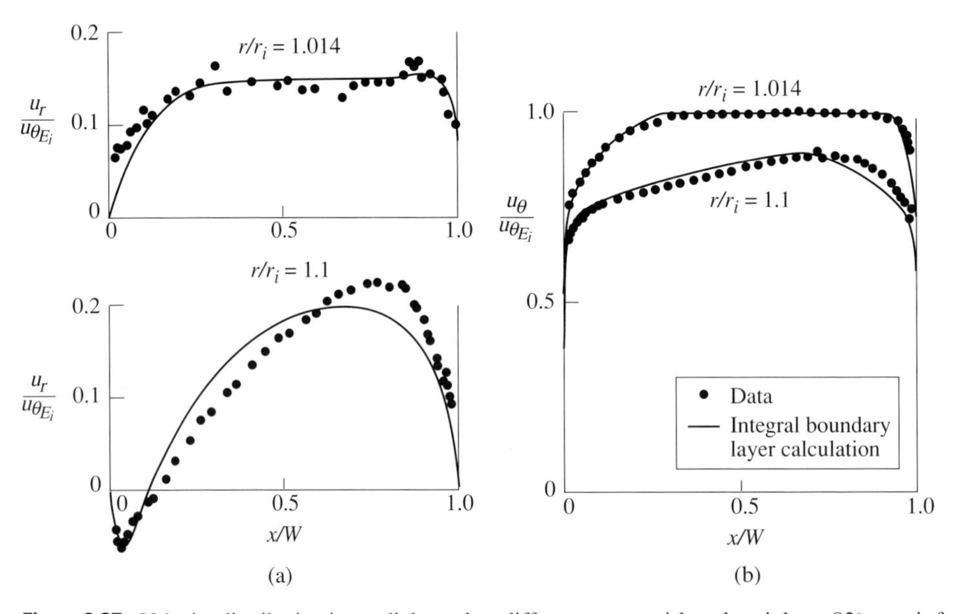

Figure 8.27: Velocity distribution in a radial vaneless diffuser, mean swirl angle at inlet = 82°, $u_{\theta_{E_i}}$ is free-stream circumferential velocity at inlet; (a) radial velocity, (b) circumferential velocity (Senoo, Kinishita and Khida, 1977).

Another view of the behavior of boundary layers in a vaneless diffuser is given in Figure 8.27, which shows the variation of radial and circumferential velocities as a function of x/W in a vaneless diffuser with an inlet average swirl angle of 82°. Data are presented for two radial stations, $r/r_i =$ 1.014 and 1.10, where r_i is the inlet radius. There is a large change in the radial velocity (and flow angle) between the two stations, with inward flow seen at $r/r_i = 1.10$. This can be contrasted with the small differences in the circumferential velocity profile between the two stations.

In terms of vaneless diffuser performance, separation in a swirling flow has generally less severe consequences than separation in two dimensions. There is increased blockage and higher loss due to

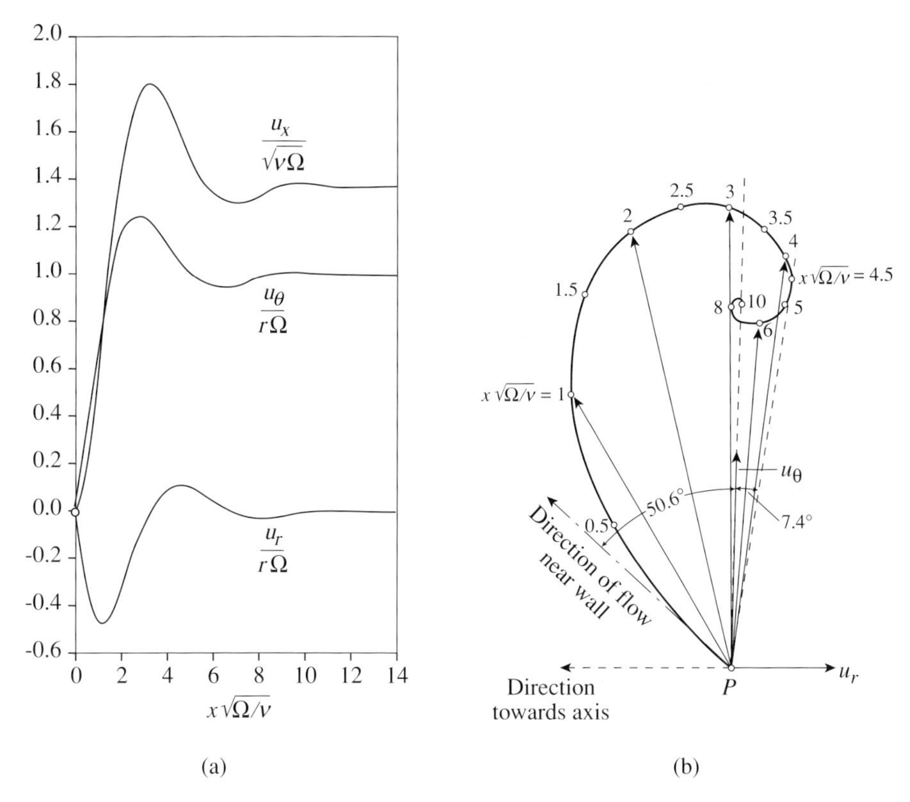

Figure 8.28: Boundary layer on a wall beneath a forced vortex flow: (a) velocity distribution in the boundary layer; (b) vector representation of the horizontal velocity component (Schlichting, 1979).

the separation, but the overall pressure rise may not be much affected because at high swirl angles pressure rise is mainly associated with the circumferential velocity component and the value, or even the direction, of the radial velocity has little influence.

Although we have focused on the problems associated with diffusing a highly swirling flow, similar considerations about cross-flow and inward motion apply to other swirling boundary layers. For example, Figure 8.28(a) shows radial, circumferential, and axial velocity components for a laminar boundary layer under a forced vortex rotating body of fluid and Figure 8.28(b) shows the corresponding angle variation through the boundary layer. In accord with the above arguments there is inward flow in the boundary layer and a substantial difference between free-steam and near wall angles.

8.9.2 Swirling flow boundary layers on rotating surfaces

The converse of the situation in which the swirling boundary layers are driven inwards by the radial pressure gradient is the outward motion of the boundary layer fluid on a rotating surface. The behavior of such flows is described in depth by Owen and Rogers (1989) and we present only the scaling for the behavior of laminar and turbulent boundary layers.

The arguments are illustrated using the specific configuration of a circular disk of radius r_o, rotating with velocity Ω, in an unbounded region of fluid with zero velocity far from the disk. As the disk rotates a layer of fluid close to the disk is set in circular motion and begins to flow outwards due to its inertia. While this occurs new fluid is brought towards the disk and in turn is pumped outwards. The resulting motion can involve considerable resistance to the disk rotation.

Following Prandtl (1952) we can develop the form of this resistance for laminar flow and then for the turbulent case. The flow across the disk, assumed parallel to the shear stress τ_w, is at an angle α to the radial direction. The velocity difference between the disk and the fluid far away from it scales as Ωr so the circumferential component of the shear stress can be represented as $\tau_w \sin\alpha \propto (\mu\Omega r/\delta)$. The radial component of the shear stress balances the centrifugal force of the fluid carried along with the disk so that $\tau_w \cos\alpha \propto \rho r\Omega^2\delta$, where δ is a representative boundary layer thickness. If α is taken independent of radius (as in the solution of Figure 8.28), the shear stress can be eliminated from these two relations to give a statement about the boundary layer thickness, namely $\delta \propto \sqrt{\nu/\Omega}$, independent of radius. The shear stress is then

$$\tau_w \propto (\rho r\Omega\sqrt{\nu\Omega}). \tag{8.9.8}$$

The total frictional torque on a disk of radius r_o is given by the product of shear stress, area, and moment arm:

$$\text{frictional torque in laminar flow} \propto \left(\rho r_o^4\Omega\sqrt{\nu\Omega}\right). \tag{8.9.9}$$

The numerical value of the constant of proportionality for a disk of radius r_o in an infinite fluid wetted on both sides has been determined as (Schlichting, 1979; Owen and Rogers, 1989)

$$C_M = \frac{2 \times \text{torque}}{\frac{1}{2}\rho\Omega^2 r_o^5} = \frac{3.87}{\sqrt{\Omega r_o^2/\nu}}. \tag{8.9.10}$$

The quantity $\Omega r_o^2/\nu$ is the disk Reynolds number.

An analogous scaling argument can be constructed for the turbulent case using the empirical relation between wall shear stress, free-stream velocity, kinematic viscosity, and boundary layer thickness introduced in Section 4.6:

$$\frac{\tau_w}{\frac{1}{2}\rho u_E^2} = 0.045\left(\frac{\nu}{u_E\delta}\right)^{1/4}. \tag{4.6.21}$$

With the substitution of Ωr for the free-stream velocity, u_E, the wall shear stress in the turbulent boundary layer on the rotating disk is expressed as

$$\tau_w \propto \left[\rho(\Omega r)^{7/4}(\nu/\delta)^{1/4}\right]. \tag{8.9.11}$$

Using the balance between the radial component of the shear stress and the centrifugal force of the fluid in the boundary layer as for the laminar boundary layer, and again making the assumption that the shear angle is constant with radius, the dependence of the turbulent boundary layer thickness is

$$\delta \propto \left[r^{3/5}(\nu/\Omega)^{1/5}\right]. \tag{8.9.12}$$

The frictional torque on the disk scales as

$$\text{frictional torque in turbulent flow} \propto \left[r_o^3\rho(\Omega r_o)^2\left(\nu/\Omega r_o^2\right)^{1/5}\right]. \tag{8.9.13}$$

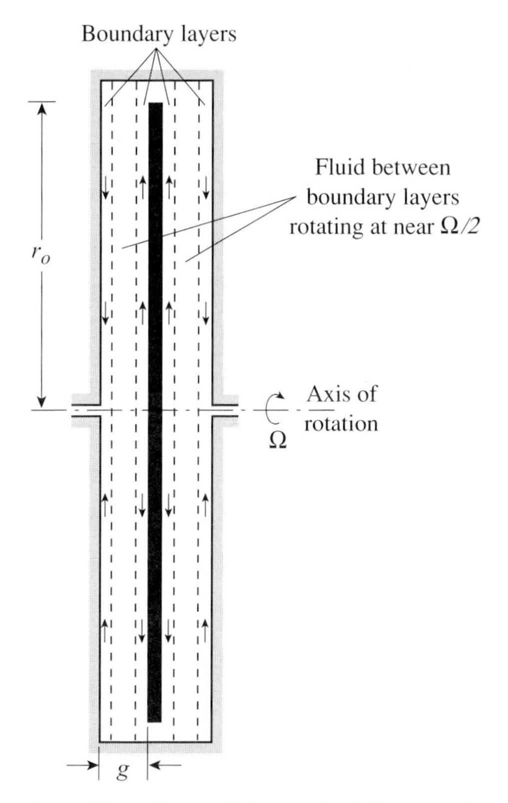

Figure 8.29: Sketch of the flow field for a rotating disk in a housing; g is the spacing.

The numerical value of the torque coefficient is (Owen and Rogers, 1989)

$$C_M = \frac{2 \times \text{torque}}{\frac{1}{2}\rho\Omega^2 r_o^5} = \frac{0.146}{\left(\Omega r_o^2/\nu\right)^{1/5}}. \tag{8.9.14}$$

8.9.3 The enclosed rotating disk

It is often the case that rotating disks are enclosed in a housing, with the disk surrounded by cylindrical and axial boundaries, as sketched in Figure 8.29. In such a configuration there is outward motion in the fluid next to the rotating disk and inward flow near the stationary boundaries. If the housing spacing is appreciably larger than the boundary layer thickness, there is a region of fluid between the two boundary layers which rotates at close to one-half the disk angular velocity. The mass of fluid set in motion by the disk loses some, but not all, of its circulation from the influence of the stationary surface, with the result that the torque is diminished by the presence of the housing.

Four behavior regimes of this configuration have been identified by Daily and Nece (1960) (see also Owen and Rogers (1989)). These are defined using coordinates of the disk Reynolds number, $Re_D(Re_D = \Omega r_o^2/\nu)$ and the spacing/radius ratio, G ($G = g/r_o$), as shown in Figure 8.30. The regimes are characterized as: (I) laminar closely spaced with no "core" between the rotating disk and the stationary housing boundary layers, (II) laminar with a core between the boundary layers,

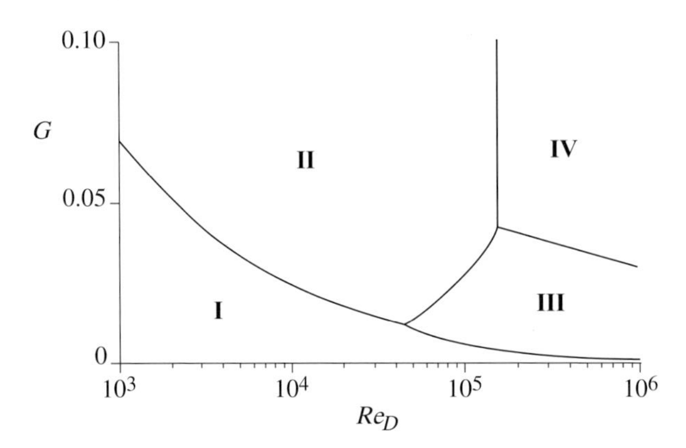

Figure 8.30: Flow regimes for a rotating disk in a housing; $G = g/r_o$, $Re_D = \Omega r_o^2/\nu$ (Daily and Nece, 1960).

(III) turbulent closely spaced, and (IV) turbulent with a core between the disk boundary layer and that on the stationary housing. Daily and Nece (1960) developed empirical expressions for the disk frictional torque in each regime as:

$$\text{Regime I: } C_M = 2\pi G^{-1} Re_D^{-1}; \tag{8.9.15a}$$

$$\text{Regime II: } C_M = 3.70 \times G^{-1/10} Re_D^{-1/2}; \tag{8.9.15b}$$

$$\text{Regime III: } C_M = 0.08 \times G^{-1/6} Re_D^{-1/4}; \tag{8.9.15c}$$

$$\text{Regime IV: } C_M = 0.102 \times G^{-1/10} Re_D^{-1/5}. \tag{8.9.15d}$$

Regime I is based on Couette flow, and Regime III on turbulent pipe flow. The moment coefficients in Regimes II and IV are those of the disk in an infinite fluid, modified by a dependence on the spacing/radius ratio, G.

Figure 8.31 presents a comparison of (8.9.15) with measured moment coefficients for rotating disks in a stationary housing as a function of disk Reynolds number, Re_D. The spacing/radius ratio, G, is 0.0255. Figure 8.32 presents computational results in Owen and Rogers (1989) and gives the radial and circumferential velocity profiles as a function of axial distance for turbulent flow in Regime IV. The overall behavior is close to the assumed two boundary layers and a core with a velocity midway between the disk speed and zero, although the core rotation rate was found to lie between 30 and 50% of the disk speed depending on Reynolds number and relative surface roughness. Further information about the parametric dependence, as well as the behavior with net flow through the chamber, can be found in Owen and Rogers (1989).

8.9.4 Internal flow in gas turbine engine rotating disk cavities

Situations exist in which several rotating disks are "stacked" together, as in the gas turbine engine axial compressor drum sketched in Figure 8.33. A problem of interest concerns the behavior of disk cooling air, which is often injected at the outer edge of the disks and enters with an initial swirl. Knowledge of the pressure drop and velocity distribution of the injected flow is critical because the velocity relative to the disk determines the heat transfer coefficient which impacts disk life.

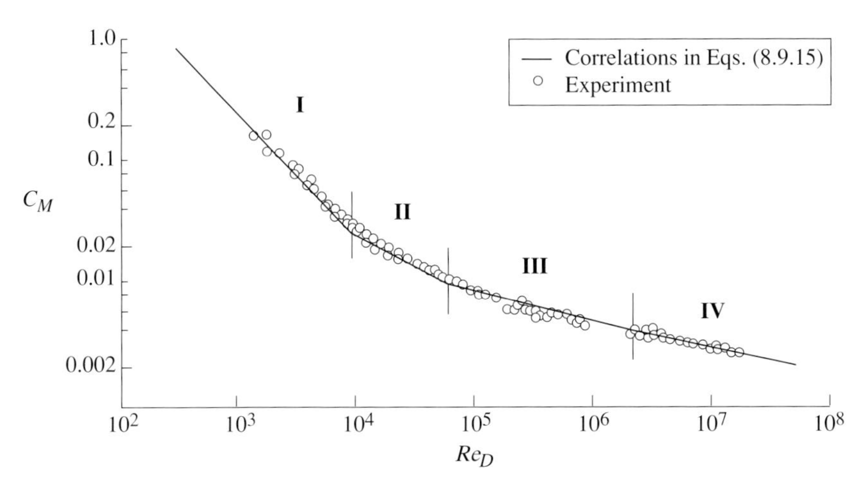

Figure 8.31: Moment coefficient for a rotating disk in a housing: $G = 0.0255$ (Daily and Nece, 1960).

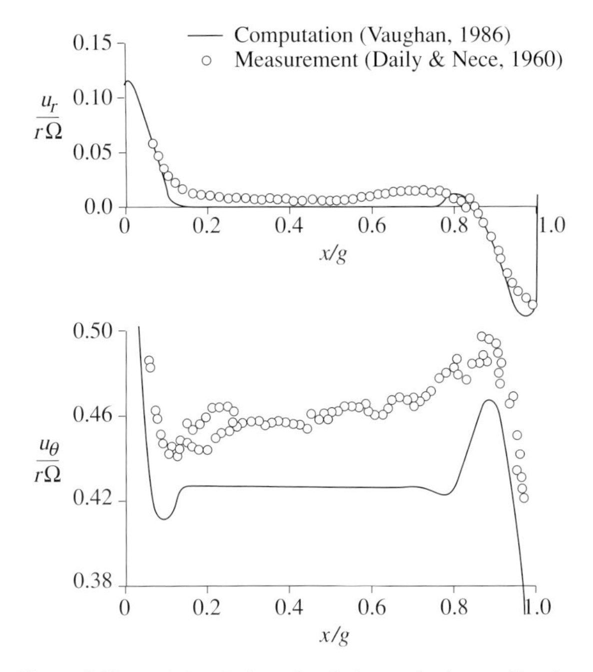

Figure 8.32: Axial variation of turbulent velocity profiles for a rotating disk in a stationary housing; $G = 0.064$, disk Reynolds number $Re_D = 4.4 \times 10^6$, radial station $= 0.765R$ (Owen and Rogers, 1989).

Three distributions of circumferential velocity versus radius are sketched in Figure 8.34. The curve labeled "free vortex" is the idealized case of circumferential velocity increasing inversely with radius. The actual velocity depends on Reynolds number and coolant flow ratio, but might be as indicated by the curve labeled "representative". The curve labeled "forced vortex" would apply if the flow were guided inward by tubes, as is the case in some designs and as shown on the right-hand side of Figure 8.33. In the first two of these situations, there is an increasing difference between the

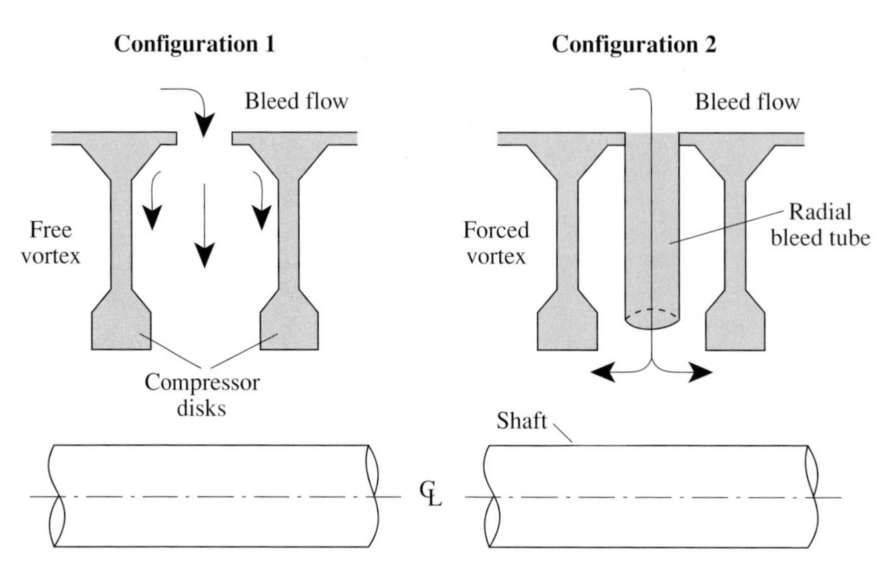

Figure 8.33: Cooling air injection in an axial compressor; shaded region is rotating (after Johnson *et al.*, 1990).

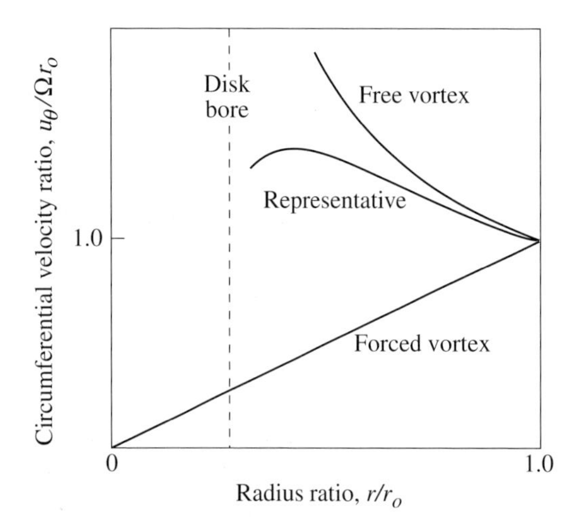

Figure 8.34: Effects of bleed configuration on circumferential velocity distribution (after Johnson *et al.*, 1990).

velocity of the disk and of the bulk fluid as radius decreases. From the perspective of an observer in a coordinate system rotating with the disks, Coriolis forces (see Sections 7.1 and 7.2) cause this change in relative circumferential velocity.

There are different pressure drops from outer radius to inner radius for the three regimes. For tube-fed flow, or for low injection rates, where shear forces from the disks have a larger opportunity to affect the incoming flow, the pressure drop is close to the value for the forced vortex, whereas for high injection rates it is more like that for a free vortex.

Measurements of the non-dimensional pressure drop in a disk cavity are given in Figure 8.35 (Johnson *et al.*, 1990). The different curves correspond to different values of a coolant flow parameter,

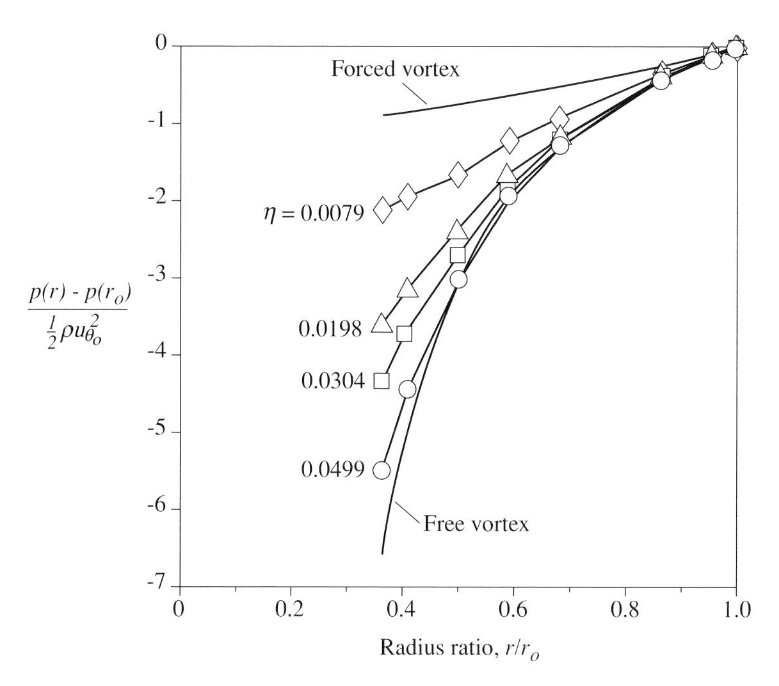

Figure 8.35: Variation of measured pressure drop as a function of radius for different coolant injection flow rates; $Re_D = 4 \times 10^6$, u_{θ_o} is the circumferential velocity at $r = r_o$ (Johnson *et al.*, 1990).

η, defined as $\eta = Re_{\dot{m}}/(Re_D)^{0.8}$ with $Re_{\dot{m}} = \dot{m}_{coolant}/(2\pi \mu r_o)$ and $Re_D = \Omega r_o^2/\nu$, and with fluid properties evaluated at the outer radius. This parameter can be written in terms of a bulk radial velocity averaged over the cavity width, $u_{r,bulk}$, as

$$\eta = \left[\frac{u_{r,bulk}}{\Omega r_o}\right]\left[\frac{W}{r_o}\right][Re_D]^{0.2}. \tag{8.9.16}$$

The experiments were carried out at constant Re_D and geometry, so an increase in η corresponds to an increase in the ratio of inflow to disk velocity and a decrease in the (non-dimensional) time over which shear forces can act. Figure 8.35 shows that as η increases the pressure distribution moves from that associated with a forced vortex towards that for free vortex swirl.

8.10 Swirling jets

Swirl can substantially increase jet mixing and entrainment and is an important design variable in devices such as combustors which depend on proper mixing to function.[8] The effects of swirl on

[8] There is a difference between swirling jets, the topic of this section, and the vortex cores discussed in previous sections. The former is a flow with no far field circulation, i.e. a jet with a circumferential velocity component injected into a stagnant fluid. The latter is a viscous vortical region embedded in a swirling freestream flow that has finite circulation far from the axis.

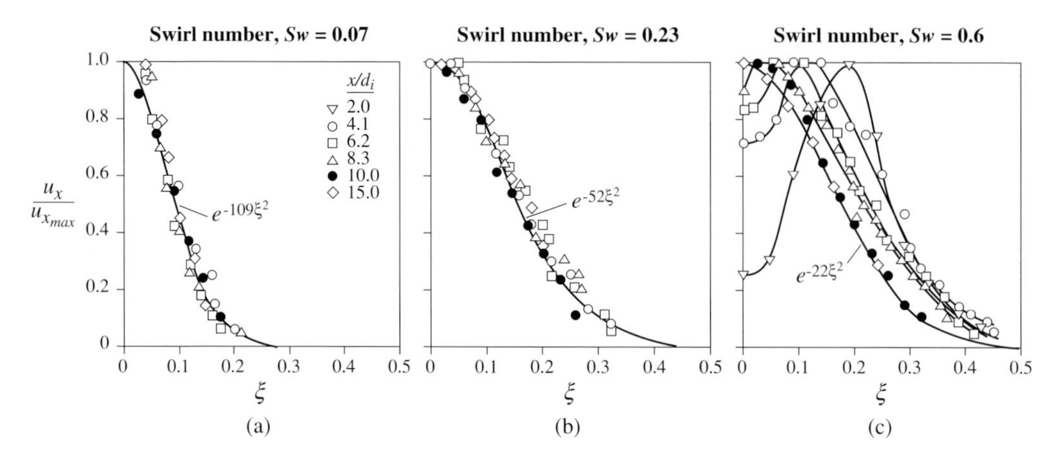

Figure 8.36: Radial distribution of axial velocity in swirling jets; $\xi = r/(x + x_v)$, x_v is the jet virtual origin, d_i is the initial jet diameter, labeled solid lines denote curve fits (Chigier and Chervinsky, 1967).

free jet behavior have been reported by Chigier and Chervinsky (1967) (see also Beer and Chigier (1972)). To characterize the swirl level those authors define a swirl number, Sw, which is the ratio of axial flux of angular momentum to r_o times the axial momentum flux, where r_o is the radius of the orifice of the swirl-producing device (and the initial radius of the jet) and p_{far} is the (ambient) pressure far away from the jet

$$Sw = \frac{\int_0^{r_o} r^2 u_\theta u_x dr}{r_o \int_0^{r_o} r \left[u_x^2 + \left(\frac{p - p_{far}}{\rho} \right) \right] dr}. \tag{8.10.1}$$

For a given experimental apparatus (implying a given distribution of flow angle and stagnation pressure coefficient) the value of Sw is sufficient to define the behavior regimes. For reference, in a jet with forced vortex rotation and uniform axial velocity, the simple radial equilibrium equation (8.2.1) may be combined with the axial momentum equation to express Sw as:

$$Sw = \frac{u_{\theta max}/2u_x}{1 - (u_{\theta max}/2u_x)^2}, \quad \text{forced vortex jet with uniform axial velocity.}$$

Measured axial velocity profiles in jets with values of Sw of 0.07, 0.23, and 0.6 are shown in Figure 8.36. The abscissa is the jet similarity parameter $\xi = r/(x + x_{vo})$, where x_{vo} is the virtual origin of the jet. The data cover downstream distances, x/d_i (d_i is the initial jet diameter), from 2 to 15.

The spreading rate and shape of the jet depend on Sw. At the highest swirl shown, there is a region of low velocity on the axis at the downstream locations and the similarity scaling no longer describes the velocity profile at all axial locations. Two different effects contribute to the increase in spreading rate. With reference to the stability discussion of Section 8.5, the circulation decreases from a maximum in the jet to zero outside the jet and the resulting instability promotes higher levels of mixing. A more important driver for the decrease in axial velocity and growth in width is the axial pressure gradient. At the initial station, where the jet issues, the static pressure on the axis is below

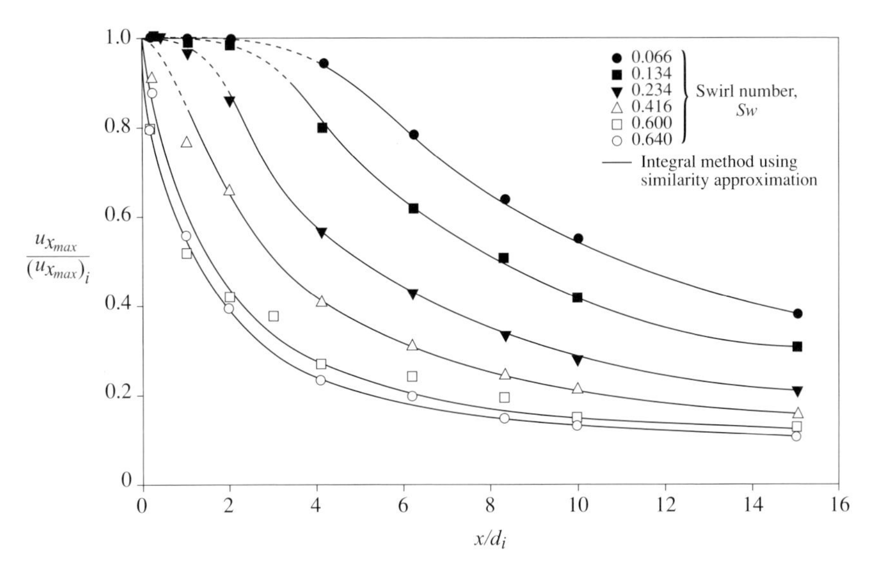

Figure 8.37: Decay of jet axial velocity maximum $u_{x_{max}}$ along the axis of the swirling jet (Chigier and Chervinsky, 1967); d_i is jet initial diameter, $(u_{x_{max}})_i$ is initial velocity maximum.

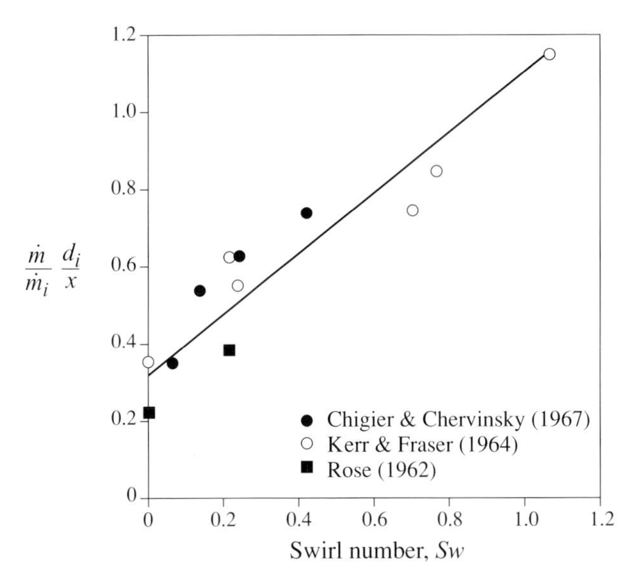

Figure 8.38: Dependence of entrainment on swirl number, Sw (Chigier and Chervinsky, 1967); d_i is the initial jet diameter, $\dot{m}_i$ is the initial jet mass flow, the straight line is $\dot{m}/\dot{m}_i = (x/d)\,(0.32 + 0.8\,Sw)$.

the value at the edge, which is at ambient conditions. Far downstream, however, the static pressure is uniform. The consequent axial pressure gradient decelerates the low stagnation pressure flow on the axis resulting in a lowered axial velocity at the center of the jet.

Figures 8.37 and 8.38 present additional features of swirling jets. Figure 8.37 shows the decay of the maximum axial velocity with downstream distance for different swirl numbers. The solid curves in the figure are derived from an integral method using similarity profiles (Chigier and Chervinsky,

1967) of the sort described in Chapter 4; this yields an expression of the maximum axial velocity in the jet as

$$\frac{u_{max}(x)}{u_{max_i}} = C_1 \left(\frac{d_i}{x + x_{vo}} \right). \tag{8.10.2}$$

An empirical expression relating C_1 and Sw is given as $C_1 \cong 1/(0.15 + Sw^2)$ (Chigier and Chervinsky, 1967; Beer and Chigier, 1972).

The increase in jet entrainment due to swirl is illustrated in Figure 8.38, which gives the ratio of the local jet mass flux to the initial mass flux, $\dot{m}_i$. The straight line represents the equation $\dot{m}/\dot{m}_i = (x/d_i)(0.32 + 0.8Sw)$. Entrainment increases roughly linearly with Sw from the value given in Section 4.9 for a non-swirling jet.

8.11 Recirculation in axisymmetric swirling flow and vortex breakdown

The presence of swirl in a diverging passage promotes the existence of stagnation or reverse flow near the axis of symmetry. Gas turbine combustor flow fields often include this type of recirculation region to enhance flame stability by providing a large enough flow residence time for fuel to mix and burn. The measured streamline patterns of Figure 8.39, derived from a simulated combustor geometry, illustrate a toroidal recirculation behind a bluff body (Beer and Chigier, 1972).

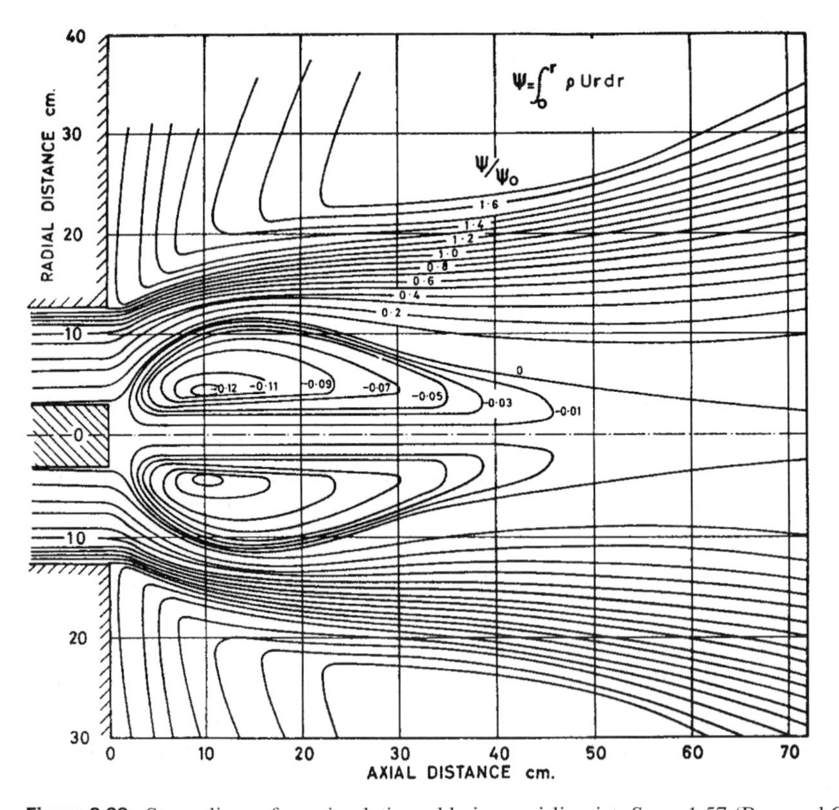

Figure 8.39: Streamlines of a recirculation eddy in a swirling jet, $Sw = 1.57$ (Beer and Chigier, 1972).

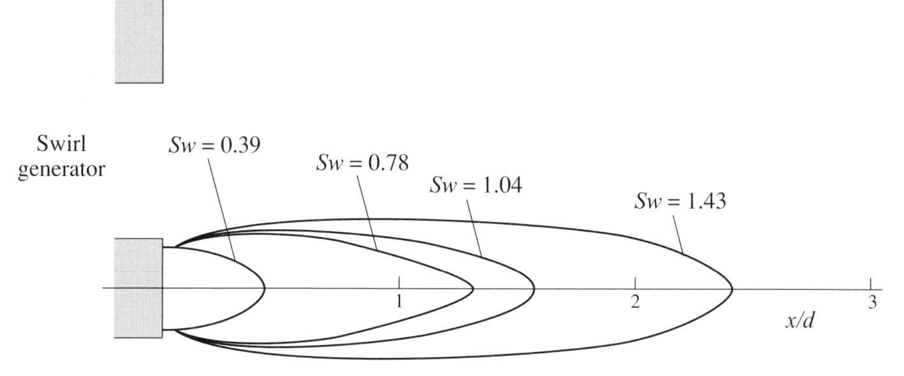

Figure 8.40: Size of the recirculation zone as a function of swirl number, Sw (Beer and Chigier, 1972).

Additional measurements of the reverse flow zone in a combustor configuration are shown for different values of the parameter Sw in Figure 8.40. These illustrate the dependence of the reverse flow extent on swirl level (Beer and Chigier, 1972; see also Gupta, Lilley, and Syred, 1984).

A generic configuration exhibiting stagnation and recirculation is a sudden expansion in a circular duct (discussed in Section 2.8 for flow without swirl). Without inlet swirl, reverse flow exists in the outer part of the duct. As the inlet swirl is increased, however, the axial velocity on the centerline decreases and reverses at high enough swirl levels.

Qualitative arguments for the deceleration on the axis can be given analogous to those presented for swirling jets in Section 8.9, although stated here in a slightly different manner. Consider a swirling flow (say, a forced vortex with angular velocity Ω_i at the inlet) which evolves, through mixing in a sudden expansion, to a downstream state with lower angular velocity and larger radius. If we approximate the swirl distribution in the throughflow as remaining a forced vortex, application of simple radial equilibrium provides an estimate for the difference between the pressure on the outer streamline of the throughflow which has a radius denoted by $r_o(x)$, and the pressure on the centerline:

$$p(r_o, x) - p(0, x) = \rho \Omega^2 r_o^2 / 2. \tag{8.11.1}$$

If angular momentum is conserved for the throughflow, Ωr_o^2 is the same at the inlet and exit. The pressure rise on the centerline is therefore larger than that on the outer edge of the throughflow region, promoting greater axial velocity decrease on the centerline than in the outer part of the flow.

The effect of swirl on the flow in a sudden expansion is illustrated in Figure 8.41. The figure shows measurements of the axial velocity at different distances downstream of an expansion of radius ratio 1.5 (Favaloro, Nejad and Ahmed, 1991). The three plots correspond to average swirl parameter, $\overline{S}$, based on average values of circumferential and axial velocity, of 0, 0.3, and 0.5. The top plot in Figure 8.41 is for no swirl and shows reverse flow in the outer part of the expansion. As the swirl is increased the axial velocity at the axis decreases. At the highest average swirl parameter, $\overline{S} = 0.5$, there is a region of reversed flow on the axis, approximately a diameter in length. At this high swirl level, in spite of the initially reversed flow on the axis, there is a more uniform axial velocity profile at the downstream station than with low swirl, presumably because of increased momentum transport in the radial direction.

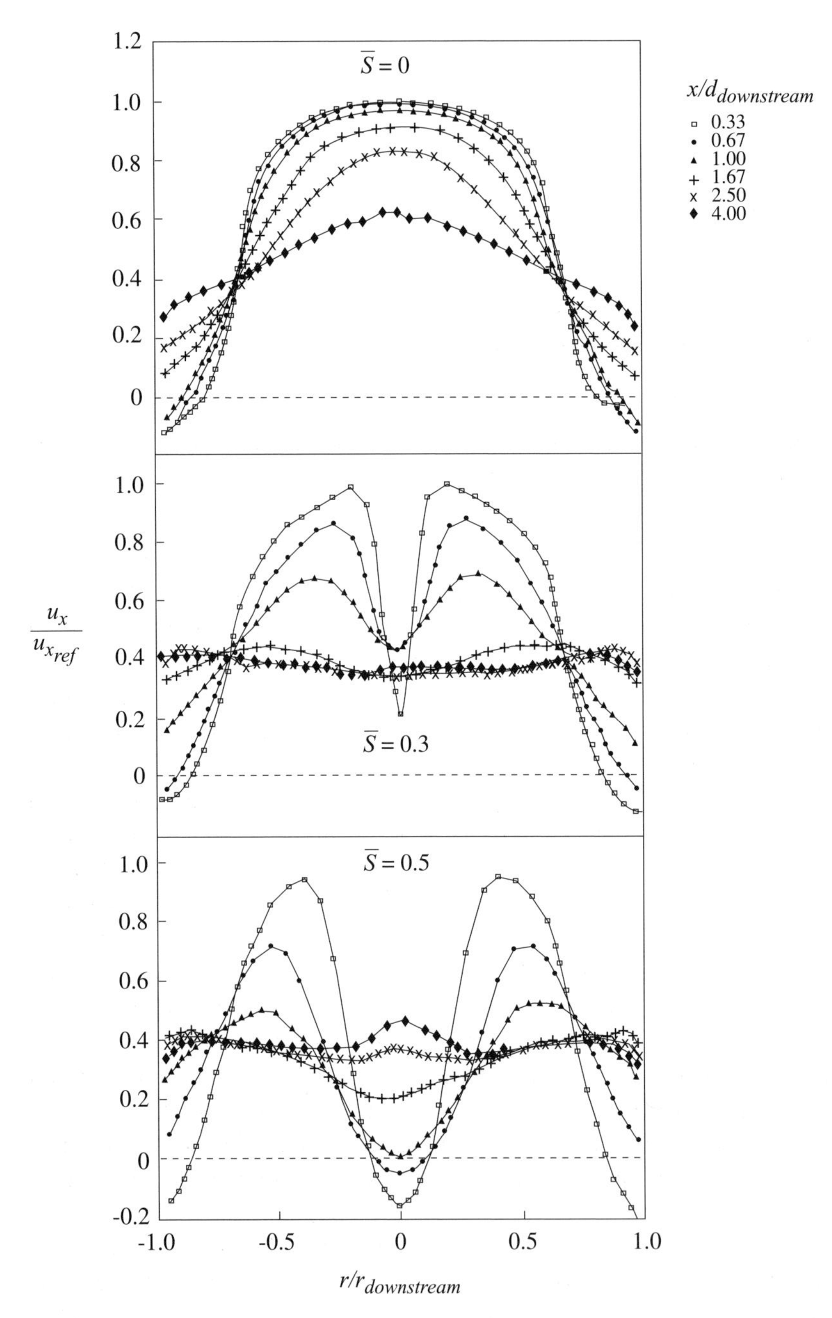

Figure 8.41: Effect of swirl on axial velocity downstream of a sudden expansion of radius ratio 1.5; $\overline{S}$ is a swirl parameter based on average circumferential and axial velocities (Favaloro *et al.*, 1991).

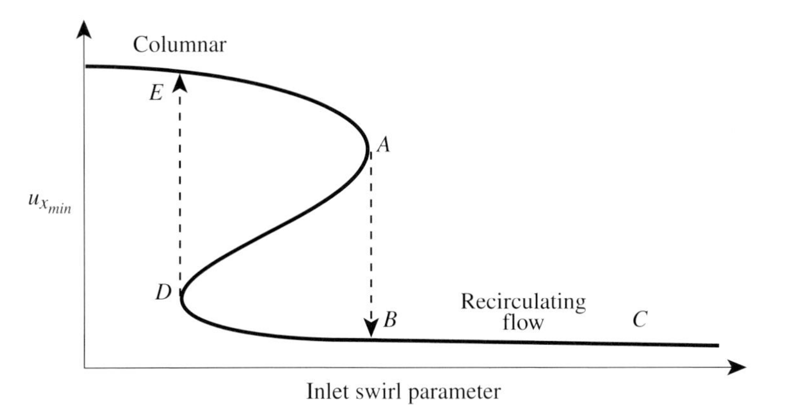

Figure 8.42: Minimum axial velocity in the vortex core, $u_{x_{min}}$, for steady solutions of the Navier–Stokes equations versus inlet swirl, S_{c_i}, as observed by Beran and Culick (1992).

The scenario for onset of recirculation in an axisymmetric swirling flow is different than without swirl. A smooth transition does not occur from a streamline pattern having no reversed flow to one in which an embedded recirculation region exists. This is in contrast to the no swirl situation, for example a wake in a diffuser (Section 4.10). In the latter if one increases the wake depth at the inlet, the wake velocity at the diffuser exit decreases to zero, with reverse flow and recirculation then encountered. With swirl, however, the transition to conditions in which stagnation and reversed flow exist (often termed vortex breakdown) is not gradual.[9]

The scenario for transition to stagnation and recirculation in a swirling flow can be introduced via the diagram of the minimum x-component of velocity on the axis versus the inlet swirl parameter in Figure 8.42. The figure is based on results of axisymmetric Navier–Stokes computations of vortex core behavior (Beran and Culick, 1992; Cary and Darmofal, 2001). As the inlet swirl parameter is increased from a value near zero to S_A, the minimum axial velocity on the axis decreases to point A, but the vortex core remains columnar with no stagnation point or region of reverse flow. Because S_A is a limit point in the solution space, however, a small increase in swirl from state A results in a jump to low or negative axial velocity at state B, which can represent a recirculation region with an axial length several times the vortex core diameter. Further increases in inlet swirl number result in quantitative changes but no further discontinuity in the qualitative form of the streamlines. A similar jump occurs when the swirl is decreased from a high value, say S_C. The recirculation region decreases in size until S reaches the limiting value S_D, when a slight decrease in swirl causes a jump back to the columnar state at E.

For axisymmetric flows the criticality of the flow can be determined using standing wave analysis (Wang and Rusak, 1997; Cary and Darmofal, 2001). The limit point at state A is signaled by the behavior of the eigenvalues associated with small amplitude disturbances, of the form $\psi' = [\exp \lambda(x/a)]f(r/a)$. Specifically, the condition at which any eigenvalue goes to zero ($\lambda \to 0$) so that disturbances no longer die away in the upstream direction corresponds to the occurrence of the limiting value of the inlet swirl number on the columnar solution branch. Figures 8.43 and 8.44

[9] Vortex breakdown also occurs in a non-axisymmetric fashion, although description of this phenomenon is beyond the scope of this text. See for example the surveys by Leibovich (1983), Escudier (1988), or Delery (1994).

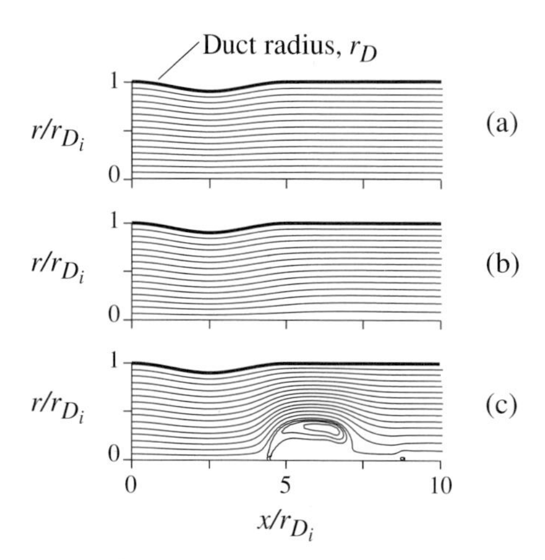

Figure 8.43: Computed stream surfaces for axisymmetric vortex core flow in a converging–diverging duct; $Re_i(u_{x_i}a/\nu) = 1000$; $S_{c_i} = (u_{\theta_{max}}/u_x)_i$, initial core area $= 0.31$ duct area: (a) $S_{c_i} = 5.6$; (b) $S_{c_i} = 1.68$ (c) $S_{c_i} = 1.69$ (Cary and Darmofal, 2001).

give results from axisymmetric Navier–Stokes computations of vortex flow in a variable-area pipe for a range of inlet swirl numbers to demonstrate this point. The inlet velocity distribution in the calculations is the Burger vortex, introduced in slightly different form in (8.7.11), which is based on a number of measurements (Delery, 1994).[10]

The numerical simulations show a transition from columnar flow to a state with stagnation and a toroidal recirculation region. This is portrayed from several perspectives in Figures 8.43 and 8.44. Figure 8.43 shows axisymmetric stream surfaces for swirling flow in the diverging region of a circular duct. Results for three inlet swirl parameters, S_{c_i} (defined as the maximum u_θ/u_x at the inlet), are presented, one at a relatively low value ($S_{c_i} = 0.56$) and the other two at the two sides of the columnar limit point ($S_{c_i} = 1.68$ and $S_{c_i} = 1.69$). Little change occurs until the critical value of inlet swirl is reached, but at this value the columnar streamline pattern changes to one with a finite length recirculation region.[11]

Figure 8.44 shows plots of: (a) minimum eigenvalue, (b) axial velocity at $r = 0$, and (c) local swirl parameter, as a function of distance along the duct for the flows in Figure 8.43. The inlet swirl parameter varies over only a small range, from 1.66 to 1.69, in increments of 0.01, but this suffices to show the changes of interest. Figure 8.44(a) shows the calculated eigenvalues versus axial distance.[12]

[10] The specific form of the Burger vortex (q-vortex) initial conditions is:

$$u_x(r/a_i, 0) = u_{x_i} = \text{constant},$$

$$ru_\theta(r/a_i, 0) = ru_{\theta_i} = S_{c_i} a u_{x_i}\{1 - \exp[-1.26(r/a_i)^2]\},$$

$$u_r(r/a_i, 0) = u_{r_i} = 0.$$

The quantity a_i is the initial vortex core radius, defined as the location of maximum u_θ/u_x.

[11] The swirl parameters have different values than in the original paper because of the different definition used here.

[12] The eigenvalue analysis described is a local approximation rather than an exact description as given by Wang and Rusak (1997). As such there is an assumption that the length scale for variation of the core is long compared to the wavelength of the disturbance (Cary and Darmofal, 2001).

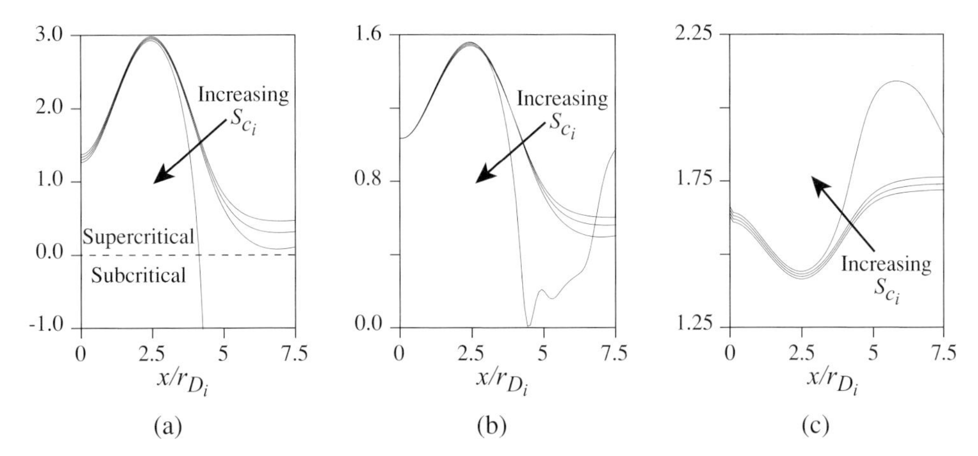

Figure 8.44: (a) Minimum eigenvalue $\lambda^2(x/a)$, (b) axial velocity at $r = 0, u_x(x/a, 0)/u_{x_i}$, and (c) swirl parameter $S(x/a)$ distributions in vortex core flow, initial conditions, $S_{c_i} = 1.66$, 1.67, 1.68, and 1.69; other conditions as in Figure 8.43 (Cary and Darmofal).

The minimum eigenvalue passes through zero as the inlet swirl parameter, S_{c_i}, increases from 1.68 to 1.69. Figure 8.44(b) shows the axial velocity on the core centerline. The axial velocity decreases as S_{c_i} increases, but is not near stagnation for S_{c_i} below the limiting inlet value. Even close to critical conditions, at $S_{c_i} = 1.68$, the velocity on the axis is far from zero. Increasing S_{c_i} to 1.69, however, gives rise to stagnation conditions and the formation of a toroidal recirculation region. Figure 8.44(c) shows the axial variation of the swirl number with, again, a large change as the value of S_{c_i} increases from 1.68 to 1.69.

In summary, transition to stagnation on the axis in an axisymmetric swirling flow is qualitatively different from flows without swirl. The former is associated with a solution limit point and occurs in a discontinuous manner as the flow parameters are altered, whereas the latter is an essentially smooth process. The distinction between the two phenomena has been stated succinctly: ". . . vortex breakdown is not simply a stagnation process, . . . rather . . . stagnation is a consequence of vortex breakdown" (Cary and Darmofal, 2001).

9 Generation of streamwise vorticity and three-dimensional flow

9.1 Introduction

In this chapter we address three-dimensional flows in which streamwise vorticity is a prominent feature. Three main topics are discussed. The first, and principal, subject falls under the general label of secondary flows, cross-flow plane (secondary) circulations which occur in flows that were parallel at some upstream station. The second is the enhancement of mixing by embedded streamwise vorticity and the accompanying motions normal to the bulk flow direction (see for example Bushnell (1992)). The third is the connection between vorticity generation and fluid impulse.

The different topics are linked in at least three ways. First, the class of fluid motions described are truly three-dimensional. Second, focus on the vortex structure in these flows is a way to increase physical insight. The perspective of the chapter is that the flows of interest are rotational and three-dimensional, and the appropriate tools for capturing their quantitative behavior are three-dimensional numerical simulations (e.g. Launder (1995)). Results from such computations, as well as from experiments, are used to illustrate the overall features. To complement detailed simulations and experiments, however, it is often helpful to have a simplified description of the motion which can guide the interrogation and scope of the computations, enable understanding of why different effects are seen, and suggest scaling for different mechanisms. The ideas about vorticity evolution and vortex structure, introduced in Chapter 3, provide a skeleton for this type of description.

9.2 A basic illustration of secondary flow: a boundary layer in a bend

9.2.1 Qualitative description

When a flow that is parallel but non-uniform in velocity or density is made to follow a curved path, the result is a three-dimensional motion with velocity components normal to the overall flow direction. Cross-flow of this type is associated with the generation of a streamwise component of vorticity and commonly referred to as *secondary flow*. The name derives from the view that one can, in many instances, identify a primary flow direction along a passage or bend and hence also specify the departures from this primary direction. Although the term secondary is in common use, it can be a misnomer because the cross-flow velocities are often a substantial fraction of the primary velocity.

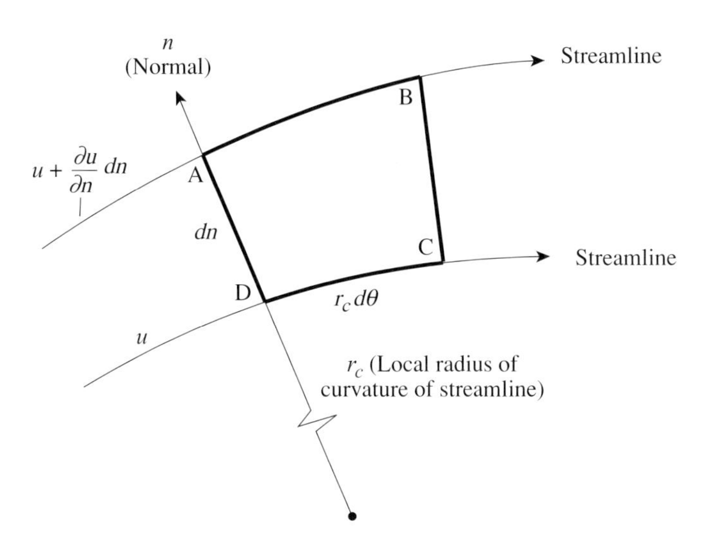

Figure 9.1: Contour ABCD for the evaluation of vorticity in a flow with curved streamlines; the circulation around contour $= (\partial u/\partial n + u/r_c)\, dnr_c d\theta$; the area of contour $= dnr_c d\theta$; the component of vorticity perpendicular to the page, $\omega_\perp = \partial u/\partial n + u_c/r_c$.

The flow in a boundary layer on the bottom of a curved passage such as a rectangular bend furnishes an illustration of the type of motions to be addressed. Generation of secondary flow can be viewed in terms of the differential convection of boundary layer vorticity through the bend (Section 3.4). Consider the flow external to the boundary layer as a two-dimensional irrotational stream. Figure 9.1 shows an elementary contour formed by two streamlines and two normals between the streamlines in a two-dimensional flow. From evaluation of the circulation around this contour, the component of vorticity perpendicular to the page, $\omega_\perp$, can be written in terms of the rate of variation of velocity in the normal (n) direction and the local streamline radius of curvature, r_c, as

$$\omega_\perp = \frac{\partial u}{\partial n} + \frac{u}{r_c}. \tag{9.2.1}$$

For a flow which is irrotational outside of the boundary layer, $\omega_\perp = 0$, (9.2.1) states that particles on the outside of a bend have a lower velocity than particles on the inside. Particles on the outside of the bend also travel a longer distance than those on the inside. If boundary layer vortex lines are convected with a velocity proportional to the local free-stream velocity, vortex lines initially normal to the flow will be tipped into the streamwise direction as they traverse the bend, as was depicted in Figure 3.10, with a resulting cross-flow as in Figure 9.2. In a plane perpendicular to the free stream (referred to as a cross-flow plane), one sees boundary layer fluid migrate toward the inside of the bend.[1]

[1] As described in Section 3.4, an alternative explanation of secondary flow can be given in terms of pressure fields and fluid accelerations. The two views of secondary flow, in terms of pressure or vorticity, are complementary and which one one adopts is a matter of choice. As should be evident by now, the authors' view is that the ability to describe fluid motions in terms of vorticity ("the sinews and muscles of fluid mechanics" (Küchemann, 1965)) provides a powerful tool for insight into internal flows.

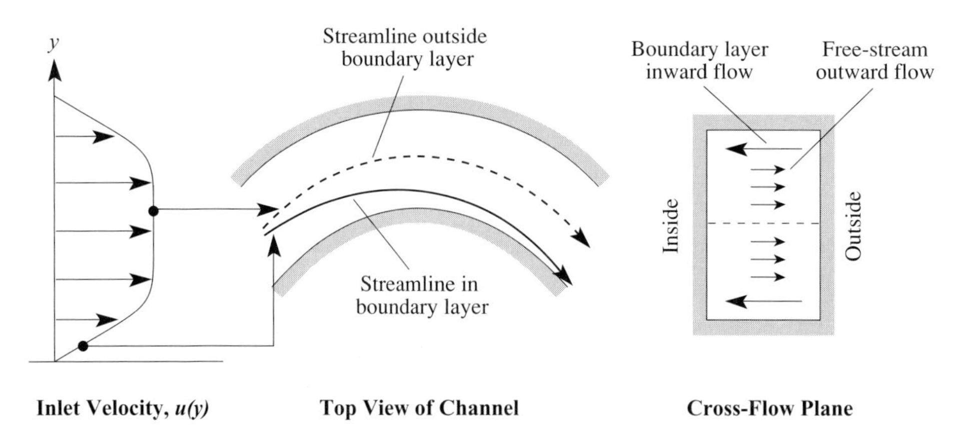

Figure 9.2: Velocity profile, streamlines, and secondary velocities in a channel. (See also Figure 3.10.)

9.2.2 A simple estimate for streamwise vorticity generation and cross-flow plane velocity components

A simple estimate for streamwise vorticity generation in a bend can be developed under the following assumptions: (i) vortex lines are convected by a background (or primary) flow which is unaffected by the secondary flow; (ii) the primary flow is irrotational and the streamlines in the bend are circular arcs; (iii) the vorticity at the inlet is normal to the free stream and parallel to the bend floor; (iv) viscous effects within the bend can be neglected.

Consider an element of a vortex line initially normal to the streamlines as it is convected through a small angle $d\theta$ in the bend.[2] A fluid cross formed from an element of a vortex line, *aa*, and an element of a streamline, *bb*, is shown in Figure 9.3 at an initial location and at a subsequent location $r_m d\theta$ further along the streamline, where r_m is the mean bend radius. The leg of the cross that is part of the streamline rotates an amount $d\theta = dl/r_m$ in the clockwise direction, where dl is the streamwise distance traveled. Since there is no net vertical vorticity, the other leg of the fluid cross, which coincides with the vortex line, must rotate an equal angle $d\theta$ in the opposite direction. If this situation is assumed to be the same all along the bend, the angle between the elements of the streamline and the vortex line thus changes from $\pi/2$ at the bend inlet to $(\pi/2 - 2\Delta\theta)$, where $\Delta\theta$ is the bend angle. As indicated on the right-hand side of Figure 9.3, if ω_i is the bend inlet vorticity which is purely normal, at the bend exit there is a streamwise component of vorticity given by

$$\omega_s \approx -2\Delta\theta\omega_i. \tag{9.2.2}$$

Within the approximations made, the streamwise vorticity generated is proportional to the product of the inlet normal vorticity and the free-stream turning angle. Equation (9.2.2), known as the Squire and Winter (1951) approximation (see e.g. Horlock (1966)), has found wide application in the estimation of streamwise vorticity generation.

[2] In Chapter 9 the local streamline angle is denoted by θ, rather than α, as previously, for two reasons. Primary is that the initial context is flow around a bend and θ is the natural variable for marking different circumferential stations, but another reason is that θ is the convention in secondary flow literature.

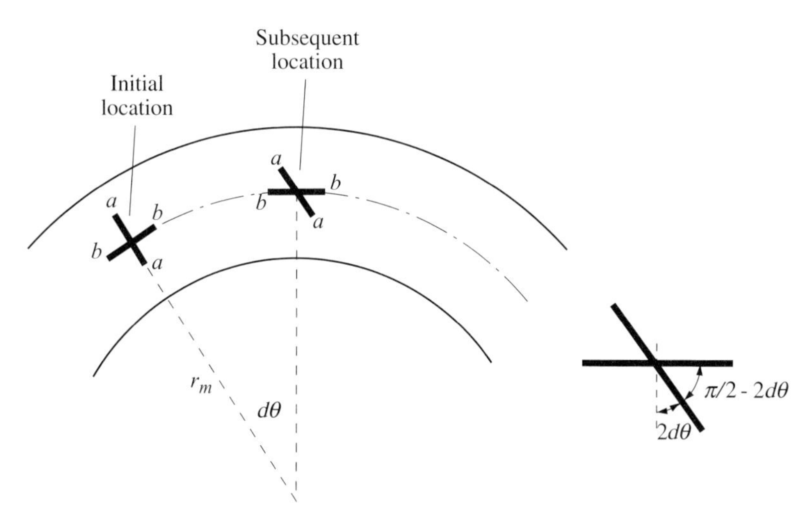

Figure 9.3: Deformation of a fluid cross moving through an incremental angle $d\theta$: leg bb aligned with the streamline, leg aa initially normal to the streamline.

Knowing the streamwise vorticity we can now find the secondary velocity field. Within the secondary flow approximation, the primary streamlines are straight downstream of the bend, the vorticity distribution is unaltered once the flow leaves the curved portion of the duct, and changes in the flow field with distance *along* the duct in this region can be neglected. Denoting the coordinate along the duct as x, the coordinate perpendicular to the bottom of the duct as y, and z as the third member of the right-handed coordinate system, with $\partial/\partial x \cong 0$, the continuity equation is

$$\frac{\partial u_y}{\partial y} + \frac{\partial u_z}{\partial z} \cong 0. \tag{9.2.3}$$

Equation (9.2.3) is a statement that the flow is (regarded as) locally two-dimensional at any x-station. This implies a relation between the streamwise vorticity and the cross-flow (y, z) velocity field in which a stream function, ψ, that satisfies (9.2.3) identically can be defined as

$$u_y = \frac{\partial \psi}{\partial z}, \quad u_z = -\frac{\partial \psi}{\partial y}. \tag{9.2.4}$$

From the definition of the vorticity, the streamwise vorticity, ω_x, is related to the stream function by the Poisson equation:

$$\frac{\partial^2 \psi}{\partial y^2} + \frac{\partial^2 \psi}{\partial z^2} = \nabla^2 \psi = -\omega_x (y, z). \tag{9.2.5}$$

The streamwise vorticity in (9.2.5) is obtained from (9.2.2). If the normal vorticity in the bottom boundary layer at the bend inlet, ω_{z_i}, is uniform in y across the passage, (9.2.5) becomes

$$\frac{\partial^2 \psi}{\partial y^2} + \frac{\partial^2 \psi}{\partial z^2} = 2\Delta\theta \times [\omega_{z_i} (y)]. \tag{9.2.6}$$

In (9.2.6) $\omega_{z_i} (= -\partial u_x/\partial y)$, the normal vorticity in the boundary layer entering the bend, is a known quantity. The boundary conditions for (9.2.6) are no normal velocity on the walls of the duct or, equivalently, that the walls of the duct are stream surfaces (surfaces of constant ψ).

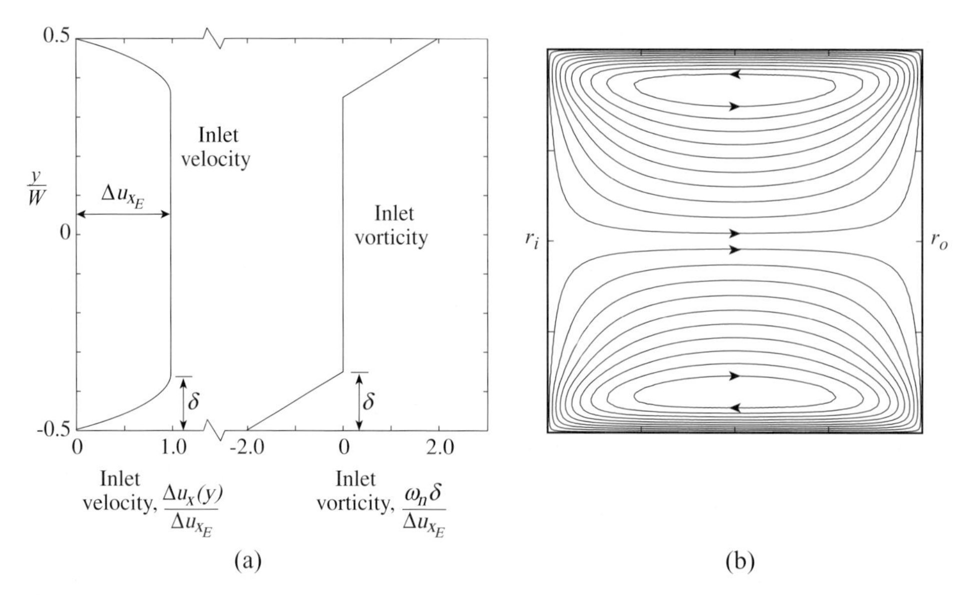

Figure 9.4: Secondary flow in a square channel of side W; linear analysis: (a) inlet velocity and vorticity distribution: velocity $u_x = u_0 + \Delta u_x(y)$, free-stream velocity $u_{x_E} = u_0 + \Delta u_{x_E}$, u_0 = constant, vorticity non-dimensionalized by Δ_{x_E}/δ; (b) streamlines in the cross-flow plane, streamline contour intervals $0.1 \Delta\theta \Delta u_{x_E} \delta$.

Figure 9.4 shows the inlet velocity and vorticity and the cross-flow plane streamlines for flow in a square bend. Figure 9.4(a) shows a linearly varying vorticity over the upper and lower 15% of the passage at the inlet with accompanying parabolic velocity profile (this is roughly representative of a thick inlet boundary layer). The streamline pattern is given in Figure 9.4(b), with r_i and r_o denoting the inner and outer radius locations. Within the approximate treatment the streamline pattern is independent of the turning angle, $\Delta\theta$, and the magnitude of the streamwise vorticity and secondary velocity at the bend exit are both linearly proportional to $\Delta\theta$.

The model described is idealized, but it shows several relevant features of three-dimensional passage flows. One is the increase in strength of the cross-flow with turning. Another is the presence of inward flow in the boundary layer and a return flow in the free stream, taking fluid from the inner radius to the outer radius. From the analysis the cross-flow plane velocity levels in the boundary layer and free stream scale roughly as the inverse of the ratio of boundary layer and free-stream heights.

There are several processes at work in the flows addressed and it is useful to comment on the effects which are taken into account by the theory. For a uniform density fluid, viscous effects at solid surfaces *generate* the vorticity, as described in Section 3.13. At the high Reynolds numbers that characterize industrial fluid devices, however, for the effects of viscosity to penetrate substantially into the stream (to have a boundary layer whose thickness is a substantial fraction of the duct dimension) requires a streamwise development length which can be from several to tens of duct dimensions. For example, at a length Reynolds number of 10^6, it takes roughly four passage heights for the edge of a turbulent boundary layer in a two-dimensional channel to grow to 10% of the channel width. Processes within fluid machines thus often occur over a length scale which is short compared to boundary layer development length, so many features can be viewed as inviscid but rotational. This difference in length scales underpins the use of secondary flow theory, which can be described as the creation of three-dimensional motions by the "processing" of inlet vorticity in a manner which is predominantly inviscid.

9.2.3 A quantitative look at secondary flow in a bend: measurements and three-dimensional computations

The sections on secondary flow have dual objectives. One is to introduce the basic ideas along with the approximations inherent in the approach. No less important, however, is to show, from computations and experiments, the connection to real three-dimensional flow situations including not only aspects which are well captured, but also limitations of the theory.

A more quantitative look at three-dimensional flow in a bend is thus given by Figures 9.5 and 9.6 which show data and viscous three-dimensional computations for laminar (9.5) and turbulent (9.6) flow in a square duct with 90° of turning (Humphrey, Taylor, and Whitelaw, 1977; Humphrey, Whitelaw, and Yee, 1981). The plots are contours of velocity along the bend, u_θ, in different cross-flow planes, with the specific stations through the bend given in the figures. The measurements are the upper half of the plots and the numerical computations the lower half. Comparison of the laminar and turbulent cases shows, as implied by the discussion of length scales in Section 9.2, a number of similarities in overall flow structure. While inward motion of the boundary layer, and the consequent return flow towards the outer radius in the midplane, seen in both figures, are in qualitative accord with the Squire–Winter result, aspects such as the vertical motions near the center of the passage are not captured by the linear analysis.

As well as indicating the ability of the computations to describe the flows under study the experiment–computation comparison helps point out where secondary flow analyses fit in. For flow round a bend the Squire–Winter analysis gives a motivation for overall trends and this carries over to other (more complex) situations later in the chapter. Comparison of Figure 9.4 with Figures 9.5 and 9.6 shows, however, that we need to go beyond the basic description to address the quantitative behavior and even to capture some qualitative features of the motions.

9.3 Additional examples of secondary flow

9.3.1 Outflow of swirling fluid from a container

We give several examples to illustrate the way in which secondary flow behavior can be determined based on the approximation of a known primary flow. The first is outflow of swirling fluid through a hole in the bottom of a circular container. Suppose the flow enters the container tangentially so that away from the boundaries (or the axis) the circumferential velocity can be approximated by $u_{\theta_E} = K/r$. Fluid near the bottom of the container has a lower velocity than this free-stream value because of viscous stresses. The resulting flow can be described by stating that the slower moving fluid in the boundary layer near the bottom is driven inward by the radial pressure gradient established by the swirling free stream. To cast the arguments in terms of streamwise (θ) vorticity generation, however, we examine the evolution of a primary flow with free vortex swirl velocity and vortex lines in the bottom boundary layer that are initially radial. If these radial vortex lines are convected by the primary flow,

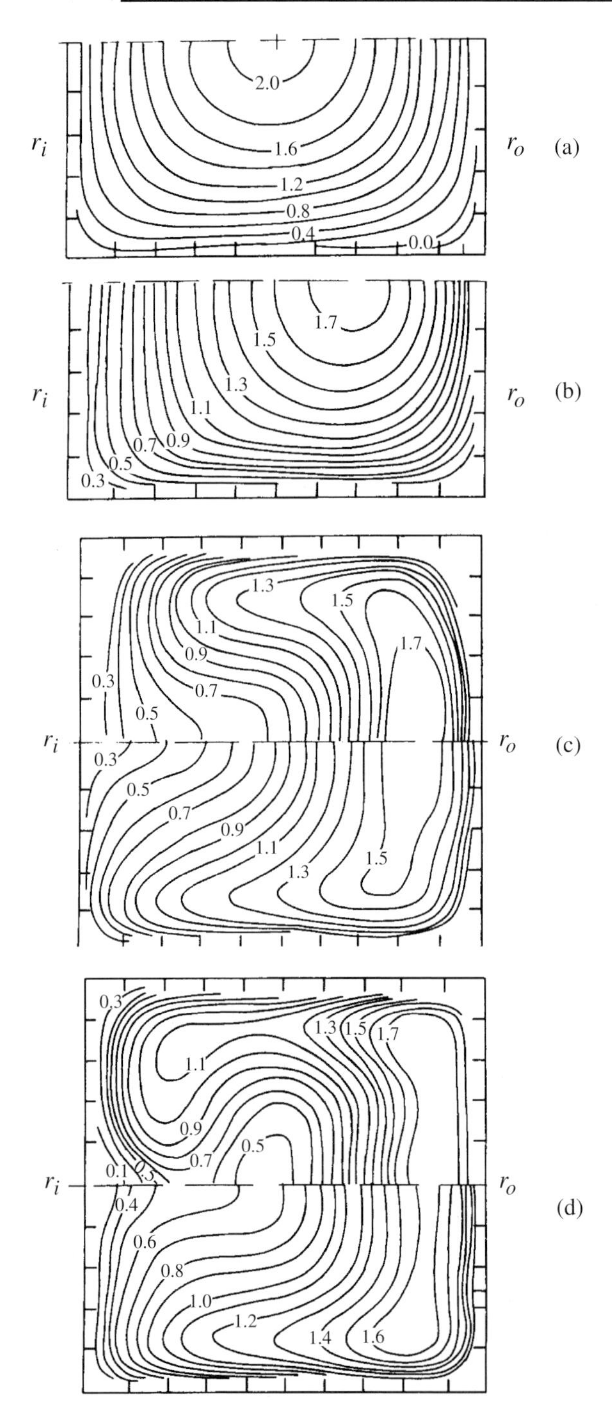

Figure 9.5: Measured and computed velocity ($u_\theta/\overline{u}$) contours in a rectangular bend, laminar flow; $\overline{u}$ is the mean velocity, $(r_o - r_i)/r_i = 0.56$: (a) 0° calculated; (b) 30° calculated; (c) 60° measured (top) and calculated (bottom); (d) 90° measured (top) and calculated (bottom) (Humphrey *et al.*, 1977).

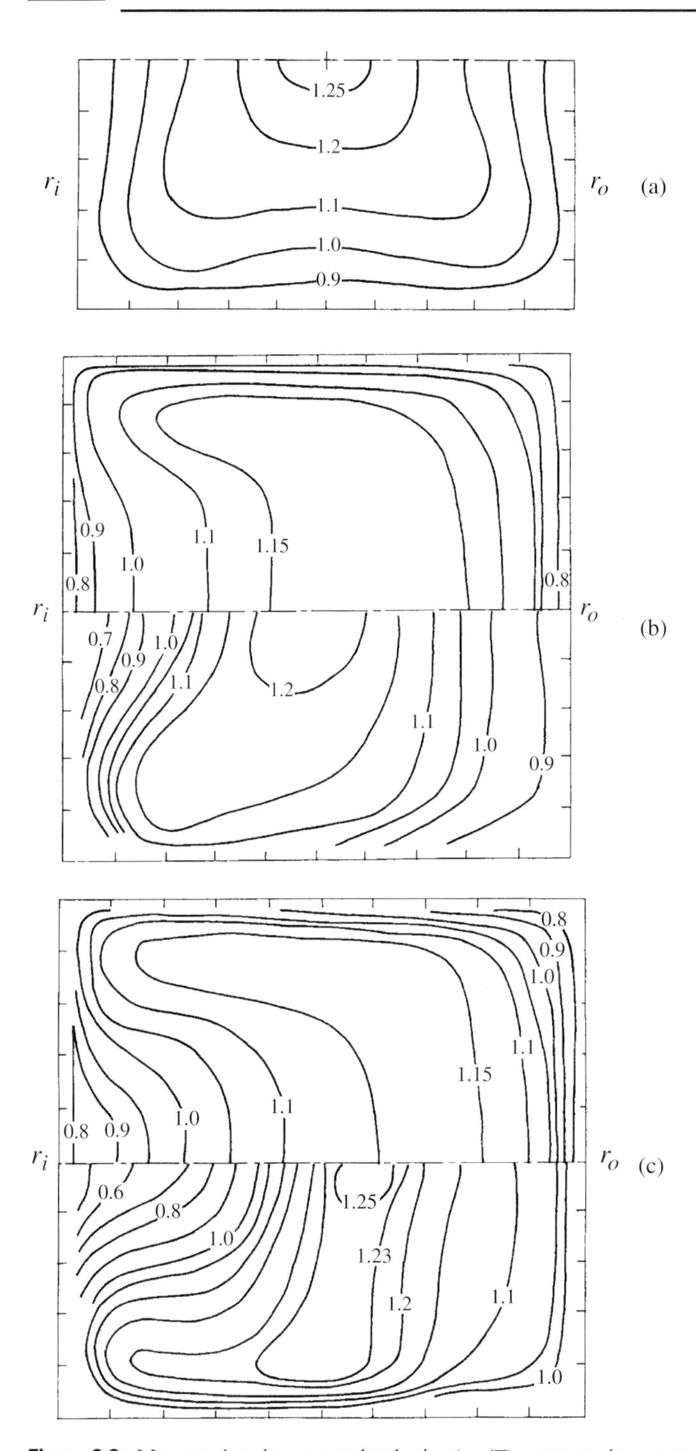

Figure 9.6: Measured and computed velocity ($u_\theta/\overline{u}$) contours in a rectangular bend, turbulent flow; $\overline{u}$ is mean velocity, $(r_o - r_i)/r_i = 0.56$: (a) upstream of bend; (b) $\theta = 71^\circ$, calculations (top) and measurements (bottom); (c) $\theta = 90^\circ$, calculations (top) and measurements (bottom) (Humphrey *et al.*, 1981).

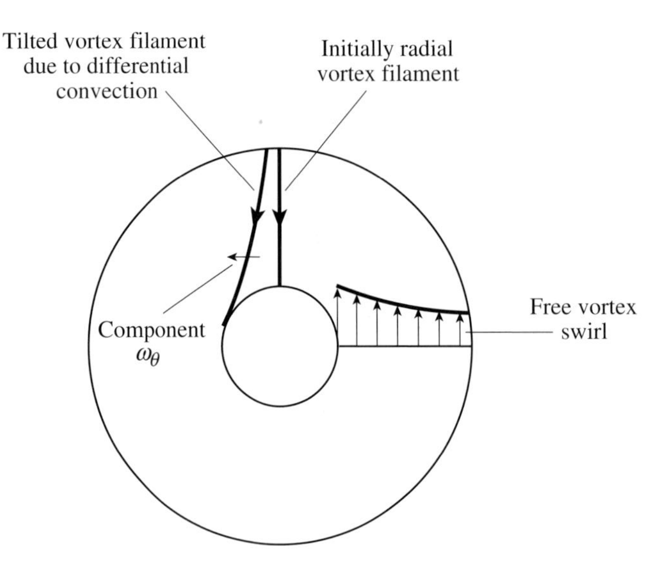

Figure 9.7: Development of the circumferential component of vorticity associated with radially inward motion in a swirling flow.

the situation is as indicated in Figure 9.7. The drawing shows an initially radial vortex line tipped into the circumferential direction, generating a θ-component of vorticity and a consequent inward radial flow.[3]

An estimate for the θ-vorticity generation can be given using secondary flow approximations. The circumferential velocity, u_θ, at any vertical (x) station is taken to be free vortex, $u_\theta = K(x)/r$, with the value of K varying along the axis of rotation (x-axis) (Horlock, 1975). The velocity components of the "primary flow" are then

$$u_x = u_r = 0; \quad u_\theta = K(x)/r. \tag{9.3.1}$$

The vorticity components of the primary flow are

$$\omega_x = \omega_\theta = 0; \quad \omega_r = -\frac{\partial u_\theta}{\partial x} = -\frac{1}{r}\frac{dK}{dx}. \tag{9.3.2}$$

Using the expression for the rate of change of vorticity in an inviscid constant density fluid ((3.4.1), $\mathbf{u} \cdot \nabla\boldsymbol{\omega} = \boldsymbol{\omega} \cdot \nabla\mathbf{u}$) the linearized expression for the generation of secondary vorticity (ω_θ) is

$$\frac{u_\theta}{r}\frac{\partial \omega_\theta}{\partial \theta} = \omega_r\left(\frac{\partial u_\theta}{\partial r} - \frac{u_\theta}{r}\right). \tag{9.3.3}$$

Because the primary velocity distribution obeys $(\partial u_\theta/\partial r) + (u_\theta/r) = 0$, (9.3.3) becomes

$$\frac{\partial \omega_\theta}{\partial \theta} = -2\omega_r. \tag{9.3.4}$$

Integrating (9.3.4) between two values of θ yields

$$\omega_{\theta_2} - \omega_{\theta_1} = -2\omega_r(\theta_2 - \theta_1)$$

analogous to (9.2.2).

[3] The arguments here are similar to those in Section 8.9.

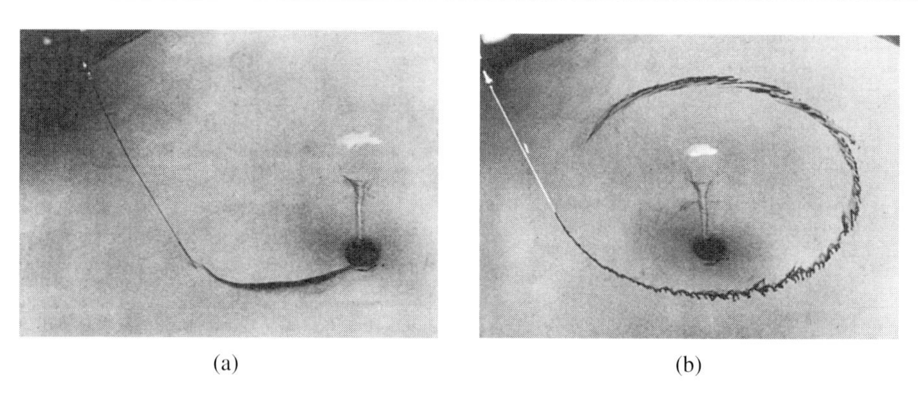

(a) (b)

Figure 9.8: Secondary flow on the endwall of a circular container with swirling flow: (a) dye trace showing radially inward secondary flow near bottom of container; (b) dye trace of circular streamlines of the primary flow away from bottom (Taylor, 1972).

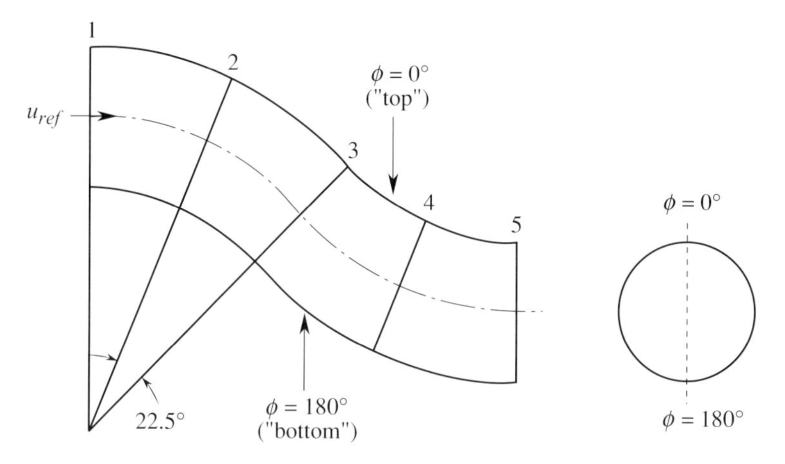

Figure 9.9: S-shaped duct (aircraft engine inlet) (Bansod and Bradshaw, 1971).

Figures 9.8(a) and 9.8(b) show the trajectory of dye injected into a swirling flow in a cylindrical container, in the boundary layer (Figure 9.8(a)) and at a location far away from the boundary (Figure 9.8(b)). There is a strong inward flow in the former, with virtually all the flow through the drain hole appearing to come from the boundary layer (Taylor, 1972).

9.3.2 Secondary flow in an S-shaped duct

Flow in an S-shaped duct, as in an aircraft engine inlet, is represented by the geometry of Figure 9.9 (Bansod and Bradshaw, 1971). Viscous effects produce the inlet vorticity essential to the existence of secondary flow, but the generation and downstream evolution of the phenomena can be predominantly inviscid. Application of (9.2.2) to the boundary layer on the duct sidewalls implies a generation of streamwise vorticity pointing in the upstream direction in the first part of the duct and hence a secondary flow in the boundary layer towards the $\phi = 180°$ location. The thickness of the region of

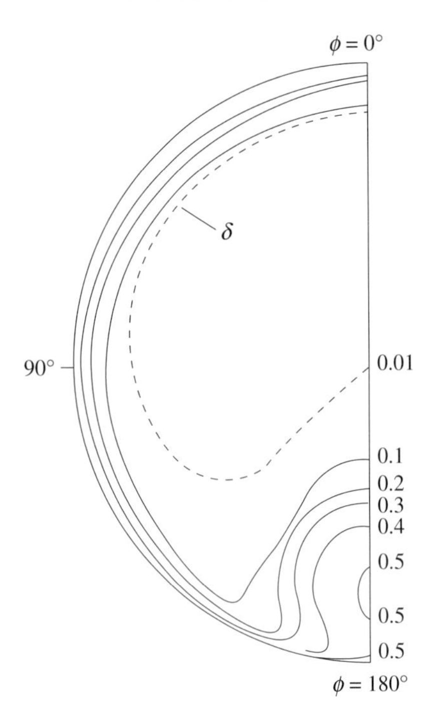

Figure 9.10: Stagnation pressure defect $[\Delta p_t/(\frac{1}{2}\rho\overline{u}^2)]$ at Station 5, for the duct shown in Figure 9.9 (Bansod and Bradshaw, 1971).

low stagnation pressure at $\phi = 180°$ increases substantially as shown by the contour plot of stagnation pressure in Figure 9.10.

Measurements indicate a pair of concentrated counterrotating vortices associated with the motion of the low stagnation pressure fluid away from the duct wall into the middle of the duct. The qualitative sense of these streamwise vortices can be understood from the secondary flow considerations, but a feature often present in actual situations is that streamwise vorticity does not remain in a layer near the wall but rather rolls up into discrete vortical structures. This concentration of vorticity results in more localized, and stronger, cross-flow plane velocities than would be the case if the vorticity remained spread along the wall.

9.3.3 Streamwise vorticity and secondary flow in a two-dimensional contraction

A third example is the secondary flow in two-dimensional contractions, such as are used in wind tunnels (Bansod and Bradshaw, 1971). Boundary layer streamlines are sketched in Figure 9.11, showing streamwise vorticity generated in the converging passage. The thickness of the boundary layer may not be greatly reduced in the contraction because the lateral convergence of the flow and the presence of the streamwise vortices lead to a thickening of boundary layer fluid. A mitigating effect is that if the contraction ratio is large the momentum integral equation shows that the momentum thickness decreases appreciably in the downstream direction, whatever the behavior of the boundary layer thickness.

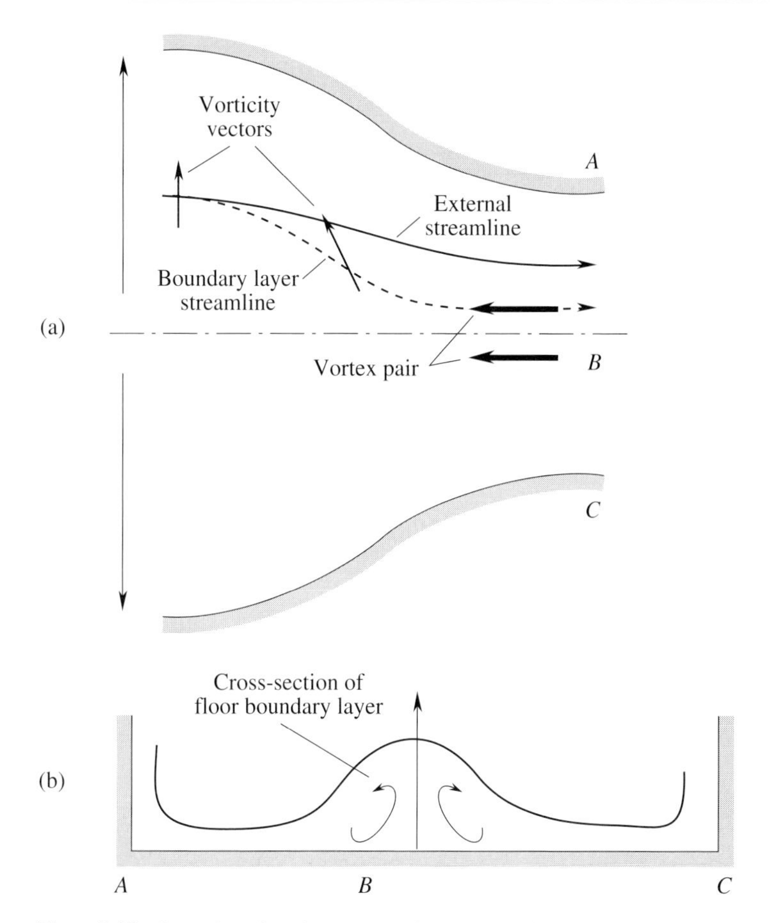

Figure 9.11: Secondary flow in a contraction: (a) top view; (b) cross-flow plane at contraction exit (Bansod and Bradshaw, 1971).

9.3.4 Three-dimensional flow in turbine passages

Three-dimensional flows play an important role in turbomachinery component performance. In a turbine blade row, for example, in addition to turning the flow, as in the bent duct, there is the formation of a horseshoe vortex round the leading edge of the airfoils (Section 3.4). Figure 9.12, from Langston (1980) is a pictorial representation of the swirling motion in a turbine vane passage. The dominant structure is the passage vortex, but also indicated is another vortex embedded near the junction of the suction surface and the endwall, smaller in area than the passage vortex, which represents the continuation of vortex filaments that thread through the passage and the horseshoe vortex structure. The formation of the passage vortex in a turbine cascade is shown in Figure 9.13 using smoke visualization (Gostelow, 1984).

Strong circulatory motions such as pictured in Figure 9.13 are associated with vortex roll up, which is inherently a non-linear process, and thus differ qualitatively from the Squire–Winter results. The area and strength of the vortex are both of interest, because they affect the overall loss as well as the flow past to the succeeding row; the vortex radial position can be important for this latter aspect.

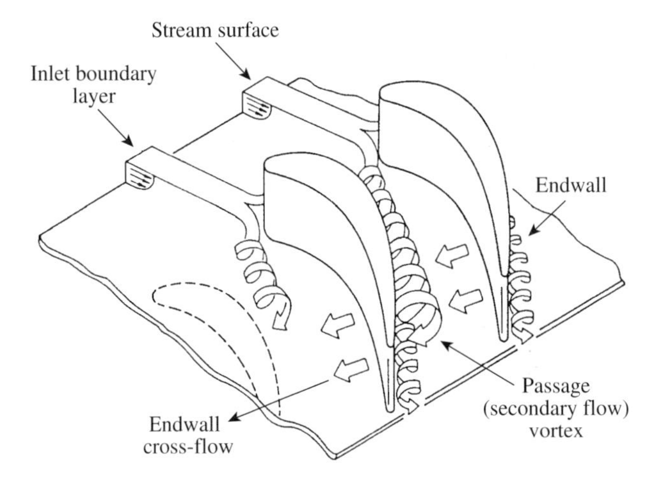

Figure 9.12: Pictorial representation of the cross-flow structure in a turbine passage (Langston, 1980).

Figure 9.13: Visualization of secondary flow near the endwall of a turbine blade cascade (Gostelow, 1984).

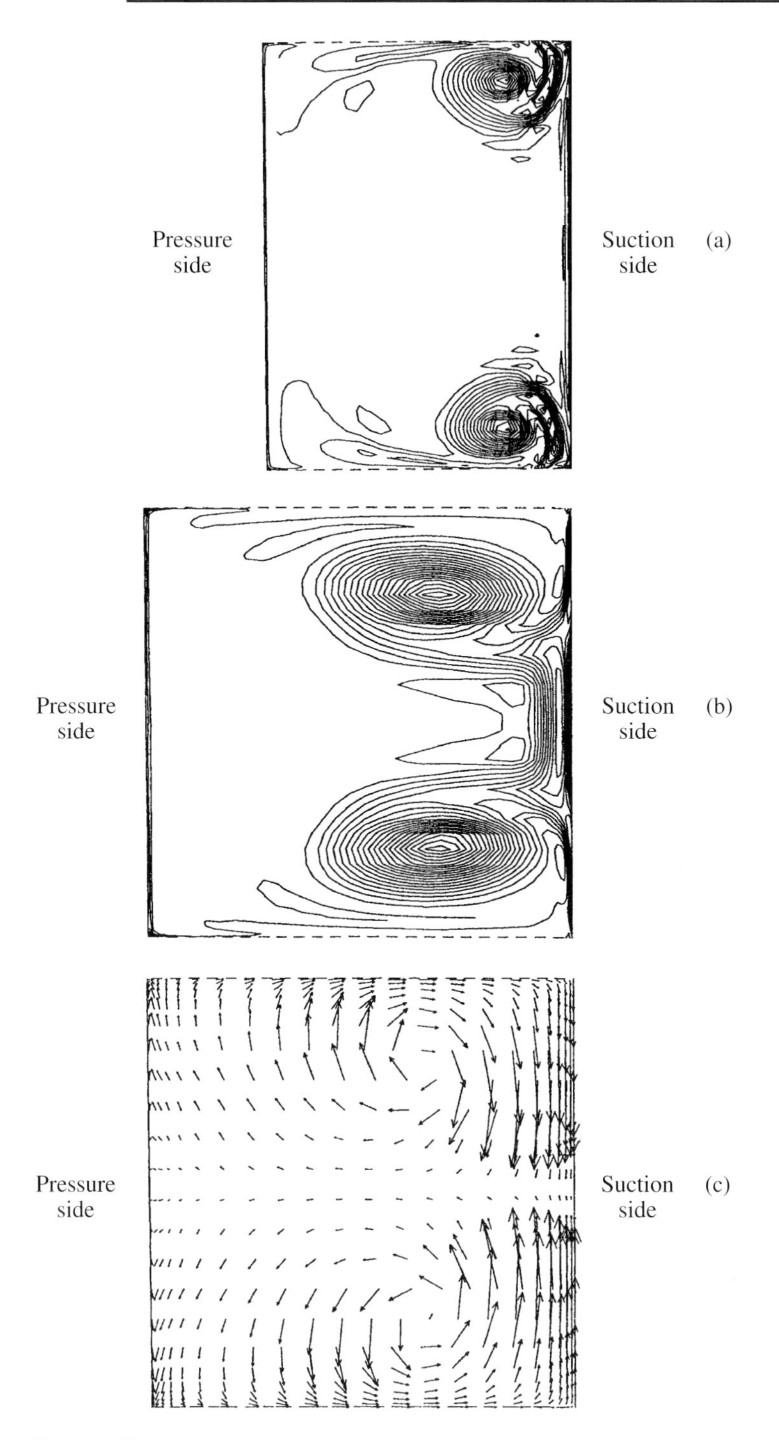

Figure 9.14: Secondary flow in a turbine cascade with aspect ratio 1.0: (a) stagnation pressure contours at midchord (contour intervals approximately $0.02\frac{1}{2}\rho\overline{u}_e^2$, $\overline{u}_e$ is mean exit velocity); (b) stagnation pressure contours at the trailing edge; (c) cross-flow plane velocity vectors at the trailing edge; inviscid computation of Denton (2000).

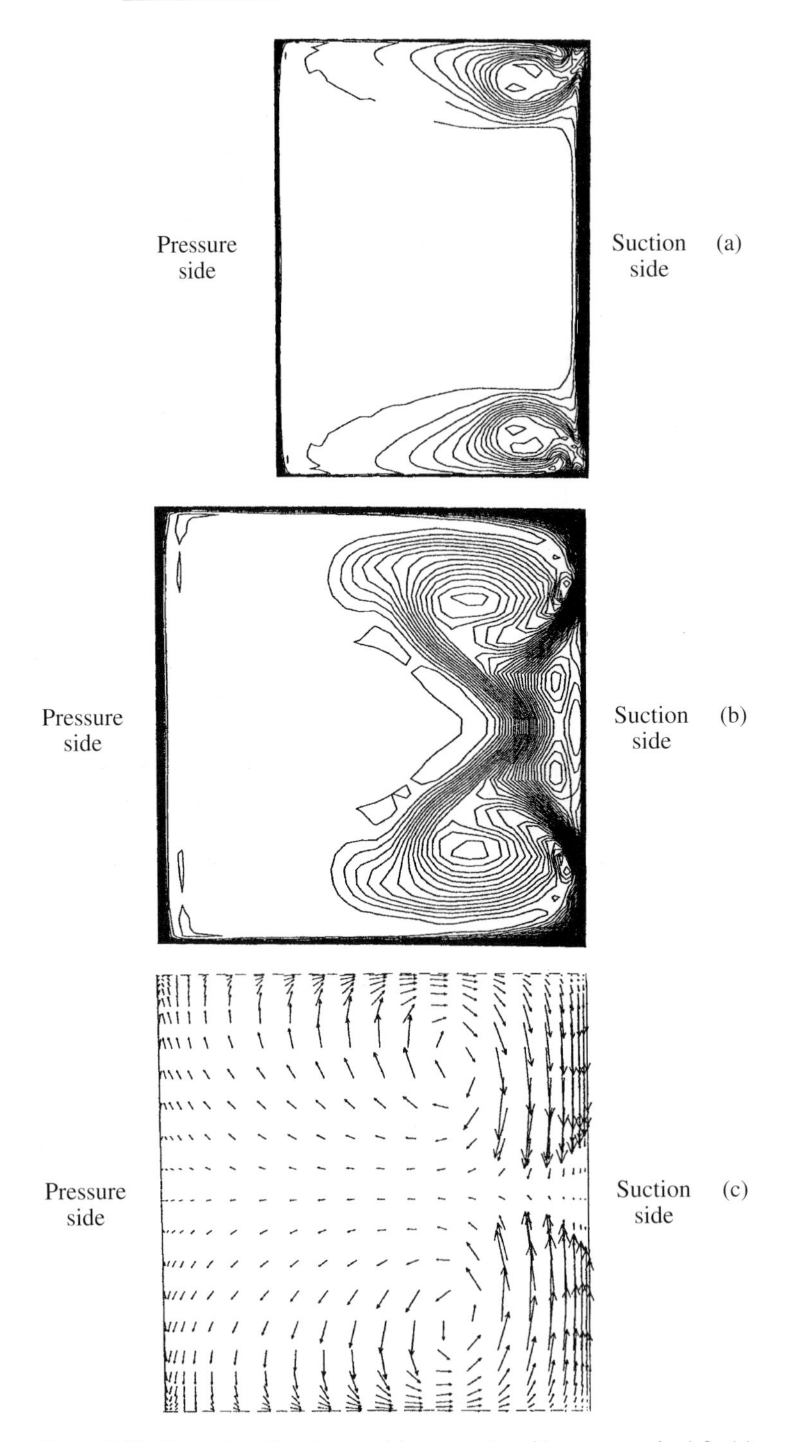

Figure 9.15: Secondary flow in a turbine cascade with aspect ratio 1.0: (a) stagnation pressure contours at midchord (contour intervals approximately $0.02\frac{1}{2}\rho\overline{u}_e^2$, $\overline{u}_e$ is mean exit velocity); (b) stagnation pressure contours at the trailing edge; (c) cross-flow plane velocity vectors at the trailing edge; viscous computation of Denton (2000).

Figures 9.14(a)–(c) and 9.15(a)–(c) show results from three-dimensional computations of flow through a turbine blade row with geometry (aspect ratio of 1.0 and turning angle of 110°) roughly similar to that in Figure 9.13 (Denton, 2000). The inlet boundary layer thickness is 15% of the half-height. Stagnation pressure contours are presented for stations at 0.5 axial chord and at the trailing edge. Cross-flow plane velocity vectors at the trailing edge are also shown. The results in Figure 9.14 are from an inviscid computation and those in Figure 9.15 from a viscous computation. The overall features, stagnation pressure defect and cross-flow plane velocity pattern, for example, are roughly similar in the two cases, although the viscous solution shows an increased region of low stagnation pressure near midspan on the suction side. The inference therefore is that much of the exit flow field structure is associated with the distortion of the inlet vorticity through the passage in a predominantly inviscid manner. Detailed comparisons of computations and measurements for this type of flow are given by Denton (1993) and Harrison (1990).

9.4 Expressions for the growth of secondary circulation in an inviscid flow

9.4.1 Incompressible uniform density fluid

A metric for the importance of secondary flow is the *secondary circulation*, the circulation in a plane normal to the velocity, about a small streamtube. If the volume flux through the streamtube is udA, where u is the magnitude of the velocity and dA is the elementary area of the tube, the secondary circulation is given by $\omega_s dA$, where ω_s is the streamwise component of the vorticity, and the ratio ω_s/u $(=(\boldsymbol{\omega}\cdot\mathbf{u})/u^2)$ is a measure of the strength of the secondary motions. This quantity, made dimensionless with an appropriate reference length, identifies regions with strong secondary flow. In the following we develop expressions for the rate of change of ω_s/u along a streamline in a uniform density inviscid flow (Hawthorne, 1951; Horlock and Lakshminarayana, 1973a).

We split the vorticity into components parallel ($\boldsymbol{\omega}_s$) and normal ($\boldsymbol{\omega}_n$) to the velocity, respectively,

$$\boldsymbol{\omega} = \boldsymbol{\omega}_s + \boldsymbol{\omega}_n = \left(\frac{\boldsymbol{\omega}\cdot\mathbf{u}}{u^2}\right)\mathbf{u} + \frac{(\mathbf{u}\times\boldsymbol{\omega})\times\mathbf{u}}{u^2}. \tag{9.4.1}$$

The vorticity field is solenoidal (i.e. $\boldsymbol{\nabla}\cdot\boldsymbol{\omega}=0$), so

$$\boldsymbol{\nabla}\cdot\left(\frac{\omega_s}{u}\right)\mathbf{u} - \boldsymbol{\nabla}\cdot\left(\frac{\mathbf{u}\times(\mathbf{u}\times\boldsymbol{\omega})}{u^2}\right) = 0. \tag{9.4.2}$$

Expanding (9.4.2),

$$\left(\frac{\omega_s}{u}\right)\nabla\cdot\mathbf{u} + \mathbf{u}\cdot\boldsymbol{\nabla}\left(\frac{\omega_s}{u}\right) - \mathbf{u}\times(\mathbf{u}\times\boldsymbol{\omega})\cdot\boldsymbol{\nabla}\left(\frac{1}{u^2}\right)$$
$$-\left(\frac{1}{u^2}\right)\boldsymbol{\nabla}\cdot[\mathbf{u}\times(\mathbf{u}\times\boldsymbol{\omega})] = 0. \tag{9.4.3}$$

A vector identity which allows simplification of the last term in (9.4.3) is

$$\boldsymbol{\nabla}\cdot[\mathbf{u}\times(\mathbf{u}\times\boldsymbol{\omega})] \equiv (\mathbf{u}\times\boldsymbol{\omega})\cdot\boldsymbol{\nabla}\times\mathbf{u} - \mathbf{u}\cdot\boldsymbol{\nabla}\times(\mathbf{u}\times\boldsymbol{\omega}). \tag{9.4.4}$$

The first term on the right-hand side of (9.4.4) contains two identical vectors, $\boldsymbol{\omega}$ and $\nabla \times \mathbf{u}$, and so is zero. The second term contains the expression $\nabla \times (\mathbf{u} \times \boldsymbol{\omega})$. The Crocco form of the momentum equation is (3.14.6)

$$\mathbf{u} \times \boldsymbol{\omega} = \frac{\nabla p_t}{\rho}. \tag{3.14.6}$$

Hence

$$\nabla \times (\mathbf{u} \times \boldsymbol{\omega}) = \frac{\nabla \times \nabla p_t}{\rho} \equiv 0. \tag{9.4.5}$$

The entire right-hand side of (9.4.4) is therefore equal to zero.

For a constant density fluid, $\nabla \cdot \mathbf{u} = 0$, and the first and last terms in (9.4.3) are zero, leaving

$$\mathbf{u} \cdot \nabla\left(\frac{\omega_s}{u}\right) = -\frac{\mathbf{u} \times (\mathbf{u} \times \omega) \cdot \nabla(u^2)}{u^4}. \tag{9.4.6}$$

Another vector identity we now make use of is

$$\nabla(u^2) \equiv 2\,(\mathbf{u} \cdot \nabla)\,\mathbf{u} + 2\mathbf{u} \times (\nabla \times \mathbf{u})\,. \tag{9.4.7}$$

Substituting (9.4.7) in (9.4.6) yields

$$\mathbf{u} \cdot \nabla\left(\frac{\omega_s}{u}\right) = -\frac{\mathbf{u} \times (\mathbf{u} \times \boldsymbol{\omega}) \cdot 2\,(\mathbf{u} \cdot \nabla)\,\mathbf{u}}{u^4}. \tag{9.4.8}$$

Equation (9.4.8) is an expression for the rate of change of the secondary circulation along a streamline.[4] The expression on the right-hand side is proportional to the product of the velocity, the vector $(\mathbf{u} \times \boldsymbol{\omega})$ which is normal to the surfaces containing the streamlines and the vortex lines (known as Bernoulli surfaces), and the acceleration, $(\mathbf{u} \cdot \nabla)\mathbf{u}$.

The interpretation of this triple product is more easily visualized using natural coordinates, l, n, b, which refer to the direction of the streamline, the outward direction of its principal normal, and the bi-normal. Figure 9.16 shows a portion of a streamline with unit vectors $\mathbf{l}$, $\mathbf{n}$, $\mathbf{b}$. The distance along the streamline is l, distance n is measured outwards along the direction of the principal normal, and distance b is measured in the bi-normal direction. The streamline lies locally in the plane defined by $\mathbf{l}$ and $\mathbf{n}$, and the vector representing the particle acceleration thus also lies in this plane. The acceleration can be resolved into two components, one along the streamline, with value $u(\partial u/\partial l)$, and the other normal to it (in the $\mathbf{n}$-direction) with the value $(-u^2/r_c)$, where r_c is the local radius of curvature of the streamline.

Employing the l, n, b coordinate system, and making use of (3.14.6), (9.4.8) is written compactly as

$$\frac{\partial}{\partial l}\left(\frac{\omega_s}{u}\right) = -\frac{2}{\rho u^2 r_c}\frac{\partial p_t}{\partial b}. \tag{9.4.9}$$

[4] A form of (9.4.8) that relates streamwise vorticity generation to gradients in the velocity magnitude along the existing vortex lines is $\mathbf{u} \cdot \nabla\,(\boldsymbol{\omega} \cdot \boldsymbol{u}) = \boldsymbol{\omega} \cdot \nabla(u^2)$.

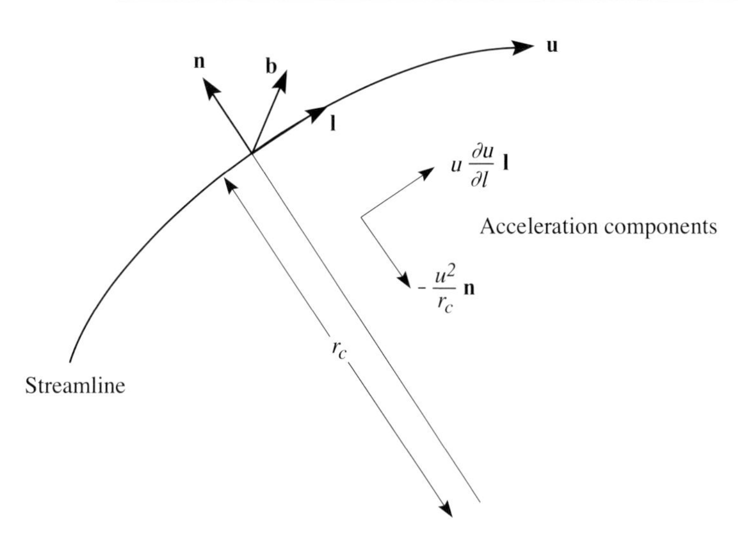

Figure 9.16: Natural coordinates for a primary flow streamline; **l**, **n**, **b** are vectors in the streamwise, normal, and bi-normal directions, **b** is out of page; $\mathbf{n} \times \mathbf{b} = \mathbf{l}$.

Equation (9.4.9) can be integrated along the streamline to give an expression for the change in secondary circulation between two locations as

$$\left(\frac{\omega_s}{u}\right)_2 - \left(\frac{\omega_s}{u}\right)_1 = -\int_1^2 \frac{2}{\rho u^2}\frac{\partial p_t}{\partial b}\, d\theta. \tag{9.4.10}$$

In (9.4.10) the integral is taken along a streamline with $d\theta$ $(= dl/r_c)$ the angle between tangents to the streamline at points arc length dl apart.

9.4.2 Incompressible non-uniform density fluid

Equations (9.4.8) and (9.4.9) have been derived for uniform density flow, but similar procedures can be carried out for an inviscid, incompressible, non-uniform density fluid, as well as (in the next section) a perfect gas (Hawthorne, 1974). In the absence of external body forces the expression for the rate of change of secondary circulation in an incompressible, non-uniform density flow is the same as for uniform density. However, stagnation pressure gradients in the bi-normal direction, which lead to production of streamwise vorticity, can now arise from gradients in density as well as velocity.

For uniform density flow the component of the stagnation pressure gradient in the binormal direction $\partial p_t/\partial b$ is

$$\frac{\partial p_t}{\partial b} = \frac{\partial p}{\partial b} + \rho u\frac{\partial u}{\partial b} = \rho u\frac{\partial u}{\partial b}. \tag{9.4.11}$$

There is no gradient of static pressure in the bi-normal direction because there is no acceleration in that direction. For an incompressible, non-uniform density flow the stagnation pressure gradient in the bi-normal direction is

$$\frac{\partial p_t}{\partial b} = \frac{u^2}{2}\frac{\partial \rho}{\partial b} + \rho u\frac{\partial u}{\partial b}. \tag{9.4.12}$$

Equation (9.4.12) shows that the effects of velocity and density non-uniformities can cancel or reinforce in creating secondary flow. Comparison of (9.4.11) and (9.4.12) also shows it is the gradient of ρu^2 in the bi-normal direction that gives rise to secondary circulation. Flows with the same distribution of ρu^2 have the same secondary circulation, and (as will be described in Chapter 10) the same streamline pattern.

The interplay between generation of secondary vorticity by differential convection of existing vorticity and production of vorticity by the pressure torque exerted on particles of non-uniform density (Section 3.5) is exhibited in a flow with constant stagnation pressure but non-constant density, i.e. inlet velocity varying such that ρu^2 is constant. For this situation, (9.4.12) implies no streamwise vorticity generation. Although there is normal vorticity at the inlet and initially normal vortex filaments are tipped into the streamwise direction, there is equal and opposite production of streamwise vorticity due to the $\nabla p \times \nabla \rho$ term and no net streamwise vorticity is produced. Further, whatever the distribution of ρ and u^2, if the value of ρu^2 is the same on corresponding streamlines for two flow fields, the same pressure distribution will maintain the two streamline patterns.

One flow corresponding to uniform ρu^2 at the inlet is a uniform density, uniform inlet velocity, irrotational flow. All flow fields with uniform ρu^2 at the inlet must therefore have no *streamwise* vorticity whatever the value of ρ or u because if this were not so the streamlines would not be the same. This cannot be said about the *normal* vorticity because there are an infinity of shear flows with constant stagnation pressure, the same streamlines, and varying density and normal vorticity distributions. In summary, for an inviscid incompressible stratified flow with no body forces, the streamline pattern is invariant with respect to the density distribution for a given inlet stagnation pressure distribution and given geometry.

9.4.3 Perfect gas with constant specific heats

For a perfect gas the expression for the rate of change of secondary circulation is (Hawthorne, 1974; Horlock and Lakshminarayana, 1973a)

$$\frac{\partial}{\partial l}\left(\frac{\omega_s}{\rho u}\right) = -\frac{2}{\rho \rho_t u^2 r_c}\frac{\partial p_t}{\partial b}. \tag{9.4.13}$$

Integration along a streamline yields

$$\left(\frac{\omega_s}{\rho u}\right)_2 - \left(\frac{\omega_s}{\rho u}\right)_1 = -\int_1^2 \frac{2}{\rho \rho_t u^2}\frac{\partial p_t}{\partial b}\, d\theta. \tag{9.4.14}$$

Equation (9.4.14) shows that the growth of streamwise circulation does not depend on the stagnation temperature distribution because only the gradient of stagnation pressure appears. Flows with only stagnation temperature gradients do not give rise to *streamwise* vorticity, whatever the distribution of *normal* vorticity. As with incompressible stratified flow this is a case of the cancellation of two equal and opposite processes for producing streamwise vorticity.

A circumstance in which this can occur is the flow of a thermally stratified fluid from a combustion chamber, say through a nozzle or a row of turbine nozzle guide vanes. Suppose the stagnation temperature is non-uniform but the stagnation pressure is approximately constant in the bi-normal direction. Equation (9.4.13) shows that no streamwise component of vorticity and no secondary circulation are generated under these conditions. There is, however, normal vorticity whose magnitude can be found from Crocco's Theorem:

$$(\mathbf{u} \times \boldsymbol{\omega}) = \nabla h_t - T\nabla s. \tag{3.14.4}$$

For constant stagnation pressure the entropy gradient is related to the stagnation enthalpy gradient by

$$T_t \nabla s = \nabla h_t. \tag{9.4.15}$$

For constant stagnation pressure (3.14.4) thus takes the form

$$\mathbf{u} \times \boldsymbol{\omega} = \frac{u^2}{2}\frac{\nabla T_t}{T_t}. \tag{9.4.16}$$

Since the vorticity is normal to the velocity, the magnitude of the vorticity is

$$\omega = \frac{u}{2}\left|\frac{\nabla T_t}{T_t}\right|. \tag{9.4.17}$$

In Chapter 10 we will see that the conclusion concerning the lack of dependence of secondary circulation on the stagnation temperature field is a special case of a more general principle concerning flow field invariance under changes in the stagnation temperature distribution.

9.5 Applications of secondary flow analyses

9.5.1 Approximations based on convection of vorticity by a primary flow

For a uniform density inviscid fluid, the growth of secondary circulation is described by (9.4.9):

$$\frac{\partial}{\partial l}\left(\frac{\omega_s}{u}\right) = -\frac{2}{\rho u^2 r_c}\frac{\partial p_t}{\partial b}. \tag{9.4.9}$$

This equation is nonlinear because the velocity field (which includes the directions in which the intrinsic coordinates l, n, and b are defined) and the stagnation pressure distribution are not known *a priori*. We can, however, make some assumptions about the behavior of the terms on the right-hand side which allow us to derive useful information from (9.4.9). In the first application we examine flow in a bend and take the vorticity associated with the secondary flow as convected along the streamlines of a background irrotational, or primary, flow. This approximation amounts to viewing the distortion of the (presumed weak) "secondary" vorticity field (and stagnation pressure non-uniformity) by a known primary flow and considering only the effects of the latter on the former.

A further approximation can be made if the bend has a radius large compared to the width so that the quantities r_c and u vary little in the bend. If so, we may take u and r_c as equal to the mean velocity across the bend ($\overline{u}$) and the mid radius (r_m) respectively and integrate (9.4.9) to give

$$\omega_s - \omega_{s_i} = -\frac{2\Delta l}{\rho \overline{u} r_m}\frac{\partial p_t}{\partial b}, \tag{9.5.1}$$

where the streamwise distance, l, is measured from the inlet station. Since $\Delta l/r_m$ is the total angle of turning of the streamline, $\Delta\theta$,

$$\omega_s - \omega_{s_i} = -\frac{2\Delta\theta}{\rho\overline{u}}\frac{\partial p_t}{\partial b}. \tag{9.5.2}$$

The normal component of vorticity at the inlet is related to the stagnation pressure derivative by the incompressible form of Crocco's Theorem (3.14.6), which can be written in terms of l, n, b coordinates as

$$\rho u\omega_n = \frac{\partial p_t}{\partial b}. \tag{9.5.3}$$

Using (9.5.3) and taking the velocity as $\overline{u}$, the expression for the streamwise vorticity generated within a bend of turning angle $\Delta\theta$ reduces to the Squire–Winter result, now extended to include an inlet component of streamwise vorticity ω_{s_i}:

$$\omega_s - \omega_{s_i} = -2\Delta\theta\omega_{n_i}. \tag{9.5.4}$$

With only normal vorticity at the inlet, (9.5.4) takes the form given previously as (9.2.2):

$$\omega_s = -2\Delta\theta\omega_{n_i}. \tag{9.2.2}$$

Flow in a bend illustrates the steps involved in the use of secondary flow approximations. The approach can be used for other geometries, with the secondary vorticity obtained by numerically integrating along known primary flow streamlines. A more complex application occurs in the "inlet-vortex" phenomenon (the generation of an intense vortex when an inlet, such as on an aeroengine, is run in proximity to the ground), where the primary flow is three-dimensional (De Siervi *et al.*, 1982). In that example the secondary flow approach enabled identification of a mechanism for vortex formation. This approximation is also used effectively in the rapid distortion theory of turbulence (i.e. Hunt (1987)).

9.5.2 Flow with large distortion of the stream surfaces

In many instances the distortion of the primary stream surfaces associated with streamwise vorticity cannot be neglected. To obtain insights into these types of flows we apply (9.4.9) with a different set of approximations. Flow in a bent pipe with a circular cross-section is examined first, because this allows views of a striking set of phenomena.

The pipe diameter is denoted by d, and the mean bend radius of curvature by r_m. The stagnation pressure at the inlet is specified as varying linearly across the pipe, so the magnitude of the stagnation pressure gradient is $\Delta p_t/d$. This magnitude is assumed to remain constant and the flow to rotate as a solid body about the pipe centerline, a major difference from the assumptions of Section 9.5.1. Figure 9.17 shows the bend and flow geometry and the nomenclature used. The angle, ϕ, which the stagnation pressure gradient makes with the bi-normal to the pipe centerline varies with the angular distance round the bend, θ. The component of the stagnation pressure gradient in the bi-normal direction is

$$\frac{\partial p_t}{\partial b} = \frac{\Delta p_t}{d}\cos\phi. \tag{9.5.5}$$

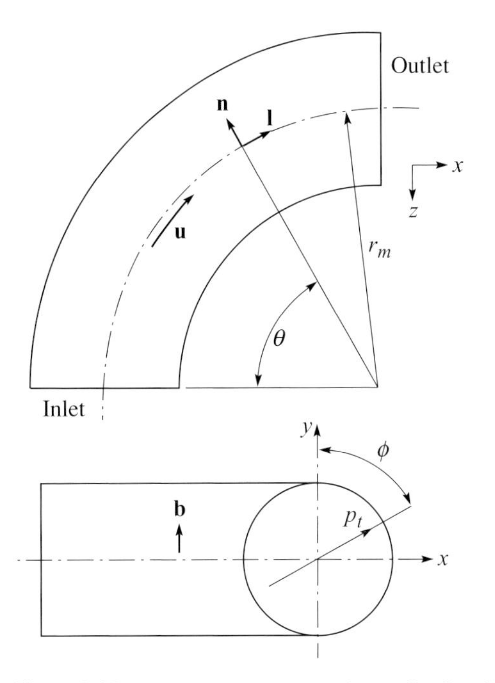

Figure 9.17: Geometry and nomenclature for flow in a curved circular pipe.

Under the assumptions described and with $\overline{u}$ denoting the mean velocity in the l-direction, (9.4.9) becomes

$$\frac{\partial}{\partial l}\left(\frac{\omega_s}{u}\right) \cong -2\frac{\Delta p_t}{d}\frac{1}{\rho \overline{u}^2}\frac{\cos\phi}{r_m}. \tag{9.5.6}$$

To integrate (9.5.6), we relate the streamwise vorticity to the orientation angle, ϕ, of the lines of constant stagnation pressure:

$$\omega_s = -2\frac{d\phi}{dt}. \tag{9.5.7}$$

The time increment, dt, is equal to $dl/\overline{u}$, or, in terms of the angle of turn in the bend, θ,

$$dt = r_m\frac{d\theta}{\overline{u}} = \frac{dl}{\overline{u}}. \tag{9.5.8}$$

Substituting (9.5.8) and (9.5.7) into (9.5.6) yields a second order ordinary differential equation for the inclination of the surfaces of constant stagnation pressure (Bernoulli surfaces):

$$\frac{d^2\phi}{d\theta^2} = \frac{\Delta p_t}{\rho \overline{u}^2}\frac{r_m}{d}\cos\phi. \tag{9.5.9}$$

For values of ϕ near $\pi/2$, i.e. for orientations in which the highest stagnation pressure fluid is near the outside of the bend, (9.5.9) reduces to the linearized form

$$\frac{d^2\phi}{d\theta^2} + \frac{\Delta p_t}{\rho \overline{u}^2}\frac{r_m}{d}\left(\phi - \frac{\pi}{2}\right) = 0. \tag{9.5.10}$$

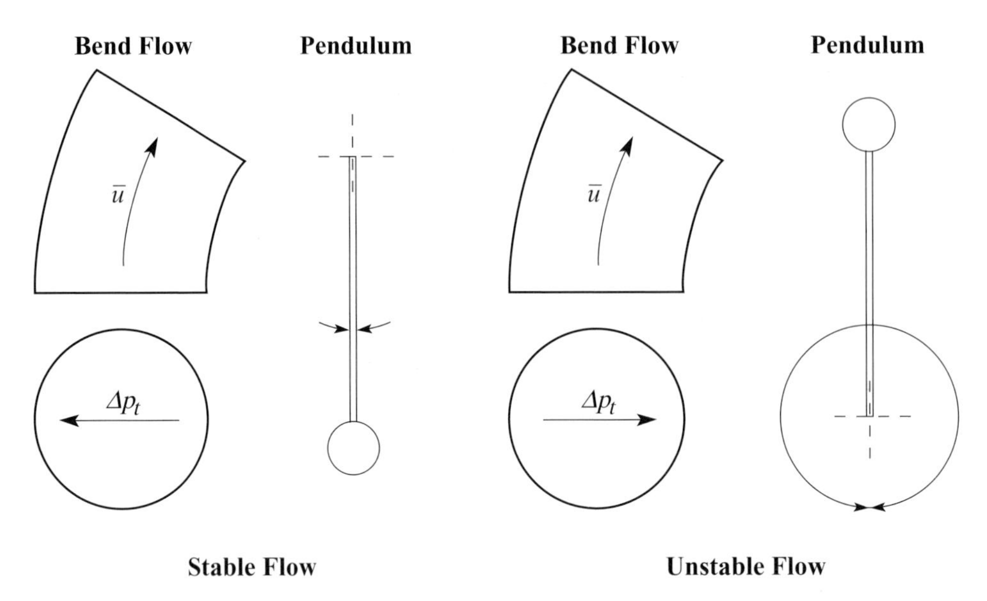

Figure 9.18: Pendulum analogy for secondary flow in a bend (Johnson, 1978).

This is the equation for simple harmonic motion about $\phi = \pi/2$ with period $2\pi\sqrt{(\rho\bar{u}^2/\Delta p_t)(d/r_m)}$.

The character of the solution to (9.5.9) can be seen by noting that this equation has the same form as the equation for the oscillations of a pendulum of length, L, about an equilibrium point:

$$\frac{L}{g}\frac{d^2\phi}{dt^2} = \cos\phi.$$

The analogy is depicted in Figure 9.18. Having the highest stagnation pressure on the *outside* of the bend is analogous to the pendulum hanging vertically downwards. If the flow is disturbed slightly from this configuration it will oscillate. Having the highest stagnation pressure fluid on the *inside* of the bend is like having the pendulum vertically above the point of suspension. Other orientations of the highest stagnation pressure fluid at the inlet to the bend result in oscillations with amplitude equal to the distance from the outside of the bend, with the period (in θ) of the oscillations $2\pi\sqrt{(\rho\bar{u}^2/\Delta p_t)(d/r_m)}$. The kinetic energy of the secondary circulation, which is analogous to the kinetic energy of the pendulum, oscillates with the same period, passing through zero after each $\pi\sqrt{(\rho\bar{u}^2/\Delta p_t)(d/r_m)}$ radians of turn.

Experimental results for flows in a 180° circular bend are given in Figure 9.19 (Hawthorne, 1951). The flow enters the bend with ϕ equal to zero, so the gradient of stagnation pressure is vertical in the plane of the paper. By the 60° location, the flow has overshot the equilibrium point and a reverse rotation is set up. In terms of the development of the streamwise vorticity, the equilibrium point with the highest stagnation pressure on the outside of the bend has zero value of $\partial p_t/\partial b$ so that the sign of the gradient, and thus the sign of the rate of change of streamwise vorticity generation, changes as the angle θ moves past this point.

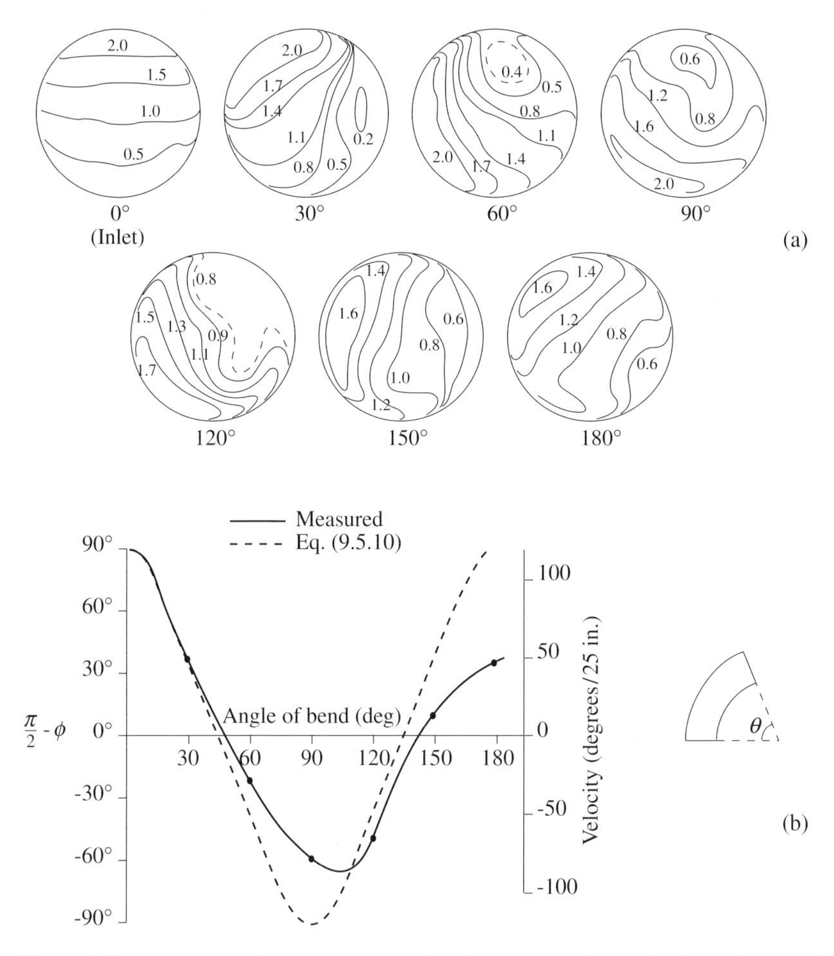

Figure 9.19: Experimental measurements of station pressure in a 180° bent circular pipe ($r_m/d = 5$): (a) stagnation pressure contours at 30° intervals around the bend (pressure measured in inches of water); (b) rotation of the maximum stagnation pressure location (Hawthorne, 1951).

9.6 Three-dimensional boundary layers: further remarks on effects of viscosity in secondary flow

Three-dimensional boundary layers can also be described within the general framework of secondary flow theory, and doing so gives a view into how viscosity affects the motion. The discussion here follows Lighthill (1963), and applies to a boundary layer which is not near separation, bounded by an irrotational free stream. The velocity gradients normal to the surface are much larger than velocity gradients along the surface. For a flat surface which coincides with the plane $y = 0$, the vorticity vector in the boundary layer is, to a very good approximation,

$$\omega = \frac{\partial u_z}{\partial y}\mathbf{i} - \frac{\partial u_x}{\partial y}\mathbf{k}, \tag{9.6.1}$$

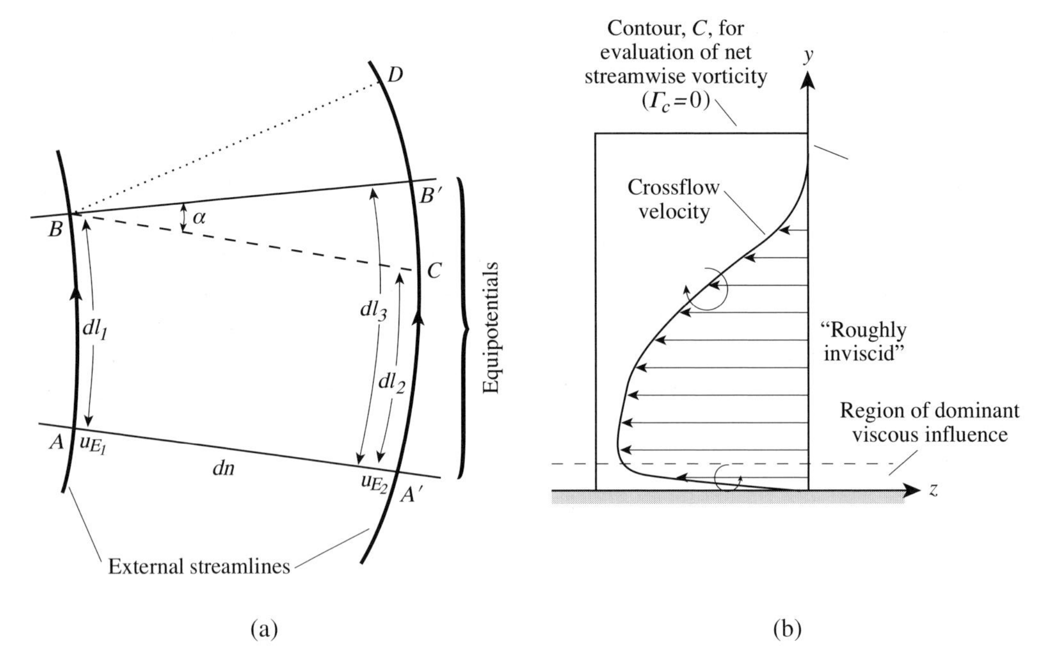

Figure 9.20: Cross-flow in three-dimensional boundary layers: (a) external flow streamlines and equipotential lines for flow outside of a three-dimensional boundary layer; (b) cross-flow in a three-dimensional boundary layer (Lighthill, 1963).

where **i** and **k** are the unit vectors in the x- and z-directions respectively. Integrating (9.6.1) from the solid surface, $y = 0$, to $y = \delta$, the edge of the boundary layer, yields

$$\int_0^\delta \omega dy = -\mathbf{u}_E \times \mathbf{j}, \tag{9.6.2}$$

where $\mathbf{u}_E$ is the free-stream velocity just outside the boundary layers and **j** is the unit vector normal to the surface (in the y-direction).

Equation (9.6.2) indicates that the mean vortex lines (the average across the layer) are perpendicular to the streamlines just outside the boundary layer. In an irrotational flow, streamlines and lines of constant velocity potential are orthogonal, so the mean vortex lines lie along the equipotentials of the external flow just outside the boundary layer. Two streamlines of the external flow, separated by a small distance dn, and two equipotentials, separated by streamwise distances dl_1, dl_3 of the same (small) magnitude, are indicated in Figure 9.20. The radius of curvature of streamline 1 is denoted as r_c, and the free-stream velocities on the two streamlines as u_{E_1} and u_{E_2}. As drawn, the center of curvature is to the left so that u_{E_2} is smaller than u_{E_1}. Because lines AA' and BB' are equipotentials, the lengths AB ($= dl_1$) and $A'B'$ ($= dl_3$) are related by $dl_1 u_{E_1} = dl_3 u_{E_2}$. In addition, the flow outside the boundary layer is irrotational and $\omega_\perp = 0$. From (9.2.1) therefore

$$\frac{\partial u}{\partial n} = -\frac{u}{r_c}. \tag{9.6.3}$$

To first order in dn/r_c, the velocity magnitude on the two streamlines is related by

$$u_{E_2} = u_{E_1}\left(1 - \frac{dn}{r_c}\right). \tag{9.6.4}$$

Consider convection of vortex lines within the boundary layer. Suppose the mean vortex line initially coincides with the equipotential line AA'. As described in Section 3.13, the average convection velocity of boundary layer vorticity is $u_E/2$. If u_{E_1} is the local free-stream velocity at point A and u_{E_2} the local free-stream velocity at point A', after a time dt the mean vortex line will be at position BC where the lengths dl_1 $(= AB)$ and dl_2 $(= A'C)$ are equal to $u_{E_1}dt/2$ and $u_{E_2}dt/2$ respectively. The mean vortex line thus makes an angle α with the equipotential BB', where α is given by $(dl_3 - dl_2)/dn$. From the condition for AA' and BB' to be equipotentials, the relation between dl_3 and dl_1 is

$$dl_3 = dl_1\left(1 + \frac{dn}{r_c}\right). \tag{9.6.5a}$$

The length dl_2 is given by $dl_2 = dl_1(u_{E_2}/u_{E_1})$ and

$$dl_2 = dl_1\left(1 - \frac{dn}{r_c}\right). \tag{9.6.5b}$$

The angle α in Figure 9.20 is thus found to be

$$\alpha = \frac{2dl_1}{r_c} = \frac{u_{E_1}}{r_c}dt. \tag{9.6.6}$$

This is the angle between the equipotential line BB' and the mean vortex line BC. The total vorticity in the boundary layer is u_E, or u_{E_1} to the order in which we have been working, so the streamwise vorticity at the exit, which has been produced in time dt over the elementary area dl_1dn, is $(u_{E_1}^2/r_c)dl_1dndt$.

The amount of vorticity convected into this area in a time dt is $(u_{E_1}^2/2)dndt$. The ratio of streamwise vorticity produced to vorticity convected in, in other words the fraction of the vorticity tipped into the streamwise direction, is just the ratio of these two quantities:

$$\frac{\text{streamwise vorticity at exiting area}}{\text{normal vorticity convected into area}} = \frac{2dl_1}{r_c}. \tag{9.6.7}$$

Since $dl_1/r_c = d\theta$, the change in angle of the external flow streamlines, (9.6.7), which accounts for the evolution of vortex lines due to inviscid effects, is another form of the Squire–Winter formula for secondary flow (Section 9.2).

We can take these considerations further using the fact that the mean vortex lines (averaged across the boundary layer) are perpendicular to the surface streamlines of the flow external to the boundary layer.[5] This means that during the time dt there must have been diffusion of streamwise vorticity into the fluid, of magnitude $u_{E_1}^2/r_c$ but opposite sign, in the elementary area shown. The new vorticity is close to the surface since it has had relatively little time to diffuse. It has the orientation shown by the dotted line BD in Figure 9.20 so the mean vorticity is lined up with the equipotentials and there is no net streamwise vorticity. This can also be seen by noting that the circulation round a contour perpendicular to the free stream with one leg on the surface and one leg outside the boundary layer

[5] The mean vortex lines that lie along the surface equipotentials of the external irrotational flow (Lighthill, 1963).

must be zero. Although there is no net streamwise vorticity, most of the fluid in the boundary layer typically has streamwise vorticity of one sign, with the vorticity due to the diffusion from the wall confined to a thin layer, as shown schematically in the right-hand side of Figure 9.20.

The vorticity production at the solid surface can be derived using the momentum equation evaluated at the surface. In terms of streamwise and normal components (to the free-stream direction) the relation between the pressure gradient on the wall and the derivative of the vorticity components is

$$\frac{1}{\rho}\frac{\partial p}{\partial l} = -u_{E_1}\frac{\partial u_{E_1}}{\partial l} = -\nu\left(\frac{\partial \omega_n}{\partial y}\right)_{\text{at the surface}}, \tag{9.6.8a}$$

$$\frac{1}{\rho}\frac{\partial p}{\partial n} = +\frac{u_{E_1}^2}{R} = \nu\left(\frac{\partial \omega_s}{\partial y}\right)_{\text{at the surface}}. \tag{9.6.8b}$$

This is a generalization of the expression for two-dimensional flow in Chapter 3.

The condition of zero net streamwise vorticity does not hold for surfaces which have a velocity in a direction normal to the free-stream velocity. For an axial free stream around a spinning body of revolution, for example, the boundary layer does have a net streamwise vorticity with the integral $\int_{r_i}^{r_i+\delta}\omega_s dr$ equal to Ωr_i, where Ω is the rotation rate and r_i is the surface radius. In line with the above discussion, however, the effect of the moving surface is often confined to a layer which is much thinner than the overall region of rotational flow and the secondary flow features can be viewed as primarily inviscid.

9.7 Secondary flow in a rotating reference frame

9.7.1 Absolute vorticity as a measure of secondary circulation

In a rotating frame, the most useful measure of secondary circulation is still the absolute vorticity. One argument for this is as follows. The relative and absolute vorticity are related by (7.5.1):

$$\boldsymbol{\omega} = \boldsymbol{\omega}_{rel} + 2\boldsymbol{\Omega}, \tag{7.5.1}$$

where $\boldsymbol{\Omega}$ is the angular velocity of the relative frame of reference. It is often the case that the rotating flow is irrotational (or nearly so) in the absolute frame at the inlet. If no vorticity is generated in a rotating passage, the relative flow will have a relative vorticity of $-2\boldsymbol{\Omega}$. In the rotating passage with the axial to radial bend sketched in Figure 9.21 the streamwise component of $\boldsymbol{\omega}_{rel}$ at the inlet is $-2\boldsymbol{\Omega}$, while the normal component is zero. At the exit, the streamwise component of $\boldsymbol{\omega}_{rel}$ is zero and the normal component is $-2\boldsymbol{\Omega}$. The change in streamwise component of $\boldsymbol{\omega}_{rel}$ is caused solely by turning of the relative velocity, $\mathbf{w}$, out of the $\boldsymbol{\Omega}$-direction, not by generation of new vorticity.

This effect is present whether or not the motion is irrotational in the absolute frame, as seen from the vorticity equation for a steady constant density inviscid flow (see Section 7.5)

$$\mathbf{w}\cdot\nabla\boldsymbol{\omega}_{rel} = ([\boldsymbol{\omega}_{rel} + 2\boldsymbol{\Omega}]\cdot\nabla)\mathbf{w}. \tag{9.7.1}$$

A change in relative vorticity occurs whenever the alignment between $\mathbf{w}$ and $\boldsymbol{\Omega}$ is changed. The conceptual breakup into primary and secondary flow is helpful when the primary flow is simple and the effect of secondary flow can be thought of as small or confined to a small region. The term $2(\boldsymbol{\Omega}\cdot\nabla)\mathbf{w}$ in the relative vorticity equation satisfies none of these criteria and it is more appropriate

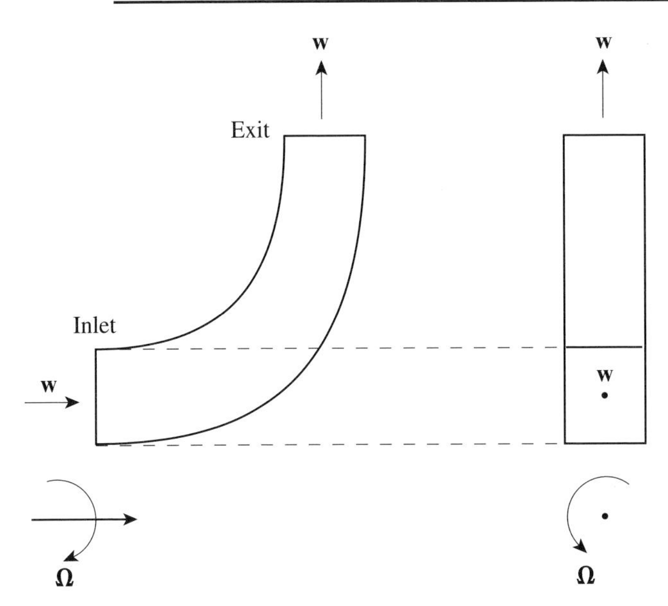

Figure 9.21: Changes in streamwise relative vorticity due to turning of the primary velocity.

to include it in the description of the primary flow. For a flow that is steady in the rotating frame, (9.7.1) can be rewritten in terms of absolute vorticity as

$$\mathbf{w} \cdot \nabla \omega = \omega \cdot \nabla \mathbf{w}, \tag{9.7.2}$$

a statement that *absolute* vorticity is convected by the *relative* flow.

9.7.2 Generation of secondary circulation in a rotating reference frame

Before deriving the general equations for generation of secondary circulation in a rotating frame, it is worth considering the consequences of the fact that the absolute vorticity is convected by the relative flow. We again examine the geometry of Figure 9.2, but now the bend is rotating about an axis perpendicular to the bend floor. The primary flow has relative streamlines which are approximately two-dimensional circular arcs in the rotating bend, and the absolute vorticity at the bend inlet is taken as normal to the relative velocity when the bend is in its initial position.

A small element of a vortex line (aa) initially normal to the streamlines, which is convected a distance dl in the bend, is shown in Figure 9.22. As the fluid cross (aa, bb) moves, the element bb rotates by an amount relative to the *rotating* frame $d\theta_{rel} = dl/r_m$. In the absolute frame the element bb rotates an amount $d\theta_{abs} = d\theta_{rel} + \Omega dt$, where dt is the time interval. The other arm of the cross, aa, which is formed by an element of a vortex line, must be rotated $-d\theta_{abs}$ since the absolute vorticity perpendicular to the floor of the bend is zero. With the assumption that this situation is the same throughout the bend, the Squire–Winter formula for the exit streamwise vorticity, (9.2.2), becomes

$$\begin{aligned}
\omega_s &= -2\Delta\theta_{abs}\omega_n \\
&= -2\left(\Delta\theta_{rel} + \Omega\Delta t\right)\omega_n \\
&= -2\left(\frac{\Delta l}{r_m} + \frac{\Omega\Delta l}{\overline{w}}\right)\omega_n .
\end{aligned} \tag{9.7.3}$$

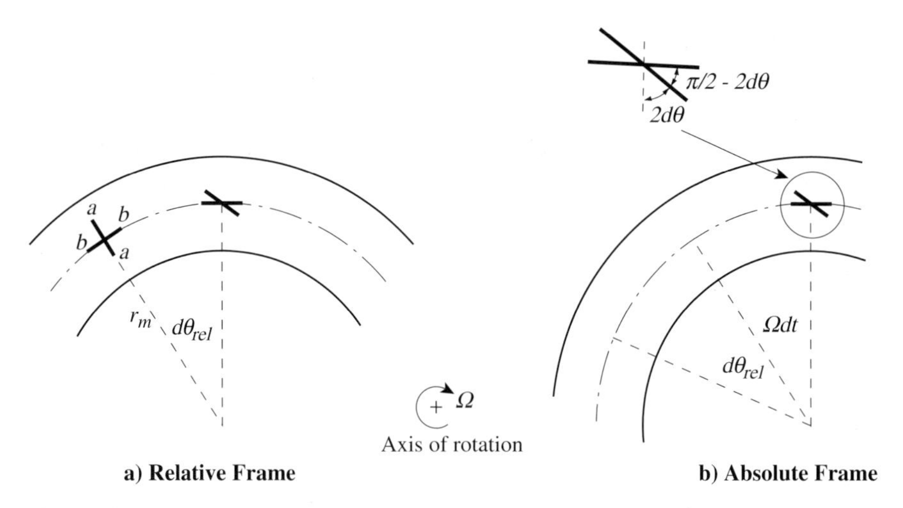

Figure 9.22: Generation of secondary streamwise vorticity in a rotating bend illustrated by fluid cross moving through incremental angle $d\theta$: (a) relative frame; (b) absolute frame (Hynes, 1991).

The movement of the line element seen from the absolute frame is the quantity to examine because it is absolute vorticity which is convected. Equation (9.7.3) indicates that generation of secondary circulation can occur even if the relative streamlines are straight, as is shown by letting $r_m \rightarrow \infty$ (see Chapter 7 for examples). The ratio of the two terms on the right-hand side of (9.7.3) is $\overline{w}/\Omega r_m$, which can be regarded as a Rossby number based on the mean bend radius as length scale. We show in the following section how (9.7.3) can be generalized.

9.7.3 Expressions for, and examples of, secondary circulation in rotating systems

The procedure to develop expressions for the change of absolute vorticity in the rotating coordinate system is similar to that used for the stationary system in Section 9.4. The vorticity is again split into two parts, one along the relative streamline and one normal to it:

$$\boldsymbol{\omega} = \boldsymbol{\omega}_s + \boldsymbol{\omega}_n = \left(\frac{\mathbf{w} \cdot \boldsymbol{\omega}}{\mathbf{w} \cdot \mathbf{w}}\right) \mathbf{w} + \frac{(\mathbf{w} \times \boldsymbol{\omega}) \times \mathbf{w}}{\mathbf{w} \cdot \mathbf{w}}, \tag{9.7.4}$$

and the divergence of the vorticity set to zero. Because we are working with a uniform density fluid, the reduced stagnation pressure, $p_{t_{red}} = p + \rho w^2/2 - \rho(\Omega^2 r^2)/2$, is used rather than the relative stagnation pressure. The Crocco form of the momentum equation for a rotating coordinate system is written in terms of reduced stagnation pressure, relative velocity, and absolute vorticity as

$$\mathbf{w} \times \boldsymbol{\omega} = \frac{\nabla p_{t_{red}}}{\rho}. \tag{9.7.5}$$

After some manipulation (Johnson, 1978; Hawthorne, 1974; Horlock and Lakshminarayana, 1973a) an equation for the rate of change of secondary circulation is obtained. This can be written to highlight the principal effects using a combination of natural and cylindrical coordinates as

$$\frac{\partial}{\partial l}\left(\frac{\omega_s}{w}\right) = \left(\frac{2}{\rho w^2}\right)\left(\underbrace{-\frac{1}{r_c}\frac{\partial p_{t_{red}}}{\partial b}}_{\substack{\text{curvature}\\ \text{term}}} + \underbrace{\frac{\Omega}{w}\frac{\partial p_{t_{red}}}{\partial x}}_{\substack{\text{rotation}\\ \text{term}}}\right). \tag{9.7.6}$$

In (9.7.6) the x-direction is along the axis of rotation consistent with the convention in Chapter 8. Equation (9.7.6) shows that absolute streamwise vorticity is generated in a rotating system whenever flow with reduced stagnation pressure varying in the bi-normal direction is taken round a bend or whenever the reduced stagnation pressure varies along the direction of the rotation axis. Introducing a Rossby number, Ro, as $Ro = w/\Omega r_c$, (9.7.6) shows the explicit dependence on Rossby number as

$$\frac{\partial}{\partial l}\left(\frac{\omega_s}{w}\right) = \left(\frac{2}{\rho w^2 r_c}\right)\left(\underbrace{-\frac{\partial p_{t_{red}}}{\partial b}}_{\text{curvature term}} + \underbrace{\frac{1}{Ro}\frac{\partial p_{t_{red}}}{\partial x}}_{\text{rotation term}}\right). \tag{9.7.7}$$

9.7.3.1 Secondary flow in a rotating straight pipe

For a straight pipe rotating around an axis normal to its length the curvature term is zero. However, gradients of reduced stagnation pressure with components normal to the axis of rotation generate absolute vorticity in the relative streamwise direction. Equations (9.7.6) and (9.7.7) exhibit the physical circumstances in which motions of this type arise.

If the pipe is long enough, the secondary flow may twist the surfaces of constant $p_{t_{red}}$ until the gradient $\partial p_{t_{red}}/\partial x$ changes sign and the flow becomes oscillatory as in the stationary bend. For a linear variation in $p_{t_{red}}$ across the pipe of $\Delta p_{t_{red}}$, analysis similar to that leading to (9.5.10) shows the equation for the angle ϕ, the angle between $\boldsymbol{\nabla} p_{t_{red}}$ and the plane of rotation, is given by

$$d^2\frac{d^2\phi}{dl^2} + \frac{\Omega d}{w}\frac{\Delta p_{t_{red}}}{\rho w^2}\sin\phi = 0. \tag{9.7.8}$$

For small ϕ (i.e. with $p_{t_{red}}$ minimum on the suction side) (9.7.8) can be approximated by

$$d^2\frac{d^2\phi}{dl^2} + \frac{\Omega d}{w}\frac{\Delta p_{t_{red}}}{\rho w^2}\phi = 0. \tag{9.7.9}$$

The angle ϕ undergoes a complete oscillation about $\phi = 0$ in a length of pipe L_c equal to

$$\frac{L_c}{d} = 2\pi\sqrt{\frac{w}{\Omega d}\frac{\rho w^2}{\Delta p_{t_{red}}}}. \tag{9.7.10}$$

The approximate formula for the streamwise vorticity in a rotating pipe, analogous to the Squire–Winter expression for the stationary bend, is obtained by setting $\phi = \pi/2$ in (9.7.8). This corresponds to the situation where $p_{t_{red}}$ is minimum on the pressure side. With l the distance along the pipe, and ω_s taken to be zero at the inlet, the result is

$$\omega_s = 2\left(\frac{\Omega l}{w}\right)\omega_n. \tag{9.7.11}$$

The relative contributions due to curvature and rotation depend on the orientation of the bend with respect to the axis of rotation. When the pipe is bent in the plane of rotation and the bend is away from the direction of rotation (Figure 9.22), the two terms in (9.7.6) are of opposite sign. (If $w/\Omega r_c = 1$, they are equal and opposite, and there is no generation of secondary circulation.) In this condition, the outside of the bend is also the suction side. Curvature results in the movement of high $p_{t_{red}}$ fluid towards the outside of the bend, and rotation produces Coriolis forces that move high $p_{t_{red}}$ fluid towards the pressure side.

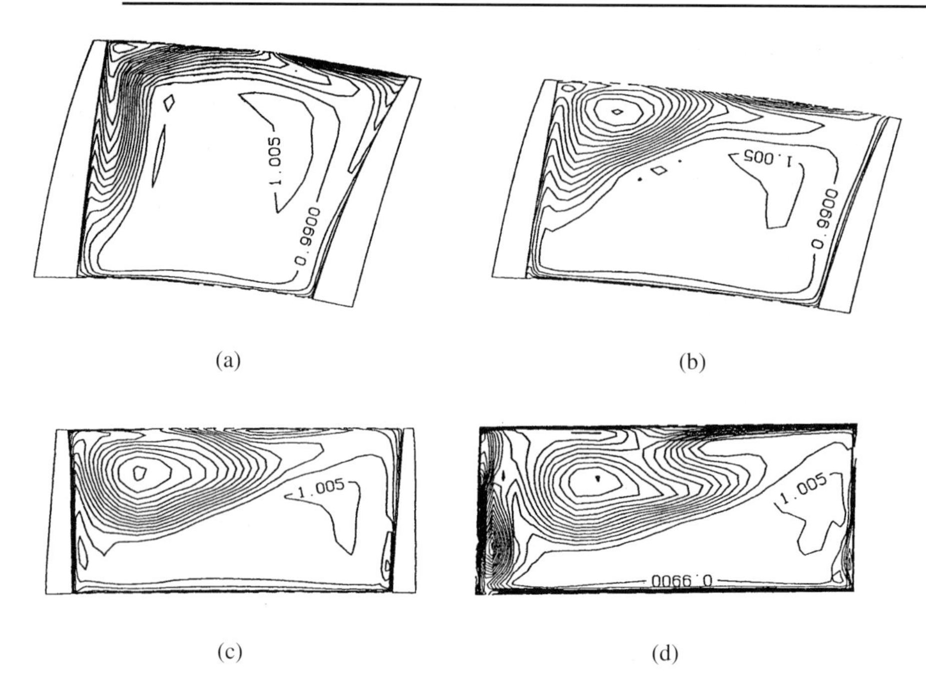

Figure 9.23: Growth of the wake through a centrifugal compressor impeller (shrouded); contours of entropy: (a) about midpassage; (b) about 3/4 of the way through the passage; (c) just before the trailing edge; (d) just after the trailing edge (Denton, 1993).

9.7.3.2 Secondary flow in a rotating axial to radial bend (impeller)

A more complex configuration is an axial to radial bend, as in a compressor impeller. Here the curvature terms can be of the same size throughout, but the rotation terms increase as the flow turns towards the radial. For this configuration impeller, three aspects of the generation of the streamwise vorticity can be identified (Johnson, 1978):

(a) In the inducer bend, the flow is turned from inlet incidence towards the axial direction. For a shrouded impeller, fluid with a low $p_{t_{red}}$ on the shroud or hub tends to be convected towards the suction surface.
(b) In the axial to radial bend, the low $p_{t_{red}}$ on the pressure and suction surfaces tends to move toward the shroud.
(c) In the rotating channel the rotation contribution increases as the flow turns to radial. Low $p_{t_{red}}$ fluid from the hub and shroud wall is moved toward the suction surface.

For many impellers Rossby numbers associated with the inducer bend are less than unity so the effect of passage rotation is more important than the effect of inducer bend on the development of secondary flow. For the axial to radial bend the Rossby number is closer to unity, and both curvature and rotation are significant. The motion described can be seen from the contours of stagnation pressure in a shrouded centrifugal impeller presented in Figure 9.23 at different locations along the flow path (Denton, 1993).

9.7.4 Non-uniform density flow in rotating passages

Although stagnation pressure gradients are necessary for the generation of secondary circulation in a stationary passage, this is not true for a rotating system. The rate of generation of secondary circulation in an incompressible stratified flow with no external body forces is given by (Hawthorne, 1974)

$$\frac{\partial}{\partial l}\left(\frac{\omega_s}{w}\right) = \left(-\frac{2}{\rho w^2 r_c}\right)\left\{\left[\frac{\partial p_{t_{red}}}{\partial b} + \left(\frac{\Omega^2 r^2}{2}\right)\frac{\partial \rho}{\partial b}\right]\right\}$$
$$+\left(\frac{2}{\rho w^2 r_c}\right)\left\{\left(\frac{1}{Ro}\right)\left[\frac{\partial p_{t_{red}}}{\partial x} + \frac{1}{2}(w^2 + \Omega r u_\theta)\frac{\partial \rho}{\partial x} - \frac{\Omega u_x}{2}\frac{\partial \rho}{\partial \theta}\right]\right\}. \quad (9.7.12)$$

For rotating systems even if the reduced stagnation pressure is uniform secondary circulation is produced by density gradients. The same considerations apply to flows of a perfect gas. In a stationary frame gradients in stagnation temperature produce no streamwise vorticity, but this is not true for a rotating frame.

9.8 Secondary flow in rotating machinery

A principal application of secondary flow in rotating systems is to turbomachinery blade rows (Smith, 1955; Horlock and Lakshminarayana, 1973b). In addition to the generation of secondary flow *within* the row, we need to consider changes in entry conditions arising from being in a rotating coordinate system. An illustration is the boundary layer entering an axial or centrifugal compressor rotor from an upstream flow with no swirl. The inlet velocity is axial and the vorticity in the boundary layer has a circumferential component only, as shown in Figure 9.24. The vorticity presented to the rotor can be resolved into inlet normal and streamwise components, both of which need to be accounted for in evaluating secondary flow.

Another example is a constant stagnation pressure, density stratified, flow through an inlet guide vane into a rotor. This might be associated with a variation in combustor exit stagnation temperature fed into a turbine or, in the low Mach number limit, with flow of an incompressible fluid of

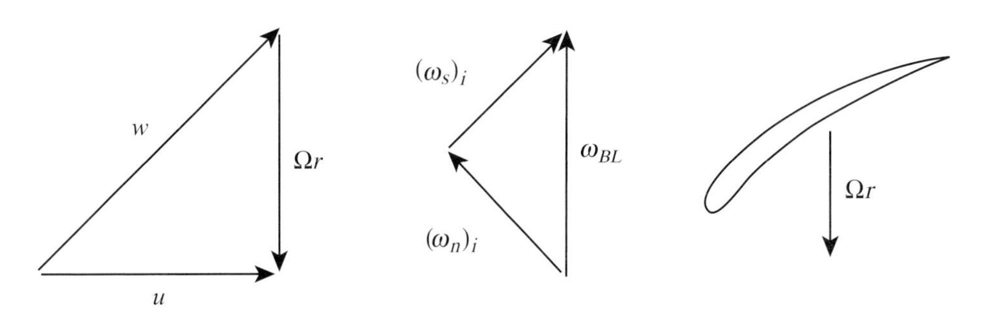

Figure 9.24: Inlet velocity and vorticity triangles for a compressor rotor with axial inlet flow; the exit streamwise vorticity consists of two contributions, one associated with the streamwise component at the inlet and the other from the processing of the normal component by the rotor.

non-uniform density. In the vane passage no secondary circulation is produced (Section 9.4) and the vorticity at the vane exit is normal to the flow.

In the rotating system, at rotor entry, part of this normal vorticity is seen as a component of absolute vorticity in the relative streamwise direction. Differential convection of the absolute vorticity along relative streamlines creates streamwise vorticity within the rotor. Density stratification, or equivalently variations in stagnation temperature, can also lead to the generation of streamwise vorticity in the rotating frame.

A third example is the flow downstream of a vane row with a circulation that varies radially. Following the arguments given in Section 3.14, if there is a radial variation in circulation there will be trailing vorticity from the vane row. For a steady, constant density, inviscid flow with uniform stagnation pressure, the Crocco form of the momentum equation, (3.14.6), becomes

$$\mathbf{u} \times \boldsymbol{\omega} = 0. \tag{9.8.1}$$

Equation (9.8.1) states that the trailing vortex lines downstream of the vane row are parallel to the absolute streamlines. In the relative system, therefore, both normal and streamwise absolute vorticity exist at rotor inlet.

To obtain qualitative features of the secondary circulation in an axial turbomachine we make the approximation that gradients of flow properties in the x (axial) or θ (circumferential) directions are small compared to those in the radial direction and the surfaces of constant $p_{t_{red}}$ and entropy remain on concentric cylinders. The binormal, $\mathbf{b}$, is then in the radial direction. Using this approach expressions for the generation of secondary circulation can be developed to show trends that are explicitly linked to blading design parameters. We consider first constant density flow and then look at the more complex case of a thermal stratification.

For constant density flow, (9.7.12) provides a basic expression relating to the generation of secondary circulation in an axial turbomachine rotor. Using the relation $p_{t_{red}} = p_t - \rho\Omega r u_\theta$ (see Section 7.3), with r_c the radius of curvature of the relative streamlines on the cylindrical surface, we find

$$\frac{\partial}{\partial l}\left(\frac{\omega_s}{w}\right) = -\frac{2}{\rho w^2 r_c}\left[\frac{\partial p_{t_1}}{\partial r} - \rho\Omega\frac{\partial}{\partial r}(r_1 u_{\theta_1})\right]. \tag{9.8.2}$$

Equation (9.8.2) displays the circumstances that lead to the generation of secondary flow in a rotor in terms of the circumferential velocity and stagnation pressure in the absolute coordinate frame.

With uniform inlet absolute stagnation pressure, (9.8.2) becomes

$$\frac{\partial}{\partial l}\left(\frac{\omega_s}{w}\right) = \frac{2\Omega}{w^2 r_c}\frac{\partial}{\partial r}(r u_\theta). \tag{9.8.3}$$

In this situation the absolute vorticity at the inlet is aligned with the absolute flow streamlines.

9.8.1 Radial migration of high temperature fluid in a turbine rotor

An example in which a number of the above effects interact occurs in an axial turbine rotor in which (as is generally the case) there is a radially non-uniform stagnation temperature and the upstream vane has a circulation that varies with radius. In this situation there will be streamwise vorticity at the rotor inlet (vane exit) as well as production of streamwise vorticity within the rotor due to

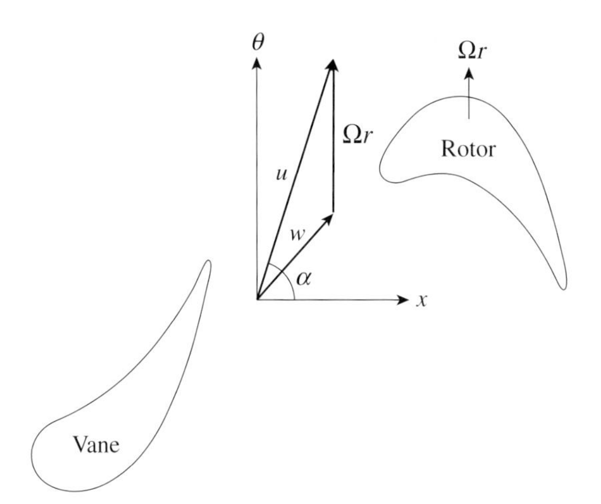

Figure 9.25: Turbine stage geometry and velocity triangle at the vane exit (Prasad and Hendricks, 2000).

both convection of normal vorticity and to density gradients. A problem associated with streamwise circulation is management of the radial transport of hot fluid to keep it away from the blade tip region which can be difficult to cool.

This problem has been analyzed using three-dimensional computations (Prasad and Hendricks, 2000), as well as secondary flow theory to display the features of the solution. The initial turbine configuration had a substantial secondary circulation, in which there was radial transport of hot fluid from the midspan location up the pressure side of the blades. A way to mitigate the radial transport is to create streamwise vorticity of the opposite sign at the rotor inlet to decrease the secondary flow. This can be done by altering the twist of the vane, changing the radial distribution of circulation and hence the inlet vorticity to the turbine.

The analysis given by Prasad and Hendricks (2000) shows that for a vane exit flow with uniform absolute stagnation pressure, starting from (9.7.12) an expression for secondary vorticity in the turbine rotor can be developed in the form

$$\frac{w^2 r_c}{2\Omega}\frac{\partial}{\partial l}\left(\frac{\omega_s}{w}\right) = (u \sin\alpha - \Omega r)\frac{r}{2\rho}\frac{\partial \rho}{\partial r} + u\cos^3\alpha\frac{\partial}{\partial r}(r\tan\alpha), \tag{9.8.4}$$

where α is the flow angle. Equation (9.8.4) shows that generation of secondary vorticity in the rotor passage arises from two effects. The first is the density gradient associated with the streak of hot fluid, which occupies a specified radial extent at the turbine vane exit. The second is vane exit absolute streamwise vorticity associated with the radial gradient of the vane exit angle. This is set by the blade twist, which can be altered to affect the size and sense of the streamwise vorticity.

For the turbine under consideration the quantity $(u \sin\alpha - \Omega r)$ in (9.8.4) is positive (see Figure 9.25 for a sketch of the geometry) and the first term on the right-hand side is of the same sign as the gradient in density. For uniform vane exit flow angle, the second term is also positive, and the two effects thus reinforce each other in the upper part of the hot streak (where $\partial\rho/\partial r > 0$) and oppose each other in the lower part of the streak where the density gradient is negative.

The quantity we seek is the radial velocity. To obtain this, we need to consider the kinematics of the circulatory flow in the rotor associated with the relative eddy (Section 7.9), in addition to

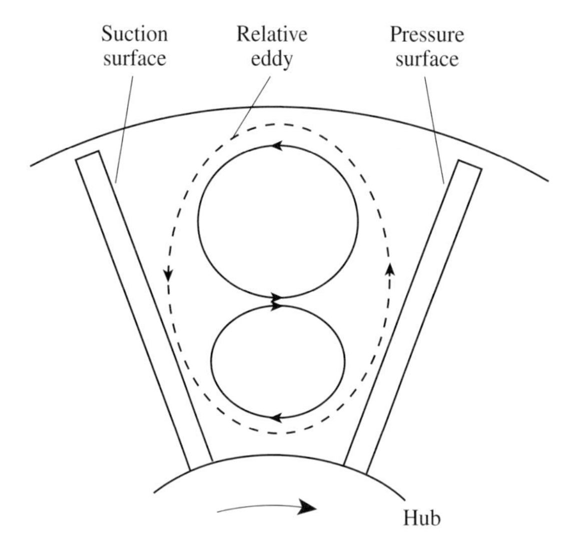

Figure 9.26: Schematic illustration of secondary flow in a turbine rotor passage due to a hot streak. The direction of the primary flow is out of the plane. In the upper part of the blade passage, the radial velocity contribution from the secondary flow (solid line) reinforces that from the relative eddy (broken line) near the pressure surface (Prasad and Hendricks, 2000).

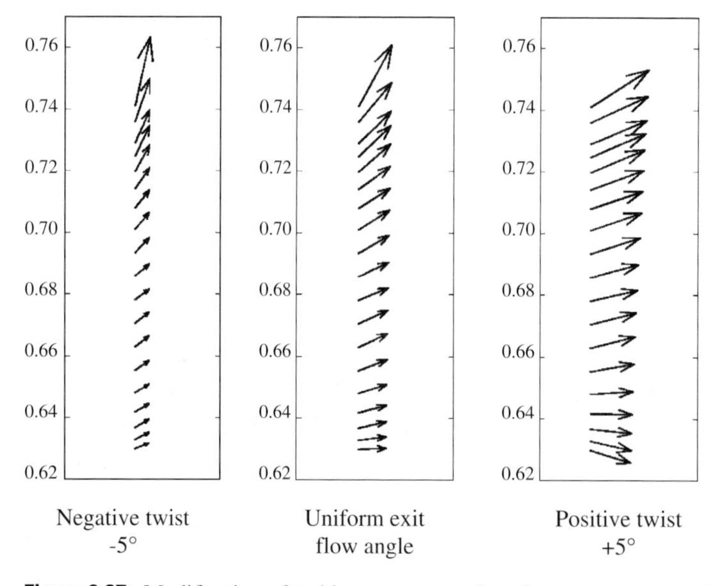

Figure 9.27: Modification of turbine rotor secondary flow using stator twist; velocity vectors from secondary flow analysis (Prasad, 1998).

the dynamical mechanisms just described. A sketch of the flow pattern in the rotor is given in Figure 9.26. From the previous discussion, the velocity field is such that close to the pressure surface the contributions to the radial velocity from the secondary flow and the relative eddy are both positive in the upper portion of the passage, while they are in opposite directions in the lower part.

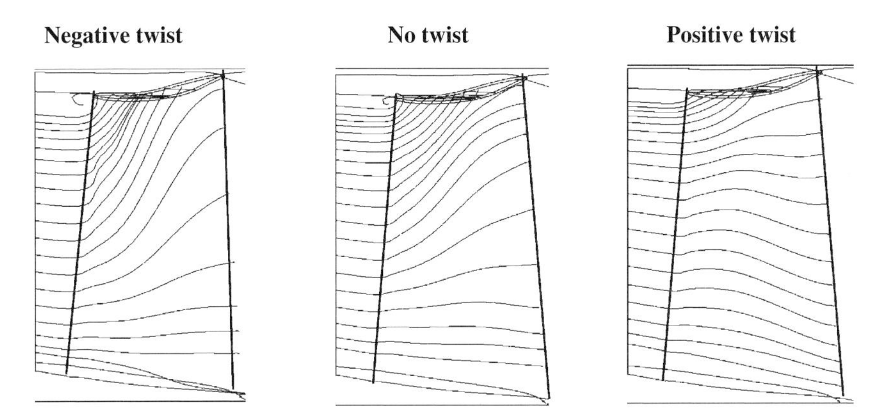

Figure 9.28: Secondary flow streamlines near the pressure surface of a turbine rotor from three-dimensional viscous computations (Prasad, 1998).

To reduce the radial transport, the radial gradient of the vane exit angle should be altered so that trailing vorticity of the opposite sign is generated. To do this the exit angle α must decrease with radius. The secondary flow velocity vectors resulting from three different blade twists are shown in Figure 9.27.

The results of three-dimensional computations for streamlines outside the boundary layer near the pressure side of the turbine rotor passage are shown in Figure 9.28. The region of high temperature on the rotor is contained on the blade (where cooling is plentiful) and the hot fluid is kept from washing over the tip using positive twist. While the secondary flow analyses are not able to capture the details of the radial migration, their application is useful not only in extracting the central fluid dynamic effects but also in indicating the pathway to solution.

9.9 Streamwise vorticity and mixing enhancement

Mixing of coflowing streams is necessary for the operation of devices such as combustors and ejectors. A powerful agent for enhancing this mixing is the introduction of embedded streamwise vortices. There are a variety of ways in which this concept has been implemented, but utilization of large-scale cross-stream circulation to augment mixing by "stirring" is inherent in all of them. In this section we describe and quantify the mechanisms contributing to the increased mixing, using the flow in a lobed mixer as the context for the discussion. Arguments are presented to illustrate the link between mixing augmentation and the strain field associated with the vortices; this strain field increases both the area available for mixing between two streams and the local gradients in fluid properties which provide the driving potential for mixing.

9.9.1 Lobed mixers and streamwise vorticity generation

A number of geometries have been used to create streamwise vortices for the purpose of increasing mixing. Vortex generators, which are seen on aircraft wings and in diffusers, are one class of configurations which can increase momentum transfer in boundary layers and prevent separation

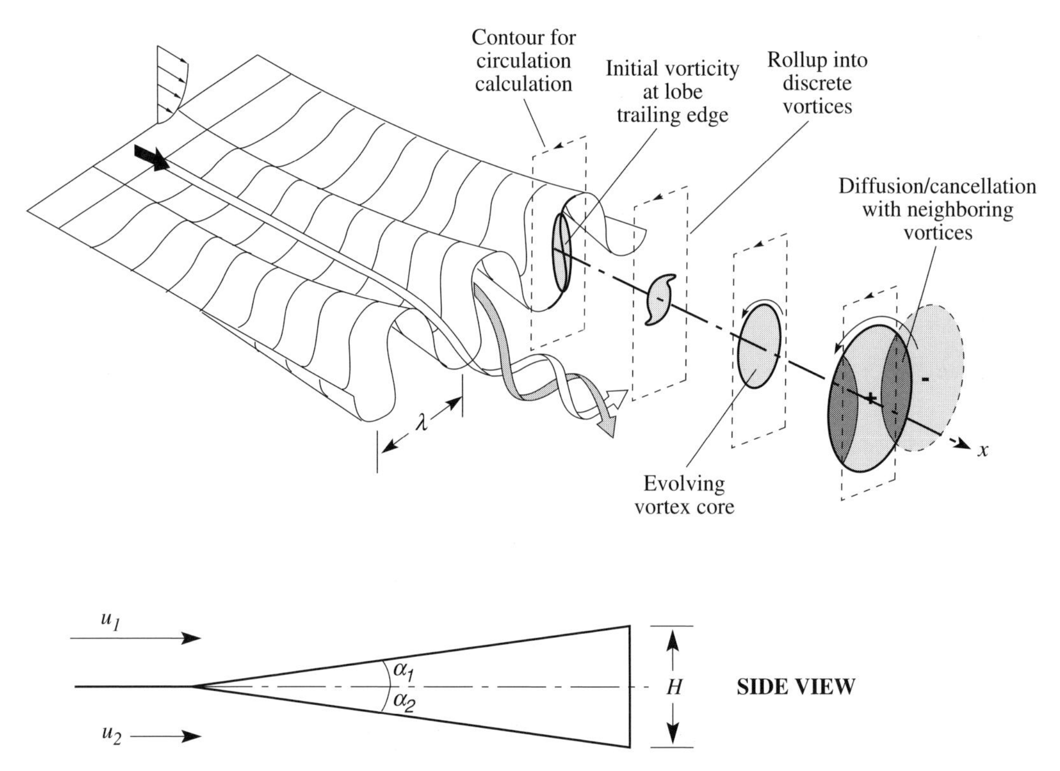

Figure 9.29: Lobed mixer; α_1 and α_2 are the top and bottom lobe penetration angles.

(see e.g., Schubauer and Spangenberg (1960), Pauley and Eaton (1988), Bushnell (1992), Lin (2002)). Vortex generator jets (cross-flow jets inclined to the oncoming stream) (Johnston and Nishi, 1990) have also been used in this application.

Another type of geometry is the *lobed mixer*. The geometry and a sketch of the downstream flow regimes for a lobed mixer are given in Figure 9.29. Lobed mixers allow the controlled introduction of streamwise vorticity along the interface between coflowing streams (Presz, Gousy, and Morin, 1986; Gutmark, Schadow and Yu, 1995; Waitz *et al.*, 1997). Ejectors with these mixers have achieved 90% of the theoretical (complete mixing) pumping value in a distance of less than two duct widths, compared to roughly five widths needed for mixing in a sudden expansion.

Figure 9.30 shows an ejector with a high pressure primary stream used to pump a secondary stream. Two different primary stream nozzles are depicted: a conventional nozzle (circular for the configuration examined) and a lobed mixer nozzle. Figure 9.31 shows pumping ratio (secondary/primary mass flow ratio) data for these two nozzles as a function of the ratio of the secondary to primary stream area at the start of the ejector mixing duct, for a supersonic ejector with a primary nozzle Mach number of 1.5. The curve marked "Ideal" in the figure is from a control volume analysis, neglecting wall friction and assuming full mixing. In the experiments the length/diameter (L/d) ratios of the mixing duct were between 1.5 and 2.3, which is considerably less than the value (5–7) needed for complete mixing with a conventional ejector. The conventional ejectors are all therefore operating far from complete mixing. With the mixer lobe configuration there is essentially complete momentum mixing in the shorter length for three of the configurations. We thus examine the lobed mixer configuration

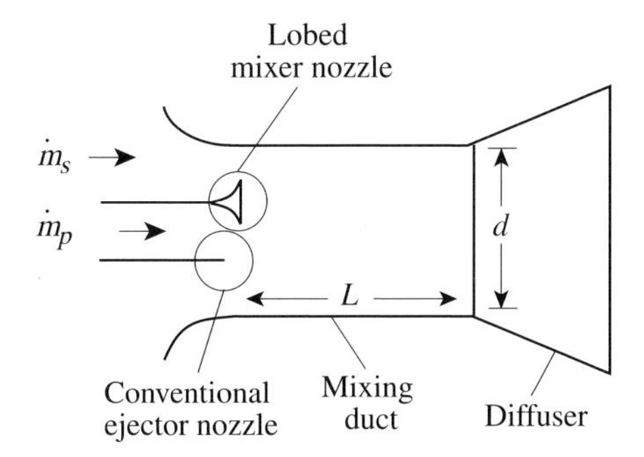

Figure 9.30: Schematic of an ejector showing nozzle configurations.

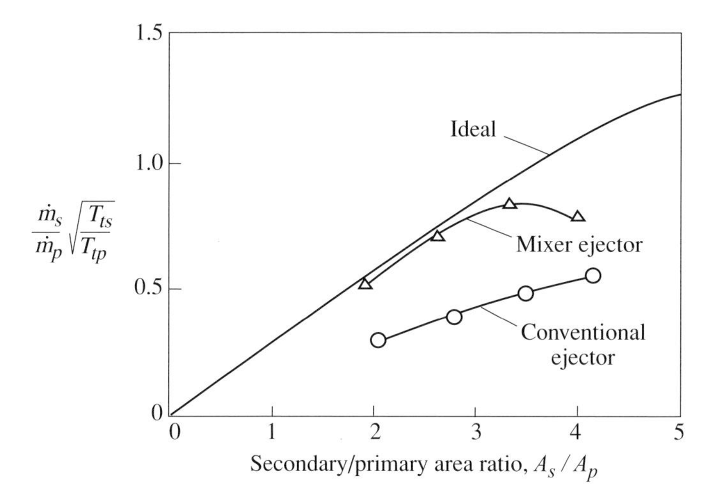

Figure 9.31: Ejector pumping performance with lobed mixer nozzles and conventional nozzles. Nozzle exit Mach number = 1.5, $T_{tp}/T_{ts} = 2.8$. The curve marked "Ideal" corresponds to control volume analysis (Tillman *et al.*, 1992).

as a specific (and effective) application, to frame the discussion of effects of streamwise vorticity on mixing.

In a lobed mixer there is a variation in aerodynamic loading (pressure difference across the solid surface) in the direction transverse to the flow. The circulation per unit axial length along the mixer, which is equal to the difference in the velocities outside the boundary layers on the two sides of the surface, thus also varies along the transverse direction. Because vortex lines cannot end in the fluid this implies the existence of streamwise vortex filaments, analogous to those which trail from a finite wing or turbomachine blade of radially varying circulation (see Section 3.14). The continuous distribution of streamwise vorticity discharged at the trailing edge evolves ("rolls up") into an array of discrete counterrotating vortices as sketched in Figure 9.29.

A useful estimate of the magnitude of the circulation just downstream of the trailing edge is obtained by approximating the fluid to exit at the lobe angle (Skebe, Paterson, and Barber, 1988). For the geometry of Figure 9.29 this yields a trailing edge circulation

$$\Gamma_{te} \approx u_{E_1} H \tan\alpha_1 + u_{E_2} H \tan\alpha_2, \tag{9.9.1a}$$

where u_{E_1} and u_{E_2} are free-stream velocities on either side of the lobe at the exit, H is the lobe height, and α_1 and α_2 are the lobe penetration angles. For $\alpha_1 = \alpha_2 = \alpha$ (9.9.1a) becomes, with $\overline{u}_E$ the average free-stream velocity, $(u_{E_1} + u_{E_2})/2$,

$$\Gamma_{te} \approx 2\overline{u}_E H \tan \alpha. \tag{9.9.1b}$$

For mixers with vertical sidewalls and penetration angles, α, up to 20°, the circulation given by (9.9.1b) has been found to be within 10% of the experimental value. For larger angles, where lobe boundary layers affect fluid exit angles appreciably, the concept can be extended using an effective lobe penetration angle, α_{eff}, and an effective lobe height, H_{eff}, to account for the influence of the boundary layers (O'Sullivan *et al.*, 1996).

If the far upstream stagnation pressures on both sides of the mixer surface are the same, there is no net transverse vorticity shed downstream. With unequal stagnation pressures, however, which is the more typical case, there is a velocity difference at the trailing edge and hence a shear layer with net transverse vorticity. The instability of the shear layer (see Section 6.5) results in vortices with axes in the transverse directions as indicated in Figure 9.32. The scale of the transverse vortices is set by the shear layer thickness. The counterrotating streamwise vortices have a larger length scale, the half-wavelength of the lobe geometry.[6] For lobed mixers in many industrial devices the shear layer thickness is small compared to the half-wavelength, at least for the downstream region (roughly 5–10 lobe wavelengths) in which much of the mixing occurs.

The difference in length scale allows the description of the interaction between the two types of vortical structures in an approximate manner in which transverse vorticity is viewed as associated with turbulent transport, as in planar shear layer mixing. In terms of a conceptual analysis of the mixing process, we consider *computing* the motions due to large scale cross-plane structures associated with streamwise vortices while (consistent with the discussion in Section 4.8) *modeling* details of the shear layers associated with transverse vorticity. Before doing this, however, we need to introduce the manner in which these embedded streamwise vortices enhance mixing.

9.9.2 Vortex-enhanced mixing

When an interface between fluids of different properties (e.g. temperature, velocity, concentration) is within the velocity field of a vortex, two related effects occur: an increase in the interfacial surface area and an increase in the magnitude of gradients normal to the interface. Both augment mixing.

The elements of the mixing enhancement processes will be demonstrated in steps. The effect of the strain rate on mixing is reviewed first, followed by an assessment of the further increase which occurs in the velocity field of a two-dimensional vortex.

9.9.2.1 Effect of the strain rate on mixing

The effect of strain rate on mixing is illustrated by the two-dimensional model problem of diffusion between two semi-infinite fluid regions separated by the x-axis, one containing fuel and one containing

[6] Horseshoe vortices can also be formed around the front of the lobes, but for representative geometries these have a circulation an order of magnitude less than either transverse or streamwise vortices and little impact on the mixing process (McCormick, 1992).

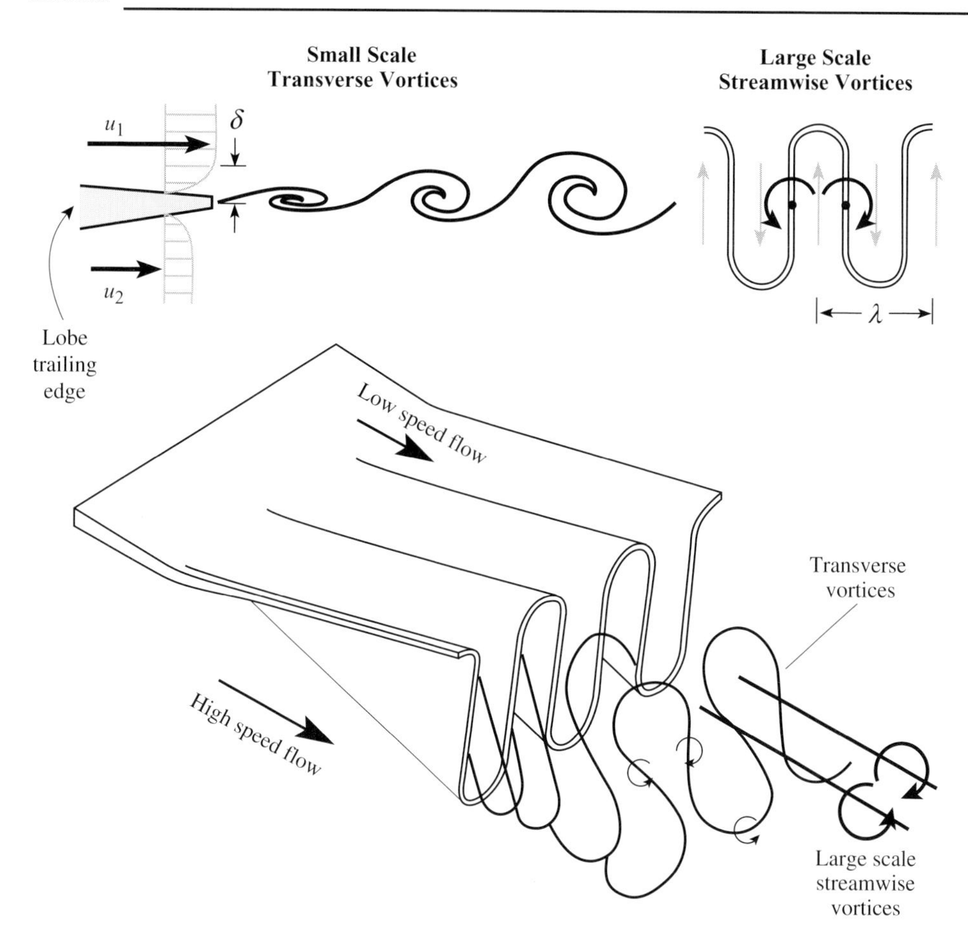

Figure 9.32: Schematic of the vortical structure about a lobed mixer (after McCormick, 1992).

oxidizer (Marble, 1985; Karagozian and Marble, 1986; Waitz *et al.* 1997). The chemical kinetics are taken as infinitely fast so that locally the reaction is diffusion controlled, the stoichiometry is taken such that equal amounts of the reactants are consumed in an infinitely thin reaction zone, the density is uniform, and effects of heat release are neglected.

The geometry is shown in Figure 9.33(a). With no strain, the mixing process is pure diffusion and the concentration field, $C(y, t)$, (the concentration of either reactant) is described by (e.g. Incropera and DeWitt (1996))

$$\frac{\partial C}{\partial t} = D\frac{\partial^2 C}{\partial y^2}, \tag{9.9.2}$$

where D is the binary diffusion constant. Note that diffusion is only one of many phenomena described by an equation similar to (9.9.2) and its subsequent modification to include the effects of strain. A range of physical problems can be modeled as diffusion processes including temporal mixing of a shear layer (momentum mixing) and heat transfer, and the ideas presented are applicable to mixing of other quantities.

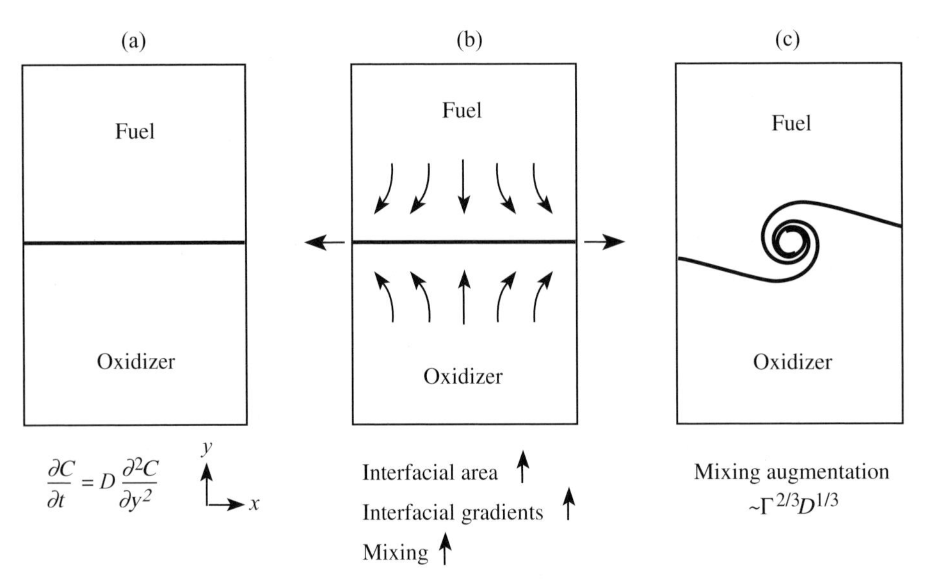

Figure 9.33: Effect of strain on mixing at an interface: (a) planar diffusion; (b) stretched interface; (c) vortex.

The solution for the concentration field of the fuel and oxidizer, C_f and C_o, can be verified by direct substitution to be

$$\begin{aligned} C_f &= \operatorname{erf}\left(y/\sqrt{4Dt}\right); \quad y > 0, \\ C_o &= -\operatorname{erf}\left(|y|/\sqrt{4Dt}\right); \quad y < 0. \end{aligned} \tag{9.9.3}$$

In (9.9.3) y is the coordinate normal to the interface and erf is the error function.[7] From (9.9.3) the thickness of the diffusion zone grows as $\sqrt{Dt}$. The reactant consumption rate, which is a direct measure of molecular mixing, can be expressed as

$$\rho D \left.\frac{\partial C}{\partial y}\right|_{y=0} = \rho \sqrt{\frac{D}{\pi t}}. \tag{9.9.4}$$

For pure diffusion the mixing rate approaches zero as $t \to \infty$.

Suppose a spatially uniform normal strain rate,[8] $\varepsilon = \partial u_x/\partial x = -\partial u_y/\partial y$, is applied as in Figure 9.33(b). The velocity normal to the interface is $u_y = -\varepsilon y$. The equation for the concentration field becomes

$$\frac{\partial C}{\partial t} - \varepsilon(t)\, y \frac{\partial C}{\partial y} = \kappa \frac{\partial^2 C}{\partial y^2}. \tag{9.9.5}$$

Equation (9.9.5) can be transformed into the form of (9.9.3) with the substitutions

$$\zeta = y \exp\left(\int_0^t \varepsilon dt_1\right), \quad \tau = \int_0^t \left[\exp\left(\int_0^{t_2} 2\varepsilon dt_1\right)\right] dt_2. \tag{9.9.6}$$

[7] The error function is defined as $\operatorname{erf}(x) = (2/\sqrt{\pi}) \int_0^x e^{-\hat{x}^2} d\hat{x}$.

[8] Shear does not affect the flow since there is no dependence on x.

The solution for the concentration field is

$$C_f = \operatorname{erf}\left(\zeta/\sqrt{4D\tau}\right); \quad y > 0$$
$$C_o = -\operatorname{erf}\left(|\zeta|/\sqrt{4D\tau}\right); \quad y < 0. \tag{9.9.7}$$

The effect of the strain rate on mixing is seen most readily for constant strain rate. In that case differentiation of (9.9.7) gives the reaction consumption rate as

$$\rho D \left.\frac{\partial C}{\partial y}\right|_{y=0} = \rho\sqrt{\frac{2\varepsilon D}{\pi}}\left(\frac{e^{2\varepsilon t}}{e^{2\varepsilon t}-1}\right)^{1/2}. \tag{9.9.8}$$

For times or strain rates such that $t\varepsilon \ll 1$ (small times or low strain rates) the interface mixing rate is similar to that for no strain. For $t\varepsilon \gg 1$ (large strain rate or long times), however, the consumption rate does not go to zero but approaches the value $(2D\varepsilon/\pi)^{1/2}$.

9.9.2.2 Mixing enhancement for a two-dimensional laminar vortex

We now describe mixing augmentation in the velocity field of a two-dimensional vortex on the interface between two reactants. There is only one non-zero component, $u_\theta(r, t)$, which is defined by (Section 1.14)

$$\frac{\partial u_\theta}{\partial t} = \nu\frac{\partial}{\partial r}\left[\frac{1}{r}\frac{\partial(r u_\theta)}{\partial r}\right]. \tag{9.9.9}$$

Equation (9.9.9) is satisfied by (Batchelor, 1967),

$$u_\theta = \frac{\Gamma}{2\pi r}\left[1 - \exp\left(\frac{-r^2}{4\nu t}\right)\right]. \tag{9.9.10}$$

Equation (9.9.10) represents a vortex with a far-field circulation Γ and a viscous core that grows with time.

Suppose the reaction and the vortex are both initiated at time $t = 0$. With increasing time the interface is deformed into a spiral, as shown in Figure 9.33(c). In the portion of the flow undergoing solid-body rotation (the vortex core) the interface is not stretched. The increase in interfacial length can therefore be obtained by examining only the region outside the viscous core. For a material element of initial length $d\ell_i$, the deformed length, $d\ell$, behaves as (Marble, 1985)

$$d\ell = \left[1 + \left(\frac{\Gamma t}{\pi r^2}\right)^2\right]^{1/2} d\ell_i. \tag{9.9.11}$$

For large values of $\Gamma t/(\pi r^2)$ (9.9.11) can be approximated by

$$d\ell = \left(\frac{\Gamma t}{\pi r^2}\right) d\ell_i. \tag{9.9.12}$$

Equation (9.9.12) shows that at large $\Gamma t/(\pi r^2)$ the local interfacial length increases as the product of time and vortex circulation. This interface stretching is the principal agent for mixing augmentation. Further, the mixing region is larger than the vortex core and much of the mixing takes place outside

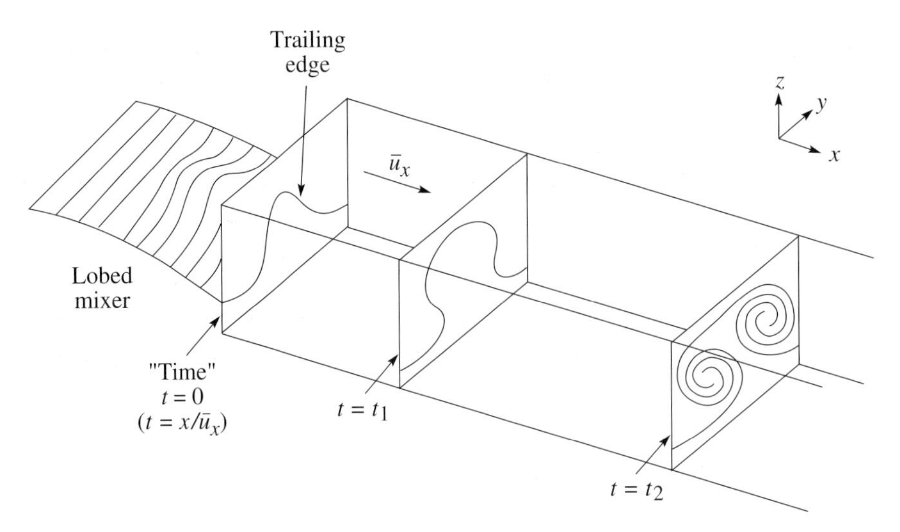

Figure 9.34: Analogy between two-dimensional unsteady flow and three-dimensional steady flow viewed by an observer traveling at $\overline{u}_x$.

the core. The behavior is thus roughly independent of the core growth rate and the ideas apply to both turbulent and laminar flow.

As with the planar interface, the area of the fully mixed fluid gives a direct measure of mixing. For the velocity field of (9.9.10), at values of Γ/D greater than roughly several hundred, the cross-sectional area of the fully mixed core has been shown to scale as (Marble, 1985)

$$A_{mixed} \cong 2\left(\frac{2}{3\pi^2}\right)^{1/3}\left(\Gamma^{2/3}D^{1/3}t\right). \tag{9.9.13}$$

The rate of core area growth in (9.9.13) is larger by a factor of $(\Gamma/D)^{2/3}$ than for radial diffusion only, where the core area would grow at a rate proportional to Dt.

9.9.2.3 Mixing enhancement for a distribution of streamwise vorticity

To enable a simplified description of mixing downstream of a lobed mixer two extensions need to be made to the analysis of mixing increase for a single vortex in a two-dimensional unsteady flow. First we have to relate the two-dimensional unsteady process to the steady three-dimensional flow in the actual geometry. Second we need to generalize the velocity field for the single vortex to that appropriate to the vorticity distribution shed from mixer lobes.

The relation between the two-dimensional unsteady flow and the three-dimensional steady flow is illustrated for the lobed mixer in Figure 9.34 (Marble *et al.*, 1990). Three-dimensional spatial development is represented by the evolution of a two-dimensional unsteady flow field. The changes along the streamwise direction (x-direction) are viewed as the changes in time seen by an observer traveling with an appropriate convection velocity, taken here as the average bulk x-velocity, $\overline{u}_x$; time is related to streamwise distance by $t = x/\overline{u}_x$.

There are several requirements which must be met in order to support this analogy between unsteady two-dimensional and steady three-dimensional motions.[9] The first is the ability to identify an appropriate convection velocity. Others are the existence of a difference in scale between transverse and streamwise motions (mentioned previously), the view of cross-stream velocities as smaller than the bulk velocity, and the idea that the length scales of the motion in the cross-plane (y, z) are smaller than the length scales over which the flow changes in the x-direction.

For a general initial distribution of vorticity specified at the lobe trailing edge, the downstream evolution of the cross-plane velocity field $[u_y(y, z, t), u_z(y, z, t)]$ can be found from solution of the two-dimensional, unsteady, Navier–Stokes equations. With the velocity field known, the distribution of any scalar quantity, ϕ (concentration, temperature), is given by (with D representing the appropriate diffusion coefficient)

$$\frac{\partial \phi}{\partial t} + u_y \frac{\partial \phi}{\partial y} + u_z \frac{\partial \phi}{\partial z} = D\left(\frac{\partial^2 \phi}{\partial y^2} + \frac{\partial^2 \phi}{\partial z^2}\right). \qquad (9.9.14)$$

Figure 9.35 shows an example of mixing enhancement due to the streamwise vorticity shed from a mixer lobe based on two-dimensional unsteady Navier–Stokes computations. The mixing is of a scalar, initially specified at $\phi = -1$ on one side of the lobe and $\phi = +1$ on the other. The different parts of the figure are for different non-dimensional times, $\tilde{t}\,(= \Gamma x/\overline{u}_x \lambda^2)$, denoting different distances downstream from the mixer lobe.[10] The features of embedded streamwise vorticity which augment mixing, the increase in the length of the interface between the two regions and the steepening of interfacial gradients, are apparent. There is a rapid growth in interface length associated with the presence of streamwise vorticity; the interface length doubles in the first two to three wavelengths downstream of the trailing edge, increasing roughly linearly with downstream distance.

Computations of the flow downstream of mixer lobes show regimes with the different characteristics depicted in Figure 9.29. A key finding, however, is that once the initial roll up of the distributed vorticity into a discrete vortex is substantially complete, the mixing process follows closely the scaling developed for the single vortex, with the increase in mixed area proportional to $\Gamma^{2/3}D^{1/3}t$ or, in physical coordinates, to $(\Gamma^{2/3} D^{1/3}x)/\overline{u}_x$. For a detailed definition of the mixing in any specific configuration, the relevant computations need to be carried out; however, it is worth noting that much of the mixing occurs in the flow regime in which the vortex has already rolled up. Because of this, the single vortex scaling gives useful guidance about vortex enhanced mixing even with more complex geometries.

9.9.2.4 Estimation of mixing enhancement for a turbulent vortex

The analysis of vortex enhanced mixing in laminar flow can be extended to provide an estimate of mixing produced by turbulent vortices. In line with the discussion of turbulent mixing in Chapter 4 we define a "turbulent transport property", diffusivity, D_t, or kinematic viscosity, ν_t.

[9] Comparisons between this approximate approach and three-dimensional Reynolds-averaged, Navier–Stokes computations are given by Waitz *et al.* (1997) in an assessment of the analogy.

[10] The scaling for time comes from the use of the lobe wavelength λ as the distance unit and Γ/λ as the cross-plane velocity unit (Waitz *et al.*, 1997).

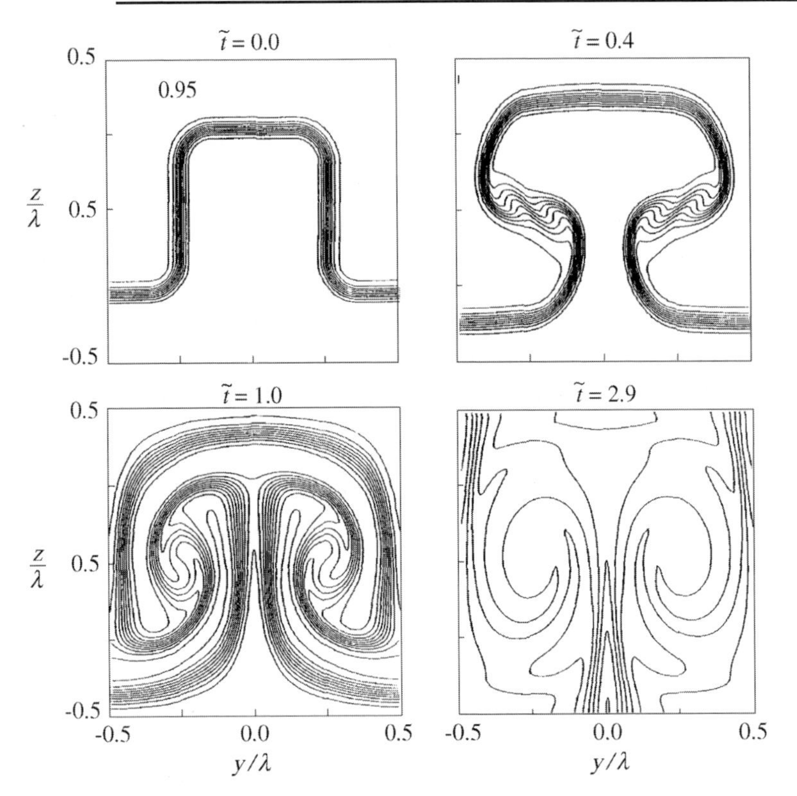

Figure 9.35: Transport of a scalar downstream of a lobed mixer, $\tilde{t}\,[= \Gamma_x/(\overline{u}_x\lambda^2)]$, λ is the mixer wavelength as defined in Figure 9.32, Γ is the initial trailing edge circulation (Qiu, 1992; Waitz *et al.*, 1997).

The θ-velocity field of a two-dimensional turbulent vortex satisfies

$$\frac{\partial u_\theta}{\partial t} = \nu_t(t)\frac{\partial}{\partial r}\left[\frac{1}{r}\frac{\partial}{\partial r}(ru_\theta)\right]. \tag{9.9.15}$$

The turbulent kinematic viscosity, ν_t, varies with time as the length scale of the turbulent core increases, and to proceed further, we need to specify the functional dependence for it and the diffusivity. Interest here is in mixing which occurs between streams of different velocities. As described in a number of texts (e.g. Schlichting (1979), White (1991); see also Section 4.8), a functional form applicable to the temporal growth of a shear layer (of thickness δ(t)) between two streams is

$$\nu_t(t) = Sc_t D_t \propto [\delta(t)](u_{E_1} - u_{E_2}) = B\left(\frac{u_{E_1} - u_{E_2}}{u_{E_1} + u_{E_2}}\right)^2 (\overline{u}_E)^2\, t, \tag{9.9.16}$$

where B is an empirical constant, Sc_t is the turbulent Schmidt number, the velocity difference is between the two free-steam values, and $\overline{u}_E$ is the mean of these two values $(u_{E_1} + u_{E_2})/2$. For temporal shear layer growth the turbulent transport coefficients grow linearly with time such that $\nu_t(t) = \beta t$:

$$\beta = B\left(\frac{u_{E_1} - u_{E_2}}{u_{E_1} + u_{E_2}}\right)^2 (\overline{u}_E)^2. \tag{9.9.17}$$

Substituting for ν_t in (9.9.15) the circumferential velocity is found as

$$u_\theta = \left(\frac{\Gamma}{2\pi r}\right)\left[1 - \exp\left(\frac{-r^2}{4\beta t^2}\right)\right]. \tag{9.9.18}$$

Analysis of the vortex/diffusion flame problem using this velocity field reveals that the ratio of mixing (growth of the mixed area) due to the vortex, compared to that due to diffusion only, has the same form as in the laminar case, except that the laminar diffusion coefficient is replaced by the turbulent version, D_t (Waitz *et al.*, 1997). The ratio of the growth of the mixed region for a turbulent vortex compared to that for "pure" diffusion is thus $\Gamma^{2/3}/D_t^{2/3}$. Since the turbulent diffusion is a function of the velocities on either side of the lobe (9.9.17), and the circulation is related to the lobe geometry by (9.9.1), the ratio can readily be connected to overall (and readily known) flow quantities.

9.9.3 Additional aspects of mixing enhancement in lobed mixers

Part of the increase in mixing for a lobed mixer is from the increased trailing edge length compared to a conventional mixing nozzle (basically a flat splitter plate between the streams). To differentiate this effect from the increase associated with streamwise vorticity, we examine the two geometries in Figure 9.36. One is a lobed mixer. The other, referred to as a convoluted plate, has the same trailing edge geometry but a parallel extension of the trailing edge. The extension allows a return towards parallel flow and hence a reduction in the magnitude of trailing streamwise vorticity. The three-dimensional Navier–Stokes calculations show that the convoluted plate geometry pictured has a trailing streamwise circulation almost an order of magnitude lower than the lobed mixer.

Figure 9.37 shows the measured (using a chemical reaction technique) molecular mixedness of two coflowing streams as a function of axial distance downstream of the trailing edge for a lobed mixer, a convoluted plate, and a flat splitter plate. In the region in which the shear layers from adjacent lobes have not yet merged the ratio of mixing with the convoluted plate compared to that with the flat splitter plate is very nearly the ratio of the geometric trailing edge lengths of the two. The mixer lobe introduces streamwise vorticity and causes additional mixing for this geometry, roughly the same as that for the increased trailing edge length.

The results of wind tunnel measurements of momentum interchange between two coflowing streams are shown in Figure 9.38 for a lobed mixer, a convoluted plate, and a flat splitter plate. The tunnel has constant area so the rise in static pressure is directly indicative of the change in momentum flux and hence the momentum interchange between streams. The horizontal axis is the downstream distance from the lobe exit, non-dimensionalized by the lobe wavelength. The vertical axis is the static pressure rise to the tunnel exit at a given location, non-dimensionalized by the ideal static pressure rise for two-stream constant area mixing. (Corrections for wall friction have been incorporated, but these are less than 10% of the ideal pressure rise). There is essentially full mixing with the lobed mixer and convoluted plate but not with the flat splitter plate. The impact of streamwise vorticity is thus shown by the difference between the lobed mixer and the convoluted plate.

One way to characterize the mixing rate is in terms of the length it takes for mixing to occur. We define a pressure rise length, L_p, as

$$\frac{L_p}{\lambda} = \frac{\int \Delta p d(x/\lambda)}{(\Delta p)_{trailing\ edge}}, \tag{9.9.19}$$

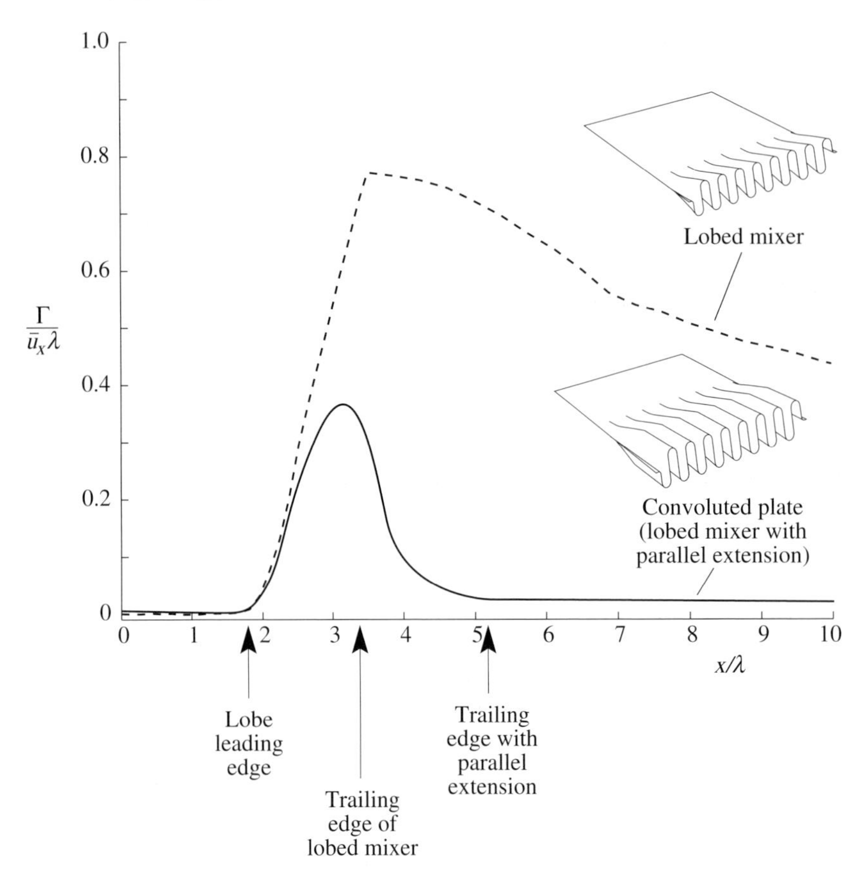

Figure 9.36: Streamwise circulation as a function of axial distance for a lobed mixer ($\alpha = 22°$, free-stream velocity ratio $u_{E_2}/u_{E_1} = 0.6$) and a convoluted plate ($\alpha = 22°$, $u_{E_2}/u_{E_1} = 0.53$). Results of three-dimensional Navier–Stokes computation (Waitz *et al.*, 1997).

where Δp is the pressure difference between a given station and the mixing duct exit. The data of Figure 9.38, as well as other data at velocity ratios of 0.1 and 0.2 (Waitz *et al.*, 1997), show the pressure rise length is 5–6 lobe wavelengths for the convoluted plate and roughly half that for the lobed mixer.

A final aspect of the discussion of lobed mixers concerns the extension of the ideas and scaling estimates to compressible flow. Two main questions need to be answered. The first is the influence of compressibility on the distribution of shed circulation in the downstream region. The second is the effect on shear layer mixing, or, within the context of the above discussion, the dependence of the diffusivity on Mach number.

For many mixer nozzle configurations the cross-stream Mach numbers are subsonic for axial Mach numbers of 2 or less. For this regime the trends concerning the magnitude and evolution of the streamwise circulation with velocity ratio do not appear to change greatly. Three-dimensional computations carried out at Mach numbers of 0.5 and 2.0 (for a flow with uniform stagnation pressure on both sides of the mixer) indicate only small changes in the growth of the mixing interface in the cross-flow plane, which represents the effect of the vortical structure.

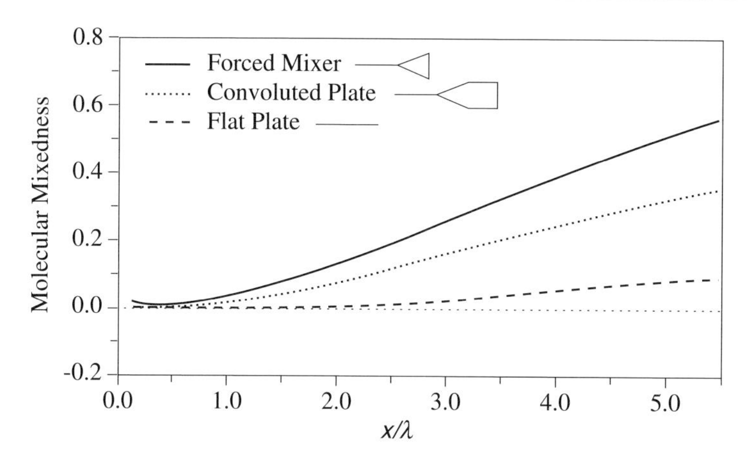

Figure 9.37: Molecular mixedness for a lobed mixer, a convoluted plate, and a flat splitter plate; free-stream velocity ratio $u_{E_2}/u_{E_1} = 0.67$, measurements of Manning (1991).

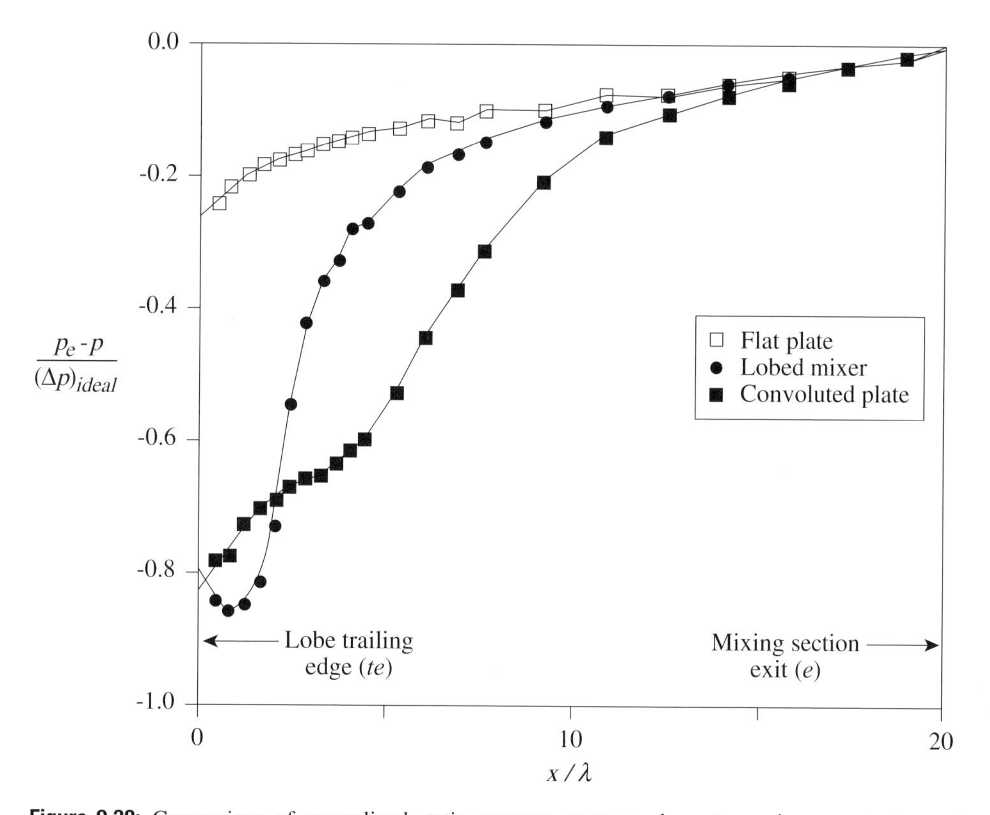

Figure 9.38: Comparison of normalized static pressure recovery downstream in a constant area tunnel for lobed mixer, convoluted plate, and flat splitter plate. Free-stream velocity ratio $u_{E_2}/u_{E_1} = 0.31$, $\alpha = 20^\circ$, height/wavelength = 2.0, $Re_\lambda = \overline{u}_x \lambda / \nu = 10^5$, $(\Delta p)_{ideal}$ from control volume analysis (Waitz *et al.*, 1997).

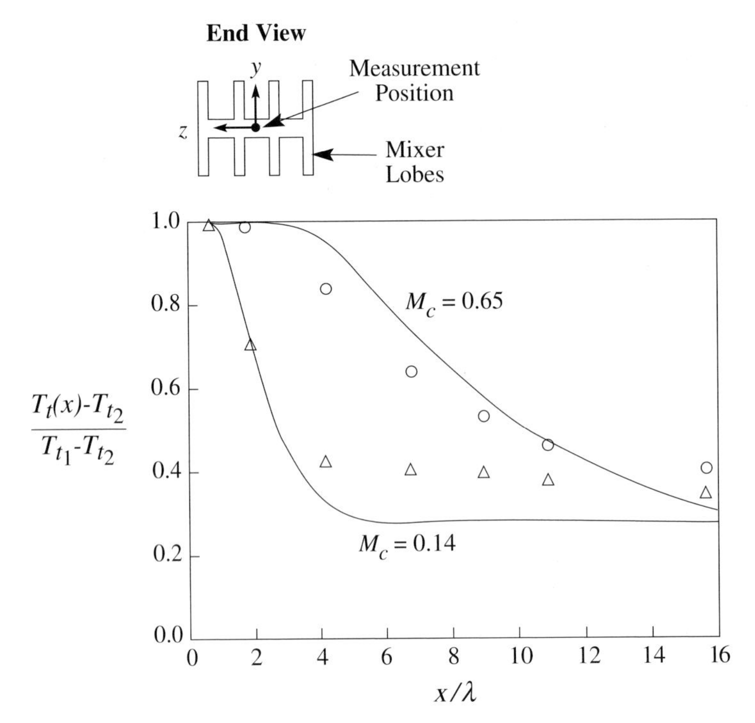

Figure 9.39: Effect of the convective Mach number on temperature mixing in an ejector; the location of measurements is shown in end view; free-stream stagnation temperature and velocity ratios at ejector inlet, $T_{t_2}/T_{t_1} = 2.8$, $u_{E_2}/u_{E_1} = 0.23$; solid lines from unsteady two-dimensional computations (Waitz *et al.*, 1997).

As discussed in Section 4.8, however, shear layer mixing rates decrease by a factor of almost 5 as the convective Mach number increases from the incompressible regime. The approximate analyses that have been developed imply that the decrease in vortex mixing rate should scale with the decay in diffusion coefficient to the one-third power. Figure 9.39 shows data for temperature mixing in an ejector at primary stream Mach numbers of 0.5 and 1.5. The primary to secondary velocity ratios were 4-to-1 at both of these conditions, corresponding to convective Mach numbers of 0.14 and 0.65. For the sparse data that exist, the decrease in vortex mixing rate with convective Mach number is in qualitative accord with this conceptual picture (Tillman, *et al.*, 1992; Waitz *et al.*, 1997).

9.10 Fluid impulse and vorticity generation

The concept of fluid impulse connects the vorticity field to the spatial distribution of external forces which act on the flow. The fluid impulse, $\mathcal{I}$, is defined (von Karman and Burgers, 1963) as the integral over time of the external force, $\mathcal{F}_{ext}$, applied to the fluid:

$$\mathcal{I} = \int_{t_1}^{t_2} \mathcal{F}_{ext} dt. \tag{9.10.1}$$

To develop the link between impulse and vorticity we consider a distribution of external forces applied to an inviscid, incompressible, uniform density fluid during a short interval of time, dt. If the changes in velocity during this interval are of the same order as the velocities that characterize the motion, denoted as U, and the characteristic length scale of the motion is L, the ratio of local accelerations (the $\partial\mathbf{u}/\partial t$ term in the momentum equation) to convective accelerations (the $(\mathbf{u}\cdot\boldsymbol{\nabla})\mathbf{u}$ term) is $L/(Udt)$. For forces which are applied over time intervals dt much smaller than L/U, the convective accelerations can be neglected during the time the force acts. With $\mathbf{F}_{ext}$ the external force per unit mass acting on the fluid, the momentum equation thus becomes

$$\frac{\partial\mathbf{u}}{\partial t} = -\frac{\boldsymbol{\nabla}p}{\rho} + \mathbf{F}_{ext}. \tag{9.10.2}$$

Taking the curl of this equation ($\boldsymbol{\nabla}\times$ (9.10.2)) gives

$$\frac{\partial\boldsymbol{\omega}}{\partial t} = \boldsymbol{\nabla}\times\mathbf{F}_{ext}. \tag{9.10.3}$$

For a flow that is irrotational before the forces act, integration of (9.10.3) over dt yields an expression for the vorticity in terms of the impulse per unit mass, $\mathbf{I}$,

$$\boldsymbol{\omega} = \boldsymbol{\nabla}\times\int_t^{t+dt}\mathbf{F}_{ext}\,dt = \boldsymbol{\nabla}\times\mathbf{I}. \tag{9.10.4}$$

Three applications of fluid impulse are given below. The first, creation of a vortex ring by a single impulse, gives an introductory illustration of the concept. The second, airfoil lift viewed as a continuous distribution of impulses, provides a different perspective on a familiar situation in terms of the impulse imparted to the flow. The third, the generation and evolution of streamwise vorticity in a jet in cross-flow, presents a case where an explicit statement about impulse is used to develop a description of jet behavior.

9.10.1 Creation of a vortex ring by a distribution of impulses

To examine the creation of a vortex ring consider the space between planes $y=0$ and $y=H$, bounded by a cylinder of radius a with the axis parallel to the y-axis, as in Figure 9.40. Outside of this space no external forces act on the fluid. Inside of the space there are external forces directed downward (in the negative y-direction). The intensity of the forces is uniform except near the radius $r=a$, where they fall to zero in a distance small compared to a. The velocity field is irrotational before the forces are applied.

The components of (9.10.4) for this configuration are:

$$\omega_x = -\frac{\partial I_y}{\partial z}, \tag{9.10.5a}$$

$$\omega_y = 0, \tag{9.10.5b}$$

$$\omega_z = \frac{\partial I_y}{\partial x}. \tag{9.10.5c}$$

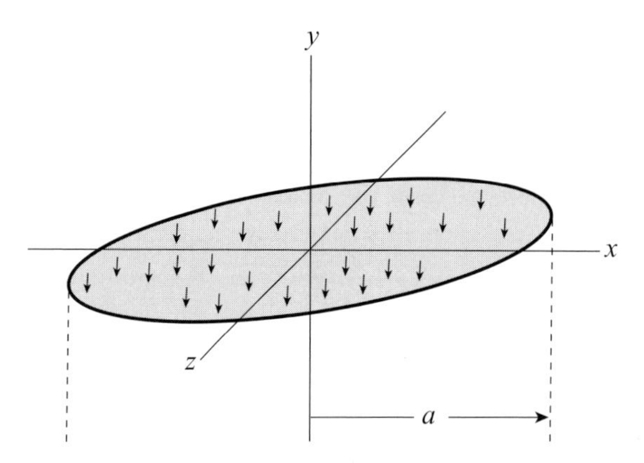

Figure 9.40: Vortex ring produced by distribution of external forces acting over a circular region of $\sqrt{x^2 + z^2} < a$ height H, $H \ll a$.

Vorticity is generated only in the region where the force (and hence the impulse) has a spatial variation. If the height H is small compared with the radius the vortex lines can be regarded as a single vortex ring, which is the heavy line in Figure 9.40. The circulation is

$$\Gamma = \int_0^H \left(\int_{a-H}^{a+H} \omega_z dx \right) dy = \int_0^H I_y dy. \tag{9.10.6}$$

For an impulse which is independent of y over the small distance H

$$\Gamma = HI_y. \tag{9.10.7}$$

The quantity ρHI_y is the y-component of the impulse per unit area for the area enclosed by the ring. Denoting this quantity, referred to as the *impulsive pressure*, by Π (Π has dimensions of (force $\times$ time)/area),

$$\Gamma = \frac{\Pi}{\rho}. \tag{9.10.8}$$

Equation (9.10.8) states that a uniform impulse applied over an area will generate a vortex ring on the bounding contour around the area. This applies not only to circular planforms but also to any area; we could break up a non-planar area into facets and consider the impulse on each of these.

The relation between circulation and impulsive pressure is most useful when applied not to a single impulse but to flows set up by the action of continuous forces, which can be viewed as the limit of a series of small impulses following immediately after one another. As an example which makes the effect of the impulse apparent, consider a flow with uniform velocity, say $\mathbf{u}_0$, to which the impulse is applied, with the background motion much stronger than the motion associated with the impulses (in other words the velocity created by the impulses is much less than u_0). If this condition is met, we can neglect the velocity field of the vortices in describing the vortex generation process. While such a linearized view carries restrictions, it enhances interpretation of the process.

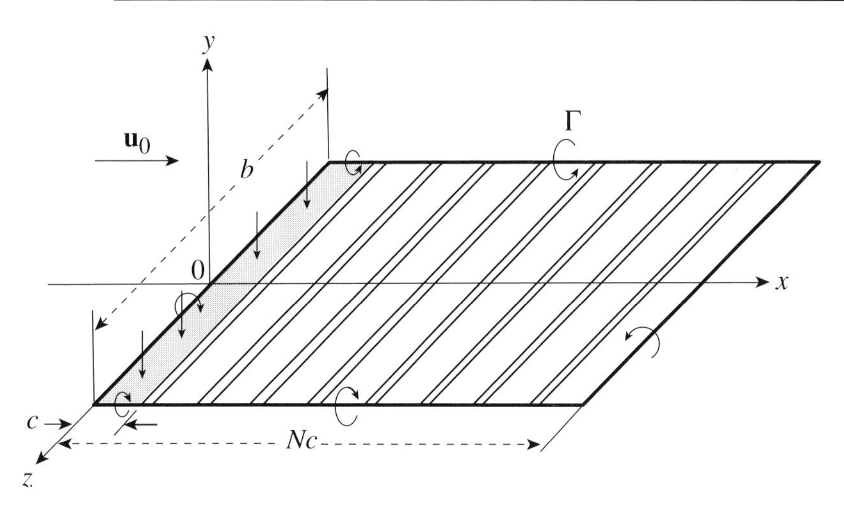

Figure 9.41: Rectangular vortex configuration for an airfoil; the airfoil occupies the shaded region.

Suppose the background velocity is along the y-axis, with the force distribution as given above. The vortices generated are confined to a thin annulus of outer radius a, effectively on a cylindrical vortex sheet of radius a. The circulation generated per unit distance along (in the y-direction) this sheet is

$$\text{circulation per unit distance} = \frac{\Pi}{\rho}. \tag{9.10.9}$$

In (9.10.9) Π is the impulsive pressure corresponding to a pressure difference Δp across the disk of radius a over a time (unit distance$/u_0$). This is an axisymmetric model for a propeller with uniform aerodynamic loading across the propeller disk. The circulation per unit distance on the cylindrical vortex sheet is the velocity difference, which we denote by Δu, between particles that have passed through the propeller disk and those on streamlines just outside of the disk. From (9.10.9) the pressure difference across the disk and the circulation per unit distance along the vortex sheet are related by

$$\Delta p = \rho u_0 \Delta u. \tag{9.10.10}$$

9.10.2 Fluid impulse and lift on an airfoil

Consider now forces directed perpendicular to the velocity $\mathbf{u}_0$, as for the flow past an airfoil. Following von Karman and Burgers (1963) the specific example is a rectangle of dimension c in the x-direction and b in the z-direction (Figure 9.41). A series of impulses in the direction of the y-axis are given to the rectangle at time intervals of c/u_0, each with strength per unit area Π. At any individual application of the impulse a rectangular vortex is generated of circulation $\Gamma = \Pi/\rho$. If the vortices are convected by the background velocity field, in a time c/u_0 they are displaced a distance c. Because there are coinciding vortices of opposite sign which cancel each other, the complete vortex system is equivalent to the single rectangular vortex shown as the heavy line in Figure 9.41.

We now pass to the limiting case in which c is small and the model is a continuous external force on the fluid, of magnitude $\mathcal{F}_{ext}$, distributed uniformly along the z-axis in the region $-b/2 \leq z \leq b/2$

and pointing in the negative y-direction. Over the short time interval c/u_0 the integral of the force is the impulse applied over the area bc or

$$\frac{\mathcal{F}_{ext}c}{u_0} = bc\Pi. \tag{9.10.11}$$

The magnitude of the force per unit length along the airfoil is

$$\frac{\mathcal{F}_{ext}}{b} = \Pi u_0. \tag{9.10.12}$$

If the total duration of time over which the impulse has been applied is Nc/u_0, the rectangular vortex ring will have breadth b, length Nc in the flow (x) direction, and circulation

$$\Gamma = \frac{\mathcal{F}_{ext}}{b\rho u_0}. \tag{9.10.13}$$

The vortex configuration described represents the vortex system for the airfoil, including the trailing vortices and the starting vortex, with $\mathcal{F}_{ext}/b$ the magnitude of the airfoil lift per unit span. Equation (9.10.12) is a (linearized) version of the Kutta–Joukowski theorem (lift/span $=\rho u_0 \Gamma$) relating the magnitude of the lift per unit span, the circulation, and the velocity.

An important relation is that between the fluid impulse and the moment of the vorticity distribution. The total impulse given to the fluid during the time Nc/u_0 is responsible for the creation of the rectangular vortex system represented by the heavy line. The moment of a vorticity distribution, $\mathbf{M}$, is defined as

$$\mathbf{M} = \int (\mathbf{x} \times \boldsymbol{\omega})\, dV. \tag{9.10.14}$$

Only the interior of a line vortex contributes to this integral. From evaluation of the integral for the rectangular vortex in Figure 9.41, the product $\rho\mathbf{M}$ is equal to $(2\rho\Gamma bNc)\mathbf{j}$, where $\mathbf{j}$ is the unit vector in the y-direction. For the rectangular vortex, referring to Figure 9.41, it can be seen that the total impulse is related to the moment of the vorticity distribution by

$$\mathcal{I} = \frac{\rho}{2} \int (\mathbf{x} \times \boldsymbol{\omega})\, dV. \tag{9.10.15}$$

The expression on the right-hand side of (9.10.15) "can be described in simple mechanical terms as one-half of the momentum of a hypothetical force distribution with force per unit mass equal to the vorticity" (Lighthill, 1986b). The relation has been developed here only for the rectangular vortex, but it is a general result for any configuration (Lighthill, 1986a; Batchelor, 1967).[11]

Force is the time rate of change of impulse or, equivalently as we have just seen, the time rate of change of one-half the moment of the vorticity distribution. For the airfoil in uniform motion with velocity u_0 this gives

$$\frac{\rho\Gamma bNc}{Nc/u_0}\mathbf{j} = \rho\Gamma bu_0\mathbf{j} = \text{lift force on the airfoil}. \tag{9.10.16}$$

[11] Equation (9.10.15) also applies, with appropriate interpretation, to flow in which there are internal boundaries on which there is a no-slip (viscous) boundary condition. In that case, as described by Lighthill (1986b), the vorticity in the integral needs to be interpreted as the vorticity field minus "a distribution of vorticity attached to the boundary in the form of a vortex sheet allowing exactly the tangential velocity (slip) associated with the potential flow" which satisfies the normal boundary condition.

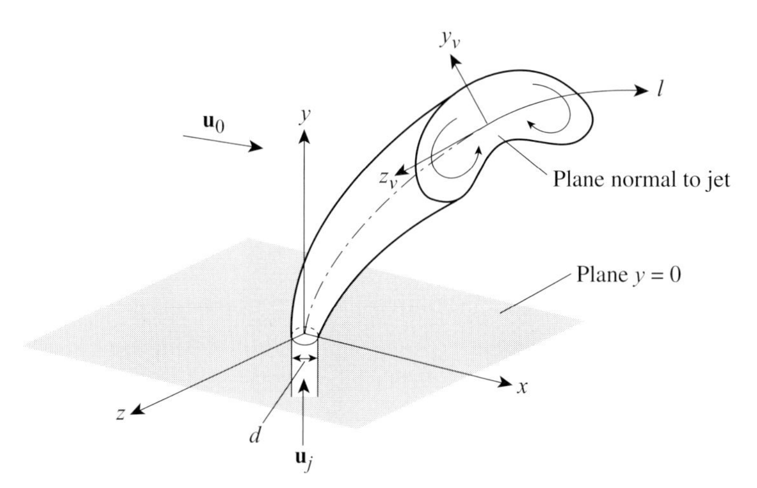

Figure 9.42: Sketch of a jet in a cross-flow emerging from a circular opening of diameter d in the plane $y = 0$. The distance along the jet is l, the coordinates in the plane normal to the jet trajectory are y_v, z_v.

Equation (9.10.16) relates the rate of change of the moment of a vorticity distribution to the force exerted on the fluid. Again, although only a restricted case has been developed, the relation between the force exerted on a fluid and the rate of change of impulse is a general one (Lighthill, 1986a).

9.10.3 Far field behavior of a jet in cross-flow

The third example in which the fluid impulse principle can be exploited is the development of an approximate description of the far field behavior of a jet in cross-flow. The configuration is sketched in Figure 9.42 in which a jet of diameter d enters the unbounded region above the plane $y = 0$ into an oncoming flow and bends into the streamwise direction (Margason, 1993). Some distance downstream[12] the time-mean jet cross-section takes the shape indicated in the figure. Central to the discussion of this "far field" jet behavior is the cross-plane motion associated with a pair of counterrotating vortices with axes at a small angle to the x-direction.[13]

We wish to illustrate several aspects of the behavior. First are general ideas concerning jet configuration and kinematics, i.e. trajectory and jet dimension, including the parametric dependence of these quantities. Second is the connection between the impulse given to fluid above the plane $y = 0$ and the streamwise vorticity in the vortex pair. The description to be given pertains to jet Reynolds numbers ($u_j\, d/\nu$) greater than roughly 10^3 where the overall features have negligible dependence on Reynolds numbers and viscosity does not enter explicitly into the considerations. It is also restricted to incompressible flow with equal densities of the jet and the incoming flow. (For information concerning unequal densities, see Margason (1993), Hasselbrink and Mungal (2001), and Karagozian (2002).)

[12] Several to ten jet diameters is a rough figure for this distance, it will be made quantitative in what follows.

[13] There are other structures associated with a jet in a cross-flow, for example the horseshoe vortex associated with the incoming boundary layer, which occurs at the upstream side of the jet, and wake vortices (Fric and Roshko, 1994; Haven and Kurosaka, 1997). These are not discussed here.

9.10.3.1 Features of velocity and vorticity fields for a jet in cross-flow

Under the above conditions dimensional analysis shows that the trajectory of the jet, as marked by features such as either the (x–y) locus of maximum velocity in the jet or the x–y curve corresponding to the centerline between the counterrotating vortices, is a function of the velocity ratio, u_j/u_0, between the jet and the oncoming flow:

$$(x/d), (y/d) = f(u_j/u_0). \tag{9.10.17}$$

To give a specific example of the parametric behavior, numerous measurements indicate that the relation between the vertical position of the maximum velocity at any cross-section and the downstream distance from the center of the emerging jet can be approximated by an equation of the form

$$\frac{y}{Rd} = A\left(\frac{x}{Rd}\right)^B, \quad R = u_j/u_0. \tag{9.10.18}$$

In (9.10.18) A and B are constants, with values roughly 1.6 and 1.3 (Hasselbrink and Mungal, 2001; Karagozian, 2002).

The main attributes of the far field velocity distribution for a jet in cross-flow are shown in Figure 9.43, which presents velocity vectors in a plane normal to the jet trajectory. The senses of the coordinates y_v and z_v in Figure 9.43, are indicated in the sketch in Figure 9.42. The origin of these y_v, z_v coordinates is at the location determined by the position of maximum upwash between the vortex pair, which is taken as the midpoint between the two vortices.[14]

The structure shown by the cross-plane velocity components corresponds to a counterrotating vortex pair. The specific data in Figure 9.43 have the origin (the point $y_v = 0$, $z_v = 0$) downstream a distance $x/d = 8.3$ from the centerline of the opening from which the jet emerged, but the qualitative configuration is characteristic for several tens (or more) diameters downstream. The contours in the figure give the ratio of the velocity along the jet trajectory to the maximum jet velocity at that plane.

The jet momentum provides the impulse responsible for the counterrotating vortex pair. The momentum flux out of the opening in the $y = 0$ plane can be regarded as a force on the fluid above the plane, analogous to that for the wing in the previous section. (The parallel being drawn may perhaps be more easily envisioned if one considers the jets to emerge from an opening with the geometry of Figure 9.41.)

Features of the vorticity field for a jet in cross-flow are depicted in Figure 9.44, which gives results from a three-dimensional (Reynolds-averaged Navier–Stokes) computation. The figure shows a perspective view of vortex lines in the region several diameters downstream of the jet entrance.[15] The vortex loops mark the locations of high vorticity in the time-mean simulation. (They are not material lines at different instants of time.)

The vortex loops in Figure 9.44 are much longer in the streamwise direction than the jet entrance diameter, d, with z-components of vorticity at the ends of the loop only. These configurations thus represent a combination of like-sign vorticity in some places and cancellation (by diffusion) of

[14] The figure is different than that in the original reference (Fearn and Weston, 1974), which had measurements that covered only $1 > z_v > -6$. In Figure 9.43 the data for $z_v < 0$ have been flipped and plotted for positive z_v to convey, assuming that left–right symmetry exists, a sense of the velocity field of the pair structure. See Smith and Mungal (1998) for comments concerning the symmetry.

[15] Detailed information concerning the generation of these counterrotating vortices in the region near the jet entrance is given by Cortelezzi and Karagorian (2001).

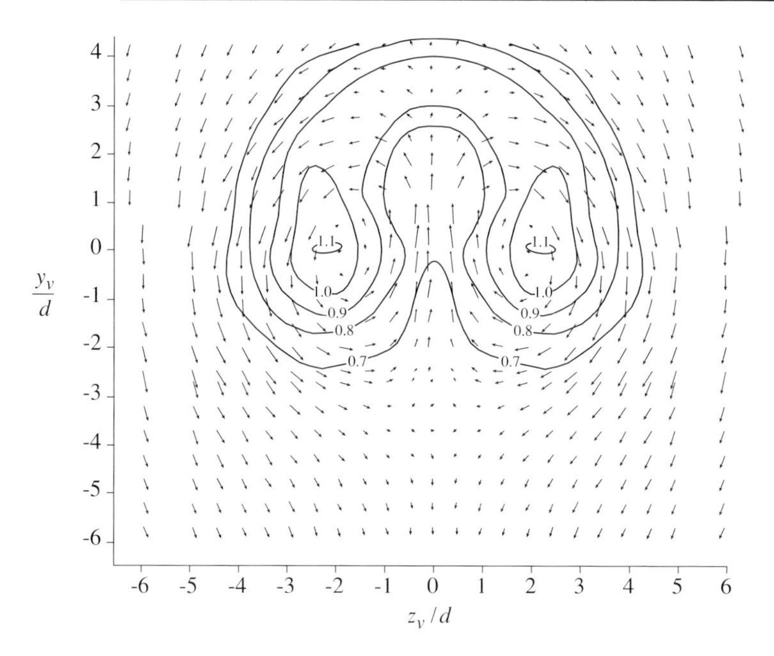

Figure 9.43: Velocity field in a plane normal to the jet trajectory; $u_j/u_0 = 8$, $x/d = 8.3$; numbers on contours refer to u/u_{max} in jet (adapted from Fearn and Weston (1974)).

opposite-sign vorticity in others. Put another way, the presence of the elongated configurations implies vorticity cancellation between (roughly circular) vortex rings which entered the region $y > 0$ at different instants of time, in a manner similar to that seen for the flow past the rectangle in the previous section. The velocity ratio used in the computations corresponds to that in the measurements of Figure 9.43. The cross-plane for the measurement location is roughly the location $x/d = 8$ marked in Figure 9.44, where x is measured from the center of the jet entrance. The vortex configuration is consistent with the observed counterrotating vortex pair.

In the counterrotating vortex pair the two vortices are close enough that there is (turbulent) diffusion of opposite-sign vorticity and the circulation of the individual vortices is not constant. For a two-dimensional vortex pair, however, even if the vortex circulation changes with time, the impulse per unit length is invariant. Application of this latter principle enables us to estimate the far field jet trajectory and circulation variation for a jet in cross-flow (Broadwell and Breidenthal, 1984).

9.10.3.2 An approximate analysis for jet kinematics and vortex pair circulation

A starting point in the development is the assumption of the far downstream flow field as due to a pair of two-dimensional counterrotating vortices nearly aligned with the x-axis. (The vortices are of finite dimension and should not be thought of as simple line vortices of infinitesimal radius.) Denoting y_{vc} as the coordinate which marks the center of the vortex pair and W as the center-to-center vortex spacing, the rate at which the vortex pair moves upward, dy_{vc}/dt, scales as

$$\frac{dy_{vc}}{dt} \propto \frac{\Gamma}{W}. \tag{9.10.19}$$

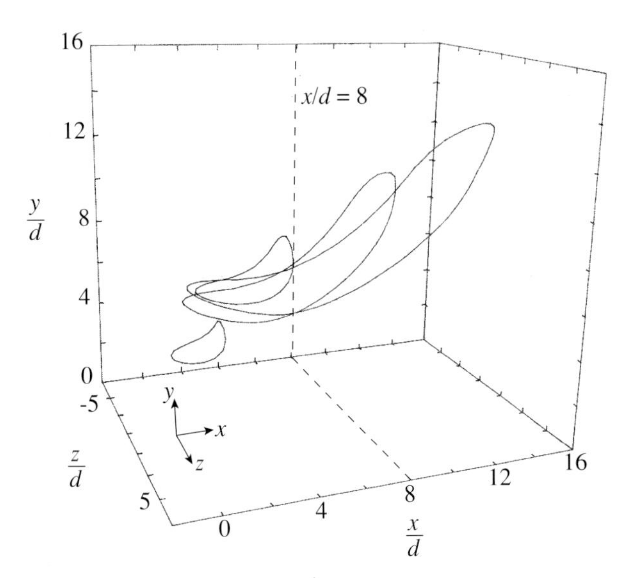

Figure 9.44: Perspective plot of the vortex lines for a circular jet in cross-flow; $\mathbf{u}_0 = u_0\mathbf{i}$, $\mathbf{u}_j = u_j\mathbf{j}$, $u_j/u_0 = 8$ (Sykes, Lewetter, and Parker, 1986).

For reference (see Section 3.15), for a two-dimensional counterrotating pair of line vortices the constant of proportionality is $1/2\pi$.

The impulse of the vortex pair per unit length in the x-direction has magnitude

$$\mathcal{I} \propto \rho \Gamma W. \tag{9.10.20}$$

For the two-dimensional line vortex pair, the associated impulsive pressure, Π, is equal to $\mathcal{I}/W$; from (9.10.8), the constant of proportionality is then unity. Combining (9.10.19) and (9.19.20) yields

$$\rho W^2 \frac{dy_{vc}}{dt} \propto \mathcal{I}. \tag{9.10.21}$$

Motivated by the experimental results (i.e. using the benefit of hindsight!) we seek a similarity solution with the spacing between the counterrotating vortices proportional to their distance from the plane ($W \propto y_{vc}$). Under this assumption (9.10.21) becomes

$$\rho y_{vc}^2 \frac{dy_{vc}}{dt} \propto \mathcal{I}. \tag{9.10.22}$$

Integrating (9.10.22) from time $t = 0$, and taking $y_{vc} = 0$ at the initial time, gives

$$y_{vc} \propto \left(\frac{\mathcal{I}}{\rho}\right)^{1/3} t^{1/3} = \left(\frac{\mathcal{I}}{\rho}\right)^{1/3} \left(\frac{x}{u_0}\right)^{1/3}. \tag{9.10.23}$$

In writing (9.10.23) the vortex pair is viewed as convecting downstream with the oncoming velocity, u_0, so that time corresponds to $t = x/u_0$.

The postulated similarity is that vortex spacing, W, and vortex pair position, y_{vc}, are proportional. This statement, the expression for fluid impulse per unit length in (9.10.20), and invariance of the fluid impulse per unit length, combine to imply that the circulation for each vortex of the pair behaves as

$$\Gamma \propto \left(\frac{\mathcal{I}}{\rho}\right)^{2/3} \left(\frac{x}{u_0}\right)^{-1/3}. \tag{9.10.24}$$

For a two-dimensional vortex pair the impulse has magnitude $\rho\Gamma Wx$ for a length x. Neglecting the contribution of the pressure forces[16] on the plane $y = 0$, the time integral of the force on the fluid above the plane $y = 0$ can be estimated as the product of the force per unit time, $\rho u_j^2 d^2$ and the time duration over which the force is applied, $t = x/u_0$. (The term $\pi d^2/4$ multiplying $\rho u_j^2 d^2$ has been omitted, consistent with other approximations made in these arguments.) Use of the connection between the impulse created and the product of force and duration of application yields an expression for the impulse in terms of jet and free-stream parameters as

$$\frac{\mathcal{I}}{\rho} \propto \frac{u_j^2 d^2}{u_0}. \tag{9.10.25}$$

Substitution of (9.10.25) in (9.10.23) yields the desired scaling for vortex pair position, y_{vc}, and the spacing between the vortices, W, as

$$\frac{y_{vc}}{Rd}, \frac{W}{Rd} \propto \left(\frac{x}{Rd}\right)^{1/3}. \tag{9.10.26}$$

Figure 9.45 shows measurements of the vortex pair trajectory, $y_{vc}/(Rd)$, for several different values of R. The solid curve in the figure represents a best fit of the scaling implied by (9.10.25). The trajectory of the maximum velocity has a similar scaling, but the quantitative value of the vertical position of the maximum lies above (roughly 20–30% depending on R) that for the vortex pair center (Fearn and Weston, 1978).

Figure 9.46 shows the measured variation in vortex circulation versus $x/(Rd)$. The solid line in the figure has the behavior

$$\frac{\Gamma}{u_0 Rd} = \frac{\Gamma}{u_j d} \propto \left(\frac{x}{Rd}\right)^{-1/3}. \tag{9.10.27}$$

As described by Hasselbrink and Mungal (2001), the measured ratio between the vortex spacing and x has been used to provide a numerical estimate for the proportionality constant in the equation. This is given in the figure.[17] Although other aspects of the problem need to be addressed for a detailed description, Figures 9.45 and 9.46 show that the basic scaling arguments provide insight into the functional dependence for vortex position, spacing, and circulation.

For a jet in cross-flow, the vortex circulation decreases with time, but there are other vortex pair configurations in which the rate of decrease in circulation is proportionally much smaller. For a flow whose evolution can be modeled by the behavior of a two-dimensional vortex pair with vortex circulation and fluid impulse invariant we can make an immediate statement about vortex behavior.

[16] See Hasselbrink and Mungal (2001) for discussion of this assumption, which implies that R is much larger than unity.

[17] Other data (Fearn and Weston, 1978) show more scatter (almost twice as much in some cases) in the circulation measurements, although this may be because the values of x/Rd are smaller and the data not truly representative of far field behavior. An estimate for the minimum at which x/Rd far field behavior is encountered is $x/Rd > 0.2\,R$ (Smith and Mungal, 1998).

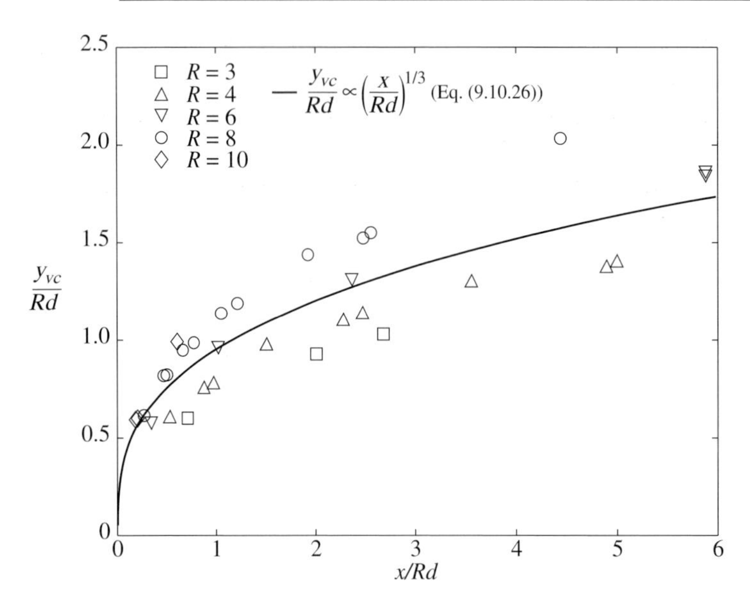

Figure 9.45: Trajectory of the center of the counterrotating vortex pair, y_{vc}, for a jet in a cross-flow, $R = u_j/u_0$. Data of Fearn (2002) after Fearn and Weston (1974).

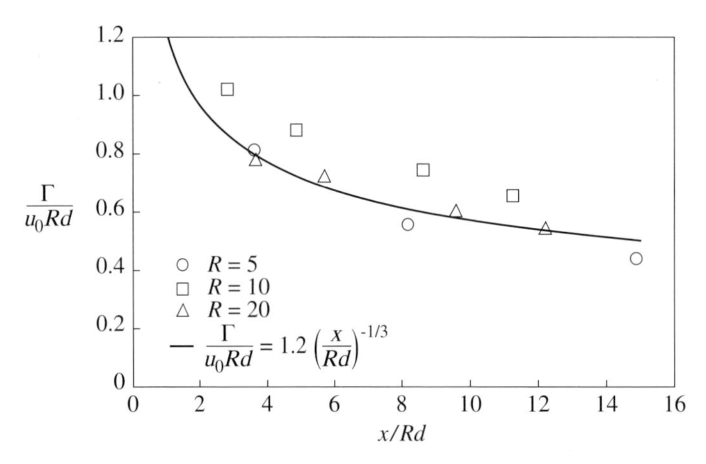

Figure 9.46: Vortex circulation versus downstream distance for a counterrotating vortex pair of jets in cross-flow, $R = u_j/u_0$; analysis of Hasselbrink and Mungal (2001).

The flow past a finite wing, where the vortex separation is large, is one example. The tip clearance flow in a turbomachine, which is characterized by a vortex which trails off the rear edge of the blade and is nearly aligned with the streamwise direction, is another.

A tip clearance vortex in proximity to a casing is equivalent to a vortex pair consisting of the vortex and its image (Section 3.15). For representative (blade-to-casing) clearances the vortex can be far enough away from the wall that the rate of change in circulation with downstream distance is small and the flow pattern in a cross-flow plane is approximated by that for a pair of two-dimensional vortices. Downstream of the blade the only forces in the direction normal to the line between the vortex and its image are the wall shear stresses and if these can be neglected the impulse of the vortex pair is constant. From (9.10.20), invariance of both vortex circulation and impulse implies the distance between the vortex pair (more precisely between the centroids of vorticity) is also invariant. For downstream distances over which these conditions are met, the distance of the vortex from the endwall is thus constant (Chen *et al.*, 1991).

10 Compressible internal flow

10.1 Introduction

Chapters 10 and 11 address flows in which substantial changes in density occur. The changes arise from processes which are dynamical (e.g. density changes from pressure variations associated with fluid accelerations) or thermodynamic (density changes primarily from bulk heat addition due to chemical reaction or phase change) or a combination of the two. This chapter focuses primarily on situations with density variations due to dynamical effects; as we saw in Section 2.2, this means flows with Mach numbers significant compared to unity. Chapter 11 discusses flows with density variations primarily due to heat addition.

Much of the material is based on quasi-one-dimensional gas dynamics. Characterization of quasi-one-dimensional analysis as "the secret weapon of the internal fluid dynamicist" (Heiser, 1995) is an apt aphorism indeed. This type of treatment enables useful engineering estimates in a wide variety of situations and is a powerful tool for providing insight into the response of compressible flows to alterations in area, addition of mass, momentum, and energy, swirl, and flow non-uniformity. This is true not only for simple duct and channel flows but also for more complex problems, for example those arising in the matching of gas turbine engine components (Kerrebrock, 1992; Cumpsty, 1998).

Many computational techniques now exist to address internal flows in complex geometries. As such, we spend little time in discussion of approximations that were necessary in the past to attack compressible flow problems. One-dimensional analysis, however, is still very much a part of modern approaches to grappling with internal flow problems, even though its use as a detailed design tool has been supplanted by more accurate computations. The analyses presented in this and the next chapter, which draw strongly on the seminal work of Shapiro (1953) and its extensions by Heiser and colleagues (e.g., Anderson, Heiser, and Jackson, 1970; Bernstein, Heiser, and Hevenor, 1967; Heiser and Pratt, 1994; Curran, Heiser, and Pratt, 1996), can be viewed as a complement to, and more importantly, an aid in interpretation of, numerical simulations in common use today.

10.2 Corrected flow per unit area

Appreciation for the overall response of a compressible flow to alterations in area, and addition of mass, momentum, and energy can be gained through consideration of the corrected flow per unit area introduced in Section 2.5. For a perfect gas with constant specific heats it was shown that the quantity $\dot{m}\sqrt{RT_t}/(Ap_t\sqrt{\gamma})$ is a function of Mach number only, denoted as $D(M)$:

$$\frac{\dot{m}\sqrt{RT_t}}{Ap_t\sqrt{\gamma}} = D(M) = \frac{M}{\left[1+\left(\frac{\gamma-1}{2}\right)M^2\right]^{\frac{(\gamma+1)}{2(\gamma-1)}}}. \tag{2.5.3}$$

The functional dependence is depicted in Figure 2.6, with $D(M)$ increasing from zero at $M = 0$ to a maximum at $M = 1$ and then decreasing at higher Mach numbers. For a specified fluid, with constant values of R and γ, the Mach number in a channel is a function of $\dot{m}\sqrt{T_t}/(Ap_t)$, where T_t and p_t are the local values of the stagnation temperature and pressure.[1]

We describe compressible flow behavior in two steps: (i) computation of the change in corrected flow per unit area, and hence Mach number, resulting from geometry variation and addition of mass, momentum, and energy, and then (ii) linkage of the changes to specific physical processes. We thus consider the corrected flow per unit area in a channel with Mach number, M_i, at an initial station i, and determine the changes between this and a downstream station if the area, mass flow, stagnation temperature, and stagnation pressure are altered from A, $\dot{m}$, T_t, p_t to $(A + dA)$, $(\dot{m} + d\dot{m})$, $(T_t + dT_t)$, $(p_t + dp_t)$. For small fractional changes in these quantities (dA/A, dT_t/T_t, etc. $\ll 1$) the change in corrected flow per unit area between the two stations is given by

$$\frac{d[\dot{m}\sqrt{T_t}/(Ap_t)]}{[\dot{m}\sqrt{T_t}/(Ap_t)]} = \left(\frac{d\dot{m}}{\dot{m}} + \frac{dT_t}{2T_t} - \frac{dA}{A} - \frac{dp_t}{p_t}\right). \tag{10.2.1}$$

Use of (10.2.1) in connection with Figure 2.6 shows several features of compressible channel flow. For both subsonic and supersonic flow, increase in physical mass flow (mass addition to the stream), increases in stagnation temperature, decreases in area, and decreases in stagnation pressure all increase the corrected flow per unit area. For *given initial conditions*, all these drive the flow towards a Mach number of unity. Further, the curve of $D(M)$ versus M flattens in the neighborhood of Mach number equal to 1 (it has zero slope at $M = 1$). A specified fractional change in corrected flow per unit area therefore implies a larger change in Mach number as conditions become closer to $M = 1$. Finally, there is a *maximum value of corrected flow per unit area*. A change to a value greater than this, for example an increase in stagnation temperature that implied a corrected flow per unit area larger than the maximum, is not possible. For fixed geometry such an increase in stagnation temperature would be accompanied by an upstream readjustment of the flow, with a consequent decrease in the Mach number at station i to accommodate the imposed stagnation temperature change. If such an upstream readjustment were not desired, the area would need to be increased to have the corrected flow per unit area not exceed the maximum allowed value.

An illustration of the last point is the operation of the variable exit nozzle on an afterburning jet engine, as sketched in Figure 10.1. At almost all flight conditions the exit nozzle is choked, and the corrected flow per unit area has its maximum value at the nozzle throat. With the afterburner unlit, the nozzle geometry has the configuration depicted as "afterburner off". When the afterburner is

[1] The term "corrected flow" (rather than corrected flow per unit area) is attached to the quantity $\dot{m}\sqrt{RT_t}/(p_t\sqrt{\gamma})$. This is typically used to describe situations in which the Mach numbers are comparable but the areas and flows are different, for example turbine engine gas generators, which are often characterized in terms of compressor inlet or exit corrected flow.

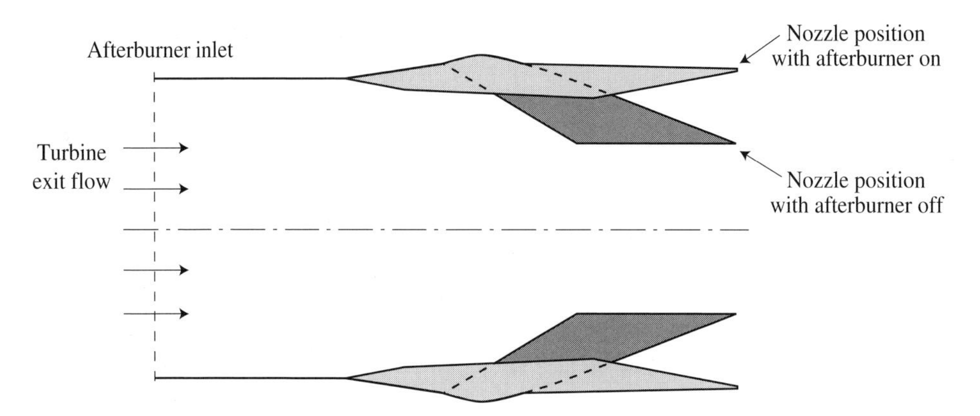

Figure 10.1: Schematic of an axisymmetric variable area nozzle on an afterburning gas turbine engine (Jane's, 1999).

lit, the stagnation temperature at the nozzle inlet can increase by a factor of 2. The workings of the turbomachinery in the gas turbine engine, however, dictate that the physical mass flow and stagnation pressure into the nozzle are roughly the same before and after the afterburner is lit. For this to be the case the nozzle throat area must be increased in proportion to the square root of temperature and, in afterburning operation, the nozzle is opened to the "afterburner on" configuration in the figure.

Changes in the flow state can also be connected to the physical processes of friction, heat addition, and shaft work (such as in turbomachinery components). The effect of friction is to lower the stagnation pressure. In both subsonic and supersonic flow, therefore, frictional effects downstream of a specified initial state cause the flow to move towards a Mach number of unity. For a constant area flow, increases in friction, as obtained for example by increases in duct length, can result in a condition in which the maximum corrected flow per unit area is reached and the flow is choked at exit.

Heat addition has two effects. The stagnation temperature increases and (as developed in Chapter 11) the stagnation pressure decreases. Both of these increase the corrected flow per unit area. As with friction, for a given initial state heat addition causes the flow to move towards a Mach number of unity in both subsonic and supersonic flow. A consequence is that if enough heat is added in a constant area duct "thermal choking" occurs in which the maximum corrected flow per unit area is reached, limiting the physical flow rate. Effects of both heat addition and friction are developed in more depth in the next sections.

Addition or extraction of shaft work, without dissipation or heat transfer, is an idealization of the process in a compressor or turbine. From the steady flow energy equation, (1.8.11), for a gas with constant specific heats, the change in stagnation temperature for an incremental amount of shaft work, dw_{shaft}, done per unit mass is

$$c_p dT_t = -dw_{shaft}. \tag{10.2.2}$$

The Gibbs equation ($Tds = dh - dp/\rho$, (1.3.19)) with $ds = 0$ allows us to relate changes in stagnation pressure for isentropic shaft work to changes in stagnation temperature as

$$\frac{dT_t}{T_t} = \frac{(\gamma - 1)}{\gamma}\frac{dp_t}{p_t}. \tag{10.2.3}$$

Equation (10.2.1) can thus be written for input or extraction of isentropic shaft work as

$$\frac{d[\dot{m}\sqrt{T_t}/(Ap_t)]}{[\dot{m}\sqrt{T_t}/(Ap_t)]} = \left(\frac{dT_t}{2T_t} - \frac{dp_t}{p_t}\right) = \left[\underbrace{\frac{2\gamma}{(\gamma-1)}\left(\frac{dw_{shaft}}{2c_pT_t}\right)}_{\text{From } p_t \text{ change}} + \underbrace{\left(\frac{-dw_{shaft}}{2c_pT_t}\right)}_{\text{From } T_t \text{ change}}\right]. \tag{10.2.4}$$

Equation (10.2.4) shows that shaft work compression processes (increases in p_t and T_t, $dw_{shaft} < 0$) decrease the corrected flow per unit area, and expansion processes increase the corrected flow per unit area.[2] In these, the change in stagnation pressure is dominant (by a factor of 7 for γ of 1.4) compared to the change in stagnation temperature so the net effect is in the direction of the former. The dominance of the stagnation pressure change is less in an actual device because the process is not isentropic, but the overall trend is valid.

The discussion in this section has been mainly qualitative. Much of the rest of the chapter is aimed at providing quantitative tools for analysis of compressible flows, with the operating regimes of inlets and wind tunnels discussed as applications. The quasi-one-dimensional description is also extended to two other important situations, compressible flow with swirl and non-uniform flow with mixed subsonic–supersonic regions. The extensions maintain much of the simplicity of the one-dimensional approach but allow examination of a much wider class of phenomena. In the last section of the chapter a flow substitution principle is introduced which enables construction of a multiplicity of solutions for different stagnation temperature distributions once a single solution has been found.

10.3 Generalized one-dimensional compressible flow analysis

One-dimensional analyses of compressible flow with mass, momentum, and energy addition can be developed starting from the conservation equations for an elementary control volume in a duct or channel shown in Figure 10.2. Within the control volume there is the possibility for mass addition, frictional forces, body forces, shaft work, and heat exchange. The resulting differential equations can be numerically integrated, but the form of the equations themselves gives much information about the direction and nature of the solution path. In the development here we take both mainstream and injected fluid to be perfect gases with constant specific heat; the derivation for varying specific heats and the injection of liquid is given by Shapiro (1953). It is convenient (and simpler) to describe the general case in two parts. Channel flows with no shaft work or work done by body forces (but all other effects) are first dealt with, followed by examination of the effects of work production.

10.3.1 Differential equations for one-dimensional flow

Referring to the control volume in Figure 10.2, conservation of mass is expressed as

$$d(\rho u A) = d\dot{m}. \tag{10.3.1}$$

[2] Note that the processes referred to are those in which shaft work is done and the stagnation quantities change; flow though a diffuser or nozzle is *not* this type of process.

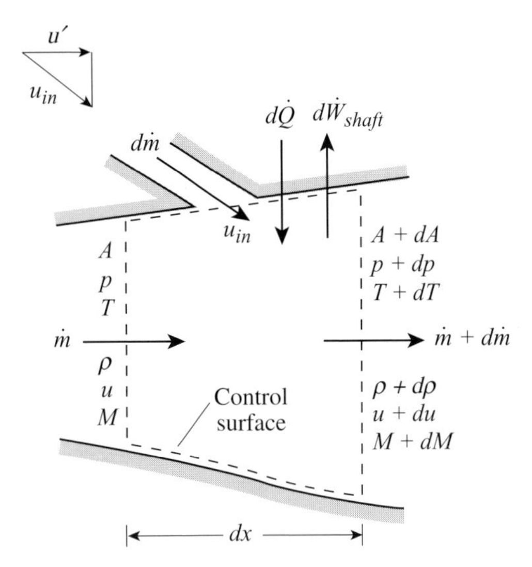

Figure 10.2: Control volume with addition of mass, momentum, and energy to a control volume.

In (10.3.1) $d\dot{m}$ represents the incremental change in mass flow across the control volume. Expanding the differential, dividing through by $\rho u A$ and keeping terms which are first order in the quantities du/u, dA/A, $d\rho/\rho$, gives

$$\frac{d\rho}{\rho} + \frac{du}{u} = -\frac{dA}{A} + \frac{d\dot{m}}{\dot{m}}. \tag{10.3.2}$$

The convention for which variables are viewed as independent (as on the right-hand side of (10.3.2)) and which as dependent is made based on application. Area, mass flow, velocity of the injected flow, shaft work, external force, and heat addition are regarded as quantities over which the designer has control and thus as independent variables.

Conservation of momentum for the control volume combined with (10.3.2) yields

$$(\rho u du + dp)\, A = -\tau_w dA_w + d\mathcal{F}_D + u_{inj} d\dot{m}. \tag{10.3.3}$$

In (10.3.3) τ_w is the shear stress at the wall due to friction, dA_w is the wetted area around the perimeter of the control volume, $d\mathcal{F}_D$ represents drag forces on the fluid such as those associated with the presence of struts or screens, for example, and u_{inj} is the velocity of the injected mass in the main stream direction. (In this equation, and in the subsequent development, to make the levels of notation less complicated we drop the notation "đ( )" for quantities such as work, heat, friction forces, which are not properties.) Dividing through by ρu^2 and employing $\rho u^2 = \gamma p M^2$,

$$\frac{du}{u} + \frac{dp}{\gamma M^2 p} = -\frac{\tau_w}{\gamma M^2 p}\frac{dA_w}{A} + \frac{d\mathcal{F}_D}{\gamma M^2 p A} + \frac{d\dot{m}}{\dot{m}}\left(\frac{u_{inj}}{u} - 1\right). \tag{10.3.4}$$

The wetted area can be related to the flow-through area using the hydraulic diameter, defined as $d_H = 4A/(\text{wetted perimeter})$. As described by Schlichting (1979) or White (1991) the use of the hydraulic diameter allows estimation of wall shear stress for ducts of arbitrary cross-section in terms of the diameter of a corresponding circular duct, for which the behavior of the wall shear stress is well documented. Writing the wall shear stress as $\tau_w = C_f(\rho u^2/2)$, where C_f is the skin friction coefficient, (10.3.4) becomes

$$\frac{du}{u} + \frac{dp}{\gamma M^2 p} = -2C_f \frac{dx}{d_H} + \frac{d\mathcal{F}_D}{\gamma M^2 pA} + \frac{d\dot{m}}{\dot{m}}\left(\frac{u_{inj}}{u} - 1\right). \tag{10.3.5}$$

Application of conservation of energy for the control volume, plus use of $dh_t = c_p\, dT_t$, gives

$$c_p dT_t = \frac{d\dot{m}}{\dot{m}} c_p (T_{t_{inj}} - T_t) + dq. \tag{10.3.6}$$

In (10.3.6) $T_{t_{inj}}$ is the stagnation temperature of the injected gas and dq is the heat addition to the stream per unit mass. In non-dimensional form

$$\frac{dT_t}{T_t} = \frac{d\dot{m}}{\dot{m}}\left(\frac{T_{t_{inj}}}{T_t} - 1\right) + \frac{dq}{c_p T_t}. \tag{10.3.7}$$

The conservation equations must be supplemented by the equation of state and the definition of stagnation temperature. In differential form these are:

$$\frac{dp}{p} - \frac{d\rho}{\rho} - \frac{dT}{T} = 0, \tag{10.3.8}$$

$$\frac{dT}{T}\left[\frac{1}{1 + \dfrac{\gamma - 1}{2} M^2}\right] + \frac{du}{u}\left[\frac{(\gamma - 1)M^2}{1 + \dfrac{\gamma - 1}{2} M^2}\right] - \frac{dT_t}{T_t} = 0. \tag{10.3.9}$$

Equations (10.3.2), (10.3.5), (10.3.7), (10.3.8), and (10.3.9) are five equations for the dependent variables:[3] du/u, dT/T, dp/p, $d\rho/\rho$, dT_t/T_t. The stagnation temperature variation, dT_t/T_t is given in terms of independent quantities in (10.3.7), so we use it directly as an independent variable to reduce the number of dependent quantities to four: du/u, dT/T, dp/p, $d\rho/\rho$. Once these are obtained, other derived quantities such as entropy, stagnation pressure, or Mach number can be determined as:

$$\frac{ds}{c_p} = \frac{dT}{T} - \frac{(\gamma - 1)}{\gamma}\frac{dp}{p}, \tag{10.3.10}$$

$$\frac{dp_t}{p_t} = \frac{dp}{p} + \left[\frac{\gamma M^2/2}{1 + \left(\dfrac{\gamma - 1}{2}\right) M^2}\right]\frac{dM^2}{M^2}, \tag{10.3.11}$$

$$\frac{dM^2}{M^2} = \frac{2du}{u} - \frac{dT}{T}, \quad \left(\text{or } \frac{dM}{M} = \frac{du}{u} - \frac{dT}{2T}\right). \tag{10.3.12}$$

As developed in Section 5.5, the entropy change ds in (10.3.10) is the change in stream entropy (per unit mass) across the control volume. The irreversible entropy creation within the control volume is

[3] The solution of the equations could of course proceed with other choices of independent and dependent variables, for example use of the area ratio as a dependent variable to determine the area change needed to create a given pressure or velocity increment.

given by (5.5.28):

$$ds_{irrev} = ds - (s_{inj} - s)\frac{d\dot{m}}{\dot{m}}. \qquad (5.5.28)$$

10.3.2 Influence coefficient matrix for one-dimensional flow

The solution for all the dependent quantities can be set up as a matrix of *influence coefficients* (Shapiro, 1953), as in Table 10.1. The top row of the table lists the independent variables and the column on the left the dependent variables. The groupings in the top row come from the conservation equations and are presented in this fashion to show where the terms arise. To use Table 10.1, one multiplies the influence coefficient in each column with the form of the dependent variable in the row at the top of the columns. The change in Mach number, for example, is given by

$$\frac{dM^2}{M^2} = \left[\frac{-2\left(1+\frac{\gamma-1}{2}M^2\right)}{1-M^2}\right]\frac{dA}{A} + \left[\frac{(1+\gamma M^2)\left(1+\frac{\gamma-1}{2}M^2\right)}{1-M^2}\right]\frac{dT_t}{T_t}$$
$$+\left[\frac{\gamma M^2\left(1+\frac{\gamma-1}{2}M^2\right)}{1-M^2}\right]\left(4C_f\frac{dx}{d_H} + \frac{2d\mathcal{F}_D}{\gamma pAM^2} - 2\frac{u_{inj}}{u}\frac{d\dot{m}}{\dot{m}}\right)$$
$$+\left[\frac{2(1+\gamma M^2)\left(1+\frac{\gamma-1}{2}M^2\right)}{1-M^2}\right]\frac{d\dot{m}}{\dot{m}}. \qquad (10.3.13)$$

All the terms in square brackets in (10.3.13) are the influence coefficients, which are the partial derivatives of the variable in the left-hand column with respect to the variable in the top row. We will use these influence coefficients in the next section to illustrate the influence of friction and heat addition to a compressible flow.

10.3.3 Effects of shaft work and body forces

Shaft work and body forces are not included as independent variables in Table 10.1. We now assess their effects, carrying out the analysis for the case of no mass addition only. To demonstrate the roles of shaft work and body forces separately we split the total force on the fluid, over and above friction and pressure forces, into three parts: body forces, drag forces, and forces associated with shaft work. For the incremental control volume of area A and length dx the contribution of the body force per unit mass, X, is $\rho AXdx$. The drag force due to stationary objects in the stream, denoted by $d\mathcal{F}_D$, has already been described. The force on the fluid within the control volume associated with the shaft work per unit mass, dw_{shaft}, is denoted as $d\mathcal{F}_{shaft}$. The relation between this force and the shaft work per unit mass is

$$\dot{m}dw_{shaft} = ud\mathcal{F}_{shaft}. \qquad (10.3.14)$$

Table 10.1 *Influence coefficients for compressible channel flow: constant specific heat and molecular weight (Shapiro, 1953)*

	$\frac{dA}{A}$	$\frac{dT_t}{T_t}$	$4C_f \frac{dx}{d_H} + \frac{2d\mathcal{F}_D}{\gamma p A M^2} - 2\frac{u_{inj}}{u}\frac{d\dot{m}}{\dot{m}}$	$\frac{d\dot{m}}{\dot{m}}$
$\frac{du}{u}$	$-\frac{1}{1-M^2}$	$\frac{1+\frac{\gamma-1}{2}M^2}{1-M^2}$	$\frac{\gamma M^2}{2\left(1-M^2\right)}$	$\frac{1+\gamma M^2}{1-M^2}$
$\frac{dT}{T}$	$\frac{(\gamma-1)M^2}{1-M^2}$	$\frac{\left(1-\gamma M^2\right)\left(1+\frac{\gamma-1}{2}M^2\right)}{1-M^2}$	$-\frac{\gamma(\gamma-1)M^4}{2\left(1-M^2\right)}$	$-\frac{(\gamma-1)M^2\left(1+\gamma M^2\right)}{1-M^2}$
$\frac{dp}{p}$	$\frac{\gamma M^2}{1-M^2}$	$-\frac{\gamma M^2\left(1+\frac{\gamma-1}{2}M^2\right)}{1-M^2}$	$-\frac{\gamma M^2\left[1+(\gamma-1)M^2\right]}{2\left(1-M^2\right)}$	$-\frac{2\gamma M^2\left(1+\frac{\gamma-1}{2}M^2\right)}{1-M^2}$
$\frac{d\rho}{\rho}$	$\frac{M^2}{1-M^2}$	$-\frac{1+\frac{\gamma-1}{2}M^2}{1-M^2}$	$-\frac{\gamma M^2}{2\left(1-M^2\right)}$	$-\frac{(\gamma+1)M^2}{1-M^2}$
$\frac{dp_t}{p_t}$	0	$-\frac{\gamma M^2}{2}$	$-\frac{\gamma M^2}{2}$	$-\gamma M^2$
$\frac{dM^2}{M^2}$	$-\frac{2\left(1+\frac{\gamma-1}{2}M^2\right)}{1-M^2}$	$\frac{\left(1+\gamma M^2\right)\left(1+\frac{\gamma-1}{2}M^2\right)}{1-M^2}$	$\frac{\gamma M^2\left(1+\frac{\gamma-1}{2}M^2\right)}{1-M^2}$	$\frac{2\left(1+\gamma M^2\right)\left(1+\frac{\gamma-1}{2}M^2\right)}{1-M^2}$
$\frac{ds}{c_p}$	0	$1+\frac{\gamma-1}{2}M^2$	$\frac{(\gamma-1)M^2}{2}$	$(\gamma-1)M^2$
$\frac{d\mathfrak{I}}{\mathfrak{I}}$	$\frac{1}{1+\gamma M^2}$	0	$-\frac{\gamma M^2}{2\left(1+\gamma M^2\right)}$	0

Note: Each influence coefficient represents the partial derivative of the variable in the left-hand column with respect to the variable in the top row; for example:

$$\frac{dp_t}{p_t} = -\left(\frac{\gamma M^2}{2}\right)\frac{dT_t}{T_t} - \left(\frac{\gamma M^2}{2}\right)\left(4C_f\frac{dx}{d_H} + \frac{dX}{\frac{1}{2}\gamma p A M^2} - 2\frac{u_{inj}}{u}\frac{d\dot{m}}{\dot{m}}\right) - \left(\gamma M^2\right)\frac{d\dot{m}}{\dot{m}}.$$

The control volume form of the momentum equation with these additional forces and no mass addition ($d\dot{m} = 0$) is

$$\frac{du}{u} + \frac{dp}{\gamma M^2 p} = -2C_f \frac{dx}{d_H} + \frac{Xdx}{u^2} + \frac{d\mathcal{F}_D}{\gamma M^2 pA} - \frac{dw_{shaft}}{u^2}. \tag{10.3.15}$$

The energy equation must be modified from the form given as (10.3.7) because stagnation temperature changes now arise not only from heat addition, but also from shaft work and from work done *by* body forces *on* the flow; the last of these has magnitude uX per unit mass. Application of the steady-flow energy equation yields

$$\dot{m}c_p dT_t = \dot{m}c_p dq - \dot{m}dw_{shaft} + \rho uXAdx.$$

Dividing through by $\dot{m}c_pT_t$,

$$\frac{dT_t}{T_t} = \frac{dq}{c_pT_t} - \frac{dw_{shaft}}{c_pT_t} + \frac{Xdx}{c_pT_t}. \tag{10.3.16}$$

To recap, the differential equations describing compressible channel flow with area change, friction, shaft work, body forces, heat addition, and drag are: continuity ((10.3.2) with $d\dot{m}/\dot{m}$ equal to zero), momentum (10.3.15), energy (10.3.16)), state (10.3.8), definition of stagnation temperature, (10.3.9), and the equations for derived quantities such as entropy, stagnation pressure, and Mach number.

Table 10.2 shows the influence coefficient matrix for this situation, again for a perfect gas with constant specific heats. A difference from the results in Table 10.1 is that the stagnation temperature is now a dependent variable, which can be altered not only from heat addition, but also from shaft work and the work of body forces. Further, these increases in stagnation temperature are not necessarily accompanied by increases in entropy and decreases in stagnation pressure, as is the case for the flow described by Table 10.1. For example if there is reversible work addition alone (adiabatic, frictionless flow with shaft work), as discussed in Section 10.2, the stagnation temperature is increased, the stagnation pressure is increased, and the entropy is unchanged. For this situation the influence coefficients from Table 10.2, which describe an isentropic compression or expansion, are shown as brackets in (10.3.17) below:

$$\frac{dT_t}{T_t} = [-1]\frac{dw_{shaft}}{c_pT_t}, \tag{10.3.17a}$$

$$\frac{dp_t}{p_t} = \left[-\frac{\gamma}{\gamma-1}\right]\frac{dw_{shaft}}{c_pT_t}, \tag{10.3.17b}$$

$$\frac{ds}{c_p} = [0]. \tag{10.3.17c}$$

An example of a fluid machinery situation with body forces is the flow in a rotating radial channel. The centrifugal body force results in a variation in stagnation quantities (as measured in the rotating channel reference frame) with radius. General results for body forces can be obtained from the influence coefficient table, but for the special case of isentropic rotating channel flow the ideas can be developed from a slightly different perspective to illustrate some overall features of the general case.

The geometry is shown in Figure 10.3, with a control volume extending from location r to $r + dr$. As developed in Section 7.3 (or as obtained from integration of the expression for the stagnation

Table 10.2 *Influence coefficients for compressible channel flow with body forces and shaft work (no mass addition, constant specific heat and molecular weight) (Fitzgerald, 2002)*

	$\frac{dA}{A}$	$\frac{dq}{c_p T_t}$	$\frac{dw_{shaft}}{c_p T_t} - \frac{X dx}{c_p T_t}$	$4C_f \frac{dx}{d_H} + \frac{2d\mathcal{F}_D}{\gamma p A M^2}$
$\frac{du}{u}$	$-\frac{1}{1-M^2}$	$\frac{1+\frac{\gamma-1}{2}M^2}{1-M^2}$	$\frac{1+\frac{\gamma-1}{2}M^2}{(1-M^2)(\gamma-1)}$	$\frac{\gamma M^2}{2(1-M^2)}$
$\frac{dT}{T}$	$\frac{(\gamma-1)M^2}{1-M^2}$	$-\frac{\left(1+\frac{\gamma-1}{2}M^2\right)(\gamma M^2-1)}{1-M^2}$	$-\frac{\left(1+\frac{\gamma-1}{2}M^2\right)}{1-M^2}$	$-\frac{\gamma M^4(\gamma-1)}{2(1-M^2)}$
$\frac{dT_t}{T_t}$	0	1	-1	0
$\frac{dp}{p}$	$\frac{\gamma M^2}{1-M^2}$	$-\frac{\gamma M^2\left(1+\frac{\gamma-1}{2}M^2\right)}{1-M^2}$	$-\frac{\gamma\left(1+\frac{\gamma-1}{2}M^2\right)}{(1-M^2)(\gamma-1)}$	$-\frac{\gamma M^2\left[1+(\gamma-1)M^2\right]}{2(1-M^2)}$
$\frac{d\rho}{\rho}$	$\frac{M^2}{1-M^2}$	$-\frac{1+\frac{\gamma-1}{2}M^2}{1-M^2}$	$-\frac{1+\frac{\gamma-1}{2}M^2}{(1-M^2)(\gamma-1)}$	$-\frac{\gamma M^2}{2(1-M^2)}$
$\frac{dp_t}{p_t}$	0	$-\frac{\gamma M^2}{2}$	$-\frac{\gamma}{\gamma-1}$	$-\frac{\gamma M^2}{2}$
$\frac{dM^2}{M^2}$	$-\frac{2\left(1+\frac{\gamma-1}{2}M^2\right)}{1-M^2}$	$\frac{\left(1+\frac{\gamma-1}{2}M^2\right)(\gamma M^2+1)}{1-M^2}$	$\frac{\left(1+\frac{\gamma-1}{2}M^2\right)(1+\gamma)}{(1-M^2)(\gamma-1)}$	$\frac{\gamma M^2\left[1+\frac{\gamma-1}{2}M^2\right]}{1-M^2}$
$\frac{ds}{c_p}$	0	$1+\frac{\gamma-1}{2}M^2$	0	$\frac{(\gamma-1)M^2}{2}$
$\frac{d\Im}{\Im}$	$\frac{1}{1+\gamma M^2}$	0	$-\frac{\gamma\left(1+\frac{\gamma-1}{2}M^2\right)}{(1+\gamma M^2)(\gamma-1)}$	$-\frac{\gamma M^2}{2(1+\gamma M^2)}$

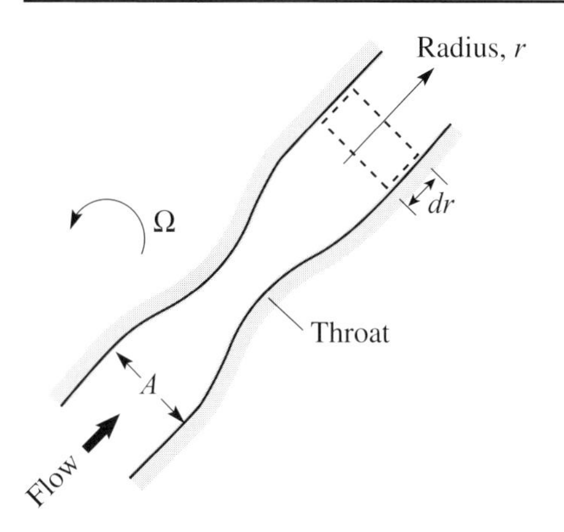

Figure 10.3: Flow in a rotating channel as an example of one-dimensional flow with body forces.

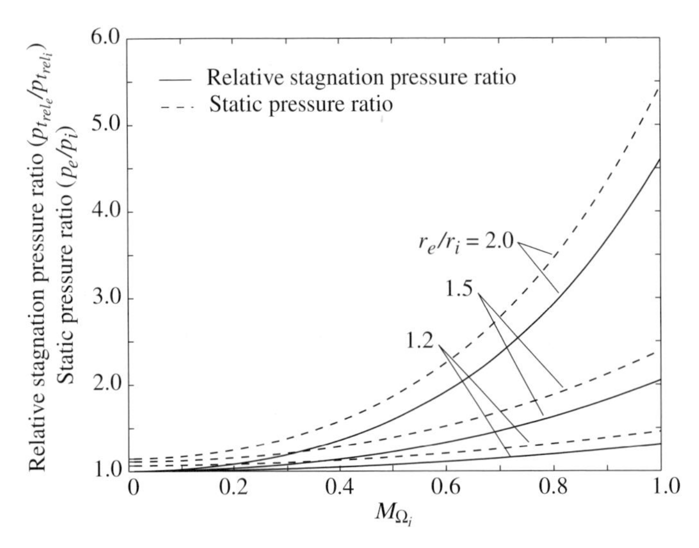

Figure 10.4: Relative stagnation pressure ratio and static pressure ratio in a radially bladed, rotating passage as a function of $M_{\Omega_i} = \Omega r_i/\sqrt{\gamma T_i}$; inlet Mach number, M_{rel_i}, = 0.5.

temperature change in Table 10.2) the rothalpy, $I_t = h + w^2/2 - \Omega^2 r^2/2$, is constant along a relative streamline. Constant rothalpy means the relative stagnation temperature, and hence the relative stagnation pressure, increases in the outward direction. Figure 10.4 shows static pressure and relative stagnation pressure ratios in a radially bladed passage for different values of exit/inlet radius ratio, r_e/r_i, as a function of the inlet rotational Mach number, $M_{\Omega_i} = \Omega r_i/\sqrt{\gamma R T_i}$. The relative inlet Mach number, M_{rel_i} is equal to 0.5. For a stationary passage ($M_\Omega = 0$) there is no change in stagnation pressure, but the relative stagnation pressure ratio increases strongly with rotational Mach number.

A further feature of flows with body forces can be seen from the one-dimensional continuity and momentum equations for a rotating channel:

$$\frac{d\rho}{\rho} + \frac{dw}{w} = -\frac{dA}{A}, \tag{10.3.18}$$

$$w dw + \frac{1}{\rho} dp = \Omega^2 r dr. \tag{10.3.19}$$

For isentropic flow, changes in pressure and density are related by $dp/d\rho = a^2$, where a is the local speed of sound (Section 1.15). Using this to eliminate dp in the momentum equation and substituting the resulting expression for $d\rho/\rho$ in the continuity equation (as described in the introduction to one-dimensional flow in Section 2.5) yields a relation between changes in velocity and changes in area and radius:

$$\frac{dw}{w} = \frac{-\frac{dA}{A} - M_\Omega^2 \frac{dr}{r}}{1 - M^2}. \tag{10.3.20}$$

In (10.3.20), two Mach numbers appear: the flow-through relative Mach number, $M = w/a$, and the rotational Mach number M_Ω ($= \Omega r/a$). The independent variables are area and radius. As discussed in Section 2.5, the condition for the velocity to smoothly increase through the sonic point ($M = 1$)

necessitates that the numerator in (10.3.20) goes to zero. The flow does not become sonic at the geometric throat ($dA/dr = 0$), but rather when $(dA/A)/(dr/r) = -M_\Omega^2$. This would occur in the converging part of the channel at a smaller radius than the throat, because pressure and density both increase with radius. The movement of the sonic point from the throat is associated with the presence of the centrifugal body force, $\rho\Omega^2 r$ per unit volume, aligned with the motion in the channel. The idea that body forces have an effect on sonic point location is general, although the specifics (and even the sign) of the sonic point displacement depend on the particular force.

10.4 Effects of friction and heat addition on compressible channel flow

We illustrate the methodology of Section 10.3 through comparison of the effects of friction and heat addition on compressible flow in a constant area duct. (Isentropic flow in ducts of varying area is discussed in Sections 2.5 and 2.7 and is thus not treated here.) Expressions for differential changes in Mach number and stagnation pressure are presented using the influence coefficients to show several physical effects. Numerical results are given for the two cases.

10.4.1 Constant area adiabatic flow with friction

For a constant area adiabatic flow with friction, combining the terms from Table 10.1 leads to an expression relating Mach number and distance, dx, along the channel

$$\frac{dM^2}{M^2} = \frac{\gamma M^2\left(1+\frac{\gamma-1}{2}M^2\right)}{1-M^2}\left(\frac{4C_f dx}{d_H}\right). \tag{10.4.1}$$

Equation (10.4.1) shows that $dM^2 > 0$ for $M < 1$ and $dM^2 < 0$ for $M > 1$; for a given initial state friction *always* drives the Mach number towards unity. If the friction coefficient, C_f, is constant, (10.4.1) can be integrated between an initial Mach number, M_i, and $M = 1$ to provide an expression for the channel length, L^*, that will cause the flow to choke

$$\frac{4C_f L^*}{d_H} = \frac{1-M_i^2}{\gamma M_i^2} + \frac{\gamma+1}{2\gamma}\ln\left[\frac{(\gamma+1)M_i^2}{2\left(1+\frac{\gamma-1}{2}M_i^2\right)}\right]. \tag{10.4.2}$$

The differential expression for stagnation pressure change in constant area adiabatic flow with friction is

$$\frac{dp_t}{p_t} = -\frac{\gamma M^2}{2}4C_f\frac{dx}{d_H}. \tag{10.4.3}$$

For constant C_f (10.4.3) can be integrated to give the stagnation pressure ratio between an initial station and the choking location:

$$\frac{p_{t_i}}{p_t^*} = \frac{1}{M_i}\sqrt{\left[\frac{\left(2+(\gamma-1)M_i^2\right)}{\gamma+1}\right]^{\left(\frac{\gamma+1}{\gamma-1}\right)}}. \tag{10.4.4}$$

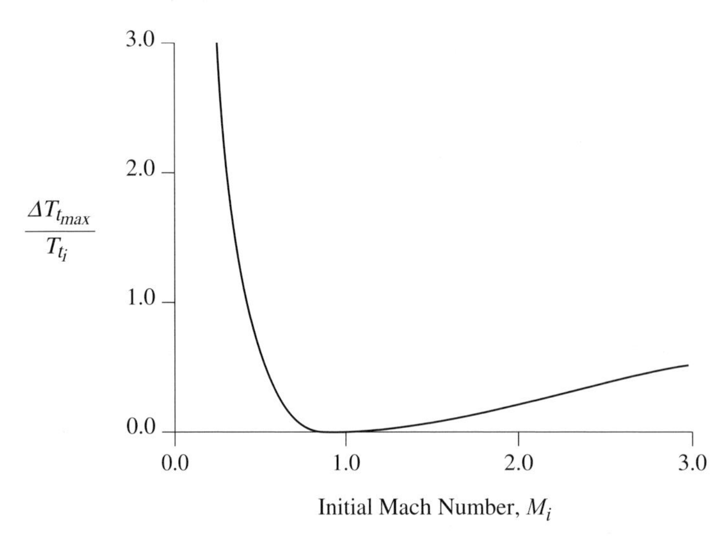

Figure 10.5: Maximum heat addition in a constant area duct in terms of initial Mach number, $\gamma = 1.4$.

10.4.2 Constant area frictionless flow with heat addition

Heat addition in a constant area channel is another important model process in the study of compressible flow. From Table 10.1 the change in Mach number for constant area heat addition is

$$\frac{dM^2}{M^2} = \frac{(1+\gamma M^2)\left[1+\frac{(\gamma-1)}{2}M^2\right]}{1-M^2}\left(\frac{dT_t}{T_t}\right). \tag{10.4.5}$$

Equation (10.4.5) states that, for any given initial flow state, heat addition drives the Mach number towards unity.[4] Further, the influence of a given fractional change in stagnation temperature increases as the initial conditions become closer to $M = 1$.

The maximum amount of heat that can be added to a stream with initial Mach number M_i in a constant area channel can be found by writing (10.4.5) as exact differentials for M^2 and T_t and integrating from the initial Mach number and stagnation temperature to $M = 1$:

$$\frac{T_t^* - T_{t_i}}{T_{t_i}} = \frac{\Delta T_{t_{max}}}{T_{t_i}} = \frac{\left(1+\gamma M_i^2\right)^2}{2(\gamma+1)M_i^2\left[1+\left(\frac{\gamma-1}{2}\right)M_i^2\right]} - 1. \tag{10.4.6}$$

In (10.4.6) T_t^* denotes the stagnation temperature at which the Mach number becomes unity. The quantity $\Delta T_{t_{max}}/T_{t_i}$ is plotted in Figure 10.5 versus initial Mach number, M_i. The small amount of heat that can be added as the Mach number nears unity is striking. To give numerical values, the maximum percentage increase in stagnation temperature is 4% at an inlet Mach number of 0.8 (subsonic flow) or 1.25 (supersonic flow), and only 2% at Mach numbers of 0.85 and 1.17.

As shown by the influence coefficients, this sensitivity near $M = 1$ is also true for area changes. For heat addition greater than the above values, the upstream conditions must alter. With the condition

[4] Note that the changes in stagnation temperature described in this section are due to heat addition only rather than to the effects of shaft work input which also include stagnation pressure changes.

of $M = 1$ occurring at the highest stagnation temperature in the constant area channel, this means the Mach number at the initial station will be moved further away from unity compared to the original value M_i.

An interesting aspect of heat addition to a flowing fluid concerns changes in static temperature. The relation from Table 10.1 to describe this is

$$\frac{dT}{T} = \frac{(1-\gamma M^2)\left[1+\dfrac{(\gamma-1)}{2}M^2\right]}{1-M^2}\left(\frac{dT_t}{T_t}\right). \tag{10.4.7}$$

For low subsonic Mach numbers ($M^2 \ll 1$), dT and dT_t have the same sign. For M between $1/\sqrt{\gamma}$ and 1, however, (10.4.7) shows they have opposite sign. In this regime, heat addition reduces the static temperature of the gas. For $M > 1$, heat addition again corresponds to an increase in static temperature.

Appreciation of this behavior is obtained by examining the portion of heat added which appears as static enthalpy per unit mass ($dh = c_p dT$) and that which appears as kinetic energy per unit mass. From Table 10.1 the two effects can be separated as

$$dT_t = \underbrace{\frac{(1-\gamma M^2)}{(1-M^2)}dT_t}_{\text{static enthalpy increase}} + \underbrace{\frac{(\gamma-1)M^2}{(1-M^2)}dT_t}_{\text{kinetic energy increase}}. \tag{10.4.8}$$

At subsonic speeds, the fraction of heat input associated with the kinetic energy change increases with M. At $M = 1/\sqrt{\gamma}$ all the heat added goes into increasing the kinetic energy and there is no change in static temperature. If M is between $1/\sqrt{\gamma}$ and unity, the change in kinetic energy of the flow is larger than the energy derived from heat addition and an additional amount must come from the gas stream, with a consequent decrease in static temperature (Zukoski, 1985).

In supersonic flow the effect of heat addition is to decrease the gas velocity and hence the kinetic energy. Heat addition thus causes an increase in static enthalpy in excess of the amount of heat added and the static temperature increase in supersonic flow is greater than the stagnation temperature increase.

Changes in flow properties with heat addition are conventionally shown using curves on a temperature–entropy diagram, as in Figure 10.6. A useful way to view this curve is in parametric form, with Mach number as the parameter. The rate of entropy change with Mach number can be found by eliminating dT_t/T_t from the expression in Table 10.1 for entropy and Mach number, yielding

$$\frac{ds/c_p}{dM^2} = \frac{M^2(1-M^2)}{1+\gamma M^2}. \tag{10.4.9}$$

The entropy is a maximum at $M = 1$.

10.4.3 Results for area change, friction, and heat addition

Effects of area change, friction, and heat transfer are displayed to scale on a T–s diagram in Figures 10.7(a) and 10.7(b) (Hill and Peterson, 1992), in which paths from an initial Mach number to the sonic condition are shown. The line corresponding to constant area flow with friction is known as the

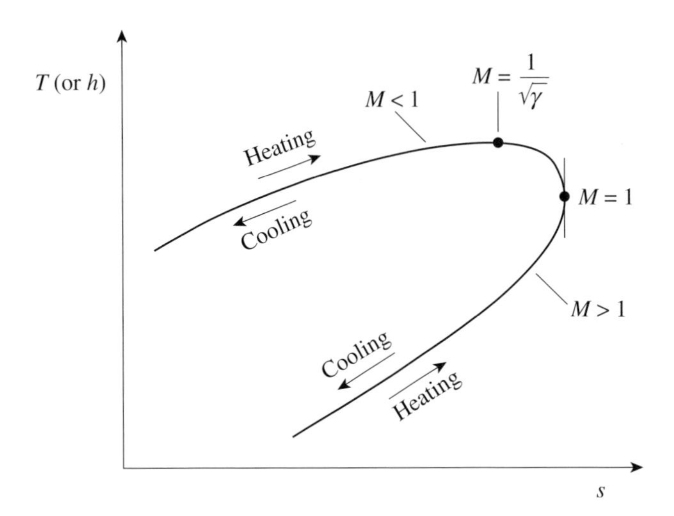

Figure 10.6: Rayleigh line for heat addition in a constant area channel.

Fanno line and that for frictionless constant area heat addition as the Rayleigh line. The upper figure corresponds to subsonic initial conditions ($M_i = 0.4$) and the lower to supersonic initial conditions ($M_i = 1.9$). Both Fanno and Rayleigh lines have $M = 1$ at the nose of the curve, the location of highest entropy. The symbols s^*, f^*, and h^* (and the corresponding stagnation values $s^*,\ f_t^*, h_t^*$) indicate the different Mach number unity states reached from the initial state by isentropic area change, constant area flow with friction, and constant area heat addition, respectively. The subsonic states are above the nose of the Fanno and Rayleigh lines and the supersonic states are below the nose. The heat addition process as modeled is reversible, so travel in both directions along the Rayleigh line is possible and cooling can move conditions away from $M = 1$. In contrast, the frictional process is irreversible, the entropy must increase along the flow path, and travel is possible only in the direction shown.

The incremental Mach number and stagnation pressure changes due to area change, friction, and heat addition in combination are summarized below from Table 10.1:

$$\frac{dM^2}{M^2} = \frac{\left(1 + \frac{\gamma - 1}{2} M^2\right)}{1 - M^2} \left[\frac{-2dA}{A} + (1 + \gamma M^2)\frac{dT_t}{T_t} + \frac{4\gamma M^2 C_f dx}{d_H} \right] \tag{10.4.10a}$$

$$\frac{dp_t}{p_t} = \left(-\frac{\gamma M^2}{2}\right)\left(\frac{dT_t}{T_t} + \frac{4C_f dx}{d_H}\right). \tag{10.4.10b}$$

From (10.4.10a) we can connect the sonic condition with the local channel geometry. At the condition $M = 1$, the quantity in square brackets must be zero for continuous variation of the Mach number through the sonic point. If so, the area change must approach the following limit as the sonic point is approached:

$$\frac{1}{A}\frac{dA}{dx} \rightarrow \frac{(1+\gamma)}{2}\frac{1}{T_t}\left(\frac{dT_t}{dx}\right) + \frac{2\gamma C_f}{d_H}, \quad M \rightarrow 1. \tag{10.4.11}$$

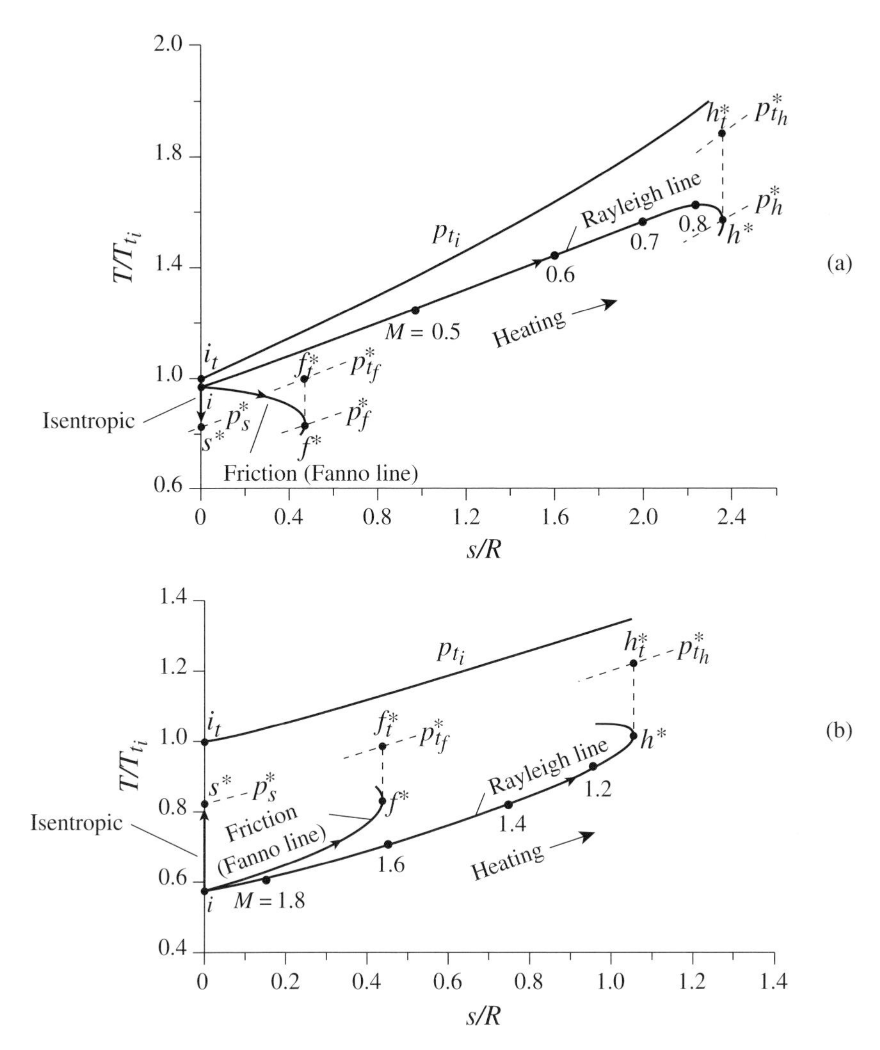

Figure 10.7: Fanno and Rayleigh lines on a temperature–entropy plane showing the choking process: (a) subsonic flow, $M_i = 0.4$; (b) supersonic flow, $M_i = 1.9$, $\gamma = 1.4$; i denotes the initial static state, i_t denotes the initial stagnation state (Hill and Peterson, 1992).

Equation (10.4.11) shows that with friction and (or) heat addition, the sonic point does not occur at the throat. Friction tends to move the sonic point downstream of the throat. For heat transfer, which side of the throat the sonic location is on depends on the sign of the heat transfer term. For the common situation with heat input from combustion, the heat addition term is large and positive and the $M = 1$ location occurs in a section of the channel with a substantial divergence rate. At the sonic conditions both the numerator and denominator of (10.4.10a) go to zero and, as described by Hill and Peterson (1992), L'Hopital's rule must be used to find the value of the derivative of the Mach number to integrate the solution through this point.

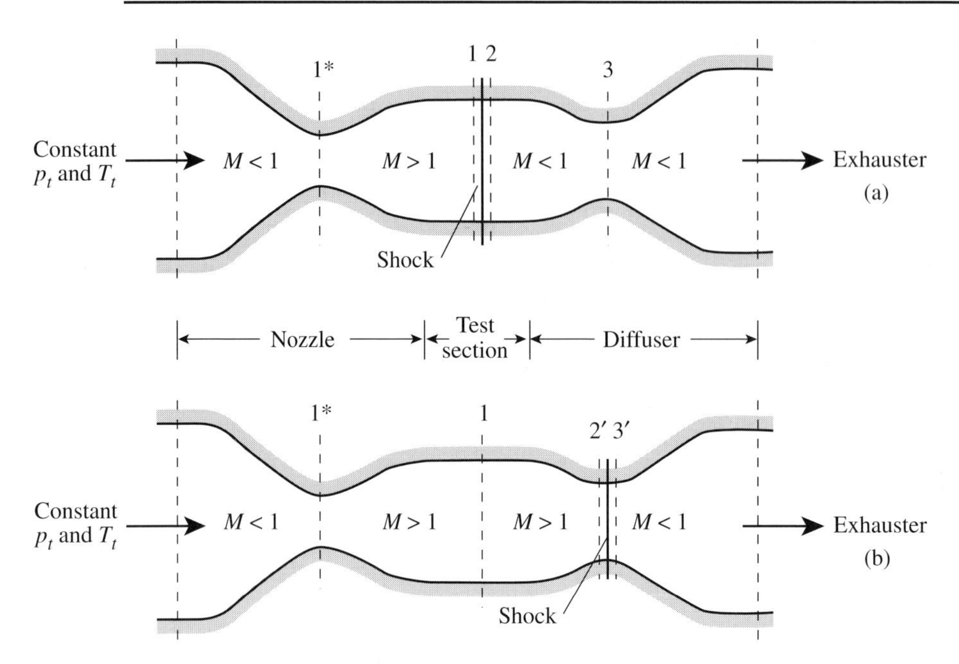

Figure 10.8: Starting of a supersonic wind tunnel diffuser (Shapiro, 1953): (a) most unfavorable starting condition; (b) best operating condition.

10.5 Starting and operation of supersonic diffusers and inlets

10.5.1 The problem of starting a supersonic flow

The relations developed for one-dimensional flow describe steady-state operation at given conditions, but for actual devices (as well as for some computational procedures) it is often necessary to address how one would attain the condition starting from an initial condition of no flow. Definition of the path to the desired condition brings out a critical problem associated with supersonic devices (Kerrebrock, 1992; Heiser and Pratt, 1994). To see this consider the supersonic wind tunnel in Figure 10.8(a) which has a converging–diverging nozzle, a constant area test section, a converging–diverging diffuser with a throat larger than the nozzle throat, and a downstream exhauster (often a centrifugal or axial compressor). The tunnel is fed by a reservoir at a specified upstream pressure and stagnation temperature, p_t and T_t, and the exhauster can be thought of as creating a specified static pressure at the exit.

At levels of exit static pressure close to the inlet stagnation pressure the flow throughout the tunnel is subsonic. As described in Section 2.7, lowering the static pressure causes the flow to become sonic at the nozzle throat, with a shock in the divergent part of the nozzle. In this mode of operation, the throat passes the maximum mass flow for the upstream p_t and T_t. Across the shock the mass flow, area, and stagnation temperature are constant so that the areas at sonic conditions (A^*) corresponding to the stagnation states upstream and downstream of the shock (denoted by 1 and 2, respectively; see

Figure 10.8(a) for station numbering for this section) are related by

$$\frac{A_1^*}{A_2^*} = \frac{p_{t_2}}{p_{t_1}}. \tag{10.5.1}$$

To pass the flow with a shock in the nozzle, the diffuser throat must be larger than the nozzle throat by the ratio given in (10.5.1). As the exit pressure is lowered and the shock moves downstream in the divergent section of the nozzle, the stagnation pressure ratio across the shock decreases, and the diffuser area required to pass the flow becomes larger. The strongest shock, and the largest required diffuser throat area, occur when the shock is in the test section as sketched in Figure 10.8(a). At this condition the diffuser throat (Station 3) must have an area at least

$$\frac{A_3}{A_1^*} = \frac{A_{\textit{diffuser throat}}}{A_{\textit{nozzle throat}}} = \frac{p_{t_1}}{p_{t_2}} = F(M_1). \tag{10.5.2}$$

At this condition, the flow in the test section downstream of the shock is subsonic and the flow at the diffuser throat is sonic.

If the shock moves into the test section it can be at any location in the constant area channel. Further, the shock will not stay in the test section. If the shock should undergo a momentary motion into the converging section of the diffuser the shock Mach number will be lowered and the downstream stagnation pressure increased. This will increase the mass flow through the diffuser throat, lowering the density and the static pressure downstream of the shock. To accommodate this, the shock must move further down the converging section. From these arguments there is no location in the converging section at which the shock will be stable so the shock will move through the throat. If no adjustments are made in conditions downstream of the diffuser, the shock will move to a location in the diverging section of the diffuser at an area corresponding to the test section area, where it will then be stably positioned. This process is known as *swallowing the shock*. Once it occurs the shock can be positioned by changing the operating conditions of the exhauster. To have the diffuser in the most efficient operating condition (the lowest stagnation pressure loss across the shock) the shock can be moved so it is at the lowest Mach number, which occurs with the shock positioned at the diffuser throat at station 2′, as in Figure 10.8(b). At this condition the stagnation pressure loss is less than that for starting. Figure 10.9 shows the stagnation pressure recovery $(\pi_d)_{max} = p_{t_3'}/p_{t_2'}$, the diffuser throat Mach number, M_2', and the area ratio between test section and diffuser throat, A_1/A_2', as functions of test section Mach number. The test section/diffuser throat area ratio A_1/A_3 which would occur with isentropic flow is also indicated.

In practice the shock must be maintained somewhat downstream of the diffuser throat because the shock is unstable in the converging part of the diffuser. If the shock moves upstream slightly, the shock Mach number increases, increasing the stagnation pressure loss and decreasing the mass flow capacity of the diffuser throat. Arguments similar to the above imply the shock would move further upstream, through the test section, until it reached a condition in the diverging part of the nozzle at which the stagnation pressure loss in the system matched the exhaust pressure of the system. In practice also, because of boundary layer growth and non-one-dimensional effects, the throat area needs to be slightly larger than the value given by (10.5.2) to ensure a supersonic diffuser of fixed geometry will start.

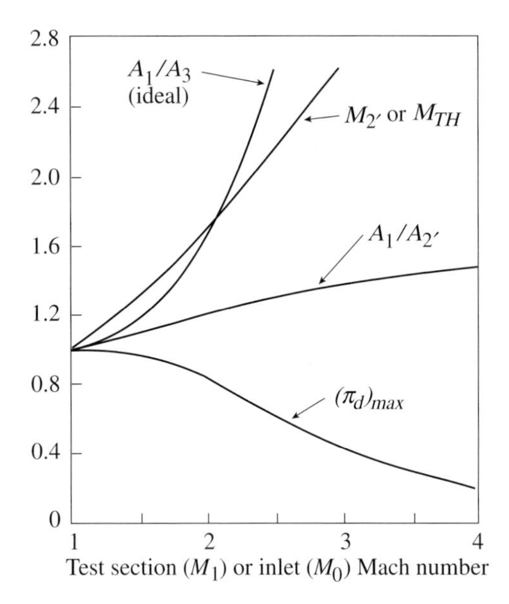

Figure 10.9: Characteristics of a fixed geometry diffuser (or internal compression inlet, see Figure 10.11), including best pressure recovery $(\pi_d)_{max}$, area ratios, and throat Mach number, M_2' (for diffuser) or M_{TH} (for inlet), $\gamma = 1.4$; nomenclature refers to Figures 10.8 and 10.11 (Kerrebrock, 1992).

10.5.2 The use of variable geometry to start the flow

The starting process can be facilitated using variable geometry, as illustrated in Figure 10.10 for a system consisting of a converging–diverging nozzle, a constant area test section which we wish to operate at supersonic conditions, and a downstream section designed to function as a subsonic diffuser. Consider the process of coming to the test condition from no flow, through the operation of a downstream exhauster, which changes the back pressure to which the diffuser discharges.

Suppose the nozzle and the diffuser initially have equal areas, as shown by the shaded contour in Figure 10.10. A small (compared to the upstream stagnation pressure) decrease in diffuser exit pressure results in a pressure distribution in the nozzle, test section, and diffuser as in curve 1 with the flow everywhere subsonic. As the exit pressure is reduced further, curve 2 results, with sonic conditions at both throats and subsonic flow in the test section. Dropping the exit pressure below the value corresponding to that for curve 2 will cause supersonic flow downstream of the diffuser throat and a shock wave in the diverging section of the diffuser, but will not affect the flow anywhere upstream of the shock.

Suppose the area of the diffuser throat is now increased to that denoted by *ab* in the figure. Reducing the exit pressure until the flow at area *ab* is sonic results in a lowered pressure in the test section (curve 3) with the formation of a shock (denoted by the dash-dot line) and then subsonic deceleration in the diverging section of the nozzle upstream of the test section. Opening the throat area to *cd* (curve 4) and decreasing the exit pressure creates a stronger shock further down the nozzle. At some level of opening, *ef*, corresponding to curve 5, the shock enters the test section (curve 5A). It can be at any location in the test section, for example the downstream end (curve 5B). In this condition the diffuser is still wholly subsonic.

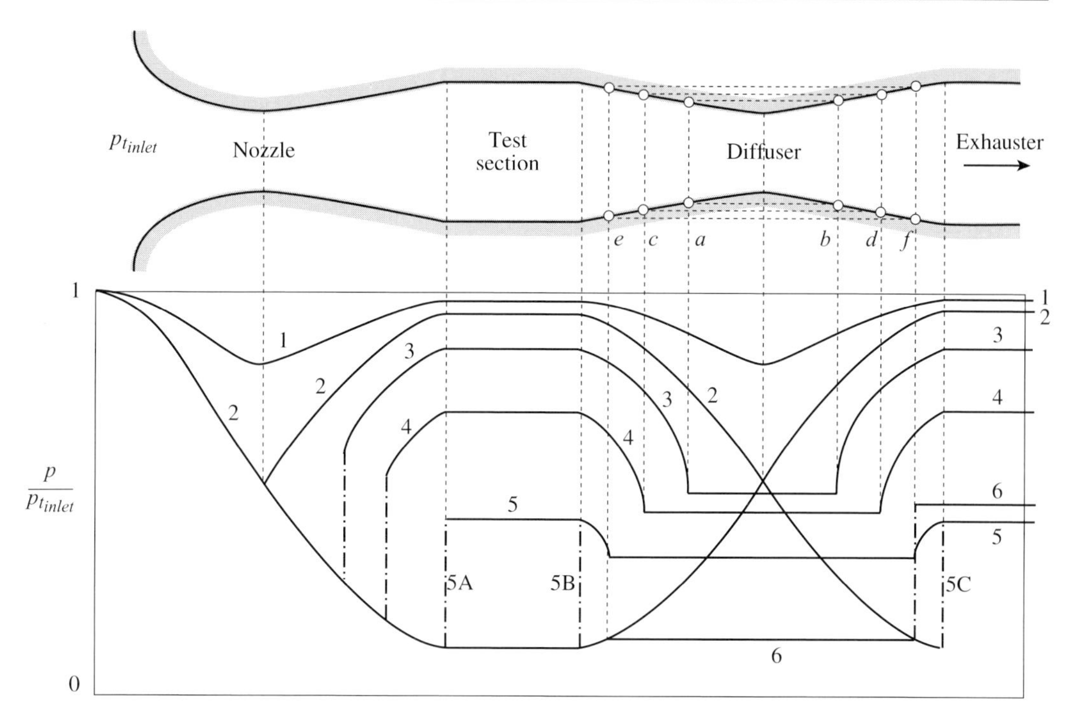

Figure 10.10: Starting sequence of a supersonic wind tunnel for various areas of the diffuser throat; dash-dot (–·–·–·) lines denote static pressure jumps across shocks (after Crocco, 1958).

The arguments given for the fixed geometry diffuser, however, imply that the condition when the shock is in the test section is unstable. If the shock moves slightly into the convergent passage, the motion will continue until the shock settles at the end of the divergent passage in the diffuser where the area is the same as the test section area (curve 5C). At this condition the shock strength and the final stagnation pressure are the same as when the shock was in the test section and the shock position is stable with respect to motion into the diffuser.

With area *ef*, therefore, the shock can be swallowed through the diffuser throat. Once this occurs, adjustment of the downstream conditions (increases in the back pressure) can move the shock to a location upstream in the diverging section of the diffuser where the Mach number is lower. (For area *ef* this would correspond to curve 6, although for reasons mentioned earlier, the shock would actually be positioned downstream.) In addition, with variable geometry the throat area can be reduced, decreasing the diffuser throat Mach number and hence the Mach number at which the shock occurs. In the ideal case this would allow sonic conditions at the diffuser throat although, again, for stability the shock would be placed downstream of the throat.

10.5.3 Starting of supersonic inlets

The starting problem is also encountered in inlets and turbomachinery blading in which entry Mach numbers are supersonic, because these devices must reach their design Mach numbers through operation at lower speeds. Consider a fixed geometry inlet designed for shock-free operation at

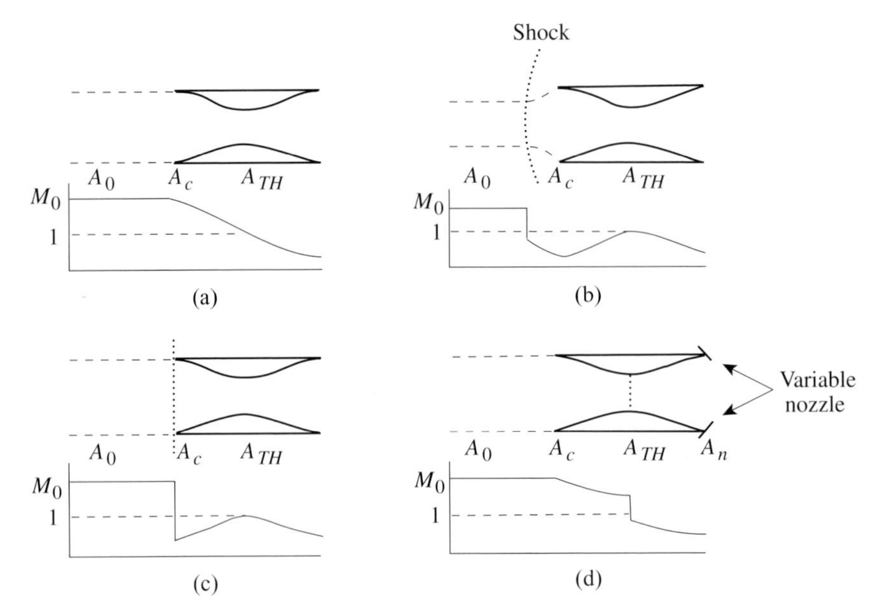

Figure 10.11: Schematics of the internal compression diffuser, showing: (a) ideal isentropic diffusion from M_0 through unity to $M < 1$; (b) operation below the critical (starting) Mach number; (c) operation at the critical M_{0_c}, but not started; and (d) operation at the critical M_{0_c} and started, with shock positioned at the throat, $\gamma = 1.4$ (Kerrebrock, 1992).

supersonic Mach number, say M_{0_c}, as sketched in Figure 10.11(a). At supersonic Mach numbers below the design value the inlet cannot pass the flow in the upstream streamtube and the excess must be diverted around the inlet. This cannot occur, however, if the flow is supersonic all the way to the inlet. A shock therefore stands in front of the inlet, as in Figure 10.11(b). As the Mach number is increased towards M_{0_c} the corrected flow per unit area of the incoming stream decreases, reducing the flow that must be spilled round the inlet, and allowing the shock to move closer to the inlet.

At the design Mach number, the shock will sit on the inlet lip. In this position it is unstable, because a small perturbation that moves it into the inlet causes a decrease in shock Mach number, an increase in the mass flow that the throat can pass, and a transient reduction in density and pressure between shock and throat. Following the arguments associated with shock swallowing in the diffuser, the consequence of the transient is shock motion through the throat to a downstream position determined by the variable nozzle. To achieve the best recovery, the nozzle is adjusted to position the shock at the throat, as in Figure 10.11(d). Figure 10.9, which shows diffuser characteristics, also applies to the internal compression inlet, with test section area A_1 corresponding to upstream streamtube area A_0 and with A_3 (or A'_2, as appropriate) corresponding to A_{TH}. The figure gives the ideal contraction ratio, the throat Mach number, and the inlet/throat area ratio, along with the area ratio for isentropic flow as a function of inlet Mach number.

Off-design behavior of a fixed geometry inlet designed for operation at Mach number $M_{0_c} = 3$ is shown in Figure 10.12 (Kerrebrock, 1992). As the Mach number increases from unity, the inlet pressure recovery corresponds to a normal shock until the shock is swallowed and the inlet is started (at a Mach number of 3), with a consequent increase in inlet pressure recovery. Further increase in

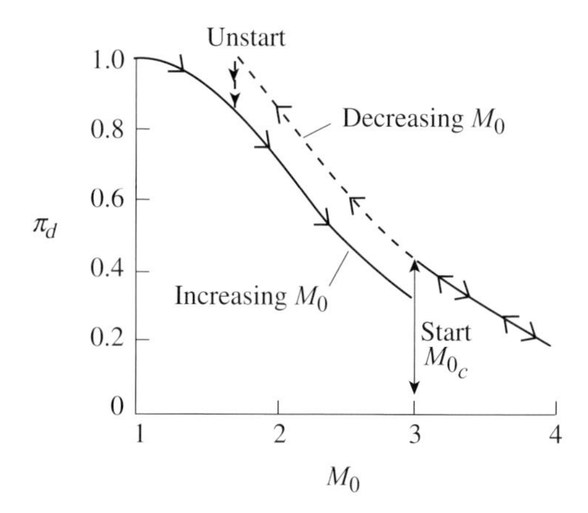

Figure 10.12: Off-design behavior (with Mach number) for the internal compression diffuser of Figure 10.9, showing hysteresis connected with starting, $\gamma = 1.4$ (Kerrebrock, 1992).

Mach number means the throat Mach number increases and the pressure recovery drops. Hysteresis behavior exists with a fixed geometry inlet in that once it is started the Mach number can be reduced with the shock still able to be positioned at the throat by adjusting the downstream conditions. This behavior continues until the throat Mach number reaches unity (at an inlet Mach number of 1.72 for the inlet shown). At this condition the corrected flow per unit area at the throat starts to decrease with Mach number so the throat can no longer pass the flow. A spill shock therefore forms upstream of the inlet, with the decrease in pressure recovery shown in Figure 10.12.

Variable area inlets enable more control over the starting process. If a shock stands ahead of the inlet, by increasing the throat area the shock is brought closer to the inlet and, when the shock reaches the inlet lip, it can be swallowed. The throat area can then be reduced (in combination with the adjustment of the exhaust pressure) so that, as with the wind tunnel, the inlet can (ideally!) be run with a Mach number of unity at the throat in a condition free of shocks.

10.6 Characteristics of supersonic flow in passages and channels

It is useful to describe features of actual flows in ducts, channels, and other passages in which shock waves play a key role. One example is turbomachinery blading. Three-dimensional viscous computations are used in much of the design work for turbomachinery components, and the goal here is not only to illustrate features of the flow but also to show the capability of computations to capture flow structure.

10.6.1 Turbomachinery blade passages

Figure 10.13 shows the results of three different calculation procedures plus measurements for an aeroengine fan rotor with supersonic relative flow. The calculations are shown for a constant radius surface. The three-dimensional numerical simulations are, from left to right and in increasing order of sophistication, an inviscid Euler calculation, a Euler calculation in which the blockage effect of the

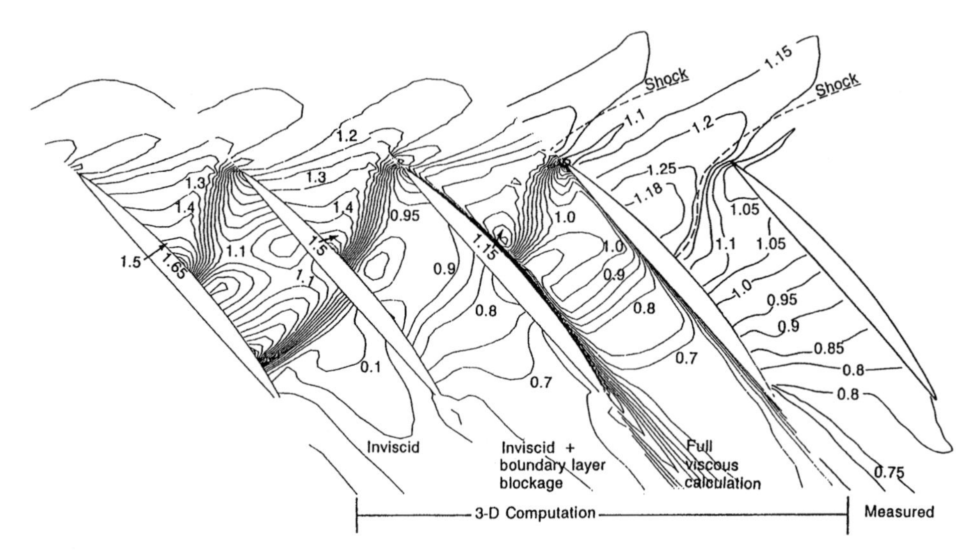

Figure 10.13: Mach number contours near the hub of an aeroengine fan rotor with supersonic relative inflow (calculations by Denton, measurements from Chima and Strazisar (1983)) (Cumpsty, 1989).

boundary layer is included and, a viscous calculation.[5] Noticeable (and important) is the similarity of the viscous calculation and the inviscid calculation including blockage, not only with one another but also with the measurements. The purely inviscid calculation predicts an entirely different flow, in which a second strong shock occurs near the trailing edge. The inclusion of a realistic estimate of blockage in a transonic calculation such as this is essential, a result that is not surprising in view of the strong influence of area changes for flow with Mach numbers near unity noted in previous sections.

Losses due to shock waves are an important part of accurate prediction methods for fan or compressor blade performance. The problem is compounded by the shock–boundary layer interaction on the suction surface which can involve a region of separated flow. Although sometimes more than a single shock is involved and a system of shocks and expansions occurs, numerical procedures based on the Reynolds averaged Navier–Stokes equations have been found to capture shock-related loss well in these configurations. As remarked by Cumpsty (1992), the methods are implemented in conservation form and can capture features such as shock losses, which are set by overall constraints in basically a control volume fashion.

10.6.2 Shock wave patterns in ducts and shock train behavior

The flow field depicted for the fan in Figure 10.13 has a single shock, but in ducts which are much longer (compared to their width) a normal, or near normal, shock wave is not generally the mechanism of transition from supersonic to subsonic conditions. The observed configuration is more complicated, with a gradual transition rather than the sharp discontinuity of a shock wave. More specifically, if duct boundary layers are thin and the shock is weak a normal shock does extend over much of

[5] A Reynolds averaged Navier–Stokes (RANS) computation, in which the turbulent transport process is modeled.

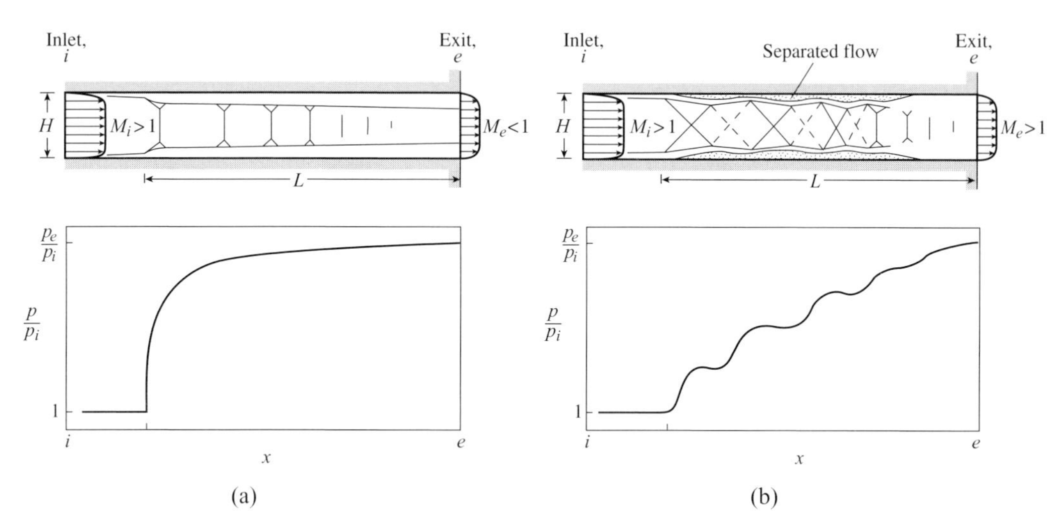

Figure 10.14: Sketch of the flow pattern and static wall pressure distribution in normal and oblique shock trains (a) normal shock train; (b) oblique shock train (Heiser and Pratt, 1994).

the stream, as sketched in Figure 10.14(a) (Heiser and Pratt, 1994; Shapiro, 1953). If the boundary layers are thick and the shock is strong (higher Mach numbers) the boundary layers separate from the walls of the duct with a series of oblique shocks and a pressure distribution as in Figure 10.14(b) results.

A one-dimensional treatment disregards the non-uniformities deriving from the presence of the boundary layers, replacing the non-uniform flow by a fictitious uniform mean flow. Suppose we have a central core of uniform flow with boundary layers near the wall. A discontinuity separating two regions with uniform values of pressure is possible in the core but is incompatible with the boundary layers where the decreased Mach number does not allow the realization, through a shock, of the same pressure ratio as in the core. Moreover, near the walls the velocities are subsonic and shocks do not exist. A normal shock across the whole duct is therefore not compatible with the presence of boundary layers.

Experiments show that if the boundary layers are removed by suction, a shock can be established across almost all of the duct width and it is possible to obtain, by proper adjustment, a practically normal shock (Shapiro, 1993), although the normal shock cannot be established right to the walls. If the shock strength is small (shock pressure ratio near unity), however, the boundary layer can adjust itself to this pressure increase in a sufficiently short length without undergoing separation so the corresponding gradual process, made possible by viscous forces, occurs in a small fraction of the duct width.

There is a limit to the possibilities for boundary layer adjustment without separation, and above a pressure ratio of roughly 2 for turbulent boundary layers, separation occurs. A consequence is that the foot of the normal shock present in the central part of the duct bifurcates near the wall and is replaced by two more or less straight oblique shocks. In this situation the configuration of the shock pattern is found to depend on the shape of the section.

In summary, only for low supersonic velocities and thin boundary layers is a quasi-normal shock possible in a duct. Otherwise, a non-one-dimensional pattern is produced. The pattern contains

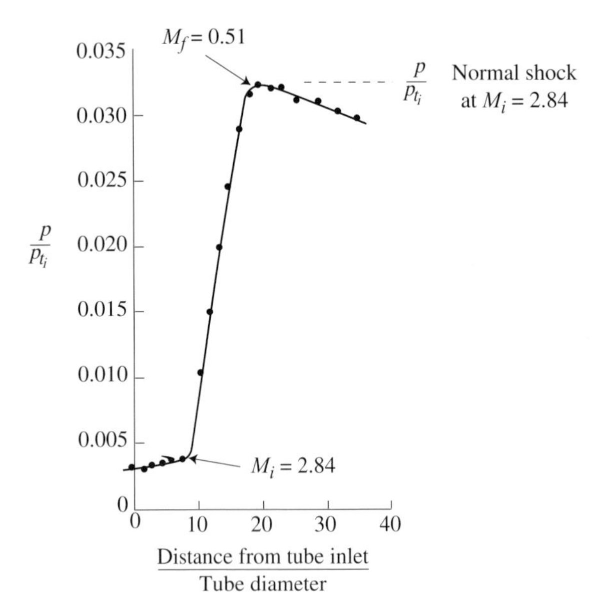

Figure 10.15: Measured wall static pressure in shock train (air); M_f is Mach number after the pressure rise (Shapiro, 1953).

oblique shocks crossing in the center of the duct and reflected back and forth in the central supersonic region of flow, while the region adjacent to the wall adjusts gradually to the corresponding pressure increase with, presumably, strong exchange of momentum due to the high turbulence level. The subsonic regions spread into the supersonic region with downstream distance until the latter disappears completely and the shock pattern terminates, with the subsequent process one of adjustment of the subsonic velocity distribution without appreciable pressure increase.

Throughout the whole process frictional effects are small. For a constant area duct, if wall friction can be neglected the relations between the initial (supersonic) and the final (subsonic) conditions are the same as those across a normal shock. The pressure rise and overall entropy increase must therefore also be closely similar to those in a normal shock. In this shock system or, as it is often referred to, *shock train*, pictured in Figure 10.14, only a part of the entropy increase is produced by the shock pattern, with the rest due to the dissipative processes present in the turbulent regions adjacent to the walls.

The length of the region occupied by the shock train is a function of the flow characteristics. Figure 10.15 (Shapiro, 1953) shows the measured wall static pressure for a constant area duct with supersonic initial conditions and varying back pressure. The shock train occurs over several diameters but, because the wall shear stress is small, the pressure rise is close to that of a normal shock at the initial Mach number.

To illustrate this further, values of the ratio p_f/p_i between the pressure maximum p_f (the "final" pressure) and the initial pressure p_i, versus Mach number M_i in the starting section, are shown in Figure 10.16 for a constant area duct. The solid curve is the theoretical result for a normal shock, and there is only a small difference between this and the measured shock train pressure ratios. The

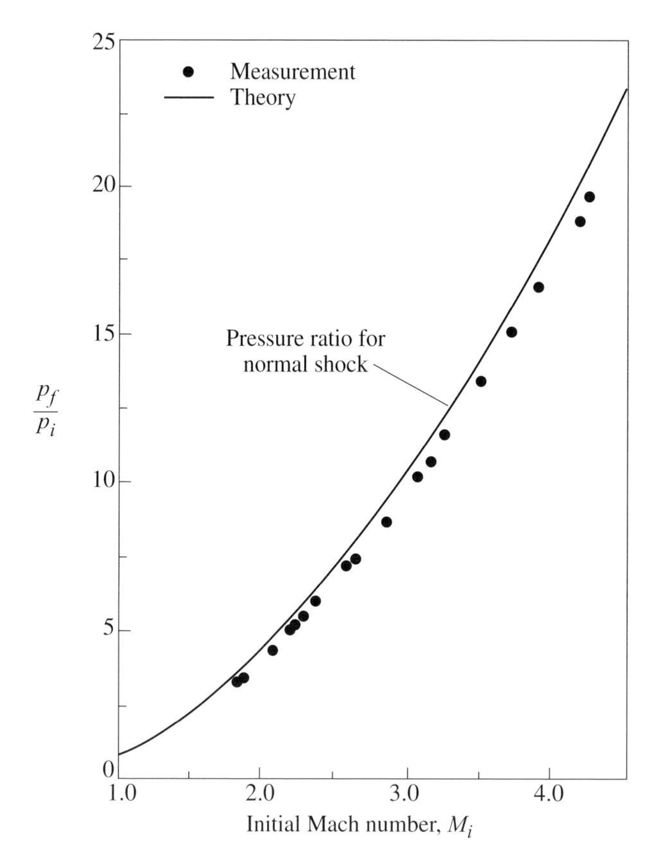

Figure 10.16: Measured and theoretical pressure ratios for a complete normal shock train in air as a function of initial Mach number (Crocco, 1958; data of Neumann and Lustwerk, 1949).

lengths over which these shock systems exist are illustrated by Figure 10.17 as a function of the initial Mach number M_i.

Shock trains can exist either in the "complete" form, illustrated by the data above, or in a truncated fashion. Which occurs depends on back pressure as indicated by the computational results of Figure 10.18. Figure 10.18(a) shows computed wall static pressure versus downstream distance in a two-dimensional channel at an inlet Mach number of 3.0. The shock train consists of a series of oblique shocks. The different curves correspond to different values of back pressure p_B/p_i. The normal shock pressure ratio is also indicated. As the overall pressure ratio is lowered, less and less of the shock train stays in the channel; the shock train can be said to slide along the channel to accommodate the appropriate pressure rise. Figure 10.18(b) shows the data of Figure 10.18(a) with the origins shifted so that the start points of the shock trains are at the same location. Other computations in the paper demonstrate that the part of the shock train structure that remains within the channel is little changed by exit conditions. A range of pressure ratios thus can be delivered by a shock train, and this accommodation is useful in high speed propulsion systems such as dual mode scramjets. Further analysis of these devices is given in Heiser and Pratt (1994) and experimental information can be found in Waltrup and Billig (1973).

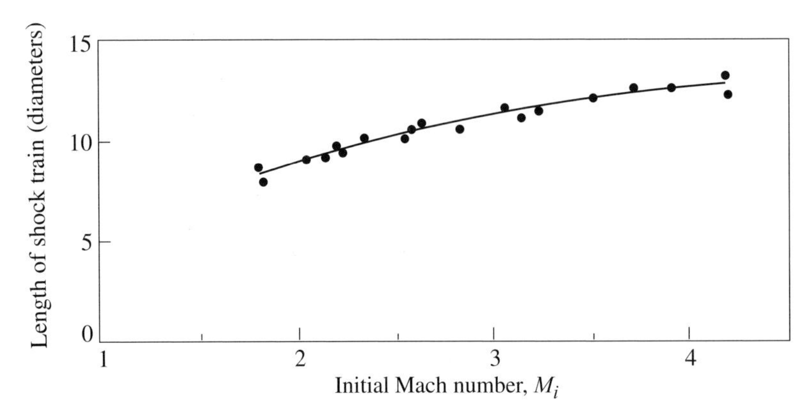

Figure 10.17: Shock train length in a circular duct as a function of the initial Mach number (air) (Crocco, 1958; data of Neumann and Lustwerk, 1949).

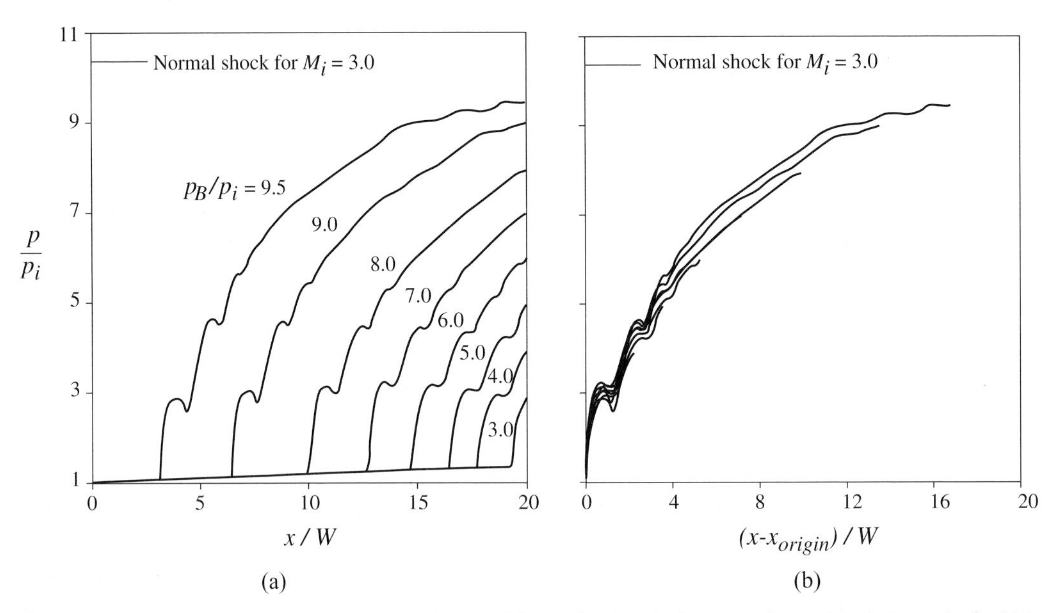

Figure 10.18: Computed wall static pressure for an oblique shock train in a two-dimensional channel of width W, $M_i = 3.0$; (a) actual distribution; (b) distribution with the shock train position shifted so origins coincide, $\gamma = 1.4$ (Lin *et al.*, 1991).

10.7 Extensions of the one-dimensional concepts – I: Axisymmetric compressible swirling flow

The swirling flows addressed in Chapter 8 had uniform density. There are many situations, however, in which both strong swirl and Mach numbers near unity exist, and the flow can no longer be considered incompressible. The combination of swirl and compressibility leads to features beyond those encountered in the examination of either phenomenon separately. These can be illustrated by extending the one-dimensional treatment of Section 10.3 to compressible swirling flows. As

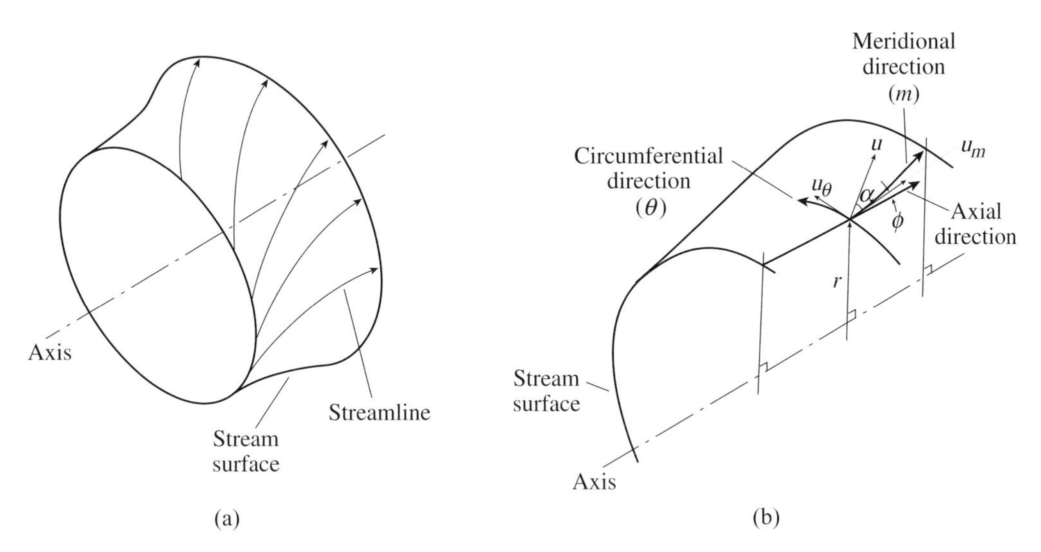

Figure 10.19: (a) Axisymmetric stream surface; (b) coordinate system for axisymmetric flow.

previously, we obtain expressions for the variation in properties associated with imposed changes in geometry (which now includes radius) as well as mass, momentum, and energy input. Even without integration the explicit form of these differential equations yields information about behavior of interest (Anderson *et al.*, 1970).

10.7.1 Development of equations for compressible swirling flow

The class of flows to be addressed are axisymmetric and steady. The fluid is a perfect gas with constant specific heat. Property values normal to stream surfaces are neglected and property values along the stream assumed continuous. To deal with friction forces, it is assumed that any friction forces act directly opposite to the local stream direction.

As in Section 10.3, there is a choice of which variables are chosen as independent (specified) and which are dependent. The perspective taken is analysis and design of fluid machinery. As such, variables chosen as independent are: (1) streamtube area; (2) streamtube radial location; (3) drag forces; (4) energy addition; (5) mass addition; and (6) wall friction. The first five can be regarded as under the control of the designer. Drag forces and wall friction are not physically independent variables, in the sense that one cannot freely vary them, but they are classified here as independent in order that their effects can be readily assessed.

The procedure followed is similar in concept to that in Section 10.3, but the control volume is annular. Further, there is an additional conservation equation (angular momentum) as well as additional kinematic relations linking the velocity components. It is therefore useful to begin the development afresh. The streamlines and streamsurfaces and the nomenclature are given in Figure 10.19 and Figure 10.20 gives a view of an elementary control volume. The *meridional component* of velocity (i.e. the projection of the velocity on an r–x plane) is denoted by u_m, with the circumferential (θ) component labeled as u_θ and the velocity magnitude as u. The streamtube area, A, is

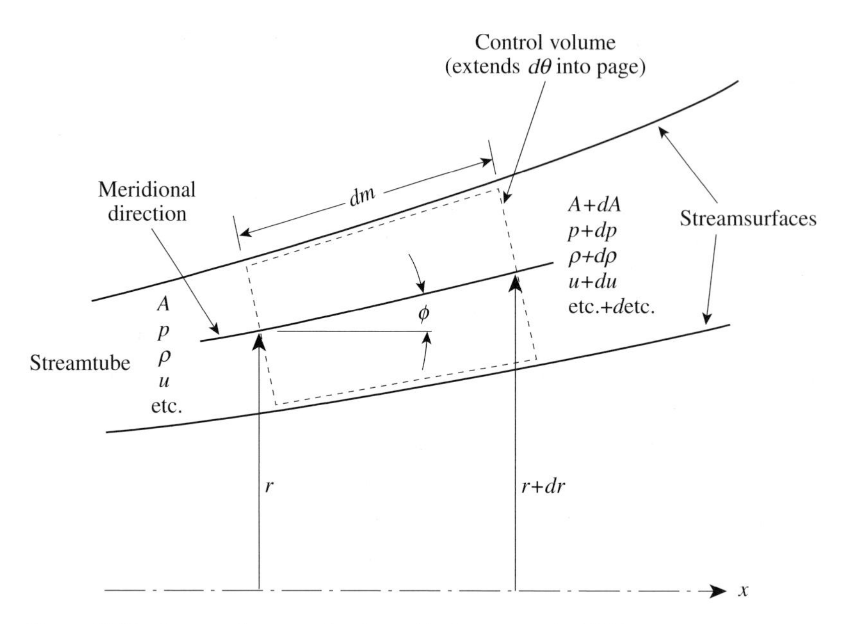

Figure 10.20: Streamtube control volume (Anderson *et al.*, 1970).

normal to the meridional direction (see Figure 10.20) and dm denotes distance along the meridional direction.

The mass flow in an axisymmetric annular streamtube is given by

$$\dot{m} = \rho u_m A. \tag{10.7.1}$$

The change in mass flow across the control volume is given by

$$\frac{d\dot{m}}{\dot{m}} = \frac{d\rho}{\rho} + \frac{du_m}{u_m} + \frac{dA}{A}.$$

Rearranging the independent (on the right) and dependent (on the left) variables,

$$\frac{d\rho}{\rho} + \frac{du_m}{u_m} = \frac{d\dot{m}}{\dot{m}} - \frac{dA}{A}. \tag{10.7.2}$$

The net flux of angular momentum out of the control volume is equal to the sum of the torques on the volume ((1.8.8) or see Section 2.8)

$$\sum r d\mathcal{F}_{total_\theta} = (\dot{m} r u_\theta)_{out} - (\dot{m} r u_\theta)_{in}. \tag{10.7.3}$$

In (10.7.3) $\sum r d\mathcal{F}_{total_\theta}$ includes all torques (drag forces and friction). We denote the force additional to wall friction in the θ-direction by $d\mathcal{F}_{D_\theta}$, with the torque $r d\mathcal{F}_{D_\theta}$. The friction force can be expressed as τdA_w, where τ is the local wall shear stress and dA_w is the wetted area, i.e. the surface area of the control volume. The wall shear stress is again defined in terms of the friction coefficient C_f as

$$\tau = C_f \tfrac{1}{2} \rho u^2.$$

In analogy with the one-dimensional usage, a hydraulic diameter, d_H, can be defined as

$$d_H = \frac{4A}{dA_w/dm} = \frac{4\,(\text{flow through area})}{\text{perimeter}}.$$

The frictional force is thus $2C_f\rho u^2\ Adm/d_H$, pointing in a direction opposite to the velocity. The θ-component is this quantity times $\sin\alpha$, where α is the angle between the streamline and the meridional direction (see Figure 10.19) so the frictional torque is $2C_f\rho u^2 r\sin\alpha\ Adm/d_H$.

The net angular momentum flux out of the control volume is evaluated as follows. The angular momentum flux out of the volume is $(\dot{m}+d\dot{m})(u_\theta+du_\theta)(r+dr)$. The angular momentum flux in is $(\dot{m}ru_\theta + ru_{inj_\theta}d\dot{m})$, where u_{inj_θ} is the circumferential component of velocity at which the injected fluid enters the control volume. (If $d\dot{m}$ is negative, u_{inj_θ} is equal to u_θ, the local circumferential velocity.) The *net* outflow of angular momentum from the control volume is thus, to first order in small changes, $ru_\theta d\dot{m} + \dot{m}rdu_\theta + \dot{m}u_\theta dr - ru_{inj_\theta}d\dot{m}$. Equating this quantity to the total torque exerted on the fluid within the control volume, dividing by $\dot{m}ru_\theta$, and invoking the relationships $u_m = u\cos\alpha$ and $u_\theta = u\sin\alpha$ and the definition of the Mach number, $M^2 = u^2/(\gamma p/\rho)$, yields the desired angular momentum relationship:

$$\frac{du_\theta}{u_\theta} = -\frac{dr}{r} - \left(1 - \frac{u_{inj\theta}}{u_\theta}\right)\frac{d\dot{m}}{\dot{m}} - \frac{2}{\sin 2\alpha}\frac{d\mathcal{F}_{D_\theta}}{\gamma pAM^2} - \frac{2}{\cos\alpha}C_f\frac{dm}{d_H}. \tag{10.7.4}$$

To derive the expression for conservation of meridional momentum, we examine the control volume in Figure 10.21 which occupies a circumferential angle, $d\theta$. The forces that act on the fluid within this elementary volume are pressure, friction, and drag forces, denoted by $\mathcal{F}_{D_m}$. Summing up the contributions, again working to first order in the small changes across the volume, the total force is $Adp + 2AC_f\rho u^2\cos\alpha\ (dm/d_H) + d\mathcal{F}_{D_m}$, where A represents the normal area for the volume in Figure 10.21(c).

These forces are balanced by the net flux of meridional momentum out of the control volume. There is a momentum flux out of both the end faces and the side faces. Through the former (sides c and d in Figure 10.21(b)), there is $u_m\dot{m} + u_{inj_m}d\dot{m}$ in and $[(u_m+du_m)(\dot{m}+d\dot{m})]$ out, where u_{inj_m} is the meridional component with which any injected mass enters.

The flow of meridional momentum through the sides of the element can be seen with reference to Figures 10.21(c) and 10.21(d). The circumferential velocity, u_θ, has a component in the meridional direction given by $-u_{\theta_p}\sin\phi$, where u_{θ_p} is the projection of u_θ in the meridional direction and ϕ is the angle between the axial and meridional directions. The velocity u_{θ_p} is given by (see Figure 10.19(d))

$$u_{\theta_p} = u_\theta\sin\frac{d\theta}{2} \approx u_\theta\frac{d\theta}{2}$$

since $d\theta \ll 1$. A flow of meridional momentum thus takes place *into* the control volume of $-\rho(u_\theta(d\theta/2)\sin\phi)\ dm \times u_\theta dn$. The flow of meridional momentum out of the volume has the same magnitude but is opposite in sign, so that the net outflow of meridional momentum through the side faces of the volume is $\rho u_\theta^2\sin\phi d\theta dmdn$. Noting that A, the area normal to the meridional direction, is given by $A = rdnd\theta$, and using $\sin\phi = dr/dm$, the net meridional momentum flux through the sides can be expressed as $-\rho u_\theta^2(dr/r)\,A$, which is the familiar "centrifugal force" term.

The last steps in the development of the meridional momentum equation consist of equating the sum of the forces in the meridional direction to the net outflow of meridional momentum and dividing

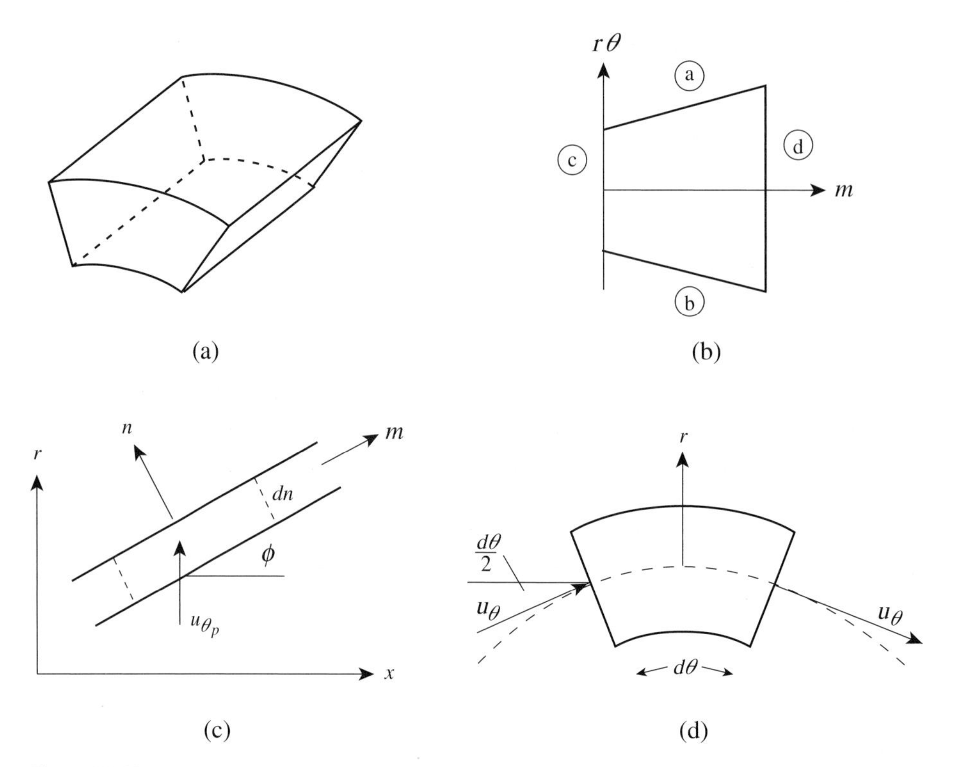

Figure 10.21: Control volume for derivation of meridional momentum equation: (a) perspective view of element; (b) projection of element; (c) meridional projection of element; (d) end-on view of element.

both sides of the equation by pA to give

$$\frac{du_m}{u_m}\gamma M^2\cos^2\alpha + \frac{dp}{p} = \gamma M^2\sin^2\alpha\frac{dr}{r} - \gamma M^2\cos^2\alpha\left(1 - \frac{u_{inj_m}}{u_m}\right)\frac{d\dot{m}}{\dot{m}}$$
$$-\frac{d\mathcal{F}_{D_m}}{pA} - 2\gamma M^2\cos\alpha C_f\frac{dm}{d_H}. \tag{10.7.5}$$

Energy exchange to the control volume is through heat addition and is specified through defining changes in stagnation temperature, T_t. Variations in static temperature and velocity are related to stagnation temperature variations through the definition of stagnation temperature:

$$T_t = T + \frac{u^2}{2c_p}. \tag{10.7.6}$$

Taking differentials of (10.7.6) and dividing by T_t,

$$\frac{dT}{T} + (\gamma - 1)M^2\frac{du}{u} = \left[1 + \frac{(\gamma - 1)M^2}{2}\right]\frac{dT_t}{T_t}. \tag{10.7.7}$$

Six unknowns have been introduced: du_m/u_m, $d\rho/\rho$, dp/p, du/u, dT/T, and du_θ/u_θ, with four equations, (10.7.2), (10.7.4), (10.7.5), and (10.7.7) to connect them. The two additional equations needed to close the system are the equation of state for a perfect gas in differential form:

$$\frac{dp}{p} - \frac{d\rho}{\rho} - \frac{dT}{T} = 0, \tag{10.7.8}$$

and the relation between the velocity magnitude, u, and the two components:

$$u^2 = u_m^2 + u_\theta^2. \tag{10.7.9}$$

Differentiating (10.7.9) and dividing by u^2

$$\frac{du}{u} - \cos^2\alpha \frac{du_m}{u_m} - \sin^2\alpha \frac{du_\theta}{u_\theta} = 0. \tag{10.7.10}$$

Equations (10.7.7) and (10.7.10) can be used to eliminate du_m and du_θ from the remaining four equations, although it is useful to maintain the grouping of the independent terms as they appear in (10.7.2), (10.7.4), (10.7.5), and (10.6.7). The equations are linear so effects of independent variables may be combined.

Other quantities of interest include the changes in: Mach number, M; meridional Mach number (the meridional component of Mach number), M_m; flow angle, α; and stagnation pressure, p_t. These can be related to the six quantities listed:

Mach number

$$\frac{dM^2}{M^2} = 2\frac{du}{u} - \frac{dT}{T}; \tag{10.7.11}$$

meridional Mach number

$$\frac{dM_m^2}{M_m^2} = 2\frac{du_m}{u_m} - \frac{dT}{T}; \tag{10.7.12}$$

flow angle

$$d\alpha = \left(\frac{du}{u} - \frac{du_m}{u_m}\right) \cot\alpha; \tag{10.7.13}$$

stagnation pressure

$$\frac{dp_t}{p_t} = \frac{dp}{p} + \left(\frac{\frac{\gamma M^2}{2}}{1 + \frac{\gamma - 1}{2} M^2}\right) \left(\frac{dM^2}{M^2}\right). \tag{10.7.14}$$

10.7.2 Application of influence coefficients for axisymmetric compressible swirling flow

As in Section 10.3, the set of linear equations for the differential quantities can be displayed in a table of influence coefficients (Table 10.3). The influence coefficients are now functions of two quantities, Mach number, M, and swirl angle, α, but the principle is the same as for the channel flow situation. The four columns in Table 10.3 are the groupings of the independent terms which arise naturally on the right-hand sides of (10.7.2), (10.7.4), (10.7.5), (10.7.7). The term $(1 - M^2 \cos^2\alpha)$ $(= (1 - M_m^2))$ has been attached to the dependent variables to simplify the table and to emphasize that it is the

Table 10.3 *Influence coefficients for axisymmetric compressible swirling flow (Anderson et al., 1970)*

Origin of source terms	Mass conservation	Conservation of energy	Conservation of azimuthal moment of momentum	Conservation of meridional momentum
Source term / *Dependent variable*	$\frac{d\dot{m}}{\dot{m}} - \frac{dA}{A}$	$\left(1 + \frac{\gamma - 1}{2} M^2\right) \frac{dT_t}{T_t}$	$\frac{dr}{r} + \left(1 - \frac{u_{inj\theta}}{u_\theta}\right) \frac{d\dot{m}}{\dot{m}} + \frac{2}{\sin 2\alpha} \frac{d\mathcal{F}_{D\theta}}{\gamma p A M^2} + \frac{1}{\cos\alpha} \frac{2C_f dm}{d_H}$	$-\gamma M^2 \sin^2\alpha \frac{dr}{r} + \gamma M_m^2 \left(1 - \frac{u_{injm}}{u_m}\right) \frac{d\dot{m}}{\dot{m}} + \frac{\gamma M^2 d\mathcal{F}_{Dm}}{\gamma p A M^2} + \gamma M^2 \cos\alpha \frac{2C_f dm}{d_H}$
$\frac{dM^2}{M^2}(1 - M_m^2)$	$2\left(1 + \frac{\gamma - 1}{2} M^2\right) \cos^2\alpha$	$(1 + \gamma M^2)\cos^2\alpha - \sin^2\alpha$	$-2\left(1 + \frac{\gamma - 1}{2} M^2\right) \cos^2\alpha\,(1 - \gamma M_m^2)$	$2\left(1 + \frac{\gamma - 1}{2} M^2\right) \cos^2\alpha$
$\frac{dM_m^2}{M_m^2}(1 - M_m^2)$	$2\left(1 + \frac{\gamma - 1}{2} M_m^2\right)$	$1 + \gamma M_m^2$	$(\gamma - 1) M^2 \cos^2\alpha\,(1 + \gamma M_m^2)$	$2\left(1 + \frac{\gamma - 1}{2} M_m^2\right)$
$\frac{dp}{p}(1 - M_m^2)$	$-\gamma M_m^2$	$-\gamma M_m^2$	$-(\gamma - 1) \frac{\gamma M^4}{4} \sin^2 2\alpha$	$-[1 + (\gamma - 1) M_m^2]$
$\frac{dT}{T}(1 - M_m^2)$	$-(\gamma - 1) M_m^2$	$1 - \gamma M_m^2$	$(\gamma - 1) M^2 \sin^2\alpha\,(1 - \gamma M_m^2)$	$-(\gamma - 1) M_m^2$
$\frac{d\rho}{\rho}(1 - M_m^2)$	$-M_m^2$	-1	$-(\gamma - 1) M^2 \sin^2\alpha$	-1
$\frac{du}{u}(1 - M_m^2)$	$\cos^2\alpha$	$\cos^2\alpha$	$-(1 - \gamma M_m^2) \sin^2\alpha$	$\cos^2\alpha$
$\frac{du_\theta}{u_\theta}$	0	0	-1	0
$\frac{du_m}{u_m}(1 - M_m^2)$	1	1	$(\gamma - 1) M^2 \sin^2\alpha$	1
$d\alpha\,(1 - M_m^2)$	$-\left(\frac{\sin 2\alpha}{2}\right)$	$-\left(\frac{\sin 2\alpha}{2}\right)$	$[1 + (\gamma \sin^2\alpha - 1) M^2] \frac{\sin 2\alpha}{2}$	$-\left(\frac{\sin 2\alpha}{2}\right)$
$\frac{dp_t}{p_t}$	0	$\frac{-(\gamma M^2/2)}{[1 + (\gamma - 1) M^2/2]}$	$-\gamma M^2 \sin^2\alpha$	-1
$\frac{ds}{c_p}$	0	1	$(\gamma - 1) M^2 \sin^2\alpha$	$\frac{\gamma - 1}{\gamma}$

meridional Mach number that plays the central role in determining the condition at which the flow regime changes and at which choking can occur. Applications of the influence coefficients are given in the next several sections.

10.7.2.1 Behavior of stagnation pressure

The first application concerns the variation in stagnation pressure:

$$\frac{dp_t}{p_t} = -\frac{\frac{\gamma M^2}{2}}{1+\frac{\gamma-1}{2}M^2}\left[\left(1+\frac{\gamma-1}{2}M^2\right)\frac{dT_t}{T_t}\right]$$
$$-\gamma M^2\sin^2\alpha\left[\frac{dr}{r}+\left(1-\frac{u_{inj_\theta}}{u_\theta}\right)\frac{d\dot{m}}{\dot{m}}+\frac{2}{\sin 2\alpha}\frac{d\mathcal{F}_{D_\theta}}{\gamma pAM^2}+\frac{2C_f}{\cos\alpha}\frac{dm}{d_H}\right]$$
$$-\left[-\gamma M^2\sin^2\alpha\frac{dr}{r}+\gamma M_m^2\left(1-\frac{u_{inj_m}}{u_m}\right)\frac{d\dot{m}}{\dot{m}}+\frac{d\mathcal{F}_{D_m}}{pA}+2\gamma M^2\cos\alpha\, C_f\frac{dm}{d_H}\right]. \quad (10.7.15)$$

Rearranging the terms in (10.7.15), the change in stagnation pressure can be written

$$\frac{dp_t}{p_t} = -\gamma M^2\left[\left(1-\frac{u_{inj}}{u}\cos\chi\right)\frac{d\dot{m}}{\dot{m}}+\frac{d\mathcal{F}_D}{\gamma pAM^2\cos\alpha}+\frac{2C_f}{d_H}\left(\frac{dm}{\cos\alpha}\right)+\frac{dT_t}{2T_t}\right]. \quad (10.7.16)$$

Equation (10.7.16) shows that changes in r and A do not affect the stagnation pressure. The term $d\mathcal{F}_D$ in (10.7.16) is the drag force in the streamwise direction, defined by $d\mathcal{F}_D = d\mathcal{F}_{D_m}\sin\alpha + d\mathcal{F}_{D_\theta}\cos\alpha$. Only forces locally parallel to the stream affect the stagnation pressure. The angle χ is the angle between the injected flow and the mainstream. The term $\cos\alpha$ in the denominator reflects the effect of extending the residence time within dm.

The behavior of the injection term u_{inj}/u indicates that stagnation pressure is affected by the components of momentum locally parallel to the stream. For a given injection velocity ratio, u_{inj}/u, stagnation pressure can best be preserved by injecting the flow so that $\chi = 0$, i.e. so that injection takes place in the direction of the mainstream.

The stagnation pressure change due to heat addition, reflected in the dT_t/T_t term, is independent of flow direction.

For the wall friction term, the quantity $\cos\alpha$ appears in the denominator because $dm/\cos\alpha$ is the actual distance that the flow must cover for a given dm. As discussed in Section 8.9, vaneless diffusers have highly swirling flows with values of α that are large, 75° or more under some conditions. If so, the actual distance traveled can be a factor of 5 or more larger than the meridional distance with a corresponding increase in frictional losses.

10.7.2.2 Planar isentropic swirling flow

A second application is planar isentropic compressible swirling flow, which can be considered an idealization of the situation in a constant axial width vaneless diffuser or a radial inflow nozzle. In this situation the meridional Mach number, M_m, becomes the radial Mach number, M_r, and the flow angle, α, is the angle between the radial direction and the velocity. The only non-zero independent variables are dA/A and dr/r, with the relation between the two given by $dA/A = dr/r$. Under these conditions the equations from the influence coefficient matrix for the variation in Mach number and swirl angle are:

$$\frac{dM^2}{M^2} = \frac{-2[1+(\gamma-1)M^2/2]}{[1-(M^2\cos^2\alpha)]}\left(\frac{dr}{r}\right) = \frac{-2[1+(\gamma-1)M^2/2]}{\left[1-M_r^2\right]}\left(\frac{dr}{r}\right) \tag{10.7.17a}$$

$$d\alpha = \frac{-[(M^2\sin 2\alpha)/2]}{[1-(M^2\cos^2\alpha)]}\left(\frac{dr}{r}\right) = \frac{-[(M^2\sin 2\alpha)/2]}{\left[1-M_r^2\right]}\left(\frac{dr}{r}\right). \tag{10.7.17b}$$

The implication from (10.7.17) is that, analogous to the situation in compressible channel flow where $M = 1$ at the minimum *area*, the condition $M_r^2 = 1$ is associated with a minimum *radius*.

To bring out the features of this compressible swirling flow, we examine the solution of (10.7.17). This can be done by integrating the differential equations, but for this simple case it is instructive to work directly with the integral forms of the conservation statements.

The axisymmetric form of the continuity equation is

$$\rho r u_r = G\rho_t = \text{constant}, \tag{10.7.18}$$

where G is a (dimensional) constant proportional to the mass flow per unit axial depth. The circumferential velocity distribution is

$$r u_\theta = \frac{\Gamma}{2\pi} = K. \tag{10.7.19}$$

The energy equation plus the condition of isentropy yields

$$u_r^2 + u_\theta^2 = u_{max}^2\left(1-\frac{T}{T_t}\right) = u_{max}^2\left[1-\left(\frac{\rho}{\rho_t}\right)^{\gamma-1}\right]. \tag{10.7.20}$$

In (10.7.20) u_{max}^2 is a reference velocity corresponding to the condition of temperature (or density) equal to zero. Combining (10.7.18)–(10.7.20) provides a relation between the non-dimensional radius, $r/(G/u_{max})$, and the static/stagnation density ratio as (Howarth, 1953)

$$\frac{r}{(G/u_{max})} = \left\{\frac{(K/G)^2(\rho/\rho_t)^2+1}{(\rho/\rho_t)^2[1-(\rho/\rho_t)^{\gamma-1}]}\right\}^{1/2}. \tag{10.7.21}$$

The denominator of the term on the right-hand side of (10.7.21) vanishes, corresponding to infinite radius, when $\rho = \rho_t$ and $\rho = 0$. The former corresponds to the limit of Mach number of zero, the latter to the limit of Mach number of infinity. At some intermediate Mach number and hence density, therefore, a minimum radius exists. The minimum radius is obtained by differentiation of the expression on the right-hand side of (10.7.21). It is found to occur when the radial component of the velocity is equal to the local speed of sound, in other words when the radial Mach number is unity. At these conditions the mass flow per unit circumferential area is a maximum. (As pointed out by Howarth (1953), from considerations of how one might generate such a flow, the field of flow cannot physically extend exactly to the minimum radius.) As implied by the influence coefficients in Table 10.3, transition of the total Mach number through unity no longer has the special significance it did

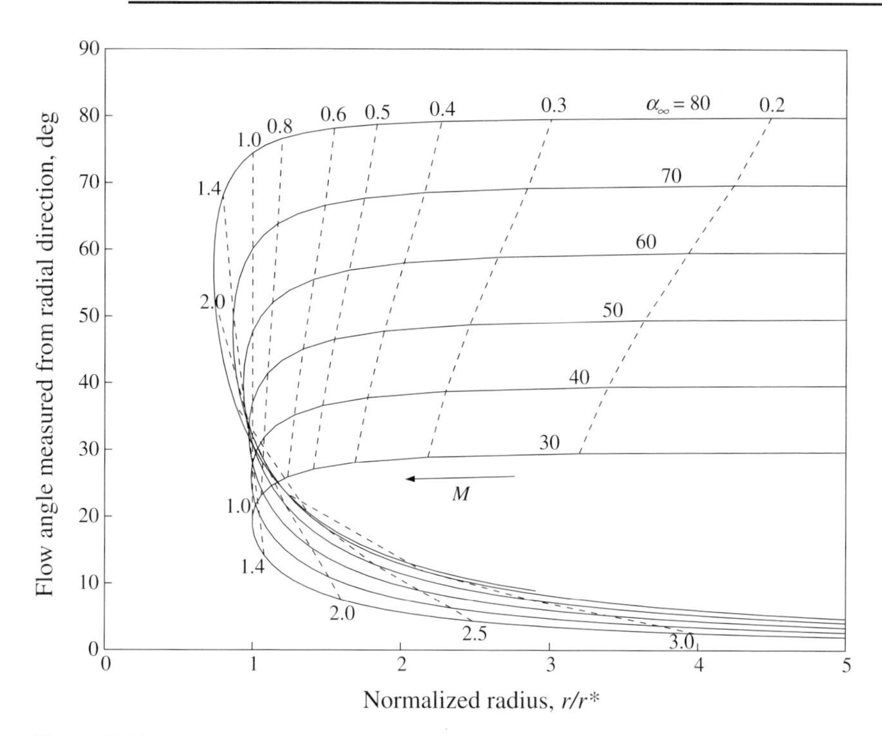

Figure 10.22: Flow angle versus radius for planar, isentropic, axisymmetric compressible swirling flow; r^* is the radius at which the Mach number is unity, α_∞ is the flow angle at radii $r/r^* \gg 1$. Solid lines mark the solution trajectory, dashed lines are loci of constant Mach number; $\gamma = 1.4$.

in one-dimensional channel flow; it is the radial Mach number (which corresponds to the meridional Mach number in the general case) that plays the important role.

The parameter K/G is an invariant for the flow. It can be viewed as the tangent of the swirl angle α when $\rho \cong \rho_t$, i.e. $K/G \to \tan\alpha$ in the regime where the Mach number is much less than unity. For a subsonic entry radial inflow swirl nozzle this would be the swirl angle at large radius, denoted by α_∞. One could impose a specified value of α_∞ for such a nozzle by setting the exit angle on a row of guide vanes located at a radius several times greater than the minimum radius.

There are two possible solution regimes for (10.7.21). In one the density tends to ρ_t at large radius and the radial Mach number is everywhere less than (or equal to at a minimum radius) unity. Where Mach numbers are low enough so the density is essentially constant, the ratio $u_\theta/u_r = \tan\alpha$ is also constant and the streamlines are spirals with angles that are uniform with radius. For the other solution, the density tends to zero at large radius, with the radial Mach number everywhere greater than (or equal to, at the minimum radius) unity. One could (conceptually at least) pass from one regime to the other using a radially subsonic inflow nozzle connected to a radially supersonic outflow nozzle section by an axial section of constant radius, which had a throat at which the meridional (axial) Mach number was unity.

Features of a planar, axisymmetric, isentropic, compressible swirling flow are illustrated in Figure 10.22. The curves show flow angle, α, versus radius, for different values of $\tan\alpha_\infty = K/G$. The radius is normalized by r^*, the radius at which the Mach number, $M(=\sqrt{M_\theta^2 + M_r^2})$, is equal to unity. Lines of constant Mach number are also indicated. Zero swirl ($\alpha^* = \alpha = 0$) is a line on the horizontal axis. For zero swirl the minimum radius is r^*. As the swirl angle, α_∞, increases, the radial Mach number at which the Mach number is unity decreases and the minimum radius becomes a smaller fraction of r^*. In the regime in which radial Mach numbers are larger than unity the radial velocity increases with radius (similar to the velocity increase in supersonic flow in a diverging channel) while the tangential velocity decreases with radius, with a consequent strong decrease in swirl angle.

10.7.2.3 Pressure distribution in a swirling flow

The static pressure distribution along the meridional path is of concern in connection with boundary layer behavior. To examine the impact of swirl on pressure distribution consider the situation with no friction, heating, drag forces, or mass addition, so only area and radius vary. Changes in static pressure are then given by

$$\frac{dp}{p} = \frac{\gamma M_m^2}{1 - M_m^2}\frac{dA}{A} + \frac{\gamma M_\theta^2}{1 - M_m^2}\frac{dr}{r}. \tag{10.7.22}$$

An interpretation of (10.7.22) is that meridional velocity may be "exchanged" for pressure largely through streamtube area variation, while circumferential velocity may be exchanged for pressure largely through variation of streamline radial location (Anderson *et al.*, 1970). Diffusers for axial turbomachines generally make use of the former of these exchanges while diffusers for radial machines employ the latter. For radial machines, the swirl velocity is often dominant as a contributor to pressure variation and virtually the same pressure rise is achieved whether the radial flow is in the desired direction or is reversed.

A related application concerns the influence of width variation and wall friction on radial vaneless diffuser pressure rise. The diffuser area at any radius is $A = 2\pi rW$, and the incremental area change is $dA/A = dW/W + dr/r$. The meridional distance dm is dr, and the hydraulic diameter, d_H, is $2W$. Substituting these in the influence coefficients of Table 10.3, the incremental pressure rise is given by

$$\frac{dp}{p} = \left(\frac{\gamma M^2}{1 - M_m^2}\right)\left\{\cos^2\alpha\frac{dW}{W} + \frac{dr}{r} - C_f\frac{r}{W}\cos\alpha\left[1 + (\gamma - 1)M^2\right]\frac{dr}{r}\right\}.$$

Non-dimensionalizing by the dynamic pressure, $\rho u^2/2$,

$$\frac{dp}{\rho u^2/2} = \left(\frac{2}{1 - M_m^2}\right)\left(\cos^2\alpha\frac{dW}{W} + \left\{1 - C_f\frac{r}{W}\cos\alpha\left[1 + (\gamma - 1)M^2\right]\right\}\frac{dr}{r}\right). \tag{10.7.23}$$

Equation (10.7.23) shows that at high values of swirl angle ($\cos^2\alpha \ll 1$) wall divergence (or convergence) has little effect on static pressure rise.

Equation (10.7.23) provides a rough estimate for the decrease in vaneless diffuser pressure rise due to wall friction. Diffusers with small width/radius ratios (large r/W) suffer the greatest decrease. Figure 10.23 shows the normalized rate of pressure rise, $[2dp/(\rho u^2)/(dr/r)]$, in a parallel wall vaneless diffuser as a function of skin friction coefficient, C_f, for different inlet swirl angles. The upper value of friction coefficient, $C_f = 0.025$, is taken from the information in Japikse (1984). (See also Cumpsty (1989) for information about vaneless diffuser performance.) For $C_f = 0$ the result is that for isentropic swirling compressible flow. A lower impact of wall friction as C_f increases is

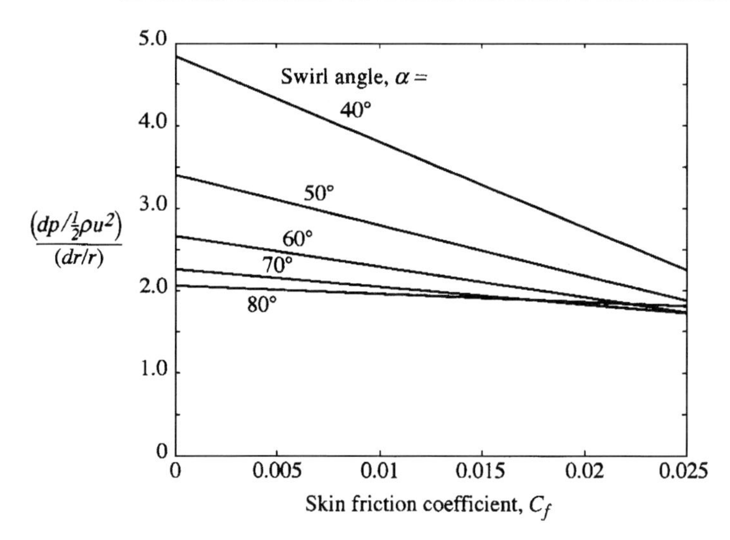

Figure 10.23: Incremental rate of vaneless diffuser static pressure rise versus wall skin friction coefficient for one-dimensional swirling flow, $M_i = 1$, $r_i/W = 20$, $\gamma = 1.4$.

shown for high swirl angles. At these angles the pressure rise is due mainly to the $\rho u_\theta^2/r$ term, and the frictional force is modeled as acting opposite to the direction of the velocity with a smaller radial component for higher values of α.

The rate of change of ru_θ ($= K$), the angular momentum per unit mass about the axis, is obtained from Table 10.3 as

$$\frac{d(ru_\theta)}{ru_\theta} = -\frac{r}{W}\frac{C_f}{\cos\alpha}\frac{dr}{r}. \tag{10.7.24}$$

Equation (10.7.24), which can be derived directly from application of conservation of mass and conservation of momentum to an axisymmetric control volume spanning the diffuser at radii r and $r + dr$, shows that increases in wall friction, r/W, and swirl angle enhance the rate at which the angular momentum decreases with radius. While the discussion in Section 8.9 shows that there are aspects of the flow in a vaneless diffuser which cannot be captured by the one-dimensional description, the approach enables extraction of qualitative (and often quantitative) parametric trends in a simple manner.

10.7.2.4 Choking in a swirling flow

The condition at which choking occurs determines the mass flow capacity. The table of influence coefficients (Table 10.3) indicates that in a swirling flow, unlike a purely axial flow, all quantities pass without difficulty through $M = 1$. The critical condition is when the meridional Mach number ($M_m = M\cos\alpha$) reaches unity. The meridional Mach number variation for a flow with only area and radius change is

$$\frac{dM_m^2}{M_m^2} = -\frac{2\left(1+\dfrac{\gamma-1}{2}M_m^2\right)}{1-M_m^2}\frac{dA}{A} - \frac{(\gamma+1)\,M_\theta^2}{1-M_m^2}\frac{dr}{r}. \tag{10.7.25}$$

The analogy with (non-swirling) one-dimensional channel flow is that one does not expect the behavior to alter qualitatively as long as axisymmetric pulses, moving at the speed of sound, can make themselves felt upstream. Conclusions from channel flow analysis concerning transition through the choking point are essentially unaltered, but they must now be applied about the condition of M_m, the meridional Mach number, equal to unity.

Equation (10.7.25) indicates that in an isentropic swirling flow it is possible to pass through the choking condition by a combination of variations in area, A, and radius, r. For example, when only r can vary, (10.7.25) becomes

$$\frac{dM_m^2}{M_m^2} = -\frac{(\gamma+1)\,M_\theta^2}{1-M_m^2}\frac{dr}{r}. \tag{10.7.26}$$

For variations in radius to have an effect on flow state, (10.7.26) shows that swirl must be present.

For constant radius isentropic flow the flow per unit area normal to the flow direction $(\dot{m}/(A\cos\alpha))$ is maximum when $M = 1$. The area normal to the flow direction must therefore increase as M increases beyond 1, even though the meridional streamtube area continues to decrease until the condition $M_m = 1$. From Table 10.3, the expression describing this is

$$\frac{d(A\cos\alpha)}{A\cos\alpha} = \cos^2\alpha\left(\frac{1-M^2}{1-M_m^2}\right)\frac{dA}{A}. \tag{10.7.27}$$

In the regime where $M > 1$ but $M_m < 1$, the flow swings toward meridional rapidly enough to allow the area normal to the flow to increase while the axisymmetric streamtube area continues to decrease.

Finally, the reduction in maximum flow rate per unit area due to swirl can be found by taking the ratio of choking mass flow parameters for fixed γ, p_t, T_t, r, and A

$$\frac{\dot{m}_{\alpha\neq 0^\circ}}{\dot{m}_{\alpha=0^\circ}} = \left[\frac{(\gamma+1)\cos^2\alpha}{\gamma+\cos 2\alpha}\right]^{\frac{\gamma-1}{2(\gamma-1)}}. \tag{10.7.28}$$

With swirl and a given stagnation pressure and temperature, at any meridional velocity the density is lower and the Mach number is higher than in axial flow. The mass flow per unit area is thus decreased compared to the situation without swirl.

10.7.3 Behavior of corrected flow per unit area in a compressible swirling flow

Figure 10.24 shows corrected flow per unit area normal to the meridional direction as a function of meridional Mach number, for different values of swirl velocity and constant radius (Millar, 1971). Curves of constant Mach number and constant circumferential velocity, non-dimensionalized by the speed of sound based on stagnation conditions $(u_\theta/\sqrt{\gamma RT_t})$, are indicated in the figure. The curve for $u_\theta/\sqrt{\gamma RT_t} = 0$ is the plot of corrected flow per unit area versus Mach number given previously in Figure 2.6.

The lines of constant Mach number in Figure 10.24 are straight lines between the origin and the corresponding points on the $u_\theta/\sqrt{\gamma RT_t} = 0$ curve. To see this, we write the corrected flow per unit area as

$$\frac{\dot{m}\sqrt{T_t R/\gamma}}{Ap_t} = \sqrt{\frac{T}{T_t}}\frac{\rho}{\rho_t}\frac{u_m}{\sqrt{\gamma RT}} = M_m f(M),$$

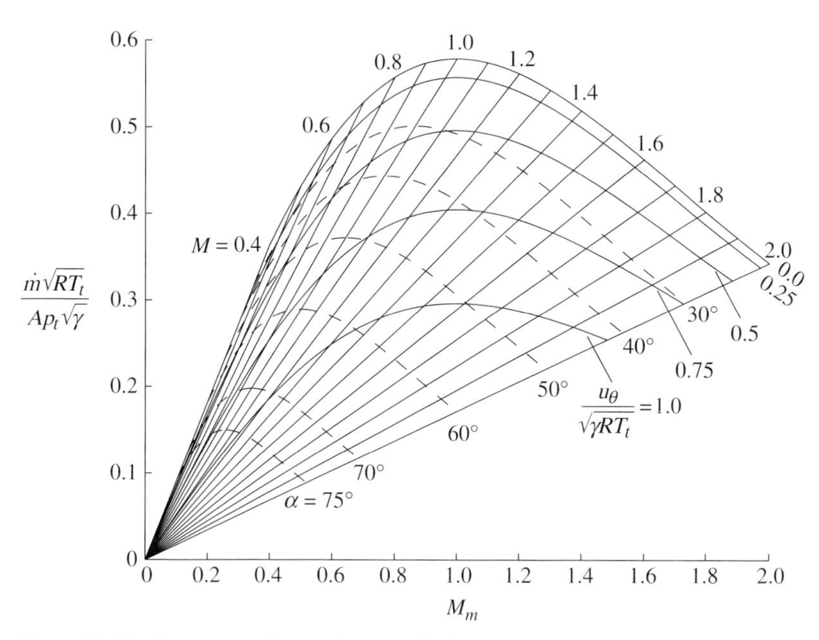

Figure 10.24: Constant radius, axisymmetric, isentropic compressible swirling flow in an annulus: corrected flow per unit annulus area as a function of meridional Mach number, M_m, for different values of total Mach number, M, swirl angle, α (dashed lines), and swirl velocity, $u_\theta/\sqrt{\gamma RT_t}$, $\gamma = 1.4$ (Millar, 1971).

where $f(M)$ denotes the ratio $(\rho/\rho_t)\sqrt{T/T_t}$. For constant Mach number, the corrected flow per unit area is linearly proportional to the meridional Mach number.

Several other features in Figure 10.24 are of interest. For an inviscid duct flow with no radius change, Kelvin's Theorem implies that the swirl velocity is fixed. As area is changed, the flow must follow a line of constant $u_\theta/\sqrt{\gamma RT_t}$, with the maximum flow per unit annulus area at a meridional Mach number of unity. In accord with the discussion of (10.7.25), the maximum corrected flow per unit annulus area decreases as the swirl velocity increases.

Figure 10.24 shows there can be two values of meridional Mach number associated with a given value of corrected flow per unit annulus area at either a specified angle or a specified non-dimensional circumferential velocity. The lines of constant flow angle have peaks along the line $M = 1.0$. For a constant radius geometry a torque must be applied to the flow in the annulus (such as that created by turbomachinery blading) to produce changes in angle. If the stream is subsonic, increases in Mach number (accelerating flow) at constant annulus area result in a higher swirl angle, so the flow swings away from the meridional direction. If the stream is supersonic, increases in Mach number result in a decreased flow angle as the flow swings towards the meridional direction. Both these trends were seen in Figure 10.22. The distance between the 60° line and the horizontal line ABC in Figure 10.25 (which shows a part of Figure 10.24) exhibits this behavior.

The vertical line DBE in Figure 10.25 shows a trajectory of constant meridional Mach number, with the total Mach number going from roughly 0.5 to 1.4, a behavior qualitatively representative of an axial turbine. To achieve this acceleration for isentropic flow, the area must increase so the corrected flow per unit annulus area decreases. During the process, the flow angle increases from 0 to approximately 68°.

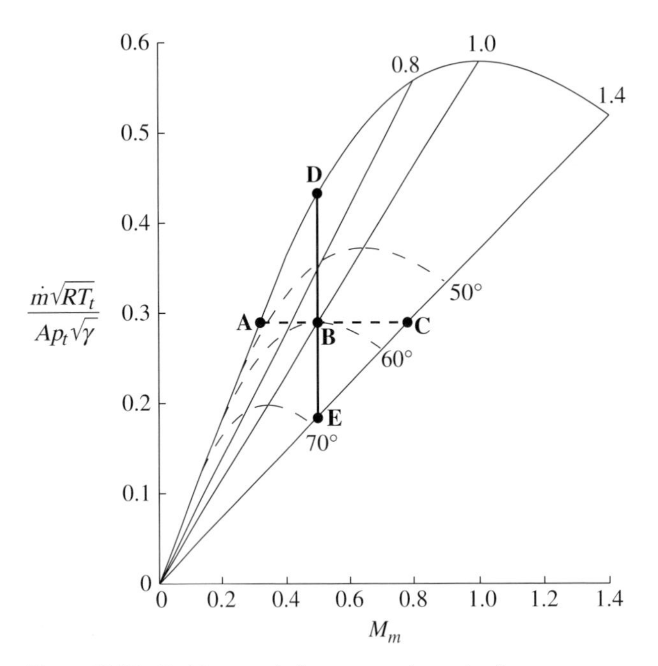

Figure 10.25: Turbine nozzle flow state trajectories for constant corrected flow per unit area (A→C) and constant meridional Mach number (D→E), $\gamma = 1.4$.

10.8 Extensions of the one-dimensional concepts – II: Compound-compressible channel flow

A configuration often encountered in fluid engineering is a duct or other flow system fed by two or more streams of different stagnation pressure or temperature. In turbofan engines, for example, core and fan streams with different stagnation temperatures and pressures can be ducted through a common nozzle with some fraction of the stream supersonic and some fraction subsonic. Computational methods exist for analyzing such flows, but considerable insight can be obtained using an extension of concepts developed for single stream quasi-one-dimensional flow to a multiple stream framework. In this section, following the treatment given by Bernstein *et al.* (1967), we develop and show applications of this idea for understanding the behavior of non-uniform compressible internal flows.

10.8.1 Introduction to compound flow: two-stream low Mach number (incompressible) flow in a converging nozzle

The principles that underpin compound-compressible nozzle flow analysis are familiar from one-dimensional channel flow, but application to even two streams leads to situations that are parametrically complex. It is thus helpful to present first some introduction to compound flow behavior. To do this with minimum complexity, we examine the low Mach number (incompressible flow) limit of an inviscid, two-stream, compound flow in a converging nozzle, noting that the important topic of

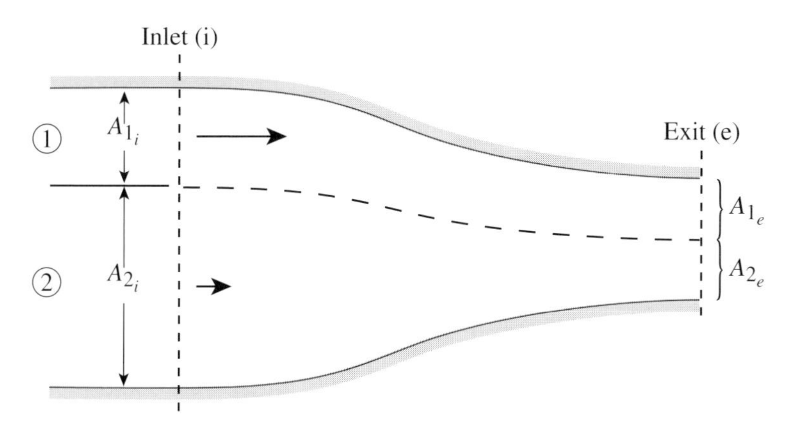

Figure 10.26: Two-stream compound flow for a converging nozzle: $A_i = A_{1_i} + A_{2_i}$, $A_e = A_{1_e} + A_{2_e}$.

choking in a compound flow will be dealt with in subsequent sections. The nozzle is considered to be fed by two reservoirs having specified stagnation conditions.

Figure 10.26 shows the nomenclature to be used. In the notation in this section the subscripts "1" and "2" refer to the two streams. The quantities which need to be specified to determine the nozzle flow are the difference between the stagnation pressure and the exit static pressure in both streams, $\Delta p_{t_1} = p_{t_1} - p_e$ and $p_{t_2} = p_{t_2} - p_e$ (or, equivalently, one of these differences and the stagnation pressure difference between streams 1 and 2, $\Delta p_{t_{1-2}} = p_{t_1} - p_{t_2}$), the density of each stream, ρ_1 and ρ_2, and a description of the nozzle geometry from inlet to exit. For one-dimensional flow the geometry description is given by any two area ratios which enable definition of the ratio of the two inlet areas (A_{2_i}/A_{1_i}) and the overall nozzle contraction ratio (A_i/A_e).

Desired information about the nozzle performance is the mass flow ratio of the two streams:

$$\text{mass flow ratio} = \frac{\dot{m}_2}{\dot{m}_1} = \frac{\rho_2 u_{2_e} A_{2_e}}{\rho_1 u_{1_e} A_{1_e}}.$$

The difference between the stagnation and the exit static pressure is known for each stream so the velocities u_{1_e} and u_{2_e} are known (from Bernoulli's equation) as $u_{1_e} = \sqrt{2\Delta p_{t_1}/\rho_1}$ and a similar equation for u_{2_e}. Since the density in each stream is constant, finding the mass flow ratio amounts to determining the exit area ratio of the two streams, A_{2_e}/A_{1_e}.

The exit area ratio can be found as follows. There are four unknowns in the description of the flow in the nozzle, the two velocities at the inlet and the two stream areas at the exit: $u_{1_i}, u_{2_i}, A_{1_e}, A_{2_e}$, respectively. Four relations between these quantities are required. The equation of continuity in the two streams furnishes two of these:

$$u_{1_i} A_{1_i} = u_{1_e} A_{1_e}, \tag{10.8.1a}$$

$$u_{2_i} A_{2_i} = u_{2_e} A_{2_e}. \tag{10.8.1b}$$

Writing Bernoulli's equation between the inlet and exit for each stream and using the fact that the static pressure is uniform across the nozzle at a given axial station in this one-dimensional flow leads to a third relation:

$$\frac{\rho_2}{\rho_1}\left(u_{2_e}^2 - u_{2_i}^2\right) = \left(u_{1_e}^2 - u_{1_i}^2\right). \tag{10.8.2}$$

The final relation is the statement that the sum of the exit areas of the two streams is the nozzle exit area, expressed non-dimensionally as

$$\frac{A_e}{A_i} = \frac{A_{1_e}}{A_i} + \frac{A_{2_e}}{A_i}. \tag{10.8.3}$$

Equations (10.8.1)–(10.8.3) can be combined into a quadratic equation for the exit area of stream 1 in terms of the variable A_{1_e}/A_{1_i}, from which the desired exit area ratio can be found

$$\left(\frac{A_{1_e}}{A_{1_i}}\right)^2\left[1-\frac{\Delta p_{t_2}}{\Delta p_{t_1}}\left(\frac{A_{1_i}}{A_{2_i}}\right)^2\right]+2\left(\frac{A_{1_e}}{A_{1_i}}\right)\left[\left(\frac{A_e}{A_{1_i}}\right)\frac{\Delta p_{t_2}}{\Delta p_{t_1}}\left(\frac{A_{1_i}}{A_{2_i}}\right)^2\right]$$
$$-\left\{1+\left(\frac{\Delta p_{t_2}}{\Delta p_{t_1}}\right)\left[\left(\frac{A_{1_i}}{A_{2_i}}\right)^2\left(\frac{A_e}{A_{1_i}}\right)^2-1\right]\right\}=0. \tag{10.8.4}$$

Equation (10.8.4) states that the exit stream area ratio, and thus the stream mass flow ratio, depends on three non-dimensional parameters. Two are associated with the nozzle geometry: the ratio of the exit area to the inlet area of stream 1 (A_e/A_{1_i}) and the ratio of the stream areas at the nozzle inlet, (A_{2_i}/A_{1_i}). The third parameter is the ratio of stagnation/static pressure differences in the two streams ($\Delta p_{t_2}/\Delta p_{t_1}$).

The density ratio does not appear in (10.8.4). The exit area ratio is therefore independent of the density ratio between the streams, although the velocities and mass flows are not. For given stagnation pressure differences, the exit velocities in each stream, j, scale as $u_{j_e} \propto 1/\sqrt{\rho_j}$. Altering the density in a given stream causes the mass flow in the stream to change as $\sqrt{\rho_j}$. The density ratio is proportional to the inverse of the temperature ratio between the two streams (see Section 1.17), and the above can also be interpreted as a statement about the independence of the streamline pattern on the temperature distribution. This is in fact a special case of a general principle for any steady, inviscid perfect gas with constant specific heats, namely that the streamline pattern is independent of the stagnation temperature distribution. This idea will be developed in Section 10.10.

To facilitate interpretation of the compound-flow nozzle operation it is helpful to cast the two geometry parameters in (10.8.4) into more conventional nozzle quantities. We thus present results for the density-weighted mass flow ratio $[(\dot{m}_2/\sqrt{\rho_2})/(\dot{m}_1/\sqrt{\rho_1})]$ in terms of the equivalent parameter set: (i) nozzle contraction ratio A_i/A_e, and (ii) nozzle area ratio at inlet (A_{2_i}/A_{1_i}).

Figure 10.27 shows the behavior of a nozzle with inlet area ratio $A_{2_i}/A_{1_i} = 3.0$ as a function of the contraction ratio for several values of stagnation pressure difference ratio ($\Delta p_{t_2}/\Delta p_{t_1}$). The curves for $\Delta p_{t_2}/\Delta p_{t_1} = 1$ correspond to a uniform inlet flow, in which the mass flow ratio mirrors the inlet area ratio. The difference between the mass flow ratio at a contraction ratio of unity (constant area duct) and the mass flow ratio at other contraction ratios represents streamtube area adjustment within the nozzle. Larger values of nozzle contraction ratio (smaller values of exit area/inlet area) mean larger static pressure drops from the inlet to the exit. In the nozzle there is greater velocity change for the low stagnation pressure stream (stream 2) than for the high stagnation pressure stream (stream 1). The trend towards equalization of the velocity at the exit as nozzle pressure drop increases implies a decrease (from inlet to exit) in the fractional area occupied by the low stagnation pressure stream, and thus in the mass flow ratio, as the contraction ratio increases (see also Section 5.9). For higher values of contraction ratios than shown the mass flow ratio goes to zero. This corresponds to the

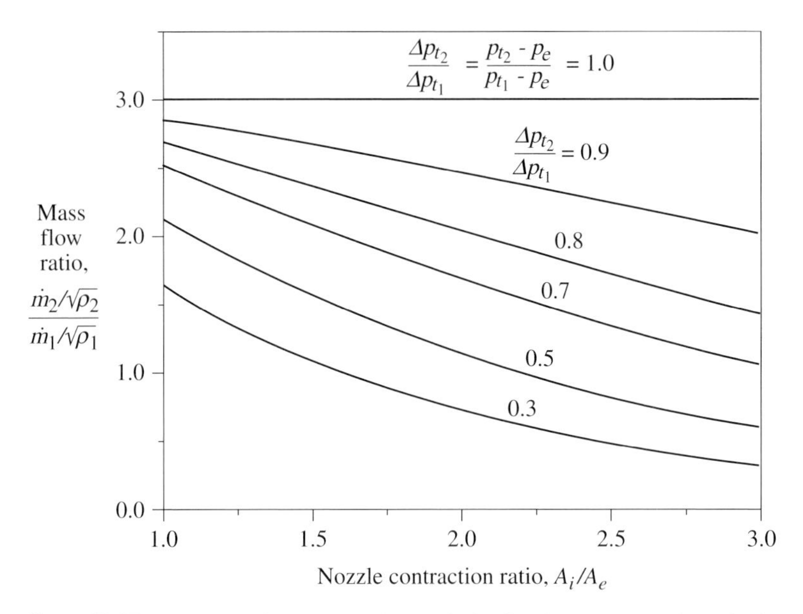

Figure 10.27: Density-weighted mass flow ratio for flow in a compound nozzle; incompressible flow, $A_{2_i}/A_{1_i} = 3.0$.

forward flow limit of stream 2, in other words the condition at which the static pressure at the nozzle inlet rises to the value of the stagnation pressure p_{t_2}.

There are two main points to be taken from the incompressible analysis. First is that flow behavior is defined once the nozzle geometry and the ratio of the differences between reservoir stagnation pressure and exit static pressure in the two streams are specified. Second is the independence of the exit stream area on the density ratio, which implies the nozzle performance is independent of the stagnation temperature ratio. These concepts carry over in a qualitative fashion to the compressible case, examined next, which exhibits the additional phenomena of compound choking.

10.8.2 Qualitative considerations for multistream compressible flow

The treatment of multistream nozzle compound flow is similar to that presented for one-dimensional flow in a single stream (Sections 2.7 and 10.3). The static pressure variation normal to the flow direction is neglected, i.e. the static pressure at a given axial location is taken as uniform across the channel. Each stream is considered as a steady, isentropic flow of a perfect gas with constant specific heats. Mixing effects are thus excluded, although comments are given later on the influence mixing is likely to have. The configuration referred to as a *compound-compressible flow* is shown schematically for three streams in a converging–diverging nozzle in Figure 10.28.

An explicit difference from the approach used for single streams is that static pressure is used as an independent variable because it varies only *along* the duct, whereas all other fluid properties can vary from stream to stream *across* the duct.

Before quantitatively examining the flow, we give a qualitative description of flow regimes and the behavior to be expected; this also aids in framing the questions to be addressed. Consider the two

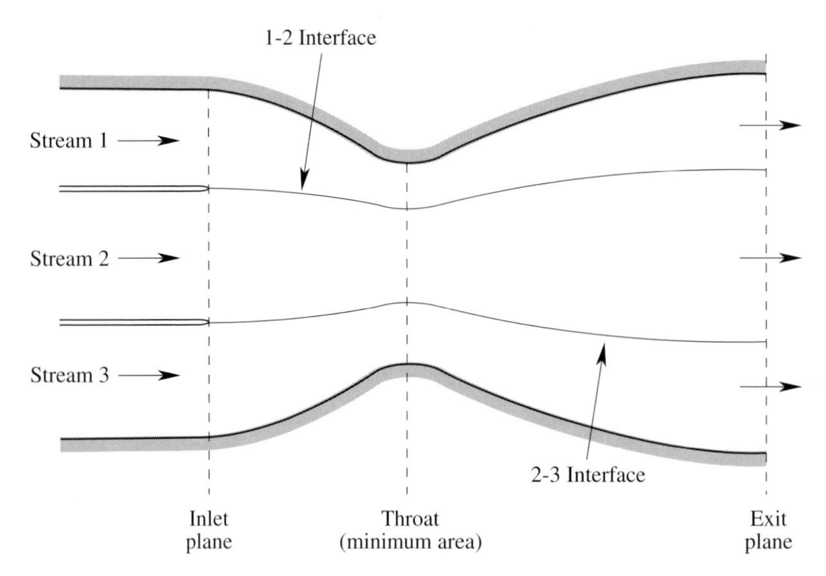

Figure 10.28: Schematic of a three-stream compound-compressible nozzle flow.

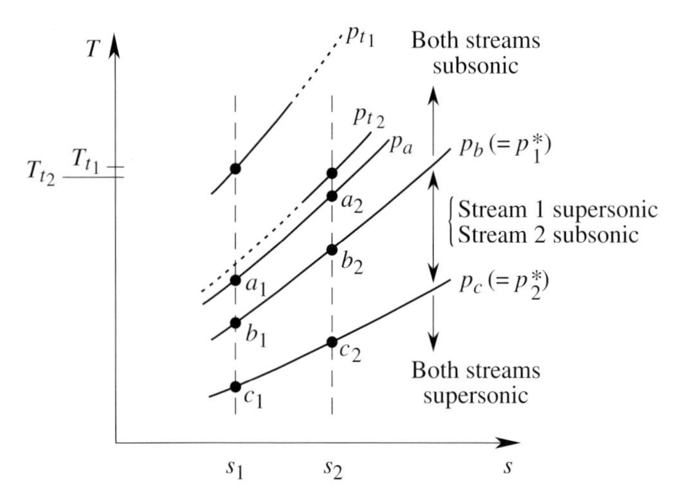

Figure 10.29: Flow regimes in two-stream compressible flow.

coflowing streams referred to by the subscripts "1" and "2", which are depicted in the temperature versus specific entropy plot of Figure 10.29. Each stream has properties along its respective constant entropy line as though the other stream were not present. In the example shown, stream 1 has a higher stagnation pressure and lower entropy than stream 2.

Constant pressure lines are indicated in Figure 10.29, corresponding to the two stagnation pressures, p_{t_1} and p_{t_2}, and to pressure levels representing local static pressure at different positions along the channel. For a given static pressure, the Mach number in either stream, j, is

$$M_j^2 = \left(\frac{2}{\gamma_j - 1}\right)\left[\left(\frac{p_{t_j}}{p}\right)^{\frac{\gamma_j - 1}{\gamma_j}} - 1\right], \tag{10.8.5}$$

which depends only on the ratio of the stream stagnation pressure to the local static pressure.

At a static pressure equal to p_{t_2}, the Mach number and velocity in stream 2 are zero, and only stream 1 will flow. When the static pressure is p_a, where p_a is close to p_{t_2}, the streams are at conditions a_1 and a_2, with both streams subsonic and a large Mach number ratio between the streams. At lower levels of static pressure, the Mach number ratios approach unity (qualitatively similar to the low Mach number results of Section 10.8.1), although M_1 is always larger than M_2.

When the pressure level falls to a value, p_b $(= p_1^*)$, at which $M_1 = 1$, the two streams are at conditions represented by points b_1 and b_2, with stream 1 sonic and stream 2 subsonic. At a still lower pressure, p_c $(= p_2^*)$, equal to the sonic value for stream 2, $M_2 = 1$ and stream 1 is supersonic and the two streams are at the conditions represented by c_1 and c_2 in Figure 10.29. At pressures below p_c, both streams are supersonic.

The above discussion implies there is a range of static pressures between p_b and p_c in which one stream is subsonic and the other supersonic. Configurations with more streams of intermediate stagnation pressures could also exist with several streams supersonic and several subsonic. A relevant issue, therefore, is whether a generalization of the ideas concerning single stream subsonic and supersonic channel flow exists which can be applied to compound-compressible flow. More specifically, several questions arise directly from their importance for the single stream situation:

(a) How does one characterize the regimes in which the compound flow behaves as a supersonic flow and as a subsonic flow?
(b) Is there an indicator whose value serves as a guide to whether the compound flow will act in a supersonic or subsonic manner?
(c) Does a nozzle throat play the same role in compound-compressible flow as in single stream compressible flow?
(d) What is the connection between the transition from subsonic to supersonic behavior, the velocities in the various streams, and the speed of propagation of small amplitude, one-dimensional disturbances in the channel?

10.8.3 Compound-compressible channel flow theory

To address these issues, we turn to the description of an n-stream compound-compressible channel flow developed by Bernstein *et al.* (1967). At any position in the channel, x, with A the total flow area and A_j the area of the jth stream,

$$A = \sum_{j=1}^{n} A_j, \quad \text{and} \quad \frac{dA}{dx} = \sum_{j=1}^{n} \frac{dA_j}{dx}. \tag{10.8.6}$$

The differential forms of the continuity and momentum equations, and the isentropic relation for each of the streams, j, follow directly from single stream analysis and are, respectively:

$$\frac{dA_j}{A_j} + \frac{d\rho_j}{\rho_j} + \frac{du_j}{u_j} = 0, \tag{10.8.7}$$

$$u_j du_j = -\frac{dp}{\rho_j}, \tag{10.8.8}$$

$$\frac{dp}{d\rho_j} = \frac{\gamma p}{\rho_j}. \tag{10.8.9}$$

Combining (10.8.7), (10.8.8), (10.8.9), and dividing by dx, the area variation along the channel for each stream is

$$\frac{dA_j}{dx} = \frac{A_j}{\gamma_j}\left(\frac{1}{M_j^2} - 1\right)\frac{1}{p}\frac{dp}{dx}. \tag{10.8.10}$$

There is no subscript on the pressure because it is uniform across the channel. Summing (10.8.10) over all streams and using (10.8.6) yields a relation between variations in the channel area and the static pressure:

$$\frac{1}{p}\frac{dp}{dx} = \frac{\dfrac{dA}{dx}}{\sum\limits_{j=1}^{n}\dfrac{A_j}{\gamma_j}\left(\dfrac{1}{M_j^2} - 1\right)} = \frac{1}{\beta}\frac{dA}{dx}. \tag{10.8.11}$$

Equation (10.8.11) defines a variable β as

$$\beta = \sum_{j=1}^{n}\frac{A_j}{\gamma_j}\left(\frac{1}{M_j^2} - 1\right). \tag{10.8.12}$$

Examination of β will be seen to be useful in answering questions (a)–(d).

We now address the problem of computing the compound-compressible flow in a fixed geometry nozzle. At the inlet, the areas occupied by different streams are known (since we know the geometry of the ducts that supply the nozzle). The stagnation temperature, pressure, and gas properties of each stream are also known, because the streams can be considered to be supplied from upstream reservoirs. To close the problem, one other flow variable must be specified. One generally cannot control the static pressure or mass flow at the inlet, and the back pressure, p_B, to which the nozzle discharges is the known variable (Section 2.7). It is conceptually more straightforward, however, to describe the nozzle behavior assuming we do know the nozzle inlet static pressure. There is no inconsistency in proceeding in this manner, and once the ideas concerning flow regimes are set forth we can then address the procedure for computing the compound flow based on exit conditions.

With static pressure specified at the nozzle inlet, the Mach numbers and hence the corrected flow per unit area in all the streams are known. Equation (10.8.11)[6] could in principle be integrated from inlet to exit to obtain the static pressure along the nozzle, and thus all other flow quantities. (We would, in fact, be finding the back pressure which must be set so the nozzle could operate with the prescribed inlet static pressure.) The result of doing this for different inlet static pressure levels is shown schematically in Figure 10.30. The stagnation pressure level of the stream with the lowest p_t is indicated by the horizontal line as $p_{t_{min}}$. Suppose the stream stagnation pressures are such that for inlet static pressure p_a, the quantities $[(p_{t_j} - p)/p_{t_j}]$ are everywhere small compared to unity. If so the Mach numbers for all streams are much below unity, $1/M_j^2$ is large, and β is positive. In this situation (10.8.11) shows that static pressure and area changes have the same sign throughout the nozzle. The static pressure has its smallest value, and the Mach numbers in all streams have their

[6] Plus the relation between the Mach number and the static/stagnation pressure ratio

$$M_j^2 = \frac{2}{\gamma_j - 1}\left[\left(\frac{p_{t_j}}{p}\right)^{(\gamma_j-1)/\gamma_j} - 1\right].$$

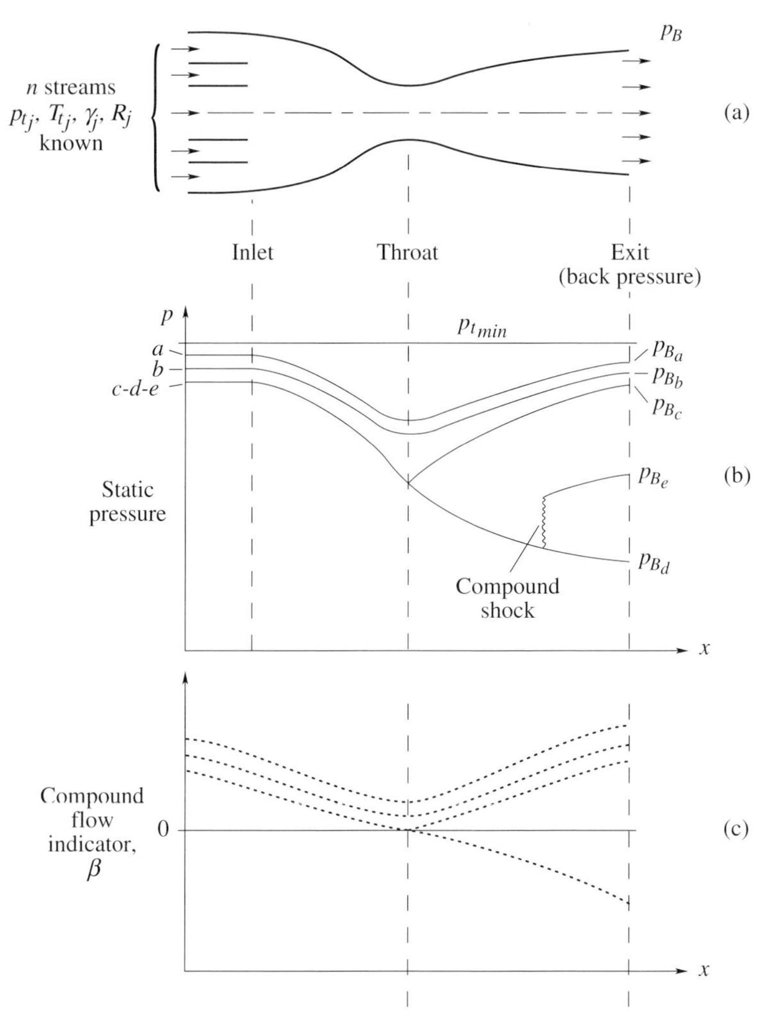

Figure 10.30: (a) Geometry, (b) static pressure, and (c) compound flow indicator, β, for compound-compressible nozzle flow.

largest values, at the nozzle throat. The back pressure, p_{B_a}, corresponding to this condition is equal to the nozzle exit pressure.

As the inlet static pressure is decreased, the inlet Mach numbers increase so that β decreases. Differentiating β with respect to static pressure gives

$$\frac{d\beta}{dp} = \sum_{j=1}^{n} \frac{A_j}{p\gamma_j^2 M_j^4} \left[\left(1 - M_j^2\right)^2 + 2\left(1 + \frac{\gamma_j + 1}{2} M_j^2\right) \right]. \tag{10.8.13}$$

The quantity $d\beta/dp$ is always positive and changes in β and p are in the same direction for all values of M_j. For inlet static pressures "a" and "b", therefore, β has a minimum at the throat. This can also be inferred from the fact that Mach numbers in all streams are maximum at this location.

As the inlet pressure is decreased further, the value of β at the throat decreases to zero. At this condition, (10.8.11) is indeterminate, but the static pressure gradient at the throat can be obtained using L'Hopital's rule (differentiating numerator and denominator) as

$$\frac{1}{p}\left(\frac{dp}{dx}\right)_{\beta=0} = \pm\left\{\frac{\dfrac{d^2A}{dx^2}}{\sum_{j=1}^{n}\dfrac{A_j}{\gamma_j^2M_j^4}\left[\left(1-M_j^2\right)^2+2\left(1+\dfrac{\gamma_j+1}{2}M_j^2\right)\right]}\right\}^{1/2}. \quad (10.8.14)$$

Equation (10.8.14) holds at the throat, where d^2A/dx^2 is positive. Just as for a single stream, there are two isentropic solutions with inlet pressure, p_c. The solution with the higher back pressure, p_{B_c} (see Figure 10.30) is qualitatively similar to curves a and b. From knowledge of the single stream behavior, however, we expect the flow regime downstream of the throat for the solution with lower back static pressure, p_{B_d}, to be qualitatively different than a, b, or c. As in the single stream case, there are no back pressures between p_{B_c} and p_{B_d} for which isentropic solutions exist. With a back pressure of p_{B_e}, say, we would expect a "compound shock" in the diverging section of the nozzle.

Consider the behavior along the curve with exit pressure p_{B_d}. Because β and the static pressure change in the same direction, $d\beta/dx$ is negative at the throat (where $\beta = 0$). The value of β therefore changes from positive to negative through the throat and decreases further, along with the static pressure, in the diverging part of the nozzle.

Figure 10.30 also shows the effect of back pressure p_B, on inlet conditions. Back pressures higher than p_{B_c} influence the inlet flow, but once the back pressure is lowered past this level further changes have no effect on the region upstream of the throat. This condition, referred to as *compound choking*, is now examined in more detail.

10.8.4 One-dimensional compound waves

For one-dimensional single stream channel flow, choking is associated with the condition at which the fluid velocity becomes equal to the propagation rate of small disturbances and information does not travel upstream. To determine whether a corresponding condition exists for compound-compressible channel flow consider the propagation of one-dimensional small amplitude disturbances on a compound stream, as in Figure 10.31. The upper part of the figure shows the behavior for one individual streamtube, while the lower part shows the streamtubes in aggregate. For one-dimensional disturbances to exist the static pressure must be uniform across the channel, implying the pressure difference across the disturbance is the same in each stream.

Suppose the disturbance velocity in the upstream direction is v_{wave} in a stationary coordinate system. In a coordinate system fixed to the disturbance, the relative velocity, w, is $u + v_{wave}$, as on the top part of Figure 10.31. We can obtain an expression relating the changes in area of the jth stream, dA_j/A_j, and the pressure difference across the disturbance, dp/p, by combining the one-dimensional continuity, momentum, and isentropic equations ((10.8.7), (10.8.8), (10.8.9)) written in the disturbance fixed coordinate system as

$$\frac{dA_j}{A_j} = \frac{dp}{\gamma_j p}\left[\frac{1}{\left(\dfrac{w_j}{a_j}\right)^2} - 1\right]. \quad (10.8.15)$$

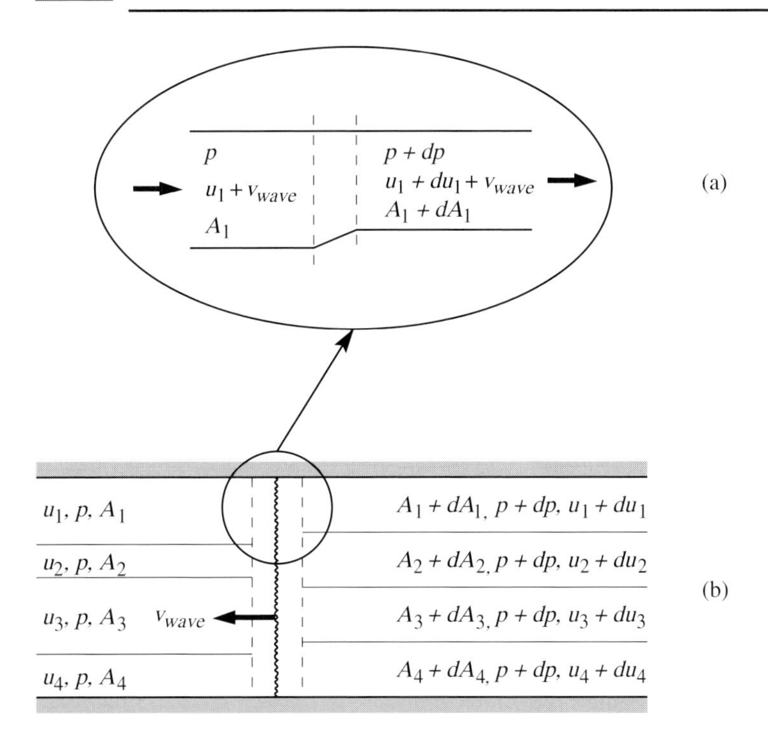

Figure 10.31: One-dimensional compound small amplitude disturbance wave; (a) stream 1 disturbance viewed in a relative coordinate system moving with disturbance velocity v_{wave}; (b) all streams viewed in an absolute system (Bernstein *et al.*, 1967).

Transforming back to the stationary coordinate system, (10.8.15) becomes

$$\frac{dA_j}{A_j} = \frac{dp}{\gamma_j p}\left[\frac{1}{\left(\frac{v_{wave}}{a_j} + M_j\right)^2} - 1\right]. \tag{10.8.16}$$

If the pressure difference occurs over a distance short compared to that over which there are appreciable variations of area in the channel,

$$\sum_{j=1}^{n} dA_j = dA = 0. \tag{10.8.17}$$

Combining (10.8.16) and (10.8.17) yields an equation for the compound disturbance wave velocity, v_{wave}:

$$\sum_{j=1}^{n} \frac{A_j}{\gamma_j}\left[\frac{1}{\left(\frac{v_{wave}}{a_j} + M_j\right)^2} - 1\right] = 0. \tag{10.8.18}$$

Positive values of v_{wave} correspond to disturbance waves propagating upstream in the absolute system, and negative values correspond to the disturbances being swept downstream by the flow. The former denotes compound subsonic flow, the latter compound supersonic flow, and the situation $v_{wave} = 0$ corresponds to compound sonic flow.

Equation (10.8.18) may be used with the definition of β to yield an alternative expression for β which illustrates a key feature of this variable,

$$\beta = \sum_{j=1}^{n} \frac{A_j}{\gamma_j} \left[\frac{1}{M_j^2} - \frac{1}{\left(\frac{v_{wave}}{a_j} + M_j \right)^2} \right]. \tag{10.8.19}$$

Equation (10.8.19) shows that the velocity of the compound disturbance, v_{wave}, and β always have the same sign. Therefore, $\beta > 0$ corresponds to compound subsonic flow, $\beta < 0$ to compound supersonic flow, and $\beta = 0$ to compound sonic flow. The variable β is thus an indicator of the compound flow regime.

Before applying these results, we recap the conclusions reached so far concerning compound-compressible flow regimes:

(1) Compound choking occurs at a nozzle throat if the compound flow indicator, β, equals zero. At this condition, some of the streams have Mach numbers greater than unity and some less than unity; the individual Mach numbers for the different streams are generally not equal to unity. The different streams influence β in proportion to their areas.
(2) For compound subsonic flow not every stream must have $M_j > 0$. For compound supersonic flow not every stream must have $M_j > 0$.
(3) The compound subsonic and compound supersonic flow regimes defined by β are analogous to the corresponding flow regimes in single stream nozzles; the compound-compressible flow results reduce to the single stream results when $n = 1$.

10.8.5 Results for two-stream compound-compressible flows

10.8.5.1 Conceptual solution procedure and non-dimensional parameter specification

The compound-compressible flow equations can be solved numerically for any number of streams, but the important points concerning nozzle behavior are displayed using two streams with the same gas properties ($\gamma_1 = \gamma_2 = \gamma$ and $R_1 = R_2 = R$). The difference between compound-compressible and single stream channel flows is brought out in the basic problem of flow through a converging nozzle of specified geometry, fed by two reservoirs considered to be at fixed levels of p_{t_1}, p_{t_2}, T_{t_1}, T_{t_2}. We examine the behavior as p_B/p_{t_1} is decreased, starting from the highest back pressure at which there is forward flow in stream 2.

The unchoked regime is similar to the low Mach number situation of Section 10.8.1 in that specifying the back pressure $p_B/p_{t_1} (= p_e/p_{t_1})$ fixes the exit Mach numbers in each stream. The nozzle geometry also needs to be specified, both the contraction ratio and the stream area ratio at the inlet. Two quantities are necessary because the stream area variation from the inlet to the exit (due to the unequal velocity change of the two streams) depends on both. For a given contraction ratio there are an infinite number of solutions corresponding to different ratios of inlet stream areas. Conversely, for a given inlet stream area ratio there are an infinite number of solutions corresponding to different contraction ratios.

The solution procedure is as follows. The relation between corrected flow per unit area in each stream ($\dot{m}_j\sqrt{T_{t_j}}/(A_j p_{t_j})$) and the ratio of static to stagnation pressure p/p_{t_i} is

$$\frac{\dot{m}_j\sqrt{T_{t_j}}}{A_j p_{t_j}} = \left(\frac{p}{p_{t_j}}\right)^{\frac{1}{\gamma}} \left\{\frac{2\gamma}{R(\gamma-1)}\left[1-\left(\frac{p}{p_{t_j}}\right)^{\frac{\gamma-1}{\gamma}}\right]\right\}^{\frac{1}{2}} = f\left(\frac{p}{p_{t_j}}\right). \tag{10.8.20}$$

Summing the stream areas at the exit:

$$\frac{\dot{m}_1\sqrt{T_{t_1}}}{p_{t_1} f_{1_e}} + \frac{\dot{m}_2\sqrt{T_{t_2}}}{p_{t_2} f_{2_e}} = A_e, \tag{10.8.21}$$

where f denotes the function of p/p_{t_j} defined in (10.8.20).

The mass flows in the streams at the nozzle inlet are

$$\dot{m}_1 = (A_{1_i} p_{t_1} f_{1_i})/\sqrt{T_{t_1}}, \tag{10.8.22a}$$

$$\dot{m}_2 = (A_{2_i} p_{t_2} f_{2_i})/\sqrt{T_{t_2}}. \tag{10.8.22b}$$

Substitution of (10.8.22) in (10.8.21) yields a relation between the inlet and exit values of the function f and the known nozzle area ratios:

$$\left(\frac{A_{1_i}}{A_e}\right)\frac{f_{1_i}}{f_{1_e}} + \left(\frac{A_{2_i}}{A_e}\right)\frac{f_{2_i}}{f_{2_e}} = 1. \tag{10.8.23}$$

Because the stagnation pressure ratio, p_{t_2}/p_{t_1} is known, (10.8.23), which contains both p/p_{t_2} and p/p_{t_1}, is an implicit equation for p_i/p_{t_1}, the only unknown. Once p_i/p_{t_1} is found (10.8.22) provide the temperature corrected mass flow ratio $\dot{m}_2\sqrt{T_{t_2}}/\dot{m}_1\sqrt{T_{t_1}}$, at which the nozzle is operating. The area ratio changes as p_e is decreased until the choking condition is reached after which all quantities remain fixed.

There are five non-dimensional quantities in the description of the nozzle flow (in addition to the specific heat ratio, γ): the stagnation pressure ratio p_{t_2}/p_{t_1}, the temperature corrected mass flow ratio $\dot{m}_2\sqrt{T_{t_2}}/\dot{m}_1\sqrt{T_{t_1}}$, the nozzle pressure ratio p_{t_1}/p_B, and the nozzle area ratios A_{1_i}/A_e and A_{2_i}/A_e. These last two are equivalent to the inlet/exit nozzle contraction ratio and the inlet area ratio. Four of the five may be specified. Which four are chosen depends on the way in which the nozzle is run but a situation encountered in practice is a nozzle fed by two reservoirs with specified stagnation pressures and temperatures and a controlled back pressure. In that case, (10.8.23) can be regarded as an equation to determine the temperature corrected mass flow as a function of the ratio of the nozzle stagnation pressure to the back pressure, p_{t_1}/p_B, for given p_{t_2}/p_{t_1} and nozzle geometry.

10.8.5.2 Converging nozzle

Figure 10.32 shows the behavior of a converging nozzle with an inlet area ratio of 3.0 ($A_{2_i}/A_{1_i} = 3.0$) and a contraction ratio of 2.0 (this is the same geometry as in Figure 10.27). The temperature corrected mass flow ratio is given as a function of p_{t_1}/p_B for different stagnation pressure ratios. The mass flow ratio goes to zero when the back pressure is such that the inlet static pressure is equal to the stagnation pressure in stream 2, as mentioned in the incompressible example of Section 10.8.1.

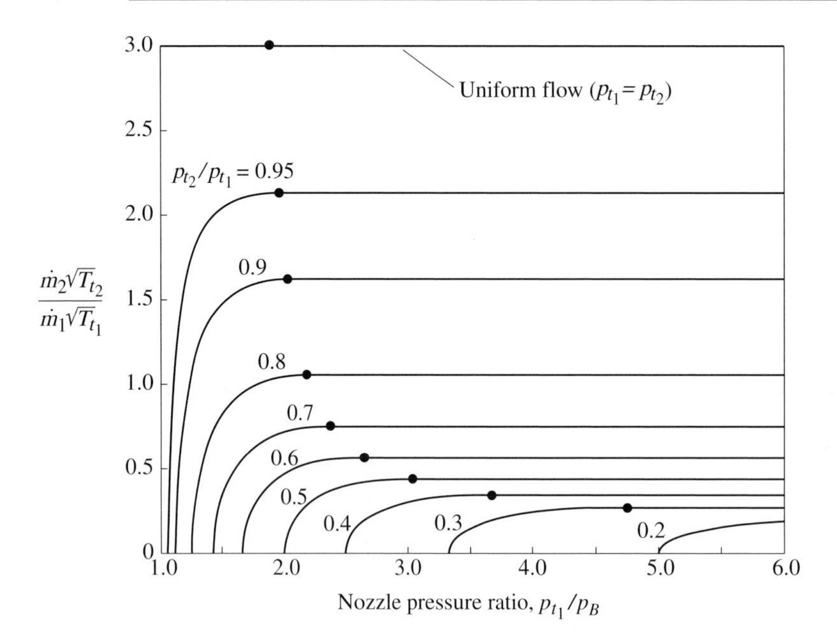

Figure 10.32: Temperature corrected mass flow ratio as a function of the nozzle pressure ratio for two-stream compound-compressible flow in a converging nozzle. Inlet area ratio $A_{2_i}/A_{1_i} = 3.0$, nozzle convergence ratio $A_i/A_e = 2.0$, $\gamma = 1.4$; solid circles indicate choke conditions.

The temperature corrected stream mass flow ratio[7] initially increases with nozzle pressure ratio but then, at a value which depends on the stagnation pressure ratio, ceases to change. The occurrence of this maximum mass flow ratio marks the condition of compound choking, where the flow upstream of the throat (the throat is at the nozzle exit) no longer responds to alterations in downstream conditions. The locus of the compound choking condition is marked by the solid symbols in Figure 10.32. The figure portrays both non-choked and choked behavior and can thus represent compound-compressible flow in the converging nozzle for any reservoir conditions and nozzle pressure ratio.

10.8.5.3 Converging–diverging nozzle

For a converging–diverging nozzle there is another geometric parameter, the nozzle area ratio from the throat to the exit, A_e/A_{TH}. For a given geometry and reservoir stagnation pressure ratio, as the nozzle pressure ratio increases the pressure distributions correspond to a, b, c in Figure 10.30. Once the nozzle pressure ratio reaches $p_{t_1}/p_B = p_{t_1}/p_B$, the condition at which the compound-compressible flow parameter, β, is equal to zero at the throat, the mass flow ratio in the nozzle does not change further.

At any location (x) in the nozzle the area, A, can be written as

$$\frac{\dot{m}_1\sqrt{T_{t_1}}}{A_e p_{t_1} f_1(x)} + \frac{\dot{m}_2\sqrt{T_{t_2}}}{A_e p_{t_2} f_2(x)} = \frac{A}{A_e}, \tag{10.8.24}$$

[7] This is often abbreviated to "mass flow ratio" below.

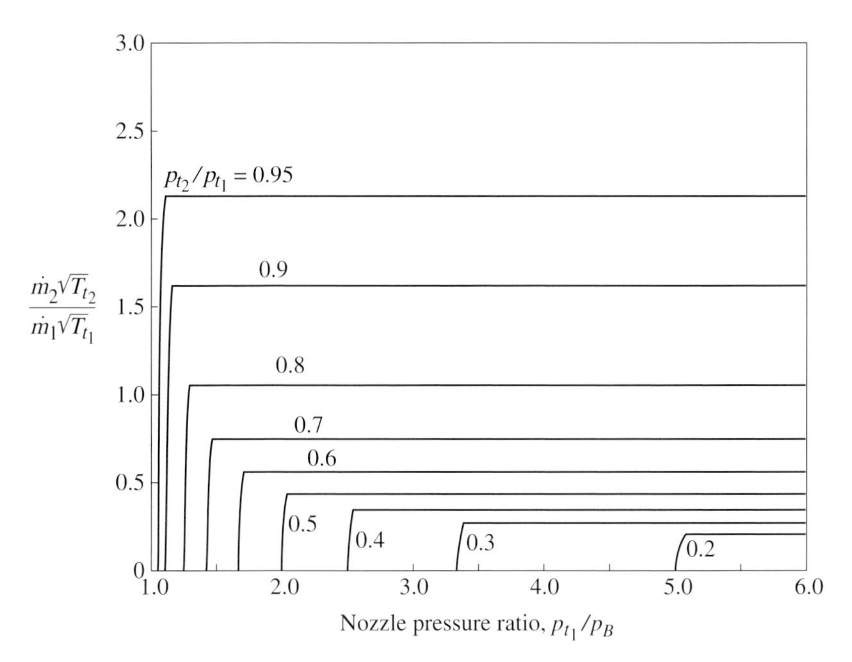

Figure 10.33: Temperature corrected mass flow ratio as a function of the nozzle pressure ratio for two-stream compound-compressible flow in a converging–diverging nozzle. Nozzle area ratios: $A_{2_i}/A_{1_i} = 3.0$, $A_i/A_{TH} = 2.0$, $A_e/A_{TH} = 2.0$ (A_{TH} = throat area), $\gamma = 1.4$.

where $f_j(x)$ denotes the corrected flow per unit area at location x, (10.8.20). From (10.8.24) the quantities $p(x)/p_{t_1}$ and $p(x)/p_{t_2}$, and hence the Mach numbers, stream areas, and value of β, can be found. For a specified nozzle geometry the nozzle pressure ratio corresponding to $\beta = 0$ can therefore be determined. At this condition, the nozzle will be compound-choked and the mass flow ratio will not change as the exit pressure is lowered further.

Figure 10.33 shows the temperature corrected mass flow ratio for a converging–diverging nozzle as a function of the nozzle pressure ratio, p_{t_1}/p_B. The geometry of the converging section is the same as in Figure 10.32, the ratio of the exit to the throat area is 2, and the results are for the same stagnation pressure ratios (p_{t_2}/p_{t_1}) as in Figure 10.32.

The nearly vertical sections of the solution curves denote the unchoked regime. These correspond to temperature corrected stream mass flow ratios for the unchoked operation of a converging nozzle with the same inlet and exit geometry as that of the specified converging–diverging nozzle. The horizontal lines denote choked operation, which is encountered at conditions corresponding to those depicted for back pressure level c (back pressure p_{B_c}) in Figure 10.30.

The converging and the converging–diverging nozzles of Figures 10.32 and 10.33 have the same ratios of inlet stream areas and inlet area to throat area. For choked flow at a given ratio of p_{t_2}/p_{t_1} the mass flow ratio in the converging–diverging nozzle is the same as that in the converging nozzle, as seen from comparing the levels of the horizontal lines in the two figures. Further, at the points of slope discontinuity in Figure 10.33 the ratio of the stagnation pressure in stream 1 to the static pressure at the throat (p_{t_1}/p_{TH}) corresponds to that for the choke points (solid circles) in Figure 10.32.

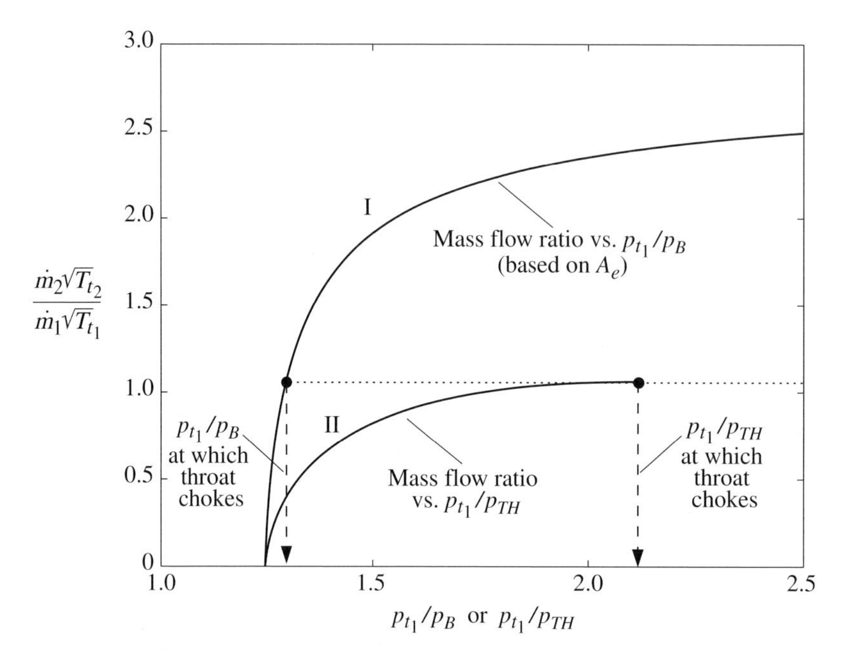

Figure 10.34: Choking behavior of a converging–diverging compound nozzle; $p_{t_2}/p_{t_1} = 0.8$, $A_{2_i}/A_{1_i} = 3.0$, $A_i/A_{TH} = A_e/A_{TH} = 2.0$, $\gamma = 1.4$.

For the converging–diverging nozzle the presence of the diverging section means there is a static pressure rise downstream of the throat, so that p_{t_1}/p_B is smaller than p_{t_1}/p_{TH}. As is the case for a single stream channel flow, therefore, the onset of choking occurs at a lower nozzle pressure ratio for a converging–diverging nozzle than for a purely converging nozzle.

These concepts are illustrated in Figure 10.34, which corresponds to the results for a stagnation pressure ratio (p_{t_2}/p_{t_1}) of 0.8 in Figure 10.33. (Only a portion of the range of Figure 10.33 is shown.) The two curves in Figure 10.34 correspond to: (I) the temperature corrected stream mass flow ratio versus the nozzle pressure ratio (p_{t_1}/p_B) and (II) the temperature corrected stream mass flow ratio versus the stagnation/throat pressure ratio (p_{t_1}/p_{TH}). The mass flow ratio increases as the nozzle pressure ratio increases along the unchoked portion of curve I. At any mass flow ratio p_{t_1}/p_{TH} is larger than p_{t_1}/p_B. When p_{t_1}/p_{TH} reaches the value for choking, the mass flow ratio is fixed and exhibits no further change with increase in nozzle pressure ratio. The solution trajectory must thus shift from moving along curve I to moving along a horizontal line at the level set by the choke point in curve II.

Several aspects of the choking behavior of a compound-compressible flow can be noted. One is the temperature corrected stream mass flow ratio which is shown in Figure 10.35 as a function of reservoir stagnation pressure ratio for different values of the inlet area ratio. In the limiting value of equal stagnation pressures the mass flow ratio and the inlet area ratio are the same.

A second aspect is the Mach number in the individual streams at the choking conditions. Figure 10.36 shows the nozzle inlet and throat Mach numbers as a function of the temperature corrected stream mass flow ratio for compound choked two-stream flow. The Mach number is not equal to unity in either stream, with M_2 being less than unity and M_1 greater than unity. This behavior is

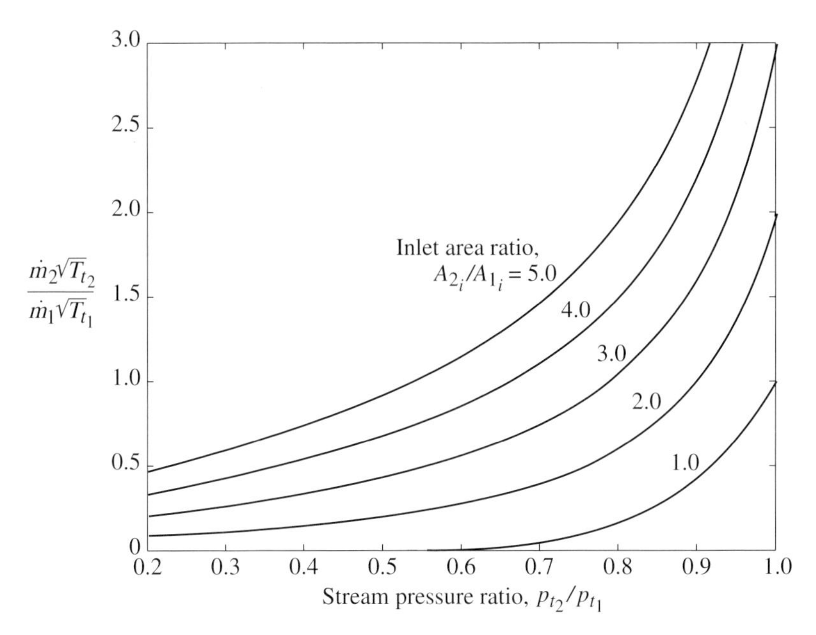

Figure 10.35: Temperature corrected mass flow ratio versus the stream stagnation pressure ratio (p_{t_2}/p_{t_1}) for compound choked two-stream flow. Nozzle contraction ratio $A_i/A_{TH} = 2.0$, $\gamma = 1.4$.

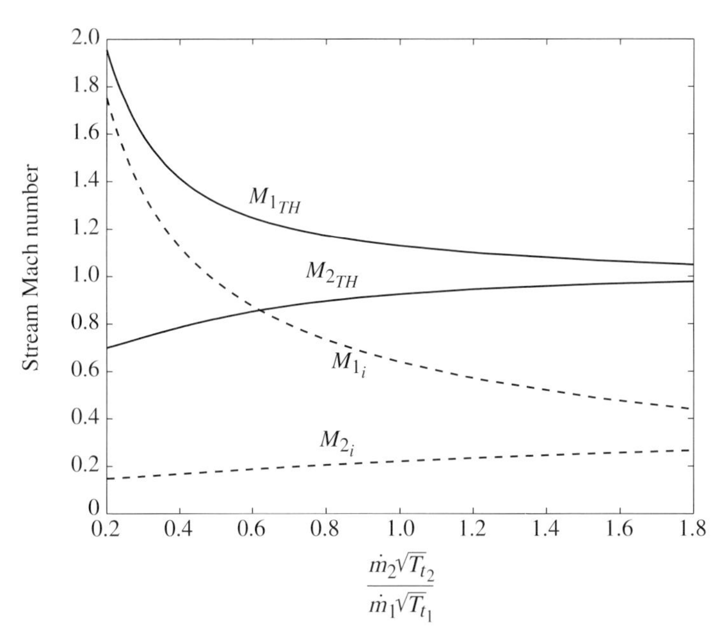

Figure 10.36: Inlet and throat Mach numbers in each stream versus the temperature corrected mass flow ratio for compound choked two-stream flow. Nozzle inlet area ratio $A_{2_i}/A_{1_i} = 3.0$, $\gamma = 1.4$.

characteristic of compound choking. As the mass flow ratio approaches the nozzle inlet area ratio (3 for this geometry) the stagnation pressures in the two streams approach one another, resulting in the trend towards equal Mach number that is seen as the mass flow ratio increases.

10.8.5.4 An alternative specification for compound-compressible nozzle operation

The discussion so far has assumed that reservoir conditions and inlet areas are prescribed, but there are other ways to specify the nozzle operating conditions. One of these is to regard the corrected mass flow per unit area at the inlet in one stream (say stream 1) as known. This was the case, for example, in the experiments described below. It would be convenient to view the problem in this manner, for example, if one of the streams that fed the nozzles was choked and delivered a constant corrected flow. In such a situation it is useful to define a reference sonic area, A_1^*, for the choked stream (stream 1) as

$$A_1^* = \frac{\dot{m}_1\sqrt{T_{t_1}}}{p_{t_1}}\sqrt{\frac{R}{\gamma}\left(\frac{\gamma+1}{2}\right)^{\left[\frac{\gamma+1}{\gamma-1}\right]}} = \frac{\dot{m}_1\sqrt{T_{t_1}}}{p_{t_1}}C, \tag{10.8.25}$$

where C is the known function of R, γ (constant for a given gas) defined in (10.8.25). This allows a simpler solution for the nozzle behavior. Summing the two stream areas at the exit and using the definition of A_1^* leads to an explicit expression for the temperature corrected mass flow ratio:

$$\frac{\dot{m}_2\sqrt{T_{t_2}}}{\dot{m}_1\sqrt{T_{t_1}}} = \frac{CA_e}{A_1^*}\frac{p_{t_2}f_{2_e}}{p_{t_1}} - \frac{p_{t_2}f_{2_e}}{p_{t_1}f_{1_e}}. \tag{10.8.26}$$

All the quantities on the right-hand side of (10.8.26) are known and the mass flow ratio can be found directly. This occurs because of the additional constraint from the imposition of known corrected flow in stream 1. Specification of A_1^* means that the nozzle inlet areas do not need to be, indeed *cannot* be, prescribed.

10.8.5.5 Experimental results for compound-compressible nozzle flows

Bernstein *et al.* (1967) give a detailed development of the behavior of a compound-compressible nozzle flow in terms of the reference sonic area, and we present results from their analysis as an aid in interpreting the experimental results that are given below. The experiments were run with fixed values of A_1^*/A_e and $\dot{m}_1\sqrt{T_{t_1}}/\dot{m}_2\sqrt{T_{t_2}}$. To correspond with this, Figure 10.37 shows compound-compressible nozzle behavior as a function of p_{t_1}/p_B for different values of A_1^*/A_e and the exit area to throat area, A_e/A_{TH}. The dependent variable here is the stagnation pressure ratio, p_{t_2}/p_{t_1}, and Figure 10.37 can thus be viewed as giving the stagnation pressure ratio required to operate at a given nozzle pressure ratio for specified area ratios and temperature corrected mass flow ratio.

The unchoked behavior, to the left of the dotted line in Figure 10.37, can be discussed with reference to a given line of A_1^*/A_e, say 0.4. As the nozzle pressure ratio, p_{t_1}/p_B, is increased holding $\dot{m}_1\sqrt{T_{t_1}}/\dot{m}_2\sqrt{T_{t_2}}$ and A_1^*/A_e constant, the stagnation pressure ratio necessary to operate at these conditions drops. The line of unchoked operation continues until the point at which the exit area equals the throat area, $A_e = A_{TH}$, at point B. Alternatively, for values of $A_e > A_{TH}$, unchoked operation can terminate at the intersection with a horizontal line corresponding to that value of A_e/A_{TH}, as at

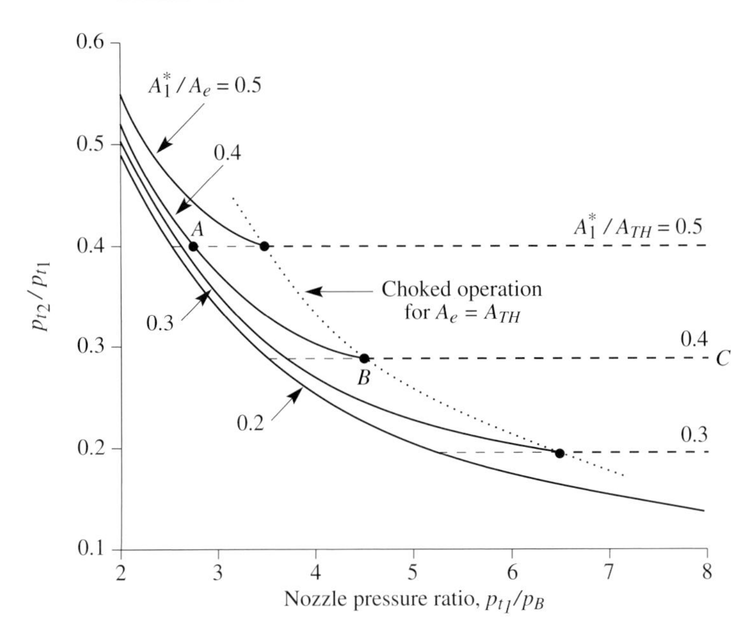

Figure 10.37: Relationship of flow parameters in a converging–diverging nozzle; $\gamma = 1.4$, $R = 288$ J(kg k), $\dot{m}_1\sqrt{T_{t_1}}/\dot{m}_2\sqrt{T_{t_2}} = 0.315$ (Bernstein *et al.*, 1967).

point A for a ratio $A_e/A_{TH} = 1.25$. For nozzle pressure ratios larger than this, operation upstream of the throat is unaffected. The complete behavior of a compound nozzle with specified A_1^*/A_e and A_1^*/A_{TH} is thus given by (a part of) the relevant solid curve plus the appropriate horizontal line. As in the previous discussion, there is a slope discontinuity in the operating point solution trajectory at the choke conditions.

With this as background, we examine the comparison of compound-compressible analysis to experiment in Figure 10.38. A sketch of the nozzle and a list of the nozzle parameters are given in the figure. Data for four temperature corrected mass flow ratios are shown, with nozzle pressure ratios, p_{t_1}/p_B, from 2 to 10. The experiments were run with the primary (stream 1) independently choked and slightly underexpanded at the inlet plane, but the conditions were close to isentropic flow. The mass flow ratio was fixed for each series of experiments.

Choked and unchoked behavior are exhibited over differing portions of the nozzle pressure ratio range. The nozzle had a straight centerline and the geometry was such that radial velocity components were small and the quasi-one-dimensional approximation appropriate. The nozzle was also short enough such that mixing, either of velocity or temperature, did not play a significant role.

In the compound choked flow regime in Figure 10.38 (the horizontal lines), the one-dimensional theory shows good agreement with experiment. This should be expected because wall boundary layers are thin because of favorable pressure gradients as are shear layers between the streams. Also, and perhaps more importantly, mixing effects downstream of the nozzle throat cannot exert any influence on the overall behavior, so the effective mixing length during choked operation is the distance from the inlet plane to the throat. For unchoked operation, mixing downstream of the nozzle throat can affect overall behavior, because the effective mixing length is the entire length of the nozzle (although many nozzle applications do have a large portion of their length supersonic).

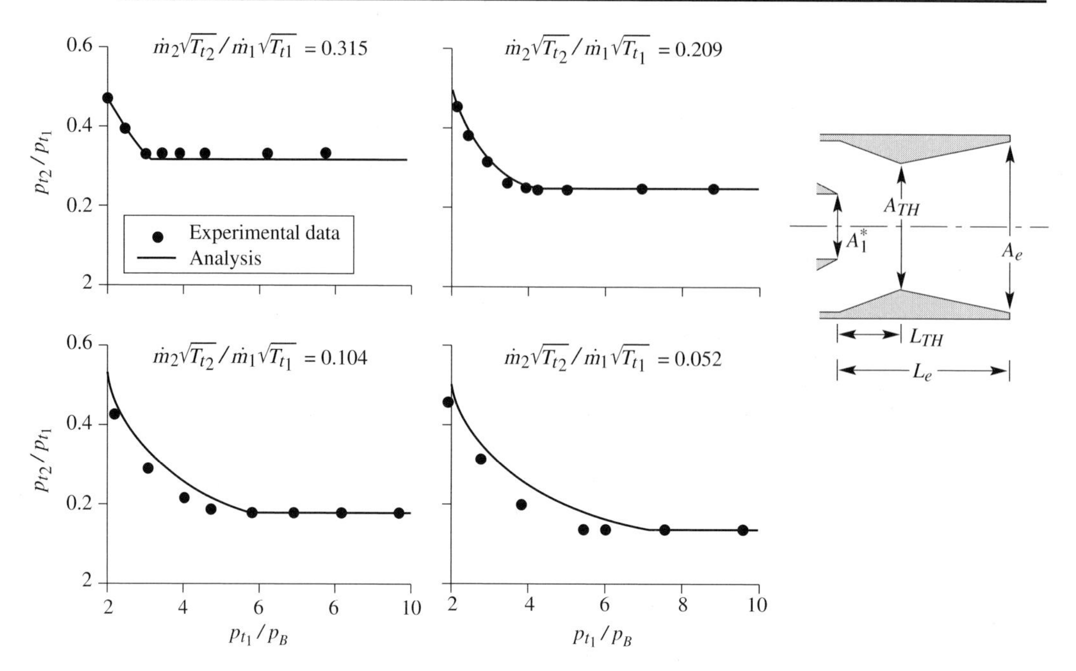

Figure 10.38: Comparison of compound-compressible flow analysis with experimental results. Nozzle dimensions: $L_{TH}/d_{TH} = 0.37$, $L_e/d_e = 0.70$, $A_1^*/A_{TH} = 0.43$, $A_1^*/A_e = 0.23$; $\gamma = 1.4$, $R = 288$ J(kg k) (Bernstein *et al.*, 1967).

Mixing has little effect at the higher mass flow ratios, but it exerts an increasing influence as mass flow ratio decreases. The effect of mixing is to increase the stagnation pressure of stream 2, in effect pumping the low velocity flow and reducing the required value of inlet p_{t_2}/p_{t_1}.

Compound-compressible flow analysis has been generalized to deal with streams having a continuous variation in properties (Decher, 1978; Lewis and Hastings, 1989). The latter reference presents an application to hypersonic propulsion (particularly the role of the forebody boundary layer on the flow inside the engine) and demonstrates the strong impact that a low-speed streamtube can have on channel flow with high freestream Mach number. The analysis has also been extended to account, in an approximate manner, for the effects of mixing in the context of a procedure for designing mixer nozzle noise suppressors (Tew, Teeple, and Waitz 1998).

10.9 Flow angle, Mach number, and pressure changes in isentropic supersonic flow

Much of this chapter has focused on flows which can be described within the quasi-one-dimensional framework, neglecting pressure differences normal to the stream. In this section we address the effect of normal pressure gradients on flow angle changes in a supersonic stream. The causal link between pressure gradient and streamline deflection is unchanged from subsonic flow, in that the deflections are in the direction of the low pressure, but the manner in which the flow angle change occurs is different in supersonic and subsonic regimes.

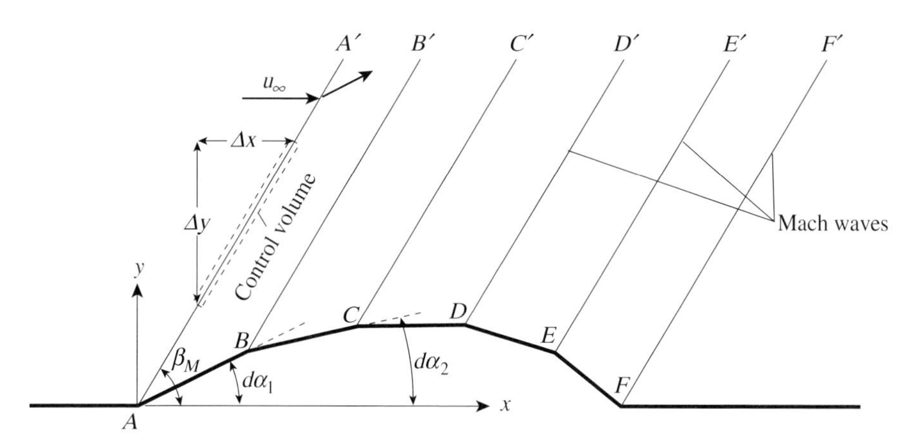

Figure 10.39: Supersonic flow around a straight-line segment bump; Δx, Δy are projections of the control volume on the x- and y-axes, u_∞ = undisturbed velocity.

10.9.1 Differential relationships for small angle changes

The issue can be illustrated through examination of the relation between pressure and turning at a bounding solid surface, confining the discussion to two dimensions. Consider the geometry in Figure 10.39 with a flat wall and a "bump" made up of straight-line segments. If the angles the segments make with the wall are small, the disturbances will also be small, the flow field will be described by a linear analysis, and the flow pattern can be built up by a superposition of small disturbances emanating from the boundary.

In Figure 10.39 a uniform parallel supersonic stream of velocity u strikes the first non-parallel element of the surface, AB, whose inclination to the stream direction is $d\alpha_1$. Two effects occur: the direction of the stream is changed by the angle $d\alpha_1$, and a pressure rise, dp_1, is produced. We wish to calculate the magnitude of dp_1 if the Mach number and the deflection $d\alpha_1$ are known.

For a supersonic flow, as described in Section 2.8, the region of influence of a disturbance is felt only behind the Mach lines, which are at an angle to the flow of $\beta_M = \sin^{-1}(1/M)$. (This can be seen by consideration of the propagation velocity of sound waves compared with the fluid velocity; sound waves are convected downstream as they propagate outwards from a disturbance.) The line AA' in Figure 10.39 denotes a Mach line with the pressure rise dp_1 behind it. In a two-dimensional flow every particle passing through AA' suffers the same deflection, $d\alpha_1$, and is subjected to the same pressure rise, dp_1.

We now apply the momentum theorem to the dashed control volume which encloses part of the Mach line AA'. There is no y-component of momentum flux entering the control volume. The y-component of momentum flux leaving the control volume is, to first order in the disturbance quantities, $(\rho u du_{y_1})\Delta y$, where non-subscripted quantities refer to undisturbed flow conditions and Δy is the control volume extent in the y-direction. The y-direction force on the fluid is $(dp_1)\Delta x$, where Δx is the extent of the control volume in the x-direction. Equating force to net outflow of momentum yields

$$\rho u du_{y_1} = dp_1 \left(\frac{\Delta x}{\Delta y}\right). \tag{10.9.1}$$

Two more substitutions are needed to achieve the desired relation. First, the flow angle, $d\alpha_1$, downstream of the Mach wave is linked to the y-component of velocity by

$$\tan(d\alpha_1) = \frac{du_{y_1}}{u + du_{x_1}}.$$

For small deflections of the stream the angle can be approximated by $ud\alpha_1 = du_{y_1}$. Second, the ratio $\Delta x/\Delta y$ for the control volume is the cotangent of the Mach angle, β_M, $\Delta x/\Delta y = \cot\beta_M = \sqrt{M^2-1}$. Substituting the expressions for du_{y_1} and $\Delta x/\Delta y$ into (10.9.1) gives the relation between the pressure change across the Mach wave and the flow deflection as

$$dp_1 = \frac{\rho u^2 d\alpha_1}{\sqrt{M^2-1}}. \tag{10.9.2}$$

The calculation can be repeated at a point farther back along the bump. If the angle of inclination of the succeeding element, BC, is $d\alpha_2$, the pressure change is $dp_2 = \rho u^2 d\alpha_2/\sqrt{M^2-1}$. Since $d\alpha_2$ is smaller than $d\alpha_1$, dp_2 is smaller than dp_1. The stream is thus accelerated in passing through the Mach line BB', experiencing a decrease in pressure equal to

$$dp_1 - dp_2 = \frac{\rho u^2(d\alpha_1 - d\alpha_2)}{\sqrt{M^2-1}}. \tag{10.9.3}$$

The pressure rise relative to the undisturbed flow decreases as we proceed downstream along the different straight-line segments. It is proportional to the local angle of inclination of the surface element and therefore remains positive until we reach the element whose inclination is zero. Past this point the angle of inclination becomes negative and the pressure falls below the undisturbed pressure level.

The above conclusions are not altered if the number of segments is increased indefinitely so the boundary has a smooth surface. If α is the angle of inclination between the tangent at a point on the surface and the undisturbed stream direction, and p denotes the undisturbed pressure, the pressure is constant along a Mach line emanating from the point and has the value $p + \rho u^2\alpha/\sqrt{M^2-1}$. Hence the pressure acting on the front part of the bump is higher than ambient, and the pressure acting on the rear is lower than ambient. There is a net pressure force in the free-stream direction, i.e. a drag. This mechanism for drag creation has no parallel in subsonic motion.

The distinction between subsonic and supersonic flow regimes is made explicitly in Figure 10.40, which portrays velocity vectors at a plane control surface, surface pressure distributions, and forces in the direction of the undisturbed velocity (thrust or drag) for isentropic flow past a slender bump on a flat surface. Three flow regimes are depicted: incompressible, subsonic ($M \cong 0.7$), and supersonic ($M \approx 1.4$). In the first two, the forces in the axial direction cancel and there is no net x-direction force. The velocity vectors are sketched to indicate no net vertical flux of x-momentum across the control surface (note the relation of the horizontal and vertical velocity components) in accord with this. For supersonic flow, the surface pressure is different than in the free stream only over the bump, with the pressures over the front and rear of the bump adding to give a drag. The velocity vectors at the plane control surface imply a net flux of x-momentum across the dashed control surface, consistent with the existence of this drag.

For low speed airfoil sections a blunt nose is often used; the main requirement is the sharp trailing edge. For supersonic speeds the blunt nose is disadvantageous because of the large angle of inclination

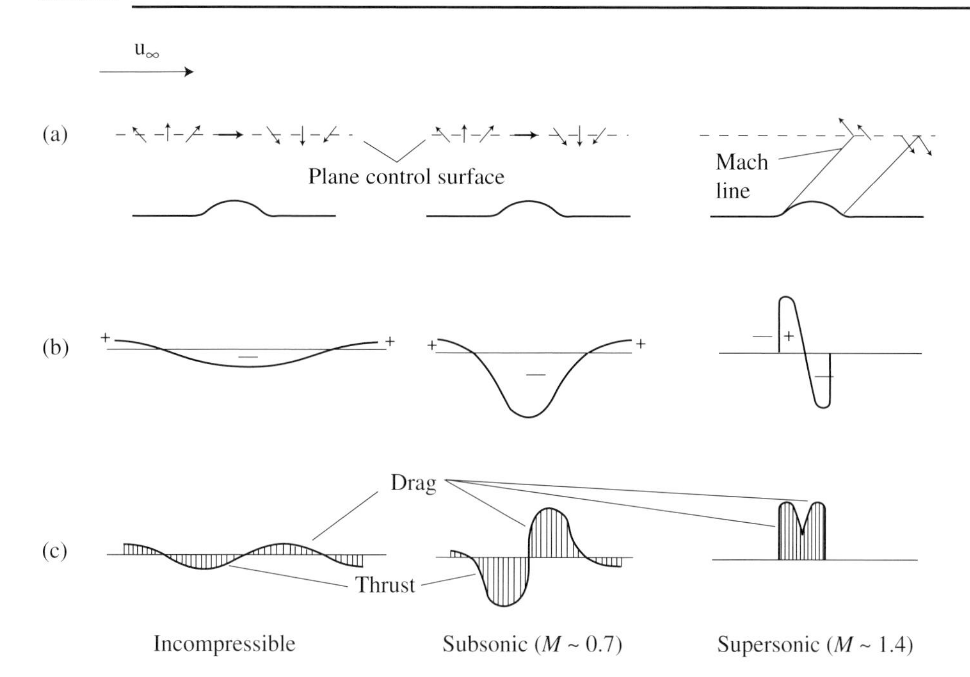

Figure 10.40: Flow over a bump in subsonic and supersonic regimes: (a) geometry, control plane, and velocity disturbances; (b) static pressure along the wall; (c) force in the x-direction (Von Karman, 1954).

it involves. A sharp training edge does not help much because negative pressure at the rear portion of the section cannot be avoided. The essential requirement for supersonic airfoil sections is a small thickness ratio, i.e. a small ratio between the maximum thickness and the chord length.

10.9.2 Relationship for finite angle changes: Prandtl–Meyer flows

The foregoing considerations concerning streamline deflection and pressure distribution apply to both compression and expansion waves. For the former, however, the Mach lines that define the compression waves can converge so the compression can become stronger. Figure 10.41 taken from Liepmann and Roshko (1957), shows the situation with compression waves converging into an oblique shock wave. It is possible to have compression without a shock, as shown in Figure 10.41(c), where a boundary has been introduced in the flow at a location at which the gradients are still small enough so that the flow is isentropic. (The upper boundary needs to be closer than the location at which Mach waves focus substantially.) For a turn in which the flow is deflected in the other direction so there are expansions, the Mach waves diverge and shock formation does not arise.

For the expansion waves, and for isentropic compression processes, the relation between the static pressure and the flow angle can be used to obtain an expression for the changes in Mach number associated with finite flow deflection. The linearized momentum equation for inviscid flow is

$$\frac{du}{u} = -\frac{dp}{\gamma M^2 p}. \tag{10.9.4}$$

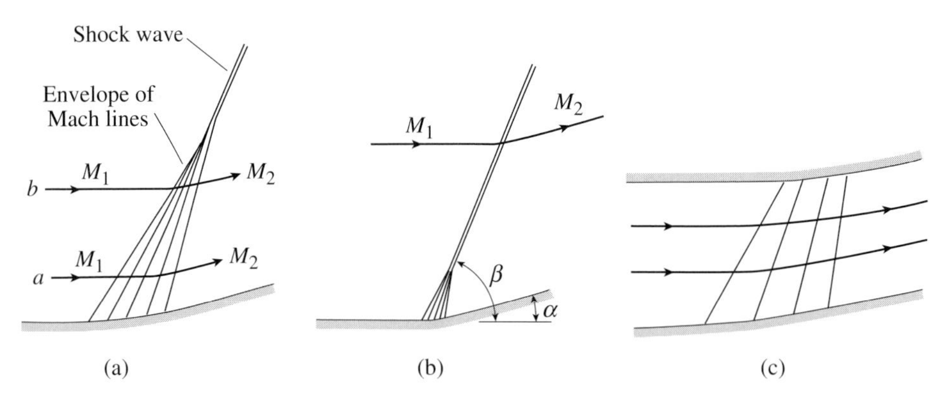

Figure 10.41: Illustrating Mach line convergence in a compression: (a) shock formation; (b) shock formation shown on a smaller scale; (c) channel conforming to streamlines of a smooth compression (Liepmann and Roshko, 1957).

Using the definition of Mach number ($du/u = dM/M + dT/2T$) and invoking (10.9.2) we obtain

$$d\alpha = -\frac{\sqrt{M^2-1}\, dM}{M\left(1+\frac{\gamma-1}{2}M^2\right)}. \tag{10.9.5}$$

Equation (10.9.5) can be integrated between two values of flow angle to find the initial (M_i) and final (M_f) Mach numbers or, more conveniently, between two Mach numbers to find the deflection between the initial and final flow angles $\alpha_f - \alpha_i$:

$$\alpha_f - \alpha_i = -\int_{M_i}^{M_f} \left[\frac{\sqrt{M^2-1}\, dM}{M\left(1+\frac{\gamma-1}{2}M^2\right)}\right] = \nu(M_i) - \nu(M_f). \tag{10.9.6}$$

In (10.9.6) $\nu(M)$ is defined as:

$$\begin{aligned}\nu(M) &= \int_{1}^{M} \left[\frac{\sqrt{M^2-1}\, dM}{M\left(1+\frac{\gamma-1}{2}M^2\right)}\right] \\ &= \sqrt{\frac{\gamma+1}{\gamma-1}}\tan^{-1}\left[\sqrt{\frac{\gamma+1}{\gamma-1}(M^2-1)}\right] - \tan^{-1}\sqrt{M^2-1}.\end{aligned} \tag{10.9.7}$$

The quantity $\nu(M)$ is a unique function of Mach number known as the Prandtl–Meyer function which is shown in Figure 10.42. The total deflection angle $|\alpha_f - \alpha_i|$ between two Mach numbers is the difference in Prandtl–Meyer function at those Mach numbers, as indicated in Figure 10.43.

This short discussion only touches the topic of non-one-dimensional effects in compressible flows. Liepmann and Roshko (1957) give additional insights on ways in which oblique shocks and expansions can be pieced together to describe more complex situations. Examples and applications of compressible flow in turbomachinery and propulsion systems are described by Cumpsty (1989),

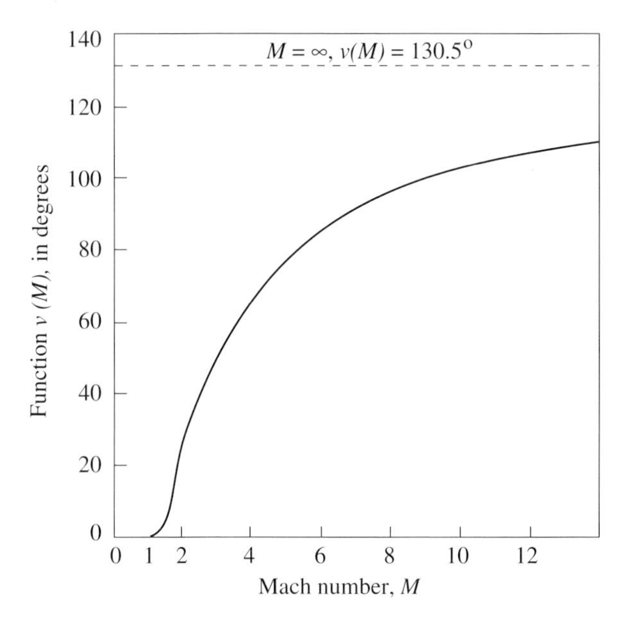

Figure 10.42: The Prandtl–Meyer function as a function of the Mach number for $\gamma = 1.4$ (Sabersky *et al.* 1989).

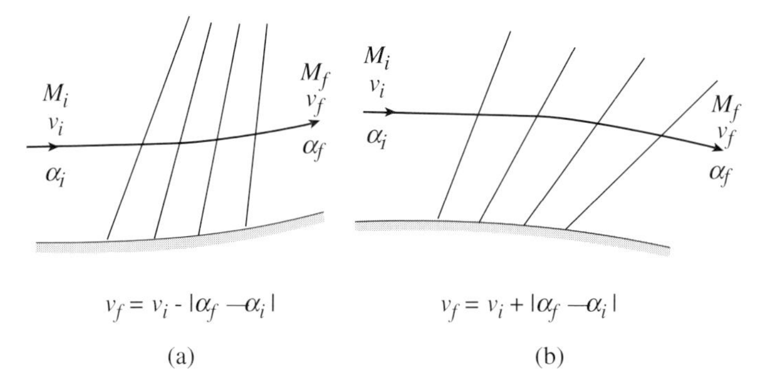

Figure 10.43: Relation of ν and α in isentropic turns (a) compression; (b) expansion (Liepmann and Roshko, 1957).

Kerrebrock (1992), Adamczyk (2000), and Heiser and Pratt (1994), and these can be consulted for more information. In this context an AGARD publication (AGARD, 1994) has as its theme the use of computational fluid mechanics for turbomachinery design in the compressible flow regime.

10.10 Flow field invariance to stagnation temperature distribution: the Munk and Prim substitution principle

Substitution principles are scaling rules that allow use of a known flow field to construct a solution for the flow in a different physical situation. In this section we demonstrate a substitution principle that allows, from knowledge of a flow field with one stagnation temperature distribution, a direct

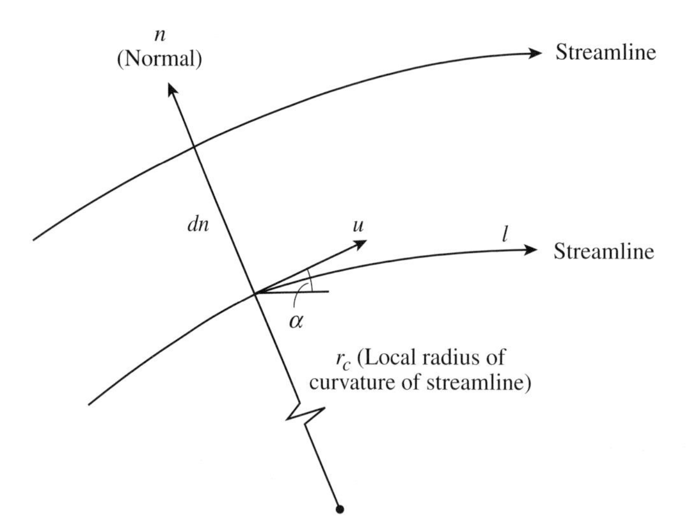

Figure 10.44: Streamlines in natural coordinates.

scaling to produce the flow field corresponding to any other stagnation temperature distribution (Munk and Prim, 1947; Tsien, 1958). We first illustrate the concepts for two-dimensional flow and then present the general three-dimensional results. Although the flows considered are inviscid and non-heat-conducting, we will see in Chapter 11 that the ideas extend, in an approximate manner, to flows with mixing and heat transfer.

10.10.1 Two-dimensional flow

For two-dimensional flow, we frame the discussion in terms of the natural coordinates introduced in Section 2.4. The coordinate normal to the streamlines is n, that along the streamlines is l, and the radius of curvature of the streamlines is r_c. The nomenclature is shown in Figure 10.44. In natural coordinates the equations of frictionless, non-heat-conducting, compressible flow are:

continuity: $\rho u dn = \text{constant along a streamtube};$ (10.10.1)

streamwise momentum: $\rho u \dfrac{\partial u}{\partial l} = -\dfrac{\partial p}{\partial l};$ (10.10.2)

normal momentum: $\rho \dfrac{u^2}{r_c} = \rho u^2 \dfrac{\partial \alpha}{\partial l} = \dfrac{\partial p}{\partial n};$ (10.10.3)

energy: $T_t = T + u^2/2c_p = \text{constant along a streamtube.}$ (10.10.4)

These are supplemented by the perfect gas equation of state, $p = \rho RT$, (1.4.1). In (10.10.3) the variable α is the angle the streamline makes with a reference direction as shown in Figure 10.44.

There are five equations for the quantities, u, α, p, ρ, T. Suppose we know a solution of these equations, i.e. we have a set of quantities u_1, α_1, p_1, ρ_1, T_1 which are functions of n and l, and which satisfy (10.10.1)–(10.10.4) and (1.4.1) for a certain boundary geometry. We ask whether there is another set of quantities with different velocity, density, and temperature u_2, α_1, p_1, ρ_2, T_2 which has the same streamline pattern and static pressure field as the first solution and is thus also a solution of the above equations.

From (10.10.3) a requirement for this to occur is

$$\rho_1 u_1^2 = \rho_2 u_2^2, \tag{10.10.5}$$

because r_c, α (the streamline pattern), and $\partial p/\partial n$ (the static pressure field) have been defined to be the same in the two cases. Using the equation of state in (10.10.5) gives the following relation between velocity and temperature fields in the two flows:

$$\frac{u_1^2}{RT_1} = \frac{u_2^2}{RT_2}. \tag{10.10.6a}$$

Equation (10.10.6a) is a statement that $M_1 = M_2$, in other words that local Mach numbers in both flows are equal. The static pressure distributions are the same and hence the stagnation pressure distributions must also be the same. Equation (10.10.6a) can be written in terms of stagnation, rather than static, temperature as

$$\frac{u_1^2}{RT_{t_1}} = \frac{u_2^2}{RT_{t_2}}. \tag{10.10.6b}$$

Equation (10.10.6b) is useful because the stagnation temperature is constant along a streamline.

The static pressure distribution and streamline configurations are the same for the two flows, so that

$$\rho_1 u_1 \frac{\partial u_1}{\partial l} = \rho_2 u_2 \frac{\partial u_2}{\partial l}. \tag{10.10.7}$$

The mass flow in a given streamtube is not the same for the two flows. The mass flow in a streamtube is $\rho u dn$, dn is the same in both flows, and the ratio of local mass flows is

$$\frac{\dot{m}_1}{\dot{m}_2} = \frac{\rho_1 u_1}{\rho_2 u_2} = \sqrt{\frac{T_{t_2}}{T_{t_1}}}. \tag{10.10.8}$$

The preceding discussion can be summarized as follows. Two steady, inviscid flow fields will have the same streamline shapes and static pressure distribution if the stagnation pressure distribution is the same, irrespective of the stagnation temperature distribution. Because the stagnation pressure is constant along streamlines, equality of the stagnation pressure distribution along a normal coordinate line that spans the flow domain ensures the stagnation pressure distribution is the same throughout. The velocities and the mass flows in a given streamtube for the two situations are different. The local velocity ratio between the two flows scales as $u_2/u_1 = \sqrt{T_{t_2}/T_{t_1}}$ and the local mass flow in any streamtube as $\dot{m}_2/\dot{m}_1 = \sqrt{T_{t_1}/T_{t_2}}$.

The results of this section can be related to the behavior of compound-compressible flow. Examining the expressions developed in Section 10.8 we find that mass flow and stagnation temperature occur everywhere only in the combination $\dot{m}\sqrt{T_t}$. A change in mass flow in any stream, coupled with a corresponding change in the stream stagnation temperature to keep the product $\dot{m}\sqrt{T_t}$ the same, leaves the behavior of the compound flow (e.g. static pressure, streamtube area, and Mach number distribution) unchanged. This can be regarded as a quasi-one-dimensional illustration of the substitution principle.

10.10.2 Three-dimensional flow

We now generalize the principle to three dimensions. As in Section 10.7 it is helpful to use the reference maximum speed, u_{max},

$$u_{max}^2 = 2c_p T_t. \tag{10.10.9}$$

The stagnation temperature, T_t, and reference speed, u_{max}, can vary normal to streamlines, but they are both constant along a streamline.

We work in terms of a non-dimensional "reduced velocity", $\mathbf{v}$, defined by

$$\mathbf{v} = \frac{\mathbf{u}}{u_{max}}. \tag{10.10.10}$$

In terms of $\mathbf{v}$, the energy equation is

$$\frac{T}{T_t} = \left(1 - \frac{u^2}{u_{max}^2}\right) = \left(1 - v^2\right). \tag{10.10.11}$$

The continuity equation is

$$0 = \boldsymbol{\nabla} \cdot (\rho \mathbf{u}) = \boldsymbol{\nabla} \cdot \left[\left(\rho_t \sqrt{2c_p T_t}\right)\left(\frac{\rho}{\rho_t}\right)\mathbf{v}\right]$$

or

$$0 = \left(\rho_t \sqrt{2c_p T_t}\right)\boldsymbol{\nabla} \cdot \left[\left(\frac{\rho}{\rho_t}\right)\mathbf{v}\right] + \left(\frac{\rho}{\rho_t}\right)\mathbf{v} \cdot \boldsymbol{\nabla}\left(\rho_t \sqrt{2c_p T_t}\right). \tag{10.10.12}$$

The derivatives of any stagnation properties along a streamline are zero. The second term in (10.10.12) is therefore zero and the equation reduces to

$$0 = \boldsymbol{\nabla} \cdot \left[\left(\frac{\rho}{\rho_t}\right)\mathbf{v}\right]. \tag{10.10.13}$$

Using the isentropic pressure–temperature relation in combination with (10.10.11) yields a form of the continuity equation that involves only $\mathbf{v}$:

$$\boldsymbol{\nabla} \cdot \left[\left(1 - v^2\right)^{1/(\gamma-1)}\mathbf{v}\right] = 0. \tag{10.10.14}$$

The momentum equation is written in terms of $\mathbf{v}$ as

$$(\mathbf{v} \cdot \boldsymbol{\nabla})\,\mathbf{v} + \frac{(\gamma - 1)}{2\gamma}\left(1 - v^2\right)\frac{\boldsymbol{\nabla} p}{p} = 0. \tag{10.10.15}$$

In obtaining (10.10.15) the invariance of stagnation properties along a streamline has been used. The Crocco form of the momentum equation ((3.14.4), $\mathbf{u} \times \boldsymbol{\omega} = \boldsymbol{\nabla} h_t - T\boldsymbol{\nabla} s$) can be written compactly as a relation between $\mathbf{v}$ and the stagnation pressure as

$$\frac{\mathbf{v} \times (\boldsymbol{\nabla} \times \mathbf{v})}{1 - v^2} = \frac{(\gamma - 1)}{2\gamma}\left(\frac{\boldsymbol{\nabla} p_t}{p_t}\right). \tag{10.10.16}$$

Reference velocities other than u_{max} can also be employed. If the velocity is divided by the local speed of sound, $\sqrt{\gamma p/\rho}$, the Mach number vector, $\mathbf{M}$, is obtained. In terms of $\mathbf{M}$ the continuity and momentum equations become

$$\nabla \cdot \left[\mathbf{M} \left(1 + \frac{\gamma - 1}{2} M^2 \right)^{-\frac{\gamma+1}{2(\gamma-1)}} \right] = 0, \tag{10.10.17}$$

$$(\mathbf{M} \cdot \nabla) \mathbf{M} - \frac{\gamma - 1}{\gamma + 1} \mathbf{M} (\nabla \cdot \mathbf{M}) - \frac{1}{\gamma - 1} \nabla \left\{ \ln \left[1 + \frac{\gamma - 1}{2} M^2 \right] \right\} + \frac{\nabla p_t}{\gamma p_t} = 0. \tag{10.10.18}$$

Equations (10.10.14) and (10.10.16) (*or* (10.10.17) and (10.10.18)) are two equations for the variables **v** and p_t (or **M** and p_t) in which the stagnation temperature does not appear. The reduced velocity, **v** (or Mach number vector, **M**) is thus invariant to changes in the stagnation temperature distribution if the far upstream stagnation pressure variation across the flow domain and the geometry of the boundaries are maintained the same. A single solution of (10.10.14) and (10.10.16) (or (10.10.17) and (10.10.18)) corresponds to a family of solutions which have different assignments of stagnation temperature to each streamline. This includes, for example, the case of uniform stagnation temperature, uniform stagnation pressure, and irrotational flow. To this irrotational solution correspond an infinite number of rotational solutions with different stagnation temperature distributions.

Changes in the stagnation temperature distribution from an original value $T_{t_{orignial}}$ to a new value T_t result in transformation of the physical velocity and the mass flow in a specified streamtube to

$$\frac{\mathbf{u}}{\mathbf{u}_{original}} = \sqrt{\frac{T_t}{T_{t_{original}}}}, \tag{10.10.19a}$$

$$\frac{\dot{m}}{\dot{m}_{original}} = \sqrt{\frac{T_{t_{original}}}{T_t}}. \tag{10.10.19b}$$

10.10.3 Flow from a reservoir with non-uniform stagnation temperature

An example of the use of the Munk and Prim substitution principle concerns a gas with varying stagnation temperature but uniform stagnation pressure which passes from a large reservoir (with negligible velocity) through a nozzle. With a uniform stagnation temperature the flow is irrotational and the velocity is uniform at nozzle exit. If the stagnation temperature is not uniform the Crocco equation shows that magnitude of the exit vorticity is

$$|\omega| = \left| \frac{\mathbf{u}}{2} \right| \cdot \left| \frac{\nabla T_t}{T_t} \right|. \tag{9.4.17}$$

This exit vorticity is normal to the streamlines.

Suppose this rotational flow is then taken through a bend which is in the plane of the exit vorticity. The combined process of vorticity generation within the nozzle plus flow round the bend can be regarded as an idealized model of the flow from a gas turbine combustor through a turbine vane. A question of interest is whether secondary circulation will be generated in this situation. Even though there is normal vorticity at the bend inlet, the substitution principle immediately tells us there is no secondary circulation because one substitution flow is the irrotational velocity field. As described in Chapter 9, scrutiny of the situation shows that there is production of streamwise vorticity from two sources, the tipping of normal vorticity into the streamwise direction and the baroclinic torque (the term $\nabla T \times \nabla s$ in the vorticity evolution equation, (3.7.3)). These are equal and opposite in the flow described and no net streamwise vorticity is created. Arguing from the perspective of pressure gradients, even though the stagnation temperature is non-uniform, the quantity ρu^2 is uniform on all the streamlines and the inertial forces are the same at all levels in the bend in spite of the temperature stratification. As emphasized in Section 9.4, gradients in stagnation temperature alone do not produce secondary circulation in a stationary duct or passage; gradients in stagnation pressure are necessary for secondary circulation to be generated.

11 Flow with heat addition

11.1 Introduction: sources of heat addition

In this chapter we examine flows in which heat addition has a major effect on the fluid motion. The situations addressed have a fractional change in bulk flow temperature which is of order unity, and we thus first inquire what circumstances lead to this occurrence.

Estimates for the magnitudes of temperature differences due to heat transfer from surfaces can be made based on the Reynolds analogy between heat and momentum transfer from a solid surface to a fluid. The physical content of this analogy, as described in a number of texts (e.g. Eckert and Drake, 1972; Kerrebrock, 1992; Schlichting, 1979; Pitts and Sissom, 1977), is that magnitude of the heat flux, q, and the shear stress, τ, (the momentum transfer per unit area across a plane parallel to the wall) arise from similar transport processes. If the two are viewed as changing in a similar manner through the boundary layer, their ratio can be approximated as constant from the wall to the free stream. Using the expressions for magnitudes of heat transfer and shear stresses in terms of gradients of velocity and temperature this ratio can be written as

$$\frac{q}{\tau} \approx \left(\frac{k\partial T/\partial y}{\mu \partial u/\partial y} \right) \approx \frac{c_p}{Pr}\frac{dT}{du}. \tag{11.1.1}$$

In (11.1.1) the thermal conductivity, viscosity, and Prandtl number, $Pr\ (= \mu c_p/k)$, are interpreted as applying to a laminar or turbulent situation as appropriate.

Equation (11.1.1) can be integrated through the boundary layer, from the wall ($T = T_w$, $u = 0$) to the free stream ($T = T_E$, $u = u_E$), to yield an expression that relates wall shear stress, τ_w, and wall heat transfer rate, q_w. For a Prandtl number of unity this is

$$\frac{q_w}{c_p(T_w - T_E)} \approx \frac{\tau_w}{u_E}. \tag{11.1.2}$$

Defining q_w as the product of a heat transfer coefficient, h, and a driving temperature, $(T_w - T_E)$, and writing (11.1.2) in non-dimensional form we obtain

$$\frac{q_w}{\rho_E u_E c_p (T_w - T_E)} = \frac{h}{\rho_E u_E c_p} \approx \frac{\tau_w}{\rho_E u_E^2}. \tag{11.1.3}$$

The fraction on the left in (11.1.3) is the Stanton number, conventionally denoted by St, and the quantity on the right is one-half the skin friction coefficient, C_f. A compact form of the Reynolds analogy is therefore

$$St \approx C_f/2. \tag{11.1.4}$$

Equation (11.1.3) or (11.1.4) allows the estimation of the heat transfer coefficient if data or computations of the skin friction are available.

For flow through a duct with perimeter, P, and length, L, the bulk temperature rise can be found using the estimate for heat input ($= q_w PL$), provided by (11.1.3) in an enthalpy balance as

$$\frac{(\Delta T)_{bulk}}{T_{bulk}} \approx \left(\frac{C_f}{2}\right)\left(\frac{T_w}{T_E} - 1\right)\frac{PL}{A}. \tag{11.1.5}$$

Levels for skin friction coefficients in turbulent flow on smooth surfaces are between 0.003 and 0.006 over the Reynolds number range of 5×10^5 to 10^7 (see Section 4.6) and the quantity involving the temperature ratio is typically order unity or less. The geometric quantity PL/A depends on the application, but for many situations (combustors, afterburners, turbine blade passages) has values of order 10 or less. Upper values of the ratio $(\Delta T)_{bulk}/T_{bulk}$ are thus 0.1, i.e. bulk temperature changes associated with wall heat transfer are an order of magnitude smaller than the inlet temperature unless the passages are long.

The temperature rise due to condensation is also roughly of this order. For example, the non-dimensional stagnation temperature change $\Delta T_t/T_t$ from the condensation of all the moisture in the air (an overestimate) at 30 °C and 100% humidity is approximately 0.2. A lower range (0.01–0.1) for the temperature rise from condensation in fluid devices is cited by Zierep (1974).

Temperature rises of these magnitudes can certainly affect device behavior, for example through thermal choking (see Sections 10.4 and 11.5), but the bulk temperature change is much less than that provided by combustion. The adiabatic flame temperature of a mixture of fuel and air is given to good approximation by

$$\frac{\Delta T_t}{T_{t_i}} = \frac{f h_{combustion}}{c_p T_{t_i}},$$

where f is the mass flow weighted fuel air ratio. (Since $f \ll 1$, the heating of the mass of the fuel has been neglected.) For a hydrocarbon fuel (heat of combustion 4.3×10^8 J/kg), at the stoichiometric fuel/air ratio[1] (0.067) and a combustor inlet temperature of 1000 K the ratio $\Delta T_t/T_{t_i}$ is approximately 2.8. While typical overall fuel/air ratios are less than stoichiometric, there is still an order of magnitude ratio between the bulk temperature changes through combustion and those associated with heat transfer from solid surfaces or heat addition from condensation. Combustion processes can decrease the density by an appreciable percentage of its inlet value, coupling the heating process into the dynamics. A consequence is that the effects of heat addition are seen even at low Mach numbers where thermal choking is not an issue. In summary, while heat transfer to or from solid surfaces or phase change (condensation) do impact fluid motions, especially for near-sonic conditions, the most powerful driver is heat addition from combustion.

Several phenomena associated with heat addition to a flowing fluid are addressed below. One is the generation of vorticity in the bulk flow, through baroclinic torque (see Section 3.5). A flow with uniform velocity passed through a region of non-uniform heat addition emerges as a rotational flow. A second is the change in the stagnation pressure associated with heat addition. This was introduced in Chapter 10, but we examine it here in more depth. In Section 11.4 a method is described for the graphical portrayal of one-dimensional compressible channel flow with heat addition, using as coordinates the kinetic energy and the static enthalpy of the fluid. The interaction between swirl and

[1] The stoichiometric fuel/air ratio is the ratio at which there is no excess fuel or air.

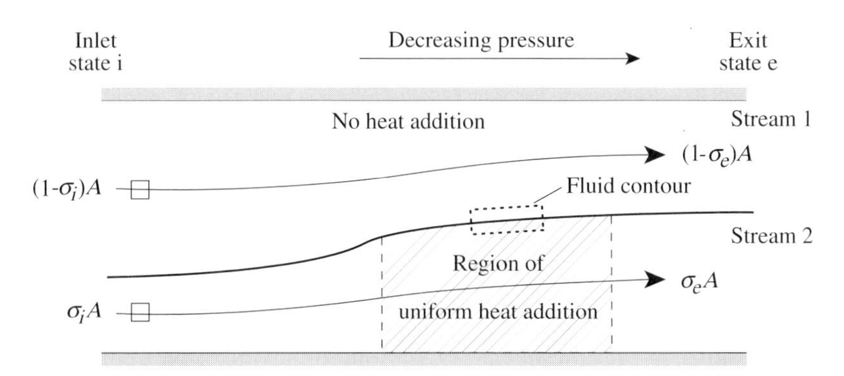

Figure 11.1: Creation of vorticity by heat addition; constant area duct with heat addition to a portion of the stream (Broadbent, 1976).

heat addition is discussed in Section 11.6, where it is seen that not only the magnitude but also the sense of the axial acceleration associated with heat addition depends on the swirl level. The final section develops an approximate extension of the Munk and Prim substitution principle of Section 10.10 to flow with heat transfer and viscous stresses, including applications of the approximate principle to mixer nozzles, ejectors, and heated jets.

11.2 Heat addition and vorticity generation

Heat addition that is non-uniform in a direction normal to a flow creates vorticity. Consider two neighboring fluid particles in an inviscid steady flow in the constant area channel sketched in Figure 11.1. The flow is parallel and uniform far upstream of the region of heat addition. The two particles are subjected to the same streamwise pressure gradient, but one receives heat and one does not. At any location along either streamline the change in velocity is given by $du = -dp/(\rho u)$. If the two particles initially have the same velocity and density, the one with heat addition has a lower density than the one with no heat addition and thus develops a higher velocity. The flow downstream of the region of heat addition is again parallel but it has non-uniform velocity and is hence rotational.

The generation of vorticity can be seen directly from Kelvin's Theorem, (3.8.9), for the rate of change of circulation (Γ) round a fluid contour:

$$\frac{D\Gamma}{Dt} = -\oint \frac{\nabla p}{\rho} \cdot d\ell = -\oint \frac{dp}{\rho}. \tag{3.8.9}$$

We apply this to the small dashed line contour in Figure 11.1. The streamwise pressure gradient is essentially the same for the top and bottom legs of the contour so the presence of the density difference between the top and bottom legs means that circulation is generated.

The effect of heat addition on vorticity generation can be defined more quantitatively using the example of quasi-one-dimensional steady flow in a constant area channel (Broadbent, 1976). The upstream flow is uniform in velocity and density, with conditions p_i, u_i, ρ_i. The flow is inviscid with no mixing, but a portion of the gas has a specified heat addition. The streamtube expansion in the region of heat addition is restricted to be small enough so the quasi-one-dimensional approximation applies and the pressure can be regarded as uniform *across* the duct. In addition, the Mach numbers are low enough such that density changes due to dynamic effects (which scale as $\Delta\rho/\rho \propto M^2$) are negligible, but density changes due to heat addition (which scale as $\Delta\rho/\rho \propto \Delta T/T \approx$ order unity) are comparable with the ambient density level. (See Section 1.17 for a discussion of this approximation.)

If the heated stream occupies a fraction σ_i of the duct at the inlet and expands to a fraction σ_e at the exit, downstream of the region of heat addition (Figure 11.1), the continuity equations in the unheated and heated streams respectively are

$$\rho_i u_i(1-\sigma_i) = \rho_i u_{1_e}(1-\sigma_e), \tag{11.2.1a}$$

$$\rho_i u_i \sigma_i = \rho_{2_e} u_{2_e} \sigma_e. \tag{11.2.1b}$$

Subscripts 1 and 2 denote exit (downstream) conditions in the unheated stream and heated stream respectively. Because of the low Mach number the density changes can be neglected in the unheated stream and we can set $\rho_i = \rho_{1_e}$.

Dividing (11.2.1) by $\rho_i u_i$ and denoting the non-dimensional properties as $\tilde{u}$, $\tilde{\rho}$, etc., the continuity equations in the two streams become

$$(1-\sigma_i) = \tilde{u}_{1_e}(1-\sigma_e), \tag{11.2.2a}$$

$$\sigma_i = \tilde{\rho}_{2_e}\tilde{u}_{2_e}\sigma_e. \tag{11.2.2b}$$

In these non-dimensional variables conservation of momentum for the duct is, noting that at the downstream location $p_{1_e} = p_{2_e}$,

$$\Delta C_p = \frac{p_i - p_{1_e}}{\left(\rho_i u_i^2/2\right)} = 2\tilde{u}_{1_e}(1-\sigma_e) + 2\tilde{\rho}_{2_e}\tilde{u}_{2_e}\sigma_e - 2. \tag{11.2.3}$$

In the unheated stream, Bernoulli's equation can be applied,

$$\Delta C_p = \tilde{u}_{1_e}^2 - 1. \tag{11.2.4}$$

Equations (11.2.2), (11.2.3), and (11.2.4) constitute four equations for the five quantities, $\sigma_e, \tilde{\rho}_2, \tilde{u}_{1_e}, \tilde{u}_{2_e}, \Delta C_p$, so that one of these must be specified. The most convenient way to solve the set of equations is to suppose the heat addition creates a known increase in the velocity of the unheated stream:

$$\tilde{u}_{1_e} = 1 + \Delta\tilde{u}. \tag{12.2.5}$$

If so the solution is

$$\tilde{u}_{2_e} = [(\Delta\tilde{u})^2 + 2\sigma_i\Delta\tilde{u} + 2\sigma_i]/(2\sigma_i), \tag{11.2.6a}$$

$$\Delta C_p = \Delta\tilde{u}(2+\Delta\tilde{u}), \tag{11.2.6b}$$

$$\tilde{\rho}_{2_e} = 2\sigma_i^2(1+\Delta\tilde{u})/\{(\Delta\tilde{u}+\sigma_i)[(\Delta\tilde{u})^2 + 2\sigma_i\Delta\tilde{u} + 2\sigma_i]\}, \tag{11.2.6c}$$

$$\sigma_e = (\Delta\tilde{u}+\sigma_i)/(1+\Delta\tilde{u}). \tag{11.2.6d}$$

For positive heat addition ($\tilde{\rho}_{2_e} = \rho_{2_e}/\rho_i = T_i/T_{2_e} < 1$) $\Delta\tilde{u}$ is positive. The velocities of the heated and unheated streams are related by

$$\tilde{u}_{2_e} = \tilde{u}_{1_e} + (\Delta\tilde{u})^2/2\sigma_i. \tag{11.2.7}$$

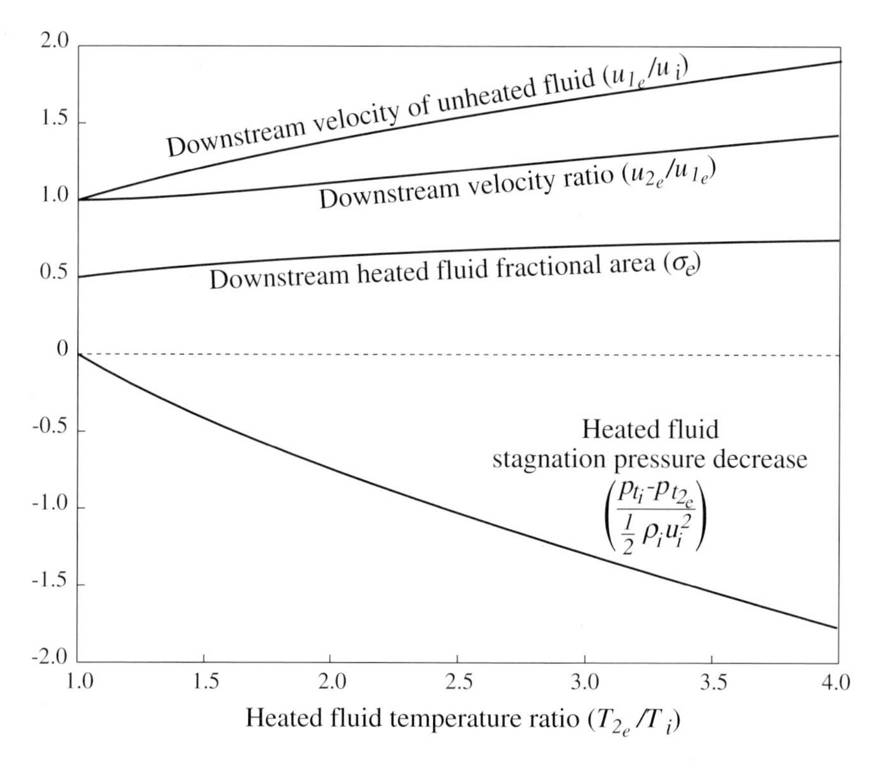

Figure 11.2: Velocities and pressures in a constant area duct downstream of a region of inviscid heating over half the duct area; u_{2_e} is the exit heated fluid velocity, u_{1_e} is the exit unheated fluid velocity, u_i is the inlet (uniform) velocity, $M^2 \ll 1$.

Equation (11.2.7) shows that the heated stream always accelerates more than the unheated stream. The change in stagnation pressure in the heated stream is

$$\frac{p_{t_i} - p_{t_{2_e}}}{\rho_i u_i^2/2} = -\frac{\Delta\tilde{u}\,(1 + \Delta\tilde{u})\,(2 + \Delta\tilde{u})}{2\,(\sigma_i + \Delta\tilde{u})}. \tag{11.2.8}$$

There is always a decrease in stagnation pressure due to heating.

Figure 11.2 shows the downstream velocity in the unheated stream, the velocity ratio between heated and unheated streams, the downstream area occupied by the heated stream, and the stagnation pressure decrease in the heated stream, all as functions of the temperature ratio in the heated stream. Large velocity and stagnation pressure non-uniformities are produced when the fluid temperature change is comparable with or greater than the inlet temperature.

11.3 Stagnation pressure decrease due to heat addition

The preceding example showed a decrease in stagnation pressure due to heat addition in a steady inviscid flow. The stagnation pressure change can be interpreted from different perspectives. For low Mach number (or incompressible) flow, stagnation pressure is a mechanical quantity. For steady

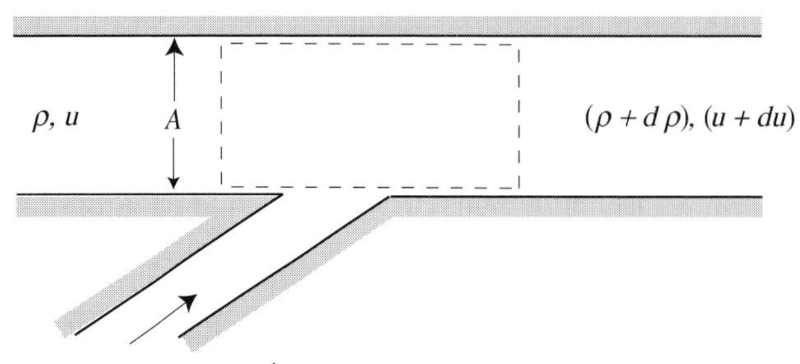

ρ_{inj}, u, volume flow rate, $d\dot{V}_{inj}$

Figure 11.3: Control volume with injection of a different density incompressible fluid into the mainstream.

inviscid flow with no body forces it is a measure of the kinetic energy of the fluid per unit volume, $\rho u^2/2$, and the work per unit volume done by the pressure to accelerate the fluid in a given stream tube.[2]

For incompressible flow, the internal energy has no dynamical significance. The impact of heat transfer on the velocity and pressure fields, and thus on the stagnation pressure, is felt only through changes in fluid density. The change in stagnation pressure resulting from small changes in pressure, velocity, and density can be written as[3]

$$dp_t = dp + \rho u du + (u^2/2)d\rho. \tag{11.3.1}$$

For inviscid flow the sum of the first two terms is zero along a streamline and the density change, represented by the last term in (11.3.1) is the only contributor. The alteration in stagnation pressure is related to the consequent change in kinetic energy per unit volume by

$$dp_t = (u^2/2)d\rho. \tag{11.3.2}$$

Equation (11.3.2) was introduced to describe a process in which heat addition causes a density change, and hence a stagnation pressure change, but the expression has a wider application. Two examples illustrate the point.

The first is the injection and mixing of two incompressible fluids of different densities in a constant area duct shown in Figure 11.3. The mainstream has density ρ, velocity u, and area A. The injected flow has density ρ_{inj}, the same velocity u as the mainstream, and a volume flow rate $d\dot{V}_{inj}$ such that $d\dot{V}_{inj}/(uA) \ll 1$. The injected flow is also taken as entering in the same direction as the mainstream. With reference to Figure 11.3 the conservation equations for this solely mechanical mixing process are:

conservation of volume flow: $du = d\dot{V}_{inj}/A$; (11.3.3)

conservation of mass: $d\rho = (\rho_{inj} - \rho)(du/u)$; (11.3.4)

conservation of momentum: $dp = (\rho_{inj} u - 2\rho u)du - u^2 d\rho$. (11.3.5)

Combining (11.3.3)–(11.3.5) yields the expression for the stagnation pressure change in (11.3.2).

[2] "The direct effect of the pressure on the energy of the material element is the same as if the element moved in a body-force field of potential energy p/ρ per unit mass" (Batchelor, 1967).

[3] In writing (11.3.1), we have neglected terms which are of order M^2 compared to those that have been retained.

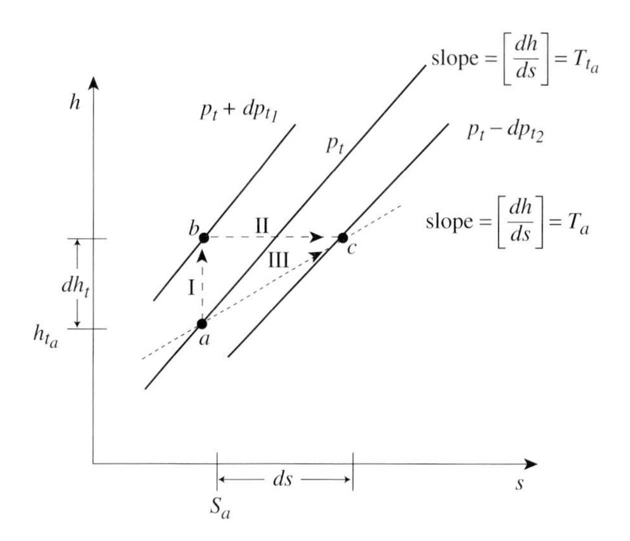

Figure 11.4: Stagnation pressure changes for a stagnation enthalpy change of dh_t and an entropy change of ds (conceptual model for heat addition to a compressible flow).

The second example is a uniform velocity, parallel, stratified flow in which there is a density variation normal to the stream direction. The kinetic energy per unit volume and stagnation pressure thus also vary normal to the stream and the connection between these two quantities is given by (11.3.2). In this situation the stagnation pressure variation can be seen to be solely mechanical in nature.

For compressible flow there is interchange between thermal and mechanical energy and discussion of stagnation pressure changes can no longer be made in terms of mechanical quantities alone. We can, however, use the characterization of stagnation pressure changes as lost work for a constant stagnation temperature process given in Chapter 5 to interpret the effect of heat transfer on stagnation pressure. The starting point is the Gibbs equation ((1.3.19), $Tds = dh - dp/\rho$), written in terms of stagnation quantities as

$$\frac{dp_t}{\rho_t} = dh_t - T_t ds. \tag{11.3.6}$$

In a general heat addition process, changes in stagnation enthalpy and entropy occur concurrently. For small increments, however, we can examine the impact of each separately to give insight into why the stagnation pressure decreases, as well as to derive an expression for the magnitude of the decrease.

The two processes ((I) – stagnation enthalpy increase at constant entropy, and then (II) – entropy increase at constant stagnation enthalpy) are shown in Figure 11.4, along with three constant pressure lines corresponding to p_t, $p_t + dp_t$, and $p_t - dp_t$. Process I, indicated by the dashed line between states a and b, is an isentropic compression resulting in a stagnation enthalpy increase dh_t and a stagnation pressure increase dp_t. Process II is at constant stagnation enthalpy. It could be achieved in several ways, for example an adiabatic irreversible process, with lost work of magnitude $T_t ds$, or a reversible isothermal expansion in which work of magnitude $T_t ds$ is received by some external agency. The system change is the same in both, it is only the interactions with the surroundings that are different. Process II is indicated by the dashed line between states b and c.

For the actual process (III) we need to combine I and II. The entropy change is $ds = dq/T = dh_t/T$. Process II results in a change in stagnation pressure which is larger than, and of opposite sign to, that in process I. The overall stagnation pressure change from state a to state c, due to heat addition, is therefore negative.

Another way to motivate this result is to note that the combined process follows along a line with slope T in the h–s plane, because $dh_t = dq = Tds$. To have no stagnation pressure change for the specified entropy change ds, however, the change in stagnation enthalpy would need to be $T_t ds$ (instead of Tds). From (11.3.6) the change in stagnation pressure resulting from processes I and II can be written as

$$\frac{dp_t}{\rho_t} = Tds\,[1 - (T_t/T)]. \tag{11.3.7}$$

This quantity will always be negative for heat addition ($ds > 0$).

Substituting the relations for T_t/T in terms of Mach number, and using the relations between ds, dq, and dh_t given just above provides a direct path to the relation between changes in stagnation pressure and stagnation temperature for steady frictionless flow with heat addition which appeared in (10.4.10b):

$$\frac{dp_t}{p_t} = -\frac{\gamma M^2}{2}\left(\frac{dT_t}{T_t}\right). \tag{11.3.8}$$

For gases at low Mach number, as mentioned in Section 11.1, the equation of state can be approximated as $\rho T \approx$ constant. Using this approximation and taking the limiting values of the quantities in (11.3.8) recovers (11.3.2), previously derived based on mechanical considerations.

11.4 Heat addition and flow state changes in propulsion devices

11.4.1 The *H–K* diagram

One reason to examine flows with heat addition is the potential for extracting thrust or power out of a thermodynamic cycle. In this context it is useful to assess the behavior of such cycles in terms of the fluid state changes that occur within cycle components. We approach this task within a quasi-one-dimensional framework, restricting discussion to perfect gases with constant specific heat.[4] The relevant equations (as seen in Chapter 10) can be integrated without difficulty and we seek a way to display these numerical results in a form which provides insight for a broad range of situations. A method for doing this is through the use of an *H–K diagram* which has the non-dimensional kinetic energy per unit mass $[u^2/(2c_pT_{t_i}) = K]$ as the abscissa and the dimensionless static enthalpy $(c_pT/c_pT_{t_i} = H)$ as the ordinate (Heiser and Pratt, 1994; Pratt and Heiser, 1993). The inlet stagnation temperature, T_{t_i}, is taken as the reference in defining H. This formulation has the advantage that several of the attributes of the flow are represented by straight lines, facilitating interpretation of processes.

[4] Comments on this aproximation are given in Heiser and Pratt (1994).

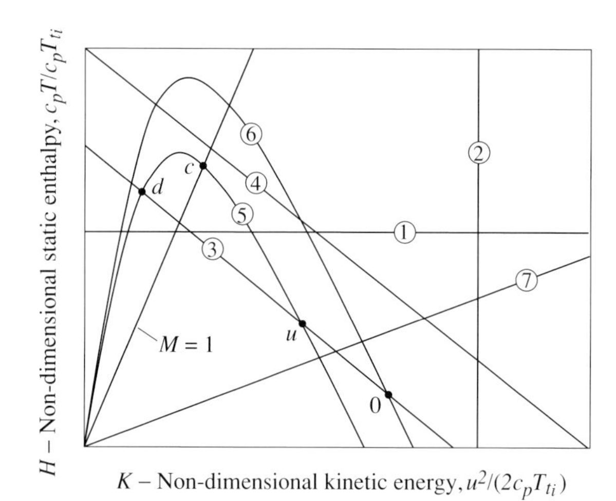

Figure 11.5: The *H–K* diagram, depicting representative constant–property isolines. Key: Point 0 = free-stream reference state. Point c = choked condition at constant impulse. Points u and d denote end states of normal shock. Circled numbers denote isolines of constant property, as follows: (1) static enthalpy, static temperature; (2) kinetic energy, velocity, pressure (for frictionless heating or cooling only); (3) total enthalpy, total temperature (adiabat), $\tau = T_t/T_{t_i} = 1$; (4) post-heat release adiabat, $\tau > 1$; (5) non-dimensional impulse function (for frictionless flow with heating or cooling only), $\Phi = \Phi_u = \Phi_d$; (6) non-dimensional impulse function, $\Phi = \Phi_0 > \Phi_u$; (7) Mach number (the line labeled $M = 1$ is also constant Mach number) (Pratt and Heiser, 1993).

In the *H–K* diagram of Figure 11.5 the horizontal line (denoted by (1)) is constant static temperature (static enthalpy) and the vertical line (denoted by (2)) is constant kinetic energy per unit mass, or equivalently constant velocity. For frictionless heating or cooling, this is also a line of constant pressure. The straight lines with a slope of −1 (denoted by (3) and (4)) are lines of constant stagnation temperature (constant stagnation enthalpy). Along such lines $u^2 + c_pT = c_pT_t$, or, in normalized form, $H + K = T_t/T_{t_i}$. These represent flow in ducts, nozzles, and diffusers, components with no shaft work and no heat addition. Line (4) corresponds to a higher stagnation temperature than line (3).

The curves denoted by (5) and (6) represent the momentum equation. The axial force between any two stations 1 and 2 in a channel is given by

$$\begin{aligned}\mathcal{F}_x &= \left(p_2A_2 + \rho_2u_2^2A_2\right) - \left(p_1A_1 + \rho_1u_1^2A_1\right)\\ &= p_2A_2\left(1+\gamma M_2^2\right) - p_1A_1\left(1+\gamma M_1^2\right).\end{aligned} \tag{11.4.1}$$

The axial force is the difference between values of the quantity $(p + \rho u^2)A$, which is denoted by $\Im$ and referred to as the impulse function. Non-dimensionalizing the impulse function by $\dot{m}\sqrt{c_pT_{t_i}}$ we obtain

$$\begin{aligned}\Phi &= \frac{\Im}{\dot{m}\sqrt{c_pT_{t_i}}}\\ &= \sqrt{\frac{1}{2}\left(\frac{2c_pT_{t_i}}{u^2}\right)}\left[2\left(\frac{u^2}{2c_pT_{t_i}}\right) + \frac{\gamma-1}{\gamma}\left(\frac{c_pT}{c_pT_{t_i}}\right)\right]\\ &= \sqrt{\frac{K}{2}}\left(2K + \frac{\gamma-1}{\gamma}H\right).\end{aligned} \tag{11.4.2}$$

The expression for Φ has been written so the quantities it contains are H and K. Curves for two values of Φ are given in Figure 11.5. For a specified value of $(p + \rho u^2)A$ the local state must be somewhere on a curve of constant Φ. In H–K coordinates, the states for frictionless constant area flow with heat addition (Rayleigh flow) would be traversed along such curves of constant Φ (i.e. constant $(p + \rho u^2)A$) with the direction towards higher stagnation enthalpy.

Finally, lines of constant Mach number in H–K coordinates are lines on which the quantity $u/\sqrt{\gamma RT}$ is constant. However, this is equivalent to the statement that H/K is constant, so lines of constant Mach number are straight lines radiating from the origin as indicated by the line marked "$M = 1$" and the line denoted by (7). The relation between H and K on a line of constant Mach number is given by

$$\text{line of constant Mach number: } H = \frac{2K}{(\gamma - 1)\, M^2}. \tag{11.4.3}$$

At point c on the line for $M = 1$ in Figure 11.5 the constant Φ line would be tangent to a constant stagnation enthalpy curve through c. This is the general condition for these two curves at $M_c = 1$.

Depiction of a shock wave is as follows. The conditions across the shock wave are constant Φ and $H + K = 1$. As such, once the upstream state (Mach number) is selected, say at point u in Figure 11.5, the solution is given by translation along the constant stagnation enthalpy straight line until it again intersects the constant Φ curve, at the downstream state indicated by point d.

The H–K diagram is used to portray different modes of heating in Figure 11.6. Frictionless constant area heating means movement towards higher stagnation enthalpy along a constant Φ curve, as indicated by the arrows. The largest possible heat addition (highest value of τ) for the specified value of Φ corresponds to the point of tangency at condition c at which the Mach number is unity and the flow is thermally choked. Adding more heat necessitates a larger value of Φ, such as would be obtained by lowering the mass flow and reducing the inlet Mach number.

Figure 11.6 also shows the regime described in Section 10.4 in which the static temperature decreases as a result of heat addition. Increasing the stagnation temperature (moving to lines of higher τ $(= T_t/T_{t_i})$ in the figure) while remaining on the $\Phi =$ constant curve corresponds to Rayleigh flow, heat addition at constant $(p + \rho u^2)A$. In the part of the constant Φ curve between $M = 1/\sqrt{\gamma} = 0.845$ (the peak) and $M = 1$, H decreases for increasing stagnation temperature.

Frictionless constant pressure heating (represented by the vertical line on Figure 11.5) and frictionless constant area heating have different effects on flow quantities. For frictionless flow, the momentum equation is $udu = -dp/\rho$, constant pressure means constant velocity and changes in static and stagnation temperature are the same ($dT = dT_t$). The relation between the change in Mach number and the static temperature is

$$\frac{dM}{M} = \frac{du}{u} - \frac{dT}{2T}. \tag{10.3.12}$$

For constant velocity this becomes

$$\frac{dM}{M} = -\frac{1}{2}\left(1 + \frac{\gamma - 1}{2} M^2\right)\frac{dT_t}{T_t}. \tag{11.4.4}$$

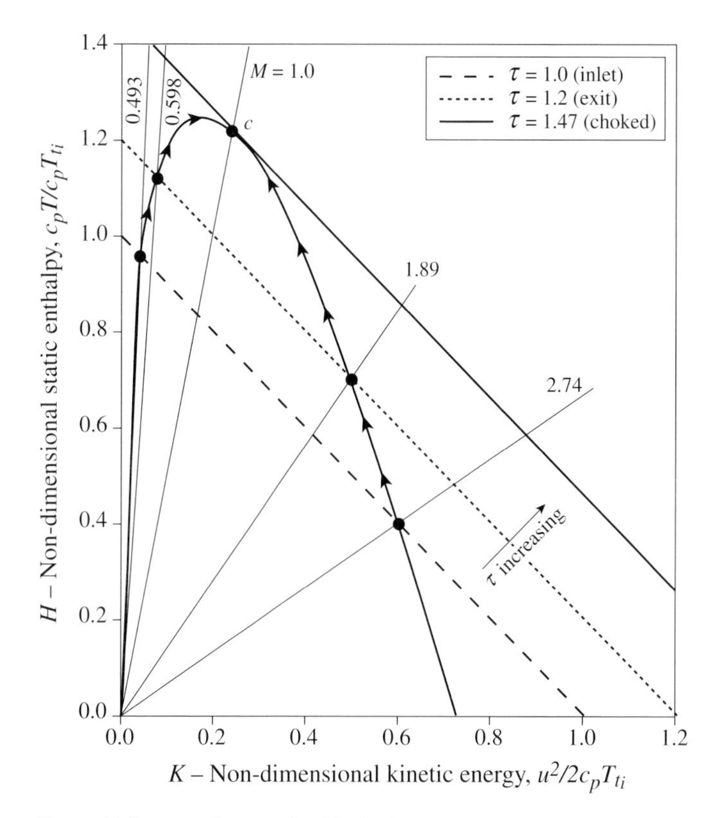

Figure 11.6: *H–K* diagram for frictionless, constant area heating; $\gamma = 1.4$, $\Phi_i = 1.2$, $\tau = T_t/T_{t_i}$. The straight lines emanating from the origin are lines of constant Mach number, increasing from left to right as indicated by their labels (Heiser and Pratt, 1994).

Integrating (11.4.4) gives the ratio of the exit Mach number to the inlet Mach number in terms of the inlet Mach number and the stagnation temperature ratio, $\tau_e = T_{t_e}/T_{t_i}$, as

$$\frac{M_e}{M_i} = \left[\tau_e\left(1 + \frac{\gamma - 1}{2}M_i^2\right) - \frac{\gamma - 1}{2}M_i^2\right]^{-\frac{1}{2}}. \tag{11.4.5}$$

The exit Mach number for constant pressure heating in a channel is shown as a function of stagnation temperature ratio in Figure 11.7(a) for four inlet Mach numbers. For supersonic inlet conditions the flow passes without difficulty through the Mach number of unity and there is no limit to the amount of heat that can be added (Heiser and Pratt, 1994).

Figure 11.7(b) shows two other aspects of constant pressure heating, the stagnation pressure ratio, p_{t_e}/p_{t_i}, and the area ratio, A_e/A_i. The former is found directly from the definition of stagnation pressure. The latter is obtained from continuity, plus the condition of constant pressure and velocity, as

$$\frac{A_e}{A_i} = \left(\frac{M_i}{M_e}\right)^2. \tag{11.4.6}$$

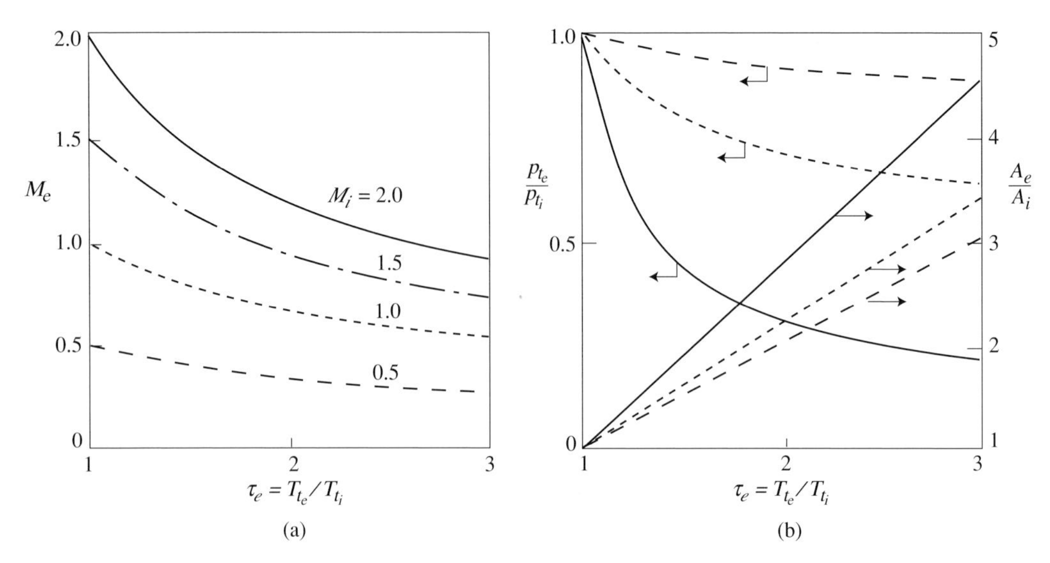

Figure 11.7: Frictionless constant pressure heat addition: (a) exit Mach number, and (b) stagnation pressure ratio and area ratio, as functions of stagnation temperature ratio τ_e and inlet Mach number M_i; $\gamma = 1.4$ (Heiser and Pratt, 1994).

Combining (11.4.5) and (11.4.6) gives the area ratio dependence on inlet Mach number and stagnation temperature ratio. For any inlet Mach number the area ratio is linear in the stagnation temperature ratio, τ_e. All these constant pressure flows occur in divergent passages.

11.4.2 Flow processes in ramjet and scramjet systems

Figures 11.8 and 11.9 give a view of the flow through a ramjet propulsion system. The overall system is sketched in 11.8. There is an initial compression, here assumed isentropic, followed by a normal shock to bring the flow to a subsonic condition. Figure 11.9 shows the process on an *H–K* diagram, starting at state 0, the free stream, to *u* and then, via the normal shock wave, to *d*, with the transition (from *u* to *d*) occurring at the same value of impulse function, Φ. Downstream of the shock there is further diffusion, with an increase in impulse function, on the constant stagnation temperature path from *d* to 1. The flow in the combustor is then (depending on design) somewhere between constant pressure and constant area flow with heat addition. In Figure 11.9 these two processes are represented by paths 1–2 or 1–3 respectively, both of which terminate on the constant stagnation enthalpy line corresponding to a combustor exit temperature of $T_t/T_{t_i} = 1.40$. For this situation the flow in the combustor is not thermally choked. The expansion through the nozzle is then along the constant stagnation enthalpy line to point 4. At the exit condition the Mach number is lower than the inlet value, but the velocity is higher than the inlet value (as necessary for positive thrust!).

Figures 11.10 and 11.11 show a view of the supersonic heat addition process in a scramjet, a situation in which interaction between heat input and area variation is critical (Heiser and Pratt, 1994). A schematic of a scramjet combustion system is given in Figure 11.10 illustrating the inlet, the (essentially constant area) "isolator", the combustor, and the nozzle. The flow is decelerated in the inlet. The isolator affords the possibility for the existence of a shock train (see Section 10.6) to accommodate downstream pressure increases. The flow exits from the nozzle.

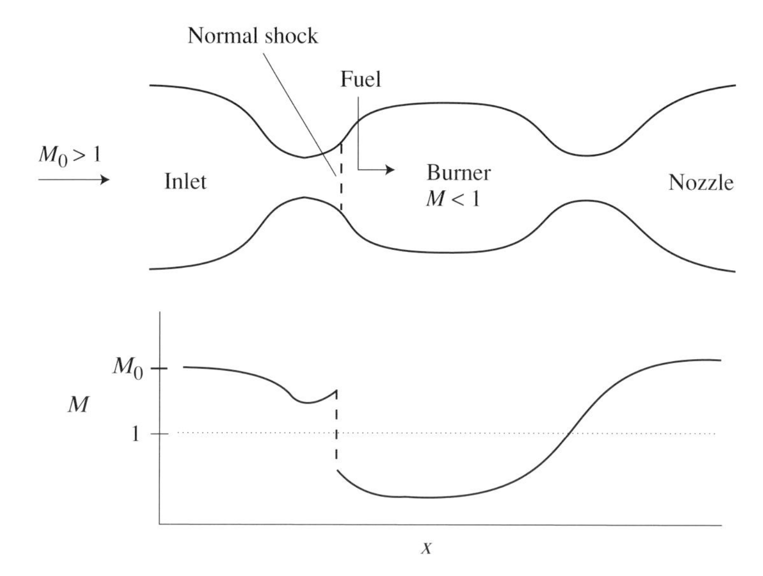

Figure 11.8: Schematic of a ramjet showing the Mach number distribution (Pratt and Heiser, 1993).

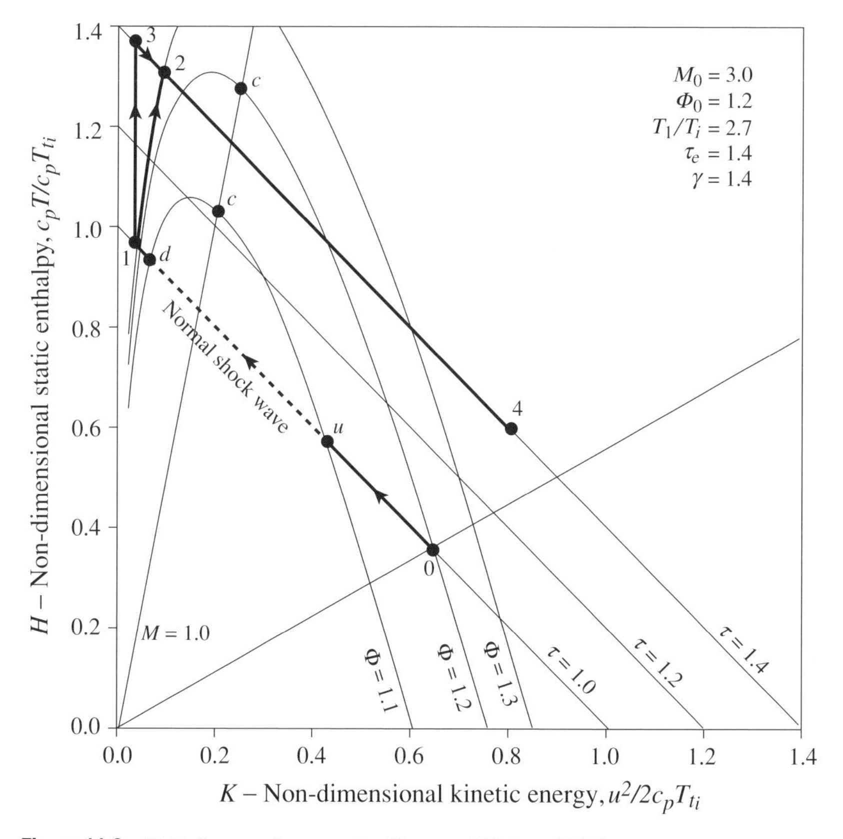

Figure 11.9: H–K diagram for a ramjet (Pratt and Heiser, 1993).

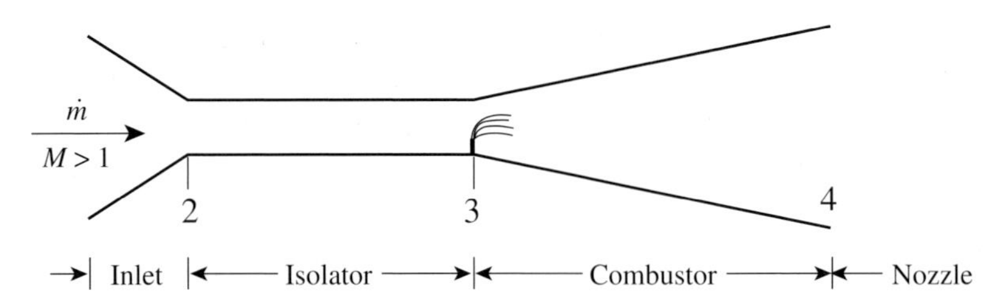

Figure 11.10: Schematic of a scramjet combustion system (Pratt and Heiser, 1993).

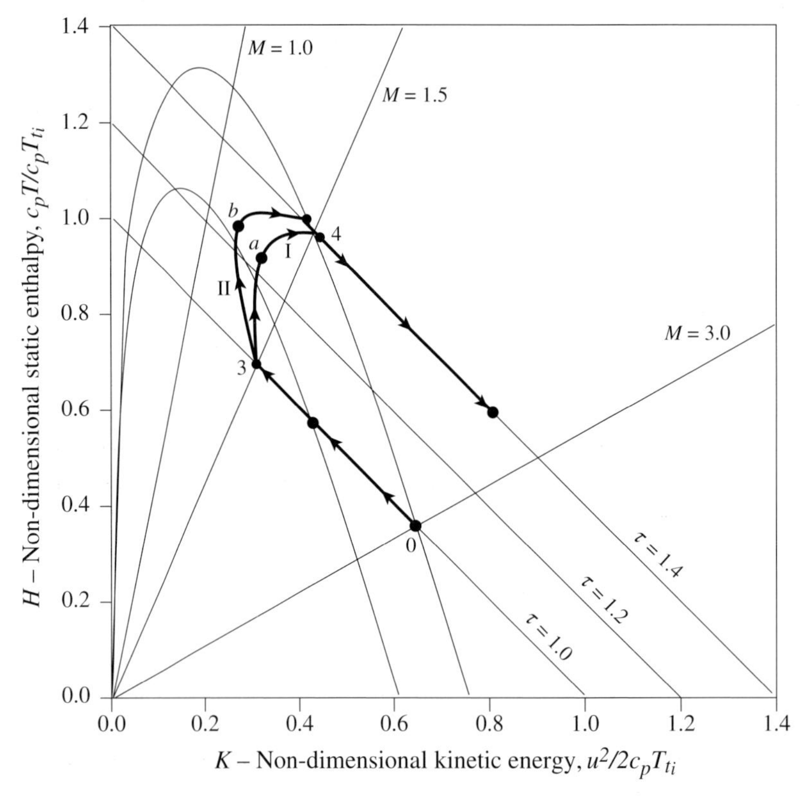

Figure 11.11: Supersonic combustion process path in H–K coordinates; $M_3 = 1.5$, linear area variation $A_4/A_3 = 2.0$ for path I, $A_4/A_3 = 1.73$ for path II, $\tau_e = 1.4$ (Curran *et al.*, 1996).

Two design problems can be posed with reference to the influence coefficients of Table 10.1. For frictionless supersonic flow with heat addition the incremental static pressure and Mach number changes are given in that table as

$$\frac{dp}{p} = \left[\frac{\gamma M^2}{M^2 - 1}\right]\left[\frac{-dA}{A} + \left(1 + \frac{\gamma - 1}{2}M^2\right)\frac{dT_t}{T_t}\right], \tag{11.4.7}$$

$$\frac{dM^2}{M^2} = \left[\frac{\left(1 + \frac{\gamma - 1}{2}M^2\right)}{M^2 - 1}\right]\left[\frac{2dA}{A} - \left(1 + \gamma M^2\right)\frac{dT_t}{T_t}\right]. \tag{11.4.8}$$

If there were no area change, heat addition ($dT_t > 0$) would cause a static pressure rise. If the pressure rise is too abrupt the boundary layer on the wall of the combustor will separate. Further, as described in Section 10.4 and implied by (11.4.8), for given inlet conditions heat addition moves the downstream Mach number towards unity, with a possibility for choking and mass flow limitation. To alleviate these problems the combustor area must increase in the downstream direction. The combination of the effects of area change and heat addition in a supersonic flow is the subject of this example.

Figure 11.11 shows the combustion system in an H–K diagram. The point 0 (corresponding to $M = 3$) denotes conditions upstream of the scramjet. The flow is decelerated to $M = 1.5$ at the inlet of the combustor (station 3) for this example. The combustor exit stagnation condition (station 4) is at a stagnation temperature ratio of 1.4. During the constant stagnation temperature deceleration from 0 to 3 the incremental change of K, the non-dimensional kinetic energy, is related to the pressure changes by

$$\frac{dK}{K} = \frac{2du}{u} = -\frac{2}{\gamma M^2}\frac{dp}{p}. \tag{11.4.9}$$

Two combustion processes are plotted in Figure 11.11, corresponding to two combustor exit area ratios, A_4/A_3. The area ratio corresponding to path I is 2.0 and that for path II is 1.73. For both paths the specified heat addition rate is largest in the entry region of the combustor and the combination of heat addition and area change provides nearly a constant pressure (and thus constant K) process in the initial part of the burner. The rate of heat addition decreases in the rear part of the combustor and the effect of area change thus becomes relatively stronger, increasing K and decreasing pressure. As a result of the heat addition the Mach number in the combustor decreases from the inlet value, going through a minimum for path I at point a, in this case $M = 1.33$. The axial location of the minimum is referred to as the thermal throat, by analogy to the physical throat in a converging–diverging nozzle. As with the physical throat, a thermal throat can exist without being choked.

For the conditions of Figure 11.11 the larger combustor area ratio corresponding to path I causes the combustor exit pressure and temperature to be lower than desired, decreasing the thermal efficiency and also making the combustion less efficient. Solution path II corresponding to a decreased combustor area ratio relieves the difficulty somewhat. For this path the minimum Mach number, at point b, drops to 1.19. Reduction of the exit area could be taken further to increase the pressure and temperature in the combustor. If the exit area is reduced enough, there will be thermal choking at the thermal throat, with the consequence that the combustor inlet flow becomes subsonic and a shock train forms in the isolator. In that situation the combustion system operates in a ramjet mode, similar to that shown in Figure 11.9. Heiser and Pratt (1994) and Curran, Heiser, and Pratt (1996) give further information on the behavior of scramjet combustion systems.

The H–K diagram allows the explicit display of many of the important quantities in propulsion systems. It is especially instructive for high speed propulsion. Heiser and Pratt (1994) note that it is not a thermodynamic state diagram, because only one axis is a thermodynamic property and there is no necessary relation between a point on the diagram and other intensive thermodynamic properties such as static pressure. They remark, however, that under many frequently encountered conditions (for example one-dimensional flow with known $\dot{m}$, A, and T_{t_i}) the H–K formulation does in fact give state information.

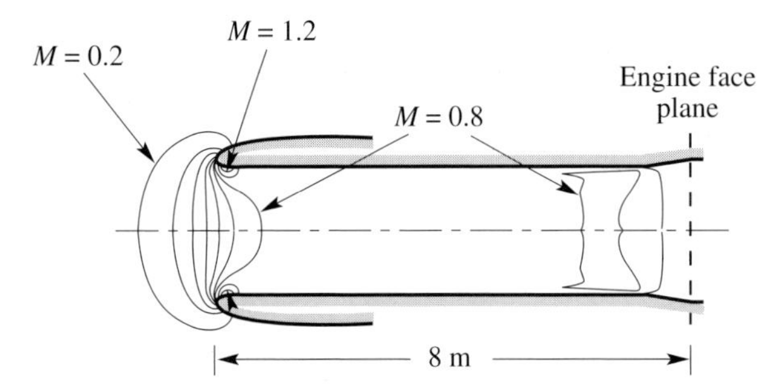

Figure 11.12: Geometry of a long aeroengine intake; contours show Mach number (Young, 1995).

11.5 An illustration of the effect of condensation on compressible flow behavior

Temperature rises due to condensation are less than those from combustion, but the former can have appreciable impact when Mach numbers are near unity. For example, problems stemming from condensation have occurred in sea level testing of gas turbine engines under conditions of high relative humidity. A methodology for assessing whether condensation is an issue, using the one-dimensional formulation, was put in place in Chapter 10; those ideas have been applied to the analysis of conditions in a jet engine static test (Young, 1995; Cumpsty, 1992). The engine had a long, straight inlet extending approximately 8 m from inlet lip to engine face. The Mach number of the axial flow along the duct was close to 0.8 over most of the length, as indicated in Figure 11.12 which shows calculated Mach number contours for dry air. Measurements during the test showed a decrease in stagnation pressure along the inlet well away from the walls (and the boundary layers).

The stagnation temperature change between two stations 1 and 2 due to the condensation of an amount of liquid of mass flow $\dot{m}_{liquid}$, with heat of vaporization h_{fg} per unit mass, is

$$\dot{m}c_p\left(T_{t_2} - T_{t_1}\right) = \dot{m}_{liquid}h_{fg}. \tag{11.5.1}$$

The mass fraction of water vapor in the air is proportional to the relative humidity, which is a function of temperature. For a relative humidity of 70%, the mass fraction of water vapor is about 0.5% at 10 °C, rising to about 1.9% at 30 °C. As the air is accelerated into the inlet, the static temperature falls and, for the conditions of sea level engine tests, typically drops below the saturation temperature. Given adequate time to condense and a supply of condensation nucleation sites, the supersaturated water vapor will condense, although this does not necessarily occur in (shorter) underwing inlets because of the shorter residence time. The long inlet provides increased time for the condensation process to occur and is thus more favorable to condensation.

The conditions at which choking occurs due to condensation are shown in Figure 11.13, based on treating the air as a dry gas with heat addition. The figure gives the Mach number that will produce thermal choking for a given inlet temperature, as a function of relative humidity, based on equilibrium conditions. The dashed line is at $M = 0.8$, which was the dry air value over much of the duct. For choking to occur with an incoming dry air Mach number of 0.8 and heat addition corresponding

Table 11.1 *Limiting conditions for thermal choking from condensation (Cumpsty, 1992)*

Inlet stagnation conditions		Ahead of condensation (1)	After condensation (2) $M_2 = 1$	
T_0 (°C)	Relative humidity	M_1	u_2/u_1	% Condensed
20.0	48	0.80	1.16	66
25.3	52	0.78	1.19	58
18.8	72	0.77	1.20	67
		Out of equilibrium	In equilibrium	

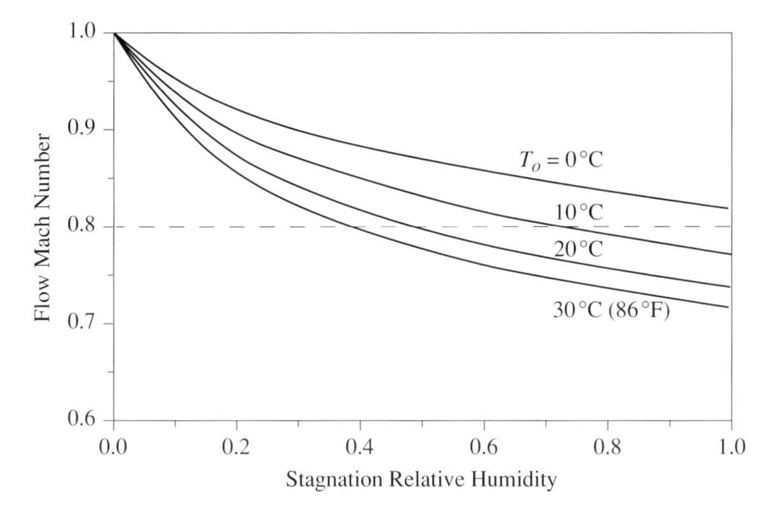

Figure 11.13: Upstream Mach number required to produce thermal choking if water vapor condensation takes place to a new equilibrium condition, air at stagnation pressure of 1 atm, and stagnation temperature T_0 (Young, 1995).

to condensation to equilibrium conditions an initial relative humidity of 40% would be needed at 30 °C, rising to 75% at 10 °C. Table 11.1 shows parameters at three test conditions at which engine fan overspeeding was encountered. Assuming the Mach number, M_2, after condensation is unity there is a large increase in axial velocity (the axial velocity ratio, after vs before, is denoted by u_2/u_1) which led to this overspeeding.

Whether condensation occurs depends on the degree of supercooling (the difference between the temperature of the supersaturated gas and the saturation temperature), the time available for condensation, and whether there are suitable dust particles to act as nucleation sites. Around the lips of the intake, the Mach numbers are roughly 1.2 and at this condition cooling is large enough for condensation to occur even without dust. For a Mach number of 0.8, dust of the correct size must be present, but many engine test facilities are in industrial areas where dust is plentiful. The paper by Young (1995) provides an instructive example of the use of one-dimensional analysis in identifying the phenomena responsible for the observed behavior as well as defining the magnitude of the effects.

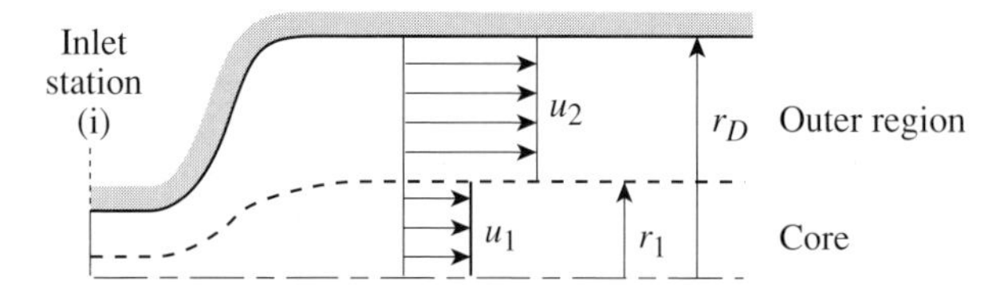

Figure 11.14: Quasi-one-dimensional model of a vortex core in a circular duct.

11.6 Swirling flow with heat addition

Many combusting flows have a strong swirl velocity. To illustrate the features and regimes encountered in such situations we extend the "simple radial equilibrium" treatment of vortex cores in Chapter 8 to include effects of heat addition. The physical mechanisms that characterize the interactions between swirl and heat addition are described, including the different influences on recirculation zone formation at low and high values of swirl.

The overall context is combustor primary zones, in which a generic configuration is swirling flow in a variable area duct with heat release. In the regimes of interest a toroidal recirculation zone exists which promotes mixing and anchors the flame. Although the one-dimensional vortex core analysis does not deal explicitly with recirculation, many features of the motion are illustrated by swirl flow influence coefficients which enable the effects of swirl, heat release, mixing, and area change to be quantified.

The vortex core analysis is an extension of that presented in Sections 8.6–8.8 to include heat addition and density change in the core (inner) stream. Several observed aspects of the flows to be addressed allow simplifications in the analysis. First, Mach numbers are low enough so we can neglect changes in density due to pressure changes compared to those arising from heat addition (temperature changes). As in Section 11.2, the state equation can be represented to order M^2 by $\rho T \approx$ constant. Second, kinetic energy changes are small compared to static enthalpy changes, so the difference between static and stagnation temperature changes can be neglected (again to order M^2). Third, only heat addition in the core stream is considered here so the temperature and density of the outer stream are constant. Finally, mixing and heat transfer between the vortex core and the outer stream is neglected; as shown by Underwood, Waitz, and Greitzer (2000) these have a quantitative influence, but do not change the results in any qualitative manner.

Application of conservation of mass, momentum, and energy, and the state equation, to the vortex core and to the outer region in the duct (see Figure 11.14) yields a system of differential equations which describe the flow evolution.[5] Conservation of mass for stream 1 (core) is

$$\frac{d\rho_1}{\rho_1} + \frac{du_{x_1}}{u_{x_1}} + \frac{dA_1}{A_1} = 0. \tag{11.6.1}$$

In stream 2 the density is constant. The area of stream 2 is related to the duct area and the area of stream 1 by $A_2 = A_D - A_1$. Conservation of mass for stream 2 is thus

[5] The notation is different than in Chapter 8 because we need to be able to denote density and temperature differences between the core and the outer stream.

$$\frac{du_{x_2}}{u_{x_2}} + \frac{d(A_D - A_1)}{(A_D - A_1)} = 0. \tag{11.6.2}$$

Denoting p_{cl} as the pressure on the centerline and S_c as the core swirl parameter, $S_c = u_\theta / u_{x_1}$, evaluated at the core radius, conservation of axial momentum for stream 1 can be written as

$$\frac{d\rho_1}{\rho_1} + \frac{dp_{cl}}{\rho_1 u_{x_1}^2} - \frac{1}{2} S_c^2 \frac{dA_1}{A_1} = 0. \tag{11.6.3}$$

Conservation of axial momentum for the outer flow, stream 2, is

$$\frac{du_{x_2}}{u_{x_2}} + \frac{(\rho_1/\rho_2)}{V^2}\frac{dp_{cl}}{\rho_1 u_{x_1}^2} - \frac{1}{2}\left[\left(\frac{\rho_1}{\rho_2}\right) + 1\right]\frac{S_c^2}{V^2}\frac{dA_1}{A_1} + \frac{1}{2}\left(\frac{\rho_1}{\rho_2}\right)\frac{S_c^2}{V^2}\left(\frac{1}{1-\sigma}\right)\frac{d\rho_1}{\rho_1} = 0. \tag{11.6.4}$$

In (11.6.4) there are three non-dimensional parameters in addition to S_c: the outer stream/vortex core axial velocity ratio, $V = u_{x_2}/u_{x_1}$, the ratio of core area to duct area, $\sigma = A_1/A_D$, and the core/outer stream density ratio, ρ_1/ρ_2. The first two were introduced in connection with the constant density vortex cores discussed in Chapter 8, but the last is new. The terms containing area and density differentials in (11.6.3) and (11.6.4) are absent for a flow with no swirl. From continuity and Kelvin's Theorem the core swirl parameter at any location is related to the inlet value by

$$\frac{S_c}{S_{c_i}} = \frac{\rho_1}{\rho_{1_i}}\sqrt{\frac{A_1}{A_{1_i}}}. \tag{11.6.5}$$

With dq_1 the increment of heat addition per unit mass, the energy equation for the core is

$$\frac{dT_1}{T_1} = \frac{dq_1}{c_p T_1}. \tag{11.6.6}$$

The equation of state is

$$\frac{d\rho_1}{\rho_1} + \frac{dT_1}{T_1} = 0. \tag{11.6.7}$$

The influence coefficients resulting from (11.6.1)–(11.6.7) are summarized in Table 11.2. They reduce to the results of Shapiro (1953) and Anderson *et al.* (1970) (given in Tables 10.1 and 10.3) for a single stream at the appropriate limits. The independent variables are taken here as the duct area, A_D, and heat release per unit mass in the core, $dq_1/(c_p T_1)$. Although the latter cannot actually be specified, the view is rather that we are examining the effect of a chosen heat addition distribution on the behavior of the swirling flow.

As an example of the use of the influence coefficients consider the change in core axial velocity (du_{x_1}/u_{x_1}) with core heat addition. There are two reasons to focus on this aspect. First, acceleration or deceleration of the vortex core is directly related to the potential for recirculation zone formation. Second, this situation illustrates a result where the behavior changes qualitatively as the swirl level is altered.

From Table 11.2 the relation between core axial velocity and heat addition is

$$\frac{du_{x_1}}{u_{x_1}} = \frac{1 - \dfrac{1}{2}\dfrac{S_c^2}{V^2}\left[\left(\dfrac{(\rho_1/\rho_2)+1}{\sigma}\right) - 1\right]}{Y}\frac{dq_1}{c_p T_1}. \tag{11.6.8}$$

Table 11.2 *Influence coefficients for vortex core flow with heat addition: $S_c = u_\theta/u_{x_1}$ at core edge, $V = u_{x_2}/u_{x_1}$, $\sigma = A_1/A_D$, Y defined in (11.6.9) (Underwood* et al., *2000)*

Coefficient	$\frac{dA_D}{A_D}$	$\frac{dq_1}{c_p T_1}$
$\frac{dp_{cl}}{\rho_1 u_{x_1}^2}$	$\frac{(1+S_c^2/2)}{Y\sigma}$	$-\frac{1-\left(\frac{S_c^2}{2\sigma V^2}\right)\left[(1-\sigma)+\frac{\rho_1}{\rho_2}\left(2+\frac{S_c^2}{2}-\sigma\right)\right]}{Y}$
$\frac{du_{x_1}}{u_{x_1}}$	$\frac{-1}{Y\sigma}$	$\frac{1-\left(\frac{S_c^2}{2\sigma V^2}\right)\left(\frac{\rho_1}{\rho_2}+1-\sigma\right)}{Y}$
$\frac{du_{x_2}}{u_{x_2}}$	$-\frac{(\rho_1/\rho_2)-(S_c^2/2)}{Y\sigma V^2}$	$\left(\frac{\rho_1/\rho_2}{YV^2}\right)\left[1+\frac{S_c^2}{2(1-\sigma)}\right]$
$\frac{dT_1}{T_1}$	0	1
$\frac{d\rho_1}{\rho_1}$	0	-1
$\frac{dA_1}{A_1}$	$\frac{1}{Y\sigma}$	$\left(\frac{\rho_1/\rho_2}{Y\sigma V^2}\right)\left(1-\sigma+\frac{S_c^2}{2}\right)$
$\frac{dp_{t_1}}{\rho_1 u_{x_1}^2}$	0	$-\frac{(1+S_c^2)}{2}$

The quantity Y is defined as

$$Y = 1 + \frac{(1-\sigma)}{\sigma V^2}\left(\frac{\rho_1}{\rho_2} - \frac{S_c^2}{2}\right). \tag{11.6.9}$$

Examining the denominator of the influence coefficient in (11.6.8), the critical swirl number that yields $Y = 0$ is

$$S_{c_{crit}} = \sqrt{2\left[\frac{\rho_1}{\rho_2} + \frac{\sigma V^2}{(1-\sigma)}\right]}. \tag{11.6.10}$$

In the discussion only swirl parameters below this critical value are considered.

The swirl parameter at which the numerator of the influence coefficient in (11.6.8) is zero denotes the condition at which the sign of the influence coefficient changes. The swirl parameter at which this occurs marks the boundary between the two regimes, denoted as S_{c_a} in Table 11.3, is

$$S_{c_a} = \sqrt{\frac{2\sigma V^2}{[(\rho_1/\rho_2)+1]-\sigma}}. \tag{11.6.11}$$

Passage through this value of S_c corresponds to a reversal in the effect of adding heat to the core. For low swirl, adding heat accelerates the core, as inferred from the discussion (for zero swirl condition) in Section 11.2. For high swirl, heat addition decelerates the core.

An explanation for the change in behavior with swirl level can be given by considering the cases of no swirl and strong swirl (e.g. swirl parameter close to critical) for a constant area duct flow.

Table 11.3 *Local trends from influence coefficients (Underwood* et al. *2000)*

Variable	Duct area change ($dA_D > 0$)	Heat release in core
p_{cl}	$\uparrow$	$\downarrow, S_c < S_{c_b}$
		$\uparrow, S_c > S_{c_b}$
u_{x_1}	$\downarrow$	$\uparrow, S_c < S_{c_a}$
		$\downarrow, S_c > S_{c_a}$
u_{x_2}	$\downarrow, S_c < S_{c_c}$	$\uparrow$
	$\uparrow, S_c > S_{c_c}$	
A_1	$\uparrow$	$\uparrow$
A_2	$\uparrow, S_c < S_{c_c}$	$\downarrow$
	$\downarrow, S_c > S_{c_c}$	
p_{t_1}	No effect	$\downarrow$

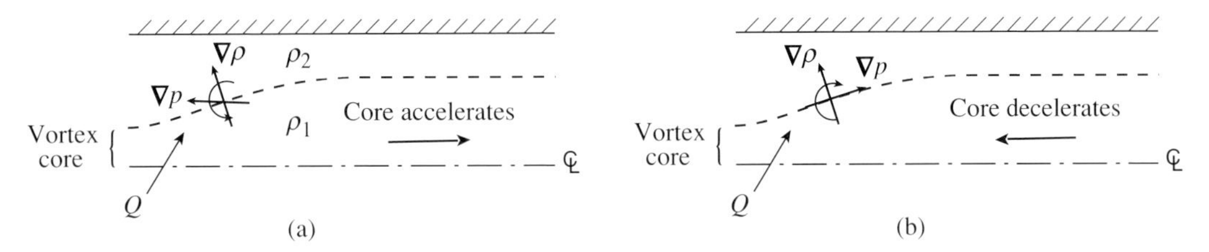

Figure 11.15: Circulation generation due to baroclinic torque for (a) no (or low) swirl and (b) high swirl.

The core and outer region are taken to have the same axial velocity upstream of the region of heat addition, although this is not a necessary part of the arguments.

With no swirl heat addition lowers the core density and causes the core to expand, as discussed in Section 11.2 and shown in Figure 11.15(a). The streamtube area in the outer region contracts, the axial velocity in the outer region increases, and the static pressure decreases. The static pressure is uniform across the duct and there is acceleration in the core.

Now consider a high level of swirl so that axial velocity changes have a weak effect on static pressure compared to the effect of the circumferential velocity (through simple radial equilibrium). The core expands in the region of heat addition as indicated in Figure 11.15(b). At the outer radius of the duct the circumferential velocity and the stagnation pressure do not vary along the duct. Simple radial equilibrium implies that in the axial region in which core expansion (heat addition) occurs the static pressure on the interface between the two streams increases with downstream distance. There is a thus a net upstream pressure force on the fluid in the vortex core in this region and hence a deceleration.

The change in behavior with swirl level can also be viewed in terms of vorticity dynamics. Changes in the circumferential vorticity (vortex lines oriented in the θ-direction, $\omega_\theta \approx -\partial u_x/\partial r$) at the edge of the core are set by a balance between the production of (negative) circumferential vorticity from stretching of existing circumferential vortex lines and the production of circumferential vorticity due to baroclinic torque and tipping of existing axial vorticity into the circumferential direction (see Figure 8.11). Production of positive circumferential vorticity is associated with core acceleration and production of negative circumferential vorticity with core deceleration. With no or low swirl the pressure on the interface between the two streams decreases in the streamwise direction. With high swirl the pressure increases in the streamwise direction (see Figures 11.15(a) and 11.15(b)). In the latter situation, negative circumferential vorticity is produced on the interface from the baroclinic torque. Negative circumferential vorticity is also produced from tipping of axial vorticity into the circumferential direction within the core.

There are two additional swirl parameters at which influence coefficients change sign. These two, denoted as S_{C_b} and S_{C_c}, define the boundaries between low swirl and high swirl behavior for the centerline static pressure change due to core heat addition and for the outer region velocity change due to duct area change respectively. The swirl parameters S_{C_b} and S_{C_c} are defined as

$$S_{C_b} = \sqrt{\frac{F - [(1-\sigma) + (\rho_1/\rho_2)(2-\sigma)]}{(\rho_1/\rho_2)}}, \tag{11.6.12}$$

where F is

$$F = \sqrt{\left[(1-\sigma) + \frac{\rho_1}{\rho_2}(2-\sigma)\right]^2 + 4\left(\frac{\rho_1}{\rho_2}\right)\sigma V^2}$$

and

$$S_{C_c} = \sqrt{\frac{2\rho_1}{\rho_2}}. \tag{11.6.13}$$

Table 11.2 also shows (as seen in Section 8.8) that an increase in duct area causes a decrease in core axial velocity and a tendency towards the formation of a recirculation zone. The local trends implied by the influence coefficients are summarized in Table 11.3.

11.6.1 Results for vortex core behavior with heat addition

The local trends given by the influence coefficients can be viewed for the whole flow field by integrating (11.6.1)–(11.6.7) and following the evolution of the swirl parameter compared to the values of S_{C_a}, S_{C_b}, S_{C_c}, and $S_{C_{cirt}}$. Figure 11.16 shows an example result at a relatively high value of inlet swirl (Underwood *et al.*, 2000).[6] The heat release profile and duct geometry are given in Figure 11.17. Six regions are defined, labeled 1–6 in Figure 11.16. Their boundaries are the swirl parameters S_{C_a}, S_{C_b} and S_{C_c}, at which the influence coefficients change sign, defined by (11.6.11)–(11.6.13) and labeled in the figure. The other boundary on Figure 11.16 is the critical swirl ratio $S_{C_{cirt}}$ defined by (11.6.10).

Behavior at any axial location (x/r_{D_i}) is determined by which region the solution (S_c) lies in at that point. Region 1 is below all the boundaries, and the behavior of the flow is qualitatively

[6] Figure 11.16 is based on calculations with a model which included additional effects, for example mixing between the core and outer stream. There are thus some quantitative differences between the solution trajectories shown and those which would be obtained from the use of (11.6.1)–(11.6.6). The discussion and conclusions concerning parametric behavior, however, are directly applicable.

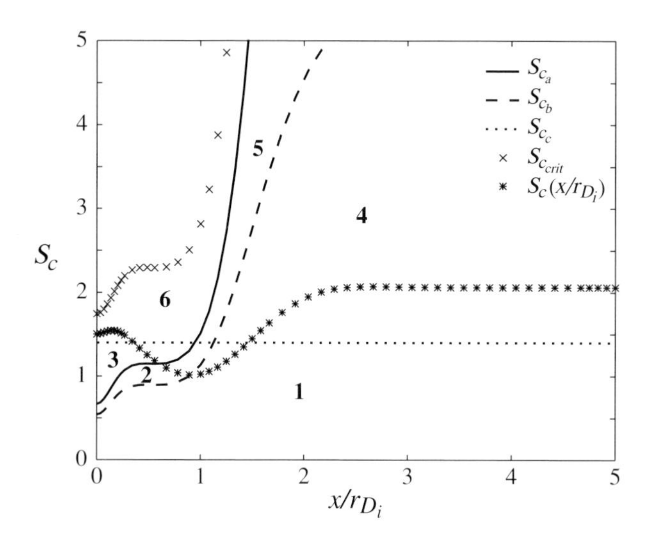

Figure 11.16: Flow regime map for vortex cores with heat addition ($S_{c_i} = 1.5$), and high heat release ($\phi = 0.8$); r_{D_i} is the duct radius at the inlet station; heat release is $dh = 0.8\, dh_{max}$; dh_{max} is based on a measured methane–air reaction heat addition profile; ϕ is the equivalence ratio (Underwood *et al.*, 2000).

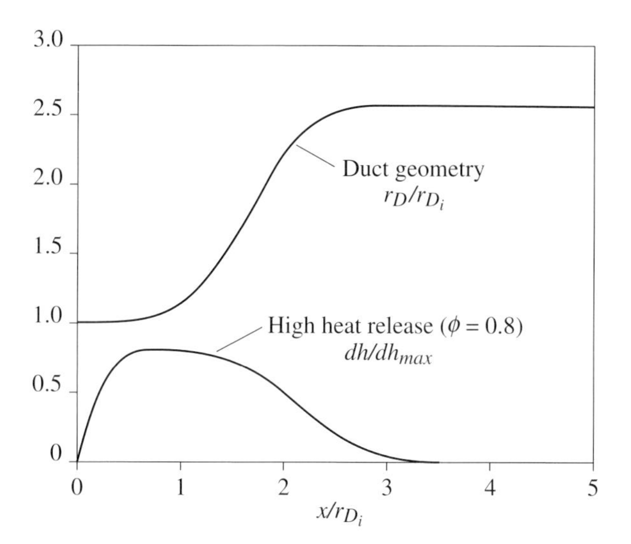

Figure 11.17: Heat release and duct radius profiles; $dh = \phi dh_{max}$; dh_{max} is based on a measured methane–air reaction heat addition profile; ϕ is the equivalence ratio (Underwood *et al.*, 2000)

similar to that for zero swirl. Region 2 is above the S_{c_b} line, the local core swirl parameter (S_c) is greater than S_{c_b}, and, as shown in Table 11.2 (or 11.3), adding heat to the core increases the core velocity. Region 3 lies above both the S_{c_a} and S_{c_b} lines. so adding heat to the core decelerates the core.

In region 4, adding heat to the core accelerates the core and decreases the centerline static pressure as in the zero swirl case. However, increasing the duct area increases the axial velocity of the outer stream and decreases the area of the outer stream, counter to the zero swirl case. In region 5, adding

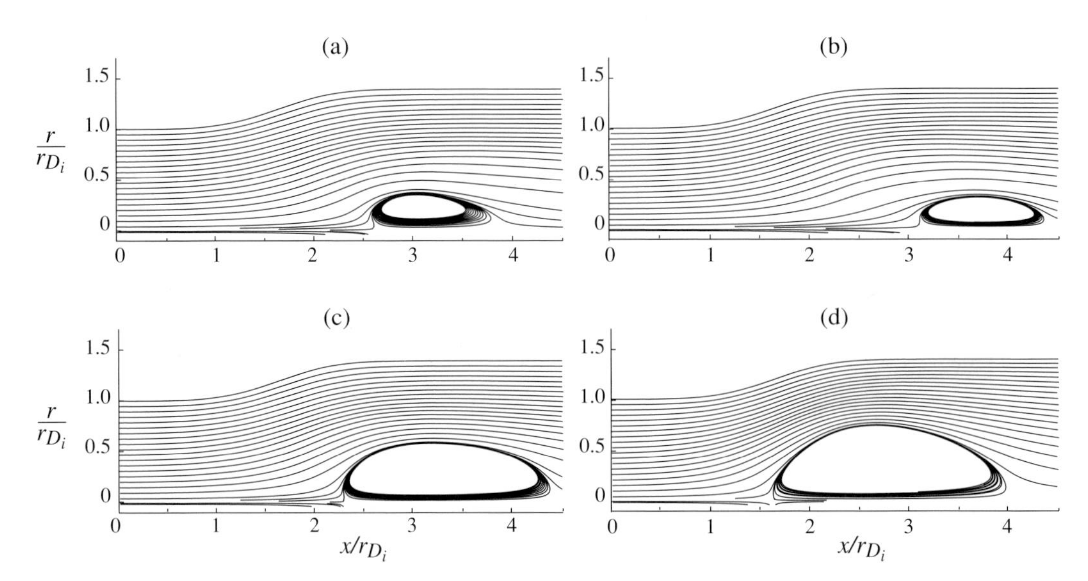

Figure 11.18: Navier–Stokes computations of vortex core flow with low and high swirl and varying heat release: (a) low heat release, law swirl; (b) high heat release, low swirl; (c) low heat release, high swirl; (d) high heat release, high swirl (Underwood *et al.*, 2000).

heat to the core increases the centerline static pressure, and increasing the duct area increases the outer stream axial velocity and decreases the outer stream area. Finally, in region 6, adding heat to the core increases the centerline static pressure and decreases the core axial velocity, whereas increasing the duct area increases the outer stream axial velocity and decreases the outer stream area. These effects are summarized in Table 11.3. Examination of the relation of the solution trajectory to the boundaries for trend reversal is helpful in understanding the parametric dependence.

Several summary comments can be made about vortex core behavior with heat addition. First, the initial core radius computation is one-half the initial duct radius and the likelihood of recirculation zone formation is increased for smaller initial area ratios (Darmofal *et al.*, 2001). In general for low swirl, heat release in the core hinders recirculation zone formation, whereas for high swirl, heat release in the core enhances recirculation zone formation. Mixing mitigates against recirculation zone formation.

The applicability of the one-dimensional vortex core analysis in providing guidance concerning the onset of recirculation can be seen by examination of results from an axisymmetric Navier–Stokes code for reacting flows (Wake, Choi, and Hendricks, 1996) in Figure 11.18. Streamlines showing the size and location of the recirculation zone for two different levels of heat release and two levels of swirl are plotted.

A low swirl ($S_{c_i} = 0.5$) case with zero and high heat release is presented in Figures 11.18(a) and 11.18(b), respectively and a high swirl ($S_{c_i} = 0.8$) case in Figures 11.18(c) and 11.18(d) for the same non-dimensional heat addition. With low swirl, heat release causes acceleration on the centerline, resulting in a weaker, further downstream, recirculation zone. With high swirl, heat release causes deceleration on the centerline, resulting in a stronger, further upstream, recirculation zone, in qualitative accord with the trends from the influence coefficients.

11.7 An approximate substitution principle for viscous heat conducting flow

Section 10.10 described a substitution principle for steady isentropic (but not necessarily uniform entropy) flow by which the behavior of a class of motions can be inferred from a single solution. This is an exact statement for inviscid flow with no heat transfer. There is no equivalent *exact* statement when heat transfer and momentum interchange through viscous stresses are present, but the idea of a substitution principle can be extended in an *approximate* manner to situations in which mixing has a strong role. (Greitzer, Paterson, and Tan, 1985).

The concept to be developed can be stated as follows. In flows with the same upstream stagnation pressure distribution but different stagnation temperature distributions, the Mach number and stagnation pressure distributions along given streamlines are approximately independent of the stagnation temperature distribution even if substantial heat and work transfer between streamtubes exist. The local streamtube area, however, is altered approximately in proportion to the square root of the local temperature.

In the next sections we discuss features of flows with heat and work exchange between streamtubes with reference to this approximate similarity. It is shown there is a relation between heat transfer and viscous work which, if satisfied, would lead to the desired result. Several model problems relating to the mixing of two streams are then defined to identify parameter regimes in which the relation is satisfied and the approximation applies. Finally, data from mixer nozzles, jets, and ejectors are presented to demonstrate concept application.

11.7.1 Equations for flow with heat addition and mixing

As a framework for the development of the ideas, we examine the equations for the steady flow of a perfect gas with viscous stresses and heat transfer. For now discussion is restricted to Mach numbers $M^2 \ll 1$, for which the equations are, with $\mathbf{F}_{visc}$ the viscous force per unit mass and $\mathbf{q}$ the heat flux vector,

$$\nabla \cdot (\rho \mathbf{u}) = 0, \tag{11.7.1a}$$

$$\mathbf{u} \times \omega = \frac{1}{\rho}\nabla p + \nabla\left(\tfrac{1}{2}u^2\right) - \mathbf{F}_{visc}, \tag{11.7.1b}$$

$$\rho c_p \mathbf{u} \cdot \nabla T = \nabla \cdot \mathbf{q}. \tag{11.7.1c}$$

For the situations of interest the Reynolds numbers are typically high enough that $\mathbf{F}_{visc}$ and $\mathbf{q}$ primarily result from turbulent rather than molecular diffusive processes. Equations (11.7.1) are supplemented by the equation of state, which, in accord with the discussion of Section 11.2 can be written as $\rho T = \rho_i T_i + 0(M^2)$, with ρ_i and T_i reference (taken here as initial) values of density and temperature. In this form of the equations of motion the terms neglected compared to those retained are of order M^2; the neglected terms contribute to dynamical processes only to order M^4. To this same order the static and stagnation temperatures are interchangeable. In the motions described the stagnation temperature can vary along a streamline as well as from streamline to streamline.

Because we are looking at conditions under which there is Mach number similarity, it is useful to cast (11.7.1) in terms of the Mach number vector, $\mathbf{M} = \mathbf{u}/a$, where a is the local sound speed:

$$\nabla \cdot \mathbf{M} = (\mathbf{M} \cdot \nabla T)/2T, \tag{11.7.2a}$$

$$\mathbf{M} \times \omega_M = \frac{\nabla p_t}{\rho_i a_i^2} + \mathbf{M}\frac{(\mathbf{M} \cdot \nabla T)}{2T} - \frac{\mathbf{F}_{visc}}{a^2}, \tag{11.7.2b}$$

$$\mathbf{M} \cdot \nabla T = \frac{(\nabla \cdot \mathbf{q})}{(\rho a c_p)}. \tag{11.7.2c}$$

The quantity ω_M is defined as (Hicks 1948)

$$\omega_M = \nabla \times \mathbf{M}. \tag{11.7.3}$$

Equations (11.7.2) contain terms with the appearance of 'sources', represented by the term $(\mathbf{M} \cdot \nabla T)/2T$ in the continuity equation, and 'body forces', represented by the term $\mathbf{M}(\mathbf{M} \cdot \nabla T)/2T$ in the momentum equation. The body-force term can be regarded as the force needed to accelerate the source mass flux, which appears at zero velocity, to the local velocity (Broadbent, 1976). Both source and body force are directly proportional to the rate of change of temperature along the streamline.

Any vector field is specified by its curl, divergence, and the boundary conditions imposed by geometry (e.g. Plonsey and Collin (1961), Batchelor (1967)). For the same boundary conditions, (11.7.2) imply there cannot be complete Mach number field equivalence between isentropic flows (which have $\nabla \cdot \mathbf{M} = 0$) and those with temperature variation along a streamline (for which, as shown by (11.7.2a), $\nabla \cdot \mathbf{M}$ is non-zero). In spite of this, for parameter regimes encountered in fluid machinery and propulsion devices, there are aspects of mixing flows that remain substantially independent of the stagnation temperature distribution.

Because many of the flows of interest have low stagnation pressure loss along streamlines it is pertinent to inquire when the stagnation pressure is invariant along a streamline if heat transfer and viscous forces exist. To see this, we make use of (11.7.2c) to write (11.7.2b) as

$$\mathbf{M} \times \omega_M = \frac{\nabla p_t}{\rho_i a_i^2} + \frac{\mathbf{M}(\nabla \cdot \mathbf{q})}{2\rho a c_p T} - \frac{\mathbf{F}_{visc}}{a^2}. \tag{11.7.4}$$

The stagnation pressure does not change along a streamline if

$$\frac{\mathbf{u}(\nabla \cdot \mathbf{q})}{2\rho c_p T} = \mathbf{F}_{visc}. \tag{11.7.5}$$

If (11.7.5) is satisfied, the equations that describe the motion[7] are (11.7.2a), (11.7.2c), and

$$\mathbf{M} \times \omega_M = \frac{\nabla p_t}{\rho_i a_i^2}. \tag{11.7.6}$$

[7] Circulation round a fluid contour is not conserved for flows obeying these equations. However, a 'reduced circulation' (based on the Mach number field) is conserved for contours composed of fluid particles that drift with a velocity proportional to the local Mach number. A formal analogy exists between the reduced vorticity in isentropic and non-isentropic flows and the actual vorticity in inviscid constant density and inviscid compressible flows respectively (Greitzer *et al.*, 1985).

The important implication of (11.7.5) is that there is a possible condition in which two competing effects cancel; the stagnation pressure change from viscous effects is balanced by that resulting from heat transfer. The point can be made specific using the example of two parallel coflowing streams in a channel with the same inlet stagnation pressure. If the streams have the same stagnation temperature, the velocities are equal and there is no heat transfer, viscous stress, or change in stagnation pressure along the flow direction. Suppose, however, that one stream has a higher inlet stagnation temperature than the other. Equality of inlet stagnation pressures means equality of the quantities ρu^2 or, because the static pressure is uniform in the parallel streams, u^2/T. The hot stream thus has a higher velocity than the cold stream. In the subsequent mixing that takes place (because of the unequal velocities and temperatures) the stagnation pressure in the cold stream tends to decrease due to heat addition (see Section 11.3) but tends to increase because of viscous work transfer from the higher speed hot stream. In the hot stream the effects are the converse. Heat extraction causes a stagnation pressure increase and the viscous stresses from the slower moving cold stream cause a decrease. We will see that a close to complete balance between the two effects exists over a wide range of conditions. As a consequence (over the defined range of conditions) stagnation pressure changes along streamlines in a mixing flow are almost independent of alterations in the inlet stagnation temperature even if large velocity and temperature gradients exist.

11.7.2 Two-stream mixing as a model problem – I: Constant area, low Mach number, uniform inlet stagnation pressure

To present the basic ideas and illustrate the degree to which the approximate scaling is valid we examine the model problem of mixing of two low Mach number streams in a constant area duct. The initial examples are for equal inlet stagnation pressures in the two streams, which is not only the simplest situation but also gives upper bounds on the departure from exact similarity. For non-uniform inlet stagnation pressures (the more usual case) the changes in stagnation pressure and Mach number resulting from temperature field alterations are found to be smaller than for uniform inlet stagnation pressures.

The inlet conditions are specified by two parameters, the temperature ratio (hot stream to cold stream) and fractional area occupied by the cold stream (stream 1) at the inlet, σ, defined as

$$\sigma = \frac{A_{1_i}}{A_{1_i} + A_{2_i}} = \frac{A_{1_i}}{A}. \tag{11.7.7}$$

As stated previously the difference between the stagnation and static temperature ratios is of order Mach number squared, and hence negligible at low Mach number, i.e. the two ratios are interchangeable in this regime. To avoid subscripts we therefore use the static temperature in the temperature ratio. In Figure 11.19, subscript 1 refers to the initially cold stream, subscript 2 to the initially hot stream, and the subscripts i and e denote inlet and exit stations. The exit condition is fully mixed. Holding σ constant corresponds to varying the inlet temperature ratio, $TR = T_{2_i}/T_{1_i}$ with geometry constant, as would generally be the case in a mixer nozzle experiment. The mass flow ratio, $\dot{m}_1/\dot{m}_2$, is related to σ and temperature ratio by

$$\frac{\dot{m}_1}{\dot{m}_2} = \frac{\sigma\sqrt{TR}}{(1-\sigma)}. \tag{11.7.8}$$

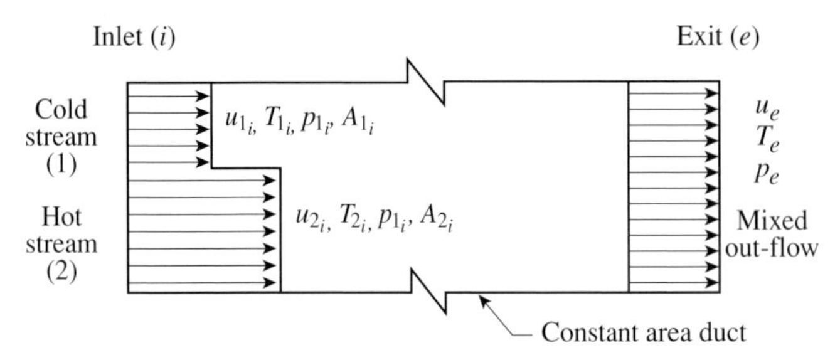

Figure 11.19: Control volume for two-stream mixing; $\alpha = A_{1_i}/A$.

The results for the mixed out states are obtained from the control volume statements.[8] Conservation of mass and energy serve to determine the mixed out velocity and temperature respectively. The momentum equation then gives the static pressure difference, inlet to exit, from which the stagnation pressure change can be obtained. The results for non-dimensional stagnation pressure change and the exit/inlet Mach number ratio are:

$$\text{stagnation pressure change: } \frac{p_{t_i} - p_{t_e}}{\frac{1}{2}\left(\rho_1 u_1^2\right)_i} = \sigma(1-\sigma)\left(\sqrt{TR} + \frac{1}{\sqrt{TR}} - 2\right), \tag{11.7.9}$$

$$\text{exit Mach number/inlet Mach number: } \frac{M_e}{M_i} = \left[1 + \frac{\left(p_{t_i} - p_{t_e}\right)}{\frac{1}{2}\left(\rho_1 u_1^2\right)_i}\right]^{\frac{1}{2}}. \tag{11.7.10}$$

Equation (11.7.9) shows that to first order in the non-dimensional inlet temperature difference, $(T_{2_i} - T_{1_i})/T_{1_i}$, the effects of viscous work and heat transfer are equal and opposite and the stagnation pressure change is zero. The cancellation is not exact for larger temperature differences, but the numerical results indicate that it is a good approximation in the range of temperature ratios considered.

Figure 11.20 presents changes in non-dimensional density, mass flow per unit area, and velocity for the initially cold stream, as a function of the inlet temperature ratio, *TR*, for equal hot and cold flow inlet areas ($\sigma = 0.5$). Substantial changes occur in velocity and density between the inlet and the exit for the higher temperature ratios. Figure 11.21, however, indicates that the change in stagnation pressure is a much smaller fraction of the inlet dynamic pressure (roughly 12% at a temperature ratio of 4 and $\sigma = 0.5$).

For the situation considered the approximate similarity proposed is that constant Mach number is also a good description of flow with non-uniform inlet temperature. The ratio of exit/inlet Mach number (the inlet Mach numbers are the same for both streams) is shown in Figure 11.22 for $\sigma = 0.5$, the situation in which the largest difference from unity occurs. Even at an inlet temperature ratio of 4 the departure from constant Mach number is only 6%.

[8] The control volume calculations give overall results and do not provide information about the local balance between *rates* of heat and momentum transfer. However, these have been examined (Greitzer *et al.*, 1985) and the utility of the approximation assessed on a local basis. This point will also be addressed by data to be shown subsequently.

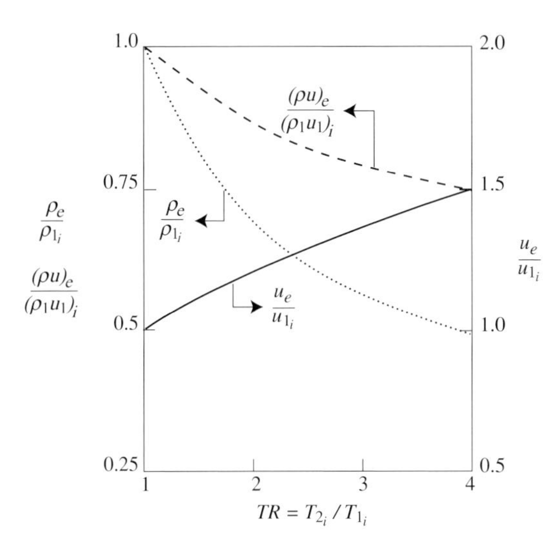

Figure 11.20: Mixing of hot and cold streams in a constant area duct. Mixed out conditions relative to the cold stream inlet conditions (subscript 1) as a function of the inlet temperature ratio of the two streams, $\sigma = A_{1_i}/A = 0.5$, $p_{t_{1i}} = p_{t_{2i}}$, $M^2 \ll 1$ (Greitzer *et al.*, 1985).

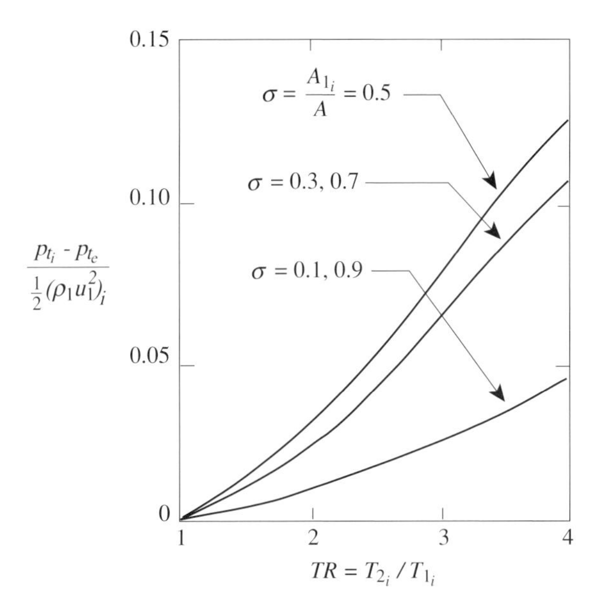

Figure 11.21: Stagnation pressure change due to mixing of hot and cold streams in a constant area duct, $p_{t_{1i}} = p_{t_{2i}}$, $M^2 \ll 1$ (Greitzer *et al.*, 1985).

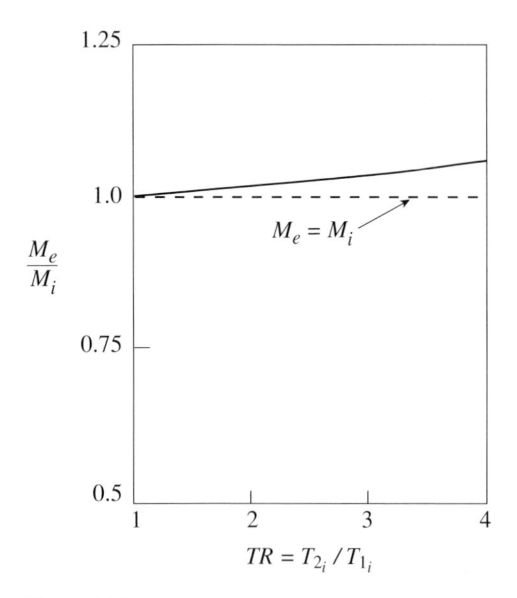

Figure 11.22: Mach number ratio for mixing of hot and cold streams in a constant area duct, $p_{t_{1i}} = p_{t_{2i}}$, $\sigma = A_{1_i}/A = 0.5$, $M^2 \ll 1$ (Greitzer *et al.*, 1985).

11.7.3 Two-stream mixing as a model problem – II: Non-uniform inlet stagnation pressures

In the more general situation in which the inlet streams have unequal stagnation pressures another non-dimensional parameter, denoted here by χ, must be specified to characterize the stagnation pressure non-uniformity at the inlet:

$$\chi = \frac{p_{t_{1i}} - p_{t_{2i}}}{\frac{1}{2}\left(\rho_1 u_1^2\right)_i}. \tag{11.7.11}$$

The velocity ratio at the inlet, u_{2_i}/u_{1_i} can be written in terms of the temperature ratio and χ as

$$\frac{u_{2_i}}{u_{1_i}} = \sqrt{TR(1-\chi)}. \tag{11.7.12}$$

Applying the control volume analysis for constant area two-stream mixing, the non-dimensional stagnation pressure change in the cold stream is found as

$$\frac{p_{t_{1i}} - p_{t_e}}{\frac{1}{2}\left(\rho_1 u_1^2\right)_i} = [\{\sigma + (1-\sigma)\sqrt{(1-\chi)/TR}\}\{\sigma + (1-\sigma)\sqrt{(1-\chi)TR}\} - 1 + 2\chi - 2\sigma\chi]. \tag{11.7.13}$$

It is instructive to examine several limiting cases. For an inlet temperature ratio of unity ($TR = 1$) and a small inlet stagnation pressure non-uniformity, $\chi \ll 1$, (11.7.13) reduces to

$$\frac{p_{t_{1i}} - p_{t_e}}{\frac{1}{2}\left(\rho_1 u_1^2\right)_i} = (1-\sigma)\chi. \tag{11.7.14}$$

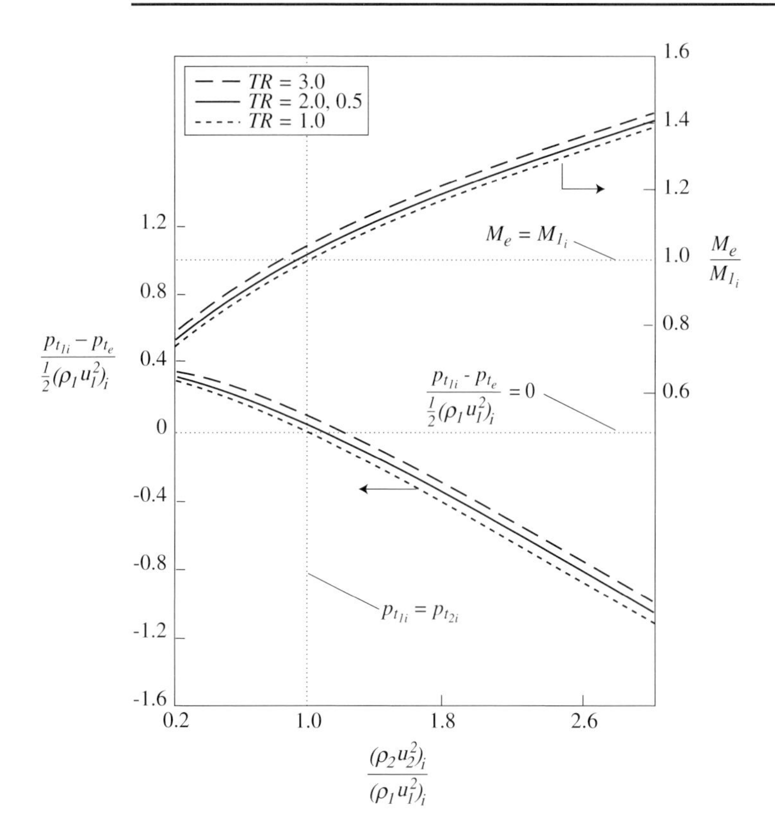

Figure 11.23: Effect of the inlet stagnation pressure and stagnation temperature differences on the stagnation pressure and Mach number changes in constant area mixing; $\sigma = A_{1_i}/A = 0.5$, $M^2 \ll 1$ (Presz and Greitzer, 1988).

If χ is negative $p_{t_{2i}} - p_{t_{1i}} > 0$, the stagnation pressure in stream 2 increases from inlet to exit, whereas if χ is positive the stagnation pressure in stream 2 decreases. The cause is work transfer between the two streams. At inlet temperature ratios near unity ($TR = 1 + \Delta T/T_{1_i}$ and $\Delta T/T_{1_i} \ll 1$) the stagnation pressure change from inlet to exit is unchanged to first order in $\Delta T/T_{1_i}$ from (11.7.14), depending only on the inlet stagnation pressure difference.

Figure 11.23 shows results for the stagnation pressure change and Mach number ratio (exit to inlet) in the cold stream as a function of the inlet stagnation pressure difference. The temperature ratios are from 0.5 to 3.0. For a specified inlet *temperature ratio*, there is a strong effect of stagnation pressure difference on Mach number ratio and stagnation pressure change. For a specified inlet *stagnation pressure difference*, however, the temperature ratio has only a small effect on the Mach number ratio and stagnation pressure change. The approximate scaling described amounts to the neglect of the effect of temperature.

11.7.4 Effects of inlet Mach number level

We now consider the effect of the inlet Mach number on similarity in two-stream mixing in a constant area duct. For an arbitrary (not small) Mach number, the exit conditions are functions of four

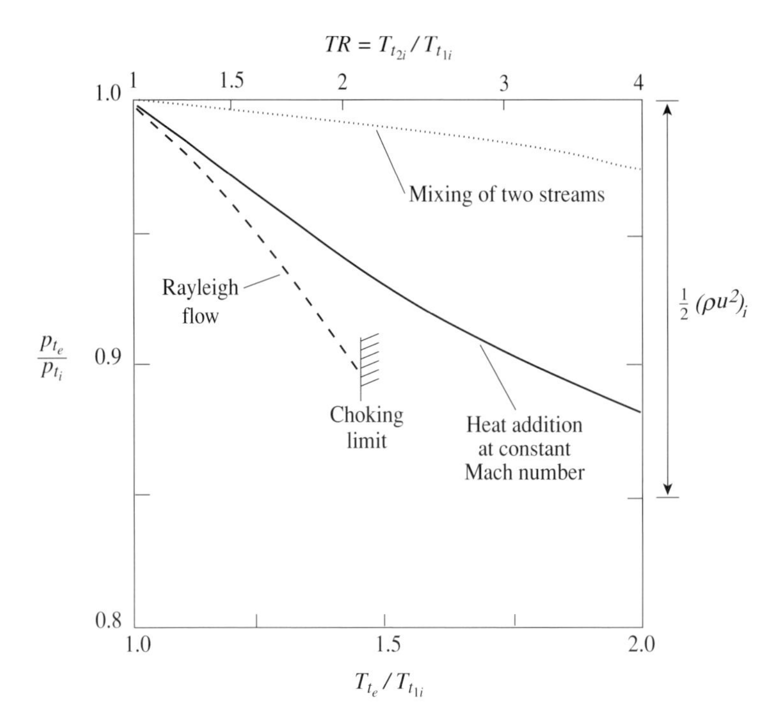

Figure 11.24: Stagnation pressure change due to: (i) mixing of hot and cold streams with equal stagnation pressure, (ii) frictionless constant area flow with heat addition (Rayleigh flow), and (iii) frictionless flow at constant Mach number with heat addition; $M_i = 0.5$ (Greitzer *et al.*, 1985).

parameters: the inlet area ratio of the two streams, the inlet stagnation temperature and stagnation pressure ratios of the two streams, and the inlet Mach number of either stream (or equivalently the ratio of stagnation to static pressure in either stream). The uniform stagnation pressure, $\sigma = 0.5$ case yields the largest departure from similarity and only results for that configuration are presented.

Figure 11.24 shows the ratio of the exit stagnation pressure to the inlet stagnation pressure for three situations: (i) constant area mixing of two streams at different inlet stagnation temperatures and the same inlet stagnation pressure; (ii) constant area frictionless flow with heat addition (Rayleigh flow) and (iii) constant Mach number flow with heat addition. The inlet conditions are $M_i = 0.5$ and, for case (i), $\sigma = 0.5$. The abscissa is the ratio of exit/inlet stagnation temperatures; for two-stream mixing it is the ratio of the mixed out exit stagnation temperature to the inlet cold stream stagnation temperature, $T_{t_e}/T_{t_{1i}}$. The stream-to-stream stagnation temperature ratio at the inlet corresponding to this mixed out value is indicated on the scale at the top of the figure. For reference the inlet dynamic pressure is shown on the right-hand side of the figure.

For the same exit/inlet temperature ratio the stagnation pressure decrease for constant area heat addition to a single stream is more than an order of magnitude larger than for the cold stream during two-stream mixing. The stagnation pressure decrease for heat addition at constant Mach number is also much larger than in two-stream mixing, even though the heat addition in the mixing flow occurs at locally higher Mach numbers. With the two-stream flow there must be an effect which acts to

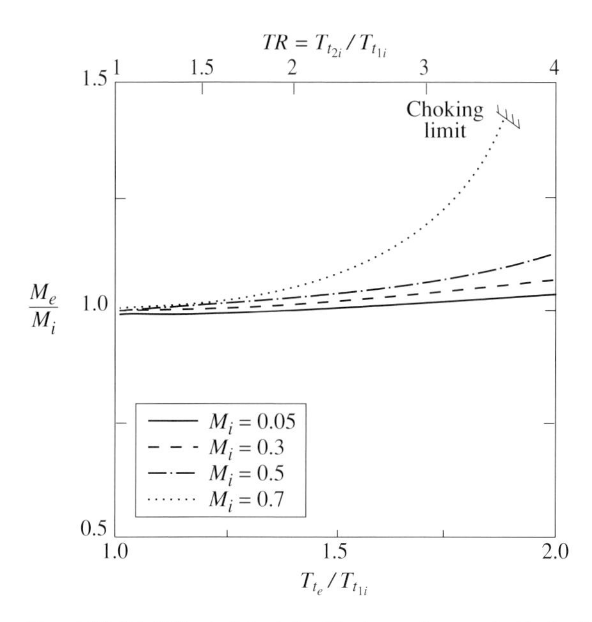

Figure 11.25: Effect of the inlet Mach number on similarity in two-stream mixing; $\sigma = 0.5$, uniform stagnation pressure (Greitzer *et al.*, 1985).

increase the stagnation pressure of the cold stream, and, as argued, this is the work transfer from shear stress exerted by the faster hot stream.

The effect of the Mach number level on the exit/inlet Mach number ratio for two-stream constant area mixing is shown in Figure 11.25. As the inlet Mach number approaches unity, the departure from the proposed scaling increases (at $M_i = 0.7$ and $TR = 2$, the error is 6%). As discussed below, however, constant area mixing is a severe test and the approximate scaling principle is more applicable than implied by Figure 11.25.

11.8 Applications of the approximate principle

11.8.1 Lobed mixer nozzles

Applications of the approximate scaling are demonstrated below for turbofan lobed forced-mixer nozzles, jets, and ejectors. The geometry for a lobed mixer nozzle (see Section 9.9 for a description of the flow structure in these devices) is given in Figure 11.26. Figure 11.27 compares measured Mach number profiles for uniform and non-uniform inlet stagnation temperature at the axial and circumferential locations of Figure 11.26. There is close correspondence between Mach numbers in the two situations even though the individual velocities and temperatures are quite different. The Munk and Prim principle should apply at the duct inlet, where effects of mixing are confined to thin shear layers emanating from the lobe surfaces. This, however, cannot be said for the conditions at the intermediate plane and nozzle exit, where measurements show that the flow is well mixed.

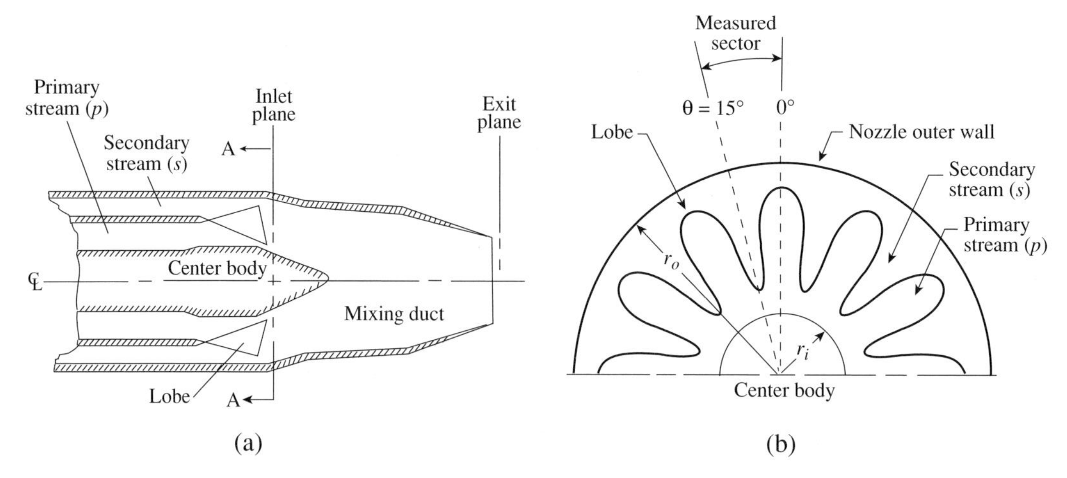

Figure 11.26: (a) Cross-section through a mixer nozzle showing axial measurement planes: mixing duct inlet plane (lobe exit) and nozzle exit plane; (b) nozzle at the mixing duct inlet measurement plane (section A–A of) (Paterson, 1982).

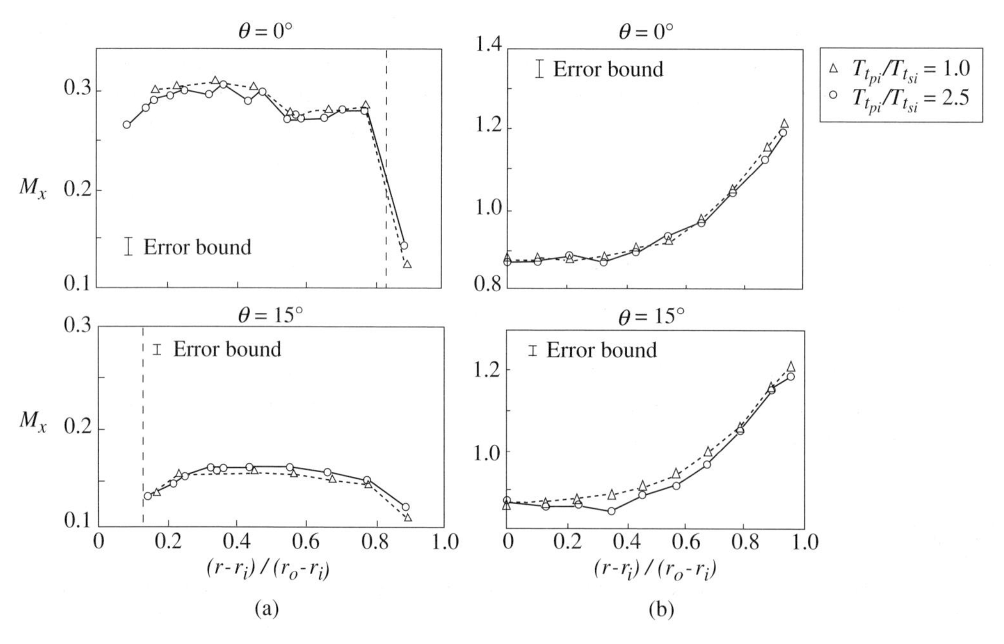

Figure 11.27: Radial distributions of axial Mach number at different circumferential positions at (a) the inlet and (b) the exit of a turbofan mixer nozzle mixing duct; vertical lines in (a) give the radial position of the lobe trailing edge (Greitzer *et al.*, 1985).

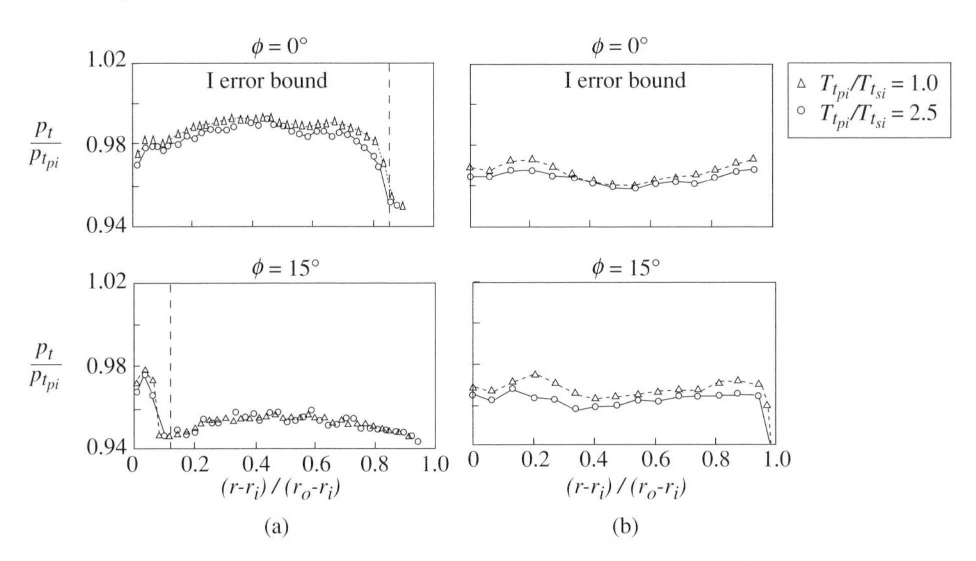

Figure 11.28: Radial distributions of stagnation pressure $p_t/p_{t_{pi}}$ for different circumferential positions at (a) the inlet and (b) the exit of a turbofan mixer nozzle mixing duct, $p_{t_{pi}}$ is primary stream inlet stagnation pressure; vertical lines in (a) give radial position of lobe trailing edge (Greitzer *et al.*, 1985).

Inlet and exit stagnation pressures in the nozzle are given in Figure 11.28. As for the Mach number profiles the exit stagnation pressure distributions are similar for the two sets of experiments.

To put the approximate principle in context it is helpful to categorize the flows addressed according to the processes that are important in setting the overall behavior. One limiting case is where pressure forces dominate, there is then little mixing, and most of the flow can be regarded as essentially isentropic. For this situation the basic Munk and Prim principle of Section 10.10 is a good descriptor. Flows in which both pressure and viscous shear forces (with the accompanying heat transfer) are important are another category, represented here by mixer nozzle flows. In these the conditions for the Munk and Prim principle are not satisfied and there is mixing throughout the flow. The approximate similarity principle, however, allows accurate scaling of isothermal results to non-isothermal situations with considerable accuracy, within the limitations that have been shown. The applicability of the proposed approximate scaling is better for a mixer nozzle than for a constant area duct because pressure forces are more important.

11.8.2 Jets

Another limiting case is a flow in which the dominant forces are viscous, with small or no pressure forces. The two-stream mixing flow and a jet issuing into a still fluid are motions of this type and we thus now assess the similarity for the jet. Figure 11.29 gives jet centerline stagnation pressure distributions (for one jet of a tandem jet configuration) at an exit Mach number of 0.6. Data are shown for ratios of jet initial stagnation temperature to ambient temperature of 1.0 and 2.7. Over an axial extent of 40 jet nozzle diameters, the cold and hot centerline stagnation pressures can be seen to be in close agreement. Radial distributions, given in Figure 11.30 for two downstream positions, show

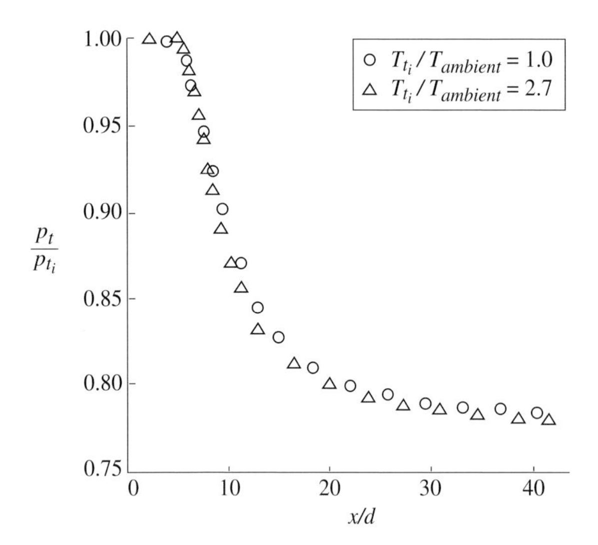

Figure 11.29: Centerline stagnation pressure distribution in a tandem jet experiment with heated and unheated jets; nozzle exit Mach number $= 0.6$, $T_{t_i}/T_{ambient}$ is the jet initial stagnation temperature/ambient temperature ratio, d is jet nozzle diameter (Greitzer *et al.*, 1985).

similar results. (It can be noted that the scaling is not valid for all aspects of the mixing. For example, the length of the potential core does depend on the temperature ratio (Lau, 1981).) Lepicovsky (1990) gives additional experimental information showing the application of the approximate substitution principle for heated jets.

11.8.3 Ejectors

A schematic of an ejector nozzle is given in Figure 11.31, showing the inflow conditions, a constant area mixing duct, and an exit diffuser.[9] Station 1 is at the start of the mixing duct, station 2 is at the end of the mixing duct, and station 3 is at the diffuser exit. A_p and A_s are the primary and secondary stream areas at station 1. Secondary (or outside) fluid is brought into the nozzle, mixes with a primary flow (for example in a jet engine) and is exhausted to ambient. Uses include plume attenuation, thrust augmentation, and noise reduction. An ejector relies on mixing for operation and its behavior provides another good test of the ideas presented concerning the independence of the Mach number and stagnation pressure on the stagnation temperature distribution (Presz and Greitzer, 1988).

The low Mach number analysis is again helpful for displaying overall trends. Ejectors are designed to have essentially complete mixing so the behavior is well described through a control volume treatment. We thus extend the analysis of Section 2.8 to include non-uniform temperatures. With reference to Figure 11.31 both primary and secondary streams are assumed isentropic from known stagnation conditions to the start of the mixing duct, with each stream uniform at this station. Mixing

[9] The notation used for the ejector is different than for the control volume, but it makes use of conventional ejector terminology including the mnemonically attractive p and s for primary and secondary streams.

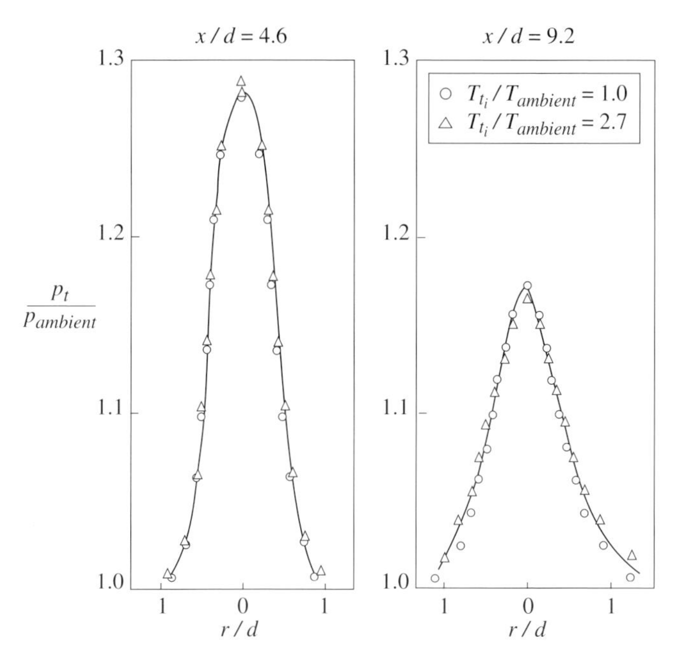

Figure 11.30: Radial stagnation pressure distribution in heated and unheated jets; nozzle exit Mach number = 0.6, d is jet nozzle diameter (Simonich and Schlinker, 1983).

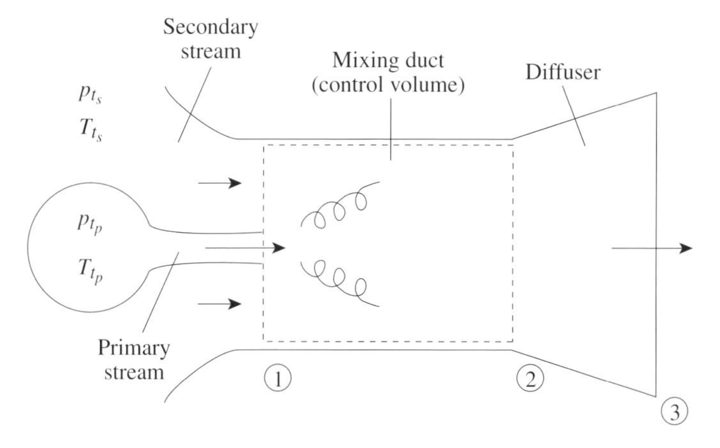

Figure 11.31: Schematic of an ejector showing primary and secondary streams and the stations used in analysis.

is taken as complete between stations 1 and 2 so the flow into the diffuser is uniform. The diffuser pressure rise performance is taken as isentropic. As previously, for $M^2 \ll 1$ the equation of state used is $\rho_p T_p = \rho_s T_s$, where the subscript p refers to the primary stream and the subscript s the secondary stream.

The necessary equations are the control volume statements for conservation of mass, momentum, and energy, plus the descriptions of lossless flow upstream and downstream of the control volume.

These can be combined into a single equation for the temperature corrected ejector pumping ratio (temperature corrected secondary/primary mass flow ratio), $\dot{m}_s\sqrt{T_s}/\dot{m}_p\sqrt{T_p}$,

$$\frac{T_s}{T_p}\left(\frac{\dot{m}_s}{\dot{m}_p}\right)^2\left[\left(\frac{A_p}{A_s}\right)^2+\left(\frac{A_2}{A_3}\right)^2\right]+\sqrt{\frac{T_s}{T_p}}\left(\frac{\dot{m}_s}{\dot{m}_p}\right)\left[1+\left(\frac{A_2}{A_3}\right)^2\right]\left[\sqrt{\frac{T_p}{T_s}}+\sqrt{\frac{T_s}{T_p}}\right]$$
$$+\left[\left(\frac{A_2}{A_3}\right)^2-1-2\left(\frac{A_s}{A_p}\right)\right]=0. \quad (11.8.1)$$

In (11.8.1) to order M^2 the stagnation and static temperature ratios are the same.

If we write $T_p = T_s + \Delta T$, to first order in $\Delta T/T_s$ the sum of the temperature ratio and its inverse, which appears in the middle term in (11.8.1) is

$$\sqrt{\frac{T_p}{T_s}}+\sqrt{\frac{T_s}{T_p}}=2. \quad (11.8.2)$$

Although this is true only for $\Delta T/T_s \ll 1$, the numerical results show that use of the approximation $\sqrt{T_p/T_s}+\sqrt{T_s/T_p}\approx 2$ has little effect on calculated ejector performance. Making use of this approximation in (11.8.2) yields a quadratic equation for the temperature-corrected ejector pumping ratio which is independent of primary/secondary temperature ratio and only depends on ejector geometry:

$$\left(\frac{\dot{m}_s}{\dot{m}_p}\right)^2\frac{T_s}{T_p}\left[\left(\frac{A_p}{A_s}\right)^2+\left(\frac{A_2}{A_3}\right)^2\right]+2\left(\frac{\dot{m}_s}{\dot{m}_p}\right)\sqrt{\frac{T_s}{T_p}}\left[1+\left(\frac{A_2}{A_3}\right)^2\right]$$
$$+\left[\left(\frac{A_2}{A_3}\right)^2-1-2\left(\frac{A_s}{A_p}\right)\right]=0. \quad (11.8.3)$$

Equation (11.8.3) is another example of the approximate similarity.

Figure 11.32 shows the pumping ratio for ejector primary/secondary temperature ratios of 0.5, 1, and 2.0. The upper part of the figure shows individual curves based on the control volume equations as in (11.7.14). The lower part of the figure shows the pumping parameter normalized by the square root of the temperatures, as suggested by (11.8.3), with the results nearly collapsed into a single curve.

Figure 11.33 presents data on ejector thrust augmentation (defined as the thrust of the ejector divided by the thrust of the primary stream expanded to the exit static pressure) for different primary/secondary temperature ratios. Eight different ejector tests are represented, with nozzle pressure ratios from 1.01 to 3.0. The abscissa is the stagnation temperature ratio of the primary/secondary stream and the ordinate is the thrust augmentation[10] normalized by the thrust augmentation for a temperature ratio of unity. The approximate principle implies no change in thrust augmentation with temperature ratio. There is a slight decrease in the thrust augmentation ratio with the temperature ratio, but the scaling does provide a good estimate for the ejector performance with a heated primary stream.

[10] Thrust augmentation is the total divided by the thrust of the primary stream expanded isentropically to exit conditions.

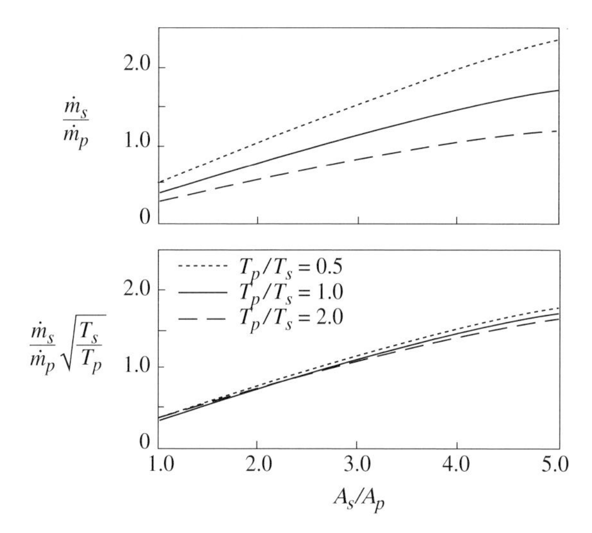

Figure 11.32: Ejector pumping as a function of secondary/primary pumping area ratio, A_s/A_p; $M^2 \ll 1$.

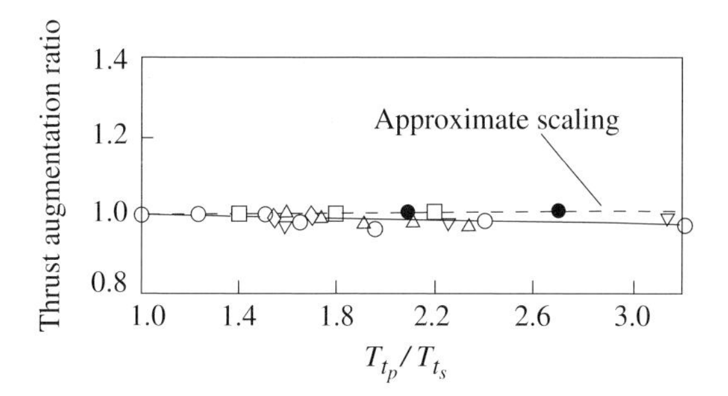

Figure 11.33: Ejector thrust augmentation ratio for different primary/secondary stagnation temperature ratios. See reference for original data sources (Presz and Greitzer, 1988).

11.8.4 Mixing of streams with non-uniform densities

The approximate similarity principle depends for its operation on the balance between the competing effects of heat transfer and shear stress work. There is an analogy between this balance and the processes that occur in the mixing of two streams of the same temperature but different densities. Consider for example coflowing streams of nitrogen and hydrogen, mixing in a constant area duct and having the same inlet stagnation pressure and temperature. Both streams have the same dynamic pressure so the hydrogen stream has a velocity which is $(28/2)^{1/2} = 3.7$ times larger than the nitrogen stream. The viscous work transfer from the high speed stream increases the stagnation pressure of the low speed nitrogen stream. However, the overall density of the stream which was initially pure nitrogen decreases, with the effect of (as implied by the arguments in Section 11.3) lowering the stagnation pressure. Control volume analysis for low Mach number flow constant area mixing shows the correspondence between the behavior with non-uniform inlet temperature and with non-uniform inlet density. The implication, namely that the approximate scaling ideas can be applied to describe flows with uniform temperature, but non-uniform density or a combination of non-uniform temperature and density, is borne out by the data shown by Presz and Greitzer (1988) for ejectors operating over a range of primary/secondary flow densities from 3.5 to less than 0.2.

11.8.5 Comments on the approximations

Several final points should be made concerning the application of these approximate scaling ideas. The first is that there are other analogies that can be made between the types of mixing processes which, for example, enable one to carry out experimental evaluation of a given process in a simpler fashion than studying the original phenomenon. An illustration is the use of mass addition to simulate the effect of heat addition to a supersonic flow. The degree to which the flow similarity for the two processes (suitably interpreted) holds has been demonstrated for both constant area and non-constant area ducts (Heiser *et al.*, 1995, 1996).

Second, although the approximate substitution principle provides an avenue to obtain information about device behavior with heated flow from knowledge of basically isothermal situations, it is important to understand the limitations of the statements that are made about flow similarity. Most evident are the limits on the temperature ratio range (or density ratio range) over which useful scaling applies; the way in which the approximation loses validity as the temperature ratio increases has been exhibited in a number of figures. Further even though there is approximate similarity in local rates of momentum and energy transfer, applications which focus on global quantities such as those for the mixer nozzles and ejectors, tend to be most appropriate. As commented on earlier, the details of mixing layers or of heated jets (both steady and unsteady) do not follow the approximate scaling. Finally, much of the information shown concerning limits of applicability has been for Mach numbers in the low supersonic regime or below. Information relating to the behavior of the stagnation pressure in shear layer mixing in a higher speed supersonic regime (convective Mach numbers up to 3.0) is given by Papamoschou (1994).

12 Non-uniform flow in fluid components

12.1 Introduction

In this chapter the discussion of fluid component and system response to disturbances, begun in Chapter 6, is extended to a broader class of flow non-uniformities. Whereas Chapter 6 considered primarily one-dimensional disturbances, that restriction is now dropped and we address more general (two- and three-dimensional) non-uniformities with variations transverse to the bulk flow direction. Examples of interest are turbomachines subjected to circumferentially varying inlet conditions and the behavior of components with geometry generated non-uniformity, such as is caused by a contraction or a bend in close proximity.

Three important issues relating to these situations can be identified. One is the effect of the fluid component on the flow non-uniformity, or *distortion*: how are the non-uniformities altered by passage through the component? A second is the effect of the non-uniformity on the component: how does the distortion modify the component performance? The approaches needed to address these two questions are fundamentally different. For the former, qualitative aspects, and even many quantitative features, can be resolved within the framework of a linearized description. For the latter, however, the problem is inherently nonlinear and a different level of analysis is needed. Beyond component performance there is a third issue. Because fluid components typically occur as part of an overall system, what changes in interactions with the rest of the system arise due to the non-uniformity?

Several integrating themes thread through the different applications discussed. The first is that fluid components do not passively accept non-uniform flow but play a major role in modifying the velocity distribution. A second concerns the type of non-uniformities to be discussed. Inlet conditions for internal flow components such as diffusers or nozzles are never actually uniform. Even with a uniform velocity core there are boundary layers on the bounding solid surfaces. As discussed in Chapter 4, there is a well-developed methodology for determining the effects of boundary layers on the performance of many fluid devices. The non-uniformities addressed in this chapter, however, have to do with variations more appropriately described as occurring outside the boundary layers. The length scales associated with these variations are typically on the order of the duct dimension (or the mean radius, for a device functioning in an annular region), and hence much larger than the boundary layer thickness. As such, features of these rotational flows can often be captured to a good approximation by an inviscid description, although we also examine situations in which this is not true and viscous effects must be considered throughout the flow domain.[1] The third theme is that because the distance over which the component's upstream influence is felt is set by the disturbance length scale (see Section 2.3), there can be strong coupling of fluid components in the presence of these non-uniformities.

A final theme concerns the level of approach to the physical situations dealt with in this chapter. In considering problems of non-uniform flow in fluid machinery, there is a large range of length and time scales, and consequently an immense amount of detail. In this type of situation it is often useful to develop simplified models which help focus on essential mechanisms.[2] Such modeling offers an effective approach to capturing features of engineering interest. Discussion in the chapter is thus aimed at providing not only insight into modeling but, more broadly, a conceptual framework for dealing with these problems including the rationale which underpins the relevant approximations.

We illustrate the above ideas with examples covering a range of fluid components: screens, diffusers, nozzles, and turbomachines. The treatments are based on constant density, but the analyses can be extended to the compressible regime.

12.2 An illustrative example of flow modeling: two-dimensional steady non-uniform flow through a screen

The overall concepts are introduced in the context of defining the effect of a screen or perforated plate on a steady, two-dimensional, velocity non-uniformity. This example is addressed in some depth to provide a framework for discussion of more complex fluid components. From examination of the flow through the screen one can see explicitly, and without lengthy description of specialized detail, key model elements and assumptions. Further, while specifics differ for other components, the approach carries through the range of applications so the features of the velocity and pressure fields for the non-uniform flow through a screen have relevance in the discussion of other devices. Finally, the problem allows assessment of the degree of fidelity of different levels of modeling including the influence of nonlinear effects, a subject we return to later in discussing turbomachinery behavior in non-uniform flow.

The first problem posed is to define desirable characteristics for a screen that is placed across (and normal to) a duct to reduce a velocity non-uniformity generated at some far upstream location. The geometry is depicted in Figure 12.1 in which the streamlines shown are plotted to scale for the conditions given in the figure, as described below.

It can be seen that there is flow redistribution in y, from upstream to downstream, caused by the interactions of the velocity non-uniformity and the screen. An argument for this redistribution can be made using the information about screens given in Section 6.6, namely that the pressure drop

[1] The viscous processes that are often responsible for the *creation* of a given non-uniformity involve a small force acting over a large distance. In this context, as argued in Chapter 9, many fluid components are short. On the length scale of the component in the flow direction, therefore, viscous effects have much less influence on the *modification* of the velocity field than do effects that are basically inviscid in nature.

[2] As stated by Anderson (1977): "Very often . . . a simplified model throws more light on the real workings of nature than any number of "ab initio" calculations of individual situations, which even where correct often contain so much detail as to conceal rather than reveal reality. It can be a disadvantage rather than an advantage to be able to compute or to measure too accurately, since often what one measures or computes is irrelevant in terms of mechanism. After all, the perfect computation simply reproduces Nature, does not explain her."

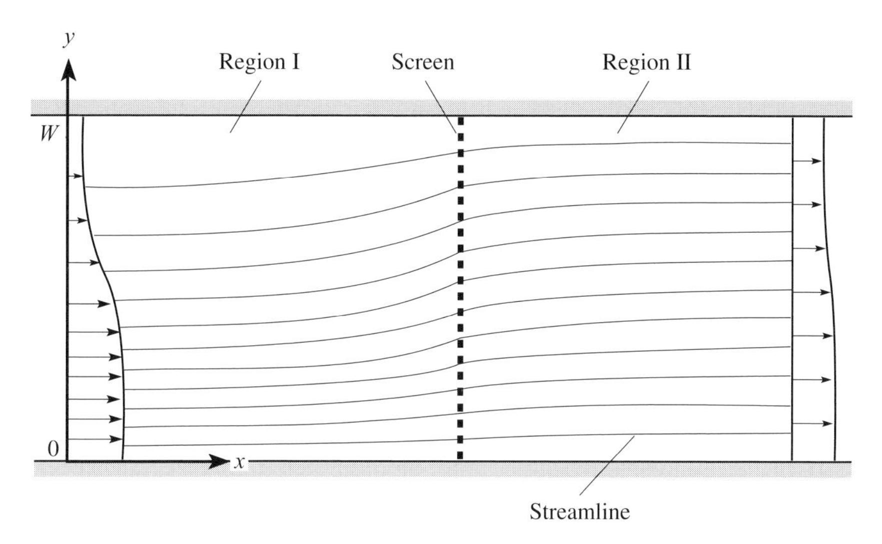

Figure 12.1: Non-uniform flow through a screen. Streamlines shown at equal values of ψ. Linearized analysis; $u_x(-\infty, y) = \overline{u}_x + A\cos(\pi y/W)$, $A = 0.67$, screen pressure drop coefficient $\mathcal{K} = 3$.

through the screen scales with the dynamic pressure of the incoming flow. Locally high velocity implies a higher pressure drop across the screen and thus a higher (than the mean) static pressure at the screen upstream face. Similarly, lower static pressure exists at the screen upstream face in locations of low inlet velocity. The resulting static pressure gradient normal to the duct centerline causes the streamline curvature shown in Figure 12.1.

To quantitatively define the screen geometry desired, a relationship is needed between screen properties and the reduction in the velocity non-uniformity. The approach to establish this is conceptually: (i) divide the flow field into regions upstream and downstream of the screen, (ii) develop descriptions of the velocity and pressure fields in each region, and (iii) apply "matching" conditions at the screen to insure compatibility of the two regions consistent with the local screen performance. In the initial examples we adopt a linearized description and restrict the discussion to incompressible flow. Although both of these constraints are readily removed, their adoption here allows explicit display of the generic flow features and resulting scaling laws.

12.2.1 Velocity and pressure field upstream of the screen

As in Figure 12.1 x- and y-coordinates denote distance along and normal to the duct respectively. The velocity components are written as a uniform "background flow" plus a disturbance:

$$u_x = \overline{u}_x + u'_x, \tag{12.2.1a}$$

$$u_y = u'_y. \tag{12.2.1b}$$

In (12.2.1) $\overline{u}_x$ is the x-component of velocity for the uniform background flow, u'_x and u'_y are the disturbance velocities (the non-uniformity), and $\overline{u}_y = 0$. The linearization to be described is based on $(u'_x/\overline{u}_x)$, $(u'_y/\overline{u}_x) \ll 1$. This allows the neglect of terms in the equations of motion which involve products of the disturbance velocities. In accord with the discussion of Section 12.1, the flow in the duct is taken as inviscid, with this assumption examined in Section 12.2.5.

The linearized equations of motion for the steady two-dimensional disturbance flow are:

$$\frac{\partial u'_x}{\partial x} + \frac{\partial u'_y}{\partial y} = 0, \tag{12.2.2a}$$

$$\overline{u}_x \frac{\partial u'_x}{\partial x} = -\frac{1}{\rho}\frac{\partial p'}{\partial x}, \tag{12.2.2b}$$

$$\overline{u}_x \frac{\partial u'_y}{\partial x} = -\frac{1}{\rho}\frac{\partial p'}{\partial y}. \tag{12.2.2c}$$

The solution sought has a specified velocity distribution at the far upstream station:

$$u_x(-\infty, y) = \overline{u}_x + u'_x(-\infty, y) = \text{known}. \tag{12.2.3}$$

The far downstream boundary condition (to be examined in the next section) is that the flow is parallel. From (12.2.2), which also hold in the downstream region, this means that the far downstream static pressure is uniform. The remaining boundary condition is the constraint of zero normal velocity at the upper and lower walls of the duct:

$$u_y(x, 0) = u_y(x, W) = 0; \quad -\infty < x < \infty. \tag{12.2.4}$$

For this two-dimensional problem, it is convenient to define a disturbance stream function, ψ_u (with the subscript u denoting the upstream region), as

$$u'_x = \frac{\partial \psi_u}{\partial y}, \qquad u'_y = -\frac{\partial \psi_u}{\partial x}. \tag{12.2.5}$$

The continuity equation is satisfied identically by this choice. There is only a z-component of vorticity, given by

$$\omega(x, y) = -\nabla^2 \psi_u. \tag{12.2.6}$$

The equation for ψ_u can be found by substituting (12.2.5) into (12.2.2b) and (12.2.2c), differentiating (12.2.2b) with respect to y and (12.2.2c) with respect to x, and subtracting the latter from the former, or by using (12.2.6) directly in the linearized two-dimensional form of the equation for the evolution of the vorticity. In either case we obtain,

$$\overline{u}_x \frac{\partial}{\partial x}(\nabla^2 \psi_u) = 0. \tag{12.2.7}$$

Equation (12.2.7) has the immediate integral

$$\nabla^2 \psi_u = f(y). \tag{12.2.8}$$

The physical content of (12.2.7) or (12.2.8) is that vorticity associated with the disturbance velocity field is viewed as convected unchanged along the background (or "mean") streamlines and as a function of y only. From Crocco's Theorem the stagnation pressure non-uniformity is also approximated as convected along the mean streamlines and as a function of y only.

The form of (12.2.8) implies the disturbance stream function is composed of two parts, the particular and the complementary solutions to the differential equation. The former (denoted as ψ_{u_p}) represents a rotational velocity field invariant with x and the latter (denoted[3] as ψ_{u_h}) represents an irrotational velocity field which is a function of both x and y:

$$\psi_u = \psi_{u_h} + \psi_{u_p}. \tag{12.2.9}$$

The two parts are governed by

$$\nabla^2 \psi_{u_h} = 0, \tag{12.2.10a}$$

$$\nabla^2 \psi_{u_p} = f(y), \tag{12.2.10b}$$

respectively, with their sum giving (12.2.8).

We now invoke the boundary condition. To minimize algebraic complexity, we choose a far upstream velocity profile anti-symmetric about the duct midplane, $y = W/2$:

$$u'_x = \sum_{k=1}^{\infty} A_k \cos\left(\frac{k\pi y}{W}\right). \tag{12.2.11}$$

The description is linear and the behavior of each Fourier component can be examined separately. For a given single (k^{th}) harmonic, the rotational component of the stream function (which is the part that is non-zero far upstream) derived from (12.2.11) and (12.2.5) is

$$\psi_{u_p} = \frac{A_k W}{k\pi}\left[\sin\left(\frac{k\pi y}{W}\right)\right]. \tag{12.2.12}$$

The velocity field represented by (12.2.11) and (12.2.12) is a parallel shear flow with no streamline curvature and uniform static pressure, which satisfies the boundary conditions at the top and bottom walls of the duct.

The irrotational component of the stream function must have a Fourier series form similar to the rotational part for the two to interact. From Section 2.3, or from substitution of the form $\psi_{u_h} = \sum_k B_k(x)\sin(k\pi y/W)$ in Laplace's equation (12.2.10a), ψ_{u_h} is found as

$$\psi_{u_h} = \sum_{k=1}^{\infty}\left(C_k e^{k\pi x/W} + D_k e^{-k\pi x/W}\right)\sin\left(\frac{k\pi y}{W}\right),$$

where C_k and D_k are constants to be determined. The velocity associated with ψ_{u_h} must be bounded everywhere, so the constants D_k are all zero and the form of ψ_{u_h} is

$$\psi_{u_h} = \sum_{k=1}^{\infty} C_k e^{k\pi x/W}\left[\sin\left(\frac{k\pi y}{W}\right)\right]. \tag{12.2.13}$$

The stream function ψ_{u_h} describes the alteration of the velocity profile from far upstream to the screen.

[3] The subscript "h" is used because the complementary solution is that for the homogeneous equation.

12.2.2 Flow in the downstream region

A similar procedure can be carried out to find the form of the stream function, $\psi_d \, (= \psi_{d_h} + \psi_{d_p})$ which describes the flow downstream of the screen. The boundary condition of uniform static pressure far downstream can be transformed into a condition on velocity using (12.2.2). The result is that $u'_y(y, \infty) = 0$. Applying the boundary condition we thus obtain

$$\psi_d = \sum_{k=1}^{\infty} \left[\left(E_k e^{-k\pi x/W} + F_k \right) \sin\left(\frac{k\pi y}{W} \right) \right]. \tag{12.2.14}$$

The term involving E_k is associated with the irrotational part of the downstream stream function while F_k is associated with the rotational part.

12.2.3 Matching conditions across the screen

Obtaining a description of the overall flow field involves solving for the three sets of constants, C_k, E_k, and F_k. To do this we need to develop the matching conditions that link the upstream and downstream stream functions. Three matching conditions are needed for each value of k. The first is mass conservation across the screen. For a uniform density flow this is the statement that the local x-components of velocity just upstream and just downstream of the screen are equal. With u'_{x_u} and u'_{x_d} denoting the x-components of velocity upstream and downstream of the screen:

$$u'_{x_u}(0, y) = u'_{x_d}(0, y). \tag{12.2.15}$$

The argument $(0, y)$ means that evaluation is carried out either just upstream or just downstream of the screen as appropriate, as indicated by the subscript. In terms of stream functions (12.2.15) is:

$$\frac{\partial \psi_u(0, y)}{\partial y} = \frac{\partial \psi_d(0, y)}{\partial y}. \tag{12.2.16}$$

The background velocity, $\overline{u}_x$, satisfies the relationship automatically and does not enter this matching condition.

The second matching relation is a statement about the screen characteristics which can be described as a refraction condition. The screen exerts a y-component force on the fluid, altering the flow angle as fluid moves through the screen. The local y-component of velocity at the screen exit is therefore less than that at the inlet as given by

$$u'_{y_d}(0, y) = \eta [u'_{y_u}(0, y)]. \tag{12.2.17}$$

In (12.2.17) η (< 1) is a property of the screen geometry and is independent of the disturbances. In this regard it is analogous to the screen pressure drop coefficient, $\mathcal{K}$, introduced in Section 6.6:

$$\Delta p = \mathcal{K} \tfrac{1}{2} \rho u^2. \tag{6.6.14}$$

There is considerable experimental information (see Laws and Livesy (1978)) which provides relationships between screen refraction and the pressure drop coefficient and the latter can be derived if the screen geometry is known. A relation between η and $\mathcal{K}$ for round wire screens suggested by

Laws and Livesy (based on data from Gibbings (1973)) is:

$$\eta = \left(\sqrt{\frac{\mathcal{K}^2}{16} + 1}\right) - \frac{\mathcal{K}}{4}. \tag{12.2.18}$$

In terms of disturbance stream functions the refraction matching condition is,

$$\frac{\partial \psi_d(0, y)}{\partial x} = \eta \left[\frac{\partial \psi_u(0, y)}{\partial x}\right]. \tag{12.2.19}$$

The two matching conditions developed so far are kinematic, but the third condition involves the pressure field and is dynamic. The physical statement is that the local pressure drop across the screen is proportional to the local upstream dynamic pressure. If $\mathcal{K}$ is the constant of proportionality (screen pressure drop coefficient), (6.6.14) becomes

$$p_u(0, y) - p_d(0, y) = \mathcal{K}\left\{\tfrac{1}{2}\rho[u_x(0, y)]^2\right\}. \tag{12.2.20}$$

The x-component of velocity is used in (12.2.20) because experiments show for inlet angles of up to 45° the pressure drop scales with inlet dynamic pressure based on the velocity component normal to the screen (Schubauer, Spangenberg, and Klebanoff, 1950).

To put (12.2.20) in terms of disturbance stream functions we write the pressure as the sum of the background value (independent of y and varying only across the screen) and the disturbance,[4] $p = \overline{p} + p'(x, y)$, and expand the right-hand side of (12.2.20) in terms of $u_x = \overline{u}_x + u'_x$. Doing this and neglecting terms which are second order in the disturbance quantities yields an equation that contains both the background pressure drop and the pressure drop associated with the disturbance. The former represents the pressure drop for a uniform flow, which is unaltered whatever the disturbance characteristics as long as linearity is maintained. The uniform flow pressure drop can thus be subtracted from the linearized pressure drop condition to provide the required relation between disturbance quantities only[5] as

$$p'_u(0, y) - p'_d(0, y) = \mathcal{K}\rho\overline{u}_x[u'_x(0, y)]. \tag{12.2.21}$$

No subscript is needed on the disturbance x-velocity component because the values just upstream and downstream of the screen are equal.

To complete the conversion to stream function form we differentiate (12.2.21) with respect to y. This can be done because the expression holds at every point across the screen. Incorporating the definition of the stream function in the y-component of the linearized momentum equation (12.2.2c) written as $(\partial p'/\partial y = \rho\overline{u}_x\partial^2\psi/\partial x^2)$ gives the third matching condition as

$$\frac{\partial^2 \psi_u(0, y)}{\partial x^2} - \frac{\partial^2 \psi_d(0, y)}{\partial x^2} = \mathcal{K}\left[\frac{\partial^2 \psi_u(0, y)}{\partial y^2}\right]. \tag{12.2.22}$$

Equations (12.2.16), (12.2.19), and (12.2.22), plus the prescription of the screen properties $\mathcal{K}$ and η, constitute the desired equations for the set of coefficients C_k, E_k, and F_k. Because the equations are linear, the equation for any Fourier component is independent of all other components and (12.2.16),

[4] In this incompressible flow, $\overline{p}$ is measured as the difference from a reference pressure, for example $\overline{p}_u$ or $\overline{p}_d$.

[5] This can also be argued by noting that in a linearized analysis the background flow, which is uniform in y, will not be coupled with any of the disturbance quantities, which all vary with y.

(12.2.19), and (12.2.22) are a set of three equations for each k. The solution is:

$$\psi_{u_k} = A_k\left(\frac{W}{k\pi}\right)\left(1 - \frac{\mathcal{K}}{1+\mathcal{K}+\eta}e^{k\pi x/W}\right)\sin\left(\frac{k\pi y}{W}\right), \tag{12.2.23a}$$

$$\psi_{d_k} = A_k\left(\frac{W}{k\pi}\right)\left(\frac{1+\eta-\eta\mathcal{K}}{1+\mathcal{K}+\eta} + \frac{\eta\mathcal{K}}{1+\mathcal{K}+\eta}e^{-k\pi x/W}\right)\sin\left(\frac{k\pi y}{W}\right). \tag{12.2.23b}$$

12.2.4 Overall features of the solution

The features exhibited by the solutions (12.2.23) are common to a variety of situations involving non-uniform flow in fluid components. The upstream disturbance consists of a rotational non-uniformity, convected unchanged by the mean velocity, plus an irrotational (or potential) disturbance which is the reaction of the screen to this rotational flow. Substituting ψ_u into the momentum equation it is seen that only the irrotational disturbance gives rise to deceleration, streamwise curvature, and static pressure variation. There is a decrease in the velocity in the high velocity region and an increase in velocity in the low velocity region, from far upstream to the screen, both of which are related (through Bernoulli's equation) to the pressure changes. The streamlines in Figure 12.1, which are from calculations for $\mathcal{K}$ of 3.0, show the streamline curvature in both upstream and downstream pressure fields, as well as the discontinuity in flow angle across the screen.[6]

The velocity at the screen for the kth Fourier component is given by

$$u'_{x_u}(0, y) = u'_{x_d}(0, y) = \frac{\partial \psi_u}{\partial y}(0, y) = A_k\left(\frac{1+\eta}{1+\eta+\mathcal{K}}\right) \tag{12.2.24}$$

and is composed of both rotational and irrotational disturbances.

The velocity far downstream is associated only with the rotational part of the downstream stream function. The ratio of the velocity non-uniformity far downstream to that far upstream is

$$\frac{u'_{x_d}(\infty, y)}{u'_{x_u}(-\infty, y)} = \frac{1+\eta-\eta\mathcal{K}}{1+\eta+\mathcal{K}}. \tag{12.2.25}$$

Figure 12.2 shows the ratio of downstream/upstream non-uniformity (12.2.25) plotted as a function of screen pressure drop, along with data from different sources. The data are for disturbances which are not sinusoidal but, because all Fourier components behave the same way, non-uniformities are unchanged in shape by the screen and the subscript k has been dropped. At a pressure drop coefficient of roughly 2.8 there is zero downstream non-uniformity; above that value the region of upstream velocity deficit has a velocity excess downstream.

The behavior is illustrated in more detail in Figures 12.3(a) and 12.3(b). In Figure 12.3(a), the value of $\mathcal{K}$ is such that the velocity defect is reduced, while in Figure 12.3(b) $\mathcal{K}$ is larger than 2.8 so that the far upstream velocity defect produces a far downstream velocity excess.

The reversal in profile shape is associated with downstream flow redistribution. Consider the behavior with an upstream profile as in Figure 12.1 for values of $\mathcal{K}$ much higher than 2.8. In this situation the axial velocity non-uniformity at the screen is small. This implies a substantial transverse velocity component just upstream of the screen and, because η is non-zero, just downstream of the

[6] A large disturbance amplitude (0.67) has been chosen to make the trends more apparent.

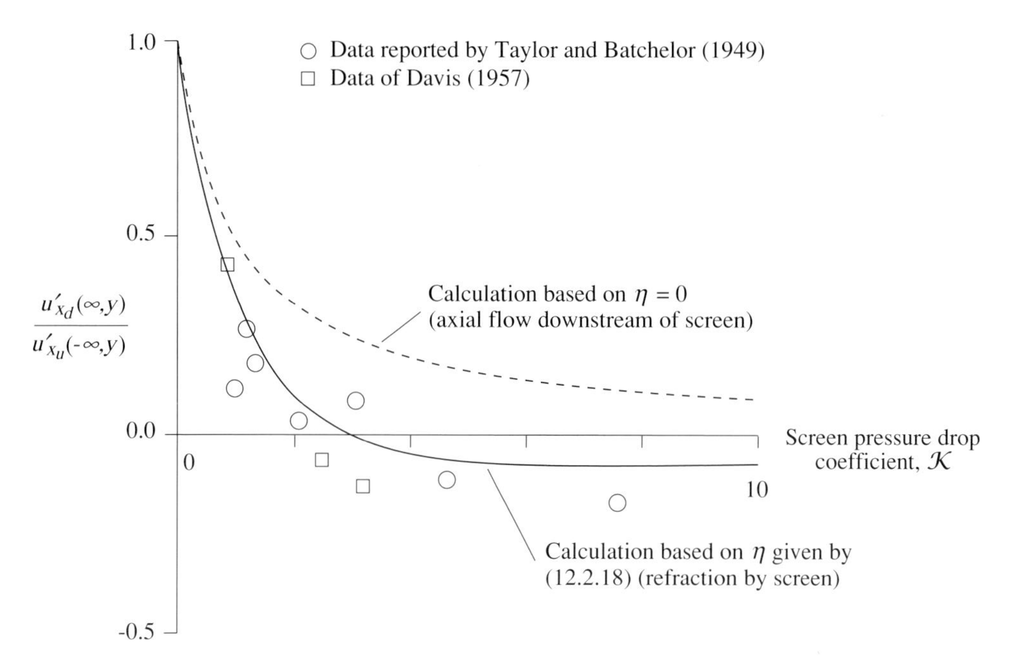

Figure 12.2: Ratio of upstream and downstream velocity non-uniformity for flow through a screen; linearized analysis.

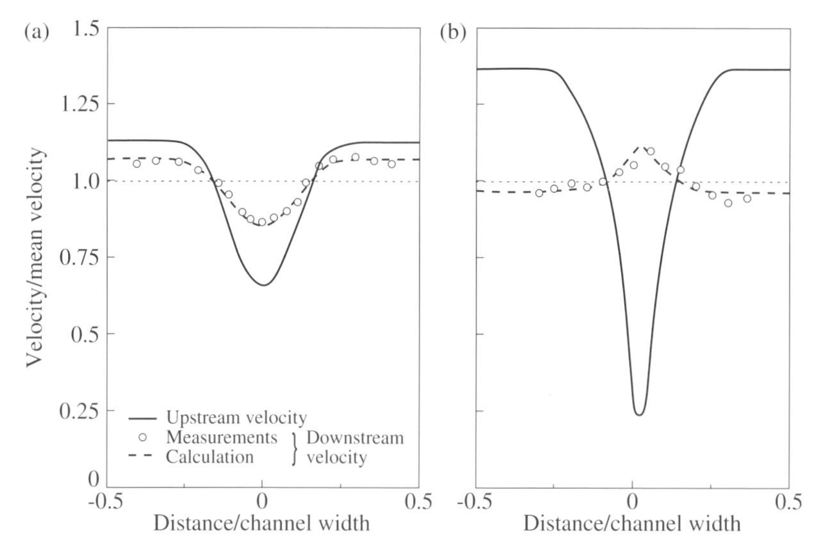

Figure 12.3: (a) Attenuation of velocity non-uniformity by screen; $\mathcal{K} = 0.88$, $\eta = 0.29$: (b) reversal of velocity non-uniformity by screen; $\mathcal{K} = 3.2$, $\eta = 0.34$ (Davis, 1957).

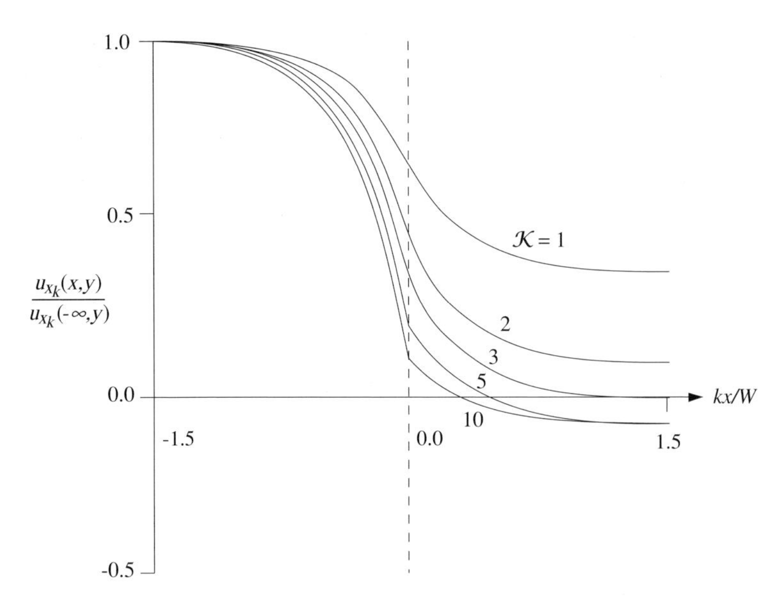

Figure 12.4: Region of influence of a screen; velocity at given y as a function of axial distance for cosinusoidal velocity non-uniformity ($u'_{x_k}(-\infty, y) = A_k \cos(k\pi W/y)$) through a uniform screen for different screen pressure drop coefficients $\mathcal{K}$; linearized analysis.

screen as well. The consequent adjustment downstream to parallel flow results in a low x-velocity region near the bottom wall and a high x-velocity region near the upper wall, relative to conditions at the screen. Achieving a zero far downstream velocity implies a balance between the effects of x-velocity attenuation upstream of the screen and the velocity adjustment that takes place downstream of the screen.

For a screen with a honeycomb or other flow straightening device as its downstream part, the refraction coefficient, η, is zero, and the behavior is different. There is no downstream redistribution and the axial velocity at the screen is the same as that far downstream. In this situation the velocity ratio between far downstream and far upstream decreases monotonically to zero with screen pressure drop, $\mathcal{K}$,

$$\frac{u'_{x_d}(+\infty, y)}{u'_{x_u}(-\infty, y)} = \frac{1}{1+\mathcal{K}} \quad \text{(for } \eta = 0\text{)}. \tag{12.2.26}$$

The refraction coefficient, η, thus controls the proportion of flow readjustment upstream and downstream. If $\eta = 0$, there is no irrotational downstream flow disturbance, no downstream static pressure variation, and no streamline curvature; all flow redistribution occurs upstream of the screen. This result finds important application in analysis of non-uniform flow in compressors.

The axial extent over which the screen affects the flow is also of interest. From the form of the solution (12.2.23), and as discussed in Section 2.3, for the kth Fourier component the region of influence is roughly $2W/(\pi k)$. In this distance the irrotational disturbance amplitude drops to e^{-2} of its value at $x = 0$. In many circumstances, the harmonic of most concern is the first so the region of influence is $|x| < W/(\pi/2)$. This is illustrated in Figure 12.4, which shows the magnitude of the velocity non-uniformity versus axial distance for different values of screen pressure drop coefficient,

$\mathcal{K}$. The degree to which the velocity is altered from far upstream to downstream varies strongly with $\mathcal{K}$ but the extent over which the flow field is affected by the screen is similar for the different values of $\mathcal{K}$.

The stagnation pressure non-uniformities upstream and downstream of the screen can be found from the velocity information. Far upstream and downstream the static pressure is uniform and the axial velocity non-uniformity corresponds directly to the stagnation pressure non-uniformity, as given by the linearized expression

$$p_t' = \rho \overline{u}_x u_x'. \tag{12.2.27}$$

Differences between far upstream and far downstream velocities reflect differences in stagnation pressure variation associated with the non-uniform loss generated across the screen.

12.2.5 Nonlinear effects

To assess the applicability of the linearizations that have been made we now examine the influence of nonlinearity. The two kinematic boundary conditions, no change in axial velocity across the screen and $u'_{y_d}(0, y) = \eta u'_{y_u}(0, y)$, are not altered, but the dynamic boundary condition is changed. One form of this boundary condition can be expressed as

$$\begin{aligned} \Delta p_{across\ screen} &= p_u(0, y) - p_d(0, y) \\ &= \mathcal{K} \times \text{upstream dynamic pressure } (0, y). \end{aligned} \tag{12.2.28}$$

Including the quadratic terms in the dynamic pressure, which involve transverse velocity components, means the static pressure drop and the stagnation pressure drop across the screen are no longer equal. The nonlinear problem also raises an issue which did not exist in the linear analysis concerning the appropriate measures of pressure drop and dynamic pressure to be employed. Relevant modeling questions are whether the pressure drop should be interpreted as the difference in stagnation or static pressure and whether the inlet dynamic pressure used to scale the pressure drop should include the y-velocity component, $u_y(0, y)$.

The first of these choices is basically a matter of convenience. The second choice, whether to include (u_y^2) in the non-dimensionalizing parameter, is a different matter. This is an issue that goes to the heart of what is the correct representation of the physical properties of the screen or, for that matter, any device, because defining the matching conditions is crucial in the modeling process. For a screen, the data of Schubauer *et al.* (1950) show the pressure drop scales with the dynamic pressure based on the component of velocity normal to the screen, up to approach angles of 45° from normal. This additional information underpins the adoption of $[\frac{1}{2}u_x^2(0, y)]$ as the relevant dynamic pressure[7] (with the proviso that the computed approach angles be examined to see whether they are in this range).

One figure of merit for the assessment of the importance of nonlinearity is the overall static pressure drop, from far upstream to far downstream, through the screen. In the linear approximation, the mean

[7] Three matching conditions are sufficient in a two-dimensional incompressible flow. For three-dimensional flow, another condition involving the third component of the velocity is needed. For compressible flow, a matching condition on a thermodynamic variable (in addition to pressure) is also necessary. For a screen, an appropriate condition would be that stagnation temperature is conserved across the screen.

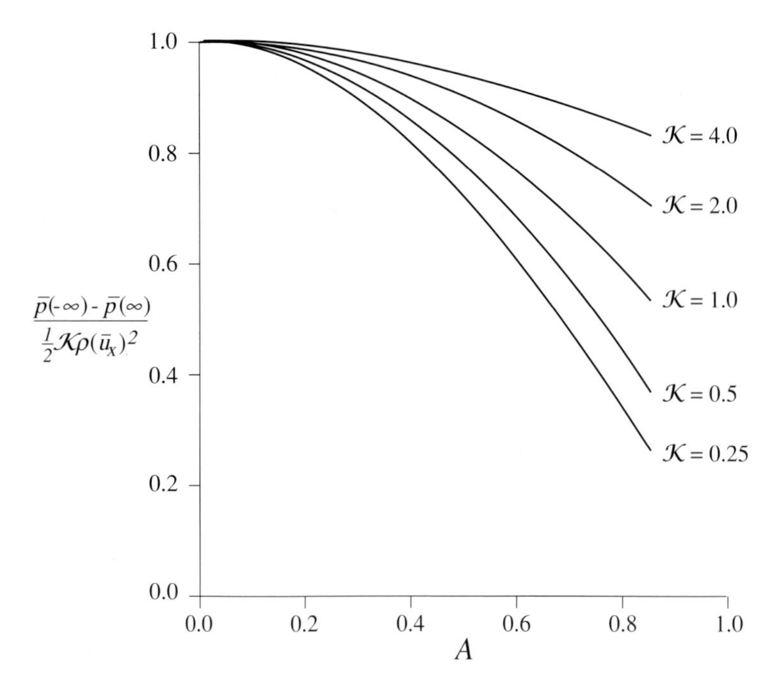

Figure 12.5: Effect of amplitude (A) on far upstream to far downstream static pressure drop $[\overline{p}(-\infty) - \overline{p}(\infty)]$ through a screen; two-dimensional channel with non-uniform flow; $u_x(-\infty, y) = \overline{u}_x + A\cos(\pi y/W)$.

static pressure at any axial location is just equal to the value for uniform flow and the static pressure difference from far upstream to far downstream is

$$\Delta\overline{p} = [\overline{p}(-\infty) - \overline{p}(\infty)] = \mathcal{K}\left(\tfrac{1}{2}\rho\overline{u}_x^2\right). \tag{12.2.29}$$

Nonlinearity increases the overall static pressure drop and the departure from the value given by (12.2.29) provides a measure of nonlinear effects.

Figure 12.5 shows a plot of the normalized pressure drop $[\overline{p}(-\infty) - \overline{p}(+\infty)]/(\frac{1}{2}\rho\overline{u}_x^2\mathcal{K})$ versus the amplitude of the non-uniformity, A, for different values of $\mathcal{K}$. This was obtained from numerical solutions for the two-dimensional steady flow through a screen. The geometry is that of Figure 12.1. A value of $A = 1.0$ implies a cosinusoidal upstream velocity distribution which goes from a minimum value of $u_x/\overline{u}_x = 0$ to a maximum of $u_x/\overline{u}_x = 2$. Even for relatively large amplitudes (up to $A = 0.4$, say), the departure from linear behavior in this problem is small.

12.2.6 Disturbance length scales and the assumption of inviscid flow

The velocity profile immediately downstream of the screen includes length scales that are associated with flow through the individual screen mesh elements. The desired "far downstream" profile, however, often has a characteristic length scale many times the screen mesh. It is because of the difference in length scales that we are able to simplify the level of flow description. The approximation that is made is portrayed in Figure 12.6, which gives a sketch of the local behavior just downstream of a round wire screen. Viewed in detail, the flow at the exit consists of high velocity jets plus wakes of near-zero velocity, as in the sudden expansion example described in Section 2.8. In terms of static

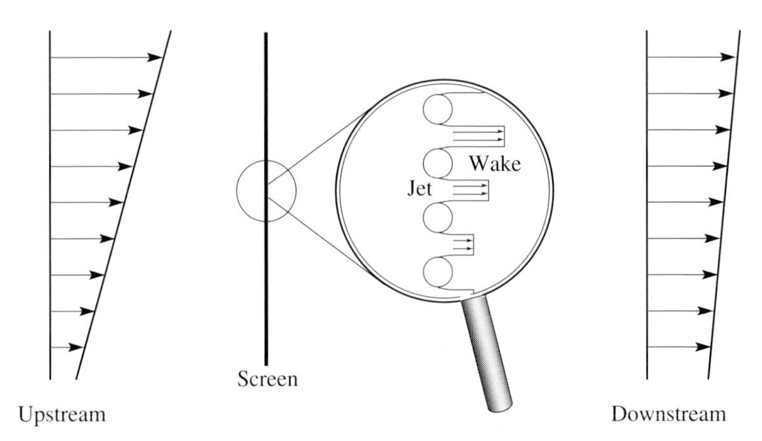

Figure 12.6: Length scales for flow through a screen; upstream and downstream large length scale flow non-uniformities and small length scale non-uniformities near the screen.

pressure the results of Section 2.8 imply that mixing is essentially complete in a distance roughly five times the mesh centerline spacing. At much larger downstream distances, therefore, there should be little trace of non-uniformity on the mesh length scale. More explicitly, if the screen mesh length scale is much smaller than the length scale of the upstream non-uniformity only phenomena with the latter scale need be computed. The details of the small scale jet mixing process can be modeled, for example (as here) lumped into the overall pressure and flow angle changes across the screen. Outside of the region in which the individual jets mix out, the details of the jet mixing process can be ignored. In terms of Figure 12.6, these would be the velocity non-uniformities labeled "upstream" and "downstream".

This approach applies to a wide variety of internal flow devices operating in non-uniform flow. It amounts to treating the flow through the device by smearing out small scale details with the implication that the local performance of the screen can be regarded as equivalent to the performance in a uniform flow at the local value of the inlet conditions. This, plus the concept that some flow field properties change discontinuously across the device, is known as the actuator disk (or, for a two-dimensional geometry, actuator strip) approximation. It has had wide application in the analysis of the behavior of screens, diffusers, nozzles, and turbomachines (see Section 6.8).

Suppose the length scale characterizing the variation in upstream velocity is some appreciable fraction of the duct width, W, say $0.1W$. To use an actuator disk approach, this length scale must be much larger than that associated with the screen mesh (or perforated plate hole spacing, or honeycomb cell size, etc.). The latter thus sets the lower limit on the length scale of motions the actuator disk approach can describe.

The assumption that viscous effects can be neglected in describing the long length scale non-uniformities also needs to be examined. To do this we compare the ratio of viscous and/or turbulent stresses to inertia forces. For industrial devices, laminar viscous forces are generally negligible compared to turbulent stresses. If the velocity non-uniformity is Δu and the transverse length scale of velocity variation is L_v, the magnitude of turbulent stresses is $\nu_t(\Delta u/L_v^2)$, where ν_t is a representative eddy viscosity. As discussed in Section 4.8 the eddy viscosity for a shear layer of velocity difference Δu and thickness L_v can be estimated as $0.014L_v\Delta u$ (Schlichting, 1979), so the ratio of turbulent stresses to inertia forces $\tau_{turb}/(\rho u \partial u/\partial x)$ is $0.014\ (W/L_v)(\Delta u/u)$. For a velocity length scale of

$0.1W$, this is roughly $0.1\Delta u/u$. If the quantity $\Delta u/u$ is less than unity, turbulent stresses are an order of magnitude or more smaller than inertial forces and the approximation of regarding the flow outside the screen as inviscid is valid.

12.3 Applications to creation of a velocity non-uniformity using screens

12.3.1 Flow through a uniform inclined screen

We can amplify the above ideas concerning non-uniform flow in fluid devices by examining two related applications. These are the creation of specified velocity or stagnation pressure non-uniformities (as might be done to assess the impact of inlet separation on compressor performance or of a wake or thick boundary layer on diffuser performance) and the suppression of diffuser separation by screens.

Section 12.2 implies that one way to create a specified velocity profile is with a screen that has a spatially varying pressure drop coefficient across the duct. Far upstream and far downstream of such a screen, the static pressure is uniform across the duct, but the downstream stagnation pressure and velocity are non-uniform. One configuration of this type is a screen that covers (or blocks) only part of the duct; particles that pass through the blocked section suffer a loss in stagnation pressure while those passing through the unblocked section suffer no loss. Another example is furnished by the flow through a uniform inclined screen spanning a duct, as in Figure 12.7. The screen creates a diversion of the flow in the upstream region, with the flow per unit screen area varying across the duct. Lower velocity through the screen, and hence lower stagnation pressure loss, exists in the lower part of the duct, with higher velocity and higher loss in the upper part. The resulting far downstream velocity profile is as sketched on the right-hand side of the diagram.

A quantitative description of this flow can be obtained by viewing the screen as an actuator disk and neglecting details of the flow on the length scale of individual meshes. The velocity field is again taken to be a uniform flow, with the constant x-component $\overline{u}_x$, plus a perturbation with components

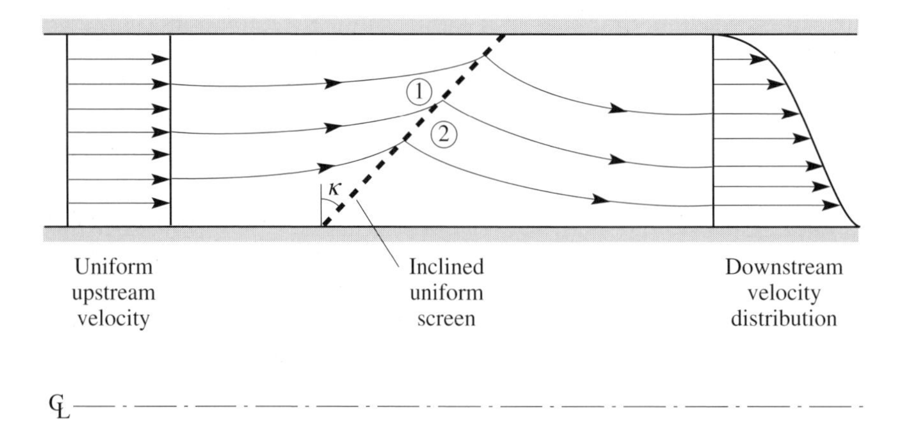

Figure 12.7: Flow field with an inclined screen in the duct; stations 1 and 2 refer to conditions just upstream and just downstream of the screen.

u'_x and u'_y respectively, which are small enough such that a linearized description can be adopted. The quantity that is the cause of the velocity non-uniformity is the inclination of the screen. Consistent with the sizes of the velocity non-uniformities, the angle of inclination of the screen, κ, can be assumed small and (analogous to the approximation of thin airfoil theory) the boundary conditions applied at $x = 0$.

The velocity components normal and tangential to the screen can be written in terms of x- and y-components as

$$u_n = u_x \cos\kappa - u_y \sin\kappa, \tag{12.3.1a}$$

$$u_t = u_x \sin\kappa + u_y \cos\kappa, \tag{12.3.1b}$$

where κ is the angle of screen inclination (see Figure 12.7). The linearized matching conditions across the screen are

$$u'_{n_u} = u'_{n_d}, \tag{12.3.2a}$$

$$\eta u'_{t_u} = u'_{t_d}, \tag{12.3.2b}$$

$$p'_u - p'_d \approx \mathcal{K}\rho\overline{u}_n u'_n, \tag{12.3.2c}$$

which can be expressed in terms of $\overline{u}_x$ and the small quantities κ, u'_x, and u'_y These are applied at $x = 0$.

The velocity and pressure field can be found by expanding as a Fourier series and solving term by term numerically, or one can use an elegant solution due to Elder (1959). The velocity u'_x as a function of y is plotted along with experimental data in Figure 12.8 for a range of screen inclinations from 10° to 45°. In qualitative accord with the remarks of Section 12.2, the agreement is excellent even for screen inclination angles one would not think of as small.

12.3.2 Pressure drop and velocity field with partial duct blockage

A common configuration is a screen or other blockage which covers a fraction of a duct as indicated in Figure 12.9 which shows a screen occupying half of a two-dimensional straight duct. The quantities of interest are the downstream velocity profile and the overall pressure drop.

In comparison to the problems described up to now there is a new aspect in that the conditions at the plane of the screen change discontinuously. For values of screen pressure drop, $\mathcal{K}$, larger than unity, and consequent large downstream non-uniformities, one would expect a linearized treatment not to be adequate.

The central idea, however, is as before. Flow fields upstream and downstream are defined and then matched across the plane at $x = 0$. The matching conditions are now explicitly functions of y; in the lower half of the duct they are such as to locally represent the flow though a screen of the appropriate porosity, while in the upper half of the duct pressures and velocities are continuous across $x = 0$. A stream function approach can still be used although the equations need to be solved numerically.

Solutions obtained for a screen that blocks half the duct are shown in Figure 12.9, which gives the velocity profile downstream of a screen of pressure drop $\mathcal{K} = 3.0$. Experimental results of Koo and James (1973) are also indicated. The solid line is a numerical computation employing the actuator disk approximation and using the relation between $\mathcal{K}$ and η given in (12.2.18). The far downstream

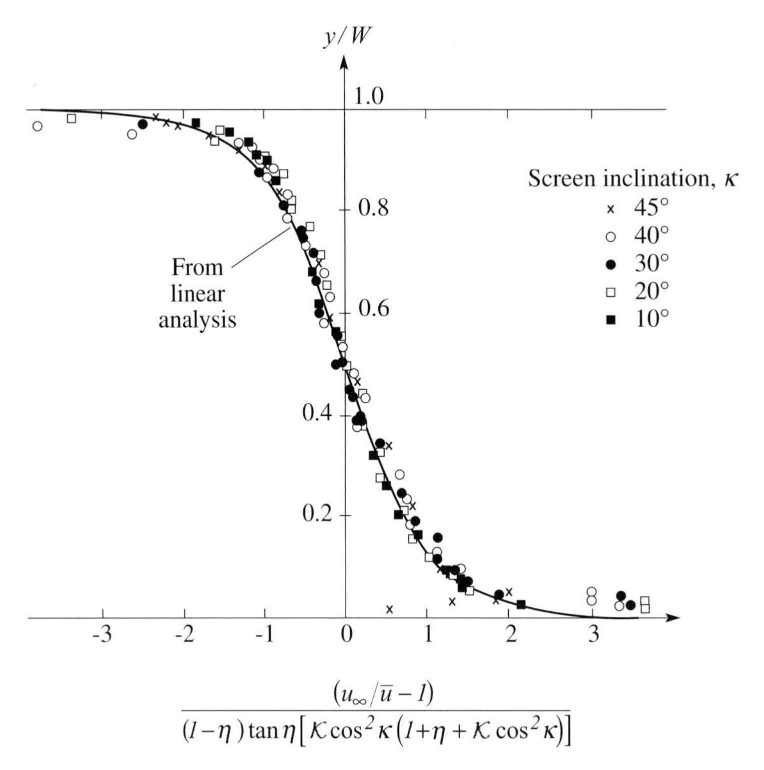

Figure 12.8: Velocity profile downstream of a uniform screen inclined to a uniform flow; $\mathcal{K} = 2.2$, $\eta = 0.78$ (Elder, 1959).

velocity field is essentially two streams of different velocity magnitudes and to a good approximation can be characterized by two quantities: (i) the ratio of far downstream velocity in the unblocked area u_∞ to the uniform far upstream velocity $\overline{u}_x$ and (ii) the ratio of the velocity in the blocked area, u_{b_∞}, to the far upstream velocity. The ratio of the duct area occupied by flow that has passed through the screen to the screen geometrical area, Λ_b = wake width/duct width, can be found from continuity as

$$\Lambda_b \frac{u_{b_\infty}}{\overline{u}_x} + (1 - \Lambda_b)\frac{u_\infty}{\overline{u}_x} = 1. \tag{12.3.3}$$

The streamline curvature generated by the presence of the screen causes the screen wake to be larger than the geometrical blockage; this can be important, for example, when considering placement of measurement stations.

The static pressure drop between far upstream and far downstream can be obtained by applying Bernoulli's equation to the stream which has not passed through the screen:

$$\frac{p_{-\infty} - p_\infty}{\frac{1}{2}\rho(u_{-\infty})^2} = \left(\frac{u_\infty}{\overline{u}_x}\right)^2 - 1.$$

There is a drop in stagnation pressure for particles that pass through the screen, so Bernoulli's equation cannot be used *across* the screen. As discussed, the viscous processes are characterized

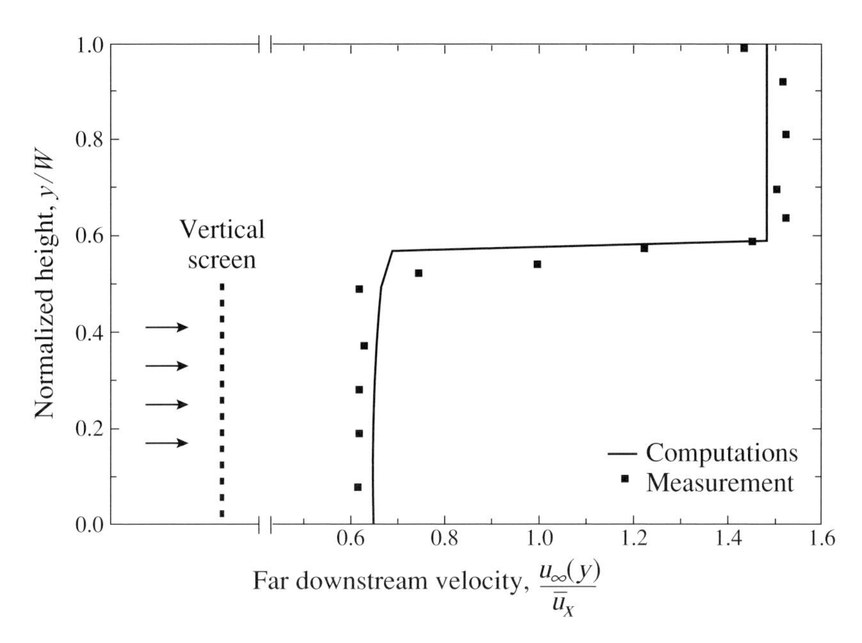

Figure 12.9: Velocity profile downstream of a partial blockage screen; screen extent/duct height = 0.5, screen pressure drop, $\mathcal{K} = 3.0$; measurements of Koo and James (1973).

as occurring "within" the screen (i.e. very near in a non-dimensional sense), and are not explicitly accounted for outside the screen.

Figure 12.10 shows computations and measurements of downstream velocities in the blocked and unblocked regions as a function of the screen pressure drop, for the geometry of Figure 12.9. As in the previous sections, the idea of an inviscid, rotational flow, coupled to an actuator disk model of the screen, gives a good description of the velocity profile, even with local regions of high turbulent stress (in the shear layer). As discussed in Section 12.2, the argument for this devolves on the issue of length scale; turbulent processes are important in setting internal features of the shear layer but the precise profile for the shear layer is not an important feature in this problem. Much further downstream (i.e. tens of duct heights), however, shear layer spreading would be felt across most of the duct and would need to be included in the description.

12.3.3 Enhancing flow uniformity in diffusing passages

Features of flow through screens lend themselves to application in enhancing velocity uniformity in passages with rapid diffusion. The context is one of creating a near-uniform velocity in a diffusing passage over a distance short enough that unseparated conditions cannot be maintained in the diffuser by itself. This would occur if the area and length of the diffusing passage were in the fully stalled regimes on the diffuser flow map of Figure 4.3 or in a sudden expansion.

The use of single or multiple screens in the diffuser can prevent or greatly suppress separation. The means by which this occurs can be inferred from Figure 12.1, interpreting the low velocity region as the boundary layer and the high velocity region as the free stream. The normal pressure gradient

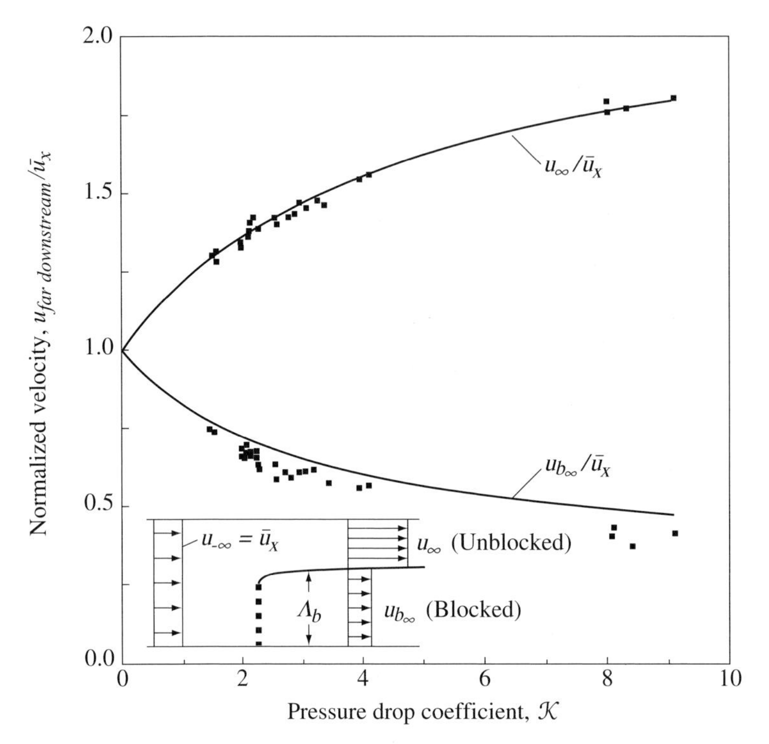

Figure 12.10: Velocity in the blocked, u_{b_∞}, and unblocked, u_∞, regions downstream of a partial blockage screen as a function of screen pressure drop, $\mathcal{K}$. Symbols are the measurements of Koo and James (1973); the solid line is a computation based on a nonlinear actuator disk analysis.

set up by the non-uniform flow through the screen causes streamline curvature and a convergence of the boundary layer streamlines. The boundary layer momentum thickness is thus decreased, the shear stress is increased, and the flow is enabled to remain attached even for diffusers that would otherwise be fully stalled. The effect can even be obtained if there is separated flow upstream of the screen. The suppression of separation does not come for free, because there is a stagnation pressure loss associated with the flow through the screen, but there are situations in which it is attractive to trade pressure drop for increased velocity uniformity.

Figures 12.11 and 12.12 show the results of experiments on a conical diffuser of area ratio 4 and a half-angle of approximately 45° (Schubauer and Spangenberg, 1948). Figure 12.11 pertains to the situation with no screen. Figure 12.11(a) shows the geometry, the measurement stations, and the measured streamlines, and indicates the proportion of total flow enclosed at different radii. Figure 12.11(b) gives the measured dynamic pressure (u^2/u^2_{ref} for this low Mach number flow, with u_{ref} a reference velocity) versus radius at the inlet station and at three locations within the diffuser. There is little change in the dynamic pressure distribution through the diffuser.

Figure 12.12 shows data for the same diffuser geometry but with three screens having pressure drop coefficients of 1.7, 1.9 and 0.9 respectively inserted in the diffuser normal to the mean flow direction at the indicated locations. The streamlines in Figure 12.12(a) show that the flow at the diffuser

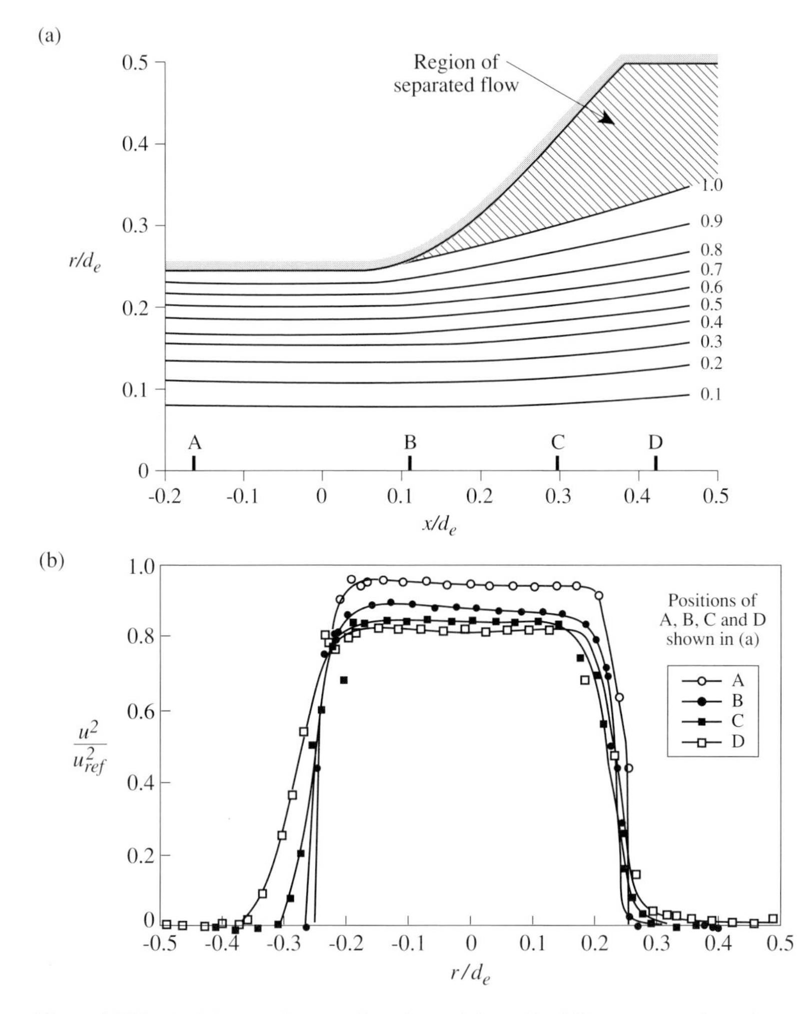

Figure 12.11: (a) Measured streamlines in a wide angle diffuser, area ratio = 4; numbers refer to the proportion of the total flow, letters refer to the location of measurement stations; (b) measured dynamic pressure versus radius in the diffuser (Schubauer and Spangenberg, 1948).

exit essentially fills the duct, and the plots of dynamic pressure versus radius in Figure 12.12(b) indicate a correspondingly much more uniform velocity than without the screen. The static pressure at the exit for the three-screen assemblage was roughly 0.2 inlet dynamic head below the diffuser inlet pressure.

The reference by Schubauer and Spangenberg (1948) gives information about flow behavior as the number of screens and the individual screen pressure drops are altered, as well as some comments on the general properties of the screen–diffuser combination. One finding emphasized is that multiple screens of low pressure drop give better results than a single screen of high pressure drop. Further guidelines for use of screens in wide angle diffusers, including the use of other than plane screens, have been developed by Mehta (1977), who presents a summary of design rules for their use.

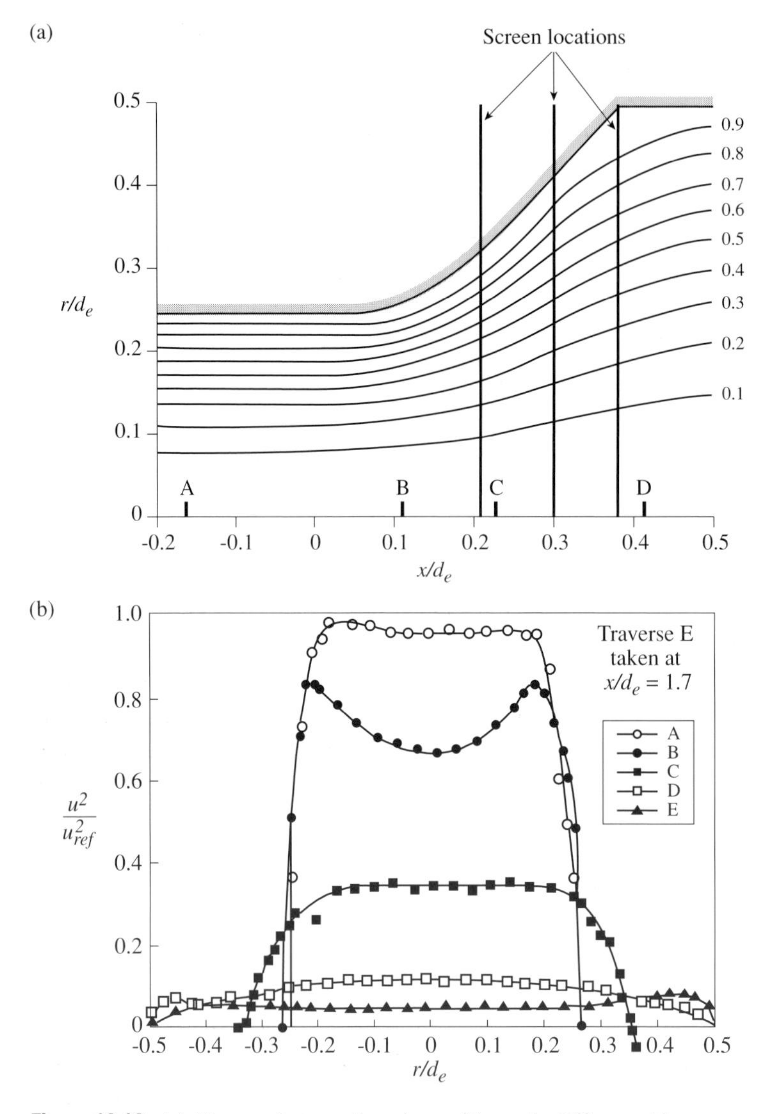

Figure 12.12: (a) Measured streamlines in a wide angle diffuser with a screen, area ratio = 4; numbers refer to the proportion of the total flow, letters refer to the location of measurement stations; (b) measured dynamic pressure versus radius in the wide angle diffuser–screen configuration of (a) (Schubauer and Spangenberg, 1948).

12.4 Upstream influence and component interaction

For incompressible flow without appreciable swirl, the upstream region of influence has been seen (Sections 2.3 and 12.2) to scale with the largest transverse length scale of the non-uniformity (here the duct height). The distance that components should be spaced so there is no upstream interaction is set by this length scale. From the linearized analysis a pressure non-uniformity decays with upstream

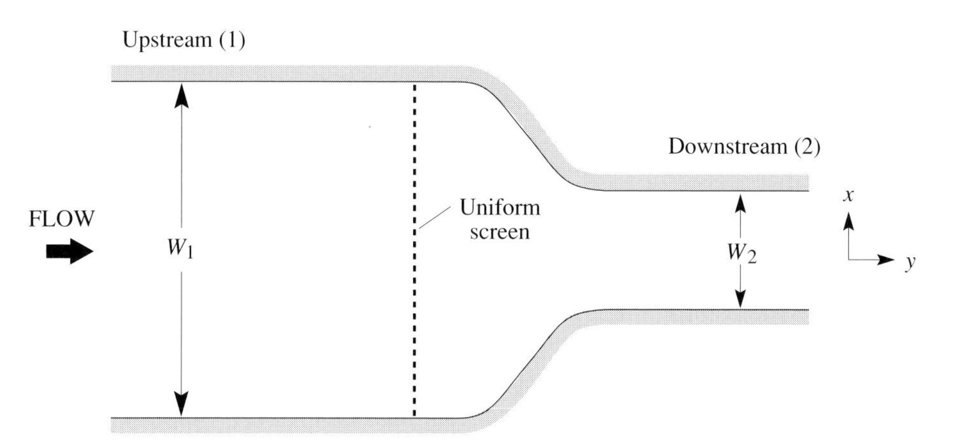

Figure 12.13: A uniform screen upstream of a two-dimensional contraction: $W_1 = 3W_2$.

distance as $(p' - \overline{p}) \propto e^{-\pi x/W}$. A spacing equal to duct width thus implies a factor of 20 decrease in the static pressure non-uniformity.

The freedom to space components far enough apart so there is negligible interaction often does not exist in fluid machinery and the magnitude and consequences of upstream influence when spacing is limited are relevant issues. An example is the design of inlet ducts for wind or cascade tunnels. Screens are used in the large area upstream sections to reduce the level of the test section velocity non-uniformity and a typical geometry is a screen, or series of screens, followed by a contraction. A two-dimensional configuration of this sort is shown in Figure 12.13, a duct with a screen in it followed by a 3:1 contraction. A design question is how close can the screen and contraction be placed without the occurrence of a substantial interaction.

At axial locations close to the contraction, the velocity near the top and bottom of the duct is lower than the velocity near the centerline. If the screen is near enough to the contraction, there will be a non-uniform normal velocity over its face and hence a non-uniform stagnation pressure drop. Screen locations where there is low velocity will have a lower stagnation pressure drop than locations with high velocity and the stagnation pressure in the downstream duct will be higher near the walls than in the center. The non-uniformity will be most severe if the screen is placed right against the contraction (as is essentially the case in some commercial air handling devices), and will decrease as the distance between the contraction and the screen is increased.

We analyze this problem in two steps, first examining the upstream influence of the contraction (because this is a geometry representative of a practical situation) and then determining the interaction between the screen and contraction. The computed static pressure field upstream of the contraction is shown in Figure 12.14, which gives the static pressure variation at different axial stations (the locations indicated by the arrows) versus duct height, non-dimensionalized by the dynamic pressure based on the mean axial velocity downstream of the contraction, $\overline{u}_{x_2}$.

The extent of appreciable variation is, as for the linearized analysis, roughly $2W/\pi$. The form of the static pressure distribution is less sinusoidal (more peaked) at axial locations near the contraction than at locations far away. The reason can be seen if one recalls that the velocity potential in the straight duct upstream of the contraction can be regarded as composed of different Fourier components, each of which obeys Laplace's equation, with the magnitude of the kth component given

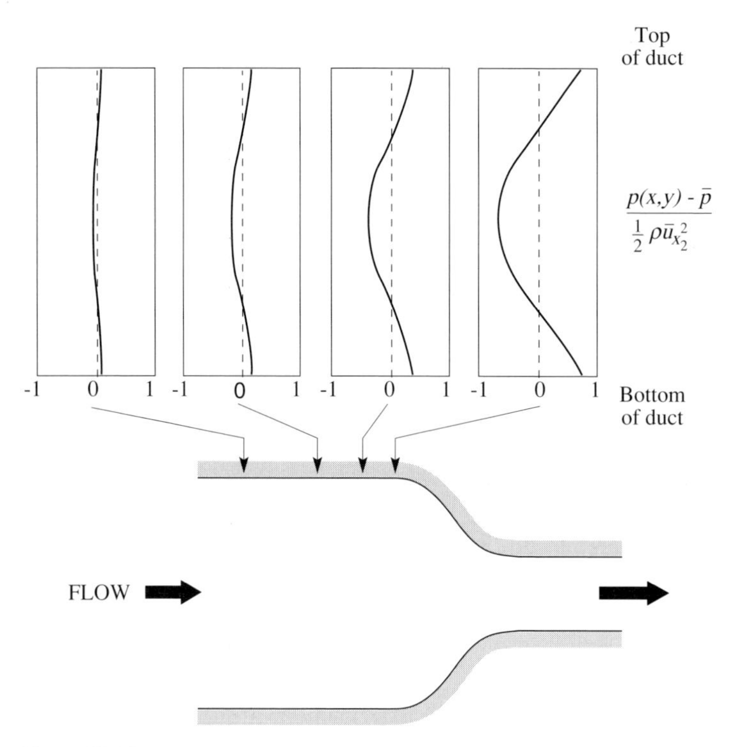

Figure 12.14: Static pressure non-uniformity upstream of a 3:1 contraction. The pressure distribution for the locations indicated: $x/W_1 = 0, -0.125, -0.25, -0.5$; $\bar{p}$ is the mean pressure at the indicated location.

by $|\phi_k| \sim |A_k e^{-k\pi/W}|$. The higher harmonics present near the contraction decrease more rapidly with distance and the influence seen farthest upstream is only the first harmonic. The pressure does not obey Laplace's equation for large amplitude variations, but the same qualitative considerations apply, with the farther upstream pressure profiles tending to purely sinusoidal form.

To illustrate the interaction between the screen and the contraction nonlinear actuator disk computations have been carried out for the screen–contraction configuration of Figure 12.14. The flow is not irrotational downstream of the screen because the stagnation pressure, and hence the vorticity (from Crocco's Theorem) is non-zero. The matching conditions used at the screen were those in (12.2.15), (12.2.17), and (12.2.28).

Figure 12.15 shows the velocity and stagnation pressure distribution in the "far downstream" uniform static pressure region of the downstream duct versus the distance across the duct for different screen locations. The velocity is non-dimensionalized by the mean axial velocity in the downstream duct, $\bar{u}_{x_2}$, and the stagnation pressure variation by the dynamic pressure based on this velocity. For screen locations closest to the contraction there is a substantial departure from uniformity in velocity and stagnation pressure with the non-uniformity decreasing to a small value as the screen locations move toward $W/2$. The profile corresponding to the screen position at $x = 0$ is discernibly non-sinusoidal, because the non-uniformity represents the effects of more than a single harmonic.

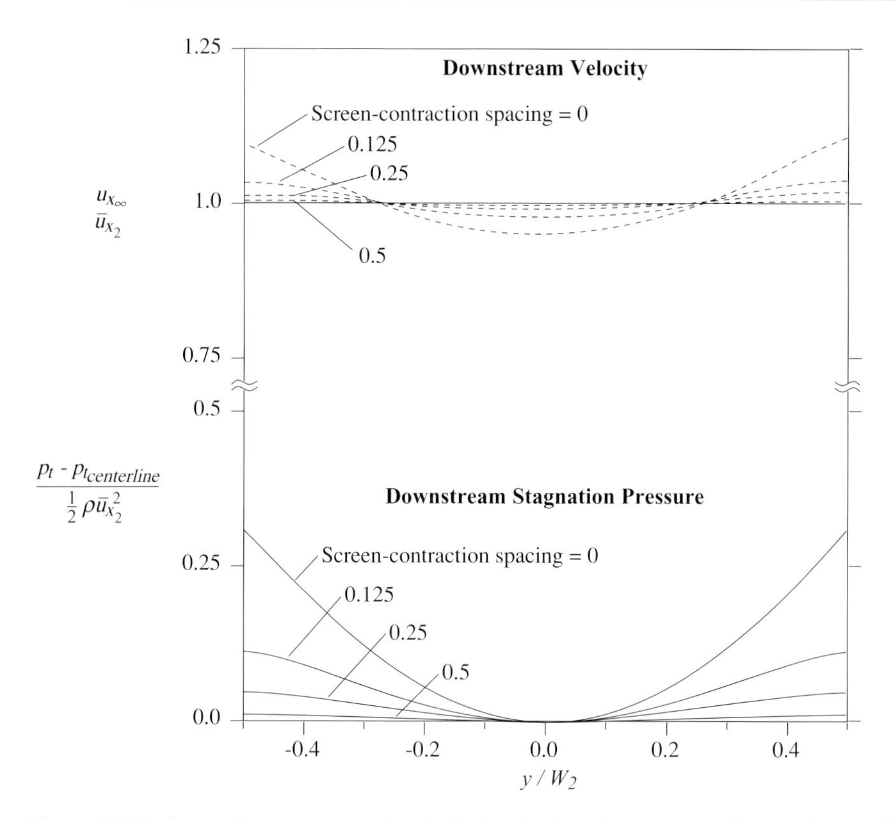

Figure 12.15: Stagnation pressure and velocity in the duct downstream of screen/contraction configuration. Curves refer to screen–contraction spacings in terms of upstream duct height, W_1, as in Figure 12.14.

12.5 Non-axisymmetric (asymmetric) flow in axial compressors

The concepts developed for the analysis of non-uniform flow through a screen can be extended to deal with the problem of turbomachinery operation in steady circumferentially non-uniform flow. If we restrict the geometries considered to a high enough hub/tip radius ratio so a two-dimensional approach can be adopted, the flow domain is as sketched in Figure 12.16, which shows an "unrolled" axial compressor and the upstream and downstream regions. As with the screen, descriptions of the flow in the regions outside the compressor are developed, with appropriate relations across the compressor to link the two descriptions and, unless specified, the flow regime has a low enough Mach number so an incompressible description can be used.

In axial compressors, there are often many pairs of rotor–stator combinations, or stages. Figure 12.16 gives a roughly scaled representation of the axial extent occupied by a compressor of five or so stages. Although the compressor is several times as wide as it is long, it is not obvious from the figure that it can be regarded as an actuator disk in the same sense that flow through a screen has been analyzed. Within the compressor, however, most of the axial length is taken up by blading

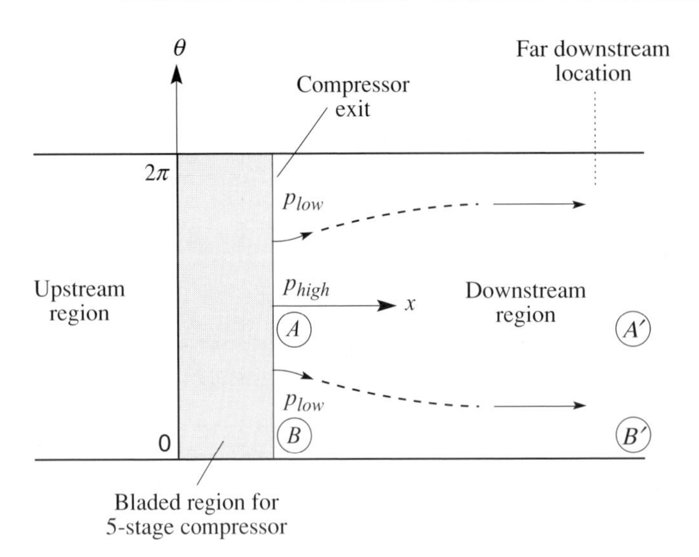

Figure 12.16: An "unrolled" compressor showing the argument for uniform static pressure downstream of axial compressor with a uniform exit flow angle. Note: only the downstream flow field is indicated.

which suppresses virtually all large-scale *circumferential* flow redistribution *within* the compressor (Cumpsty, 1989; Longley and Greitzer, 1992). The shape of the velocity distribution, as a function of θ, at the inlet of the compressor is therefore nearly the same as at the exit.[8]

Compressor response to a circumferentially non-uniform stagnation pressure is an important problem. For a stagnation pressure distribution far upstream of the compressor we wish to know the velocity field at the compressor face, because that velocity is what the blades actually experience. From the discussion of flow through screens we expect upstream flow redistribution and an alteration of the velocity field between far upstream and the compressor face, and the goal is to connect this alteration to compressor performance parameters. In the following sections we examine the flow in the upstream and downstream regions and develop appropriate matching conditions to do this.

12.5.1 Flow upstream of the compressor

Upstream of the compressor, the background flow is taken as axial, as is generally the case in practice. Viscous effects are neglected outside the compressor blade rows. The non-uniformities are viewed, for this initial discussion, as small enough that a linearized description can be adopted. There is no computational bar to analyzing nonlinear problems, but the linear analysis suffices for many situations and shows the general behavior simply. Discussion of nonlinear effects is given later in the chapter.

[8] To make this plausible, it may be helpful to think of the compressor blading as composed of a large number of passages, with no axial gaps between the rows. The difference in spatial extent of an axial velocity defect of large circumferential extent, say 90°, can only be of order one passage width. If the passage width is much less than the mean radius of the machine, the axial velocity distribution, $u_x(\theta)$, will be essentially the same at inlet and outlet.

The upstream flow field is similar to that described in Section 12.2, but it is useful to take a slightly different perspective and work directly in terms of the stagnation pressure non-uniformity which is a variable of great interest in turbomachinery. The stagnation pressure is convected unchanged along streamlines and enters naturally into the compressor boundary conditions.

The linearized equation that describes the behavior of the stagnation pressure non-uniformity, denoted by p_t', in the upstream flowfield is

$$\frac{Dp_t'}{Dt} = \overline{u}_x \frac{\partial p_t'}{\partial x} = 0. \tag{12.5.1}$$

Equation (12.5.1) states that stagnation pressure is convected unchanged along the background flow streamlines from far upstream, so $p_t' = p_t'(\theta)$. The variable θ is kept in these sections as a reminder that the transverse coordinate represents the position around the circumference, with the distance in the circumferential direction given by $r_m\theta$, where r_m is the turbomachine mean radius. We leave until later the further specification of the upstream flow field and turn now to the description of the downstream flow.

12.5.2 Flow downstream of the compressor

Within the compressor, the flow can be regarded as well guided by the rows of airfoils, with the relative angle at exit of each row varying little over a range of operating conditions. In particular, the flow angle at exit of the machine is taken as uniform around the circumference. This condition (which is analogous to the case of $\eta = 0$ for a screen) allows considerable simplification in the problem description because it implies the downstream static pressure is uniform. This was discussed in Section 12.2.4, but it is helpful to argue it afresh along lines similar to those for the subsonic nozzle in Section 2.5.

Consider a flow at the compressor exit which has a non-uniform velocity but a uniform leaving angle. Suppose the static pressure were non-uniform with a high pressure p_{high} over some part of the circumference, as indicated in Figure 12.16. If so, the streamline curvature near the compressor exit would be as sketched. Far downstream, however, there is nothing to cause streamline curvature, the streamlines are parallel, and the static pressure uniform. This scenario, however, is not self-consistent because to reach the supposed far downstream state the velocity would decrease from A to A' and increase from B to B', resulting in a static pressure difference between B' and A' that is larger than that between B and A, i.e. the far downstream static pressure non-uniformity would be larger than at the compressor exit. The only way out of the dilemma is to withdraw the supposition concerning non-uniform static pressure at the compressor exit. The conclusion developed is thus that the downstream static pressure is uniform if the flow angle leaving the compressor is uniform.

The condition of uniform static pressure at the compressor exit is a substantial simplification for the analysis and it is worthwhile to emphasize that it is applied to a steady two-dimensional flow with a uniform angle at the "entrance" to the downstream region and a downstream duct which is straight and has a constant area. We will see later in the chapter that asymmetric flow with a diffuser or nozzle downstream of a compressor can lead to circumferentially non-uniform static pressure, even with a uniform flow angle at the compressor exit.

12.5.3 Matching conditions across the compressor

Because the downstream static pressure is uniform there is no irrotational downstream disturbance and only two matching conditions are needed across the compressor. The first is that the background axial velocity and the axial velocity non-uniformity are the same at the inlet and exit:

$$\overline{u}_{x_i} = \overline{u}_{x_e}, \tag{12.5.2a}$$

$$u'_x(\theta)_i = u'_x(\theta)_e. \tag{12.5.2b}$$

The second relation concerns the compressor pressure rise. Several levels of sophistication are possible but the model used here (at least initially) is a compressor with a row of inlet guide vanes of fixed leaving angle so the compressor pressure rise is a function of axial velocity only. The axial velocity and the compressor pressure rise vary around the circumference. The approximation made concerning compressor performance is similar to that for the screen, namely that the local pressure rise of the compressor is the pressure rise corresponding to the local value of the axial velocity.[9] As in Section 12.2, its application is appropriate for non-uniformities with transverse length scales much larger than the scale of a blade passage.

The compressor pressure rise ($\Delta P_{t/s}$) is represented as the difference between the *exit static* pressure and the *inlet stagnation* pressure:

$$\Delta P_{t/s}(\theta) = p_e - p_{t_i}. \tag{12.5.3}$$

This representation is convenient because it makes explicit use of the uniformity in exit static pressure (Stenning, 1980; Longley and Greitzer, 1992).

In a linearized description, the pressure rise across the compressor consists of a (circumferentially-averaged) mean, denoted by ($\overline{\ }$), plus a perturbation. For a small amplitude non-uniformity in axial flow, u'_x, the compressor pressure rise can be written as

$$\Delta P_{t/s}(\theta) = \overline{\Delta P_{t/s}} + \left(\frac{\overline{d\Delta P_{t/s}}}{du_x}\right) u'_x(0, \theta). \tag{12.5.4}$$

In (12.5.4) ($\overline{\Delta P_{t/s}/du_x}$) is the slope of the curve of pressure rise versus axial velocity for a uniform flow, evaluated at a given mean operating point, and $u'_x(0, \theta)$ is the local axial velocity at the compressor inlet face, $x = 0$. Equation (12.5.4) corresponds to a Taylor series expansion about the mean conditions of pressure rise and axial velocity, $\overline{\Delta P_{t/s}}$ and $\overline{u}_x$.

If we subtract the mean flow quantities, (12.5.3) and (12.5.4) can be combined into a compact and useful relation between non-uniformities in the axial velocity at the compressor face, the exit static pressure and the compressor inlet stagnation pressure:

$$p'_e - p'_{t_i} = \left(\frac{\overline{d\Delta P_{t/s}}}{du_x}\right) u'_x(0, \theta). \tag{12.5.5}$$

[9] This implies that the rotor blades respond quasi-steadily to the unsteady flow field they experience. This approximation will be removed in Section 12.6.

The exit static pressure non-uniformity is zero, however, so (12.5.5) gives the compressor face velocity as

$$u'_{x_i}(\theta) = -\frac{p'_{t_i}(\theta)}{\left(\dfrac{d\Delta P_{t/s}}{du_x}\right)} = -\frac{p'_{t-\infty}(\theta)}{\left(\dfrac{d\Delta P_{t/s}}{du_x}\right)}. \tag{12.5.6}$$

Equation (12.5.6) can be put in non-dimensional form using as reference velocity the mean rotor speed,[10] Ωr_m. Defining the pressure rise coefficient $\Psi_{t/s} = \Delta P_{t/s}/[\rho(\Omega r_m)^2]$ and mean and disturbance flow coefficients $\Phi = \overline{u}_x/\Omega r_m$ and $\Delta\Phi = \phi' = u'_x/\Omega r_m$,

$$\frac{u'_x(0,\theta)}{\Omega r_m} = -\frac{\left[p'_{t-\infty}/\rho(\Omega r_m)^2\right]}{\left[\dfrac{d\Delta P_{t/s}/\rho(\Omega r_m)^2}{du_x/\Omega r_m}\right]} \tag{12.5.7a}$$

or

$$\phi'(0,\theta) = -\frac{\left[p'_{t-\infty}/\rho(\Omega r_m)^2\right]}{\left(\dfrac{d\Psi_{t/s}}{d\Phi}\right)}. \tag{12.5.7b}$$

Equation (12.5.7) gives the velocity non-uniformity at the compressor in terms of the upstream stagnation pressure non-uniformity and the slope of the non-dimensional compressor pressure rise characteristic, $\Psi_{t/s}$.

12.5.4 Behavior of the axial velocity and upstream static pressure

A graphical representation of the above solution for the compressor inlet velocity non-uniformity is shown in Figure 12.17. The abscissa is the flow coefficient, Φ, and the ordinate is the non-dimensional pressure rise, inlet stagnation pressure to exit static pressure. For a given far upstream stagnation pressure non-uniformity, $p'_{t-\infty}(\theta)$, the axial velocity at the compressor face is given by the horizontal distance *a–b*. If the slope of the constant speed compressor pressure rise characteristic (known as a "compressor speedline") is steep, there is a smaller velocity non-uniformity at the compressor face (C_1) than if the pressure rise versus flow curve is flat (C_2). For two compressors with the same pressure rise but different slopes, the one with the steeper slope has the smaller velocity non-uniformity. For a given compressor, the characteristic flattens out near stall, so as stall is approached there is also less attenuation of the far upstream velocity non-uniformity.

As with non-uniform flow through a screen, the change in axial velocity from far upstream to the compressor face is associated with an upstream static pressure variation. The compressor face static

[10] Although Ωr_m is not explicitly in the problem as formulated, the non-dimensional parameters defined are in common use in compressor aerodynamics.

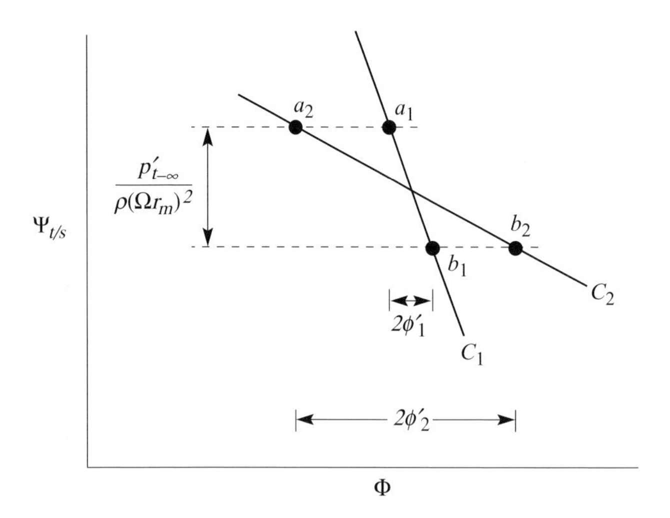

Figure 12.17: Velocity non-uniformity at the compressor for a specified (far upstream) stagnation pressure non-uniformity $p'_{t-\infty}$. The increased slope of the compressor characteristic C_1 versus C_2 implies a decreased velocity non-uniformity at compressor.

pressure is given by

$$\frac{p'(0,\theta)}{p'_{t-\infty}} = 1 + \frac{\Phi}{\left(\dfrac{d\Psi_{t/s}}{d\Phi}\right)}. \tag{12.5.8}$$

There is an upstream circumferential velocity associated with the irrotational part of the flow. For a stagnation pressure non-uniformity of the form $p'_{t-\infty}(\theta) = A_k \sin k\theta$ the kth harmonic of the circumferential velocity at the inlet can be obtained from the linearized θ-momentum equation as

$$u'_\theta(0,\theta) = A_k \cos k\theta \left[1 + \frac{\Phi}{\left(\dfrac{d\Psi_{t/s}}{d\Phi}\right)}\right]. \tag{12.5.9}$$

As in Section 12.2, the static pressure and the irrotational part of the velocity field obey Laplace's equation with exponential upstream decay.

Figure 12.18 presents measurements of the magnitude of the static pressure variation upstream of a gas turbine engine subjected to a stagnation pressure non-uniformity created by a screen of 180° circumferential extent. There is good agreement between the measured decay in amplitude of the static pressure variation and the exponential decay curve based on the first Fourier component response of the simple analysis.

For incompressible flow the inlet and exit axial velocity distributions are the same. Since the exit static pressure is uniform, the amplitude of the inlet axial velocity non-uniformity determines the stagnation pressure non-uniformity at the compressor exit. Figure 12.19 shows theoretical and experimental results for the magnitudes of stagnation and static pressure non-uniformities at the inlet,

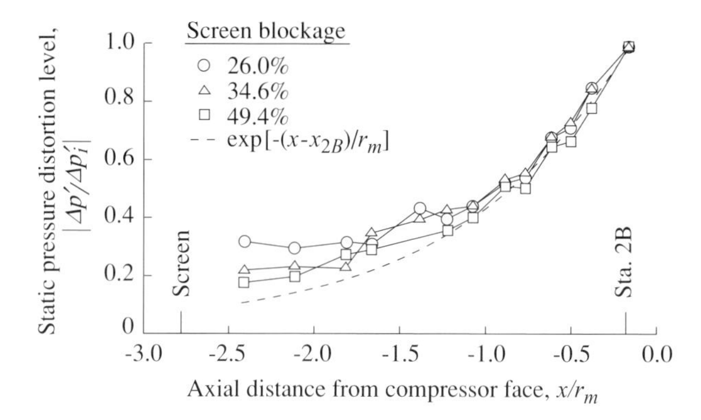

Figure 12.18: Variation of static pressure non-uniformity magnitude with the distance upstream of an axial compressor; 180° inlet total pressure distortion (Soeder and Bobula, 1979).

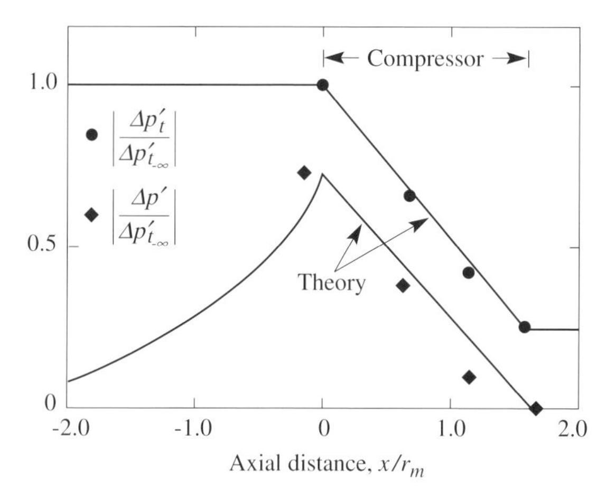

Figure 12.19: Variation of stagnation and static pressure magnitudes upstream and through a three-stage compressor (Stenning, 1980).

the exit, and inside a three-stage compressor. The stagnation and static pressure non-uniformities march in step through the compressor and essentially zero static pressure variation exists at the compressor exit as argued earlier.

The ideas presented can be adapted to a simple one-dimensional analysis of the response of a turbomachine to a stagnation pressure non-uniformity of arbitrary amplitude. Consider the behavior of the compressor when operating with two streams of differing stagnation pressures, as could be achieved using a screen that partially blocks the annulus (sketched on the left-hand side of Figure 12.20). The right-hand side of the figure shows a graphical solution for the axial velocity variation at the compressor inlet, with a specified mean flow, indicated by the solid circle, and a specified magnitude of the stagnation pressure variation. The local representation we have been using implies that the compressor can be viewed as operating at two different points, one at low flow and one at high flow, indicated by the two open circles. The exit static pressure is uniform, so the difference in the ordinate of the two operating points is equal to the difference in the inlet stagnation

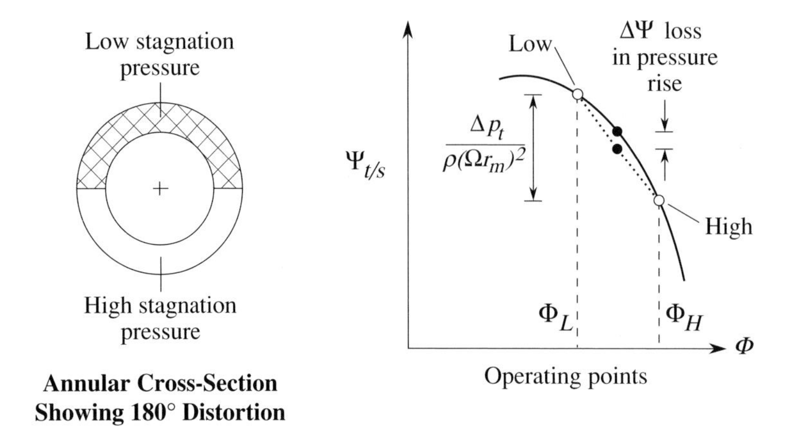

Figure 12.20: Parallel compressor model for the compressor response to circumferential stagnation pressure distortion.

pressure. The region of the compressor with low inlet stagnation pressure provides a higher pressure rise than the region with high stagnation pressure.

Figure 12.20 shows an aspect of compressor behavior with non-uniform flow which a linearized analysis does not capture. Because of the curvature of the compressor pressure rise characteristic the mean pressure rise is below that achieved with uniform flow at the mean flow rate. This is indicated in the figure by the quantity $\Delta\Psi$ which denotes the loss in pressure rise capability due to the distortion. This one-dimensional approach, which views the compressor as two (or more) compressors which exhaust to a uniform static pressure, is often referred to as the parallel compressor approximation. The compressible flow extension of this model has been widely used to assess compressor response to stagnation pressure distortion (Cumpsty, 1989; Longley and Greitzer, 1992).

12.5.5 Generation of non-uniform flow by circumferentially varying tip clearance

The local quasi-steady description of the compressor response just developed can be applied to other problems of asymmetric flow in turbomachines. One of these is the effect of circumferentially non-uniform tip clearance, as might occur from case ovalization or rotor non-concentricity. Tip clearance has a strong effect on compressor performance. As an example, Figure 12.21 shows data from a six-stage axial compressor at different values of rotor tip clearance (Cumpsty, 1989). The abscissa is mass flow (axial Mach number) and the two variables on the ordinate are the overall pressure ratio p_{t_e}/p_{t_i} and the adiabatic efficiency. At constant speed (say 100% of design) the peak stagnation pressure rise delivered by the compressor, p_{t_e}/p_{t_i}, decreases by roughly 25% for an increase in rotor tip clearance of 2.5% of the chord of the last rotor in the machine. There is also a substantial decrease in efficiency. The magnitude of the decrease varies with the specific machine geometry, and can be less than that shown, but values of 3–6% in peak pressure rise for each 1% increase in rotor clearance to chord are quoted as representative (Wisler, 1998).

If the variation in compressor performance with a tip clearance in an axisymmetric situation is known, we can use the approach for analyzing inlet distortion to predict compressor performance with a tip clearance that varies around the circumference. Denoting the non-dimensional tip clearance

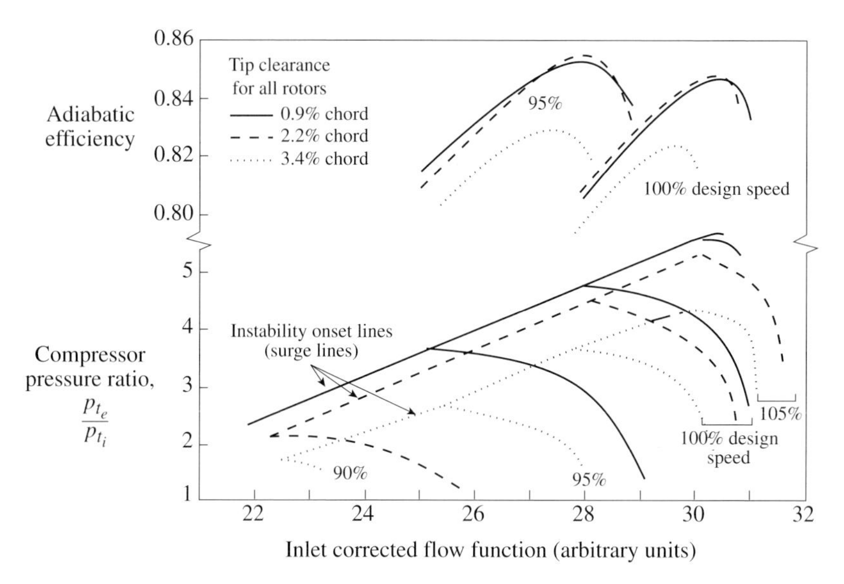

Figure 12.21: Effect of tip clearance on the pressure ratio, the stall line and the efficiency of a six-stage, high speed compressor; last stage rotor chord 32 mm, aspect ratio 1.0. Data of Freeman (1985) as quoted by Cumpsty (1989).

(e.g. clearance/blade height) as ε $(= \overline{\varepsilon} + \varepsilon'(\theta))$, the compressor pressure rise (outlet static minus inlet stagnation pressure) can be written as a function of tip clearance and compressor inlet axial velocity as

$$(p_e' - p_{t_i}') = \left(\frac{\partial \overline{\Delta P_{t/s}}}{\partial u_x}\right)_{\varepsilon} u_x'(0, \theta) + \left(\frac{\partial \overline{\Delta P_{t/s}}}{\partial \varepsilon}\right)_{u_x} \varepsilon'(\theta). \tag{12.5.10}$$

In (12.5.10) the subscripts on the partial derivatives indicate the quantity held constant during the differentiation. The second partial derivative on the right-hand side is the change in pressure rise with tip clearance, evaluated in a circumferentially uniform flow.

Suppose the clearance varies sinusoidally around the circumference with the smallest clearance at $\theta = 0$ and the largest at $\theta = 180°$, as in Figure 12.22. From the quasi-steady arguments the pressure rise of the compressor (which is the axial force exerted on the fluid) is greatest at $\theta = 0$ and least at $\theta = 180°$. The axial velocity in the compressor will thus vary around the circumference, and a flow which is irrotational upstream of the compressor exits from the compressor with a non-uniform stagnation pressure and a radial component of vorticity. A quantitative treatment of this problem is given in Section 12.7.

12.6 Additional examples of upstream effects in turbomachinery flows

12.6.1 Turbine engine effects on inlet performance

The length scales for upstream influence (Sections 2.3, 12.2) with stagnation pressure distortion imply that fluid components in a gas turbine engine are often closely coupled. If so the behavior

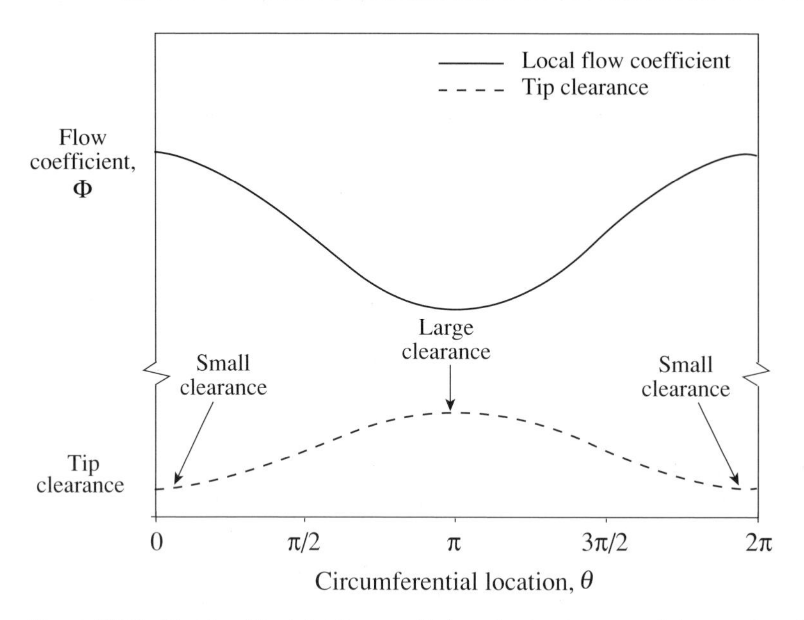

Figure 12.22: Sketch of the inlet flow coefficient distribution around the annulus of the compressor with non-axisymmetric tip clearance.

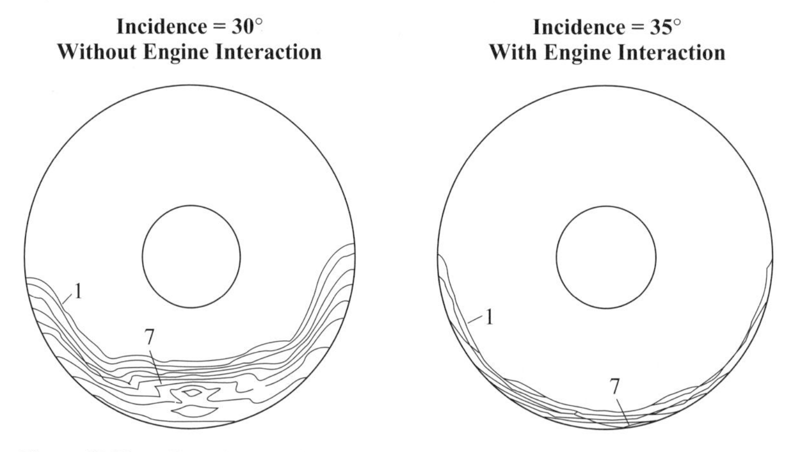

Figure 12.23: Effect of the engine presence on the total pressure distribution in a short pitot inlet (Hodder, 1981).

of the overall system can differ from that predicted from examination of individual components separately (Section 12.4 showed another illustration of this situation). An example of this is the effect of a gas turbine engine on inlet behavior.

Aircraft inlets sometimes operate at high angles of attack, leading to separation and flow non-uniformity into the engine. Civil aircraft inlet lengths are usually short compared to their diameter. The arguments in this chapter imply that the engine can substantially affect the static pressure field within the inlet and it is not necessarily correct to assess inlet performance without simulating the presence of the engine. This is demonstrated in Figure 12.23 which shows the contours of stagnation pressure in a cylindrical inlet at incidence, with and without engine influence (Hodder, 1981). In the latter case, the inlet duct was followed by a constant area duct long enough so engine upstream influence was negligible at the measurement station. At 30° incidence a large region of low stagnation pressure existed associated with separation. With an engine present, even at 35° incidence, the region of low stagnation pressure is much smaller because the compressor acts to equalize the velocities and suppress separation. Simulations (either theoretical or experimental) of internal flow devices need to include such interaction (Hsiao *et al.*, 2001); the ideas presented show that a screen can provide a useful simulation of the desired effect.

12.6.2 Strut-vane row interaction: upstream influence with two different length scales

Another example of upstream influence concerns the response of turbomachine blade rows to downstream struts (Barber and Weingold, 1978; Chiang and Turner, 1996). There are often a number of struts (say, eight) downstream of a compressor or fan. The compressor rotor which passes through the upstream pressure field of such struts experiences an eight per revolution disturbance. This is a forcing function for vibratory stresses and the magnitude of the velocity non-uniformity is of concern. The strut–stator configuration presents a new aspect, the existence of two upstream length scales, the spacing between the stator blades and the (much larger) spacing between the struts.

A representative configuration is portrayed in Figure 12.24, which shows a strut and the stator blades ahead of it. There are six stator blades per strut gap. The upper part of the figure gives measurements and analysis of the static pressure distribution at a location one-half stator chord upstream of the stator row leading edge. The computations were carried out using a vortex method of the type described in Section 3.15 and the dots show the location of the vortices used.

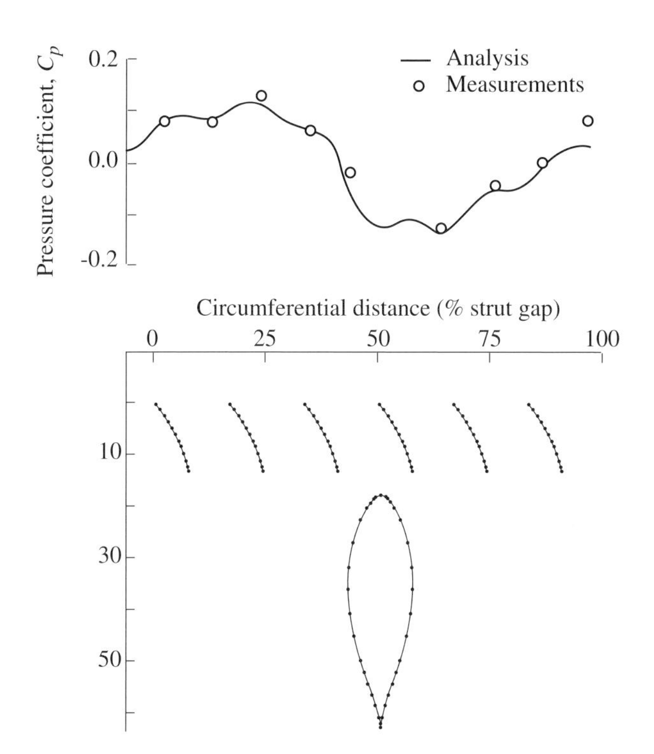

Figure 12.24: Static pressure non-uniformity upstream of the stator–strut configuration (Barber and Weingold, 1978).

The influence of the two length scales can be seen in the figure. The static pressure variation associated with the stators has decayed to a low value at the axial station where the data were taken, but the variation on the scale of the strut-to-strut spacing has not. If we looked at a station another half-stator chord upstream, the small scale ripple on the static pressure profile would be absent, whereas the magnitude of the larger scale variation would be little altered.

12.7 Unsteady compressor response to asymmetric flow

The quasi-steady model of compressor behavior in asymmetric flow has proved useful in a number of problems, but quantitative application to a range of situations (including self-excited propagating instabilities in compressors) is considerably increased if we extend the description to include effects of unsteady flow in the blade rows. Unsteady blade passage flows occur when there is a circumferential variation in velocity and also when the flow in the engine fixed coordinate system is unsteady.

We consider unsteady flow in stators first for simplicity. The passage is idealized as a channel that is much less wide than it is long, with the flow treated in a one-dimensional manner. For the moment, viscous effects are neglected. We wish to develop a relation between the stagnation pressure difference along the blade channel and the velocity. The approach is analogous to that of Section 6.3, with the starting point the one-dimensional momentum equation cast as an expression for the spatial rate of change of stagnation pressure. With u the velocity magnitude in the channel and l the streamwise distance, this is

$$\rho \frac{\partial u}{\partial t} = -\left[\frac{\partial (u^2/2)}{\partial l} + \frac{\partial p}{\partial l} \right] = -\frac{\partial p_t}{\partial l}. \tag{12.7.1}$$

Integrating (12.7.1) from channel inlet to exit:

$$(p_t)_i - (p_t)_e = \rho \int_{inlet}^{exit} \frac{\partial u}{\partial t} \, dl. \tag{12.7.2}$$

For a constant area channel, $\partial u / \partial t$ is constant and (12.7.2) becomes

$$(p_t)_i - (p_t)_e = \rho b_{stator} \frac{\partial u_i}{\partial t}, \tag{12.7.3}$$

where b_{stator} is the stator chord.[11] There is a stagnation pressure change from the inlet to the exit because of the local acceleration of the fluid in the blade channel.

The stagnation pressure difference in (12.7.3) is in addition to any difference associated with viscous losses. The latter can be estimated using the quasi-steady analysis as implied by (12.5.4) so the viscous losses have the form of a mean plus a variation of the form

$$\text{quasi-steady loss variation} = \left(\frac{d(\overline{\text{loss}})}{d(\text{inlet conditions})} \right) (\text{variation in inlet conditions}).$$

[11] To account for non-constant area, an "effective chord" can be introduced following Section 6.3.

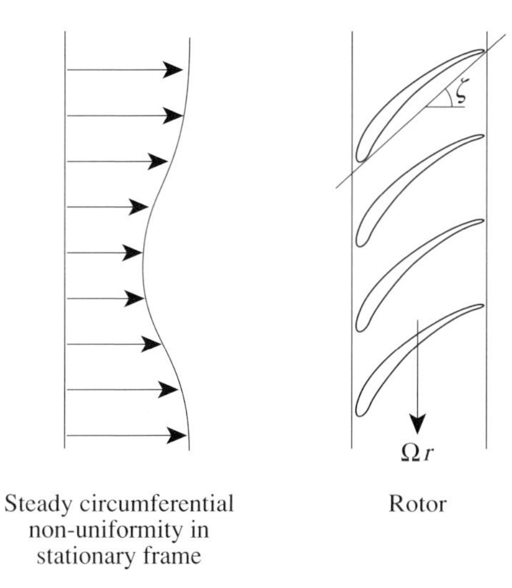

Figure 12.25: Compressor rotor blades passing through a circumferentially non-uniform flow which is steady in the stationary (absolute) frame see an unsteady flow field; ζ = stagger angle.

The overall difference in stagnation pressure between the blade row inlet and exit is thus viewed as the sum of viscous losses and local acceleration effects. For a stator blade row,

$$(p_t)_i - (p_t)_e = \rho b_{stator}\frac{\partial u_i}{\partial t} + \text{quasi-steady losses.} \tag{12.7.4}$$

For a rotor a similar statement about unsteadiness can be written in terms of the time rate of change of the relative velocity (w) in the passage and the stagnation pressure, $p_{t_{rel}}(= p + \frac{1}{2}\rho w^2)$, measured in the rotating coordinate system:

$$\rho\frac{\partial w}{\partial t} = -\frac{\partial p_{t_{rel}}}{\partial l}. \tag{12.7.5}$$

Time variations in the rotor arise in two ways: from unsteadiness in the absolute (engine fixed) system and from blades passing through a steady, circumferentially varying, flow. The latter is shown in Figure 12.25. The change in fluid properties in the rotor frame of reference is:

$$\left[\frac{\partial()}{\partial t}\right]_{rotor\ frame} = \left[\frac{\partial()}{\partial t} + \Omega\frac{\partial()}{\partial \theta}\right]_{absolute\ frame}. \tag{12.7.6}$$

The change in relative stagnation pressure from inlet to exit of the rotor passage is

$$(p_{t_{rel}})_i - (p_{t_{rel}})_e = \rho b_{rotor}\left[\frac{\partial w}{\partial t} + \Omega\frac{\partial w}{\partial \theta}\right]_{absolute\ frame}. \tag{12.7.7}$$

An expression similar to (12.7.4) can be developed for viscous losses through the rotor. Further, within the one-dimensional description the direction of the flow in the blade passage is constrained by the blading. The relative and axial velocities are thus related by, for stator and rotor respectively,

$$u = u_x / \cos\zeta_{stator}; \quad w = u_x / \cos\zeta_{rotor}, \tag{12.7.8}$$

where ζ is the stagger angle (the angle the chord of the blade makes with the x-axis) as indicated in Figure 12.25).

In this one-dimensional description the two mechanisms (viscous losses and local accelerations) of stagnation pressure change are viewed as operating in series. The blade channel can be thought of as an impedance consisting of an actuator disk at the front of the passage (a resistance-like element which provides quasi-steady loss variations in response to the varying inlet conditions) and the unsteady flow in a constant area channel (an inertance). For this reason the combination is sometimes referred to as a "semi-actuator disk" approximation. The modeling can also be carried further to include a description of the unsteady viscous response (Haynes, Hendricks, and Epstein 1994; Longley, 1994; Hendricks, Sabnis, and Feulner, 1997), but the representation here is sufficient to illustrate a number of aspects of the blade row unsteady response.

To use the expression for rotor stagnation pressure change in a description of the overall compressor, the relative stagnation pressure change must be cast in terms of quantities in the absolute coordinate system. From the definition of the relative stagnation pressure ($p_{t_{rel}} = p + \frac{1}{2}\rho w^2$) and the vector velocity triangles between absolute and relative velocity, and using the subscript "i/e" to denote the change from inlet to exit,

$$(\Delta p_{t_{rel}})_{i/e} = (\Delta p_{t_{abs.}})_{i/e} - (\Omega r_m \Delta [u_\theta])_{i/e}. \tag{12.7.9}$$

Substituting (12.7.8) and (12.7.9) in the expressions for rotor and stator stagnation pressure changes gives an expression for the change in absolute stagnation pressure across a complete compressor stage (rotor plus stator) as

$$\begin{aligned} p_{t_e} - p_{t_i} = & \underbrace{\rho\Omega r_m (u_{\theta_e} - u_{\theta_i})_{rotor} - Loss_{rotor} - Loss_{stator}}_{\text{Quasi-steady stagnation pressure change due to inlet condition variations}} \\ & \underbrace{-\rho\left[\frac{\partial u_x}{\partial t}\left(\frac{b_{stator}}{cos\zeta_{stator}} + \frac{b_{rotor}}{cos\zeta_{rotor}}\right) + \Omega\frac{\partial u_x}{\partial\theta}\left(\frac{b_{rotor}}{cos\zeta_{rotor}}\right)\right]}_{\text{Stagnation pressure change due to local accelerations in channel}}. \end{aligned} \tag{12.7.10}$$

The first set of quantities on the right-hand side of (12.7.10) is the quasi-steady response and includes both axisymmetric and asymmetric terms. The second set is the change in stagnation pressure associated with local accelerations. For a multistage compressor the effects in the different stages can be added and, after some manipulation, the pressure rise written in terms of the difference between exit static pressure and the inlet stagnation pressure non-uniformities as

$$\begin{aligned} (p'_e - p'_{t_i})_{compressor} = & \left(\overline{\frac{d\Delta P_{t/s}}{du_x}}\right) u'_x(0,\theta,t) \\ & - \rho r_m \left(\mathcal{I}_R \Omega \frac{\partial u'_x(0,\theta,t)}{\partial\theta} + \mathcal{I}_{RS}\frac{\partial u'_x(0,\theta,t)}{\partial t}\right). \end{aligned} \tag{12.7.11}$$

In (12.7.11) $\mathcal{I}_R$ and $\mathcal{I}_{RS}$ are inertia parameters for the compressor (with j rotors and k stationary blade rows; j stators plus an inlet guide vane, and possibly an outlet guide vane) defined as:

$$\mathcal{I}_{RS} = \sum_{l}^{All\ j+k\ rows} [b_l/(r_m \cos \zeta_l)], \tag{12.7.12a}$$

$$\mathcal{I}_R = \sum_{l}^{j\ rotors} [b_l/(r_m \cos \zeta_l)]. \tag{12.7.12b}$$

Non-dimensionalizing velocities by Ωr_m and pressures by $\rho(\Omega r_m)^2$, (12.7.11) becomes

$$\left(\frac{p'_e - p'_{t_i}}{\rho \Omega^2 r_m^2}\right) = \left(\frac{d\overline{\Psi_{t/s}}}{d\Phi}\right) \phi'(0,\theta,t) - \left[\left(\mathcal{I}_R \frac{\partial \phi'(0,\theta,t)}{\partial \theta}\right) + \left(\frac{\mathcal{I}_{RS}}{\Omega}\frac{\partial \phi'(0,\theta,t)}{\partial t}\right)\right]. \tag{12.7.13}$$

Equations (12.5.10) (the quasi-steady description of circumferentially non-uniform tip clearance) and (12.7.13) are both linear and can be superposed to give an expression that accounts for unsteadiness and asymmetric clearance as

$$\left(\frac{p'_e - p'_{t_i}}{\rho \Omega^2 r_m^2}\right) = \left(\frac{\partial\overline{\Psi_{t/s}}}{\partial\Phi}\right)_{\varepsilon} \phi'(0,\theta,t) + \left(\frac{\partial\overline{\Psi_{t/s}}}{\partial\varepsilon}\right)_{\Phi} \varepsilon'(\theta) - \left[\left(\mathcal{I}_R \frac{\partial \phi'(0,\theta,t)}{\partial \theta}\right) + \left(\frac{\mathcal{I}_{RS}}{\Omega}\frac{\partial \phi'(0,\theta,t)}{\partial t}\right)\right]. \tag{12.7.14}$$

We will apply (12.7.14) to three situations connected with asymmetric flow in compressors: conditions for the onset of self-excited disturbances in axial compressors, generation of flow non-uniformities due to circumferentially varying tip clearance, and response to unsteady inlet distortion. In the remainder of Section 12.7 the treatment is in the context of linear analysis in order to introduce a single one of the three effects of the behavior. In Section 12.8 a nonlinear version of the theory is used to address situations in which the different effects interact.

12.7.1 Self-excited propagating disturbances in axial compressors and compressor instability

Determining the conditions for compressor instability to propagating disturbances follows the approach laid out in Sections 6.5 and 6.6, namely examining the behavior of small disturbances to a steady flow to see whether they grow or decay. The steady flow now is the specified mean axial velocity parameter, Φ, which determines the compressor operating point and the slope of the characteristic. Defining appropriate forms of the disturbance flow fields upstream and downstream of the compressor and linking them through matching conditions leads to an equation for the eigenvalues (complex frequencies) which define disturbance growth or decay.

Expressing the compressor characteristic as the exit static pressure minus the inlet stagnation pressure is convenient here because both are directly related to axial velocity perturbations. For self-excited disturbances the velocity field is irrotational ahead of the compressor, the perturbations decay upstream, and the conditions far upstream are steady and uniform. The disturbances of interest propagate round the circumference and must be of the form $\exp[ik(\theta - \omega_k t)]$, where k is the harmonic number and ω_k is the complex frequency. As in Section 6.8 this implies that the general solution for the upstream disturbance velocity potential, φ, obeying Laplace's equation is

$$\varphi = \sum_{\substack{k=-\infty,\\ k\neq 0}}^{\infty} A_k e^{k[(x/r_m)+i(\theta-\omega_k t)]}.$$

The relation between stagnation pressure variations and the disturbance velocity potential is given by the first integral of the momentum equation for irrotational flow as derived in Section 6.4:

$$\frac{\partial \varphi}{\partial t} + \frac{p_t}{\rho} = f(t). \tag{6.4.3}$$

The far upstream flow is uniform and steady and the function $f(t)$ is thus a constant. Equation (6.4.3) for the stagnation pressure variation (the stagnation pressure is a function of x, θ, and t) applied to the stagnation pressure perturbation is

$$\frac{p_t'}{\rho} = -\frac{\partial \varphi}{\partial t}. \tag{12.7.15}$$

Equation (12.7.15) shows the upstream stagnation pressure variation must have the same form as the velocity potential, φ, so

$$\frac{p_t'}{\rho} = \sum_{\substack{k=-\infty,\\ k\neq 0}}^{\infty} B_k e^{k[(x/r_m)+i(\theta-\omega_k t)]}. \tag{12.7.16}$$

The flow at the compressor exit emerges from the last vane row at an essentially constant angle and can be taken to be axial. The downstream static pressure, however, is not uniform, as would be the case in a steady flow. The static pressure obeys Laplace's equation ($\nabla^2 p' = 0$) (see Section 2.3)[12] and the downstream static pressure perturbation has the form

$$p' = \sum_{\substack{k=-\infty,\\ k\neq 0}}^{\infty} C_k e^{-k[(x/r_m)+i(\theta-\omega_k t)]}. \tag{12.7.17}$$

Using the continuity equation, the x-momentum equation can be written as (with $\overline{u}_y = 0$)

$$\frac{\partial u_x'}{\partial t} + \overline{u}_x \frac{\partial u_x'}{\partial x} = \frac{\partial u_x'}{\partial t} - \left(\frac{\overline{u}_x}{r_m}\right)\frac{\partial u_\theta'}{\partial \theta} = -\frac{1}{\rho}\frac{\partial p'}{\partial x}. \tag{12.7.18}$$

The condition on the compressor exit flow angle means that u_θ' is zero at the compressor exit. Equation (12.7.18) evaluated at the compressor exit thus reduces to

$$\left(\frac{\partial u_x'}{\partial t} = -\frac{1}{\rho}\frac{\partial p'}{\partial x}\right)_{\text{compressor exit}}. \tag{12.7.19}$$

There is a static pressure non-uniformity at the compressor exit due to unsteadiness.

[12] This can be seen by differentiating the x-component of the momentum equation with respect to x and the y-component of the momentum equation with respect to y, adding them and making use of the continuity equation.

Equations (12.7.13), (12.7.15), and (12.7.19), plus the mass conservation matching condition across the compressor ($u'_{x_i}(\theta, t) = u'_{x_e}(\theta, t)$) and the functional forms of the velocity potential, upstream stagnation pressure, and downstream static pressure, can be combined into a single equation for the complex frequency, $\omega_k = \omega_{k_{real}} + i\omega_{k_{imaginary}}$. In terms of real and imaginary parts for the kth Fourier component this is

$$\left(\frac{\omega_{k_{imaginary}}}{\Omega}\right)_k = \frac{1}{\rho\bar{u}_x}\left(\frac{\overline{d\Delta P_{t/s}}}{du_x}\right) = \left(\frac{1}{\Phi}\frac{\overline{d\Psi_{t/s}}}{d\Phi}\right), \tag{12.7.20a}$$

$$\left(\frac{\omega_{k_{real}}}{\Omega}\right)_k = \left(\frac{\mathcal{I}_R}{2/k + \mathcal{I}_{RS}}\right), \quad k \neq 0 \tag{12.7.20b}$$

Equation (12.7.20a) states that propagating disturbances grow, and the flow through the compressor becomes unstable, when the compressor operates with a positive slope of the total-to-static pressure rise characteristic. The instability criterion is thus that the operating point has reached the peak of this characteristic. This is useful as a guideline, although it has been found that there are also compressors which exhibit propagating disturbances on the negatively sloped portion of the compressor characteristic (Day 1993; Camp and Day, 1998). Such behavior appears to be linked to a different, and nonlinear, mechanism (Gong *et al.*, 1999).

Equation (12.7.20b) states that the disturbance will propagate round the circumference at a speed that is a fraction of, and scales with, the rotor speed. For a first harmonic disturbance ($k = 1$) and a compressor having rotors and stators with identical geometries, $\mathcal{I}_{RS}$ is approximately twice $\mathcal{I}_R$ and the propagation speed is

$$\omega_{1_{real}} \approx \left(\frac{\mathcal{I}_R}{1 + \mathcal{I}_R}\right)\frac{\Omega}{2}. \tag{12.7.21}$$

As the number of stages (and $\mathcal{I}_R$) increases, the propagation speed approaches $\Omega/2$ for a many-stage machine, in accord with multistage compressor data on rotating stall, which is the mature form of the propagating instability (Cumpsty and Greitzer, 1982).

The mechanism that feeds energy into the disturbances can be described with reference to the slope of the compressor characteristic, in a manner similar to that given in Section 6.6 for compression system oscillations. On a local (in θ) basis, the product of the perturbation velocity and the perturbation in pressure rise, which is a quadratic energy-like quantity, is positive when the mean operating point is in the positively sloped region, corresponding to energy fed into the disturbance flow field, and negative for operation in the negatively sloped region. The basic ideas carry over from the instability onset description in Section 6.6, although here the eigenmodes vary with θ as well as with x and t. Further information on the link between the growth of small amplitude traveling waves and the onset of compressor rotating stall, including the use of (wave behavior based) feedback control as a means of enhancing the compressor stable flow range, is given by Paduano, Greitzer, and Epstein (2001).

12.7.2 A deeper look at the effects of circumferentially varying tip clearance

As mentioned in Section 12.5 asymmetric flow can be generated by circumferentially non-uniform compressor tip clearance. In this situation, the flow is steady in the absolute system, with the compressor inlet stagnation pressure and exit static pressure both uniform. Equation (12.7.14) therefore reduces to a non-homogeneous differential equation (in θ) for the axial velocity variation around the compressor:

$$\left(\mathcal{I}_R \frac{\partial \varphi'(0,\theta)}{\partial \theta}\right) - \left(\frac{\partial \overline{\Psi_{t/s}}}{\partial \Phi}\right)_{\varepsilon} \varphi'(0,\theta) = \left(\frac{\partial \overline{\Psi_{t/s}}}{\partial \varepsilon}\right)_{\Phi} \varepsilon'(\theta). \tag{12.7.22}$$

The behavior can be illustrated using a non-dimensional sinusoidal clearance variation of the form $\varepsilon' = \varepsilon_k \sin k\theta$. The corresponding axial velocity variation at the compressor inlet is

$$\phi_k'(0,\theta) = \frac{\left(\dfrac{\partial \overline{\Psi_{t/s}}}{\partial \varepsilon}\right)_{\Phi}}{\sqrt{\left(\dfrac{\partial \overline{\Psi_{t/s}}}{\partial \Phi}\right)_{\varepsilon}^{2} + k^2 \mathcal{I}_R^2}} \left\{(\varepsilon_k) \sin[k(\theta + \Theta_k)]\right\}. \tag{12.7.23a}$$

The circumferential phase shift between the clearance minimum and the axial velocity maximum is given by

$$\tan k\Theta_k = \frac{k\mathcal{I}_R}{\left(\dfrac{\partial \overline{\Psi_{t/s}}}{\partial \Phi}\right)_{\varepsilon}}. \tag{12.7.23b}$$

Equations (12.7.23) exhibit a dependence of the axial velocity disturbance on: (1) the harmonic content of the clearance variation, k, (2) the rotor fluid inertia parameter, $\mathcal{I}_R$, (3) the sensitivity of the compressor pressure rise to axisymmetric clearance changes, $\partial \overline{\Psi_{t/s}}/\partial \varepsilon$, and (4) the slope of the compressor pressure rise characteristic, $\partial \overline{\Psi_{t/s}}/\partial \Phi$. More negatively sloped compressor characteristics, larger inertia parameters, a higher harmonic content of the clearance variation and a decreased sensitivity to clearance all promote smaller variations in axial velocity. In addition, as the slope of the compressor pressure rise characteristic decreases (for example as a given compressor is throttled toward stall) the axial velocity variation increases in magnitude and shifts in phase with respect to the clearance variation. At the peak of the compressor pressure rise characteristic ($\partial \overline{\Psi_{t/s}}/\partial \Phi = 0$) the axial velocity and clearance variation are predicted to be in quadrature, with a 90° phase shift.

12.7.3 Axial compressor response to circumferentially propagating distortions

The third situation is the response of compressors to imposed propagating non-uniformities as can occur in multishaft gas turbine aircraft engines in which a disturbance created in an upstream compressor is fed to the downstream one. If the disturbance has an angular velocity $f\,\Omega$, where Ω is the angular velocity of the rotor, the unsteady upstream and downstream disturbances given previously are augmented by an unsteady stagnation pressure perturbation[13] of the form

$$\frac{p'_{t_{-\infty}}(x,\theta,t)}{\rho \Omega^2 r_m^2} = \sum_{\substack{k=-\infty,\\ k\neq 0}}^{\infty} A_k e^{ik(\theta - f\Omega t)}. \tag{12.7.24}$$

[13] The disturbances are not coupled in a uniform background flow, and the interaction is through the matching conditions across the components.

The stagnation pressure disturbance in (12.7.24) is associated with the rotational part of the velocity field and does not obey Laplace's equation (compare with the form in (12.7.16)). The pressure matching condition across the compressor is again (12.7.14), with the compressor inlet stagnation pressure, p'_{t_i}, specified from (12.7.24). For a given (k^{th}) harmonic component the velocity disturbance at the compressor face is

$$\phi'_k(0, \theta, t) = \frac{A_k e^{ik(\theta - f\Omega t + \Theta_k)}}{\sqrt{(d\Psi/d\Phi)^2 + [k\mathcal{I}_R - f(k\mathcal{I}_{RS} + 2)]^2}}. \qquad (12.7.25)$$

The phase angle, Θ_k, is given by

$$\tan k\Theta_k = \frac{k\mathcal{I}_R - (k\mathcal{I}_{RS} + 2)f}{\left(\dfrac{d\Psi}{d\Phi}\right)}. \qquad (12.7.26)$$

In (12.7.25) and (12.7.26) we again see the (by now familiar) dependence on the slope of the compressor characteristic and the harmonic content of the non-uniformity. A new consideration is that as the angular frequency of the imposed distortion passes through the natural frequency of self-excited perturbations, the velocity disturbance in response to the forcing can become large. In essence, (12.7.25) shows resonance behavior, where the analog of the damping coefficient is the slope, $d\Psi/d\Phi$, of the compressor pressure rise characteristic curve. Near the peak of the pressure rise characteristic where $d\Psi/d\Phi$ is small, the axial velocity non-uniformity in the compressor will be large when the forcing is near the natural disturbance propagation speed ($f = \mathcal{I}_R/[(2/k) + \mathcal{I}_{RS}]$).

12.8 Nonlinear descriptions of compressor behavior in asymmetric flow

The foregoing descriptions of compressor behavior in asymmetric flow can be developed into a nonlinear treatment which allows examination of the important effect of circumferential distortion on compressor stability. The two problems, instability onset and the response to inlet distortion, have been addressed up to now from a linear perspective, where they can be regarded *independently* as the forced and free response of the compressor flowfield. To assess their interaction, a nonlinear analysis is needed.

In contrast to the stability problem of Section 12.7, the background flow upon which the small disturbances are now superposed is that associated with a finite amplitude stagnation pressure distortion and is non-uniform in θ. The eigenmodes are thus not sinusoidal and, if expressed as sums of sines and cosines, exhibit a rich harmonic content. Further, the background flow is not simply defined by the choice of mean flow. The θ-distribution of the axial velocity at the compressor face must be solved for because the local compressor pressure rise is determined by the local slope of the compressor characteristic and this varies around the circumference.

Discussion of the conceptual approach to solving this problem highlights the type of choice between modeling and computation which is inherent in many fluids engineering problems. The description of the flow domain can be separated into three parts: upstream, internal to the compressor, and downstream. In the linearized problems solved previously all these are treated in terms of small perturbations to a uniform flow. In addressing the nonlinear problem we can define aspects which are essential to describe in a nonlinear manner as well as those which are well approximated by a linear treatment. The compressor, which has a pressure rise characteristic whose slope can vary from strongly negative to positive as the axial velocity varies by $\pm 10\%$ of the mean, is the most important nonlinearity. Not only the magnitude but also the sign of the slope can change. The behavior of the upstream and downstream regions, however, is little affected by nonlinearities for the parameter range of interest. Put another way, what is needed for problem closure is a relation between the axial velocity and the pressure at the compressor inlet and exit, i.e. the impedance of the upstream and downstream regions. These quantities are much less dependent on nonlinear effects than the compressor response which provides the damping (either positive or negative) that drives or damps the instability.[14]

The above arguments imply that the background flow can be usefully modeled with a linear description of the upstream and downstream impedances coupled to a nonlinear description of the compressor behavior. It cannot be emphasized too strongly that there is no computational barrier to adding the upstream and downstream nonlinearities; the point is whether the gain in fidelity is worth the complication. The decision of what to model and what to compute is especially important when fluid components are part of a complex system or when there is an application (such as control or optimization) where a description in terms of a relatively small number of states is desired.

With an asymmetric background flow the Fourier components are coupled and the system and compressor responses can no longer be treated separately; a propagating sinusoidal disturbance on a circumferentially non-uniform background flow causes a time varying annulus average flow.[15] Because of this, the model must include all the other system components, as pictured in Figure 12.26 with reference to the problem of flow stability for non-uniform tip clearance. (For stagnation pressure distortion there would be a distortion-generating screen or other component in the upstream flow field.)

Figures 12.27 and 12.28 show two features of the compressor velocity field due to non-uniform tip clearance, based on a nonlinear model applied to a four-stage axial compressor (Graf *et al.*, 1998). Figure 12.27 gives the compressor inlet axial velocity distribution as a function of θ for a "one-lobed" clearance variation. As the compressor is throttled and the operating point approaches the peak pressure rise, the axial velocity variation increases in magnitude and changes shape. The phase also shifts towards being in quadrature with the clearance variation.

The effect of the disturbance Fourier component on flow coefficient variation is given in Figure 12.28, where behavior for a two-lobed clearance non-uniformity is compared to that for a one-lobed clearance variation. There is a difference in magnitude with operating condition and also an effect of harmonic number. The latter is basically an effect of increasing reduced frequency (consider the unsteady flow seen by the rotor) in decreasing the compressor response to non-uniformity.

These two figures show only the behavior of the steady background flow field. Conditions for the growth of self-excited small disturbances have also been assessed and the computations of instability onset show results in good accord with experiments.

[14] The argument can be presented with respect to the computation of the static pressure in a situation with a known velocity disturbance and uniform stagnation pressure. The static pressure non-uniformity is $\overline{p} + \Delta p = -\rho(\overline{u}^2/2 + \overline{u}\Delta u + (\Delta u)^2/2)$, where the notation Δp and Δu has been used to emphasize that neither of these is small compared to the background values. For a 10% (of the mean) velocity non-uniformity, nonlinear effects give rise to a 1% change in the static pressure non-uniformity and the main contribution is the linear term which gives a 20% variation. However, a 10% change in velocity can be enough to change the sign of the slope of the compressor characteristic.

[15] Quadratic nonlinear interaction of two Fourier components that are of the form $\sin\theta$, will result in sum and difference disturbances of the form $\sin 2\theta$ as well as disturbances that are uniform in θ.

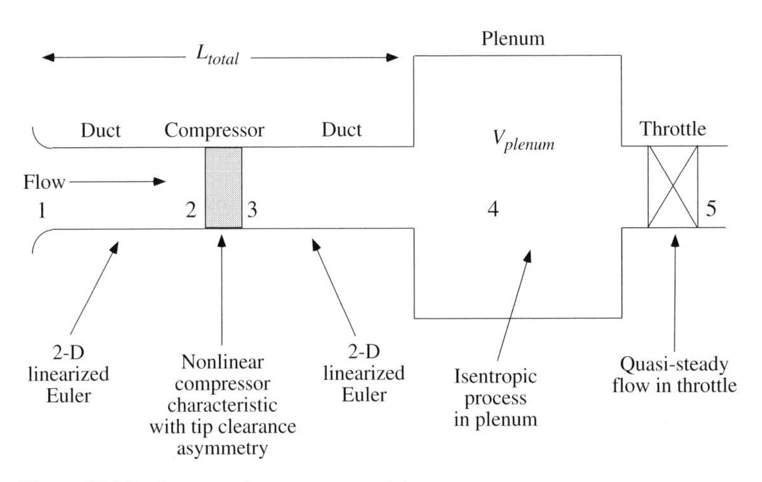

Figure 12.26: Compression system model.

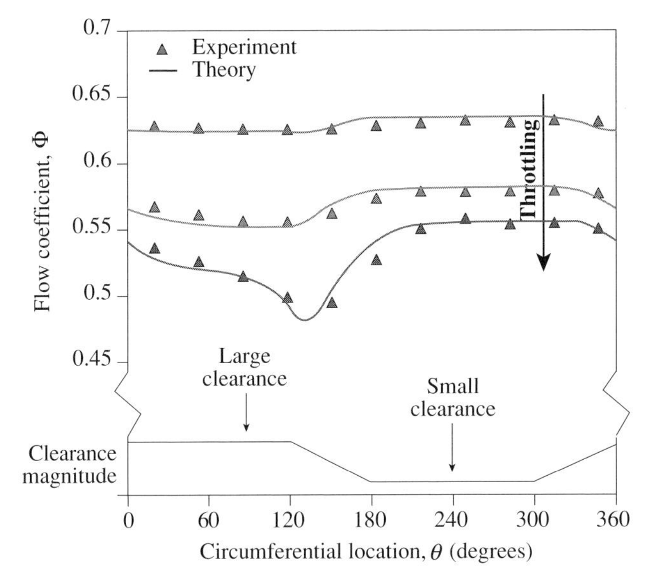

Figure 12.27: Flow coefficient distribution with one-lobed asymmetry; large clearance = 4.4% of blade height, small clearance = 1.4% of blade height (Graf *et al.*, 1998).

Another application of the nonlinear analysis is to rotating (propagating) distortions. As implied by (12.7.25) there is a strong response of the system at certain frequencies and there can be a substantial effect on compressor stability. Figure 12.29(a) shows theoretical and experimental results for the annulus average flow coefficient at instability onset as a function of the distortion rotation frequency. The experiments were conducted with a rotating screen in front of a four-stage compressor. The nominal design point of the compressor is at an axial velocity coefficient of approximately 0.62. The instability point with a stationary inlet distortion occurs at roughly 0.51. The flow range is defined as the increment between instability onset and the design point. The effect of the disturbance rotation rate is to move the instability point to 0.60, decreasing the flow range by roughly a factor of 7. To

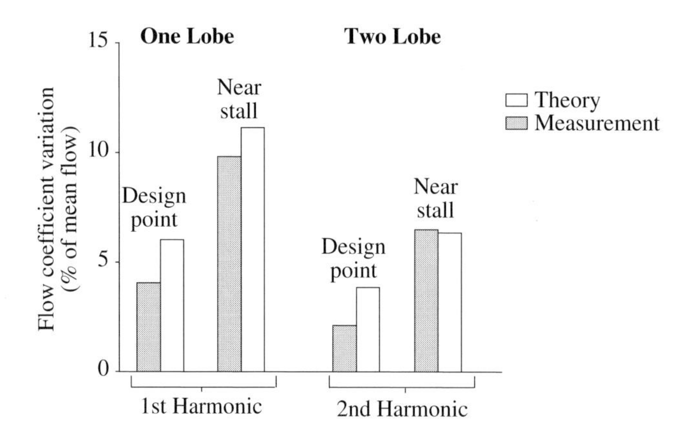

Figure 12.28: Circumferential variation in flow coefficient due to asymmetric tip clearance; spatial harmonics of steady flow non-uniformity for single-lobed and two-lobed clearance variations (Graf *et al.*, 1998).

give an idea of the range of flow accessed in practice compared to the change in stability point the compressor characteristic curve with uniform flow is plotted in Figure 12.29(b).

The previous arguments concerning the mechanical energy input leading to instability onset can be generalized to apply to the situation with inlet distortion. With non-uniform background flow the local operating point of the compressor and the local slope of the compressor characteristic vary around the circumference. Some circumferential locations have energy added to the disturbances, whereas others have energy extracted. Figure 12.30 shows computations of background flow and disturbance quantities for a three-stage compressor geometry subjected to a stationary inlet stagnation pressure distortion. Figures 12.30(a) and 12.30(b) show the background axial velocity distribution and the local slope of the compressor characteristic, at the instability point, as a function of θ. The extent of the far upstream stagnation pressure defect is also indicated and it is seen that the axial velocity at the compressor inlet and the far upstream stagnation pressure (and axial velocity defect) are roughly in quadrature.

Figure 12.30(c) shows the net value (averaged over a period $2\pi/\omega$) of the product of the non-dimensionalized local pressure rise $((d\Psi_{t/s}/d\Phi)\phi')$, and local velocity perturbation, ϕ'. Although the average (in θ) slope of the compressor characteristic is negative, there is a positive net flux of disturbance mechanical energy. For this to occur the square of the axial velocity perturbation must be larger in the region of positive slope (roughly $\theta = \pi$ to $\theta = 2\pi$) than in the region of negative slope. The square of the axial velocity perturbations is depicted in 12.28(d) and it is seen that this is the case. Chue *et al.* (1989) give additional information on the interactions between axisymmetric and non-axisymmetric harmonic components which lead to this situation.

12.9 Non-axisymmetric flow in annular diffusers and compressor–component coupling

Examination of circumferentially asymmetric flow in annular diffusers brings into prominence the effect of component length or, more precisely, the ratio of length to disturbance wavelength. As

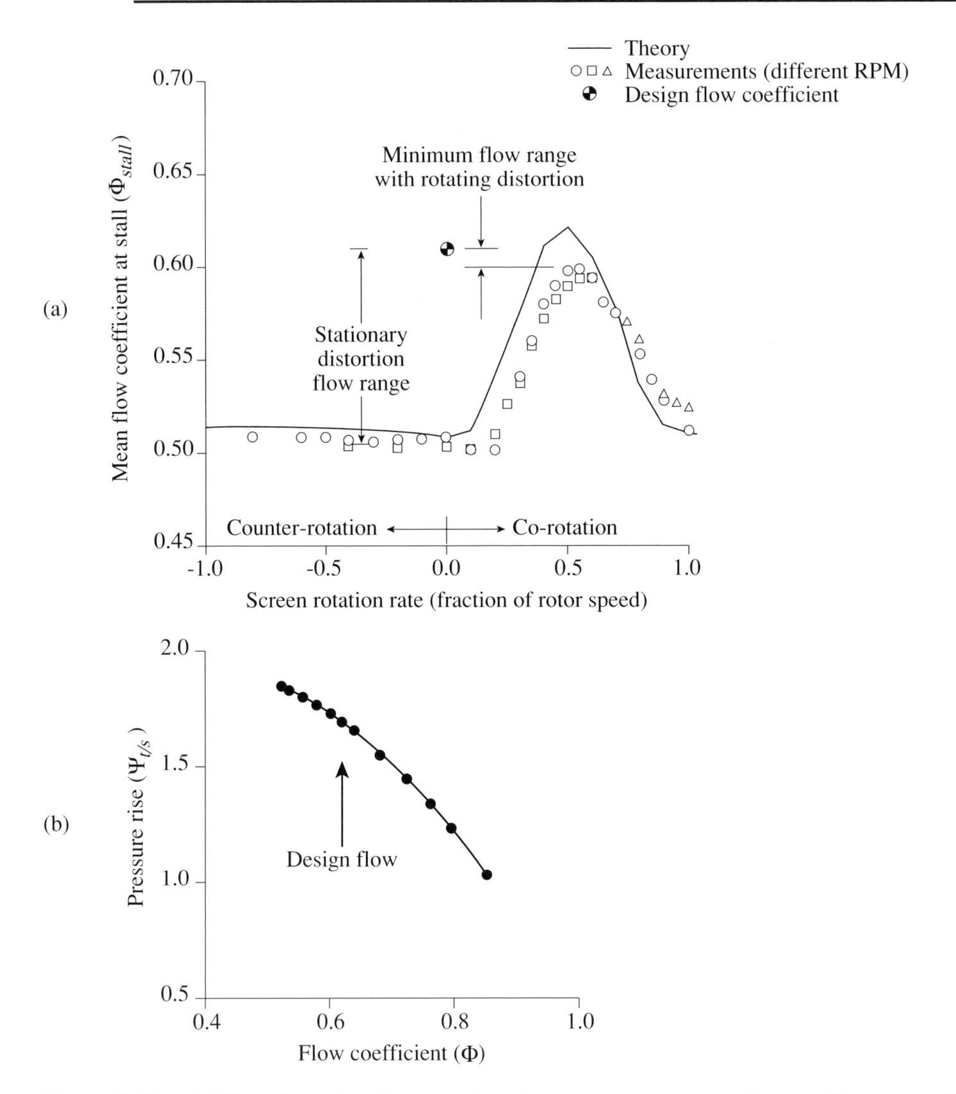

Figure 12.29: (a) Effect of rotating distortion for a four-stage compressor; flow coefficient at stall versus distortion rotation rate. The design point is indicated. (b) Uniform flow pressure rise versus flow characteristic for the compressor used in (a) (Longley *et al.*, 1996).

in the description of asymmetric flow in compressors, for the basic description here the annulus hub/tip radius ratio is regarded as high enough that flow quantities can be represented by radially-averaged conditions. The annulus is thus "unrolled" to give the flow domain on the right-hand side of Figure 12.31 with all quantities periodic in θ.

For a constant area straight duct the compressor exit static pressure in a steady flow was shown to be circumferentially uniform if the exit flow angle was uniform. We now wish to inquire whether this is the case with a downstream diffuser or nozzle and, if not, what determines the static pressure non-uniformity.

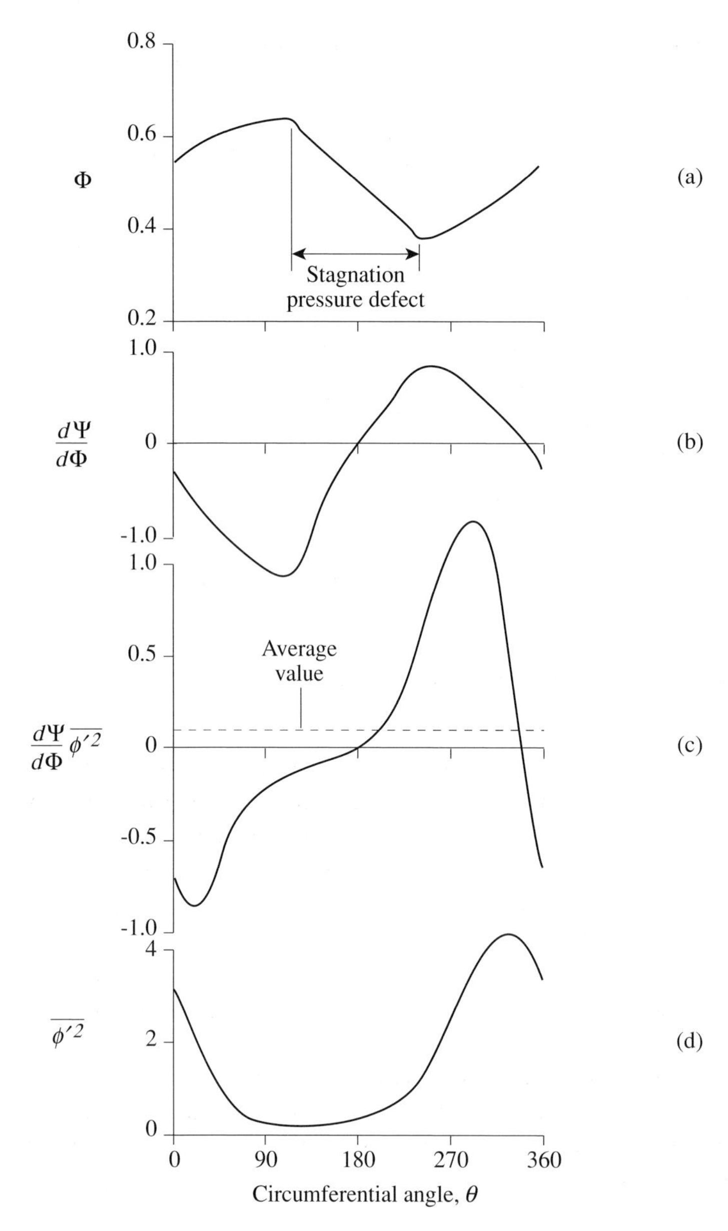

Figure 12.30: Circumferential distributions of mean and perturbation variables at a neutral stability point near resonance; $\mathcal{I}_R = 1, \mathcal{I}_{RS} = 2$, 120° extent distortion, $\Delta p_t/\rho(\Omega\, r_m)^2 = 0.2$: (a) background axial velocity at the compressor inlet; (b) local slope of the compressor characteristics; (c) net mechanical energy flux from the compressor; (d) mean square axial velocity perturbation (Chue *et al.*, 1989).

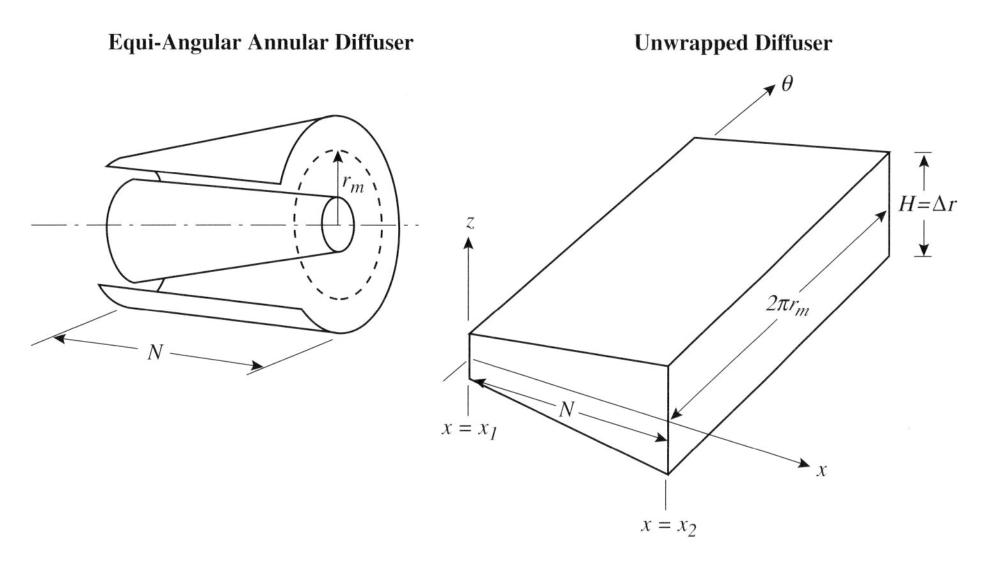

Figure 12.31: Flow field geometry for an equi-angular annular diffuser and an unwrapped diffuser.

12.9.1 Quasi-two-dimensional description of non-axisymmetric flow in an annular diffuser

We analyze the problem in terms of small amplitude asymmetric disturbances superposed upon an axisymmetric, axial, background flow, which varies in the x-direction. Looked at in detail, the flow in the diffuser is three-dimensional with x-, θ-, and r-components. Diffusers for turbomachinery applications, however, are often of small divergence angle to avoid separation, radial differences in static pressure are small, and we can work in terms of radially-averaged quantities which are functions of x and θ only (Greitzer and Griswold, 1976). The approach is analogous to the approximations in one-dimensional channel flow and can be referred to as quasi-two-dimensional. The diffuser core flow is treated as incompressible and inviscid, but boundary layer blockage can be accounted for through use of an effective area ratio.

In the linearized treatment adopted, the background axisymmetric flow and the asymmetric perturbations can be addressed separately. For the background flow, which has only an x-velocity component, the continuity and momentum equations can be written in terms of the radially averaged axial velocity and static pressure, denoted by overbars, as

$$\frac{d}{dx}(\overline{u}_x H) = 0, \tag{12.9.1}$$

$$\overline{u}_x \frac{d\overline{u}_x}{dx} + \frac{1}{\rho}\frac{d\overline{p}}{dx} = 0. \tag{12.9.2}$$

The quantity $H(x)$ is the local diffuser height. For an annular diffuser, it is the difference in radius (Δr) between the inner and outer walls.

To develop equations for the asymmetric velocity and pressure fields, the most direct approach is to take the perturbations (denoted by primes) as invariant with radius and carry out mass and momentum balances on a control volume with the elementary area $rdxd\theta$ which spans the diffuser.[16] The continuity, x-momentum, and θ-momentum (again using $r_m\theta$ to represent distance around the circumference) equations thus obtained are

$$\frac{\partial u'_x}{\partial x} + \frac{1}{r_m}\frac{\partial u'_\theta}{\partial \theta} + \frac{u'_x}{H}\frac{dH}{dx} = 0, \tag{12.9.3a}$$

$$\overline{u}_x \frac{\partial u'_x}{\partial x} + u'_x \frac{d\overline{u}_x}{dx} = -\frac{1}{\rho}\frac{\partial p'}{\partial x}, \tag{12.9.3b}$$

$$\overline{u}_x \frac{\partial u'_\theta}{\partial x} = -\frac{1}{\rho r_m}\frac{\partial p'}{\partial \theta}. \tag{12.9.3c}$$

A stream function, ψ, can be defined for the asymmetric velocity components as

$$u'_x = \frac{1}{Hr_m}\frac{\partial \psi}{\partial \theta}, \tag{12.9.4a}$$

$$u'_\theta = -\frac{1}{H}\frac{\partial \psi}{\partial x}. \tag{12.9.4b}$$

Equations (12.9.4) satisfy the continuity equation identically.

The radial (z-component) vorticity, ω_z, can be defined in terms of ψ as

$$\frac{\partial^2 \psi}{\partial x^2} - \left(\frac{1}{H}\frac{dH}{dx}\right)\frac{\partial \psi}{\partial x} + \frac{1}{r_m^2}\frac{\partial^2 \psi}{\partial \theta^2} = -H\omega_z. \tag{12.9.5}$$

An equation for the rate of change of the vorticity can be obtained by eliminating the pressure from (12.9.3b) and (12.9.3c)[17]

$$\frac{\partial \omega_z}{\partial x} = \left(\frac{1}{H}\frac{dH}{dx}\right)\omega_z. \tag{12.9.6}$$

Equation (12.9.6) has an immediate integral for the z-component of vorticity at any x location in terms of the vorticity, ω_{z_1}, at the diffuser inlet station (denoted by subscript 1), which corresponds to $x = x_1$:

$$\frac{\omega_z(x,\theta)}{H(x)} = \frac{\omega_{z_1}(\theta)}{H_1}. \tag{12.9.7}$$

Equation (12.9.7) is a quasi-two-dimensional form of Helmholtz's Theorem stating that radial vortex filaments are stretched as they are convected along the diffuser with a consequent increase in vorticity for a given fluid particle.

Equations (12.9.5) and (12.9.7) can be combined into a single equation for ψ:

$$\frac{\partial^2 \psi}{\partial x^2} - \left(\frac{1}{H}\frac{dH}{dx}\right)\frac{\partial \psi}{\partial x} + \frac{1}{r_m^2}\frac{\partial^2 \psi}{\partial \theta^2} = -\left(\frac{\omega_{z_1}}{H_1}\right)H^2. \tag{12.9.8}$$

[16] An alternative method is to formally average the three-dimensional equations from the inner to the outer radius, employing the condition of flow tangency at the bounding surfaces.

[17] This can also be obtained from the linearized form of the vorticity evolution equation: $\overline{u}\partial\omega_z/\partial x = -\omega_z d\overline{u}_x/dx$ combined with (12.9.1).

To proceed further, specific boundary conditions and diffuser shapes are needed. In the context of the problem described, a diffuser downstream of a compressor in a circumferentially non-uniform flow, we prescribe the stagnation pressure distribution at the diffuser inlet (which is the compressor exit) and take the flow angle at compressor exit to be circumferentially uniform and axial. As is often the case in practice, the diffuser is taken as straight-walled (H increasing linearly with x). Finally, the region downstream of the diffuser is regarded as a large volume, for example a combustion chamber or collector. Conditions at the diffuser exit are thus similar to those at the discharge of a subsonic duct into a large chamber described in Section 2.7, with the pressure at the diffuser exit equal to the chamber pressure and circumferentially uniform.

The solution of (12.9.8) can be represented as a Fourier series in θ, with each Fourier component independent. The equation for the kth harmonic is

$$\frac{\partial^2 \psi}{\partial x^2} - \frac{1}{x}\frac{\partial \psi}{\partial x} - \frac{k^2 \psi}{r_m^2} = -\left(\frac{\omega_{z_1}}{H_1}\right) H^2. \tag{12.9.9}$$

The three boundary conditions can be written in terms of the stream function and the vorticity. Crocco's Theorem gives a relation between the diffuser inlet stagnation pressure gradient and vorticity:

$$\omega_{z_1}(\theta) = -\frac{1}{\rho \bar{u}_x r_m}\frac{\partial p_t'(x_1,\theta)}{\partial \theta}. \tag{12.9.10}$$

The condition of axial velocity at the diffuser inlet ($x = x_1$) is written in terms of the stream function as

$$\left(\frac{\partial \psi}{\partial x}\right)\bigg|_{x=x_1} = 0. \tag{12.9.11}$$

The circumferentially uniform static pressure at the diffuser exit ($x = x_2$) is written in terms of the stream function as

$$\left(\frac{1}{x}\frac{\partial \psi}{\partial x} - \frac{\partial^2 \psi}{\partial x^2}\right)\bigg|_{x=x_2} = 0. \tag{12.9.12}$$

12.9.2 Features of the diffuser inlet static pressure field

Solution of (12.9.10) subject to the given boundary conditions can be expressed numerically or in terms of Bessel functions. In the context of the questions initially posed, the most important quantity is the static pressure variation at the diffuser inlet. With the stagnation pressure variation expressed as $p_t'(\theta) = \sum A_k \cos k\theta$, the static pressure at the inlet is (Greitzer and Griswold, 1976)

$$p'(x_1,\theta) = \sum A_k \left[F_k\left(\frac{kx_1}{r_m}, \frac{kx_2}{r_m}\right)\right] \cos k\theta. \tag{12.9.13}$$

In (12.9.13) the function F_k is a combination of modified Bessel functions of the first and second kind which is always negative. For any harmonic, therefore, the static and total pressure distortions at the diffuser inlet are out of phase. The magnitude of the static pressure variation decreases from the value at the diffuser inlet to zero at the diffuser exit.

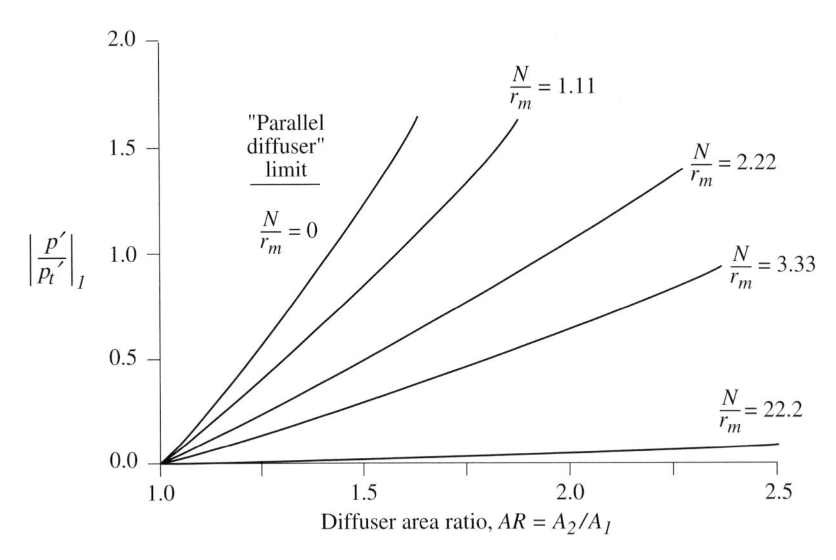

Figure 12.32: Diffuser inlet (station 1) static pressure distortion; cosinusoidal stagnation pressure distribution ($p_t' = \varepsilon \cos\theta$), N is the diffuser axial length, r_m is the mean radius (Greitzer and Griswold, 1976).

Figure 12.32 shows the magnitude of the diffuser inlet static pressure asymmetry normalized with respect to the magnitude of the inlet stagnation pressure asymmetry versus the diffuser effective area ratio ($AR = A_2/A_1$) for several non-dimensional diffuser lengths, N/r_m. There are two relevant length scales. One is associated with the circumferential non-uniformity and is given by r_m/k for the kth harmonic. The other is the axial length of the diffuser, N. For a given area ratio, the shorter the diffuser (in terms of kN/r_m) the larger the static pressure non-uniformity. For a given non-dimensional length, the larger the effective area ratio the larger the static pressure non-uniformity.

The uppermost curve, that for $N/r_m = 0$, which corresponds to the limiting case of a very short diffuser, provides a vehicle to discuss the relative phases of the static and stagnation pressures. At the diffuser inlet, the circumferential velocity is zero. For a short diffuser it remains small because there is not enough length for the streamlines to deflect appreciably. If the circumferential travel of a fluid particle is negligible, the streamtube divergence from inlet to exit at *any* θ location is set by the diffuser area ratio only. Any two circumferential locations in the annular diffuser can therefore be viewed as two equal area ratio diffusers operating in parallel, as sketched in Figure 12.33, with the local (in θ) diffuser static pressure rise depending only on the local inlet dynamic pressure. Regions of high velocity (high diffuser inlet stagnation pressure) are associated with high diffuser static pressure rise, and regions of low velocity (low stagnation pressure) with low static pressure rise. Because the diffuser exit static pressure is uniform, the former corresponds to low diffuser inlet static pressure and the latter to high inlet static pressure. The diffuser inlet static and stagnation pressure variations are therefore out of phase.

Figure 12.34 shows the decrease in static pressure non-uniformity with harmonic number for an annular diffuser geometry typical of aeroengine compressor exit diffusers. The figure shows the ratio of the magnitude of the inlet static pressure non-uniformity to the stagnation pressure non-uniformity for three situations: an inlet distortion sector 180° in extent, two 90° sectors, and four 45° sectors. The situations with the shortest *distortion* length scale have, as mentioned earlier, the lowest static pressure non-uniformity.

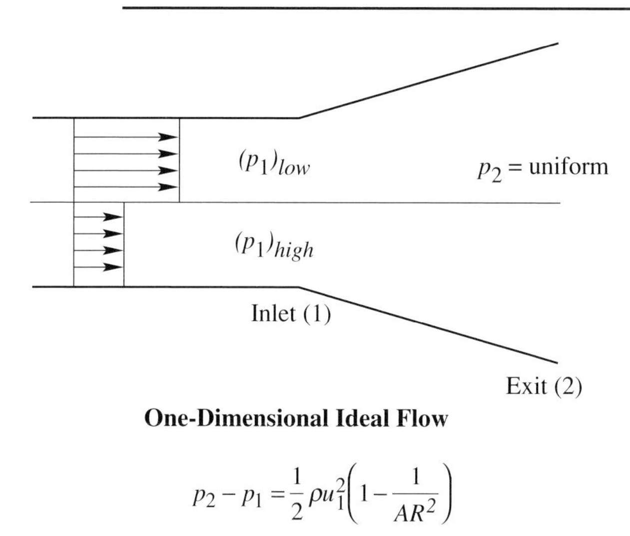

Figure 12.33: Diffusers in parallel (limiting case of short diffuser).

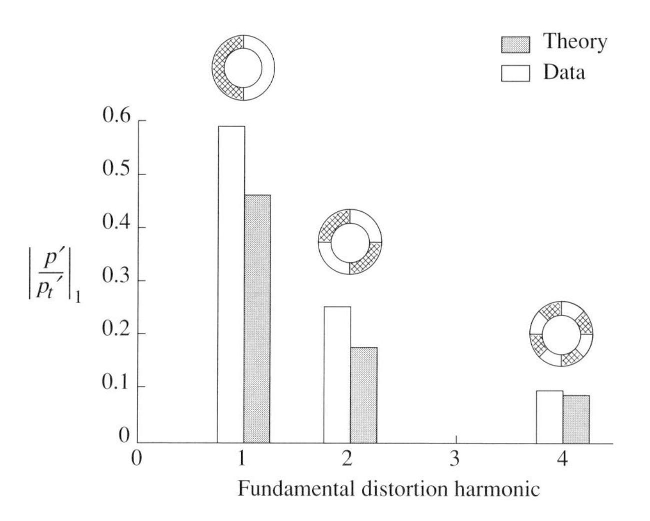

Figure 12.34: Effect of the harmonic number on the diffuser inlet static pressure non-uniformity (distortion screen configurations as indicated in figure); diffuser area ratio = 1.5, $\Delta r/r_m = 0.22$, $N/r_m = 1.1$ (Greitzer and Griswold, 1976).

Arguments in terms of local area change provide a way to interpret these results. As the harmonic number, k, increases, the diffuser becomes longer in terms of the disturbance length scale. The circumferential velocity set up by the circumferential pressure gradient tends to increase streamtube divergence in the high static pressure region and decrease it in the low static pressure region, compared to the situation with the short diffuser. This streamtube area variation evens out the local static pressure rise around the circumference, reducing the diffuser inlet static pressure non-uniformity.

As a consequence of the dependence of diffuser response on harmonic number, the diffuser acts as a low pass filter. The "output", which is the static pressure non-uniformity at the inlet, has a distribution with less higher harmonics than the "input" stagnation pressure non-uniformity. This can be seen in Figure 12.35, where the stagnation pressure profile is much less sinusoidal than the

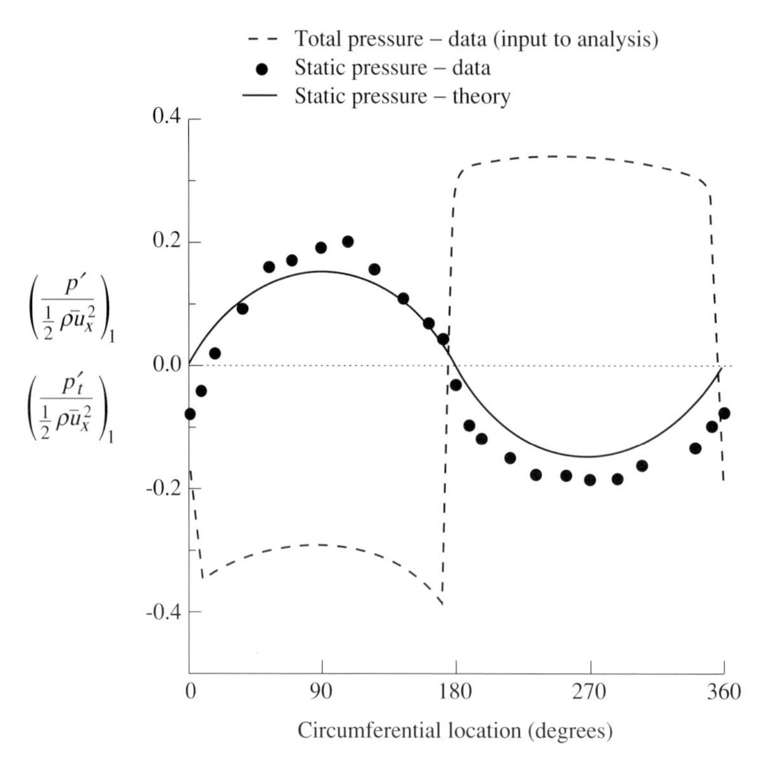

Figure 12.35: Circumferential variation in the static and stagnation pressure at the diffuser inlet (station 1); area ratio $= 1.5$, $\Delta r/r_m = 0.22$, $N/r_m = 1.11$ (Greitzer and Griswold, 1976).

static pressure because higher harmonic components have smaller proportional inlet static pressure non-uniformities. The stagnation pressure non-uniformity was created with a sharp edged screen and honeycomb combination and has many harmonics. The diffuser inlet static pressure non-uniformity is closer to a single sinusoid with the fundamental wavelength.

12.9.3 Compressor–component coupling

The description of diffuser behavior that has been developed provides a tool to assess the coupling between a compressor and a downstream diffuser or nozzle. The overall behavior can be described qualitatively using the short diffuser (or parallel diffuser) representation. Figure 12.36, which is a more inclusive version of Figure 12.33, shows the situation. Static and stagnation pressure variations at the diffuser inlet are out of phase. In the low stagnation pressure stream (denoted by subscript L), there is a high static pressure at the diffuser inlet (compressor exit) and conversely. For a *nozzle*, on the other hand, the diffuser inlet static and stagnation pressure non-uniformities are in phase.

The consequences on compressor performance are sketched in Figure 12.37, which compares operation in asymmetric flow for a compressor with a constant area downstream annulus, with an exit diffuser, and with an exit nozzle. For the same mean flow and inlet stagnation pressure distortion, the local working points of the compressor are altered by the downstream component with the low flow side pushed nearer to stall by the presence of the diffuser.

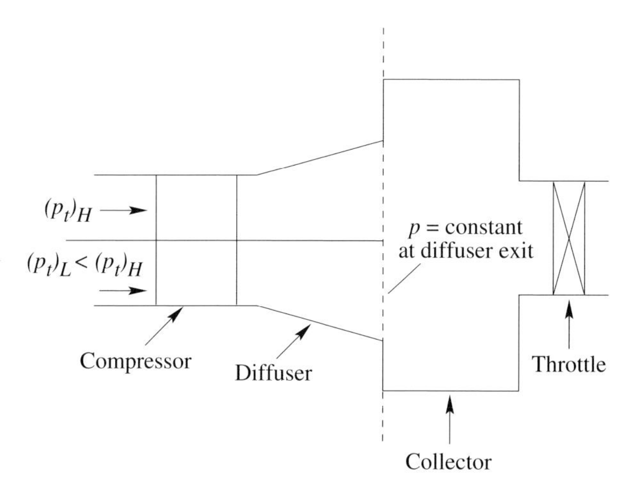

Figure 12.36: Parallel diffuser model for compressor–component coupling; subscripts L and H denote regions of low and high inlet stagnation pressure.

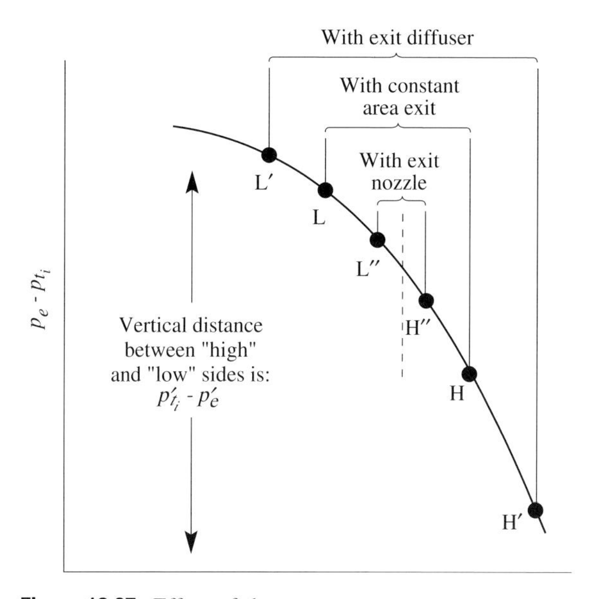

Figure 12.37: Effect of downstream components on compressor performance with inlet stagnation pressure asymmetry; subscripts L and H (and primed versions) refer to regions of low and high inlet stagnation pressure, subscripts i and e refer to compressor inlet and diffuser exit.

A more quantitative view of this phenomenon is shown in Figure 12.38, which gives measurements and calculations of the circumferential distribution of static pressure at the exit of a three-stage compressor run with the three different geometries (Greitzer, Mazzaway, and Fulkerson, 1978). The computations are based on a description of the whole compression system (upstream region, compressor, and downstream component flow models) which included a nonlinear representation of the compressor behavior and a linearized analysis for upstream and downstream flows. The mean flow and inlet stagnation pressure distortion were the same for all three tests. As suggested by the

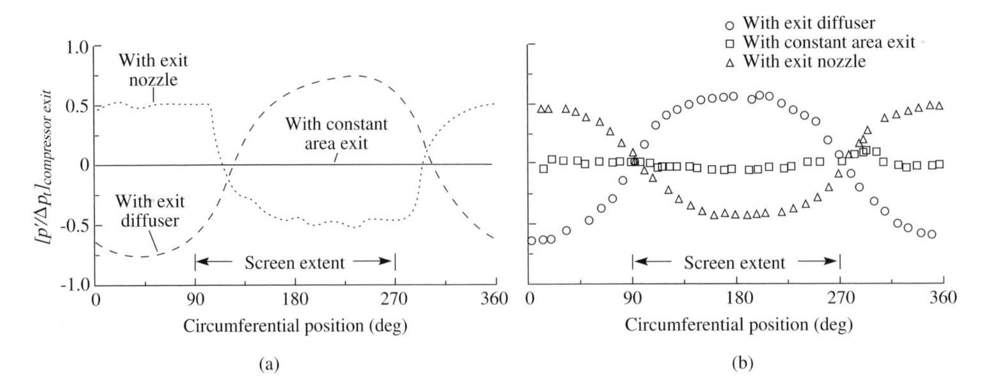

Figure 12.38: Effect of downstream components on static pressure non-uniformity at the exit of a three-stage compressor: (a) theory, (b) experiment; Δp_t is magnitude of stagnation pressure variation (Greitzer *et al.*, 1978).

foregoing arguments, the static pressure non-uniformity at the compressor exit is in phase with the stagnation pressure for the nozzle, out of phase for the diffuser, and virtually zero for the constant area annulus.

12.10 Effects of flow non-uniformity on diffuser performance

In much of the discussion of viscous effects, the approximation has been made that they were confined to thin regions (boundary layers, shear layers, or wakes). In this section, we give an introductory discussion of flow fields in fluid devices which are long enough so viscous forces become important in the core[18] regions and there are no regions which can be considered effectively inviscid. The context in which the discussion is set is the behavior of a diffuser with a non-uniform core velocity at the inlet. Examination of this type of flow not only shows the interplay of viscous and pressure forces in shaping the velocity profile, but also provides perspective on the relative roles the two forces play in determining the pressure rise as a function of the geometric parameters.

Figure 12.39 shows velocity profiles (velocity/mean velocity versus distance across the channel/channel width) for two two-dimensional diffusers with the same area ratio, $AR = 1.5$, but different non-dimensional lengths, $N/W_1 = 3$ and 6, where N is the diffuser axial length and W_1 is the inlet width (Wolf and Johnston, 1969). The inlet profile consists of a uniform shear (linear variation in velocity) over most of the channel plus thin boundary layers near each wall. The dashed line labeled "calculated core slope at exit" represents an exit velocity profile with the same vorticity as at the inlet; the apparent slope increase is because the exit channel width is 1.5 times the inlet channel width. The regions occupied by retarded flow in the boundary layers at the exit are several times as large as at the inlet, with the boundary layer growth particularly marked on the low velocity (left) side of the channel. The velocity in the boundary layer of the shorter diffuser also falls below that for the longer diffuser on the low velocity side.

[18] In Section 12.10, the term "core" refers to the flow outside of the diffuser wall boundary layers.

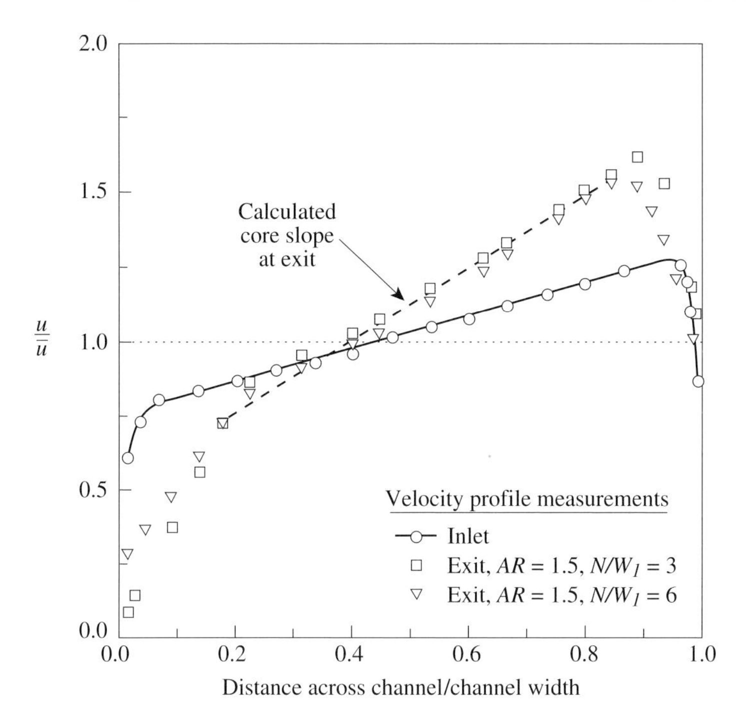

Figure 12.39: Inlet and exit velocity profiles for two two-dimensional diffusers with uniform shear flow in the core; N is the diffuser length, W_1 is the inlet width (Wolf and Johnston, 1969).

Figure 12.39 encapsulates a number of aspects of the behavior of a shear flow in a pressure gradient. First, the response of the core to the pressure field is essentially inviscid, but the size of the core is affected by viscous effects. Second, in an adverse pressure gradient, boundary layers associated with low core velocity regions grow substantially more than those associated with regions of high velocity. Third, viscous effects can improve aspects of performance through momentum transfer in the boundary layer if the device is long enough. With the longer diffuser, the exit boundary layer blockage is decreased, even though (as shown below) the pressure rise is higher than that in the shorter diffuser.

Pressure rise coefficients, $C_p = [(p_2 - p_1)/(\frac{1}{2}\rho\overline{u}_1^2)]$, for two-dimensional diffusers operating with the inlet uniform shear profile of Figure 12.39 are given in Figure 12.40 as a function of diffuser area ratio. Data are included for two non-dimensional diffuser lengths, N/W_1, 3 and 6. Also included are a curve showing the measured pressure rise with uniform core flow and the same inlet boundary layer blockage, and two ideal flow curves, one for uniform inviscid flow ($C_{p/uniform}$) and one for inviscid uniform shear ($C_{p/shear}$). The small arrow indicates the area ratio at which the inviscid forward flow limit occurs (see Section 5.9) for the shear profile; the $C_{p/shear}$ curve has been extended at this pressure rise for larger area ratios.

The difference between the two ideal curves is due to distortion of the core velocity profile as it enters a region of higher static pressure. The actual pressure rises are less than the ideal, but the difference between the two ideal curves represents the major part of the difference between the uniform core and shear flow results, especially for unstalled diffusers (area ratios of 1.8 or less in the figure).

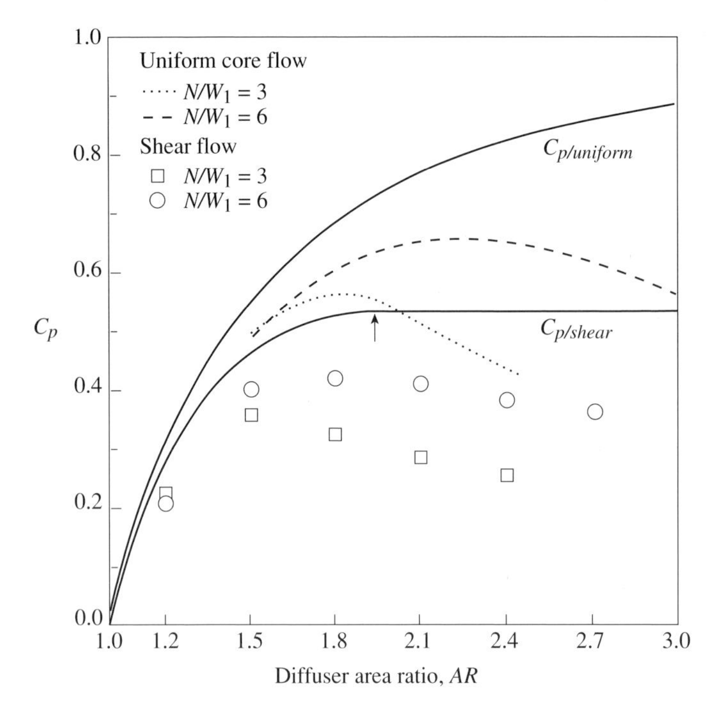

Figure 12.40: Static pressure rise coefficient for two-dimensional diffusers in shear flow, $u_{max}/u_{min} = 1.73$. Symbols denote measured performance with the inlet shear profile of Figure 12.39, dashed lines denote the measured performance with uniform core flow; boundary layer blockage at inlet = 0.017. Solid lines denote ideal performance, with the arrow indicating the forward flow limit for shear flow at inlet (Wolf and Johnston, 1969).

For these diffusers profile effects in the core, rather than effects directly associated with boundary layers, are responsible for the major part of the decreased pressure rise with uniform inlet shear flow.

These points are further illustrated through comparison of the behavior of diffusers that operate with the two non-uniform inlet flows shown in Figure 12.41. The "jet" profile has a region of high velocity in the center of the channel and low velocity near the walls, and the "wake" profile has the opposite. Inlet boundary layer blockage for the two profiles is the same.

Measured pressure rise coefficients as a function of diffuser area ratio with jet and wake inlet velocity profiles and with uniform core flow are shown in Figure 12.42 for non-dimensional lengths of 3, 6, and 12. The longer diffusers have a higher pressure rise. The diffusers operating with the wake profile have a higher performance than those with the jet profile, because the boundary layers of the high velocity stream are more capable of negotiating a given pressure rise.

Figures 12.39–12.42 represent situations in which both viscous forces (or rather forces due to gradients of turbulent stresses) and pressure forces need to be accounted for, because it is the competition between the two which results in accentuation or attenuation of the non-uniformity in velocity. This competition is exhibited by examination of differential changes in stagnation pressure, static pressure and velocity along a streamline:

$$\frac{du}{u} = \frac{1}{\rho u^2}(dp_t - dp). \tag{12.10.1}$$

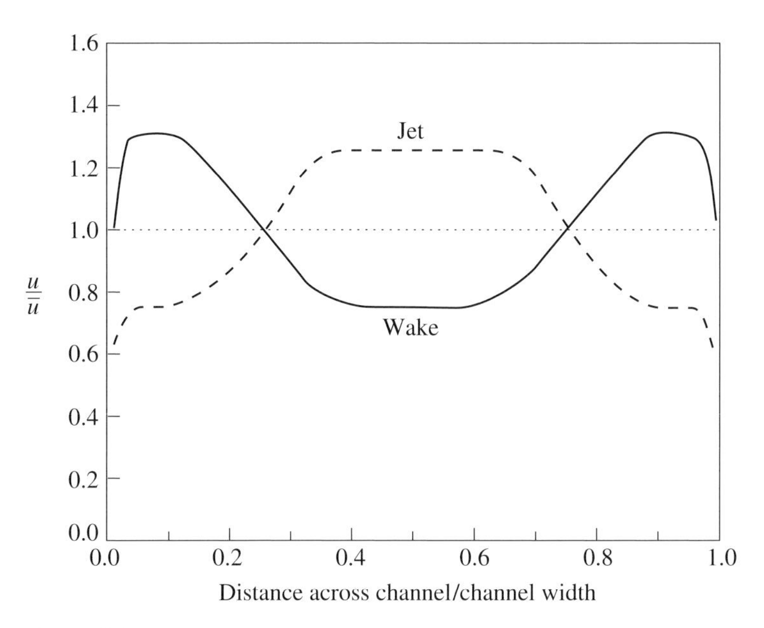

Figure 12.41: Jet and wake inlet velocity profiles for a two-dimensional diffuser (Wolf and Johnston, 1969).

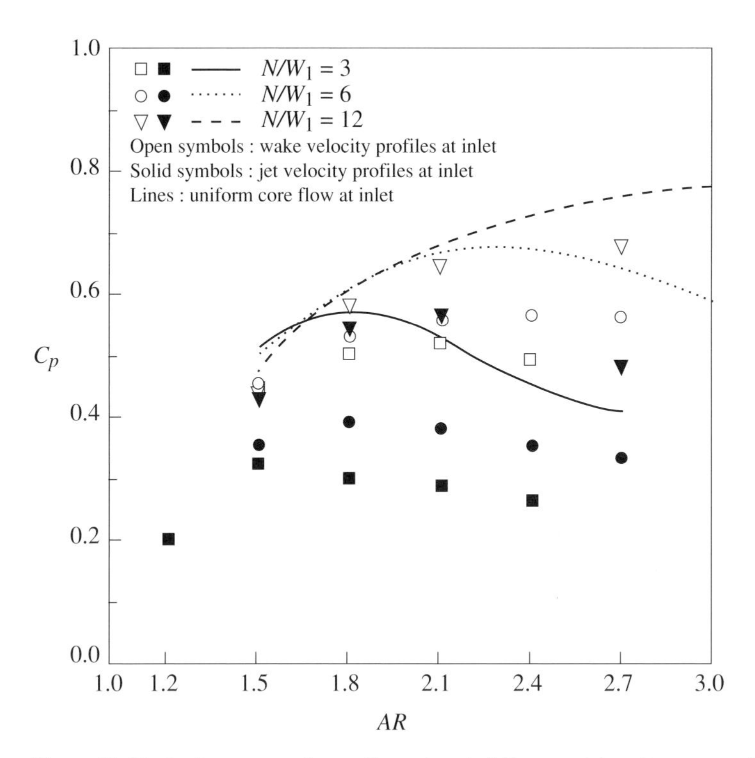

Figure 12.42: Performance of two-dimensional diffusers with uniform core flow at the inlet and with jet and wake inlet profiles; inlet boundary layer blockage = 0.012 (Wolf and Johnston, 1969).

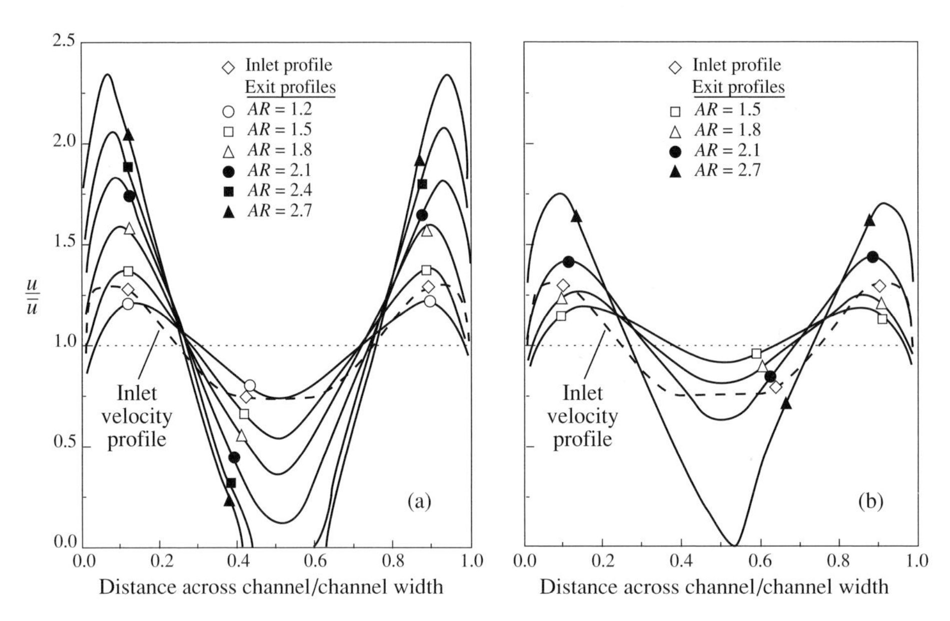

Figure 12.43: Inlet and exit velocity profiles in a two-dimensional diffuser; (a) $N/W_1 = 6$, (b) $N/W_1 = 12$ (Wolf and Johnston, 1966).

For a diffuser, dp is positive. Depending on the streamline and amount of mixing, dp_t can be negative, zero, or positive. For a streamtube with a lower velocity than its neighbors, the effect of mixing is to increase the stagnation pressure. In terms of the fractional velocity non-uniformity $u/\overline{u}$ (where $\overline{u}$ is the mean velocity at a given axial station and $\overline{u}W =$ constant):

$$d\left(\frac{u}{\overline{u}}\right) = \left(\frac{u}{\overline{u}}\right)\left(\frac{dp_t - dp}{\rho u^2} + \frac{dW}{W}\right). \tag{12.10.2}$$

The content of (12.10.2) can be seen in the data of Figures 12.43 and 12.44, which give inlet and exit velocity distributions for two-dimensional diffusers with a wake-type inlet velocity profile. Figure 12.43(a) shows the behavior in diffusers with N/W_1 equal to 6 and area ratios from 1.2 up to 2.7. At the lowest area ratio, the relative velocity non-uniformity is decreased through the diffuser because of the proportionally large influence of viscous forces, with dp_t larger than dp. (Because the velocities are referred to the mean, and the mean velocity at the exit is less than at the inlet, the actual, non-normalized, exit velocity non-uniformity is less than the numerical value in the figure.)

As the diffuser area ratio is increased, pressure forces become more important, resulting in a decrease in normalized wake velocity and, at area ratios of 2.4 and 2.7, reverse flow in the central portion of the wake region. Viscous forces have an effect, indicated by the rounding of the wake profile, but cannot overcome the influence of pressure forces. In this situation it is reverse flow in the center of the channel, rather than at the wall, which is responsible for the limits on pressure rise.

Lengthening the diffuser increases the influence of viscous forces (through increasing the opportunity for mixing) and Figure 12.43(b) shows results for diffusers with non-dimensional length $N/W_1 = 12$. For a given area ratio, the exit velocity non-uniformity is reduced compared to Figure 12.43(a). Even for the largest area ratio of 2.7, there is no flow reversal with the longer diffuser, in spite of the fact that the static pressure rise is higher than that for the shorter diffuser by roughly fifteen percent.

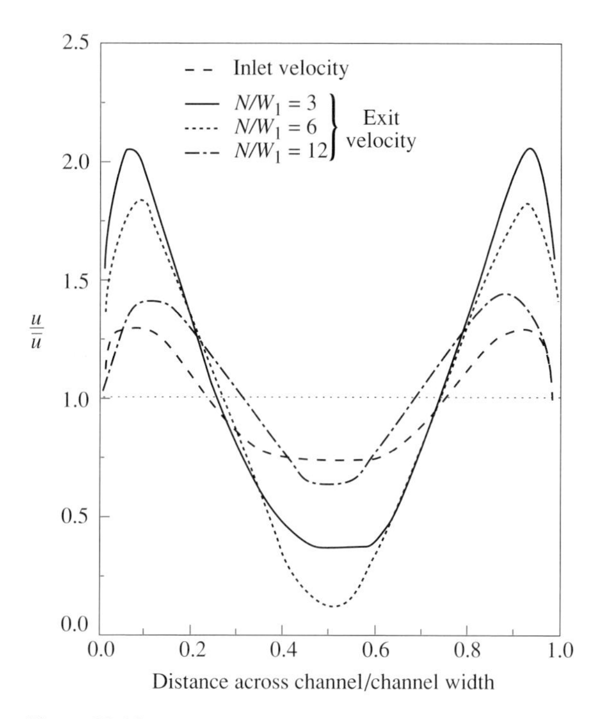

Figure 12.44: Inlet and exit velocity profiles in a two-dimensional diffuser showing the effect of viscous forces on the profile; area ratio = 2.1 (Wolf and Johnston, 1966).

Figure 12.44 gives another view of the data. The figure shows measured inlet and exit velocity profiles for diffusers of area ratio, $AR = 2.1$, and different non-dimensional lengths, N/W_1, 3, 6, and 12. The exit velocity non-uniformity for the $N/W_1 = 6$ diffuser is more severe than that for the $N/W_1 = 3$ diffuser, consistent with Figure 12.43, which shows a larger static pressure rise for the longer diffuser. For these two diffusers the stagnation pressure rise in the middle of the wake was found to be negligible, implying little impact of viscous forces. A different situation occurs when comparing the $N/W_1 = 6$ and $N/W_1 = 12$ diffusers. Viscous forces in the latter result in a stagnation pressure increase of $0.20\ (1/2\rho u^2)$ at the exit compared to the exit of the $N/W_1 = 6$ diffuser, with a consequent decreased velocity non-uniformity.

Section 12.10 can be summarized as follows. The distorting effect of pressure forces on a core velocity non-uniformity can be a major contributor to the decreased performance of fluid components relative to the performance with uniform core flow. The location of the low velocity regions in the core is also important; keeping these regions away from solid surfaces has a favorable effect. Finally, the influence of viscous forces in transferring momentum to low velocity streams tends to mitigate the effect of pressure forces and, if the device is long enough, can result in improved performance.

12.11 Introduction to non-axisymmetric swirling flows

The non-uniformities treated so far have been two-dimensional. In flows with swirl additional features enter the problem which are inherently three-dimensional. Such conditions are found downstream of

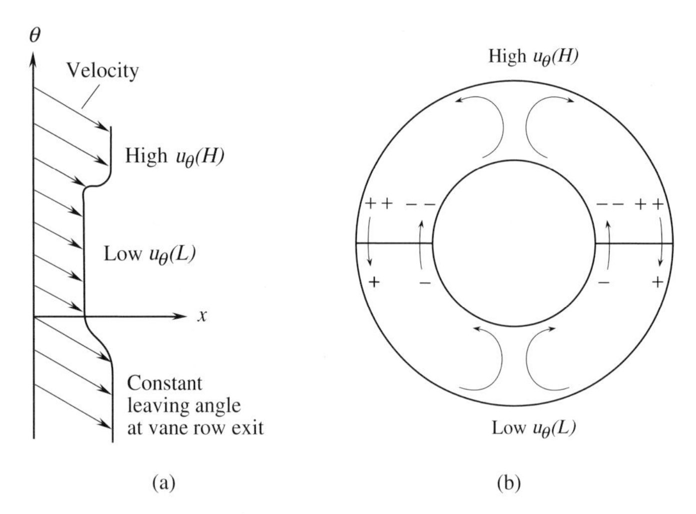

Figure 12.45: Radial flow due to a circumferentially non-uniform velocity in a swirling flow behind a row of stator vanes: (a) the velocity distribution at the vane exit looking radially inward; (b) the flow pattern due to static pressure imbalance (+ and − denote relative levels of static pressure; arrows show direction of secondary flow).

a turbomachinery blade row where wakes are superposed upon a flow with mean swirl. They are also seen in turbomachines operating with circumferential inlet distortion; once the circumferentially non-uniform flow passes through one or more blade rows, it generally has appreciable swirl. To illustrate the phenomena associated with these situations we examine a simple generic problem (incompressible, steady, asymmetric swirling flow in a constant area annulus) in which the salient features are demonstrated explicitly.

With a mean swirl the different types of flow disturbances developed in Chapter 6 (pressure and vorticity) are coupled. The mechanism for the coupling can be introduced through consideration of the circumferentially non-uniform stagnation pressure flow behind a row of constant exit angle vanes in an annulus. There is a circumferential region of stagnation pressure defect and corresponding low velocity so the vane exit velocity at a given radius as sketched in Figure 12.45(a).

Without swirl, the shear (vortical) disturbances would be convected unchanged and the static pressure would be uniform. This, however, is not the case if there is a mean swirl, as seen from the following arguments. Suppose:

(1) the low velocity (L) region has θ-velocity component u_{θ_L}, and the high velocity region (H) has u_{θ_H};
(2) the velocities at any θ-location are roughly the same at all radii;
(3) the streamlines are cylindrical, as they would be for purely convected disturbances.

Under these conditions, the radial pressure gradient can be approximated from simple radial equilibrium as

$$\left(\frac{dp}{dr}\right)_H = \rho\frac{u_{\theta_H}^2}{r}$$

in the H (high velocity) region, and

$$\left(\frac{dp}{dr}\right)_L = \rho \frac{u_{\theta_L}^2}{r}$$

in the L (low velocity) region.

Since $u_{\theta_H} > u_{\theta_L}$, the radial variation in pressure is larger in the "H" region than in the "L" region and at the outer radius of the annulus, the pressure is higher in the H region than in the L region. The resulting circumferential pressure gradient causes flow from high velocity to low velocity regions. Similar arguments apply at the inner radius, with the direction of motion from the low velocity to the high velocity region. Figure 12.45(b) is a view of the annulus looking upstream, which illustrates the situation. The plus and minus signs indicate relative static pressure levels, compared to other circumferential locations at the same radius. The arrows show the direction of the motion that arises because of the pressure gradients. Purely convected disturbances cannot occur because circumferential velocities will be generated as well as (from continuity) outward radial velocities in the H region and inward radial velocities in the L region.

12.11.1 A simple approach for long length scale non-uniformity

A basic quantitative description of steady circumferentially non-uniform swirling flow can be developed to demonstrate some generic three-dimensional features (Greitzer and Strand, 1978). The approximations are: (1) non-uniformities small enough so a linearized analysis is applicable, (2) background flow with a free vortex velocity distribution and uniform axial velocity, (3) a characteristic length scale of the disturbance much larger than the annulus height ($(\Delta r)/r_m \ll 1$), and (4) inviscid, constant density flow.

The coordinate system and nomenclature used is given in Figure 12.46. The relevant equations are derived by linearizing the cylindrical coordinate form of the continuity and momentum equations (see Section 1.14). The background flow has $\overline{u}_\theta = K/r$, $\overline{u}_x$ = constant and $\overline{u}_r = 0$. The resulting

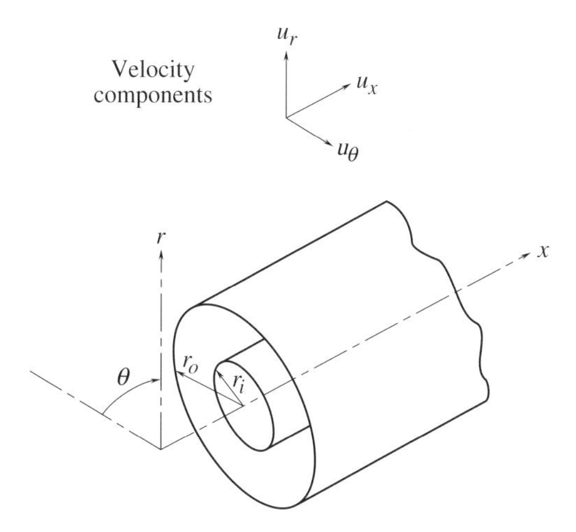

Figure 12.46: Coordinate system for an annular flow field.

equation set is

$$\frac{1}{r}\frac{\partial}{\partial r}(ru'_r) + \frac{1}{r}\frac{\partial u'_\theta}{\partial \theta} + \frac{\partial u'_x}{\partial x} = 0, \tag{12.11.1a}$$

$$\frac{\overline{u}_\theta}{r}\frac{\partial u'_r}{\partial \theta} + \overline{u}_x\frac{\partial u'_r}{\partial x} - \frac{2\overline{u}_\theta u'_\theta}{r} = -\frac{1}{\rho}\frac{\partial p'}{\partial r}, \tag{12.11.1b}$$

$$\frac{\overline{u}_\theta}{r}\frac{\partial u'_\theta}{\partial \theta} + \overline{u}_x\frac{\partial u'_\theta}{\partial x} = -\frac{1}{\rho r}\frac{\partial p'}{\partial \theta}, \tag{12.11.1c}$$

$$\frac{\overline{u}_\theta}{r}\frac{\partial u'_x}{\partial \theta} + \overline{u}_x\frac{\partial u'_x}{\partial x} = -\frac{1}{\rho r}\frac{\partial p'}{\partial x}. \tag{12.11.1d}$$

In (12.11.1), u'_x, u'_θ, u'_r, and p' are the x, θ, and r disturbance velocity components and the pressure, and $\overline{u}_x$ and $\overline{u}_\theta$ are the background (circumferentially uniform) components.

The central idea to be pursued can be expressed as follows. With no swirl, purely convected vorticity perturbations exist with uniform static pressure. With swirl this is not the case. To highlight the behavior differences we split the disturbance velocity components into purely convected perturbations (denoted by subscript c) and a "secondary flow" representing the departure from purely convected disturbances (denoted by subscript s). The former have $u'_r = 0$ and are invariant along the background streamlines. For the latter u'_r is not necessarily zero. The velocity components are:

$$u_r = u'_{r_s} \qquad r\text{-component}; \tag{12.11.2a}$$

$$u_\theta = \overline{u}_\theta + u'_{\theta_c} + u'_{\theta_s} \qquad \theta\text{-component}; \tag{12.11.2b}$$

$$u_x = \overline{u}_x + u'_{x_c} + u'_{x_s} \qquad x\text{-component}. \tag{12.11.2c}$$

The purely convected disturbances, u'_{x_c}, u'_{θ_c}, satisfy

$$L_c\begin{bmatrix} u'_{\theta_c} \\ u'_{x_c} \end{bmatrix} = \left(\frac{\overline{u}_\theta}{r}\frac{\partial}{\partial r} + \overline{u}_x\frac{\partial}{\partial x}\right)\begin{bmatrix} u'_{\theta_c} \\ u'_{x_c} \end{bmatrix} = 0. \tag{12.11.3}$$

L_c is the background flow convective operator defined in (12.11.3). For an asymmetry with θ-dependence of the form $\exp(in\theta)$, the velocity components u'_{x_c}, u'_{θ_c} are:

$$\begin{bmatrix} u'_{\theta_c} \\ u'_{x_c} \end{bmatrix} = A\overline{u}_x\begin{bmatrix} \tan\overline{\alpha} \\ 1 \end{bmatrix}\exp\left\{in\left[\theta - \left(\frac{x}{r}\right)\tan\overline{\alpha}\right]\right\}. \tag{12.11.4}$$

In (12.11.4) A is a non-dimensional disturbance amplitude and $\overline{\alpha}$ is the background flow angle, $\tan\overline{\alpha} = \overline{u}_\theta/\overline{u}_x$.

To simplify (12.11.1) we examine the magnitudes of terms in the momentum equations for a disturbance with a wavelength equal to the circumference ($n = 1$). The continuity equation implies that velocity components scale as

$$\left(\frac{u'_{\theta_s}}{r_m}, \frac{u'_{x_s}}{r_m}\right) \approx \frac{u'_{r_s}}{\Delta r/2}. \tag{12.11.5}$$

Because $\Delta r/2r_m$ is small,

$$\left(\frac{u'_{r_s}}{u'_{\theta_s}}, \frac{u'_{r_s}}{u'_{x_s}}\right) \ll 1. \tag{12.11.6}$$

If the secondary flow velocities are comparable to, or smaller than, the convected disturbances (arguments addressing this are given below) the scaling in (12.11.6) means the term involving u'_r in (12.11.1b) can be neglected and the equation written as an expression of *local* simple radial equilibrium:

$$2\frac{\overline{u}_\theta}{r}\left(u'_{\theta_c} + u'_{\theta_s}\right) = \frac{1}{\rho}\frac{\partial p'}{\partial r}. \tag{12.11.7}$$

For annuli of high hub/tip radius ratio ($r_i/r_o \to 1$), (12.11.7) can be evaluated at the mean radius to obtain an approximation for the pressure distribution as

$$p'(r,\theta,x) - p'(r_m,\theta,x) = 2\rho\overline{u}_\theta\left(u'_{\theta_c} + u'_{\theta_s}\right)\Big|_{r=r_m}\left(\frac{r-r_m}{r_m}\right). \tag{12.11.8}$$

Another simplification stems from the constraint that a vane row puts on the flow angle at the row exit. At this location which we take as the upstream end of the domain ($x = 0$), there is no secondary flow and u'_{θ_s} and u'_{x_s} are zero. The earlier qualitative picture of secondary flow driven by the presence of convected disturbances implies that u'_{θ_s} increases from zero and for some distance downstream is smaller than u'_{θ_c}. If so the former can be neglected in (12.11.8), which becomes a statement linking the three-dimensional pressure field to the convected velocity disturbance at the mean radius:

$$p'(r,\theta,x) - p'(r_m,\theta,x) = 2\rho\overline{u}_\theta[u'_{\theta_c}(r_m,\theta,x)]\left(\frac{r-r_m}{r_m}\right). \tag{12.11.9}$$

Circumferential and axial pressure gradients therefore exist at any radii away from the mean.

12.11.2 Explicit forms of the velocity disturbances

Using the form of the convected velocity components from (12.11.4) in (12.11.9), substituting in (12.11.1c) and (12.11.1d), and applying (12.11.3) yields a pair of ordinary differential equations that describe the growth of the secondary flow disturbances:

$$L_c\begin{bmatrix} u'_{\theta_s}(r,\theta,x) \\ u'_{x_s}(r,\theta,x)\end{bmatrix} = -2\frac{\overline{u}_\theta(r-r_m)}{r_m}\begin{bmatrix}(\partial u'_{\theta_c}/r\partial\theta) \\ (\partial u'_{\theta_c}/\partial x)\end{bmatrix}_{r=r_m}. \tag{12.11.10}$$

Integration of (12.11.10) provides the secondary velocity components. With $\overline{\alpha}_m$ as the background flow angle evaluated at the mean radius,

$$\begin{bmatrix} u'_{\theta_s} \\ u'_{x_s}\end{bmatrix} = 2Ai\overline{u}_x\begin{bmatrix} -1 \\ \tan\overline{\alpha}_m\end{bmatrix}\left\{\left(\frac{r-r_m}{r_m}\right)\left(\frac{x}{r_m}\right)\right.$$

$$\left.\left(\tan^2\overline{\alpha}_m\right)\exp\left(i\left[\theta - \left(\frac{x}{r_m}\right)\tan\overline{\alpha}_m\right]\right)\right\}. \tag{12.11.11}$$

12.11.3 Flow angle disturbances

The flow angle, α, is of interest in turbomachinery applications. Expanding the quantities in the definition $\tan\alpha = u_\theta/u_x$ as mean values plus perturbations and keeping first order terms yields

$$\alpha' = \cos\overline{\alpha}_m\sin\overline{\alpha}_m\left[\left(\frac{u'_{\theta_c}}{\overline{u}_\theta} - \frac{u'_{x_c}}{\overline{u}_x}\right) + \left(\frac{u'_{\theta_s}}{\overline{u}_\theta} - \frac{u'_{x_s}}{\overline{u}_x}\right)\right]. \tag{12.11.12}$$

Because the purely convected disturbances represent parallel streamlines at the background flow angle the first term in parenthesis is zero and

$$\alpha' = \cos\overline{\alpha}_m \sin\overline{\alpha}_m \left(\frac{u'_{\theta_s}}{\overline{u}_\theta} - \frac{u'_{x_s}}{\overline{u}_x} \right). \tag{12.11.13}$$

The flow angle perturbation is therefore associated with the departure from purely convected disturbances.

12.11.4 Relations between stagnation pressure, static pressure, and flow angle disturbances

The static pressure and flow angle disturbances can be cast in terms of the stagnation pressure non-uniformity, since this quantity is typically specified or known. For the convected disturbance,

$$p'_t/\rho = \overline{u}_x u'_{x_c} + \overline{u}_\theta u'_{\theta_c}. \tag{12.11.14}$$

In terms of the amplitude, A, of the convected velocity disturbances (12.11.4), the magnitude of the stagnation pressure non-uniformity is defined as $\Delta p_t = A\rho\overline{u}^2$, with $\overline{u}^2 = \overline{u}_x^2 + \overline{u}_\theta^2$. The static pressure and flow angle distributions can be written in terms of the amplitude of stagnation pressure distortion, Δp_t, the background swirl angle, $\overline{\alpha}_m$, (both evaluated at $r = r_m$), and the radial position as

$$\frac{p'(r,\theta,x)}{\Delta p_t} = 2\sin^2\bar{\alpha}_m \left(\frac{r - r_m}{r_m} \right) \exp\left\{ i\left[\theta - \left(\frac{x}{r_m} \right) \tan\overline{\alpha}_m \right] \right\}, \tag{12.11.15}$$

$$\frac{\alpha'(r,\theta,x)}{\Delta p_t/\rho\overline{u}^2} = -2i\tan^2\overline{\alpha}_m \left(\frac{r - r_m}{r_m} \right) \left(\frac{x}{r_m} \right) \exp\left\{ i\left[\theta - \left(\frac{x}{r_m} \right) \tan\bar{\alpha}_m \right] \right\}. \tag{12.11.16}$$

12.11.5 Overall features of non-axisymmetric swirling flow

Features of the asymmetric swirling flow are illustrated below for a single lobed sinusoidal stagnation pressure distortion ($n = 1$). Two non-dimensional parameters characterize a given configuration: $\overline{\alpha}_m$ and the annulus hub/tip radius ratio, r_i/r_o. The dependence of static pressure and flow angle non-uniformities on these is seen in (12.11.15) and (12.11.16): the radial coordinate, $(r - r_m)/r_m$, can take on larger values as the hub/tip radius ratio decreases leading to larger values of p' and α' and the strength of the radial pressure non-uniformity depends on the swirl angle as $\sin^2\overline{\alpha}_m$.

Two additional features are of note. For the regime in which the analysis applies, the static pressure distortion is in effect "locked in" to the stagnation pressure distortion; both are convected round the annulus at the background flow angle, a phenomenon not seen in the absence of swirl. In addition, flow angle perturbations increase linearly with downstream distance ((12.11.16)) because the deflection of a given particle is proportional to the static pressure gradient it experiences and the length of time (i.e. distance) over which the static pressure gradient is applied.

Figure 12.47 gives the (normalized) magnitude of static pressure variation as a function of background swirl angle, for three hub/tip radius ratios. The quantity shown corresponds to conditions at the hub or tip, which is the maximum variation in static pressure round the annulus.

Figures 12.48–12.50 present a more detailed look at the three-dimensional asymmetric swirling flow in an annulus with background swirl angle of 48° and hub/tip radius ratio 0.43. Features of the

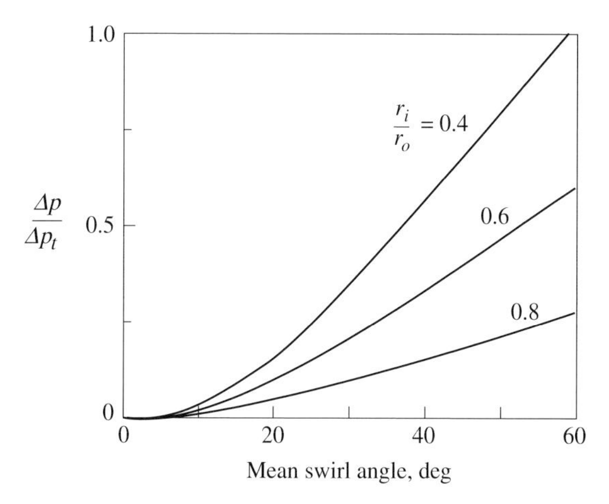

Figure 12.47: Effect of the background flow angle, $\overline{\alpha}_m$, and the hub/tip radius ratio (r_i/r_o) on static pressure non-uniformity in an annular swirling flow.

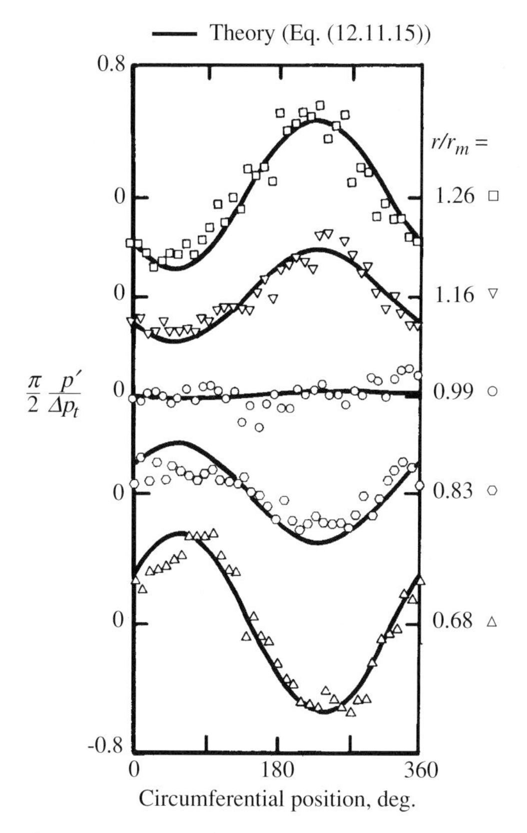

Figure 12.48: Static pressure distribution at different radii in an annulus; $x/r_m = 0.87, \overline{\alpha}_m = 48°, r_i/r_o = 0.43$ (Greitzer and Strand, 1978).

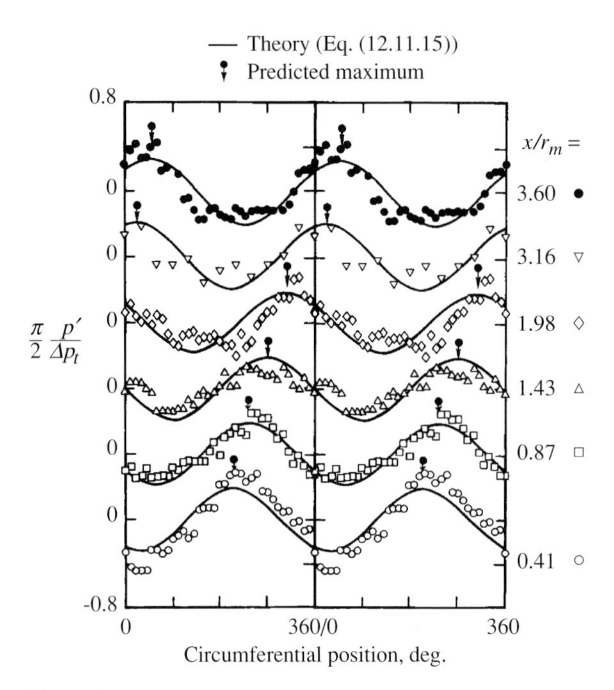

Figure 12.49: Static pressure distribution at different axial stations in an annulus; r/r_m =1.16, $\overline{\alpha}_m = 48°$, $r_i/r_o = 0.43$ (Greitzer and Strand, 1978).

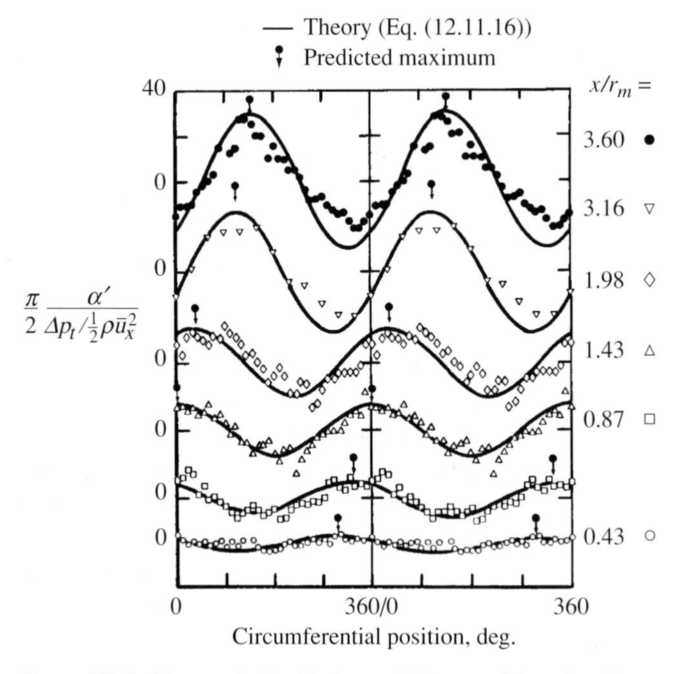

Figure 12.50: Flow angle distribution at different axial stations in an annulus; $r/r_m = 1.16$, $\overline{\alpha}_m = 48°$, $r_i/r_o = 0.43$ (Greitzer and Strand, 1978).

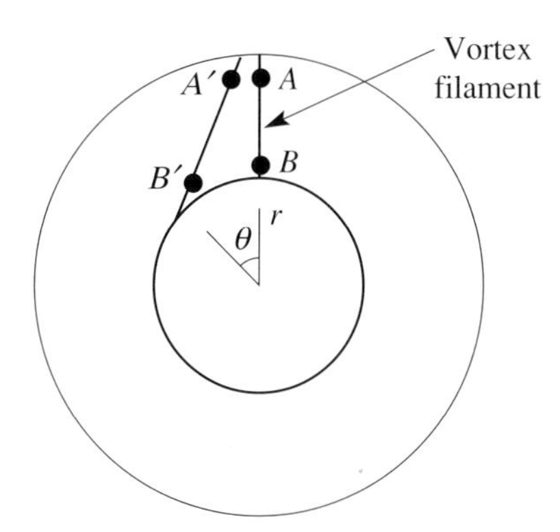

(a) Convection of perturbation vortex filaments by mean flow

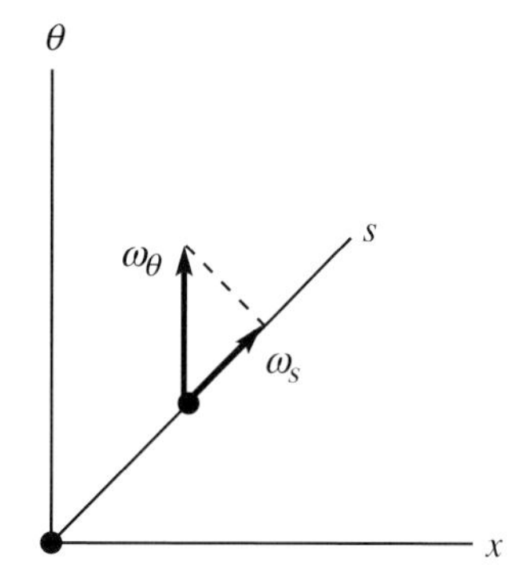

(b) Circumferential and streamwise vorticity components

Figure 12.51: Generation of streamwise vorticity in an asymmetric swirling flow: (a) vortex filaments at the exit of stator vanes (AB) and further downstream $A'B'$; (b) projection of vortex filaments on the θ–x plane.

static pressure and flow angle fields are illustrated one by one. Figure 12.48 highlights the local simple radial equilibrium approximation. The static pressure variation is shown versus θ (circumferential position) at a given axial location for five radii, two on each side of the mean radius and one near the mean. The solid lines represent the approximate analysis, (12.11.15). In accord with simple radial equilibrium the static pressure perturbations are approximately zero at the mean radius, with their magnitude increasing linearly with distance from this mean. Further, at a given circumferential location the perturbations have opposite signs depending on whether the radial location is inward or outward of the mean radius.

The static pressure distribution at constant radius as a function of downstream distance is given in Figure 12.49. The abscissa is circumferential position, the ordinate is non-dimensional static pressure perturbation, and the solid curves again show the approximate analysis. To exhibit the trends more clearly two cycles of data have been plotted side by side, with the vertical arrows indicating the position of the predicted maximum in static pressure. The rotation of the static pressure disturbance round the annulus is evident.

Finally, the flow angle variation with downstream distance is shown in Figure 12.50. The large increase in the magnitude of the flow angle with downstream distance can be remarked.

Flow angles and secondary velocities are both predicted to increase linearly with distance, similar to the secondary flow analyses in Chapter 9, but similar to the secondary flow results there are restrictions on the regime in which the theory applies. As the fluid moves downstream, particles initially at the same circumferential location but different radial locations develop larger angular position differences due to the differences in angular velocity at different radii. The condition of particles at one circumferential location all having the same total pressure (or pure shear velocity perturbation) thus becomes less valid as one moves further downstream. The radial pressure differences also decrease as particles at the different radii shift in phase relative to one another, limiting the growth of the secondary flow. The basic analysis thus cannot be used to give the behavior of the flow field at far downstream locations or, most notably, for disturbances with length scales comparable with (or smaller than) Δr, i.e. large values of harmonic number, n.

12.11.6 A secondary flow approach to non-axisymmetric swirling flow

The three-dimensional motions that arise out of the interaction of vortical disturbances with a background swirl have been referred to as secondary flow disturbances. Figure 12.45 shows these motions are characterized by the existence of streamwise vorticity. We now make a link to the secondary flow analyses for generation of streamwise vorticity, described in Chapter 9.

Consider steady flow with a circumferentially non-uniform stagnation pressure perturbation superposed on a free vortex, constant axial velocity background velocity. Crocco's Theorem tells us there is an associated vorticity field. At $x = 0$ we specify the vortex lines to be purely radial, as they might be at the exit of a vane row. The evolution of these vortex lines is shown schematically in Figure 12.51(a), where the points A and B are fluid particles on the vortex line at $x = 0$, and the points A' and B' represent the same fluid particles after the vortex line has moved some distance downstream. In the context of secondary flow analysis the perturbation vorticity is convected with the background flow and the axial velocity of the two particles is the same. However, the inner particle has a higher angular velocity so that the vortex line becomes inclined to the radial direction, providing a component of vorticity in the θ-direction. The projection of the vortex line $A'B'$ on an x, θ surface (a cylindrical surface at constant radius), drawn in Figure 12.51(b), shows that the presence of a θ-component of vorticity implies streamwise vorticity and hence secondary circulation.

References

Abbott, M. M., and Van Ness, H. C., 1989, *Theory and Problems of Thermodynamics*, Second Edition, Schaum's Outline Series, Mc-Graw-Hill Publishers, New York, NY.

Abernathy, F. H., 1972, Fundamentals of Boundary Layers, in *Illustrated Experiments in Fluid Mechanics*, National Committee for Fluid Mechanics Films, MIT Press, Cambridge, MA.

Abu-Ghannam, B. J., Shaw, R., 1980, Natural Transition of Boundary Layers – The Effects of Turbulence, Pressure Gradient, and Flow History, *J. Mech. Engineering Science*, **22**, No. 3, 213–228.

Ackeret, J., 1967, Aspects of Internal Flow, in *Fluid Mechanics of Internal Flow*, Sovran, G. (ed.), Elsevier Publishing Company, Amsterdam.

Adamczyk, J. J., 2000, Aerodynamic Analysis of Multistage Turbomachinery Flows in Support of Aerodynamic Design, *ASME J. Turbomachinery*, **122**, 189–217.

AGARD, 1994, Turbomachinery Design Using CFD, *AGARD Lecture Series 195*, Advisory Group for Aerospace Research and Development, Neuilly Sur Seine, France.

Anderson, J. D., 1990, *Modern Compressible Flow with Historical Perspectives*, McGraw-Hill Publishers, New York, NY.

Anderson, L. R., Heiser, W. H., Jackson, J. C., 1970, Axisymmetric One-Dimensional Compressible Flow – Theory and Applications, *ASME J. Applied Mechanics*, **37**, 917–923.

Anderson, P., 1977, in *Nobel Lectures – Physics, 1971–1980*, S. Lundqvist (ed.), World Scientific, Singapore.

Andreopoulis, Y., Agui, J. A., Briassulis, G., 2000, Shock Wave-Turbulence Interaction, in *Annual Review of Fluid Mechanics*, Volume 32, Annual Reviews, Inc., Palo Alto, CA, 309–346.

Aris, R., 1962, *Vectors, Tensors and the Basic Equations of Fluid Mechanics*, Prentice-Hall, Englewood Cliffs, NJ.

Bansod, P., and Bradshaw, P., 1971, The Flow in S-shaped Ducts, *Aeronaut. Quart.*, **23**, 131–140.

Barber, T. J., Weingold, H. D., 1978, Vibratory Forcing Functions Produced by Non-Uniform Cascades, *ASME J. Eng. Power*, **100**, 82–88.

Bark, F. H., 1996, Rotating Flows, in *Handbook of Fluid Dynamics and Fluid Machinery*, Volume 1, *Fundamentals of Fluid Dynamics*, Schetz, J. A. and Fuhs, A. E. (eds.), John Wiley & Sons, New York, NY.

Batchelor, G. K., 1964, Axial Flow in Trailing Line Vortices, *J. Fluid Mech.*, **20**, 645–658.

Batchelor, G. K., 1967, *An Introduction to Fluid Dynamics*, Cambridge University Press, Cambridge, England.

Beer, J. M., Chigier, N. A., 1972, *Combustion Aerodynamics*, Halstead Press Division, John Wiley & Sons, Inc., New York, NY.

Bejan, A., 1988, *Advanced Engineering Thermodynamics*, Wiley-Interscience Publications, John Wiley & Sons, New York, NY.

Bejan, A., 1996, *Entropy Generation Minimization, The Method of Thermodynamic Optimization of Finite-Size Systems and Finite-Time Processes*, CRC Press, Boca Raton, FL.

Beran, P. S., Culick, F. E., 1992, The Role of Non-uniqueness in the Development of Vortex Breakdown in Tubes, *J. Fluid Mech.*, **242**, 491–527.

Berger, S. A., Talbot, L., Yao, L.-S., 1983, Flow in Curved Pipes, in *Annual Review of Fluid Mech.*, Volume 15, Annual Reviews, Inc., Palo Alto, CA, 461–512.

Bernstein, A., Heiser, W. H., Hevenor, C., 1967, Compound-Compressible Nozzle Flow, *ASME J. Applied Mech.*, **34**, 548–554.

Betchov, R., Criminale, W. O., 1967, *Stability of Parallel Flows*, Academic Press, New York, NY.

Bo, T., Iacovides, H., Launder, B. E., 1995, Developing Buoyancy – Modified Turbulent Flow in Ducts Rotating in Orthogonal Mode, *ASME J. Turbomachinery*, **117**, 474–484.

Bradshaw, P., 1969, The Analogy Between Streamline Curvature and Buoyancy in Turbulent Shear Flow, *J. Fluid Mech.*, **77**, 153–176.

Bradshaw, P., 1996, Turbulence Modeling with Application to Turbomachinery, *Prog. Aerospace Sci.*, **32**, 575–624.

Bradshaw, P., Koh, Y. M., 1981, A Note on Poisson's Equation for Pressure in a Turbulent Flow, *Phys. Fluids*, **24**, 777.

Bragg, S. L., Hawthorne, W. R., 1950, Some Exact Solutions of the Flow Through Annular Actuator Disks, *J. Aeronautical Sciences*, **17**, 243–249.

Brennen, C. E., 1994, *Hydrodynamics of Pumps*, Concepts ETI, Inc., Wilder, VT and Oxford University Press, Oxford, UK.

Broadbent, E. G., 1976, Flows With Heat Addition, *Prog. Aerospace Sci.*, **17**, 93–108.

Broadwell, J. E., Briedenthal, R. E., 1984, Structure and mixing of a transverse jet in incompressible flow, *J. Fluid Mech.*, **148**, 405–412.

Brown, G. L., Lopez, J. M., 1990, Axisymmetric vortex breakdown Part 2: Physical mechanisms, *J. Fluid Mech.*, **221**, 553–576.

Brown, G. L., Roshko, A., 1974, On Density Effects and Large Structure in Turbulent Mixing Layers, *J. Fluid Mech.*, **64**, 775–816.

Bushnell, D. M., 1992, Longitudinal Vortex Control – Techniques and Applications, the 32nd Lanchester Lecture, *Aeronautical J. Roy. Aeronautical Soc.*, **96**, 419–431.

Camp, T. R., Day, I. J., 1998, A Study of Spike and Modal Stall Phenomena in a Low-Speed Axial Compressor, *ASME J. Turbomachinery*, **120**, 393–401.

Carlson, J. J., Johnston, J. P., 1965, Effects of Wall Shape on Flow Regimes and Performance in Straight Two-Dimensional Diffusers, Report PD-11, Thermosciences Division, Mechanical Engineering Department, Stanford University, Stanford, see also Carlson, J. J., Johnston, J. P., Sagi, C. J., 1967, Effects of Wall Shape on Flow Regimes and Performance in Straight Two-Dimensional Diffusers, *ASME J. Basic Engineering*, **89**, Series D, No. 1, 151–160.

Carta, F. O., 1967, Unsteady Normal Force on an Airfoil in a Periodically Stalled Inlet Flow, *J. Aircraft*, **4**, 416–421.

Cary, A. W., Darmofal, D. L., 2001, Axisymmetric and Non-Axisymmetric Initiation of Vortex Breakdown, in *Proceedings of Research and Technology Organization Applied Vehicle Technology (AVT-072) Symposium on Advanced Flow Management: Vortex Flows and High Angle of Attack – Military Vehicles*, Loen, Norway.

Casey, M. V. 1994, Computational Methods for Preliminary Design and Geometry Definition in Turbomachinery, in *Turbomachinery Design Using CFD*, AGARD Lecture Series 195, Advisory Group for Aerospace Research and Development, Neuilly Sur Seine, France.

Cebeci, T., Bradshaw, P., 1977, *Momentum Transfer in Boundary Layers*, Hemisphere Publishing Co., New York, NY.

Chen, G-T., Greitzer, E. M., Tan, C. S., Marble, F. E., 1991, Similarity Analysis of Compressor Tip Clearance Flow Structure, *ASME J. Turbomachinery*, **113**, 260–271.

Chiang, H-W. and Turner, M. G., 1996, Compressor Blade Forced Response Due to Downstream Vane-Strut Potential Interaction, *ASME J. Turbomachinery*, **118**, No. 1, 134–142.

Chigier, N. A., Chervinsky, A., 1967, Experimental Investigations of Swirling Vortex Motions in Jets, *ASME J. Applied Mech*, **34**, 443–451.

Chima, R. V., Strazisar, A. J., 1983, Comparison of Two and Three-Dimensional Flow Computations with Laser Anemometer Measurements in a Transonic Compressor Rotor, *ASME J. Eng. Power*, **105**, 596–605.

Cho, D., 1995, Effect of Vortex Core Stagnation Pressure on Tip Clearance Flow Blockage in Turbomachines, S. M. Thesis, Dept. of Aeronautics and Astronautics, MIT, Cambridge, MA.

Chue, R, Hynes, T. P, Greitzer, E. M., Tan, C. S., Longley, J. P., 1989, Calculations of Inlet Distortion Induced Compressor Flow Field Instability, *Int. J. Heat and Fluid Flow*, **10**, No. 2, 211–223.

Clauser, F. H., 1956, The Turbulent Boundary Layer, in *Advances in Applied Mechanics*, Volume IV, Academic Press, New York, NY.

Coles, D. E., 1972, Channel Flow of a Compressible Fluid, in *Illustrated Experiments in Fluid Mechanics*, National Committee for Fluid Mechanics Films, MIT Press, Cambridge, MA.

Coles, D. E., 1985, Dryden Lecture: The Uses of Coherent Structure, AIAA Paper 85–0506, American Institute of Aeronautics and Astronautics, Washington, DC.

Coles, D., Hirst, E. A., 1968, Computation of Turbulent Boundary Layers, 1968 AFOSR-IFP-Stanford Conference, *Proc. 1968 Conf.* Volume 2, Stanford University, Stanford, CA.

Cornell, W. G, 1958, Losses in Flow Normal to Plane Screens, *ASME Transactions*, **80**, 791–799.

Cortelezzi, L., Karagozian, A.R., 2001, On the Formation of the Counter-Rotating Vortex Pair in Transverse Jets, *J. Fluid Mech.*, **446**, 347–373.

Cravalho, E. G., and Smith, J. L., 1981, *Engineering Thermodynamics*, Pitman Publishing Inc., Marshfield, MA.

Crocco, L, 1958, One-Dimensional Treatment of Steady Gas Dynamics, in Emmons, H. W., *Fundamentals of Gas Dynamics*, Volume III in the Princeton Series High Speed Aerodynamics and Jet Propulsion, Princeton University Press, Princeton, NJ.

Cumpsty, N. A., 1989, *Compressor Aerodynamics*, Longman Group UK Ltd., London, England.

Cumpsty, N. A., 1992, Aerodynamics Of Aircraft Engines – Strides And Stumbles, Gas Turbine Laboratory Report #213, Massachusetts Institute of Technology, Cambridge, MA.

Cumpsty, N. A., 1998, *Jet Propulsion*, Cambridge University Press, Cambridge, England.

Cumpsty, N.A., Greitzer, E.M., 1982, A Simple Model for Compressor Stall Cell Propagation, *ASME J. Eng. Power*, **104**, 170–176.

Cumpsty, N.A., Horlock, J. H., 1999, Averaging for Non-uniform Flow of an Ideal Gas, internal memo, Whittle Laboratory, Cambridge University, Cambridge, England.

Curran, E. T., Heiser, W. H., Pratt, D. T., 1996, Fluid Phenomena in Scramjet Combustion Systems, in *Annual Review of Fluid Mechanics*,Volume 28, Annual Reviews, Inc., Palo Alto, CA, 323–360.

Daily, J. W., Nece, R. E., 1960, Chamber Dimension effects on Induced Flow and Frictional Resistance of Enclosed Rotating Disks, *ASME J. Basic Engineering*, 217–232.

Darmofal, D. L, 2002, personal communication.

Darmofal, D. L., Khan, R.Greitzer, E. M., Tan, C. S., 2001, Vortex Core Behaviour in Confined and Unconfined Geometries: A Quasi-One-Dimensional Model, *J. Fluid Mech.*, **449**, 61–84.

Davis, G. de Vahl, 1957, The Flow of Air Through Wire Screens, Ph.D. Thesis, Department of Engineering, University of Cambridge, England.

Day, I. J., 1993, Stall Inception in Axial Flow Compressors, *ASME J. Turbomachinery*, **115**, 1–9.

Dean, R. C., 1959, On the Necessity of Unsteady Flow in Fluid Machines, *ASME J. Basic Engineering*, **81**, 24–28.

Decher, R., 1978, Nonuniform Flow Through Nozzles, *J. Aircraft*, **15**, 416–421.

Delery, J. M., 1994, Aspects of Vortex Breakdown, *Prog. Aerospace Sci.*, **30**, 1–59.

Delery, J. M., 2001, Robert Legendre and Henri Werle: Toward the Elucidation of Three-Dimensional Separation, in *Annual Review of Fluid Mechanics*, Volume 33, Annual Reviews, Inc., Palo Alto, CA, pp. 129–155.

Denbigh, K, 1981, *The Principles of Chemical Equilibrium*, Fourth Edition, Cambridge University Press, Cambridge, England.

Den Hartog, J. P., 1948, *Mechanics*, Dover Publications, Inc., New York, NY.

Denton, J. D., 1990, Entropy Generation in Turbomachinery Flows, Cliff Garret Award Lecture, paper SP-846, SAE International Aerospace Technology Conference and Exposition, Long Beach, CA.

Denton, J. D., 1993, Loss Mechanisms in Turbomachines – The 1993 IGTI Scholar Lecture, *ASME J. Turbomachinery*, **115**, 621–656.

Denton, J. D., 2000, personal communication.

De Siervi, F., Viguier, H. C., Greitzer, E. M., Tan, C. S., 1982, Mechanisms of Inlet-Vortex Formation, *J. Fluid Mech.*, **124**, 173–207.

Dimotakis, P. E., 1986, Two-Dimensional Shear-Layer Entrainment, *AIAA Journal*, **24**, No. 11, 1791–1796.

Dimotakis, P. E., 1991, Turbulent Free Shear Layer Mixing and Combustion, in *High-Speed Flight Propulsion Systems*, Progress in Aeronautics and Astronautics, Volume 137, American Institute of Aeronautics and Astronautics, Washington, DC, pp. 265–340.

Dou, H-S, Mizuki, S., 1998, Analysis of the Flow in Vaneless Diffusers with Large Width-to-Radius Ratios, *ASME J. Turbomachinery*, **120** 193–202.

Dowling, A. P., Ffowcs Williams, J. E., 1983, *Sound and Sources of Sound*, Ellis-Horwood Publishers, Chicester, England.

Drazin, P. G., Reid, W. H., 1981, *Hydrodynamic Stability*, Cambridge University Press, Cambridge, England.

Drela, M., 1989, Integral Boundary Layer Formulation for Blunt Trailing Edges, AIAA Paper 89-2166-CP, American Institute of Aeronautics and Astronautics, Washington, DC.

Drela, M. 1998, Assorted Views on Teaching of Aerodynamics, AIAA paper 98–2792, American Institute of Aeronautics and Astronautics, Washington, DC.

Drela, M., 2000, personal communication.

Drela, M., Giles, M. B., 1987, Viscous-Inviscid Analysis of Transonic and Low Reynolds Number Airfoils, *AIAA J.*, **25**, No. 10, 1347–1355.

Eckert, E. R. G., 1984, Experiments on Energy Separation in Fluid Streams, *Mechanical Engineering*, **106**, 58–65.

Eckert, E. R. G., 1987, Cross Transport of Energy in Fluid Streams, *Warme- und Stoffubertragung*, **21**, 73–81, 1968–1979.

Eckert, E. R. G., Drake, R. M. Jr, 1972, *Analysis of Heat and Mass Transfer*, McGraw-Hill, Inc., New York, NY.

Ekaterinaris, J. A., Platzer, M. F., 1997, Computational Prediction of Airfoil Dynamic Stall, *Prog. Aerospace Sci.*, **33**, 759–846.

Elder, J. W., 1959, Steady Flow Through Non-Uniform Gauzes of Arbitrary Shape, *J. Fluid Mech.*, **5**, 355–363.

Escudier, M., 1987, Confined Vortices in Fluid Machinery, in *Annual Review of Fluid Mechanics*, Volume 19, Annual Reviews, Inc., Palo Alto, CA, 27–52.

Escudier, M., 1988, Vortex Breakdown: Observations and Explanations, *Prog. Aerospace Sci.*, **25**, 189–229.

Favaloro, S. C., Nejad, A. S., and Ahmed, S. A., 1991, Experimental and Computational Investigation of Isothermal Swirling Flow in an Axisymmetric Dump Combustor, *AIAA J. Propulsion and Power*, **7**, 348–356.

Fearn, R., 2002, personal comminication.

Fearn, R. Weston, R. B. 1974, Vorticity Associated with a Jet in Crossflow, *AIAA J.*, **12**, 1666–1671.

Fearn, R. Weston, R. B. 1978, Induced Velocity Field of a Jet in Crossflow, NASA Technical Paper 1087.

Feynman, R. P., 1985, *Surely You're Joking, Mr. Feynman*, W. W. Norton and Company, New York, NY.

Fink, D. A., Cumpsty, N. A., Greitzer, E. M., 1992, Surge Dynamics in a Free-Spool Centrifugal Compressor System, *ASME J. Turbomachinery*, **114**, 321–332.

Fitzgerald, N. 2002, personal communication.

Freeman, C., 1985, Effects of Tip Clearance Flow on Compressor Stability and Engine Performance, Lecture Series 1985–05, Von Karman Institute for Fluid Dynamics, Rhode-St-Genese, Belgium.

Fric, T. F., Roshko, A., 1994, Vortical Structure in the Wake of a Transverse Jet, *J. Fluid Mech*, **279**, 1–47.

Fried, E., Idelchik, I. E., 1989, *Flow Resistance: A Design Guide For Engineers*, Hemisphere Publishing Co., New York, NY.

Gibbings, J. C., 1973, The Pyramid Gauze Diffuser, *Ingenieur-Archiv*, **42**, 225–233.

Gibbs, J. W. 1901, *Vector Analysis* (based on lectures of J. W. Gibbs, book by E. B. Wilson), Yale University Press, New Haven, CT.

Goldstein, H., 1980, *Classical Mechanics*, Second Edition, Addison-Wesley Publishing Company, Reading, MA.

Goldstein, M. E., 1978, Unsteady Vortical and Entropic Distortions of Potential Flows Round Arbitrary Obstacles, *J. Fluid Mech.*, **89**, 433–468.

Goldstein, S., 1960, *Lectures on Fluid Mechanics*, Interscience Publishers Ltd, London, England.

Gong, Y., Tan, C. S., Gordon, K., Greitzer, E. M., 1999, A Three-Dimensional Computational Flow Model for Short Length Scale Instability in Multistage Axial Compressors, *ASME J. Turbomachinery*, **121**, No. 4, 726–734.

Gostelow, J. P., 1984, *Cascade Aerodynamics*, Pergamon Press, New York, NY.

Graf, M. B., Wong, T. S., Greitzer, E. M., Marble, F. E., Tan, C. S., Shin, H-W., Wisler, D. C., 1998, Effects of Nonaxisymmetric Tip Clearance on Axial Compressor Performance and Stability, *ASME J. Turbomachinery*, **120**, 648–661.

Green, S. I., 1995, Wing Tip Vortices in Green, S. I. (ed.) *Fluid Vortices*, Kluwer Academic Publishers, Dordrecht, The Netherlands.

Greenspan, H. P., 1968, *The Theory of Rotating Fluids*, Cambridge University Press, Cambridge, England.

Greitzer, E. M., 1981, The Stability of Pumping Systems – The 1980 Freeman Scholar Lecture, *ASME J. Fluids Engineering*, **103**, 193–243.

Greitzer, E. M., Griswold, H. R., 1976, Compressor–Diffuser Interaction with Circumferential Flow Distortion, *J. Mech. Eng. Sci.*, **18**, No. 1, 25–38.

Greitzer, E. M., Strand, T. 1978, Asymmetric Swirling Flows in Turbomachine Annuli, *ASME J. Eng. for Power*, **100**, 618–629.

Greitzer, E. M., Mazzawy, R. S., Fulkerson, D. A., 1978, Flow Field Coupling Between Compression System Components in Asymmetric Flow, *ASME J. Eng. for Power*, **100**, 66–72.

Greitzer, E. M., Paterson, R. W., Tan, C. S., 1985, An Approximate Substitution Principle for Viscous Heat Conducting Flows, *Proc. Roy. Soc. London*, A **401**, 163–193.

Gutmark, E. J., Schadow, K. C.,Yu, K. H., 1995, Mixing Enhancement in Supersonic Free Shear Layers, in *Annual Review of Fluid Mechanics*, Volume 27, Annual Reviews, Inc., Palo Alto, CA, 375–417.

Gupta, A. K., Lilley, D. G., and Syred, N., 1984, *Swirl Flows*, Abacus Press. Tunbridge Wells, Kent, England.

Gysling, D. L., 1993, Dynamic Control of Rotating Stall in Axial Flow Compressors using Aeromechanical Feedback, Gas Turbine Laboratory Report No. 219, Massachusetts Institute of Technology, Cambridge, MA.

Hall, M. G., 1966, The Structure of Concentrated Vortex Cores, *Prog. Aeronautical Sci.* **7**, 53–110.

Hall, M. G., 1972, Vortex Breakdown, in *Annual Review of Fluid Mechanics*, Volume 4, Annual Reviews, Inc., Palo Alto, CA, 195–218.

Hall, W. B., Orme E. M., 1955, Flow of a Compressible Fluid Through a Sudden Enlargement in a Pipe, *I. Mech. E. Proc.*, **169**, 1007–1020.

Hansen, K. E., Jorgensen, P., Larsen, P. S., 1981, Experimental and Theoretical Study of Surge in a Small Centrifugal Compressor, *ASME J. Fluids Eng.*, **103**, 391–395.

Harrison, S., 1990, Secondary Loss Generationin a Linear Cascade of High Turning Turbine Blades, *ASME J. Turbomachinery*, **112**, 618–624.

Hasselbrink, E. F., Jr, Mungal, M. G., 2001, Transverse Jets and Jet Flames. Part 1. Scaling Laws for Strong Transverse Jets, and Transverse Jets and Jet Flames. Part 2. Velocity and OH Field Imaging, *J. Fluid Mech.*, **443**, 1–25 and 27–68.

Haven, B. A., Kurosaka, M., 1997, Kidney and Anti-Kidney Vortices in Cross-Flow Jets, *J. Fluid Mech.*, **352**, 27–64.

Hawthorne, W. R., 1951, Secondary Circulation in Fluid Flow, *Proc. Roy. Soc. London*, A, **206**, 374–387.

Hawthorne, W. R., 1957, Some Aerodynamic Problems of Aircraft Engines, *J. Aeronautical Sci.*, **24**, No. 10, 713–730.

Hawthorne, W. R., 1974, Secondary Vorticity in Stratified Compressible Fluids in Rotating System, CUED/A-Turbo/TR 63, Dept of Engineering, University of Cambridge, England.

Haynes, J. M., Hendricks, G. J., Epstein, A. H., 1994, Active Stabilization of Rotating Stall in a Three-Stage Axial Compressor, *ASME J. Turbomachinery*, **116**, 226–239.

Heiser, W. H., 1995, personal communication.

Heiser, W. H., Pratt, D. T., 1994, *Hypersonic Airbreathing Propulsion*, American Institute of Aeronautics and Astronautics, Washington, DC.

Heiser, W. H., McClure, W. B., Wood, C. W., 1995, Simulating Heat Addition Via Mass Addition in Constant Area Compressible Flows, *AIAA J.*, **33**, 167–170.

Heiser, W. H., McClure, W. B., Wood, C. W., 1996, Simulating Heat Addition Via Mass Addition in Variable Area Compressible Flows, *AIAA J.*, **34**, 1076–1078.

Hendricks, G. J., Sabnis, J. S., Feulner, M. R., 1997, Analysis of Instability Inception in High-Speed Multistage Axial Flow Compressors, *ASME J. Turbomachinery*, **119**, 714–722.

Hicks, B. L., 1948, Diabatic Flow of a Compressible Fluid, *Quart. Appl. Math.*, **6**, 221–237.

Hill, P. G., Peterson, C. R., 1992, *Mechanics and Thermodynamics of Propulsion*, Second Edition, Addison-Wesley Publishing Company, Reading MA.

Hill, P. G., Schaub, U. W., Senoo, Y., 1963, Turbulent Wakes in Pressure Gradients, *ASME J. Appl. Mech.*, **30**, Series E, No. 4, 518–524.

Hodder, B. K., 1981, An Investigation of Engine Influence on Inlet Performance, NASA CR-166136.

Horlock, J. H., 1966, *Axial Flow Turbines: Fluid Mechanics and Thermodynamics*, Butterworth and Company Ltd; London, England (also 1973 reprint, Robert E. Krieger Publishing Co. Inc, NY).

Horlock, J. H., 1975, Secondary Flows, Von Karman Institute Lecture Series 72: *Secondary Flows in Turbomachinery*, Von Karman Institute, Rhode St-Genese, Belgium.

Horlock, J. H., 1978, *Actuator Disk Theory*, McGraw-Hill, New York, NY.

Horlock, J. H., 1992, *Combined Power Plants Including Combined Cycle Gas Turbine (CCGT) Plants*, Pergamon Press, Oxford, England.

Horlock, J. H., Lakshminarayana, B., 1973a, Generalized Expressions for Secondary Vorticity Using Intrinsic Coordinates, *J. Fluid Mech.*, 59, pp. 97–115 (see also Horlock, J. H. Lakshminarayana, B. 1991, Corrigendum, *J. Fluid Mech.*, **226**, 661–663).

Horlock, J. H., Lakshminarayana, B., 1973b, Secondary Flows: Theory, Experiment and Application in Turbomachinery Aerodynamics, in *Annual Review of Fluid Mechanics*,Volume 5, Annual Reviews, Inc., Palo Alto, CA, 247–280.

Hornung, H. G., 1988, Vorticity Generation and Secondary Flows, AIAA Paper 88-3751-CP, American Institute of Aeronautics and Astronautics, Washington, DC.

Howard, L. N., 1963, Fundamentals of the Theory of Rotating Fluids, *ASME J. Applied Mech.*, **30**, 481–485.

Howarth, L., 1953, *Modern Developments in Fluid Dynamics: High Speed Flow*, Clarendon Press, Oxford, England.

Hsiao, E., Naimi, M., Lewis, J. P., Dalbey, K., Gong, Y., Tan, C. S., 2001, Actuator Duct Model of Turbomachinery Components for Powered-Nacelle Navier–Stokes Calculations, *AIAA J. Propulsion and Power*, **17**, No. 4, 919–927.

Humphrey, J. A. C., Taylor, A. M. K., Whitelaw, J. H., 1977, Laminar Flow in a Square Duct of Strong Curvature, *J. Fluid Mech.*, **83**, 509–527

Humphrey, J. A. C., Whitelaw, J. H., Yee, G., 1981, Turbulent Flow in a Square Duct with Strong Curvature, *J. Fluid Mech.*, **103**, 443–463.

Hunt, J. C. R., 1987, Vorticity and Vortex Dynamics in Complex Turbulent Flows, *Trans. CSME*, **11**, No. 1, 21–35.

Hynes, T., 1991, personal communication.

Incropera, F. P., De Witt, D. P., 1996, *Fundamentals of Heat and Mass Transfer*, Fourth Edition, John Wiley & Sons, New York, NY.

Ishigaki, H., 1994, Analogy Between Laminar Flows in Curved Pipes and Orthogonally Rotating Pipes, *J. Fluid Mech.*, **268**., 133–145.

Ishigaki, H., 1996, Analogy Between Turbulent Flows in Curved Pipes and Orthogonally Rotating Pipes, *J. Fluid Mech.* **307**, 1–10.

Jacobs, J. W., 1992, Shock-Induced Mixing of a Light-Gas Cylinder, *J. Fluid Mech.*, **234**, 629–649.

Jane's Aero Engines, 1999, W. Gunston (ed.), Jane's Information Group Limited, Coulsdon, Surrey, England.

Japikse, D., 1984, *Turbomachinery Diffuser Design Technology*, Concepts ETI, Inc, Wilder, VT.

Johnson, B. V., Daniels, W. A., Kawecki, E. J., Martin, R. J., 1990, Compressor Drum Aerodynamic Experiments and Analysis with Coolant Injected at Selected Locations, ASME paper 90-GT-151.

Johnson, M. W., 1978, Secondary Flow in Rotating Bends, *ASME J. Eng. for Power*, **100**, 553–560.

Johnston, J. P., 1978, Internal Flows, in *Turbulence*, Bradshaw, P. (ed.), Springer Verlag, Berlin, 109–172.

Johnston, J. P., 1986, Boundary Layers in Internal Flow – Performance Prediction, in *Advanced Topics in Turbomachinery Technology*, Japikse, D. (ed.), Concepts ETI Press, Wilder, VT.

Johnston, J. P., 1997, Diffuser Design and Performance Analysis by Unified Integral Methods, AIAA Paper 97–2733, American Institute of Aeronautics and Astronautics, Washington, DC.

Johnston, J. P., Nishi, M., 1990, Vortex Generator Jets – Means for Flow Separation Control, *AIAA J.*, **28**, No. 6, 989–994.

Johnston, J. P., Halleen, R. M., Lezius, D. K., 1972, Effects of Spanwise Rotation on the Structure of Two-Dimensional Fully Developed Turbulent Channel Flow, *J. Fluid Mech.*, **56**, 533–557.

Karagozian, A. R., 2002, Background on and Applications of Jets in Crossflow, in *Manipulation and Control of Transverse Jets*, Karagozian, A. R., Cortelezzi, L., and Soldati, A. (eds.), Springer-Verlag, Berlin.

Karagozian, A. R., Marble, F. E., 1986, Study of a Diffusion Flame in a Stretched Vortex, *Combustion Sci. Tech.*, **45**, 65–84.

Katz, J. and Plotkin, A., 2001, *Low Speed Aerodynamics*, Cambridge University Press, Cambridge, England.

Kerrebrock, J. L., 1992, *Aircraft Engines and Gas Turbines*, MIT Press, Cambridge, MA.

Kerwin, J. E., Kinnas, S. A., Lee, J.-T., Shih, W.-Z., 1987, A Surface Panel Method for the Hydrodynamic Analysis of Ducted Propellers, The Society of Naval Architects and Marine Engineers Annual Meeting, No. 4, New York, NY.

Kestin, J., 1979, *A Course in Thermodynamics*, Volume One, Hemisphere Publishing Co., New York, NY.

Khalak, A., 2000, Personal communication.

Khalid, S. A., Khalsa, A. S., Waitz, I. A., Tan, C. S., Greitzer, E. M., Cumpsty, N. A., Adamczyk, J., Marble, F. E., 1999, Endwall Blockage in Axial Compressors, *ASME J. Turbomachinery*, **121**, No. 3, 499–509.

Kheshgi, H. S., Scriven, L. E., 1985, Viscous Flow Through a Rotating Square Channel, *Phys. Fluids*, **28**, No. 10, 1968–1979.

Kline, S. J., Johnston, J. P., 1986, Diffusers – Flow Phenomena and Design Layers in Internal Flow – Performance Prediction, in *Advanced Topics in Turbomachinery Technology*, Japikse, D. (ed.), Concepts ETI Press, Wilder, VT.

Kline, S. J., Bardina, J. G., Strawn, R. C., 1983, Correlation of the Detachment of Two-Dimensional Turbulent Boundary Layers, *AIAA J.*, **21**, No. 1, 68–73.

Koo, J.-K., James, D. F., 1973, Fluid Flow Around and Through a Screen, *J. Fluid Mech.*, **60**, 513–538.

Köppel, P, Roduner, C., Kupferschmied, P., Gyarmathy, G., 1999, On the Development and Application of the FRAP (Fast-Response Aerodynamic Probe) System in Turbomachines – Part 3: Comparison of Averaging Methods Applied to Centrifugal Compressor Measurements, ASME Paper 99-GT-154, presented at International Gas Turbine & Aeroengine Congress, Indianapolis, IN.

Krasny, R., 1986, Desingularization of Periodic vortex Sheet Roll-Up, *J. Comp. Physics*, **60**, 513–538.

Kristoffersen, R., Andersson, H. L., 1993, Direct Simulation of Low-Reynolds-Number Turbulent Flow in a Rotating Channel, *J. Fluid Mech.*, **256**, 163–197.

Küchemann, D., 1965, Report on the I. U. T. A. M. Symposium on Concentrated Vortex Motion in Fluids, *J. Fluid Mech.*, **21**, 1–20.

Küchemann, D., 1978, *The Aerodynamic Design of Aircraft*, Pergamon Press, Oxford, England.

Küchemann, D, Weber, J. 1953, *Aerodynamics of Propulsion*, McGraw-Hill, New York, NY.

Kurosaka, M., Gertz, J. B., Graham, J. E., Goodman, J. R., Sundaram, P., Riner, W. C., Kuroda, H., Hankey, W. L., 1987, Energy Separation in a Vortex Street, *J. Fluid Mech.*, **178**, 1–29.

Lagerstrom, P. A., 1996, *Laminar Flow Theory*, Princeton University Press, Princeton, NJ (originally published 1964 in *Theory of Laminar Flows*, Moore, F. K. (ed.), Princeton University Press).

Lamb, H., 1945, *Hydrodynamics*, Sixth Edition, Dover Books, New York, NY.

Landahl, M. T., Widnall, S. E., 1971, Vortex Control in *Aircraft Wake Turbulence and its Detection*, A. Goldburg, J. H. Olsen, M. Rogers (eds.), Plenum Press, New York, NY.

Landau, L. D., Lifshitz, E. M., 1987, *Fluid Mechanics*, Second Edition, Pergamon Press, Oxford, England.

Langston, L. S., 1980, Crossflows In A Turbine Cascade Passage, *ASME J. Eng. Power*, **102**, No. 4, 866–874.

Lau, J. C., 1981, Characteristics of Mach Number and Temperature on Mean-Flow and Turbulence Characteristics in Round Jets, *J. Fluid Mech.*, **105**, 193–218.

Launder, B. E., 1995, Modeling the Formation and Dispersal of Streamwise Vortices in Turbulent Flow, the 35th Lanchester Lecture, *Aeronautical J. Roy. Aeronautical Soc.*, **99**, 419–431.

Laws, E. M., Livesy, J. L., 1978, Flow Through Screens, in *Annual Review of Fluid Mechanics*, Volume 10, Annual Reviews, Inc., Palo Alto, CA, pp. 247–266.

Lee, J. F., Sears, F. W., 1963, *Thermodynamics*, Addison Wesley Publishers, Reading MA.

Lei, U., Hsu, C. H., 1990, Flow through Rotating Straight Pipes, *Phys. Fluids*, **A2**, No. 1, 63–75.

Lei, U., Lin, M. J., Sheen, H. J., Lin, C. M., 1994, Velocity Measurements of the Laminar Flow through a Rotating Straight Pipe, *Phys. Fluids*, **6**, No. 6, 1972–1982.

Leibovich, S., 1983, Vortex Stability and Breakdown: Survey and Extension, *AIAA J.*, **22**, , No. 9, 1192–1206.

Leibovich, S., Kribus, A., 1990, Large-Amplitude Wavetrains and Solitary Waves in Vortices, *J. Fluid Mech.*, **216**, 459–504.

Leonard, A. 1985, Computing Three-Dimensional Incompressible Flows with Vortex Elements, in *Annual Review of Fluid Mechanics*, Volume 17, 523–59 Annual Reviews, Inc., Palo Alto, CA, pp. 523–559.

Lepicovsky, J., 1990, Total Temperature Effects on Centerline Mach Number Characteristics of Freejets, *AIAA Journal*, **28**, 478–482.

Lewis, M. J., Hastings, D. E, 1989, Application of Compound-Compressible Flow to Nonuniformities in Hypersonic Propulsion Systems, *J. Propulsion and Power*, **5**, 626–635.

Lewis, R. I., 1991, *Vortex Element Methods for Fluid Dynamic Analysis of Engineering Systems*, Cambridge University Press, London, England.

Lezius, D. K., Johnston, J. P., 1976, Roll cell instabilities in rotating laminar and turbulent channel flows, *J. Fluid Mech.* 77, pp. 153–176.

Liepmann, H. W., Roshko, A., 1957, *Elements of Gas Dynamics*, John Wiley & Sons, New York, NY.

Lighthill, M. J., 1954, The Response of Laminar Skin Friction and Heat Transfer to Fluctuations in the Stream Velocity, *Proc. Roy. Soc. London*, A **224**, 1–23.

Lighthill, M. J., 1958, On Displacement Thickness, *J. Fluid Mech.*, **4**, 383–392.

Lighthill, M. J., 1962, Physical Interpretation of the Mathematical Theory of Wave Generation by Wind, *J. Fluid Mech*, **14**, 385–398.

Lighthill, M. J., 1963, Introduction; Boundary Layer Theory, in *Laminar Boundary Layers*, Rosenhead, L. (ed.), Clarendon Press, Oxford, England (reprinted 1988 by Dover Press, New York, NY).

Lighthill, M. J., 1966, Dynamics of Rotating Fluids: a Survey, *J. Fluid Mech.*, **26**, 411–431.

Lighthill, M. J., 1978, *Waves in Fluids*, Cambridge University Press, Cambridge, England.

Lighthill, M. J., 1986a, *An Informal Introduction to Theoretical Fluid Mechanics*, Clarendon Press, Oxford, England.

Lighthill, M. J., 1986b, Fundamentals Concerning Wave Loading on Off-Shore Structures, *J. Fluid Mech.*, **173**, 667–681.

Lin, J. C., 2002, Review of Research on Low-Profile Vortex Generators to Control Boundary-Layer Separation, *Prog. Aerospace Sci.*, **33**, 389–420.

Lin, P., Rao, G. V. R., and O'Connor, G. M., 1991, Numerical Investigation on Shock Wave/Boundary-Layer Interactions in a Constant Area Diffuser at Mach 3, AIAA Paper 91–1766, American Institute of Aeronautics and Astronautics, Washington, DC.

Livesey, J. L., 1972, Duct Performance Parameters Considering Spatially Non-Uniform Flow, AIAA Paper 72–85, American Institute of Aeronautics and Astronautics, Washington, DC.

Livesey, J. L. Hugh, T. 1966, 'Suitable Mean Values' in One-Dimensional Gas Dynamics, *J. Mech. Eng. Sci*, **8**, No. 4, 374–383; see also the Communication on the paper by Ward Smith, A. J., 1967, *J. Mech. Eng. Sci.* **9**, No. 3, 241–245.

Longley, J. P., 1994, A Review of Nonsteady Flow Models for Compressor Stability, *ASME J. Turbomachinery*, **116**, 202–215.

Longley, J. P., Greitzer, E. M., 1992, Inlet Distortion Effects in Aircraft Propulsion System Integration, in *Steady and Transient Performance Prediction of Gas Turbine Engines*, AGARD Lecture Series 183, Advisory Group for Aerospace Research and Development, Neuilly Sur Seine, France.

Longley, J. P., Shin, H-W., Plumley, R. E., Silkowski, P. D., Day, I. J., Greitzer, E. M., Tan, C. S., Wisler, D. C., 1996, Effects of Rotating Inlet Distortion on Multistage Compressor Stability, *ASME J. Turbomachinery*, **118**, 181–188.

Lucas, M. J., Noreen, R. A., Sutherland, L. C., Cole III, J. E., Junger, M. C., 1997, *Handbook of the Acoustic Characteristics of Turbomachinery Cavities*, ASME Press, New York, NY.

Lumley, J. L., 1967, The Applicability of Turbulence Research to the Solution of Internal Flow Problems, in *Fluid Mechanics of Internal Flow*, Sovran, G. (ed.), Elsevier Publishing Company, Amsterdam.

Lyrio, A. A. Ferziger, J. H., 1983, A Method of Predicting Unsteady Turbulent Flows and Its Application to Diffusers with Unsteady Inlet Conditions, *AIAA J.*, **21**, 534–540.

Lyrio, A. A., Ferziger, J. H., Kline, S. J., 1981, An Integral Method for the Computation of Steady and Unsteady Turbulent Boundary Layer Flows, Including the Transitory Stall Regime in Diffusers Report PD-23, Thermosciences Division, Department of Mechanical Engineering, Stanford University, CA.

MacFarlane, I., Joubert, P. N., Nickels, T. B., 1998, Secondary Flows and Developing Turbulent Boundary Layers in a Rotating Duct, *J. Fluid Mech.*, **373**, 1–32.

Mahesh, K., Lele, S. K., Moin, P., 1997, The Influence of Entropy Fluctuations on the Interaction of Turbulence with a Shock Wave, *J. Fluid Mech.*, **334**, 353–379.

Manning, T., 1991, Experimental Studies of Mixing Flows with Streamwise Vorticity, MS Thesis, Dept of Aeronautics and Astronautics, Massachusetts Institute of Technology, Cambridge, MA.

Marble, F. E., 1985, Growth of A Diffusion Flame in the Field of a Vortex, in *Recent Advances in the Aerospace Sciences*, Corrado Casci (ed.), Plenum Publishing Corp, New York, NY, 395–413.

Marble, F. E., Candel, S, M., 1977, Acoustic Disturbance from Gas Non-Uniformities Convected Through a Nozzle, *J. Sound and Vibration*, **55**, 225–243.

Marble, F. E., Zukoski, E. E., Jacobs, J. W., Hendricks, G. J., Waitz, I. A., 1990, Shock Enhancement and Control of Hypersonic Mixing and Combustion, AIAA paper 90–1981, American Institute of Aeronautics and Astronautics, Washington, DC.

Margason, R. J., 1993, Fifty Years of Jet in Cross Flow Research, AGARD Conference Proceedings 534, Advisory Group for Aerospace Research and Development, Neuilly Sur Seine, France.

Marshall, J. S., 1991, A General Theory of Curved Vortices With Circular Cross-Section and Variable Core Area, *J. Fluid Mech.*, **229**, 283–307.

Marshall, J. S., 1993, The Effect of Axial Pressure Gradient on Axisymmetrical and Helical Vortex Waves, *Phys. Fluids*, A **5**, 588–599

Maxworthy, T. M., 1988, Waves on Vortex Cores, *Fluid Dynamics Res.*, **3**, 52–62.

Mayle, R. E., 1973, personal communication.

Mayle, R. E., 1991, The Role of Laminar-Turbulent Transition in Gas Turbine Engines – The 1991 IGTI Scholar Lecture, *ASME J. Turbomachinery*, **113**, 509–537.

McCormick, D. C., 1992, Vortical and Turbulent Structure of Planar and Lobed Mixer Free-Shear Layers, PhD Thesis, University of Connecticut, Storrs, CT.

McCroskey, W. J., 1977, Some Current Research in Unsteady Fluid Dynamics – The 1976 Freeman Scholar Lecture, *ASME J. Fluids Engineering*, **99**, 8–38.

McCroskey, W. J., Philippe, J. J., 1975, Unsteady Viscous Flow on Oscillating Airfoils, *AIAA Journal*, **13**, No. 1, 71–79.

McCroskey, W. J., Pucci, S. L., 1982, Viscous–Inviscid Interaction on Oscillating Airfoils in Subsonic Flow, *AIAA J.*, **20**, No. 2, 167–174.

Mehta, R. D., 1977, The Aerodynamic Design of Blower Tunnels with Wide-Angle Diffusers, *Prog. Aerospace Sci.*, **18**, 59–120.

Millar, D. A. J., 1971, A Note on Axisysmmetric Isentropic Compressible Flow in an Annular Duct, *Canadian Aero. Space Inst. Trans.*, **4**, 145–146.

Miller, D. S., 1990, *Internal Flow Systems*, British Hydraulic Research Association (BHRA) Information Services, Cranfield, Bedfordshire, England.

Moin, P., 1998, Numerical and Physical Issues in Large Eddy Simulation of Turbulent Flows, *JSME International Journal*, Series B, **41**, No. 2, 91–100.

Moin, P., 2002, Advances in Large Eddy Simulation Methodology for Complex Flows, *Int. J. Heat and Fluid Flow*, **23**, 710–220.

Moin, P., Mahesh, K., 1998, Direct Numerical Simulation: A Tool in Turbulence Research, in *Annual Reviews of Fluid Mechanics*, Volume 30, Annual Review, Inc, Palo Alto, CA, pp. 539–578.

Moore, J., 1973a, A Wake and an Eddy in a Rotating, Radial Flow Passage, Part I: Experimental Observations, *ASME J. Eng. for Power*, **95**, 205–212.

Moore, J., 1973b, A Wake and an Eddy in a Rotating, Radial Flow Passage, Part II: Flow Model, *ASME J. Eng. for Power*, **95**, 213–219.

Morris, P. J., Giridharan, M. G., Lilley, G. M., 1990, On the Turbulent Mixing of Compressible Free Shear Layers, *Proc. Roy. Soc. London*, A, **431**, 219–243.

Morris, W. D., 1981, *Heat Transfer and Fluid Flow in Rotating Coolant Channels*, Research Studies Press Division, John Wiley & Sons, New York, NY.

Morse, P. M., Feshbach, H., 1953, *Methods of Theoretical Physics*, Part I, Chapter 1, McGraw-Hill Book Company, New York, NY.

Morse, P. M., Ingard, K. U., 1968, *Theoretical Acoustics*, Princeton University Press, Princeton, NJ.

Munjal, M. L., 1987, *Acoustics of Ducts and Mufflers*, John Wiley & Sons, New York, New York.

Munk, M., Prim, R., 1947, On the Multiplicity of Steady Gas Flows Having the Same Streamline Pattern, *Proc. Nat. Acad. Sci.*, **33**, 137–141.

Neumann, E. P., Lustwerk, F., 1949, Supersonic Diffusers For Wind Tunnels, *ASME J. Applied Mech.*, **16**, 195–202.

Ng, S. L., Brennen, C. E., 1978, Experiments on the Dynamic Behaviour of Cavitating Pumps, *ASME J. Fluids Engineering*, **100**, 166–176.

Nitsche, M., Krasny, R., 1994, A Numerical Study of Vortex Ring Formation at the Edge of a Circular Tube, *J. Fluid Mech.*, **276**, 139–161.

Ogata, K., 1997, *Modern Control Engineering*, Prentice-Hall, Upper Saddle River, NJ.

O'Sullivan, M. N., Waitz, I. A., Greitzer, E. M., Tan, C. S., Dawes, W. N., 1996 A Computational Study of Viscous Effects on Lobed Mixer Flow Features and Performance, *J. Propulsion and Power*, **12**, No. 3, 449–456.

Owen, J. M., Rogers, R. H., 1989, *Flow and Heat Transfer in Rotating-Disc Systems*, Volume 1 – *Rotor-Stator Systems*, Research Studies Press, Ltd., Taunton, England, John Wiley & Sons, Inc, New York, NY.

Paduano, J. D., Greitzer, E. M., Epstein, A. H., 2001, Compression System Stability and Active Control, *Annual Review of Fluid Mechanics*, Volume 33, Annual Reviews Publishers, Palo Alto, CA.

Pagan, D., 1989, Contribution a L'Etude Experimental et Theoretique de L'Eclatement Tourbillonnaire en Air Incompressible, Thesis de Doctorat de L'Universite Paris VI, Paris, France.

Panton, R. L., 1984, *Incompressible Flow*, John Wiley & Sons, New York, NY.

Papamoschou, D., 1994, Thrust Loss Due to Supersonic Mixing, *J. Propulsion and Power*, **10**, No. 6., 804–809.

Papamoschou, D., Roshko, A., 1988, The Compressible Turbulent Shear Layer: An Experimental Study, *J. Fluid Mech.*, **197**, 453–477.

Patel, M. H., 1975, On Laminar Boundary Layers in Oscillatory Flow. *Proc. Roy. Soc. London*, A **347**, 99–123.

Paterson, R. W., 1982, Turbofan Forced Mixer-Nozzle Internal Flowfield: I. A Benchmark Experimental Study, NASA Contractor Report No. NASA CR 3492.

Paterson, R. W., Weingold, H. D., 1982, Experimental Investigation of a Simulated Compressor Airfoil Trailing-Edge Flowfield, Pratt & Whitney Report FR-15859, Pratt & Whitney Aircraft Group, Government Products Divsion, West Palm Beach, FL.

Paterson, R. W., Weingold, H. D., 1985, Experimental Investigation of a Simulated Compressor Airfoil Trailing-Edge Flowfield, *AIAA J.*, **23**, 768–775.

Pauley, W. R. Eaton, J., K., 1988, Experimental Study of the Development of Longitudinal Vortex Pairs Embedded in a Turbulent Boundary Layer, *AIAA J.*, **26**, No. 7, 816–823.

Pianko, M., Wazelt, F., 1983, Propulsion and Energetics Panel Working Group 14 on Suitable Averaging Techniques in Non-Uniform Internal Flows, AGARD Advisory Report, No. 182, Advisory Group for Aerospace Research and Development, Neuilly Sur Seine, France.

Pitts, D. R., Sissom, L. E., 1977, *Theory and Problems of Heat Transfer,* McGraw-Hill Publishers, New York, NY.

Plonsey. R., Collin R. E., 1961, *Principles and Applications of Electromagnetic Fields*, McGraw-Hill Publishers, New York, NY.

Poinsot, T. J., Trouve, A. C., Veynante, D. P., Candel, S. M., Esposito, E. J., 1987, Vortex-Driven Acoustically Coupled Combustion Instabilities, *J. Fluid Mech.*, **177**, 265–292.

Prandtl, L., 1952, *Essentials of Fluid Dynamics*, Blackie and Son Limited, London, England.

Prasad, D., 1998, private communication.

Prasad, D., Hendricks, G. J., 2000, A Numerical Study of Secondary Flow in Axial Turbines with Application to Radial Transport of Hot Streaks, *ASME J. Turbomachinery*, **122**, 667–673.

Pratt, D. T., Heiser, W. H., 1993, Isolator–Combustor Interaction in a Dual-Mode Scramjet Engine, AIAA Paper AIAA-93–0358, American Institute of Aeronautics and Astronautics, Washington, DC.

Preston, J. H., 1961, The Non-steady Irrotational Flow of an Inviscid, Incompressible Fluid, with Special Reference to changes in Total Pressure through Flow Machines, *Aeronautical Quart.*, **12**, 343–360.

Presz, W., Gousy, R., Morin, B., 1986, Forced Mixer Lobes in Ejector Designs, AIAA paper AIAA-86–1614, American Institute of Aeronautics and Astronautics, Washington, DC.

Presz Jr, W. M., Greitzer, E. M., 1988, A Useful Similarity Principle for Jet Engine Exhaust System Performance, AIAA Paper AIAA-88–3001, American Institute of Aeronautics and Astronautics, Washington, DC.

Qiu, Y. J., 1992, A Study of Streamwise Vortex-Enhanced Mixing in Lobed Mixer Devices, PhD Thesis, Dept of Aeronautics and Astronautics, Massachusetts Institute of Technology, Cambridge, MA.

Reneau, L. R., Johnston, J. P., Kline, S. J., 1967, Performance and Design of Straight Two-Dimensional Diffusers, *ASME J. Basic Engineering*, **89**, Series D., 151–160.

Reynolds, W. C., Perkins, H. C., 1977, *Engineering Thermodynamics*, McGraw-Hill Book Company, New York, NY.

Ricou, F. P., Spalding, D. B., 1961, Measurement of Entrainment by Axisymmetrical Turbulent Jets, *J. Fluid Mech.*, **11**, 21–32.

Roberts, Q. D., and Denton, J. D., 1996, Loss Production in the Wake of a Simulated Subsonic Turbine Blade, ASME Paper 96-GT-421, IGTI Gas Turbine Congress.

Roshko, A., 1954, On the Drag and Shedding Frequency of Two-Dimensional Bluff Bodies, NACA Technical Note 3169.

Roshko, A., 1976, Structure of Turbulent Shear Flows: A New Look, *AIAA J.*, **14**, 1349–1357.

Roshko, A., 1993a, Instability and Turbulence in Shear Flows, in *Theoretical and Applied Mechanics 1992*, S. R. Bodner, J. Singer, A. Solan, Z. Hashin (eds.), Elsevier Science Publishers, Amsterdam, The Netherlands.

Roshko, A., 1993b, Free Shear Layers, Base Pressure, and Bluff Body Drag, *Proceedings of the Symposium on Developments in Fluid Dynamics and Aerospace Engineering*, December 1993, S. M. Des hand (ed.) Interline Publishers, Bangalore, India.

Roshko, A. 1993c, Perspectives on Bluff Body Aerodynamics, *J. Wind Eng. Industrial Aerodynamics*, **49**, 79–100.

Rothe, P. H., Johnston, J. P., 1976, Effects of System Rotation on the Performance of Two-Dimensional Diffusers, ASME Paper 76-FE-15.

Ryan, L. F., 1951, Experiments on Aerodynamics Cooling, Thesis for Dr. degree, Swiss Federal Institute of Technology, Zurich, (as cited in Eckert, E. R. G, 1987, Cross Transport of Energy in Fluid Streams, *Warme-und Stoffubertragung*, **21** 73–81).

Sabersky, R. H., Acosta, A. J., Hauptmann, E. G., 1989, *Fluid Flow: A First Course*, Third Edition, Macmillan Publishing Company, New York, NY.

Saffman, P. G., 1981, Dynamics of Vorticity, *J. Fluid Mech.*, **106**, 49–58.

Saffman, P. G., 1992, *Vortex Dynamics*, Cambridge University Press, Cambridge, England.

Sandham, N. D., Reynolds, W. C., 1990, Compressible Mixing Layer: Linear Theory and Direct Simulation, *AIAA J.*, **28** No. 4, 618–624.

Sarpkaya, T. 1988, Computational Methods With Vortices – The 1988 Freeman Scholar Lecture, *ASME J. Fluids Eng.* **11**, 5–52.

Sarpkaya, T, 1994, Vortex Element Methods for Flow Simulation, *Adv. App. Mech.*, **31**, 113–247.

Schetz, J. A., 1980, *Injection and Mixing in Turbulent Flow*, Progress in Aeronautics and Astronautics, Volume 68, Summerfield, M. (ed.), American Institute of Aeronautics and Astronautics, Washington, DC.

Schlichting, H., 1968, *Boundary Layer Theory*, Sixth Edition, McGraw-Hill Book Company, New York, NY.

Schlichting, H., 1979, *Boundary-Layer Theory*, Seventh Edition, McGraw-Hill, New York, NY.

Schubauer, G. B., Spangenberg, W. G., 1948, Effect of Screens in Wide-Angle Diffusers, NACA Technical Note 1610.

Schubauer, G. B. Spangenberg, W. G., 1960, Forced Mixing in Boundary Layers, *J. Fluid Mech.*, **8**, 10–32.

Schubauer, G. B., Spangenber, W. G., Klebanoff, P. S., 1950, Aerodynamic Characteristics of Damping Screens, NACA Technical Note 2001.

Schwind, R. G., 1962, The Three-Dimensional boundary Layer Near a Strut, Gas Turbine Laboratory Report No. 67, Massachusetts Institute of Technology, Cambridge, MA.

Senoo, Y., Kinishita, Y., Ishida, M., 1977, Asymmetric Flow in Vaneless Diffusers of Centrifugal Blowers, *ASME J. Fluids Engineering*, **99**, 104–114.

Sforza, P. M., Mons, R. F., 1978, Mass, Momentum, and Energy Transport in Turbulent Free Jets, *Int. J. Heat and Mass Transfer*, **21**, 371–384.

Shapiro, A. H., 1953, *The Dynamics and Thermodynamics of Compressible Fluid Flow*, Volume 1, The Ronald Press Company, New York, NY (now John Wiley & Sons, New York, NY), 1975 reprinted by The Ronald Press Company, New York, NY.

Shapiro, A. H., 1972, Pressure Fields and Fluid Accelerations, in *Illustrated Experiments in Fluid Mechanics*, National Committee for Fluid Mechanics Films, MIT Press, Cambridge, MA.

Sherman, F. S., 1990, *Viscous Flow*, McGraw-Hill, New York, NY.

Simon, J. S., Valavani, L., 1991, A Lyapunov Based Nonlinear Control Scheme for Stabilizing a Basic Compression System Using a Close-Coupled Control Valve, in *Proceedings of the 1991 American Control Conference*, Institute of Electrical and Electronics Engineers, New York, NY.

Simonich, J. C., Schlinker, R. H., 1983, personal communication; see also Simonich, J. C., 1986, Isolated and Interacting Round Parallel Heated Jets, AIAA Paper 86–0281, American Institute of Aeronautics and Astronautics, Washington, DC.

Simpson, R. L., 1996, Aspects of Turbulent Boundary-Layer Separation, *Prog. Aerospace Sci.*, **32**, 457–521.

Skebe, S. A., Paterson, R. W., Barber, T. J., 1988, Experimental Investigation of Three-Dimensional Forced Mixer Lobed Flow Fields, AIAA paper AIAA 88–3785, American Institute of Aeronautics and Astronautics, Washington, DC.

Smith, L. H., 1955, Secondary Flow in Axial-Flow Turbomachinery, *ASME Trans.*, **77**, 1065–1076.

Smith, L. H., 1966a, The Radial-Equilibrium Equation of Turbomachinery, *ASME J. Eng. Power*, **88**, 1–12.

Smith, L. H., 1966b, Wake Dispersion in Turbomachines, *ASME J. Basic Eng.*, **88**, 688–690.

Smith, L. H., 1970, Casing Boundary Layers in Multistage Axial Compressors, in *Flow Research on Blading*, L. S. Dzung (ed.), Elsevier Publishing Company, Amsterdam, The Netherlands.

Smith, L. H., 1993, Wake Ingestion Propulsion Benefit, *J. Propulsion and Power*, **9**, No. 1, 74–82.

Smith, L. H., 2001, Personal communication.

Smith, S. H., Mungal, M. G., 1998, Mixing, Structure and Scaling of the Jet in Crossflow, *J. Fluid Mech.*, **357**, 83–122.

Soeder, R. H., Bobula, G. A., 1979, Effect of Steady-State Pressure Distortion on Flow Characteristics Entering a Turbofan Engine, NASA Technical Memorandum 79134 and AVRADCOM Technical Report 79–19.

Sommerfeld, A., 1964, *Mechanics of Deformable Bodies*, Academic Press, New York, NY.

Sonntag, R. E., Borgnakke, C., Van Wylen, G. J., 1998, *Fundamentals of Thermodynamics,* Fifth Edition, John Wiley & Sons, New York, NY.

Sovran, G., Klomp, E. D., 1967, Experimentally Determined Optimum Geometries for Rectilinear Diffusers with Rectangular, Conical, or Annular Cross-Section, in *Fluid Mechanics of Internal Flow*, Sovran, G. (ed.), Elsevier Publishing Company, Amsterdam, The Netherlands.

Squire, H. B., Winter, K. G., 1951, The Secondary Flow in a Cascade of Airfoils in a Non-Uniform Stream, *J. Aeronautical Sciences*, **18**, 271–277.

Stenning, A. H., 1980, Inlet Disortion Effects in Axial Compressors, *ASME J. Fluids Engineering*, **102**, 1, 7–14.

Stetson, H. D., 1984, Designing for Stability in Advanced Turbine Engines, *Int. J. Turbo and Jet Engines*, **1**, 235–245.

Stoker, J. J., 1950, *Nonlinear Vibrations*, Interscience Publishers, New York, NY.

Strawn, R. C., Ferziger, J. H., Kline, S. J., 1984, A New Technique for Computing Viscous–Inviscid Interactions in Internal Flows, *ASME J. Fluids Engineering*, **106**, 79–84.

Strogatz, S. H., 1994, *Nonlinear Dynamics and Chaos*, Addison-Wesley Publishers, Reading, MA.

Stuart, J. T., 1963, Unsteady Boundary Layers in *Laminar Boundary Layers*, L. Rosenhead (ed.), Clarendon Press, Oxford, England (reprinted 1988 by Dover Press, New York, NY).

Sykes, R. I., Lewellen, W. S., and Parker, S. P., 1986, Vorticity Dynamics of a Turbulent Jet in a Crossflow, *J. Fluid Mech.*, **168**, 393–413.

Tannehill, J. C., Anderson, D. A., Pletcher, R. H., 1997, *Computational Fluid Mechanics and Heat Transfer*, Taylor & Francis, Philadelphia, PA.

Tatro, P. R., Mollo-Christensen, E. L., 1967, Experiments on Ekman Layer Instability, *J. Fluid Mech.*, **28**, 531–543.

Taylor, E. S., 1971, Boundary Layers, Wakes and Losses in Turbomachines, Gas Turbine Laboratory Report No. 105, Massachusetts Institute of Technology, Cambridge, MA.

Taylor, E. S., 1972, Low-Reynolds-Number Flows, in *Illustrated Experiments in Fluid Mechanics*, National Committee for Fluid Mechanics Films, MIT Press, Cambridge, MA.

Taylor, G. I., Batchelor, G. K., 1949, The effect of a gauze on small disturbances in a uniform stream, *Quart J. Mech. Appl. Math*, **2**, 1–29.

Teeple, B. S., 1995, personal communication.

Telionis, D. P., 1979, Review – Unsteady Boundary Layers, Separated and Attached, *ASME J. Fluids Engineering*, **101**, 29–43.

Telionis, D. P., Romaniuk, M. S., 1978, Velocity and Temperature Streaming in Oscillating Boundary Layers, *AIAA J.*, **16**, 488–495.

Tennekes, H., Lumley, J. L., 1972, *A First Course in Turbulence*, MIT Press, Cambridge, MA.

Tew, D. E., Teeple, B. S., Waitz, I. A., 1998, Mixer-Ejector Noise-Suppressor Model, *J. Propulsion and Power*, **14**, 941–950.

Thompson, P. A., 1984, *Compressible-Fluid Dynamics*, Maple Press.

Thwaites, B., 1960, *Incompressible Aerodynamics*, Clarendon Press, Oxford, England (reprinted 1988, Dover Publications, New York, NY).

Tillman, T. G., Paterson, R. W., Presz, W. M., 1992, Supersonic Nozzle Mixer Ejector, *J. Propulsion and Power*, **8**, 513–519.

Tobak, M., Peake, D. J., 1982, Topology of Three-Dimensional Separated Flows, in *Annual Review of Fluid Mechanics*, Volume 4, Annual Reviews, Inc., Palo Alto, CA, pp. 61–85.

Tritton, D. J., Davies, P. A., 1981, Instabilities in Geophysical Fluid Dynamics, in *Hydrodynamic Instabilities and the Transition to Turbulence*, Topics in Applied Physics, Swinney, H. L., Gollub, J. P., (eds.), Volume 45, Springer Verlag, Berlin, 229–269.

Tritton, D. J., 1988, *Physical Fluid Dynamics*, Clarendon Press, Oxford, England.

Truckenbrodt, E., 1952, A Method of Quadrature for Calculation of the Laminar and Turbulent Boundary Layer in Case of Plane and Rotationally Symmetric Flow, translated as NACA TM 1379.

Tsien, H. S., 1958, The Equations of Gas Dynamics, in *Fundamentals of Gas Dynamics*, Emmons, H. W. (ed.), Princeton University Press, Princeton, NJ.

Turner, J. S., 1986, Turbulent Entrainment: The Development of the Entrainment Assumption, and Its Application to Geophysical Flows, *J. Fluid Mech.*, **173**, 431–471.

Underwood, D. S., Waitz, I. A., Greitzer, E. M., 2000, Confined Swirling Flows with Heat Release and Mixing, *AIAA J. Propulsion and Power*, 16, No. 2, pp. 169–177.

Valkov, T., Tan, C. S., 1999, Effects of Upstream Rotor Vortical Disturbances on the Time-Averaged Performance of Axial Compressor Stators: Part 1 – Framework of Technical Approach and Rotor Wake-Stator Blade Interactions, Part 2 – Rotor Tip Vortex/Streamwise Vortex-Stator Blade Interactions, *ASME J. Turbomachinery*, **121**, No. 3, 377–397.

Van Zante, D. E., Adamczyk, J. J., Strazisar, A. J., Okiishi, T. H., 2002, Wake Recovery Performance Benefit in a High-Speed Axial Compressor, *ASME J. Turbomachinery*, **124**, No. 2, 275–284.

von Karman, T. 1954, On the Foundation of High Speed Aerodynamics in *General Theory of High Speed Aerodynamics*, W. R. Sears (ed.), Princeton University Press, Princeton, NJ.

von Karman, T., Burgers, J. M., 1963, General Aerodynamic Theory – Pefect Fluids, in *Aerodynamic Theory*, Volume II, Durand, W. R. (ed.), Dover Publications, NY.

Waitz, I. A., Qiu, Y. J., Manning, T. A., Fung, A. K. S., Elliot, J. K., Kerwin, J. M., Krasnodebski, J. K., O'Sullivan, M. N., Tew, D. E., Greitzer, E. M., Marble, F. E., Tan, C. S., Tillman T. G., 1997, Enhanced Mixing with Streamwise Vorticity, *Prog. Aerospace Sci.*, **33**, No. 5–6, 323–351.

Wake, B. E., Choi, D., Hendricks, G. J., 1996, Numerical Investigation of Pre-Mixed Step-Combustor Instabilities, AIAA Paper 96–0816, American Institute of Aeronautics and Astronautics, Washington, DC.

Waltrup, P. J., Billig, F. S., 1973, Structure of Shock Waves in Cylindrical Ducts, *AIAA J.*, **11**, 1404–1408.

Wang, S., Rusak, Z., 1997, The dynamics of a swirling flow in a pipe and transition to axisymmetric vortex breakdown, *J. Fluid Mech.*, **340**, 177–223.

Ward-Smith, A. J., 1980, *Internal Fluid Flow*, Clarendon Press, Oxford, England.

White, F. M., 1991, *Viscous Fluid Flow*, McGraw-Hill, Inc., New York, NY.

Wisler, D. C., 1998, Axial-Flow Compressor and Fan Aerodynamics, in *Handbook of Fluid Dynamics*, R. W. Johnson (ed.), CRC Press, Boca Raton, FL.

Wolf, S., Johnston, J. P., 1966, Effects of Nonuniform Inlet Velocity Profiles on Flow Regimes and Performance in Two-Dimensional Diffusers, Report PD-12, Thermosciences Division, Stanford University, Stanford, CA.

Wolf, S., Johnston, J. P., 1969, Effects of Nonuniform Inlet Velocity Profiles on Flow Regimes and Performance in Two-Dimensional Diffusers *ASME J. Basic Engineering*, **91**, 462–474.

Yang, J., Kubota, T., Zukoski, E. E., 1994, A Model for Characterization of a Vortex Pair formed by Shock Passage over a Light-Gas Inhomogeneity, *J. Fluid Mech.*, **258**, 217–244.

Yang, K.-Y., Kim, J., 1991, Numerical Investigation of Instability and Transition in Rotating Plane Poiseuille Flow, *Phys. Fluids*, **A 3**, 633–641.

Young, A. D., 1989, *Boundary Layers*, BSP Professional Books, Oxford, England.

Young, J. B., 1995, Condensation in Jet Engine Intake Ducts During Stationary Operation, *ASME J. of Engineering for Gas Turbines and Power*, **117**, No. 2, 227–236.

Young, J. B., Wilcock, R. C., 2001, Modeling the Air-Cooled Gas Turbine Part 2 – Coolant Flows and Losses, ASME Paper 01-GT-0392, ASME International Gas Turbine Institute Turbo Expo, New Orleans, LA.

Zhuang, M., Kubota, T., Dimotakis, P. E., 1990, Instability of Inviscid, Compressible Free Shear Layers, *AIAA J.*, **28**, 1728–1733.

Zierep, J., 1974, Theory of Flow in Compressible Media with Heat Addition, AGARDdograph No. 191, Advisory Group for Aerospace Research and Development, Neuilly Sur Seine, France.

Zukoski, E. E., 1985, Afterburners, in *Aerothermodynamics of Aircraft Engine Components*, G. C. Oates (ed.), AIAA Education Series, American Institute of Aeronautics and Astronautics, Washington, DC.

Supplementary references appearing in figures

Ashajaee, J., Johnston, J. P., Kline, S. J., 1980, Subsonic Turbulent Flow in Plane Wall Diffuser: Peak Pressure Recovery and Transitory Stall, Report PD-21 Department of Mech. Engineering, Stanford University, also *ASME J. Fluids Engineering*, **102**, 275–282, 1980.

Bennett, H. W., 1953, An Experimental Study of Boundary Layer Transition, Report, Kimberly-Clarke Corporation, Research and Development Laboratories.

Brown, B., Burton, R. C., 1977, The Effects of Free-Stream Turbulence Intensity and Velocity distribution on Heat Transfer to Curved Surfaces, ASME Paper, 77-GT-48.

Cham, T.-S., Head, M. R., 1970, Calculation of the Turbulent Boundary Layer in a Vortex Diffuser, Aeronautical Research Council Reports and Memorandum 3646.

Cousteix, J., 1979, Couches Limits en Ecoulement Pulse, ONERA Report No. NT-1979-1.

Didden, N., 1979, On the Formation of Vortex Rings: Rolling Up and Production of Circulation, *Z. Agnew Math. Phys.*, **30**, 101–116.

Fraser, H. R., 1958, The Turbulent Boundary Layer in a Conical Diffuser, *J. Hydraulic Division, Proc. ASCE*, **84**, No. H43, 1684.

Gardow, E. B., 1958, The Three-Dimensional Turbulent Boundary Layer in a Free Vortex Diffuser Gas Turbine Laboratory Report #42, Massachusetts Institute of Technology, Cambridge, MA.

Hewson, C. T., 1958, Growth and Separation of a Turbulent Boundary Layer, in *ASME Symposium on Stall*, American Society of Mechanical Engineers, New York, NY.

Hislop, G. S., 1940, The Transition of a Laminar Boundary Layer in a Wind Tunnel, Ph. D. Thesis, Cambridge University.

Karlsson, S. K. F., 1959, An Unsteady Turbulent Boundary Layer, *J. Fluid Mech.*, **5**, 622–636.

Kerr, N. M., Fraser, D., 1964, Back Flows in Rotating Fluids Moving Axially Through Expanding Cross Sections, *Am. Inst. Chem. Engrs J.*, **10**, 83–88.

Ludwieg, H., Tillman, W., 1949, Untersuchungenn uberdie Wandschubspannung in turbulenten Reibungsschichten, *Ing.-Arch*, **17**, 288–299. (English translation in NACA TM 1285, 1950).

Martin, B. W., Brown, A., Garrett, S. E., 1978, The Effects of Free-Stream Turbulence Intensity and Velocity Distribution on Heat Transfer to Curved Surfaces, *Proc. Inst. Mech. Engrs*, **192**, 225–235.

Moses, H. L., 1964, The Behavior of Turbulent Boundary Layers in Adverse Pressure Gradients, Gas Turbine Laboratory Report 73, Massachusetts Institute of Technology, Cambridge, MA.

Murphy, J. S., 1955, The Separation of Axially Symmetric Turbulent Boundary Layers, Part I, Douglas Aircraft Company Report ES 17513.

Newman, B. G., 1951, Some Contributions to the Study of the Turbulent Boundary Layer Near Separation, Department of Supply, Aeronautical Research Consultative Committee, Australia, Report No. ACA-53.

Rose, W. G., 1962, A Swirling Round Turbulent Jet, *ASME J. Applied Mechanics*, **29**, 616–625.

Rotta, J. 1951, Schubspannungsverteilung un Energiedissipation bei turbulenten Grenzschichten, *Ing. Arch.*, **20**, 195–207.

Sandborn, V. A., 1953, Preliminary Experimental Investigation of Low-Speed Turbulent Boundary Layers in Adverse Pressure Gradients, NACA TN 3031.

Schubauer, G. B., Klebanoff, P. 1950, Investigation of Separation of the Turbulent Boundary Layer, NACA TN 2133 (also NACA TR 1030, 1951).

Schubauer, G. B., Skramstad, H. K., 1948, Laminar Boundary Layer Oscillations and Transition on a Flat Plate, NACA TR 909.

Simpson, R. L., Strickland, J. H., Barr, P. W., 1974, Laser and Hot-Film Anemometer Measurement in a Separating Turbulent Boundary Layer, Southern Methodist University Tech. Report WT-3 (see also *J. Fluid Mech.* **79**, 553–594, 1977).

Index

Simpson, R. L., Chew, Y.-T., Shivaprasad, B. G., 1980, Measurement of a Separating Turbulent Boundary Layer, Southern Methodist University Tech. Report SMU-4-PU.

Stratford, B. S., 1958, An Experimental Flow with Zero Skin Friction throughout its Region of Pressure Rise *J. Fluid Mech.*, **5**, 17–35.

Vaughan, C., 1986, A Numerical Investigation into the Effect of an External Flow Field on the Sealing of a Rotor-Stator Cavity. D. Phil. Thesis, University of Sussex, Sussex, England.

Von Doenhoff, Tetervin, 1943, Determination of General Relations for the Behavior of Turbulent Boundary Layers, NACA TN 772.

Wells, C. S., 1967, Effect of Free-Stream Turbulence on Boundary-Layer Transition, *AIAA J.* **5**, 172–174.